MILITARY PROJECT MANAGEMENT HANDBOOK

Editors

David I. Cleland
Ernest E. Roth Professor and Professor of Engineering Management
University of Pittsburgh

James M. Gallagher, PMP
Wright State University

Ronald S. Whitehead
Vought Aircraft Company

McGRAW-HILL, INC.

New York San Francisco Washington, D.C. Auckland Bogotá
Caracas Lisbon London Madrid Mexico City Milan
Montreal New Delhi San Juan Singapore
Sydney Tokyo Toronto

Library of Congress Cataloging-in-Publication Data

Military project management handbook / David I. Cleland, James M. Gallagher, Ronald S. Whitehead, editors.
p. cm.
Includes index.
ISBN 0-07-011330-0
1. United States—Armed Forces—Procurement—Management—Handbooks, manuals, etc. I. Cleland, David I. II. Gallagher, James M. III. Whitehead, Ronald S.
UC263.M463 1993
355.6′212′0973—dc20 93-20463
CIP

1 2 3 4 5 6 7 8 9 0 DOC/DOC 9 9 8 7 6 5 4 3

ISBN 0-07-011330-0

The sponsoring editor for this book was Larry S. Hager, the editing supervisor was Stephen M. Smith, and the production supervisor was Pamela A. Pelton. It was set in Times Roman by Carol Woolverton, Lexington, Massachusetts, in cooperation with Warren Publishing Services, Eastport, Maine.

Printed and bound by R. R. Donnelley & Sons Company.

This book is printed on acid-free paper.

Whenever "man," "men," or their related pronouns appear in this work, either as words or parts of words (other than with obvious reference to named male individuals), they have been used for literary purposes and are meant in their generic sense.

The views expressed in this work are those of the authors and do not necessarily represent the official position of the U.S. Government or any of its departments, agencies, or other organizational elements.

CONTENTS

CONTRIBUTORS

James A. Abrahamson *Oracle Corporation, Redwood Shores, California* (Chap. 11)

Fred Abrams *Modern Technologies Corporation, Dayton, Ohio* (Chap. 18)

Norman R. Augustine *Martin Marietta Corporation, Bethesda, Maryland* (Introduction)

Bud Baker *Wright State University, Dayton, Ohio* (Chap. 37)

Robert R. Barthelemy *Wright-Patterson Air Force Base, Dayton, Ohio,* (Chap. 15)

Alan W. Beck *Defense Systems Management College, Ft. Belvoir, Virginia* (Chap. 5)

Michael J. Browne *Defense Systems Management College, Ft. Belvoir, Virginia* (Chap. 24)

Wallace T. Bucher *Wright-Patterson Air Force Base, Dayton, Ohio* (Chap. 30)

Tom Caudill *Wright-Patterson Air Force Base, Dayton, Ohio* (Chap. 9)

Virginia Caudill *Wright-Patterson Air Force Base, Dayton, Ohio* (Chap. 9)

Jeffrey D. Cerny *Redstone Arsenal, Huntsville, Alabama* (Chap. 16)

John A. Ciucci *Logistics Management Institute, Bethesda, Maryland* (Chap. 33)

David I. Cleland *University of Pittsburgh, Pittsburgh, Pennsylvania* (Chap. 38)

Charles B. Cochrane *Defense Systems Management College, Ft. Belvoir, Virginia* (Chaps. 12, 31)

Curtis R. Cook *Educational Services Institute, Falls Church, Virginia* (Chap. 42)

Earl D. Cooper *Florida Institute of Technology, Alexandria, Virginia* (Chaps. 28, 43)

Margaret G. Cunningham *Xavier University, Cincinnati, Ohio* (Chap. 27)

A. J. DiMascio *Florida Institute of Technology, Alexandria, Virginia* (Chap. 10)

James H. Dobbins *Defense Systems Management College, Ft. Belvoir, Virginia* (Chaps. 8, 17, 36)

Vernon J. Edwards *Educational Services Institute, Falls Church, Virginia* (Chap. 42)

C. Michael Farr *Air Force Institute of Technology, Dayton, Ohio* (Chap. 41)

Donald S. Fujii *Defense Systems Management College, Ft. Belvoir, Virginia* (Chap. 20)

James M. Gallagher *Wright State University, Dayton, Ohio* (Chap. 25)

Kevin P. Grant *Air Force Institute of Technology, Dayton, Ohio* (Chap. 22)

Michael E. Heberling *Air Force Institute of Technology, Dayton, Ohio* (Chap. 35)

Norah H. Hill *Air Force Institute of Technology, Dayton, Ohio* (Chap. 44)

Thomas C. Hone *Defense Systems Management College, Ft. Belvoir, Virginia* (Chap. 19)

Timothy J. Kloppenborg *Xavier University, Cincinnati, Ohio* (Chap. 27)

Michael G. Krause *Principal, Krause Associates, Silver Spring, Maryland* (Chap. 26)

Jerome G. Lake *Defense Systems Management College, Ft. Belvoir, Virginia* (Chap. 2)

J. Gerald Land *Defense Systems Management College, Ft. Belvoir, Virginia* (Chap. 23)

James Lis *Douglas Aircraft Company, Albuquerque, New Mexico* (Chap. 4)

Herbert W. McCarthy *Formerly, Deputy Assistant Secretary for Logistics and Material Management, Department of Defense, Washington, D.C.* (Chap. 1)

William T. Motley *Defense Systems Management College, Ft. Belvoir, Virginia* (Chap. 7)

John F. Phillips *Wright-Patterson Air Force Base, Dayton, Ohio* (Chap. 39)

Helmut H. Reda *Wright-Patterson Air Force Base, Dayton, Ohio* (Chap. 15)

Daniel G. Robinson *Defense Systems Management College, Ft. Belvoir, Virginia* (Chap. 21)

Benjamin C. Rush *Defense Systems Management College, Ft. Belvoir, Virginia* (Chap. 3)

Jerry D. Schmidt *Wright-Patterson Air Force Base, Dayton, Ohio* (Chap. 40)

Steven M. Shaker *Global Associates, Ltd., Arlington, Virginia* (Chap. 29)

Carl R. Templin *Air Force Institute of Technology, Dayton, Ohio* (Chap. 34)

Edward J. Trusela *Defense Systems Management College, Ft. Belvoir, Virginia* (Chap. 14)

Ernst Peter Vollmer *Defense Systems Management College, Ft. Belvoir, Virginia* (Chap. 13)

Daniel R. Vore *Air Force Institute of Technology, Dayton, Ohio* (Chap. 44)

John W. Warner *U.S. Senate, Washington, D.C.* (Introduction)

Rita Lappin Wells *Industrial College of the Armed Forces, National Defense University, Washington, D.C.* (Chap. 6)

David A. Yosua *The Analytic Sciences Corporation (TASC), Fairborn, Ohio* (Chap. 32)

PREFACE

The U.S. defense industry is currently facing one of its greatest challenges—to survive in the face of the government's increasingly stringent defense budgets. The changing political systems of the world, the movement toward the end of the Cold War, and the need to deal with an unpredictable economy in the United States have put many defense contractors in perilous times.

The President's proposals for a reduction in nuclear arms should bring further welcome reductions in the nuclear threat to the world. The changing times in the nations formerly integrated under the Soviet Union pose both problems and solutions for the future of defense systems in the military arsenals of the United States and its allies. Even as this book is being finalized, the political and economic changes in the world will impact the strategic management of military systems technology.

The critical question for any military systems developer or manufacturer is simply, "What's left?" Is there going to be enough left in the future to ensure an effective deterrent to any attack upon the United States and its allies from any quarter? If the United States gives up any of its military capabilities, we must make sure that no power can come against us. Even though the President of the United States calls for continued support for essential military preparedness programs, it is not clear what Congress will do. Defense cuts are already hitting far more than most people expected, with nearly 400,000 defense-related jobs lost since the beginning of the recession in July 1990. If Congress and the President agree to slice tens of billions out of the Pentagon budget over the next five years, the economic recovery facing the nation is likely to be delayed. The optimists say that over the long run, the resources freed up should boost innovation, support new industries, create new jobs, and put the U.S. economy on a higher growth track. But, of course, no one has been able to offer any specifics in these matters.

Defense spending has amounted to a covert industrial policy for the United States. Federal funds have been sent to prime military contractors who in turn moved these funds on to second- and third-tier contractors. Economists estimated that for every job cut by a prime contractor, an equal number of jobs are lost by the company's subcontractors and suppliers. The defense mobilization is hitting almost every region; no one state seems exempt. State and local governments are just starting to develop job programs and grants to redirect businesses into nondefense work. The dramatic demobilization of the U.S. defense economy is devastating large portions of industrial America. Civilian manufacturers will enjoy a bounty of highly trained professionals and managers, and research and development spending should be more oriented toward commercial products and services. Indeed, for the first time since 1940, the U.S. economy will have to find ways to grow without a major boost from defense spending. The defense industry is in a retreat to a lower and permanent plateau after over 40 years of supporting a U.S. strategic policy emphasizing a technological edge in military systems. And the military services are also cutting the size of their forces. With the political changes in the breakup of the former Soviet Union, the need to develop and rush military systems into the field has been diminished. Instead, the Pentagon will spend more resources developing and testing military systems without any commitment to full production. Former Defense Secretary Cheney has stated: "We will focus on research and development to create a storehouse of technology, which we can use when needed." Whether or not a defense industry can survive in a meaningful way by just building prototypes remains to be seen.

The terms project management, program management, and systems management are sometimes used interchangeably in the text. Usually these terms mean the same thing and should be interpreted accordingly.

These national policies are sending shock waves through the military systems developers, suppliers, and users. It is in this environment that the editors have chosen to publish this *Military Project Management Handbook.* It is our hope that this handbook will make a real contribution to a more effectively managed military systems development, procurement, and production process.

In the fast-changing military project management business, there is a need for a common base of theory and practice to provide developers and users a sense of permanence in a changing environment. This handbook has been designed to provide a litany of practical explanations and descriptions by leading authors and experts in the field of the concepts, strategies, tools, and techniques of military projects. Useful concepts and approaches have been drawn from the abundance of Department of Defense (DoD) documentation and literature in the defense project management field.

It is the purpose of this handbook to provide a practical guide to the theory and practice of military project management for DoD policy makers, military commanders, military project managers, industrial general managers, and functional managers, as well as for professionals in the field. This handbook will also be useful to other people in the many defense-related organizations in the nation who have a stake in the management of military projects. Thus in such a rapidly changing field as is found in the defense business, even experienced people are faced with the challenge of keeping up with current developments and management strategies. The handbook is organized so that it can be used either as a reference or for sustained study. Not only will practitioners find the handbook valuable, students attending formal courses as well as those who wish to pursue a self-study professional reading program will find the book of value. This handbook can serve as a valuable touchstone for those people dedicated to managing projects in the most effective and efficient manner.

We have given the authors considerable latitude in the content and style with which they have presented their material in terms of format, breadth and depth of coverage, and scope for each chapter. We also welcome their personal judgment in the interpretation of the cause and effect of the DoD documentation—to fill in the spaces between the lines of formal DoD policy—tempered by their judgment as survivors in the military project management arena.

The chapter authors have, in some instances, used material from DoD public domain sources. We deeply appreciate the opportunity to use material from such sources to add to the authenticity of the handbook material.

We accept that it is a time of major change in the process of managing military projects. Indeed, the changes in the last two years have been as substantial as any since those caused by World War II. The publication of the handbook was delayed so that the chapter authors could incorporate the provisions of the new DoD 5000 series policy documentation, as well as recent changes in the Federal Acquisition Regulations in their writings. This small delay has been beneficial as it provided the opportunity to publish a handbook that reflects contemporary DoD strategies and policies in the management of military projects.

Whatever its value to the users, the *Military Project Management Handbook* reflects the knowledge, skills, and attitudes of a remarkable gathering of professional people concerning the key pivotal factors and forces surrounding the management of military projects. Five interdependent areas of military project management are developed and presented in the following order:

1. *Functional Management.* Project management is described in this part in terms of the many disciplines that must be pulled together and integrated in order to develop, design, produce, and put military systems into an operational environment.

2. *The Environment of Military Project Management.* This part provides an overview of the major characteristics of the environment in which project management is planned and implemented in DoD–defense industry circumstances.

3. *The Human Element of Military Project Management.* People and the many factors and forces that influence the performance of the human subsystem are reviewed in this part of the handbook.

4. *Military Project Management Framework.* An important consideration in the theory and practice of project management is the basis on which the supporting environment and the management functions are carried out. This part provides an overview of such issues.

5. *Strategic Outcomes.* Three special topics that can influence the strategic outcome of military projects are presented in this final part.

We believe that the change that is impacting the U.S. Department of Defense is just beginning. This handbook reflects some of that change—the changing theory and practice of project management in the defense establishment. We hope that the basic philosophies and practices outlined in this book will help the readers to better understand how best to manage their defense projects. If this hope is realized, the creation of this book will have been worth the effort put forth by the editors and authors. We wish the readers good reading—as they peruse this first-of-its-kind publication.

David I. Cleland
James M. Gallagher, PMP
Ronald S. Whitehead

ACKNOWLEDGMENTS

There are many persons who have given support to the development and publication of this handbook. Special credit is given to the chapter authors whose excellent credentials are listed on the title page of each chapter. A perusal of these credentials should convince the reader that truly a gathering of experts have contributed their knowledge and skill in creating this handbook, the first of its kind. We thank these chapter contributors.

We thank many notables, too numerous to name, in the U.S. Department of Defense, in the U.S. defense industry, and in academia, who have provided an environment for the authors to carry out the creative work required to write the chapters.

We are deeply indebted to Claire Zubritzky of the Industrial Engineering Department of the University of Pittsburgh, who in her usual sustained superior style, managed the administrative details involved in the design, development, and production of this handbook. Claire kept us on schedule and within budget on this important project.

A special note of thanks is extended to Edward Trusela, Air Force Chair, Executive Institute of the Defense Systems Management College, Fort Belvoir, Va., for his able counsel in the selection of many of the authors and in much of the design of the strategy for the handbook.

We wish to recognize Victoria Sievert, who must share major responsibility in getting the coeditors together to create this book. Victoria also assisted us in identifying and selecting many of the authors.

We also thank Dr. Harvey W. Wolfe, Chairman of the Industrial Engineering Department, and Dr. Charles A. Sorber, Dean of the School of Engineering, of the University of Pittsburgh, who provided us with the appropriate resources and the intellectual environment to pursue this strategic project.

At Wright State University we thank Dr. Waldemar M. Goulet, Dean of the College of Business and Administration, and Ms. Marsha L. Adams, Assistant Dean and Executive Director of the Organizational Services Group, who provided the environment and a considerable amount of university resources to pursue this venture. Thanks also go to Ms. Susan Young, Secretary, Organizational Services Group, who contributed state-of-the-art word processing and administrative skills in support of the project.

Special thanks goes to Kaye Cramblit of Vought's Dayton office for her administrative prowess in helping develop the book.

Finally we thank our families, who provided us with a suitable cultural environment at home to pursue this strategic initiative.

INTRODUCTION

Currently, U.S. defense contractors are facing their greatest challenge—to survive in an intensely competitive market in the face of continuing reductions in Department of Defense (DoD) budgets. Downsizing, restructuring, diversification, and the conversion to nondefense product development and production are some of the alternatives available to these companies. The sobering recognition is that many of the small and midsized defense contractors that have grown to depend on a flow of contracts from prime defense contractors may not survive. Even some of the prime contractors may not survive without the development and implementation of innovative survival strategies.

New product and process technologies have outdated some time-proven manufacturing management strategies. Product design teams, quality circles, production teams, and project management teams will be required to deal with the realities of computer-integrated manufacturing (CIM), MRP, JIT, TQM, computer-aided design (CAD), CAM, CAE, and flexible manufacturing systems (FMS), all pieces of a development-production medley that ties modern innovations and the changing defense factories together. At the same time the reduction in DoD budgets, the growing complexity of doing business in the defense industry, and the continued growth of DoD procurement regulations add to the challenges of doing business in the defense market, yet still earning a reasonable return on the resources that have been invested.

In the material that follows two leaders from the defense industry environment present an overview of their perceptions of the changing defense business. First, Sen. John W. Warner of Virginia presents a legislative viewpoint. This is followed by a defense industry perspective by Norman R. Augustine, Chairman and Chief Executive Officer of the Martin Marietta Corporation.

LEGISLATIVE PERSPECTIVE

John W. Warner

John W. Warner is the ranking Republican on the Senate Armed Services Committee and a member of the Select Committee on Intelligence. He holds a B.S. degree in basic engineering from Washington and Lee University and a law degree from the University of Virginia. He served in World War II and in the Korean War. In 1956 he was appointed an assistant U.S. Attorney; from 1969 to 1972 he served as Under Secretary of the Navy and from 1972 to 1974 as Secretary of the Navy. Before entering the Senate, Senator Warner carried out special assignments for the United States in government-to-government negotiations on military, scientific, economic, educational, and cultural matters, which gave him a background in national security and foreign affairs.

"To ye who shall see these presence, greetings: Know ye that reposing special trust and confidence in the patriotism, valor, fidelity and abilities of 'The Military Project Manager'... (he or she is hereby expected to now be all things to all people at all times!)." I will assume that the reader of this handbook is already familiar with the words in the first part of the foregoing quotation, which constitute the first official greetings a newly commissioned, recently promoted, or award-deserving officer hears or reads to commemorate the respective event. These words are not only as familiar to the lexicon of the American fighting men and women as "pass in review" or "KP," but contain the embedded essence of the authority to exercise leadership: "trust and confidence."

In my years of service in both the Navy and the Marine Corps, and later as the service secretary for both, I observed the increasing responsibilities imposed on our military officers as technology created expanding management challenges. Exercising special trust and confidence is no longer a matter of simple upward and downward loyalty between troops and commanders. The complexities facing today's acquisition professionals require the same personal application of leadership (trust and confidence) as good company commanders, platoon leaders, squad sergeants, and fire-team leaders have traditionally invoked. To today's project manager, your program *is* your command in every classic sense. As with troops, you must care for its welfare, feed and clothe it, and train it to improve its endurance for the time it will be called upon to fight. You must also discipline it when required. Robert E. Lee once stated that "discipline is training which makes punishment unnecessary." Nowhere within the Department of Defense is General Lee's sage proclamation more applicable than within the program management arena. Driven in part by declining budgets and in part by the continuing need to avoid wasteful expenditure of tax dollars, Congress has increasingly demanded program "discipline" from all the services.

The price of failing to exercise disciplinary (and penal, when required) aspects of the special trust and confidence charter has been demonstrated in program "casualties" like the Sergeant York, the P-7 aircraft, and the A-12 medium attack stealth bomber. On the other hand, Congress has warmly embraced well-disciplined, trim and fit, ready to fight programs such as the advanced tactical fighter (ATF), now known as the F-22.

I may seem to be suggesting that Congress is the enemy, and I know that many readers reaching for this handbook for the first time might already agree with that premise. In a sense, it is true, and if a program maintains or gains health and strength from its manager adopting such a posture, then the system is working. Just remember, however, that unlike a classic military enemy, unlike a true threat to democracy, you *must not* classify your "battle plans" as top secret when dealing with Congress. Smooth and open communication to and from Capitol Hill is, in fact, one of the most effective instruments of battle in your quiver. I will not insult the reader by suggesting that a new program manager spend all of his or her time "learning to work with Congress." There are far too many political facets, which change almost hourly, for a program manager to manipulate externally or to form tactics to compete against. But the "intelligence" you can obtain from your various congressional liaison offices can be most helpful in the need for increased "assaults" of program information. Let the liaison offices work at what they do best: carrying information under the white flag; and you do what you do best: exercising your special trust and confidence by devoting all your energies to the care and feeding (*and the disciplining*) of your program.

One of the more important concerns that has evolved from the congressional perspective is the concept of "fly before you buy." While I admit that, as a risk-reduction effort, this catchy phrase has significant political appeal, I also strongly endorse it as an ingredient for any successful future program acquisition recipe. It is this very quality that has led to the upsurge of congressional support for the ATF and, I predict, will engender that same level of backing for the jointly developed interdiction aircraft (without regard to its eventual common name: A-X, FA-X, VFAX, or AFX). The congressionally mandated "competitive prototype" phase for A-X not only enhances the "fly before you buy" effort seen in the ATF program, but further serves to strengthen the defense industrial base at a time when budgetary constraints appear to be attacking this base from all sides. Embracing concepts such as "competitive prototyping" and "fly before you buy" is therefore all the more important since the military project managers of today will be asked, if not tasked, to oversee the change of a *defense* industrial base into an industrial base. A certain amount of this type of change, better titled defense conversion, is appropriate to realistically and wisely apportion valuable tax dollars in a changed world environment. Nonetheless you, as military project officers, must help to ensure that we do not "convert" to an extent that would preclude a rapid reversal for defense needs, should world conditions reverse again as quickly as we have experienced in the early 1990s.

With this thought in mind, let us add one more concept to your growing "quiver." The budgetary limitations of the foreseeable future will force rethinking the acquisition strategies of new-technology, expanded-capability, high-cost weapon systems. The choice simplifies to: develop and procure less capable (and thus less expensive) systems in numbers large enough to provide a mass defense approach; or develop and produce the most advanced systems of which we are capable, but in smaller, "silver bullet," numbers. I would lean toward the latter, augmented with the concept of "high-low mix." This means that while an arsenal of more expensive but highly capable weapon systems (such as the F-22 and the A-X) can only be afforded in small numbers, their force-multiplying effect should then be bolstered by low-end, lower-cost, and lower-capability systems which can be procured in larger numbers (such as the F-18 and F-16 aircraft). The resultant "high-low mix" offers both volume and capability.

I have attempted in this introduction to identify certain "arrows" which may be of greatest use to a project manager from a congressional perspective. This handbook fully supplies your "arrow" collection with hard learned knowledge and well-directed suggestions designed to fend off the most tenacious challenges from virtually every conceivable source. This handbook *is* your "quiver." Use the "arrows" which it provides, together with the leadership required of one given special trust and confidence, and not only will your program win the acquisition battle, but our soldiers, sailors, airmen and women, and Marines who will use your system will win the *real* battles to defend democracy and keep us safe. (It will not do your fitness report, performance evaluation, or OER any harm either.)

INDUSTRY PERSPECTIVE

Norman R. Augustine

Norman R. Augustine is Chairman and Chief Executive Officer of Martin Marietta Corporation, a major defense contracting firm. He has held numerous high-level posts in government and private industry, including Assistant Secretary of the Army for Research and Development, Assistant Director of Defense Research and Engineering in the Office of the Secretary of Defense, Chairman of the Defense Science Board, and President of the American Institute of Aeronautics and Astronautics. He is the author of the bestselling book *Augustine's Laws* and co-author of *The Defense Revolution.* He is a former program manager.

The job of project manager is among the most important and most difficult assignments in America's peacetime military. The attendant responsibilities include the control of substantial financial resources coupled with the challenge of meeting major technological goals—all in an environment of intense external scrutiny. Consequences of the project manager's decisions can be of the highest order, possibly measured in the preservation of lives and the maintenance of freedom.

The job of project manager is also potentially a career buster.

Performed properly, the project management role can make enormous contributions and can even affect the course of history. This was vividly demonstrated during the Persian Gulf War, as it has been on various other occasions in the past; for example, a half-century earlier in the development of the atomic bomb during the Manhattan Project. In truth, America's armed forces in the Persian Gulf were so well trained and motivated that they would probably have won with the other side's equipment. But it was to a considerable extent equipment that was the product of nearly four decades of effort by U.S. project managers that resulted in a favorable 1000 to 1 equipment loss ratio between combatants. There are few who today would disagree that modern technology can save lives of our military personnel when called upon to go in harm's way. It is the job of the project manager to take technology off the shelves of our nation's laboratories and place it in the hands of our military forces in the field.

Truly, the stakes are greater than ever as world stability continues to erode and the likelihood of a reduction in U.S. military procurement by two-thirds and research and development by one-third in constant dollar terms becomes increasingly real.

The need for perfection in project management has perhaps never been greater, particularly in the management of cost.

Two challenges of military project management are particularly noteworthy. The first is inherent in technology itself. In the effort to obtain the maximum possible advantage over an adversary, military equipment is generally designed at the very edge of the state of the art. But as one former Assistant Director of Defense Research and Engineering noted in a moment of frustration over the repeated inability of advanced electronic systems to meet specified mean time between failure goals, "airborne radars are not responsive to enthusiasm." In short, managerial adrenalin is not a substitute for managerial judgment when it comes to transitioning technology from the laboratory to the field.

Second, the military acquisition environment should never be confused with the free-enterprise system. Most of the aspects of a multibuyer, multiseller marketplace which have made the free-enterprise system so enormously successful are simply not present in defense acquisition. This is very fundamental. Not only is there but a single buyer (or at least a single buyer with the authority to approve or disapprove sales to other potential buyers), but in many cases there is also but a single seller. The latter is likely to become increasingly the case in the years ahead as the defense industry drastically consolidates and downsizes. The normal incentives of the free market are therefore not present in defense procurement, which at times can be characterized as a monopoly wrapped in a monopsony. Thus the essence of military project management is to find synthetic substitutes for the market forces which exist naturally in the commercial free-enterprise system.

The nation's record at finding these means of emulating free-market forces has to date not been strikingly successful—even in the most basic matter of establishing suitable contracting mechanisms. For over four decades military project managers have sought an acceptable means of balancing risk and reward between buyer and seller when carrying out the highly unpredictable tasks of research and development. But whatever may have been the factors which have troubled the acquisition process over the years, consistency has certainly not been among them. The military acquisition process has seen policy on contract types swing back and forth like a pendulum: first embracing cost-reimbursable-type contracts which place most of the risk on the buyer, and then shifting to fixed-price-type contracts which place most of the risk on the seller, and then back again.

General Eisenhower wrote that "The fullest utilization by the Army of the civilian resources of the nation cannot be procured merely by prescribing the military characteristics and requirements of certain types of equipment. Scientists and industrialists are more likely to make new and unsuspected contributions to the development of the Army if detailed directions are held to a minimum." In short, contractors should be provided considerable flexibility. But less than a decade later, Secretary of Defense McNamara stated that engineering development should not even begin until the "... technology needed is sufficiently in hand" and a basis exists for a "firm fixed price or fully structured incentive contract for Engineering Development." A half-dozen years later, Deputy Secretary of Defense Packard observed that "it is not possible to determine the precise production cost of a new complex defense system before it is developed ... Cost-type prime and subcontracts are preferred where substantial development effort is involved." But shortly thereafter, under the leadership of Secretary of the Navy Lehman, the policy was once again reversed: "A Systems Commander will not ... proceed with Full Scale Engineering Development until he is satisfied that advanced development has reduced risks sufficiently to enable the contractors to commit to a fixed price type contract that includes not-to-exceed prices or priced production options." More recently, Secretary Cheney instructed that "the Under Secretary of Defense/Acquisition will strictly limit the use of cost-sharing contracts for systems development and the use of fixed-price type contracts for high risk development."

Thus although it is (currently) the stated *policy* of the Department of Defense to conduct research and development under cost-reimbursable-type contracts, in practice such "imaginative" new contracting concepts have been introduced as cost-reimbursable contracts with ceilings, cost-reimbursable contracts with share lines, cost-reimbursable contracts excluding provisions for economic inflation, cost-reimbursable development contracts with prespecified fixed-price production options, cost-reimbursable contracts with prescribed cancellation ceilings, and so on. One is reminded of the baboon that P. T. Barnum once exhibited as "the most remarkable gorilla in captivity." Responding to a skeptical zoologist who argued that "gorillas don't have tails," Barnum is said to have enjoined, "yes, but this one does. That's what makes him so remarkable."

Further complicating the task of the military project manager is that he or she is required to operate in an environment characterized by highly constrained management flexibility. This derives from the diffusion of acquisition authority among large numbers of governmental entities and individuals, involving many of the various branches of government. In one recent case, a judge demanded to see a request for proposal before it was released to potential contractors. It is standard practice for the Office of the Secretary of Defense to review all major requests for proposals, and the detailed program management guidance emanating from Congress is legion.

The solution to acquisition problems encountered in the past has too often been simply to promulgate regulations insisting that whatever problem has occurred never ever occur again. The resulting body of procurement "law" has therefore been evolutionary—rather than the product of a carefully considered, "zero-based" effort to determine how best to manage military research, development, procurement, and support successfully.

Also relatively unique to military project management is the fact that most often the independent variable which a manager has available to use as a control mechanism is schedule. This contrasts sharply with commercial project management practice wherein near-term funding is most often the control variable. Commercial project managers apply funds expeditiously so as to avoid costly delays; military project managers seldom have the flexibility to provide additional funding in the short term—and therefore adjust schedule to comport with a rigid close-in funding profile. The consequences of this practice can, over the long term, be punitive indeed.

In short, the successful project manager must learn to unhorse "the four horsemen of the apocalypse" as they relate to military procurement. The purpose of this book is to help do exactly this.

First among the horsemen, conditions must be studiously avoided which may preclude other than absolutely the "finest" (most honest, motivated, competent, and experienced) individuals from holding responsibilities in the project management structure.

Second, practices must be avoided which prevent "margins" from being established to accommodate uncertainties in development activities—margins in cost, schedule, product performance, staffing, alternative technical approaches, and so forth. One of Augustine's widely unremembered laws states that the difference between a good program manager and a great program manager is "reserves!" And "reserves" is *not* a four-letter word, in spite of its treatment as such in the government procurement process.

Third, steps must be taken to contain those aspects of the military procurement process that produce turbulence—turbulence in people, funding, schedule, goals, and so on. This is critical: "turbulence" *is* a four-letter word.

And, fourth, the "grass is greener" syndrome must be avoided at all cost. This tendency has over the years been one of the greatest enemies to successful procurement. It embodies the notion that every time a substantial problem is encountered in a development or production program (which is almost always, if significant advances are being sought, even in the best managed programs), the project should be terminated and an altogether new effort undertaken "which has no problems." This is of course the epitome of self-delusion. One simply cannot afford to be unduly risk-averse. If the goal is never to let anything go wrong, it is likely that nothing great will ever go right.

The author of this introduction has, based on over three decades of first-hand observation, prepared a "checklist for an acquisition adventure" (or, in more pragmatic terms, "a formula for failure") of military development projects. The list, published herein for the first time, is a tried and proven set of actions which represent the accumulation of years of scar tissue by a large number of very able individuals at a financial cost overwhelming that of the world's finest business school education. The list comes with a guarantee: if a project manager studiously follows these policies, he or she will be virtually assured of failure! Thus a score of 0 makes success unavoidable. A score of 1 to 5 results in success *sometimes* being unavoidable. A 6 to 10 score assures that only agonizing success is possible, whereas 11 to 20 guarantees a good chance of disaster. But a score of 21 to 36 virtually dictates that disaster will not be left to chance.

Augustine's Checklist for an Acquisition Adventure
or
"A Formula for Failure"

- ☐ Settle for Less Than the Best People — Reduce Payroll Cost
- ☐ Build an Adversarial Relationship Between Buyer and Seller
- ☐ Change Management Frequently — Provide Opportunities
- ☐ Avoid Evolutionary Growth to New Capabilities — Take Grand Leaps
- ☐ Continually Revise Schedule and Funding — Generate Excitement
- ☐ Include All Features Anyone Wants — Make Everybody Happy
- ☐ Allow No Margins in Funding, Schedule or Technical Approach — Nothing Will Go Wrong
- ☐ Divide Management Responsibility Among Several Individuals — Two Heads are Better Than One
- ☐ Whenever Difficult Problems are Encountered, Start All Over with a New Approach Having No (Known) Problems
- ☐ Promote Continued Debate over Goals Throughout Life of Project — Variety is The Spice of Life
- ☐ Give Reliability Low Priority — Especially Avoid Redundancy
- ☐ Develop Underlying Technology and End-Product Concurrently
- ☐ Do Not Plan Intermediate Test Milestones — Just One Glorious Display
- ☐ Create as Many Interfaces as Possible — Help People Get to Know Each Other
- ☐ Focus on the Big Picture — The Details Will Take Care of Themselves
- ☐ Disregard Seller's Track Record — The Law of Averages Will Work Out
- ☐ Cut Costs by Reducing Testing — Especially Environmental and Full-System Testing
- ☐ Ignore the Users — They Don't Understand High-Tech
- ☐ Choose Among Sellers Based on What They Promise — No One Likes a Pessimist
- ☐ Get Head-Start on Work Prior to Finalizing Goals, Schedule, and Cost — This is Especially True for Software - Which is Easy to Change
- ☐ Share Authority for Project Direction with Staff Advisors
- ☐ Eliminate Independent Checks and Balances — They Just Create Friction
- ☐ Don't Compete Potential Suppliers at Outset — Pick a Friend
- ☐ Once Underway Continue to Compete Selected Supplier with Outsiders — Change as Often as Possible to Assure "Freshness"
- ☐ Minimize Managers' Latitude for Judgment — Rely on Regulations
- ☐ Deal Harshly with Anyone Surfacing Problems — One Can't Afford Troublemakers
- ☐ Never Delegate — Hold Authority at the Top Where People Really Know What's Going On
- ☐ Maximize Individual Incentives — Teamwork is Just the Sum of the Parts
- ☐ Make Up for Schedule Slips by Overlapping Design and Build — Especially When Test Results are Disappointing
- ☐ Include at Least as Many Auditors on the Project as Workers — Reviews Give Everyone a Chance to Participate
- ☐ Do All Possible to Minimize Profits of Participating Contractors — Save the Money
- ☐ Don't Waste Time Communicating (Especially Face-To-Face) — It Takes Time; and Time is Money
- ☐ Eschew Strong Systems Engineering — It Just Complicates Decision-Making
- ☐ Delay Establishing Configuration Control Until the Last Minute — Reduce Cost of Management
- ☐ Always Pick the Low Bidder — They Must Know Something Special and are Often Courageous
- ☐ Don't Worry About the Form of Contract — Just Enforce It

Score	
0	Success Unavoidable
1-5	Success Sometimes Unavoidable
6-10	Agonizing Success Possible
11-20	Good Chance of Disaster
21-36	Disaster Not Left to Chance

With all the tribulations of military project management, it is nonetheless among the most rewarding careers one could pursue. It presents challenging work with important consequences. It involves the latest in technology. It offers the opportunity to work with a quality group of associates. And over the years, its practitioners have generated a large number of truly enormous successes.

It is noteworthy that in the case of almost every one of these successes, a few individuals—usually in government but occasionally in industry—have saved their program from unwarranted cancellation, often by laying their careers or, more important, their reputations on the line. This was true of the Sidewinder, Apache, Aegis, Polaris, SR-71, and Patriot. It was true of Tomahawk, Abrams, F-16, Phalanx, and A-10, and it was true of a whole host of other systems. The American public, and sometimes even those who have been called upon to stake their lives on the performance of these systems, often know neither the names nor the contributions of these unsung heroes who daily meet and overcome the tribulations of the military procurement process. But this circumstance in no way diminishes the profound impact that this small group of dedicated individuals has had, nor the enormous intangible rewards they enjoy. For they know who they are.

P · A · R · T · 1

FUNCTIONAL MANAGEMENT

Project management is the military acquisition process for developing new or modernizing existing weapon systems or subsystems and has long been chosen as the "best," most effective approach. In the typical systems program office work is carried out by the matrixed, shared resource functional divisions, such as engineering, manufacturing, program control, logistics, or contracts, and is supported by staff groups in higher organizational levels. It is the project manager who is selected to be given the full responsibility for all aspects of the military or industry counterpart business. Often he or she must negotiate for that support, and must rely on persuasion and influence rather than positional authority or rank to get things done. The keys to functional support and leadership of the project are interdependence and empowerment.

The project manager cannot be everywhere at once and, in fact, must often spend an inordinate amount of time competing for attention upward in the organizational channel and responding to numerous requests for information and "what if" exercises. Issues and problems do not usually come all at once, but rather piecemeal, incrementally. Rarely do these problems or issues affect only one function in a straightforward fashion. The project office is no Greek drama in which all the players maneuver according to an orderly plot moving inexorably to a predetermined outcome. Thus the project manager must lead his or her team in a way that allows open communication horizontally and vertically. All must have their eyes on the same plans and objectives and all must feel a real sense of ownership with the program. A weak, confederate matrix is never successful; management of programs must be a coordinated, involved, dedicated, and supported effort with clear, strong, central focus. The project manager needs to know how to empower the functionals at all levels in order to act and to trust them enough to share the power with those of different perceptions of roles, methods, responsibilities, and priorities. But he or she must also have the courage to act, to decide, to select the path. The project manager must be able to require functional staffs to be at once interdependent yet true to their professional requirements and the policies they are required to pursue.

But it is the project manager alone who must select the path to recommend to the higher authority, whether military or civilian, and these choices are tough and will, in today's environment, tip over not a few "rice bowls." This is not a time for rigid rules, but a time for creative, innovative, comprehensive thinking and superior performance. Consensus is not the goal of the project manager, but he or she must ensure that each functional area participates and supports. Furthermore, the project manager must not allow functions to lose sight of their role on the team and must put the emphasis on their being brought together to achieve field performance rather than simply delivering equipment. Only with a working integration of all of the technical disciplines involved, both military and industry, aligned together during the system's life cycle, can a project be considered a success.

In fact, the charge could be made that past projects have often been guilty of developing and producing systems or subsystems that provide hardware meeting the specifications but did not ultimately provide the needed operational mission capability. Each functional part worked, but as a whole it failed to be integrated into the operational manufacturing characteristics which drive costs, performance, and support. These issues are real today: the A-12 was killed, the B-2 truncated, the MultiRole Force has a customer who demands an affordable design if a new fighter is selected; LOSAT is in trouble; each service has its impetus for change.

This part treats those functional entities and gives the prospective project manager some strong indicators of how to achieve integration and empowerment. It provides perspectives of each functional area.

Herbert McCarthy presents in Chap. 1 the logistics functions, which pervade every aspect of military equipment, from cradle to grave. He provides a compelling story of the true scope of logistics considerations for the project manager and indications of its complexity.

In Chap. 2, Jerome Lake describes the processes, control, and phases of military project engineering, and explains how it has become truly interdisciplinary and is at the heart of satisfying user needs.

The broad topic of program control is covered by Benjamin Rush in Chap. 3. This subject is made more complex due to service differences and the broad spectrum of tasks covered by this important military project function. Dr. Rush provides a grand overview and detailed considerations throughout this chapter.

James Lis focuses on the test functions in Chap. 4. He discusses their use throughout the various phases of the project and provides fresh insights into current and upcoming practices, approaches, and considerations of the process.

Alan Beck and Rita Wells provide their views of project management contracting and project contract administration in Chaps. 5 and 6, respectively. For the project manager, the contract is the "heart of it all," and understanding this function is vital to project success. Alan Beck's overview of the contracting principles provides an excellent start, and Dr. Wells explores the roles and responsibilities of those charged to administer and perform buying and oversight tasks.

William Motley's Chap. 7 on manufacturing and the defense industrial base updates the reader on one of the single most changing areas of project management. Companies in the defense business are revolutionizing long established, antiquated practices to enter the world of total quality, introducing integrated process design, reducing costs, and tearing down vertical stovepipe organizational structures. This function has suddenly found itself thrust to the forefront of every military project, and is now a major consideration in every project acquisition strategy.

In Chap. 8 James Dobbins explains the configuration management function and its importance as a comprehensive part of the life-cycle success of the project.

This part concludes with Chap. 9, wherein the husband and wife team of Caudill provides a comprehensive review of the Security Assistance Program. They conclude that security assistance is an exciting, energizing, and dynamic field.

CHAPTER 1
LOGISTICS

Herbert W. McCarthy

Herbert W. McCarthy is a former Deputy Assistant Secretary of Defense for Logistics and Material Management. He served on the board of directors of the NATO Maintenance and Supply Agency, and negotiated logistics agreements with European and Pacific area countries. He has received the Secretary of Defense Distinguished Civilian Service Medal and Meritorious Civilian Service Medal, and the Navy Superior Civilian Service Medal. He earned a bachelor's degree at Boston College and a master's degree at George Washington University. He is also a graduate of the Industrial College of the Armed Forces and the Federal Executive Institute. He currently is an Adjunct Professor at the University of North Florida; George Washington University; NOVA University, Ft. Lauderdale; and Jacksonville University.

Any discussion of logistics must begin with an understanding of terms, specifically, a common definition. This is because there are so many definitions of logistics. Every author of a book on logistics has a definition, as does every dictionary. And no two of these seem to match. For our purpose, we need only to look at JCS Publ. 1 for the Department of Defense definition.

> *Logistics.* The science of planning and carrying out the movement and maintenance of forces. In its most comprehensive sense, those aspects of military operations which deal with: a. design and development, acquisition, storage, movement, distribution, maintenance, evacuation, and disposal of material; b. movement, evacuation, and hospitalization of personnel; c. acquisition or construction, maintenance, operation, and disposition of facilities; and d. acquisition or furnishing of services.[1]

It is with this definition in mind that we can proceed to a discussion of how logistics affects project management.

First, we need to appreciate the scope of the entire spectrum of logistics. Logistics does not relate only to the development of weapon systems. It is an ongoing everyday, 24-hour-a-day function. Within the Department of Defense (DoD) it involves approximately five million items of supply, in excess of $100 billion worth of inventory, and over $200 billion of equipment in use. It has annual expenditures of close to $40 billion for maintenance and another $11 billion for transportation. There are over one million people employed in logistics. Many of them are employed at the 21 inventory control points, 30 wholesale storage depots, and 33 maintenance depots currently undergoing consolidation reviews because of the defense management reviews emanating from the Office of the Secretary of Defense.

This is a very unusual time for logistics within the DoD. The outcome of the Gulf War was a logistics triumph for the United States. The start of the war was delayed, not only until sufficient forces were in place to overcome the Iraqis, but also to ensure that the logistics support would be there to back them up. Iraq, on the other hand, was logistically vulnerable. The war was to be fought in their "backyard" and they would have the advantage of short supply lines and rapid resupply and reinforcing capability while their primary adversary, the United States, had to rely on a logistics pipeline that stretched halfway around the world. But the infrastructure of Iraq, specifically the configuration of its transportation network, was such that all supplies and reinforcements had to pass through a single point. That point was Basra. By cutting off the southward flow of material and forces from Basra to Kuwait, the U.S. forces were able to make it impossible for Iraq to resupply its forces in Kuwait. In addition, the demoralized Iraqi forces within Kuwait had to face the fact that replacement or augmenting forces were not going to arrive to help them. Thus the United States was able to control the outcome of the war.

At the same time, the inability of Iraq to mount any kind of offensive either on the seas or in the air prevented them from cutting the U.S. supply lines. Thus the war was a logistics disaster for Iraq and a ringing logistics success for the United States.

During the first three weeks of that war, more people and equipment were moved than during the first three months of the Korean War. Aircraft delivered 75,000 troops and 65,000 tons of equipment in just one month. As of December 31, 1990, more than 300,000 personnel and 305,000 short tons of equipment had been airlifted to the Gulf. During this period over 9000 missions had been flown, representing over 90 percent of the available military airlift command each day. As for sealift, more than 225 ships delivered almost 2.5 million short tons of equipment. It is important to note that about 85 percent of all supplies and equipment was moved by sea.

Yet in spite of this glowing success, the DoD is facing significant cutbacks in the defense budget. This is to be expected with the current and projected world situation. What makes this situation of particular concern to the logistician is that there is a long history within the DoD of absorbing budget reductions by passing a disproportionate share to the logistics community. Part of this is due to the long held belief that the logistics "tail" is far larger than it needs to be in comparison to the combat unit "teeth," although many people are beginning to recognize that the complexity of today's equipment dictates a greater level of support. This means that the logistics organizations will have to look for improved ways of doing business, such as increased reliance on other services and agencies for support.

Secretary of Defense Richard Cheney's "Annual Report to the President and the Congress" speaks of reducing the cost of depot maintenance operations by $1.7 billion from fiscal year 1991 through fiscal year 1997 through streamlining and reducing the size of the depot maintenance infrastructure. The long-range plans which are still under development will reduce these costs by an additional $2.2 billion through increased capacity utilization, closing unneeded facilities, enhancing competition between the services and among the services and the private sector, and improved productivity.

In this same report he also discusses the possibility of future consolidations of the 21 inventory control points within the services and the Defense Logistics Agency.[2]

It is clear that, for the foreseeable future, logisticians will have to keep up the same level of support with diminished resources. The logisticians involved in project management can provide the basis for a great deal of this support by performing thorough integrated logistics support planning as early in the weapon system development cycle as possible.

Logistics has been in a long struggle for recognition in the acquisition process. Project managers all too frequently took the position that they were not going to spend money for logistics support. There was a general attitude that if the system were fielded, the money for logistics support would have to be made available from sources other than the project itself.

The money available to the project manager was then used to produce the system itself without consideration of downstream logistics costs. This was due in part to the different appropriations involved, procurement dollars providing the funds for acquisition of the system, while operation and maintenance dollars supported the system once it was deployed. Thus the motivation of the program manager was to save procurement dollars to offset cost overruns or simply to complete the project on time and within the funds available while not spending procurement dollars to increase supportability once the system was in operation.

This attitude cannot be afforded any longer. As we have seen, the funds may not be available for operational support of the system if logistics is not taken into consideration during system development. The system, when deployed, may not be supportable.

The solution to this problem began to evolve in 1964 with the publication of DoD Directive 4100.35 which, for the first time, required that integrated logistics support (ILS) development be initiated concurrently with performance requirements. It established logistics support as a design consideration. This was extremely important because a system designed without consideration of logistics support will either be inoperative once it is deployed or not recover from its first failure in the field.

In 1968 the DoD ILS planning guide was published, and this was followed in 1972 by the DoD ILS implementation guide, which was tailored to each service. Finally, on November 17, 1983, DoD Directive 5000.39 "Acquisition and Management of Integrated Logistics Support for Systems and Equipment" was issued. This made the consideration of logistics a responsibility of the program manager and required that an "adequately funded" ILS program be established. It also required that logistics be subject to separate treatment in source selection. Although this directive was canceled with the publication of DoD Directive 5000.1, "Defense Acquisition," in February 1991, these logistics requirements are carried forward in DoD Directive 5000.2, "Defense Acquisition Management Policies and Procedures," dated February 23, 1991. So the logistician within the project office has a basis for providing improved logistics throughout the life cycle of the system.

To assist the logistician in carrying out these responsibilities, the program manager may have an assistant program manager for logistics who will be responsible for ensuring that all of the logistics elements are considered throughout the acquisition of the system, depending on the size and complexity of the program. Further, if the program is sufficiently large, it may warrant deputy program managers for each of the logistics elements who will report to the assistant program manager for logistics.

In order to visualize this, it is appropriate to identify the "logistics elements," as they have been defined within the DoD.

LOGISTICS ELEMENTS

DoD Directive 5000.2, part 7, sec. A, sets forth the definition of the 10 elements that make up ILS. These elements must be addressed for both hardware and software under both war and peace conditions. They are the resources required to field and maintain a weapon system in an operationally capable condition. The ILS elements are as follows:

1. *Maintenance Planning.* The process conducted to evolve and establish maintenance concepts and requirements for the lifetime of the system.
2. *Manpower and Personnel.* The identification and acquisition of military and civilian personnel with the skills and grades required to operate and support the system over its lifetime at peacetime and wartime rates.
3. *Supply Support.* All management actions, procedures, and techniques used to determine what is needed to acquire, catalog, receive, store, transfer, issue, and dispose of secondary items. This includes provisioning for both initial support and replenishment supply support. It includes the acquisition of logistics support for support and test equipment.
4. *Support Equipment.* All equipment (mobile or fixed) required to support the operation and maintenance of the system. This includes associated multiuse end items, ground handling and maintenance equipment, tools, metrology and calibration equipment, test equipment, and automatic test equipment.
5. *Technical Data.* Scientific or technical information recorded in any form or medium (such as manuals and drawings). Computer programs and related software are not technical data; documentation of computer programs and related software are. Also excluded are financial data or other information related to contract administration.
6. *Training and Training Support.* The processes, procedures, techniques, training devices, and equipment used to train civilian and active duty and reserve military personnel to operate and support the system. This includes individual and crew training (both initial and continuation); initial, formal, and on-the-job training; and logistics support planning for training equipment and training device acquisitions and installations.
7. *Computer Resources Support.* The facilities, hardware, system software, software development and support tools, documentation, and people needed to operate embedded computer systems.
8. *Facilities.* The permanent, semipermanent, or temporary real property assets required to support the system, including conducting studies to define facilities or facility improvements, locations, space needs, utilities, environmental requirements, real estate requirements, and equipment.
9. *Packaging, Handling, Storage, and Transportation.* The resources, processes, procedures, design considerations, and methods to ensure that all system, equipment, and support items are preserved, packaged, handled, and transported properly, including environmental considerations, equipment preservation requirements for short- and long-term storage, and transportability.
10. *Design Interface.* The relationship of logistics-related design parameters to readiness and support resource requirements. These logistics-related design parameters are expressed in operational terms rather than as inherent values and specifically relate to system readiness objectives and support costs of the system.

With these definitions in mind we can take a closer look at the logistics elements to expand their definition. Then we can proceed to show how they interrelate in the logistics support analysis.

Maintenance Planning

In the DoD there is a general recognition of three levels of maintenance: organizational, intermediate, and depot. Basically they may be differentiated by the skills and equipment available at each of the levels.

Organizational maintenance is performed at the point at which the equipment is in use, the infantry company in the Army, the flight line in the Air Force, and the flight deck in the Navy. The tools available at this level are limited and the skills of the personnel at this level are dedicated to the operation of the equipment; so the maintenance generally takes the form of visual inspections, cleaning of equipment, external adjustments, some removal and replacement of components, and identification of those components that have failed and will need to be repaired at a higher maintenance level.

Intermediate maintenance is performed further away from the operational unit and involves some retrograding of the equipment to a mobile, semimobile, or fixed repair facility. At this level there is a greater range of test equipment and the maintenance-related skill levels of the personnel are greater than at the organizational level. For example, it is common at this level to isolate, remove, and replace circuit cards as well as performing microminiature repair. It is customary for one intermediate repair facility to support more than one operating unit.

Depot maintenance is capable of performing all work not within the capabilities of the intermediate level. The depot generally performs maintenance for a large number of operating units and is a fixed facility. Depot-level repair may be nonorganic, that is, it may be performed at the contractor's plant. Whether it is organic or nonorganic, the capability of depot-level maintenance includes rebuilding, overhaul, and calibration of equipment.

In any event, the levels of maintenance will undergo continued review as a result of the experience gained in the Gulf War. The technological advances in today's weaponry have some people thinking of as many as five levels of maintenance for these more complex items of equipment.

Organic versus Nonorganic Maintenance. A great deal of discussion has centered around the issue of whether repairs will be made by service personnel in-house (organic) as opposed to using contractor personnel and facilities out-of-house (nonorganic).

The complexity of today's high-technology equipment may make it prohibitive in many cases to train sufficient numbers of military personnel to maintain this equipment adequately, especially in times of a declining defense budget. Thus the alternative of contractor maintenance is an important consideration.

DoD Directive 4151.1, "Use of Contractor and DoD Resources for Maintenance of Material" (July 15, 1982), states:

> Organic depot maintenance capabilities and physical capacities established or retained within the DoD Components for support of DoD materiel shall be kept to the minimum required to ensure a ready, controlled source of technical competence and resources necessary to meet military contingencies.

Contractor maintenance may take many forms which are not mutually exclusive, but may instead serve in combination: (1) expansion of the contractor performing depot-level maintenance as the military services have experienced over the years; and (2) increased reliance on contractor maintenance personnel in the field, at both the organizational and the intermediate levels.

Nevertheless, the point to be made is that the decision as to who will perform the maintenance does not change the levels of maintenance.

Reliability and Maintainability. Over the years there has been a strong drive toward increasing the reliability of weapon systems. This has caused some unanticipated repercussions. As reliability increases, it becomes more difficult to justify budgets for maintenance and spare parts. It is not unusual for the logistician testifying before Congress on the defense budget to have to defend the dollars included in the budget for spares. Usually the argument raised by the congressional committees or by individual members will center on why a certain level of spares is needed when funds for the system have been justified based on increased reliability. The answer lies in the factors involved in maintenance planning such as reliability-centered maintenance and the failure modes, effects, and criticality analysis (FMECA).

FMECA is an analysis method that identifies possible system failures, the causes of the failures, and the effects of the failures on the system. It will also provide information as to the criticality of the failure (in terms of the effect both on the capability of the system to fulfill its mission and on safety), the frequency with which the failure may be expected to occur, and special maintenance considerations that may be required.

The criticality of the potential failure is classified into four categories: (1) catastrophic; (2) critical; (3) marginal; and (4) minor. Catastrophic means that death or the loss of a weapon system may occur. Critical means that severe injury could occur, or a major property loss, or mission failure. Marginal means that a minor injury could occur, or minor property loss, or the mission will be degraded. Minor means that unscheduled maintenance will be required. As you can see, the criticality of the failure will dictate what happens to the design of the system. FMECA is an important part of the process of designing a weapon system and should be done in conjunction with the reliability centered maintenance (RCM) analysis so that the FMECA results can be used in the RCM analysis. The FMECA results should identify the potential failures that can be corrected by design changes and those that will require some other form of corrective action. RCM analysis will provide an opportunity for establishing preventive maintenance actions in lieu of corrective (and more expensive) maintenance actions.

Manpower and Personnel

In every weapon system there is a requirement for certain skills and a certain amount of manpower to be brought together to operate the system effectively over its lifetime at peacetime and wartime rates. This element identifies those skills and their related grades.

It is not easy to fulfill this requirement because, at the time the information is called for, available data are limited. Manpower constraints are to be identified prior to program initiation and an initial estimate of manpower requirements is to be developed during the concept exploration phase. Therefore the initial efforts are based on comparing the proposed system to similar systems with the same general mission and extracting those data that seem likely to apply to the new system. There are several models available for the purpose of bringing this information into the logistics support analysis (LSA) process. For example, the Army's MANPRINT is structured to bring pertinent information regarding human factors engineering, manpower, personnel, training, safety, and health hazards into the LSA process. In that process, the MANPRINT concept is applied to each ILS element. In moving into concept exploration, the data become more refined as the alternative concepts are identified and narrowed to those from which the final system will be selected. Then, as with all of the ILS elements, the analysis continues through each stage of system development.

Supply Support

Supply support begins with the determination of requirements that will determine the range and depth of items that will be carried in the supply system. In the case of new weapon systems in development, this is done through provisioning.

There are two important factors to consider in provisioning: (1) the provisioning itself, which is the process of determining the range and depth of repair parts which are required to support an end item for an *initial* period; and (2) the provisioning technical documentation (PTD), which is the information supplied by the manufacturer of the item and used by the military services for the identification, selection, and determination of initial requirements and cataloging of support items to be procured through the provisioning process. The PTD can be in many forms: provisioning parts list, common and bulk items lists, postconference provisioning list, system configuration provisioning parts, tool and test equipment list, interim support items list, supplementary provisioning technical documentation, repairable items list, and long lead items list.

The long lead items list is very important. As you can see in Fig. 1.1, the long lead items must be identified and pursued early on if they are going to be available when the system reaches its planned material support date. The material support date is the date by which the military service is capable of supporting the system with its organic supply system.

Once the items that will be carried in the supply systems have been identified by the supply personnel involved with the program, the descriptive information is forwarded to the Defense Logistics Services Center in Battle Creek, Mich., for screening to see whether the items are already in the supply system. If not, then the items are assigned a national stock number (NSN), and upon receipt of the NSN, the service begins the procurement action, including directing where the items will be delivered. The items are then stored, awaiting requisition actions by the users. At that time the items will be shipped to the user.

Spares Acquisition Integrated with Production (SAIP). The procurement of selected spares may be combined with the procurement of identical items being procured for deployment when it is determined to be cost-effective. This not only can save dollars, but it will save time in getting spares into the system. Such spares may be bought from the contractor or from a subcontractor. SAIP requirements will be specified in the ILS plan.

Demand Development Period. Once the system is fielded, actual demand can then be recorded at the inventory control points, and replenishment planning and inventory level development can be refined. The system will then become routinely supported.

The replenishment cycle continues as long as the items are still in demand. Once there is no further demand (need) for an item, it moves to disposal.

Support Equipment

Support equipment consists of the equipment needed to support the operation and maintenance of the weapons system. It includes ground handling equipment, tools, test equipment (and the logistics support required for the test equipment), and metrology and calibration equipment.

The determination of what support equipment may be needed is accomplished in the concept exploration phase of system development. Again, equipment in use in the support of similar systems is evaluated to see whether it could perform the function for the new system. It is important to standardize equipment wherever possible in order to keep down the cost of the new system and to avoid unduly overburdening the supply system. If suitable equipment is not found within the system, then a determination must be made as to what new support equipment will be acquired.

At the time of full-scale development, the trade-off studies that were begun dur-

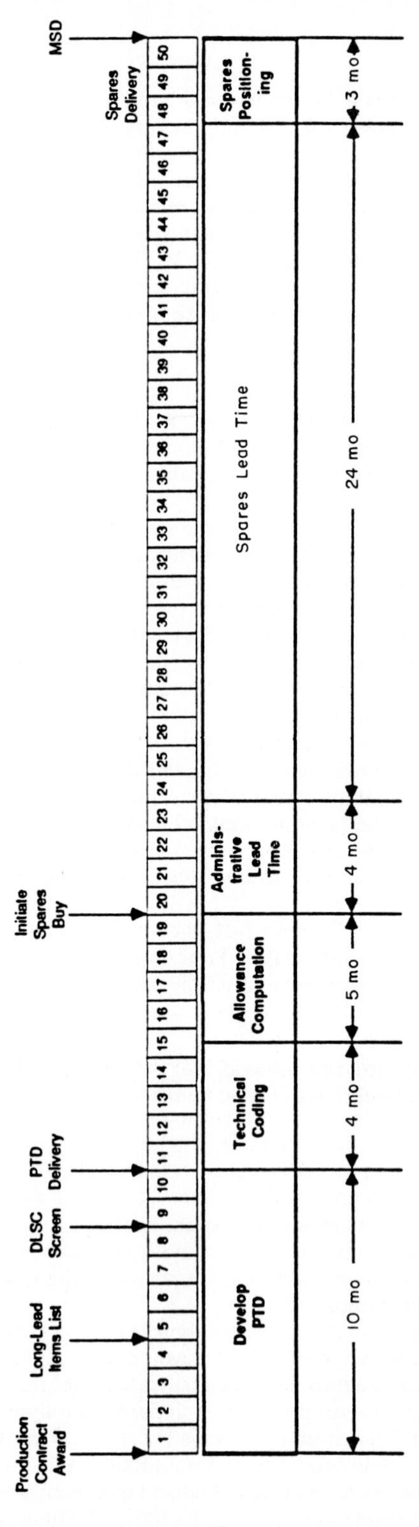

FIGURE 1.1 Provisioning cycle.

ing concept exploration should have identified both standard and special support equipment that will be needed for the new system. These analyses and their results will become input for the LSA.

The selection of automated test equipment should be explored carefully. The complex technology involved makes this a high-cost item, which should not be considered lightly. To accomplish this, within each service there has been standardization of automated test equipment or families of such equipment. This applies also to the software used to automate test procedures.

Technical Data

Technical data involve all recorded information developed by the contractor in order to build and test the system. This includes all data of technical and scientific nature relating to the weapon system, such as drawings, inspection and test procedures, calibration instructions, operating manuals, maintenance manuals, and specifications. Also included are any computer programs which are used to train or guide personnel in the operation and support of the equipment. Technical data may be found in the design documentation as well as in production instructions developed by the contractor. Technical manuals describe the levels of expertise required to operate and maintain a system, thus providing a basis for analyzing the skills and training requirements needed to support the system.

As discussed earlier, FMECA and RCM analysis provide detailed technical data with respect to corrective and preventive maintenance, respectively. All of this information eventually is documented in technical manuals as well as in the LSA documentation.

Timely delivery of technical data is essential. During the development of the weapon system, preliminary technical manuals must be available during demonstration and validation in order to perform test and evaluation and to begin training. A formal validation and verification plan must be developed prior to operational testing and evaluation.

It is important to remember that, at the time the system is fielded, the users must have in their hands all the information necessary to operate and maintain it. This means that the development of technical and operating manuals must begin during demonstration and validation and become an iterative process throughout the remainder of the system development.

The receipt of technical data upon completion of the program is dependent on the provisions of the contract. The data should, of course, be complete. But they also should have been requested in the form in which they are going to be used. It does no good to receive truckloads of data in paper form which cannot be manipulated by the user. In this respect, the future lies in the requirement for the data to be provided in digitized form in accordance with the standards being developed as part of the computer-aided acquisition and logistics system (CALS).

Training and Training Support

The training and training support element consists of the procedures, processes, techniques, training devices, and equipment used to train personnel in the operation, maintenance, and general support of the equipment. As stated in the definition, this includes individual and crew training (both initial and continuation); new equipment training; initial, formal, and on-the-job training; and logistics support planning for training equipment and training device acquisitions and installations.

Again, detailed descriptions of the current and projected skills and training resources must be developed beginning in the concept exploration phase of the program and refined over the remainder of system development. These data also feed into the LSA process in order to identify equipment and training requirements at the task level and to feed into the procedures to be used in the operational test and evaluation that will occur during full-scale development. Finally, this information will evolve into the procedures that will be used in actual operation.

The equipment and training devices identified in this element will become the object of separate procurements during the development of the system. However, these procurements must be timed so as to ensure the availability of the equipment and training devices prior to the system being fielded. The equipment and training devices must be available in sufficient time to allow appropriate training of the users before receipt of the system.

Computer Resources Support

This element includes all of the facilities, hardware, system software, software development and support tools, documentation, and people needed to operate and support embedded computer systems.

The technology available to weapon system designers today makes them capable of incorporating the most technologically sophisticated embedded computer systems in the weapon system. This growth in the numbers of embedded systems coupled with the unprecedented increase in their complexity, as well as the high cost of these systems and their related software call for configuration control and configuration status accounting of the software.

The ILS manager (and the computer resources support element manager, if one is assigned) need to ensure that diagnostic programs are fully tested and any deficiencies corrected prior to full-scale development.

Facilities

Facilities include all the permanent, semipermanent, and temporary real property assets required to support the system. This also includes the studies that are conducted to define facilities or improvements to facilities, locations, space needs, utilities, environmental requirements, real estate requirements, and equipment.

The ILS manager must ensure that the required facilities are available to allow the testing of the system, the operation of the system, and support of the system as each of these activities require. The budget cycle for military construction appropriations is such that a minimum of five years must be allowed for planning purposes. This means that the facility requirements must be determined as early as possible. If the construction is to be in a foreign country, even more time must be allowed to accommodate the additional debates and approvals that will be needed before construction is authorized.

As with the other logistics elements, it is important to analyze space and equipment needs while in the concept exploration phase so that gross facility requirements can be determined. This is particularly true of those facilities that will be needed for testing purposes during the development of the system, thus reducing the time available for planning, designing, and scheduling the construction of these facilities.

In preparing facility-related information for the analysis and eventual introduction into the LSA, all available information must be included. This includes data on existing facilities, projected space availability, facility funding constraints, and planned operational and maintenance requirements.

Packaging, Handling, Storage, and Transportation

This element includes all resources, processes, procedures, and design considerations involved with the packaging, handling, storage, and transportation (PHS&T) of the item. It includes the methods used to ensure that all system, equipment, and support items are preserved, packaged, handled, and transported properly. It also includes environmental considerations, equipment preservation requirements for short- and long-term storage, and transportability.

In planning for PHS&T, there are several factors to be considered. Among them are existing packaging standards, containers, transportability constraints, and the capabilities of existing material handling equipment and storage facilities. During concept exploration it is important to establish design constraints to be certain that the design will be compatible with the planned support system. During the demonstration and validation phase, the design of the item is reviewed for compatibility with existing packaging capabilities and to determine unique protection and handling requirements. Also, pertinent packaging and handling standards should be specified. All of this information, including shelf-life data, dimensions, special handling, and storage, are entered into the LSA process.

Design Interface

As defined, design interface is the relationship of logistics-related design parameters to readiness and support resource requirements. This is an interactive relationship and iterates throughout the development cycle. It must be emphasized that the greatest opportunities to affect design will occur in the early phases.

INTEGRATED LOGISTICS SUPPORT

Before discussing ILS, we need to understand the phases of weapon system development. Figure 1.2 shows the various phases in system development and the purposes of each phase. It is most important to note that ILS begins at program initiation with the accumulation of data for the various analyses being conducted in the concept exploration phase. The earlier the analyses are conducted, the greater the opportunity to influence design.

DoD Directive 5000.2 states that the policies and procedures contained therein establish the basis for ensuring that:

(1) Support considerations are effectively integrated into the system design; and

(2) Required support structure elements are acquired concurrently with the system so that the system will be both supportable and supported when fielded.

This is important because it speaks directly to logistics influencing design. It is critical.

The volumes of data are translated and processed into the program through the LSA required by MIL-STD-1388-1A. Figure 1.3 is a graphic portrayal of the ILS process and the part played by the LSA.

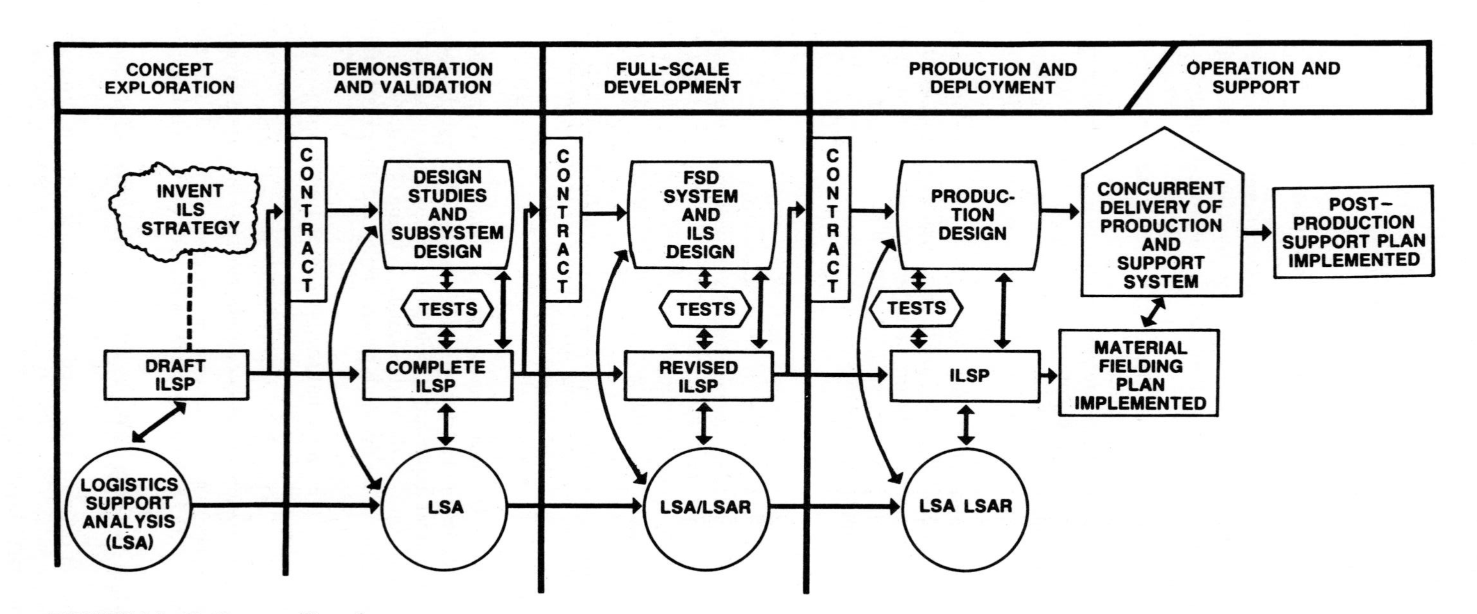

FIGURE 1.2 DoD system life cycle.

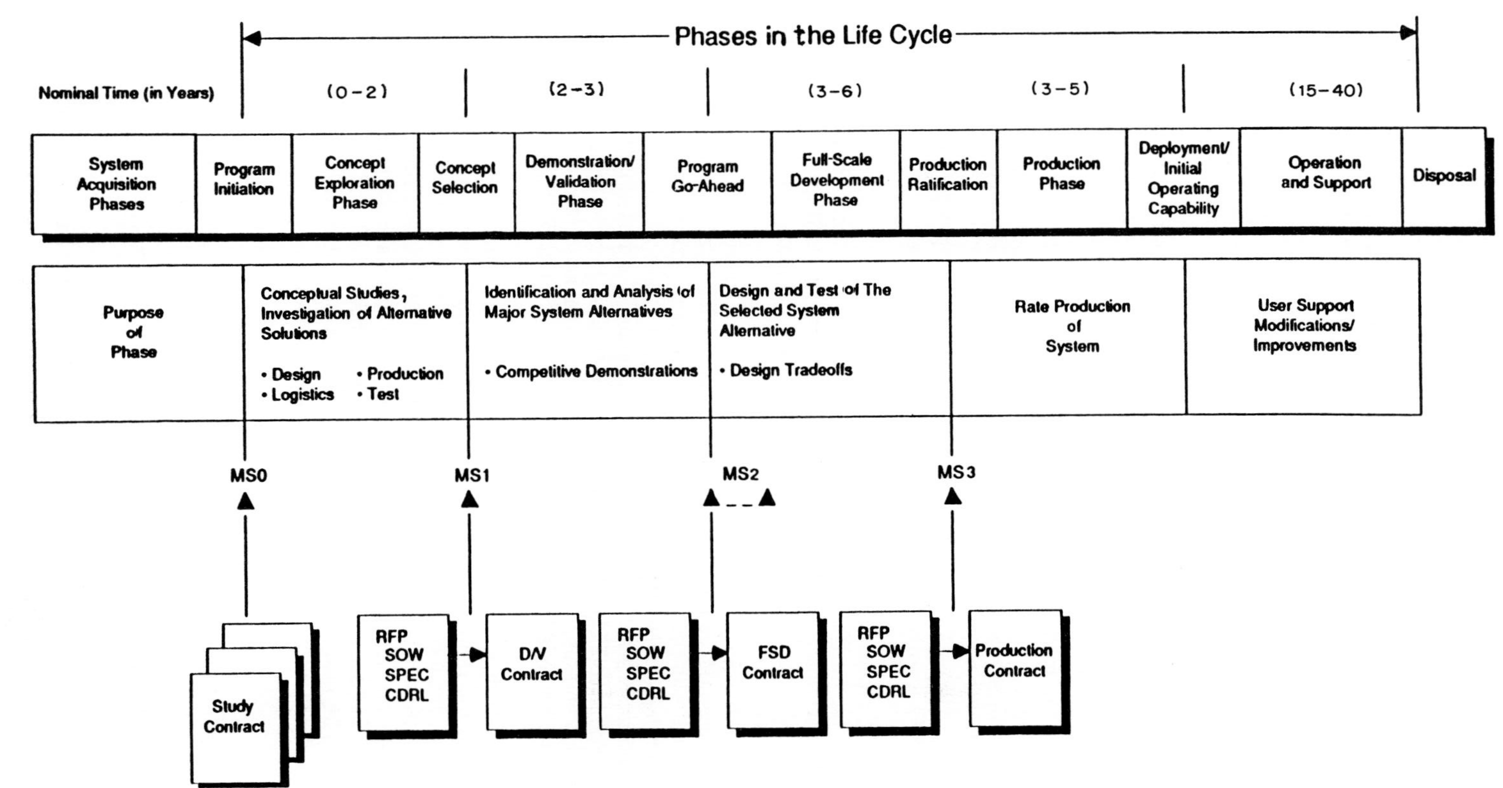

FIGURE 1.3 Integrated logistics support process. (*Source: Integrated Logistics Support Guide, Defense Systems Management College, May 1986.*)

Integrated Logistics Support Planning

Discussion of ILS should begin with mention of the DoD policy as stated in DoD Directive 5000.2, "Defense Acquisition Management Policies and Procedures":

> **a.** An effective integrated logistics support effort shall be established within each program office. Integrated logistics support shall be managed as a disciplined, unified iterative approach to the management and technical activities necessary to:
> **(1)** Developing support requirements that are related consistently to readiness, to design, and to each other,
> **(2)** Effectively integrating support considerations into the system and equipment design,
> **(3)** Identifying the most cost-effective approach to supporting the system when it is fielded, and
> **(4)** Ensuring that the required support structure elements are developed and acquired.
> **b.** Post-production support planning, a subset of the overall integrated logistics support effort, shall be accomplished to ensure continued attainment of readiness objectives with economical logistics support after cessation of production.
> **c.** Integrated logistics support efforts shall encompass the ten elements identified in attachment 1.

(These are the elements defined earlier in this chapter.)

The planning for ILS begins as soon as the need for a system is known. The centerpiece of this effort is the integrated logistics support plan (ILSP), which is the documentation of the management approach, decisions, and plans associated with the planning effort.

The plan will provide the direction for coordinating logistics planning efforts to ensure that each of the integrated logistics elements described earlier is addressed and integrated with the other elements throughout the program. It will also include planning for deployment and postproduction support.

As mentioned earlier, the database for this planning is in the tailored LSA, in accordance with MIL-STD-1388. The vehicle for recording, processing, and reporting supportability and support data is the logistics support analysis record (LSAR). The results of the analysis are documented in the logistics support analysis document (LSAD).

The contract may require the contractor to develop an integrated support plan (ISP). This is the contractor's view of how the weapon system may be maintained and operated, and it contributes to the development, by the service, of the ILSP.

Whether an overall ISP is required or not, the contract for phase II, engineering and manufacturing development, will require the contractor to include postproduction support considerations in the early trade-off studies prescribed by MIL-STD-1388.

This plan for postproduction support should be presented at ILS reviews. It also must be updated throughout system development.

LOGISTICS SUPPORT ANALYSIS

The logistics support analysis (LSA) is a single logistics database. The purpose of the analysis is to:

1. Cause logistics support considerations to influence the design of the system
2. Identify support problems and cost drivers early

3. Develop logistics support resource requirements for the life span of the system

A properly performed LSA will:

1. Quantify supportability requirements during conceptual design
2. Enable consideration of supportability and cost compared to performance attributes during trade-off analyses
3. Provide a basis for evaluating alternatives
4. Provide a structure to help integrate the voluminous data that will be generated
5. Document tradeoffs and alternatives

The guidance for performing LSA is contained in two publications. The first is MIL-STD-1388-1A, which describes the tasks that may be considered for analysis, and provides guidance as to how to tailor the analysis for the particular system. The second is MIL-STD-1388-2B, which discusses the LSAR and describes (in relational database format) the data elements, data record formats, report formats, and guidance for tailoring the records.

Figure 1.4 details where the steps of the LSA occur in the system development, Fig. 1.5 shows some of the data flow, and Fig. 1.6 illustrates the relationship within the database.

Tailoring of the LSA will be controlled by several factors, such as the amount of freedom to change the design, the availability of funds for the LSA, the estimated return on investment, schedule constraints, the availability of the data, and their relevance. The bottom line is that LSA can be expensive if it is not limited (tailored) to what is really necessary. But it is important that LSA be accomplished at its most cost-effective level. It is worth emphasizing that programs that are funded "up front" for LSA, and for ILS in general, will be more successful in supporting the system in the field. Figure 1.7 shows that, even though only a small percentage of life-cycle cost has been expended, 70 percent of life-cycle cost is already committed.

LIFE-CYCLE COST

There is a significant impact on life-cycle cost by the cost of logistics support. Figure 1.8 shows all the costs associated with a weapon system, while Fig. 1.9 shows that 60 percent of life-cycle costs are attributable to logistics. This is particularly true in the case of recent weapon system decisions, which have extended the life of systems and platforms beyond original expectations. Aircraft carriers, the M-60 tank, the B-52 bomber have all been recent examples. The operation and maintenance costs of these systems have grown dramatically because of these decisions. More of the same can be expected in the future as the defense budget declines.

The development of a weapon system may take five to six years, but when it remains in the active inventory for another 30 or 40 years, the logistics support costs are heavy.

Computer-Aided Acquisition and Logistics (CALS)

No discussion of logistics can end without a discussion of CALS because of its dramatic impact on the interchange of data. Unfortunately it is also shrouded with doubts about its future. To understand this, we need to understand the background

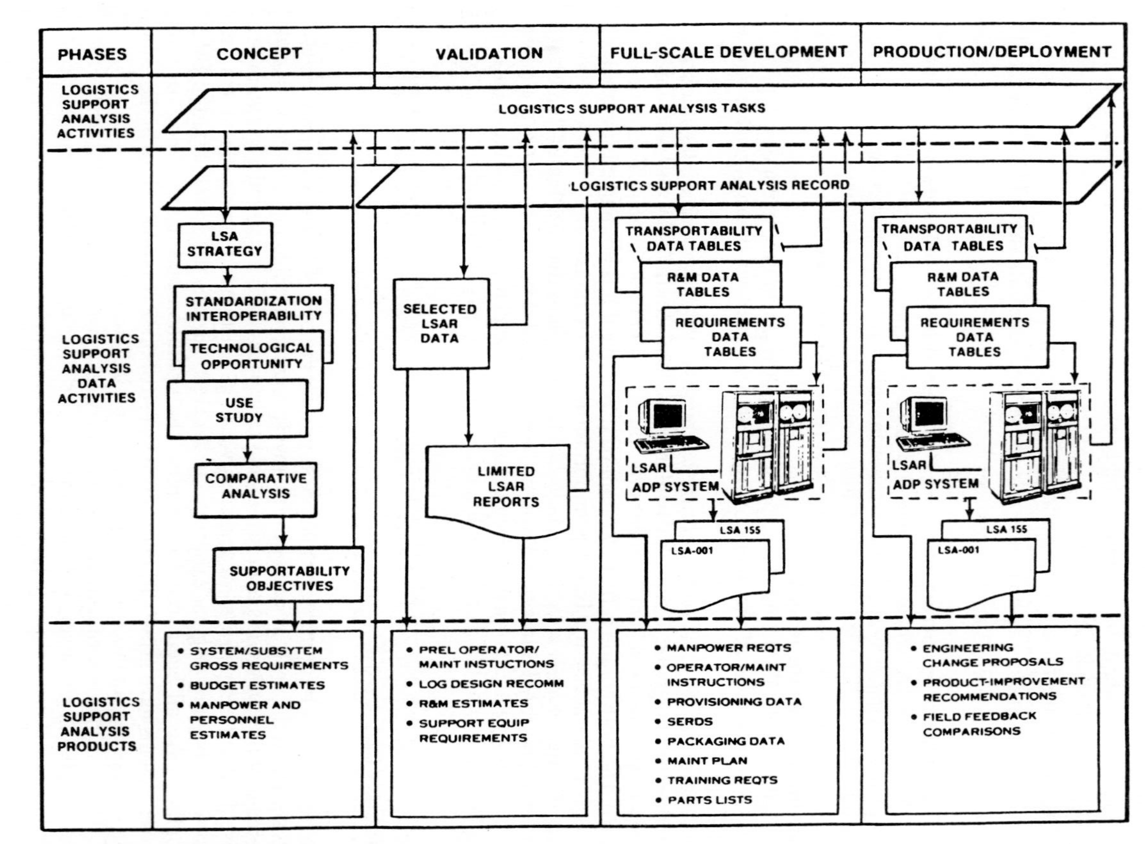

FIGURE 1.4 LSA data documentation process.

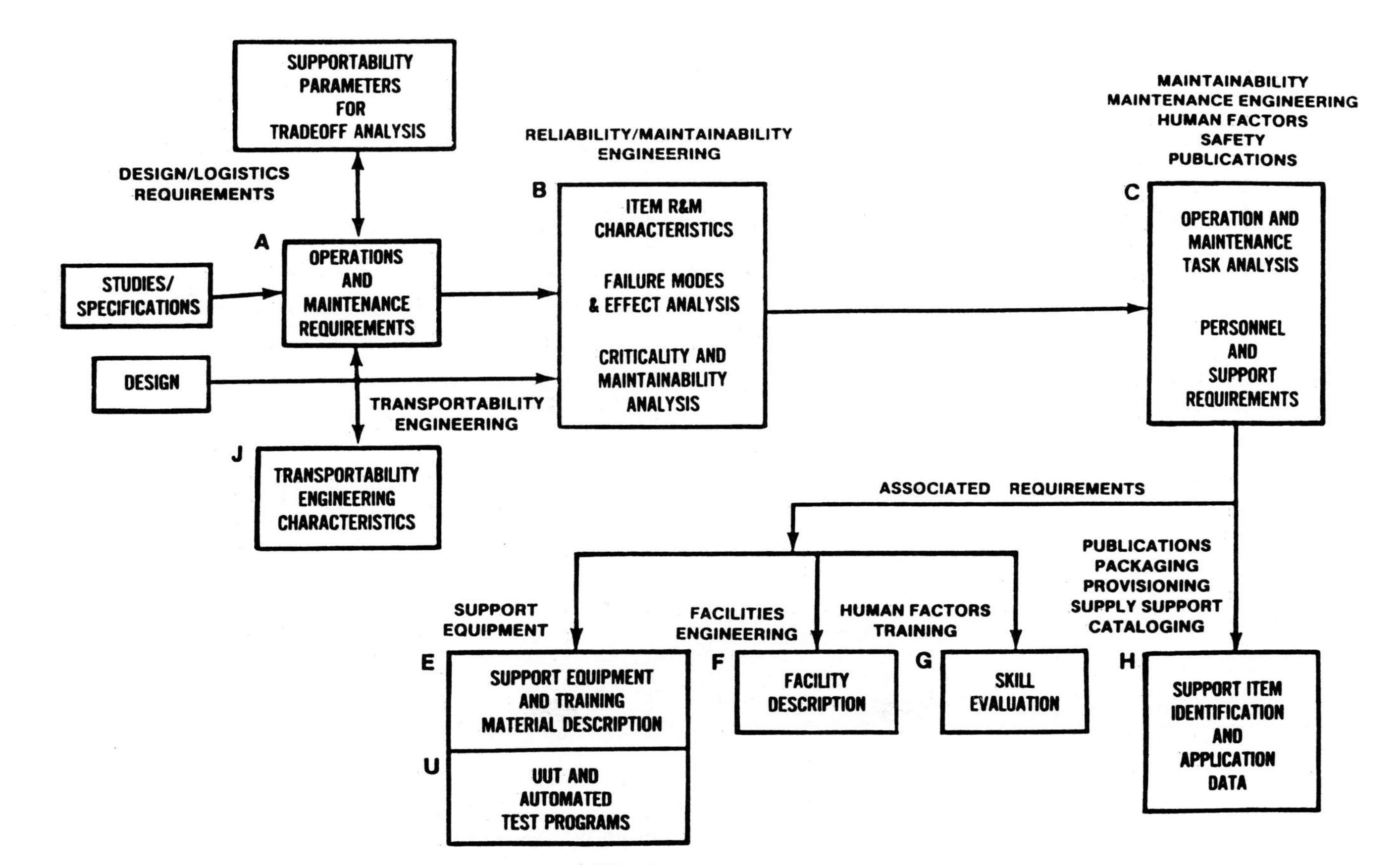

FIGURE 1.5 LSAR data flow and systems engineering interface.

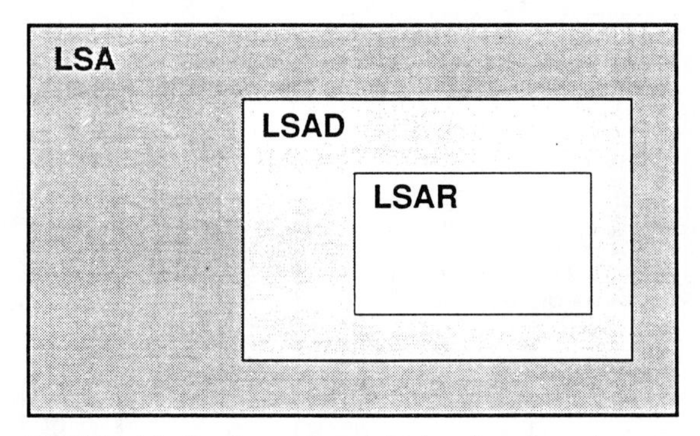

FIGURE 1.6 Database oriented. LSA = all analyses that produce logistics information; LSAD = documentation of the analyses; LSAR = portion of LSAD applicable to LSA data records.

and origin of CALS. For decades, the DoD and industry have sought ways to generate technical data in a medium that would be usable to both parties. Early efforts to provide technical data gave rise to a myriad of drawings with each weapon system and resulted in the generation of volumes of paper and aperture cards. There simply did not seem to be an effective way to prepare, handle, transfer, and store the data. In addition, the data that were received were difficult to recall because of a lack of sufficient and efficient storage media.

It began to look like this problem would be insurmountable as weapon systems became even more complex as the years went by. This would significantly reduce the government's chances to establish alternative production sources using technical

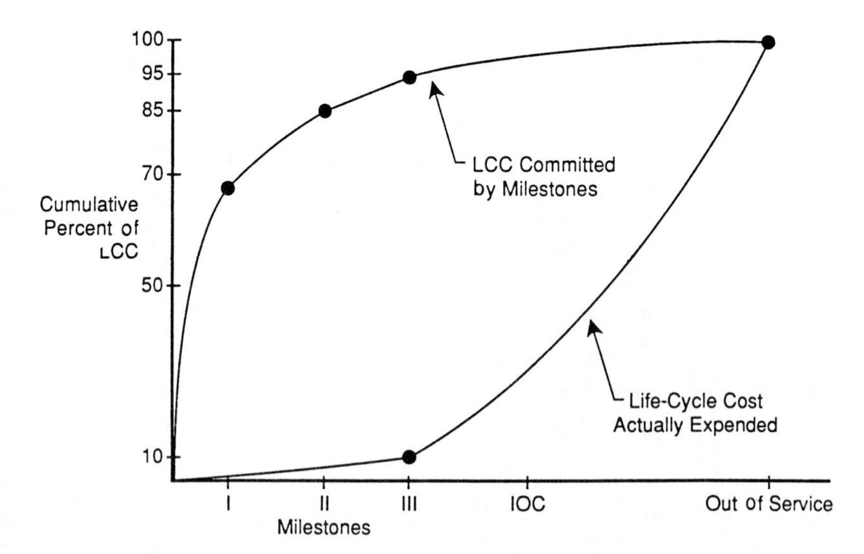

FIGURE 1.7 LSA influences life-cycle costs: 70 percent of life-cycle cost is committed in earliest phase of acquisition. (*Source: Integrated Logistics Support Guide, Defense Systems Management College, May 1986.*)

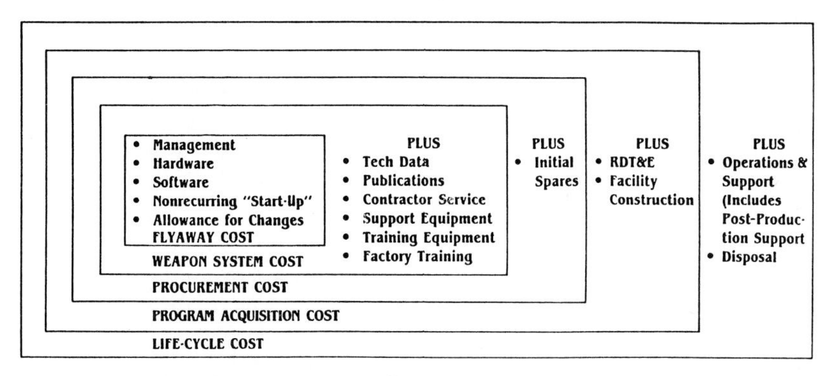

FIGURE 1.8 Life-cycle cost composition. (*Source: Integrated Logistics Support Guide, Defense Systems Management College, May 1986.*)

data supplied by the original producers of the weapon system. It also would preclude the development of reprocurement data that could be used to expand the contractor base for components of the weapon system. Such entities as the Government Industry Data Exchange Program (GIDEP) of the early 1960s were doomed to failure unless a better way of exchanging data was found.

When Congress passed the Competition in Contracting Act in 1984 it had heard a great deal of testimony about the problems faced by the government in gaining sufficient technical data to expand competition. The arguments involved the high cost of the data, the reluctance of industry to provide the data, the lack of complete data when they were bought, and the inability of the government, in some cases, to verify the data before accepting them because of their sheer volume.

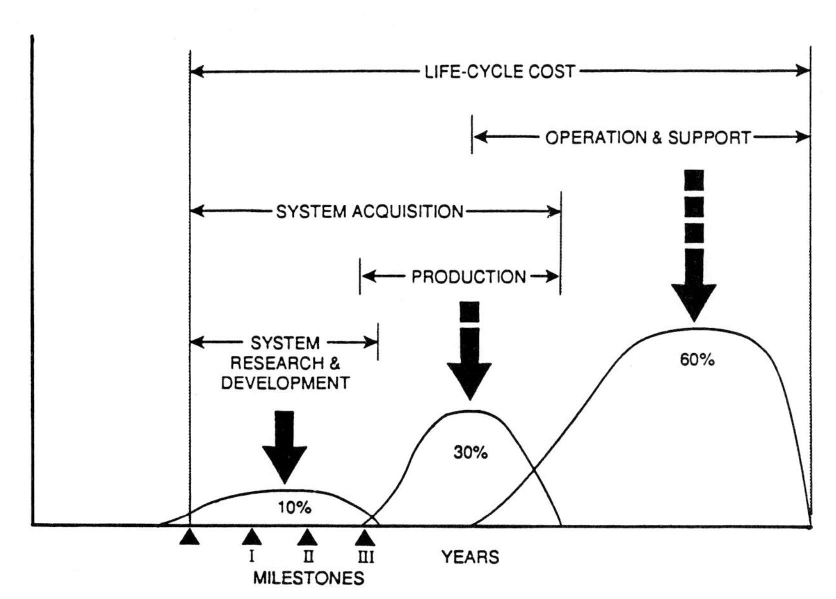

FIGURE 1.9 LSA influences life-cycle cost: up-front commitment to effective LSA controls is majority of life-cycle cost; cost of operation and support is 60 percent of life-cycle cost.

The breakthrough came with the advent of the compact disk and the capability to digitize technical data, including drawings.

In 1985 the CALS objectives were spelled out. They included reducing lead time by improving the timeliness of the data, thus shortening the acquisition cycle; reducing acquisition cost by eliminating labor-intensive data and moving to paperless information, and by sharing data among all those involved in the design and production of the weapon system; and improving quality by reducing errors, making data consistent, and building quality in the process.

CALS is an attempt to build a unified interface between industry and government for the exchange of data to (1) lower the cost of data management; (2) provide for more supportable weapon systems; and (3) provide more consistent and accessible data. This was to be accomplished initially by the development of standards which would bring about commonality in data. Then provision would be made for the automated interchange of technical data, to include digital delivery of engineering drawings. Eventually the solution lies in having a single database to which all parties to a contract provide input, extract the resultant data and utilize them for specified functions, and return them to the database for interactive use by others.

Among other things, the CALS database would be the recipient of the LSAR concerning each weapon system.

The primary concerns with respect to CALS are as follows:

1. Both industry and the military services should be educated in CALS.
2. Business elements are lagging in implementation plans.
3. Document development must be accelerated (DIDs, MILSPECS, etc.).
4. Concurrent engineering must be integrated under the CALS strategy.
5. The proprietary rights of the contractor must be protected.
6. Security, privacy, and export should be controlled.
7. Industry should be given incentives to invest in CALS.
8. How can the data be accepted and verified upon delivery?
9. Where does legal liability rest?
10. What should be done when other nations are involved (coproduction)?
11. Changes are required to the Defense Supplement to the Federal Acquisition Regulations.
12. The ADP acquisition process should be used.

Is the DoD serious about a program so far-reaching with respect to change as CALS? It appears so. In 1987 the Office of the Assistant Secretary of Defense (Production and Logistics) in a report to the House Appropriations Committee said, "Companies without CALS capabilities will not win contracts."

REFERENCES

1. JCS Publ. 1, "Department of Defense Dictionary of Military and Associated Terms," June 1, 1979.
2. "Annual Report to the President and the Congress," Jan. 1991.

CHAPTER 2

MILITARY PROJECT ENGINEERING

Jerome G. Lake

Jerome G. Lake, Ph.D., is a professor of systems engineering at the Defense Systems Management College. He has over 10 years of acquisition management experience in both government and industry. Dr. Lake is past president of the National Council on Systems Engineering and a director and fellow of the American Society for Engineering Management. His advanced degrees in aerospace engineering are from the University of Michigan and the University of Oklahoma.

Many military projects include the development, production, test, and deployment of defense systems, such as aircraft, armored vehicles, command centers, communication satellites, missiles, and ships, or subsystems of these systems. Such projects require large capital investments, span many years, involve multiple disciplines, and are engineering intensive. In this chapter a systems engineering approach for accomplishing such projects is presented in terms of a conceptual framework, process, and life-cycle application. The intent is to assist government and industry project management personnel in defining, performing, managing, and evaluating project engineering efforts.

SYSTEMS ENGINEERING FRAMEWORK

Systems engineering is considered integral to military project management.[1] It provides for the complete, integrated technical effort necessary to develop, produce, test or verify, and deploy or install a system. Systems engineering is a logical process for deriving the desired outputs and determining the inputs needed for each phase of the project; for evaluating and controlling project engineering efforts; and for ensuring that all life-cycle needs for system operation, support, training, and disposal are considered in the design.

Purpose of Systems Engineering

There is a fivefold purpose of systems engineering.

First, systems engineering is to ensure that a project is completed on time and within budget while providing a system that meets all life-cycle requirements. Such a system must have an acceptable total cost of ownership and be able to successfully perform its intended operational tasks, meet all performance objectives, be interoperable with other deployed systems, and require a minimum of support to maintain military readiness and capability over time.[2]

Second,* systems engineering is to guide the definition, redefinition, and documentation of life-cycle requirements and constraints that the system must satisfy to meet users' needs. Two new paradigms are integral to this purpose: (1) Systems engineering is applied iteratively from conception through disposal. This total life-cycle paradigm encompasses the activities involved in concept definition, engineering design, fabrication, integration, test, production, deployment or installation, and operations until final decommissioning of the system and its ultimate disposal. (2) Eight categories of users must be considered, namely, developers, producers, testers, installers or deployers, trainers, supporters, operators, and disposers of the system. Each has unique, yet potentially overlapping, requirements that must be iteratively defined and redefined for each phase of the life cycle.

Third, systems engineering is to develop balanced system product and process designs that efficiently meet the specific requirements of all users during each phase of the life cycle. This purpose recognizes that those involved with development may have specific, unique processes and products that must be provided for development tasks to be completed. Testability, producibility, and supportability requirements may also call for definition and analysis of unique products and processes to accomplish respective test, production, and support roles in the project process. The same is true for the other users. This purpose is consistent with the total quality management principle that requires satisfaction of all customers, where a customer is defined as anyone receiving a desired service or output from another.

Fourth, systems engineering is to provide documents that thoroughly describe selected designs and related functional and physical characteristics and interfaces. These must be controlled to ensure that the resulting system will satisfy specified life-cycle requirements. Such documentation is provided to developers for each successive development phase. The technical data package that evolves from development efforts must enable the production of producible products and processes.

Fifth, systems engineering must ensure that appropriate cost, schedule, performance, and risk data are provided to the government project office to permit timely preparation of decision documentation (for example, integrated program summary, test and evaluation master plan, integrated logistics support plan, and the cost and operational effectiveness analysis)† for the next project phase. To satisfy this purpose, a performance-based measurement system is required to provide exit information for making event-based, rather than calendar-based, decisions.

Definition

The definitions of systems engineering are many and varied. Most authors fall back on the definition given in MIL-STD-499A(USAF)[5]:

*The last four purposes are adapted from Lake[3] and used by permission.

†The description and details of these documents can be found in DoD 5000.2-M.[4]

> Systems engineering is the application of scientific and engineering efforts to (a) transform an operational need into a description of system performance parameters and a system configuration through the use of an iterative process of definition, synthesis, analysis, design, test, and evaluation; (b) integrate related technical parameters and ensure capability of all physical, functional and program interfaces in a manner that optimizes the total system definition and design; (c) integrate reliability, maintainability, safety, survivability, human, and other factors into the total engineering effort to meet cost, schedule, and technical performance objectives.

This definition is generally accepted as satisfactory since its focus is more on systems engineering tasks than a clear statement of meaning or philosophy. It does not, however, capture the essence of current thinking on the simultaneous development of both products and related processes and on the use of cross-discipline design teams. A better definition, one that is used at the Defense Systems Management College, is*:

> Systems engineering is an interdisciplinary approach to evolve and verify an integrated and life-cycle balanced set of system product and process solutions that satisfy stated customer needs.

Framework

The systems engineering framework for accomplishing the fivefold purpose of systems engineering is depicted in Fig. 2.1.[8] For each of the primary functions (development, production, test, deployment, operations, support, training, and disposal) listed on the left side of the cube in Fig. 2.1, input needs for the applicable phase of development are determined. The systems engineering process activities of requirements analysis, functional analysis and allocation, synthesis, and system analysis and con-

*This definition was derived by John Kordik and the author for the May 6, 1992, For Coordination Draft of MIL-STD-499B, *Systems Engineering.* It is based on incorporating the essential philosophies of concurrent engineering and systems engineering as envisioned by pioneers such as Chase[6] and Chestnut.[7]

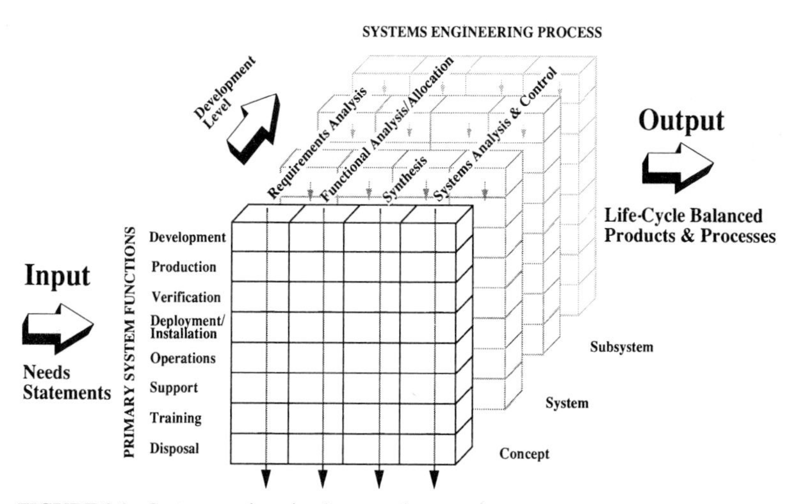

FIGURE 2.1 Systems engineering framework.

trol, shown across the top of the cube, are applied to each need individually and collectively to convert the needs into a life-cycle optimized set of outputs. Output products and processes are project and level of development dependent and used to develop event-based progress information for decision makers and as input information for the next development level of the project.

The systems engineering process activities and the phasing of project systems engineering tasks and accomplishments are discussed in detail in the sections that follow.

SYSTEMS ENGINEERING PROCESS

The top-level activities of the generic systems engineering process applied to military projects are presented in Fig. 2.2.* This process is applied iteratively to provide the progressive definition of the system, subsystems, and configuration items and their verification, as suggested in Fig. 2.1. Process activities evolve the needs of the various users into integrated and balanced end-item products and processes. The activity involved in each application of the process is determined by the level of development and the project.

Process Outputs

Process outputs provide the status of the design effort, identification of the configuration, and performance-based progress data. Three output views—operational, functional, and architectural—are initially generated and become mature throughout development with each application of the systems engineering process.

The *operational* output view fully describes system use in terms of mission, operational environment, and any applicable constraints. The government initially prepares a baseline concept document that is updated to reflect maturing of the system definition. This baseline document provides a point of departure for concept development, rather than a final statement of what the system has to be. It serves as a single, focused concept both for application of the systems engineering process, and for initial cost estimates and trade studies. The general system overview is described: initial operations concept, operational equipment and interfaces, training concepts and equipment, maintenance and supportability concept, facilities, test and evaluation concept, manufacturing concept, system deployment or installation (including assembly and checkout, where applicable) concept, and disposal concept.† It also includes pertinent program decisions and agreements. The baseline concept document is most useful during concept definition and validation. After the functional baseline is established it provides the operational view of the system. During later phases of development and during operations and support, the allocated baselines provide operational perspectives for configuration items.

The *functional* output view provides a description of required subsystems and configuration items, the tasks they perform, and how well they must accomplish each task. All functions are described in sufficient detail to define modes of operation, required inputs and outputs, and interface requirements. Functional flow block diagrams, N^2 diagrams, IDEF, and time-line analyses are effective in identifying those

*Figure 2.2 is adapted from the systems engineering process description developed by John Kordik and the author for MIL-STD-499B.

†From a baseline concept description briefing given by Bordelon.[9]

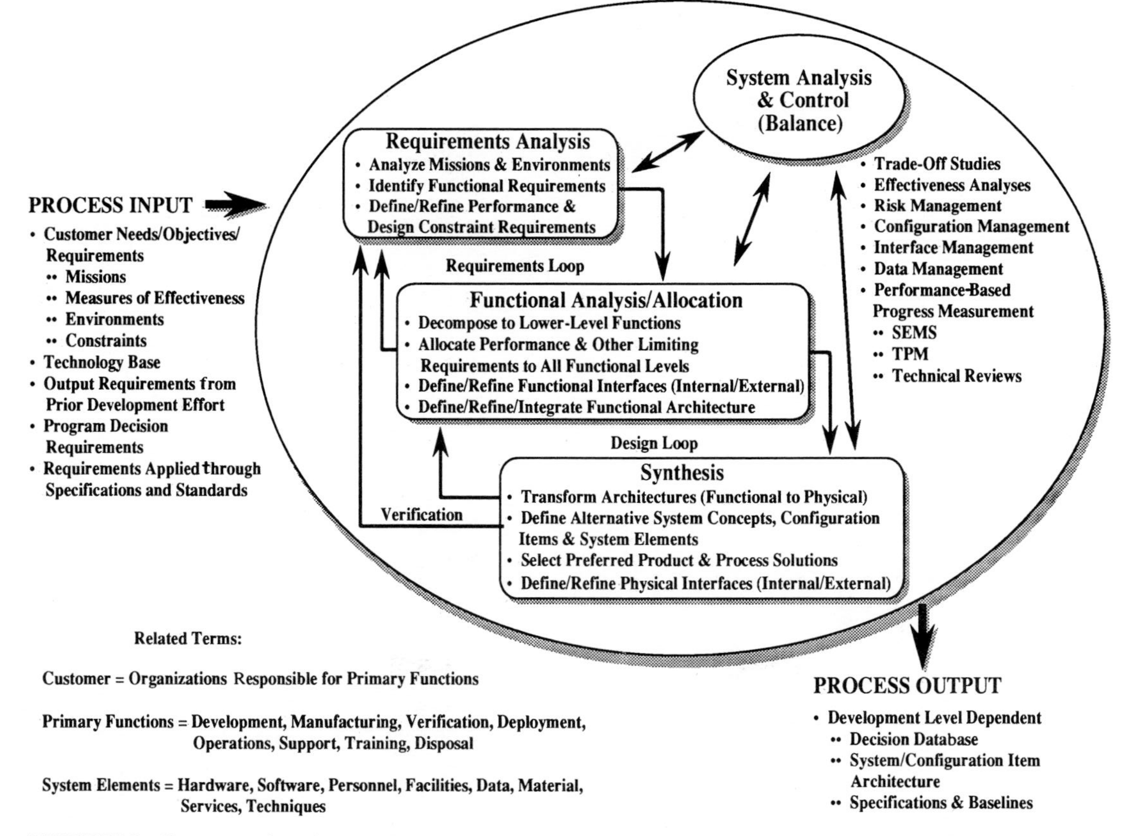
PROCESS INPUT
• Customer Needs/Objectives/ Requirements
•• Missions
•• Measures of Effectiveness
•• Environments
•• Constraints
• Technology Base
• Output Requirements from Prior Development Effort
• Program Decision Requirements
• Requirements Applied through Specifications and Standards
Requirements Analysis
• Analyze Missions & Environments
• Identify Functional Requirements
• Define/Refine Performance & Design Constraint Requirements
Requirements Loop
Functional Analysis/Allocation
• Decompose to Lower-Level Functions
• Allocate Performance & Other Limiting Requirements to All Functional Levels
• Define/Refine Functional Interfaces (Internal/External)
• Define/Refine/Integrate Functional Architecture
Design Loop
Synthesis
• Transform Architectures (Functional to Physical)
• Define Alternative System Concepts, Configuration Items & System Elements
• Select Preferred Product & Process Solutions
• Define/Refine Physical Interfaces (Internal/External)
Verification
System Analysis & Control (Balance)
• Trade-Off Studies
• Effectiveness Analyses
• Risk Management
• Configuration Management
• Interface Management
• Data Management
• Performance-Based Progress Measurement
•• SEMS
•• TPM
•• Technical Reviews
PROCESS OUTPUT
• Development Level Dependent
•• Decision Database
•• System/Configuration Item Architecture
•• Specifications & Baselines

Related Terms:

Customer = Organizations Responsible for Primary Functions

Primary Functions = Development, Manufacturing, Verification, Deployment, Operations, Support, Training, Disposal

System Elements = Hardware, Software, Personnel, Facilities, Data, Material, Services, Techniques

FIGURE 2.2 Systems engineering process.

functions which must be performed to satisfy objectives of the primary tasks. The functions identified by such tools provide functional view information. The methods for defining functional requirements are described in more detail in Blanchard and Fabrycky[10] and in the *Systems Engineering Management Guide.*[11]

The *architectural* output view describes the system and its configuration items in terms of system physical characteristics, interfaces, and processing to be performed. Diagrams and drawings provide product and process descriptions in ever-increasing detail as the project evolves.

Combinations of these three views make up system (functional) and configuration item (allocated and product) baseline descriptions and performance specifications (type A, B, C, D, and E) that are generated prior to appropriate decision points.

Additional outputs include test results, prototypes, system analysis reports, and specialty and management plans. A digital database is recommended to store and control the output views, verifications, decision criteria, trade study assessments, system and configuration item capability assessments, and other documentation generated.

Process Inputs

These include evolving outputs from the previous development phase; new or updated user requirements; new or revised directions or limitations established by the acquisition decision authority; and specifications called out in designated military standards and specifications, industry standards, or company standards. User requirement inputs include mission definition (what the system or configuration item must do); performance (how well the item must perform); constraints (such as cost, schedule, form and fit, interface, interoperability, operator skills, weight, technology, material and parts availability, standardized parts and materials, and government furnished equipment); and environmental conditions (such as geographic and atmospheric conditions under which items must perform, day or night operational expectations, and threat environments).

Process Activities

Figure 2.2 indicates top-level activities for requirements analysis, functional analysis and allocation, and synthesis. Initially these activities are accomplished at the system level, later at subsystem and configuration item levels. The approach provides for top-down definition of partitioned functional requirements for which design solutions will be derived through bottom-up design.

Functional analysis decomposes or partitions system level functions into lower-level identifiable system elements (hardware, software, personnel, facilities, data, material, service, or technique) and identifies all internal and external functional interfaces. Performance requirements are then allocated by flow down from system-level requirements or by derivation to the system elements and interfaces.

The synthesis activity produces design requirements for each system element so that derived functional requirements and all internal and external physical interfaces are satisfied. System performance requirements and functional requirements for each system element are verified by appropriate demonstration, test, simulation, or analysis of the products and processes fabricated or defined by engineering drawings derived according to design requirements.

System analysis and control balances performance requirements, functional requirements, and design requirements derived through the systems engineering process. This activity provides the assessment, progress measurement, and decision mechanism to evaluate design capabilities, determine progress in satisfying technical requirements and technical program objectives, formulate and evaluate alternative courses of action, and evolve the total system to satisfy all user needs.

System Analysis. System analysis includes trade studies, effectiveness analyses, and risk assessments. Generally, trade studies are made to identify and execute tradeoffs among stated user requirements, functional requirements, designs, project schedules, and life-cycle costs. Specific trade studies are also performed to resolve conflicts with user requirements; establish configuration items; support make-or-buy, process, rate, and location decisions; and examine alternative technologies to satisfy functional or design requirements for which moderate- to high-risk technologies are initially identified.

Continuing system and cost-effectiveness analyses are conducted throughout the project to ensure that engineering decisions are made only after considering impacts on system effectiveness and life-cycle costs.

Effectiveness analyses are conducted to balance system and subsystem designs with downstream production, test or verification, deployment or installation, operations, supportability, training, and disposal tasks. In addition, environmental and life-cycle cost analyses are conducted to ensure that system solutions are socially responsible and affordable.

Risk assessments are an integral and ongoing component of system analysis and control. Potential sources of risk are identified and then impacts on cost, schedule, and technical effort determined. For identified risks, alternatives are developed and recommendations provided as to whether the risk can be avoided, controlled, or must be accepted.

Control. Control of the systems engineering process is accomplished through configuration management, data management, and performance-based progress measurement. Configuration management maintains the physical and functional integrity of the system and its configuration items during their life cycle by controlling changes to those characteristics, and recording and reporting any changes processed and their implementation status. It also includes checking that the system or configuration item complies with its configuration identification.[8]

Control of technical data is essential to maintain evolving requirements traceability and drawing changes, and to provide a repository for engineering decisions.

Performance-based progress measurement provides control of project engineering through a systems engineering master schedule, technical performance measurement, and technical reviews. The systems engineering master schedule lists key project tasks which must be completed successfully to achieve an identified event, and the metrics and accomplishment criteria by which event completion will be determined. Events reflected in the systems engineering master schedule include technical reviews and audits, demonstration milestones, and decision points. A series of major technical reviews and audits over the project life cycle provides an in-depth technical assessment of design maturity. Key tasks include critical activities such as establishing baselines and verifications. Verification may be accomplished by test, analysis, simulation, or demonstration.[8]

Between each review the major activities of the systems engineering process are accomplished. The reviews and audits provide the government program office with the opportunity to evaluate progress, evaluate design maturity, determine if the contractor is ready to proceed with the activities toward the next event, identify and resolve issues, and evaluate readiness for the next decision point. Demonstration milestones include demonstration of the system's, subsystem's, or configuration item's readiness for test; showing that the manufacturing process is ready for full production, or that there is a validated need for a new system start or major modification to a system in production. Reviews are the basis for decisions at various oversight levels to decide whether or not to make the resource investment required to continue the project through its next development phase.[8]

The metrics of the systems engineering master schedule include technical performance parameters which are critical to the system achieving its desired operational objectives. Technical performance measurement is a method to assess compliance to requirements and the level of risk in a project. It includes analysis of the differences from planned achievements to date, actual versus planned costs, and required goals or thresholds identified as technical objectives or specifications.[8]

LIFE-CYCLE APPLICATION

Project Phase Descriptions

Figure 2.3 portrays five phases of a military project life cycle, the overall objective for each phase, and related inputs and decision milestones.[12] This portrayal represents the formal phasing that most unprecedented* military projects will follow. It is also instructive and useful for adaptation by precedented systems (not significantly different from a previous system) and by smaller military engineering projects, such as exploring a technological opportunity in a laboratory. Adaptation may include omitting, combining, or abbreviating one or more phases.

The intent of the information and decision flows of Fig. 2.3 is explained in Fig. 2.4.[12] The information flows of Fig. 2.4 apply to each phase of the project. The nature of the information, however, becomes more detailed as the definition and design mature and the investigation expands into the definition of alternative solutions for components of the total system.

*An unprecedented system, as used here, is one that differs significantly from a previous system as a result of changes in mission need, risk, or performance requirements.[13]

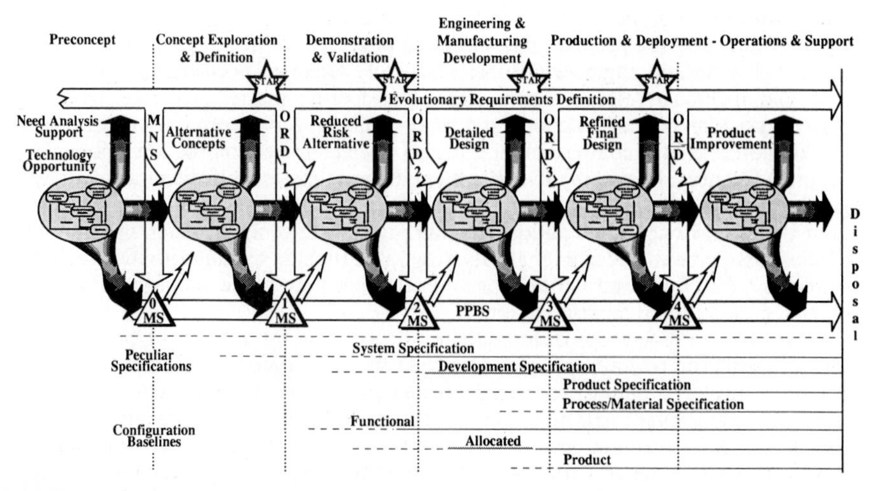

FIGURE 2.3 Systems engineering life cycle.

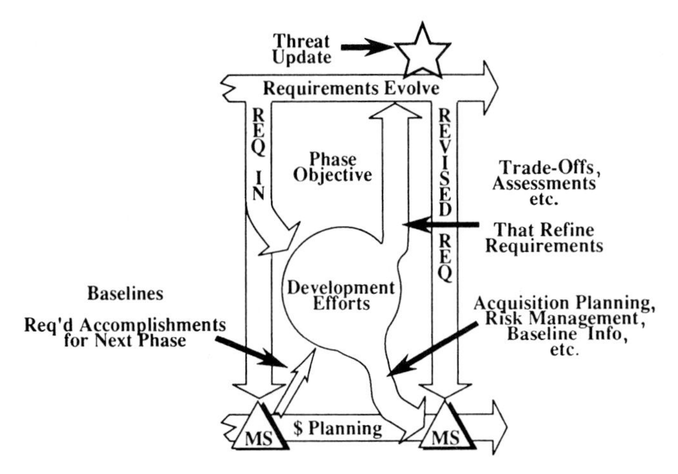

FIGURE 2.4 Input-output flow definitions.

Preconcept Exploration. Prior to the initiation of a military project for a new defense system, a valid operational need must be identified. Determination of need is a government activity generally initiated by a service operational command. The need is derived from a service-conducted mission area analysis which analyzes various approaches to satisfy the need (changes to tactical doctrine, cooperative developments, modifications to existing systems, nondevelopmental items, or a new system). If a new system start is indicated, a mission needs statement (MNS) is drafted, which describes the validated threat or need and an operational concept.

Concept Exploration and Definition. Defense system projects are generally initiated after an identified need has been validated. The go-ahead for the concept exploration and definition phase is provided by an initial set of operational requirements from the MNS, an agreed-to acquisition strategy, required accomplishments to determine satisfactory completion of the phase (exit criteria), and initial funding as established from reprogramming or prior-year program planning and budgeting system (PPBS) funds appropriated and authorized by Congress.

Although a government support activity may accomplish the efforts of this phase, it is typical that a number of parallel, short-term contractor studies are used to investigate alternative concepts and approaches.

The objective of the first phase is to select a preferred system that will satisfy the specified operational need from among alternative system concepts. Selection is made based on cost, schedule, performance, and risk trade studies. Trade studies are made to ensure that production, test, and support, as well as development, deployment, operation, training, and disposal needs are considered and implications evaluated. Effectiveness assessments are also completed to determine which system concept best satisfies identified needs while providing a value added over current systems performing similar missions.

The results of each study contractor's assessments are reviewed by the government during alternative design reviews. These reviews examine contractor understanding of system requirements, trade study results, and recommended system design. The government will generally select one system concept (for example, a manned aircraft, unmanned air vehicle, or satellite system for a reconnaissance need) from several concepts reviewed. This selected concept will be recommended for continued development.

Outputs from this phase include baseline concept definition, system requirements and technical objectives, a program work breakdown structure, technical documentation to support demonstration and validation, specific demonstration and validation phase exit criteria, and other decision support data. Data include manufacturing, support, test or verification, deployment or installation, training, and disposal process assessments for all products of the system. Also included are identification and assessments of applicability for existing and emerging technologies, identification of demonstration or prototype requirements, and a risk management approach. This information provides a basis for developing an acquisition strategy, and deciding whether or not to proceed with the next development phase. Design information for the approved concepts is provided to prospective contractors through the request for proposal (RFP) and contract.

Requirements identified in the MNS and those derived from requirements analysis of primary system functions are converted into initial technical performance objectives. The technical objectives are balanced with the derivation of and changes to operational requirements expressed in an initial operational requirements document (ORD). The threat definition peculiar to the system concept under development is also integrated into the technical objectives and operational requirements. This threat definition is documented in a system threat assessment report (STAR).

Demonstration and Validation. The demonstration and validation (D/V) phase is generally conducted by two or more contractors selected through a competitive process. This phase is normally managed by the government program office formed after the milestone I decision to proceed with development has been made.

The objectives of this phase are to identify and assess the risks associated with the selected system and to demonstrate the critical components and "improve understanding of, and confidence in, technologies and processes needed to attain the system concept,"[14] to derive system-level performance requirements, and to allocate system performance requirements to subsystem and configuration item levels.

Following decomposition of system-level functions, technology, cost, and schedule risks are balanced to determine the most efficacious combination of system subfunctions that integrate into a system. Risk assessment is accomplished by analysis, simulation, bread boarding, and prototyping. When a technical risk is considered unacceptable, but increased performance potentials are deemed to be highly beneficial, a preplanned product improvement (P^3I) strategy can be implemented.* Such high-risk capabilities should be developed in parallel to other less risky options. They can be incorporated as a block change to the system during the production or operations and support phase. This permits early production articles to be achieved with less risk.

It is essential to identify the cost drivers derived through risk assessment so that they may be managed and not adversely affect the project.

Two major design reviews are completed during this phase. The first is a system requirements review. This review allows the government to determine whether the contractor understands the system operational concept and the resulting system requirements so that risk reduction efforts may be undertaken and overall system and

*Preplanned product improvement (P^3I) was directed in 1981 by the secretary of defense to be used as an acquisition strategy on all programs. DoDI 5000.2[15] defines P^3I as a phased approach that incrementally satisfies operational requirements in order to address the cost, risk, or relative time urgency of different elements of the system being developed. With this approach, selected capabilities are deferred so that the system can be fielded (with sufficient capability to meet near-term threats) while the deferred element is developed in a parallel or subsequent effort.

subsystem specifications can be established. Prior to this review the contractor must complete at least one iteration of the systems engineering process to establish the key technologies, determine demonstration or maturation requirements, and arrive at a set of draft system specifications and a preliminary system design.

The second technical review is the system design review. The systems engineering process is applied prior to this review to help accomplish the following tasks: (1) verify that product and process technology proposed are sufficiently complete to ensure a high confidence commitment to system performance requirements, cost, and schedule, and that any remaining technical risks are acceptable or controllable; (2) establish system and system segment specifications; (3) derive draft configuration item functional performance requirements; (4) complete alternative system and system segment designs in sufficient detail to confirm system requirements; and (5) identify P^3I requirements.

Following the system design review, the functional (system) baseline and a configuration control board (to control changes to the system baseline) are normally established.

Outputs from demonstration and validation include system-level requirements and specifications, initial development specifications which incorporate all primary system functions, initial definition of processes needed for end items, contract work breakdown structures for engineering and manufacturing development (EMD), a specification tree, P^3I candidates, and required technical accomplishments and exit criteria for EMD.

For small projects this phase and the concept exploration phase are often not undertaken. Such projects may be performed within a government laboratory or within a contractor's internal research and development effort. This kind of project often exploits a technological opportunity to achieve a new capability. Such projects can often proceed directly into EMD, with the establishment of a verified need and go-ahead approval, unless there are technical risks that need to be demonstrated through an abbreviated D/V phase.

Consistent with concurrent engineering concepts, it is imperative that all of the project's early phases be accomplished using interdisciplinary teams. The purpose of such teams is to have cross-functional team members consider each aspect of the system design as it evolves. This approach ensures that primary function needs, capabilities, and limitations are considered and incorporated into the evolving design.

Engineering and Manufacturing Development. The interdisciplinary planning done in earlier phases has its payoff in the EMD phase. The objective is to complete the design and production configuration for each system element—hardware, software, data, personnel, service, material, and technique. Production configurations must be able to be effectively and efficiently manufactured or implemented. The resulting products must be able to be deployed or installed, to carry out the intended mission, and to be properly supported.

In this phase the selected contractor must ensure that all configuration items which make up the system and its segments are defined, designed, and built or procured; and that configuration items are integrated to make up the specified system. In addition, the contractor will, through appropriate test, analysis, simulation, and demonstration, verify that all system, subsystem, and configuration item specifications are met.

During this phase the government's major role is to provide oversight which ensures progress assessment through its appropriate design reviews and audits. A software system review is accomplished early in this phase to review computer software configuration item (CSCI) software specifications and interface requirement specifications. The adequacy of the specifications and the operations concept are demonstrated to permit establishment of the allocated baseline for preliminary CSCI design.[16]

At least two iterations of the systems engineering process are accomplished during EMD by each contractor developing a configuration item (CI). The first iteration develops the preliminary design; the second develops the detailed design.

After the preliminary design of the products and processes for each CI has been completed, a preliminary design review is accomplished. The purpose of this review is to[16]:

> (1) evaluate the progress, technical accuracy, and risk resolution (on a technical, cost, and schedule basis) of the selected design approach, (2) determine its compatibility with the performance and engineering specialty requirements of the Hardware Configuration Item (HWCI) development specification, (3) evaluate the degree of definition and assess the technical risk associated with the selected manufacturing method/processes, and (4) establish the existence and compatibility of the physical and functional interfaces among the configuration item and other items of equipment, facilities, computer software, and personnel.

At this review the CSCIs will also be reviewed to evaluate progress and technical adequacy and compatibility. Development specifications are authenticated and an allocated baseline is established for the CI. After this review is completed, hardware configuration item (HWCI) detailed design and software coding are initiated by the contractor.

After completing HWCI detailed design and CSCI coding, a critical design review is conducted to examine HWCI performance specifications, internal and external physical and functional interface control requirements, and all drawings to ensure readiness for the fabrication of test articles. CSCIs are reviewed to determine the acceptability of the design solution.

Fabrication and test of the CI and integration and test of the system or system segment are conducted following these reviews to verify that the as-built article meets the appropriate development (CI) and system or system segment specifications. A functional configuration audit is accomplished on each CI and the system or system segment to ensure that test results do in fact verify the approved design specifications. Other reviews accomplished during this phase include production readiness reviews and test readiness reviews. Details of these reviews are provided in MIL-STD-1521B.[16]

Technical data packages are the primary output of EMD. These packages include detailed design of all end items to be produced and descriptions of all processes needed to produce, deploy or install, and support the system, as well as those needed to dispose of related manufacturing wastes and system components which are no longer usable or required. Other EMD outputs include cost, schedule, and technical data needed to support the go-ahead decision into full rate production.

The first three phases (concept exploration, D/V, and EMD) comprise what is commonly called research, development, test, and evaluation (RDT&E). During the final two phases of the life cycle (production and operations and support) over 80 percent of total project life-cycle cost is expended. The RDT&E effort determines whether the resulting unit procurement costs and operations and support costs will be affordable.

Production—Operations and Support. The objective of production is to produce the products that will meet operational needs. Production is typically carried out by a single contractor. However, the government's desire for lower unit procurement costs has put increased emphasis on competition and pushed second-source contract arrangements. During this phase and the operations and support phase, operational site facilities are activated, required training is carried out, support systems are implemented, and completed products are deployed or installed. In addition, the final design continues to be refined through feedback of operational test and evaluation results, and succeeding system items are modified as necessary to incorporate preplanned improvements and other modifications which correct deficiencies, improve performance, extend the system lifetime, or are required by threat changes. Finally, expired product components are disposed of in such a way that environmental concerns are satisfied.

The systems engineering process is applied as necessary to meet the needs of any engineering changes initiated during production and for product improvements during operations and support.

SUMMARY

All military engineering projects have as a goal the creation of a mature design that can be produced and tested, supported properly in the field, and that will function according to system requirements. The systems engineering approach described in this chapter is intended to guide managers through the complex multidisciplinary engineering activities challenging military projects. Most projects will generally not go through each phase described. But sufficient detail is provided to help project managers plan successfully a project requiring an engineering effort.

REFERENCES

1. A. M. Frew, Jr., "System Engineering Overview," Briefing, TRW Military Space Systems Division, Redondo Beach, Calif., 1990.
2. D. M. Clemons, "An Overview of Systems Engineering," Briefing, Systems Engineering Institute, General Dynamics Space Systems Division, Nov. 1989.
3. J. G. Lake, "Systems Engineering Re-energized: Impacts of the Revised DoD Acquisition Process," in *Proceedings of the Joint Conference of the American Society for Engineering Management and National Council on Systems Engineering,* Chattanooga, Tenn., Oct. 1991.
4. DoD 5000.2-M, *Manual on Defense Acquisition Management Documentation and Reports,* Under Secretary of Defense for Acquisition, Washington, D.C., Feb. 1991.
5. MIL-STD-499A (USAF), "Engineering Management," Headquarters Air Force Systems Command, Andrews AFB, Md., May 1, 1974.
6. W. P. Chase, *Management of System Engineering,* Wiley, New York, 1974.
7. H. Chestnut, *Systems Engineering Tools,* Wiley, New York, 1965.
8. J. Kordik and J. G. Lake, *Systems Engineering,* For Coordination Draft MIL-STD-499B, Aeronautical Systems Division and Defense Systems Management College, Andrews AFB, Md., May 6, 1992.
9. L. Bordelon, "Baseline Concept Definition," Briefing to the Integrated Weapons System Management Systems Engineering/Configuration Management Process Action Team, Wright Patterson AFB, Ohio, Aug. 27, 1991.
10. B. S. Blanchard and W. J. Fabrycky, *Systems Engineering and Analysis,* 2d ed., Prentice-Hall, Englewood Cliffs, N.J., 1990, pp. 55–64.
11. *Systems Engineering Management Guide,* Defense Systems Management College, Fort Belvoir, Va., Jan. 1990, pp. 6.1–6.17.

12. J. Kordik, "Systems Engineering Process," Briefing, Aeronautical Systems Division, Wright Patterson AFB, Ohio, July 23, 1991.
13. R. L. Roe, "Coordination Meeting MIL-STD-499B," Briefing at the Defense Systems Management College, Boeing Defense and Space Group, July 30, 1991.
14. R. A. Warren and C. R. Cooper, "Science, Technology and the Program Manager," *Program Manager,* vol. XX, no. 4, DSMC 103, Defense Systems Management College, Fort Belvoir, Va., July-Aug. 1991, pp. 20–27.
15. DoDI 5000.2, "Instruction for Defense Acquisition Management Policies and Procedures," Under Secretary of Defense for Acquisition, Washington, D.C., Feb. 23, 1991.
16. MIL-STD-1521B (USAF), "Technical Reviews and Audits for Systems, Equipment, and Computer Software," Headquarters Electronics Systems Division, Hanscom AFB, Mass., June 4, 1985.

CHAPTER 3
PROGRAM CONTROL

Benjamin C. Rush

Benjamin C. Rush is presently dean of faculty at the Defense Systems Management College (DSMC). Over the last 15 years he has served as professor, department chairman, associate dean, and dean at DSMC. He has been a leader in improvements to the program management curriculum. Dr. Rush worked for McDonnell Douglas Corporation for over 10 years in several program control management positions. He has extensive government program office experience with the Navy (F-14 Program), the National Aeronautics and Space Administration (Apollo Program), and the Air Force (B-52 Program). His academic background includes doctoral and master's degrees in business administration and a bachelor of science degree in mechanical engineering.

INTRODUCTION

The subject of program control involves a broad cross section of processes and activities in weapon system acquisition. It is difficult to narrow to a lesser scope because of differences among the services within the Department of Defense (DoD) as to what is included in program control, and the integrative nature of the tasks conducted by program control. In this introduction, program control is defined and its organizational units are identified, while acknowledging the differences among services. To examine the integrative nature, a major section is devoted to the role program control plays in four major processes of weapon system acquisition. In the final section, the primary activities of program control are examined to gain insight into specific methodologies used. These primary activities of work definition, scheduling, cost estimating, budgeting, performance measurement, and program integration and reporting are a major part of effective project management.

Program control is defined as the business management of weapon system acquisition. However, the organizational terminology is far from standard among the services. While the Air Force consistently uses the title of program control to describe the business activities, the Navy and Marines with equal consistency use the term business/financial management. The Army uses a number of different terms for the business management organization, including program management, financial management, and program control. In some cases the Army changes the name of the organizational unit doing the business management functions to match the stage of the life cycle and the changing mix of activities required.

The organizational structure within program control reflects an even greater degree of variability than the name. The size of the program, the degree to which it is matrixed versus projectized, the structure of the services supporting organizations, and the stage in the life cycle of the program determine the activities required in a program control organization. A typical projectized program control organization might include program planning and scheduling, cost estimating, budgeting, and performance measurement. The program analysts working in program planning and scheduling are likely to be involved in the preparation of a number of integrated planning documents, the preparation for program reviews, as well as the preparation of program master schedules. The cost analysts doing cost estimating prepare program cost estimates for input to program reviews, source selection cost panels, program documentation, and budget inputs. Another critical activity of program control is the formulation and submittal of weapon system program budgets by budget analysts. A number of organizations within and outside the program office provide inputs to the budget analysts as they initiate the process of obtaining the budget authority necessary for program continuation. Another major role of program control is the monitoring of contractor cost performance. The analysts within the performance measurement element of program control analyze contractor cost and schedule reporting during the execution of contracts. Project management is the efficient management of cost, schedule, and technical performance. The discussion of program control emphasizes the interrelationships and the integrative role it plays in effective project management.

An important business management function, which is typically not a part of program control, is contract management. The contract management function is normally matrixed to the program office, although a large program could have a dedicated contract management office. When contract management is within the program office, it is normally not a part of program control, but a separate element within a larger business management organizational unit.

PROGRAM CONTROL AND THE PRIMARY PROCESSES OF DEFENSE ACQUISITION

There are five primary processes in defense systems acquisition management. These are requirements process, systems acquisition life-cycle process, systems engineering process, resource allocation process, and contract management process. To understand the role and activities of the program control function, it is necessary to show how program control interfaces with each of the last four processes. The requirements process of developing the mission need and operational requirements is primarily within the command which is the ultimate user of the weapon system and not as important in understanding the role of program control. Each of the other four processes is overviewed with emphasis on program control and its role in the management of and interaction with each of these processes. In many respects, the program control function serves as an integrator across these primary processes in program management of defense systems.

Systems Acquisition Life-Cycle Process

Defense programs are structured around a life cycle consisting of five phases and a set of acquisition milestones which provide the major decision reviews for conduct of the acquisition program. These milestones and phases are shown in Fig. 3.1. The fol-

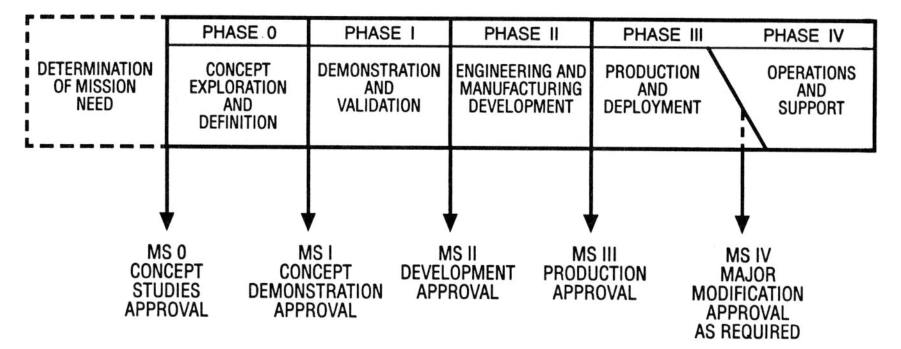

FIGURE 3.1 Acquisition milestones and phases. (*Source: Ref. 15.*)

lowing provides an overview of the milestones and phases and the key program control activities in each phase. The concept exploration and definition phase evaluates alternative system concepts that are identified and will satisfy the mission need. This is normally accomplished through competitive, parallel, short-term studies by the government or contractors. The key program control activities in this phase have to do with determining the affordability of each alternative concept through the development of preliminary cost and effectiveness estimates of the competing alternatives. These early estimates will provide the basis for funding inputs to the resource allocation process, which is discussed later in this section. During this phase contractors are doing paper studies on the feasibility of alternative concepts which provide promising solutions. Factors evaluated by the program control function are schedules, cost, budget, and overall evaluation of the program. The key product of the definition phase is the program acquisition strategy. The acquisition strategy lays out the life cycle and balances the achievement of the program performance, cost, schedule, and readiness objectives. These efforts lead to a milestone I Defense Acquisition Board (DAB) review which approves a concept baseline and provides for program initiation into the demonstration and validation phase. Program control inputs leading to a milestone I review include the development of an independent cost estimate, the conduct of an initial cost and operational effectiveness analysis (COEA), the detail acquisition strategy, schedules, and a program budget. A successful milestone review will have addressed the concerns on affordability and life-cycle cost as well as the ability to meet the initial operating capability schedule.

In the demonstration and validation phase, the primary design of the system and demonstration of the critical processes and technologies are accomplished. As the design and development testing of subsystems is accomplished, analysis will identify the areas of program risk. In the program office, with the systems engineering and program control functions leading, the emphasis is on identifying cost drivers and alternative designs. Assessments are made of the cost of various design approaches. The program office will conduct tradeoffs required to balance costs, schedule, and performance benefits to the government. The degree of competition during this phase may be limited to design competition at the subsystem or component level, or there may be full system-level competition. Program control will update life-cycle cost estimates for program development, procurement, and operations and support cost and provide inputs into the program planning and budgeting system. The demonstration and validation phase concludes with milestone II and approval of the development baseline for entering into the engineering and manufacturing development phase. With the declining DoD budget of the 1990s, the emphasis will be on prototyping in the demonstration and validation phase, with few major programs proceeding into the next phase.

The engineering and manufacturing development phase initiates a period of significant increase in expenditure of program resources. This phase will mature and finalize the selected design, and validate the manufacturing and production processes as well as providing the necessary testing to evaluate the proposed system. Control of the design is very important, and several design reviews are conducted to ensure that the detailed design satisfies performance and engineering requirements of the development specifications. In this phase the production schedule will be developed to support the initial operational capabilities (IOC) and full operational capabilities (FOC) scheduled dates. The design to average unit-production cost objective is updated and life-cycle cost estimates must be updated to reflect operations and support considerations. This phase leads to milestone III production approval to enter the production and deployment phase and the production baseline is approved. Included in the milestone III DAB review are production cost verification, affordability and life-cycle cost assessment, and production and deployment schedules. In addition there is a cost-effectiveness analysis review of the plans for production competition through two or more system contractors or dual sourcing of subsystems.

In the production and deployment phase, the objectives are (1) to establish a stable, efficient production and support base; and (2) to achieve operational capability that satisfies mission need. Program control considerations in the production and deployment phase include costing of program changes, production at an economic rate, and monitoring of contract cost and schedule performance. The Defense Contract Management Command personnel provide surveillance of the production effort at the contractors' facilities.

The operations and support phase overlaps with the production and deployment phase. Postfielding and supportability readiness reviews will be conducted, as appropriate, to identify and resolve operational and supportability programs. Milestone IV for major modification approval to modify systems still in production will be held as required. A common major modification program is a service life extension program for a system. A major modification is defined as one that exceeds the specified dollar thresholds for programs that must be reviewed at the military department or defense agency level or above.

The preceding discussion of phases and milestones indicates a fixed sequence. In reality a tailored acquisition strategy is encouraged, which uses concurrency or combinations of phases. The objective is to minimize the time and cost it takes to satisfy the identified need consistent with common sense, sound business practice, and basic management policies. In tailoring the acquisition strategy, the plans should be evaluated using the criteria of realism, stability, flexibility, balance, and risk. The program objectives and strategic approach should be realistic and sufficiently robust such that negative influences are inhibited from disrupting the process. The plan should have the flexibility necessary to accommodate changes and be sufficiently specific such that the probability of not achieving program performance, cost, and schedule goals can be estimated. Program control is heavily involved in assessing the impact of alternative strategies with regard to funding, scheduling, cost, and risk management.

Systems Engineering Process

The systems engineering process defines the design of the system and therefore is the major determinant of the development activities of the system acquisition life cycle. The systems engineering process transforms operational needs into descriptions of system performance parameters and a system configuration through the use of an iterative process of definition, analysis synthesis, design, test, and evaluation. In systems theory there is an input, transformation system, and output. This can be related to the systems engineering process, where the input is the user requirement and the output is the description of the system to meet the user requirement. The transformation process is the systems engineering process of functional analysis (what functions need to be performed), synthesis (design of how these functions are to be accomplished), evaluation (integration and balance of the functional disciplines), and description of system elements. The systems engineers integrate and balance across all of the design engineers, both specialty and traditional, while program control assists the engineering process with integration of cost, schedule, and technical performance.

The systems engineering process as shown in Fig. 3.2 is used across the system development cycle. It translates the requirements into a series of specifications which define the system. MIL-STD-490A defines five types of program specifications. The systems specification states the technical and mission requirements for a weapon system. It allocates requirements to functional areas, documents design constraints, and defines the interfaces among the functional areas. Systems specifications are referred to as type A specifications and are developed during the concept exploration and definition phase.

Development specifications state the requirements for the design or engineering development of a product during the development period. They describe the performance characteristics that each configuration item is to achieve and provide the baseline for detailed design. Development specifications are also known as type B or "design-to" specifications and are completed early in the engineering and manufacturing development phase.

Product specifications provide the form, fit, and function of a system as well as a detail design of the parts and assemblies. The product specifications are referred to as type C or "build-to" specifications, which are completed in the engineering and manufacturing development phase and enable the production of the system.

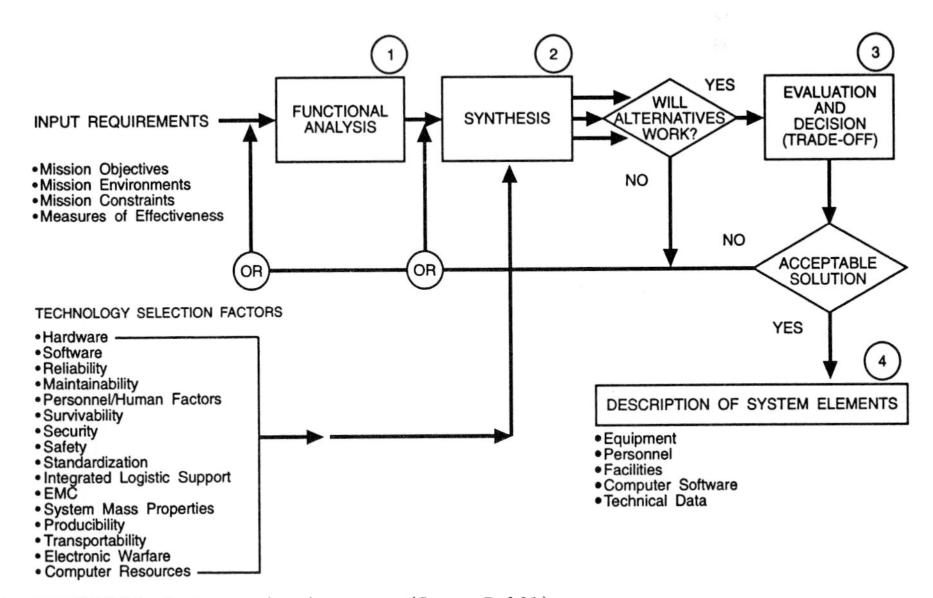

FIGURE 3.2 Systems engineering process. (*Source: Ref. 21.*)

Two additional specifications are process specifications and material specifications. The process specifications cover manufacturing techniques which require a specific procedure to achieve a required end product and are also known as type D specifications. The material specifications apply to materials used in the fabrication of the product and are known as type E specifications. The process and material specifications are also completed during the engineering and manufacturing development phase.

These specifications are developed through a series of trade-off analyses during the systems engineering process, which evaluate to obtain the best technical solution and then evaluate to obtain an acceptable solution from an affordability point of view. The final selection of the design approach must consider such factors as producibility and operational suitability as well as performance, cost, and schedule.

The principal policy statement on cost in design is provided in DoD Directive 5000.2, part 6, sec. K, "Design to Cost." This establishes cost as a design constraint early in the acquisition life cycle. This directive calls for designing to achieve life-cycle cost goals based on credible acquisition and operations and support cost parameters that are consistent with program plans and budgets and that achieve the best balance among the cost, schedule, performance, reliability, and supportability characteristics. Design-to-cost goals must be established by milestone II. A typical weapon system program might have 60 percent of the life-cycle cost in operations and support, 30 percent in production, and 10 percent in development. Because of this it is necessary to ensure that the most cost-effective concepts and design have been selected over the entire life cycle and that the accepted concept and design will not have excessive operations and support costs. While a large portion of the cost is incurred late in the life cycle, the design decisions which determine these costs are made early in the program. It has been estimated that approximately 70 percent of a system's life-cycle cost has been determined by design decisions made by milestone I. In controlling the life-cycle cost these figures show the importance of the design decisions in controlling the high but still far in the future cost of systems support.

During development a design-to-cost program seeks to balance unit-production costs against operation and support cost and requires a system to be designed and built within specific cost goals. Design-to-cost and life-cycle cost concepts should be applied early in the development cycle, when the design may be heavily influenced through the requirements generation. The life-cycle cost composition is given in Fig. 3.3. A flyaway (sailaway or rollaway) cost goal is established by the program office based on such factors as budgetary constraints, independent cost estimates, economic forecasts, and prior concept exploration studies. A certain portion of flyaway cost is allocated to the contractor as the goal for design to unit-production costs. The program office will retain the remainder of the flyaway cost to cover the internal DoD investment costs and engineering change allowances. Design-to-cost goals should be established from the knowledge of cost improvement potential for the system, together with budgetary limitations. A key to achieving design-to-cost goals is flexibility in allowing the designer freedom of choice as the trade-off analyses are conducted to achieve the design that satisfies the system requirements.

Life-cycle cost analysis (LCC analysis) is the structured study of cost estimates and design elements to identify the life-cycle cost drivers, life-cycle total cost to the government, areas of cost risk, and cost-effective changes. While the lead in LCC analysis is with the systems engineering function, the program control function provides significant input in cost estimating, budgetary constraints, schedule inputs, and risk analysis. A number of computer models are available for modeling of LCC analysis; however, models must be selected and tailored to fit the individual weapon system. The Defense Systems Management College (DSMC) has three models that can be run on personal computers: a life-cycle cost model, CASA; a quick reaction

FACTORS INCLUDED IN EACH CATEGORY OF PROGRAM COST	
Management Hardware Software Nonrecurring Production Change Allowance	= FLYAWAY, ROLLAWAY, SAILAWAY
PLUS Technical Data Publications Contractor Services Support Equipment Training Equipment Factory Training	= WEAPON SYSTEM COST
PLUS Initial Spares	= PROCUREMENT COST
PLUS RDT&E Facility Construction	= PROGRAM ACQUISITION COST

FIGURE 3.3 Life-cycle cost composition. (*Source: Ref. 10.*)

model, DPESO; and a schedule and cost risk model, VERT. Each of the models provides a means for examining cost impact.

The contractor develops a life-cycle cost plan in accordance with MIL-STD-499A, which describes the approach for integrating the life-cycle cost into the management of the program and design effort on the weapon system. The contractor provides a life-cycle cost estimate that will serve as a cost baseline for the program. The documentation for this estimate requires three basic elements: (1) the data and sources of data on which the estimate is based; (2) the estimating methods applied to the data; and (3) the results of the analysis. During the engineering and management phase the design goals established in the design-to-cost analysis are monitored to measure performance toward achieving the cost-related design goals. During the production phase the design to unit-production cost goals are statused at the summary work breakdown structure (WBS) level for each specified WBS element in terms of hours and dollars for each functional cost and for recurring and nonrecurring costs. The program acquisition cost goal is critical and tracked throughout the life cycle. Variance analysis on this datum indicates those areas in need of corrective action.

Resource Allocation Process

The resource allocation process includes development of budgets through the planning, programming, and budgeting system (PPBS) of the Department of Defense (DoD) and the congressional budget process of the U.S. Congress. An overview of the resource allocation process is shown in Fig. 3.4. The function of the PPBS is to develop a plan, formulate a program to implement the plan, and cost the program in a DoD budget request which, when approved, enables execution of the plan. The DoD budget then becomes part of the President's budget which is submitted to Congress. The PPBS is calendar driven and focuses on balancing all of the DoD requirements within DoD financial constraints. The funding of weapon system acquisition programs is only a subset within PPBS. In 1986 the PPBS changed from an annual to a biennial cycle, and now during the odd-numbered years, DoD initiates the first phase of PPBS. While two budget years are submitted to Congress, the congressional

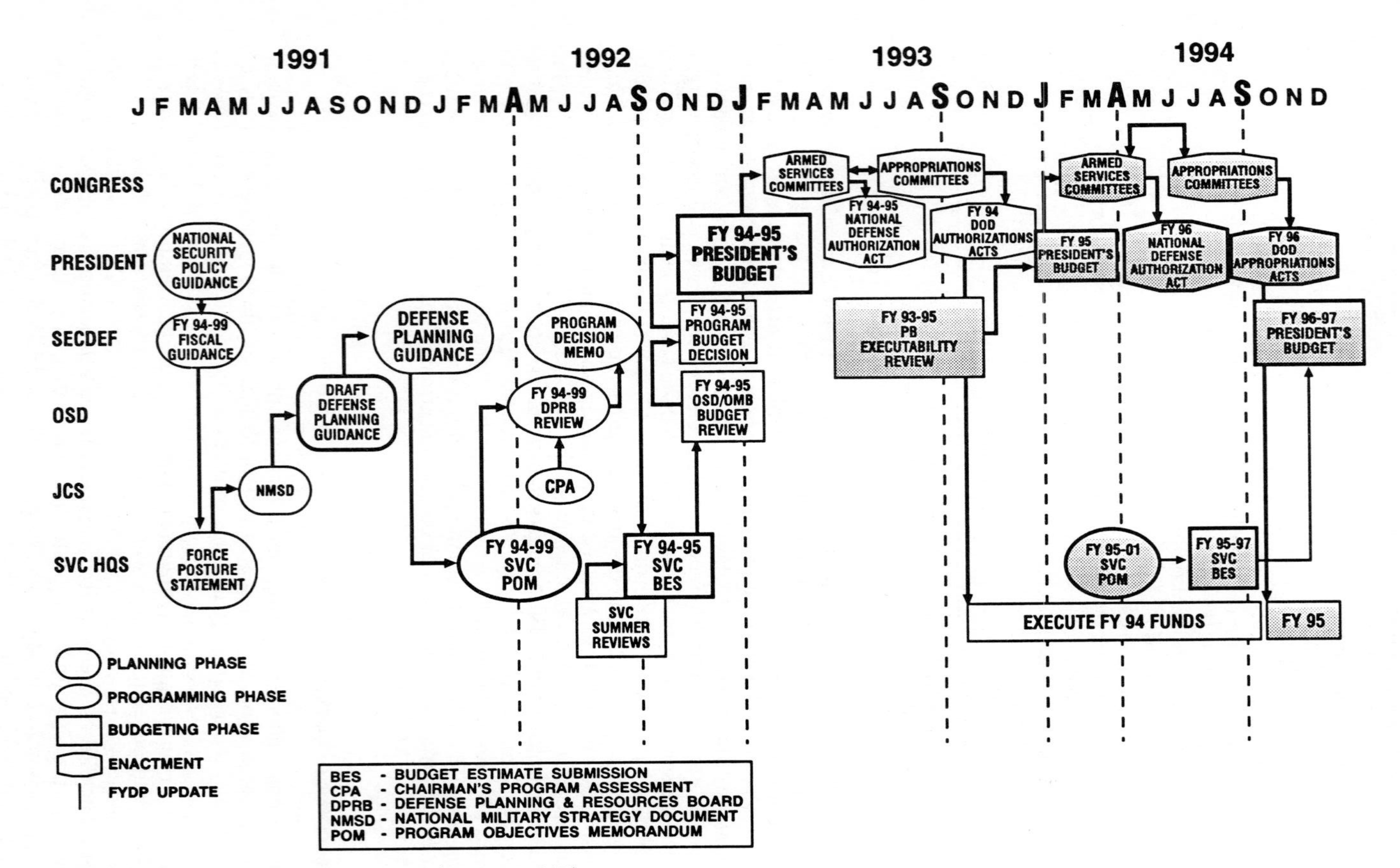

FIGURE 3.4 Resource allocation process. (*Source: Ref. 17.*)

budget system has continued to consider one year at a time and DoD is required to prepare and submit an amended second-year budget as the President's budget is prepared for the second year.

In 1961 Robert McNamara was appointed as the Secretary of Defense. The McNamara years were characterized by three key changes: (1) increasing authority of the Secretary of Defense within DoD as the decision maker rather than the arbitrator between the services; (2) a more analytical management approach to the problems of DoD with an emphasis on cost benefit analysis; and (3) greater integration of strategy, force requirements, and defense budget in the development of alternative solutions. To accomplish this, Mr. McNamara brought Charles Hitch and Alain Enthoven from RAND Corporation to develop a planning and control system for DoD. A system structured along the lines of PPBS had been used at RAND Corporation to analyze defense issues for some number of years. With Hitch as the DoD comptroller, a PPBS system was developed and implemented in less than a year. This process enabled a significant change in decision making and resource allocation within DoD.

Each budget cycle of the weapon system program has approximately three years elapse from the beginning of PPBS until congressional enactment is completed. In addition, up to five years are allowed to get the funds on contract and additional time for completion of execution of the budget. This means that there is significant overlap of the budget cycles which ensures that several cycles are simultaneously in progress for a given weapon system program. This is further complicated by the timing mismatch with the systems acquisition process, which keys on milestone decision points determined by programmatic decisions. The milestone events of the systems acquisition process and the PPBS events rarely coincide. It is up to the program office, with much of the burden falling to program control, to ensure that the decisions of a Defense Acquisition Board (DAB) review do not result in a disconnect from the resources planned in the PPBS process. The DAB decision to proceed at a milestone review certifies that the program is ready to move forward into the next stage of the life cycle and that the program has sufficient resources in the budget to be executed as recommended. This requires anticipation in the budgeting process of the alternatives that will be approved, and an assessment of affordability remains a key issue in PPBS.

The initial phase of PPBS is the planning phase. It begins more than two years in advance of the fiscal year for which the budget authority will be requested. The planning phase focuses on the military role and posture of DoD, with the major objective of planning the integrated military forces required to support the military strategy during the next two to six years. This phase is under the direction of the Under Secretary of Defense (Policy) and results in the development of a broad long-range investment plan which projects the major modernization and investment requirements. The direction necessary to program for these requirements is provided in a draft defense planning guidance document prepared using objectives from the President's National Security Council, fiscal guidance from the Office of Management and Budget, an assessment of the threat from the Defense Intelligence Agency/Central Intelligence Agency, as well as strategy inputs from the Joint Chiefs of Staff (JCS) and each of the services. The draft defense planning guidance is influenced by the national military strategy document prepared by the JCS and a review by the summer defense planning and review board. A final defense planning guidance document is completed in October or November of each odd year. The planning phase should not derive from existing programs and budgets and therefore while having inputs on fiscal guidance, does *not* operate under budgetary constraints. The final defense planning guidance provides direction on defense policy, strategy, and force and resource planning to DoD components for use in formulating the programming decisions in the next phase.

The programming phase of PPBS translates approved objectives into time-phased resource requirements. Fiscal constraints are an important part of this phase, and the available resources are matched to the approved force objectives. The programs for all of DoD are costed out in terms of financial and manpower resources six years into the future. The document containing all of these programs is called the Future Years Defense Program (FYDP). The programming phase is documented in the development of a program objective memorandum (POM). The FYDP is updated after each POM submission and contains prior year, current year, the biennial budget year, and the following four years, matching dollar and manpower resources to programs. The FYDP is considered an internal DoD working document, and data beyond the budget years are not released outside of the executive branch without permission of the DoD comptroller.

The FYDP has two basic dimensions of mission areas and appropriations. There are 11 mission areas called Major Force Programs (MFP) and numerous congressional appropriation categories. In the programming phase, the primary emphasis is on the MFP structure which enables analysis by mission area. The intersection of the mission area and the appropriation matrix is the program element which is the primary data element in the FYDP. These program elements are designed to be mutually exclusive and are the basic building blocks of the PPBS. A program element will be unique to a major weapon system or grouping of weapon systems within a specific service. The five primary appropriation categories used in systems acquisition are military personnel, RDT&E (research, development, test and evaluation), military construction, procurement, and operations and maintenance.

The programming phase starts with the issuance of the defense planning guidance (DPG) and ends with the program decision memorandum (PDM). Each of the military services and defense agencies submits a POM document which recommends force and resource levels within the specific fiscal guidance and in accordance with the DPG. The service POM submissions are reviewed by the joint staff to assure compliance with the DPG, and this review is documented in the chairman's program assessment (CPA). The Office of the Secretary of Defense staff also reviews the service POM and recommends changes in the form of POM issue papers. These documents define specific issues for review by comparing the proposed program with the objectives and requirements established in the DPG. These issues are then resolved by the Defense Planning and Resources Board (DPRB), the Assistant Secretary of Defense for Program Analysis and Evaluation (ASD-PA&E), or are deferred until the budget phase, depending on the category of issue. The ASD-PA&E is the lead for the programming phase. When all issues have been resolved and approved, the Secretary of Defense issues a PDM for each military department and defense agency. These PDMs become the basis for the budget phase.

The budget phase translates the programmatic decisions of the prior phase into resource requirements in an appropriations format. The emphasis shifts from the major force programs showing mission structure in the programming phase to the congressional appropriation structure used in budget development, review, and execution. The DoD comptroller is responsible for the budgeting phase, and the emphasis of the review is on program justification and execution, but may reconsider program alternatives discussed in the programming phase. In this phase, the primary document is the budget estimates submission (BES). Based on the issuance of the PDM (the document concluding the programming phase), the service budgets are updated in the BES and submitted in September in the even years. These submittals are also used for the second update of the FYDP in this PPBS cycle. The entire BES is reviewed to ensure conformity with the PDMs and other approved changes.

The executability of each program budget is the primary concern of the budget scrub. Budget hearings are conducted with budget analysts from the DoD comptroller and the Office of Management and Budget (OMB). The decisions to adjust the budgets are reflected in program budget decision (PBD) documents. The PBDs may be provided to the services for comment and reclamma with response required within as short a period of time as 24 hours. The services have one last opportunity to identify issues that are serious enough to warrant a major budget issue (MBI) in meetings between the Service Secretary and the Secretary of Defense. If the services are successful in the argument for a major issue, they will normally be required to provide dollars from other programs within that service to offset the buyback of the major issue. After the MBIs are resolved, the services finalize their budgets and the DoD budget is approved by the Secretary of Defense and sent to the President and OMB.

OMB is responsible for consolidating the DoD budget with that of all other executive branch budgets and the President's budget is submitted to Congress on the first Monday after January 3. The budget is for the two fiscal years commencing on the following October. The submittal to Congress of the President's budget completes the formulation of the budget and begins the political debate on Capitol Hill.

The congressional budget process is a continuation of the activities in the budgeting phase of the PPBS process, but is under the control of Congress. The President's budget is the input to the congressional enactment process (authorization and appropriation). The output of this process is the Defense Authorization Bill and the Defense/Military Construction Appropriations Bills. The first stage of the congressional enactment process is the concurrent resolution which provides a set of budget targets for the overall federal budget and for each of the appropriation bills. The concurrent resolution is not signed by the President and does not have the force of law but sets the targets toward which Congress will work during the authorization and appropriation process. The congressional budget timetable calls for the concurrent resolution by April 15. The budget resolution reflects an examination of the overall financial picture with revenue, outlay, and deficit targets for the federal government and specific budget allocations to the authorization and appropriation committees of Congress.

The authorization phase of the enactment process has evolved over time as congressional oversight has increased. The House and Senate Armed Services Committees pass authorization legislation providing maximum amounts that are believed to be justified for the development, procurement, and operations and support appropriations of specific weapon systems. An authorization bill authorizes a weapon system program and provides recommended funding levels. This bill authorizes defense budget authority, procurement quantities, end strength, and military construction. It is the recommended funding level which should not be exceeded in the appropriations bill. Hearings will be conducted by the House and the Senate Armed Services Committees followed by mark-up sessions in which they conduct a line by line (program element) review of the bill. The Senate will complete its bill after passage of the House bill and a conference will convene to discuss the differences between the House and Senate bills. A single bill is then passed by both the House and the Senate and sent to the President for signature. The final step in the enactment process is appropriation. The Senate and House Appropriations Committees generally start their hearings in March and schedule to pass these appropriations bills prior to the start of the fiscal year on October 1. The Defense Appropriations Bill and the Military Construction Appropriations Bill provide the budget authority to operate the DoD and its many programs. If the appropriations bills are not passed by October 1, Congress must provide Continuing Resolution Authority (CRA) or shut down the agencies and programs affected. Under a CRA the agencies are normally required to limit their rates of obligation and no new program starts are permitted.

The final step in the resource allocation process is budget execution. The congressional appropriations are funds made available by legislation to purchase resources for specific purposes within a given period of time. The Treasury Department then prepares an appropriation warrant that documents all appropriations by title, number, and amount, thereby making the funds available. Following appropriation there must be a delegation of budget authority to the using commands within DoD to provide them permission to enter into transactions that will result in cash flowing out of the Treasury. The distribution begins with apportionment by the Office of Management and Budget, which limits the amount of obligations that may be incurred during a specified time period. Thus the Office of Management and Budget apportions, the DoD subapportions, the military departments allocate, special operating agencies suballocate, and finally the installations and activities distribute funds by suballotment, obligational authority, and citation of funds. The program control office is responsible to tell the accounting and finance office how the funds will be utilized at the detail level.

Accounting for availability and use of funds is a critical part of the resource allocation process. It provides the connection between resource planning and execution of the funds by the program office. It also ensures that funds are not distributed in excess of the amount available. Each program office must keep an accurate accounting of funds in terms of commitments, obligations, and expenditures. These are the three stages of the accounting cycle used to track availability of funds and the purposes to which they are applied. Commitments are an administrative reservation of funds resulting from a purchase request or purchase order that enables certification that the funds are available in the amount requested, in the correct fiscal year, and in the proper appropriation. An obligation is incurred when an order is placed or a contract is awarded for delivery of goods or the performance of services. An obligation legally encumbers a specified sum of money which will require expenditures and outlays. The timing of the obligation occurs with the placement of an order by the signing of a contract or purchase order. An expenditure is a charge against available funds for work that has been performed. The expenditure represents the presentation of a check to the performer of the work. Finally an outlay is the actual cash being paid from the U.S. Treasury when the check representing the expenditure is presented to the Treasury.

In recent years there has been increasing emphasis for program control to do more detailed tracking of expenditures on a weapon system program to ensure that the budget authority is being used properly. The contract funds status reports (CFSRs) are received by program control from the contractors and provide a forecast of expenditure billings to the government for the total authorized work. The CFSRs, program schedules, and time required for bill paying are used to develop the expenditure forecast. If actual expenditures are not matching the planned profile, the budget authority may be withdrawn. Thus it is important for program control to ensure not only that the forecasted obligations are achieved by getting the activity on contract, but also that progress on the contract as reflected by expenditures is matching the forecast. Overvariances from obligation and expenditure plans require corrective action to ensure that fiscal year plans are not exceeded. Undervariances may result in potential funding sources for other programs through reprogramming if the underrun is significant.

Reprogramming is the process of using funds for purposes other than that for which they were originally appropriated. It is *not* the requesting of additional funds from Congress, but the reapplication of resources within a particular appropriation. There are several different reprogramming actions, some of which require congressional approval and others which require approval by the DoD comptroller or by the individual service or defense agency.

Funds are appropriated for one to five years, depending on the appropriation.

(The specific period for each appropriation is discussed in a later budgeting section.) After this time the appropriation funds are considered expired and no longer available for new obligations. The unobligated budget authority is withdrawn annually by the Treasury Department, but may be restored to cover valid claims against the original appropriation. Two years after an appropriation expires the appropriation lapses and payment for expenditures is made against a successor merged (M) account. The Authorization Act for FY 91 made significant changes to the use of expired appropriations. In the future there will be a five-year period after which these expired accounts will be closed and no further expenditures may be recorded or paid. In this situation, current program appropriations must be used to pay expenditures which would have been charged to expired accounts. In the face of already tight budgets, program control must avoid this situation by working with the administrative contracting officer and government auditors to expedite negotiation of unallowable costs and other factors delaying billing of contractor expenses.

Contract Management Cycle

The contract management cycle is discussed in three parts. The first is program acquisition planning, which includes the development of the acquisition strategy and addresses the entire program across the life cycle. The second part is the contract solicitation, selection, and negotiation, which is unique to the specific activities, leading to a single contract. Across the life cycle of a weapon system program, each phase of the life cycle will typically result in one or more contracts. The third part in the cycle is also contract specific and is the administration of the contract after award, including such things as contract changes and contract surveillance.

Program acquisition planning means the process by which the efforts of all personnel responsible for an acquisition are coordinated and integrated through a comprehensive plan to fulfill a requirement in a timely manner and at a reasonable cost. It includes developing the overall strategy for managing the acquisition. Those involved will include all those who are responsible for significant aspects of the acquisition such as contracting, program control, legal, and technical personnel. Earlier in the discussion of the system acquisition life-cycle process, the development of an acquisition strategy was a key product of the concept exploration and definition phase. This acquisition strategy is incorporated into an acquisition plan and is a separate entity in the acquisition strategy report. The development of the acquisition plan is under the control of the program office and is a specific responsibility of the program manager. The acquisition strategy for the weapon system program leads to functional strategies for each of the functional disciplines and specific plans within each of these disciplines which are then incorporated into the acquisition plan. Therefore the acquisition plan includes not only the weapon system program acquisition strategy but the functional plans to achieve program goals. The plan should address the acquisition background and objectives and a plan of action including the contract method. A description of typical required contents is given in part 7 of the *Federal Acquisition Regulation* (FAR).

There are two basic approaches to government contracting: sealed bids or negotiated contracts. The traditional preferred method of government contracting has been to get competitive sealed bids and award to the low bidder. Detailed formal procedures are written in FAR part 14 on exactly how to solicit and handle these sealed bids. Sealed bids are appropriate when there is sufficient definition of requirements, when an award is based on price, when it is not necessary to conduct discussions with the responding sources, and when competitive bids are reasonably expected. In the acquisition of weapon systems it is normally not possible to have sufficient definition of the system to enable the contracting officer to meet the necessary conditions for the use of sealed bids, but this method of procurement may be used for subsystems or components, particularly when commercial items can be used. The predominant method of contracting for weapon systems is negotiation, defined in FAR part 15 as "contracting through the use of competitive or other than competitive proposals and discussions." The negotiated method may be either competitive or sole-source contracting, but sole-source contracting may only be used when certain specific statutory exceptions from the general requirement to use competitive procedures are met.

With the acquisition strategy determined and the contracting approach selected, the second phase of the contract management process is the contract solicitation, selection, and negotiation. Planning for a specific contract solicitation and selection will overlap with the weapon system program acquisition planning and be included as part of the acquisition plan. In the solicitation planning there will be a synopsis of the requirement and public notice of sources being sought for this contract. A source selection plan is developed to assure fairness and timely selection of the most realistic proposal. The solicitation is called a request for proposal (RFP) and is used to communicate government requirements to perspective contractors and to solicit proposals. The FAR provides a uniform contract format that shall be used in the preparation of RFPs and provides consistency and organization to government solicitations. DoD directive 5000.2 provides principles applicable to competitive solicitations for major weapon system acquisition, which are also applicable to less than major systems. These principles include the following: (1) Evaluation criteria and their relative importance must flow from the statement of work and must be furnished to all potential offerors in the solicitation. Numerical weights will not be disclosed to either offerors or evaluators. (2) The statement of work should include a description of the mission need and should be written in terms of performance requirements rather than design requirements. (3) Solicitation should provide guidance to offerors regarding proposal page limitations, number of copies required, and the structure of the proposal. (4) The use of draft RFPs is encouraged to obtain feedback and perspective. (5) The source selection plan shall be prepared by the program manager, reviewed by the procuring contracting officer (PCO), and approved by the source selection authority before the issuance of the solicitation.

The FAR requires formal designation of a responsible official as a source selection authority (SSA), formal establishment of an evaluation organization, and preparation of a source selection plan. A source selection advisory council (SSAC) serves as staff and advisors during the source selection process to the SSA. The Source Selection Evaluation Board (SSEB) does the detail evaluation of the offeror's proposals. This board will include program control personnel involved in cost estimating and price analysis as well as personnel from a number of technical and management functions. These personnel will typically evaluate the cost, technical, and management proposals of the offerors and prepare an evaluation report which contains the evaluation criteria, detail narrative assessments of each proposal against the criteria, scoring and a summary of the significant strengths, weaknesses, and risks of each proposal.

The finalized RFP is released to industry sources after final approval from the SSA and notification of release in the *Commerce Business Daily.* Contractors receiving the RFP have a very limited time to complete their proposals and submit to the appropriate contracting office. For most weapon systems the contractors will have been preparing their proposals long before receipt of the final RFP. The government contracting office begins an initial evaluation after receipt of the proposals in order to determine the competitive range and eliminate any contractors not within that competitive range. The contracting office and program office will receive and analyze field reports and audits conducted by the Defense Contract Management Command (DCMC) and the Defense Contract Audit Agency (DCAA). The areas in which the proposals will be evaluated will vary with the individual weapon system. However, typical areas to be evaluated would be the technical area, supportability area, management capability area, and cost area. Program control personnel will probably be involved in assessing the contractors' organizational structures, management plans, and overall program schedules. Management control and information systems may also be evaluated in the management capability area. Program control personnel will be involved in assessing the cost area. A cost panel will evaluate the affordability of each offer as well as the methodology used in constructing the estimates. The proposed price will be evaluated for realism, reasonableness, and consistency. The cost evaluation will be based on an assessment of each offeror's proposed cost and the government's estimate of the most probable total life-cycle cost of each offeror. The overall evaluation of each proposal may include on-site inspections and results of preaward surveys to provide information to the SSA regarding the offeror's current and future capability to perform all aspects of the weapon system program.

During the evaluation, the SSEB members may determine that insufficient data have been presented by the offerors and further communication is required with the offerors. There are three kinds of written communication used during the source selection: process clarification, deficiencies, and amendments. Clarifications are used when contradicting information is provided and the proposal or portions of the proposal are missing. Deficiencies are used by the evaluators to record parts of the proposal which do not meet the minimum requirements set forth in the RFP. An amendment is a formal change to the RFP which all offerors receive. Fact-finding discussions can begin when these communications are completed. The fact-finding process allows the government team to completely understand each offeror's proposal during face-to-face discussions. During the fact-finding session there is to be no questioning of the subjective judgments made by the offeror, but merely an understanding of the facts, processes, and assumptions presented within the proposal and the obtaining of necessary back-up data. With the negotiated sole-source method of contracting, the negotiation begins at the completion of the fact-finding process. The negotiation is bargaining to achieve an agreement on price, schedule, technical requirements, type of contract, and other terms of a proposed contract. With the negotiated competitive method of contracting, at completion of discussions, a request for best and final offers may be made. Negotiations are completed when best and final offers are received. Upon receipt, all that should be necessary to affect a binding contract is the government's acceptance and notification of award. Best and final offers may not be used as an auctioning technique to obtain unrealistic promises of performance or costs. The proposal evaluators will update their initial evaluation with the changes made in the best and final offer and the SSEB will report its findings to the SSAC who forwards them with their recommendation to the SSA. After the SSA has made the source selection decision and informs the appropriate management level, the Procuring Contracting Officer will award the contract and ensure that appropriate notifications are made.

The final part of the contract management process is contract administration after award. The DCMC is responsible for the contract administrative services accomplished on site at the contractor's plant. Some of the contract administration functions provide critical support to program control activities. Indirect cost monitoring includes negotiation of forward pricing rate agreements, negotiation of billing and final indirect cost rates, and determination of cost allowability. Also contract administration provides surveillance of progress against production schedules as well as surveillance of a number of contractor management systems, including accounting, purchasing, subcontracting, and cost-estimating systems. As the government representatives in the contractor's facility, the contract administration personnel are key members of the acquisition team. Coordination and cooperation between the program control and contract administration personnel are important to a successful program.

During the life of a major weapon system contract it is common to have a number of changes to the contract. The preferred way to achieve the incorporation of a change is to have a negotiated agreement on all modifications to the contract signed by both parties before the work is authorized. When this has been achieved and both parties have signed, it is called a supplemental agreement. More common perhaps is the use of undefinitized changes where the price of the work has not yet been negotiated but the government under the Changes clause in the contract unilaterally directs changes in the specification, shipping, or packing. The contractor will then request an equitable adjustment to the contract cost or schedule, or both, and these changes must be negotiated and definitized into a supplemental agreement at a later time after work has already proceeded on the change. The program control office will be significantly involved in evaluating and negotiating these changes.

PROGRAM CONTROL ACTIVITIES

Work Definition

The first of the program control activities to be discussed is the development of the work breakdown structure (WBS). The WBS provides the framework for definition of the products to be developed and manufactured into the weapon system. It is essential to providing program management with the capability to structure, define, and integrate the technical, schedule, and financial aspects of the program. The primary document controlling the application of the WBS is MIL-STD-881A, "Work Breakdown Structure for Defense Material Items." A WBS is a listing or graphic display which defines the weapon system to be developed and produced and divides the system into a series of subsystems or critical elements relating each to the total end system. Each WBS element is a discrete, identifiable item of hardware, software, data, or service.

There are four WBS formats identified in MIL-STD-881A. These are the summary WBS, project summary WBS, contract WBS (CWBS), and project WBS. The summary WBSs are specified in MIL-STD-881A, providing uniform element terminology, definition, and placement in the family tree structure of the top three levels of the WBS for each of seven types of weapon system programs. The seven example systems provided in MIL-STD-881A are aircraft systems, electronics systems, missile systems, ordnance systems, ship systems, space systems, and surface vehicle systems. The summary WBSs are to be used as guidelines in the development of a project summary WBS.

The project summary WBS is tailored from the summary WBS to a specific weapon system. It is also specified to three levels of detail and is constructed by the program control and systems engineering personnel to identify the major areas of effort. This project summary WBS provides the framework for all work to be accomplished by contractors and government on this weapon system.

Within the project summary WBS a number of CWBSs may be developed. For each contract a WBS will be included in the request for proposal to define the structure of work on that contract. The CWBS provides a direct link between the contracted effort and the total project with traceable summarization of individual CWBSs into the approved project WBS. The CWBS developed for a request for pro-

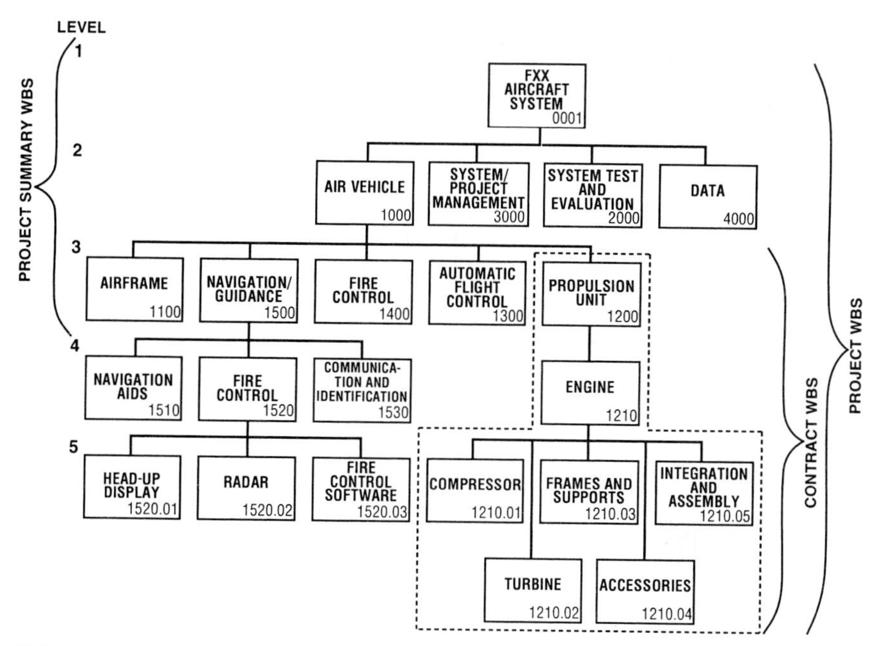

FIGURE 3.5 Components of project work breakdown structure. (*Source: Ref. 14.*)

posal will change during the solicitation and negotiation as the government and contractor determine contract activities. After contract award the CWBS will be extended by the contractor to lower levels for internal management purposes. The CWBS provides the framework for the work to be performed. The contractor's organizational structure indicates the relationship of the organizational units that will perform the work. The interrelationship of the CWBS and the organizational structure is shown in Fig. 3.5 at the cost account level, which defines functional responsibility for identified units of work. The lower levels of the extended CWBS are interrelated to contract line items, end items, configuration items, data items, and statement of work tasks. The contract will indicate the level within the CWBS at which cost accumulation shall be reported by the contractor to the government.

The project WBS defines the WBS for an entire weapon system program. It is prepared by the program office and reflects all of the CWBS elements flowing from the project summary WBS. Thus it is an extension of the project summary WBS to the levels of the CWBS plus the government efforts required for completion of the total weapon system. The project WBS is completed prior to the initiation of production and is maintained throughout the system acquisition process. Figure 3.6 shows the relationships among project summary WBS, CWBS, and project WBS.

The WBS is the foundation for (1) program and technical planning; (2) cost estimation and budget formulation; (3) schedule definition; (4) statements of work and specification of contract line items; and (5) progress status reporting and problem analysis. The WBS is developed jointly by program control, technical, and manufacturing personnel in the program office. This management technique for subdividing a whole program or project into manageable elements reflecting all products to be delivered and services to be performed was standardized by the DoD in the mid-1960s. Standardization of the WBS assists program office personnel in integrating

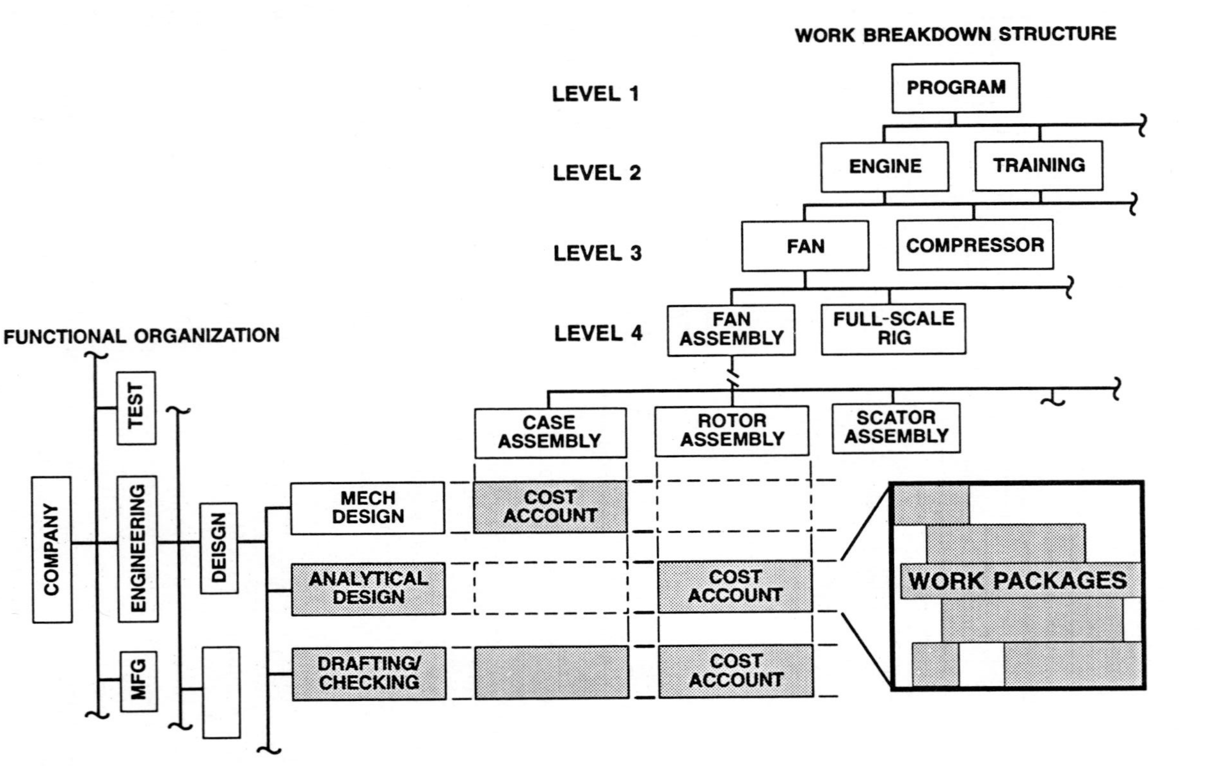

FIGURE 3.6 Responsibility assignment matrix, detail level. (*Source: Ref. 14.*)

across the weapon system program as well as comparing to other weapon system programs, such as conducting cost comparisons on proposals. While tailored to the individual weapon system, the WBS provides a consistent framework within which the program management functions occur.

When the WBS is shown graphically, each box represents a WBS element. A WBS element is an individual, discrete portion of the WBS representing an identifiable item of hardware, data, or service. These WBS elements are organized into a hierarchy of levels beginning with level 1, which represents the entire weapon system. An indented listing of the number and names of each WBS element at all the different levels is called a WBS index. The technical and cost content associated with each WBS element at its lowest level of detail is provided in a WBS dictionary. The dictionary is a sequential listing of project or contract WBS element descriptions where each page provides WBS title; element number; revision number, authorization, and description of changes; element task description; specification number and title; contract line item; contract end item and quantity; cost content and description; and contractor and subcontractor names. The WBS dictionary is provided by contractors in response to an RFP.

Scheduling

Scheduling is the developing of information on how long a job should take, who is responsible for doing it, relating it to the other jobs required to deliver the product, and laying out the data in a specific format to show when they must be done to comply with all of the program constraints. A weapon system program schedule is a documented plan of work to be done within a specified timetable. The system acquisition life cycle and all of its phases and milestones are documented in an acquisition strategy. This acquisition strategy is the starting point for the development of a program master schedule that integrates all major program activities, major decision points, and officially directed milestone dates. The development of the program master schedule provides the schedule baseline for the program and, when combined with the cost and technical baselines, provides the three critical facets of program management. The schedule baseline should be totally integrated with the cost and technical baseline, and a change to any one of these three requires each to be reviewed for possible impact. The key to a successful planning process is to define the relationships linking technical performance, cost, and schedule and to continue to work the plan as the program progresses.

To achieve good schedule management, a waterfall of schedules by the contractors and program office provides progressively more detail. These schedules are structured to the WBS. The contractor's detail schedule should flow up to a contract master schedule, which should integrate with the program master schedule. A good scheduling system has the configuration control discipline to ensure that this integration continues as the schedules are changed or statused over the life cycle of the program.

Program control usually has the responsibility for integrating schedule requirements from the various functional areas. Program analysts must be familiar with the various scheduling techniques and the sources of data for developing and integrating program schedules. Sources of scheduling information include program office documentation and documentation from other government offices and from government contractors. An important document in the program office is the program management directive (PMD), which typically includes a number of directed schedule dates, including the required submission of various plans, and the initial operating capability (IOC) date for the system. An integrated program summary is a statement of how the program office intends to complete the program and is a source of data as well as a place where the integrated master program schedule is documented.

The most important sources of schedule information are the contractors that design, develop, and produce the weapon systems. Some of this information will be delivered to the program office as specified in the contract data requirement list (CDRL). The CDRL should specify schedules that will report by WBS item. The program office must determine which areas of the program present the greatest risk and request schedule data at lower WBS levels for areas that present the greatest risk. It is not a good practice to specify schedules at the same level of the WBS for the entire program. As shown in the responsibility assignment matrix in work definition, the WBS and the contractor's functional organization are integrated at the cost account level. Care should be taken that the development of schedules and schedule reporting is oriented to the WBS and not just to the functional organization. The contractor's detail planning should be reviewed periodically during visits to the contractor's facility to ensure that the information being summarized and reports submitted to the government flow appropriately from the detail to the summary level and are properly integrated with the technical and cost control systems.

The following provides the major scheduling techniques used to display program schedules graphically. It is important to remember that one of the functions of a schedule is communication, and the technique should be selected which allows for optimum exchange of information within the involved government offices and between the program office and its contractors. The scheduling techniques discussed are Gantt charts, milestone schedules, networks, and line of balance. A final technique discussed is the process flowchart.

The most basic of the scheduling techniques is the Gantt chart, or the bar chart as it is sometimes called. On this chart the vertical axis typically shows tasks, organizations, or people, and the horizontal axis is used to portray time, with a bar showing the length of a particular activity. Progress is shown by shading in the bar to represent actual completion versus planned work. This shaded area is contrasted with a vertical time-now line, indicating the status date. The advantage of the Gantt chart is its simplicity, and it is a good technique for showing summary levels of a total program. A simple example is given in Fig. 3.7. While simple relationships may be indicated between the activities, this technique does not enable complex interrelationships to be shown.

Another technique which is quite simple is the milestone technique, which shows when an event should occur if the particular activity is to proceed as planned. The

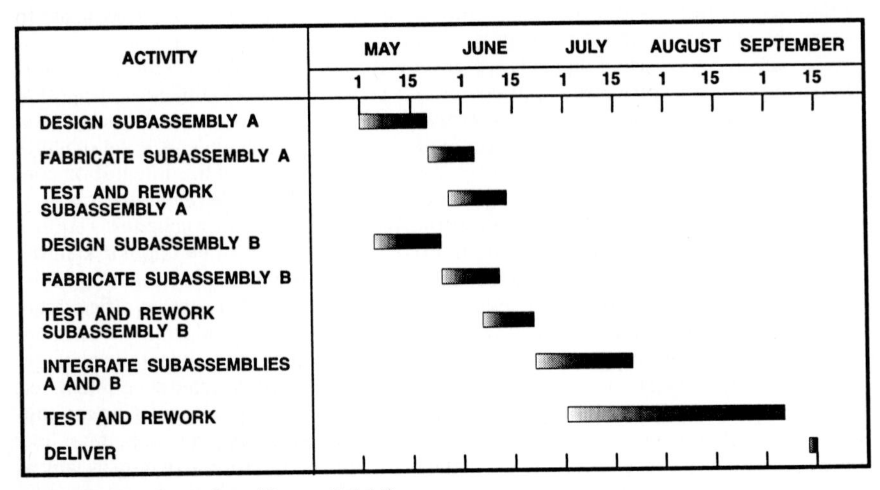

FIGURE 3.7 Gantt chart. (*Source: Ref. 14.*)

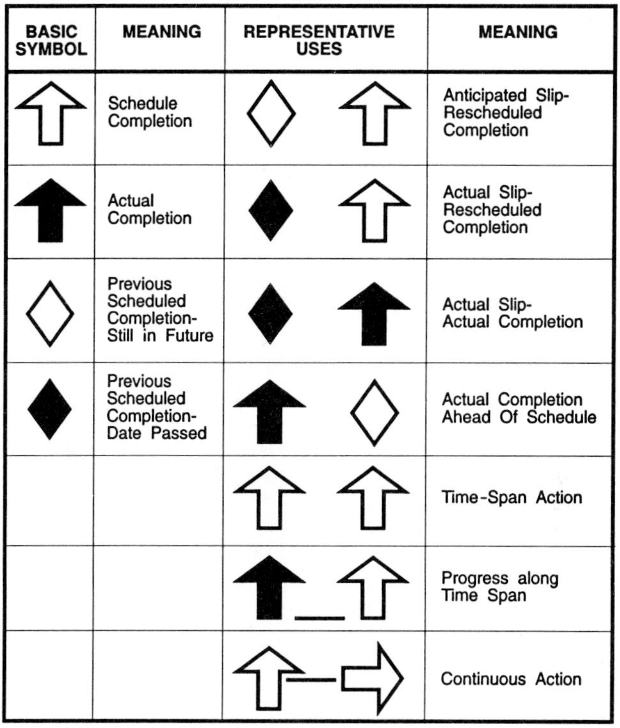

BASIC SYMBOL	MEANING	REPRESENTATIVE USES	MEANING
	Schedule Completion		Anticipated Slip-Rescheduled Completion
	Actual Completion		Actual Slip-Rescheduled Completion
	Previous Scheduled Completion-Still in Future		Actual Slip-Actual Completion
	Previous Scheduled Completion-Date Passed		Actual Completion Ahead Of Schedule
			Time-Span Action
			Progress along Time Span
			Continuous Action

FIGURE 3.8 Milestone chart symbols. (*Source: Ref. 14.*)

program milestones are provided on the vertical axis with the horizontal axis again displaying time. The milestone chart uses a relatively standard set of symbols to display the scheduled and actual completion dates, as shown in Fig. 3.8. An example of the milestone chart is given in Fig. 3.9. Milestone charts are one of the most common techniques used in a program office and are also a common way for a contractor to submit the program schedule contract data item. The symbology is relatively standard and simple to use and allows statusing against a baseline plan as well as a current plan. As with the Gantt chart, a major weakness of the milestone technique is the lack of a mechanism to show interdependencies. It can be used in conjunction with the network technique by showing major milestones with the data determined from a network.

Network analysis shows the interdependencies of events and activities and their predecessor and successor relationships. This type of schedule provides the logic and structure to analyze a great number of activities and their interrelationships. Also the process of constructing the network is a valuable experience for program control personnel. Their research to develop a program network will necessitate interfacing with all major functions working on the program, and interrelationships of the activities will frequently be determined by the disciplined analysis necessary to construct the network. A key characteristic of networks is the ability to identify activities or events for the most time-consuming sequence of activities (critical path). The time for tasks on the critical path determines the total duration of the schedule. The length of time which activities on the noncritical paths could be delayed without going beyond the critical-path end date is called slack time. By determining the crit-

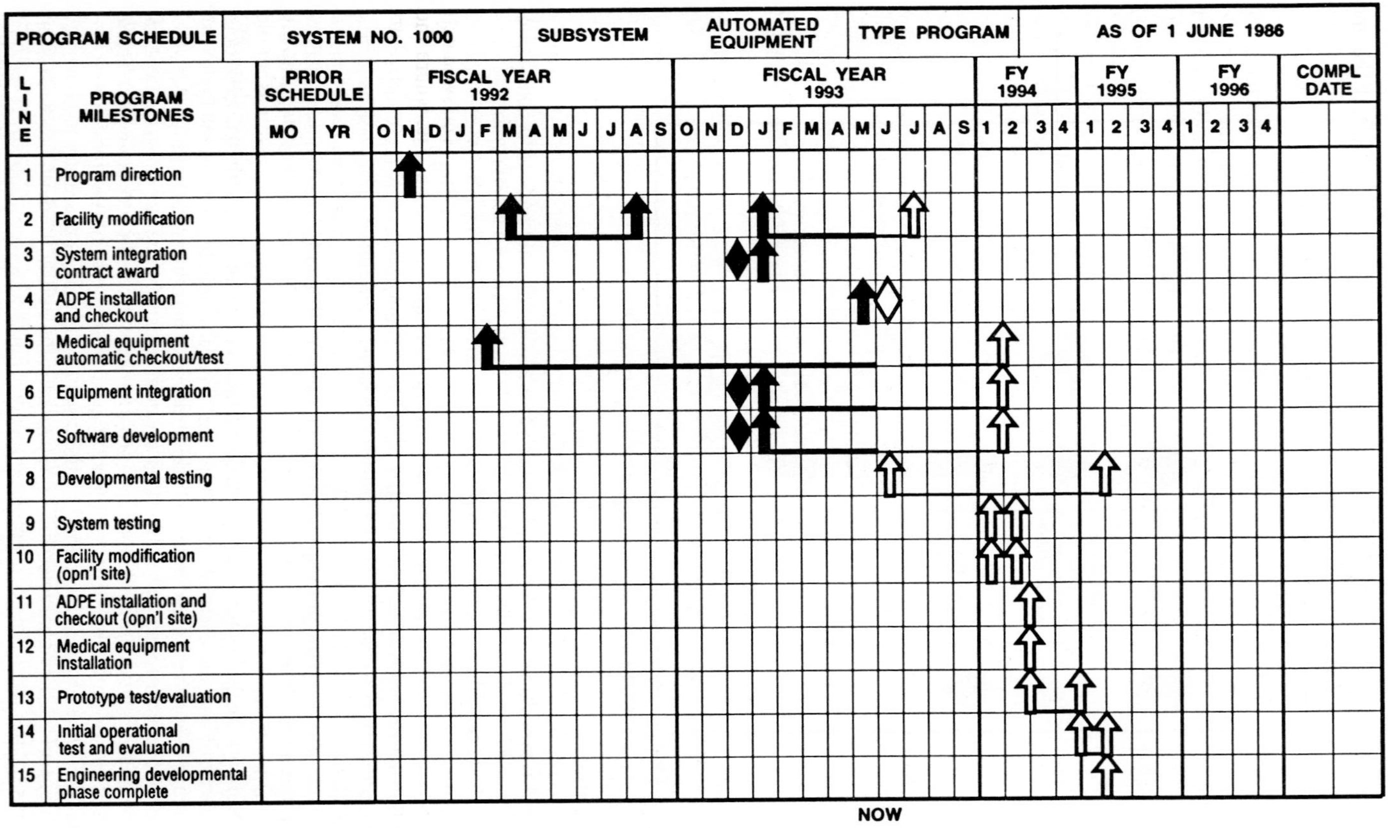

PROGRAM SCHEDULE | SYSTEM NO. 1000 | SUBSYSTEM AUTOMATED EQUIPMENT | TYPE PROGRAM | AS OF 1 JUNE 1986

LINE | PROGRAM MILESTONES | PRIOR SCHEDULE (MO, YR) | FISCAL YEAR 1992 (O N D J F M A M J J A S) | FISCAL YEAR 1993 (O N D J F M A M J J A S) | FY 1994 (1 2 3 4) | FY 1995 (1 2 3 4) | FY 1996 (1 2 3 4) | COMPL DATE

1 Program direction
2 Facility modification
3 System integration contract award
4 ADPE installation and checkout
5 Medical equipment automatic checkout/test
6 Equipment integration
7 Software development
8 Developmental testing
9 System testing
10 Facility modification (opn'l site)
11 ADPE installation and checkout (opn'l site)
12 Medical equipment installation
13 Prototype test/evaluation
14 Initial operational test and evaluation
15 Engineering developmental phase complete

NOW

FIGURE 3.9 Milestone chart. (*Source: Ref. 19.*)

ical path, management can focus its attention on those tasks which affect the end date of the program. For a large-scale network, the critical-path and slack-time calculations are typically done by computer by what are called forward-pass and backward-pass operations along all paths of the network. The forward pass determines the earliest expected completion dates for each event by adding the expected duration of each activity to reach that point. The backward pass starts with the final event and establishes the latest allowable completion for each event earlier in the network by subtracting the expected duration of each activity from the succeeding event. The difference between the earliest expected and the latest allowable completion dates at each point is the slack time at that point. This process lends itself to automated schedule analysis, enabling hundreds and even thousands of activities to be analyzed within a single schedule. An automated critical-path analysis provides identification of critical activities and a schedule which can easily accommodate changes and be reprinted. The automated systems also facilitate sensitivity analysis and provide the ability to summarize the network data into other scheduling techniques such as a milestone chart or a Gantt chart.

The initial use of networking for large-scale development programs was in the 1950s when the critical-path method (CPM) and the program evaluation and review technique (PERT) were developed. The PERT network is event-oriented while the CPM network is activity-oriented, although in the use of these networks this distinction is not critical. The most difficult aspect of either technique is to estimate the expected duration for a specific activity. In the CPM technique, the estimate is typically a single estimate providing the most likely value of the expected duration. In PERT the expected time is derived from a three-point estimate for the activity, based on (1) the most optimistic time a; (2) the most likely time m; and (3) the most pessimistic time b. The PERT technique assumes a beta probability distribution function with the expected time approximated by the following formula:

$$T_e = \frac{a + 4m + b}{6}$$

The difficulties in obtaining and using three-point estimates for each event has gradually caused the use of the PERT network to be reduced in favor of the critical-path method. The network technique is an excellent way of scheduling a large number of activities or events, but the cost of maintaining this schedule must be considered.

One variation of the traditional CPM network is the swan network. The swan network shows activities as horizontal bars against a horizontal time line, where the length of the bar represents the length of the activity. The vertical lines on this chart represent "fences," which indicate that the activity preceding that vertical line must be completed prior to successor activities starting. A comparison of a simple CPM network and its swan equivalent is shown in Fig. 3.10. By using a time dimension in the display of the activities in the network, the swan network pictorially presents the slack and critical path, with the slack shown as dots and the critical path as a bold horizontal line. Note that in this example there are two critical paths, activities BFI and CJK. It is possible to combine a version of the Gantt chart and the information from a network. This is done by showing the activity bar in a succession of small bars to represent individual days. The outline of these small daily bars can be varied from a normal line to a dotted line to represent slack time or a bold line to represent critical path. While this is too troublesome for hand drafting, it is easily portrayed by computer programs.

While these network scheduling techniques were primarily oriented to nonrepetitive activities such as those in the development phase of a weapon system, another technique, called the line of balance (LOB), is useful for repetitive activities such as those in the production phase. The line-of-balance technique consists of

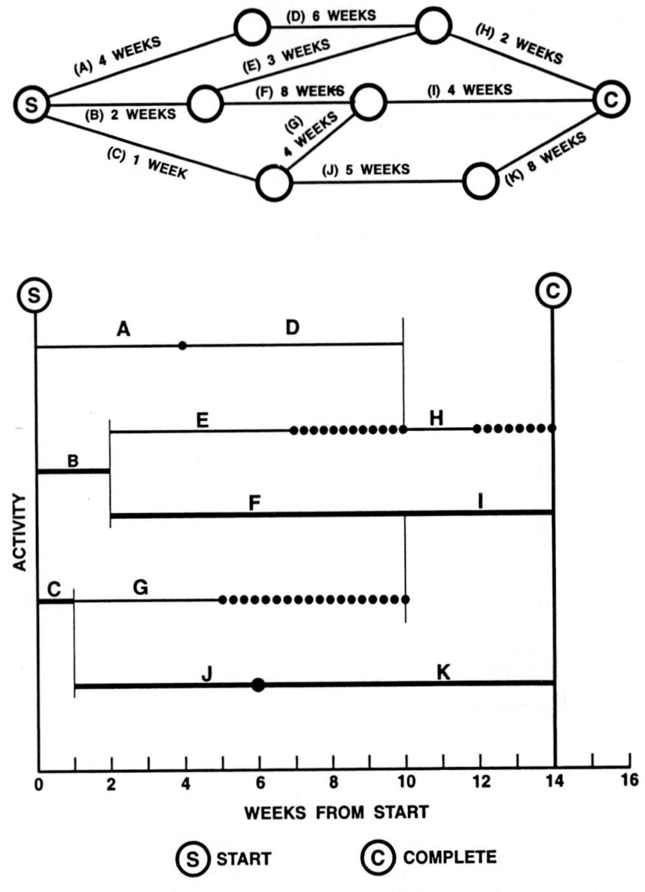

FIGURE 3.10 CPM and swan equivalent networks. (*Source: Ref. 19.*)

three charts: (1) objectives of the program; (2) production plan; and (3) current program status. A line-of-balance chart reflecting each of these portions is shown in Fig. 3.11. The objective chart shows cumulative units on the vertical scale and dates of delivery on the horizontal scale. The contract schedule line shows the cumulative units which the contractor is committed to deliver over the life of the program. The production plan is a lead time chart, indicating the setbacks required to complete all of the required events in order to complete the system in time for final delivery. The production plan can be for any level of assembly, whether it be the total system, subsystem, or subassembly. Each event on the production plan chart is numbered, with number 1 assigned to the longest lead time event. These events are called steps, and the production plan chart shows the interrelationships and the sequence of the major steps, including the lead times. The production status chart shows each step, providing the cumulative number of units through that step as compared to the required number to support the contract schedule. The required number to support the contract schedule is represented as a line of balance on the program status chart as of a given status date. This shows which steps in the production plan are behind or ahead of schedule to meet the contract delivery schedule. This technique, while not a

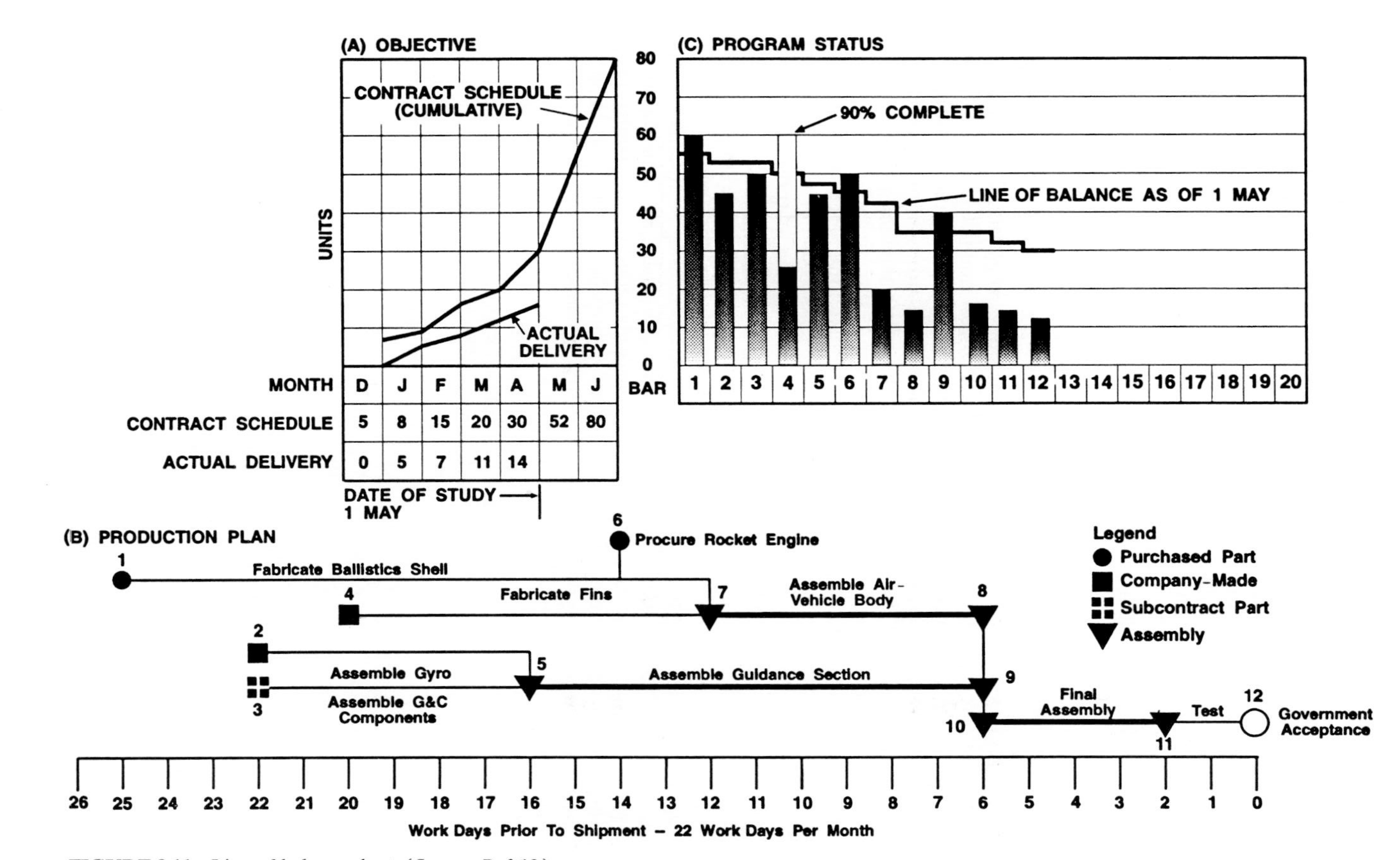

FIGURE 3.11 Line-of-balance chart. (*Source: Ref. 19.*)

great planning tool, is excellent for monitoring the progress of a repetitive production operation.

While not a true scheduling technique, a valuable tool to program control is the process flowchart. Used to depict standard operating procedures (SOPs) or work-cycle processes, it does not usually have a time line. The process flowchart is used to show the key players in each step of a process and the appropriate sequence. For SOPs a fairly standard set of symbols shows whether the step is a decision point, an activity, or a document product. For work-cycle processes the American Society of Mechanical Engineers defines a set of symbols to show operations, transportation, inspection, and storage. The pictorial display of a process flowchart can normally be much more effective in communicating a process than just a narrative description.

Cost Estimating

Cost-estimating activities support a number of decisions in each of the major acquisition processes reviewed earlier. The accuracy of the cost estimates is critical to the success of each of these processes. To better understand the development of cost estimates the following reviews the cost-estimating process, different methods of cost estimating, and different uses of the estimate. The cost-estimating process can be viewed in four primary steps: (1) task identification and estimating methodology; (2) obtaining and evaluating data; (3) developing the estimate; and (4) review and documenting the estimate. The first step, task identification, goes back to work definition of the effort to be costed and should be keyed to the WBS. The purpose of the estimate must be known as well as the ground rules, constraints, and assumptions. The estimating methodology is selected and each step of the process is detailed in a formal estimating plan. The second step is the search for data, and the ability to find the appropriate data may require changes in the selected estimating methodology. The estimator researches, collects, and analyzes the data to be used. Here an important factor is to ensure that the historical data appropriately relate to the planned task. The third step is forming the estimate structure for purposes of data collection as well as actually developing the estimate. The WBS provides the framework for the estimate structure. The formation of the estimate will also depend on whether the estimate is for a development or a production task and the level of detail required of the various elements of cost. The estimating methodology is then used to project the cost to accomplish the specific task being estimated. The final step is review and documentation of the estimate. The usefulness of the estimate is greatly influenced by the quality of the documentation. The quality of the documentation is greatly enhanced by having the format established at the beginning, with all those involved continually contributing to the documentation while the estimate is being developed. The documentation should be sufficiently complete that the estimate can be replicated by another analyst.

A number of cost-estimating methods are available, and the selection of a particular method will be guided by several considerations: (1) availability of historical data; (2) availability of time to do the estimate; (3) adequacy of task definition; (4) level of detail required and purpose of the estimate; and (5) resources available to accomplish the estimate. Depending on these considerations, one or more of the following methodologies is used. There are four major cost-estimating techniques: (1) analogy; (2) parametric; (3) extrapolation from actuals; and (4) detailed engineering.

The analogy method is used when historical data are available on a single system which is comparable to the system being estimated. The cost analyst makes a subjective judgment relative to the similarities between the weapon system being estimated and the system for which historic data are available. If the two systems are totally comparable, then they only have to be adjusted for inflation reflecting differences in time. This is accomplished by utilizing data in terms of a given base year and factoring these data for escalation by an appropriate price index. It is also possible to factor the data from a historical system for differences in complexity. If the system being estimated is considered to be 25 percent more complex, then the estimate for the new system could be as simple as a 125 percent of the cost for the historical system. Obviously the utilization of an analogy whenever the systems are not totally similar will result in an estimate with significant subjectivity in developing the estimate. One way to adapt the analogy estimate when a complete comparable system is not available for comparison is to develop analogy estimates for each of the subsystems within the total system. This may be an excellent choice when the new system is primarily a new combination of existing subsystems for which recent and complete historical data are available. In general we must remember that the analogy estimate is based on a single historical cost data point and there can be considerable risk in basing an estimate on so little data. It is also likely that as the complexity ratio increases, the quality of the analogy estimate decreases. Any complexity factors above 2 could be very suspect unless a convincing rationale were available to support the quantitative determination of the difference between the old system and the new one. In any case, the basis for complexity factors must be clearly documented in a way that is logical and persuasive. The cost analyst may work with a technical specialist who understands both the prior and the new systems and can think of complexity as a number directly related to cost differences.

Analogies must sometimes consider miniaturization factors where the new system will be expected to operate within tighter weight and volume constraints. This must be evaluated in terms of the impact on the expected cost of the new system. It is also possible to have a factor which reduces costs for significant productivity improvements between production of the old and new systems. Here a judgment is required as to the cost ratio of producing a new system using the anticipated manufacturing technology as compared with that of the old system. Because analogy cost estimates require factor value judgments which significantly affect the final results, they should be reviewed with another cost analyst experienced in analogy cost estimating, and it is desirable to perform sensitivity analysis to show how sensitive the total estimate is to a key complexity value. An analogy estimate is most appropriate early in the life cycle of the system when insufficient definition of the system is available to use more detailed methods. It is a valued method when appropriate historical data from a comparable system are available.

The parametric cost-estimating method is a statistical approach in which a cost-estimating relationship (CER) is developed between cost, the dependent variable, and an explanatory variable of a physical or performance parameter of the new system. Thus the CER might have an independent variable of weight which is used to predict the dependent variable of weapon system cost. Because the CER is a statistical relationship, there must be sufficient historical data which are appropriate to the weapon system being estimated to enable a statistically valid predictor. The objective of constructing the equation is to use the independent variables about which information is available, or can be obtained, to predict the value of the dependent variable which is unknown. The selection of an independent variable which has a logical causal relationship to cost is the first step in the determination of the CER. The most critical aspect then becomes the development of a database which has data from systems that are truly comparable to the system being estimated. Because a statistical relationship is being developed, the greater the number of data points, the greater will be the confidence level. The tradeoff for the cost analyst is having a greater number of systems in the database, and therefore a potentially higher confidence level, versus having systems in the database that are not truly comparable to the system being estimated. Given the cost analyst has selected the appropriate database for the particular system being estimated, the next step is the development of the CER equation through the use of statistical regression. This may be a simple linear relationship obtained through linear regression or a more complex curvilinear function using nonlinear regression techniques.

The use of the parametric method is more appropriate early in the program life cycle when the total system cost is being estimated. However, parametrics are used throughout the life cycle where statistical relationships can be developed at lower and lower levels of detail within the WBS. As more becomes known about the system to be estimated, more detail methods of costing become appropriate. When conceptual studies are being done early in system development, one of the important benefits of CERs is the ability to do trade-off and sensitivity analysis with minimum effort. In the source selection process, CERs can also serve the government cost analysts as checks on the reasonableness of the detailed cost proposals of the contractor. There are a number of parametric cost models in use for almost every type of weapon system. However, the cost analyst is best served, if the time is available, to develop his or her own CER utilizing the data most appropriate to the system being estimated.

The engineering method of cost estimating is the most detailed of all the techniques and the most costly to implement. Other names for this method are "bottoms up" or "grass roots," which reflect the technique used in this methodology of estimating at the lowest level within the WBS of definable work. This methodology is normally used by contractors in cost proposals to the government in response to an RFP. The cost proposal will be substantiated by cost and pricing data from the contractor, providing details of how each cost element was estimated.

In the development of cost proposals the contractor will define the lowest level within the WBS that discrete packages of work can be estimated. These estimates can be summarized up the WBS to obtain cost for contract line items. The estimate is also structured to the contractor's organization by the estimation of direct labor hours for each of the labor rate categories involved in the work, typically using several labor rate categories for both engineering and manufacturing. An example for engineering is systems engineer, design engineer, and draftsperson.

In addition to direct labor the other major direct cost elements are direct material and other direct charges. The direct material includes such things as raw material, purchased parts, and subcontracts. Other direct charges include travel, computers, and consultants. The indirect cost is then included in the contractor's cost proposal through the use of overhead and general and administrative rates, which typically have been negotiated in a forward pricing rate agreement with the administrative contracting officer.

The estimation of the direct labor reflects the reason for the name of engineering cost-estimating method. For direct labor within the manufacturing area the estimates are developed by industrial engineers using labor standards, typically developed by time and motion studies. For the engineering labor the estimates are done by the engineers doing a job and reflect their best subjective estimate of the task. The direct material is developed by engineers from the design drawings and results in a bill of material which typically is then priced by personnel responsible for material procurement.

A number of factors, analogies, and parametric relationships are used in determining direct labor, direct material, and other direct charges. In estimating the engineering labor it is common to use a simple parametric relationship between number of drawings and average number of hours to produce a single drawing. Some labor categories such as planning, tooling, and quality control are typically estimated using a percentage relationship to the number of manufacturing hours. These factors are standard factors used by the contractor in all cost proposals for a given year. Factors are sometimes negotiated as a part of the forward pricing rate agreement between the contractor and government. Analogies, parametrics, and actual vendor quotes are common methods for pricing the detailed listing of the bill of material.

The indirect cost of the contractor is recovered through the application of overhead and general and administrative (G&A) rates that are forecasted for each fiscal year over the life of the proposed contract. Indirect costs are defined as those not directly traceable to a single contract. Understanding the contractor's indirect costs is very important, considering that as much as two-thirds of total in-plant costs are indirect. The contractor provides a disclosure statement required by the government to be filed by any contractor having to negotiate defense business within a given fiscal year of greater than $10 million. The disclosure statement provides sufficient detail to ensure that the contractor's cost accounting system is adequately defined, showing which detail elements of cost are direct versus indirect, and for indirect cost, showing the methods of accumulation into the overhead and G&A pools and allocation to contracts. It is a valuable tool for program control personnel required to understand the contractor's cost proposal.

The costing of the direct elements of cost and the application of overhead and G&A are shown schematically in Fig. 3.12. The sequence of activities shown in the waterfall indicate the importance of the initial estimate for a specific unit at a specific point in time, and the necessity to use learning curve theory to develop estimates for all units to be produced. It also shows the requirement to time phase all of the direct cost elements in order that the appropriate escalation can be included to reflect the cost at time of performance.

While the contractor develops a cost proposal using the detail engineering method for each contract throughout the life cycle, the government may also be required to develop a detailed estimate using the engineering method. This is typically possible in production where the government has sufficient data for the cost analyst

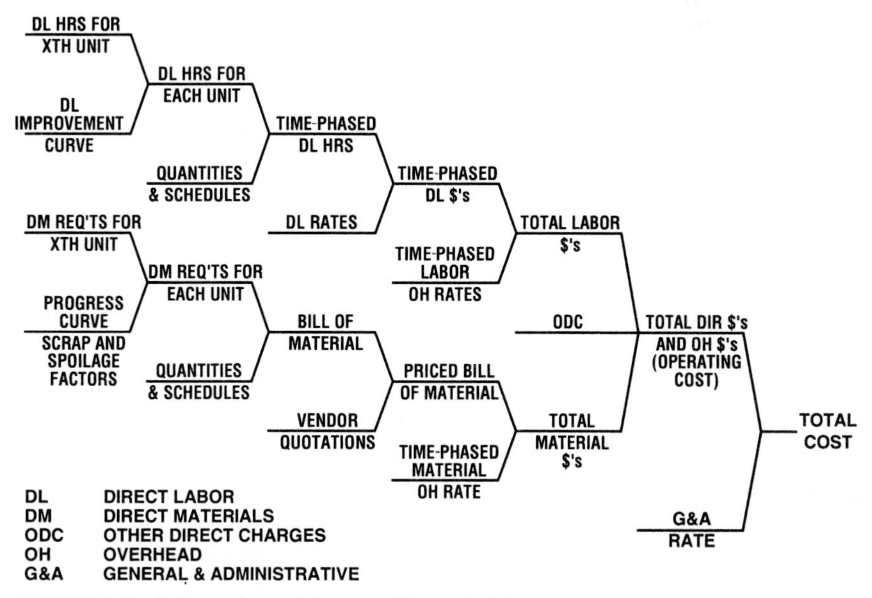

FIGURE 3.12 Labor and material costing. (*Source: Ref. 16.*)

to develop his or her own direct labor and material estimates at a low level within the contract WBS. Typically these government cost estimates will not be at as low a level in the WBS as those performed by the contractor. There is a special type of government engineering cost estimate called a should-cost estimate. The should-cost estimate is developed by a team of government personnel working on site at the contractor's facility. They develop a detailed estimate using the contractor's processes, procedures, and data. A should-cost estimate is normally performed on large sole source production contracts. This is because of the significant cost to the government of using an extensive multidisciplinary team at the contractor's facility for an extended time. In a large sole source procurement, competition is not available to ensure appropriate market prices, and it may be appropriate for the government to use a should cost team early in the production phase.

The last of the estimating methods is extrapolation from actuals. This technique is based on data from previous units of the same system. It is typically used in the production phase after the initial production contract has been completed. This is a very accurate method of forecasting contract costs when the production processes and the weapon system have not changed. To the extent that there have been changes which will affect the learning curve or the extent of work, the extrapolation method must be adjusted. This method relies on the use of a learning curve to extrapolate the historical data. The slope of the learning curve is a major issue in negotiation and is a function of how well the historical data from this weapon system fit a given slope.

In the definition of life-cycle costs it was clear that the cost analysts must be able to conduct all types of cost estimates, ranging from life-cycle costs in the development phases to the flyaway cost for a specific aircraft system to the contract cost for the next phase in the acquisition life cycle. No single estimating method meets all of these requirements, and it is sometimes necessary to use a combination of methods in order to develop the required estimate. The purpose of the estimate, whether it be for estimating a program line item in the budget, or evaluating a contractor's proposal for a specific contract, requires the analyst to draw on all the sources of data and all of the estimating methods appropriate to the individual problem.

Budgeting

The budgeting process for a government project office is critical to ensure that sufficient budget authority is received in the resource allocation process for efficient execution of the program. To be successful, the program must have developed reasonable and accurate cost estimates which reflect the program baseline, containing objectives for key cost, schedule, and technical performance parameters. The budget must reflect the complete set of financial requirements and the structure to ensure that the entire project is included as shown in the project WBS. The remaining step in the budget is to ensure the appropriate time phasing, and this starts with the program master schedule. Utilizing these inputs, budget analysts then formulate budget requests for input to the PPBS process.

For a given weapon system program it is likely that several congressional appropriation categories will be necessary in obtaining budget authority. The appropriations of research development test and evaluation (RDT&E), procurement, operations and maintenance, military personnel, and military construction each have specific rules as to what and how budget inputs are made. Each service has an RDT&E appropriation which pays for expenses incurred in support of research, development, test, and evaluation effort performed during the year for which the funds are appropriated. This is an incremental funding policy which requires dividing all the development effort on the weapon system project into fiscal year increments.

Thus a funding profile for development efforts is developed from the cost expected to be incurred in each fiscal year. The profile represents the tasks in the program WBS that are RDT&E funded.

There are a number of procurement appropriations for each service. The procurement appropriation provides budget authority for all costs expected to be incurred in producing and delivering a weapon system that has entered production. The procurement appropriations are governed by the full funding concept that requires that each year's request contain the funds estimated to cover the total cost to be incurred in completing delivery of a given quantity of usable end items. The intent of this is to ensure that the total weapon system is delivered and that the procurement is not broken up into a number of subsystems. There are exceptions to the full funding policy such as advance procurement, which allows long lead time components to be purchased with funds from a fiscal year prior to the entire weapon system. The full funding concept does not mean that the total service requirement for the weapon system over its entire production run is funded at one time. Rather it means that one year's production of that total requirement is funded, all of which must be delivered as completed units. A one-year production run means 12 months from delivery of the first unit until delivery of the last unit and recognizes that for large complex weapon systems these deliveries may not begin until two or three years after contract award. Procurement funding is available for new obligations (contract awards) for three years beginning with the year for which it is appropriated.

In the operations and maintenance category there are again a number of appropriations for each of the services. These appropriations are for operational expenses as opposed to the procurement of investment assets as in the procurement appropriations. Examples are government civilian salaries, expenses for operation of military forces, training, and education, and base operations support. These appropriations are available for new obligations only in the year for which appropriated and for this reason are referred as having one-year money.

The military personnel appropriations of each of the services are for the cost incurred in compensation for active and retired military personnel and national guard. In addition, these accounts pay for such expenses as those associated with permanent change of duty station, temporary duty, and recruiting.

The last appropriation category is military construction and includes costs to carry out major military construction projects, including the cost of construction design and property acquisition cost. The military construction appropriations are considered an investment account and subject to the full funding concept just as the procurement appropriation. The military construction funds are available for new obligations for five years beginning with the year for which they are appropriated.

An important concept for program control is to ensure that the budget submitted considers the likely contingencies in the weapon system program. There are likely to be changes over time, and these must be anticipated in the budgeting process. The project office should not be in the position of having to request additional funds for effort that might reasonably have been anticipated. The project office must include a reserve in the budget that is commensurate with the risk. In the budget structure there can be no line item called "management reserve," thus these dollars must be included in the cost estimates for the specific tasks.

In the development of the program cost estimates, the historical data and the weapon system estimate were initially in constant-year dollars that enabled comparison of dollars from one year to the next and eliminated the impact of inflation. The budget submittals must be in current-year (then-year) dollars. The following describes the process of converting constant-year dollars into then-year dollars for a specific appropriation budget submittal. It must be recognized that procurement appropriation budget authority being requested will not become an outlay in the year in which budget authority is requested. An outlay profile represents the rate at which an appropriation is spent, expressed as a percentage of the appropriation by fiscal year. Because in the procurement appropriations the actual expenditures may take place over a period of five to seven years, the budget analyst must utilize outlay rates which will spread the budget authority of a given fiscal year over a period of time in which the actual outlay occurs. The outlay rates are provided by the Office of the Secretary of Defense (OSD) and represent the historical averages of programs within that procurement appropriation. While the outlay profile may not be an accurate representation of when the dollars will be expended for a specific weapon system program, the values have been shown to be a good approximation over all of the DoD weapon system programs in that appropriation.

With the budget dollars time-phased in the year in which they will be expended, it is a straightforward calculation for the budget analyst to apply a price level index to reflect inflation as forecasted for the year in which the expenditure takes place. The price level index is called a compound index and relates the price level for each fiscal year to a base year. Thus the compound index shows the incremental price increase from one year to the next and enables conversion of the constant-year dollars for a given base year to then-year dollars which are submitted in the budget. The application of outlay rates and the compound index is converted into a single-step process by calculating a weighted index where the outlay rate for a specific year is multiplied by the price index for that year and summed across all of the years in the outlay profile. This sum of the price escalated outlays is called a weighted index, or composite index. The budget analyst then applies this composite index to the constant-year dollar cost estimates to obtain then-year budget dollars requested for that specific appropriation for that year.

The indices used by DoD are provided by the Office of Management and Budget (OMB). The estimate of inflation which these indices provided is an administration forecast, which typically understates the inflation that will actually be experienced. In part this is because officially published inflation predictions tend to become self-fulfilling prophecies because they are used in determining the budget available for all of the purchases by the government. Typically the pressure is on the administration to hold inflation at lower levels, and their optimistic predictions are typically lower than those developed by the use of econometric models in the private sector. This understatement of inflation can be a major cause of weapon system program cost overruns during periods of high inflation.

The program control budget analysts are responsible for submitting the RDT&E and procurement exhibits which support the budget estimate submissions (BESs) by the services to OSD and updated for the President's budget submission to Congress. The RDT&E exhibits provide information by program element line item and define a development effort providing detailed financial and justification information. The procurement exhibits, or P-forms, are by procurement line item and provide the budget and quantity profile with narrative description and justification for the procurement appropriations. These forms are exceedingly important to the project as errors or inconsistencies may well cause the loss of budget in the review process.

Performance Measurement

Subsequent to contract award the emphasis shifts to analysis of performance in the execution of the contract. The primary responsibility for execution of the contract and analysis of performance while work is in progress is with the contractor. In the government role of monitoring contractor performance, the DoD has developed and implemented a criteria approach which sets forth the characteristics and capabilities which should be inherent in an effective cost and schedule control system of the contractor. The cost/schedule control systems criteria (C/SCSC) are not a management system or report but a set of standards for measuring the adequacy of the contractor's management control systems. While it is necessary that the contractor's management systems reflect the unique needs of their organization and management philosophies, the systems must provide the data and capabilities specified in the criteria in order to be acceptable to DoD.

A critical aspect of the criteria approach is that a contractor's management systems satisfy both the internal needs of the contractor and the information needs of DoD. Performance measurement systems provide measurement of the work compared to the consumption of resources in a way that accurately measures the status of the weapon system program. The system must provide valid and timely data and enable analysis in a way that provides insight into future problems. The C/SCSC performance measurement system provides a framework in which the work is defined, scheduled, and budgeted to the lowest level of an organization responsible for its accomplishment. It must also provide an accurate way to measure progress against a performance baseline.

The introduction of C/SCSC in the late 1960s incorporated the very important concept of earned value. Earned value is an objective measure of how much work has been accomplished or obtained by summing the budget planned for those tasks now completed. Prior to the earned value only two elements were used in measuring cost performance: (1) the plan which is the budgeted cost of work scheduled (BCWS); and (2) the actual cost of work performed (ACWP). The earned-value concept added the budgeted cost of work performed (BCWP). Without the earned-value concept it was not possible to recognize which portion of the planned work had been accomplished. This concept enables the variance to be divided into cost and schedule variances. The cost variance is the difference in the amount of budget earned for work completed and the cost actually incurred. It is expressed as BCWP – ACWP. A negative cost variance is a true variance in dollars between the actual cost and the budgeted cost for the same set of tasks. The schedule variance is equal to BCWP – BCWS. This is an interpretation of the schedule variance in terms of dollars, and a negative schedule variance indicates a behind schedule condition that is going to take at least the variance amount of dollars to get back on schedule. Accurate analysis of the schedule condition in terms of time requires utilization of the program network or milestone schedules. Cost and schedule variances are displayed graphically in Fig. 3.13, which shows the variances to date on a given contract.

C/SCSC has become widely used within the government and has been implemented in the Department of Energy, the National Aeronautics and Space Administration, and the Federal Aviation Agency as well as in the Department of Defense. There is a C/SCSC joint implementation guide (JIG), dated October 1987, which provides a common approach to industry by each of the services within DoD. The criteria provided in the JIG are divided in five major sections: organization, planning and budgeting, accounting, analysis, and revisions. The organization criteria ensure that the contract work is defined appropriately using the contract work breakdown structure (CWBS) while integrated with the contractor's planning, scheduling, budgeting, work authorization, and cost accumulation systems. The integration of the CWBS with the contractor's functional organizational structure is required such that cost and schedule performance measurement can be managed in both these frameworks. The planning and budgeting criteria ensure that the authorized work is scheduled in a way that the sequence of work, task interdependencies, and program milestones can be measured at the lowest level within the CWBS. The criteria require time-phased budgets by task and organization to enable the identification and control of both direct and indirect costs. Direct cost budgets must identify

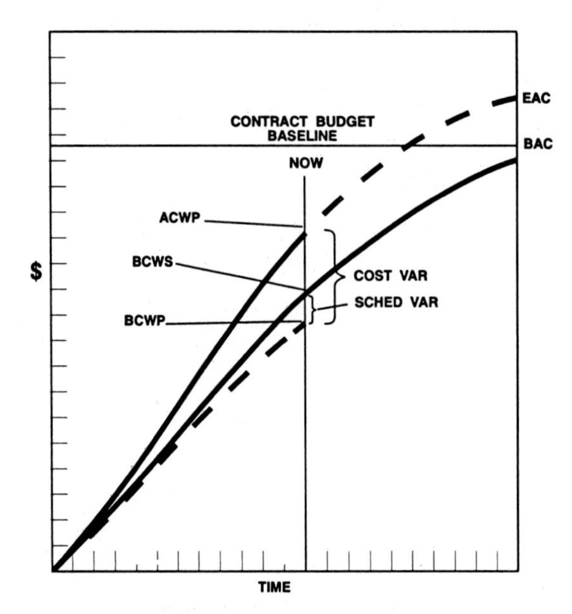

FIGURE 3.13 Cumulative performance. (*Source: Ref. 2.*)

management reserves and ensure contractor reconciliation of contract budgets with authorized work. The accounting criterion requires that the contractor's cost accounting system will summarize direct cost without allocation of a single cost account into two or more WBS elements or two or more organizational elements. The accounting criterion also ensures that the material accounting system enables accurate determination of cost and variances. The analysis criterion requires that the accounting system will provide BCWS, BCWP, and ACWP at the cost account level on a monthly basis and that the reason for variances can be determined. A final criterion section on revisions and access to data requires timely inclusion of contract changes such that the integrity of the relationship between work authorized, scheduled, and budgeted is maintained and reconcilable to the original baseline. Without this baseline integrity during changes, variance analysis is meaningless.

A critical step in performance measurement is establishing the budget baseline. In the transition from contract negotiations to contract award, budgeting to develop a baseline is tedious at best. The negotiations internal to the contractor's organization between contractor program management and functional elements may be just as intense as the contract negotiations between government and contractor. The necessary result is a timely, well-planned, and realistic baseline for controlling internal performance and reporting contract status to the government.

The summation of all the time-phased budgets from the lowest level in the CWBS up to level I, the total contract level, provides the performance measurement baseline for the contract. The performance measurement baseline represents the cost, schedule, and technical content of the work to be performed, not including management reserve. Management reserve is an amount of budget withheld for management control purposes rather than being allocated as budget for a specific task. Management reserve is budget set aside for future unknown, unexpected work effort which may arise during the contract performance and can only be used for work which is not in the scope of a given cost account. It cannot be used to cover up over-

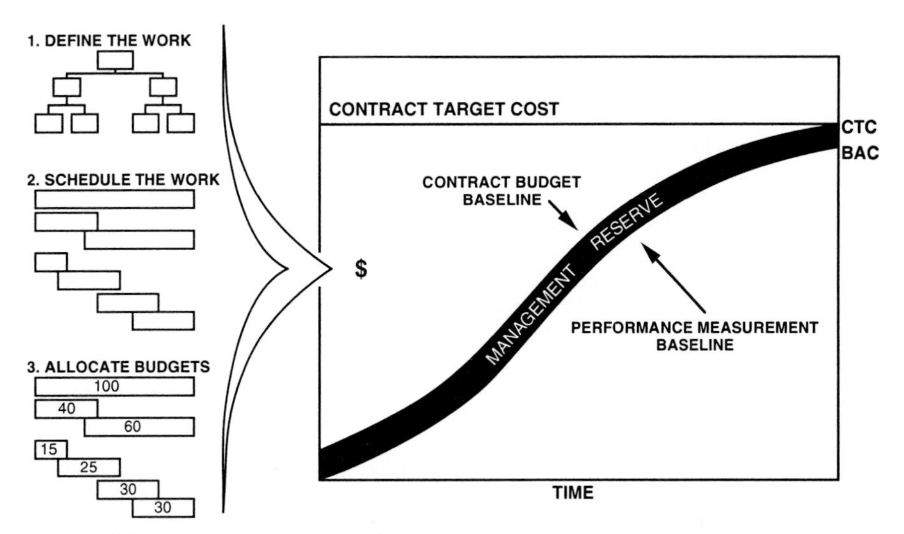

FIGURE 3.14 Establishing the baseline (an interactive three-step process). (*Source: Ref. 17.*)

runs. To use management reserve for eliminating overrun variances would defeat the purpose of the system, which is to provide identification of problem areas. The time-phased performance measurement baseline totals to the budget at completion (BAC). When management reserve is added to the time-phased performance measurement baseline, it becomes the contract budget baseline, which sums to the contract target cost (CTC) at the projected completing date. The process of establishing the performance measurement baseline through defining the work, scheduling the work, and allocating budgets to specific tasks is shown in Fig. 3.14.

To this point, work below the cost account level has not been discussed. The planning and budgeting criteria require that all authorized work assigned to cost accounts be time-phased in segments of work. All work must be categorized into one of three different types of effort: (1) discrete tasks which are identified as work packages; (2) level of effort activity which is measured through the passage of time such as sustaining engineering; and (3) apportioned effort which is related in direct portion to the performance of discrete tasks. All contract work must eventually be planned in one of these three categories. The work packages are short-term detail tasks identified by the contractor for accomplishing work under the contract. They represent units of work distinguishable from all other work packages and are assigned to a single organizational element. Each work package will have a budget and scheduled start and completion dates. At the beginning of the contract it is frequently not possible to plan in detail the discrete work packages of work across the entire contract life cycle. To aid in this, a planning package is utilized which is a logical aggregation of work within the cost account that can be identified and budgeted, but is not yet divided into work packages. These planning packages are then divided into discrete tasks during the life of the contract when greater definition is available. The division of the planning packages should be accomplished no later than six months before the work is accomplished.

A goal of C/SCSC is to maximize the amount of discrete work effort in the cost accounts so that more objective measurements can be used in determining contract progress. To implement the earned-value concept, techniques must be utilized which enable objective calculation of BCWS and BCWP for work packages. For discrete work packages there are a number of different techniques, an example of which is the weighted milestone technique using the 50/50 rule. Using the milestones for work package start and completion and the 50/50 rule, half of the budget would be earned when the work package is opened, with the other half earned on completion of the work package. Other weightings could be 0/100 or 100/0. To calculate BCWS, 50 percent of the budget for each work package schedule to start before time now is added to 50 percent of the budget for each work package which is scheduled to complete before time now. To calculate BCWP, 50 percent of each work package which has already started is added to 50 percent of each work package that has already been completed. In calculating BCWS and BCWP for apportioned work, the budget value for the apportioned effort would be based on the percentage of work completed in the related work package. For level of effort work, time is the only measurable parameter; thus the earned value is equal to the planned value of work performed, and BCWS and BCWP are always the same. In this circumstance, no schedule variance can occur, although cost variances may appear if actual expenditures vary from the level of effort planned.

Reporting of performance measurement data by the contractor to the government is accomplished using either the cost performance report (CPR) or the cost/schedule status report (C/SSR). In general the CPR is required on major contracts (greater than $60 million for development or $250 million for production in FY 90 constant dollars), and the C/SSR is used on small nonmajor contracts of over $2 million and a duration of over one year. Depending on the size and type of contract, either the CPR or the C/SSR must be identified and made a requirement in the contract data requirements list (CDRL). Without inclusion in the CDRL, no performance reporting is required, although the contractor's management control systems must be in compliance with the criteria if they meet the above values shown for CPR reporting.

The CPR summarizes cost and schedule status for all authorized work under a contract and is reported monthly to the program office. Although contractors may use their own reporting formats as long as they include the required data, there are five specified report formats for the CPR: format 1, work breakdown structure—current and cumulative cost and schedule status by WBS elements, typically reported at level 3; format 2, functional categories (Fig. 3.15)—current and cumulative cost and schedule status by organizational entities; format 3, baseline report provides changes to the performance measurement baseline from contract inception; format 4, manpower loading—actual and planned direct manpower by functional category; format 5, problem analysis—a narrative explanation of costs, schedule, and other performance measurement problems. The C/SSR shown in Fig. 3.16 provides contract cost and schedule performance data structured by WBS elements, including a narrative explanation summarizing problems and corrective actions. Further the C/SSR does not require current period reporting and only cumulative data are presented.

The cost analysts generally receive the CPR or C/SSR within 25 days after the close of the contractor's accounting month. The information provided is the prime source of cost performance data for use by the cost analyst. There are a number of techniques available for analysis of current contract status using performance indices, trend analysis, and the forecasting of a contract estimate at completion (EAC).

The calculation of cost variance (BCWP – ACWP) and schedule variance (BCWP – BCWS) is the starting point for analysis of current program status. A cost performance index (CPI) measures the cost efficiency with which work has been accomplished and is calculated by the formula CPI = BCWP/ACWP. A CPI of 1.0 indicates that for every dollar of actuals, a dollar's worth of work was accomplished. An index of 0.9 would indicate 90 percent efficient performance. Also required is analysis of the projected contract variance at completion, which is calculated by subtract-

COST PERFORMANCE REPORT - FUNCTIONAL CATEGORIES

CONTRACTOR: LOCATION: RDT&E ☐ PRODUCTION ☐	CONTRACT TYPE/NO.:	PROGRAM NAME/NUMBER.:	REPORT PERIOD	SIGNATURE, TITLE & DATE	FORM APPROVED OMB NUMBER 22R0280

QUANTITY	NEGOTIATED COST	EST COST AUTH, UNPRICED WORK	TGT PROFIT/FEE %	TGT PRICE	EST PRICE	SHARE RATIO	CONTRACT CEILING	EST CONTRACT CEILING

ORGANIZATIONAL OR FUNCTIONAL CATEGORY	CURRENT PERIOD					CUMULATIVE TO DATE					REPROGRAMING ADJUSTMENTS		AT COMPLETION		
	BUDGETED COST		ACTUAL COST WORK PERFORMED	VARIANCE		BUDGET COST		ACTUAL COST WORK PERFORMED	VARIANCE		COST VARIANCE	BUDGET	BUDGETED	LATEST REVISED ESTIMATE	VARIANCE
	WORK SCHEDULED	WORK PERFORMED		SCHEDULE	COST	WORK SCHEDULED	WORK PERFORMED		SCHEDULE	COST					
(1)	(2)	(3)	(4)	(5)	(6)	(7)	(8)	(9)	(10)	(11)	(12)	(13)	(14)	(15)	(16)
COST OF MONEY															
GEN AND ADMIN															
UNDISTRIBUTED BUDGET															
SUBTOTAL															
MANAGEMENT RESERVE															
TOTAL															

FORMAT 2

FIGURE 3.15 CPR format 2, functional categories. (*Source: Ref. 17.*)

CONTRACTOR: LOCATION: RDT&E ☐ PRODUCTION ☐	COST/SCHEDULE STATUS REPORT			SIGNATURE, TITLE & DATE	FORM APPROVED OMB NUMBER 2250327
	CONTRACT TYPE/NO.	PROGRAM NAME/NUMBER	REPORT PERIOD:		

Contract Data

(1) ORIGINAL CONTRACT TARGET COST	(2) NEGOTIATED CONTRACT CHANGES	(3) CURRENT TARGET COST (1) + (2)	(4) ESTIMATED COST OF AUTHORIZED, UNPRICED WORK	(5) CONTRACT BUDGET BASE (3) + (4)

Performance Data

WORK BREAKDOWN STRUCTURE	CUMULATIVE TO DATE					AT COMPLETION		
	BUDGETED COST		ACTUAL COST WORK PERFORMED	VARIANCE		BUDGETED	LATEST REVISED ESTIMATE	VARIANCE
	Work Scheduled	Work Performed		Schedule	Cost			
(1)	(2)	(3)	(4)	(5)	(6)	(7)	(8)	(9)
GENERAL AND ADMINISTRATIVE								
UNDISTRIBUTED BUDGET								
MANAGEMENT RESERVE								
TOTAL								

FIGURE 3.16 Cost/schedule status report. (*Source: Ref. 17.*)

ing the government estimate at completion (EAC) or the contractor's latest revised estimate at completion (LRE) from the budget at completion (BAC). The projected EAC is typically determined based on a cumulative to-date cost performance index.

The schedule performance index (SPI) indicates the efficiency with which the planned work has been done on time and is given by the formula SPI = BCWP/BCWS. An index of 1.0 means that performance is on schedule, while any value less than 1.0 indicates the contract is behind schedule and values above 1.0 indicate an ahead of schedule condition. The schedule variances are most meaningful when calculated using cumulative data and when accompanied by a detailed review of the program network schedules.

Figure 3.17 provides a graphic display of both SPI and CPI, where values below 1 indicate a behind schedule condition and a cost overrun. Calculation of an EAC based on performance factors relates past performance to the budgeted cost of work remaining. A commonly used formula for this is $EAC = BAC/CPI_{CUM}$. An alternative to the CPI as a performance factor is the use of a cumulative CPI and cumulative SPI combined. The CPI-based performance factors become more useful with the quantity of data increasing as the quantity of work remaining decreases. An SPI, however, must be used with caution and is more helpful as an indicator early in the contract. A number of techniques are available and cost analysts must use their best judgment in determining the best EAC method. There are two important software programs which can assist the cost analyst in evaluating contract status. These are the contract appraisal system (CAPPS) available through the Defense Systems Management College and the performance analyzer program developed by the Air Force.

C/SCSC implementation involves both validation of the contractor's system and surveillance of that system through subsequent application reviews. The *Joint Implementation Guide* provides a checklist and guidance for the evaluation of contractor's proposals concerning C/SCSC implementation and conducting demonstration reviews. The phases of a typical review cycle include evaluations preliminary to contract award, an implementation visit after contract award, readiness assessment after the implementation visit, and a demonstration review leading to official acceptance. The contractor provides a C/SCSC system description which details the system's compliance with the criteria.

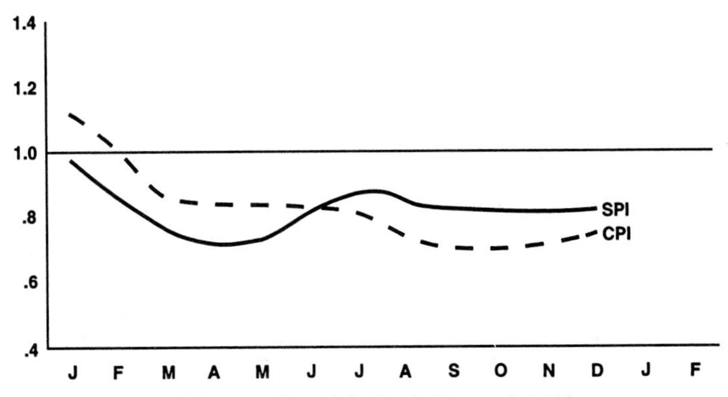

FIGURE 3.17 Variance trend graph (indices). (*Source: Ref. 17.*)

Program Integration and Reporting

Program control provides the organizational focal point for integration of the weapon system program activities. This integration is required in the key processes of system acquisition and in the documentation of those processes. There are three key integrated program reports which describe and rank the program within DoD and to the Congress. These are the integrative program summary (IPS) which is required at each milestone in the acquisition life cycle; the defense acquisition executive summary (DAES) report program which provides status to the Under Secretary of Defense for Acquisition [USD(A)]; and the selective acquisition report (SAR) which reports program status to Congress. The following provides an overview of each of these reports as described in DoD 5000.2-M, "Defense Acquisition Management Documentation and Reports."

The IPS is the primary decision document provided to acquisition executives at each of the milestones in the acquisition life cycle. It provides a comprehensive summary of program structure, status, assessment, plans, and recommendations by the program manager and the program executive officer. The functions of the IPS report include: (1) providing program execution status for cost, schedule, and performance; (2) assessing program alternatives including the rationale for those felt to be most promising; (3) assessing program risk and documenting plans for risk reduction; and (4) information on program cost, schedule, and performance necessary to establish the acquisition program baselines. The format of the IPS and its supporting indexes as provided in 5000.2-M is shown in Fig. 3.18. Following submittal of the IPS to the acquisition executive, an integrative program assessment will be prepared by the appropriate defense acquisition board committee and submitted to the acquisition executive. This concept applies to all milestone decision authorities, but is streamlined for smaller programs.

The second major integrative report is the DAES. This report provides quarterly reporting to the Under Secretary of Defense for Acquisition [USD(A)], who is the defense acquisition executive for all major acquisition category (ACAT I) programs. The intent of the report is to provide advance notice of potential problems on the theory that improved communication and recognition of problems can lead to early resolution. The DAES report format is provided in 5000.2-M with eight reporting sections, as shown in Fig. 3.19. The program manager will assess program status in nine program assessment areas. These are performance characteristics test and eval-

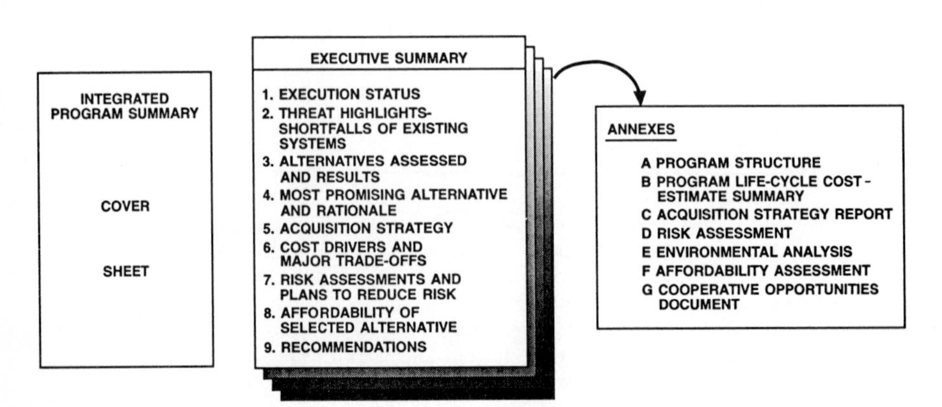

FIGURE 3.18 Integrated program summary. (*Source: Ref. 10.*)

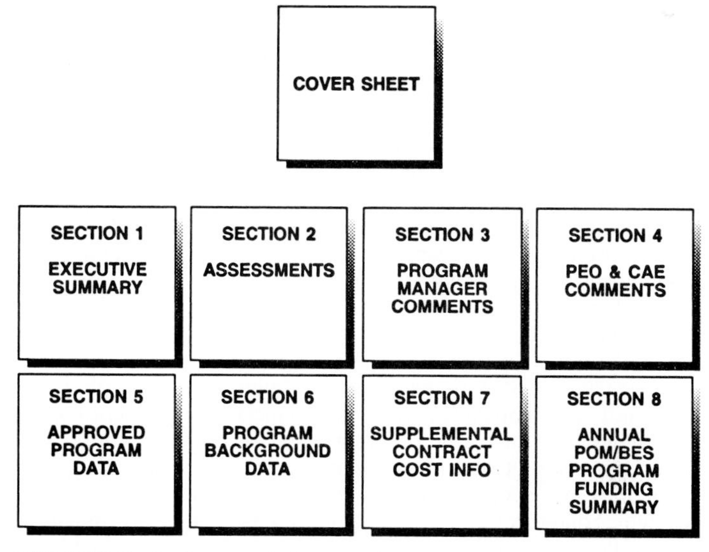

FIGURE 3.19 Defense acquisition executive summary report structure. (*Source: Ref. 10.*)

uation, logistics and readiness, cost, funding, schedule, contracts, production, and management structure. The program manager will provide a narrative on each of these, including status of corrective actions and changes since the last report. The program executive officer (PEO) comments on the significance of the problems reported by the program manager and the risk to the acquisition program baseline. The report also includes summary cost and funding data, including unit-cost data, data by appropriation and fiscal year, contract performance data, and production quantities. These are all compared to the acquisition program baseline.

The SAR is a comprehensive summary of cost schedule and performance data that is provided by DoD to Congress. The SAR establishes baselines for unit-cost reporting and provides the current estimate of total program acquisition cost, schedule, and performance data. The SAR is submitted annually for the period ending December 31. Additional quarterly reports are required when there has been a 15 percent or more increase in program acquisition unit cost or current production unit costs or a six-month or greater delay in the current estimate of any scheduled milestone. These unit-cost SAR baselines reflect the estimates of the program at the last milestone decision point and remain in effect until the next milestone decision point.

These milestone cost estimates entitled planning estimate prior to milestone 2, development estimate prior to milestone 3, or production estimate after milestone 3 are provided in both base-year and then-year dollars. Cost variances are classified according to the following categories: inflation, quantity, schedule, engineering, estimating, other, and support. The cost variances are identified and reported according to these categories. The importance of this report and of having to report cost variances through DoD executives to Congress is clear. It also provides an example of the degree of oversight with which the program manager of a major program must live.

The preceding reports as well as the activities and processes previously described make clear the requirement for an automated information system to support the needs of the program office and particularly program control. No single automated system is currently in use that provides a centralized database and automated analysis and reporting within the context of the DoD 5000.1 and 2 series. A great deal of software is available to assist in specific activities of program control, but not an integrated system specifically for the government weapon system program office.

A new system is under development at the Defense Systems Management College, entitled "Program Manager's Support System" (PMSS). This system is an application of decision support systems technology to the specific problems of the program manager in a government program office. PMSS utilizes a centralized database with distributed processing on computers capable of running UNIX or centralized processing for microcomputers not capable of running UNIX. The system provides overview of the project, impact assessment, scheduling, and financial management. The scheduling and financial management components are the basic building blocks of the system with capability to easily manage both the funding and the schedule information at the task level. The scheduling component provides both network and Gantt chart capability, while the financial management component addresses PPBS submissions by the program office as well as execution analysis. The project overview component enables summary analysis in any of eight functional categories in measuring cost or schedule performance. The impact assessment component evaluates budgets, cost, and schedules, and provides assessments and recommendations. The system utilizes expert systems software to provide analysis of variances to plan. The system is task-oriented and utilizes the program WBS as the integrative framework for cost and schedule data.

PMSS 2.0 completed for initial site installation in mid-1992 and is available to all government program offices. This integrated software system will continue to need added components to address such features as cost estimating and technical performance measurement. However, the utilization of a common centralized database to assist in analysis and reporting can be a major productivity improvement. The needs of the government program office, and those of program control in particular, require the continued development of software tools to assist in information management.

BIBLIOGRAPHY

1. *AFSC Financial Management Handbook,* 2 vols., Air Force Systems Command, Directorate of Programs and Budget, 1990.
2. *The AFSE Cost Estimating Handbook Series,* 2 vols., Analytic Sciences Corp., 1987.
3. Arnavas, D. P., and Ruberry, W. J., *Government Contract Guidebook,* Federal Publications, 1987.
4. Brown, J. W., Fleming, Q. W., and Humphreys, G. C., *Project and Production Scheduling,* Probus Publishing, 1987.
5. Cheslow, R. T., and Nelson, J. R., *The Executive Workshop on Cost/Performance Measurement,* Institute for Defense Analysis, Oct. 1989.
6. *Contract Management for Program Managers Course,* Defense Systems Management College, 1992.
7. *Cost Realism Handbook,* Navy Office for Acquisition Research, 1985.
8. *Cost/Schedule Control Systems Criteria Joint Implementation Guide,* Department of Defense, Oct. 1987.
9. Department of Defense, *DoD Directive* 5000.1, Feb. 1991.
10. Department of Defense, *DoD Instruction* 5000.2, Feb. 1991.

11. Department of Defense, DoD Manual 5000.2M, *Defense Acquisition Management Documentation and Reports,* Feb. 1991.
12. Fleming, Q. W., *Cost/Schedule Control Systems Criteria,* Probus Publishing, 1983.
13. McNaught, W., *Defense Requirements and Resource Allocation,* National Defense University, Aug. 1989.
14. *Program Control Handbook,* vols. I–IV, Electronic Systems Division, Hanscom AFB, Mass., Feb. 1983.
15. *Program Management Course* (Acquisition Policy and Environment, Principles of Program Management, and International Program Management), Defense Systems Management College, 1992.
16. *Program Management Course* (Contract Management and Contractor Finance), Defense Systems Management College, 1992.
17. *Program Management Course* (Funds Management, Cost Schedule Control, and Program Management), Defense Systems Management College, 1992.
18. *The Program Manager's Notebook,* Defense Systems Management College, Mar. 1989.
19. *Scheduling Guide for Program Managers,* Defense Systems Management College, Jan. 1990.
20. *Systems Acquisition Funds Management Course,* Defense Systems Management College, 1992.
21. *Systems Engineering Management Guide,* Defense Systems Management College, 1990.

CHAPTER 4

TEST AND EVALUATION*

James Lis

Chief Master Sergeant James Lis is currently cargo systems integration specialist, C-17A, Douglas Aircraft Company. He was formerly a strategic aircraft systems test manager at the Air Force Operational Test and Evaluation Center, Kirtland Air Force Base, N. Mex. Chief Lis has over 25 years of experience and training in the aircraft loadmaster field.

INTRODUCTION

Within the Department of Defense (DoD) test and evaluation (T&E) is one of the essential functions of U.S. government system acquisition as well as a critical component of successful system development, production, and deployment. Formal planning, effective testing, and accurate reporting are the three key ingredients to effective test management. These critical elements are controlled by management tools which are highly structured, extremely dynamic, and evolutionary in nature. This chapter is structured to provide an overview of how acquisition and T&E relate to and affect one another throughout a system's life cycle.

ACQUISITION AND T&E MANAGEMENT

The collective DoD process of system acquisition and T&E is directed through the Office of the Secretary of Defense (OSD) and managed or overseen by a senior executive staff consisting of both acquisition and T&E principals. This important body establishes policies and directives which control T&E management activities throughout DoD. Figure 4.1 illustrates this structure and quite notably demonstrates the independence in management oversight between research and engineering and operational T&E. This separateness allows the T&E process to effectively focus on technical and operational criteria independently and in relative terms.

The forum which governs selected major defense acquisition program decisions is

*This chapter represents the views of the author and does not necessarily reflect the official opinion of the Department of Defense or any U.S. government agency which directs the development and implementation of acquisition or test and evaluation policies and procedures.

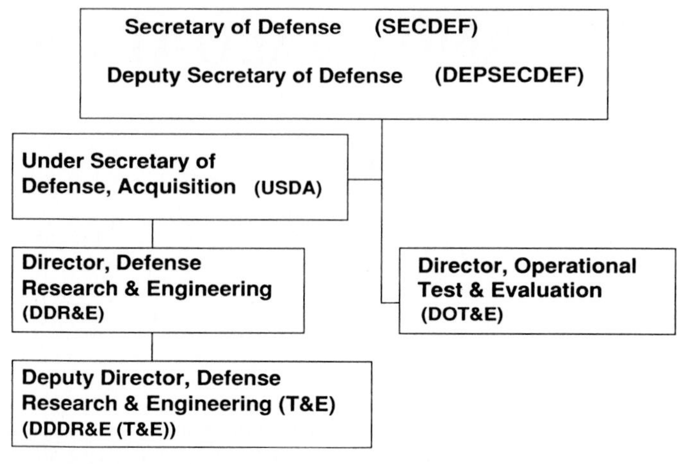

FIGURE 4.1 DoD T&E principals.

the Defense Acquisition Board (DAB). Inasmuch as national defense planning strategies, politics, and the U.S. Congress ultimately determine the fate of a major defense system, or any government-procured system for that matter, the DAB's decisions are generally respected and supported by Congress. The DAB convenes at major program decision points (milestones) to review the progress of a particular program and, based on evaluations from service acquisition executives, acquisition committees, and associate board members, recommend to the Under Secretary of Defense for Acquisition whether to proceed into the next phase of acquisition, redirect a program's evolution, or terminate acquisition efforts altogether. Figure 4.2 demonstrates the composition of the DAB and, in particular, shows the relationship the directors of T&E (DDR&E and DOT&E) have with this conclave. Assessments and recommendations made by the directors are based on independent evaluations conducted by their respective subordinate military department acquisition and test organizations. The formal DAB process is primarily reserved for major acquisition

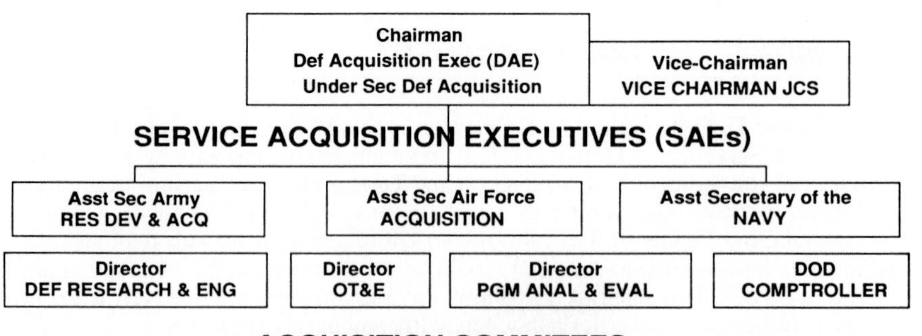

FIGURE 4.2 Defense Acquisition Board (DAB).

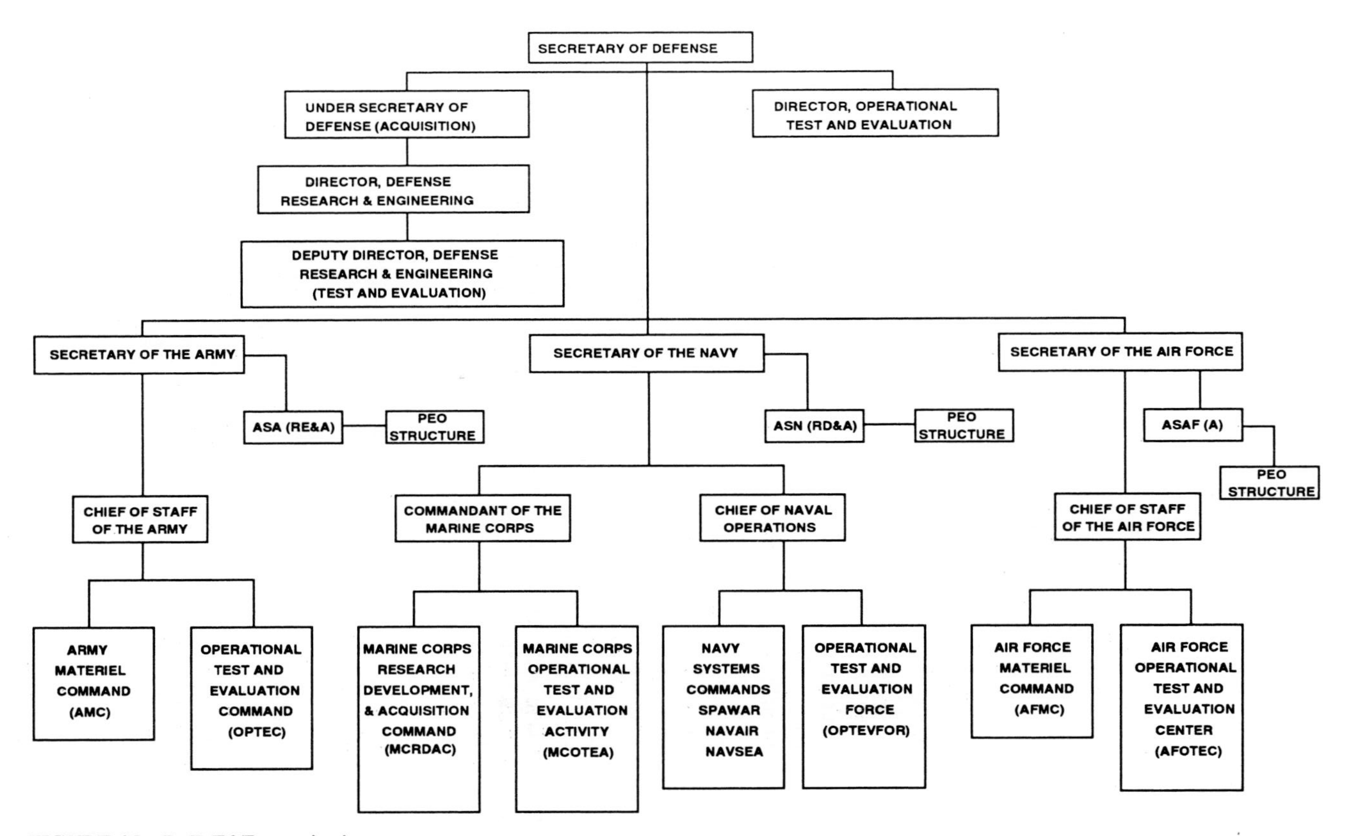

FIGURE 4.3 DoD T&E organization.

programs (high cost) or for programs which are considered important enough (high risk) to warrant oversight at the OSD level. Programs which are less critical in both cost and risk are managed at the DoD component level by designated service acquisition executives or ancillary milestone decision authorities (MDAs). Typically both major and nonmajor programs follow the same paths and are managed in the same fashion, but there are exceptions.

The mechanics of functionally supporting all system acquisition and T&E has been delegated by the Under Secretary of Defense for Acquisition through the service secretaries and service chiefs to designated subordinate acquisition and operational test commands or agencies. Figure 4.3 depicts the overall DoD T&E management structure and further demonstrates the separateness between developmental and operational test agencies.

Taking this a step further are the recently revised DoD 5000 series defense acquisition management policies and procedures. These changes substantially alter and align system acquisition and test management practices so that they:

1. Expand and standardize test management activities so that they occur early in the acquisition life cycle
2. Increase the level of formal test planning activities so that acquisition, user, and test agencies more closely align themselves throughout the life of a particular program
3. Initiate formal testing, evaluating, and reporting activities earlier in the system acquisition life cycle

This formalized relationship between T&E and acquisition also provides test managers with direct responsibility and accountability for both the conduct and the completeness of T&E. Moreover, the resultant level of success will be directly proportional to how well testing is planned, executed, and documented. When properly conducted, T&E:

1. Provides senior decision makers with information necessary to identify, assess, and reduce acquisition risks
2. Verifies to what degree a system's design quantitatively or qualitatively meets technically stated performance and operationally stated mission objective criteria
3. Verifies that the system is the right system for its intended use
4. Most importantly, complements and supports a sound acquisition decision

THE FUNDAMENTALS OF T&E

Understanding and managing T&E is a challenging process. It is complicated by the fact that it involves contractors, users, developers, and support agencies who tend to view T&E in abstract terms. Figure 4.4 humorously illustrates this in a remarkably accurate fashion.

What Is T&E?

There are two fundamental categories of T&E, formally known as developmental test and evaluation (DT&E) and operational test and evaluation (OT&E). *Test* in the T&E can be defined as a technically driven (developmental test) or operationally oriented (operational test) activity whose goals are to quantitatively or qualita-

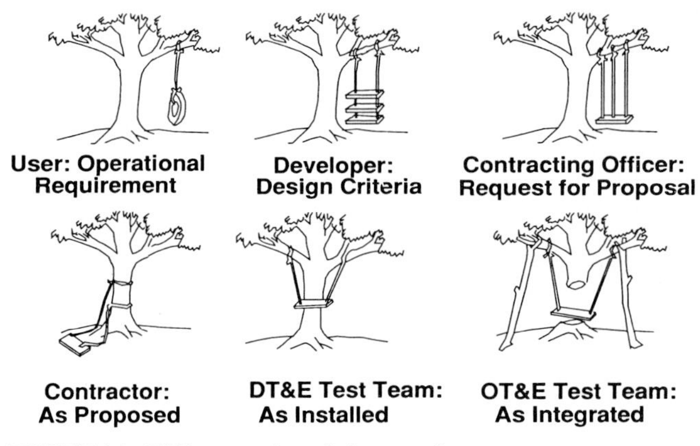

FIGURE 4.4 T&E as seen through the eyes of . . .

tively gather data which clearly identify to what degree a test article or system meets specified design criteria and operational mission objectives, respectively. *Evaluation* is the formal review, analysis, and assessment process supported by contractors, acquisition agencies, users, and supporting test organizations. Collectively, T&E accumulates data, formulates test results, and postulates on how well the system meets design specifications (DT&E) and how well it performs, or will perform (OT&E), in its intended operating environment throughout its expected life cycle.

Why Do We Test?

Within DoD formalized developmental tests are conducted to verify contract specification compliance and design maturity of particular systems, subsystems, equipment, hardware, and software while formalized operational tests are performed to learn how effective (how well does the product or system perform its intended mission?) and suitable (how reliable, maintainable, and supportable is it?) those items are when measured against formally stated standards. As stated earlier, testing is also performed to improve the T&E process and the product itself, and to support the overall acquisition decision process.

How Do We Test?

Each service tests a product or system by "shaking it down," evaluating the results, assessing its merits, and reporting on the findings. Operational test strategies are particularly protected from unnecessary management intervention which could, and in some cases would, influence test results or test findings through independent and direct reporting responsibilities delegated by each service chief to specialized subordinate acquisition and test organizations. Figure 4.3 illustrates that arrangement along with the interservice bond, which becomes critically important when programs involve multiple services. When directed, when practical, or when joint testing requires it, these service acquisition and test agencies work together to develop, acquire, test, and deploy systems which maximize system utility while minimizing acquisition risks and test costs.

THE RELATIONAL ASPECTS OF DEVELOPMENTAL AND OPERATIONAL TESTING

DT&E generally begins during the early development phases of acquisition, although it may occur at any point in a system's life cycle. In addition, this technically focused form of testing normally precedes OT&E. Although this sequential approach to testing often affects program schedules adversely and increases acquisition costs, Congress and the DoD began to require independent operational testing prior to production decisions for major programs in the early 1970s. The purpose of this policy was to focus on both the system's design as well as its ability to perform its mission in operational environments. OT&E usually concludes with a dedicated operational testing phase, which only begins after the implementing command (program manager) certifies, through DT&E results, that the system is ready for operational testing. Although the processes of DT&E and OT&E are separate and highly distinctive, resources needed to conduct and support the two are often quite similar, as many of the test data generated by one are beneficial and critical to the other. Consequently, a formally conducted combined DT&E/OT&E has, until recently, been considered appropriate by some test agencies so long as the results of combined testing remained clearly separable and contractor OT&E involvement was limited to that which would be normally employed during peacetime and combat operations. Simply stated, if the contractor does not support the system during actual combat operations, then the contractor does not participate in OT&E. When planning to conduct an approved combined test effort, necessary test conditions and data requirements are formally integrated and accomplished in three distinct segments. During the first segment, critical technical and engineering (DT&E) test events assume priority. OT&E personnel participate, when practical, to gain an operational insight into the system and preview any development test data which can, or will, support dedicated OT&E later. The second segment consists of combined test events consisting of selectively shared DT&E and OT&E data which, when viewed independently, clearly support either. The last segment consists of dedicated OT&E or separate operational test events which solely support operational mission objectives. Dedicated OT&E is managed and conducted by the operational test agency (OTA). The OTA, along with the implementing (lead acquisition) command, ensures that combined tests are planned and executed in such a manner as to provide unbiased operational test information. The OTA actually conducting the testing formulates an independent evaluation of the dedicated OT&E results and is ultimately responsible for the achievement of all planned OT&E objectives.

When DT&E and OT&E are combined, data collection, evaluations, and mandated test reports are administered independently. This is directly attributable to public law, the Federal Acquisition Regulation, DoD directives, and service regulations and policies.

WHY DISTINCTIVE TEST AND EVALUATION?

DT&E

DT&E measures the technical qualities of a product against a design specification whose criteria were derived from operationally defined user requirements. The results of DT&E are most important to the lead acquisition agency as they are used to:

1. Verify that system engineering development is complete and alternative design solutions, if appropriate, have been identified and pursued
2. Support the identification of potential technical (system design) and operational (system requirement) cost-performance tradeoffs
3. Substantiate that contractually identified critical technical parameters* are met and design risks minimized
4. Formally certify that the system is ready for OT&E

An alternative to formal DT&E is a technical testing method commonly identified as qualification test and evaluation (QT&E). QT&E is usually conducted in lieu of DT&E on programs which encompass system upgrades or modifications. Examples include situations where the existing overall system design remains intact, where off-the-shelf equipment only requires minor modifications in order to support system integration with other similar systems, where off-the-shelf commercial systems only require minor modifications to support the user's mission objectives, or where items require no development. QT&E lends itself to development testing and essentially is managed and conducted in the same manner as DT&E.

Another form of DT&E is preproduction qualification testing (PPQT) and production qualification testing (PQT). Both are normally conducted by the contractor and usually occur during the system engineering development phase. PPQT is a series of formal contractual tests conducted on preproduction hardware, components, subsystems, or systems fabricated to proposed production design specifications and drawings to ensure the integrity of the design over a specified operational and environmental range. These tests and their results are critical to production release on programs which involve volume acquisition. PPQT also includes tests which evaluate contractually defined reliability and maintainability criteria. PQT is conducted on production hardware, components, subsystems, or systems to verify the integrity of the manufacturing process, equipment, and procedures. PQT is accomplished on a random sample of the initial production lot, and is repeated if the process changes significantly and when a second or alternate source is brought on line.

OT&E

OT&E is accomplished to evaluate (or to refine estimates of) a system's operational qualities with regard to effectiveness, suitability, reliability, maintainability, availability, and supportability, and to identify any operational and logistics support deficiencies. The results of OT&E are most important to the user and are used to make determinations similar to those made during DT&E. The OT&E test environment must be structured such that:

1. Threats must be as representative as practical.
2. The test team must consist of typical users who will operate and maintain the system under realistic peacetime as well as combat stress conditions.
3. Production or production representative articles must be used for that portion of OT&E which supports a production or procurement decision.

*Critical technical parameters are derived through systems engineering technical efforts. The lead acquisition agency works through their engineering staff with user and test agencies to translate validated user requirements into detailed design (system, product, or end-item) specifications. Collectively, critical design characteristics and minimum performance criteria are established and verified through demonstrations and tests by DT&E and OT&E agencies.

4. Direct contractor involvement during the dedicated phase of OT&E must be limited to that which will be provided during peacetime and combat operational system deployment.
5. OT&E test data must be collected, compiled, and reported independently, without any contractor involvement.*

OT&E focuses on two specific areas:

1. Operational effectiveness
2. Operational suitability

Operational effectiveness is determined by how well a system performs its mission when used by representative personnel in the planned operational environment relative to organization, doctrine, tactics, survivability, vulnerability, and electronic, nuclear, or nonnuclear threats.

Operational suitability is determined by the degree to which a system can be placed satisfactorily in the field with consideration given to availability, compatibility, transportability, interoperability, reliability, wartime usage rates, maintainability, safety, human factors, manpower and logistics supportability, documentation, and training requirements.

Both operational effectiveness and operational suitability go hand in hand and are virtually inseparable, as elements within one are readily inclusive in either test area.

As previously stated, OT&E is conducted under conditions which are as operationally realistic as possible and must be representative of both combat stress and peacetime operating conditions foreseen throughout the system life cycle. Through OT&E the test team measures the system against operational thresholds ("need to have") and objectives ("nice to have, but still important") outlined in program documentation formally developed by military departments or service component operations and support commands.

OT&E results provide information on organizational structure, personnel requirements, support equipment, doctrine, training, and tactics. OT&E also provides data to the field which validate operating instructions, maintenance procedures, computer and other documentation, training programs, publications, and handbooks. As much as is practical, OT&E teams comprise personnel with the same type of skills and qualifications as those who will operate, maintain, and support the system when deployed. This assures credibility of OT&E results and findings.

Although each service categorizes OT&E with its own acronyms, the process includes the general phases listed below. These activities overlap, but they usually occur in the following sequence during the life cycle of a particular system:

1. Early operational assessments (EOAs)
2. Operational assessments (OAs)
3. Initial operational test and evaluation (IOT&E)
4. Follow-on operational test and evaluation (FOT&E)
5. Qualification operational test and evaluation (QOT&E)

EOAs and OAs are conducted during system definition and predevelopment acquisition phases and consist of system maturity assessments and full-scale development readiness evaluations along with estimations of the system's ability to begin and

*For major defense system acquisition programs as defined in DoD Instruction 5000.2, this independence is mandated and explained by USC Title 10, sec. 2399. Nonmajor programs are controlled by service regulations which generally encourage similar compliance when practical.

complete initial operational testing. EOAs support the early acquisition process by providing decision makers with preliminary test data which compare the proposed system design with user-stated requirements and specify their completeness, clarity, sufficiency, priority, rationale, and testability. EOAs support the testing process by providing test managers with information which helps them determine the adequacy of planning along with potential risks associated with completed or pending DT&E or OT&E activities. OAs are similar to EOAs, but provide a more detailed examination since they commonly occur later in the acquisition process and are conducted to support development, and in some cases, production approval decisions.

IOT&E usually occurs during the engineering development phase of acquisition and is structured to provide acquisition managers with inputs throughout the remaining program decision milestones. The maturity of articles tested during IOT&E is considered and evaluated and is accomplished by service OT&E commands or component OTAs as designated by the appropriate acquisition executive or milestone decision authority (MDA). Testing is conducted using prototypes, preproduction articles, or pilot production items. IOT&E must be completed prior to the full-rate production decision milestone. This assures senior decision makers that all test results and issues have been thoroughly addressed and resolved.

FOT&E is normally conducted after the system is well into production and operationally deployed. It is accomplished by the designated OT&E command or service component OTA. As the system matures and mission objectives and goals evolve, incremental FOT&Es can and do occur, as needed, throughout the remainder of the system's life cycle. Their purpose is to further refine operational effectiveness and suitability estimates, identify operational deficiencies, evaluate system changes, and reevaluate the overall system utility against changing operational mission objectives.

QOT&E is common with nonmajor systems and is conducted in lieu of IOT&E where there have been no preceding research, development, and test and evaluation (RDT&E) funded development efforts such as those described earlier. QOT&E is usually accomplished prior to acceptance of the first production article. However, under certain circumstances (low-risk programs or one-of-a-kind production systems) acceptance and production go-ahead can occur prior to the completion of QOT&E.

DT&E and OT&E have their unique qualities and, as management tools, are genuinely important. However, without the integration and interaction of the formal acquisition process, neither T&E nor acquisition can function adequately or effectively. This relationship is influenced and guided by different forces. For example, it can be rationalized that acquisition focuses on and is driven by events (milestones), whereas T&E is influenced and guided by processes (acquisition phases). This can be further hypothesized.

HOW ACQUISITION INFLUENCES AND GUIDES T&E

T&E, like any management activity, begins with planning (in this case test planning), which starts at the outset of an acquisition and evolves throughout and in parallel with a program's life cycle. As our national security interests and strategies change, so do our military regional plans for supporting strategic and tactical mission objectives. In concert with the National Command Authority, the Joint Chiefs of Staff (JCS) and the service commanders in chief (CINCs) develop defense and service mission objective planning guidance. The major commands (MAJCOMs) review these plans for their missions and objectives and with support from service compo-

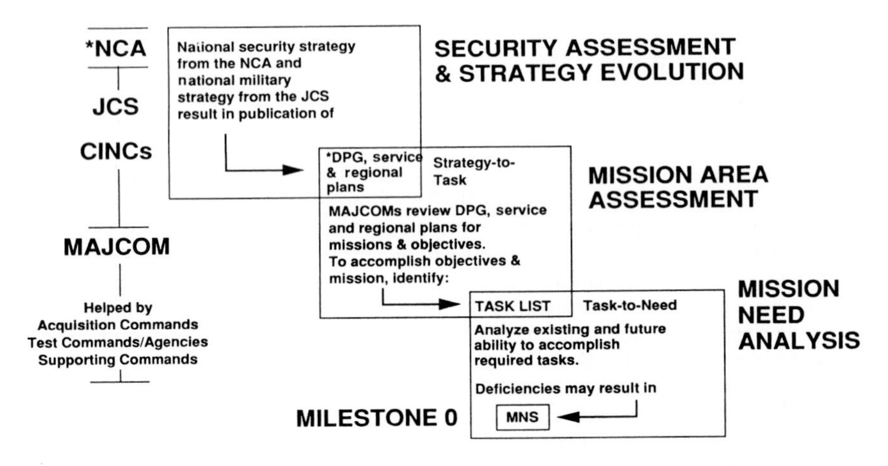

FIGURE 4.5 Mission need determination process.

nent, acquisition, logistics, test, and component support organizations develop a mission objective task listing. The task listing is reviewed as a task-to-need analysis and compared with a particular service's current and future planned ability to accomplish those mission objectives. The results of a specific mission area analysis may well lead to a deficiency identification. After operational or system deficiencies have been identified and documented by a particular military service, they are evaluated and either resolved within the service's regulatory and fiscal capabilities or processed as unresolved deficiencies. If a particular identified deficiency is validated, it becomes a mission need and is formally processed as a mission need statement (MNS). Figure 4.5 demonstrates this process and how it associates with the first acquisition decision milestone.

Once an MNS is developed, categorized, validated, and approved by either a unified or a specified command, a military department, or the Office of the Secretary of Defense (OSD), it becomes the baseline requirement document for succeeding program documentation to include future test plans. The MNS's level of approval is predicated on associated values and risks. This program identification and approval process establishes and categorizes a program into one of four acquisition categories (ACATs). Figure 4.6 defines these categories and includes the decision authorities. All acquisition programs require an MNS and are placed into one of the four ACATs. Although cost, schedule, and risks are different for major and nonmajor ACAT programs, acquiring and testing processes are identical.

After a program has been validated, categorized, and approved by the milestone decision authority, the acquisition evolution begins and programmatic events occur which set the stage for program initiation. Typically, system acquisition programs follow a single path, which encompasses five milestones and five phases. Figure 4.7 demonstrates this process along with milestones and phases.

Milestone 0: Concept Studies Approval

After this first milestone is initiated, the appropriate defense or service acquisition executive designates a principal lead acquisition command or agency who in turn es-

PROGRAM DESIGNATION

BASIC NEED DOCUMENT	TYPE OF SYSTEM	THRESHOLDS	DECISION REVIEW BODY	LEVEL OF APPROVAL = MDA
MISSION NEED STATEMENT	ACAT I	$300M RDT&E (FY 90) $1.8B PROCUREMENT (")	DAB	SECDEF
	ACAT II	$115M RDT&E $540M PROCUREMENT	AS DIRECTED	SAF
	ACAT III/IV	AS DIRECTED		AS DIRECTED

FIGURE 4.6 Acquisition category (ACAT) thresholds table. DAB = defense acquisition board; MDA = milestone decision authority.

tablishes a program management activity known as a system program office (SPO) or program office (PO). In addition, a principal operating service or command is identified, supporting services (joint service programs) or commands are designated, and supporting test agencies are established. Formal direction, which makes all this happen, is provided via a formal program directive commonly termed program management directive (PMD). During the succeeding acquisition phase (phase 0, concept exploration) all participants work together with the implementing and operating commands to formalize program management and operational requirement documents. Most importantly the principal user refines the MNS by developing an operational requirements document (ORD). This living program document lists and identifies critical operational issues (show stoppers), minimum acceptable operational performance requirements ("absolutely need-to-haves"), mission scenarios,

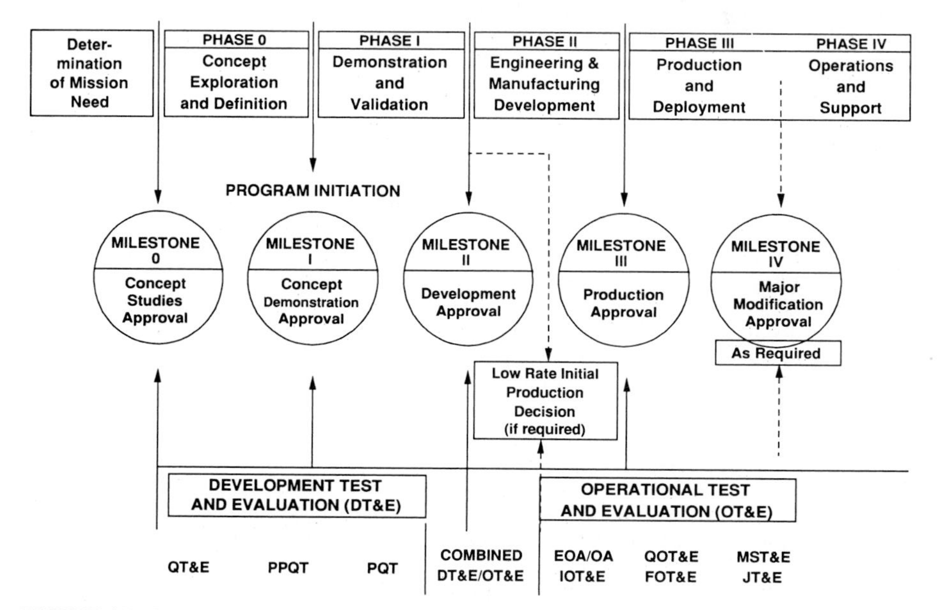

FIGURE 4.7 Program milestones and test phases.

and so on. Concurrently, corollary test and user agencies support the implementing command's SPO test directorate through a joint development of the preliminary test planning and control document called the test and evaluation master plan (TEMP).

Milestone I: Concept Demonstration Approval

Although management resources are in place, formally recognized program efforts only begin after milestone I approval. Once initiated, test planning activities branch out to include early T&E (EOAs, OAs, etc.). The ORD now becomes the principal baseline for the system or product specifications. In addition, test documents, including the TEMP, are further developed, updated, and expanded to match program direction and user-refined operational mission objectives stated within the ORD. Although the ORD supports and baselines the TEMP, neither is complete until the contract system or product specifications are formally defined and contractually implemented. During these early stages of acquisition both DT&E and OT&E agencies institutionalize their relationship by establishing a formal test working group, known by such names as the test planning working group (TPWG), test integration working group (TIWG), or test working group (TWG). Since this controlling test working group is usually a mid- to upper-level management activity, subgroups often form to deal with and support detailed test planning activities. The test working group is established and chaired by the implementing command, through the program office, and as such becomes the responsible agency for TEMP development, coordination, and approval. Since the TEMP is a key high-level management tool, supplemental detailed test plans must be developed to support actual testing. For example, the contractor, working with both DT&E and OT&E agencies, develops detailed test procedures from system or product specifications and identifies preliminary test schedules for DT&E and OT&E activities, test resource requirements, test schedules, verification and validation schedules, and so on. Also, during this period OT&E agencies further ameliorate their test planning efforts by working with the user to refine the ORD to ensure testability. These preliminary T&E planning activities begin to take on somewhat of a chaotic air about the time the program meets the next milestone.

Milestone II: Development Approval

DT&E activities move into high gear after approval of this milestone, and it is also about this time that OT&E plans begin to fully evolve. In terms of importance to the test community, this milestone, and the phase it leads into (phase II, engineering and manufacturing development), is most critical to both. Milestone II and its associated events prove the worth of a new system or product. That worth, however, is directly affected by the quality and thoroughness of T&E which supports acquisition. A sound and well-structured relationship between program office, user, and test agencies will result in timely and accurate test results.

Milestone III: Production Approval

After approval, role reversal begins to occur. T&E activities transition from development testing into operational testing. OT&E activities continue, as necessary, to allow the OT&E test teams the opportunity necessary to adequately evaluate tactics, assess alternative solutions, and evaluate interoperability with other systems. As time passes and as threats change, strategic and tactical mission objectives evolve accordingly. National security and defense planning strategies inevitably evolve to the extent that a system must be upgraded, modified, or replaced. Since the latter is the most expensive solution, system upgrades or modifications are usually the preferred and most cost-effective approaches. Consequently, if timely and cost-effective upgrades or modifications are to occur, an orderly process must be used similar to that which is used during major defense system acquisitions. The next milestone in the evolution of a system allows for just those activities to occur in both logical and practical fashion.

Milestone IV: Major Modification Approval

The modification cycle begins with the identification of the first deficiency that warrants a change to an existing fully operational system. During and throughout the remainder of the system's useful life cycle the user further refines system capabilities through permanent or temporary modifications, expanded or modified use of the existing system, or system integration. Enhancing the capability of an existing system without actually modifying it usually falls into the category of FOT&E. Modifying a system with new equipment or system upgrades requires regeneration of the MNS or initiation of the modification proposal process which, in either case, restarts the formal acquisition life cycle and recirculates the program back through the developmental and operational test community. Usually modifications fall into ACAT II through IV categories and are managed the same as large development programs. Many modification programs lack milestone definition since most have little or no R&D associated with them, or they consist of existing off-the-shelf purchases. Those that do have formal milestones tend to become major or high-visibility programs rather than simple system modifications. The smaller modification program acquisition and test life cycles are generally two to four years in length and require fewer program team members, whereas major programs tend to take longer (7 to 10 years), require program staffs of as many as several hundred personnel, and include numerous inter- and intraservice agencies. In some cases, modification programs are managed by a single system program manager (SPM). In addition, milestone decision authority is delegated to the lowest practical level which, in some cases, includes the SPM. Since specific system acquisition and test management requirements may vary from one modification program to another as well as one service component to the other, the overall process is regulated and controlled by the same DoD policies as are major programs, with minor exceptions:

1. Acquisition and T&E are accomplished on a remarkably smaller scale.
2. Program management functions with fewer resources.
3. Funding is limited and at times may be virtually nil.
4. Development, test, and production schedules are highly compressed.

HOW T&E SUPPORTS SYSTEM ACQUISITION

Throughout a system acquisition life-cycle testing is conducted to assure decision makers as well as users (customers) that the system or product they are acquiring provides them with the most "bang for their buck" and does what it is intended to do. With the advent of new DoD directives and policies focusing on earlier testing as

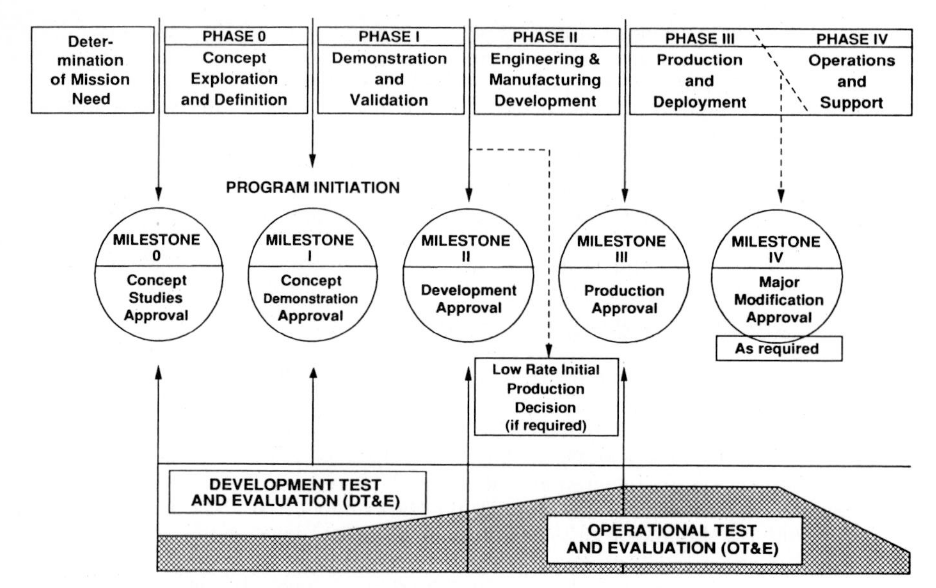

FIGURE 4.8 T&E and its influence on acquisition.

well as reporting, T&E and its association with the system acquisition process becomes more critical than ever. Together, DT&E and OT&E agencies work throughout the test execution cycles to:

1. Improve the test process
2. Evaluate or assess a system's operational capability
3. Support the acquisition decision process with quantitatively or qualitatively developed test data

Figure 4.8 is an expanded view of the acquisition life cycle and shows how T&E timing ties in to and essentially supports the acquisition process.

Phase 0: Concept Exploration and Definition

During phase 0, and prior to milestone I approval, system concept or technology testing is accomplished to determine whether preferred alternative concepts, technologies, and designs exist which might alleviate the need for a new system altogether. Test resources for subsequent T&E such as ranges, targets, and threat simulators are formulated and identified so that successive testing can be done under operationally realistic conditions. This also allows sufficient lead time for the test planning and procurement of any new or modified resources. If any formalized testing occurs, it generally consists of DT&E and encompasses component, laboratory, and, occasionally, field tests of "similar to" components or systems which determine and enhance cost, schedule, and performance alternatives, technical and operational performance criteria established by concept studies, MNSs, and initial operational requirement documents.

Phase I: Demonstration and Validation

During phase I DT&E is more likely to occur and is conducted to determine a preferred technical approach, identify design risks, and demonstrate feasible solutions or alternatives. DT&E further enhances the system acquisition process by supporting design trade-off decisions which will satisfy user-identified operational requirements cost-effectively. This is accomplished through formal testing of subsystems, components, system prototypes, or preproduction articles which help to establish and justify a development approval decision.

If tradeoffs and preliminary design changes are proposed, they are assessed by lead and supporting agencies. In addition, T&E resource estimates are refined and long-lead procurement (such as raw materials, forgings, tooling) actions initiated. Systems which are intended to operate in an electromagnetic threat environment are analyzed and assessed for potential vulnerabilities and susceptibilities prior to entering the next phase of acquisition.

Should OT&E activities occur, they usually consist of EOAs or OUEs to help acquisition managers forecast risk and system readiness for dedicated OT&E.

Phase II: Engineering and Manufacturing Development

During phase II DT&E is conducted to ensure that engineering development is reasonably complete and significant design problems are identified or solved. The accomplishment of IOT&E and QOT&E is structured to provide operational assessment for the low-rate initial production (LRIP) acquisition decision and for mission-oriented testing necessary to evaluate a system's operational effectiveness and suitability before advancing beyond LRIP.

Phase III: Production and Deployment

During phase III DT&E is conducted to evaluate product improvements which occur as a result of design refinements, approved engineering change proposals (ECPs), proposed system modifications which meet forecast threat changes, along with changes which reduce system life-cycle costs and improve reliability, maintainability, and availability (RM&A). OT&E is conducted during this phase to further evaluate the system's operational effectiveness and suitability and examine the operational impact any completed or ongoing production cycle design changes may have on the overall system.

Phase IV: Operations and Support

Phase IV is an extension of milestone III and, as stated previously, begins when the first major modification occurs. During this phase incremental DT&E and OT&E continue as programmed until either the system's life cycle is complete or mission objectives change beyond that which allows cost-effective support. Thus the genesis of a new deficiency and renewal of the acquisition life cycle.

The aforementioned relationship between acquisition and T&E, as depicted in Fig. 4.8, summarizes the fact that acquisition and test are virtually inseparable. Although strategies and methodologies for actively managing T&E are well defined and formally structured, orchestrating this requires a fundamental document which logically and chronologically drives T&E to its desired end: credible test completion. That document, or management tool, is the test and evaluation master plan (TEMP).

TEST AND EVALUATION MASTER PLAN

The TEMP is the key to successful acquisition and is the essential working tool for all T&E. It documents the overall structure and objectives of the T&E program while providing a framework within which to generate detailed test plans, substantiate test schedules, assess resource implications, and define planned DT&E and OT&E activities. Further, the TEMP associates the overall program schedule, management structure, test strategy, and required resources to:

1. Critical operational issues (operational show stoppers)
2. Critical technical parameters (design show stoppers)
3. Minimum acceptable operational performance requirements ("absolutely need-to-haves")
4. Evaluation criteria (how system will be tested)
5. Acquisition decision points (milestones)

How the TEMP Evolves

The TEMP is developed during phase 0 and approved by the appropriate acquisition executive or authority prior to milestone I approval. TEMPs for all major defense system programs or other programs which, because of their importance or levels of risk, are placed on the OSD T&E oversight list and approved by both the director of operational test and evaluation and the deputy director of defense research and engineering for T&E (see Fig. 4.3). TEMPs for all other acquisition category programs are approved by the DoD component milestone decision authority as designated by service-generated program management documentation.

As system acquisition and employment costs increase, multiservice or jointly sponsored programs become more common. When this does occur, the designated lead service develops a single TEMP, coordinates all testing requirements, and constructs a single T&E report summarizing the operational effectiveness and suitability test results for each participating service. Testing is planned according to the lead service regulations. If, however, a supporting service has unique test requirements, testing is planned, funded, and conducted according to that service's regulations.

The TEMP must be coordinated and approved by all principal service and component agencies before testing begins.

Updating and Revising the TEMP

The TEMP is revised or updated when program milestones occur and transition to the next phase of acquisition, when the program baseline (cost, performance, schedule) has been breached, or whenever significant program changes occur, usually as a result of formal redirection. The designated acquisition executive or milestone decision authority will determine the appropriate TEMP revision criteria for programs without milestones, such as modification programs.

Terminating a TEMP

TEMPs are formally terminated via the PMD, which normally occurs when system development is complete and all critical operational issues are satisfactorily resolved. This includes the verification and correction of deficiencies discovered during testing. Examples include:

1. A fully deployed system which has no planned operationally significant product improvements or block modification efforts
2. A system whose production is ongoing and whose fielding has been initiated with no significant deficiencies observed as a result of production qualification test results
3. A partially fielded system, in early production, which has successfully completed all developmental and operational test objectives
4. A system wherein planned T&E is only a part of routine aging and surveillance testing, service-life monitoring, or tactics development
5. A system where no further DT&E, OT&E, or live fire testing is required by any DoD user component
6. A system where future testing requirements (product improvements or block upgrades) have been incorporated in an unrelated separate TEMP

Guidelines for Developing a TEMP

As previously stated, DoD guidelines for developing the TEMP have been instituted to ensure a common approach to T&E for all services and component agencies involved with T&E. The formally stated process is described in DoD 5000.2-M and is mandatory for all major (ACAT I) programs along with those programs on the OSD oversight list. Although other (ACAT II, III, and IV) programs are exempt from mandatorily following the format, at the discretion of the MDA, a wise program or test manager will, at the least, ensure that each segment of the TEMP outline is assessed and formally addressed. This guarantees three things:

1. It enables the test manager to thoroughly document all program issues, technical criteria, and operationally critical issues associated with a particular program.
2. It allows program and test managers to effectively determine what must be tested as well as when and how.
3. It assures the ultimate decision authority that test planning, execution, and reporting have been wholly and adequately addressed.

Constructing the TEMP

The TEMP is constructed in five parts and includes significant subheadings. Since it is an important management tool, it must be accurate and readily available for use. Classified TEMPs should be avoided as much as is practical. Any classified matters which are critical to T&E should be addressed in separate annexes or appendixes if possible. What follows is a synopsis of each section with the critical aspects of each explained. Figure 4.9 represents the TEMP outline along with critical subheadings.

Part I: System Introduction

Mission Description. This segment consists of a brief system mission description as it is described in the MNS and keys on how the system will perform when operationally deployed. The description is written in nontechnical terms and high-

TEST & EVALUATION MASTER PLAN OUTLINE (FORMAT)

PART I	SYSTEM INTRODUCTION (2 pages suggested - add annexes if necessary)
	MISSION NEED DESCRIPTION
	SYSTEM THREAT ASSESSMENT
	MINIMUM ACCEPTABLE OPERATIONAL REQUIREMENTS
	SYSTEM DESCRIPTION
	CRITICAL TECHNICAL PARAMETERS
PART II	INTEGRATED TEST PROGRAM SUMMARY (2 pages suggested)
	INTEGRATED TEST PROGRAM SCHEDULE
	MANAGEMENT
PART III	DEVELOPMENTAL TEST AND EVALUATION OUTLINE (10 pages suggested)
	DEVELOPMENTAL TEST AND EVALUATION OVERVIEW
	DEVELOPMENTAL TEST AND EVALUATION TO DATE
	FUTURE DEVELOPMENTAL TEST AND EVALUATION OVERVIEW
	SPECIAL DEVELOPMENTAL TEST AND EVALUATION TOPICS
	LIVE FIRE TEST & EVALUATION
PART IV	OPERATIONAL TEST AND EVALUATION OUTLINE (10 pages suggested)
	OPERATIONAL TEST AND EVALUATION OVERVIEW
	CRITICAL OPERATIONAL ISSUES
	OPERATIONAL TEST AND EVALUATION TO DATE
	FUTURE OPERATIONAL TEST AND EVALUATION
PART V	TEST AND EVALUATION RESOURCE SUMMARY (6 pages suggested)
	SUMMARY
	TEST ARTICLES
	TEST SITES AND INSTRUMENTATION
	TEST SUPPORT EQUIPMENT
	THREAT SYSTEMS/SIMULATORS
	TEST TARGETS AND EXPENDABLE
	OPERATIONAL FORCE TEST SUPPORT
	SIMULATIONS, MODELS AND TESTBEDS
	SPECIAL REQUIREMENTS
	TEST AND EVALUATION T&E FUNDING REQUIREMENTS
	MANPOWER/TRAINING
	KEY RESOURCES
Appendix A	BIBLIOGRAPHY
Appendix B	ACRONYMS
Appendix C	POINTS OF CONTACT

FIGURE 4.9 TEMP outline.

lights planned operational missions, expected performance objectives, and general capabilities.

System Threat Assessment. This is also an overview section and includes a brief description of each threat which must be encountered and, if appropriate, countered throughout the system's life cycle. This includes a generalized summary of the operational, physical, and technological environment in which the system will perform. Source material for this section is derived from the formal system threat assessment report (STAR), which is developed either by service intelligence commands (ACAT I programs) or by DoD component intelligence commands or agencies (ACAT II,

III, IV programs). This requirement may seem somewhat extreme for smaller nonmajor programs which may only involve the acquisition of a component or software upgrade, but if the component is being integrated into a combat system, then that component essentially faces the same threat as the overall system itself. As such, it is imperative that the overall threat be considered and identified, if only to assure the decision authority that a system being upgraded or modified will survive in its intended operational environment.

Minimum Acceptable Operational Performance Requirements. This segment includes critical operational effectiveness and suitability (discussed earlier) parameters derived from the ORD, or other appropriate program documentation. Included are constraints such as manpower allocations, training, software, computer resources, and transportability. This section can be either in statement or, as is more common, in matrix format and addressed as a figure or attachment to part I. Figure 4.10 is an example of a matrix summary and includes characteristics along with thresholds and objectives necessary to perform the system's operational mission. The user or the user's representatives work with the implementing command to develop this segment and ensure its adequacy.

System Description. A brief summary of the system's design follows, which addresses:

1. Critical features and subsystems, both hardware and software (such as architecture, firmware, data bus interfaces), which allow the system to perform its operational mission.
2. Critical interfaces with existing or planned systems which are needed for mission accomplishment. This includes both developmental and nondevelopmental (off-the-shelf) systems.

	PARAMETER	THRESHOLD
OPERATIONAL EFFECTIVENESS	GROUND OPERATIONS	
	BACKUP CAPABILITY WITH 160,000-lb PAYLOAD TO FLY 1,000 NM, SEA LEVEL, 90-DEGREE F DAY	1.5% GRADE
	COMBAT OFFLOAD CAPABILITY	INDIVIDUAL OR LINKED PALLETIZED CARGO AND AIRDROP PLATFORMS
	180-DEGREE TURN	IN 90 FT WITH APPROXIMATELY 3 MANEUVERS
	NORMAL CREW COMPLEMENT	2 PILOTS AND 1 LOADMASTER
	LAPES (LOW-ALTITUDE PARACHUTE EXTRACTION SYSTEM) AIRDROP OPERATIONS	LOADS UP TO 60,000 lbs, SINGLE OR SEQUENTIAL PLATFORMS IN TRAINS UP TO 60 FT LONG
OPERATIONAL SUITABILITY	RELIABILITY	AT 9,000 FLEET FLYING HOURS (FFH)
	MISSION COMPLETION SUCCESS PROBABILITY	.84
	AIR VEHICLE MEAN TIME BETWEEN MX, CORRECTIVE (HRS)	.60
	AIR VEHICLE MEAN TIME BETWEEN MX, INHERENT (HRS)	1.24
	AIR VEHICLE MEAN TIME BETWEEN REMOVAL (HRS)	2.13
	MAINTAINABILITY	
	MX MANHOUR PER FLIGHT HOUR	30.13
	MEAN MANHOURS TO REPAIR	3.38
	AVAILABILITY	
	FMC RATE	72.6
	PMC RATE	80.4
	MX DOWNTIME PER SORTIE (HRS)	7.8

FIGURE 4.10 Minimum acceptable operational performance requirements.

3. Critical system characteristics or unique support concepts which will result in special test and analysis requirements (such as postdeployment software support; chemical, biological, or nuclear hardening; development of new threat simulation, simulators, or targets).

Critical Technical Parameters. This segment is constructed in matrix format and includes design characteristics which have been, or will be, evaluated during the remaining phases of DT&E. The matrix specifically includes critical technical parameters derived from the system design specifications supported by the ORD and critical system characteristics extracted from program documents. Adjacent to each technical parameter there are the appropriate accompanying objectives and thresholds. Highlighted are critical data (exit criteria) which must be evaluated and demonstrated before entering the next acquisition or operational test phase. Included are any actual values or test results which have been demonstrated to date. A sample critical technical parameters matrix, derived from an approved TEMP, is displayed in Fig. 4.11. The parameters and thresholds listed are those that would need to be demonstrated prior to the full-rate production decision milestone. It is important to note here that the technical parameters listed within the matrix must be those that are absolutely critical to the needs of the program. Listing superfluous or extraneous noncritical parameters could quite possibly complicate or even undermine what is otherwise a sound and credible program. This same approach should also be taken when listing the minimum acceptable operational performance requirements pre-

CRITICAL TECHNICAL PARAMETERS	TOTAL EVENTS	TECHNICAL THRESHOLD FOR EACH TEST EVENT	LOCATION	SCHEDULE	MILESTONE DECISION SUPPORTED	DEMO VALUE
PERFORMANCE						
PAYLOAD/RANGE (UNERFUELED, 2.25G LOAD FACTOR STD. DAY, PLUS RESERVE)	15	160,000/2400 NM	EDWARDS AFB (EAFB)	08/93	III	
MAXIMUM DESIGN PAYLOAD	15	172,200 lbs	EAFB	08/93	III	
TAKEOFF CRITICAL FIELD LENGTH	05	7,600 FT	EAFB	10/92	III	
MAXIMUM EFFORT LANDING FIELD LENGTH	15	2,650 FT	EAFB	10/92	III	
CRUISE SPEED	15	.77 MACH	EAFB	08/92	III	
BACKUP CAPABILITY	01	2% GRADE	EAFB	08/93	III	
FLYING QUALITIES						
NORMAL STATES	03	LEVEL I WITHIN OPERATIONAL FLIGHT ENVELOPE LEVEL II WITHIN SERVICE FLIGHT ENVELOPE	EAFB CF[1]/EAFB	08/93	III	
FAILURE STATES	08	LEVEL III		08/93	III	
MISSION SYSTEMS						
SINGLE-LOAD AIRDROP, INCLUDING LAPES	6 LVAD[2] 4 LAPE[3]	60,000 lbs RIGGED	EAFB	08/93	III	
SEQUENTIAL-LOAD AIRDROP	2 LVAD	110,000 lbs	EAFB	08/93	III	
STRUCTURAL STRENGTH	05	SHORT-FIELD LANDING AT 15 FT/SEC MAX SINK RATE	EAFB	08/93	III	
SOFTWARE	50[4]	INTEGRATED TO ALLOW SAFE AND EFFECTIVE OPERATIONS DURING ALL MISSIONS	EAFB	08/93	III	

1 CONTRACTORS FACILITY
2 LOW-VELOCITY AIRDROP
3 LOW-ALTITUDE PARACHUTE EXTRACTION
4 SORTIES TO ASSESS MISSION COMPUTER, IRUs & GPS

CH-4

FIGURE 4.11 Critical technical parameter matrix

viously discussed. This can and will only happen if all involved acquisition and test agencies work together with adequate program documentation.

Part II: Integrated Test Program Summary

Integrated Test Program Schedule. The most important aspect of this graphically represented schedule is the inclusion of critical T&E phases along with test events, related activities, and planned cumulative expenditures by funding appropriation (that is, 3600, 3010). Included are event dates such as milestone decision points, operational assessments, test article availability, software version releases, live fire test and evaluation, operational test and evaluation schedule, low-rate initial production deliveries, full-rate production deliveries, initial operational capability, and full operational capability. A single schedule should be provided for multiservice, joint, or capstone* TEMPs showing all DoD component system event dates. Obviously, a one-page schedule which includes all this will be rather busy. Nevertheless, it is important to remind the reader that the TEMP is an important management tool which is developed to stand alone and be as complete as possible; the more detail, the better. Figure 4.12 is an example of an integrated schedule and demonstrates a stand-alone capability.

Management. Management and management support responsibilities for all participating organizations (developers, testers, evaluators, users) are described

*A capstone program is one that involves the acquisition and testing of a system comprising a collection of "stand-alone" component systems which function collectively to achieve the objectives of the total system.

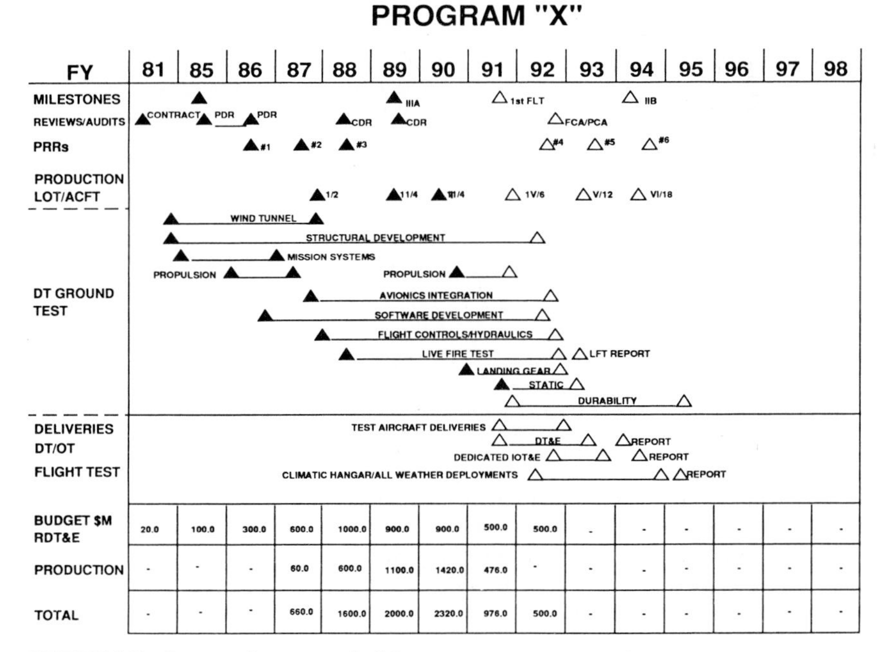

FY	81	85	86	87	88	89	90	91	92	93	94	95	96	97	98
BUDGET $M RDT&E	20.0	100.0	300.0	600.0	1000.0	900.0	900.0	500.0	500.0	-	-	-	-	-	-
PRODUCTION	-	-	-	60.0	600.0	1100.0	1420.0	476.0	-	-	-	-	-	-	-
TOTAL	-	-	-	660.0	1600.0	2000.0	2320.0	976.0	500.0	-	-	-	-	-	-

FIGURE 4.12 Integrated program schedule.

within this segment. Minimally, this segment should specifically identify responsibilities associated with the implementing command, principal test working group, supporting test organizations or agencies, operating commands, participating services (joint test programs), contractor, and supporting logistics commands or agencies. Depending on the complexity and magnitude of the program, an organizational chart might be appropriate.

Supplementally included is the fiscal quarter date when the decision to proceed beyond low-rate initial production* is planned. Low-rate initial production quantities (production systems or kits for modification programs) required for OT&E must be identified for approval by the director of OT&E prior to milestone II for ACAT I programs or other ACAT programs designated for OSD T&E oversight. Again, for nonmajor programs where the decision process has been delegated to a component decision authority, that authority will determine what constitutes a development approval decision point.

Lastly this segment identifies known operational issues along with vulnerability and live fire test requirements which will be incomplete prior to the low-rate initial production decision milestone.

Part III: Developmental Test and Evaluation Outline

Developmental Test and Evaluation Overview. This section details how DT&E will verify the status of engineering and manufacturing development progress. It also clarifies how design risks have been minimized to date, and substantiates the achievement of contract technical performance requirements. Specifically, this section identifies any system technology or component design which has yet to demonstrate its ability to contribute to system performance and, ultimately, fulfill contract requirements. Quantitatively addressed, this is how well system hardware and software designs have stabilized so as to reduce manufacturing and production uncertainties. Results from any PPQT or PQT would be included here.

Developmental Test and Evaluation to Date. Any completed DT&E listed in the critical technical parameters matrix which revealed marginal or unsatisfactory results should be amplified and explained here. Conversely, if the system exceeded expected results to a remarkable degree, then that should be addressed here as well.

Future Developmental Test and Evaluation. Of all the sections within the TEMP this is the most dynamic. Starting with the initial TEMP and concurrent with successive revisions this section evolves accordingly. This segment identifies all planned DT&E, beginning with the date of the current approved TEMP (to include revisions), and extends through production completion. Emphasis is placed on the next phase of testing and at the least includes the following:

1. A summary of the configuration description to include functional system capabilities and how they will differ from the production system.
2. Critical technical parameter objectives, which must be confirmed prior to entering the next acquisition phase. This should also include any specific technical parameters which the milestone decision authority has designated as exit criteria (thresholds) or directed to be demonstrated during this phase of testing.
3. A synopsis of test events, test scenarios, and the test layout concept. This is quantified where practical and, at the least, incorporates the number of test hours, test events, and test firings (if appropriate). In addition, any specific threat systems,

*For nonmilestone programs such as those consisting of system modifications or upgrades, this event could be substituted with event dates such as trial (prototype) installation or proof (production) installation dates.

surrogates, countermeasures, components of subsystem testing, and test beds whose use would be critical in determining whether DT&E objectives are met, are identified and explained. If an outside agency will be conducting any significant parts of testing in lieu of the principal test agency, the methods of evaluation will be explained here. If simulation and modeling are used, an explanation of their use along with the associated logic is included here. In addition it is described how testing will evaluate the system's performance, interoperability, and compatibility with other similar systems when operated in natural environmental conditions representative of the intended area of operations (such as temperature, pressure, humidity, fog, precipitation, clouds, blowing dust and sand, icing, wind conditions, steep terrain, wet soil conditions, high sea state, storm surge, and tides).

4. A review of any test limitations which might significantly affect the evaluator's ability to draw objective conclusions, including their impact along with potential solutions.

Live Fire Test and Evaluation. If live fire T&E is directed or required, this segment will address and highlight the overall live fire T&E strategy for the system or critical components. This will include all critical live fire T&E issues such as required levels of system vulnerability or lethality, management of the live fire T&E program, schedules, funding plans and requirements, related prior and future live fire T&E efforts, evaluation plan and shot-axis selection process, and known major test limitations.

Part IV: Operational Test and Evaluation Outline

Operational Test and Evaluation Overview. This section lends itself to the creative writing skills of the author. However, when developed, this section should clearly preview upcoming OT&E and be consistent with preceding sections.

Critical Operational Issues. Critical operational issues (COIs), by definition, are operational issues which must be examined during OT&E to evaluate or assess the system's capability to perform its mission. Correctly stated COIs identify essential capabilities, risks, or uncertainties which must be explored before the overall worth of the new or modified system can be estimated. They are critically important to the acquisition process since their outcome directly affects the decision which determines whether or not a system should advance into the next acquisition phase. COIs are those operational characteristics most essential to the system's effectiveness and suitability. They should address the user-identified critical mission requirements and areas of operational risk that affect mission performance most significantly. Failure to achieve these identified characteristics would render the system unresponsive to the user's needs. The user develops COIs with support of the implementing command test management organization. COIs evolve just as do mission objectives and as such, COIs which were appropriate during early phases of OT&E (IOT&E) may have less application during later phases (FOT&E). COIs form the basis for OT&E objectives and are typically phrased as questions which must be answered qualitatively in order to properly evaluate operational effectiveness (such as, "How well does the system meet user requirements for intertheater and intratheater range and payload?") and operational suitability (such as, "How well does the system's mission reliability permit completion of assigned missions?"). Some COIs have both technical and operational performance parameters. However, individual achievement of these attributes does not guarantee that a COI will be resolved favorably. The judgment and findings of the operational test agency are used by the decision authority to determine whether the COI has been resolved favorably. If every COI is resolved favorably, then the system should be operationally effective and operationally suitable when employed in its intended environment by typical users.

Operational Test and Evaluation to Date. This segment is arranged in chronological fashion to identify and date test reports which detail the currency and results of testing and operational assessments. This also includes COIs which have been resolved (favorably or unfavorably), partially resolved, or unresolved at the completion of each phase of testing.

Future Operational Test and Evaluation. The planned remaining phases of OT&E are addressed herein separately and include the following:

1. A summary of the configuration's description, which includes the system that will be tested during each remaining test phase. Any known differences between the tested system and the system that will be fielded are also identified. Included, where applicable, are software maturity values and the extent of integration with other systems with which it must function. The system should be characterized in terms relative to its physical configuration at the time of testing (such as, prototype, engineering development model, and production model).

2. Clearly stated OT&E objectives, which include minimum acceptable operational requirements and COIs addressed by type of OT&E along with the milestone decision reviews each supports. OT&E which supports the full-rate (or similar) production decision should have test objectives which examine operational effectiveness and suitability in total.

3. A concise summary of mission scenarios which include among other items events to be conducted, types of resources to be used, threat simulators and simulations to be employed, type of representative personnel who will operate and maintain the system, status of logistics support, operational and maintenance documentation that will be used, environment under which the system will be employed and supported during testing, plans for interoperability and compatibility testing with other U.S. or allied weapon and support systems as applicable. Also included is the identification of planned information sources (such as DT&E, testing of related systems, modeling, simulation) which may be used by the operational test agency to supplement a particular phase of OT&E. Whenever models and simulations are used, an explanation along with the rationale for their credible use is included here. If OT&E will be conducted or completed during this iteration and the outcome will be an operational assessment in lieu of an evaluation, this should be clearly stated and explained.

4. A narrative on test limitations which should include threat resource availability, limiting operational (military, climatic, nuclear) and support environments, maturity limitations associated with tested system, safety, or any specific limitation which may impact the resolution of affected COIs. If limitations are addressed, their impact on the ability to resolve and formulate conclusions regarding operational effectiveness and operational suitability must be discussed thoroughly. After discussing each limitation, each affected COI is highlighted.

Part V: Test and Evaluation Resource Summary. This final section of the TEMP consists of a detailed summary (preferably in table or matrix format) of all key T&E resources. It includes both government and contractor assets which will be used during the course of the acquisition program. Specifically, the following test resources must be addressed:

1. *Test Articles.* Identified are the actual number of test articles along with the timing requirements. This also includes key support equipment and technical information (user and maintenance manuals) required for testing during both DT&E and

OT&E. If key subsystems (components, assemblies, subassemblies, or software modules) are to be tested individually and prior to being tested in the final system configuration, then either section III or section IV must include the identity of the key subsystems along with the quantities required. Clarify when and how prototype, engineering development, preproduction, or production models will be used.

2. *Test Sites and Instrumentation.* Specific test ranges or facilities required for both DT&E and OT&E are listed here. This includes a comparison of test range or facility requirements with their availability and capability to support testing during planned DT&E and OT&E periods. Highlighted are any major shortfalls, such as inability to test under representative natural environmental conditions. Test instrumentation requirements are addressed here and include all equipment (strain gages, high-speed photographic equipment) necessary to support the entire test program. This segment also identifies who is responsible for providing, maintaining, and operating the test instrumentation and associated test support equipment. If necessary, any existing or proposed memoranda of agreements (MOAs) might also be appropriate here.

3. *Test Support Equipment.* This segment also consists of a list of all test support equipment necessary to maintain the entire test program. Support equipment is addressed in terms of what it is, where it is needed, and what testing it actually supports.

4. *Threat Systems or Simulators.* Herein the type, number, availability, and fidelity requirements for all threat systems or simulators are listed. When developing this segment, it is important to compare the requirements for threat systems or simulators with known available and projected assets along with their capabilities. Major shortfalls are highlighted. Prior to use, each threat simulator must be subjected to validation procedures, which establish and document a baseline comparison with its associated threats and ascertain the extent of the operational and technical performance differences between the two throughout the simulator's life cycle.

5. *Test Targets and Expendables.* This section and its depth will depend on the combat role the system will have. If applicable, type, number, and availability requirements for all targets, flares, chaff, sonobuoys, smoke generators, acoustic countermeasures, and so on, which are required for each phase of testing, are identified along with any special support requirements. These might include ordinance disposal or suppression systems necessary to protect the tested system. If shortfalls are known or expected, then these must be identified along with their impact on testing.

6. *Operational Force Test Support.* For each test and evaluation phase (DT&E and OT&E) it is critical and necessary to identify the type and timing of planned aircraft flying hours, ship steaming days, and other critical operating force support requirements. An example might include safety or chase aircraft, vessels, or vehicles needed for dedicated IOT&E to support field tests.

7. *Simulations, Models, and Test Beds.* With testing costs equal to or greater than some acquisition costs use of simulation and modeling is becoming more acceptable as an alternative to actual conducted tests. This is primarily a result of modern simulation technology which can be as realistic as, or in some cases more realistic than, the system itself. Substitution, however, is decided case by case, and when this alternative appears to be viable, this segment will include and identify the system simulations required, including computer-driven simulation models and hardware-in-the-loop test beds. The rationale for their credible usage or application must be explained and validated before their use. This should also include the impact on T&E should this method be unsuccessful.

8. *Special Requirements.* This segment identifies requirements for any significant noninstrumentation capabilities and resources such as special data processing or databases, unique mapping, charting, or geodesy products, extreme physical environmental conditions or restricted or special-use air, sea, or landscapes.

9. *Test and Evaluation Funding Requirements.* Herein the estimates are given by fiscal year and appropriation line number (program funding elements), of the funding required to pay the direct costs of planned testing. The funding currently appearing in those lines (program elements) is to be stated by fiscal year, and any major shortfalls must be identified.

10. *Manpower and Training.* This section identifies existing and projected manpower and training requirements along with any known shortfalls which may also be known. Any adverse impact these shortfalls may have should also be explained and potential alternative solutions included, such as extending the test schedule window.

Resource Summary. The initial preliminary test and evaluation master plan should project the key resources necessary to accomplish demonstration and validation testing along with any planned early operational assessments. The preliminary (initial) TEMP should estimate, to the degree known at milestone I, key resources necessary to accomplish DT&E, live fire test and evaluation (if required), and OT&E. Test range and facility requirements should be stated and their capabilities to support testing included. In addition, any unique instrumentation, threat simulators, and targets should also be included. As system acquisition progresses, the preliminary test resource must be reassessed and refined while subsequent TEMP updates reflect any changed system concepts, resource requirements, or updated threat assessments. Any resource shortfalls which introduce significant test limitations should be discussed and include planned corrective action or potential work around.

Appendix A: Bibliography. This final portion of the TEMP includes and references all documents referred to in the TEMP. It should also identify all reports, including their completion dates, which document technical testing or OT&E. This might include workload studies, PPQT or PQT results, demonstrations, and so on.

Appendix B: Acronyms. This section lists and defines all acronyms used throughout the TEMP.

Appendix C: Points of Contact. Listed here are all the guilty parties.

Annexes or Attachments. Many TEMPs are quite complex and require more space than is allocated by DoD guidance. There are no limits to the number and depth of supplemental annexes or attachments. This section should be used judiciously, however, and only support sections within the TEMP which require amplification. An example might include an annex which details criteria for exiting or entering another phase of testing.

SUMMARY

T&E is a process which, like any management technique, will always be open to criticism and subject to change. It has, however, moved to the forefront of acquisition and has become an equal partner in the system acquisition process. As stated at the beginning of this chapter, effective T&E is predicated on how well it is planned, conducted, and documented. The impact T&E and acquisition have on each other can be realized when one looks at the DoD-directed documentation required to support a major system. Of the no less than 23 primary documents which directly affect any major program six are T&E specific, eight are acquisition specific, and the remainder are consolidated documents which depend on inputs from both. It may be difficult to convince some that 23 documents streamline a program, but consider this: newly revised DoD 5000 series directives have resulted in the elimination or consolidation of 36 Air Force regulations encompassing Air Force acquisition management. The process will continue to evolve, and as new policies unfold and imbed themselves, it will further improve the acquisition and T&E processes. Streamlining can also be noted by the advent of the newly established DoD acquisition corps. Establishing a universal T&E corps is just as possible, and with the right approach can be just as effective and successful. These new approaches to system acquisition and T&E may not solve all our government procurement ills, but they are a big step in the right direction.

BIBLIOGRAPHY

DoD Directive 3200.11, "Major Range and Test Facility Base," Sept. 29, 1980.

DoD Instruction 5000.2, "Defense Acquisition Management Policy and Procedures," Feb. 23, 1991.

DoD 5000.2-M, "Defense Acquisition Management Documentation and Reports," Feb. 23, 1991.

DoD 5000.3-M-4, "Joint Test and Evaluation Procedures Manual," Aug. 1988.

"New Direction for Acquisition and Testing," *MRTFB Gazette,* no. 4, p. 1, July 1991.

E. Rodriguez, "Operational Suitability Guide," vol. 2, *Templates,* Office of the Director of Operational Test and Evaluation, May 1990.

Title 10, U.S. Code, sec. 138, "Director of Operational Test and Evaluation."

Title 10, U.S. Code, sec. 2350a.(g), "Side-by-Side Testing."

Title 10, U.S. Code, sec. 2362, "Testing Requirements: Wheeled or Tracked Armored Vehicles."

Title 10, U.S. Code, sec. 2366, "Major Systems and Munitions Programs: Survivability Testing and Lethality Testing Required before Full-Scale Production."

Title 10, U.S. Code, sec. 2399, "Operational Test and Evaluation of Defense Acquisition Programs."

Title 10, U.S. Code, sec. 2400, "Low-Rate Initial Production of New Systems."

Title 10, U.S. Code, sec. 2457, "Standardization of Equipment with North Atlantic Treaty Organization Members."

U.S. Air Force Regulation 800-1, "Air Force Acquisition System," Oct. 1, 1991.

U.S. Air Force Approved C-17 Test and Evaluation Master Plan, rev., July 31, 1991.

CHAPTER 5
PROJECT MANAGEMENT CONTRACTING

Alan W. Beck

Alan W. Beck has had experience preparing, negotiating, and managing major system contracts as an Air Force major systems contracting officer and as a systems contracting division chief. Since 1980 he has been teaching about systems acquisition contracting at the Defense Systems Management College, the Industrial College of the Armed Forces, and American University. He has published numerous articles on acquisition contracting in *Program Manager, Contract Management,* and *The Bureaucrat.*

Contracting is the process by which the government acquires or "procures" items and services from contractors, and it is how contractors buy from subcontractors or suppliers. The contract is the formal agreement which sets the terms by which the contractor promises to deliver goods or services according to some specification, at an agreed time, for some equitable amount of money or other consideration. The major activities which occur before contract award are planning, preparing contract documents for the solicitation or contract, evaluating contractor offers to select contractors, and awarding the contract.

When a contract is awarded, the period of contract performance begins. This is a time for monitoring compliance, paying for progress or deliveries, and (in complex project management procurements) for making changes to the contract. The government procurement process has thousands of pages of rules and regulations; many professionals are needed to interpret and comply. This management team includes contracting officials, technical or project management personnel, financial managers, logisticians, and various internal and external auditors and reviewers.

Because people are what makes the system work, understanding the activities in the contracting process calls for a look at the players on the contracting team. Both industry and government have contracting officers who are the authorized agents to enter into contracts for goods and services. The government agent awarding contracts is called the procuring contracting officer (PCO). After a contract has been awarded, the management of the contractual activity for a company is coordinated by a contract administrator and, for the government, by an administrative contracting officer (ACO). Should the government decide to halt a procurement in process, a specialized individual may serve as terminating contracting officer (TCO).

For less complex and lower-dollar-value procurements, the contracting officer buys what is needed with little interaction with the requirements people. As requirements grow in complexity, more technical involvement is needed; and both government and industry turn to a project management team concept. Successful functioning of this concept hinges on the relationship of the project or program manager and the contracting officer.

Both program or project managers and contracting officers are given responsibility in acquiring goods and services. The differences in their job responsibilities, backgrounds, authority, and reward system can contribute to tension; yet, the successful PMs and PCOs learn to function well as a team.

Responsibility. Program managers have the overall responsibility for managing systems acquisitions, subject, of course, to the various controls of government, Department of Defense (DoD), and service policy and regulation. Contracting officers are responsible for planning, preparing, obtaining, and documenting contracts and for managing or administering contractor performance.

Authority. Program managers get their authority from a charter—a written document giving their mission requirements or "taskings." Contracting officers get their authority from a certificate or warrant, which appoints them as a contracting officer and specifies their authority to sign contracts.

Background and Training. Usually contracting officers have background in business areas and professional courses related to contracting. Their professional schooling involves general courses in awarding and administering contracts and specific courses in contract law, pricing, and contract negotiation. Program or project managers often come from an engineering or technical background. The program manager typically has broad management experience and program management education.

Rules. The program manager in the DoD operates under DoD's 5000 series directives; the contracting officer's activity is primarily governed by the Federal Acquisition Regulation and supplements. Commercial contracts are governed by the Uniform Commercial Code.

Reward System. The program manager is rewarded for getting systems on schedule, within cost, and meeting technical requirements. The contracting officer may be scored on schedule but is more likely scored by some evaluation of contract quality (low noted errors in reviews) or numbers and dollars of contractual actions.

Contracting officers are in charge of negotiation, sign official correspondence to contractors, and issue unilateral changes pursuant to the contract's changes clause. They may unilaterally stop work or terminate the contract and may make final decisions on any disputes with the contractor. Because contractors may appeal a contracting officer's final decision to the Board of Contract Appeals or to the U.S. Claims Court, contracting officers are generally cautious and get legal help with contentious issues. In systems acquisition, the contracting officer works with a project management team in developing acquisition strategy and in preparing and maintaining the acquisition plan. Sometimes contracting officers delegate a part of their authority to specific representatives, known as contracting officer's representative (COR) or contracting officer's technical representative (COTR). Since official correspondence relating to the contract flows through the contracting officer (or representative), this job has been compared to being in the middle of a double-ended funnel—with large numbers of people at both the buying and the selling ends of the funnel.

In spite of the differences in the two roles, some of the job challenges are similar. Both must manage and coordinate the activities of others to get their job done.

Eighty to ninety percent of the calendar time to "get on contract" is often time for other functions not directly under the control of the contracting officer. Pushing (or "pulling" as General George Patton said in comparing leadership with moving spaghetti) contract actions through these various steps makes leadership a delicate balancing act for the contracting officer who must balance the desire for fast action and satisfactory results. It may be like the old proverb, "if you want it bad, you get it bad." Rushing auditors, legal reviews, pricing personnel, or technical evaluators may result in a bad deal for the government. Failure to adequately review or prepare proposed contract wording may be very costly downstream.

Within individual purchasing offices, contracting officers are supported by professional procurement or contracting personnel known as contract negotiators, contract specialists, or buyers. Other specialty functions of pricing, contract writing, and policy compliance review may be performed by either the contracting officer's buyers or other specialists within the procurement organization. It is the contracting officer who coordinates the activity of these assigned and supporting specialists.

The contracting officer's perspective of the contractor's requirements and the government's desires makes him or her a potentially valuable resource to program managers. Program managers who understand, cultivate, and tap this resource as advisor or confidant can improve the overall teamwork and efficiency of the program. Regular "open-door" communication, automatic invitation, and sincere welcome at meetings generate a feeling that the contracting officers (PCOs and ACOs) are "on the team." This can pay big dividends.

Unfortunately there is often a scarcity of contracting personnel, so workload and continued training needs restrict their time for attending many program office meetings. With increased political attention on procurement, managers at all levels—particularly program managers—need to seek and maintain effective contracting support.

THE CONTRACT AWARD PROCESS

The contracting process before award involves gathering contract requirements, writing down what the contractor is to do, selecting the contractual method, determining the contract type, planning criteria for selecting contractors, developing the contractual documents, obtaining contractor offers, and reaching agreement on the specific price and terms of the contract.

Planning

The planning phase begins with the identification of needs and funding. The Federal Acquisition Regulation (FAR) Part 7 (with supplements) prescribes the requirements for preparing acquisition plans for proposed acquisitions. This overall planning sums up or leads to more detailed planning needed for specific activities and documents required to get on contract. Key items needed to plan for the contract solicitation include the technical requirements, the type of contract, and the method of procurement.

Establishing Contract Technical Requirements

To transition from planning to getting paperwork ready to issue a contract requires careful translation of needs into several documents. The statement of work (SOW) describes the work to be done. The specifications describe the requirements that the product or service must meet. The contract data requirements list consolidates deliverable data requirements. A purchase request package consolidates and transfers all of the requirements documents plus funding information to the contracting specialists. It all starts with the basic question of *what* the contractor should do. Structure for the various elements of the project is provided by the work breakdown structure (WBS).

Work Breakdown Structure

The WBS is an organized way to break down a project into logical subdivisions at lower and lower levels of detail. A pictorial example of part of a WBS is shown in Fig. 5.1.

WBS is useful in organizing and managing a project. Contractors use it to plan and track individual work "package" effort at the lowest levels. Government cost monitors track the cost and schedule accomplishment using these work packages. WBS also serves as a basis for planning the contractual statement of work, data requirements list, and specifications.

Specification guidance is provided in Part 10 of the FAR, "Specifications, Standards, and Other Purchase Descriptions." This policy calls for specifications to state only the government's actual minimum needs and is designed to promote full and open competition. More detailed guidance on the format and contents of specifica-

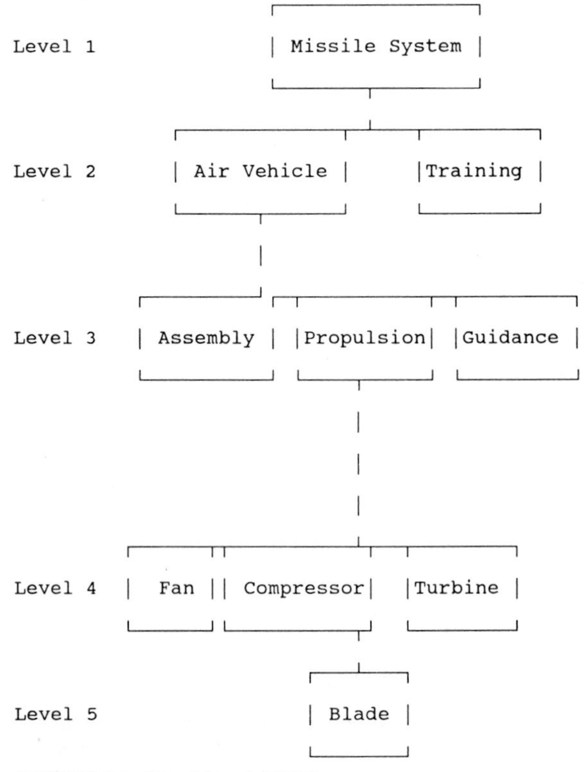

FIGURE 5.1 Sample partial WBS.

tions for DoD is provided in MIL-STD 490A, "Specification Practices." This military standard provides uniform practices to prepare specifications in five specified types:

1. *Type A: System Specifications.* They define mission and technical requirements, allocate functional areas, and define interfaces to provide a functional baseline of specification requirements.
2. *Type B: Development Specifications.* They describe performance characteristics of each configuration item, providing an "allocated baseline" of specification requirements.
3. *Type C: Product Specifications.* They define form, fit, function and performance, and test requirements for acceptance contributing to the product baseline, which also includes the process and material specifications.
4. *Type D: Process Specifications.*
5. *Type E: Material Specifications.*

The specifications logically become more detailed as a program progresses from early stages of broad general type A system specifications to the definitive product specifications used in production for the rigid quality and conformance inspections on delivered hardware. Sometimes it makes sense to provide only type C product specifications with functional requirements so that industry has some flexibility in methods of process and material. The preference is for stating "what is necessary" instead of "how to." The nature of the requirement in the specification and the WBS combine to determine the level of detail and general thrust desired in the SOW preparation.

Statement of Work Preparation

The SOW is prepared to establish and communicate nonspecification requirements and identify the desired contractor work effort. The SOW is a key to effective proposal preparation, evaluation, selection of contractors for award, and contractor performance on the contract. As the work is performed, information that may be required for retention is developed. Since the SOW is not the contractual vehicle to require data delivery, the preparer must concurrently identify and list data requirements.

Contract Data Requirements

We "buy" or require delivery of data by listing the requirements on a contract data requirements list (CDRL). This list tells contractors what data to deliver, when and how data will be accepted, where to look for preparation instructions, where in the contract (SOW reference paragraph) the preparation effort is required, and other information.

As the SOW is prepared, the preparers should keep track of tasks which produce data that may be contracted for delivery. Data requirements will need special identification.

Data Call and Review

CDRL candidate items are developed by persons with data needs in response to the program manager's "data call." Candidate data items are reviewed by a data review board or individual to check for need, duplication, or potential "nice-to-have" but not essential cost drivers. The scrubbed down data requirements list is finally prepared and double checked with its associated SOW to ensure correct cross-reference paragraph numbers. The SOW, specifications, and CDRL then form the backbone of the package which the contracting officer will need to prepare and award a contract. This contract package accompanies a funding document called a purchase or procurement request (PR), and the entirety is called a PR package. For the contracting officer to procure the items or services requested in the purchase request package, the officer must determine the appropriate type of contract and method of contracting. While these decisions may officially belong to the contracting officer, in complex project managed acquisitions, the contracting officer coordinates with the project or program manager's team.

DETERMINING THE TYPE OF CONTRACT

The type of contract determines how the risk for cost of performance is shared between the parties. It makes little sense to ask an inventor to conceive, design, and produce a new complex product on a firm fixed-price contract if no one can reasonably guesstimate the cost of performance. Conversely, there is less motivation to cut performance costs on a repeat production contract if the contract is cost-reimbursable. If, as with a space system, reliability and quality are far more important than cost, then appropriate incentives (such as award fee or performance incentives) should be used.

Categories of Contracts. The two basic categories, sometimes called families, of contracts are cost-reimbursement (where the government pays the cost, subject to limitations on allowability, allocability, and reasonableness) and fixed-price (where the government pays a price, subject to some fixed maximum ceiling amount if a sharing incentive is used). Within these categories are other common types such as firm fixed price (FFP), fixed-price incentive (FPI), cost plus incentive fee (CPIF), cost plus award fee (CPAF), and cost plus fixed fee (CPFF).

The Contractual Promise. Because the contractor promises to deliver on time and to meet the contract specifications, fixed-price contracting is serious business. If a contractor is late or if the product does not meet the specifications for acceptance, the government may terminate the contract for default and not pay. In some circumstances the government will reprocure the items from another source and, if they are more expensive, charge the first contractor the difference. Clearly, in fixed-price contracting, the contractor bears the cost risk. The higher the contractor perceives the cost risk, the higher he or she will price the contract.

In cost-reimbursement contracts, the government promises to pay all (allowable, allocable, and reasonable) costs incurred on the contract. In return, contractors promise to exert their best efforts to perform the desired work. If the work turns out to need more money than originally estimated, the contractor notifies the government and may stop work when the money runs out (or perhaps earlier if so directed or authorized).

Incentive Sharing. In either fixed-price or cost-reimbursement contracts the parties may agree to sharing costs over or under preset targets. This concept of cost incentive means that the contractor's profit (if fixed-price) or fee (if cost-reimbursement) is increased or decreased by a predetermined share (percentage) of the overrun or underrun.

Fixed-Price Incentive. Fixed-price category contracts with incentives (cost-only or multiple) are called fixed-price incentive (FPI). The most common FPI contract is FPIF (fixed-price incentive firm target), which means that there is a firm target established from which to compute the overrun or underrun later. A rarely used and more sophisticated variant, fixed-price incentive with successive target (FPIS), has successive targets which are firmed up later based on some initial contract performance cost data.

A top limit on sharing on an FPI contract is logical if one considers the nature or promise of fixed-price contracting. At some set price, the contractor must deliver on time and on specification. Therefore a ceiling price, which is the maximum amount the government will pay, is set. If performance costs the contractor more, the contractor loses.

Cost plus Incentive Fee. CPIF contracts allow similar over- or underrun sharing of costs via a predetermined formula for fee adjustments which apply to incentives for cost category contracts. Within these guidelines and the basic concept of the government paying all costs for a cost contract, the limits for a CPIF contract become those of maximum and minimum fees.

Award Fee. Award fee contracts permit a subjective unilateral decision on how much fee to award at each award period. The subjective award fee is generally used with a set base fee on a cost-plus-award-fee (CPAF) contract. However, the subjective fee award approach is also used as an add-on incentive with other types of contracts.

SELECTING CONTRACT TYPES

The contract type determines how cost risk is passed to the contractor. In a move from cost-reimbursement toward fixed-price contracts, the contractor assumes most of the cost risk.

Matching the cost risk to the appropriate contract type is basic to selecting the contract type. Given that the expected costs of performance are uncertain, the issue becomes "how uncertain." If the work is easy to price with high confidence that the actual prices will be within a few percent of the estimates, the contract should probably be FFP. If the work is very hard to estimate and the probability of the price being close to the estimate is risky, then a CPFF or CPAF may be appropriate. Early in the developmental life cycle one may expect more cost-type contracts, with a move toward CPIF, FPIF, and FFP as items are refined and begin production.

Incentive Contract Pricing

Cost uncertainty is expected in incentive contracting, and the expected range of probable cost outcomes is a major determinant in selecting the appropriate contract type. The pricing arrangement should provide for a fair profit or fee for expected performance with reasonable increases or decreases in profit or fee for exceeding or failing to achieve incentive goals. This is the reason for an incentive contract.

Developing an effective incentive structure with a fair pricing formula begins early in the acquisition strategy discussions and is further refined as the acquisition plan is developed and the solicitation prepared. Industry comments in response to draft solicitations may provide further refinements in the proposed contract structure for final solicitation.

If the technical experts agree that the preliminary cost estimates are very likely accurate within a narrow (±10 percent) range, an FFP contract—the strongest incentive—is probably appropriate. If cost estimates are less certain, the question shifts to whether to use a performance incentive or an award fee. Basically a management decision, this will be reflected in the solicitation. Since fair, equal competition is a key policy, the contracting officer treats all offerors equally.

When multiple incentives are considered in addition to cost incentives, price interrelationships get even more complicated. Having too many incentives can dilute their value and lessen their effectiveness. Where performance emphasis may be desired in several different areas and may even vary through the contract life, an award-fee provision may be more appropriate for a flexible, subjective incentive.

Pricing involves estimating and uncertainty; one must remember that there is a range of incentive effectiveness (RIE) of possible or probable cost outcomes. The term *cost overrun* is used for costs over target, but should not necessarily have a negative connotation so long as costs are within the contemplated RIE. The target price is, by definition, imprecise. Managers must realize that these aggregate imprecise numbers are the summation of many other imprecise estimates.

Management of Incentive Contracts

Depending on the type, an incentive contract may require or permit varying management. Cost-reimbursement contracts generally require monitoring costs to ensure that sufficient funds are obligated for the work. On CPIF contracts increasing overruns may mean less fee will be due—so the contracting officer may need to reduce or suspend fee payment.

While cost contracts lead to closer government management, fixed-price contracts lead to a more hands-off type management. Award-fee contracts lead to the closest relationship as periodic performance evaluations are made (and hopefully discussed) as a basis for the government's award-fee decision.

Changes to incentive contracts may reopen the negotiation of the appropriate targets, limits, and even sharing formula. The fair pricing adjustments for each change must be calculated according to that change's estimated costs and relative risk.

Incentive contracting, properly applied and with a contractor who is motivated by the incentives, can be a powerful tool to challenge industry to superior performance. Pragmatic determination of strategy, sound selection and structuring of incentives, and careful management after award can result in success. The novice and the experienced acquisition person alike will want to continue to study the techniques and seek out lessons learned to do the best job, appropriate for the goals, risks, and uncertainties of each program.

DETERMINING THE METHOD OF PROCUREMENT

The choice of how to buy goods and services is related to the contract type. Just as goods and services which cannot be precisely priced in advance should have other than a fixed-price contract, the requirement that needs some negotiation discussions before award should have other than a sealed-bid method. These subjective decisions are made by the contracting officer working with the project management team. What are the government policy and principles on methods of procurement?

Fundamental Policy—Fair. Because government officials buying goods or services are dealing with the public trust, the government procurement laws and regulations have a fundamental assumption reflected in the thousands of pages of regulations on how to buy for the government. This assumption, or ethical imperative, is fairness. The government should pay a fair and reasonable price and should be fair in selecting contractors for government contracts. Whether the government uses price or technical competition or demands contractor cost data for analysis before negotiating in noncompetitive situations, the fundamental policy is a fair price for the government and a fair opportunity for the contractor.

The methods of procurement differ based on the dollar value and the availability of competition by price or technical merit. A look at the basic methods—sealed bid, two-step sealed bid, competitive negotiated, and sole source—will help understand the many ways to procure goods and services.

Sealed Bidding. The traditional preferred method of government procurement (called contracting within DoD) has been to get competitive sealed bids and award to the low bidder. Precise sealed-bid rules are purposely written and interpreted to give all firms a fair chance at government business.

The basic concept and process include the following steps:

1. *Publicizing.* The government purchasing offices are to publicize proposed procurements, including an official synopsis, in the Department of Commerce's publication, *Commerce Business Daily,* prior to issuing the formal invitation for bids (IFB) solicitation document.

2. *Public Opening.* The sealed bids are held for a public bid opening with subsequent award to the responsive and responsible low bidder, based on price and other factors (such as transportation cost). Responsive means that the bidder complies with all IFB requirements; responsible means that the bidder is capable of performing the work on schedule. Responsibility is a subjective opinion of the government after evaluating such things as financial capability, personnel or manpower, scheduled workload, and production capability. These responsibility factors are checked by what is called a preaward survey.

The sealed-bid method is the preferred way to buy items where conditions are appropriate and where it makes good business sense. One of our early astronauts joked in space that he felt uneasy riding on a product that was the result of low bids on government contracts. Perhaps he knew that NASA contracts were often cost-reimbursable with quality and reliability the primary concerns, but the general public and, therefore, some public officials do not realize or appreciate the need to award some government business on factors other than low bids. Just as private consumers often prefer to make purchases with quality, long-term life, or potential repair or support cost as criteria, so does this also make sense for our government.

Sealed bids are appropriate when there is enough time, when award is based on price competition, when it is not necessary to conduct discussions with the responding sources, and when there is an adequate government specification for an FFP contract. When these circumstances do not exist, the contracting officer should use another method. For most major project management work, this is to solicit competitive proposals for "negotiated" contracts.

Negotiation as a Method (Not Sealed Bid). Negotiation as a method of contracting occurs when the sealed-bid method is not used. It may involve face-to-face negotiations, as happens in negotiating a sole-source contract, or it may involve competition between several offerors, with the award based on some preestablished criteria (such as technical excellence) instead of low price.

Since Congress prefers competition, where possible, to promote fair opportunity for winning government contracts, special approval is required for sole-source contracting. Where competition exists, the competitive negotiated method does not need special approval. Negotiation is a confusing term with dual meanings. Actually, Negotiation (capital N) as a method merely means using some way other than sealed bid. Negotiated procurements range from small-purchase procedures (which save paperwork for items less that $25,000, but still require competition where possible) up to complex research and development acquisition (which may have competition in technical areas rather than price).

The confusion over Negotiation as a contracting method and negotiation as a way to seek agreement on terms and price leads many to misunderstandings on how the negotiation method works under competition. Because the term used to describe this method is "negotiated," many people think that there has to be a price negotiation during this process. In a competitive negotiated acquisition, competing contractors submit their prices in initial proposals or (if the government elects to discuss proposals to resolve questions) as a best and final offer. These competitively priced offers provide the price.

Sole-Source Approval. Proposed sole-source contracts require special approvals. For federal contracts, the law generally requires competitive procedures, but provides specific statutory exceptions when there is only one responsible source and no other type of property or service will do, or when there are special circumstances, such as a special law, international treaty, or national security requirement.

Two-Step Sealed Bidding. A newcomer to a study of procurement methods may think that two-step is a dance that was in vogue years ago. An old hand in procurement may be saying, "What happened to two-step?" Two-step is a variant of the sealed-bid procurement method. It is a way to get sealed bids after an initial step to discuss, reject, or accept technical proposals.

Two-step is appropriate when there is either a loose "functional" specification or a tight restrictive specification. It implies a willingness to go to the low technically acceptable bidder. Firms are allowed to submit one or more technical (not price) proposals which are then evaluated against criteria indicated in the solicitation. Next, all satisfactory technical approaches are allowed to submit bids in response to an IFB, which uses each individual's approach as the contractual specification with a sealed-bid award process to the low responsive and responsible bidder.

Two-step is a way to allow technical "go/no-go" evaluation but still use sealed bids for competitive award based on low bid rather than using the detailed scoring of a formal negotiated method source selection. Since two-step requires award to the low bidder in the second step, the two-step technical evaluator's slogan may be "when in doubt, rule them out."

Selecting the Method. To select the appropriate method of procurement, a series of questions must be asked in early acquisition strategy discussions about the nature of the procurement. The thrust of public policy is to increasingly seek alternative solutions or functional specifications to capture the innovative competitive potential of industry. Selecting the right method is the basis of a good start toward a reasonable contract.

PREPARING THE SOLICITATION

To put either type of solicitation together, the contracting officer needs to collect the information required by the uniform contract format as specified in the FAR. The format will look like this:

Section	Title
Part I: The Schedule	
A	Solicitation/contract form
B	Supplies or services and prices/costs
C	Description/specification/work statements
D	Packaging and marking
E	Inspection and acceptance
F	Deliveries and performance
G	Contract administration data
H	Special contract requirements
Part II: Contract Clauses	
I	Contract clauses
Part III: List of Documents, Exhibits, and Other Attachments	
J	List of attachments
Part IV: Representations and Instructions	
K	Representations, certifications, and other statements of offerors or quoters
L	Instructions, conditions, and notices to offerors or quoters
M	Evaluation factors for award

Although they are major inputs, the specifications, SOW, and CDRL provide only part of the needed information for a contract. A brief study of the uniform contract format will show that it outlines other requirements.

Section B is a list of deliverable supplies or services. There can be subtle strategy in whether items are grouped or separately listed and in how items are priced. At the solicitation stage, Section B will have blanks for prices; the contractor fills in these prices when sending the offer to the government.

Sections D, E, and F provide information of particular relevance to the field contract administration personnel involved in contract acceptance and payment. Sections C, H, and J provide information guiding contractor performance during the contract.

The RFP

Part IV of the uniform contract format entitled "Representations and Instructions" provides the guidance needed to tell contractors how to respond (Section L), how the offers will be evaluated (Section M), and the various administrative requirements such as socioeconomic program compliance (Section K).

The instructions and the evaluation factors will affect the ability of the government to select the best contractor for the job. Source selection planning must be coordinated with the phase of RFP development to ensure an effective evaluation of well-organized proposal information.

Sections H and I: Contract Clauses and Special Requirements

Most technically oriented people focus on the SOW or contract specifications to see what the contractor is required to do. While these parts of the contract are vital to describing the basic work to be done, the experienced person will also want to check the contract clauses and special contract requirements sections of the contract.

Although the contract clauses are standardized approved wording and are generally included by reference instead of full text, particular clauses used do vary from contract to contract. The contract includes certain clauses which are required depending on the type and purpose of the contract; it also has optional clauses which may be included as applicable. Sometimes these standardized clauses call for implementing language or "fill in the blank" type information, which will probably be included in the contract clauses section. Special non-FAR standard clauses may be included in Section I or, if particularly tailored for your contract, will more likely be found in Section H, "Special Contract Requirements."

What Language Is Most Important? Significant language to check in contract clauses or special contract requirements is that concerning payment and performance responsibility.

Payment. Standard payment clauses for fixed-price contracts cover payment on delivery (and acceptance), with optional clauses for earlier progress payments of a percentage of incurred cost for satisfactory progress under the contract. Progress payments are commonly used for larger dollar value contracts (over $1 million in large business; $100,000 in small), where there will be a significant delay period (four months for small business, six months for large) before the contractor will be able to bill for performance.

Standard payment clauses for cost-reimbursement contracts provide for the contractor to be reimbursed regularly as costs are incurred during contract performance. To ensure that the government has enough money obligated on the contract to pay for the cost of performance, a standard "fill in the blank" clause is included to state how much money is allotted to the contract, to limit the government's responsibility to pay (or the contractor's responsibility to perform), and to require the contractor to submit advance notice when funds are running out and more will be needed. These clauses are entitled, "limitations of cost" or "limitation of funds." If the government fails to provide the additional money for performance, the contract becomes in effect self-terminating because contractors are not expected or required to work without money. Government personnel can get in serious trouble if they request (or perhaps even condone) contractors to perform for the government when funds are not properly approximated and available for payment.

Contracting over Several Years. The congressional budget appropriation procedures lead to limitations on contracting over several years. The contract must be structured to accommodate the peculiar rules for the use of different appropriations and the availability of funds.

Fixed-price contracts frequently require performance over several years. If production items are being bought and sufficient money is available to obligate at the award of the contract, then delivery may take place in later years. If sufficient production money is not available in the first year (year of award) for all desired quantities, then the government may structure future years as options, or (with necessary approval) write a multiyear contract to provide for funding for additional quantities in succeeding years. The advantage of the multiyear contract, instead of options, is that the contractor can plan for the full quantity, making appropriate capital investment decisions, long-term subcontract agreements, and economic order-quantity purchases of components. The disadvantage to the government is the need to pay the costs of canceling a multiyear contract (up to an agreed ceiling) if the government changes its mind and does not fund subsequent years. The advantage of options is that they are optional and the government is not bound to order the extra quantities; however, this also means that the contractor cannot count on the whole deal.

Fixed-price contracts are occasionally, but not typically, used for research and development efforts extending over several years. In this case a clause is inserted into the contract, similar to the limitations of the cost clause used in cost contracts, to allow the government to fund the contract as the annually appropriated funds become available and to provide that the contractor does not have to perform beyond the dollars obligated if the government fails to fund later years.

Performance Responsibility. Performance milestones are sometimes highlighted in special contract requirements with language to tie specific contractual consequences to performance events of "milestones." These features are often called demonstration milestones, as the contractor must demonstrate particular performance. Various contractual actions can be tied effectively to milestone events instead of calendar dates. Test reports can be made due *X* days after test. Option validity dates can be *X* days after successful completion of final test. With events driving actions, there is less confusion when contractual dates slip.

Various special requirements clauses have been developed to control responsibility when complex systems involve several contractors. A total system performance responsibility (TSPR) clause is sometimes written to give a prime contractor total responsibility for the performance of a system, even including responsibility for performance of government-furnished property (GFP) once this GFP is accepted by the prime contractor. This clause involves a large degree of risk and thus may cost a lot. To help reduce the cost of a TSPR clause, try to ease the uncertainty or cost risk on the prime contractor. One way is to give the contractor more control over the acceptability of the GFP.

Total system integration responsibility (TSIR) clauses are sometimes used to specify the prime contractor's responsibility as a system integrator. This responsibility is less than performance responsibility but can include responsibility to coordinate all interfaces and ensure that system design in no way degrades the performance of GFP.

Interface control working group (ICWG) clauses are sometimes written to establish and require participation in groups involving contractors and government representatives. Inserting implementing ICWG language in each participating contractor's contract can provide a structural and procedural mechanism to help ensure effective coordination.

Getting a Good Contract. The special contract requirements, together with the standard contract clauses, govern much of the contractor's effort. To implement the desired acquisition strategy effectively, technical and requirement personnel need to work effectively with contracting specialists to get the right special language for the request for proposal (RFP). A dialogue with industry using a draft RFP may reveal an unexpected problem in proposed language. A seemingly small change in the wording of a special requirement could have a significant effect on the risk to industry and thus the price of that requirement.

Other pre-RFP activities include identification of any GFP requirements, obtaining solid promises of resources the government must provide, drafting any special contract requirements language for Section H, and selecting appropriate preprinted contract clauses from the FAR. Clause selection varies by the nature of the requirement (supplies, services, construction) and the type of contract. Because contract type is the key element controlling cost risk to the contractor, it must be selected carefully to meet the circumstances. Much advance thought on contract type is needed during acquisition planning and SOW drafting.

OBTAINING COMPETITION

Competition, where possible and practicable, is the policy on contracting. Planning for competition is a mandatory consideration in writing the acquisition plan.

Requirements should be stated carefully so as not to restrict competition. Stating goals or minimal requirements in functional terms allows for maximum contractor innovation in competing ways to meet government requirements.

Before a solicitation is released, the potential buy is publicized by printing a synopsis in the *Commerce Business Daily* (CBD) to notify interested firms of the nature of the buy and where to ask for the solicitation. For major systems, the potential prime firms will know of the requirement and will be working on proposal strategy before the CBD synopsis. However, for many subcontractors or suppliers and for less complex buys, competing sources are obtained from the CBD synopsis.

Selection of a source can vary in procedure. The relatively simple sealed-bid method simply awards to the low-price (FFP) bidder who is responsive (meets the requirements) and responsible (can do the work).

For more complex acquisition, where the award criteria may be other than just price, a detailed source-selection evaluation procedure may be used to carefully evaluate and score each contractor's proposal against the evaluation criteria in the RFP.

Where a logical WBS has led to a clear and meaningful SOW, specification, and overall requirement package, and where the type of contract is appropriately selected according to the cost risk of the expected contract, the source selection has the best chance of obtaining a good contract to achieve the government's objectives.

CONTRACT AWARD

Changes resulting from discussions with contractors during source selection are incorporated in revised offers to the government. Before final contract award, the contracting officer will initiate a preaward survey to verify the capability of the contractor to perform. Contract award takes place when the contracting officer has signed and distributed the contract to the contractor.

PROTESTS

During the preaward period, a potential contractor may feel unfairly treated and may elect to protest. All individuals involved in the acquisition process may have a chance to cause or prevent protests. Protests indicate a lack of satisfaction by a contractor in the government's contract award actions. The frustrated feelings or suspicions of unfair treatment can lead to a formal protest that may cost a lot in time, administrative effort, and legal services.

Let us review some basic information and rules on handling protests to learn how to limit protests and their impact on programs.

What Is a Protest? In acquisition terminology, a protest concerns the award or proposed award of a contract. Traditionally this is contrasted with a claim relating to an existing (or implied) contract. Contractors or potential contractors considered to be interested parties may protest when they feel that government action concerning the solicitation and award process is unfair.

The protest, or objection to the award process actions, either can be relatively informal, even verbal, or it can be formally written. Often the implied or actual threat of a protest, or simply someone's perception that a protest could result, is sufficient cause for the government to change requirements or procedures. An overly restrictive specification is a common example of this area where the government may decide to "loosen" the specification or use a functional specification in order to enhance competition while preventing a potential protest.

Perhaps the possibility of a formal protest and the resultant program delay help keep the government personnel concerned to be as fair as possible to all offerors. Thus the fact that government actions concerning the award of contracts are subject to challenge and review tends to provide protection to contractors against unfair treatment by the government.

Who Gets the Protest? Normally protests are directed to the contracting officer responsible for solicitation and award of the contract. Many formal and informal protests can be settled easily at that level by the contracting officer with simple coordination and communication to resolve the actual or perceived problem. If the problem cannot be resolved or if the contractor does not feel inclined to even ask the contracting officer first, the protest may be presented to some higher authority in the service, directly to the Comptroller General of the United States General Accounting Office (GAO), or directly to the U.S. Claims Court for judicial injunctive relief. In addition to being taken to the contracting officer, some higher-level official in the agency, or the GAO, some protests are taken to the courts or, in the case of automated data processing (ADP) protests, to the General Services Board of Contract Appeals.

When protests occur before award, GAO regulations require notification to GAO if the head of the contracting activity decides to proceed with award without waiting for a ruling on the protest. Although awarding the contract while a protest is pending is discouraged and requires special approval, it is not uncommon.

Besides schedule delay, other program impacts of a formal protest are the extensive paperwork necessary to support or defend the case and the higher-level visibility or scrutiny of the issues. The potential need for data to resolve protests demands cautious documentation of fair and equitable treatment and procedures. The detailed procedures and rules are sometimes frustrating and always time-consuming but are aimed at preserving the integrity and fairness of the source-selection process.

Winners and Losers. Sometimes we hear of winning or losing a protest. Too often all parties lose. The protest drains resources from all participants and costs valuable time as both sides defend their positions and wait for a ruling. Many protests are withdrawn before decision, perhaps indicating either agreement to change by the government or new understanding which caused the contractor to drop the case. Which types of protests are successful and which a waste of time? Subjective decisions of government contracting officers will generally not be "second guessed" by GAO unless the contracting officer was obviously unfair by not following required procedures or by being grossly arbitrary. GAO may sustain a protest based on objective issues. Late bids are late and a protester who thinks the government should have let him be a few minutes late will find no help from the GAO. Similarly, GAO's filing times are rigid and a protest may be dismissed, regardless of its merit, if it is not timely.

Procurement managers often say that if everything is done professionally, fair to all parties with justified (documented) government actions, there is no reason to fear losing a protest. More significantly, if actual fairness is communicated by actions, procedures, and tone of letters, there should be little reason to have a protest. The challenge is to maintain a professional environment that is fair to all.

CHANGES

Changes are needed whenever the original contract agreement is no longer adequate to document what needs to be done. While relatively simple or technologically stable goods or services are procured in contracts which may have no changes, contracts for complex items developed in an arena of technological uncertainty may require thousands of changes.

There are two major ways to put a change on a contract. The preferred way is to get full and final negotiated (definitized) agreement on work, cost, and schedule with both parties signing the change. When both parties sign the change, it is called a supplemental agreement. The less preferred, but sometimes appropriate way if time is critical, is to have an undefinitized change.

Undefinitized changes are not as desirable because the price remains to be negotiated. While it is often undesirable to have someone (such as a home builder) working with a blank check, it sometimes makes sense to have work started before firm prices are established (when the gas station hears, "My car is running rough, please fix it."). To protect against the unlimited potential blank-check cost, contracting officers normally obtain a not-to-exceed price agreement to put a cap on the yet to be negotiated price for the change (like telling the gas station, "Fix it if it's not over $200.").

Change orders may direct undefinitized changes in contract specifications, shipping, or packaging. Because complex work is often likely to require change, the FAR has standard change clauses for government contracts. Since the change may well impact the costs and schedule of the contract, the clause allows for the contractor to request an equitable adjustment in the contract cost or schedule. Settling this price adjustment, normally by supplemental agreement, is called definitizing the change order.

How does the change order process work? Basically the contracting officer needs technical detail of what is to be changed in the requirements, funding to cover anticipated costs, and authority to issue the change order. The greatest of these hurdles may well be getting authority to issue the change order. Generally, contractual authority under the FAR changes clause is not the difficulty. The issue is more often a management concern for limiting undefinitized work. Many organizations have additional review procedures requiring special justification and approval before their contracting officers are allowed to issue undefinitized changes. Even Congress has gotten into the management of undefinitized contractual actions through legislation limiting overall percentages on new awards.

The contracting officer wants the contractor to agree to a not-to-exceed price to limit the government's potential cost exposure for the change. When there are uncertainties in the pricing, a rational contractor will increase the not-to-exceed price slightly over "best guess" estimates. When the contractor is rushed to provide a price without time to talk to subcontractors or to do detailed cost estimating, the word may be "if you want it bad, you get it bad." Unusual circumstances might dictate contractual direction without waiting to agree on a not-to-exceed price, but this is very rare and not good practice. Given management approval, money, requirements, and an agreed not-to-exceed price, the easy part is to make the contract change. It may be as simple as a one-page contract change form with a fund citation and a remarks block entry which simply says, "Engineering change proposal 336 is incorporated, pursuant to the changes clause, at not-to-exceed $800,000.00." The contracting officer may need to get internal legal or other review and may want to include by reference the contractor's letter offer of the not-to-exceed price or may ask the contractor to also sign the undefinitized change to formally record agreement to the not-to-exceed price.

The preferred way to change contract requirements is with a definitized supplemental agreement, which both parties sign, documenting full agreement on what is changed and the price adjustments, if any, resulting from the change. This is the way the vast majority of changes are incorporated into contracts. Undefinitized change orders are relatively rare, but sometimes justified, because the process of issuing a supplemental agreement can take many months. A look at the key steps will show what is done and why it sometimes takes six months or more to complete the action.

1. *The contractor prepares and submits a change proposal.* The contractor's proposal preparation time depends on the complexity of the pricing. Where there is extensive subcontracting, the pricing may be more accurate if the contractors have time to get detailed price quotes from subcontractors rather than quick estimates. Even a relatively simple change may take a month for the contractor to get all the prices estimated, summed, and approved as a proposal.

2. *The government evaluates and audits the proposal.* To determine the right terms, technical effort, and price for the change requires careful proposal review and cost analysis. This review typically takes two major actions—technical evaluation and audit. The technical evaluation reviews the proposed materials, labor, and work requirements. The materials review may cover technical characteristics, scrap estimates, and quantities required. The labor review includes both levels of each skill required and the hours for each level. The work requirements review would include overall evaluation on any changes in the SOW or technical specifications. The audit (for larger changes) reviews three major areas: the contractor's labor costs for each proposed level of skill; the indirect overhead cost rates for labor or material; and the costs for proposed materials. The audit may take six to eight weeks, with some additional time for inputs by the contract administration team.

3. *The government and contractor prepare for negotiations.* Negotiation preparation involves identification of any issues to be resolved and detailed cost analysis to estimate the appropriate pricing for the effort. Fact finding discussions may be required to provide a basis for negotiation. Negotiation strategy for the entire package, including any changes to the pricing structure or contract incentives, may then be reviewed for prenegotiation clearance according to local procedure. This pricing and negotiation preparation process may take a few days or a couple of weeks.

4. *The agreement is negotiated.* Negotiation of the contract agreement may be quick or lengthy. How long it takes depends on the complexity of the change and the participants' interaction. Negotiation is the art of obtaining agreement on what is to be done and on what terms. Anyone can deadlock; skilled negotiators forge agreements through an understanding of different opinions and causing compromise.

5. *The contract language is drafted and reviewed.* Controversial new language will probably be worked out at the negotiation table, but after negotiation, the exact wording and format of the planned change have to be drafted. Depending on the dollar value and management interest, the proposed change package will then be subject to review by legal officials and a special contract review activity according to local procedure. It is hoped that none of these reviews will require reproposal and renegotiation.

6. *The contractor reviews and signs.* The proposed contract modification is mailed to the contractor for signature. The contractor may require internal reviews before signing the proposed change. This could take a few hours if the change was relatively simple or anticipated; it may take weeks if the change was serially coordinated from organization to organization with no special priority.

7. *The contracting officer reviews and signs.* After receipt of the contractor-signed proposed contract modification, the contracting officer will make a final review and may have requirements for other internal reviews or clearance before signature. This step may take several days, particularly if someone in headquarters wants to ask more questions.

8. *Reproduction and mailing.* Reproduction of copies for the contract mailing list takes additional time. The supplemental agreement does not become effective on signature, but rather when it is distributed. The time for each step in the process may vary depending on the complexity of the action and the efforts of the people working the problems. Sometimes changes such as a disagreement or a change in the requirements late in the process cause a return to earlier steps.

Other Formal Changes

In addition to the change order and the supplemental agreement, there are some other areas in which the PCO can issue changes. Administrative (no-cost) changes may be issued by the contracting officer unilaterally to make minor (noncontroversial) changes such as changing the name of a government representative or obligating additional funding for an incrementally funded contract. A provisioned item order (PIO) is used to order spare parts through the provisioning process. These orders may be added unilaterally to a production contract to permit timely (concurrent) manufacture of the initial spare parts while the production parts are being made. PIOs are an administratively convenient way to order spares. They may cite estimated prices (undefinitized) with the ACO negotiating final prices. Prompt ordering using undefinitized PIOs to allow concurrent production with end items may provide schedule and cost benefits which outweigh the usual government preference to wait for negotiation of definitized prices before ordering.

Managing Change

The manager's challenge is how to limit or eliminate nonessential changes while seeing that necessary changes are made in a timely manner within the constraints of law and policy. Frequently changes are needed when (or before) they are suggested, so the users or requiring activities may want the change made as soon as possible. This increases the pressure on the manager for rapid approval technically, for finding existing money rather than waiting for justification in a budget submission, and for wanting the contractor to start work quickly rather than wait for a definitized change. Managers are paid for making things happen; however, the system inserts pragmatic checks to be sure no mistakes are made. The preference for getting a definitized fair price before work starts is often frustrating to contractors and government managers who want changes to be made quickly.

Undefinitized work authorization may make good sense when the costs of delay in implementing change are weighed against the benefits of earlier change. Managers have to decide when it makes good business sense to order changes without firm negotiated prices. As senior managers (even Congress) have perceived problems with undefinitized work authorization, working-level management's flexibility to work with procurement personnel to determine how to contract for change has become more limited. Management by quotas tends to take over with managers and procurement offices scored by how they minimize "undesirable" (undefinitized) actions. As these controls change from watching trends to management by ultimatum such as "There will be no more undefinitized changes in my organization," harmful effects can result from well-meaning attempts to cope by contractors and government managers.

What happens as senior officials limit the flexibility of contracting officers to issue formal change orders to promptly initiate desired changes? One insidious result is the cost of schedule delay. Where a production effort is under way, delay in change implementation can mean costly out-of-station work or even field rework or retrofit. Delay in the completion of items may mean a delay in fielding needed operational capability. A second action, sometimes resulting in negative impact, misunderstanding, and claims, occurs when a contractor senses the need and proceeds with the change "at his or her own risk" without authorization. This means that the contractor is no longer performing in accordance with the written contract. If the change is never approved and added to the contract, the contractor might be unable to deliver and thus not be paid for the work. A third negative action could be government liability if the government knows that the contractor is performing the changed work, and government personnel encourage or even do not expressly discourage the action. This could result in a board or court ruling that the contractor was doing work for the government and thus should be paid. Those who ask for work without proper funding and authority risk personal liability for violating federal law and regulation. A fourth adverse impact of delay in contract change approval is in suboptimal more costly "work arounds" if a contractor stops work in the area to be changed (perhaps as a result of a government stop work order).

Constructive Changes

Sometimes actions or inactions by government personnel in authority, or simply by circumstances, cause a contractor to perform work differently than required by the written contract. This is called a constructive change. A contractor who feels that he or she has a constructive change may file a claim for equitable adjustment in the contract. To prevent constructive change claims, many acquisition managers and contracting officers have implemented educational programs, included disclaimers for correspondence and meetings, and put special requirements into contracts to require timely notification of any circumstances where the contractor feels that a change has been directed. While many people understand that a constructive change occurs when a contractor is directed to do something beyond contract requirements, there are other situations which cause constructive changes, some of which may not be as preventable. Extra or different unexpected work might be caused by unforeseen problems in vague, insufficient, or defective specifications. A lack of timely government action might cause the contractor to have delay cost (called constructive deceleration). More work requirements than contemplated, such as additional work requests or excessive government reviews, could result in a claim of constructive acceleration for schedule relief and more money.

Improving Change Management

Solutions to improve management of the change process encompass both contracting techniques and management effort to improve the people side of the business.

Contracting techniques include writing contracts with provisions to help manage change as well as taking timely and professional action to change contracts when needed. Contractual language which helps manage changes includes incentives such as the optional "swing" clause which incorporates contractor-initiated engineering change proposals below a certain threshold at no change in contract cost. This clause saves administrative time in evaluating prices while also motivating the contractor to propose only essential changes. Specially tailored language in the special contract requirements section may also help to limit changes. A notification of changes clause may help limit constructive changes and should promote timely action. Managing change with timely action requires appropriate consideration of the pros and cons of undefinitized change orders, followed by careful planning and management of the change-definitization process. When there is a heavy volume of changes, the contracting officer may group several changes in process into one contract modification to avoid duplicative paperwork and review cycles. Sometimes technical managers also do this by holding several nonurgent changes for "block" change or even a new model designation for configuration management and logistics support reasons. Once the change gets into the contract definitization process described, effective management planning and follow-up can make a big difference. Over 80 percent of the change definitization time may be taken up by activities outside the contracting officer's office. The effective contracting professional is one who uses overall management skill in coordinating and influencing work done by a variety of other activities. When many changes are pending at the same time, the follow-up actions required to shepherd actions through the evaluation, audit, pricing, negotiation, writing, review, and approval process is a management coordination challenge.

Good change management requires top-quality professionals who understand the rules and the process and who have a dedication to getting the job done in a professional timely manner. These managers need authority to make decisions and act as they see in the government's best interest without multiple layers of management review and second guessing to lengthen the process. With sufficient high-quality dependable people to work with requirements people and managers, contracts and changes can accurately reflect and promote agreement.

CONTRACT ADMINISTRATION

When the contract is awarded, the action begins. This chapter has focused on the major activities in preparing for award. During the contract performance period, the government contract administration team monitors the performance, implements changes where necessary, evaluates contractor systems for compliance with contract requirements, accepts completed goods or services, and manages the payment process. Teamwork is required between the government buying and program office people and the rest of the team involved in administering the contract. Complex systems acquisition contracts run for several years and involve numerous contract changes as they progress.

CHAPTER 6
PROJECT CONTRACT ADMINISTRATION

Rita Lappin Wells

Rita Lappin Wells is a Professor of Acquisition at the Industrial College of the Armed Forces (ICAF), National Defense University, Washington, D.C. Dr. Wells was previously on the faculty of the Air Force Institute of Technology, Wright-Patterson Air Force Base, Ohio. Before that, she was employed at the Naval Avionics Center, Indianapolis, Ind., as the Procuring Contracting Officer for the terminal guidance system for the TOMAHAWK cruise missile. She is the author of several articles on innovative management practices in the contracting field and frequently lectures on contract financing and profit policy.

CONTRACT ADMINISTRATION—THE JOURNEY TO PROJECT COMPLETION

Tremendous energy and resources are used by both industry and the government to reach the point of contract award. Acquisition strategy is carefully planned, funds are committed, statement of work and specifications are drafted, solicitation is prepared, delivery schedules are determined, government-furnished property is identified, a determination is made that the price is fair and reasonable, and source selection is completed. The end result of all these efforts—signing a contract—is really the beginning of a journey to project completion. The success of the journey—and of the project—depends a great deal on the quality of the contract administration efforts of all those involved.

This chapter provides an overview of the activities involved in contract administration. The role of the Defense Contract Management Command (DCMC) of the Defense Logistics Agency (DLA) is addressed. New policies and initiatives affecting contract administration will also be discussed.

REGULATORY GUIDANCE

The regulations covering federal contract administration are contained in the Federal Acquisition Regulation (FAR) Part 42—Contract Administration.[1] Additional regulations covering defense contracts are contained in the Defense Federal Acquisition Regulation Supplement (DFARS) Part 42—Contract Administration.[2] These regulations change frequently due to changes in laws and policies. Therefore it is important to check the most recent editions before making any major contract management decisions.

DEFINITIONS

Administrative Contracting Officer (ACO). A key person in the contract administration office who holds a warrant which authorizes action as an agent of the government in administering contracts.*

Buying Activity. As used here, all the government personnel—contracting, technical, legal, financial, and program specialists—who work in the organization which is responsible for the award of contracts. Many different terms are used by military organizations to designate such offices, but for purposes of clarity the generic term "buying activity" will be used here.

Contract Administration Office. A government office which performs contract administration functions in accordance with FAR 42.302, the contract terms, and applicable regulations of the servicing agency.†

Contracting Officer (CO). A key person in the buying activity who holds a warrant which authorizes that individual to act as an agent of the government in awarding contracts.‡ Although the term is no longer used in the regulations, contracting officers in buying activities are often referred to as procuring contracting officers (PCOs).

ASSIGNMENT OF CONTRACT ADMINISTRATION RESPONSIBILITIES

Once the government buying activity has made a contract award, the contract may be assigned to another government organization, a contract administration office (CAO), for postaward management. The December 1991 edition of the DFARS tightened the rules for assignment of contracts by stating that "DoD [buying] activities shall not retain any contract for administration that requires performance of any contract administration function at or near contractor facilities."§

Exceptions to Assignment of a Contract to a CAO

Base, post, camp, or station contracts, which normally require performance on a military installation, are exceptions to this rule (that is, these buying activities may retain administration of their contracts). Other exceptions are shown in Table 6.1, but a significant number of defense contracts do not fall under these categories. There-

*FAR 1.602, Contracting Officers.
†FAR 42.3, Contract Administration Office Functions.
‡FAR 1.602, Contracting Officers.
§DFARS 242.203(a)(i), Retention of Contract Administration.

TABLE 6.1 From DFARS 42.203, Retention of Contract Administration

DoD activities shall not retain any contract for administration that requires performance of any contract administration at or near contractor facilities, except contracts for—

(A)	The National Security Agency
(B)	Research and development with universities
(C)	Flight training
(D)	Consultant support services
(E)	Mapping, charting, and geodesy services
(F)	Base, post, camp, and station purchases
(G)	Operation or maintenance of, or installation of equipment at radar or communication network sites
(H)	Communications services
(I)	Installation, operation, and maintenance of space track sensors and relays
(J)	Dependents Medicare program contracts
(K)	Stevedoring contracts
(L)	Construction and maintenance of military and civil public works, including harbors, docks, port facilities, military housing, development of recreational facilities, water resources, flood control, and public utilities
(M)	Architect-engineer services
(N)	Airlift and sealift services (Military Airlift Command and Military Sealift Command may perform contract administration services at contractor locations involved solely in performance of airlift or sealift contracts)
(O)	Subsistence supplies
(P)	Ballistic missile sites (contract administration offices may perform supporting administration of these contracts at missile activation sites during the installation, test, and checkout of the missiles and associated equipment)
(R)	Purchase orders issued via written telecommunications

fore a substantial number of defense contracts are, in fact, assigned to CAOs for administration.

Contract Administration Functions Automatically Delegated to the CAO

When a contract is assigned to a CAO, responsibility is automatically delegated to the CAO for 67 specific functions listed in FAR 42-302(a) and eight functions listed in DFARS 242.302. In addition, the buying activity which assigns the contract may also delegate to the CAO any of the 10 optional contract administration functions listed in FAR 42-402(b). These 75 mandatory and 10 optional contract administration functions are listed in the Appendix to this chapter.

SELECTION OF THE APPROPRIATE CAO

When it is proper for a buying activity to assign a contract to a CAO, the buying activity must identify the appropriate CAO. The *DoD Directory of Contract Administration Services Components*[3] is the basic guide used to locate the cognizant CAO.

CAOs are identified by the geographic location of the contractor. Very large defense contractors will have separate listings for their own dedicated on-site CAOs. For example, the *DoD Directory of Contract Administration Services Components* identifies one CAO as having cognizance for all defense contractors in Ft. Wayne, Ind., with the exception of Magnavox Electronics Systems Co., which is so large that it has its own dedicated CAO (Ref. 3, p. II-20).

Secondary Delegations of Contract Administration Services

Sometimes a CAO is assigned a contract to administer which requires contract administration services at another location. For example, a contract might be assigned to a CAO in Florida where the contractor's primary production facility is located. However, part of the work will be performed at another of the contractor's facilities in New York. Therefore quality assurance and production surveillance will be required in New York as well as in Florida. In cases like this, the Florida CAO to whom the contract is assigned will issue a secondary delegation to a CAO in New York for quality assurance and production. One of the automatic delegations from the buying activity is to "(57) Assign and perform supporting contract administration." Therefore secondary delegations are generally made by the CAO, not by the buying activity.

DEFENSE CONTRACT MANAGEMENT COMMAND

History

Today the majority of CAOs are a part of the Defense Contract Management Command (DCMC) of the Defense Logistics Agency. In 1990 DCMC was organized as the single activity for defense contract administration. The birth of DCMC represented a massive reorganization and consolidation of existing defense contract administration activities. This consolidation was one of the basic goals stated by Secretary of Defense Dick Cheney in his 1989 "Defense Management Report to the President" (DMR):[4]

> In addition, all DoD contract administration services (CAS), including those currently performed in the Defense Logistics Agency (DLA) and the Military Departments, will be consolidated under a newly-created Defense Contract Management Agency (DCMA), which will report to the USD/A [Undersecretary of the Defense for Acquisition] and be charged with more efficiently and effectively performing the CAS function. . . . This plan will, among other things, seek to streamline existing CAS organizations, promote uniform interpretation of acquisition regulations, improve implementation of DoD procurement policy, and upgrade the quality of the CAS workforce while eliminating overhead and reducing payroll costs. The plan should make appropriate provision for continued technical and other support to program offices. It should also preserve the existing regulatory division of responsibilities between those of administrative contracting officers, to be exercised within the DCMA, and those of procuring contracting officers, which will continue to be exercised within the Military Departments.

In the DMR, Secretary Cheney stated his goals for implementing recommendations which had been made earlier in the 1986 Packard Commission report.[5] Such recommendations for a consolidation of contract administration services organizations actually were first made in the early 1960s.

Prior to 1962, each military service administered its own contracts. If a contractor had contracts with all three military services, the contractor had to deal with three different sets of contract administration employees, each with a different way of doing things. This caused duplication of effort, confusion, and needless aggravation for many contractors. In 1962 Secretary of Defense Robert McNamara established Project 60[6] to:

(1) develop a plan for establishing a uniform system of contract management within all three services and

(2) develop several alternate plans for placing the entire DoD contract management function under one DoD agency.

As a result of Project 60, an organization was formed in 1965, the Defense Contract Administration Services (DCAS), to consolidate defense contract administration. DCAS was part of the Defense Supply Agency, the forerunner of the Defense Logistics Agency (DLA). DCAS developed into an organization with nine geographic regions reporting to DLA headquarters at Cameron Station, Alexandria, Va. Under each region there were Defense Contract Administration Services Plant Representative Offices (DCASPROs) and Defense Contract Administration Services Management Areas (DCASMAs). DCASPRO employees were located at the plants of major defense contractors. DCASMA employees were not located at plants, but were responsible for the administration of all other contracts awarded to defense contractors within designated geographic areas. DCAS employees administered *all* military contracts awarded to the contractors under their cognizance; however, not all contractors were under their cognizance.

Project 60 created DCAS, but it permitted the military services to maintain their own contract administration offices in the plants of the largest defense contractors. These military contract administration offices were called Air Force Plant Representative Offices (AFPROs), Navy Plant Representative Offices (NAVPROs), and Army Plant Representative Offices (ARPROs). They were organized so that no more than one PRO would be located at any one plant, and that PRO would administer all defense contracts awarded to the contractor. For example, the NAVPRO at the McDonnell Douglass plant in St. Louis would administer Air Force, Army, and DoD contracts as well as Navy contracts.

When DCMC was established in January 1990, all the existing DCAS offices and service PROs were disestablished. Defense policy makers decided it would be less disruptive and more efficient to create a new *command* under DLA instead of developing an entirely new *agency* or letting one of the existing CAO organizations simply absorb the others.[7]

Current Posture of DCMC

Today DCMC is organized into five districts which report to DCMC headquarters at DLA headquarters, Cameron Station, Alexandria, Va. The DCMC commander, a major general, is also the deputy director for acquisition management at DLA. The district headquarters are in Atlanta, Chicago, Philadelphia, Boston, and Los Angeles. DCMC Defense Plant Representative Offices (DPROs) are located at major contractors' plants, many of them at the sites of the former service PROs. The DCASMAs—offices based on geographic boundaries—have been replaced by Defense Contract Management Area Operations (DCMAOs).

DCMC also consolidated overseas contract administration. In October 1990, the Defense Contract Management Command International (DCMCI) was chartered with headquarters at Wright-Patterson Air Force Base, Ohio. Contract administration offices at diverse locations outside the United States report to DCMCI, which in turn reports to DCMC headquarters.

DCMAO and DPRO Organization

The basic organization of a CAO under DCMC contains a team of specialists who are dedicated to contract management. This team includes administrative contracting officers (ACOs), contract administrators, price analysts, property specialists, and transportation and packaging specialists. A similar team within the CAO is dedicated to quality assurance. Government quality assurance representatives are primarily concerned with making sure that products accepted by the government conform to the requirements of the contracts.

When DCMC was established, a new directorate was established for program and technical support. This was done to assure the program offices that they would get the same program support from DCMC that they had enjoyed under the service PROs. One of the goals stated in the DMR was to "make appropriate provision for continued technical and other support to program offices."[4]

This goal is also reflected in one of the automatic delegations to a CAO: "(67) Support the program, product, and project offices regarding program reviews, program status, program performance and actual or anticipated program problems."* Engineers and industrial specialists within the program and technical support directorate are charged with this responsibility.

DCMC has adopted a "best practices" approach to the consolidation. The new organization has been actively seeking the best, most effective contract administration practices which existed under DCAS and the service PROs.

The DCMC consolidation and reorganization occurred as the defense budget was shrinking. With fewer defense contracts awarded, fewer contract administration employees were needed. The contract administration work force has been shrinking since DCMC was established, partly because of the downward trend in defense contracting and partly because of the economies of the organizational consolidation.

DEFENSE FINANCE AND ACCOUNTING SERVICE

Each of the nine DCAS regions had a contract finance center which was responsible for making payments to contractors within the region. In the late 1980s an effort was begun to consolidate all the DCAS payment offices into one payment office in Columbus, Ohio. In January 1991, the consolidation was taken a step further—a new Defense Finance and Accounting Service (DFAS) was formed.[8] The primary DFAS payment office is located in Columbus, Ohio, but DFAS is no longer a part of DLA.

INTERRELATIONSHIP OF CONTRACT ADMINISTRATION RESPONSIBILITIES OF THE BUYING ACTIVITY, DFAS, AND THE CAO

Figure 6.1 illustrates the interrelationship of contract administration functions among the buying activity, DFAS, and the CAO. All three organizations have important roles in project contract administration. Figure 6.1 shows that even when contracts are assigned to a CAO, the two other government organizations—the buying activity and DFAS—still have significant contract administration responsibilities.

*FAR 42.302.

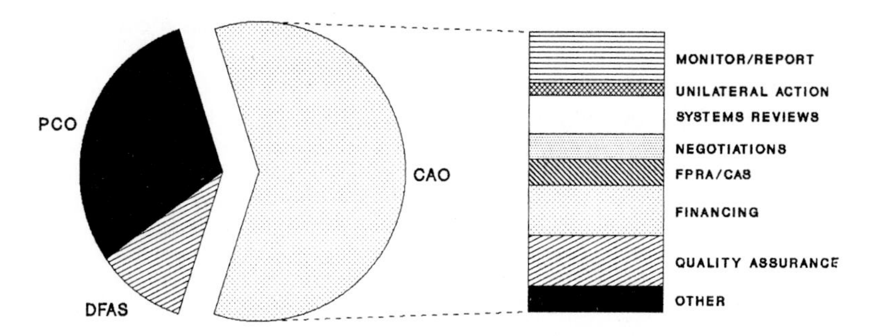

FIGURE 6.1 Distribution of contract administration responsibilities. PCO = buying activity.

Also shown are the many different types of contract administration functions that are included in the assignment of contract administration responsibilities to the CAO. Note that the proportions shown in Fig. 6.1 are for illustration purposes only. The proportion of contract administration responsibilities handled by the buying activity, DFAS, and the CAO will vary from one project to another. Likewise, the proportion of CAO responsibilities for specific functions such as negotiation or financing will vary from one contract to another.

ROLE OF THE BUYING ACTIVITY IN PROJECT CONTRACT ADMINISTRATION

In the buying activity, a team of interdisciplinary specialists is often needed to completely define the government requirements prior to issuance of a solicitation and award of a contract. This interdisciplinary team may include contracting officers, contract specialists, price analysts, engineers, attorneys, and project specialists, among others. After award, many of these specialists often are involved in the many decisions—schedule, technical, and financial—which must be made to bring the contract to completion.

If the contract is not assigned to a CAO, the buying activity will be responsible for all postaward contract administration. Even if the contract is assigned to a CAO for contract administration, the buying activity remains involved until the contract has been completed and closed out. The buying activity is responsible for anything that is not specifically called out in the FAR and DFARS as CAO contract administration functions (see Appendix).*

The following are among the most important responsibilities which are not automatically delegated to the CAO and, thus, are generally retained by the buying activity:

Funding. The buying activity has the ultimate responsibility for providing funding for any work not already funded by the contract. Changes which increase the con-

*FAR 42.302(c).

tract price cannot be made unless the buying activity agrees to provide the necessary funding.*

Design Changes. If the contractor or the government wishes to make design changes, review and approval by the buying activity is required. Technical specialists within the buying activity must evaluate suggested changes, such as engineering change proposals (ECPs), to determine whether the end item use would be adversely affected.

Schedule Changes. One of the automatic delegations to the CAO is:†

> **(64)** Negotiate and execute one-time supplemental agreements providing for the extension of contract delivery schedule up to 90 days on contract within an assigned Criticality Designator of C (see 42.1105). Notification that the contract delivery schedule is being extended shall be provided to the contracting office. Subsequent extensions on any individual contract shall be authorized only upon concurrence of the contracting office.

All other requests for schedule changes must be approved by the buying activity.

Other Contract Provisions. Any significant changes in contract provisions—other than clerical or spelling errors—require action by the contracting officer in the buying activity.

ROLE OF DFAS IN PROJECT CONTRACT ADMINISTRATION

When the consolidation of payment organizations has been completed, all payments of contracts assigned to DCMC will be made by DFAS. Payment responsibilities will vary according to specific contract provisions. However, in general, DFAS will be responsible for payment of interim and final vouchers on cost contracts, financing including progress and advance payments, and invoice payments including both partial and final payments.

Payments cannot be made by DFAS until specific actions are completed by individuals in other organizations. This can be time-consuming and sometimes confusing, especially for the novice. Contractors must be careful to prepare vouchers and invoices correctly. Final vouchers on cost contracts cannot be paid until reviewed by the Defense Contract Audit Agency (DCAA) and approved by the ACO. Invoice payments generally cannot be made until DFAS receives proof of acceptance by the government. Proof of acceptance may be completed by the quality assurance representatives in either the CAO or the buying activity. Financing requests require initial approval and periodic review by the ACO. And finally, all payments require complete and current information in the contract database used by the DFAS mechanized payment system. Clerical and data entry errors must be corrected before payments can be made.

*FAR 1.602-2(a).
†FAR 42.302.

ROLE OF THE CAO IN PROJECT CONTRACT ADMINISTRATION

Figure 6.1 shows several categories of CAO responsibilities: monitoring and reporting status, accomplishing unilateral actions, performing systems reviews, negotiating settlements, completing forward pricing rate agreements (FPRAs), administering cost accounting standards (CAS), managing financing, performing quality assurance, and other duties. Not all these categories apply to all contracts. For example, contracts with small businesses are exempt from cost accounting standards (CAS).* Therefore the CAO would have no CAS responsibilities for these contracts.

Monitor, Report, and Assist. Several of the contract administration functions already have been discussed, which are automatically delegated when a contract is assigned to a CAO. Many of the automatic delegations to the CAO are stated in the form of "monitor, review, analyze, report, advise" and "assist the buying activity." In performing these functions, the CAO employees become the "eyes and ears" of the buying activity. The CAO employees are located at or near the contractor's place of business, they perform periodic progress reviews, and they evaluate requests from the contractor. Information gathered by the CAO is transmitted to the buying activity for decision and action. The authority to take action in these situations is generally retained by the buying activity.

For example, when a contract is assigned, the CAO automatically receives a delegation to "(49) Monitor the contractor's value engineering program." However, if the contractor actually submits a value engineering change proposal, the CAO will forward it to the buying activity for a decision as to whether to accept or reject the proposal.

Production surveillance is one of the key monitoring services provided to the buying activity. Depending on the criticality of the items being produced, the CAO production specialists may be required to evaluate regular production status reports submitted by the contractor throughout the life of the contract. Or, for routine low-priority items, the CAO production specialists may be asked only to follow up if the items have not been shipped within one month after the due date.

Unilateral Actions. Several of the automatic delegations, however, give the CAO the authority and responsibility for action, *independent of the buying activity*. The CAO generally will notify the buying activity of planned action. However, the ultimate responsibility rests with the CAO, not the buying activity. These unilateral actions range from relatively minor administrative actions to significant responsibilities which affect contractor costs and processes. FAR 42.302 contains several delegations which permit the CAO to take unilateral actions on relatively minor administrative matters. Examples are:

- **(59)** Issue administrative changes, correcting errors or omissions in typing, contractor address, facility or activity code, remittance address, computations which do not require additional contract funds, and other such changes (see 43.101).
- **(63)** Cancel unilateral purchase orders when notified of nonacceptance by the contractor. The ACO shall notify the contracting officer when the purchase order is canceled.

*FAR 30.201-1, CAS applicability.

Examples of significant CAO unilateral actions affecting costs and processes are automatic delegations to:

(7) Determine the allowability of costs suspended or disapproved as required (see Subpart 42.8), direct the suspension or disapproval of costs when there is reason to believe they should be suspended or disapproved, and approve final vouchers.

(50) Review, approve or disapprove, and maintain surveillance of the contractor's purchasing system (see Part 44).

Systems Reviews. Several automatic delegations require the CAO to monitor the contractor's overall company plans, programs, and systems for compensation, insurance, estimating, property management, material management and accounting systems, labor relations, traffic operations, value engineering, purchasing, small business subcontracting, and a drug-free workplace. Specialists within the CAO monitor these systems on a regular basis. Systems reviews focus on the *contractor,* not just on a specific *contract.* Consequently, CAO systems reviews benefit all defense contracts awarded to the contractor under review.

Negotiations. The CAO negotiation responsibilities are twofold. (1) CAOs support negotiations by the buying activity through review and analysis of contractors' proposals, and (2) when authorized by the buying activity, CAOs may also conduct negotiations themselves. Many of the optional delegations in FAR 42.302(b) give the CAO authority to negotiate on behalf of the buying activity. Examples are:

1. Negotiate or negotiate and execute supplemental agreements incorporating contractor proposals resulting from change orders issued under the changes clause. Before completing negotiations, coordinate any delivery schedule change with the contracting office.
2. Negotiate prices and execute priced exhibits for unpriced orders issued by the contracting officer under basic ordering agreements.
3. Negotiate or negotiate and execute supplemental agreements changing contract delivery schedules.

When CAOs conduct negotiations which result in increased contract prices, the funding must be furnished by the buying activity.

FPRA/CAS. Forward pricing rate agreements (FPRAs) are agreements by a contractor and the government as to the cost rates to be used on all contracts awarded to that contractor.* The ACO is given the authority to "(5) Negotiate forward pricing rate agreements." By negotiating an FPRA, the ACO is predetermining a large part of the costs for every contract awarded to the contractor. Negotiating an FPRA is a tremendous amount of work and responsibility for the ACO, but will save time and effort for all contracting officers in buying activities which negotiate with the contractor.

When contractors are subject to CAS, the ACO is responsible for all actions required to administer CAS:†

(11) In connection with Cost Accounting Standards (see Part 30)—

- **(i)** Determine the adequacy of the contractor's disclosure statements;
- **(ii)** Determine whether disclosure statements are in compliance with Cost Accounting Standards and Part 31;

*FAR 15.809.
†FAR 42.302(a).

(iii) Determine the contractor's compliance with Cost Accounting Standards and disclosure statements, if applicable; and

(iv) Negotiate price adjustments and execute supplemental agreements under the Cost Accounting Standards clauses at 52.230-3, 52.230-4, and 52.230-5.

Administration of CAS affects the ways in which a contractor classifies, accumulates, and allocates costs on all covered contracts. As with FPRAs, an ACO's actions with regard to administration of CAS affect contracts awarded by all buying activities to a specific contractor.

Financing. One of the most important responsibilities delegated to the CAO is to "(12) Review and approve or disapprove the contractor's requests for payments under the Progress Payments clause." Progress payments, a form of financing, are important to contractors because they provide cash flow—the money needed to pay salaries and other bills as work progresses.

A contractor does not receive final payment from DFAS until after (1) goods have been produced or services rendered; (2) the government has accepted those products or services; (3) the contractor has submitted a proper invoice; and (4) the government payment office has processed the invoice. Depending on the nature of the goods to be produced or services rendered, a substantial amount of time may pass before the contractor receives final payment.

The contractor will need money to pay bills and meet payroll before the final payment is received. If a contractor does not have the funds to pay the bills, vendors may refuse to deliver materials and employees will look for work elsewhere. Contractors must carefully plan cash flow to make sure that production is not delayed—or stopped completely—due to lack of funding to pay the bills.

There is wide variation in the financial and technical capabilities of government contractors. Therefore it is not surprising to find that there is corresponding variation in the frequency, timing, and extent of Progress Payment reviews. Guidance given to the ACO in the FAR states:*

> The extent of Progress Payment reviews should vary inversely with the contractor's experience, performance record, reliability, quality of management, and financial strength, and with the adequacy of the contractor's accounting system and controls. Supervision shall be of a kind and degree sufficient to provide timely knowledge of the need for, and timely opportunity for, any actions necessary to protect government interests.

Progress Payments may be paid first and reviewed after payment. Or a review of each Progress Payment may be considered necessary before payment. Even if Progress Payments are made without a prepayment review, the CAO must schedule regular reviews to keep itself informed of the status of the contractor's overall performance and financial conditions.

If problems develop on contracts containing the Progress Payment clause, the ACO can take several actions with regard to the Progress Payments. The ACO may suspend Progress Payments, reduce Progress Payments, increase the rate of liquidation, or take a combination of these actions.

The decision made by the ACO in these circumstances is one of the most difficult decisions to be made in the administration of government contracts. Cash flow is the life blood of government contracts. To cut off—or significantly reduce—Progress Payments is like a medic applying a tourniquet to an arm or a leg. The tourniquet may stop the dangerous loss of blood, but the arm or leg itself also may be lost—because of the tourniquet. If a contractor already has serious financial or technical dif-

*FAR 32.503-2, Supervision of Progress Payments.

ficulties, cutting off the cash flow from Progress Payments may incapacitate the firm to the point that it threatens performance under the contract.

On the other hand, the ACO is responsible for public funds which have been obligated under a contract for a specific purpose. If the contractor's situation has deteriorated to the extent that performance in accordance with the terms of the contract is doubtful, the ACO must act to protect the government's interests.

Quality Assurance. Overall responsibility for controlling the quality of the product required by a contract rests with the contractor.* However, one of the automatic delegations to the CAO is to "(38) Ensure contractor compliance with contractual quality assurance requirements." The responsibilities of the CAO quality assurance representatives (QARs) vary according to the quality assurance requirements in the contract. Some contracts give QARs the authority to sign for acceptance by the government of the contractor's finished products. Other contracts retain the authority for acceptance by the buying activity; the QAR is authorized only to observe the contractor's in-process production and inspection methods.

The buying activity decides which of three different levels of quality assurance requirements to put in the contract before award. The level of quality assurance requirements has a significant impact on the quality assurance actions performed by the CAO.† The most basic level of quality relies on inspection by the contractor; no government quality assurance actions are performed at the location where the item is manufactured.

The second level of quality requires compliance with the Standard Inspection clause. The contractor must have an inspection system acceptable to the government. The Standard Inspection clause also gives the CAO quality assurance representative (QAR) access to the contractor's facility. In the event that the contractor presents material for inspection and acceptance which does not conform to the contract requirements, the Standard Inspection clause permits the government to assess the contractor reinspection costs.

The third level of quality requires compliance with one of two higher-level quality requirements: MIL-I-45208A, "Inspection System," or MIL-Q-9858A, "The Quality Program." These higher-level quality requirements specify that the contractor have a formal, documented inspection system or program. The QAR is responsible for reviewing the system or program documentation and procedures as well as the end products.

The DCMC has recently adopted a new approach to quality assurance called the In-Plant Quality Evaluation (IQUE) program.[9] The program applies the principles of total quality management (TQM) to government quality assurance. Teamwork is emphasized between the contractor and the government QAR. Under IQUE, emphasis is placed on the contractor-government team, improving the contractor's processes so that they yield conforming products—instead of the QAR simply looking for errors in the final products.

ROLE OF THE CONTRACTOR IN PROJECT CONTRACT ADMINISTRATION

The primary responsibilities of the contractor are to comply with the requirements stated in the contract and deliver the correct quantity of conforming products to the

*FAR 46.105, Contractor Responsibilities.
†FAR 46.2, Contract Quality Requirements.

designated destination point by the scheduled delivery date. However, project contract administration may entail much more for the contractor than just "getting the product out the door." There is great variation among defense contract requirements, but most major defense projects require the contractor to maintain specific systems, processes, and documentation. For example, in addition to delivery of hardware, the contract may also require submission of cost and schedule reports during the performance period, performance of special tests, and delivery of data with the final products.

On cost and incentive type contracts, the contractor must maintain an acceptable accounting system which permits the government to audit costs being reimbursed under the contract. If the contractor wishes to receive financing, such as Progress Payments from the government, the contractor must submit detailed requests and permit the government to have access to financial records which support the request.

Government contractors are expected to support socioeconomic programs such as equal employment opportunity (EEO) and to maintain a drug-free workplace. They also are expected to understand and respect ethics rules regarding the conduct of government procurement, and to refrain from retaliation against whistle blowers—employees who report misconduct on government contracts.

If government property is furnished on a contract, the contractor is expected to safeguard it and only use it for the designated contract work. If classified information is involved in the performance of a contract, the contractor must adhere to stringent security rules. If ammunition or hazardous materials are handled in the performance of a contract, the contractor must follow safety regulations for the handling, transportation, and disposal of the material.

Military project contract administration requires contractor employees to have a thorough understanding of government contract administration practices as well as knowledge of the product or service being provided. It also requires coordination with government specialists in the buying activity, CAO, and DFAS.

How Does the Contractor Know Whom to Talk To?

After award of a contract, contractors frequently find themselves working with two different organizations—the buying activity and the CAO. The same titles may be given to employees in both offices. For example, both the buying activity and the CAO may have employees with the titles contracting officer, contract specialist, contract administrator, industrial specialist, price analyst, or engineer. This can cause much misunderstanding about routing of correspondence and authority to make decisions.

Contractor correspondence is generally sent to the CAO. Depending on the subject of the correspondence, the CAO may review it and forward it to the buying activity or may take action. This can be confusing for contractors. Contractor employees who are involved in project contract administration need to become familiar with the list of automatic delegations contained in FAR 42.302(a) and DFARS 242.302 (see Appendix). Unclear areas should be discussed with both offices or addressed during a postaward orientation conference.

Postaward Orientation Conference

A postaward orientation conference (PAOC) frequently is held after the award of a contract to a new defense contractor or a contract which contains particularly complex requirements.* One of the automatic delegations to a CAO is "(3) Conduct

*FAR 42.500, Post Award Orientation.

post-award orientation conferences." The PAOC is an excellent opportunity for the contractor to meet with employees of the CAO and the buying activity. Contract requirements are reviewed, roles defined, and procedures discussed.

CONCLUSION

Military contract administration is complex and challenging. This chapter has discussed the roles of employees of the buying activity, DFAS, the CAO, and the contractor. Everyone involved in contract administration needs to have a good understanding of their role and their relationship to the other players. They are key people who can, by working together, guide the project to successful completion.

APPENDIX

FAR 42.302(a): Contract Administration Office Functions

The following are the normal contract administration functions to be performed by the cognizant CAO, to the extent they apply, as prescribed in 42.202:

(1) Review the contractor's compensation structure.

(2) Review the contractor's insurance plans.

(3) Conduct post-award orientation conferences.

(4) Review and evaluate contractors' proposals under Subpart 15.8 and, when negotiation will be accomplished by the contracting officer, furnish comments and recommendations to that officer.

(5) Negotiate forward pricing rate agreements (see 15.809).

(6) Negotiate advance agreements applicable to treatment of costs under contracts currently assigned for administration (see 31.109).

(7) Determine the allowability of costs suspended or disapproved as required (see Subpart 42.8), direct the suspension or disapproval of costs when there is reason to believe they should be suspended or disapproved, and approve final vouchers.

(8) Issue Notices of Intent to Disallow or not Recognize Costs (see Subpart 42.8).

(9) Establish final indirect cost rates and billing rates for those contractors meeting the criteria for contracting officer determination in Subpart 42.7.

(10) Prepare findings of fact and issue decisions under the Disputes clause on matters in which the administrative contracting officer (ACO) has the authority to take definitive action.

(11) In connection with Cost Accounting Standards (see Part 30):

- **(i)** Determine the adequacy of the contractor's disclosure statements;
- **(ii)** Determine whether disclosure statements are in compliance with Cost Accounting Standards and Part 31;
- **(iii)** Determine the contractor's compliance with Cost Accounting Standards and disclosure statements, if applicable; and
- **(iv)** Negotiate price adjustments and execute supplemental agreements under the Cost Accounting Standards clauses at 52.230-3, 52.230-4, and 52.230-5.

(12) Review and approve or disapprove the contractor's requests for payments under the Progress Payments clause.

(13) Make payments on assigned contracts when prescribed in agency acquisition regulations (see 42.205).

(14) Manage special bank accounts.

(15) Ensure timely notification by the contractor of any anticipated overrun or underrun of the estimated cost under cost-reimbursement contracts.

(16) Monitor the contractor's financial condition and advise the contracting officer when it jeopardizes contract performance.

(17) Analyze quarterly limitation on payments statements and recover overpayments from the contractor.

(18) Issue tax exemption certificates.

(19) Ensure processing and execution of duty-free entry certificates.

(20) For classified contracts, administer those portions of the applicable industrial security program designated as ACO responsibilities (see Subpart 4.4).

(21) Issue work requests under maintenance, overhaul, and modification contracts.

(22) Negotiate prices and execute supplemental agreements for spare parts and other items selected through provisioning procedures when prescribed by agency acquisition regulations.

(23) Negotiate and execute contractual documents for settlement of partial and complete contract terminations for convenience, except as otherwise prescribed by Part 49.

(24) Negotiate and execute contractual documents settling cancellation charges under multiyear contracts.

(25) Process and execute novation and change of name agreements under Subpart 42.12.

(26) Perform property administration (see Part 45).

(27) Approve contractor acquisition or fabrication of special test equipment under the clause at 52.245-18, Special Test Equipment.

(28) Perform necessary screening, redistribution, and disposal of contractor inventory.

(29) Issue contract modifications requiring the contractor to provide packing, crating, and handling services on excess Government property. When the ACO determines it to be in the Government's interests, the services may be secured from a contractor other than the contractor in possession of the property.

(30) In facilities contracts:

- **(i)** Evaluate the contractor's requests for facilities and for changes to existing facilities and provide appropriate recommendations to the contracting officer;
- **(ii)** Ensure required screening of facility items before acquisition by the contractor;
- **(iii)** Approve use of facilities on a noninterference basis in accordance with the clause at 52.245-9, Use and Charges;
- **(iv)** Ensure payment by the contractor of any rental due; and
- **(v)** Ensure reporting of items no longer needed for Government production.

(31) Perform production support, surveillance, and status reporting, including timely reporting of potential and actual slippages in contract delivery schedules.

(32) Perform pre-award surveys (see Subpart 9.1).

(33) Advise and assist contractors regarding their priorities and allocations responsibilities and assist contracting offices in processing requests for special assistance and for priority ratings for privately owned capital equipment.

(34) Monitor contractor industrial labor relations matters under the contract; apprise the contracting officer and, if designated by the agency, the cognizant labor relations advisor, of actual or potential labor disputes; and coordinate the removal of urgently required material from the strikebound contractor's plant upon instruction from, and authorization of, the contracting officer.

(35) Perform traffic management services, including issuance and control of Government bills of lading and other transportation documents.

(36) Review the adequacy of the contractor's traffic operations.

(37) Review and evaluate preservation, packaging and packing.

(38) Ensure contractor compliance with contractual quality assurance requirements (see Part 46).

(39) Ensure contractor compliance with contractual safety requirements.

(40) Perform engineering surveillance to assess compliance with contractual terms for schedule, cost, and technical performance in the areas of design, development, and production.

(41) Evaluate for adequacy and perform surveillance of contractor engineering efforts and management systems that relate to design, development, production, engineering changes, subcontractors, tests, management of engineering resources, reliability and maintainability, data control systems, configuration management, and independent research and development.

(42) Review and evaluate for technical adequacy the contractor's logistics support, maintenance, and modification programs.

(43) Report to the contracting office any inadequacies noted in specifications.

(44) Perform engineering analyses of contractor cost proposals.

(45) Review and analyze contractor-proposed engineering and design studies and submit comments and recommendations to the contracting office, as required.

(46) Review engineering change proposals for proper classification, and when required, for need, technical adequacy of design, producibility, and impact on quality, reliability, schedule, and cost; submit comments to the contracting office.

(47) Assist in evaluating and make recommendations for acceptance or rejection of waivers and deviations.

(48) Evaluate and monitor the contractor's procedures for complying with procedures regarding restrictive markings on data.

(49) Monitor the contractor's value engineering program.

(50) Review, approve or disapprove, and maintain surveillance of the contractor's purchasing system (see Part 44).

(51) Consent to the placement of subcontracts.

(52) Review, evaluate, and approve plant or division-wide small and small disadvantaged business master subcontracting plans.

(53) Obtain the contractor's currently approved company—or division-wide—plans for small business and small disadvantaged business subcontracting for its commercial products, or, if there is no currently approved plan, assist the contracting officer in evaluating the plans for those products.

(54) Assist the contracting officer, upon request, in evaluating an offeror's proposed small business and small disadvantaged business subcontracting plans, including documentation of compliance with similar plans under prior contracts.

(55) By periodic surveillance, ensure the contractor's compliance with small business and small disadvantaged business subcontracting plans and any labor surplus area contractual requirements; maintain documentation of the contractor's performance under and compliance with these plans and requirements; and provide advice and assistance to the firms involved, as appropriate.

(56) Maintain surveillance of flight operations.

(57) Assign and perform supporting contract administration.

(58) Ensure timely submission of required reports.

(59) Issue administrative changes, correcting errors or omissions in typing, contractor address, facility or activity code, remittance address, computations which do not require additional contract funds, and other such changes (see 43.101).

(60) Cause release of shipments from contractor's plants according to the shipping instruction. When applicable, the order of assigned priority shall be followed; shipments within the same priority shall be determined by date of instruction.

(61) Obtain contractor proposals for any contract price adjustments resulting from amended shipping instructions. ACOs shall review all amended shipping instruction on a periodic, consolidated basis to assure that adjustments are timely made. Except when the ACO has settlement authority, the ACO shall forward the proposal to the contracting officer for contract modification. The ACO shall not delay shipments pending completion and formalization of negotiations of revised shipping instructions.

(62) Negotiate and/or execute supplemental agreements, as required, making changes in packaging sub-contractors or contract shipping points.

(63) Cancel unilateral purchase orders when notified of nonacceptance by the contractor. The ACO shall notify the contracting officer when the purchase order is canceled.

(64) Negotiate and execute one-time supplemental agreements providing for the extension of contract delivery schedules up to 90 days on contract with an assigned Criticality Designator of C (see 42.1105). Notification that the contract delivery schedule is being extended shall be provided to the contracting office. Subsequent extensions on any individual contract shall be authorized only upon concurrence of the contracting office.

(65) Accomplish administrative closeout procedures (see 4.804-5).

(66) Determine that the contractor has a drug-free workplace program and drug-free awareness program (see Subpart 23.5).

(67) Support the program, product, and project offices regarding program reviews, program status, program performance and actual or anticipated program problems.

FAR 32.302(b)

The CAO shall perform the following functions only when and to the extent specifically authorized by the contracting office:

(1) Negotiate or negotiate and execute supplemental agreements incorporating contractor proposals resulting from change orders issued under the Changes clause. Before completing negotiations, coordinate any delivery schedule change with the contracting office.

(2) Negotiate prices and execute priced exhibits for unpriced orders issued by the contracting officer under basic ordering agreements.

(3) Negotiate or negotiate and execute supplemental agreements changing contract delivery schedules.

(4) Negotiate or negotiate and execute supplement agreements providing for the deobligation of unexpended dollar balances considered excess to known contract requirements.

(5) Issue amended shipping instructions and, when necessary, negotiate and execute supplemental agreements incorporating contractor proposals resulting from these instructions.

(6) Negotiate changes to interim billing prices.

(7) Negotiate and definitize adjustments to contract prices resulting from exercise of an economic price adjustment clause (see Subpart 16.2).

(8) Issue change orders and negotiate and execute resulting supplemental agreements under contracts for ship construction, conversion, and repair.

(9) Execute supplemental agreements on firm-fixed-price supply contracts to reduce required contract line item quantities and deobligate excess funds when notified by the contractor of an inconsequential delivery shortage, and it is determined that such action is in the best interests of the Government, notwithstanding the default provisions of the contract. Such action will be taken only upon the written request of the contractor and, in no event, shall the total downward contract price adjustment resulting from an inconsequential delivery shortage exceed $250.00 or 5 percent of the contract price, whichever is less.

(10) Execute supplemental agreements to permit a change in place of inspection at origin specified in firm-fixed-price supply contracts awarded to nonmanufacturers, as deemed necessary to protect the Government's interests.

DFARS 242.302: Contract Administration Functions

(a)(4) Also, review and evaluate:

(A) Contractor estimating systems (see FAR 15.811); and

(B) Contractor material management and accounting systems under Subpart 242.72.

(8) Monitor contractor costs under Subpart 242.70.

(11) The DoD lead negotiating agency under FAR Subpart 41.20 (i.e., the tri-service contracting office) shall make all determinations under FAR 42.302 (a)(11)(ii) and (iii) related to Cost Accounting Standard 420 for contractors requiring advance agreements for independent research and development/bid and proposal costs.

(19) Also negotiate and issue contract modifications reducing contract prices in connection with the provisions of paragraph (b) of the clause at FAR 52.225-10, Duty-Free Entry, and paragraph (c) of the clause at 252.225-7008, Duty-Free Entry—Qualifying Country End Products and Supplies.

(33) Also perform industrial readiness and mobilization production planning field surveys and negotiate schedules.

(39) See 223.370 for safety requirements on contracts for ammunition and explosives.

(41) In contracts with cost schedule control system requirements (see 234.005-70):

(A) Perform postaward surveillance of contractor progress in demonstrating that its cost schedule control systems meet the cost schedule control systems criteria;

(B) Provide assistance in the review and acceptance of the contractor's cost schedule control systems; and

(C) After acceptance of the systems, perform surveillance to monitor their continuing acceptable operation.

(b)(S-70) Issue, negotiate and execute orders under basic ordering agreements for overhaul, maintenance and repair.

REFERENCES

1. *Federal Acquisition Regulation,* prescribed jointly by the Secretary of Defense, the Administrator of General Services, and the Administrator, National Aeronautics and Space Administration; issued as Chapter 1 of Title 48, Code of Federal Regulations (CRF), U.S. Government Printing Office, Washington, D.C., 1984.
2. Secretary of Defense, *Defense Federal Acquisition Regulation Supplement,* U.S. Government Printing Office, Washington, D.C., 1991.
3. *DoD Directory of Contract Administration Services Components,* DLAH 4105.4, Defense Logistics Agency, Cameron Station, Alexandria, Va., Jan. 1991.
4. R. Cheney, "The Defense Management Report to the President," Office of the Secretary of Defense, Washington, D.C., July 1989, p. 17.

5. D. Packard, "A Quest for Excellence; Final Report by the President's Commission on Defense Management," Washington, D.C., 1986.
6. F. E. Bartlett, "DOD Project 60 and the Activation of the Air Force Contract Management Division," AFSC Historical Publ. Ser. 66-80-1, Apr. 1966, p. 3.
7. "Defense Contract Management Command," Fact Sheet, Office of Public Affairs, Defense Logistics Agency, Alexandria, Va., Jan. 1990, p. 2.
8. "Contractor Payment Information," Defense Finance and Accounting Service, Columbus, Ohio, 1991, p. 1.
9. *Defense Logistics Agency Handbook* (DLAH), 8200.5.

CHAPTER 7

MANUFACTURING AND THE DEFENSE INDUSTRIAL BASE

William T. Motley

William T. Motley is chairperson of the Manufacturing Management Department, United States Defense Systems Management College, Fort Belvoir, Va. He has been the director for both the Defense Manufacturing Management course and the Total Quality Management course. Mr. Motley previously worked for ASEA-Flakt, a firm specializing in the design and manufacture of electromechanical controls and structural components for the nuclear and fossil-fuel powerplant industries.

Manufacturing* is a conversion process which transforms raw material into a finished product. Manufacturing includes the functions necessary to organize and execute this transformation process. The transformation process comprises factors of manpower, machinery, work methods, metrology, and material. Manufacturing is generally viewed as including the functions of planning, scheduling, manufacturing engineering, industrial engineering, material control, fabrication and assembly, test and inspection, and tooling fabrication and control. Effective manufacturing requires that all of these functions be organized and coordinated with the specific objectives of providing a quality product, on time, and at the lowest possible total cost. Schonberger outlines six true measures of manufacturing effectiveness:[1]

- Increasing quality
- Increasing throughput
- Increasing plant capacity
- Decreasing inventory levels
- Decreasing total costs
- Decreasing cycle time

*In this chapter the terms manufacturing and production will be used synonymously. Some purists refer to one-off or small-volume production as manufacturing, while production is used to denote the high-volume manufacture of standardized products.

These factors should indicate to the reader the synergistic nature of manufacturing. Improvements or problems in one area of the manufacturing process almost always result in improvements or problems appearing in different and sometimes unexpected areas of the manufacturing process. The quality of the produced product is one of these obvious synergistic areas. It always requires more time to repair or rework a product than to produce it correctly the first time. Once rework or repair is eliminated, the rework time and material become free resources and capacity. Quality products also result in required inventory levels being lowered with a corresponding reduction in working capital and required floor space. Such synergisms and interdependencies provide great opportunities for the alert manager and, correspondingly, great traps for the unsuspecting. There are almost never independent problems or independent solutions to be found in the manufacturing environment.

THE ROLE OF MANUFACTURING

Manufacturing is moving from a traditionally reactive function, normally subservient to design engineering, toward a discipline integrated with design and having substantial influence on the design process. The role of manufacturing during acquisition phases I, II, and III can be stated as (1) influencing the design process, (2) planning for production, and (3) executing the plan. This is not a role completely accepted by all managers, but it is certainly a philosophy that will require a cultural change among many engineering and program managers.

In many organizations, manufacturing is considered as a necessary but strategically trivial task that consumes valuable organizational resources that could be better used elsewhere. These manufacturing organizations have evolved around a reactive philosophy of "damage control." Manufacturing engineers and their managers are routinely directed to solve impossible manufacturing problems stemming from unproducible designs, changing requirements, poorly written statements of work, ambiguous specifications, and unrealistic production schedules. These types of problems must be addressed by the entire organization early in the life cycle of the product or system. The new dynamic nature of manufacturing can be seen in the role manufacturing is playing in the philosophy of concurrent engineering.[2] Early involvement of the manufacturing process in engineering design means beginning at milestone I and continuing through deployment.

It is sometimes difficult for managers to understand how manufacturing can be of value during concept exploration. Manufacturing involvement during concept exploration is one of determining manufacturing feasibility and related risk. In many cases this is a materials question. What material characteristics are needed? Does this material exist? Does the material exist in the shape and forms needed? What is the cost? How has this material been used in the past? Are firms available that can manufacture this material? What is the yield of this process? Is the yield repeatable? Are the material characteristics repeatable? Questions such as these may lead to the conclusion that there will not only be necessary a product development process, but a manufacturing process development as well.

The manufacturing manager can make many feasibility assessments based on probable materials, the estimated production rate, and the total production buy. The strategic implications of production rate and total procurement cannot be overstated. Rate and quantity affect capital equipment amortization and overhead allocations. Rate and quantity also affect manufacturing process selection. For example, the selection of automated equipment over manual equipment will affect decisions such as manpower loading, inspection methods, operator training, required inventory levels, line-flow balancing, process sheets, work sequencing, scheduling, and tooling.

Modern manufacturing is focused on the process of manufacturing rather than solely on the outcome. By definition, if our process factors (manpower, machinery, work methods, metrology, material) are correct and repeatable, then our end product will be correct and repeatable. Uniformity of product characteristics is a result of a successful manufacturing process. Uniform product implies statistical control of processes and, as a result, consistent product performance during operation. Another outcome of controlled processes is defect-free product. Defect-free does not imply perfection or necessarily require the use of the latest technologies. It denotes only that the product has met its functional and physical requirements. The defect-free nature of the product results in lower product cost. Thus the new technical manufacturing definition of quality is focused on meeting or exceeding requirements at lower cost. In contrast to the traditional definition of quality as "conformance to requirements," a more definitive, dynamic definition of quality has evolved. Products are deemed to possess quality if they are uniform, defect-free, and meet all of the user's needed physical and functional requirements.

The role of manufacturing management during the actual transformation process can be stated as (1) achieving design intent by "building to print"; (2) developing repeatable manufacturing processes; and (3) improvement of all manufacturing processes through variability reduction.

PROCESS CLASSIFICATION

A description of the Department of Defense (DoD) manufacturing environment is complicated by the high degree of variety in the types of products and manufacturing systems that are encountered. Typically, defense products run the gamut from small high-precision parts such as would be seen in gyroscopes or laser range-finding assemblies, through extremely complex parts such as advanced applications for electronics countermeasures or target classification, to extremely large weapons such as aircraft and ship assemblies. A weapon system development program typically includes many of these types of products. If a major weapon system is decomposed into its component parts, it is not unusual to find most of the major product types and technologies represented. This inherent diversity and resultant complexity present many problems unique to the defense industry in terms of component integration, the availability of capable manufacturers, developing repeatable manufacturing processes, and ensuring that the manufactured product reflects design intent at an affordable cost.

This complexity is amplified by the increasing percentage of piece parts and assembled subassemblies purchased by large defense contractors who are continuing the trend toward systems integration. In the last 10 years it has not been unusual to find 60 to 90 percent of the unit production cost of major systems to have been purchased from vendors or subcontractors. Thus subcontractor management and material control are two of the government program office's and prime contractor's major management tasks preceding and during production.

Manufacturing systems also have certain characteristics which affect their ability to produce at given rates and quantities. A job shop process comprises many physically unconnected machines or processes that possess great process flexibility but cannot produce at great rates. They are characterized by the need for very skilled operators and high setup costs. Satellite manufacture is an example of a process usually conducted in this manner.

The process may be an automated connected line-flow process where a standardized, identical product is produced in large quantity. Automobile production is a very advanced form of line-flow process. This process would be relatively rare within the defense environment because of the limited quantities of systems typically produced. It may be encountered in some areas of ordnance acquisition and in some of the piece parts manufacturing to support a large-quantity electronics program.

More typical of the defense environment is batch manufacturing. Typical products include aircraft and missiles. This type of manufacturing system produces similar parts or end items in defined lots with a discrete start and stop point for each lot. The production of lots is sometimes spread irregularly over time. Machining is a process very often associated with batch production.

The fourth type of manufacturing system is referred to as specialty manufacturing or construction. This type of system is generally found in extremely large fabrications where there is a fixed position for final assembly. Construction of facilities and ship construction projects fall into this category. This type of production is heavily capital- and labor-intensive.

THE CLASSIC TRADEOFF IN MANUFACTURING

Typically defense weapon system manufacturing can be characterized as a mix of various kinds of products fabricated through various kinds of manufacturing systems. The challenge is to organize the manufacturing process in a manner that yields the required end product, delivered at the point in time, that is required and possessing the physical and functional characteristics necessary to meet the system requirements. Proper selection of the manufacturing process to match the characteristics of the product is a critical task. This matching process is heavily driven by the anticipated production rate and total production run. High-volume standardized components cannot be produced quickly and economically in a job shop. Unique, one-of-a-kind items cannot be produced quickly or economically on an automated assembly line. This matching of product with process develops into what may be termed the classic manufacturing trade-off issue. Do we wish to have technological flexibility, or do we wish to have the capacity to produce large volumes of product very quickly? The ideal manufacturing facility would possess both traits. However, given the current state of manufacturing technologies, this is, for most products, not possible. The manufacturing strategy must usually focus on the high-volume production of standardized components (ammunition), or the extremely low-rate production of unique, one-off items such as satellites, or attempt a compromise with batch manufacturing.

Current technological thrusts such as flexible manufacturing and computer-aided engineering and manufacturing have as one of their goals the enhancement of batch manufacturing. These enhancements would in theory provide one group of machines or a facility the rapid ability to economically switch back and forth between the production of one-of-a-kind part numbers and the high-rate production of standardized parts. The goal is to be able to produce any component, economically, in lots of one or 1000. This flexibility would, in great part, be provided by robotics and the use of software that would automatically generate process steps and thereby eliminate setup and programming costs. Much of the current research in DoD manufacturing is focused on flexible manufacturing systems in machining because of the high preponderance of machined components produced in batch fashion which are used in weapon systems.* As will be discussed later under industrial base issues, the future

*A very good example is the Automated Manufacturing Research Facility of the National Institute of Standards and Technology.

promise of such systems in both electronic and mechanical manufacturing is one of the technological foundations of the concept of "virtual swords" or the "long shadow" concept.

The tradeoff between flexibility and capacity has been a factor in manufacturing since the Industrial Revolution. Manufacturing tolerances were introduced during the Industrial Revolution in order to facilitate standardization. Standardization became a key to mass production. One of the first known uses of tolerances was by Eli Whitney in 1798 in order to facilitate the mass production of 10,000 muskets for the U.S. government.[3] Mr. Whitney wished to produce muskets from a standardized pattern rather than to the designs of individual artisans. Mass production became the centerpiece of what was later to become known as the "American method" of manufacturing. Mass production brought many benefits, but it also brought with it a lack of flexibility.

Many consider Henry Ford to be the father of mass production. He demonstrated his genius with the 60 percent learning curve achieved during production of the Model T automobile. Ford achieved this efficiency through brilliant production planning, heavy standardization of product and process, and organizing almost completely vertically. However, his competitor, Alfred Sloan, so successfully changed the nature of automobile competition that Ford was forced to shut down his River Rouge plant for five years in order to retool for the Ford Model A. Sloan changed the nature of competition by competing with the flexibility of his products rather than with the low production costs brought about by the standardization of product and process. The flexibility Sloan introduced was in the form of different body styles, colors, internal heaters, internal starters, and roll-up windows. Ford was so focused on capacity that he could not respond to the threat.

The same conflict between flexibility and capacity was illustrated dramatically during World War II with the production of aircraft for the U.S. Army Air Corps.[4] The B-24 was an aircraft at one time planned for production at 1000 per month. This rate was never achieved due to the large number of design changes and modifications which were driven by changes in enemy tactics and equipment. One of the solutions that was developed, and is still used today, is the block change. This is a scheduling solution to the conflict between flexibility and capacity. Army Air Corps aircraft product changes were very disruptive to the concept of adapting automobile-type transfer lines to aircraft production. These changes were especially disruptive to tooling.

The block change concept requires design changes or modifications to be stopped as of a certain date. After this date only accepted changes will be put into production for a certain number of aircraft, for example, 100. After this block had been produced, all other incoming changes would be incorporated into the next block of 100 or perhaps 500 aircraft. This concept remains in effect today, resulting in groups of aircraft differing in configuration rather than having each individual aircraft possess a unique configuration.

Another example from World War II illustrates a success based on maximizing the capacity of a production system based on design standardization. More Liberty freighters were produced in four years, 2708, than in all of shipbuilding history.[5] Largely responsible for this success were the West Coast shipyards of Henry Kaiser. The efficiency of Kaiser's Richmond, Calif., shipyard was such that the *Robert E. Peary* was built and launched in just over four days during November 1942.

Kaiser's shipyards achieved this type of mass-production efficiency by using a standardized design, changing the organization of the typical shipyard, and utilizing the latest technologies to improve the existing shipyard manufacturing processes. These changes to the manufacturing process included the use of welding instead of riveting and the utilization of very large cranes. Both of these technologies increased the speed of assembly and allowed the use of prefabricated hull sections. Many naval architects believe that the great successes the Japanese have scored in commercial shipbuilding stem in great part from the Japanese study and implementation of Henry Kaiser's ideas.

PROGRAM OFFICE CONCERNS

Every effort should be made by the program office to develop realistic, stable estimates of production rate and total buy. It is recognized that in defense acquisition such funding issues are usually not controlled by the program office. However, the fact remains that unstable production rates and unstable total buys are some of the great disrupters of any production line. Commercial manufacturers of capital equipment such as automobiles will go to great lengths to stabilize production runs due to their very high fixed costs. These efforts often include offering special financing packages to prospective buyers that are below the prevailing market rate.

The accurate prediction of production rates and total quantity allow the contractor to best match the product to the manufacturing process and minimize the impacts of the flexibility versus capacity tradeoff. The program office must be aware of any inconsistencies between acquisition strategy, manufacturing strategy, and funding profiles. Examples of such an inconsistency would be the not unheard of plan to produce at high volumes, in order to reduce unit-production costs, while delaying the design freeze until the end of engineering and manufacturing development. This is an extremely difficult if not impossible scenario to execute.

A primary management tool of the program office is the manufacturing plan. The prime guidance for the manufacturing plan is found in MIL-STD-1528, "Manufacturing Management Program." The manufacturing plan should be a required contract data item and should be clearly referenced against the tasking given for manufacturing planning in the statement of work.

The manufacturing plan is meant to be a dynamic, evolving document providing guidance and coordination for both the government and the contractor. Initial manufacturing planning should begin during the demonstration and validation phase. The plan must be comprehensive enough to ascertain whether the contractor has adequately planned for production and it should also assist in the monitoring of the contractor's efforts to ensure the timely execution of the production program. The plan should cover at least five main areas.

1. Identification of manufacturing processes and methods and the related requirements for tooling, capital equipment, and plant facilities.
2. Definition of schedule. Will resources be available when needed and are product delivery dates achievable?
3. Identification of needed skills, number of personnel, training and certification programs, and the availability of needed personnel.
4. Establishment of a make-or-buy plan, with justifications. This establishes the distribution of effort between the prime contractor and subcontractors, a very critical manufacturing decision.
5. Facilities plan. Material control and flows within the plant. Evaluation of plant capacity versus projected schedule. Plant requirements, including power, special test and handling equipment, clean rooms, and the storage, handling, and disposal of hazardous or explosive materials.

This standard, as should all military standards, must be tailored to suit the individual requirements of a particular program.

VISIBILITY

In planning for and executing the production phase there are a number of significant issues which involve the program manager. The first issue is cost sensitivity. Cost is of concern to every manager, but there is a unique aspect for the manufacture of defense products. In a life-cycle context, production costs typically comprise at least 30 percent of a system's total life-cycle cost. Life-cycle costs are a large sum, but are usually overlooked largely due to the fact that most organizations do not have a good method for collecting or evaluating costs over 20 or 30 years. Because life-cycle costs are poorly perceived, manufacturing costs (problems) appear magnified in size to Congress and the public. This magnification effect also occurs because production costs are substantially greater than the preceding research and development costs and because production funds have the greatest expenditure per unit of time. Such expenditure rates bring with them increased scrutiny from both within and without the government.

COST DRIVERS

The set of processes and the manufacturing approaches chosen for the program have significant impact on cost. Often there are alternative manufacturing techniques available to produce the same product. Each of these options has associated with it a set of costs. By a careful selection of processes, significant cost savings can be achieved. To obtain the optimum cost impact, manufacturing process selection needs to be treated early in the design process.

The reader need not be reminded that the greatest program cost levers occur early in the life cycle. This phenomenon has been explored by numerous firms, academic researchers, and the DoD. The general consensus is that 80 to 90 percent of the final life-cycle cost of a system has been determined once the design is released for production. Changes in manufacturing methods, organization, and administrative processes have little effect on total costs once production begins. Early program decisions have the greatest impact and possess the greatest risk.

As specific design approaches are evaluated, the manufacturing cost implications of each design approach must also be taken into account. By identifying cost to manufacture as a key design parameter, the design team will be sensitized to cost and the probability of developing a low-cost design greatly increased. Historical data indicate that the three largest cost drivers creating upward pressure on original manufacturing cost estimates are engineering changes, unstable production rates, and unstable total quantities. These three factors create a loss of learning and increase cost through changes to manufacturing processes and work methods, tooling modifications, and changes to overhead allocation. Clearly, the program office must be sensitive to the potential impact of these factors. Success-oriented programs subject to evolving designs and unstable funding have almost no chance of succeeding. In a perfect world manufacturing managers desire a frozen design and a frozen production schedule.

The second area of cost sensitivity relates to the planning of the manufacturing system. By failing to invest in production planning as part of the demonstration and validation effort or earlier, the program manager runs the risk of initiating production without a defined plan. A typical situation, after approval to start production, is that a maximum compression of the schedule is required due to political pressures not to delay initial operational capability (IOC). The result is that the initial units are produced with marginal, and in some cases minimal, planning. As a result, there are tremendous inefficiency and wasted resources.

The schedule is also sensitive to the manufacturing plans generated. There is a historical tendency to arbitrarily assign fixed points in the acquisition schedule, such as milestone III and the IOC, and to create a schedule of manufacturing events to fit between the two dates. This situation is sometimes exacerbated by an almost fanatical reluctance by the buying service to delay IOC, regardless of the difficulties experienced during product or process development. IOC dates are often established and frozen as much or more by political concerns as they are by real or perceived military threats. Unrealistic schedules continue as a major contributor to technical difficulties on many acquisition programs.[6]

While the arbitrary assignment of milestones may be necessary to meet program objectives, these actions almost always fail to address the question of the actual effort that must be exerted to achieve the compressed schedule. By looking at the manufacturing process unconstrained by program events the most realistic time requirements can be identified. If the unconstrained manufacturing schedule is inadequate to meet deployment commitments, then attention can be directed to those areas which must be compressed and action taken to identify the extraordinary management actions required to achieve that compression.

There is a set of schedule penalties that are paid when planning is insufficient. By not clearly defining what must be done prior to starting and not identifying required resources, the time required to accomplish these tasks increases.

PRODUCIBILITY

The product is a result of both the design and the manufacturing processes which execute that design. The processes to be utilized must be considered against the design demands to determine the likelihood that those processes can repeatedly produce products which have the requisite physical and functional characteristics. By selecting processes which have inherent capabilities less than the design demands, the contractor can enter a situation in which the performance or the quality of the items produced will not meet requirements.

Producibility must be a key design driver. Producibility is a shared engineering function between design and manufacturing, directed toward achieving a design which is compatible with the realities of the existing or feasible manufacturing processes. Producibility may be defined in various ways:

- The relative ease of producing an item or system which is governed by the characteristics and features of a design that enable economical fabrication, assembly, inspection, and testing using available production technology.
- A design accomplishment that enables manufacturing to repeatedly fabricate hardware which satisfies both functional and physical requirements at an economical cost.

The compatibility of the design with the realities of the shop floor has large potential impacts on product quality, unit-production cost, the ability to increase production rates rapidly, and life-cycle operation and support costs. High unit-production costs, high maintenance costs, and poor field reliability often have their causes in insufficient attention having been given to producibility engineering. Products which are inherently difficult to fabricate, assemble, and test in the controlled environment of the production facility are potentially very difficult to regularly operate and maintain in the field.

Unfortunately engineering design activities that are necessary for product development are often treated as a discrete functional activity, with little or no involvement of the other plant functions, such as manufacturing, quality assurance, or production engineering. Particular projects are often compartmentalized within a multiprogram organization. This approach to product development stresses performance and gives little attention to producibility considerations. As a result, the product's design meets performance specifications at the completion of development, but does not allow for the limitations of manufacturing processes and procedures found on the factory floor. Hence the apparently mature product configuration does not survive rate production without performance degradation, and significant redesign is required for efficient production. Many systems are, unfortunately, developed more than once in order to be made producible.

Producibility is an integral part of the systems engineering process. Therefore the producibility effort must be performed by a team of specialists from across the program and supporting functions. One individual cannot possibly accomplish the total producibility effort without assistance from other functional areas. Considering the number of new processes and materials that are being developed, materials specialists should be brought into the areas of manufacturing, test, and evaluation and to the design process. Personnel from the various disciplines are necessary so that a detailed interaction can occur between the product designers and the production personnel who have specific knowledge of the available manufacturing technologies and their relevant costs.

Producibility is often identified as one of the items to be covered in a design review, but is not discussed as one of the major cost drivers in the transition from development to production. Several DoD directives and military standards discuss the topic of production design, but provide very little direction or guidance. Producibility, as a subset of product design, is usually not a major concern during the design review activity. As a result, it is not given sufficient attention to impact the design process in the early development phases. Producibility and general manufacturing issues need to be required agenda items at all major reviews, and not restricted to production readiness reviews.

The first step in the design process is to review the functional requirements. After the design has been scrubbed for unnecessary requirements and reviewed for completeness and clarity, ideas are formulated on how to meet the cited requirements. Here producibility is considered as part of the design criteria to be evaluated for cost-effectiveness and ease of manufacture versus the degree of compliance with the functional requirements. Preliminary analyses should be made tentatively to select components, configuration, materials, and processes without selecting any initial design concepts. These initial selections merely provide a basis for the designer to evaluate the concept. With a number of possibilities to consider, analysis is required to choose the approach that shows the greatest promise. As a minimum, design alternatives, function versus cost, schedule versus cost, and component design versus manufacturing capability must be considered.

As in any situation where resources and time are constrained, managers must have a method of focusing on high-leverage areas. With regard to producibility the program office must have a system to focus its efforts and the contractor's efforts. A contractor's producibility efforts should be explicitly called out in the contract statement of work. In addition to the normal data items required to document the management and analysis of producibility, the program office should seriously consider the use of DoD Standard 2101, "Classification of Characteristics." This military standard establishes management practices for the selection, classification, and identification of essential design characteristics. If one believes the old engineering adage that 20 percent of your components contribute to 80 percent of the cost or problems, this military standard forces the contractor to identify these vital 20 percent of the components. Design characteristics are classified as:

- *Critical,* a characteristic that if defective is likely to create or increase a hazard to human safety, or to result in the failure of a weapon system or major system to perform a required mission.
- *Major,* a characteristic that is not critical but is likely, if defective, to result in the failure of an end item to perform a required mission.
- *Minor,* a characteristic that is significant to product quality but is not likely, if defective, to impact the mission performance of the item.

Such design classifications should also be required of all subcontractors and component suppliers. Classification not only assists in focusing producibility efforts, but aids in quality assurance procedures and tracking technical progress. Tracking engineering drawing releases or engineering change proposals of the critical and major design characteristics, especially if they lie on the program's critical path, is a powerful management tool for the program office.

The achievement of production phase objectives usually requires the use of the most efficient, shop-proven processes for material transformation. These two process descriptors, efficient and shop-proven, often tend to be mutually exclusive. New processes and approaches to manufacturing, such as computer-aided manufacturing, often do not have extensive shop experience. The challenge is to maintain maximum efficiency of manufacture within the risk levels deemed acceptable for a specific program. It is important to recognize that advanced manufacturing technologies generally bring certain levels of risk to a program, along with the potential benefits of improved efficiency. However, advanced manufacturing technologies often can be avoided by the use of standardization, statistical process control, tolerance analysis, and designing for assembly.

Contractor design policies and guides should be established which specifically outline the technical considerations to be implemented during the production design process. Management participation in design and producibility reviews is critical to its success. Colocation of design engineers, production engineers, quality assurance, and purchasing personnel greatly encourages cooperation, communication, and participation in early reviews of the design to assure its eventual producibility at rate. Producibility must be confirmed before the production decision to assure that a stable mature design is transitioned to the factory. In addition, it is mandatory for low risk that proof-of-manufacturing models be required and that all processes be proofed to assure that the design is indeed consistent with production processes and capabilities. Very often proof-of-manufacturing models are not provided or required, which results in tooling and process problems not being totally resolved before production. As a consequence, many producibility issues are not discovered until production, and depending on the severity of these problems, rate production may be impacted severely. Retooling, rework, new equipment, considerable redesign, exotic manufacturing processes, and the like are often required—at great additional expense and time—to achieve required production quantities.

The contractor should be encouraged to utilize statistical process control rather than relying on the mass inspection of the finished product. Once production begins, statistical process control should be implemented as soon as practical for all critical and major manufacturing processes. Statistical process control is necessary to control variability in manufacturing and ensure that original design specifications are met during production. Control of manufacturing processes and the resulting consistency of product is impossible without statistical process control. Planning for the use of statistical process control should begin during engineering and manufacturing development.

The government's acquisition strategy and contracting approach have major impacts on a system's producibility. The government's approach and emphasis will determine how well the contractor incorporates producibility into the end product. The program office must be aware of the impact of the contract type on the contractor's motivation during production. The contract must be written so as to make clear to the contractor what is important, and the financial terms of the contract must be evaluated to determine whether they could encourage dysfunctional contractor behavior. Fixed-price development programs or cost-type development programs with fixed-price production options should be used with care. Manufacturing risks increase if the contractor is forced to make future financial and schedule commitments based on unproven technologies or unproven manufacturing processes.

Proposed production rates and quantities affect the contractor's cost structure, business plans, facilities, and selected manufacturing methods and processes. Stable rates and quantities are a prerequisite for production quality and declining unit-production cost. The government's funding and tracking of producibility engineering and planning (PEP) activities force discipline and visibility for the early consideration of producibility. Several incrementally phased production readiness reviews (PRR) should be used to manage the PEP effort and to identify production risks. In order to manage the PRR and PEP efforts, the program office must staff itself with individuals appropriately trained in manufacturing and industrial engineering.

PRRs are used to manage the producibility effort and assess production preparation. They evaluate the risk associated with proceeding into low-rate initial production (LRIP) and full-rate production. The primary focus of the PRR is on producibility of the design, control over production processes, and identification of required resources. Risks should be estimated for processes, costs, schedule, and supportability. Incremental production readiness reviews aid in making the producibility effort a proactive, integrated activity during the design process.

Producibility has both economic and engineering aspects. It requires coordination and cooperation between government and contractor. Failure to consider producibility may require costly and time-consuming modifications to the design and to the manufacturing process in order that performance and quality requirements can be met.

Levels of the required producibility effort by a contractor in order to ensure a producible system are hard to estimate. However, one study by the Logistics Management Institute (LMI) provides rules of thumb based on empirical data derived from the U.S. Air Force F-16 fighter and air-launched cruise missile programs.[7] The LMI study indicated that for a major program, such as an aircraft or missile development, using advanced, but not necessarily leading-edge technologies, an acceptable level of producibility effort would be 10 percent of all engineering manhours during both concept exploration and demonstration and validation. Twenty-five percent of all engineering manhours during engineering and manufacturing development should be devoted to manufacturing and producibility issues. These are estimates based on only two programs from a particular service, but they provide a reasonable starting point to evaluate contractors' proposals. Such heuristics need to be modified to suit a particular product's technological maturity, process maturity, and a contractor's skills and past experience with similar programs.

As discussed earlier, the manufacturing program has significant impacts on the program budget. Often the lowest-cost approach to production involves substantial front-end investments in special tooling, special test equipment, and facility layout. By planning for these investments significant positive impacts on the recurring manufacturing cost of the system can be achieved. However, the difficulty that arises is that early program budgets must be sufficient to sustain these investments. Where the front-end budget is insufficient the potential manufacturing structure for the weapon system cannot be established. This lack of structure is later reflected in missed schedules, cost overruns, and problems with product quality.

DoD Instruction 5000.2[8] tasks the program manager as follows:

- At Milestone I, Concept Demonstration Approval, manufacturing feasibility and industrial base capability assessments will be presented. Areas of production risk and manufacturing technology or industrial modernization efforts to reduce that risk will be identified. Design to unit procurement cost objectives should be established. Tradeoffs should be used to minimize strategic or critical materials use.
- A producibility program will be established during Phase I, Demonstration and Validation. This program will be an integral part of the systems engineering effort.
- At Milestone II, Development Approval, the producibility of the emerging product design, risk reduction efforts undertaken, and plans for proofing new or critical manufacturing processes will be specifically assessed. Updated manufacturing feasibility and defense industrial base capability assessments must also be presented.
- At Milestone III, Production Approval, the production decision will be supported by a production readiness review.

It should be noted that DoD acquisition policy only requires one PRR before milestone III. One PRR is not sufficient for most programs. Modern manufacturing management desires to integrate manufacturing into the design process and manage the producibility effort proactively. Therefore it is recommended that at least four PRRs be conducted incrementally during the engineering and manufacturing development phase. One PRR will only identify problems, leaving no time for corrective action before the milestone decision is due.

SUMMARY

Schedule, cost, and quality problems are not independent occurrences. They are symptoms of technical problems. The severity of many technical problems has, historically, first been recognized during the transition to production. The total impact of these factors suggests the need for early visibility to the program office of the types of manufacturing approach that are being considered for the development and production program. By bringing these issues into the design trade process and establishing a formal process for planning, the program office increases the likelihood of schedule attainment.

Planning for the design and execution of the manufacturing system is, in many cases, as significant a design and planning task as the design and planning for the weapon system itself. Tooling design and fabrication is often a major program within itself. These facts are frequently not recognized. In the DoD the term development is almost always associated with products and not processes. The transition from engineering and manufacturing development into production has historically been the greatest risk period for any program. This period of time reveals whether prototype systems can be scaled up and produced repeatedly at accelerated rates. Fundamental plans in design concepts or deficiencies in manufacturing processes become glaringly evident.

Certain facts have become evident over the years in terms of how a program should prepare to succeed in manufacturing and successfully transitioning to production. The GAO has studied defense acquisition for several years, and in one particularly good study has produced valuable insights.[9]

The study evaluated two "successful" programs, successful being defined as having met user requirements at or near estimated cost and having been deployed at or near the planned initial operational capability. The two successful programs were the U.S. Air Force F-16 fighter aircraft and air-launched cruise missile (ALCM). The programs deemed less than successful were the U.S. Army Copperhead projectile, the Army Blackhawk helicopter, the U.S. Navy high-speed antiradiation missile, and the Navy Tomahawk cruise missile.

The GAO study identified several discriminators between successful and less than successful programs. The successful programs were identified as having possessed the following characteristics:

- Stable funding during development and production
- A programmatic balance between performance and manufacturing concerns
- Adequate numbers of suitably trained government program office personnel to plan, coordinate, and evaluate manufacturing plans
- The use of four incremental production readiness reviews during the engineering and manufacturing development phase rather than relying on only one review, as is required by DoD acquisition policy

Successful manufacturing management strategies are quite logical and are based on well-proven experience. However, due to political or financial exigencies they are often ignored or curtailed. The principles of good manufacturing management are deceptively simple and obvious. The fundamental difficulty in managing the manufacturing process arises in funding, implementing, and sustaining these management principles.

THE U.S. INDUSTRIAL BASE

> The defense technology and industrial base is the combination of people, institutions, knowledge, material and facilities used to design, develop, manufacture and maintain the weapons and supporting defense equipment needed to meet the U.S. national security objectives.*

> The industrial base is . . . that part of the total privately and government owned industrial production and maintenance capacity of the U.S. government, territories and possessions as well as the capacity located in Canada, expected to be available during emergency to manufacture and repair items required by the armed forces.†

The study of the industrial base is a mammoth undertaking. Such a study includes the financial markets, economics, political and legal systems, natural resources, educational system, and technology base of a nation. Such a broad scope tends to overwhelm individuals or organizations that hope to influence such a system. This brief

*Congressional Office of Technology Assessment.
†U.S. Department of Defense.

section is obviously not intended to be an exhaustive study of the industrial base. It is intended to present an overview of the current status of the industrial base, projected trends, and some of the latest thinking on possible restructuring of the base.

Today within the DoD there is an ongoing debate about the health and future path of the U.S. industrial base. Basically the debate is focused on the issue of whether the industrial base is of concern to the DoD and whether the DoD could or should try to use its buying power to influence the industrial base's structure.

Conventional wisdom, as will be discussed in the remainder of this chapter, dictates that the decline of the domestic industrial base will have negative implications not only for national defense but for national economic competitiveness as well. Many in the Pentagon and industry believe that the United States is already involved in a global economic war. Raw military power and physical confrontations will play a decreasingly influential role in the new world order.

Another equally convinced group believes that globalization of the industrial base is inevitable and cannot be stopped. Today nearly one-third of the largest 50 suppliers of electronics to the DoD are foreign firms.[10] These individuals believe that the erosion of on-shore manufacturing capabilities and the technology base have progressed beyond correction. This group believes that the departments of defense, state, and commerce must work toward multinational agreements and coproduction arrangements so that needed materials and components will always be readily available from our economic and military allies. The human skills required to be self-sufficient in all or most of the critical technologies and the cost of capitalizing the required production facilities are beyond the capabilities of any one nation.

Any attempts by the DoD to improve the industrial base are hampered by the fact that it is an unstated policy of the Bush administration that the United States will not have an industrial policy. The allocation of national resources will be controlled by free market forces. This results in no national policy that integrates critical technologies and manufacturing capabilities with the needs of national defense. This lack of industrial base structural planning and the resultant systemic problems have been well documented by Gansler in his excellent book, *Affording Defense*:[11]

> The problems of U.S. industry and the potential corrective actions are beyond the control of the Defense Department. For America's overall industry to be positively affected, broad changes in education, industrial management and government policy are required.

It appears that even with the sweeping changes occurring in the former Soviet Union and Eastern Europe, the United States will remain globally engaged and will thus continue to require significant military forces. Since World War II the military strategy of the United States has been based on technological superiority, deterrence, flexible response, and a concept of forward defense. In order to respond to a changing global environment, the DoD must rethink its military strategies and the corresponding procurement and industrial base strategies.

The congressional Office of Technology Assessment has characterized future U.S. forces as possessing the following characteristics:[12]

- Small active and ready reserve forces
- Less forward basing, greater strategic mobility
- Continuing weapons performance advantage
- Substantial nuclear capability
- Chemical and biological defense capabilities
- Greater dependence on mobilization

The defense industrial base, as a result of the democratization of Eastern Europe, will in all likelihood be affected by potentially large reductions in the defense budget. The Office of Technology Assessment expects that overall defense spending will decline from a 1985 peak of 6.4 percent of gross national product to approximately 3.8 percent of gross national product by 1996 (Ref. 12, p. 3). Thus defense spending will be the smallest proportion of the gross national product since before World War II. These funding drops will almost assure a drastic decline in major system new starts and an increasing emphasis on modifications, retrofits, and service life extension programs.

Reconstitution as a national defense strategy is receiving increased study as a means of responding to changing threats and declining budgets. Reconstitution is the national ability to rebuild military strength within the time made available by strategic warning. Such a strategy would require accurate in-depth intelligence sources, armed forces prepared to absorb heavy augmentation, trained manpower, material resources, and the availability of current technologies and manufacturing processes and facilities.

Besides the decline in defense spending and a need for the ability to reconstitute forces, the U.S. industrial base faces continuing problems which have been developing since the Viet Nam conflict. These include a decline in the number of defense firms at the second and third tier levels.

The Center for Strategic and International Studies reported that the number of manufacturing firms providing goods to the DoD declined from more than 118,000 in 1982 to approximately 38,000 in 1987 (Adelman and Augustine,[10] p. 161). This occurred despite a major defense buildup during this time period. Of major concern are the smaller firms that provide specialty items such as fasteners, seals, ball bearings, optics, pumps, forgings, and castings. There is also an increasing dependence on foreign technology in terms of components and manufacturing equipment.

A study by the National Academy of Engineering found that one U.S. missile system would require more than one year to replace all foreign piece parts with domestically produced hardware (Adelman and Augustine,[10] p. 110). In another air-to-air missile it was found that the missile contained 16 foreign components, any one of which if not available could delay production for up to 18 months.

Many of the accusations of poor management and structural impediments associated with the steel and automobile industries have also been directed at the major defense contractors. These problems include a managerial focus on short-term results, excess capacity, and old capital equipment. A GAO study recently concluded that two-thirds of all contractor plant equipment costing over $50,000 could not be considered state of the art, and over 50 percent of this equipment was over 15 years old.

In all fairness, these firms have had to survive within a monopsonistic environment embedded within a heavily regulated industry. This monopsony has resulted in a market dominated by a relatively few specialized firms, each possessing excess capacity, engaged in fierce competition in a market with extremely high barriers to entry and exit. It is estimated that the top 100 defense contractors perform 75 percent of all defense-related work (Gansler,[11] p. 245). The situation is further exacerbated by an unstable market driven by uneven government funding and by the decline of smaller subcontractors and vendors. According to a report by the DoD Technology Assessment Team:[13]

> National security rests on U.S. economic and military leadership of the free world. Both are tightly coupled to the strength of American manufacturing. Since the 1960s, manufacturing has been implicitly de-emphasized by the Department of the Defense and, coincidentally, for different reasons, by the civil sector. While a variety of factors, including

the emergence of many more nations as manufacturing competitors, have played a role in our declining international competitiveness, the de-emphasis of manufacturing as a strategic element in corporate planning coupled with a lack of national leadership and often conflicting policies are the most significant. Similarly, the de-emphasis of manufacturing and manufacturing research by the Department of Defense in favor of advanced device technology is responsible for many of the problems of cost, long development time, and quality that plagues major weapons systems acquisition.

The manufacturing problems faced by the Department of Defense are exacerbated by the fact that the Defense Industrial Base cannot necessarily turn to the commercial sector for its manufacturing technology. Department of Defense requirements often exceed the capabilities of current commercial production particularly in light of the decline of many commercial sectors. Policies that directly affect the viability of these commercial sectors generally do not acknowledge that linkage to the national security.

The Office of Technology Assessment foresees the following as the most probable trends in the U.S. industrial base ("Redesigning Defense,"[12] p. 65):

- Extensive but declining research and development capability
- A continuing surplus production capacity in prime contractors
- A continuing decline in the number of subtier subcontractors and suppliers
- Continued limited access to civilian technology
- Increasing costs of production
- Consolidation of maintenance and repair capabilities
- Increasing globalization of the base

It is also reasonable to assume that the industry may see fewer prime contractors and more teaming and coproduction agreements between prime contractors. These teaming or coproduction arrangements can be expected to include both Japanese and European firms. Foreign military sales offer potential sources of income and technology during a period of declining U.S. defense budgets.

Given these uncertainties in global ideologies and in defense funding, and the systemic weaknesses in the industrial infrastructure, the Office of Technology Assessment outlines desirable characteristics of the future industrial base ("Redesigning Defense,"[12] p. 8):

- Advanced research and development capability
- Ready access to civilian technology
- Continuous design and prototyping capability
- Limited, efficient peacetime engineering and production capabilities in key defense sectors
- Responsive production of ammunition, spares, and consumables
- Healthy, mobilizable civilian production capacity
- Robust maintenance and overhaul capability
- Good, integrated management

LONG SHADOWS AND VIRTUAL SWORDS

Wagner, Gold, and other defense strategists believe the current defense acquisition system lacks the flexibility and speed of response needed to deal with uncertainties in the defense budget and potential threats. These innovative thinkers have evolved an acquisition and industrial base strategy which addresses flexibility and speed of procurement:[14]

> What is required is the selective use of a flexible acquisition strategy designed to be responsive to, and to help channel a drastically changing high technology threat. This is extremely important to keep design teams alive, and this could be done by prototyping, possibly skipping a generation if the threat allowed and if the life expectancy of the present system could be both upgraded and extended—all within the context of significantly reduced budgets while at the same time protecting and enhancing the technology base. This could be accomplished by utilizing the acquisition milestone system in such a way that selected systems are started *without* a commitment to go beyond the prototype stage. Some systems would be produced, others with very limited production, and some not produced but the capability to produce maintained. Others may be deliberately kept several years away from production while the technology is recycled and updated until their production is required. Finally, selected systems would not be produced and would represent a "long shadow" of technology for our adversary; that is, demonstrating the technological ability to develop a system could in itself act as a deterrent.
>
> . . . a virtual deployment is a capability brought within sometime before actual deployment—months to several years—and then put on "hold" to be maintained at that (or a time varying) state of future deployability, introducing new technology as it becomes available.

The author believes that such innovative strategies should be encouraged and advanced. This concept would appear to fall within the purview of the Defense Advanced Research Projects Agency (DARPA). Any practical implementation of virtual deployment would rely heavily on continuing advances in flexible manufacturing and in computer-aided engineering and manufacturing software. These technological advances will occur, no doubt. However, defense strategists and acquisition executives must recognize that virtual deployments assume that the very hazardous transition to production is accomplished successfully—this is a very large assumption and should be given due consideration.

DUAL-USE TECHNOLOGIES

Jacques Gansler has developed a concept of reducing defense costs while maintaining a viable defense industrial base. This concept appears to be infinitely more feasible within the existing infrastructure than the "long shadow" concept of Gold and Wagner. Gansler's approach is based on what he calls civil/military integration or dual-use technologies.

The concept rests on the premise that many industries could use the same technology, personnel, research, production facilities, and administrative procedures for both commercial and military products. The results would be lower costs and higher quality of defense products, and a larger industrial base, which would permit greater economies of scale. An added benefit would be that the producers would not be dependent on the government for their economic survival:[15]

> Such an integration approach would not have been possible in the past, when defense technology was usually more advanced in performance and was designed for more severe environments than commercial equivalents, and when commercial production required high volumes to achieve efficiency. Today, however, ruggedized commercial product technologies and flexible manufacturing process technologies have advanced to the point where such civil/military integration is both possible and desirable.

The dual use of technologies would require changes to government procurement policy and incentives for commercial manufacturers to become involved with the DoD. Basically, Gansler recommends four areas of change:[15]

Accounting requirements—limit the use of unique Department of Defense accounting requirements which were developed for use by firms doing primarily defense business.

Specifications and standards—reverse the governmental bias towards the use of military standards and specifications. The Department of Defense needs to identify commercial substitutes at the subsystem, component and raw material levels.

Technical data rights—the government must constrain its demands for unlimited rights in both hardware data and software.

Unique contract requirements—exempt commercial suppliers and their products from contractual delegations that are not consistent with the Uniform Commercial Code.

Federal laboratory systems—legislation should be enacted which mandate an increased transfer of technology from the federal laboratories to private business.

THE JAPANESE EXPERIENCE

The Japanese industrial experience since World War II is of great interest in economic, military, and manufacturing circles. The Japanese are of interest to economists, defense strategists, and program managers for at least three main reasons:[16]

1) The obvious economic and manufacturing success of the Japanese in building high quality, low cost products. Simultaneously moving from a producer of low technology commercial goods to a designer and producer of high technology commercial and military products. How were these successes planned and managed? Can the U.S. emulate the Japanese style of management?
2) The increasing reliance of U.S. weapons systems upon Japanese produced components.
3) The threat of the Japanese winning an economic war and the resultant possibility that the U.S. will decline as both an economic and military superpower.

There can be little argument that Japan has developed an industrial establishment of unparalleled efficiency which is dominating targeted technologies and markets at an astounding pace. Their economic power is based on a shared national vision for world economic domination. Japan has become a world economic superpower because of their ability to exert extraordinary control over its destiny.

. . . A serious analysis of emerging economic trends indicates that the U.S. is in trouble as a power in the transnational economy of 2000 and beyond. Yet there is no popular consensus among U.S. citizens that these problems exist or that prompt action to stem U.S. economic decline must be taken.

. . . Japan will be able, because of American ignorance and lack of insight, to launch an economic sneak attack, from which the U.S. may not recover.

Regardless of the program manager's feelings about the Japanese economic threat, it is obvious that U.S. managers can learn a great deal about successful manufacturing and strong industrial bases by studying the Japanese experience.

Recommendations of the DoD Technology Assessment Team on Japanese manufacturing technology were:[13]

1. *Manufacturing and the National Security.* We recommend the Department of Defense carefully consider how its industrial base policies can be coordinated with, and influence, those of other parts of government to achieve maximum benefit for the Defense Industrial Base.

2. *Recognition of Manufacturing as a Strategic Factor.* We recommend that the Department of Defense move toward an acquisition strategy acknowledging the quality of the production process as a major consideration of the Defense Acquisition Board and other like decision-making bodies and as a major element of competition.

3. *Investment.* We recommend establishment of increased incentives to reduce the financial risks to the Defense Industrial Base associated with the introduction of new manufacturing equipment and techniques.

4. *Concurrent Design.* Department of Defense should move with dispatch to develop policies and procedures to actively encourage, but not mandate, the implementation of concurrent design.

5. *Quality.* We recommend that the Department of Defense create policies which explicitly acknowledge that quality is designed in, not inspected in. Incentives should be provided to companies that improve quality through product and process design and process control. Department of Defense should emphasize and encourage the widespread use of systematic product and process design techniques to identify and reduce sources of process or material variation and to achieve quality in the face of remaining variability.

6. *Continuous Manufacturing Process Improvement.* Department of Defense should establish an explicit RDT&E (Research, Development, Test and Evaluation) design and manufacturing research budget line in DDR&E (Director, Defense Research and Engineering) which is a significant percentage of funds allocated for device research and weapons system development. Similar line items should be created in the service RDT&E budgets.

7. *Human Resources.* The Department of Defense should take an active role in ensuring the presence of well-educated personnel for the Defense Industrial Base trained to support the requirements of new process technologies and practices. Consideration should be given to establishing a program similar to the National Defense Education Act of 1958 to provide manufacturing engineers for the Defense Industrial Base. Classroom work in such a program should be supplemented by practicum in a manufacturing facility analogous to a "teaching hospital" operated jointly by an industry/university team with government/industry support.

8. *Vendors.* We recommend that policies be established which permit and encourage prime contractors to establish stable, long-term relations with their vendor base. Quality should receive more emphasis in vendor selection. Vendors should be involved early in product and process design as members of the concurrent design team. Less reliance should be placed on rigid, detailed specifications in communicating with vendors and more reliance placed on vendor participation in the concurrent design process.

9. *Technology Assessment Team (Japanese Manufacturing Technology)* We recommend that Technology Assessment Team visits to study Japanese manufacturing technology be made on a periodic basis to maintain currency on Japanese manufacturing practices and to identify opportunities for transfer of Japanese manufacturing technology to the Defense Industrial Base. Visits should also be made to other countries which have particular centers of manufacturing excellence. We recommend that the Department of Defense determine how opportunities to observe and study Japanese and other manufacturing can be made more broadly available to those in the Department of Defense and the Defense Industrial Base concerned with the manufacturing improvement and in particular to major weapon system program managers and their industrial counterparts.

PROGRAM OFFICE CONCERNS

Regardless of the program manager's feelings concerning how industrial base issues might be resolved, it is clear that the issues concerning the base must be addressed.

Industrial base planning should begin during concept exploration and proceed through the engineering and manufacturing development phase. DoD Instruction 5000.2 tasks the program manager as follows:[8]

> The industrial base implications of proposed acquisition program objectives during peacetime, surge and mobilization to include conflicts with other Department of Defense or commercial programs, shall be addressed at each milestone decision.
>
> Program planning shall identify and minimize the potential impact of foreign dependencies and diminishing manufacturing sources and material shortages on production and support objectives. If such items must be used, plans must describe actions to ensure the availability of the items/materials during production and support and under surge and mobilization conditions.
>
> Industrial base parameters will be included as part of the Acquisition Strategy Report as part of the Integrated Program Summary.
>
> The acquisition strategy will include an analysis of the industrial base's ability to develop, produce, maintain and support the program and, if applicable, the strategy to make production rate and quantity changes in the program in response to surge and mobilization objectives.

The majority of industrial base decisions are beyond the scope of the program office's scope and influence. It is critically important, however, that the program office be sensitized to the potential impact of industrial base problems and be prepared to elevate these issues to higher echelon decision makers within the DoD's acquisition hierarchy.

REFERENCES

1. R. J. Schonberger, *Japanese Manufacturing Techniques,* Free Press, New York, 1982.
2. "The Role of Concurrent Engineering in Weapons System Acquisition," Rep. R-338, Institute for Defense Analysis, Dec. 1988.
3. *DataMyte Handbook,* DataMyte Corp., Minnetonka, Minn., 1984, pp. 1–2.
4. "Production," in *Buying Aircraft: Material Procurement for the Army Air Forces, United States Army in World War II,* chap. 20.
5. "The Ships that Broke Hitler's Blockade," *Invention and Technology,* pp. 26–41, Winter 1988.
6. "Unguided Missiles," *Military Forum,* pp. 16–24, Apr. 1988.
7. "Program Management Guidelines for Producibility Engineering and Planning (PEP)," Logistics Management Institute, Bethesda, Md., Jan. 1985.
8. DoD Instruction 5000.2, "Defense Acquisition Management Policies and Procedures."
9. "Why Some Weapon Systems Encounter Production Problems While Others Do Not: Six Case Studies," GAO/NSIAD-85-34, May 1985.
10. K. L. Adelman and N. R. Augustine, *The Defense Revolution,* ICS Press, San Francisco, Calif., 1990, p. 110.
11. J. Gansler, *Affording Defense,* M.I.T. Press, Cambridge, Mass., 1989, pp. 240–242.

12. "Redesigning Defense," OTA-ISC-500, Congress of the United States, Office of Technology Assessment, U.S. Government Printing Office, Washington, D.C., July 1991, p. 8.
13. Report on Japanese Manufacturing Technology, U.S. Department of Defense Technology Assessment Team, June 1989.
14. R. L. Wagner, Jr., and T. S. Gold, "Long Shadows and Virtual Swords: Managing Defense Resources in the Changing Security Environment," in *Science and International Security: Responding to a Changing World,* edited by E. H. Arnet, American Association for the Advancement of Science, Washington, D.C., 1990, p. 62.
15. J. S. Gansler, "A Future Vision of the Defense Industrial Base," Testimony before the Senate Armed Services Committee, Apr. 16, 1991.
16. "Japan 2000," Rochester Institute of Technology, Rochester, N.Y., Feb. 11, 1991.

CHAPTER 8

CONFIGURATION MANAGEMENT ON DoD PROGRAMS

James H. Dobbins

James H. Dobbins is an internationally recognized specialist in software quality and reliability with over 28 years of experience in these and related disciplines. He is currently Professor of System Acquisition Management at the Defense Systems Management College, Ft. Belvoir, Va. He is an attorney-at-law licensed to practice in Virginia and various federal courts.

INTRODUCTION

Configuration management is a subject often ignored in practice, little understood at the implementation level, and the source of many program failures. It is a subject which is easy to give surface acknowledgment to and to spend considerable funds accomplishing, without realizing until too late that those activities and expenditures are ineffective. But configuration management, given the proper level of attention and understanding from both the government and contractors, can and should be accomplished with a relative degree of certitude and effectiveness.

The primary objectives of configuration management are (Ref. 1, sec. 3.6.2, p. 6):

a. To ensure that all personnel involved on a project know:
  (1) What is supposed to be built or produced,
  (2) What is being built or produced, and
  (3) What has been built or produced.

b. To ensure that the product being designed, manufactured, tested, and delivered by the contractor is exactly what was specified.

c. To ensure requirements can be traced into the final product.

d. To effectively and efficiently manage engineering changes.

e. To ensure the final product can be logistically supported and reproduced, as necessary.

There are six keys to effective configuration management. They are (Ref. 1, sec. 3.7, pp. 6–7):

a. *Early Planning and Involvement by Configuration Management.* Contractors must implement CM early to exercise control over the developing product so that the government can exercise meaningful CM over approved requirements and other baselines.

b. *Early involvement by logistics support.* Early involvement by logistics support personnel is necessary to ensure that Integrated Logistic Support (ILS) requirements are incorporated into and are reflected in the contract.

c. *Management Support.* Without management support in delegating authority and providing support, the tasks of CM will not be effectively realized.

d. *Training.* CM training, at a minimum in rudimentary form, must be provided for all personnel. Personnel cannot be expected to respond to CM needs if they do not understand the rudiments of CM objectives.

e. *Centralized Direction.* A centralized CM program allows resource sharing and standardization, and helps avoid repeated mistakes.

f. *Simplicity.* The CM plan should be easily understood and followed.

SOURCES

The sources of regulations for configuration management are found at both the Department of Defense (DoD) level as well as at the service level, and these are currently being updated. The updates, even though subject to modification, will be referred to herein to provide the reader with the benefit of the latest thinking on the subject. At the DoD level, there are standards such as MIL-STD-480B,[2] dated July 15, 1988, which superseded DoD-STD-480A, dated April 12, 1978, and DoD-STD-2167A.[3] There is also the new DoD Instruction 5000.2, Part 9, secs. A and B.[4] There is the draft MIL-STD-973,[5] dated April 22, 1991. A corresponding draft handbook, MIL-HDBK-61,[1] dated May 13, 1991, has been produced to aid in the implementation of draft MIL-STD-973. Draft MIL-STD-973, when published in its final form, "is intended as the principal DoD standard for configuration management." (Ref. 5, Foreword, p. ii.)

At the service level, each service has implemented configuration management directives. In this chapter one of the service directives, AFSC pamphlet 800-7,[6] will be referred to as an example of what can be found at that level.

Other federal and private agencies and organizations have implemented, or are in the process of implementing, configuration management regulations and standards such as the NASA technical memorandum 85908,[7] the draft NASA change authorization process standard, dated April 18, 1991, and the IEEE Std. 828-1983,[8] dated June 24, 1983.

In several of these references there is current activity in place to update standards and regulations which are about 10 to 15 years old. This really says two things. (1) The environment for both hardware and software has changed enough to require the change. (2) The practice of configuration management is not so volatile that change in the process is required every year or so.

WHAT IS CONFIGURATION MANAGEMENT?

Among the 17 pages of definitions contained in draft MIL-STD-973, configuration management is defined as follows (Ref. 5, par. 3.19, p. 11):

1. A discipline applying technical and administrative direction and surveillance to:
   a. Identify and document the functional and physical characteristics of configuration items.
   b. Audit configuration items to verify conformance to specifications, interface control documents, and other contract requirements.
   c. Control changes to configuration items and their related documentation.
   d. Record and report information needed to manage configuration items effectively, including the status of proposed changes and the implementation status of approved changes.

This is in contrast to configuration control, which is defined as "The systematic proposal, justification, evaluation, coordination, approval or disapproval of proposed changes, and the implementation of all approved changes in the configuration of a configuration item after formal establishment of its baseline." (Ref. 5, par 3.15, p. 10.) Configuration control is a generalized process. Configuration management is a set of processes which have specific product relationships.

Both of these definitions refer to configuration items. A configuration item is defined as "An aggregation of hardware, firmware, or computer software or any of their discrete portions, which satisfies an end-use function and is designated by the Government for separate configuration management. Configuration items may vary widely in complexity, size, and type, from an aircraft, electronic, or ship system to a test meter or a round of ammunition. Any item required for logistic support and designated for separate procurement is a configuration item." (Ref. 5, par. 3.18, p. 11.) Note that the examples given in the preceding definition for configuration item are all hardware-related since it is easy to visualize hardware, but the definition is just as applicable to software in its final form or any of its subcomponents.

DoD-STD-2167A does not define configuration management precisely, but states that "the contractor shall perform configuration management in compliance with the following requirements." (Ref. 3, secs. 4.5, 4.5.1, 4.5.2, 4.5.3, p. 17.) It then follows with three sections, one each on configuration identification, configuration control, and configuration status accounting. This standard describes configuration control (Ref. 3, sec. 4.5.2, p. 17) as those activities required to:

1. Establish a development configuration for each CSCI
2. Maintain current copies of the deliverable documentation and code
3. Provide the contracting agency access to documentation and code under configuration control
4. Control the preparation and dissemination of changes to the master copies of deliverable software and documentation that have been placed under configuration control so that they reflect only approved changes.

Note how DoD-STD-2167A combines the concepts of configuration management and configuration control as described in draft MIL-STD-973, and presents a greater degree of concern with contracting agency interface and access than does draft MIL-STD-973.

Other definitions are found in the related government publications such as Air Force Regulation AFR 65-3, which defines configuration management as a discipline which encompasses concepts, policies, and procedures for managing functional and physical characteristics of an item and its documentation.

There is no fundamental disagreement with these older definitions, but the newer definitions as given here tend to elaborate and clarify the meaning in such a way as to not only identify in detail what configuration management activities are to encompass, but also when the process of configuration management should begin.

When Does the Configuration Management Process Begin?

Perhaps the proliferation of definitions has been a source of some concern and confusion, and may have led some government program offices and some contractors to assume that the practice of configuration management does not begin until after the functional configuration audit (FCA) and the physical configuration audit (PCA) are completed. Part of this confusion may be related to definitions of configuration management such as that found in draft MIL-STD-973, par. 3.19, subpar. a, wherein there is a reference to functional and physical characteristics of configuration items. This confusion should be dispelled. Configuration management, for both hardware and software, begins at the time of contract award. In par. 4.2 of draft MIL-STD-973 it is clearly stated that configuration management applies to the technical data (documentation) and then extends to the product itself. Further, in par. 4.2.2, configuration management planning is required as an early step in each phase of the program or project. If the configuration management process is not applied continuously during the development process, there is no way that a product will be brought under configuration management successfully for the first time after completion of FCA/PCA. This is equally true for software and hardware, but it is even more crucial for software since, given the absence of a separate production phase, the first article produced is the final product. The rest are just electronic copies of that first article.

In managing the process and activities of configuration management, it is necessary that the requirements and provisions of the contractually imposed standards be incorporated into subcontractor statements of work (SOW), including all elements of a comprehensive configuration management program. Configuration management activities are not considered peripheral to the overall development effort. Draft MIL-STD-973, par. 4.2.4, requires that contractor configuration management personnel participate in (not just attend) all technical reviews conducted, and goes further to elaborate on the specific duties to be performed at the technical reviews, including evaluation of the quality of the configuration identification (the documentation establishing the functional, allocated, and product baselines and the developmental configuration for each configuration item), ensuring that the configuration item has met development milestone requirements, and that it is under formal configuration control. This set of requirements seems to overlap somewhat with the responsibilities of the software quality organization, and the program manager must work out with the contractor how this will be handled, especially on contracts where DoD-STD-2167A, DoD-STD-2168, and the finally approved MIL-STD-973 are all invoked.

Contractor Configuration Change Control Processes

The configuration change control process used by the contractor must:

1. Ensure effective control of all configuration items and their approved configuration identifications.
2. Propose engineering changes to configuration items, request deviations or waivers for pertinent items, prepare software change notices (SCNs), and prepare notices of revision (NOR) on the appropriate form (currently, DD form 1695). The NOR is only used for proposing changes to documentation (specifications, drawings, associated lists, or other referenced documents) which require revision after engineering change proposal (ECP) approval.
3. Establish a developmental configuration for each software configuration item.

4. Maintain current copies of deliverable documentation and code.
5. Provide the contracting agency access to documents and code under configuration control.
6. Control the preparation and dissemination of changes to items under configuration control so that they reflect only approved changes.

Configuration Item Changes

Changes to configuration items which have been placed under configuration control are accomplished through deviations, waivers, NORs, and specification change notices (SCNs) (Ref. 5, secs. 4.4.5–4.4.8). Requests for deviations are submitted by the contractor prior to manufacture. They are submitted when the contractor considers it necessary to depart from the mandatory requirements of the specified configuration identification. Note that this only applies to documentation. Requests for waivers are submitted when the contractor, through error during manufacture or development, does not conform to the configuration identification. In such a case, without approval of the waiver, the government is prohibited from accepting the product. The NOR is used where documents are not controlled by the ECP preparing activity. In cases where the specifications are controlled by the ECP preparing activity, the SCN is used to describe the exact changes made to the affected configuration item specifications if the related ECP is approved.

Interfaces

Interface management, especially on large systems, may be handled by a separate interface or integration contractor. The prime development contractor's configuration management plan is required to describe either the contractor's interface identification and control, or the separately administered interface management program.

CONFIGURATION MANAGEMENT PLANNING

It should come as no surprise that effective configuration management requires early planning. The government's configuration management plan should be developed early in the concept exploration life-cycle phase, and should be updated at the beginning of each succeeding phase to reflect changing requirements. The contractor's configuration management plan is usually required as a separate CDRL item to be delivered during the early phase of the contract. All configuration management plans should be reviewed regularly to determine whether an update is necessary. Configuration management plans should never be one of the documents which are produced and delivered and never looked at again. They should be referred to continuously as a management tool. Whether or not updates to the contractor configuration management plan should be deliverable is a matter for negotiation. The information can be briefed during, and provided alternatively through minutes of, regular program review meetings.

The government configuration management plan and the contractor configuration management plan should, as a minimum, be coordinated. Initially, the configuration management plans contain only planning and procedural information. As the task progresses into subsequent life-cycle phases, the content expands to include historical information, such as results of various key activities, as well as expanded or modified planning and procedural information.

Contractor's Configuration Management Plan

A contractor which is process mature will have a companywide or facilitywide configuration management process or configuration management Procedures Manual which is applied to each contract, perhaps in a tailored form, and followed during the entire life-cycle process leading to delivery and operational support. This plan or procedures document will include a reference to the standards used to create the document, the applicable contractor policies, procedures, and directives, a description of the resources to be employed, and the organizational structure within which the configuration management activity will be implemented. Configuration management plans for a specific project should include a description of the confirmation item, the procedures by which configuration management will be applied, how the configuration management program or standards should be tailored, the tasks and participants, and the events to be conducted.

The contractor's configuration management plan CDRL for the identified development effort should identify the contract, the customer, organizational elements to include any subcontractor relationships, the authority and responsibility of each member of the organization, the items to which the configuration management plan will be applied, and any special configuration management problems on this contract. Also included should be any special facilities, including hardware and software, to be utilized. The product to which configuration management is applied should be examined for possible inclusion of privately developed items, government-furnished items, or other unique considerations requiring special configuration management attention.

The contractor configuration management plan should have separate sections detailing the contractor's processes and procedures for configuration identification, configuration control, status accounting, subcontractor/vendor control, configuration audits, and life-cycle phasing considerations. For those contractor plans which include computer software configuration items (CSCIs), it should also describe the software configuration management organization, the software identification documents, the software change control process (SCCB), and how it interfaces with the system change control process (CCB), procedures for marking, storage, and handling of software-related documentation, the establishment, control, and maintenance of software libraries, and CSCI configuration status accounting records.

Because the configuration identification and control of software has consistently plagued government program managers and contractors as well, every individual software module (CSU) must be configuration-identified and uniquely identified by revision level. This job is usually done using automated tools, and is almost never done effectively or successfully using manual methods. The quality of the configuration identification process is absolutely crucial to effective configuration management of software. If ever there is a condition where two different versions of the same CSU are present with the same revision level, configuration management has been lost. The large number of products which have been developed to handle software configuration control and identification, both commercial and home-grown, are testimony to the recognized need for such control, but the quality and capabilities of these configuration management products vary widely. Some have built-in limits, possibly not even known to the developers, in terms of the project size and number of configuration items which can be managed. Most commercial products are not built and tested on programs of the size and complexity found in today's

DoD weapon systems, which may have 10 to 20 million lines of code and 100 or more different CSCIs. Some of these products are generalized, while others, such as CCC,* have been specifically built to meet the requirements of government standards. Taken alone, the tasks required for configuration identification of software are an indication of the functionality the software configuration management tools must have. The software configuration identification tasks include:

1. Identifying the documentation, and the associated computer software media (tape, disks, and so on) containing the code or the documentation, or both, which are placed under configuration control.
2. Identifying each computer software configuration item, its corresponding components (CSCs), and the associated CSUs by name and revision level, so that any one or any combination of these can be uniquely identified and reproduced.
3. Identifying all identification details of each deliverable item, including the version, release, revision level, change status, and any other identification detail.
4. Identifying the correspondence between each version of each software configuration item, CSC, or CSU, and its associated documentation.
5. Identifying the specific version of software contained on any specified delivery medium, including all changes incorporated since the most previous release.

The contractor configuration management plans are delivered as a CDRL item subject to government approval. Sometimes a generic facilitywide configuration management plan is requested to be included with the proposal and a more detailed, contract-specific plan is required after contract award. The configuration management plan format and contents may vary depending on what requirements are imposed. DoD-STD-2167A has an associated DID which describes the requirements of the plan. Draft MIL-STD-973[5] describes the plan requirements within the body of the standard (sec. 5.22, pp. 29–30). In any instance, what the government approves is what it gets. Configuration management is far too important to both the government and the contractor for the government to go through a rubber-stamp approval process.

One of the specific standard requirements for a configuration management plan (Ref. 5, sec. 5.2.2(d), pp. 29–30) is to address, as a minimum, the configuration control, interchangeability, and interoperability for all nondevelopmental items (NDIs). This is a growing concern for software development efforts, given the focus on reusability and the encouragement being given for the use of commercial off-the-shelf software (COTS) and other NDI in software development. This focus has grown in parallel with the push for Ada in development, and many development contractors are reporting a growth in reuse of 30 percent per year as Ada becomes more widely used. Although COTS is usually a purchased item, NDI may be a purchased off-the-shelf item, or may be software developed under another contract which is proposed for reuse, or reuse with modification, on the current development effort. NDI may also be a product developed by a contractor using overhead or internal research and development funds, and proposed for inclusion in the CSCI to be developed. The configuration management and configuration control issues, when these multiple kinds of products from multiple sources, some of which may be government-furnished (GFE or GFI), each being governed by one or more kinds of data rights, are all used and integrated on one development effort, are far more complicated than the issues encountered when every software product is developed by one

*CCC is a product of SofTool Corporation.

contractor. When the issues noted here are compounded through the use of subcontractors, who may all have their own combinations and mixes of products and types of data rights, the need for extensive and effective configuration management planning becomes paramount.

Should the program management office choose to utilize industry standards rather than DoD or MIL standards, there are a series of IEEE and ANSI standards available, such as the IEEE Std. 828-1983 for software configuration management plans. This standard defines in detail the industrywide minimum acceptable requirements for a software configuration management plan. It requires six sections, defined as introduction, management, software configuration management (SCM) activities, tools, techniques and methodologies, supplier control, and records collection and retention. This IEEE standard is consistent in purpose and intent with the government standards, and should be invoked in the absence of other standards. This is particularly applicable on most information resources management (IRM) programs where the usual DoD 5000 series documents and associated standards, such as DoD-STD-2167A, are generally not invoked.

DoDI 5000.2 Requirements

DoDI 5000.2, dated February 23, 1991, went into effect on all new contracts signed after August 23, 1991. Part 9, secs. A and B, define the requirements for (A) configuration management and (B) technical data management. Section A replaces DoD Directive 5010.19, DoD Configuration Management. Section A is only three pages long, compared to draft MIL-STD-973, which is 79 pages long, plus notes and formats of various forms. DoDD 5000 and DoDI 5000.2 set policy to be followed by the program manager, whereas the standards establish the requirements for implementing that policy. Standards, not DoD directives and instructions, are imposed on contractors as contractually binding requirements documents to the extent specified in the acquisition documents.

DoDI 5000.2, sec. 15, Definitions, p. 15-3, defines configuration management and configuration item as follows:

- *Configuration Management.* The technical and administrative direction and surveillance actions taken to identify and document the functional and physical characteristics of a configuration item; to control changes to a configuration item and its characteristics; and to record and report change processing and implementation status.*
- *Configuration Item (CI).* An aggregation of hardware, firmware, or computer software or any of their discrete portions, which satisfies an end-use function and is designated by the government for separate configuration management. Configuration items may vary widely in complexity, size, and type, from an aircraft, electronic, or ship system to a test meter or round of ammunition. Any item required for logistics support and designated for separate procurement is a configuration item.

DoDI 5000.2 does not define configuration control. Note that the DoDI 5000.2 definition of configuration management is similar in many respects to that contained in draft MIL-STD-973, and the definition of configuration item in DoDI 5000.2 is

*This definition is quoted in, and incorporated into, draft MIL-HDBK-61,[1] the implementing companion handbook to draft MIL-STD-973,[5] confirming the consistency between DoDI 5000.2 and draft MIL-STD-973.

virtually identical to that in draft MIL-STD-973. Therefore there should not be any major disparity between the current direction provided in DoDI 5000.2 and that which will be contained in the finally approved MIL-STD-973. This is important since DoD directives and instructions are generally not imposed on contractors, but standards are. Therefore the imposition of the finally approved MIL-STD-973 should be consistent with the obligation placed on the government program office by DoDI 5000.2.

DoDI 5000.2 takes a comprehensive life-cycle view of configuration management. It focuses on a configuration management program which results from a configuration management general process. It describes the requirements of a configuration management program as (Ref. 9, Part 9A, par. 3(a), subpar. (1), (2), (3), p. 9-A-2):

> **(1)** Procedures that must be tailored consistent with the complexity, criticality, quantity, size, and intended use of the CIs. It requires the development of standard processes, and requires that these processes be used, but requires that they be tailored by means of the application of relevant military standards which are adapted to specific program characteristics.
>
> This is consistent with the modern philosophy expressed within the body of the more recent standards, such as DoD-STD-2167A and DoD-STD-2168, which advocate the tailoring of standards for every program and discourage the blanket imposition of a set of standards on a program.
>
> **(2)** Configuration management activities conducted by the government Program Managers during the acquisition program. It requires that these activities, initially conducted by the Program Manager, transfer to the various service systems, logistics, or materiel commands when the CI is transferred from the control of the Program Manager.
>
> Inherent in this requirement is the need to transfer, or separately procure for the receiving service or command, the configuration program support environment, including automated support, necessary for the continued application of the configuration management activities.
>
> **(3)** The processes and procedures established by mutual agreement, and documented by the lead DoD Component when more than one DoD Component is involved in the acquisition, modification, or support of the CI.
>
> This requires the government to establish a chain of control and authority over the CI configuration, and to do so in ways which each of the other participants can accept. This assures a smooth and orderly transition of CIs and responsibilities between the different DoD components.

DoDI 5000.2 (Part 9, sec. A, par. 3(b)(1)–(3), p. 9-A-2) requires that individual configuration items be directly traceable to the work breakdown structure (WBS). How this is done is not readily evident for software, and so the WBS handbook[10] is currently being upgraded to reflect software more clearly, and in a way that the requirement as expressed in DoDI 5000.2 can be implemented with relative ease. DoDI 5000.2 also clarifies an issue which has been the subject of much question and debate for several years. It states (Part 9, sec. A, par. 3(b)(3), p. 9-A-2) that "Computer software will be treated as computer software configuration items throughout the life of the program, regardless of how the software is stored (e.g., read-only memory devices, magnetic tape or disc, compact discs, nonvolatile random access memory)."

Recognizing the role that various DoD and military standards play in the acquisition process, DoDI 5000.2 (Part 9, sec. 3(c)(1), p. 9-A-2) requires that the functional, allocated, and product configuration baselines be identified and documented in accordance with MIL-STD-483 or MIL-STD-490 or both. When draft MIL-STD-973 is approved and released, this section of DoDI 5000.2 may be upgraded to reflect the addition of that standard. Part 9 of DoDI 5000.2 does not define these three baselines, preferring to accept the definitions as contained in the relevant DoD and military standards, such as those in draft MIL-STD-973 (sec. 3, definitions 3.5.1, 3.5.2, and 3.5.3, p. 8).

At this point it will be helpful to understand the differences between the three principal baselines. A functional baseline (FBL) is defined (sec. 3.5.1, p. 8) as the initial approved documentation describing a system's or item's functional characteristics and the verification required to demonstrate the achievement of those specified functional characteristics. An allocated baseline (ABL) is (sec. 3.5.2) the initially approved documentation describing an item's functional and interface characteristics that are allocated from those of a higher-level configuration item, interface requirements with interfacing configuration items, additional design constraints, and the verification required to demonstrate the achievement of those specified functional and interface characteristics. A product baseline (PBL) is (sec. 3.5.3) the initially approved documentation describing all of the necessary functional and physical characteristics of the configuration item, and required joint and combined operations interoperability characteristics of a configuration item (including a comprehensive summary of the other service(s) and allied interfacing configuration items or systems and equipments), and the selected functional and physical characteristics designated for production acceptance testing and tests necessary for support of the configuration item.

One of the requirements of DoDI 5000.2, Part 9, sec. 3(e), is the establishment of a change control board to control and approve changes to approved baselines and approved configuration identification, and advise the program manager as to configuration status. It further requires the grouping of engineering changes for implementation to reduce the total number of different changes supported by the field. It likewise requires that all documentation associated with the changes be upgraded to reflect the condition as reflected in the changes. When a change is made to a configuration item which has been fielded, everything necessary to implement the change, including the documentation, will be provided as one complete kit. All changes to configuration items and configuration identifications will be maintained and the status accounting will be accomplished in accordance with the tailored application of MIL-STD-483, MIL-STD-482, and DoD-STD-2167A, as appropriate. DoDI 5000.2 also requires configuration audits in accordance with MIL-STD-1521B and DoD-STD-2167A.

Configuration Item Numbering

Because each configuration item requires traceability, each requires a unique name and number. Each hardware configuration item shall be identified by a unique part number in accordance with DoD-STD-100.[11] Although there is no comparable numbering standard for software, and the applicable standard (Ref. 5, sec. 5.3.3.1(g), p. 34) states that the computer software configuration item shall be identified in accordance with the provisions of the contract, many contractors apply a part numbering scheme to software just as is done for hardware. The software part number drawing will list each individual CSU by name and revision level, which go to make up the part number being described. Although most software deliveries are made as copies of the bonded master, the part number drawing tells the contractor exactly which components are necessary to again reproduce that specific part number should that ever be necessary (for example, the bonded master becomes defective or the media is compromised).

Documents are also numbered. Military specifications are numbered in accordance with MIL-STD-961. Program-peculiar specifications are numbered in accordance with MIL-STD-490 or, when authorized, in accordance with the contractor's numbering system. Although identification labels with part numbers or serial numbers are generally affixed directly to the hardware, identification numbers for computer software shall be both embedded in the machine readable code and also contained on a label affixed directly to the media containing the computer software, the specification for that software, and the related manuals for that configuration item. Upgraded versions are identified by suffixes to the base number.

Change Classifications

Changes to baseline products are made via engineering change proposals (ECP), DD form 1692, each of which is classified by the originator as either a class I or a class II change. A class II change is any change which is not class I.

A change is class I if (Ref. 5, sec. 5.4.2.16, pp. 41–42)

a. The Functional Configuration Identification (FCI) or Allocated Configuration Identification (ACI), once established, is affected to the extent that any of the following requirements would be outside specified limits or specified tolerances:

- **(1)** Performance
- **(2)** Reliability, maintainability, or survivability
- **(3)** Weight, balance, moment of inertia
- **(4)** Interface characteristics
- **(5)** Electromagnetic characteristics
- **(6)** Other technical requirements specified in the FCI or ACI or PCI.

b. A change to the physical configuration identification (PCI), once established, will affect the ACI or FCI as described above, or will impact one or more of the following:

- **(1)** GFE
- **(2)** Safety
- **(3)** Deliverable operational, test, or maintenance computer software associated with the CI or computer software CI being changed.
- **(4)** Compatibility or specified interoperability with interfacing CIs or computer software CIs, support equipment/software, spares, trainers or training devices/equipment/software.
- **(5)** Configuration to the extent that retrofit action is required.
- **(6)** Delivered operation and maintenance manuals for which adequate change/revision funding is not provided in existing contracts.
- **(7)** Preset adjustments or schedules affecting operating limits or performance to such an extent as to require assignment of a new identification number.
- **(8)** Interchangeability, substitutability, replaceability as applied to CIs, and to all subassemblies and parts except the pieces and parts of non-repairable subassemblies.
- **(9)** Sources of CIs or repairable items at any level defined by source-control drawings.
- **(10)** Skills, manning, training, biomedical factors or human-engineering design.

c. Any of the following contractual factors are affected:

- **(1)** Cost to the government including incentives and fees.
- **(2)** Contract guarantees or warranties.
- **(3)** Contractual deliveries.
- **(4)** Scheduled contract milestones.

Those ECPs which are class I shall be assigned a priority (Ref. 5, sec. 5.4.2.16.6, pp. 46–47) as emergency, urgent, or routine. Emergency priority means that national security is affected, or a personnel hazard is present. Urgent priority means that mission effectiveness of deployed forces is compromised, or a potentially hazardous condition exists. Other reasons include the need to meet significant contract requirements, to avoid cost or schedule slips, or that the change will result in net life-cycle cost savings to the government.

Using MIL-HDBK-61

Draft MIL-HDBK-61[1] is a companion to draft MIL-STD-973, and is intended to be used in conjunction with that standard when MIL-STD-973 is approved. MIL-HDBK-61 is intended to provide general and detailed implementing guidance to government agencies, and should be used to tailor the configuration management requirements of any contract where MIL-STD-973 is invoked. Note that it is intended to aid the government, not contractors, although there is no reason why contractors should not also benefit from its use. Although the standard may be a contractual requirement, the handbook will not be contractually binding.

Arguments against Configuration Management

Draft MIL-HDBK-61 provides some insight into why some organizations resist configuration management (sec. 3.9.2, pp. 9–10). One argument is that we simply do not need configuration management. This is typical of the kind of argument one hears from contractors or government organizations who are level 1 on the SEI software process maturity scale (see Chap. 36) or the manufacturing equivalent. The typical rationale given is the lack of desire to spend resources on a discipline so little understood, rather than admitting a lack of ability to be able to exercise discipline and process control. Rather than opting for education and process improvement, the response is rejection. The solution is education. Configuration management has clear benefits, but fully understanding those benefits implies an understanding of the need for process control and standardization of practices. Another argument against configuration management is that there are no clearly visible results. On the contrary, initiating configuration management at the beginning of a project and developing a close working relationship with the project manager, will make configuration management as an activity more visible. Properly implemented, the evidence of configuration management at work is a smooth product development, documentation, distribution, upgrade, and control process. Absent effective configuration management, the out-of-control products, the lack of revision-level identification and consistency, and the lack of consistent documentation will make configuration management *very* visible, but in a highly adverse manner.

Some argue that configuration management is too burdensome; that it just gets in the way and is a bottleneck to progress. It is seen as an overhead activity consuming numerous resources required to complete forms and follow complicated procedures. Configuration management should be, and can be, implemented as a conceptually simple and effective system, and should include projectwide education on the configuration management requirements, processes, and practices, and how each part of the organization should interface with configuration management. It also helps when configuration management is provided with automated support systems. Lack of knowledge and understanding can make it appear much more complex than necessary.

Others argue that configuration management is too idealistic. To the contrary, it is a highly practical discipline, and should include clear step-by-step guides for appropriate personnel to follow. The configuration management plan should be written in such a way as to clearly delineate its role, including the role it plays in the product development. Program managers who have already been through the battle of developing and fielding a system, whether successful or not, have learned the value of configuration management. They do not have to be convinced of the disastrous results attendant to losing configuration control.

Acquisition Process Configuration Management Activities

Taking recognition of DoDI 5000.2, MIL-HDBK-61 (sec. 3.10–3.10.2.5, pp. 10–16) describes the configuration management-related activities beginning with concept exploration and going through operations and support. The handbook recognizes the changing nature and emphasis for configuration management, moving from a sometimes peripheral activity focused on documentation to a full production and deployment activity involving a full system. It is significant that the handbook takes cognizance of the need to have configuration management purview of both the product development description drawings and documents as well as the interface control drawings and documents. These interface documents may describe interfaces between similar components, and may also describe interfaces between hardware and software. The important thing is for configuration management personnel to always think in terms of a *system.*

When thinking *system,* an obvious component that is too often overlooked is software. Software, by its nature, presents the program manager with a greater level of uncertainty, for a longer period of time, than other typical system components. Software configuration is also not apparent (Ref. 1, sec. 3.13). One cannot open a drawer, or the back panel of a unit, and see that a major software component is missing or broken, or not built according to specification. Software is also a product which for the most part is generated, not manufactured. The pieces (modules, or CSUs) are generated and modified by humans at every step, and at every step are subject to inadvertent error introduction. It is often the case that different software components of a system are generated by different companies, with different software engineering environments, different quality assurance and configuration management skills and processes, and different corporate management practices. The interface requirements, in addition to the software requirements specification (SRS), govern the generation of all these diverse components.

A process of software configuration management is often difficult for hardware engineers, managers, and configuration management practitioners to grasp because it is impossible to describe software. Software has no form. When one engineer grappling with this problem asked if software was the box of Holerith punched cards he held in his hands, he was told that the software was the holes in the cards. Software has no physical form, although it is transmitted on media which do have form. When doing nondocument software configuration management, one is concerned with the information contained in or on the media, not the media itself. When MIL-HDBK-61 and MIL-STD-973 refer to the physical characteristics of software, they both mean the executable image; that distinct binary state at which it is ready to be loaded into the target system and used to satisfy its intended end purpose. This is why the configuration management control of software documentation is so very important, and why verification that the binary image and the documentation are consistent and at the same level is likewise important. When considering or evaluating changes to software, we do not look at software itself, but rather at the associated documentation.

Configuration Identification

Of all the elements of configuration management (configuration change control, configuration identification, status accounting, and FCA/PCA audits) configuration identification can easily be argued as the most important (Ref. 1, sec. 4.1, p. 21). All the others are based on configuration identification. It is impossible to control changes to a configuration not properly identified.

Configuration identification can be thought of as both a verb and a noun. As a verb, it is the activity of selecting the technical documents which describe the functional and physical characteristics of a configuration item. This activity, performed by the government, requires an intelligent and meaningful balance between the requirement for full and detailed identification of physical and functional characteristics of a configuration item and the cost of production and dissemination of the documents. As a noun, the more common usage, it is the set of documents which define the approved configuration, or part thereof, of a configuration item.

Configuration identification is accomplished at three levels. These are:

1. *Functional Configuration Identification (FCI).* The approved functional baseline plus approved changes
2. *Allocated Configuration Identification (ACI).* The approved allocated baseline plus approved changes
3. *Product Configuration Identification (PCI).* The approved product baseline plus approved changes

Configuration identification is seen in operation as a set of numbers affixed to parts or documents. These may be part numbers, revision levels, serial numbers, and the like. These numbers serve as a substitute for long and wordy descriptions of the part or document to which the inquirer is being referred. Even on small projects, several hundred drawing numbers may be issued, and individual part numbers may exceed 1500 (Ref. 1, sec. 4.5.7, p. 53). On large projects there may be 10,000 or more drawing numbers issued. It is absolutely imperative that drawing and part number format selection, composition, and assignment be performed accurately and carefully in order to avoid loss of configuration control. That means keeping the numbering scheme as simple as possible while still accomplishing the goal. It also means never changing an assigned number and never reassigning an already used number to a different element, even if the original element is discontinued. It is also necessary to follow government regulations governing numbering, such as DoD-STD-100, and to assign drawing numbers from a central source. It is also important to remember that all parts which are interchangeable, even if manufactured by different contractors, must have identical numbers.

Configuration Items

Most systems are composed of one or more configuration items. A configuration item may be hardware or software, and it may be decomposed into lower-level subelements. Configuration items are classified as prime item, software item, critical item, or noncomplex item.

Prime items are documented with type B1 specifications, commonly referred to as the prime item development specification, and type C1, commonly referred to as the prime item product specification. These specifications are documented, as appropriate, according to the requirements as defined in MIL-STD-970. A prime item is normally found at the higher levels of the overall system architecture. Prime items are usually those components which have some or all of the following characteristics:

1. They are separately accepted on DD form 250.
2. Quality conformance inspections are required.
3. Provisioning activity is required.
4. Technical and user manuals for the configuration item are required.

Software items are documented in the type B5a (software requirements), type C5 (software product) specifications, or type B5b (interface requirements) specifications. Separate software items are identified when the complexity or distinct functionality requires a separate specification to document functional capabilities. This is usually the case with embedded systems found as part of weapon systems. Software is also identified as a configuration item when it is developed or modified using government funds. This protects the government, the contractor, and the public. Software can also be identified as a configuration item when it will be widely distributed and supported, perhaps in different locations and on different platforms from other system software or other software items.

Critical items are documented with type B2 (critical item development specification) and type C2 (critical item product specification) specifications. Once beyond the development stages, critical items are documented, as appropriate, in accordance with MIL-STD-970. Critical items, although less complex than prime items and even though they are always components of a prime item, are given that designation because they are either engineering critical or logistics critical (Ref. 1, sec. 4.2.2.6, p. 25). Items which are engineering critical are those whose performance requirements must be carefully defined and controlled during development. The reasons may be founded in technical complexity, reliability requirements, or safety and health concerns. Those which are logistics critical are those whose performance or design requirements must be separately specified and controlled in order to facilitate future procurement of the critical item. The reasons may be driven by repair parts provisioning, or where the item has been designated for multisource reprocurement.

Noncomplex items, documented in accordance with MIL-STD-970, are documented with type B3 (noncomplex item development) and type C3 (noncomplex item product fabrication) specifications. These are generally simple design items.

Software Configuration Identification

Because of the difficulty in describing software in terms of physical characteristics, the linkage, or association, between a computer software configuration item (CSCI) and its configuration identification is essential. One of the unique features of software configuration identification is the possible inclusion of a definition of the development software engineering environment and software test environment (SEE/STE) (Ref. 1, sec. 4.7.2, p. 60). The configuration identification of a CSCI should always include a precise identification and description of critical SEE/STE components if those components are necessary to regenerate the CSCI in its accepted version and to evaluate the CSCI performance.

Another wrinkle which arises with software is configuration management applied to firmware. Firmware can be treated in configuration management as:

1. Nondevelopmental hardware and software
2. Nondevelopmental hardware and government-developed software

3. Government-developed hardware and nondevelopmental software
4. Government-developed hardware and software

Each of these different mixes must be understood and both identified and controlled in light of their unique nature. This may mean references to the firmware identifications in either software specifications and drawings or hardware specifications and drawings, or both.

Contract Requirements

The government must be careful to invoke the level of configuration management necessary for the contract in the contract itself. This is normally done in the statement of work (SOW). The SOW must identify the importance of the configuration management activity, and must describe the configuration management requirements in a separate and clearly identified section on the same level as other major development and management activities. Configuration management requirements should never be buried several paragraphs down under some other heading such as project management.

In the configuration management section of the SOW, all configuration management requirements should be detailed. This section should not be limited to a one-line reference to some other standard or regulation. It should reflect any unique needs or features of the contract, and should also reflect a proper and intelligently considered tailoring of any imposed standards such as MIL-STD-973 or DoD-STD-2167A.

SOW requirements should be stated explicitly. Vague or general overarching statements, including general references to a standard, provide the contractor with no guidance and the government with no yardstick for performance monitoring and measurement. SOW requirements should be specific, and should detail the exact references for each activity, and the sections of any imposed standards that do or do not apply to the contract. For each requirement, the SOW should identify the exact deliverables, the delivery dates, the work to be performed, and any milestones.

Configuration management requirements cannot be imposed through the application of a data item description (DID). A DID only specifies the format and content of a contract deliverable. A specific SOW requirement must be present which requires the activity and then prescribes the use of the DID to document that activity or result of the activity. A DID necessary to meet specific contract requirements may not be readily available and may have to be created for the contract.

The government, before imposing configuration management requirements in a SOW, should perform an estimated cost analysis of the contemplated activities. The value of the resulting configuration management activity and products must be reasonable in terms of the total contract value. Where possible, the SOW should identify the expected commitment from the contractor in terms of a percentage of total costs for a project. For example, hardware configuration management may be 0.5 to 1 percent of project cost; for software it may be 6 to 8 percent of software development costs.

The SOW should also state how the government will interact with the contractor on the configuration management program. This will include whether or not the government will chair technical review meetings, attend CCB or SCCB meetings, attend ICWG meetings, and conduct periodic audits of the configuration management system and specifications during the project.

If imposing a standard, know the contents beforehand. The government has wasted countless hours in acquisition because of the imposition of inappropriate or inapplicable standards, only to have the contractor call the matter to the government's attention and thereby cause a delay while the corrective action change notices are sent out to all bidders. It is also necessary, where multiple standards are imposed, to avoid imposing conflicting standards. In one example, MIL-STD-1679, DoD-STD-1679A, DoD-STD-2167, and draft DoD-STD-2168 were all called out, even though DoD-STD-1679A superseded MIL-STD-1679, and DoD-STD-2167 superseded DoD-STD-1679A. In addition, standards in draft form cannot be imposed on a contract.

The program office should also watch for chaining. It is often the case that a standard is imposed "to the extent specified in" another higher-level document, and that higher-level document never referred to the standard. In effect, it was not imposed at all. Sometimes this chaining can go through three or more levels before it can be determined exactly what, if anything, is actually required. The operative word is clarity. Be clear, and be specific, in articulating the requirements. MIL-HDBK-61, beginning on p. 148, provides a model SOW section for configuration management, including tailoring guidance for MIL-STD-973.

SUMMARY

Configuration management is a discipline which requires proper management awareness and attention. This will become increasingly important as the trend develops within the DoD toward fewer new program starts and more upgrades to existing systems. Many of these upgrades will be software-intensive since software provides much of the functionality for weapon and intelligence systems. Since software is considered to be the most difficult element to bring under configuration control, proper software configuration management process development and management will be increasingly important. The program manager will be faced with systems which have been in the field for five years or more, to which there may have been several upgrades already, and for which the principles of configuration management discussed herein have not be uniformly applied.

There will be instances where one contractor was the system developer, and a series of other contractors served the role of change agent over a period of time following delivery. The level of configuration management competencies may have been less than uniform among these contractors, and now the government wants to make a major upgrade to the existing system. One of the critical priorities of the program manager managing this change will be to establish configuration management over what now exists, and do so with the understanding that this task may be resource draining. If configuration management of the base is not firmly established, the probability of acquiring configuration management of the total system by default as the new upgrade is developed and integrated is zero. When a system is not under strict configuration management, the job of program manager becomes as difficult as trying to grab a handful of smoke.

REFERENCES

1. Draft MIL-HDBK-61, *Configuration Management,* Joint Logistics Commanders, May 13, 1991 (subject to modification).
2. MIL-STD-480B, *Configuration Control—Engineering Changes, Deviations and Waivers,* dated 15 July 1988.

3. DoD-STD-2167A, "Defense System Software Development," 29 February 1988.
4. DoD Instruction 5000.2, Part 9, secs. A and B, "Configuration Management and Technical Data Management."
5. Draft MIL-STD-973, *Configuration Management,* dated 22 April 1991. Prepared by Joint Logistics Commanders. Subject to modification.
6. AFSC Pamphlet 800-7, Department of the Air Force, "Acquisition Management, Configuration Management," dated 1 December 1977.
7. NASA Technical Memorandum 85908, "Software Control and System Configuration Management: A Systems-Wide Approach," August 1984.
8. IEEE Std. 828-1983, "IEEE Standard for Software Configuration Management Plans," June 24, 1983.
9. DoD Instruction 5000.2, "Defense Acquisition Management Policies and Procedures," Feb. 23, 1991.
10. MIL-HDBK-WBS.SW (2d draft), being prepared by Army CECOM Center for Software Engineering, Oct. 1, 1991.
11. DoD-STD-100, "Engineering Drawing Practices."

CHAPTER 9

THE SECURITY ASSISTANCE PROGRAM*

Virginia and Tom Caudill

Virginia Caudill is an Assistant Professor at the Defense Institute of Security Assistance Management at Wright-Patterson Air Force Base, Dayton, Ohio, where she specializes in FMS process, financial management, and logistics. She has over 15 years experience working in several aspects of security assistance programs including FMS program management.

Tom Caudill is currently Director of Arabian Programs at the Air Force Security Assistance Center at Wright-Patterson Air Force Base, Dayton, Ohio. He has a strong background in security assistance which included chief of the plans and policy office at the International Logistics Center. Mr. Caudill spent several years working in acquisition logistics for the Air Force and is able to incorporate that experience into FMS program management.

Tom and Virginia met and married while they were in the Peace Corps as teachers in Thailand.

INTRODUCTION

The conduct of foreign policy and the influence of diverse players, all acting in a manner destined to secure their own needs (which may or may not coincide with our perception of their needs) requires tools as diverse as the circumstances leading to those tools' creation. One of these foreign-policy tools used by successive presidential administrations is the Security Assistance (SA) Program. The various components of security assistance play important roles in the foreign policy of this country. The SA Program is specifically designed to enhance the foreign-policy objectives of the United States by contributing to its national security through the concept of "collective security."

*The views expressed herein are those of the authors and do not necessarily represent the official position of the Department of Defense or any of its agencies, departments, or other organizational elements.

The SA Program is divided into five major components. The first component, the economic support fund (ESF), promotes economic assistance and political stability by providing for balance of payment support, infrastructure and other technical development efforts, as well as health, education, agricultural, and family planning projects in the developing countries throughout the world. The ESF is managed by the Department of State (Samelson,[1] p. 44). Peace keeping operations (PKO) is the second component of the SA Program. PKO provides for the multinational force and observers, United Nations forces in Cypress, and other programs designed specifically to maintain peace in various regions of the world. PKO is also managed by the Department of State (Samelson,[1] p. 44). The third element of the SA Program is international military education and training (IMET). Under IMET, technical training is provided to foreign personnel on a grant basis (Samelson,[1] pp. 42–43). IMET is managed by the Department of Defense (DoD). Fourth, the foreign military financing program (FMFP) provides for credits and loan repayment guarantees to foreign governments to purchase defense articles, services, and training from the U.S. government (Samelson,[1] pp. 38–39). Finally, foreign military sales (FMS) and foreign military construction sales allow foreign governments to purchase defense articles, services, and training directly from the U.S. government on a cash basis. All costs associated with this type of FMS sale are paid by the foreign government. Like the IMET program, FMS, including the FMFP and foreign military construction sales, are managed by DoD. In addition to the five components of the SA Program, a foreign government may also purchase defense articles, services, and training from a U.S. contractor through a direct commercial contract.

By far the largest components of the DoD SA Program are FMS and direct commercial sales. Consequently this chapter is dedicated to the role of the project manager vis-à-vis these components. Many of the concerns, actions, organizations, and management tools and techniques used in a “typical” DoD project manager's environment are applicable to the SA environment as well, such as program control, funds management, and network analysis. Moreover, the acquisition rules for a “normal” DoD program (such as DoD 5000 series documents, Federal Acquisition Regulation clauses, military service acquisition regulations) apply to the SA Program as well. As you will see in the second part to this chapter, we have concentrated on those features unique to the SA environment with special emphasis on FMS.

PURPOSE

The SA Program is advantageous to the United States in that it allows for the maintenance of internal order in the receiving countries, and generally increases regional stability in areas of the world critical to U.S. foreign-policy objectives. This stability may reduce the need and potential for direct U.S. military involvement. The SA Program enhances the standardization of materiel, doctrine, and training among U.S. allies and maintains cooperation and closer relations with recipients through arms transfers, resupply, logistics, and training. The successful implementation of Desert Shield and Desert Storm represents the value of the SA Program to the United States and its friends and allies. Among the numerous elements contributing to the success of the allied effort, standardization and interoperability of command, control, communications, weapon systems, strategy, and tactics rank high. Other benefits gained include maintaining our arms production base, which can be converted to U.S. use in an emergency. It is often asserted that other benefits of the SA Program include increasing U.S. employment, assisting in resolving the balance of payments problem, reducing unit costs of equipment to the U.S. military services through larger production runs, and partially recovering research and development costs. In addition, by assuring friendly and allied nations a dependable source of conventional arms for their security needs, the requirement to build their own armament industries or to develop their own nuclear capabilities is removed.[2]

Security assistance is, however, "... far more than an economic occurrence, a military relationship, or an arms control challenge—arms sales are foreign policy writ large" (Pierre,[3] p. 3).

Security assistance must be viewed in essentially political terms, that is, the relationship created by the transfer of arms between the supplier and the recipient. The world is in a state of flux where political, economic, and military power is being diffused from the industrial nations to the developing nations of the Third World (Pierre,[3] pp. 3–4):

> Arms are a major contributing factor to the emergence of regional powers such as Israel, Brazil, South Africa, or until recently, Iran; their purchase makes a deep impact upon regional balances and local stability. The diffusion of defense capabilities contributes at the same time to the erosion of the early post-war system of imperial or hegemonic roles played by the major powers around the globe. Thus the superpowers, and even the medium-sized powers such as Britain and France, are losing the ability to "control" or influence events in those former colonies or zones of special influence.

The various components of the SA Program assist this nation's efforts to meet the international challenges facing it in a variety of ways. In general, this element of our foreign policy is designed to[4]:

1. Bolster the military capabilities of our friends and allies, permitting them in some cases to undertake responsibilities which otherwise we ourselves might have to assume.
2. Contribute to the broad cooperative relationships we have established with many nations which permit either U.S. facilities on their territory or access by U.S. forces to their facilities in time of threat to mutual interests. (*Note:* Of the many facets contributing to the success of the allies in the 1990–1991 Gulf crisis, one of paramount importance was the agreement of the Saudi Arabian government to allow foreign military forces to use its territory as a staging and supply point against Iraqi forces in Kuwait.)
3. Help our friends and allies provide for their own defense and furnish tangible evidence of our support for their independence and territorial integrity, thus deterring possible aggression.
4. Provide a means of demonstrating U.S. constancy and willingness to stay the course in support of nations whose continued survival constitutes a basic purpose of our foreign policy.
5. Help alleviate the economic and social causes of instability and conflict, particularly in countries whose necessary military expenditures would otherwise impose severe strains on their economies.

Many countries receiving security assistance are situated in a geographic proximity to the natural resources necessary for our economic continuance. Regional allies possess knowledge of events of an historical, cultural, or socioeconomic nature beyond our appreciation, and can best understand these events and use their knowledge to influence these events to remain within the parameters of political control. Finally, many of these states have military and social development forces trained and experienced in operating in these unique geographic regions[5]:

> As we strengthen these states, we strengthen ourselves and we do so more effectively and at less cost. Friendly states can help to deter threats before they escalate into world-shaking crises. The issue is not whether a local state can single-handedly resist a Soviet

assault. Rather, it is whether it can make that assault more costly, more complicated, and, therefore, potentially less likely to occur.

[*Note:* The allied experience in Desert Shield and Desert Storm reemphasized the importance of having defense partners in critical areas of the world capable of providing the infrastructure (such as airfields, port facilities, command, control and communications, interoperable weapon systems and tactics) necessary to delay or halt an aggressor's assault.]

One important argument for security assistance is the influence the supplier has in its relationship with the recipient. Assistance can become a symbol of support and friendly relations resulting in influence. "Arguments for the sale of weapons to China have been based not so much on the need to enhance its military capabilities against the [former] Soviet Union, for the Chinese will remain comparatively weak under any circumstances, as to demonstrate American friendship and further the normalization of relations" (Pierre,[3] p. 14). The same can be said for continued sale of weapon systems and support to Saudi Arabia now that Desert Storm has concluded (Pierre,[3] p. 15):

> Arms may provide access to political and military elites. This has been the traditional justification for many of the U.S. military assistance programs to Latin American nations, where often there was no serious military threats or needs for arms. The continuing contacts between defense establishments, which accompany arms transfers through training missions and the sending of Latin American military officers to U.S. military schools, is thought to be important because of the political role played by the military on the continent. . . .

Perhaps the strongest political benefit for the United States in its application of security assistance is the leverage gained over the recipient countries' foreign policy decisions. The sale of F-15 aircraft to Israel was used as a bargaining point in convincing the Israeli leadership to accept the 1975 Sinai disengagement agreement. Further, "the Carter administration's decision in 1978 to sell F-5E fighters to Egypt was strongly influenced by the need to buoy up Anwar Sadat in order to dissuade him from breaking off the peace negotiations after the initiative he had launched seemed to be going nowhere" (Pierre,[3] p. 14). More recently, U.S. deployment of Patriot missiles to protect Israeli cities from Iraqi SCUD missile attacks during Desert Storm helped preclude the Israelis from attacking Iraq and thus threatening the alliance among the Arab states.

The most traditional argument for security assistance is to help meet the security needs of allied and friendly nations. Following World War II and continuing until the mid-1970s, most U.S. security assistance took the form of grant aid. Meeting the security needs of allies was the principal reason for providing assistance to NATO, Japan, South Korea, and South Vietnam. The Nixon doctrine emphasized the need for security assistance to build up local indigenous forces as a replacement for U.S. military personnel (Pierre,[3] p. 19).

Other strategic concerns affect our decisions to offer security assistance to various nations and international organizations. These include the authority to establish a military presence in the recipient country (Pierre,[3] p. 21):

> Following World War II, as the U.S. came to rely upon a global network of overseas bases for its bombers and for monitoring the Soviet Union, arms were often transferred as a *quid pro quo* for the availability of bases. The U.S. received base rights in Pakistan, Ethiopia, and Libya, and naval facilities in Spain and the Philippines with such understanding. More recently, after the Soviets moved into Afghanistan, Washington promised arms to Oman, Somalia, and Kenya in exchange for access to bases. . . .

Another advantage for the United States, popular with the DoD, is the argument that U.S. developed weapon systems are more likely to be tested in combat by the recipient than by the United States. The success for U.S. supplied antitank weapons and fighter aircraft used by Israel in the Yom Kippur War and lessons learned from that encounter "...had a profound impact upon military planners thereafter" (Pierre,[3] p. 22).

Likewise, the striking superiority of U.S. technology over other countries' technology during the Gulf crisis has altered military thinking in many client states previously predisposed to purchase weapon systems from France, Britain, Brazil, or other suppliers.

A final argument for security assistance is the economic advantage arms transfers are asserted to bring. Arms sales are said to help in the balance of trade problems and earn foreign exchange for the U.S. Treasury (Pierre,[3] p. 24):

> Arms sales are also thought to provide significant employment in the defense industries of the producers. In addition, the export of arms is seen as an excellent way to create economics of scale, thereby reducing the per-unit costs of arms to be manufactured for the armed forces of the producer country. Exports are also a way of spreading out, or recouping, some of the research and development expenses.

In reality, however, during the 1980s arms exports amounted to only 4 or 5 percent of total U.S. exports. In 1988, the latest figure available, arms exports represented 4.4 percent of total U.S. exports.[6] The Defense Security Assistance Agency estimates that an annual FMS program of $10.5 billion generates between 265,000 and 317,000 worker-years of direct employment. The advantages are not spread evenly across the states, but are concentrated in those states with large defense industries.[7] According to Pierre[3] (p. 21):

> As to the savings generated through enlarged production runs, and recoupment of research and development expenses, a Congressional Budget Office study has estimated that arms exports of $8 billion per year produce savings of only $560 million.... A study prepared by the Department of Defense found that "there is only a loose relationship between production readiness and cost economics on the one hand, and the total dollar volume of transfers on the other."

It appears then that the political and strategic advantages for the United States vis-à-vis the SA Program assume more importance than the economic gains. However, economic benefits cannot be totally ignored.

When judging the economic and political value of these programs, one must recognize the connection between today's assistance from us to our allies and tomorrow's needs provided by our allies to us. For in the final analysis, a state's foreign policy is neither moral nor immoral; it is amoral. Policy must be designed to achieve objectives deemed in the interest of the nation and the people it was created to serve. Political realities in today's multipolar world and U.S. interests foreclose the extreme options for the SA Program, particularly arms transfers, which some proponents purport. A total embargo on arms exports and an unrestricted policy allowing arms sales to any country are extreme options which are neither feasible nor expedient. Since the end of the Gulf crisis, there has been renewed interest in banning U.S. transfer of major defense equipment to the Middle East with a view to preclude "undermining regional stability." This measure was passed by the House of Representatives on July 10, 1991 and sent to the Senate for action.[8]

A variety of other approaches to the transfer of arms exports is, instead, available.

1. *Weapons Category Restrictions.* Numerous factors, such as the cost of the weapon, research and development costs, the most probable use of the weapon (offensive or defensive, tactical or strategic), and the sophistication of the system, can be used to control arms transfers.

2. *Sophistication.* Weapons can be transferred on the basis of the highest level of sophistication that the recipient nation can operate and maintain. We can transfer items only to a level of sophistication already introduced into a given region.

3. *Costs.* Arms sales can improve the U.S. balance of payments posture, reduce unit costs of defense equipment, and improve defense production capabilities and efficiency by selling large numbers of expensive items. The defense needs and the economic conditions of recipient countries can be considered, and the United States can encourage these allies to procure less expensive systems. The United States can prohibit sales of expensive defense equipment to developing nations that cannot afford such expenditures.

4. *Country Category Restrictions.* Transfer decisions can be made on the basis of whether a proposed client is an ally or an adversary, or whether it is threatened by others or is threatening to others.

5. *Linkage.* Arms transfers can be linked to other U.S. foreign-policy objectives: the international actions or the domestic practices of the client state.

6. *Ceilings.* A ceiling on the total volume of arms transfers, in terms of either dollar value or number of items or systems, can be established. A ceiling can also be placed on transfers to specific countries or regions.

7. *Transfers with Conditions.* The United States can establish conditions for future transfers predicated on the use of the equipment by an ally or on the client's restraint of arms purchases and use. If necessary, transfers could be approved only if U.S. personnel monitor the use of certain weapons. By withholding the technology necessary to produce or repair weapons and by withholding a sufficient supply of spare parts, ammunition, or some operational component, the United States could maintain some control of the weapon system after it arrives in a country.[9]

A well-balanced arms transfer policy must use several of these approaches to maintain a dynamic role in U.S. foreign policy. No single approach will suffice for all the contingencies challenging this nation. Our perception of arms transfers must not be single-dimensional, that is, it should not be viewed in purely military terms. The SA Program is a political barometer throughout the international community. Its most significant impact is not in its military nor economic influence worldwide; its impact is most significant in its political effect of this nation upon another. Arms transfers affect the political relationship between the supplying nation and the client state. This bilateral impact may well affect the political alignment of other countries within the same geographical area. It can, and often does, restructure the global alignment of nations and changes their international positions and strategies (Pierre,[3] pp. 14–19).

The global strategy of the United States during the latter part of this decade is based on seven broad policy goals:

1. Promotion of peace in the Middle East
2. Enhancement of cooperative defense and security
3. Deterrence and combat of aggression
4. Promotion of regional stability
5. Promotion of key interests through FMS cash sales and commercial military exports

6. Promotion of democratic values
7. Countering transnational dangers, such as environmental degradation, narcotics trafficking, and terrorism

These goals, individually and collectively, support the United States' broad strategy to assist friendly and allied nations to secure the means to defend themselves while complementing the rebuilding of U.S. military strength and increasing the human and material resources available for the defense of our interests (Ref. 10, pp. 3–7).

First, the Middle East continues to be of vital interest to the United States both strategically and economically. Seventy percent of the noncommunist world's known oil reserves comes from this area. Its location, astride the sea lanes linking Europe and Asia, has had profound influence on the western world's historical development. The conflict besetting the Middle East today threatens the economic and political well-being of nations far removed from the region (Ref. 10, pp. 3–7).

Consequently, peace between Israel and its Arab neighbors still remains the primary goal of U.S. foreign policy in the Middle East. To be realistic, this goal requires (Ref. 11, p. 14):

1. The recognized legitimacy of Israel and the maintenance of its security.
2. The self-government of West Bank Palestinians in association with Jordan.
3. Direct negotiations between the concerned parties if any possible settlement is to be implemented.

U.S. arms transfers to the region are intended to enhance our friends "... self-defense capabilities, facilitating political support of the peace process" while at the same time laying the "ground work for broader political and security cooperation with Israel, Egypt, Jordan, and other responsible states in the area." (Ref. 11, p. 14.)

Second, the United States depends on the cooperation of many nations to contribute to the concept of coalition defense. The SA Program has as the pivotal point in its basic conceptual design, the strengthening of recipient nations economically and militarily to ensure full partnership capabilities as much as possible (Ref. 11, p. 14). The SA Program builds the necessary confidence that friends and allies and the United States can rely on one another when a crisis arises by increasing the resources and capabilities available for cooperative efforts. The recent allied coalition in the Gulf crisis serves to emphasize this goal (Ref. 10, p. 5):

> Through FMS and commercial sales, the United States has built strong security relationships with friendly countries in the Persian Gulf as well as other countries in the United Nations coalition participating in Operation Desert Storm. Moreover, many allied military personnel have experienced U.S. training under FMS or the IMET program, thereby enhancing compatibility in language, military doctrine, and technical proficiency. The fact that these countries have built inventories of U.S. equipment with the accompanying training has greatly eased the difficulties faced by our forces in fighting as part of the multinational coalition. Due in part to the security assistance program, the oft-cited goal of interoperability has been made a reality in Operation Desert Storm.

The third goal of security assistance is to deter aggression. Since World War II, a basic challenge to U.S. interests has been "externally supported aggression and subversion" (Ref. 11, p. 15). This challenge has been met through the SA Program with adequate military and economic assistance to deter and combat such aggression. U.S. security assistance is designed to eliminate "inequities that can be exploited by external forces ..." and "... promote the political and economic reforms necessary for the safeguarding of internationally recognized human rights for the development of viable democratic institutions and for economic and social progress" (Ref. 11, p. 16).

Fourth, even without having to confront and defend themselves against externally supported aggression and subversion, many nations in the Third World face endemic poverty and a lack of political and economic opportunities which are root causes of both national and regional instability. Such instability is a threat to "... the orderly conduct of international, political, and economic relations and thus to U.S. interests." The economic support fund set up under the SA Program assists recipients "... to alleviate systemic causes of poverty to promote economic development.... Directed toward demonstrating that static economic policies are not only ineffective but harmful, these programs encourage structural economic reform, diversification, individual enterprise, improved productivity, and the sustained growth of recipient economics" (Ref. 11, p. 16).

Fifth, many nations purchase defense articles and services through the FMS program or through direct commercial transactions with U.S. contractors. As is true with USG funded security assistance transfers, these FMS and commercial transactions are subject to U.S. foreign-policy decisions (Ref. 11, p. 17). These types of sales make possible improvements in the defense capabilities of NATO partners, Australia and Japan especially. "They contribute to the military and economic strength and political cohesiveness of the free world" (Ref. 11, p. 18).

The sixth broad policy goal is to promote professional military relationships through grant training. To date more than 500,000 foreign officers and enlisted [personnel] have benefitted from the IMET program since 1950 (Ref. 10, p. 19). IMET takes place in the United States and overseas at military training facilities and at schools and research institutions.

The IMET program encourages mutually beneficial relations between U.S. military services and those of friends and allies. It increases the self-reliance of participating countries and improves their ability to utilize their resources effectively, including defense equipment and services obtained from the United States. Equally important, the IMET program promotes broad mutual understanding and the awareness of foreign nationals of the basic issues which drive U.S. policy, especially the concern with internationally recognized human rights (Ref. 11, p. 16). As of fiscal year 91 IMET training was expanded to include training for international military personnel and civilians in (Ref. 10, p. 19)

> ... managing and administering military establishments and budgets, and in creating and maintaining effective military judicial systems and military codes of conduct...
>
> A key part of this expanded IMET training will consist of training of foreign military and civilian government officials (including civilian personnel from ministeries other than defense) in order to: contribute to responsible defense management; foster greater respect for and understanding of the principle of civilian control of the military; and improve military justice systems and procedures in accordance with internationally recognized human rights.

The seventh and final broad goal is to counter transnational dangers such as narcotics trafficking and terrorism by using local military forces equipped and trained under the SA Program as reconnaissance and interdiction units.

The SA Program is one diplomatic tool available to us which can help establish the infrastructure necessary to foster the initiatives of others. Selective application of the various components of the SA Program allows the flexibility essential to our foreign policy. Its successful implementation has occurred in part because of its integration into the mainstream of DoD and State Department policies and procedural processes.

SECURITY ASSISTANCE PROCESS WITHIN DOD

Program Planning for Foreign Military Financing Program (FMFP)

The FMS components of security assistance initially involve many government agencies beyond the military services. In planning for security assistance to be funded from the FMFP, the same budgetary process found in the program objective memorandum is followed. Figure 9.1 shows some of the key players in this process.

The State Department and the DoD require input from various sources prior to preparing a request to Congress for funding each year. The funding request and the detailed justification to support the proposed programs are prepared annually in the congressional presentation document (CPD). This document accompanies the Executive Branch's annual proposed authorization and appropriations legislation. The information used in the CPD is a compilation of planning information from the U.S. embassies worldwide, matched with an analysis of current national policy goals. The annual security assistance funding cycle is completed with legislative enactment of the authorization and appropriations bills (Samelson,[1] pp. 102–105).

Project Management in Foreign Military Sales (FMS)

Project management in the FMS arena of security assistance incorporates all the tools of project management in a military environment with the international and political elements involved with dealing with a foreign purchaser. The FMS project manager in DoD is able to tap into the same resources that are available to all military project managers, along with special assistance to incorporate unique FMS requirements. One of the major attractions of FMS to foreign customers is the knowledge that DoD resources and expertise will be used to support their requirements.

A letter of offer and acceptance (LOA) is the primary document used as authority and guidance to process foreign sales in DoD. The LOA is a contractual sales agreement between the United States and an eligible foreign country or international organization. Each LOA will have a unique identification code, an assigned

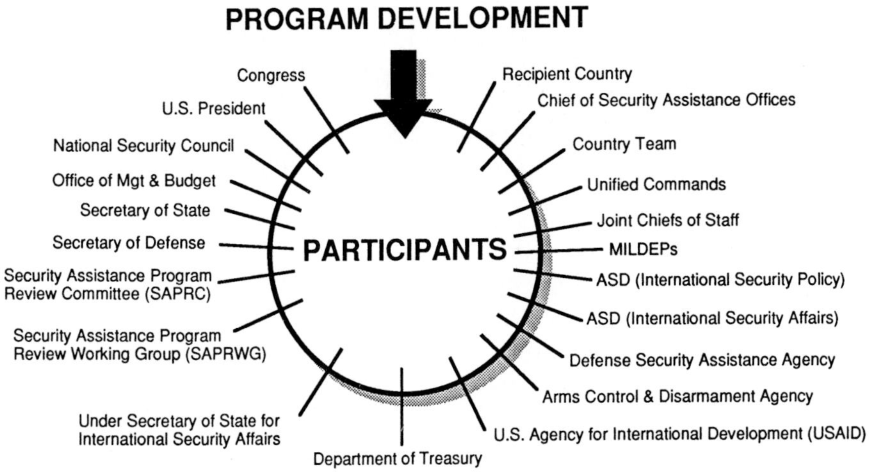

PROGRAM DEVELOPMENT

PARTICIPANTS

Congress
U.S. President
National Security Council
Office of Mgt & Budget
Secretary of State
Secretary of Defense
Security Assistance Program Review Committee (SAPRC)
Security Assistance Program Review Working Group (SAPRWG)
Under Secretary of State for International Security Affairs
Department of Treasury
U.S. Agency for International Development (USAID)
Arms Control & Disarmament Agency
Defense Security Assistance Agency
ASD (International Security Affairs)
ASD (International Security Policy)
MILDEPs
Joint Chiefs of Staff
Unified Commands
Country Team
Chief of Security Assistance Offices
Recipient Country

FIGURE 9.1 Program development.

case designator (which will indicate the country, U.S. military service responsible for the transfer, and type of equipment or service provided) and will be used on all subsequent documentation related to the sale. An FMS case, as defined by the case designator, will have a designated case manager to manage the project from cradle to grave. According to the *Security Assistance Management Manual* (SAMM), the case manager is assigned by the military service as "that individual who is designated to accomplish the task of integrating functional inter- and intra-organizational efforts directed towards the successful performance of an FMS Case" (Ref. 12, p. 704-1).

The SAMM further indicates that "the case manager should have the authority to take actions to task inter- and intra-organizational areas relating to financial, logistics, procurement, and administration matters in the day-to-day operation of a case." The following is a list of specific management aspects delineated in the SAMM under the responsibilities of the case manager (Ref. 12, p. 704-2):

a. Establish initial and long-range goals and objectives for case execution.
b. Prepare a case master plan.
c. Develop a financial and logistics management plan.
d. Approve plans of execution, scope and schedule of work.
e. Review and verify funding/program requirements.
f. Integrate the program and logistics financial plan with the execution of the case.
g. Initiate requirements.
h. Validate that costs are accurate and billed.
i. Respond to requirements of counterpart managers, functional activities, and other supporting agencies in the resolution of interface or operating problems.
j. Initiate, when necessary, working agreements with supporting activities when appropriate.
k. Analyze case performance in relation to required performance specifications.
l. Maintain a complete chronological history (significant events and decisions).
m. Provide status, progress, and forecast reports.
n. Develop and execute a case closure plan.
o. Ensure DIFS [Defense Integrated Financial System] and DoD Component case records are in agreement.
p. Ensure that records are retained in accordance with DoD 7290. 3-M [*Foreign Military Sales Financial Management Manual*] and this manual.
q. Ensure that schedules are timely and accurate.

The FMS case manager is a key player in the overall project management of an FMS sale. Depending on the size and intensity of an anticipated foreign military sale, the case manager will work closely with a program manager from the primary office responsible for the weapon system to be sold to accomplish the project management objectives.

THE INITIAL REVIEW PROCESS

The FMS program begins in the foreign country when, after careful evaluation of its military requirements and an analysis of the marketplace, a conscious decision is made to buy a military article or service from the United States. The country then has the option of going directly to a contractor for that purchase, or coming through the FMS system managed by DoD. If a purchaser decides to buy from a U.S. defense contractor without using the DoD FMS system, appropriate export licenses must be obtained from the State Department, Center for Defense Trade, Bureau of Politico-Military Affairs. The State Department is responsible for setting the policy guidelines for commercial defense trade in accordance with the International Traffic in Arms Regulation (Ref. 10, pp. 43–44).

Key political decisions which will impact the execution of the FMS program are made as soon as a foreign country decides to make a purchase through the DoD FMS system. The review process involves close scrutiny of the proposed transfer of military articles or services to a foreign country to ensure that such transfer is in the national best interest of the United States and conforms to our laws and policy directions. While the project manager may not be actively involved in the initial review process, the decisions that are made in these preliminary stages will impact on project execution and decision making during the implementation of the program.

An official letter of request (LOR) from a foreign purchaser must be reviewed by several offices before the sale can be approved. The level and intensity of the reviews depend on the type of equipment being sold and the dollar value, along with the regional political issues for any specific country. Routine military sales of follow-on support items seldom require more than a military department cursory review to ensure a country is eligible and has not been suspended for any political or economic reason.

Sales for significant military equipment, as defined in the International Traffic in Arms Regulation as items with significant military utility, and major defense equipment, as defined by the level of investment [research, development, test, and evaluation (RDT&E) of more than $50 million or total production exceeding $200 million], require an intensive and time-consuming review process. The reviews involve State Department concerns along with the DoD issues. Before a DoD project gets the "go ahead," the State Department must concur. In addition to the State Department and DoD reviews, FMS cases which reach a certain dollar threshold or which are politically sensitive are also submitted for congressional review.*

The first review of an LOR for significant military equipment comes through the U.S. embassy of the country of origin. The ambassador is required to look at several areas of concern as listed in the SAMM (Ref. 12, p. 700-6):

1. The reason the nation desires the article and/or services.
2. How the items will affect the recipient's force structure and how it would affect the recipient's capability to contribute to mutual defense or security goals.
3. The anticipated reaction of neighboring nations.
4. The ability of the purchaser to operate, maintain and support the article. Training required either in country or in the U.S. and the impact of any in country U.S. presence that might be required as a result of providing the article.
5. The source of financing and the economic impact of the proposed acquisition.
6. Relevant human rights considerations that might bear on the proposed acquisition.
7. Whether the U.S. Government should approve transfer of the article and the reasons therefore.

Although the DoD project manager may not be involved in the decision-making process that goes on at the State Department and embassy levels, it is important to

*Section 36(b)(1) of the Arms Export Control Act (AECA) requires congressional review of any LOA over $50 million total value of articles and services; $200 million design and construction services; or $14 million major defense equipment. Congress must adopt a joint resolution within 30 days opposing the sale, otherwise it will proceed. In addition, DSAA has agreed to provide 20 days advance notification, prior to the formal submission, in order to allow Congress sufficient time to review the information.

be aware of the evaluations and analyses submitted by the different agencies that may impact project management decisions further down the line.

The DoD may get involved with the proposed FMS sale at several different levels. Initially the military staff at the embassy in the country may have to contribute to the ambassador's comments. In addition, the security assistance officer (SAO) located in the country may also have inputs to the initial request evaluation. At the same time the commander-in-chief of the regional unified command may be looking at the request from a regional military perspective. Within the United States, the principal office within DoD for receiving, processing, and doing initial evaluation of FMS letters of request is the Defense Security Assistance Agency (DSAA).

DSAA has been designated as the agency that performs administrative management, program planning, and operations functions for the U.S. security assistance programs at the DoD level under the policy direction of the Assistant Secretary of Defense, International Security Affairs. In this capacity, DSAA acts as the focal point for the execution of FMS programs, even though it may direct one of the military services to actually "take the lead" to fulfill the requirements of any specific request. If congressional notification is required, DSAA will act as the centralized location to gather data from the proponent military service and prepare the congressional presentation material.

Concurrent with the State Department and congressional reviews, the proponent military department will begin processing a formal LOA. The primary offices for processing the initial request for major systems within the military services are:

Secretary of the Air Force, International Affairs (SAF/IA)

Navy International Programs Office (Navy IPO)

United States Army Security Assistance Command (USASAC)

The military service having primary responsibility for a piece of equipment within DoD will be assigned responsibility for the FMS case. For example, a request for aircraft will go to the Air Force; tanks to the Army; and ships to the Navy. Each service will then assign a program manager to a project who will be responsible for fulfilling the customer's requirements. The program manager will work closely with the case manager to ensure timely delivery of the customer's requirements. The first task of the program manager is to develop a milestone plan that will lead up to program implementation. The initial major milestones include collection of pricing and leadtime information from various sources, identification and integration of country-specific requirements, and preparation and coordination of the LOA.

LETTER OF OFFER AND ACCEPTANCE (LOA) DEVELOPMENT

FMS project management begins at the service level when the preliminary review process has been completed and it is time to finalize the official letter of offer. The LOA will become part of the congressional package if congressional review is required and will not be sent to the country until the congressional review has been completed. The program must pull together all the major elements of a DoD purchase and integrate the specific country requirements into a cohesive program to be offered to a foreign customer on the LOA.

It is the official policy of DoD to propose a total package sale to the foreign purchaser. This means that when a major weapon system is sold, all the elements of an integrated logistics support plan with all the appropriately trained personnel are in place to operate and maintain that system. The key to having such a program is to have a properly developed and detailed LOA.

The LOA is a country-to-country legal agreement which gives the United States the military authority to fulfill a program on behalf of a foreign government or organization and, as such, acts as the logistics management and funding authority on any given program. Once a foreign purchaser signs the LOA, the military program manager in DoD has in effect been given the authority to proceed to execute the program as spelled out, and use the funds that are provided on the LOA. A sample LOA is given in Appendix A. The terms and conditions on the LOA provide additional conditions of sale and expand on the requirements of the AECA governing all military sales.

The lack of a complete and well-written LOA will cause delays and confusion during program implementation. Anytime it is necessary to go back to a country for a decision on something that was not included in the LOA, an automatic delay in the program will be caused, especially if additional funding is required. A foreign government may have to staff any changes in the program through different levels for approval. An LOA for a major weapon system should contain all the elements for an integrated logistics support plan. Introducing and activating a new major weapon system in a foreign country requires a complete analysis of the country's capabilities in order to have an operational system. Along with the obvious requirement for identification of the end item and the delivery schedule, it is also necessary to identify the operational, maintenance, and supply concepts in order to determine the levels of spares and support equipment required, in addition to making a determination on the manpower and training requirements and the facilities that are available or have to be built. In other words, all the major considerations taken into account when activating a new weapon system in the DoD must be applied to the foreign country. Checklists for major weapon systems are included in the service regulations pertaining to FMS. Appendix B gives a sample checklist from Air Force Regulation 130-1 to be used for an aircraft system sale.

If the required information is not provided in the LOR, and the details are not readily available, a site survey may be necessary in order to consolidate the data. The program manager can put together a survey team of experts to go in-country and analyze the facilities and requirements of a country in order to put together a comprehensive LOA. The site survey report would take into account all the elements required to put together a total package sale, along with any specialized requirements to support the country. All data acquired during the site survey should be represented in the LOA.

In some countries unique requirements may include transportation pipelines for maintenance support, manpower strength available to meet training requirements, and a facilities review to determine construction requirements. If the purchasing nation has agreed to perform part of the necessary functions (such as construction or upgrade of certain facilities), those requirements will have to be built into the master plan and should be clearly annotated on the LOA.

In projecting milestones and planning a program, the program manager must be careful to include sufficient administrative lead time to process a comprehensive LOA and get it signed by the appropriate authorities in-country. The normal lead time for country review and signature is 60 days; however, some countries may require additional time to process the LOA. Frequently a country will request an LOA presentation which involves the program manager and key members of the LOA development team to travel in the country and give a briefing on the program as outlined on the LOA.

LOA development for a major program may be a time-consuming process, but timely execution of the program may depend on how well the LOA is written. Figure 9.2 is a sample time line for an LOA which shows the key participants.

In the development of a total package sale, the program manager must make a projection of the dedicated manpower (U.S. military and civilian) that may be required to implement the program. The program manager must also project the expenses for the program team to attend regularly scheduled meetings along with unanticipated trips that may be necessary to ensure program integration. These costs are included in a line on the LOA to be used for funding throughout the life of the program.

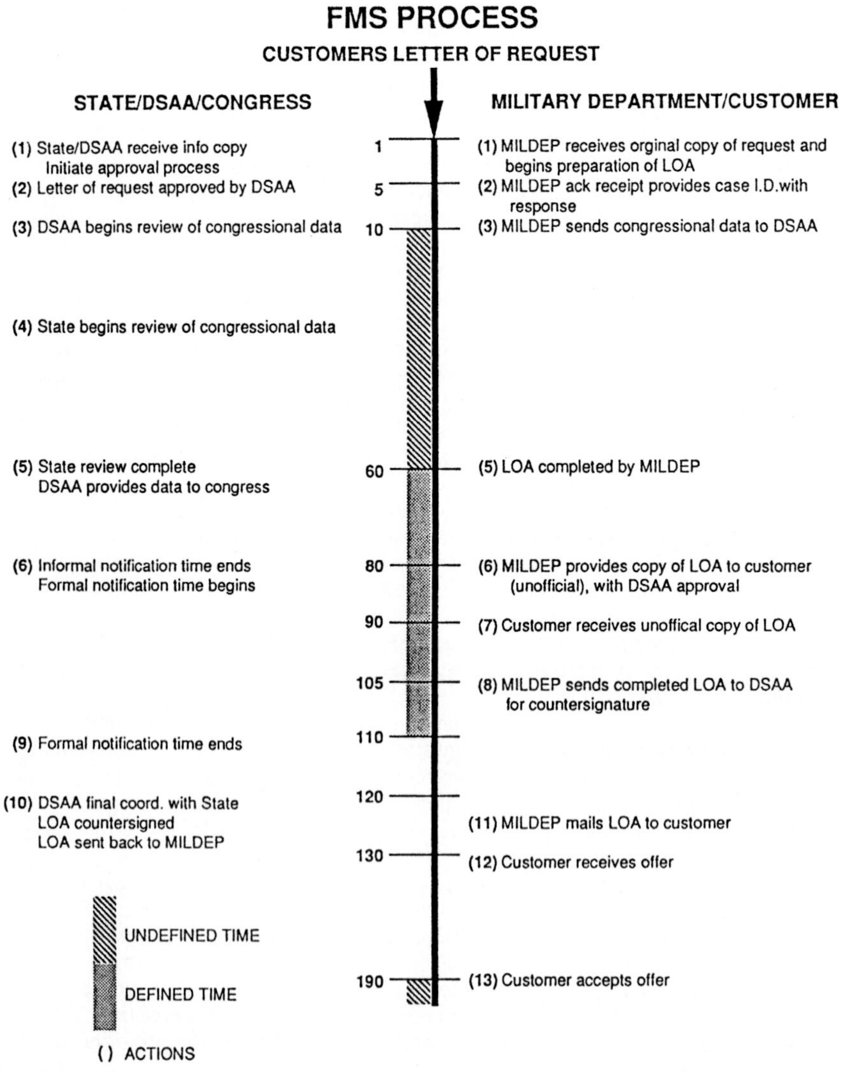

FMS PROCESS

CUSTOMERS LETTER OF REQUEST

STATE/DSAA/CONGRESS		MILITARY DEPARTMENT/CUSTOMER
(1) State/DSAA receive info copy Initiate approval process	1	(1) MILDEP receives orginal copy of request and begins preparation of LOA
(2) Letter of request approved by DSAA	5	(2) MILDEP ack receipt provides case I.D.with response
(3) DSAA begins review of congressional data	10	(3) MILDEP sends congressional data to DSAA
(4) State begins review of congressional data		
(5) State review complete DSAA provides data to congress	60	(5) LOA completed by MILDEP
(6) Informal notification time ends Formal notification time begins	80	(6) MILDEP provides copy of LOA to customer (unofficial), with DSAA approval
	90	(7) Customer receives unoffical copy of LOA
	105	(8) MILDEP sends completed LOA to DSAA for countersignature
(9) Formal notification time ends	110	
(10) DSAA final coord. with State LOA countersigned LOA sent back to MILDEP	120	(11) MILDEP mails LOA to customer
	130	(12) Customer receives offer
	190	(13) Customer accepts offer

UNDEFINED TIME

DEFINED TIME

( ) ACTIONS

FIGURE 9.2 FMS process. (*Source: The Management of Security Assistance, 11th ed., April 1991.*)

During LOA development it is important for the program and case managers to work closely with the prime contractor in order to develop a cohesive program. The contractor support team will be key performers in the implementation of the program. The degree of support to be expected from or provided by the contractor should be planned early in the program so that there are no surprises down the line. Contracted services such as interim contractor support or contractor training are important considerations in putting together the total package.

Another important contact for the program manager during the development phase is the SAO. Depending on the size of the security assistance programs in a country, the SAO may be one person, or a large group of U.S. military personnel located in a foreign country with assigned responsibilities for carrying out the security assistance management functions. The larger SAOs are called by a variety of names such as MILGRP (U.S. military group), JUSMAG (joint military assistance group), OMC (office of military cooperation), or ODC (office of defense cooperation). The personnel in the SAO have access to in-country military data, which are invaluable when developing an FMS program. For example, their assistance before and during the site survey can help identify specific areas for review. In addition, as members of the ambassador's team in the country, they can give insight into the impacts of the sale on not only the country's military position, but the political and economic situation as well. This puts the SAO in the enviable position of knowing the benefits and obstacles to the sale along with the expectations of the foreign purchaser.

The officers assigned to the SAO may already have worked with the prime contractor when the system was being marketed in the country. In a July 10, 1990, message then acting Secretary of State Lawrence S. Eagleburger sent a cable to all U.S. ambassadors worldwide which emphasized the role of the embassy in support of U.S. defense trade. In it he wrote: "It is the policy of the United States that our diplomatic posts abroad should support the marketing efforts of U.S. companies in the defense trade arena as in all other spheres of commercial activity."

Through experience and personal contacts, the SAO is also aware of the country's logistics system capabilities. The program manager can work with the SAO in developing support for areas of weakness that may impact program execution. Once the program begins, the SAO will serve as the local representative of the program manager to assist in evaluating how the program is going, and in providing quick fixes as needed.

PROGRAM EXECUTION

After the LOA has been signed by the country, and the required funds to cover the initial deposit as specified in the LOA have been deposited with the Defense Finance and Accounting Service, Denver Center, the program manager has the authority to proceed with program execution. A definitization conference is usually necessary to define the requirements not delineated on the LOA such as spares, publications, and training requirements. The milestones developed for the LOA must now be expanded into a detailed program management plan (PMP), which provides information on how the program elements will be satisfied, who the key players are, and the specific time frames for the completion of the actions involved in the program. The PMP should also include any contractor information that is relevant to program execution and incorporate all elements of an integrated logistics support plan.

At this point the program manager begins to function as an integrator of the DoD, contractor, and country assets and requirements to bring a program to fruition.

A typical FMS major system sale may take 48 to 60 months to deliver the end items. During the years prior to delivery, the program manager must oversee contract negotiations to ensure delivery time frames. At the same time it is necessary to integrate all elements of the program: monitor training; order, track, and deliver spares and support equipment; make arrangements to order publications and technical orders that will be current at the time of delivery; and monitor configuration changes. All of the efforts for a major sale will require the program manager to have a team—commonly identified as the management action team—comprised of individual specialists at different locations which meets regularly to evaluate the status of the program. A key member of the team is the financial advisor who will monitor the funds.

Financial management of an FMS program is very different from a DoD program. The availability of funds for an FMS program is controlled by the financial annex of the LOA and must be balanced by line item. As a basic rule, U.S government funds cannot be used to purchase items directly for a foreign purchaser. Therefore country funds must be collected in advance. The authority to obligate funds and their limits are set by the LOA; the LOA acts as a financial management document and as a logistics guide.

The collection of the funds on the LOA, the subsequent billings, and the accounting of those funds are performed for all military services at the Defense Finance and Accounting Service, Denver Center. The funds on the LOA may come from a variety of sources: a country's national funds, a private commercial agreement, U.S. DoD appropriated funds through the FMFP, a third country, or any combination of these. The funds are controlled in a treasury account at the Denver Center and maintain their integrity to a particular country and FMS case.

At the time the LOA is prepared, the financial data are based on the planned implementation of the program, including a quarterly schedule of payments that the foreign purchaser must make. The payments are scheduled so as to collect funds prior to anticipated expenditures. Periodic financial reviews need to be accomplished in order to make sure that the funds are being collected properly. As soon as the contract for the major item has been finalized and fully negotiated and accepted, a detailed financial assessment should be accomplished to reevaluate the funding requirements. It is essential that the program manager oversee all financial operations and be aware of the financial status of the program at all times. At the same time there must also be an awareness of the fiscal policies of the country purchasing the items and any legal restraints that may impact program execution. As the deliveries are performed and the bills paid, the financial documents must be in balance and a final bill issued to the customer.

Another key element in FMS program management is configuration management. If a customer requests a configuration that is not the standard DoD configuration, special efforts will have to be made to incorporate the unique features. Detailed integration plans may require specialized attention. In an ideal program, the configuration to be released to a foreign customer is approved during the LOA process and the integration is incorporated into the contract production schedule. In the real world, however, frequent changes in configuration from the customer side, and engineering change proposals within DoD, require constant attention.

It is necessary to monitor any changes in the DoD production configuration in order to arrange for incorporation into the foreign customer's asset. Many times this may mean establishing a retrofit plan and reevaluating support equipment, training, and spares. The number of assets a country is purchasing will affect its decision on whether to retrofit or continue with the previous configuration. Releasability issues may also affect configuration changes.

An important configuration concern, which was highlighted during the Gulf crisis, involves environmental or climatic issues. Special protection against the fine sand in the Arabian desert had to be developed to maintain the operability of the equipment, especially aircraft engines and tank guns. The foreign customer will have concerns over operations in unique climatic conditions and may require special configuration requirements. Some of these requirements may not be readily identified at the onset of a program and will have to be dealt with as they arise.

The composition of the team supporting the program manager and the number of people in it will depend on the level of effort required to support a specific sale. Other key personnel may include a support equipment manager, a munitions monitor, a maintenance oversight team, a facilities construction overseer, engineers, and anyone else required to execute a complete program. Many of the manpower positions required for support of the sale will be funded directly from the program management line on the LOA and therefore will require early identification and justification in the planning stages. The program management line will also fund for travel expenses of U.S. personnel in support of the program. Additional support, on a limited basis, is available from FMS personnel funded through the general administrative fund. Reevaluation of the manpower requirements is an ongoing process and, like any other manpower and travel budget, requires constant attention.

Although the program manager is immediately concerned with putting together an initial sale, one of the key elements of the total package sale is a detailed plan for follow-on support. A country will need a continuing life-of-system support plan in effect that not only covers the life-cycle costs but also the logistics requirements for continuing operations.

As much planning that may go into an FMS program and preparation of an LOA, the program manager may still end up in a reactive mode. Not only do changes occur within DoD that may impact the program as it develops, but there may also be policies and laws in the foreign country that may cause delays and problems.

INTERNATIONAL CONSIDERATIONS

A foreign military sale is by definition an international program, and elements of cross-cultural communication may become vital to a program's success. It is not sufficient for an FMS program manager to merely know the FMS process and its legal restrictions, the logistics systems, the financial flow of FMS funds, the budget process for credit funds, and the dynamic DoD acquisition system. It is also crucial that he or she be aware of these same elements within the foreign country being represented. Frequently a program may have to be tailored to conform to a specific country's legal or fiscal requirements.

It is also imperative that a program manager be intensely aware of the culture of the purchasing country along with the history and geography which contributed to the development of that culture. Communication with foreigners involves more than just language. The total behavioral aspects of nonverbal communication can have a major impact on smooth program implementation. In analyzing the thought patterns of another culture, the program manager can work more effectively with international counterparts in executing a program.

Learning some of the spoken language and social customs is a good beginning. However, to function in the international environment it is equally important to learn the unspoken language. Words are only the beginning of the communication link. They must be received and understood in the same manner as the speaker intended. Verbal communication makes up approximately 10 percent of the pattern; the behavioral nonverbal feedback is what is required to close the link.[13] Since behavior patterns are so varied between cultures, even though the same words are being spoken, communication may be totally blocked.

Cultural sensitivity is a vital component of the FMS project. Cross-cultural communication can make the difference between a program that is on schedule and one that is plagued with delays and cost overruns due to misinterpretation of requirements. FMS project management must include understanding and coping with the cultural difference of an international customer.

SUMMARY

Security assistance is an exciting and dynamic field. The political direction or redirection and impacts of the program on the international arena can be quite frustrating at times, but the rewards of working out unique solutions make the experience exciting. You may be watching the evening news one night and see a report on new administration policy or congressional mandates that will impact your program the following morning. An FMS project manager is more than the title implies; he or she must also be a diplomat and an international relations specialist. The AECA and the policies stated in the SAMM give firm direction to the management of FMS programs while, at the same time, allowing for flexibility for the individual military service to decide the best way to implement its programs.

APPENDIX A

United States of America
Letter of Offer and Acceptance (LOA)

[Case Identifier]

Based on [XXXXXXXXXXX]

Pursuant to the Arms Export Control Act, the Government of the United States (USG) offers to sell to **[the Government of XXXXXXXXXX]** the defense articles or defense services (which may include defense design and construction services) collectively referred to as "items," set forth herein, subject to the provisions, terms, and conditions in this LOA.

This LOA is for **[XXXXXX]**

Estimated Cost: **[$XXXXX]** Initial Deposit: **[$XXXXX]**
Terms of Sale: **[XXXX]**

This offer expires on **[Date]**. Unless a request for extension is made by the Purchaser and granted by the USG, the offer will terminate on the expiration date.

This page through page [#], plus Letter of Offer and Acceptance Standard Terms and Conditions attached, are a part of this LOA.

The undersigned are authorized representatives of their Governments and hereby offer and accept, respectively, this LOA:

_______________	_______	_______________	_______
U.S. Signature	Date	Purchaser Signature	Date

_______________	_______________
Typed Name and Title	Typed Name and Title
_______________	_______________
Implementing Agency	Agency

_______________	_______
DSAA	Date

Information to be provided by the Purchaser:

Mark For Code ________, Freight Forwarder Code________, Purchaser Procuring Agency Code________, Name and Address of the Purchaser's Paying Office

____

____

Explanations for acronyms and codes, and financial information, may be found in attached "Letter of Offer and Acceptance Information."

Items to be Supplied (costs and months for delivery are estimates):

(1) Itm Nbr	(2) Description/Condition	(3) Qty, Unit of Issue	(4) Costs (a) Unit	(4) Costs (b) Total	(5) SC/MOS/ TA Notes	(6) Ofr Rel Cde	(7) Del Trm Cde

(8) Net Estimated Cost
(9) Packing, Crating, and Handling
(10) Administrative Charge
(11) Transportation
(12) Other (specify; e.g., supply support arrangement)
(13) Total Estimated Cost

To assist in fiscal planning, the USG provides the following anticipated costs of this LOA:

ESTIMATED PAYMENT SCHEDULE

Payment Date	Quarterly	Cumulative

LETTER OF OFFER AND ACCEPTANCE STANDARD TERMS AND CONDITIONS

Section

1 Conditions – United States Government (USG) Obligations
2 Conditions – General Purchaser Agreements
3 Indemnification and Assumption of Risks
4 Financial Terms and Conditions
5 Transportation and Discrepancy Provisions
6 Warranties
7 Dispute Resolution

1 Conditions – United States Government (USG) Obligations

1.1 Unless otherwise specified, items will be those which are standard to the US Department of Defense (DoD), without regard to make or model.

1.2 The USG will furnish the items from its stocks and resources, or will procure them under terms and conditions consistent with DoD regulations and procedures. When procuring for the Purchaser, DoD will, in general, employ the same contract clauses, the same contract administration, and the same quality and audit inspection procedures as would be used in procuring for itself, except as otherwise requested by the Purchaser and as agreed to by DoD and set forth in this LOA. Unless the Purchaser has requested, in writing, that a sole source contractor be designated, and this LOA reflects acceptance of such designation by DoD, the Purchaser understands that selection of the contractor source to fill requirements is the responsibility of the USG, which will select the contractor on the same basis used to select contractors for USG requirements. Further, the Purchaser agrees that the US DoD is solely responsible for negotiating the terms and conditions of contracts necessary to fulfill the requirements in this LOA.

1.3 The USG will use its best efforts to provide the items for the dollar amount and within the availability cited.

1.4 Under unusual and compelling circumstances, when the national interest of the US requires, the USG reserves the right to cancel or suspend all or part of this LOA at any time prior to the delivery of defense articles or performance of defense services. The USG shall be responsible for termination costs of its suppliers resulting from cancellation or suspension under this section. Termination by the USG of its contracts with its suppliers, other actions pertaining to such contracts, or cessation of deliveries or performance of defense services is not to be construed as cancellation or suspension of this LOA itself under this section.

1.5 US personnel performing defense services under this LOA will not perform duties of a combatant nature, including duties relating to training and advising that may engage US personnel in combat activities outside the US, in connection with the performance of these defense services.

1.6 The assignment or employment of US personnel for the performance of this LOA by the USG will not take into account race, religion, national origin, or sex.

1.7 Unless otherwise specified, this LOA may be made available for public inspection consistent with the national security of the United States.

2 Conditions – General Purchaser Agreements

2.1 The Purchaser may cancel this LOA or delete items at any time prior to delivery of defense articles or performance of defense services. The Purchaser is responsible for all costs resulting from cancellation under this section.

2.2 The Purchaser agrees, except as may otherwise be mutually agreed in writing, to use the defense articles sold hereunder only:
2.2.1 For purposes specified in any Mutual Defense Assistance Agreement between the USG and the Purchaser;
2.2.2 For purposes specified in any bilateral or regional defense treaty to which the USG and the Purchaser are both parties, if section 2.2.1 is inapplicable; or,
2.2.3 For internal security, individual self-defense, or civic action, if sections 2.2.1 and 2.2.2 are inapplicable.

2.3 The Purchaser will not transfer title to, or possession of, the defense articles, components and associated support material, related training or other defense services (including plans, specifications, or information), or technology furnished under this LOA to anyone who is not an officer, employee, or agent of the Purchaser (excluding transportation agencies), and shall not use or permit their use for purposes other than those authorized, unless the written consent of the USG has first been obtained. The Purchaser will ensure, by all means available to it, respect for proprietary rights in any items and any plans, specifications, or information furnished, whether patented or not. The Purchaser also agrees that the defense articles offered will not be transferred to Cyprus or otherwise used to further the severance or division of Cyprus and recognizes that the US Congress is required to be notified of any substantial evidence that the defense articles sold in this LOA have been used in a manner which is inconsistent with this provision.

2.4 To the extent that items, including plans, designs, specifications, technical data, or information, furnished in connection with this LOA may be classified by the USG for security purposes, the Purchaser certifies that it will maintain a similar classification and employ measures necessary to preserve such security, equivalent to those employed by the USG and commensurate with security agreements between the USG and the Purchaser. If such security agreements do not exist, the Purchaser certifies that classifed items will be provided only to those individuals having an adequate security clearance and a specific need to know in order to carry out the LOA program and that it will promptly and fully inform the USG of any compromise, or possible compromise, of US classified material or information furnished pursuant to this LOA. The Purchaser further certifies that if a US classified item is to be furnished to its contractor pursuant to this LOA: (a) items will be exchanged through official government channels, (b) the specified contractor has been granted a facility security clearance by the Purchaser at a level at least equal to the classification level of the US information involved, (c) all contractor personnel requiring access to such items have been cleared to the appropriate level by the Purchaser, and (d) the Purchaser will assume responsibility for administering security measures while in the contractor's possession. If a commercial transportation agent is to be used for shipment, the Purchaser certifies that such agent has been cleared at the appropriate level for handling classified items. These measures will be maintained throughout the period during which the USG may maintain such classification. The USG will use its best efforts to notify the Purchaser if the classification is changed.

3 Indemnification and Assumption of Risks

3.1 The Purchaser recognizes that the USG will procure and furnish the items described in this LOA on a non-profit basis for the benefit of the Purchaser. The Purchaser therefore undertakes to indemnify and hold the USG, its agents, officers, and employees harmless from any and all loss or liability (whether in tort or in contract) which might arise in connection with this LOA because of:
3.1.1 Injury to or death of personnel of Purchaser or third parties, or
3.1.2 Damage to or destruction of (a) property of DoD furnished to Purchaser or suppliers specifically to implement this LOA, (b) property of Purchaser (including the items ordered by Purchaser pursuant to this LOA, before or after passage of title to Purchaser), or (3) property of third parties, or
3.1.3 Infringement or other violations of intellectual property or technical data rights.

3.2 Subject to express, special contractual warranties obtained for the Purchaser, the Purchaser agrees to relieve the contractors and subcontractors of the USG from liability for, and will assume the risk of, loss or damage to:
3.2.1 Purchaser's property (including items procured pursuant to this LOA, before or after passage of title to Purchaser), and
3.2.2 Property of DoD furnished to suppliers to implement this LOA,
to the same extent that the USG would assume for its property if it were procuring for itself the items being procured.

4 Financial Terms and Conditions

4.1 The prices of items to be procured will be billed at their total cost to the USG. Unless otherwise specified, the cost of items to be procured, availability determination, payment schedule, and delivery projections quoted are estimates based on the best available data. The USG will use its best efforts to advise the Purchaser or its authorized representatives of:
4.1.1 Identifiable cost increases that might result in an overall increase in the estimated costs in excess of ten percent of the total value of this LOA,
4.1.2 Changes in the payment schedule, and

4.1.3 Delays which might significantly affect estimated delivery dates. USG failure to advise of the above will not change the Purchaser's obligation under all subsections of section 4.4.

4.2 The USG will refund any payments received for this LOA which prove to be in excess of the final total cost of delivery and performance and which are not required to cover arrearages on other LOAs of the Purchaser.

4.3 Purchaser failure to make timely payments in the amounts due may result in delays in contract performance by DoD contractors, claims by contractors for increased costs, claims by contractors for termination liability for breach of contract, claims by USG or DoD contractors for storage costs, or termination of contracts by the USG under this or other open Letters of Offer and Acceptance of the Purchaser at the Purchaser's expense.

4.4 The Purchaser agrees:
4.4.1 To pay to the USG the total cost to the USG of the items even if costs exceed the amounts estimated in this LOA.
4.4.2 To make payment(s) by check or wire transfer payable in US dollars to the Treasurer of the United States.
4.4.3 If Terms of Sale specify "Cash with acceptance", to forward with this LOA a check or wire transfer in the full amount shown as the estimated Total cost, and agrees to make additional payment(s) upon notification of cost increase(s) and request(s) for funds to cover such increase(s).
4.4.4 If Terms of Sale specify payment to be "Cash prior to delivery", to pay to the USG such amounts at such times as may be specified by the USG (including initial deposit) in order to meet payment requirements for items to be furnished from the resources of DoD. USG requests for funds may be based on estimated costs to cover forecasted deliveries of items. Payments are required 90 days in advance of the time DoD plans such deliveries or incurs such expenses on behalf of the Purchaser.
4.4.5 If Terms of Sale specify payment by "Dependable undertaking", to pay to the USG such amounts at such times as may be specified by the USG (including initial deposit) in order to meet payments required by contracts under which items are being procured, and any damages and costs that may accrue from termination of contracts by the USG because of Purchaser's cancellation of this LOA. USG requests for funds may be based upon estimated requirements for advance and progress payments to suppliers, estimated termination liability, delivery forecasts, or evidence of constructive delivery, as the case may be. Payments are required 90 days in advance of the time USG makes payments on behalf of the Purchaser.
4.4.6 If Terms of Sale specify "Payment on delivery", that bills may be dated as of the date(s) of delivery of the items, or upon forecasts of the date(s) thereof.
4.4.7 That requests for funds or billings are due and payable in full on presentation or, if a payment date is specified in the request for funds or bill, on the payment date so specified, even if such payment date is not in accord with the estimated payment schedule, if any, contained in this LOA. Without affecting Purchaser's obligation to make such payment(s) when due, documentation concerning advance and progress payments, estimated termination liability, or evidence of constructive delivery or shipment in support of requests for funds or bills will be made available to the Purchaser by DoD upon request. When appropriate, the Purchaser may request adjustment of any questioned billed items by subsequent submission of discrepancy reports, Standard Form 364.
4.4.8 To pay interest on any net amount by which it is in arrears on payments, determined by considering collectively all of the Purchaser's open LOAs with DoD. Interest will be calculated on a daily basis. The principal amount of the arrearage will be computed as the excess of cumulative financial requirements of the Purchaser over total cumulative payments after quarterly billing payment due dates. The rate of interest paid will be a rate not less than a rate determined by the Secretary of the Treasury taking into consideration the current average market yield on outstanding short-term obligations of the USG as of the last day of the month preceding the net arrearage and shall be computed from the date of net arrearage.
4.4.9 To designate the Procuring Agency and responsible Paying Office and address thereof to which the USG will submit requests for funds and bills under this LOA.

5 Transportation and Discrepancy Provisions

5.1 The USG agrees to deliver and pass title to the Purchaser at the initial point of shipment unless otherwise specified in this LOA. With respect to items procured for sale to the Purchaser, this will normally be at the manufacturer's loading facility; with respect to items furnished from USG stocks, this will normally be at the US depot. Articles will be packed, crated, or otherwise prepared for shipment prior to the time title passes. If "Point of Delivery" is specified other than the initial point of shipment, the supplying US Department or Agency will arrange movement of the articles to the authorized delivery point as a reimbursable service but will pass title at the initial point of shipment. The USG disclaims any liability for damage or loss to the items incurred after passage of title irrespective of whether transportation is by common carrier or by the US Defense Transportation System.

5.2 The Purchaser agrees to furnish shipping instructions which include Mark For and Freight Forwarder Codes based on the Offer/Release Code.

5.3 The Purchaser is responsible for obtaining insurance coverage and customs clearances. Except for articles exported by the USG, the Purchaser is responsible for ensuring that export licenses are obtained prior to export of US defense articles. The USG incurs no liability if export licenses are not granted or they are withdrawn before items are exported.

5.4 The Purchaser agrees to accept DD Forms 645 or other delivery documents as evidence that title has passed and items have been delivered. Title to defense articles transported by parcel post passes to the Purchaser at the time of parcel post shipment. Standard Form 364 will be used in submitting claims to the USG for overage, shortage, damage, duplicate billing, item deficiency, improper identification, improper documentation, or non-shipment of defense articles and non-performance of defense services and will be submitted promptly by the Purchaser. DoD will not accept claims related to items of $200. or less for overages, shortages, damages, non-shipment, or non-performance. Any claim, including a claim for shortage (but excluding a claim for nonshipment/nonreceipt of an entire lot), received after one year from passage of title to the article or from scheduled performance of the service will be disallowed by the USG unless the USG determines that unusual and compelling circumstances involving latent defects justify consideration of the claim. Claims, received after one year from date of passage of title or initial billing, whichever is later, for nonshipment/nonreceipt of an entire lot will be disallowed by the USG. The Purchaser agrees to return discrepant articles to USG custody within 180 days from the date of USG approval of such return.

6 Warranties

6.1 The USG does not warrant or guarantee any of the items sold pursuant to this LOA except as provided in section 6.1.1. DoD contracts include warranty clauses only on an exception basis. If requested by the Purchaser, the USG will, with respect to items being procured, and upon timely notice, attempt to obtain contract provisions to provide the requested warranties. The USG further agrees to exercise, upon the Purchaser's request, rights (including those arising under any warranties) the USG may have under contracts connected with the procurement of these items. Additional costs resulting from obtaining special contract provisions or warranties, or the exercise of rights under such provisions or warranties, will be charged to the Purchaser.
6.1.1 The USG warrants the title of items sold to the Purchaser hereunder but makes no warranties other than those set forth herein. In particular the USG disclaims liability resulting from infringement or other violation of intellectual property or technical data rights occasioned by the use or manufacture outside the US by or for the Purchaser of items supplied hereunder.
6.1.2 The USG agrees to exercise warranties on behalf of the Purchaser to assure, to the extent provided by the warranty, replacement or correction of such items found to be defective, when such materiel is procured for the Purchaser.

6.2 Unless the condition of defense articles is identified to be other than serviceable (for example, "As is"), DoD will repair or replace at no extra cost defense articles supplied from DoD stocks which are damaged or found to be defective in respect to material or workmanship when it is established that these deficiencies existed prior to passage of title, or found to be defective in design to such a degree that the items cannot be used for the purpose for which they were designed. Qualified representatives of the USG and of the Purchaser will agree on the liability hereunder and the corrective steps to be taken.

7 Dispute Resolution

7.1 This LOA is subject to US Federal procurement law.

7.2 The USG and the Purchaser agree to resolve any disagreement regarding this LOA by consultations between the USG and the Purchaser and not to refer any such disagreement to any international tribunal or third party for settlement.

LETTER OF OFFER AND ACCEPTANCE INFORMATION

1. **GENERAL.** This provides basic information pertaining to the LOA for US and Purchaser use. Additional information may be obtained from the Security Assistance Management Manual, DOD 5105.38-M, the in-country Security Assistance Office, the DSAA Country Director, or from the implementing agency.

2. **INFORMATION ENTERED BY THE USG.**

a. **Terms of Sale**, and Purchaser responsibilities under those Terms, are described on the LOA. A list of all Terms of Sale, with explanations for each, are shown in DOD 5105.38-M.

b. **Description/Condition**. The item description consists of coding for use in US management of the LOA (starting with Generic/MASL and MDE "(Y)" or non-MDE "(N)" data such as that in DOD 5105.38-M, Appendix D) plus a short description of what is to be provided. When items are serviceable, Code "A" (new, repaired, or reconditioned material which meets US Armed Forces standards of serviceability) may be used; otherwise, Code "B" (unserviceable or mixed condition without repair, restoration, or rehabilitation which may be required) may be used. In some instances, reference to a note in the Terms and Conditions may complement or replace these codes.

c. The **Unit of Issue** is normally "EA" (each, or one; for example, 40 EA) or blank (unit of issue not applicable; for example, services or several less significant items consolidated under one LOA Item Number). When blank, a quantity or Unit Cost is not shown.

d. The **Source Code** (SC) in the Articles or Services to be Supplied Section is one or more of the following:

S - Shipment from DoD stocks or performance by DoD personnel
P - From new procurement
R - From rebuild, repair, or modification by the USG
X - Mixed source, such as stock and procurement, or undetermined
E - Excess items, as-is
F - Special Defense Acquisition Fund (SDAF) items

e. Availability leadtime cited is the number of months (MOS) estimated for complete delivery of defense articles or performance of defense services. The leadtime starts with Acceptance of this Offer, including the conclusion of appropriate financial arrangements, and ends when items are made available to transportation.

f. **Type of Assistance** (TA) Codes are as follows:

3 - Source Code S, R, or E; based on Arms Export Control Act (AECA) Section 21(b).
4 - Source Code X; AECA Sections 21(b), 22(a), 29, or source undetermined.
5 - Source Code P; AECA Section 22(a).
6 - Source Code S, R, or E, payment on delivery; AECA Section 21(d).
7 - Source Code P, dependable undertaking with 120 days payment after delivery; AECA Section 22(b).
8 - Source Code S, R, or E, stock sales with 120 days payment after delivery; AECA Section 21(d).
M - MAP Merger; Foreign Assistance Act (FAA) Section 503(a)(3).
N - FMS Credit (Nonrepayable); AECA Sections 23 or 24.
U - Source Code P; Cooperative Logistics Supply Support Arrangement (CLSSA) Foreign Military Sales Order (FMSO) I.
V - Source Code S; CLSSA FMSO II stocks acquired under FMSO I.
Z - FMS Credit; AECA Sections 23 or 24.

g. **Training notes:** AP - Annual training program; SP - Special training designed to support purchases of US equipment; NC - This offer does not constitute a commitment to provide US training; SC - US training concurrently being addressed in separate LOA; NR - No US training is required in support of this purchase.

h. **Offer Release Codes** (Ofr Rel Cde) and Delivery Term Codes (Del Trm Cde) below may also be found in DOD 4500.32-R, MILSTAMP, Appendix M, Figure M-1. The following Offer Release Codes also pertain to release of items for shipment back to Purchaser on repair LOAs:

A - Freight and parcel post shipments will be released automatically by the shipping activity without advance notice of availability.

Y - Advance notice is required before release of shipment, but shipment can be released automatically if release instructions are not received by shipping activity within 15 calendar days. Parcel post shipments will be automatically released.

Z - Advance notice is required, before release of shipment. Shipping activity will follow-up on the notice of availability until release instructions are furnished. Parcel post shipments will be automatically released.

X - The Implementing Agency (IA) and country representative have agreed that the:

-- IA will sponsor the shipment to a country address. Under this agreement, the Freight Forwarder Code must also contain X and a Customer-within-Country (CC) Code must be entered in the Mark For Code on the front page of the LOA. The MAPAD must contain the CC Code and addresses for each type of shipment (parcel post or freight).

-- Shipments are to be made to an assembly point or staging area as indicated by clear instructions on exception requisitions. Under this agreement, the Freight Forwarder Code must contain W. A Mark For Code may be entered in the Mark For Code space on the front page of the LOA and the MAPAD must contain the Mark For Code if the Mark For Address is to be used on the shipment to the assembly point or staging area.

i. For the following Delivery Term Codes, DoD delivers:

2 - To a CONUS inland point (or overseas inland point when the origin and destination are both in the same geographic area)

3 - At the CONUS POE alongside the vessel or aircraft

4 - Not applicable (Purchaser has full responsibility at the point of origin. Often forwarded collect to country freight forwarder.)

5 - At the CONUS POE on the inland carrier's equipment

6 - At the overseas POD on board the vessel or aircraft

7 - At the overseas inland destination on board the inland carrier's equipment

8 - At the CONUS POE on board the vessel or aircraft

9 - At the overseas POD alongside the vessel or aircraft

Delivery Term Codes showing DoD transportation responsibility for repair LOAs are shown below. The LOA will provide a CONUS address for each item identified for repair. The customer must assure this address is shown on all containers and documentation when materiel is returned.

A - From overseas POE through CONUS destination to overseas POD on board the vessel or aircraft

B - From overseas POE through CONUS destination to CONUS POE on board the vessel or aircraft

C - From CONUS POD on board the vessel or aircraft through CONUS destination to CONUS POE on board the vessel or aircraft

D - From CONUS POD on board the vessel or aircraft through the CONUS destination to overseas POD on board the vessel or aircraft

E - Not applicable (Purchaser has complete responsibility.)

F - From overseas inland point through CONUS destination to overseas inland destination

G - From overseas POE through CONUS destination to overseas POD alongside vessel or aircraft

H - (For classified items) From CONUS inland point to CONUS POE alongside vessel or aircraft

J - (For classified cryptographic items) From CONUS inland point to overseas inland destination

3. **INFORMATION TO BE ENTERED BY THE PURCHASER. Mark For and Freight Forwarder Codes** are maintained in the Military Assistance Program Address Directory (MAPAD), DOD 4000.25-8. The **Purchaser Procuring Agency** should show the code for the Purchaser's Army, Navy, Air Force, or other agency which is purchasing the item(s). The **Name and Address of the Purchaser's Paying Office** is also required.

a. **Mark For Code.** This Code should be entered for use in identifying the address of the organization in the Purchaser country which is to receive the items. This includes return of items repaired under an LOA.

(1) This address will be added by the US DoD to the Ship To address on all freight containers. It will also appear on items forwarded by small parcel delivery service, including parcel post. The address should include the port of discharge name and designator (water or air); country name, country service name, street, city, state or province, and (if applicable) in-country zip or similar address code.

(2) Shippers are not authorized to apply shipment markings. If codes and addresses are not published, containers will be received at the freight forwarder or US military representative in-country unmarked for onward shipment with resultant losses, delays, and added costs. The USG will sponsor shipment of this materiel to FOB US point of origin.

b. **Freight Forwarder Code.** When Offer Release Code X applies, Code X or W, discussed under Offer Release Code X above, must be entered.

4. **FINANCIAL.**

a. The method of financing is shown in the LOA, Amendment, or Modification. The initial deposit required with Purchaser signature of the LOA is an integral part of the acceptance.

b. LOA payment schedules are estimates, for planning purposes. DFAS (SAAC) will request payment in accordance with the payment schedule unless DoD costs, including 90-day forecasted requirements, exceed amounts required by the payment schedule. When this occurs, the US will use its best efforts to provide a new schedule via LOA Modification at least 45 days prior to the next payment due date. The Purchaser is required to make payments in accordance with quarterly DD Forms 645 issued by DFAS regardless of the existing payment schedule.

c. The DD Form 645 serves as the bill and statement of account. An FMS Delivery Listing, identifying items physically or constructively delivered and services performed during the billing period, will be attached to the DD Form 645. DFAS forwards these forms to the Purchaser within 45 days before payments are due and Purchasers must forward payments in US dollars to the USG in time to meet prescribed due dates. Costs in excess of amounts funded by FMF agreements must be paid by the Purchaser. Questions concerning the content of DD Forms 645 and requests for billing adjustments should be submitted to the Defense Finance and Accounting Service (SAAC/FS), Lowry AFB, CO 80279-5000.

d. The preferred method for forwarding cash payments is by bank wire transfer to the Department of the Treasury account at the Federal Reserve Bank of New York using the standard federal reserve funds transfer format. Wire transfers will be accepted by the Federal Reserve System (FRS) only from banks that are members of the FRS, therefore, non-US banks must go through a US correspondent FRS member bank. The following information is applicable to cash payments:

Wire transfer--
United States Treasury
New York, New York
021-030-004
DFAS/SAAC
Agency Code 3801
Payment from (country or international organization) for
Letter of Offer and Acceptance (Identifier at
the top of the first page of the LOA)

Check mailing address--
Defense Finance and Accounting Service (DFAS)
DE/SAAC/F
Denver, CO 80279-5000

e. To authorize payments from funds available under FMF loan or grant agreements, the Purchaser may be required to submit a letter of request to the Defense Finance and Accounting Service (DFAS/DE-FCC), Denver, Colorado 80279-5000. Purchasers should consult applicable FMF agreements for explicit instructions. Questions pertaining to the status of FMF financing and balances should be directed to DSAA-COMPT-FMD.

f. Payments not received by DFAS (SAAC) by the due date may be subject to interest charges as outlined in paragraph 4.4.8 of the LOA Standard Terms and Conditions.

g. The values on the LOA are estimates. The final amount will be equal to the cost to the USG. When deliveries are made and known costs are billed and collected, SAAC will provide a "Final Statement of Account" which will summarize final costs. Excess funds will be available to pay unpaid billings on other statements or distributed as agreed upon between the Purchaser and the Comptroller, DSAA.

h. The Purchaser may cancel this LOA upon request to the implementing agency. An administrative charge that equals one-half of the applicable administrative charge rate times the ordered LOA value, which is earned on LOA acceptance, or the applicable administrative charge rate times the actual LOA value at closure, whichever is higher, may be assessed if this LOA is cancelled after implementation.

5. **CHANGES TO THE LOA.** Changes may be initiated by the USG or by requests from the Purchaser. After acceptance of the basic LOA, these changes will take the form of Amendments or Modifications.

a. Amendments encompass changes in scope, such as those which affect the type or number of significant items to be provided. Amendments require acceptance by the USG and the Purchaser in the same manner as the original LOA.

b. Modifications include changes which do not constitute a change in scope, such as increases or decreases in estimated costs or delivery schedule changes. Modifications require signature only to acknowledge receipt by the Purchaser.

c. When signed, and unless alternate instructions are provided, copies of Amendments and Modifications should be given the same US distribution as the basic LOA.

d. Requests for changes required prior to acceptance by the Purchaser should be submitted to the implementing agency for consideration. See DOD 5105.38-M, section 70105.M.2.

6. **CORRESPONDENCE.** Questions or comments regarding this LOA should identify the Purchaser request reference and the identification assigned by the implementing agency within DoD.

APPENDIX B

Checklist for an Aircraft System Sales Request

The following checklist is to be used by the FMS purchaser for all aircraft. Include this checklist with the initial request for new U.S. Air Force inventory and excess U.S. Air Force aircraft. When preparing the checklist, each item must be addressed and an entry made. Enter "NA," if not applicable.

1. *Purchaser—Project Security Classification.* (USG normally handles all requests as unclassified on congressional notification or LOA).
2. *Purpose.* Request for LOA, planning and availability (P&A) or planning and review (P&R) data (circle one).
3. *Aircraft Model, Designation or Series (MDS).*
4. *Quantity.*
5. *Basic Configuration:*
   - **a.** Additions to basic (attach list).
   - **b.** Deletions to basic (attach list).
   - **c.** Option items to be separately priced.
   - **d.** Changes to configuration:
     - **(1)** Included in aircraft cost.
     - **(2)** Optional item.
   - **e.** Specific computer program identification number, if known.
6. *Source Data:*
   - **a.** Inventory aircraft:
     - **(1)** Prepare for one time flight.
     - **(2)** Serviceable, reconditioned, or rehabilitated according to AFR 400-6 and chapter 7.
   - **b.** Production.
   - **c.** Development.
7. *Delivery Data (Schedule):*
   - **a.** First aircraft at plant.
   - **b.** Desired monthly production rate.
   - **c.** Method of delivery (ferry, surface, or airlift).
   - **d.** Delivery by USG or purchaser?
   - **e.** Desired in-country delivery rate (how many per month).
8. *Missiles, Bombs, Ammunition, or Electromagnetic Communications (EC) Systems:*
   - **a.** Type.
   - **b.** Quantity.
   - **c.** Initial spares.
   - **d.** Support equipment (standard or developmental).
   - **e.** Furnish definitive list of make line item subject to provisioning conference.
   - **f.** Developmental system requirements.
9. *Anticipated LOA Acceptance.*
10. *Operational Concept:*
    - **a.** Role of aircraft:
      - **(1)** Primary.
      - **(2)** Secondary.

b. Number of squadrons:

(1) Number of aircraft per squadron.

(2) Anticipated monthly flying hours per aircraft.

c. Number of main operating bases (MOB). (Number of squadrons at each MOB.)

d. Number of forward operating bases (FOB):

(1) Number of aircraft to be supported at each FOB.

(2) Estimated time aircraft will be supported at each FOB.

(3) Mission to be performed at FOB.

(4) Will support be prepositioned at FOB?

11. *Maintenance Concept (See Note 1):*

a. Organizational and intermediate level:

(1) Number of organizational support equipment sets.

(2) Number of intermediate support equipment sets.

b. Depot level:

(1) Number of depot-level support equipment sets.

(2) Identify systems to be supported.

c. Level and amount of required technical data.

d. Assumptions regarding present maintenance capability and availability of existing facilities or equipment.

e. If software support is required for embedded computer support (ECS) complete appropriate parts of attachment 23.

12. *Supply Concept (See Note 2):*

a. Number of years initial spares should cover.

b. Anticipated special requirements (identify).

c. Planned flying hours—each aircraft, each month.

13. *Contractor Engineering and Technical Services (CETS) (See Note 3):*

a. Number of persons required.

b. Specialty required (for example, air frame, engine, avionics, or supply).

c. Time required for each person.

14. *Weapon Systems Logistics Officer (WSLO) or System Acquisition Officer (Country Should Provide a Statement):*

a. Number required.

b. Time required for each.

15. *Training Concept:*

a. Number or type aircrew requiring continental United States (CONUS) training (pilot, navigator, electronic warfare officer, weapon system officer, flight engineer):

(1) Student background (type aircraft flown, number of hours, etc).

(2) English language capability.

(3) Type mission to be qualified for: air-to-air, air-to-ground, all weather intercept, ferry, etc.

(4) Physiological training qualification.

(5) Date CONUS training to be completed.

b. Number or type of maintenance personnel who require CONUS training (breakout by Air Force specialty code AFSC). (Identify variances between purchaser Air Force and U.S. Air Force specialties; for example, weapon mechanic performs egress; avionics communications technician performs avionic navigation; host does not have fire control specialties).

(1) Student background (type aircraft or system).

(2) Training level desired: organizational, intermediate or depot. (Identify desires for contractor or U.S. Air Force training).

(3) Required CONUS completion date.

(4) English language capability.

c. U.S. Air Force instructor aircrew and maintenance mobile training teams (MTTs) desired for training in-country:

(1) Number of aircrew MTT and duration.

(2) Description of flying training facilities, ranges, navigation aids.

(3) Number, composition, and duration of maintenance MTT.

(4) Language qualification of host students.

(5) Availability of interpreters, if required.

(6) Training start date.

d. Training devices:

(1) Quantity.

(2) Weapon system simulators.

(3) Mobile training sets (MTS) (maintenance).

(4) Other (attach description).

e. Security training to adequately protect U.S. Air Force classified equipment and information.

16. *Insurance:* Purchaser will arrange own insurance unless extenuating circumstances exist that justify a Defense Security Assistance Agency (DSAA) exception to policy.

17. *Quality Assurance:*

a. Air Force.

b. Other services.

c. Consultants.

18. *Test Measurement Diagnostic Equipment (TMDE) Calibration Services.*

19. *Requirements for In-Country Surveys:*

a. Preliminary to LOA development.

b. After LOA acceptance to help define support requirements.

20. *Preservation or Packaging Requirements.*

21. *Requirements for In-Country Contractual Support:*

a. Type of support.

b. Period of support.

22. *Facilities Beddown Requirements:*

a. Definition of requirements.

b. Design and construction schedules.

c. Information (as built) on facilities and utilities being modified.

23. *Requirements for Participation in U.S. Air Force Programs.*

24. *Automated Logistics and Maintenance System.*

25. *Other Pertinent Remarks.*

Notes:

1. Baselines for maintenance procedures are in AFM 66-1.

2. Baselines for supply procedures are in AFM 67-1.

3. CETS requests must include information required for a "G" case.

GLOSSARY

AECA	Arms Export Control Act.
FMFP	Foreign military sales financing program
FMS	Foreign military sales
LOA	Letter of offer and acceptance, DD Form 1513
LOR	Letter of request (for an FMS case)
SA	Security assistance
SAO	Security assistance officer or organization; the DoD representative in a foreign country responsible for carrying out security assistance functions.

REFERENCES

1. L. J. Samelson (ed.), *The Management of Security Assistance,* Defense Institute of Security Assistance Management, Wright-Patterson AFB, Ohio, 1991.
2. "United States Arms Transfer and Security Assistance Programs," report prepared for the Subcommittee on Europe and the Middle East of the Committee on International Relations, U.S. House of Representatives, p. 6, Mar. 21, 1978.
3. A. J. Pierre, *The Global Politics of Arms Sales,* Princeton University Press, Princeton, N.J., 1982.
4. J. L. Buckley, "Testimony of Under Secretary of State for Security Assistance, Science, and Technology before the Subcommittee on International Security and Scientific Affairs, Committee on Foreign Affairs, House of Representatives," pp. 3–4, Mar. 19, 1981.
5. A. M. Haig, Jr., "Statement before the House Committee on Foreign Affairs," p. 7, Mar. 18, 1981.
6. "World Military Expenditures and Arms Transfers, 1989," U.S. Arms Control and Disarmament Agency, p. 111, Oct. 1990.
7. W. D. Bajusz and D. J. Louscher, *Arms Sales on the U.S. Economy,* Westview Press, Boulder, Colo., 1988, p. 53.
8. "Fascell Reacts to Five-Power Talks on Mideast Arms Control," Committee on Foreign Affairs, U.S. House of Representatives News Release, July 10, 1991.
9. "Changing Perspectives on U.S. Arms Transfer Policies," Report prepared for the Subcommittee on International Security and Scientific Affairs of the Committee on Foreign Affairs, House of Representatives, pp. 95–99, Sept. 15, 1981.
10. Congressional Presentation [Document] for Security Assistance Programs (CPD), Fiscal Year 1992.
11. Congressional Presentation [Document] for Security Assistance Programs (CPD), Fiscal Year 1988.
12. *Security Assistance Management Manual,* DoD 5105.38-M, Department of Defense, Defense Security Assistance Agency, Washington, D.C., Mar. 1, 1991.
13. E. T. Hall, *The Dance of Life,* Anchor Books/Doubleday, New York, 1983, p. 4.

PRIMARY SOURCES FOR SECURITY ASSISTANCE

Arms Export Control Act (AECA) of 1976, as amended.

Foreign Assistance Act (FAA) of 1961, as amended.

U.S. Department of State and U.S. Department of Defense, Defense Security Assistance Agency (DSAA), "Congressional Presentation for Security Assistance Programs, Fiscal Year 1992."

U.S. Department of Defense, *Security Assistance Management Manual (SAMM),* DoD 5105.38-M.

U.S. Department of Defense, *Foreign Military Sales Financial Management Manual,* DoD 7290.3-M.

U.S. Department of the Air Force Regulation (AFR) 130-1, "Security Assistance Management."

U.S. Department of the Army Regulation (AR) 12-8, "Foreign Military Sales Operations/ Procedures."

U.S. Department of the Navy, Naval Supply (NAVSUP) System Command Publication 526, "Foreign Military Sales Customer Supply System Guide."

P · A · R · T · 2

THE ENVIRONMENT OF MILITARY PROJECT MANAGEMENT

Part 2 presents an overview of the major aspects of the environment of the project manager. Today's military project manager, whether within the industrial sector or within the military, must be acutely aware of each and all interdependent entities surrounding the project and interacting with it. The military project shares with its commercial endeavors an increasing emphasis on the "customer" and "stakeholder." The military project manager's environment includes not only the integration of company or organizational divisions, departments, and functions of the prime contractor, subcontractors, suppliers, vendors, and others, but it must also interact with military requirements, acquisition processes, regulations, desires, and, of course, stringent contractual provisions. The military project managers in government and industry are obverse of the same coin, serving a common end.

The military project manager, both in industry and in the military, has no real counterpart in the commercial sector. Witness the major differences of the project managers of the McDonnell-Douglas MD-80 and of the Douglas C-17A airlifter. The job of the military project is to translate requirements of a user into an acquisition program, to establish a specification baseline for the system, to lead the government and industry team, to evaluate alternatives or trades, to recommend choices among capabilities, schedules, risks, and costs for the user to select, and, finally, to assess government and industry performance against the established baseline. A weapon system must meet user needs without surprises. The industry project manager manages the design, competitive proposal, manufacture, subcontractor or vendor, and support according to the contract, rules, regulations, and needs of the government as given through the military project manager.

The diversity of the practice of military project management can be shown throughout this handbook. The authors of these chapters include military officers and NCOs; senior civilians in government; aerospace, military, and naval project managers; industry managers, executives, staff, and functional specialists; and academicians. All are represented in different specialties and a variety of sizes and complexities in the projects or program experiences. And what they can agree upon is far more instructive than what they find different. To be a successful military project manager, in industry or in the government, requires fundamental situational awareness. The project manager who does not understand the Pentagon and Capitol Hill in terms of decision making and roles and processes, for example, will have a very difficult time. Similarly, it is necessary to be aware of other environmental concerns, such as the flow of money downward or laterally into the project or the flow of technology or ecological impacts affecting the project. Future projects will require new thinking to achieve built-in quality requirements, affordability, and a deeper understanding of key characteristics of the design, test, support, and training environment. Each functional element is operated in vertical tunnels. The services can ill afford compartmentalization of the functional entities and arbitrary distinctions between system life cycles. The emphasis in the future will be on customer and supplier processes and how business is conducted, and on holding the entire team accountable. This means empowering the team to act as one. Lines between design, production, and support will be redrawn or eliminated, walls between functions will be lowered.

Chapter 10 presents A. J. DiMascio's anatomy and descriptive model of the project throughout its life within the military environment. He depicts the phases of the system and explains the unique features of each. This chapter sets the stage for an overall summary understanding of the military project as practiced in today's U.S. Department of Defense environment.

In Chap. 11, James A. Abrahamson provides a personal perspective, based on a wealth of experience, on who and what a project manager is and how he or she can influence the environment to achieve successful execution. His brief advice contains a wealth of wisdom and words that can have meaning for all future project managers.

In the next three chapters we gain a rare perspective. Charles B. Cochrane in Chap. 12, Ernst Peter Vollmer in Chap. 13, and Edward J. Trusela in Chap. 14 provide overviews of the approaches of project management within the U.S. Army, Navy, and Air Force, respectively. Here we see close up the many similarities, differences, and distinct features of each service as they evolved even from the same set of DoD directives.

In Chap. 15, Robert R. Barthelemy and Helmut H. Reda begin a discussion of perhaps the most difficult form of military project management. "Traditional" projects take on innumerable forms, but at least the management approaches are predictable and follow evolved formats. Nontraditional military projects, such as the National Aerospace Plane, must strike out into unchartered territory, and create lateral thought processes rather than following old methods. Nothing is taken for granted: resources, support, established contractual provisions, communication, direction, or commitment. Each of the nontraditional alternatives, joint government programs, consortia, contractor teaming, and Skunk Works is discussed and provides insight into the way of future efforts.

As a second part to this important environmental topic, in Chap. 16 Jeffrey D. Cerney provides an in-depth look into a joint project office, the unmanned ground vehicle. He explores the environmental background of the joint program and provides some clues to the prospective joint project manager as to how to maintain a "fragile system" in a balanced way.

James H. Dobbins, in Chap. 17, explores the factors critical to successful project management, that is, his environmental topic relates to the metrics of measurement of success. This is a topic that has gone largely unexplored until now and provides a springboard for deliberate thought and practice. How do we know when and how a project is considered a success or a failure? What devices measure the project? What is meaningful and useful, and how does the project manager select?

Next the discussion of the environment takes on a unique consideration. In Chap. 18, Fred Abrams introduces the U.S. Air Force approach to integrated weapon system management. The Army and the Navy have never really separated project management into development, production, and logistics support as has the Air Force. In 1992 the Air Force merged Air Force Systems Command and Air Force Logistics Command. This will result in effectively redesigning long-standing management practices and, in many ways, starting over from scratch. Thus these Air Force leaders are able to design some aspects of project management from a clean slate, and thus this important topic of evolving a total comprehensive weapon system management approach is included in this section. The chapter explains the approach used to define and mature the integrated weapon system concept and explores the uses made of total quality management precepts.

Finally the discussion of the environment of the military project manager explores the political process. Thomas C. Hone, in Chap. 19, provides an historical and practical primer in the whys and wherefores of the Congress and the budget process, and how they affect the project manager.

CHAPTER 10
THE PROJECT CYCLE

A. J. DiMascio

A. J. DiMascio is a recognized expert in engineering management, system engineering, and system acquisition management. In addition he is an authority on product assurance and has significant expertise in organizational development. He is a member of the faculty of Florida Institute of Technology, teaching in the graduate programs of logistic management, acquisition management, and system management. He has extensive technical and management experience in design engineering, test and evaluation, system engineering, system acquisition, and program in the Navy Department and in the private sector. He was the Director of the Office of Naval Acquisition Support (ONAS) prior to leaving the Navy Department in 1986. Immediately prior to his appointment as Commander ONAS, he was Deputy Commander of the Naval Air Systems Command. Dr. DiMascio received the B.S. and M.S. degrees from Drexel Institute of Technology and the doctorate from George Washington University in 1979. He has authored several papers and articles on technical and management subjects.

ANATOMY OF A PROJECT

A project can be generally characterized as a set of interdependent activities or tasks, which are integrated to accomplish a specific set of goals and objectives within a specified time period. Nominally a project is a relatively major undertaking of an enterprise involving the commitment and expenditure of relatively large amounts of resources. This characterization indicates that a project has a degree of uniqueness requiring customized planning and control. It also implies that there is a significant level of risk, and, therefore, risk management is a critical consideration.

Given the general characterization of uniqueness, goal specificity, and resource requirements, the life of a nominal project is determinate—a project has a finite time frame. Time is an uncontrollable variable in this context. Accordingly, planning and control are critical project management functions. The project activities must be carefully planned and scheduled. The scheduling will be dependent on the technical content of the activities and tasks, the resource requirements and availability considerations, and the risk management framework. The scheduling must be compatible with the associated resource availability.

The technical context of most project activities and tasks is usually the critical consideration; the technical performance requirements in most projects are generally the pacing considerations. In addition the technical aspects are relatively complex. The project activities and tasks must be accomplished by multifunctional and multidiscipline teams. The orchestration and the integration of the efforts of these teams require a system management approach and comprehensive project planning and control. Risk management is a critical component of the prescribed system management approach.

The resource availability will always be subject to the general economic decision calculus—the methods for allocation of scarce resources to satisfy several competing needs within a priority preference network. There will be resource limitations and constraints that must be accommodated prior to the initiation of a project and during project execution. Accordingly, affordability is a strategic factor for general management and project management. Affordability is the state of being able to bear the cost. However, one must recognize that this state of affordability is a complex function of needs and their strategic criticality and priority, cost, and the overall availability of resources. General management and project management decisions and plans must reconcile the tensions that are generated by operational needs, technical opportunities, and limited resources.

The general characteristics of a project can be succinctly summarized as follows:

- Relative degree of uniqueness within the corporate framework
- Definitive requirements (purpose and desired results)
- Explicit beginning and ending
- Necessity for multifunctional or multidisciplinary teaming
- High degree of interaction and interdependencies
- Limitations and constraints on resources (affordability is a critical consideration)
- Dedicated, intensive, and parochial management
- An evolutionary progression through a series of differentiated phases—the project cycle.

These attributes reflect a distinctive interrelationship among project activities, technical effort, resource requirement, and time. This distinctive project framework is referred to as the project cycle. It represents a descriptive model for time phasing the activities and tasks and the proposed means for accomplishing the tasks in accordance with some risk management and affordability context. An example of an affordability analytical framework is illustrated schematically in Fig. 10.1. In this model the hierarchy of priorities is defined in the strategic planning and programming processes. Project performance and technical requirements are defined to satisfy the basic needs and operational requirements. The performance and technical requirements should be defined in terms of minimum acceptable values (threshold values) and goals in order to provide a logical boundary for cost-effective trades. These performance and technical requirements dictate the technical activities and tasks and, therefore, are the drivers of the project phasing and resource requirements. The project budget is derived from the phasing or scheduling and the resource requirements determination. As indicated in Fig. 10.1, the actual resource allocation or phasing of the resources is conditioned by the affordability reconciliation. It is significant to note the iterative nature of this model. The iterations indicated (such as project requirements versus affordability and budgeting versus affordability) are demanded in order to prevent "requirements versus budgeting" decoupling. This decoupling would occur when the project plan and the actual phasing of resources are not reconciled. Accordingly, the project planning and control functions must be congruent with the practical realities of the affordability imperative.

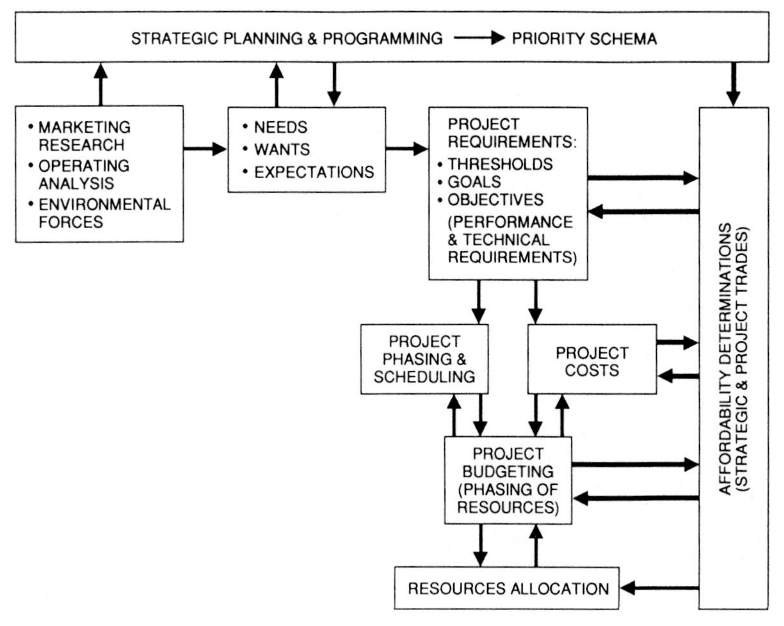

FIGURE 10.1 Project affordability analysis.

An understanding of project cycle considerations and risk management requirements, combined with the affordability logic illustrated in Fig. 10.1, inherently improves project management discipline. However, no two projects are exactly alike. Further, there is no prescriptive or normative model of a generic project cycle which is universally acceptable. A variety of practical nominal descriptive models of project cycle are extant in the literature. Projects are different in general and in specifics. There are differences related to the industry (such as defense versus commercial), the firm or enterprise (such as public versus private, profit versus not for profit), and the technical environment. The descriptive model must be tailored to account for the relevant project cycle considerations and the project-decision logic.

Some fundamental considerations which pertain to most project planning and control situations are:

- Scope (project and activities or tasks) or work content
- Performance and quality
- Schedule
- Cost
- Risk

The scope or work content is prescribed to satisfy all of the project goals and objectives. The scope determines the technical and business approaches and is generally embodied in a work breakdown structure. The performance parameters and related technical requirements are defined to satisfy the project goals and objectives. Quality is an abstract concept which represents a set of characteristics and attributes (physical, functional, and perpetual). In a total quality context, quality is a composite set of characteristics and attributes through which a product will meet the needs and expectations of the customers or users. The customer satisfaction orientation is the critical aspect. We can describe quality, from an external perspective, as the conformance to the requirements as explicitly and implicitly defined by the customers. In the internal perspective we can describe quality in terms of the efficiency and productivity of operations and processes employed to satisfy the needs of the customers.

Schedule refers to the time dimension of each activity, task, and work package within the project. Cost refers to the resource expenditures in fiscal terms. There are functional relationships which exist among cost, schedule, and performance. Therefore these fundamental considerations are subject to trades during the project evolution. Trade studies and trade-off compromises are elemental tools of the system management approach.

PROJECT MANAGEMENT DISCIPLINE

Programs or projects are established within a corporate framework to accomplish specific goals and objectives which have significant impact on the achievement of the strategic goals and objectives of the overall corporation, enterprise, or agency. The required program or project is a relatively unique undertaking, which requires special management attention and emphasis for a relatively long, but finite period of time. This period of time is referred to as the project cycle. The special management requires unique dedication and a high degree of project parochialism relative to the general management and functional management positions. This special management discipline is called project management. Since projects are usually complex in nature, often involving new product or system development, project management is necessarily a complex discipline.

Project management is system management. It represents a system approach to dedicated, parochial, and intensive management to ensure technical and business success of a large, complex, multidiscipline, and host-cost endeavor. The basic concept requires that the authority, responsibility, and accountability for overall project success be vested in one individual—the project manager. The primary determinants of the need for this type of focused management are:

- Rigorous resource, schedular, and technical performance requirements
- Significant complexity and scope
- Multifunctional, multidisciplinary, and multiorganizational dimensions
- Critical impacts and risks (the benefits from success and the penalties of failure are significantly high)
- Significant interrelationships (internal and external)

These factors are the natural consequences of the fundamental characteristics of a project which were outlined previously.

The project manager can be characterized as the "master system integrator." This individual must be the active integrator of the objective system (that is, the product system—hardware or software elements and support elements), integrator of the technical and administrative process systems, and integrator of the organizational elements. This integration is facilitated through the business, administrative, and technical management of the sequence of project activities in the project cycle. Project planning and control are the keystone functions for effecting this orchestration and integration. A typical project planning framework is illustrated in Fig. 10.2. This framework represents an integrated project planning process in which all of the

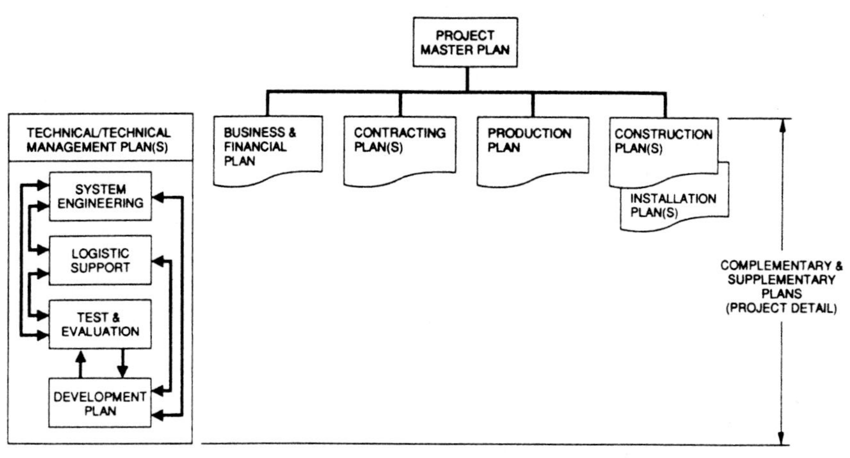

FIGURE 10.2 Generic project planning framework (integrated planning process and plan set).

functional plans are developed in a coordinated and collaborative network to complement and supplement the project master plan. These plans not only provide the strategies and detail courses of action to execute the project, but also define the standards which are used for the control process.

A typical project control process is illustrated schematically in Fig. 10.3. The project standards are defined in the applicable plans (cost, schedule, and technical performance measures). The work packages are established by the work breakdown structure. Technical performance measurement is accomplished by a review of ana-

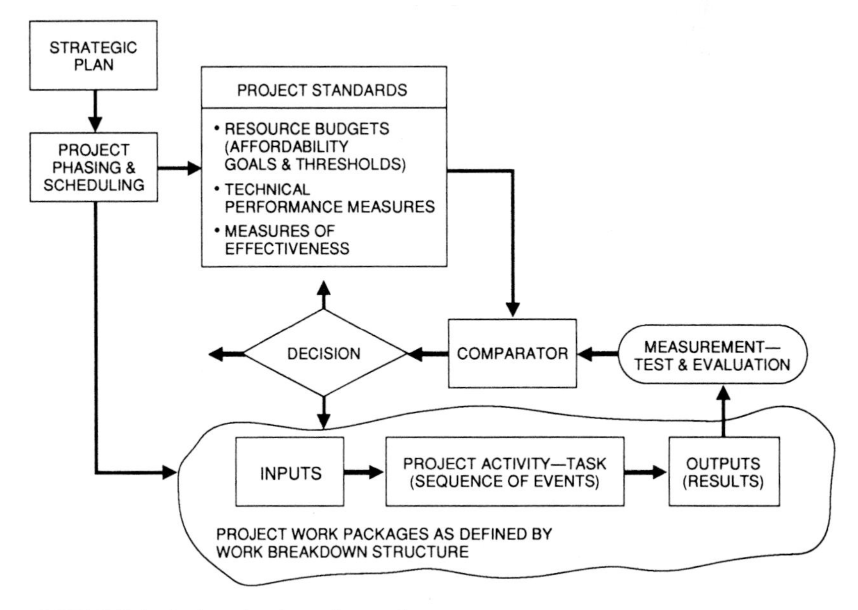

FIGURE 10.3 Project planning and control.

lytical and simulation results or predictions or by actual test and evaluation demonstration events, depending on the phase of the project cycle during the measurement. The decision subsequent to the comparison of the actual performance to the planned performance is based on demonstrated events (demonstrated analytically in the early phases and physically during the late phases) not some calendar time requirements. Major decision milestones should coincide with the points of transition from one phase to a succeeding phase in the project cycle. This is a critical step in the project evolution—a baseline for the succeeding phase is defined at this critical decision milestone. In addition achievement of the planned goals and objectives of the preceding phase is validated at this point. The integrated plan set provides the logic and mechanism progression through the project cycle and the decision-making logic. Risk management is a critical component of this logic and the progression mechanics. Risk management is a systematic process for dealing with uncertainty and the concomitant risks. It involves explicit considerations of risk exposure and contingency planning. The primary elements of a risk management methodology are risk identification and assessment, risk planning, risk analysis, and risk handling. This methodology must be integrated in the project system management approach. In the early stages of the project planning, risk planning is incorporated to develop the overall risk reduction strategy, to develop contingencies and alternative courses of action in high-risk areas, and to establish schedule and financial reserves to accommodate the risk exposure. Since some degree of uncertainty, and consequently risk, will always be inherently present in projects, risk management is a critical subset of system management.

THE PROJECT CYCLE: A NOMINAL MODEL

Each project is unique. Therefore, in each case project management must be tailored specifically to the individual project environment. This project environment will be determined by many forces and factors, including affordability considerations, technology content, and risk tolerance conditions. However, it is clear that every project evolves and progresses through several distinct phases. We can describe a nominal model of a project cycle: the set of interrelated phases through which the project passes as it matures. It is impossible to prescribe a detailed model of a single generic project cycle which defines the required stepwise evolutionary approach. However, it is possible to provide descriptive models for the various categories of projects. These models of project cycles represent basic blueprints for the formulation of project management logic. In addition they are useful as the models for project planning and control as well as coordination.[1]

The project cycle model describes the phased process of project progression. The project phases reflect a stepwise, layered approach, which permits continuation of reasonable flexibility and retention of practical options until the appropriate project decision milestone. This progression and decision convergence is dependent on the following general considerations, which must be carefully interrelated:

- Risk reduction
- Risk exposure or risk tolerance versus cost exposure
- Technical maturation or design maturation
- Affordability or resource availability

The fundamental phasing and decision-making logic dictates that the major project phases and decision milestones be defined to ensure that resource allocations are

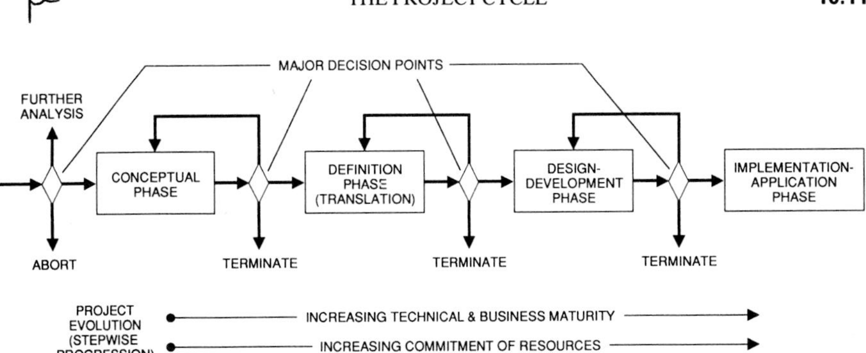

FIGURE 10.4 Generic project cycle.

contingent upon appropriate technical progress and risk reduction. The phases must be portioned by logical project baselines and major decision milestones. Decisions at these milestones must be based on a specific determination that the overall confidence in success is high enough to justify further allocation of resources at an increased level and intensity. This confidence must be derived from technical maturation—learning which clearly indicates that the knowledge, to date, or the prospects of yielding satisfactory results justify the higher levels of investment required during the next phase. This technical definition is evidenced through some combination of analyses, predictions, simulations, design analyses, and test and evaluation, depending on the phase of the project. This project management discipline is represented graphically in Fig. 10.4, which represents a generic project cycle for a typical project established to develop a system or product. Four distinct project phases can be identified in this representation case: a conceptual phase, a definition or translation phase, a design-development phase, and an implementation-application phase. Four major decision milestones are indicated representing the planned culmination of a sequence of activities and the successful achievement of phase goals and objectives. The first major decision milestone is the decision to initiate the project. The succeeding key milestones are of special significance because they are needed to provide the progressive assessment of risk reduction, technical maturation, and affordability. These decisions must be based on actual project accomplishments (events). Each phase is generally distinguishable from any other by the type of tasks, the characteristics of each phase, and the risk exposure. Each major milestone establishes a project baseline which defines the measurement standards for the succeeding phase.

The conceptual phase is characterized by tasks that define the system concepts and determine technical, economic, and environmental feasibilities. Several practical alternatives and enabling technologies are identified. A determination of the adequacy of the available technology base is critical during this phase. Resource requirements will be refined prior to the completion of this phase. The definition phase is characterized by tasks which define the complete set or system performance and design requirements. The various system interfaces are defined and a variety of risk-reduction activities are completed. At the completion of this phase realistic cost, schedule, and performance determinations should be available. The system or product is fully designed and developed during the design or development phase. A wide range of system tests and evaluation events are incorporated in this phase to assist in the development and demonstrate achievement of the project goals and objectives. The system or product design is implemented or applied in operations during the last phase. This may require production and construction activities, depending on the type of system or product in each case.

Each application of any model of a generic product cycle must be conditioned specifically by the unique project requirements—the project goals and objectives. The overall project goals and objectives as well as those established for each phase of the project cycle are the foundations for the project baselines. They define the planning and control basics. Goals and objectives are the aims or the desired results. Each goal is normally fully defined by two or more objectives. The objectives represent the means to achieve the goal and they serve to indicate greater specificity. Each objective is achieved by the appropriate sequencing of activities or tasks. The completion of such an activity or task is an event, and milestones are events of special significance in the planning and control scheme. The major decision milestones reflected in Fig. 10.4 are used in project planning and control to provide a structure and discipline for risk management. At these key decision points further commitments of resources must be justified by the risk-reduction status. These control points are also incorporated to assure that the project need and worth as well as its affordability are validated prior to higher levels of investment. These decisions establish planning, control, and coordination baselines.

The simple descriptive model of a generic project cycle presented in Fig. 10.4 can be expanded for those projects involving the development, production, and deployment of relatively complex hardware and software systems and products. Such an expanded model is illustrated graphically in Fig. 10.5. The project genesis refers to all of the activities, studies, and corporate-level programming which lead to the decision to initiate a designated project. The project needs and affordability determinations are established during this preproject stage. These determinations, supported by various studies, analyses, and strategic planning, provide the bases for the definition of the

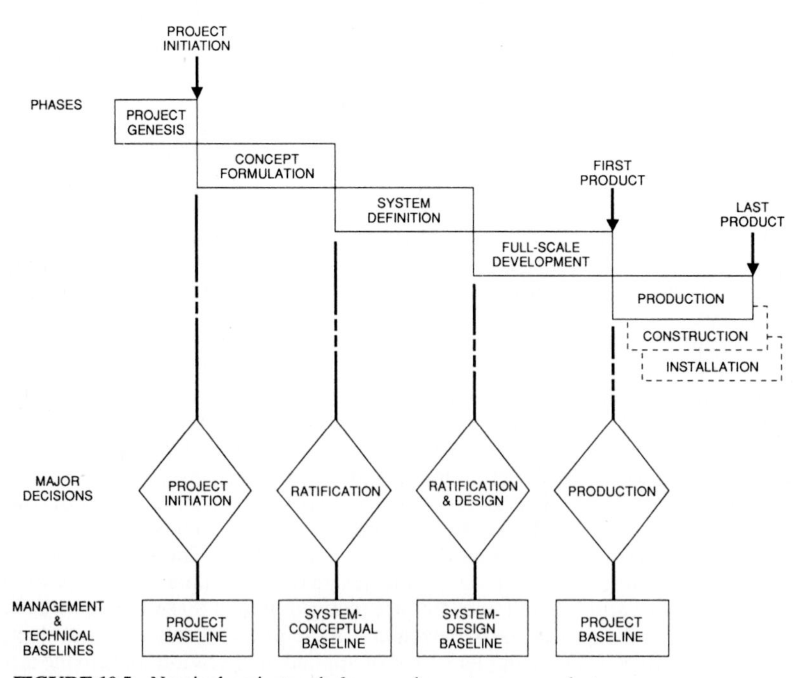

FIGURE 10.5 Nominal project cycle for complex systems or products.

first baseline—the project baseline—which establishes overall project requirements, including thresholds, goals, and objectives. The system concept is defined at the system conceptual baseline. The system and the design requirements are defined at the system design baseline. The product baseline defines the production requirements. The most critical decision milestones are the ratification and design and the production decisions. These represent critical project validations and the commitment of significant resources to complete the project. It is important to note that sound project management discipline requires strict objectivity at these decisions. None of the major decision milestones can be treated as being so rational to permit inordinate flexing because of emotional, political, or competitive pressures. The project plan and the preceding baselines define the decision logic in advance. The accomplishments must be indicated by demonstrations and analytical evidence of progress as required by the plan. The basic logic for this project planning and control discipline is presented graphically in Fig. 10.6.

The characteristic curves in Fig. 10.6 indicate relationships among the system development progression, test and evaluation process, risk management, and resource management considerations. System development is determined by increasing technical definition of both the product design and the process design. Increasing technical maturation is accompanied by decreasing uncertainty. The technical definition or maturation process is both assisted and demonstrated by test and evaluation. In the early phases experimental and other development testing is used to facilitate the design-development process. In the latter stages of full-scale development, test and evaluation is implemented to demonstrate achievement of the project goals and objectives. The risk management imperative dictates that risk will be reduced to the appropriate levels, as determined by the risk tolerance environment, prior to the progressive inputs of resources. The cost, schedule, and performance uncertainties decrease significantly as technical maturation progresses. This is demonstrated by the integrated test and evaluation events. Accordingly, the project risk decreases and risk exposure is reduced progressively. Time management is a critical aspect of this logic. Most projects characterized by the project cycle illustrated in Fig. 10.5 have significant developmental and creative contents. Creative processes take time. Appropriate timing must be accommodated in the project planning. The risk management framework must establish an environment for reasonable risk taking in order to preclude an inordinate risk-aversive decision logic or a subjectively optimistic risk posture.

A technical management view of the system development process in the project cycle is illustrated in Fig. 10.7. Complex systems and products will generally proceed through three design stages prior to establishment of the product baseline: conceptual design, preliminary design, and detail design. This reflects the stepwise layered developmental approach mentioned previously. The conceptual design establishes the system conceptually and functionally. Practical alternatives are identified and preliminary cost, schedule, and performance estimates are developed. The work in this phase, including trade studies and technology assessments, provides the conceptual baseline for further development. The preliminary design defines the preferred system and establishes the detail design requirements. Design specifications are developed during this phase. The work done in the phase generally classified as advanced development, including engineering studies, technology insertion studies, life-cycle cost analyses, and system trade studies, provides the system and design baseline for full-scale development. The primary risk-reduction efforts are accomplished during advanced development, and the system concept or conceptual design are validated. The production design is defined during the detail design phase. The product design and process designs are fully developed and documented. In addition the support system is designed during this phase. The output of this phase, the prod-

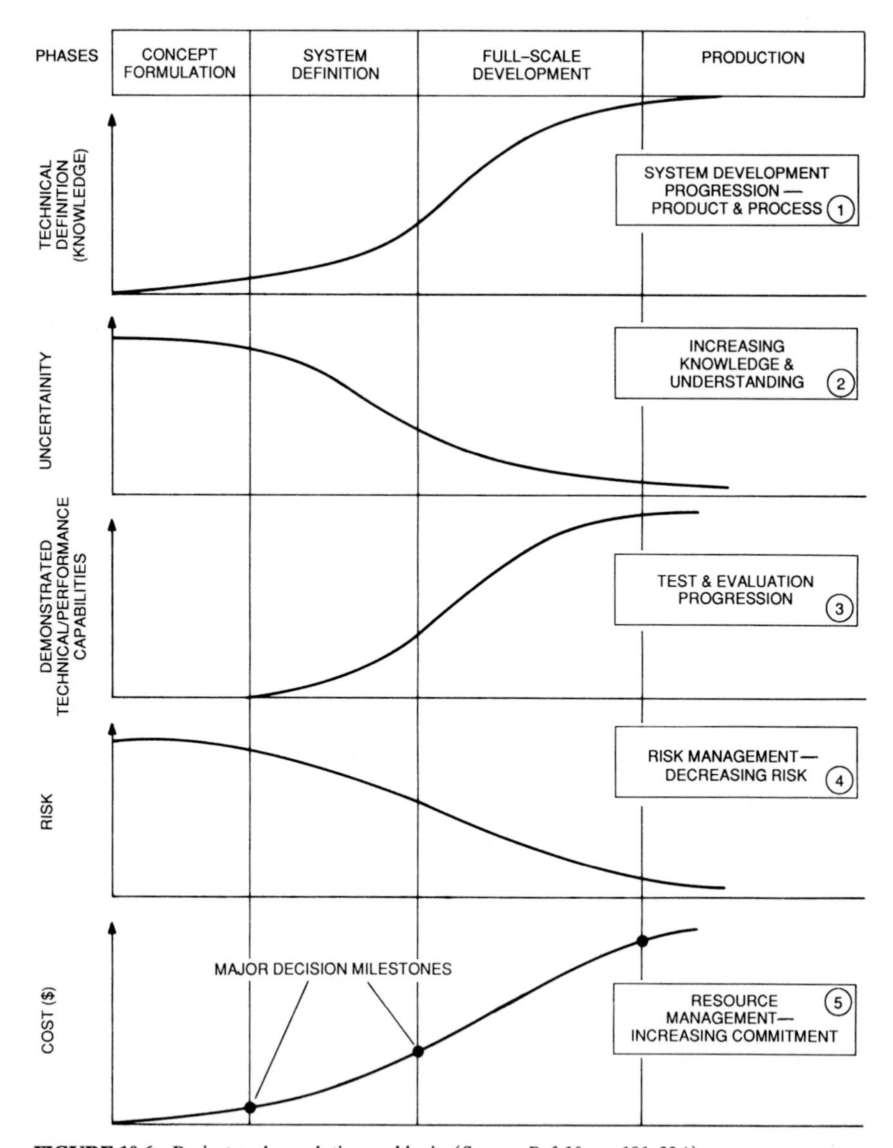

FIGURE 10.6 Project cycle, evolution and logic. (*Source: Ref. 10, pp. 191–224.*)

uct baseline, defines the system production requirements: product, process, materials, and support.

The project management discipline and project evolution logic outlined in the preceding reflect sound system management concepts. This system context should become more apparent when one places the project cycle illustrated in Fig. 10.5 within the system life-cycle perspective. One representation of the system life cycle for a typical commercial system is given in Fig. 10.8. The system life cycle has three regimes which overlap: the project cycle (assuming the system or product was devel-

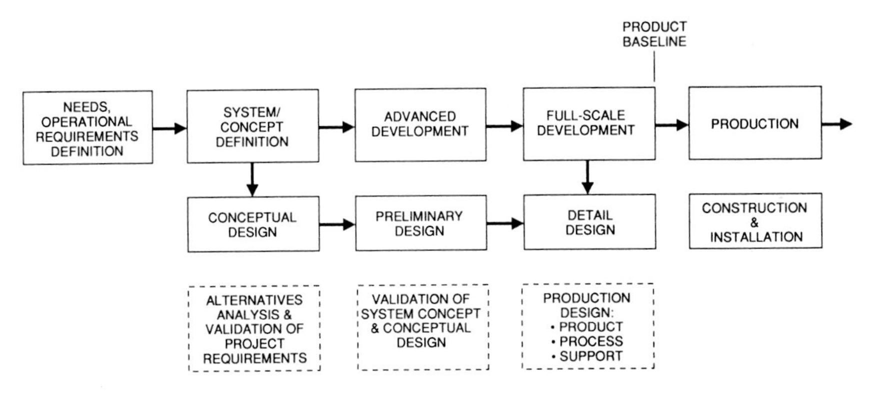

FIGURE 10.7 Nominal project cycle, technical management perspective.

oped and produced in a project framework), the service life or operational life, and the disposal or divestment period. The service-life disposal stages for typical commercial systems or products can be described by the product-market life-cycle curve, as shown in Fig. 10.8. This product-market life-cycle curve describes graphically the life-cycle stages of a commodity within a particular product market in terms of industry sales and profits. The research and development (R&D) stages of the system life cycle precede the product-market life cycle. The investment associated with the R&D and production activities is presented graphically in Fig. 10.8.

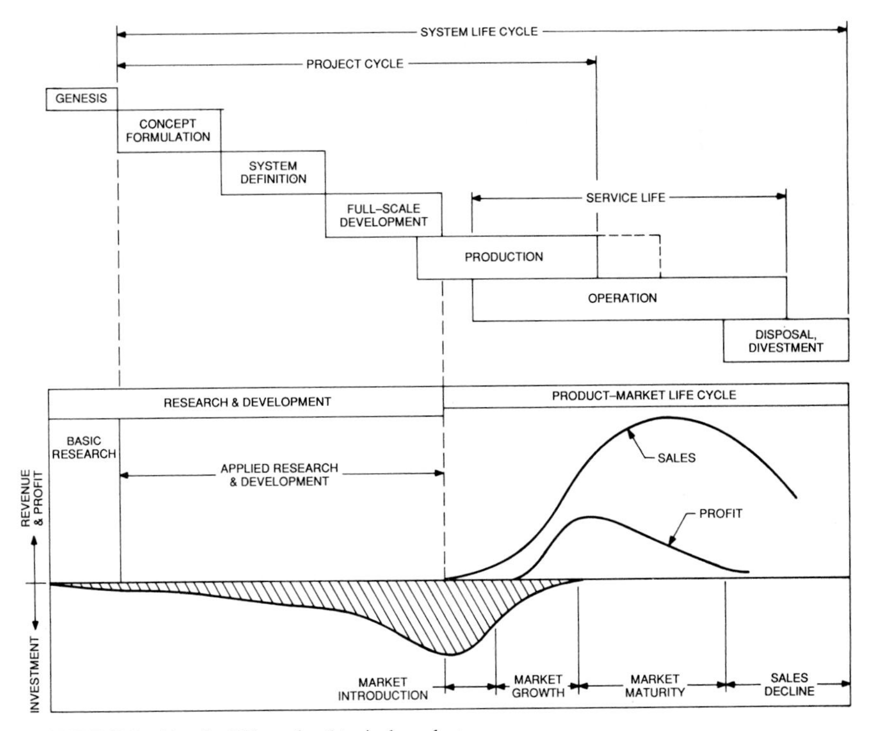

FIGURE 10.8 Nominal life cycle of typical product.

The project cycle, as described in Fig. 10.5, is shown in Fig. 10.8 as a subset incorporated into the system life cycle. The development phases of the project cycle are normally characterized by applied research and development activities, as illustrated. The basic research efforts are usually not subsumed under the project; they are generally managed as technology base activities which are unfocused. However, there are some projects that would include the full range of R&D activities.

The system life-cycle model depicted in Fig. 10.8 is a system management tool. It provides a framework for a life-cycle cost perspective. The life-cycle cost perspective is required during the project cycle in order to ensure that all of the system trades which are conducted account for life-cycle cost impacts.

DEFENSE SYSTEM LIFE CYCLE

Project management as practiced for major defense system acquisitions is generally more complex in nature than for typical commercial system projects. The technical and business complexities are the natural results of the necessarily large and complex defense bureaucracy, the complicated executive and legislative fiscal and political environments, the system operational requirements and mission needs, and the intense public visibility. These and other related considerations affect both the government project manager, the primary agent of the Department of Defense (DoD) and the military service, and the industrial counterpart project manager. The DoD or service project manager is the system acquisition manager. The project manager for the system contractor is responsible for the actual system development and production in accordance with the applicable contracts. This is the normal relationship since most major defense systems are acquired by contracting for the development and production with one or more industrial firms.

The complexity of a typical major weapon system is indicated in Fig. 10.9. This is a simplified diagram of a generic weapon system showing the major elements: a vehicle or platform (such as aircraft, ship, or tank), mission equipment, the weapons, and the support systems. The primary support system is the logistic support system made up of the logistic elements or resources (such as maintenance, supply support, support equipment, and facilities). In some cases the weapon system acquisition is managed as a program, with the major subsets (vehicle, mission equipment, weapons) as designated projects.

The system life cycle of the typical major system represented in Fig. 10.9, while

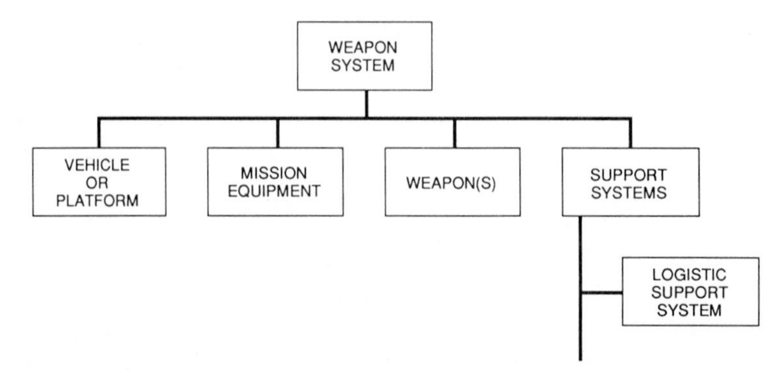

FIGURE 10.9 Generic model of defense system (weapon system).

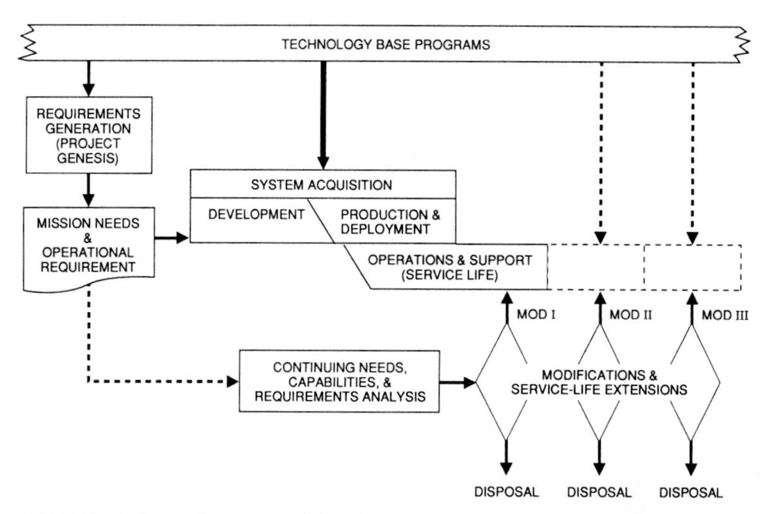

FIGURE 10.10 Defense system life cycle.

similar to that illustrated in Fig. 10.8, has several unique characteristics related to the military requirements and operations. The product-market cycle is significantly different since the military users' "market" (operations) is unlike any commercial market segment. Accordingly the model of a system life cycle shown in Fig. 10.8 requires some conceptual modification for defense system applications. Figure 10.10 provides a simplified model of a typical defense system life cycle. This reflects the DoD or service project manager's view. Programs or projects are initiated to achieve the goals and objectives which satisfy the mission needs and operational requirements defined by the user community—the operational forces. In most cases these programs and projects are initiated to acquire systems which provide the required capabilities. The project cycle in this case represents the system acquisition process. If continuing needs and operational requirements assessments establish requirements for system modifications and service life extensions, projects may be protracted and expanded, or a follow-on major modification project may be initiated.

The user community, that is, the operational forces, defines the operational requirements based on its continuing mission area analyses and threat projections. The material support establishments in the military services implement the system acquisition process and provide the full range of material support. Material support includes technology base efforts, system or equipment acquisition, and in-service support. The primary component of in-service support is logistic support services, including the provisioning of the logistic resources.

DEFENSE SYSTEM ACQUISITION MANAGEMENT

Defense system acquisition management (DSAM) and the system acquisition process are basically guided by the policies established in the Office of Management and Budget (OMB), Circular A-109.[2] The general principles and process rationale prescribed in OMB circular A-109 are illustrated schematically in Fig. 10.11. The fundamental phase evolution and decision logic reflected in this process cycle diagram

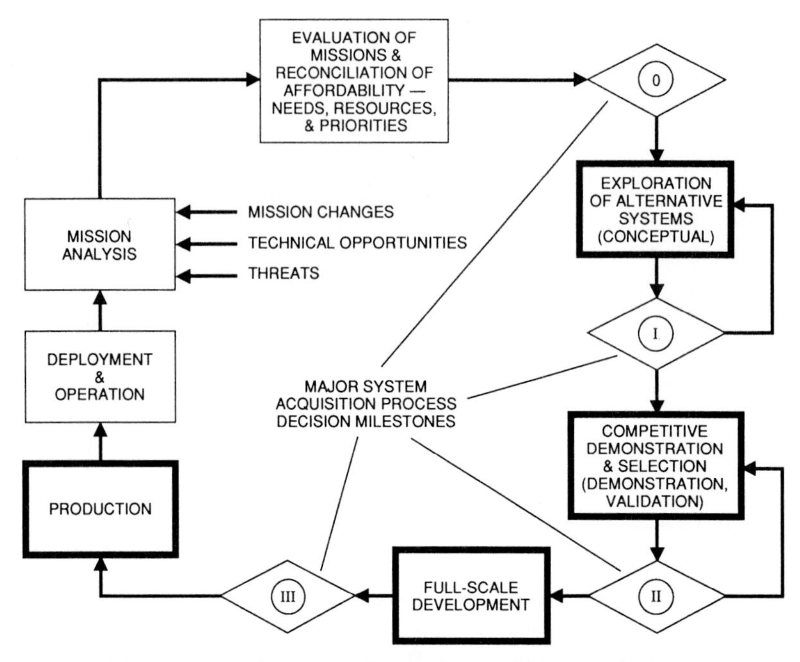

FIGURE 10.11 Major system acquisition process cycle. (*Adapted from Ref. 2.*)

are the same as those outlined in Figures 10.4, 10.5, and 10.10. It is important to note that Fig. 10.11 represents the entire system life cycle as defined in Fig. 10.10. The system acquisition process with the four major decision milestones is expanded in this schematic.

The primary precepts prescribed in OMB Circular A-109 and reflected in Fig. 10.11 are as follows:

- The ultimate users should continuously assess capabilities, capacity, operational and technical threats, technical opportunities, and mission effectiveness through structural mission analyses, which are used to define mission needs and operational requirements.
- Operational requirements and capabilities should be defined in mission need terms and not in specific system prescriptions in order to preclude unjustified precipitous convergence on a detail design.
- The decision to initiate a program or project in response to the defined mission needs and operational requirements should be carefully derived and ultimately sanctioned directly by the head of the agency (Secretary of Defense or the secretary of the military department for DoD cases).
- The decision to initiate the program or project should address and accommodate the affordability implications—the decision to initiate represents an agency commitment to provide funding for the project cycle.
- The project evolution, that is, stepwise progression, should provide for optimum competition among system and design notions in the early phases and ultimate

convergence upon the preferred system and detail design based on risk reduction, technical maturity, and cost-effectiveness.

- The project phasing should be planned and executed so as to guard against premature foreclosure of any practical system or technical option.
- The decisions at the three major decision milestones during the project cycle should be based on careful analyses and assessments of the work completed during the preceding phase—the decision to proceed to the following phase should be guided by successful demonstrable events of the preceding phase (analytical, simulations, design reviews, and test and evaluation, as appropriate).[2]

It is clear that the process indicated in Fig. 10.11 is congruent with the technical management model previously presented in Fig. 10.7. Such a technical management approach is required to comply with the precepts outlined.

Project Management

When one understands the magnitude and the complexity of the business, technical, economic, and political considerations involved with acquiring systems similar to that illustrated in Fig. 10.9 in accordance with the procedures represented in Fig. 10.11, one quickly recognizes the need for implementation of the project management approach. A major defense system acquisition is the epitome of a project as outlined previously. The project cycle and project management framework elaborations are complicated since two distinct perspectives must be explained—the defense system acquisition management (DSAM) process and the developing or producing contractors' projects.

The DSAM process is necessarily complex. In addition the organizational environment in which it is set is also complex—the DoD bureaucracy is large. The project manager must be a master integrator—the orchestrator of the process. The magnitude of this coordination and integration function establishes the necessity for a holistic approach—the system management imperative.

The nature of the problems which face the project manager in implementing the prescribed system approach can only be understood when one understands the requirements for integrating the DoD resource allocation process; the DoD research, development, test, and evaluation (RDT&E) programming; the DoD system acquisition process; and the federal contracting or procurement regulations. This set of problems is diagrammed conceptually in Fig. 10.12. The resource allocation process, the planning, programming, and budgeting system (PPBS), is the centralized system for strategic planning, program development, and resource allocations. It is designed to interface with the presidential budget development and the legislative authorization and appropriation processes. Accordingly it is a calendar-driven process which is influenced directly by political and economic forces. The contracting rubrics are established by legislative actions and the Federal Acquisition Regulations (FAR). These define the methods and procedures for instrumenting contracts in procuring products and services from industry. The DoD RDT&E programming framework establishes guidelines, procedures, and methodologies for implementing the R&D processes (technology base development through operational system development) and the T&E processes (laboratory experimentation through operational test and evaluation). This RDT&E framework provides the direction and programming rules for technology transfer and technical maturation. RDT&E is event-driven; however, the DoD RDT&E programming is driven directly by the PPBS operations. The system acquisition process incorporates RDT&E in the evolution logic illustrated

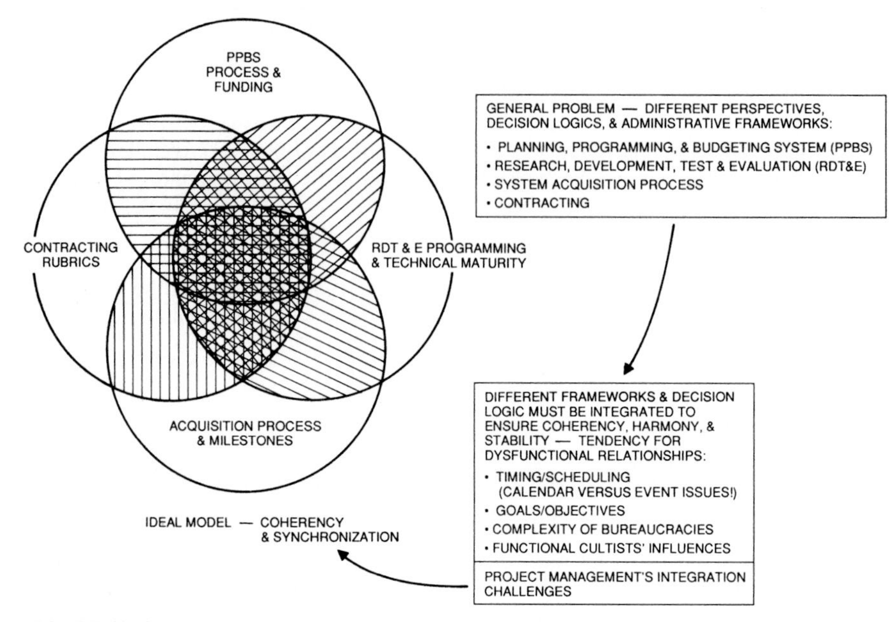

FIGURE 10.12 Defense system acquisition projects.

graphically in Figs. 10.5 and 10.7. The project cycle is largely determined by RDT&E progression, which facilitates technical maturation.

The ideal project management model for the DSAM situation is depicted by the Venn, or logic, diagram in Fig. 10.12. This reflects the need for congruency among the four major interrelated frameworks. The coherence and synchronization result from the system management efforts and ensure stability during the project cycle. However, this congruency is not easy to achieve and maintain. There are several dysfunctional relationships, as indicated in Fig. 10.12, which must be overcome. Adherence to the project cycle model can help to reduce the effects of these dysfunctional tendencies.

Integration and coordination imply balance and coherence. Balance and coherence apply to both the objective system, as illustrated in Fig. 10.9, and the acquisition process, as illustrated in Fig. 10.11. Use of the project cycle model and the project planning and control discipline facilitates the required orchestration and harmony. In addition an effective and efficient organizational project design will facilitate functional integration. The primary functional integration considerations are represented schematically in Fig. 10.13. This functional scheme reflects the necessity to accommodate the complex project environment. Risk management and configuration management are two fundamental project management functions and tool sets. The objective system configuration refers to the functional and physical characteristics as achieved in the operational system and defined in the technical documentation. Accordingly, configuration management is a project management discipline, as well as a system management tool set for:

- Technical definition and documentation of the system functional and physical characteristics (configuration identification)

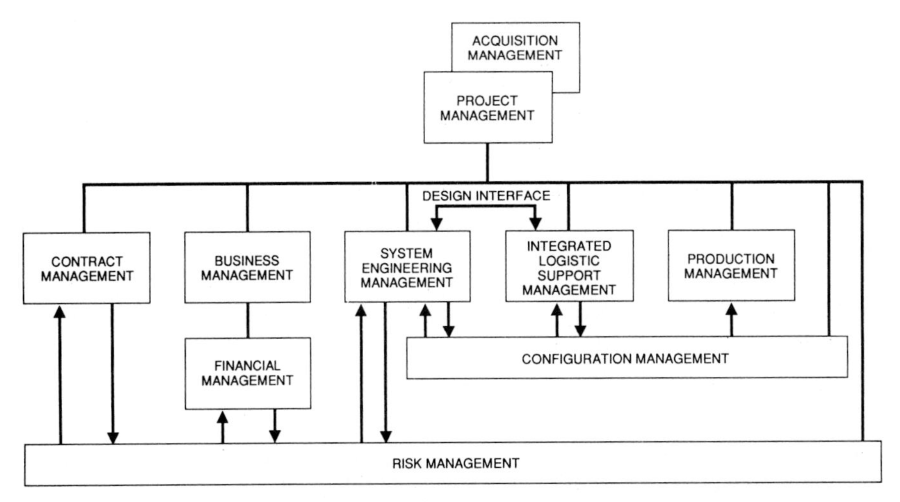

FIGURE 10.13 Defense system acquisition management, project management (function integration).

- Technical and administrative control of all changes to these characteristics and attributes (configuration control)
- Recording and reporting change processing and implementation status (configuration status accounting)

The configuration management discipline and tools are essential project components for the baseline management discussed previously and illustrated in Fig. 10.5.

System engineering management, integrated logistic support (ILS) management, and production management are the three technical system management disciplines and functions in the project management network. System engineering management is usually the lead technical management discipline. There is no universally accepted definition of system engineering. However, most practitioners would agree that it can be characterized as the practical application of scientific and engineering methods and techniques to:

- Transform mission need and operational requirement statements into a description of a system configuration which best provides the required capabilities in accordance with the specified effectiveness metrics
- Coordinate the system design definition and detail design process
- Integrate all technical considerations and assure compatibility of all physical, functional, and technical program interfaces to facilitate an optimized total system design (technical program planning and control)
- Integrate the activities of all the applicable engineering disciplines and specialties into a unified engineering team approach to facilitate an effective and efficient system design process (engineering specialties integration).[3]

ILS is the composite of all the logistic support considerations and resources required to assure effective and economic support of the system throughout its service life. It is characterized by balance, harmony, and coherence among all of the logistic elements or resources which make up the logistic system (such as maintenance planning, supply support, support equipment, and training or training devices). ILS management is a system management approach for developing and deploying the logistic support system needed to achieve the specified operational effectiveness and suitability as well as supportability. Effective ILS management designs and develops this logistic support system concurrent with the objective system development to ensure that:

- The supportability considerations influence the system design process
- ILS impacts are addressed during the system trade studies in the early project phases
- ILS options and tradeoffs can be considered before the system design is frozen or the product baseline is established
- The ultimate logistic system is an integrated system with congruence among all logistic elements

The design interface between the system engineering management function and the ILS management function illustrated in Fig. 10.13 reflects the necessity for a continuous interface and concurrent development of the objective system and its logistic support system.[4]

Production management involves the full range of activities required to produce the total system effectively and efficiently. It includes production planning and support; production engineering and industrial engineering for process, facilities, and plant equipment design, development, and integration; and production inventory management and control. The effectiveness and the efficiency of the production phase of the project cycle are dependent on the production management and the producibility designed into the objective system. In fact, production management effectiveness and producibility are interdependent. Producibility is an inherent system design attribute which characterizes its ability to satisfy the technical and performance requirements while being produced in a specified total time, at a specified production cost, with readily available materials, and using the most advantageous manufacturing processes and methods. Producibility represents the degree to which an item can be produced in the specified production line environment, given the available manufacturing technology and resources, cost and schedule constraints, quality and reliability considerations, and the technical documentation which defines the product baseline.

In recognition of the project management necessity to develop a coherent and cost-effective total system which is both producible and supportable, the concurrent engineering management approach has been articulated. This approach emphasizes the full synchronization of the system engineering management, ILS management, and production management networks and the integration of their process outputs and inputs. It is a logical extension of the design interface instrumentation. Concurrent engineering is a system management approach to the integrated and simultaneous development of all elements of the total system—the product elements and their related processes, including production and support. It is a systematic approach, which is directed at causing the developers to specifically address all components of the system life cycle from project initiation, including the acquisition process as well as operation and support. A notional concurrent engineering approach model is illustrated diagrammatically in Fig. 10.14. In this notional model we see that the ultimate standard for determining total system quality is the satisfaction of the ultimate customers' (users') needs.

Effective planning and control discipline and a well-designed project organization (including structure administrative system, communications system, and team culture) will facilitate the implementation of the concurrent engineering approach. The inte-

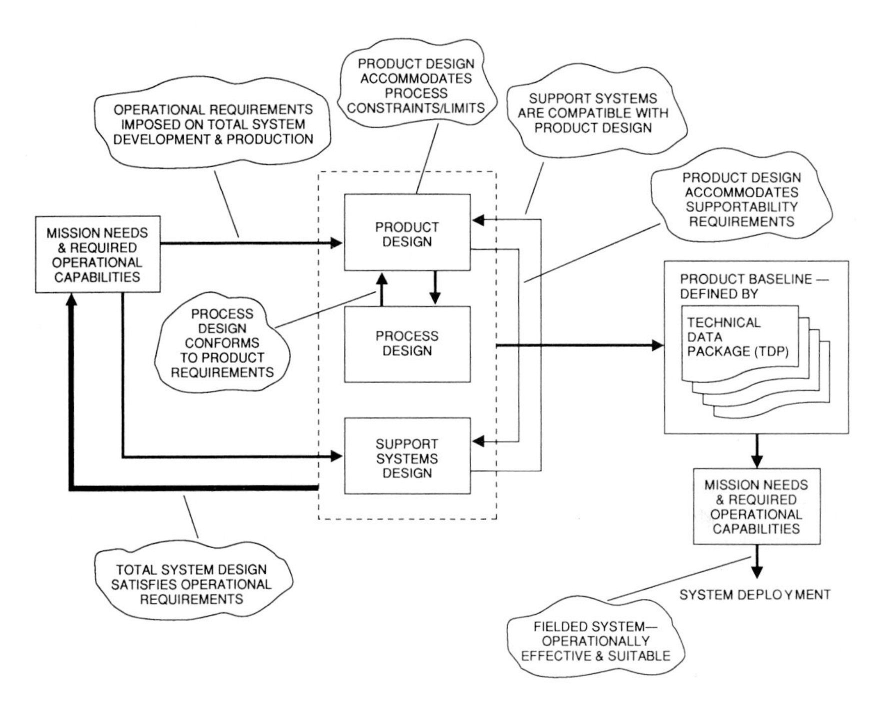

FIGURE 10.14 System development, concurrent engineering approach.

grated planning process and plan set depicted in Fig. 10.2 is a prerequisite for the successful incorporation of concurrent engineering. Incorporation of this concurrent engineering regimen ensures the attainment of a cost-effective total system design which can be replicated in an economically efficient production environment and fully supported throughout its service life. Further, it ensures that there will be no significant degradation of quality or reliability during production or operations.

A second prerequisite for successful incorporation of the concurrent engineering regimen, as well as for overcoming the dysfunctional tendencies outlined in Fig. 10.12, is implementation of a project organization which is oriented specifically toward system management. Figure 10.15 provides an illustrative example of a structure for such an organization. The structure must be complemented with the other organizational subsystems appropriately designed and matched. The nine functional or discipline components indicated in Fig. 10.15 represent the primary supporting business, administrative, and technical components of project management in a typical major DSAM project. This type of structure would accommodate the required functional integration illustrated in Fig. 10.13. It is important to note the criticality of the test and evaluation management and configuration management. These sets of functions are critical to the practical application of the project cycle management logic displayed graphically in Fig. 10.10. The planning and analysis coordinator is another critical team member. Coordination of the development, maintenance, and communication of the integrated planning process and the plan set illustrated in Fig. 10.2 is a critical project management requirement. The preeminence of the four assistant project managers is self-evident in the context of the concurrent engineering framework and the necessity to fully integrate the contract management discipline.

The final success of a defense system acquisition project can be expressed quanti-

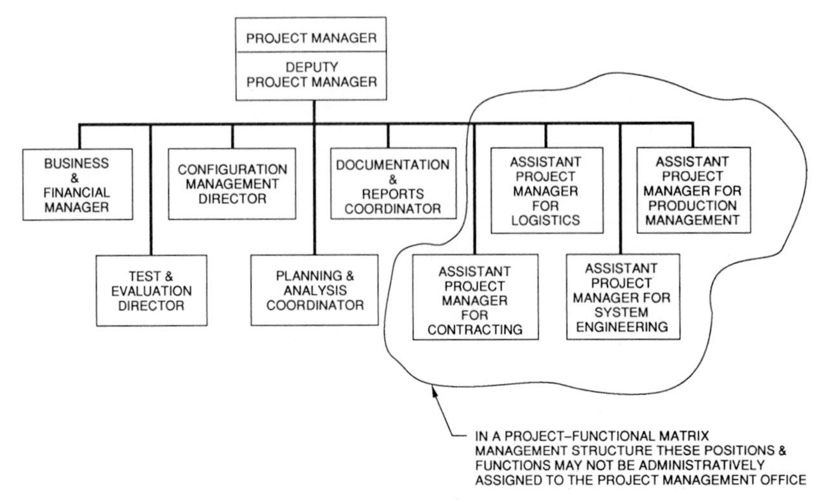

FIGURE 10.15 Project management organization, executive management team.

tatively or qualitatively in terms of three fundamental project dimensions: performance, cost, and schedule. A useful measure of the effectiveness for overall system performance is system effectiveness, which is a complex function of operational availability A_o, dependability D, and operational capability C. This comprehensive total system performance measurement was originally developed in the early 1960s by the Weapon System Effectiveness Industry Advisory Committee (WSEIAC).

Operational availability is a surrogate measure for operational readiness—a measure of the degree that a system is operable and committable at the start of a mission when the mission is initiated at a random time. The operational availability A_o is the probability that the system is operable at the start of a mission. It is a function of the inherent availability A_i, which is a design attribute, and the support system effectiveness. The inherent availability is a function of the reliability and maintainability of the system, both design characteristics.

Dependability is a measure of the system condition at one or more points during the mission. It is the probability that the system, if available at the beginning of a mission, is able to successfully remain in a satisfactory condition throughout the mission. Dependability is a complex function of reliability, maintainability, and survivability.

Capability specifically accounts for the full performance spectrum of the system. It is a function of design adequacy, which is a measure of whether the system can successfully accomplish its specified missions, given that it is operating within the design specifications.

A comprehensive measure of system cost is the system life-cycle cost (LCC). The LCC is the total cost of ownership of a system: the acquisition costs plus the operations and support (O&S) costs plus the disposal costs. The acquisition costs are the development costs and the production costs.

The system schedule relates to the project cycle. It represents the schedule for developing, producing, and ultimately deploying a militarily useful total system within the operational force structure.

Figure 10.16 is a graphic representation of this comprehensive project effectiveness and efficiency measurement schema. It is generally difficult to express all of the

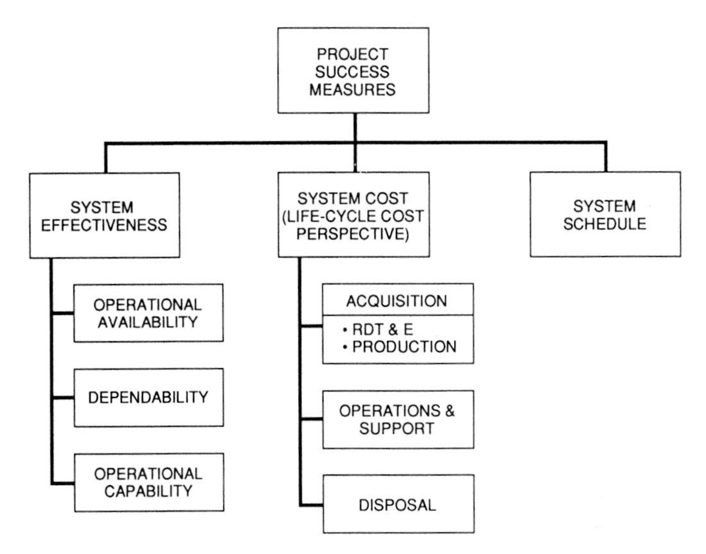

FIGURE 10.16 Project management, factors that determine project success.

indicated components in objective and quantified measurements. However, subjective quantification and qualitative treatments can be utilized to both guide and evaluate the project management processes.

The schema illustrated in Fig. 10.16 and summarized in the preceding provides the means to address the cost-effectiveness imperatives for project management and the decision logic outlined in Figs. 10.1, 10.4, and 10.11. It also provides the means to rationally implement informed trade studies during the project cycle. Cost-effectiveness can be defined as the measure of the benefits to be derived from the resources expended on a system (either for acquisition or for the life cycle). Cost-effectiveness can be expressed as a function of system effectiveness and system cost (LCC in the most comprehensive model), as shown in Fig. 10.17. Cost-effectiveness analysis is a process for developing and displaying, in a specifically tailored format, an array of system cost and system effectiveness characteristics which can facilitate evaluation of the cost-effectiveness or cost benefits of a system. This could be a framework and methodology for presenting information for either objective or subjective evaluation. Trades can be made in this context in a true design to cost mode, a mode in which logical system performance, LCC, system schedule, and supportability tradeoffs can be accomplished in a balanced approach with cross-impact visibility. This concurrent engineering and trade study model is compatible with the concurrent engineering regimen.

Project Cycle

The DoD has established a specific project cycle generic model for major defense system acquisition projects. This model is based on the OMB Circular A-109 process cycle illustrated in Fig. 10.11. However, it is elaborated and tailored for the DoD acquisition management and project management environment. DoD Directive 5000.1, DoD Instruction 5000.2, and DoD Manual 5000.2M[5–7] prescribe the model and the full range of system acquisition management policies and procedures. This

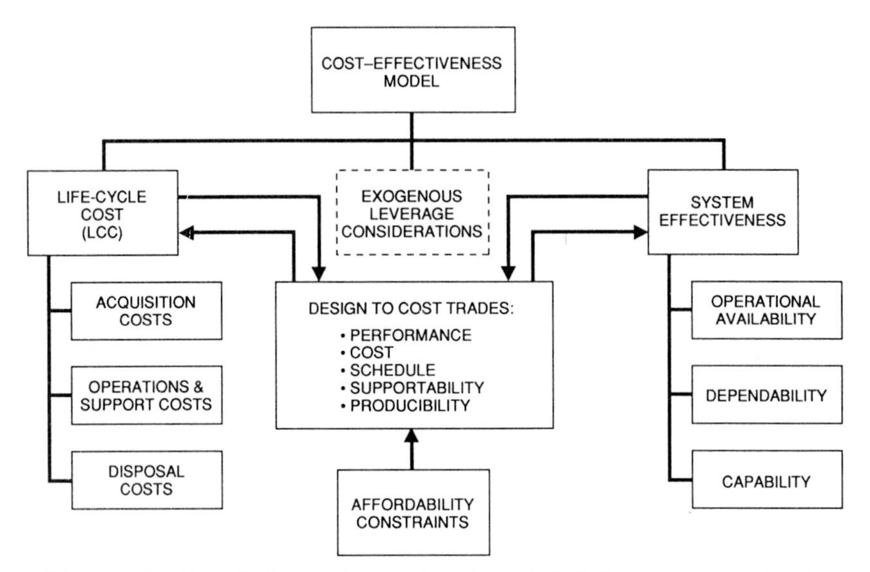

FIGURE 10.17 Cost-effectiveness framework, trade study logic for concurrent engineering management.

defense system acquisition project cycle model is summarized diagrammatically in Fig. 10.18. It is similar in structure and principle to the basic project cycle illustrated in Fig. 10.5.

There are four phases in this project cycle: concept exploration and definition, concept demonstration and validation, engineering and manufacturing development, and production and deployment. A fifth phase, operations and support, extends the concept of the project life to incorporate the full system life cycle. There are four major acquisition decision milestones: concept decision, concept demonstration decision, development decision, and production decision. A potential fifth decision, the major modification decision, is also established to address the potential need for implementation of a major modification project during the operations and support cycle. In practice more than one major modification project may be established. It is significant to note that the third phase of the project cycle, the full-scale engineering development phase, is designated as engineering and manufacturing development. This reflects the concurrent engineering emphasis.

Figure 10.18 indicates the three primary acquisition management documents: the mission need statement, the operational requirements document, and the integrated program summary (IPS). The mission need statement is required for the project initiation decision—it defines the operational requirements baseline. The mission need statement provides a summary of the mission area assessment or mission needs analyses, current programs and capabilities assessments, technology base advancement analyses, life-cycle cost reduction analyses, and defense policy and threat changes analyses. The IPD contains the program structure outline, an LCC estimate summary, the acquisition strategy report (primary project management road map), the risk assessment, the environmental analysis, the affordability assessment, and the cooperative opportunities document. The IPS is updated with each baseline definition: concept, development, and production.

The concept exploration and definition phase is a conceptual phase. During this

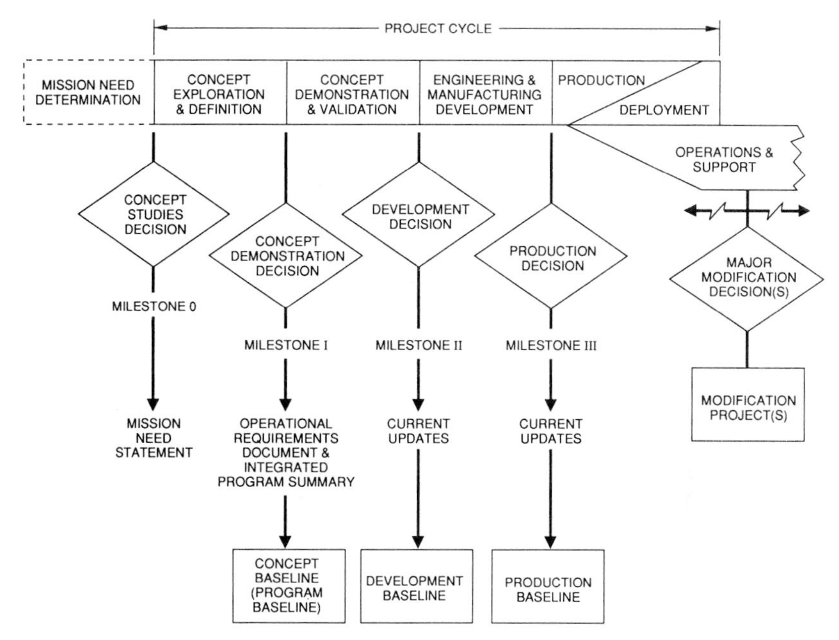

FIGURE 10.18 Defense system acquisition project phases. (*Sources: Ref. 5, pp. 2–5; and Ref. 6, pp. 3–8.*)

phase all practical system alternatives that could satisfy the mission needs and operational requirements are identified. Knowledge and information needed to select the best alternatives for system concept development and conceptual design are acquired. Alternatives which meet the required operational capabilities, cost, schedule, and supportability goals and are affordable are selected. The conceptual designs are completed during this conceptual phase. In addition the acquisition strategy, which will guide the completion of the project, is developed during this phase—the concept or program baseline is defined during this phase.

The concept demonstration and validation phase is initiated subsequent to the affirmative concept demonstration decision. This is generally an advanced development phase in which preliminary design is completed. During this phase the system concept and conceptual designs are validated. The preliminary design is verified and the feasibility (technical and economic) of competing concepts is verified. Testing, analyses, and simulations are completed to demonstrate that all risks have been reduced to acceptable levels and that only engineering development is required to complete the system development. The preferred system is selected for full-scale development, and project affordability is confirmed. In addition the detail design requirements are established. The development baseline is defined at the completion of the concept demonstration and validation phase.

The engineering and manufacturing phase is initiated subsequent to the affirmative development decision, the most critical decision milestone. This is a full-scale engineering development phase in which the total system detail design is completed. In accordance with the concurrent engineering regimen, all product, process, and support system designs are completed and verified or validated by test and evaluation. The independent operational test agency validates the system operational effectiveness and suitability through operational test and evaluation (operational evaluation). Production readiness is confirmed by production engineering and producibility assessments and pilot or limited production prototyping. The production baseline is defined by the complete technical data package at the end of the engineering and manufacturing development phase.

The production and deployment phase is initiated subsequent to the affirmative production decision. During this phase (sometimes the final stage of transition from development to production may occur in two steps—first low-rate initial production and then full-rate production) the total system is produced and deployed. The primary goal of this phase is to produce and deliver cost-effective systems which are fully supported. Production acceptance test and evaluation is conducted to ensure that system performance, reliability, and quality are not degraded as a result of production anomalies. If the transition from development to production has been continuously accommodated from conceptual design to definition of the production baseline in accordance with good concurrent engineering practice, production quality problems can be minimized to acceptable levels, and cost-effective production with high quality can be assured.

The major modification decisions will be programmed as required. The first milestone will generally occur approximately 5 to 10 years after the initial operational capability or first deployment. This decision is often a question of major system upgrade or system replacement. It is precipitated by changes in mission, changes in threat, technical opportunities, affordability problems, supportability problems, or the onset of premature aging, wearout, or fatigue.[6]

The project cycle model presented in Fig. 10.18 represents administrative process and industrial process perspectives. However, it is critical to recognize that the project manager must emphasize the industrial process content. The typical project defined by the project cycle is a technical, manufacturing, and logistic support enterprise, not merely an administrative exercise. We must acknowledge the requirements for the administrative framework—the structure for discipline, the procedures for standardization, and the tools for orchestration. However, we must assiduously guard against administrative process focus. The ultimate consequences of this focus and the resulting administrative emphasis are:

- An acquisition strategy and project execution controlled primarily by exogenous forces, often politically driven
- Blind reliance on military standards and specifications
- Extraordinary emphasis on acquisition cost and schedule and resultant uninformed decisions
- Project management control by administrative milestones instead of measured technical progress
- Subversion of the technical definition and industrial aspects of the project cycle
- Disregard of risk management discipline

This type of inordinate administrative process focus exacerbates the inherent dysfunctional tendencies problem illustrated in Fig. 10.12.

One of the most critical functional integration problems during the project cycle is that of planning, control, and coordination to integrate the system acquisition process with the defense strategic planning and resource allocation process—the planning, programming, and budgeting system (PPBS). The inherent difficulty in accomplishing this integration is indicated in Fig. 10.19. The project cycle for the acquisition process should establish a systematic, achievement-oriented procession in

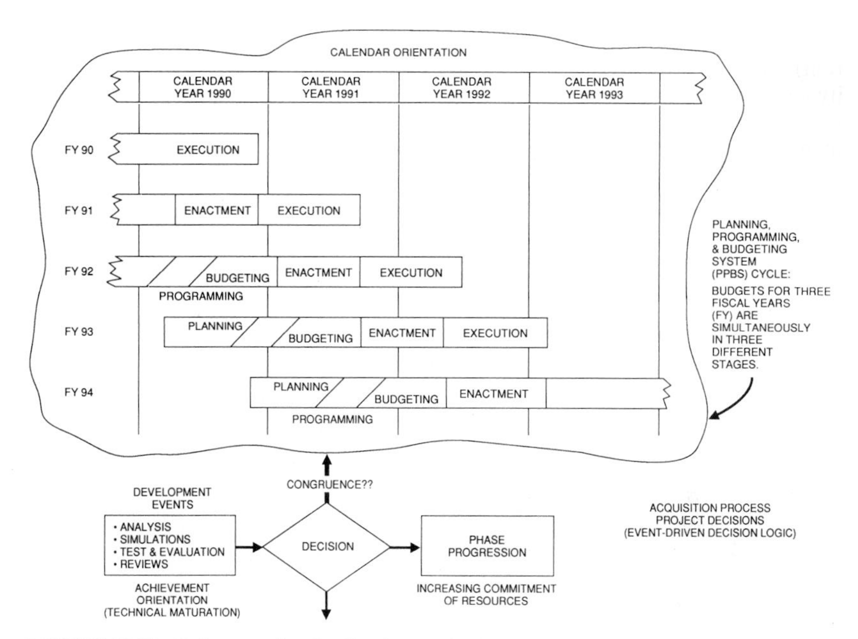

FIGURE 10.19 Defense project funding, interaction of PPBS and system acquisition process.

which the decisions are guided by a system of logical reasoning and the accomplishment of specific event objectives. Accordingly this ordered set of planned interdependent activities must be event-driven. However, the PPBS must be synchronized with the congressional annual authorization and appropriation process. It is calendar-oriented and schedule-driven. The project manager, the military service, the DoD, and the executive department must plan, control, and coordinate effectively in order to ensure that the project cycle sequencing is kept congruent with the PPBS. The major acquisition decision milestones cannot be forced to be calendar-driven. However, since phase progression is dependent on the availability of increased resources, the decision must be carefully integrated within the PPBS and budget enactment sequencing. Further, the PPBS should be used to establish and maintain the affordability context for the project.

Reconciliation of the event-driven acquisition process and the calendar-driven PPBS is a major functional integration problem for project management and general management. Figure 10.20 shows the magnitude of this integration task within the overall project and agency integrative network. The operational requirements and performance or technical requirements transformation and translation, the technical maturation and design and development, and the resource allocation must be orchestrated within the system acquisition management framework. Project management discipline and the application of a system approach facilitate the required functional integration. Implementation of concurrent engineering and effective technology transfer are the fundamental technical management components of the system approach. Planning, control, and coordination are the enabling management functions. The project manager and appropriate general managers at the military service department and the Office of the Secretary of Defense levels must plan, control, and coordinate within the project cycle, the PPBS, and military operations to en-

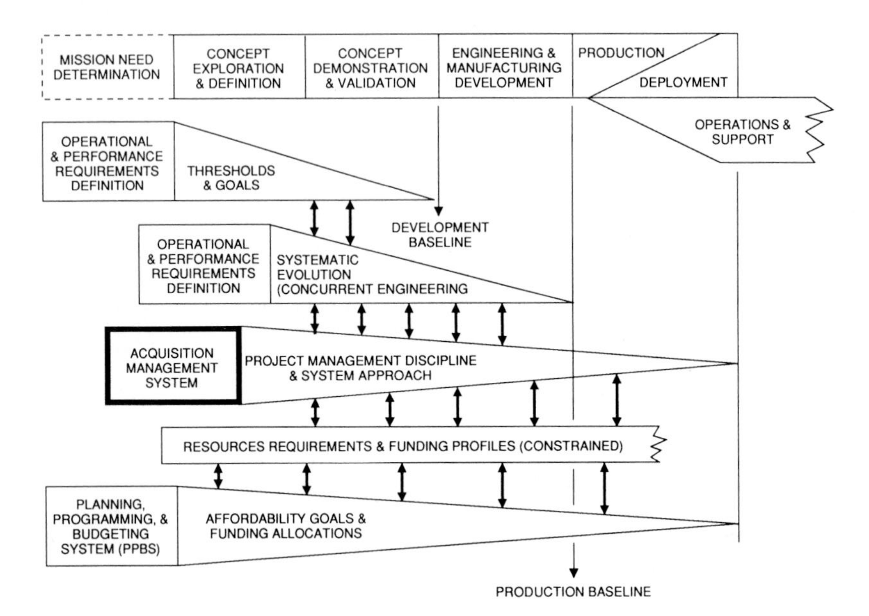

FIGURE 10.20 Defense system acquisition management, an integrated management paradigm.

sure coupling of the requirements definition and transformation aspects, system development, affordability determination, and resource allocation. Decoupling among these interrelated activities is a major cause of project instability, ineffectiveness, and failure.

Experience during the last 15 years indicates that the time required to develop, produce, and deploy a major defense system ranges between 10 to 18 years, as illustrated in Fig. 10.21. In many cases inordinate project schedule slippages or stretch-outs result in extension to the higher end of this range with the resulting consequences as indicated.

Project histories have shown that excessive slippages and stretch-outs are most often caused by some combination of the following:

- Project instability, often resulting from the uncoupling described (instability is reflexive in that slippages or stretch-outs reinforce instabilities)
- Overoptimism (such as engineering optimism)
- Ineffective planning and control
- Inadequacy of the technology base or immaturity of the enabling technologies
- Ineffective risk management discipline
- Inaccurate cost analyses and estimation
- Uncontrolled "requirements (operational, performance, or technical) creep"
- Ineffective quality management

The inordinate schedule slippages and project stretch-outs can, in turn, result in project failures. Premature technological obsolescence and the concomitant service life

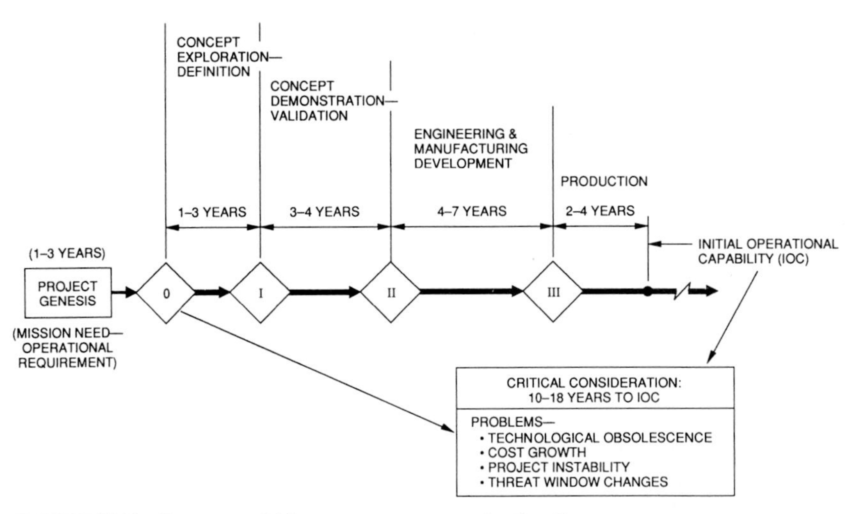

FIGURE 10.21 System acquisition process, representative time line.

supportability difficulties, excessive cost growth, and decreased military worth and utility are also resultants.

The requirements-funding decoupling phenomena are particularly insidious since these usually result in the inability to execute the planning. Effective implementation of concurrent engineering is often precluded by a lack of timely funding. In addition inappropriate funding profiles will result in schedule slippages, project stretch-outs, and reduction in the project cost-effectiveness. The design, production, and logistic discontinuities are generally never resolved adequately. False economies in the early phases of the project cycle usually result in ultimately higher acquisition and life-cycle costs, excessive project delays, and lower than projected military utility or worth.

Project planning and control provide the foundation for project success. A rational affordable acquisition strategy is the foundation for an effective defense system acquisition project plan. The basic principles of the integrated management paradigm outlined graphically in Fig. 10.20 should be incorporated and reflected in any efficient acquisition strategy.

Figure 10.22 is an illustration of nominal acquisition strategy, which involves maximum competition. In this strategy technical competition is implemented during the first two phases of the project cycle. The technical competition consists of two or more separate and distinct conceptual and preliminary design efforts by two or more competing contractor teams. During these phases emphasis is placed on appropriate risk reduction, cost-effectiveness tradeoffs, and technical maturation progression. The result of this competition is the selection of a preferred cost-effective alternative (or a combination of alternative conceptual or preliminary design results) for full-scale development and detail design. During the engineering and manufacturing development phase a single detail design approach is implemented. However, an affordable cost-effective production competition option is established and maintained (such as some type of "leader-follower" full-scale development strategy). This would include the development of a compatible technical data package (TDP) and a low-rate initial production stage in which the TDP is properly validated. The TDP

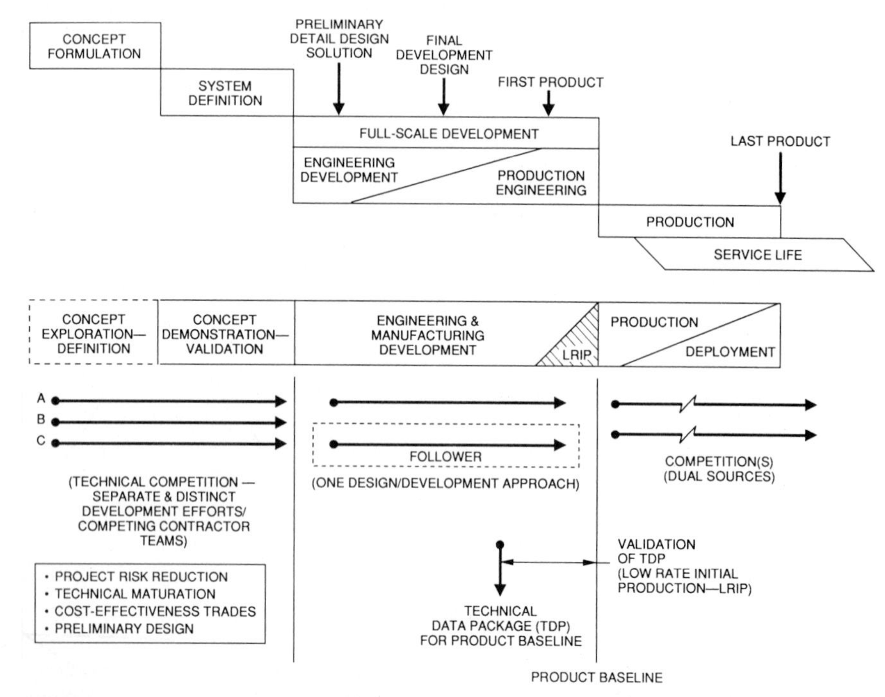

FIGURE 10.22 Nominal acquisition strategy, maximum competition. (*Source: Ref. 11, pp. 4-1 to 5-18.*)

contains detail design specifications as well as relevant production specifications (such as tooling, fixtures, and factory test equipment). Any significant requirements-funding decoupling will render this type of competitive strategy inexecutable.

DEFENSE SYSTEM ACQUISITION PROJECT CYCLE

Concurrent Engineering Perspective

The defense system life cycle has two major subsets: the system acquisition process and the service life (operations and support), as illustrated in Fig. 10.10. The typical system acquisition process is implemented in the nominal project cycle illustrated in Fig. 10.18. Figure 10.23 shows typical representative cost profiles for a major defense system over its life cycle. The life-cycle cost (LCC), or cost of ownership, includes the acquisition costs, the operations and support costs, and the disposed costs. Experience has shown that the LCC is generally distributed as illustrated in the profiles presented in Fig. 10.23—approximately 10 percent for development, 30 percent for production, and 60 percent for operations and support. These costs are actually expended as indicated by the cumulative LCC expenditures curve: approximately 40 percent is expended during the acquisition process. It is significant to note that approximately 60 percent of the LCC is expended for operations and support.

In the typical defense system acquisition project 10 percent of the LCC is spent in

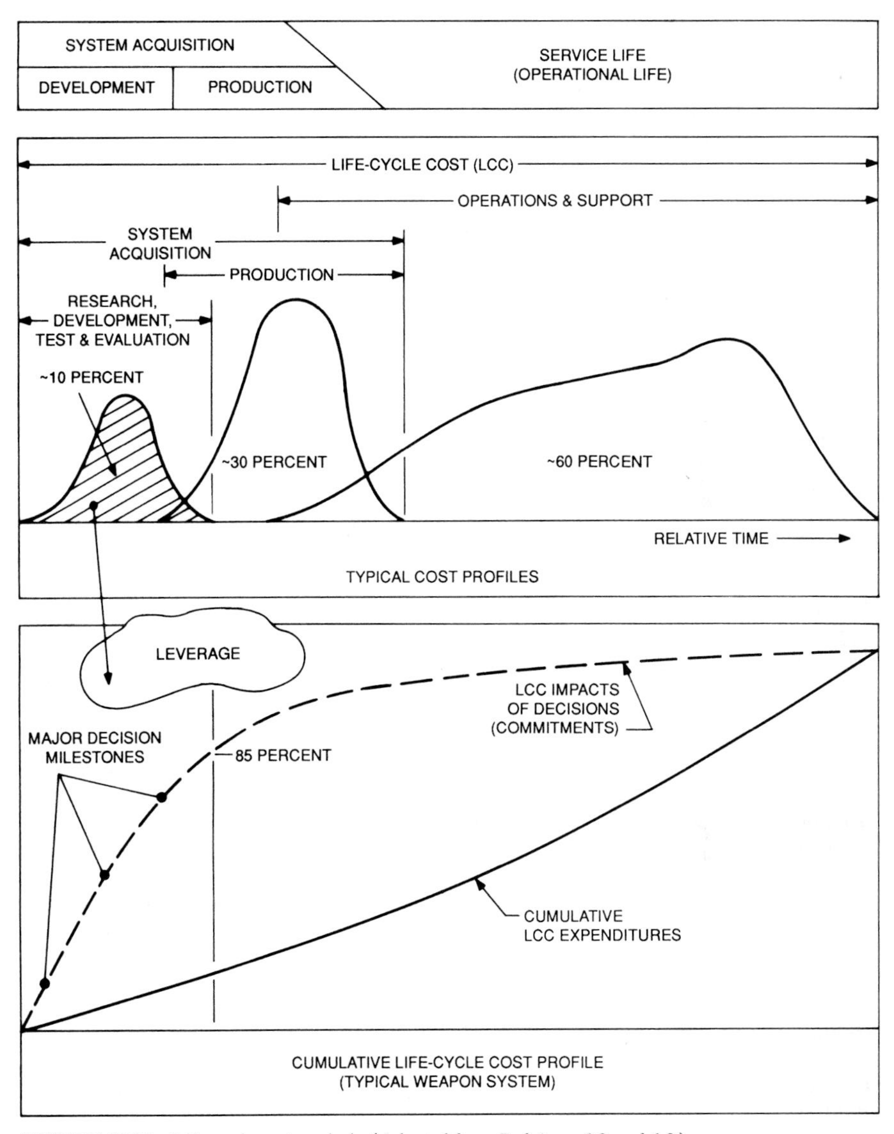

FIGURE 10.23 Life-cycle cost analysis. (*Adapted from Ref. 4, pp. 6-2 and 6-3.*)

development. However, during the development phases of the project cycle, decisions are made and implemented which directly influence approximately 85 percent of the LCC. These development decisions, product and process designs, and logistic support system design are the total system performance and cost drivers. They dictate how the system will be produced, operated, and supported during its entire life cycle. Therefore the 10 percent invested in research, development, test, and evaluation (exclusive of any relevant technology base inputs) has critical leverage effects

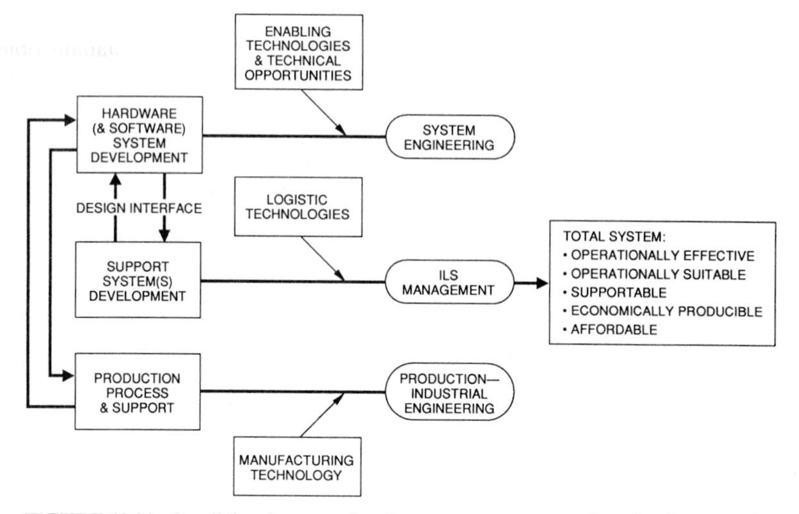

FIGURE 10.24 Parallel and process development, concurrent engineering (integrated simultaneous stepwise progression).

on the cost-effectiveness of a representative system for its entire life cycle. The appropriate investment in effective development efforts can produce significant LCC cost-effectiveness results within the context of the system approach indicated schematically in Fig. 10.17. Recognition of this LCC leverage concept further reinforces the criticality of the concurrent engineering regimen.

This concurrent engineering regimen prescribes the integrated parallel development of the objective (product) system, the manufacturing or production processes, and the logistic support system, as illustrated in Fig. 10.24. This requires that the system engineering process, the ILS process, and the production or industrial engineering process be directly interfaced from the initiation of the project cycle and continuously coordinated during all phases of the cycle. It also requires that the technology transfer or technology insertion considerations be expanded to promote and facilitate the appropriate transfer or insertion process technologies (such as manufacturing technology) during the integrated parallel development.

Unification of the concurrent engineering efforts is managed within the system engineering management framework. The primary integrating mechanism is the design interface between the ILS process and the system engineering process. This concept was previously illustrated in Fig. 10.13. Trade studies and the logistic support analysis (LSA) in this interface management provide the critical tools. The LSA provides the basic tools for ILS management and the enabling mechanics and tools for interfacing with system engineering management. Some of these tools and mechanics are overlapping and reciprocal with the system engineering methods. The LSA incorporates the utilization of a wide range of analytical methods, techniques, and models. It is actually a planned series of tasks which examine all aspects of the total system to:

- Define the supportability requirements
- Evaluate alternative system and equipment configurations
- Determine a preferred system solution
- Evaluate given design configurations

- Establish logistic element requirements
- Relate design, operational, and support characteristics to system availability objectives
- Influence the system development process

The basic system engineering process description reflects the iterative nature of the classical requirements transformation and definition, the system optimization, synthesis, or definition, and the detail design cycle. It is significant to note that the initial requirements transformation and definition procedure is a functional analysis. In addition the supportability requirements are factored directly into the trade studies and the system definition procedures.

The concurrent engineering approach is compatible with, and a prerequisite for, the quality engineering methods pioneered by Dr. Genichii Taguchi. The Taguchi methods for designing robust products and processes, which are relatively insensitive to uncontrollable variables, are effective system engineering methods.

Research, Development, Test, and Evaluation Perspective

The development phases of the project cycle involve some or all of the RDT&E continuum of activities. In most cases a system acquisition project does not require any basic research; however, it often requires applied research and development activities. A simplified model of an industrial R&D spectrum is illustrated in Fig. 10.25. This indicates the relationships between basic research, applied research, and development in the R&D continuum. The basic information and the hardware inputs and

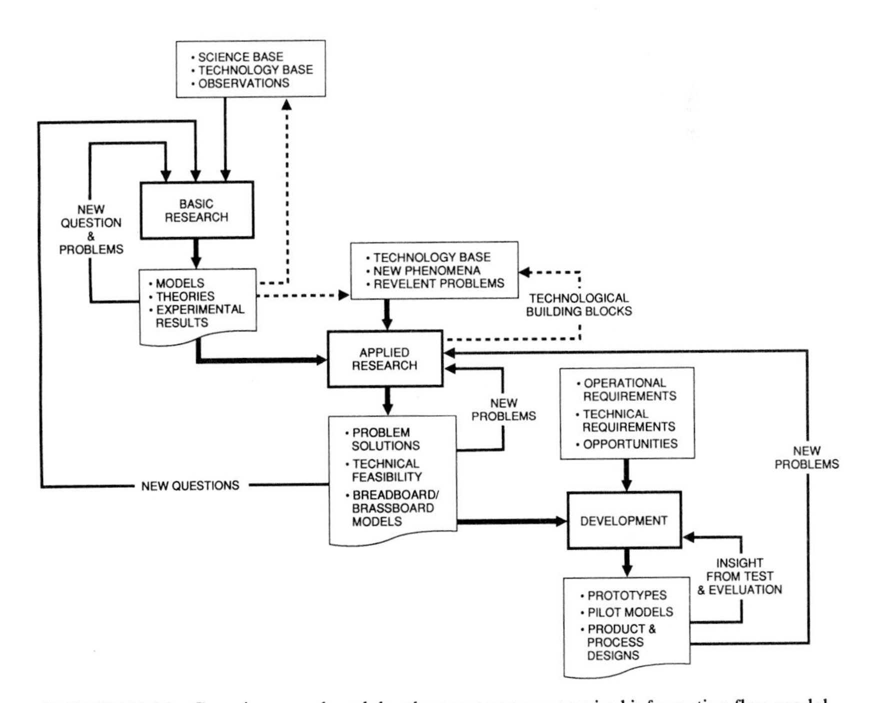

FIGURE 10.25 Generic research and development process, nominal information flow model.

INDUSTRY CATEGORY	BASIC RESEARCH	APPLIED RESEARCH			DEVELOPMENT	
DEFENSE RDT&E PROCESS (CATEGORIES)	RESEARCH (CATEGORY 6.1)	EXPLORATORY DEVELOPMENT (CATEGORY 6.2)	TECHNOLOGY-ORIENTED ADVANCED DEVELOPMENT (CATEGORY 6.3)	SYSTEM-ORIENTED	ENGINEERING DEVELOPMENT (CATEGORY 6.2)	OPERATIONAL SYSTEMS DEVELOPMENT (CATEGORY 6.6)
FUNCTIONAL PURPOSE	DEVELOPMENT OF KNOWLEDGE BASE & SCIENCE	DEVELOPMENT OF TECHNOLOGY & INDICATE TECHNICAL FEASIBILITY	COMBINE NEW TECHNOLOGIES IN TECHNOLOGICAL BUILDING BLOCKS & RISK REDUCTION		DESIGN & DEVELOP SYSTEM FOR SERVICE USE (INTEGRATION OF ENABLING TECHNOLOGIES)	REFINE SYSTEM DESIGN & CORRECT DEFICIENCIES
GENERAL ORIENTATION	SCIENTIFIC		TECHNOLOGICAL		DESIGN & SYSTEM ENGINEERING	PRODUCTION ENGINEERING
R&D PROGRAM MODE	TECHNOLOGY MODE			SYSTEM DEVELOPMENT/APPLICATION		
BASIC OUTPUTS	TECHNOLOGY STOCK: •TECHNICAL OPPORTUNITIES • ENABLING TECHNOLOGIES • TECHNOLOGICAL BUILDING BLOCKS SCIENCE STOCK			OPERATIONAL SYSTEMS		

FIGURE 10.26 Research and development program structure.

outputs of each set of activities are summarized in order to characterize the content of the efforts in each of these stages. As indicated, the basic research activities increase the science and technology base stock and, in some cases, can be an integral stage of a system development process.

Figure 10.26 outlines the DoD R&D program structure, which relates the generic R&D spectrum illustrated in Fig. 10.25 to the DoD programming framework. The three stages of R&D which are generally included in the generic industrial model are subdivided into six categories of RDT&E in the DoD process model. These categories can be classified as either technology base development or system development, as shown. All or some of the system development stages usually constitute the development phases of the defense system acquisition project cycle. In some cases the defense system acquisition project will specifically subsume some technology base activities. However, as a general rule the technology base development efforts are managed as separate program activities.

The nominal separation of the technology base development and the system development within the system acquisition process is elaborated in Fig. 10.27. Two fundamental modes of technology transfer, which can be incorporated in the project cycle, are indicated—technology transition or technology insertion. Transition, which is often driven by technology-push incentives, refers to the logical progression of some definitive equipment or system through the full spectrum of R&D stages from research or exploratory development to engineering development. In this case the technology base activities are specifically focused for the ultimate transition into system-oriented advanced development or engineering development.

Technology insertion involves the identification and integration of enabling technologies, technical opportunities, and technological building blocks, which are needed to meet operational, performance, or technical requirements. In this mode these product and process technologies are drawn from the technology base stock, which is established as a result of DoD-sponsored technology base development as well as all other technology base programs (such as independent research and development and industrial research and development). Insertion is the primary mode of technology transfer used, with a few notable exceptions. In this mode it is usually dif-

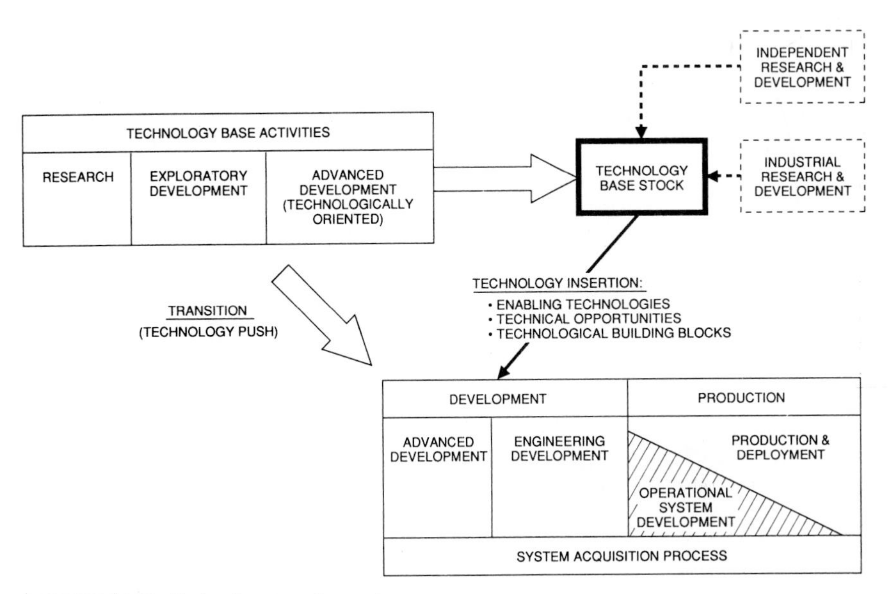

FIGURE 10.27 Technology transfer modes.

ficult to point to any single output of the technology base as the major source of system development. System development is generally the resultant of the combined effects of several enabling technologies, technological building blocks, and technical opportunities. Progress in these areas usually accumulates over long periods. Effective and efficient engineering development requires the skillful selection and integration of appropriate technological building blocks, technical opportunities, and enabling technologies in the design of a total system. The technology insertion is most often implemented in response to the requirements-pull considerations.

The requirements pull–technology push considerations for project initiation and technology transfer strategy are shown in Fig. 10.28. In most cases the project initiation decision logic is derived primarily from the mission need and operational requirements base. However, some aspects of a technology-push rationale are often factored into the decision and the acquisition strategy. Similarly the system development strategy and approach are usually based on the requirements-pull rationale. However, in most cases some degree of technology push will be incorporated in the technology transfer. In the extreme case the technology-push rationale would be the primary driver of project initiation—the project is initiated to exploit technology for some advantage.

The five summary categories of R&D in the DoD programming structure (six when the advanced development category is subdivided into the two subcategories technology-oriented and system-oriented advanced development) are summarized in Fig. 10.26. Research includes scientific efforts directed to increase the knowledge of natural phenomena and to solve basic problems in physical, engineering, environmental, and life sciences. Research involves systematic and intensive study with the purpose of developing comprehensive scientific knowledge and understanding. Exploratory development is directed to shape the basic knowledge base into technological building blocks. It also involves efforts directed to evaluate the technical

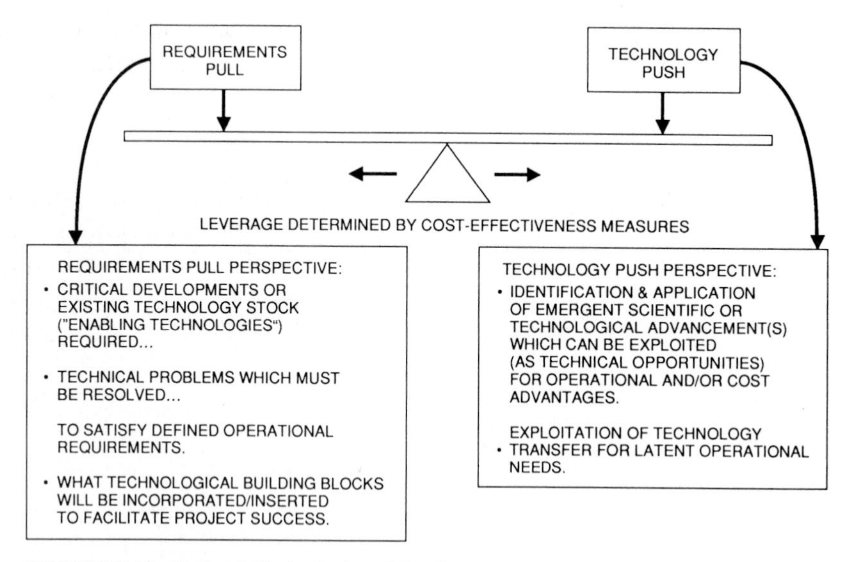

FIGURE 10.28 Project initiation logic and development approach.

feasibility of solutions. When coupled with research it serves to sustain technology base momentum for orderly progress in the state of the technological art. Advanced development involves efforts to assemble technological building blocks in some conceptual or preliminary system configuration for experimentation or validation of system concepts. In complex system developments it represents a sequential link in the development process that bridges the gap between the technology base and engineering development. In this context it is implemented to increase knowledge and technical definition while reducing the risk prior to large investments in engineering development activities. Advanced development models (ADM) or prototypes are fabricated to demonstrate the system concept (conceptual design or preliminary design) and to facilitate the engineering development activities. Engineering development is the full-scale development of the total system for production and operational service use. It includes the full range of design and development activities, culminating in the definition of the production baseline. Engineering development models (EDMs) and production prototypes are fabricated for test and evaluation prior to the establishment of the production baseline. In some cases an additional stage of development, operational system development, may be needed after release to production to refine the system design or correct deficiencies identified in operational test and evaluation. It is important to note that in each of the aforementioned R&D stages, product and process technologies and logistic technologies must be progressed and integrated to develop a total system which is operationally effective and suitable, supportable, economically producible, and affordable for its entire life cycle.

Test and evaluation (T&E) is a critical set of activities of the project cycle. It is actually a spectrum of activities ranging from early experimentation to ultimate production acceptance tests. The various T&E events are required to assist in the development process, verify achievement of the performance and technical objectives, demonstrate operational effectiveness and suitability, and validate the continuing quality of the delivered systems and equipment. Accordingly, the T&E process provides data and information to facilitate the system development, certify achievement of the project requirements (operational, performance, and technical), confirm risk reduction, provide confidence for decision authorities at the major decision milestones, and evaluate operational concepts and tactics. Testing is the execution of the actual test events or simulations, and evaluation is the analysis and assessment of the results.

In the defense system acquisition project cycle, the T&E spectrum is subdivided into three major categories: development test and evaluation (DT&E), operational test and evaluation (OT&E), and production acceptance test and evaluation (PAT&E). The DT&E process is the responsibility of the developing agency, while the OT&E process is the responsibility of the independent operational testing agency. The PAT&E process is the responsibility of the developing agency. The various DT&E events are conducted as either contractor tests or government tests, in accordance with the test and evaluation master plan (TEMP), one of the critical project management plans. The DT&E events are planned and executed to:

- Facilitate development and assist concurrent engineering
- Verify risk reduction
- Verify achievement of the performance and technical requirements (technical evaluation)
- Certify readiness for the operational evaluation, the critical stage of the OT&E process which is the major consideration for release to production (production decision)

The OT&E events are planned and executed to:

- Estimate the operational effectiveness, operational suitability, and operational readiness of the system and equipment
- Assist the project manager in selecting the preferred system alternatives during the development
- Assess the operational impacts and utility of the system
- Evaluate the operational concepts, tactics, and doctrine[8]

It is significant to note that the two most critical stages of the DT&E and OT&E processes, technical evaluation and operational evaluation, respectively, should be conducted using test articles which are representative of the actual production versions—production prototypes. In addition the total system should be available for the operational evaluation.

Figure 10.29 shows an outline of a nominal T&E program, indicating the successive phasing of the processes. The various models which are used to conduct the different T&E events during the progressive development are reflected in this simplified synopsis. These models include the mathematical models and the computer models which are used for analytical simulations during the early stages of system development and technology base development.

The general rationale for integrating the R&D process and the T&E process within the system acquisition framework is illustrated schematically in Fig. 10.30. The R&D process establishes the methods for technical maturation and risk reduction. The T&E assists in this and provides the data and information to support the decision making. It is important to note the possible options at each major decision milestone—proceed directly into the succeeding phase, reassess the requirements and objectives, or cancel the project. Canceling the project should always be retained as a practical option without stigmatizing the project team or the agency. The tendency to build a project management environment of subjective optimism based on

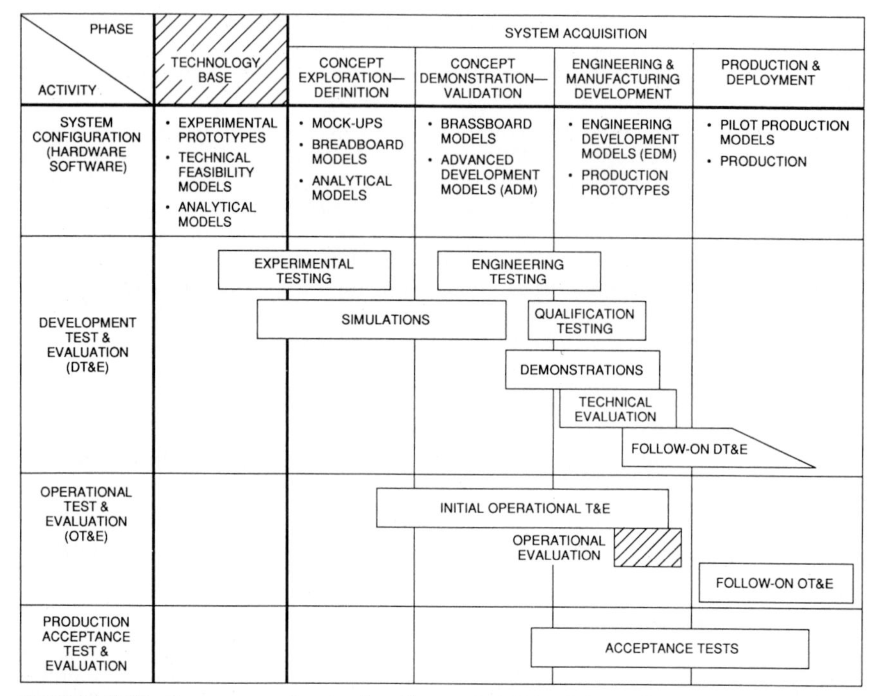

FIGURE 10.29 System test and evaluation. (*Source: Ref. 12, pp. 3-1 to 4-7.*)

inordinate zeal and advocacy must be continuously controlled. In addition the project manager must guard against creating an inappropriate technological imperative—an invitation for a technological orgy. In most cases the affordability rule will demand affordable solutions and, therefore, some design to cost logic, as illustrated in Fig. 10.17. It is a historical fact that the development of defense systems has pushed the state of the technological art in material, devices, and physical methods. However, project management caution and informed skepticism are prudent. Often the caution of Voltaire should be considered: "The best is the enemy of good enough!"

The technology transfer from the technology base to application in the operational system development, or from other extant applications, must be carefully planned and executed. Comprehensive and exhaustive technological forecasts and assessments should be the basis for the technical development plan, which identifies the technical opportunities and enabling technologies. The transition or insertions of these technological building blocks should be systematically subsumed in the design and development evolution. The technology transfer process (transition or insertions) should be planned and controlled within the concurrent engineering framework order to ensure systematic evolution and design maturation. Unplanned and uncontrolled insertions, particularly those which introduce relatively revolutionary changes in the late stages of development, result in significant perturbations, including cost growth and schedule slippages. These are usually caused by requirements creep or unbridled technology push.

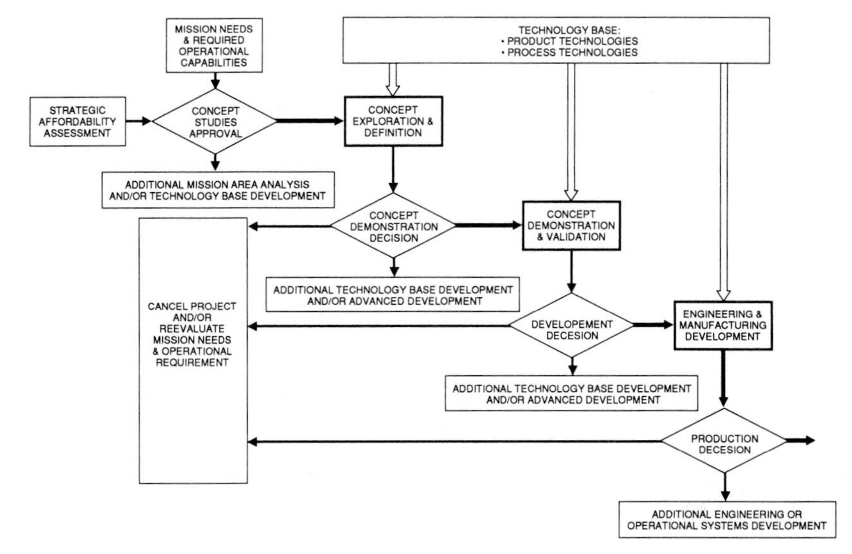

FIGURE 10.30 System development, decision events and technical maturation.

Configuration Management Perspective

Configuration management is a critical project management function. It is critical to project management in general and effective technical management because it provides the tools and discipline for preventing engineering, logistic support, and production anarchy. These tools and discipline are used to:

- Guide and control the design and development
- Guide and control the technical documentation process
- Implement baseline management

The guidance, control, and baseline management discipline facilitates and promotes the orderly definition of operational, performance, technical, design requirements, and production requirements as well as the complementary T&E requirements. However, configuration management is effective and efficient only when it is coordinated and integrated with the basic system acquisition decision milestone. The baselines must be synchronized with the project cycle phasing.

A baseline is a standard used for goal and objective setting, process calibration, and control. It is a quantitative and qualitative description formally specified. In the context of defense system acquisition projects, four mutually supportive and interdependent types of baselines can be identified:

- Program or project baselines (concept, development, production, and operational support; see Fig. 10.18)
- Technical performance baselines (parameter)
- Cost baselines
- Configuration baselines (functional, allocated, and product)

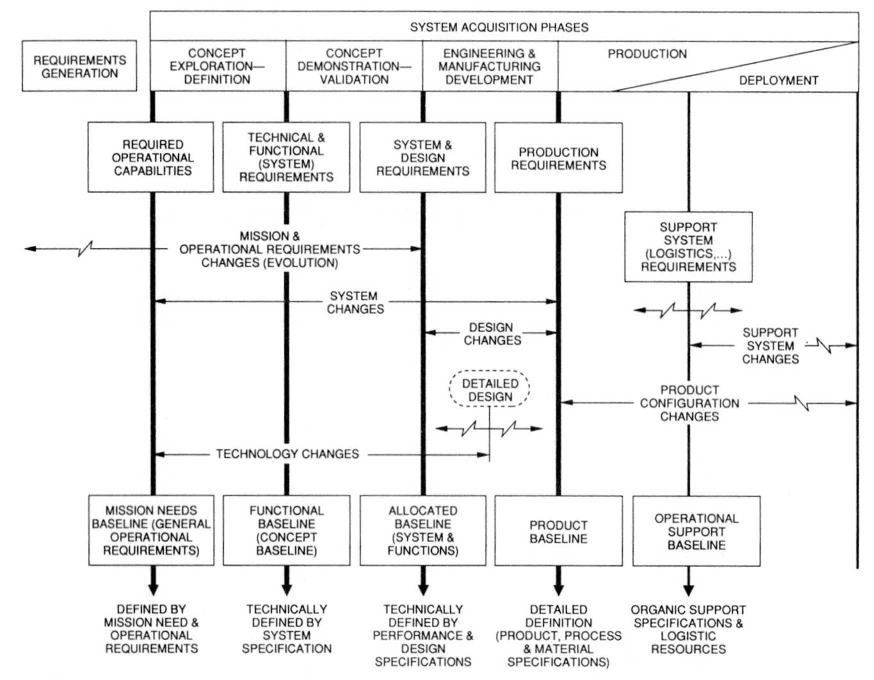

FIGURE 10.31 Project baseline management, configuration management model. (*Source: Ref. 13, pp. 6.6a–6.6e.*)

Configuration is the set of physical and functional characteristics of the system. A configuration baseline definition constitutes the current system configuration identification. It is documented by specifications and engineering drawings. The configuration baseline definitions must be fully coordinated with, and supported by, the program or project, technical performance, and cost baseline definitions.

Figure 10.31 illustrates some of the basic considerations of baseline management using configuration management tools and discipline. The configuration identification and configuration control aspects of configuration management are outlined graphically within the project cycle context. In this descriptive model five configuration baselines are progressively defined as a result of the system development and the configuration identification process. Three of these baselines are the formal configuration baselines prescribed by standard configuration management practice—the functional, allocated, and product baselines. These are defined by the system specification, the development specifications (performance and design specifications), and the system technical data package (product, process, and material specifications, including detail design drawings), respectively. Two additional baselines are also included in the model, operational requirements baseline and operational support baseline. The operational requirements baseline is formally defined by the mission need statement and the operational requirements documentation. The operational support baseline is the resultant of the ILS process, specifically logistic support analysis (LSA). It is defined by the logistic support analysis record (LSAR), the integrated logistic support plan (ILSP), and the supporting operational logistic elements or resource plans.

The ideal configuration control logic is depicted in Fig. 10.31. This notional ideal model assumes that a definitive operational requirements baseline is formally established or a result of a comprehensive and exhaustive mission area assessment or mission need analysis process. It also assumes that the technical development plan establishes the means to identify the feasible technical opportunities and enabling technologies early in the development evolution and to effect the required technology transfer within the planned system development phasing. Both of these assumptions imply the early integration of operational, technical, support, and affordability considerations to promote the progress of cost-effective and affordable project solutions. In this idealized framework, configuration control attends to the expected changes shown in Fig. 10.31 during the project phases as indicated. This change discipline accommodates the simultaneous needs for evolution flexibility and change control. Significant mission and operational requirements changes subsequent to the establishment of the allocated baseline will result in significant project performance, cost, and schedule impacts. Likewise, significant technology changes subsequent to the system design freeze will result in significant project impacts. In all cases changes to the defined configuration baselines must be controlled to prevent requirements to creep and project anarchy. The predictable consequences of requirements creep and project anarchy are continuous project instability, inability to implement concurrent engineering, cost volatility, schedule volatility, and the continuous risk of loss of the affordability commitments.[9]

The exercise of practical caution is prudent in applying the configuration identification and configuration control tools and logic. Inclusion of requirements for premature technical details in the configuration identification process must be diligently avoided. A systematic evolutionary scheme for the progressive growth of technical and production requirements detail should be incorporated in the configuration management planning. This evolution is conditioned by the growth of knowledge and understanding and the risk reduction which is actualized as reflected in Fig. 10.6. In practice, this will guide the retention of appropriate flexibility to optimize the system design during each phase of the development while assuring efficient convergence on a product baseline fully defined by a technical data package which is tailored for the approved acquisition strategy.

One of the most critical components of the configuration identification process is the design review. The system of progressive design reviews is employed to validate achievement of the planned project objectives of each phase prior to the establishment of the succeeding baseline. Management of the design review process is accomplished as a subset of system engineering management. The formal design review and audit progression for major system acquisition projects is shown in Fig. 10.32. This progression is parallel to the system development progression—the reviews and audits are components of project management or technical program control. Technical program planning and control is one of the system engineering management functional areas. However, the design review process incorporates the full range of concurrent engineering considerations—system engineering, integrated logistic support, and production management. Since the progression within the project cycle should be based on demonstrated achievement of phase objectives, the design review management must be synchronized with the T&E process.

As shown in Fig. 10.32, the definition of each baseline is immediately preceded by a formal design review (and audits in the case of the product baseline). These reviews are conducted by multidisciplinary teams, in keeping with the concurrent engineering and affordability philosophies. The system requirements review, the first major review in the project cycle, is used to determine whether the total system requirements have been fully defined. The primary focus for the review is the output of the conceptual phase system engineering or conceptual design documentation and

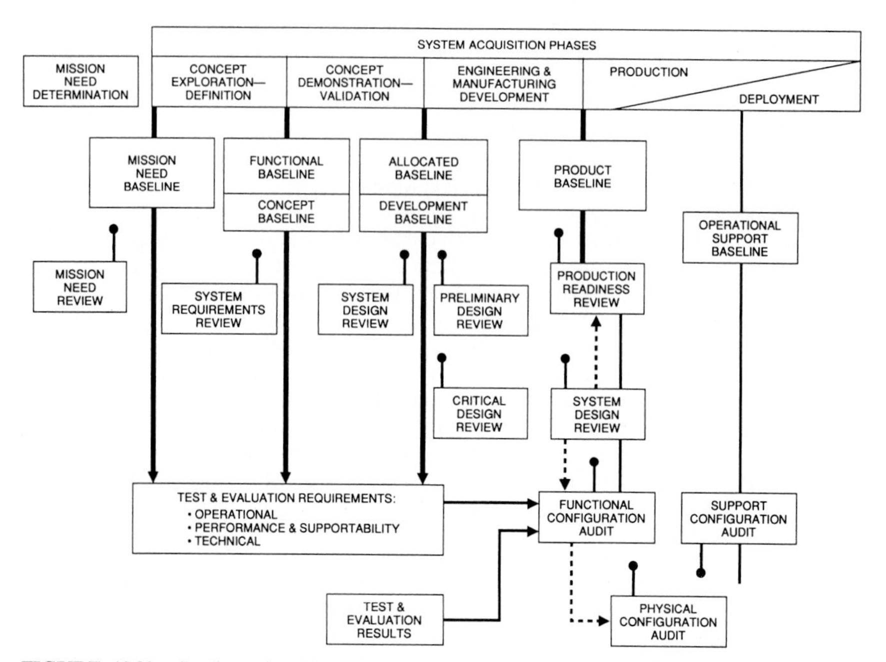

FIGURE 10.32 Configuration identification discipline, review and audit management. (*Source: DoD (Defense Systems Management College), System Engineering, Defense Systems Management College, Washington, D.C., 1986, pp. 12-1 to 12-19.*)

concept formulation reports (such as functional flow block diagrams and requirements to allocation sheets). These documents are reviewed within the context of the mission area assessment or mission need analysis requirements. The system design review, the second major design review, is used to evaluate the preliminary system design, system design optimization, and risk posture as well as the preliminary logistic support considerations. The primary focus is placed on achievement of the system specification requirements, which were defined as a result of the functional baseline. The various trade studies, preliminary logistic support analysis, cost-effective analysis, preliminary system design definition, manufacturing process assessments, and advanced development model or prototype testing reports are the major inputs for this review.

The preliminary design review is conducted subsequent to completion of the preliminary design. It is based on the prime item development specification requirements, which were defined as a result of the allocation baseline. The completeness and adequacy of the development specification are also evaluated at this time. Satisfactory completion of the preliminary design review is the necessary antecedent for initiation of the detail design stage of the development. The critical design review is conducted prior to release for production (engineering development model or productive prototype). The system critical design review is generally conducted after completion of the incremental reviews of each configuration item in the system. The draft fabrication, process, and material specifications (technical data package) are evaluated during this set of reviews.

The T&E requirements are derived from the operational requirements, functional and allocated baseline definitions. The functional configuration audit is conducted subsequent to completion of the technical evaluation stage of development, test, and evaluation to assess the completeness and adequacy of the T&E program and the results of the T&E events. The physical configuration audit is conducted subsequent to the satisfactory completion of the functional configuration audit. It is a formal examination of the as-fabricated production prototype or first production item in comparison to the fabrication, process, and material specifications (technical data package). It includes a detailed examination of the engineering drawings.

The production readiness review is conducted to determine whether the system design should be released to full-rate or limited-rate production. Producibility and quality control are the major considerations. The primary question to be addressed is whether the system is ready for efficient production in an economical production line environment.

The design review and audit management requirements illustrated in Fig. 10.32 reflect a comprehensive and exhaustive approach, one that is fully integrated in the system engineering or concurrent engineering framework. In practice the design review and audit requirements should be tailored specifically for the actual project needs, based on complexity and risk management considerations. In addition it is important to note that, for complex systems, many of the reviews and audits are implemented incrementally and progressively. In these cases the reviews represent a series of events rather than a single discrete event.

REFERENCES

1. H. Kerzner, *Project Management: A Systems Approach to Planning, Scheduling and Controlling,* 3d ed., Van Nostrand Reinhold, New York, 1989, pp. 72–88.
2. Office of Management and Budget, "Major System Acquisitions," Circular A-109, Apr. 5, 1976, pp. 5–7.
3. Department of Defense (Defense Systems Management College), *Systems Engineering Management Guide,* 2d ed., Defense Systems Management College, Washington, D.C., 1986, pp. 4-1–4-8.
4. Department of Defense (Defense Systems Management College), *Integrated Logistics Support Guide,* Defense Systems Management College, Washington, D.C., 1986, pp. 1-3–1-10.
5. DoD Directive 5000.1, "Defense Acquisition," Feb. 23, 1991.
6. DoD Instruction 5000.2, "Defense Acquisition Management Policies and Procedures," Feb. 23, 1991, pp. 2–7.
7. DoD Manual 5000.2M, "Defense Acquisition Management Documentation and Reports," Feb. 23, 1991.
8. DoD Directive 5000.3, "Test and Evaluation," Mar. 12, 1986, pp. 2–7.
9. DoD Directive 5000.45, "Baselining of Selected Major Systems," Aug. 25, 1986, pp. 3–6.
10. D. I. Cleland and W. R. King, eds., *Project Management Handbook,* 2d ed., Van Nostrand Reinhold, New York, 1988.
11. DoD (Defense Systems Management College), *Acquisition Strategy Guide,* Defense Systems Management College, Washington, D.C., July 1984.
12. DoD (Defense Systems Management College), *Test and Evaluation Management Guide,* Defense Systems Management College, Washington, D.C., 1988.
13. DoD (Defense Systems Management College), *The Program Manager's Notebook,* Defense Systems Management College, Washington, D.C., 1985.

CHAPTER 11

MILITARY PROGRAM MANAGEMENT: A PERSONAL PERSPECTIVE

James A. Abrahamson

James A. Abrahamson, Lieutenant General, USAF (Retired), is currently Chairman of the Board of Oracle Corporation. Among his many assignments during 33 years of active military service, he served as a fighter pilot, instructor, and director of the Maverick, F-16, and Strategic Defense Initiative programs. He was also the NASA associator for space flight, having overall responsibility for the Space Shuttle program and three major NASA centers.

My objective in writing this chapter is not to provide readers with a recipe for successful program management, but to provide a framework through which I might pass on thoughts and opinions which were formulated over almost three decades of program work. In effect, these thoughts and opinions are also a synopsis of the lessons I learned from the very many top-notch professionals with whom I was privileged to work over those three decades.*

During my career, I found neither a recipe for successful program management nor general agreement as to how a program manager (PM) should do his or her job, except to do it very well. Be assured, however, that program management is always interesting, exciting, and very challenging—in short, a great job! As a PM, I felt that I was contributing, and that my work was educational and usually a fun experience. I cannot remember a single day when something new and beneficial was not learned. Similarly, rare indeed were the days when we, in the program office, did not enjoy a good laugh—frequently at ourselves!

Most importantly, as Norm Augustine points out in the introduction to this handbook, the results have a direct bearing on the combat effectiveness of our nation's

*I especially wish to acknowledge Colonel Joseph Rougeau, who worked with me on this chapter and who has been a constant source of wisdom through many of the program management "experiences" mentioned here.

armed forces, and the lives of our service men and women. The lives of our service men and women are a sobering and serious dimension that must, above all other considerations, take precedence in the performance (and outcome) of the PM's job.

THE PROGRAM MANAGER

Every military PM must view himself or herself first as part of the combat team and only secondarily as a manager, engineer, or scientist. The PM's job is to acquire superior (but not necessarily complex) weapon systems and supporting equipment. As almost every PM has found out, the process of acquiring superior weapon systems and supporting equipment can prove lengthy and sometimes difficult, but it need never be unmanageable.

Ideally the PM is expected to introduce a critical system or component at or below cost, on or ahead of schedule. The PM also is expected to meet or better performance requirements. If at any time, from the conceptual to the deployment phases, these interdependent parameters do not balance, the PM, alone, is ultimately responsible for identifying the causes, sorting out the issues, and initiating aggressive corrective action—and the PM better do it. If the PM cannot do it, he or she soon will be replaced by someone who can. So being named a PM does not ensure personal job safety, nor, in my opinion, should it. Being a PM can be risky. The costs of failure can be great, but being a PM can also be the most rewarding job imaginable.

The PM must variously be counselor, engineer, designer, historian, accountant, logistician, administrator, strategist, planner, and commander. The PM also must be a student, because the old pros, and the young pros, in the program office have many invaluable lessons to pass on to the PM. Without these skills and lessons learned, the PM may find it difficult to optimize cost, schedule, and total system performance. In the course of business, the PM will have to develop and use expert judgment relative to financial management, procurement, contract administration, systems engineering, test and evaluation, production, and many other functions, including congressional and public relations. More than anything else, the PM must learn that success lies in how effectively he or she manages, motivates, and leads the people assigned to the program office.

THE PROGRAM OFFICE

During the 1961 to 1964 time frame when I was assigned to the Vela Nuclear Detection Satellite program, AFSC Manual 373-3, *System Program Office Manual,* stated that "... the real reason for creating ... program offices ... was simply to optimize the time, cost, and total system performance. ... " The compelling reason for creating a program office has not changed and remains a cogent factor in allowing PMs a measure of flexibility in organizing their program offices. In my opinion, there should not be a prescribed way to organize a program office. A program office prescription would put it at risk of becoming unwieldy and unresponsive.

When I was a PM within the Air Force Systems Command, program offices were treated as line elements that interacted with staff or support activities, but were nonetheless allowed many prerogatives. I took over the Maverick AGM program when it was in advanced development. The program office organization was already set and, apart from expanding the configuration management and program control functions, I did little to change it, even as the Maverick entered into production. There was no need to do so—the organization was good, responsive, and well adapted to the program's unique requirements.

As the F-16 PM I took over the program just as a fighter prototype competition was winding down. The F-16 program office was, in effect, a new organization, and I had the freedom to tailor the organization to meet program needs. Aside from the more traditional (but critical) program office functions, the F-16's multinational content also required the assignment of key personnel to oversee the very unusual and important functions associated with coproduction, international liaison, and currency transactions. Over time, the organization grew from approximately 30 people to over 350 people. Nonetheless, the organizational chart of Fig. 11.1, which dates back to 1976, continues to reflect largely the F-16 program office of some 15 years later.

The challenges posed by the Strategic Defense Initiative (SDI) were different than those of any program to which I had previously been assigned. Though 80 personnel were authorized, the SDI organization (SDIO) initially had a cadre consisting of only 15 people. Even 80 people could not have accomplished more than those first 15 SDIO people did in firming up what has become history's largest defense-based research and development program—and perhaps its greatest technological challenge. Nonetheless, the technical requirements and the political demands imposed upon the organization became so great, that SDIO had to grow if it was to survive. During my tenure as the SDIO director, it expanded threefold over its initially authorized manning. As the SDI program proves its worth in addressing the dramatically changing world situation, I am pleased to note that SDIO, faced with new challenges and opportunities, continues to expand.

Converting the President's vision into a reality required a different emphasis than when addressing an operational requirement. The Army, Air Force, and DARPA had all worked on aspects of the SDI program. We folded these efforts into our overall planning, but decided that they would best be continued within the services. SDIO had its comptroller and procurement offices, but the thrust within SDIO first had to be technology development and integration, and external relations (a function which we then euphemistically coined planning and development) because of the pressing need to communicate to friends and critics that the President's vision could—and would—become reality. To tap their vast potential, I also attempted to strengthen our interface with colleges and universities and forge partnerships with their research activities.

I was assigned to NASA headquarters from 1981 to 1984. NASA and the Space Shuttle program was equally a challenging, interesting, and rewarding experience. By no means do I intend to slight it, but as this is to be a chapter on military program management, I will confine my remarks to the latter. In general, however, the essential elements that I have outlined below were as relevant and as helpful to me at NASA as they were within the Department of Defense (DoD).

ESSENTIAL ELEMENTS FOR PROGRAM MANAGEMENT

No two weapon system acquisition programs are exactly alike, but I believe there are elements common to good program management everywhere. The elements are simple, obvious, and interrelated. They are neither new nor original. In my case, they are based on the common sense and the experience that came through almost 30 years

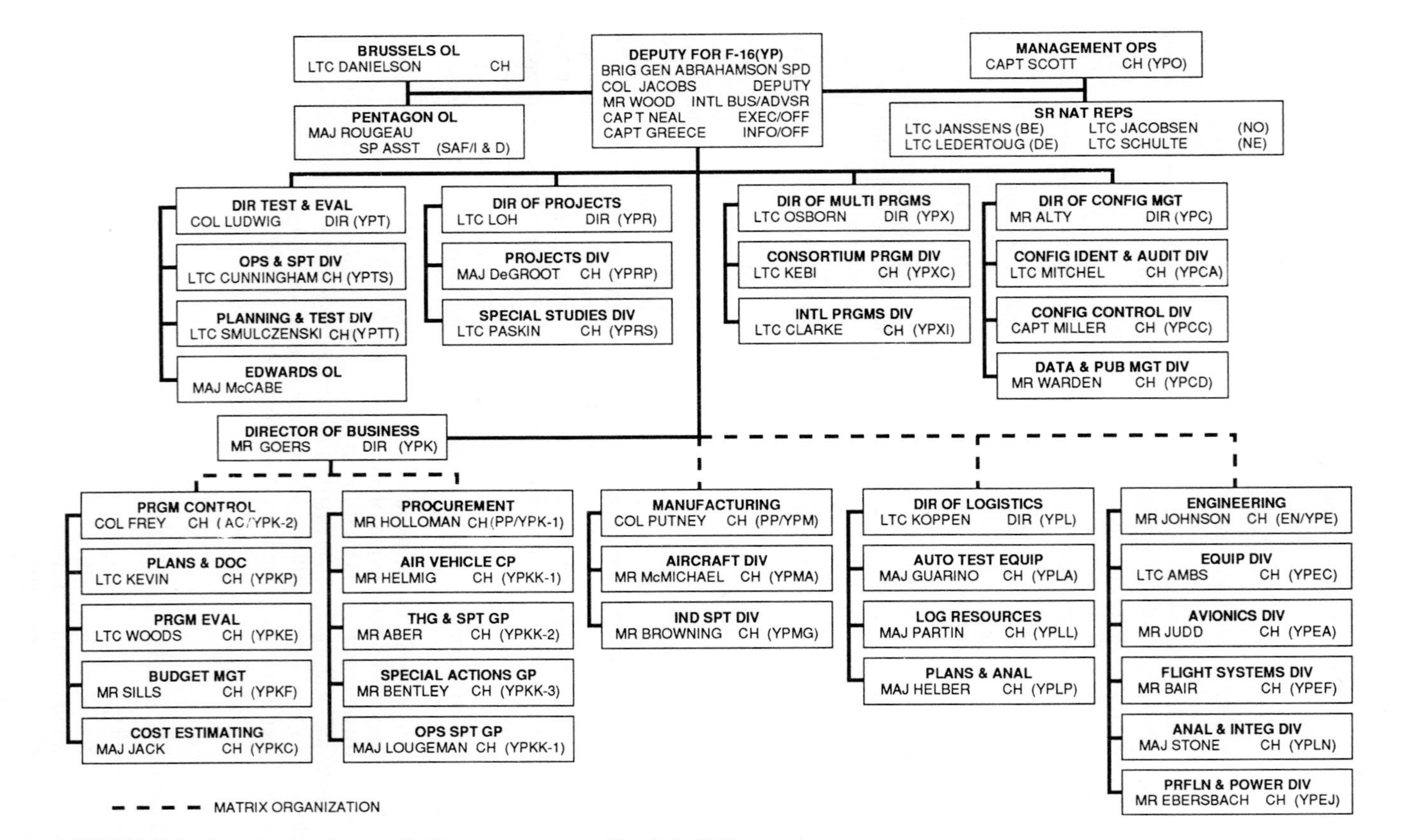

BRUSSELS OL
LTC DANIELSON CH
PENTAGON OL
MAJ ROUGEAU
SP ASST (SAF/I & D)
DEPUTY FOR F-16(YP)
BRIG GEN ABRAHAMSON SPD
COL JACOBS DEPUTY
MR WOOD INTL BUS/ADVSR
CAPT NEAL EXEC/OFF
CAPT GREECE INFO/OFF
MANAGEMENT OPS
CAPT SCOTT CH (YPO)
SR NAT REPS
LTC JANSSENS (BE) LTC JACOBSEN (NO)
LTC LEDERTOUG (DE) LTC SCHULTE (NE)
DIR TEST & EVAL
COL LUDWIG DIR (YPT)
OPS & SPT DIV
LTC CUNNINGHAM CH (YPTS)
PLANNING & TEST DIV
LTC SMULCZENSKI CH (YPTT)
EDWARDS OL
MAJ McCABE
DIR OF PROJECTS
LTC LOH DIR (YPR)
PROJECTS DIV
MAJ DeGROOT CH (YPRP)
SPECIAL STUDIES DIV
LTC PASKIN CH (YPRS)
DIR OF MULTI PRGMS
LTC OSBORN DIR (YPX)
CONSORTIUM PRGM DIV
LTC KEBI CH (YPXC)
INTL PRGMS DIV
LTC CLARKE CH (YPXI)
DIR OF CONFIG MGT
MR ALTY DIR (YPC)
CONFIG IDENT & AUDIT DIV
LTC MITCHEL CH (YPCA)
CONFIG CONTROL DIV
CAPT MILLER CH (YPCC)
DATA & PUB MGT DIV
MR WARDEN CH (YPCD)
DIRECTOR OF BUSINESS
MR GOERS DIR (YPK)
PRGM CONTROL
COL FREY CH (AC/YPK-2)
PLANS & DOC
LTC KEVIN CH (YPKP)
PRGM EVAL
LTC WOODS CH (YPKE)
BUDGET MGT
MR SILLS CH (YPKF)
COST ESTIMATING
MAJ JACK CH (YPKC)
PROCUREMENT
MR HOLLOMAN CH (PP/YPK-1)
AIR VEHICLE CP
MR HELMIG CH (YPKK-1)
THG & SPT GP
MR ABER CH (YPKK-2)
SPECIAL ACTIONS GP
MR BENTLEY CH (YPKK-3)
OPS SPT GP
MAJ LOUGEMAN CH (YPKK-1)
MANUFACTURING
COL PUTNEY CH (PP/YPM)
AIRCRAFT DIV
MR McMICHAEL CH (YPMA)
IND SPT DIV
MR BROWNING CH (YPMG)
DIR OF LOGISTICS
LTC KOPPEN DIR (YPL)
AUTO TEST EQUIP
MAJ GUARINO CH (YPLA)
LOG RESOURCES
MAJ PARTIN CH (YPLL)
PLANS & ANAL
MAJ HELBER CH (YPLP)
ENGINEERING
MR JOHNSON CH (EN/YPE)
EQUIP DIV
LTC AMBS CH (YPEC)
AVIONICS DIV
MR JUDD CH (YPEA)
FLIGHT SYSTEMS DIV
MR BAIR CH (YPEF)
ANAL & INTEG DIV
MAJ STONE CH (YPLN)
PRFLN & POWER DIV
MR EBERSBACH CH (YPEJ)
MATRIX ORGANIZATION

FIGURE 11.1 Organizational chart of F-16 system program office (July 1975).

in program management. Hopefully, that will not diminish their worth. The elements fall short of being a recipe for success. However, I believe that they are worth restating.

Well-Defined and Clearly Communicated Objectives

In all cases the PM's first and most important function is to have a "vision" of where the program is going and to clearly communicate that vision to his or her staff, key officials, and contractors.

Within the military services, the communications process is facilitated through the use of program management directives or other higher headquarters tasking documents. The quality of these documents in conveying policy and thresholds, and their balance and level of detail, are very important to the common understanding of a program's goals. When the objective is one geared to more traditional acquisitions, such as the development and production of a fighter aircraft, battle tank, or rifle, the goal is more easily understood and communicated at all levels.

However, when the system embraces a new and different capability and crosses service or operational lines, setting and communicating program goals become diplomatic as well as vital programmatic exercises. This was true for SDI and it is true for many other programs, especially C3I efforts. Their problems are exacerbated since many objectives often cannot be formally stated, sometimes because the programs are compartmentalized due to national security considerations.

Sound and Reasonably Consistent Plan

The initial program plan, with its basic acquisition strategy, is the keel of a successful acquisition program. If the plan is accomplished in a precipitous manner and fails to reflect or resolve program unknowns, be they cost, schedule, or performance related, it likely will be a prescription for failure. For example, the program plan often neglects or defers the support concept, a hazardous omission in this era of declining or highly variable budgets. An extra effort on the front end to resolve support and other issues will reap dividends for the PM, and spare the PM many future headaches.

In addition to a sound plan, a program must have reasonable consistency. No matter how risky, concurrent, or complex it may be, a program usually can be accomplished if resources are properly applied. When the plan is altered, however, the program's momentum, including that of the government and industry team, is interrupted and sometimes shattered. This happens when budget ceilings and changing priorities force an adverse funding action or because of an arbitrary, even capricious, technical or budget decision within the DoD, OMB, or the Congress. The political process is tough on acquisition programs and even tougher on PMs.

Authority Commensurate with Responsibility and Expectations

Authority commensurate with responsibility and expectations is a basic management principle and an essential ingredient for program success. However, it is a principle rarely held intact throughout the life of a program, especially if the program is dependent on matrixed organizations for support. Based on my experiences, there are two common practices that serve to subtly undermine a PM's authority.

The first practice is employment within the program of matrixed personnel over whom the PM has little or no control and whose actions may be influenced by different policies, procedures, and objectives than those guiding the PM. This creates the potential for conflict because given specified cost constraints, the PM is graded on attaining an optimal mix of performance, schedule, and support packages. In order to attain this mix, the PM may be required to introduce change orders and waive or reduce some specifications. However, the change orders and the reduced specifications may be resisted by the program's matrixed procurement and engineering support personnel, who are under specific direction, irrespective of program, to minimize change orders and effect standardization by not allowing any deviations from specifications. In the F-16 program, matrixed personnel were physically located within the program office and rated by me. This helped to ensure that all personnel were responsive to program goals and it minimized the potential for conflict. However, few PMs will be fortunate enough to have this authority.

The second practice is even more pervasive in undermining the PM's authority. It occurs when the PM's prerogatives are coopted as a result of actions or decisions taken by Congress, OMB, or higher headquarters. These decisions, sometimes effected on a macro level without knowledge of their specific impact on acquisition programs, can undo years of determined planning and work. Persuasion and documented rebuttals are often the PM's only recourse. Never, however, should the PM and program office personnel give up the fight. If the PM is to succeed, he or she must be resilient and work assiduously to meld seemingly conflicting interests and positions—by executing program direction in a responsible and creative manner in the program office and by pushing for responsible, innovative, and flexible implementation of policy guidance at higher headquarters.

Planning and Control

Budgetary limitations, long lead times, changes in program direction and scope, and the multifaceted nature of weapon system acquisition make effective planning and control an absolute necessity for every program office and PM. Without effective planning and control, the responsible and creative program office execution and the innovative and flexible headquarters implementation that I suggested will be impossible.

Consequently one of my first priorities always was to set up a control room. The control room was a valuable tool in clearly communicating program status to every one in the program office. As much as possible, we held our program reviews and staff meetings in the control room. In being able to see the whole program with all of its pieces, program office personnel became more alert to trends that signaled potential programs, and, in most cases, they were able to initiate corrective actions before the program could be adversely affected.

I insisted on a dynamic scheduling system that not only laid out program and financial goals and milestones, but also incorporated functions and feedback. The dynamic scheduling system coupled with an activity network, gave me the ability to compare and contrast program status with program goals, program costs with program budgets. It allowed the program office to build alternative strategies and courses of action; and to determine if and when they needed to be introduced. The activity network outlined the interrelationships between program activities and functional areas. It proved invaluable in isolating program unknowns and in effecting viable workarounds.

It takes a lot of work to implement effective planning and control, but it will help keep a program on a stable course. Because our planning and control system in the F-16 program provided us the requisite information, we could quickly and definitively assess the impact on program goals when program direction was changed or funding levels were reduced. These quick, but specific assessments were decisive, for example, in the F-16's often avoiding cuts that plagued many other programs.

Leadership by Example

Leadership by example is—or should be—so fundamental a precept that I hesitate to emphasize it. Nevertheless its need is sometimes overlooked because of the hectic pace of daily business and the relatively unique environment in which each program office operates. This unique environment, in fact, makes leadership by example all the more important and imposes upon each PM an obligation for selfless performance and accomplishment. Nothing can be taken for granted. Program office demands will strain the PM's perceptions, and tax his or her persuasive skills and tolerance, but lead the PM must!

The PM must not avoid the glare that comes with the program management job. The PM is responsible for the end result and must never pass the buck. Neither should the PM seek the spotlight. Recognition, first and foremost, must be of the program and the people, both government and industry, who are instrumental in making it (and the PM) a success. Discipline, morale, and leadership are integral to the fabric of progressive program management. This fabric is critical to building a motivated, productive program office team.

A Trusted Team

Industry is—and must be treated as—a key member of the program team. The acquisition process is best served when there is excellent communication and coordination between the program office and the contractors. Misunderstandings are not likely to occur if, from the outset, it is clear that it is the PM who bears ultimate responsibility for program success, and that the program office must, and will, manage contracted work with the same vigor that it manages other parts of the program. I never worked with an industry member who did not understand and appreciate my role as a PM and, hence, my responsibilities as a steward of appropriated funds.

Whenever appropriate, we held our government-contractor program reviews in the program office control room. This helped gel our relationship. In addition, it gave us the ability, when necessary, to scope and analyze problems together, even to the work breakdown structure or component level. We traced the problems and worked the solutions as a team. Because of this, the team progressively claimed ownership of the corrective actions recommended and taken. Team members invariably redoubled their efforts. Thus getting the team to own an action almost was tantamount to guaranteeing success.

As a PM I felt that maintaining the public confidence and trust was one of my most important duties. Clearly, I tried to set the example. In this vein, there was, is, and always will be no substitute for personal and professional integrity. I also did everything possible, in oral and written communications, to ensure an understanding by program office and contractor personnel of the importance of avoiding even the slightest appearance of a conflict of interest in their dealings with one another. Industry always responded with the same spirit and intent.

I never hesitated to let industry know that it was a key member of the program team. Similarly, I never hesitated to let industry know that it was critically important for the program office and its contractors to be candid, even brutally honest, with one another; that it was important for both to maintain a close working relationship; that an adversary relationship between the program office and the contractors should never be permitted to materialize. And that, above all, integrity must govern the relationship.

Building a trusted team relationship was fundamental to our success. It worked well, much to the benefit of our taxpayers and our national security.

Motivated People

Any success that I had as a PM was due to the outstanding support given me by good bosses and the great people assigned to my programs. I am serious—it was those people who made the programs the success they were.

I had no control or choice over who were to be my bosses, so I was very lucky to have had such good ones. For that matter, I did not have much more flexibility in the choice of people who worked in support of my programs. However, I knew that we had a tough job ahead of us and that we were expected to produce results. Consequently I worked long and hard to positively motivate them, be they assigned to the program office, on support staffs, or in industry. This extra effort always reaped dividends. Irrespective of the program, I found that, with few exceptions, the people wanted most to be part of a dedicated, productive team.

In my opinion, motivated people are a key element of leadership and the most important ingredient for success. My approach to people did not come out of a textbook. It developed as a natural result of my trying to treat people in much the same manner that I wanted them to treat me.

I learned early on that every person is unique—every one I know prefers to be recognized as an individual. Consequently I tried to know and address people in the program office by their first names. I encouraged the people to take pride in the programs and in their individual and collective contributions. I never lacked confidence in their ability to do the tough jobs and to respond to significant challenges—and I let them know it. Predictably, they completed the tough jobs and they responded successfully to the challenges. Most important, my trust of the people and their judgment was fully returned.

Sometime during the process, we, in the program office, all came together. We got to know and appreciate one another and our talents and strengths. More important, we became a dedicated, productive team of which we were very proud. And that dedicated, productive team, in turn, contributed significantly to our national security.

A GREAT JOB

There should be nothing inherently complex about program management, but it is becoming increasingly bureaucratic. Program management should be a continuously evolving management concept, framed by competent people, rational priorities, and clearly defined authority and responsibility—but it is not. Public distrust, unresponsive procurement practices, higher echelon micromanagement, burgeoning use of paper studies, rigid compartmentalization, and restrictive technology transfer policies sometimes make a PM's job more difficult and less satisfying than it should be. These are not new problems. They go with the job. They plagued PMs in the 1960s, 1970s, and 1980s. In all likelihood, they will plague PMs in the 1990s.

Why, then, should anyone aspire to be a PM? Based on my experiences, the question is easy to answer. Being a PM is a great job!

A good PM is *the* sine que non in the weapon system acquisition process. No matter the problems, no matter the restrictions, the PM can succeed by power of personality and creative and persuasive skills. Given some freedom of action, the PM needs no special incentives. The job, when well done, is its own reward.

CHAPTER 12

PROGRAM MANAGEMENT IN THE UNITED STATES ARMY

Charles B. Cochrane

Charles B. Cochrane is a professor of systems acquisition management, Acquisition Policy Department, Defense Systems Management College. Mr. Cochrane retired from the U.S. Army with over 25 years of active service. While on active duty, Mr. Cochrane (then Lieutenant Colonel) was the principal staff officer at headquarters, Department of the Army, for implementation of the Packard Commission recommendations for streamlining the program manager's reporting chain. He also played a key role as the acquisition member of the study group charged with implementing the DoD Reorganization Act of 1986.

THE EARLY YEARS: A BRIEF HISTORY

Prior to World War I, U.S. Army research, development, and acquisition was generally conducted by the Ordnance Corps (then called the Ordnance Department) through a network of arsenals. The first was established in Springfield, Mass., in 1777. Others, such as Watervliet Arsenal, Watervliet (then West Troy), N.Y., established in 1814, still exist today. By 1860 there were 24 of these arsenals, all east of the Mississippi River, providing an almost totally in-house research, development, and production capability for powder, guns, and armaments until World War II.

The War Department did not institutionalize policies for departmental oversight for the management of research, development, and acquisition until 1924, with the publication of Army Regulation 850-25. This regulation was the basis for a process now well known to Army program managers: a statement of the requirement; development and production of test samples; testing, to include comparison testing with fielded items; more development to correct test deficiencies; determination of production feasibility; more service testing; and finally, adoption for service use.

Prior to World War II, powerful chiefs of technical services controlled Army research, development, and acquisition. There were seven: quartermaster, ordnance, engineers, chemical warfare, signal, transportation, and the Surgeon General. General George C. Marshall, Army Chief of Staff, reorganized the War Department in

March 1942, placing supervision of these technical services under the Army Services Force (ASF), a new major army command. The Army general staff, within the office of the G-4 (logistics), exercised staff supervision over research and development. However, the chiefs of the technical services were dominant, maintaining their strong role in actually performing the mission. It was the Manhattan Project, led by the Corps of Engineers, a technical service, that pioneered the concept of program management and produced the first atomic bomb.[1]

The reign of the technical service chiefs lasted from the early 1800s until Secretary of Defense McNamara approved the consolidation of many of their functions under a new Army Materiel Command (AMC) in 1962. Two of the technical services, the Ordnance Department and the Quartermaster Corps, had performed most of the materiel acquisition functions from inception to obsolescence that are today the responsibility of AMC. During these early years, general staff supervision of the research and development process was almost nonexistent, even with the assignment of the staff function to the G-4 in 1940 and the creation of an ASF Directorate of Materiel in 1942. Both of these staff agencies had little influence over the technical services. It was a perceived lack of preparedness for the Korean War that seemed to convince the Army leadership that warfare was moving rapidly into a modern era. Significant changes in the Army's processes and organizations for the pursuit of technology and the fielding of modern weapon systems were made. Between 1952 and 1976 the Army went through a number of management oversight and structural changes, which are summarized here from the final report of the ROBUST task force[2]:

1952. Secretary of the Army Pace assigned the Deputy Chief of Staff for Plans on the Army staff general staff responsibility to ensure that research and development was integrated into operational planning.

1954–1955. The Deputy Chief of Staff for Plans and Deputy Chief of Staff for Logistics (DCSLOG) have conflicting missions: the Deputy Chief for Plans has general staff responsibility for integrating research and development into planning; however, the technical services report into the DCSLOG for general staff supervision. To correct this, the Secretary of the Army shifted responsibility for all research and development policy to a new Chief of Research and Development (CRD), and a Research and Development Directorate was created within the secretariat. Now the technical services work through two chains of authority, the primary to DCSLOG and the secondary to CRD.

1961. Army Regulation 11-25 was published, establishing a four-year objective from project initiation to production roll-off. Although this optimistic objective may not have been exactly consistent with the two chains of authority, program managers began to feel the heat concerning development cycles that were too long. (However, in the mid-1960s the Vietnam War effort consumes manpower and material resources and many major program new starts are delayed.)

1962. As one result of Defense Secretary McNamara's Project 80 reforms, a U.S. Army Materiel Development and Logistics Command (USAMLDC) was established, absorbing the functions of five of the technical services. USAMDL was almost immediately renamed Army Materiel Command (AMC). The ordnance, engineer, transportation, chemical, signal, and quartermaster branches retained their identity, but only as career specialty areas. Almost all research, development, and acquisition functions were transferred to AMC. AMC also assumed responsibility for logistics support to the fielded force. The Chief of Engineers and the Surgeon General are two technical service chiefs that were not eliminated and even today retain research and development functions.

1962–Mid-1970s. The training functions of the technical services were transferred to the Continental Army Command (CONARC), soon to become the Training and Doctrine Command (TRADOC). Requirements generation was assumed by another new field command, the Combat Developments Command (CDC). The role of CDC was assumed by TRADOC in the mid-1970s.

Mid-1970s–1986. As the Vietnam War winds down, the Army can now devote attention to a series of major program initiatives: the multiple launch rocket system (MLRS), the Patriot air defense system, the Army tactical missile system (ATACMS), Pershing II theater support missile system, Sergeant York air defense system (terminated soon after production started), Copperhead (a laser-guided artillery round), the Apache attack helicopter, the Abrams tank, and the Bradley fighting vehicles. Many of these new systems would see combat in the Persian Gulf War of 1991.

The Chief of Research and Development was assigned responsibility for the acquisition of materiel from the DCSLOG in 1973, and appointed as the Appropriations Director both for the research, development, test, and evaluation (RDT&E) and for the procurement accounts. The title of the CRD was changed to Deputy Chief of Staff for Research, Development and Acquisition (DCSRDA). The office of the DCSRDA became the operational staff for all matters pertaining to Army RDA, from technological base or laboratory activities through development and production. The Assistant Secretary of the Army for Research, Development and Acquisition [ASA(RDA)] was a small policy oversight staff within the secretariat. This would last until the implementation of the DoD Reorganization Act in 1987.

The only other change of significance during these years was the internal reorganization of AMC's major subordinate commands. A 1974 study directed by Under Secretary Staudt and General Weyand, Vice Chief of Staff, established an Army Materiel Acquisition Review Committee. The AMARC recommended separation of the management for new weapon systems and major product improvements from logistics or readiness. AMC reacted by dividing each major subordinate command (such as the Missile Command) into two commands, one for development and one for readiness.[3] It took about four years to split up the commands. One year after the last command was split, AMC abandoned the idea and started putting the MSCs back together into one command both for research and development and for readiness.

1986 TO THE PRESENT: MANAGEMENT CHALLENGES PERSIST

The Army again went through a series of agonizing changes starting in 1986, creating the research, development, and acquisition process as it is today. These changes were the result of congressional perceptions on a lack of civilian control, layering and duplication of staff functions, and the fraud, waste, and abuse issues highlighted by the media in the early to mid-1980s. The executive branch was also caught up in this media blitz, even as the Reagan defense buildup began to slow down significantly. Figure 12.1 summarizes how these changes came to pass. In 1985 President Reagan charted a blue ribbon panel on defense management, chaired by David Packard, Chairman of the Board, Hewlett Packard Company, to look into the management of the Department of Defense. A series of interim reports were issued in 1986. The first, "A Formula for Action: A Report to the President on Defense Acquisition," was is-

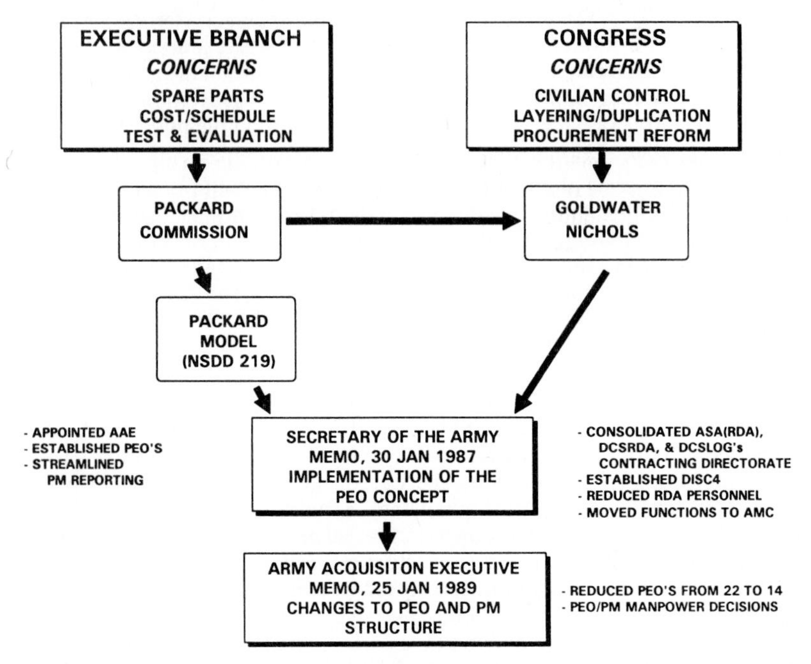

FIGURE 12.1 Army acquisition 1986 to present.

sued in April 1986. This interim report is the basis for the program management structure as it exists in the Army today. This Packard Commission indicated that defense program managers needed a streamlined reporting chain, that is, there should be no more that two levels of authority between the program manager of a major defense acquisition program and the defense acquisition executive (DAE). These two levels were to be a program executive officer (PEO) and a service or Army acquisition executive (AAE). President Reagan directed this streamlined reporting system by NSDD 219 on April 1, 1986.[4]

In mid-1986, headquarters, Department of the Army (HQDA), chartered the Army Acquisition Task Force to implement the recommendations of the Packard Commission. This task force reported through the Deputy Chief of Staff for Research, Development and Acquisition (DCSRDA) and the Assistant Secretary of the Army for Research, Development and Acquisition [ASA(RDA)] to the Under Secretary of the Army and the Vice Chief of Staff. The goal of this task force was to develop a system of PEOs and PMs reporting to the AAE outside of normal reporting channels, that is, bypassing the commanders of the materiel developing commands (primarily, AMC) and eliminating layers of authority in accordance with Packard's recommendation. This streamlined reporting concept was strongly resisted by the commander of AMC, who recommended that the CG of AMC be appointed the AAE and that the commanders of AMC major subordinate commands (MSCs) be dual-hatted as PEOs. The Navy and the Air Force also resisted this streamlining concept. The Air Force simply assigned their product division commanders a dual role as PEOs for all assigned programs. The Navy dual-hatted their

Systems Command commanders as PEOs.* Nevertheless, the Army decided to comply, and the Acquisition Task Force developed the following eleven guiding principles for the management of major Army programs, which are basically intact today:

1. No more than one layer of supervision between the PM and the AAE. This is the PEO. (Same as Packard's recommendation.)
2. The PEO should be responsible for a reasonable and defined number of similar programs. (Army added the word "similar" to Packard's recommendation.)
3. The AAE and PEOs should be devoted full time to their acquisition responsibilities. ("and PEOs" added by the Army to Packard's recommendation.)
4. PMs report directly to the PEO for program matters. (A Packard recommendation.)
5. PEOs report directly to the AAE for program matters. (A Packard recommendation.)

The following were not Packard Commission recommendations. They were added by the Army Acquisition Task Force.

6. PEOs should be "hands-on" type individuals colocated with their PMs.
7. The PEO is responsible for ensuring that the PM is properly resourced (people, dollars, facilities).
8. The PM is responsible for successful execution of the program.
9. The Army should utilize the "capstone" PMs and PEOs.†
   a. Groups interrelated programs to enhance mission area management
   b. Provides training of general officers for future acquisition assignments
10. Enhance program transition from development and initial production to fielding and sustainment.
11. Weapon systems normally transition from the PEO to the materiel command after successful fielding.

Task force recommendations for a PEO structure independent of AMC were finalized and approved by General John A. Wickham, Jr., Army Chief of Staff, in September 1986. The implementing instruction approved by the chief designated the ASA(RDA) as the AAE, and the Army's "capstone" program managers as PEOs. The ultimate number of PEOs was to have been decided during an implementation phase ending on October 1, 1987. However, this implementing instruction was overtaken by actions on Capitol Hill and was not approved by the Secretary of the Army, John O. Marsh, Jr.‡

The Department of Defense Reorganization Act of 1986 (Public Law 99-433, October 1, 1986), better known as the Goldwater-Nichols Act, directed the military de-

*The Navy and Air Force did not create a PEO structure independent from their materiel commands until required to do so by Secretary of Defense Cheney's defense management report (DMR) to the President in July 1989.

†In 1986 the Army had capstone PMs, or PMs responsible for a number of acquisition programs within the same mission area. For example, the tank systems PM, with subordinate PMs for tanks and armored vehicle programs; Air Defense Systems PM, with subordinate PMs for air defense weapons and command and control systems; and a tactical communications system PM, with subordinate PMs for tactical radios and other command, control, and communications systems.

‡Information on the activities of the Army Acquisition Task Force was derived from the author's personal archives.

partments to transfer management and oversight of research, development, and acquisition from the staff of the service chief to the office of the service secretary. Indeed, this law specifically prohibits a duplication of these staff functions within the office of the chief. Further, the law had to be fully implemented by April 1, 1987. The Army Reorganization Commission was chartered by Secretary Marsh on October 11, 1986. The commission was cochaired by the Assistant Secretary of the Army for Financial Management, Michael P.W. Stone, and Lieutenant General Max W. Noah, Comptroller of the Army. In addition to effecting a major HQDA reorganization within six months, the Army still had to comply with NSDD 219 and implement the Packard Commission's recommendation for the management of acquisition programs.

Secretary Marsh signed instructions on January 30, 1987, for implementing both the HQDA reorganization for acquisition management and the PEO concept. Secretary March's implementation memorandum directed that the HQDA offices of the ASA(RDA), DCSRDA, and the DCSLOG directorate of contracting be combined into a new office of Assistant Secretary of the Army for Acquisition [ASA(A)]. The ASA(A) was soon renamed ASA(RDA). The ASA(RDA) was appointed the AAE and provided a military deputy ASA(RDA), a lieutenant general. The implementing instructions also provided for a PEO structure using the Army's "capstone" PMs. For example, PM Tank Systems, a brigadier general, managed seven subordinate PMs. He reported to and was supported by the Tank Automotive Command (TACOM), a major subordinate command of AMC. Under the Army's implementation of Packard's recommendations, PM Tank Systems would be designated a PEO for Combat Vehicles reporting directly to the ASA(RDA), the Army acquisition executive. TACOM would continue to support this PEO with functional services such as financial management, contracting, quality assurance, and legal assistance. Consistent with Packard's recommendation, two layers of supervision had been eliminated: HQ TACOM and HQAMC.[5]

Although the ASA(RDA) had been designated the AAE, Secretary Marsh changed his mind, and on February 18, 1987, designated Under Secretary of the Army James R. Ambrose the AAE. Secretary Ambrose almost immediately issued instructions to the field establishing 22 PEOs. Nearly all Army PMs reported within this PEO structure. About 15 of these PEOs fell under the AMC for functional support, but not program management authority. The others were supported by the Information Systems Command, the Strategic Defense Command, the Chief of Engineers, the Surgeon General, and the commander of the Intelligence and Security Command. All reported to Secretary Ambrose for acquisition program management.[6]

Between February 1987 and January 1989 the Army struggled with the issue of resourcing and overseeing this large number of PEOs, all general officers or senior executive service civilians. In the meantime Assistant Secretary Stone was nominated by the Bush administration and confirmed as the replacement for Under Secretary Ambrose. As the Army acquisition executive, Secretary Stone streamlined the PEO structure and reduced the number of PEOs to 14, with about 113 PMs. Instructions to the field in January 1989 directed the implementation of this new structure and provided specific manpower allocations for the PEOs and their remaining PMs. The number of PEOs had been reduced, and many PMs were transferred back to the materiel commands.[7,8]

Today the Army structure for management of acquisition programs is basically as it was in 1989. The number of PEOs continues to be reduced, and as the Army realigns its missions pursuant to a changing global environment, the numbers and types of PMs, both within and external to the PEO chain, continue to fluctuate. As the mission areas of the PEOs grow with increased emphasis on conventional warfare and land combat, the number of acquisition programs requiring formal program management has actually increased. As of October 1992 there were 10 PEOs with 37 major and 111 related nonmajor PMs under the PEO structure. Non-PEO managed programs included one major and 53 nonmajor defense acquisition programs.[9]*

THE U.S. ARMY PROGRAM MANAGER

The Army categorizes "program managers" as program managers, project managers, or product managers.† These PMs provide intensive, centralized management of selected Army acquisition programs. The AAE is the approval authority for designating Army programs for the appropriate level of program management and for the termination of all PMs. The AAE charters all PEOs and direct reporting PMs. PEOs and commanders of materiel commands charter assigned PMs.[10]

Program. Acquisition programs that require the performance of a broad mission over a protracted period of time. These programs are normally highly complex and involve substantial resources. A program manager is normally a general officer or senior executive service civilian appointed by the AAE.

Project. This designation is based on mission criticality; urgency of need; congressional, DoD, or Army interest; organizational or technical complexity; and dollars. A project manager is a HQDA board selected colonel or GM-15 civilian.

Product. Typically applies to weapon subsystems, derivative systems, or series or models of existing systems or software. A product manager is a HQDA board selected lieutenant colonel or GM-14 civilian.

Generally, program or project management is approved for major defense acquisition programs, or acquisition category (ACAT) I programs. Project management is also applied to major systems, or ACAT II programs, reviewed by the Army Systems Acquisition Review Council (ASARC). The acquisition category refers to the level of milestone decision authority (MDA) and to the dollar value or other criteria that determine the relative importance of the program (see Fig. 12.2).

Major automated information systems, reviewed by the Major Automated Information Systems Review Council, chaired by the Assistant Secretary of Defense for Command, Control and Communications, may also be selected for project management. Other programs not specifically designated ACAT I or II or as major automated information systems may be approved by the AAE for project or product management based on a requirement for substantial RDT&E and procurement costs, the impact on the U.S. military posture, unusual organizational complexity, emerging and difficult to manage technology, or interface control.

Military project and product managers are selected centrally by a HQDA selection board. The selection is based on a review of the performance file of all officers within a zone of consideration determined by the Secretary of the Army. Colonels and lieutenant colonels selected for project or product management are considered to be serving in an assignment equivalent to that of a brigade or battalion commander, an essential step in the career development to general officer rank. Officers

*The only non-PEO managed major defense acquisition program was the air defense non-line-of-sight (NLOS) system.

†As of June 1992 there were five program managers, 90 project managers, and 108 product managers for a total of 213. Most are Army officers. However, the Army is increasing the number of civilian PMs, and 34 were civilians in the grades of GS-14 and GS-15 and three were senior executive service (SES) civilians.

ACAT ID:	• Designated by DAE • Reviewed by DAB • Decision by DAE	*$300M RDTE/ $1.8B Procurement (FY90 Constant $)*
ACAT IC:	• Designated by DAE • Reviewed by ASARC • Decision by AAE	*$300M RDTE/ $1.8B Procurement (FY90 Constant $)*
ACAT II:	• Designated by AAE • Reviewed by ASARC • Decision by AAE	*$75M RDTE/ $300M Procurement (FY80 Constant $)*
ACAT III:	• Designated by AAE • Reviewed by IPR • Decision by PEO or Materiel Command Commander	*High Visibility Special Interest*
ACAT IV:	• Designated by AAE • Reviewed by IPR • Decision by Materiel Command Commander*	*All Other Acquisition Programs*

*May be further delegated by the Materiel Command Commander

FIGURE 12.2 Army acquisition categories (ACAT).

serving as members of the Army Acquisition Corps are considered career acquisition managers and do not compete for command of combat battalions or brigades. These officers may be considered for command of acquisition-related positions, such as laboratories and ammunition plants. Army civilians normally are appointed to PM positions; however, in the near fugure the Army plans to start competing PM positions between the military and civilian members of the Army Acquisition Corps.[11] For ease of reference, the term PM will be used here to apply to any level of Army program management, military or civilian.

Army PMs may report to a program executive officer (PEO), to a commander of a materiel command, or directly to the AAE. Direct reporting PMs may be assigned the same duties and responsibilities as a PEO (to be discussed later). The duties and responsibilities of Army PMs at all levels are outlined in Army Regulation 70-1[10]:

> Plan and manage, from day to day, acquisition programs consistent with, and supportive of, the policies and procedures issued by the AAE and contained in DoDD 5000.1, DoDI 5000.2, this regulation and other appropriate regulations, policies, procedures, and standards.
>
> Develop and submit financial, manpower, matrix, and contractor support requirements to the AAE, the respective PEO or other materiel developer. Coordinate for required functional support from the appropriate materiel command.
>
> Develop, coordinate and commit to an acquisition program baseline and immediately report all imminent and actual breaches of approved baselines.
>
> Prepare and submit timely and accurate periodic program performance reports, as required.
>
> Identify critical intelligence parameters for inclusion in the System Threat Assessment Report.
>
> Conduct the logistics support analyses (including industrial base considerations) necessary to recommend a system support concept.
>
> Establish and maintain control over funds received from the AAE, the supervising PEO or other materiel developer.

Develop and coordinate the test and evaluation master plan (TEMP) for assigned systems.

Execute the MANPRINT* program, and share equally with the combat developer in continuous planning of the MANPRINT program.

Promptly and accurately record and update data required by Army management systems and data bases for all assigned programs/projects/products.

Be responsible for configuration management.

U.S. ARMY ORGANIZATION FOR ACQUISITION MANAGEMENT†

Today's Army is organized for acquisition management with headquarters, Department of the Army (HQDA), oversight, four materiel development commands, four combat development commands, a PEO structure directly reporting to the AAE, and two test and evaluation commands. Figure 12.3 illustrates how the Army generally fits into the overall DoD structure for acquisition management. As previously mentioned, HQDA has two large staffs, the Army staff and the secretariat, both of

*MANPRINT, acronym for manpower and personnel integration, is the process by which the Army integrates the full range of human factors engineering, manpower, personnel, training, health hazard assessment, and system safety throughout the development and acquisition process. AR 602-2 implements the MANPRINT program, complying with the human systems integration requirements of DoDI 5000.2,[12] Part 7.B.

†See AR 70-1 for the acquisition-related responsibilities of all HQDA staff principals. For a more complete description of all responsibilities see AR 10-5.

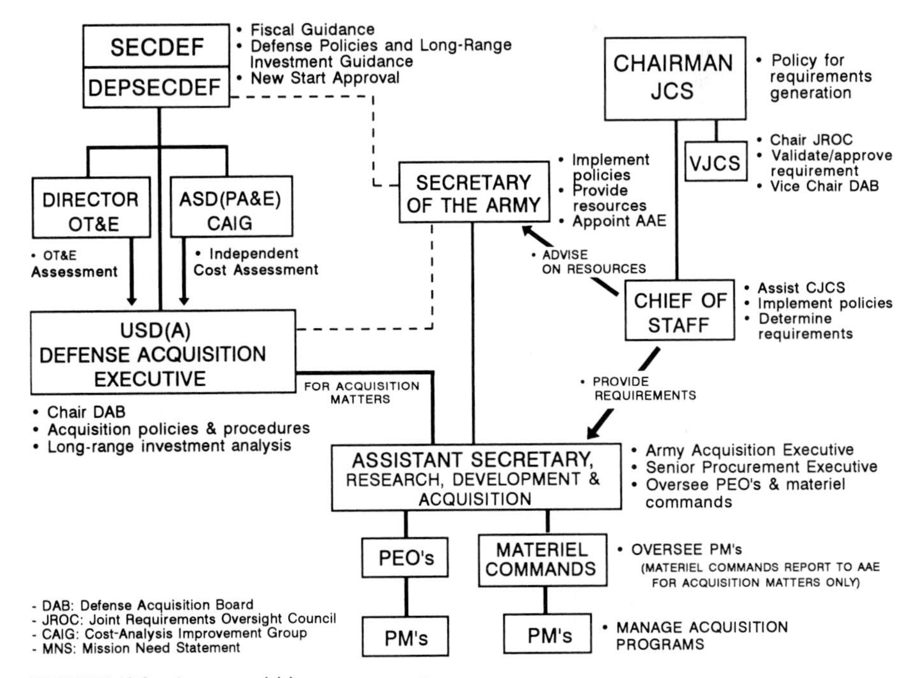

FIGURE 12.3 Army acquisition management.

which have immense influence on the daily activities of the program manager. The Army acquisition executive (AAE) reports directly to the defense acquisition executive (DAE) for acquisition management matters, with parallel reporting to the Secretary of the Army for all other matters, to include procurement. The organization of the Army secretariat is shown in Fig. 12.4. Duties and responsibilities of the major acquisition players are described in the next section.

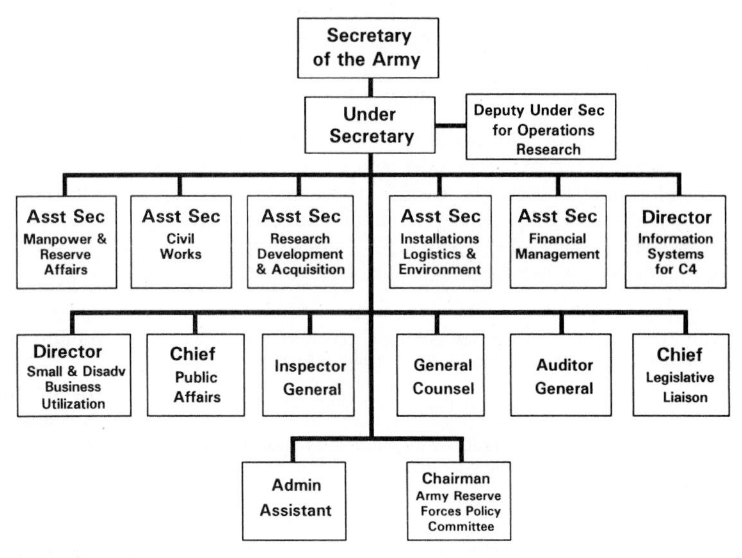

FIGURE 12.4 Army secretariat.

ARMY SECRETARIAT

Deputy Under Secretary for Operations Research. The DUSA(OR) is responsible for oversight of all Army test and evaluation associated with research, development, and acquisition. He is also responsible for software development for modeling and simulations and for software test and evaluation policy. The DUSA(OR) carries out these responsibilities in close contact and coordination with the Office of the Chief of Staff and the Deputy Chief of Staff for Operations and Plans (DCSOPS).

General Council. The GC advises the AAE and the Army Systems Acquisition Review Council (ASARC) concerning legal issues which may arise during research, development, and acquisition.

Assistant Secretary of the Army for Research, Development and Acquisition. The ASA(RDA) jointly sponsors the Army Long-Range Research, Development and Acquisition Plan (LRRDAP) with the DCSOPS. ASA(RDA) and the DCSOPS provide joint guidance for the development of the Long-Range Army Materiel Requirements Plan (LRAMRP). Both the LRRDAP and the LRAMRP are discussed later. The ASA(RDA) coordinates with the Assistant Secretary of the Army for Financial

Management [ASA(FM)], the Director for Program Analysis and Evaluation (DPAE), and the DCSOPS to develop the research, development, and acquisition portions of the Army program objectives memorandum (POM) and budget* to include the justification materiel for the Office of the Secretary of Defense and the Congress. The ASA(RDA) is also responsible for the development of long-range investment analysis that provides the structure for the timing and affordability of proposed acquisition programs.

Army Acquisition Executive. AAE is not a statutory title, nor a title reserved for a specific member of the secretariat. The Secretary of the Army designates the AAE among the members of the secretariat.† Currently the ASA(RDA) is the AAE and the Army's senior procurement executive. Under the AAE hat, the ASA(RDA) is responsible for discharging the responsibilities for component acquisition executives in accordance with DoD Directive 5000.1. The AAE supervises the performance of the entire Army acquisition system to include the workforce, the Army Acquisition Corps, and is the final authority on all matters affecting that system except as defined by public law or DoD policy. The AAE is the milestone decision review authority (MDA) for Army acquisition category (ACAT) IC and II programs, and assigns the MDA for ACAT III and IV programs. With the Vice Chief of Staff, the AAE cochairs the ASARC. He has authority to designate Army acquisition programs for centralized program management, and to review and approve all agreements to transfer management responsibility between the PEO and materiel command functional structures. PEOs report directly to the AAE.

The office of the ASA(RDA) is organized to provide the AAE with functional staff specialists for oversight of the Army acquisition system. Army PEOs and PMs of major defense acquisition programs deal with this office on an almost daily basis. Figure 12.5 shows the organization of the office of the ASA(RDA) as of October 1991. Note the broad spectrum of responsibilities listed under each major deputy. Within the office of the Deputy for Systems Management, and the military deputy DISC4, there are staff officers or PEO liaison officers, or both, assigned for each Army acquisition program. These are officers in the grade of major or lieutenant colonel, or civilians in the grade of GM-14. These staff officers were once officially known as Department of the Army systems coordinators (DASC). The term DASC (pronounced "dask"), although no longer an official title, is still used to refer to these staff officers throughout the Army acquisition community. PMs should know these staff officers. They are the eyes and ears of the PM within the halls of the Pentagon and maintain contact and provide liaison with the Army staff, Office of the Secretary of Defense (OSD), and other agencies that have an interest in the program, as well as other services and Congress. These staff officers, along with staff officers from the office of the DCSOPS (ODCSOPS), defend RDT&E and procurement programs before OSD, the White House Office of Management and Budget (OMB), and Congress.

---

*Resource requirements to fund RDT&E and procurement programs for six or more years are displayed in the Army POM submitted to the Office of the Secretary of Defense in April of even-numbered calendar years. This POM also updates the DoD six-year Future Years Defense Program (FYDP). After POM approval by the Secretary of Defense, the Army translates the first two years into a budget submission that is provided to the DoD comptroller in September of the same year.

†However, DoDD 5000.1,[13] Part 3.D, states that the military department acquisition executive will be at the assistant secretary level, and that the acquisition executive is also the senior procurement executive (SPE) pursuant to the Office of Federal Procurement Policy Act (41 U.S.C. 414). As the AAE, the ASA(RDA) reports to the USD(A) for acquisition matters. As the SPE, the ASA(RDA) is responsible to the Secretary of the Army for procurement actions and contract management.

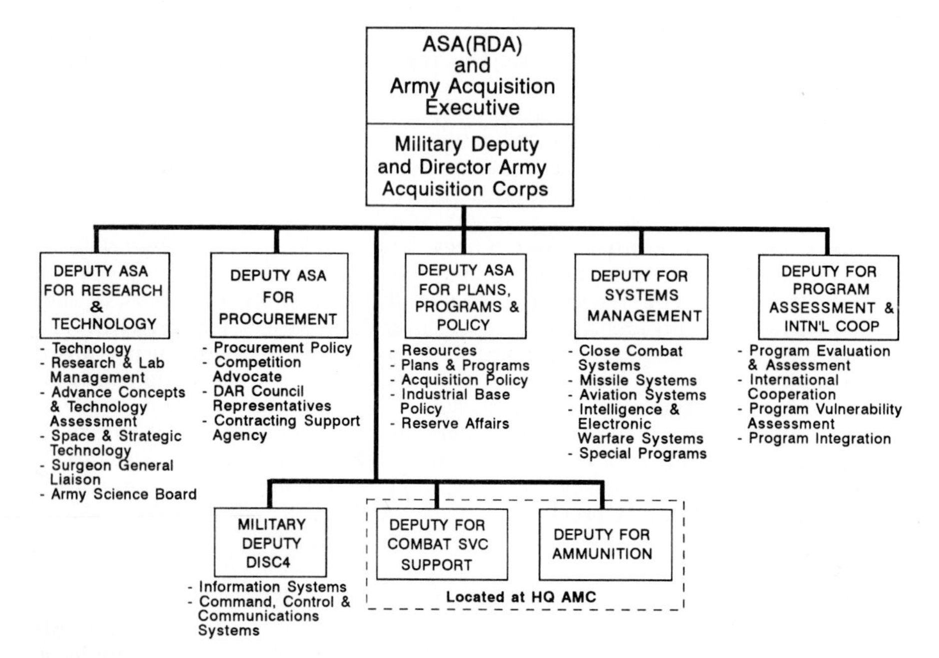

FIGURE 12.5 Office of the Assistant Secretary of the Army for Research, Development and Acquisition.

Assistant Secretary of the Army for Financial Management. The ASA(FM) has responsibility for all financial management activities and operations for appropriated funds, including the establishment, receipt, allocation, withdrawal, and redistribution of funds. He administers the Army reprogramming process and ensures materiel developer compliance with the contract cost data reporting requirements. The ASA(FM) represents the AAE on all cost and economic analysis matters relating to the acquisition process. The Comptroller and the Director of the Army Budget are located within the office of the ASA(FM).

Assistant Secretary of the Army for Manpower and Reserve Affairs. The ASA (MR&A) oversees the personnel and training considerations required to support new systems, monitors the Army manpower and personnel integration (MANPRINT) program in coordination with the Deputy Chief of Staff for Personnel (DCSPER), and reviews and transmits to the Office of the Assistant Secretary of Defense for Force Management and Personnel the manpower estimate report (from the DCSOPS) required prior to milestone decision reviews II and III.*

Assistant Secretary of the Army for Installations, Logistics, and Environment. The ASA(IL&E) has secretariat responsibility for policy and oversight of logistics management, installations and housing, environment, safety and occupational

*The statutory requirement for the manpower estimate report (MER) was repealed by sec. 801 of the National Defense Authorization Act for fiscal years 1992 and 1993. However, the report is still required by the USD(A) to support milestone II and III reviews of major defense acquisition programs.

health, and chemical demilitarization. In coordination with the DCSLOG, the ASA(IL&E) oversees logistics supportability of materiel systems, including integrated logistics support (ILS).

Director of Information Systems for Command, Control, Communications, and Computers. Unlike the assistant secretaries, who are appointed from civilian life by the President, the DISC4 is a general officer recommended by the Chief of Staff and appointed by the Secretary.* The DISC4 serves as a military deputy to the AAE for information mission area activities, including planning, programming, life-cycle management, and software policy for both automated information systems and weapon systems.[14]† The DISC4 oversees the PEOs managing command, control, communications, computer, and information systems acquisition programs. The DISC4 also provides software research and development advice and management oversight for systems subject to ASARC and Major Automated Information Systems Review Council (MAISRC) review. The DISC4 cochairs the Army MAISRC with the Deputy ASA(RDA) for Procurement.

ARMY STAFF

The Army staff is the uniformed staff that reports to the Chief of Staff.‡ The chief and the Army staff also play a heavy role in the acquisition management process. Although prohibited by law from having an office within the Army staff that can focus on research, development, and acquisition, the chief is still responsible for programming, priorities, and requirements. The Vice Chief of Staff supervises the Army staff (see Fig. 12.6). The duties and responsibilities of the Army staff members that an Army program manager must deal with are as follows.

Vice Chief of Staff. The VCSA cochairs the ASARC with the AAE.

Director of Program Analysis and Evaluation. The DPAE has Army staff responsibility for management of the Army's planning, programming, budgeting, and execution system (PPBES)§ and maintains the Army portion of the DoD Future Years Defense Program (FYDP). The DPAE supports ASARC and Defense Acquisition Board reviews with affordability assessments of Army programs and cochairs the Program Budget Committee with the director of the Army budget during the programming and budgeting phases of PPBES.

*The DoD Reorganization Act of 1986 required all information management functions to be placed in the secretariat. The Army Reorganization Commission recommended that the Assistant Chief of Staff for Information Management (ACSIM), a general officer position, be moved from the Army staff and established as the DISC4. To place all responsibility for information systems under one office, the executive oversight function from ASA(FM) and the ACSIM headquarters staff function were combined under the new DISC4.

†This MOU assigns overall responsibility for Army software policy and plans for automated information systems and for weapon systems to the DISC4 and clarifies the duties of all three offices.

‡Both active-duty military and career government civilians serve on the Army staff. The principals (deputy chiefs of staff), however, are uniformed general officers, in contrast to the secretariat, where the assistant secretaries are appointed from civilian life by the President.

§The Army added "execution" to the DoD PPBS to reflect and emphasize the execution phase of the resource allocation process. Within DoD, PPBS ends with the submission of the DoD budget. PPBS is the first of four phases of the resource allocation process. The others are *enactment* of the budget by Congress, *apportionment* of budget authority to the executive agencies by the Office of Management and Budget, and *execution,* the obligation of funds by the agencies and outlay by the treasury.

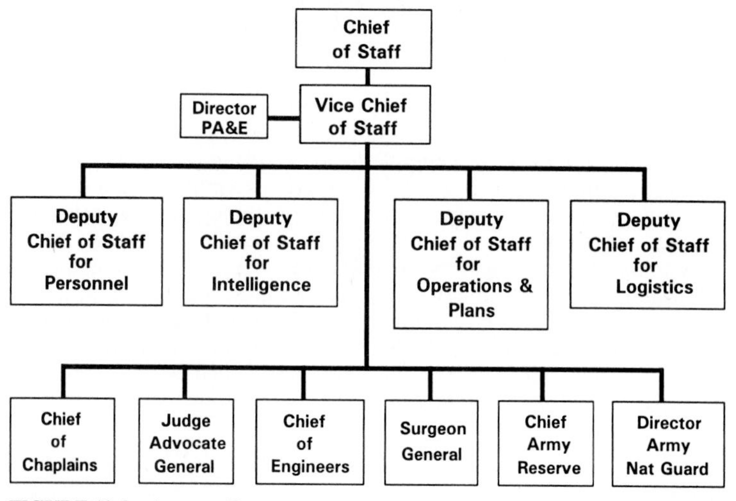

FIGURE 12.6 Army staff.

Deputy Chief of Staff for Operations and Plans. Like the ASA(RDA) in the secretariat, the DCSOPS is the key player on the Army staff for acquisition programs. This is due to the DCSOPS responsibility for the requirements process, including requirements generation, validation, integration, and approval of materiel requirements. The DCSOPS also establishes and validates Army priorities throughout the development of the Army POM and budget (the PPBES process). The DCSOPS is coproponent with the ASA(RDA) of the LRRDAP, assists ASA(RDA) in preparing program and budget documentation, prepares the manpower estimate report, assists the DUSA(OR) in establishing operational test and evaluation policy, develops policy and guidance for cost and operational effectiveness analysis (COEA), and recommends to the AAE the appropriate management level for acquisition programs. The DCSOPS also approves, for the Chief of Staff, the Army's Five Year Test Program for materiel systems.

It is the Army DCSOPS who determines the need for a special task force (STF) or special study group (SSG) at milestone 0 for all ACATs, and the DCSOPS supervises STF activities in the conduct of the concept exploration and definition phase.

Deputy Chief of Staff for Logistics. The DCSLOG has Army staff responsibility for logistical acceptability and supportability for materiel systems. The DCSLOG serves as the logistician for other than medical equipment and provides oversight over the logistics aspects of modification programs. The DCSLOG, in coordination with the ASA(RDA) and ASA(IL&E), establishes priorities for the Army logistics R&D program. He also monitors the acquisition programs to ensure that planning for organic support takes place early in the acquisition process, provides advice and analysis to materiel and combat developers to influence weapon system design for combat resilience, and ensures that spares and repair parts are in balance with the LRAMRP and LRRDAP. The Army Materiel Systems Analysis Activity (AMSAA) is responsible to the DCSLOG for performing integrated logistic support (ILS) program surveillance and independent logistics supportability assessments of all materiel programs, except those medical items that are the responsibility of the Army Medical Materiel Agency.[15]

Deputy Chief of Staff for Personnel. The DCSPER has the Army staff responsibility for the MANPRINT program and for assessing the adequacy of MANPRINT planning for ACAT I and II programs, supervises human performance RDT&E efforts, and coordinates with DCSOPS in the preparation of the manpower estimate report.

Deputy Chief of Staff for Intelligence. The DCSINT establishes and implements policy governing technology security, foreign disclosure, and automation security, establishes and implements threat support and documentation policy, designates threat integration staff officers for ACAT I and II programs, approves and validates threat documentation, and obtains Defense Intelligence Agency validation of threat documentation (normally a Systems Threat Assessment Report) to support a Defense Acquisition Board (DAB) review of acquisition programs.

Surgeon General. The Surgeon General has Army staff responsibility for medical research, development, test, and evaluation. The Surgeon General is the Army's medical materiel developer, responsible for ensuring the implementation of systems acquisition policy as it applies to combat medical systems, including the coordination of the PPBES for these efforts. A materiel command, the U.S. Army Medical Research and Development Command, with two laboratories and seven research institutes, reports to the Surgeon General.

Chief of Engineers. The COE has staff responsibility to support the AAE by overseeing the requirements and R&D necessary to provide construction design criteria, techniques, and materiel; fixed facility concealment, camouflage, and deception; and environmental quality associated with construction of R&D activities. The COE is also the Commander, U.S. Army Corps of Engineers, and as such is a materiel developer. The Corps of Engineers is discussed in the next section.

MATERIEL DEVELOPERS

The Army considers the PEOs, four Army major commands, and the Surgeon General (who supervises the Medical Research and Development Command) to be materiel developers.

Program Executive Officers (PEOs). When a number of similar acquisition programs interrelate in such a way as to warrant centralized management, the AAE may establish a PEO for management oversight of these programs. The decision as to whether a program will be placed in the PEO or materiel command structure is based on dollars, level of oversight required, and the relationship of the program to other programs within a PEO structure. The PEO structure as of October 1992 is depicted in Fig. 12.7. The duties and responsibilities of the 10 PEOs are outlined in AR 70-1[10] as follows:

> Administer assigned programs to ensure all necessary support is available to achieve programmatic goals.
>
> Support the total Army perspective of programs, rather than advocating individual programs.
>
> Be responsible to the AAE for programmatics (that is, materiel acquisition cost, schedule and total system performance), and the planning, programming, budgeting, and execution necessary to guide programs through each milestone. Immediately report baseline breaches to the AAE.

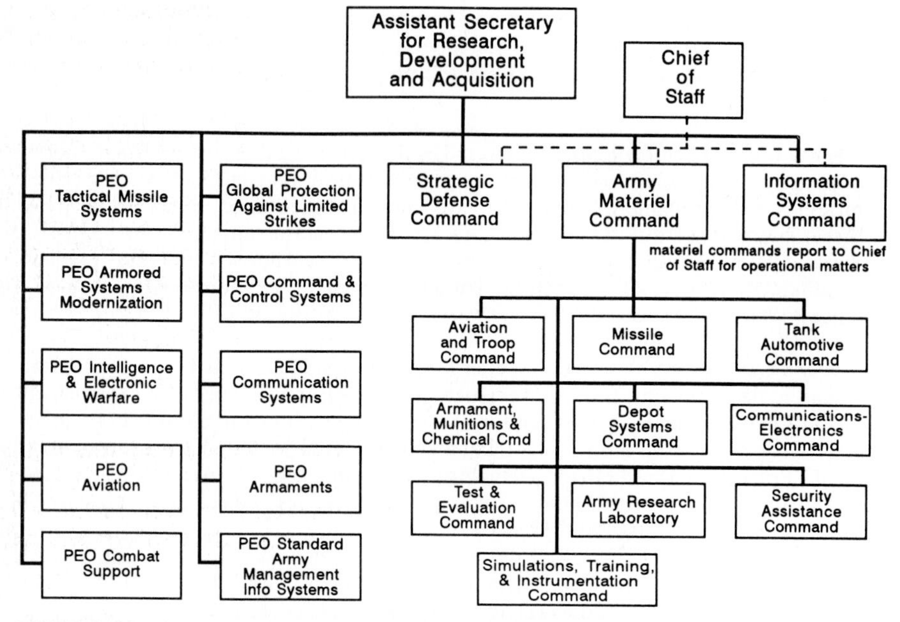

FIGURE 12.7 Army Program Executive Officers and Materiel Commands.

Provide the planning guidance, direction, control, oversight, and support necessary to minimize life-cycle cost and to field systems within cost, schedule, and performance baselines.

Provide program information to the AAE, HQDA, DoD, and the Congress, and defend assigned programs to Congress, through the Army Legislative and Budget Liaison Offices.

Participate in the development of data to support AAE programmatic decisions in the PPBES and provide development and acquisition resourcing data to TRADOC for the Long-Range Army Materiel Requirements Plan (LRAMRP).

Provide technical and functional integration across assigned programs. Ensure that functional (matrixed) support to subordinate PMs is planned and coordinated with the supporting organizations.

Charter, supervise, and evaluate assigned PMs.

Perform the fund control responsibilities of an independent general operating agency.

Approve acquisition plans for assigned programs under the provisions of paragraph 3-5.c (AR 70-1).

Serve as the safety officer for assigned systems with responsibility for proper planning and execution of system safety and environmental requirements (see AR 385-16).

MATERIEL COMMANDS

Four major Army commands—Army Materiel Command, Strategic Defense Command, Information Systems Command, and the Corps of Engineers—along with the Medical Research and Development Command (mentioned earlier), are considered to be materiel development commands. These materiel commands are responsible for all research, development, and acquisition programs that are not assigned by the AAE to the PEO structure. The commanders of these materiel commands report directly to the AAE for acquisition matters, the same as the PEOs. However, unlike the PEOs, these commanders also have an operational chain of command. All report into the office of the Chief of Staff for operational matters, including training and logistics support to fielded Army forces.*

Army Materiel Command. By far, AMC is the largest, most complex materiel command in the Army with worldwide responsibilities for research, development, acquisition, and logistics support. AMC is also responsible for Army security assistance efforts, a large part of which are foreign military sales. There are hundreds of acquisition programs managed under AMC major subordinate command's research and development centers and within the Army laboratory system. The Army's PEOs are colocated with the major subordinate commands of AMC. AMC's major subordinate commands are shown in Fig. 12.7. As of June 1992 there were 34 chartered program managers reporting under the AMC management structure, three of which report directly into HQAMC. AMC's responsibilities for acquisition management can be summarized as follows:

Equip and sustain the Army forces in the field.

Provide equipment and services to other nations through the Security Assistance Program.

Develop and acquire non-PEO managed systems.

Provide functional support to PEOs and PMs. This includes procurement, contracting, legal, finance, systems engineering, development test and evaluation, integrated logistics support, MANPRINT, and environmental.

Serve as single manager for conventional ammunition, and for the development and acquisition of targets, threat simulators, and unique test instrumentation.

Manage the Army laboratory system to define, develop, and acquire superior technologies.

Strategic Defense Command. SDC is the Army's materiel developer for strategic defense systems. The CG, SDC is the principal staff advisor to the Secretary of the Army and the Chief of Staff for matters pertaining to research, development, and acquisition for the Army Strategic Defense Program. CG, SDC is also the Army's primary point of contact with the Strategic Defense Initiative Organization (SDIO).

Information Systems Command. ISC is the materiel developer responsible for the research, development, and acquisition of Army information systems. Like AMC, ISC is responsible for providing assistance to information systems under the PEO structure and for programs that report directly into ISC. In June 1992 there were 16 PMs reporting under the ISC management structure. ISC reports into the AAE through the AAE's military deputy for DISC4 (refer to Fig. 12.4).

U.S. Army Corps of Engineers. USACE is an Army materiel developer responsible for research, development, and acquisition of engineering programs and materiel. In addition, USACE is responsible for looking at emerging systems for digital

*AR 70-1 contains detailed duties and responsibilities for both materiel and combat developers. This is a summary.

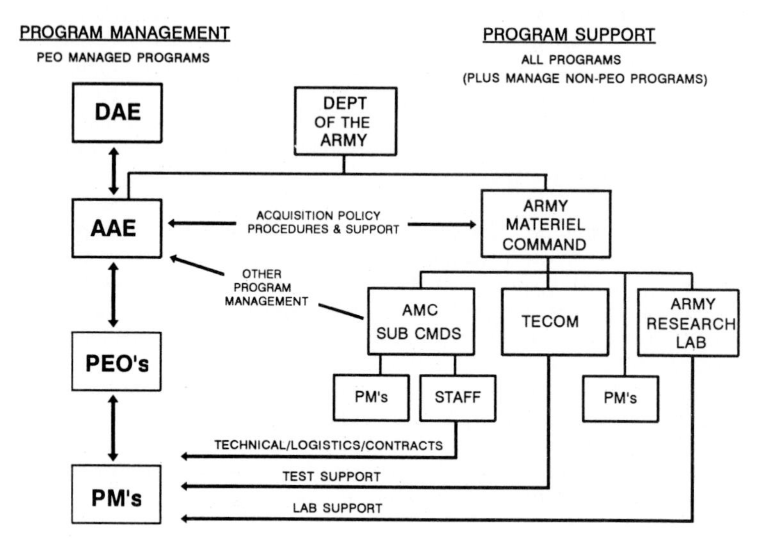

FIGURE 12.8 Army Materiel Command support for program management.

terrain data requirements and environmental effects, such as climate, terrain, and hazardous wastes. USACE also reviews mapping, charting, and geodesy requirements for materiel systems. The corps has responsibility for four Army laboratories. In 1992 there were no chartered PMs under the USACE management structure.

The relationship of the PEO and the PMs under the PEO structure to the materiel commands is critical to the success of all PEO managed programs. The management system is intended to be responsive to the needs of PMs who work for PEOs, and to those who work for the materiel command commander. This relationship is best described using the Army Materiel Command, the Army's largest materiel developer (see Fig. 12.8). The CG, AMC reports directly to the Army Chief of Staff for all operational matters and for the support of the fielded Army. He also has large major subordinate materiel commands (MSCs), a Test and Evaluation Command (TECOM), and the Army Research Laboratory (ARL), all of which must support a large number of in-house programs while providing functional matrix support to the PEO structure. These AMC organizations and their impact on the Army PM are discussed in more detail later. Not shown is the Depot Command (DESCOM), the arsenals, the ammunition plants, and other infrastructure that helps the CG, AMC fulfill the mission to support the Army in the field. Also, for acquisition matters, that is, matters dealing with the acquisition programs that fall within the AMC major subordinate command structure, the CG, AMC reports directly to the AAE, while keeping the Chief of Staff informed.

The reader must remember that program management responsibilities and the various programs themselves are divided between the PEO structure and the materiel commands. In the early stages of program initiation almost all programs fall under the materiel commands in the research and development centers or laboratories. As a program matures (soon after milestone I) the AAE must decide whether the program warrants placement under the PEO structure for management oversight. Also, as production stabilizes, and major modification efforts fall off, PEO programs may be transferred back to program or functional management under the materiel commands.

COMBAT DEVELOPERS

Only the Army and the Marine Corps have a formal management structure, separate from the field commands and the service headquarters, for developing warfighting requirements. Although the Army considers combat developments to be part of the acquisition process, the policy and procedures for most of this activity fall under the proponency of the Army staff DCSOPS, not the AAE. Army combat developers definitely have a significant impact on the Army program manager throughout the life cycle of the program. Army combat developers represent the using commands, the soldiers who must operate the equipment when fielded. It is the Army combat developer who develops the requirement. Figure 12.9 illustrates the relationship between the combat developer (user representative) and the materiel developer. The special task force (STF) or special study group (SSG) takes the larger acquisition programs through phase 0, the concept exploration and definition phase, and prepares all the documentation to support milestone I.[10] This is a combat development function under the staff supervision of the DCSOPS (for STF), or TRADOC (for SSG), supported by the materiel developers.

The combat developer develops the operational performance requirement with minimum acceptable values for both technical performance and supportability. Combat developers also define what actions are required to attain initial operational capability (IOC) and full operational capability (FOC). These parameters become the acquisition program baseline for cost, schedule, and performance that the PM, PEO, AAE, and DAE must sign.* Ironically, the combat developer does not actually sign the acquisition program baseline. However, it is the Army DCSOPS who represents the combat developer to the OSD staff, or the VCSA who represents the combat developer to the Joint Requirements Oversight Council (JROC). It is the DCSOPS, the combat developer, or the VCSA to whom the AAE and the DAE will

*See DoD 5000.2-M,[16] Part 14, for a complete description of an acquisition program baseline (APB). The APB contains quantified targets for key performance, cost, and schedule parameters of an acquisition program throughout the phases of the acquisition process.

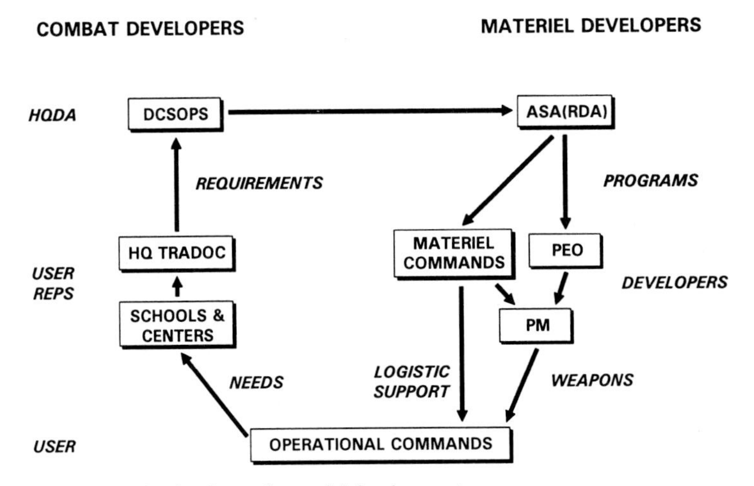

FIGURE 12.9 Combat and materiel development process.

turn for justification of performance requirements. For this reason it is important for PMs to understand the necessity of frequent and direct coordination with the combat developer to resolve performance issues that are certain to arise throughout development, production, and fielding. The role of combat developers in the Army acquisition process is summarized in the following.

Training and Doctrine Command. TRADOC is the Army's principal combat developer. The other combat developers coordinate their activities with HQ TRADOC. TRADOC is the lead for the Army's Long-Range Materiel Requirements Plan (LRAMRP). Other combat developers provide input to TRADOC for their portion of the LRAMRP. HQ TRADOC submits requirements developed by the combat developments activities of the TRADOC schools and integrating centers to HQDA, DCSOPS. TRADOC uses a process known as the concept-based requirements system (CBRS) to identify, validate, and integrate warfighting needs. These needs are listed in a battlefield development plan (BDP). (CBRS, LRAMRP, and BDP are discussed later, under "Resourcing the Program.")

Warfighting needs requiring a materiel solution are documented in a mission need statement (MNS) and forwarded to HQ TRADOC for approval. Approved MNSs are then sent by TRADOC to HQDA, DCSOPS for Army level approval and further action. The MNS is the basis for decisions by the AAE or DAE to proceed into the first phase of the weapon system life cycle. Army programs that may result in a new ACAT I program are forwarded through the JROC for review and approval to the DAE for the milestone 0 decision (Ref. 13, pp. 2–4).

Some of the specific TRADOC responsibilities to support the Army PM include preparing or updating the operational requirements document (ORD)* starting at milestone I; ensuring that critical operational issues and test criteria are developed to support the test and evaluation master plan (TEMP); participating in design reviews, ASARC/DAB preparation reviews; preparing and coordinating critical operational issues and criteria for operational test and evaluation; and conducting cost and operational evaluation analysis (COEA) of acquisition programs for the AAE.

TRADOC has established system managers to assist the PM in fielding a totally integrated system. The TRADOC System Managers (TSMs) are colocated with the combat development directorates at the schools and centers where the requirements are developed. The TSM is a counterpart to the PM, chartered by the CG, TRADOC. For example, the Army Tactical Missile System (ATACMS) has a program manager located at the Army's Missile Command in Huntsville, Ala., and reporting to the PEO for fire support systems. TRADOC established a TSM for ATACMS at the field artillery school, Fort Sill, Okla. Both the TSM and the PM are Army colonels. The TSM represents the user's requirement at the ASARC and the DAB, while the PM is responsible for the acquisition strategy to support the requirement. Generally, TRADOC establishes a TSM for each major defense acquisition program.[17] TRADOC responsibility for most systems is transferred after fielding to the school or center's functional directorate for combat developments. However, the PM office may last for many years into fielding. In 1992 TRADOC had 26 colonels assigned as TSMs.

Intelligence and Security Command. INSCOM is the Army's combat developer for strategic signal intelligence systems and in-house intelligence and electronic warfare systems. INSCOM also has responsibility for conducting development, test, and

*While the MNS describes a warfighting deficiency in broad mission terms, such as the inability to provide close fire support to defeat an enemy, the ORD describes a system-specific solution such as a multiple-launch rocket system. The MNS supports the decision to study the deficiency; the ORD describes the solution obtained from the study effort.

evaluation and serves as operational tester and evaluator for assigned classified or secure systems.

Health Services Command. HSC conducts medical combat development activities assigned by the Surgeon General and CG, TRADOC. HSC reviews requirements documents for consideration of health hazards and provides acquisition and logistics guidance for medical materiel. HSC also serves as the operational tester and evaluator for medical materiel.

Criminal Investigation Command (CIC). This is the combat developer for equipment used in criminal investigations.

The materiel and combat developers are the key players below the level of HQDA. These are the organizations that "make it happen" in Army acquisition. This is where the Army PMs, TSMs, and combat development staff officers are located. These organizations have research and development centers, laboratories, analysis agencies, and other functional support on which the PM depends for the life of the program. Figure 12.10 provides a quick reference for these key players, who they are, and what they do.

MATERIEL COMMANDS	PROGRAM EXECUTIVE OFFICERS (PEO's)	COMBAT DEVELOPERS
• Army Materiel Command • Information Systems Command • Medical Research & Development Command • Corps of Engineers • Strategic Defense Command	• Tactical Missile Systems • Armored Systems Modernization • Combat Support • Armaments • Aviation • Intelligence & Electronic Warfare • Communication Systems • Command & Control • Standard Army Management Information Systems • Global Protection Against Limited Strikes	• Training & Doctrine Command • Health Services Command • Intelligence & Security Command • Criminal Investigation Command
▸ Manage non-PEO Programs ▸ Matrix support PM's & PEO's ▸ Conduct DT&E ▸ Manage Tech Base ▸ Support the Army in the Field	▸ Manage major and selected nonmajor defense acquisition programs	▸ Develop doctrine, organizations & requirements

FIGURE 12.10 Army acquisition, key players.

*PROGRAM REVIEW FORUMS**

Army Systems Acquisition Review Council. The ASARC is the Army's equivalent of the Defense Acquisition Board (DAB). They both review the status of acquisition programs during development and production, and make recommendations to the acquisition executive as to whether or not the program should be allowed to continue. Many of the Army secretariat and staff principals as well as commanders of or-

*DoD Instruction 5000.2[12] is the reference for USD(A) review of Army ACAT ID programs. AR 70-1[10] is the reference for review of ACAT IC, II, III, and IV programs.

ganizations mentioned are members of the ASARC. The ASARC is convened for all ACAT I and II programs at each formal milestone to determine a program's readiness to proceed into the next phase of development or to production. As the milestone decision authority (MDA) for ACAT IC and II programs, the AAE makes the decision to either proceed or not proceed. The ASARC may also be convened at any time to review the status of a program, and will review all ACAT ID programs before the DAB review. Documentation to support an ASARC for all categories of programs is prescribed by DoDI 5000.2.[12] However, the AAE has discretionary authority for tailoring some of the documentation required for ACAT II programs.

ASARC review procedures tend to parallel those of the DAB, and the PM must keep in mind that when developing a program's milestone schedule, all ASARC requirements will precede those of the DAB. An ASARC preliminary review is held two to four weeks prior to the ASARC to identify the major issues and finalize the ASARC agenda. The ASARC executive secretary, located within the executive office of the ASA(RDA), is responsible for administrative control of the ASARC review as well as coordinating Army participation in DAB and DAB committee* reviews. For ASARC reviews, the executive secretary documents the results in an acquisition decision memorandum (ADM) and forwards the ADM to the AAE for signature.

The ASA(RDA) as the AAE and the VCSA cochair the ASARC. The VCSA is also the Army member of the Joint Requirements Oversight Council (JROC). The JROC will review the program after the ASARC, but before the DAB. Normally both the PM and the PEO will attend ASARC and DAB reviews. The PM must attend and brief the ASARC and the DAB, and may be required to brief the JROC on the status of the program. The following are members of the ASARC:

AAE and VCSA as cochairs

Deputy Under Secretary of the Army (Operations Research)

Assistant Secretary of the Army (Financial Management)

Assistant Secretary of the Army (Installations, Logistics, and Environment)

Assistant Secretary of the Army (Manpower and Reserve Affairs)

Commanding General, Army Materiel Command

Commanding General, Training and Doctrine Command

General Counsel

Director, Information Systems for Command, Control, Communications, and Computers

Deputy Chief of Staff for Logistics

Deputy Chief of Staff for Operations and Plans

Deputy Chief of Staff for Personnel

Deputy Chief of Staff for Intelligence

Chief, Army Reserve

Chief, National Guard Bureau

Chief, Legislative Liaison

*See DoD Instruction 5000.2,[12] Part 13, for a description of the DAB review process. There are three committees that support the DAB. A program will go through one or more committee reviews prior to the DAB review. The committee has overall responsibility for program oversight, and the committee chair makes a recommendation to the DAB chair [the USD(A)] as to whether or not the program should be allowed to continue in development or production.

Military Deputy to the ASA(RDA)

Director, Program Analysis and Evaluation

Comptroller of the Army

Commander, Operational Test and Evaluation Command

Others may be invited if there is a significant issue within their area of responsibility

Major Automated Information Systems Review Council (MAISRC).* The MAISRC is the equivalent to a Defense Acquisition Board (DAB) or ASARC review for automated information systems (AISs). These are systems acquired under the provisions of Army Regulation 25-3, "Life Cycle Management of Information Systems," and DoD Directive 7920.1, "Life Cycle Management of Automated Information Systems." All AISs except for those embedded in weapon systems and those used exclusively for cryptologic activities or information systems acquired under the National Foreign Intelligence Program for operational support of intelligence and electronic warfare systems, are managed in accordance with these directives.

The OSD MAISRC reviews AISs when total program costs exceed $100 million, or AISs having program costs in excess of $25 million in any one year. The Army MAISRC reviews AISs with total program costs of $10 million or more. PMs for these systems fall under the Army Information Systems Command or under the PEO for Standard Army Management Information Systems. The DISC4, as a military deputy to the ASA(RDA), cochairs the MAISRC with the Deputy ASA(RDA)(Procurement). Voting members of the Army MAISRC are DISC4, Deputy ASA (RDA)(Procurement), ASA(FM), ASA(IL&E), ASA(MRA), DCSLOG, DCSOPS, DCSPER, Director PA&E, and the Operational Test and Evaluation Command (OPTEC). Both the DISC4 and the Deputy ASA(RDA)(Procurement) sign the system decision memorandum documenting the results of the MAISRC review.

In-Process Reviews. The IPR is the equivalent of the ASARC review for ACAT III or IV programs and for MAISRC programs costing less than $10 million or delegated to the functional proponent, PEO, or major command. The milestone decision authority (MDA) is designated by the AAE, and that MDA chairs the IPR. The MDA would normally be a PEO or an equivalent commander of a materiel command. Procedures are basically the same as for the ASARC for weapon systems or the MAISRC for information systems.

Members of an IPR will include the MDA as chair, the combat developer, the logistician, the trainer (if different from the combat developer), the operational test independent evaluator (normally OPTEC), the developmental test independent evaluator (normally AMSAA), the materiel command's functional support organization, and others as determined by the MDA. Documentation to support an IPR may be tailored by the MDA, but should be consistent with the documents required by the ASARC, MAISRC, or DAB.†

---

*A letter of instruction (LOI)[18] provides instructions for conduct of the MAISRC and the documentation required for both the Army and OSD level review.

†DoDI 5000.2,[12] Parts 11-C and 11-D, list all documents and reports required for all acquisition categories (ACAT).

RESOURCE MANAGEMENT AND THE ARMY PROGRAM MANAGER

Army Modernization Plans. DoD Directive 5000.1 requires the military departments to develop long-range investment and modernization plans to support the OSD Planning, Programming, and Budgeting System (PPBS) (Ref. 13, pp. 2–9). The Army developed a series of modernization plans to provide guidance for force modernization well into the next century.* This series of plans were consolidated into one Army Modernization Plan (AMP) in the fall of 1992. The AMP will include guidance for the near-term (2 years for the budget), mid-term (6 years for the POM), and long-term (out to 20 years for planning). The AMP is constrained to the resources and force structure expected to be available, and is published by October 31 annually to support the PPBS process. ODCSOPS is the primary author of the AMP and will coordinate it with the Army staff, ASA(RDA), TRADOC, and AMC.

The AMP is preceded by the TRADOC Army modernization architecture (AMA). The AMA will replace the concept of battlefield functional mission areas (BFMAs) under the current TRADOC Concept-Based Requirements System (CBRS) (to be discussed later). The AMA is functional mission area specific focused on needs and/or requirements identified by the CBRS, and covers the same time frame as the AMP; however, the AMA is not limited to available resources (unconstrained). Each annex to the AMP is supported by a research, development, and acquisition program plan written by the office of the ASA(RDA). Program plans focus on new system development, are resource-constrained, and are PPBS related. Program plans are coordinated with the PM/PEO structure, approved by the ASA (RDA), and attached as an appendix to each AMP annex.

Many PMs have programs whose future depends on the AMP. Currently the AMP is expected to have the following annexes:†

Close Combat—Heavy
Close Combat-Light
Engineer/Mine Warfare
Theater Missile Defense
Fire Support
Air Defense
Command, Control, and Communications
Medical
Combat Service Support
Tactical Wheeled Vehicles
Soldier
Aviation
Nuclear, Biological, and Chemical
Information Mission Area

*See statement by the ASA(RDA) to the first session of the 102d Congress,[19] and "Weapon Systems."[20] "Materiel for Winning," jointly published on an annual basis by AMC, TRADOC, and ISC, contains additional investment strategy information for all the Army's mission areas.[21]

†Specific guidance on AMP content and format is published by the office of the DCSOPS. The latest guidance was forwarded to the Army staff by an internal ODCSOPS memorandum dated 30 June 1992. It is expected that the first AMP will be published in fall 1992, or very early in CY 1993. The AMP will be unclassified and made available throughout DoD and to the defense industry.

Intelligence/Electronic Warfare

Training

Concept-Based Requirements System. TRADOC Regulation 11-15 governs the CBRS process. CBRS is the process TRADOC uses to develop concepts, identify and prioritize mission needs and solutions, and obtain and synchronize changes caused by solutions. CBRS is guided by Army Long-Range Planning Guidance (ALRPG) issued by the DCSOPS. The ALRPG provides a common basis for the development of long-range plans by providing basic planning assumptions, characteristics of the future Army, and long-range planning goals. TRADOC is in the process of enhancing CBRS to make it more supportive of the ALRPG and of the AMP process (discussed earlier).

The CBRS consists of three phases conducted over 30 months. The three phases are cyclic, that is, they are ongoing activities based on a biennial cycle that coincides with PPBS. These three phases are (1) review concepts (capabilities required in the future) (April to October); (2) identify needs as warfighting deficiencies, opportunities, and obsolescence (October to June); (3) identify solutions in terms of changes to doctrine, training, leader development, organizations, and materiel (June to October). Implementation from identified solutions is to obtain solutions from the functional processes of doctrine, training, leader development, organizations, and materiel and then synchronize delivery, that is, ensure that all necessary resources for employment of a change are in the right place at the right time.

Warfighting needs are identified as part of an ongoing branch* (TRADOC schools) planning process. TRADOC major subordinate commands eliminate redundancies, group like needs, and do initial prioritization. Final prioritization is accomplished by HQ TRADOC. The final prioritized list of needs is published in May of even-numbered years as the Battlefield Development Plan (BDP). The BDP is distributed to the Army, other services, and allies, and is available to industry through the Defense Technical Information Center.

The two key major subordinate commands (MSCs) in this process are the Combined Arms Center (CAC) for combat and combat support and the Combined Arms Support Command (CASCOM) for combat service support needs. Based on guidance from HQ TRADOC and these MSCs, the branches (schools) develop solution candidates (doctrine, training, leader development, organizations, and materiel changes) to satisfy the needs listed in the BDP. The MSCs group the recommended changes into capability packages and battlefield functional mission areas (BFMAs) for integration and prioritization.

BFMAs represent the following battlefield functions: maneuver, fire support, air defense, mobility and survivability, intelligence, command and control, and combat service support. Each BFMA has associated capability packages. For example, the fire support BFMA includes capability packages for counterfire, suppression of enemy air defenses, attack of uncommitted forces, and attack of emitters. BFMA master plans precede the development of the DCSOPS modernization plan.

The Army modernization memorandum (AMM) is the key TRADOC product out of CBRS, listing changes to doctrine, training, organizations, leader development, and materiel identified as solutions to warfighting needs listed in the Battlefield Development Plan. The AMM has annexes for each of these areas. The AMM is published by October 1 of odd-numbered years (1991, 1993, etc.). Materiel solutions are

---

*The term branch refers to the combat and support arms: infantry, armor, field artillery, air defense artillery, ordnance, transportation, signal, quartermaster, and others. The proponents for these branches are the associated TRADOC schools. For example, the proponent for field artillery is TRADOC's Field Artillery School at Fort Sill, Okla.

identified in the Long-Range Army Materiel Requirements Plan (LRAMRP) annex.

Long-Range Army Materiel Requirements Plan. The LRAMRP depicts future Army materiel requirements, prioritized for resourcing. It is a document prepared by TRADOC with input by the other combat and materiel developing commands. HQ TRADOC directs the process, has final approval authority, and submits the LRAMRP to HQDA. The LRAMRP drives the HQDA development of the Long-Range Research, Development, and Acquisition Plan (LRRDAP) discussed next. The LRAMRP covers the PPBES six-year program (for example, FY 1994–1999) and is extended nine years beyond the POM. PMs must develop and provide programmatic input for this 15-year program in terms of dollars and quantities.

The LRAMRP process starts in February of each odd-numbered year, the planning year under the biennial PPBES process. A series of reviews are held throughout the spring and summer, and a final LRAMRP is submitted to HQDA by October.* The LRAMRP is the starting point for the HQDA LRRDAP process.

Since the LRAMRP prioritizes warfighting mission area deficiencies, it is very important for the PM to ensure that accurate and timely data are submitted. The PM may want to attend these reviews personally when appropriate, particularly if the PMs program is within a mission area in turmoil. Figure 12.11 illustrates the LRAMRP process.

*The LRAMRP replaced the field LRRDAP for the FY 1994–2008 planning period. TRADOC is considering an enhanced CBRS (ECBRS) to support the FY 1996–2010 period. The LRAMRP would be replaced by the Army Modernization Plan (AMP) to feed development of the LRRDAP. The AMP is to be derived from the Army Modernization Architecture (AMA), which replaces the AMM. The BDP would be eliminated. Also, TRADOC created six "battle labs" in the summer of 1992 to define capabilities, identify requirements, and help determine priorities. This concept is still maturing; however, it appears that the timelines for PMs to influence any enhanced process will be about the same, as the PPBS has not changed.

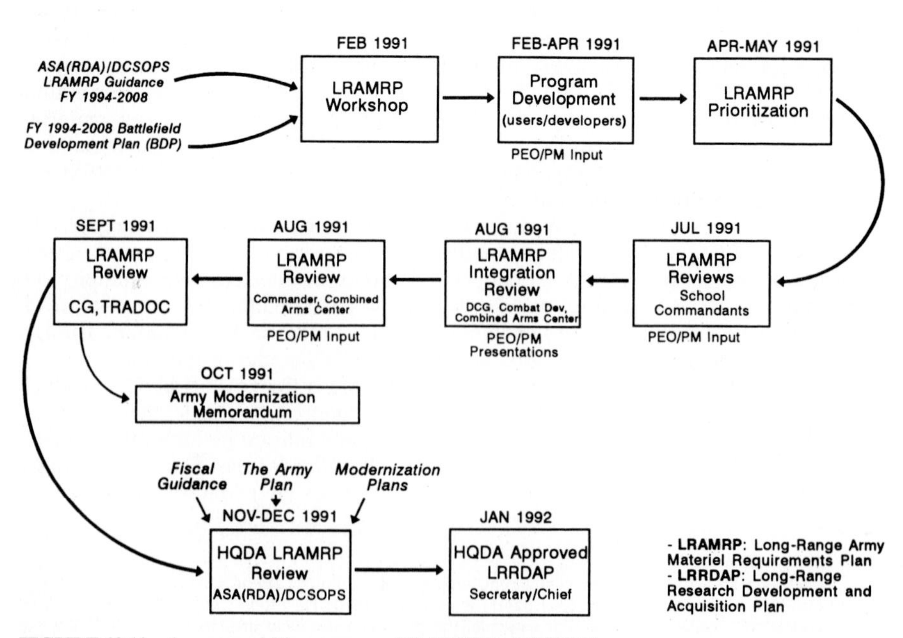

FIGURE 12.11 Army acquisition program, LRAMRP to LRRDAP.

Programming and Budgeting. On receipt in October of odd-numbered years, HQDA translates the LRAMRP into a Long-Range Research, Development, and Acquisition Plan (LRRDAP). The office of the DCSOPS and the office of the ASA(RDA) jointly sponsor development of the LRRDAP. Working closely together and with the director for program analysis and evaluation (DPAE), OASA(RDA) and ODCSOPS convert the LRAMRP into the LRRDAP, a HQDA prioritized listing of technology and development programs which will feed the development of the Army POM.

The LRAMRP-to-LRRDAP conversion process is heavily influenced by the modernization plan mentioned earlier, and the Army Plan (TAP) issued by the DCSOPS. TAP summarizes national military strategy and security policy for the Army, and develops and articulates the Army's priorities within expected resource levels. The resulting program represents the continuance of the Army's modernization program.

Throughout late fall and winter, and again in the spring of the following year, the Army PM will be asked to update the data submitted previously for the LRAMRP, and to answer questions on the impact of funding decrements and changes in HQDA priorities.

The Army acquisition program must survive a series of POM and budget review panels led by colonels, a program budget committee led by generals, and finally the select committee, after which the secretary and the chief will approve the POM for submission to the Assistant Secretary of Defense for Program Analysis and Evaluation on April 1 of each even-numbered year (April 1992, 1994, etc.). The PM must not lose sight of the fact that the program is only one of hundreds, and that all acquisition programs are competing for resources among themselves and with the other accounts (such as operations and support, military construction, and military pay).

After the Army submits the POM in April there is a POM review cycle conducted by the OSD and joint staffs. The PM must be available to answer questions on programmatic issues raised by these agencies. At the same time, the Army's Assistant Secretary for Financial Management now has the responsibility to translate the six-year POM into a two-year budget. So the PM is responding to program analysts from OSD [ASD(PA&E)], action officers from the DAE and other OSD staffs, and the joint staff. At the same time the PM is preparing to defend the first two years of the POM to the director of the Army budget in OASA(FM), and after the Deputy Secretary of Defense approves the POM around June or July, to the comptroller of the DoD. Figure 12.12 depicts the LRRDAP-to-POM approval process.

The PM will work with the director of the Army budget from POM approval in July or August until the Army submits its two-year budget to the DoD comptroller on September 15. The PMs budget must survive a review cycle similar to that of the POM. Further, the PM must stand by to answer questions that will come up during the program budget decision (PBD) cycle as staff officers from the office of the DoD comptroller and the Office of Management and Budget revisit the Army's requirements for many programs. DoD's goal is to complete the budget and provide it to OMB before Christmas so that the President can submit it to Congress in January. Figure 12.13 illustrates the Army POM-to-budget development process.

The Army PM should keep in mind the nature of the biennial cycle for development of the LRAMRP, LRRDAP, POM, and the budget. For example, the budget submitted to OSD by the Army in September 1992 may or may not reflect the warfighting needs, or the acquisition programs identified in the Battlefield Development Plan issued in February 1991, or the LRAMRP sent to HQDA in October 1991. Also, under the DoD biennial budget cycle, even though the POM is submitted every two years, there is still a budget cycle every year. The PM must play in both the biennial POM cycle and the annual budget cycle. The key to success during this entire resource allocation process is to keep in touch with the TRADOC systems man-

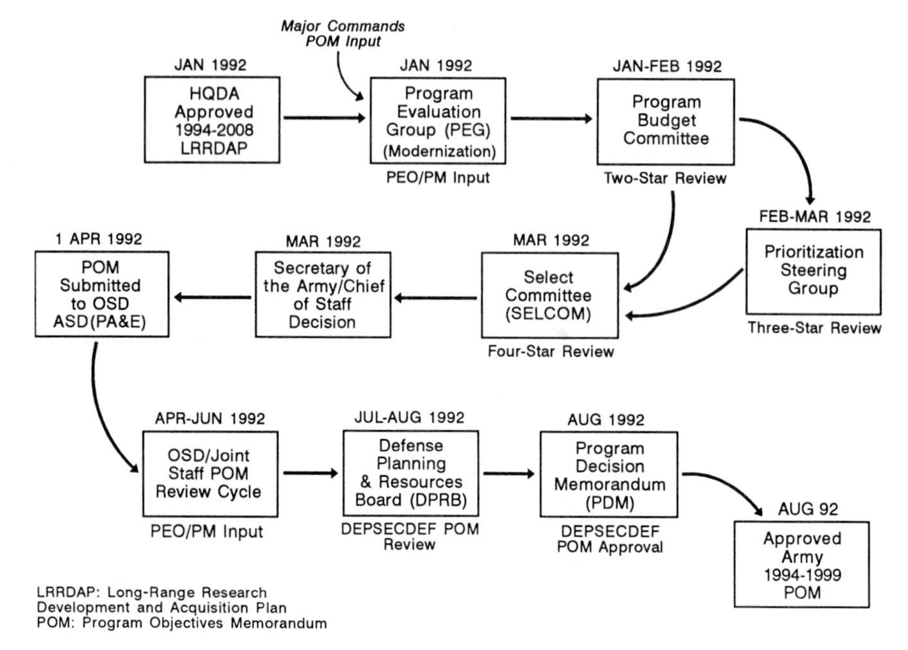

FIGURE 12.12 Army acquisition program, LRRDAP to POM.

ager or combat developer, and the staff officers that represent the PMs program in the offices of the ASA(RDA) and DCSOPS. The most likely points at which the PM may intervene during this complex resource allocation are shown in Figs. 12.11, 12.12, and 12.13. Also, the PM must continue to manage the program.

Funds Management. PEOs and materiel commands provide budget justification documents for assigned programs directly to HQDA. Funds control responsibility for PEO-managed programs resides solely in the PEO structure, separate from the materiel commands. Reprogramming* authority of RDT&E and procurement funds, within established thresholds, resides with the PEO for assigned programs and with the materiel command commander for non-PEO managed programs. Reprogramming of funds between PEOs, between materiel commands, or between a PEO and a materiel command must have HQDA approval. PEOs and commanders of materiel commands also have discretionary authority for reprogramming of operations and maintenance, Army (OMA) funds within assigned programs. Exceptions

*Congress allows DoD to shift funds from one program to another within the same appropriation. Congress has agreed with DoD that funds may be shifted under certain circumstances without notifying Congress. This is called "below-threshold" reprogramming authority. For a below-threshold reprogramming action which moves funds into a RDT&E program element, the limit is $4 million; for moving funds out of an RDT&E program element, the limit is $4 million or 10 percent, whichever is greater. For procurement the limit is an increase of $10 million or more; or a decrease of $10 million or more or 20 percent of the procurement line item (whichever is greater). Other rules apply. See DoD "Budget Guidance Manual," May 1990, sec. 3, Reprogramming. Also, refer to House of Representatives Report 102-328 (FY 92 Conference Report on appropriations).

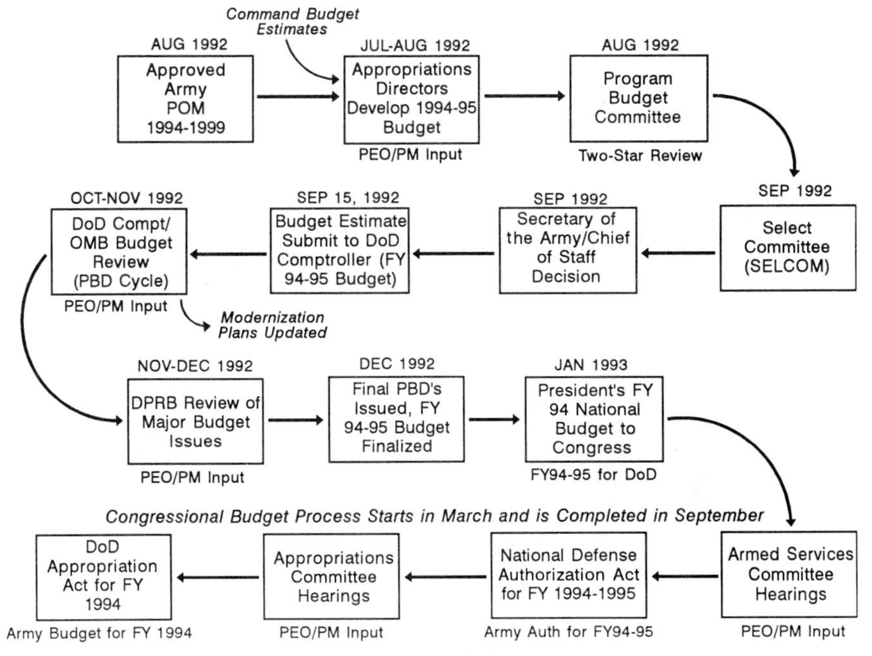

FIGURE 12.13 Army acquisition program, POM to budget.

are OMA dollars for intelligence or depot maintenance (hardware) programs, which require HQDA approval (Ref. 13, pp. 62–63).*

Matrix Support for PEO or Direct Reporting PMs. PEOs and PMs must negotiate a memorandum of understanding (MOU) with the materiel commands for support required to execute the program. The MOU is considered to be a living document, updated as changes occur. The PEO or PM will identify the tasks to be accomplished and the materiel command will identify the functional resources required. A funding schedule for these required resources must be in the MOU. The PEO and the commander of the materiel command or major subordinate command sign the MOU (Ref. 13, p. 59).

Personnel Management. Matrix support arrangements between the PEOs and the materiel commands vary by location. Some programs have matrix support personnel collocated with the program office or the PEO office. Other program offices are supported by personnel that remain in the materiel command's functional directorate. Still other matrix support is provided on a temporary-duty travel status when there are great distances between the program office and the materiel command.

The AMC matrix support model provides for a PM core office, a resident direct support matrix, and a nonresident support matrix, which are described below.†

*AR 37-1 specifies funds control responsibilities in detail.

†This model was developed immediately after the PEO structure was established in 1987. However, it is not formally endorsed by HQDA, and implementation is spotty at best. It is presented here as an AMC method to manage the matrix support.

PM Core. Manpower approved by ASA(RDA) and permanently assigned to the PEO or PM office. These personnel are those required regardless of the phase of the life cycle the program is in. They include program management, financial management, cost analysis, and program evaluation personnel.

Direct Support Matrix—Resident. Personnel required to support a specific acquisition phase or phases. Personnel are assigned to the materiel commands. However, they are collocated in the PEO or PM office. They include systems engineering, product assurance, test planning and execution, configuration management, logistics, and (for some PMs) security assistance.

General Support Matrix—Nonresident. These are personnel normally not required on a full-time basis. They are assigned to the materiel commands and justified based on the overall mission of the command, not just acquisition programs. Examples are procurement, legal, resource management, logistics assistance, and other functional areas.

Starting in FY 1992, the PEOs and assigned PMs were provided the funds necessary to reimburse the materiel commands for all matrix support.[23] The materiel commands are given the opportunity to meet the PEO or PM support requirements, basically through the MOU negotiation process previously discussed. If the materiel command cannot provide the required functional support, the PEO or PM has authority to use contractor (private industry) support. HQDA provides funds for all sustainment efforts not provided to the PEO or PM directly to the materiel commands.

Performance Evaluations. The PEOs and PMs must be in the rating chain for matrix support personnel. For direct support matrix—resident personnel, the PEO or PM is in the rating chain for military evaluation reports, and is a designated reviewing official for civilian performance appraisals. For general support matrix—nonresident personnel, the PEO or PM will provide letter input to the performance evaluations for both civilian and military personnel.

A process has been developed by AMC, whereby PEOs and PMs rate the materiel commands on quality and completeness of matrix support. This is accomplished on a quarterly basis. The PM provides a report to the PEO, rating each functional area against five categories: activity expertise, timely response, service quality, management involvement, and manpower provided. The PEO summarizes the reports of assigned PMs for each category with a single rating. The PEO provides this summary report to the commander of the materiel command. The materiel command forwards the report to HQAMC with appropriate comments.[24]

The Burden of Proof. Recognizing that the authority and responsibility for program management is vested in PMs is not always easy for the functional matrix. Conflicts are bound to arise between PMs and organizations outside of the programmatic chain that provide advice to decision makers in the chain. AR 70-1 clearly shifts the "burden of proof" as to the value added by functional requirements placed on PMs to the Army functional organization that developed the requirement. The AR states that "when these organizations place functional requirements on the PM which are not required by statute, Federal Acquisition Regulation (with supplements), or DoD directives and instructions, the organizations must justify the value of the added requirements to the PM—they are not applied automatically." PMs may decide not to comply. Disagreements that cannot be resolved will be raised to the milestone decision authority, who has the authority to enforce the functional request or exempt the program from the requirement.

Test and Evaluation. The program manager is concerned with two major areas of test and evaluation during the development of a weapon system. First, developmental testing is conducted to ensure that the system meets the contract specification. The Army uses the term "technical testing," or TT, for this process. Technical testing is conducted at various Army research and development centers and test ranges and proving grounds throughout the United States. Live fire test and evaluation to assess the vulnerability of systems to enemy action, or the lethality of U.S. systems against enemy systems, is also considered developmental, or technical, testing.

The second type of testing is operational testing to ensure that the system meets the user's requirement. The Army refers to operational testing as "user testing," or UT. User testing is conducted with typical user troops from operational units. Unlike technical testing, user testing is conducted under realistic combat conditions at Army installations that can best provide this environment.*

The Army PM is concerned with both contractor and government test and evaluation. From a resource standpoint, the contract for the system will normally provide for contractor testing prior to systems being delivered to the government for further testing. The PM must also identify funding to support testing. The test agency budgets funds for test ranges, instrumentation, and so on, according to its test plans.

A test integration working group (TIWG) (pronounced "tee wig") will be established and chaired by the PM to coordinate and resolve the issues and criteria for testing, including the measures of effectiveness. TIWG membership will include the development and operational testers and the independent evaluators: AMC's Test and Evaluation Command (TECOM), the Operational Test and Evaluation Command (OPTEC), and the Army Materiel Systems Analysis (AMSA) activity. Also represented on the TIWG will be the TSM or combat developer, the trainer, the logistician, and the threat system officer.

Other participants on the TIWG may include the integrated logistic support management team, the contractor (when appropriate), live fire test representatives, HQDA and major command staffs, and others who serve in a monitor role (such as the Surgeon General representative for health-related aspects associated with testing). The TIWG should be established as early as possible, during concept exploration and definition, but no later than the demonstration and validation phase of development.

The Army has three major agencies that the PM must work with to ensure a coherent test and evaluation master plan.

Test and Evaluation Command. TECOM is an AMC major subordinate command that functions as the Army's development test agency. TECOM is the technical or development tester for all Army materiel systems except those assigned to the Information Systems Command or to the Medical Research and Development Command. TECOM is also responsible for a number of development test ranges, such as White Sands missile range in Texas and New Mexico, and Aberdeen proving grounds in Maryland.

Army Materiel Systems Analysis Activity. AMSAA is an independent agency reporting directly to HQAMC that provides an independent assessment of the developmental testing conducted by TECOM. The TECOM test report and the AMSAA independent evaluation provide the PM and the milestone decision authority with the results of developmental testing during the development effort and at milestone reviews.

---

*AR 73-1[25] provides policy for testing all systems acquired under the auspices of AR 70-1 and information mission area systems acquired under the AR 25 series.

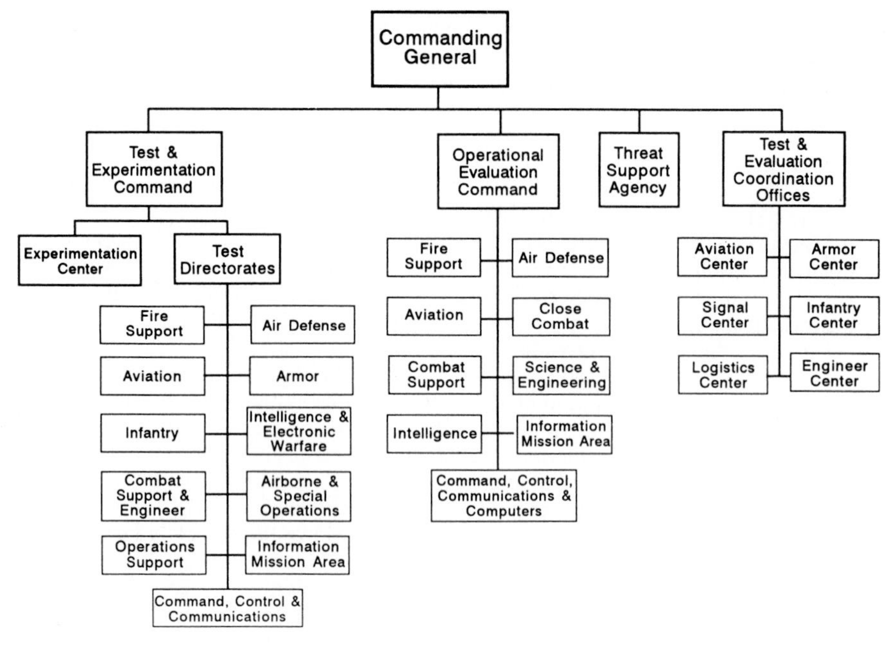

FIGURE 12.14 Army Operational Test and Evaluation Command.

U.S. Army Operational Test and Evaluation Command. OPTEC is the Army's independent operational tester, reporting directly to the Army Vice Chief of Staff. OPTEC consists of the Test and Experimentation Command, the Operational Evaluation Command, test and evaluation coordination offices, and the OPTEC Threat Support Agency* (see Fig. 12.14).

The Text and Experimentation Command (TEXCOM) is organized with a series of test directorates located at major Army installations. These test directorates conduct operational testing of all new materiel systems, except those assigned to the Health Services Command, the Intelligence and Security Command, and the Chief of Engineers. This also includes all information mission area systems subject to review by the Major Automated Information Systems Review Council and joint or multiservice systems. The Operational Evaluation Command (OEC), located near Washington, D.C., is the independent evaluator of results from TEXCOM testing. TEXCOM conducts the operational test, writes a test report, and forwards the report to OEC. The independent evaluator from OEC observes the test, reviews the test report, and writes an independent evaluation report. Both the test report and the independent evaluation report are forwarded to the Army acquisition executive prior to milestone decision reviews.

*Prior to 1991 the Test and Experimentation Command (TEXCOM) was a TRADOC command with test boards at the major TRADOC schools and centers. These test boards conducted the operational testing, with the U.S. Army Operational Test and Evaluation Agency (OTEA) providing an independent evaluation of test results. In late 1990 TEXCOM and OTEA were combined into one command, the U.S. Army Operational Test Command (OPTEC).

CONCLUSION

The policies and procedures for program management in the U.S. Army have stabilized since the reorganization of the Army's system for acquisition management in accordance with the DoD Reorganization Act of 1986. The Army was the first service to establish a program executive officer structure independent of the materiel development commands. The Army is unique among the military departments in that it has provided the program managers and the program executive officers with streamlined procedures for resource allocation. Both funding and manpower are provided directly from HQDA.

The Defense Acquisition Workforce Improvement Act (Title XII, Public Law 101-510) required the services to establish an "acquisition corps." Currently the Army Acquisition Corps (AAC) consists of about 3000 military officers and civilian managers. This corps is not a separate organization; the members retain their branch and career field identities. The corps is a management structure that formalizes the requirements for education, training, and experience for its members. All program managers, deputy program managers, and other key personnel with a program office are members of the AAC. The AAC is an elite organization of professionals tasked with providing effective and reliable materiel systems to the Army. As the military strategy of the nation changes, the mission of the AAC will be of critical importance to the readiness of the Army.

REFERENCES

1. Association of the United States Army, "A Primer on Research and Development in the United States Army," Special Report, Aug. 1990, pp. 2–7.
2. Office of the Chief of Staff, Army, "Comprehensive Review of Active Component and Reserve Component Table of Distribution and Allowances (TDA)," Memorandum, Dec. 16, 1988, pp. 17-1–17-5, 32-2–32-10; final study report of the Redistribution of Base Operations Unit Strength within TDA (ROBUST) Task Force.
3. "Transmittal of Report of the AMARC Study," Memorandum for the Secretary of the Army, Apr. 1, 1974; final report of the Army Materiel Acquisition Review Committee (AMARC).
4. "A Quest for Excellence," Appendix C, Final Report by the President's Blue Ribbon Commission on Defense Management, National Security Decision Directive 219, Apr. 1, 1986; White House Summary, June 1986.
5. Secretary of the Army, "Implementation of the PEO Structure," Jan. 30, 1987.
6. Office of the Under Secretary, "Implementation of the Program Executive Officer Concept, Memorandum, Apr. 29, 1987.
7. Office of the Under Secretary, "Program Executive Officer Organization," Memorandum, Aug. 4, 1988.
8. "Changes to the PEO and PM Structure," Army Acquisition Executive Memorandum, Jan. 25, 1989.
9. Army Acquisition Executive Support Agency, "Program Executive Officers and Program, Project, Product Managers Listing," June 1992.
10. Army Acquisition Executive Interim Operating Instruction for Army Acquisition Policy, Army Regulation 70-1, "Army Acquisition Policy—Research, Development and Acquisition," April 17, 1992.
11. "Army Acquisition Corps," Information Paper, Total Army Personnel Center, Oct. 10, 1990.

12. DoD Instruction 5000.2, "Defense Acquisition Management Policies and Procedures," Feb. 23, 1991.
13. DoD Directive 5000.1, "Defense Acquisition," Feb. 23, 1991.
14. Memorandum of Understanding (MOU) Among the Director for Information Systems, Command, Control and Computers (DISC4), the Assistant Secretary of the Army for Research, Development and Acquisition [ASA(RDA)], and the Deputy Under Secretary of the Army for Operations Research [DUSA(OR)], "Headquarters, Department of the Army, Software Management Responsibilities," Mar. 8, 1991.
15. Memorandum of Understanding between Deputy Chief of Staff for Logistics, U.S. Army Materiel Command, U.S. Army Operational Test and Evaluation Command, U.S. Army Materiel Systems Analysis Activity, and U.S. Army Logistics Evaluation Agency, "Transfer of the Department of the Army Independent Logistician Integrated Logistics Support (ILS) Mission and Functions," Sept. 30, 1990.
16. DoD Manual 5000.2-M, "Defense Acquisition Management Documentation and Reports," Feb. 23, 1991.
17. TRADOC Regulation 71-12, "TRADOC Systems Management," Nov. 19, 1990.
18. "Letter of Instruction (LOI) for Conduct of Major Automated Information Systems (AIS) Reviews," Office, Director of Information Systems for Command, Control, Communications, and Computers, Department of the Army, June 15, 1991.
19. "Equipping the U.S. Army," Statement to the Congress, FY 92 Army RDT&E and Procurement Appropriations, Steven K. Conver, Assistant Secretary of the Army (Research, Development and Acquisition), and Lieutenant General August M. Cianciolo, Military Deputy to ASA(RDA), Mar. 1, 1991.
20. "Weapon Systems, United States Army 1992," Department of the Army publication available through the Government Printing Office, Washington, D.C., March 1, 1992.
21. U.S. Army Materiel Command, U.S. Army Training and Doctrine Command, and U.S. Army Information Systems Command, "Materiel for Winning," Nov. 1990.
22. TRADOC Regulation 11-15, "Concept Based Requirements System," Aug. 1, 1989.
23. "Matrix Support Policy for Program Executive Officer Managed Systems," Army Acquisition Executive Policy Memorandum 91-4, Feb. 28, 1991.
24. Headquarters, Army Materiel Command, "PEO/PM Evaluation Program for Matrix Support Activities," Memorandum, May 20, 1991.
25. Army Regulation 73-1, "Test and Evaluation Policy," Oct. 15, 1992.

CHAPTER 13
PROGRAM MANAGEMENT IN THE U.S. NAVY

Ernst Peter Vollmer

Ernst Peter Vollmer is presently chair of the Acquisition Policy Department at the Defense Systems Management College, Ft. Belvoir, Va. A decorated, retired Naval officer with significant operational aviation experience, he served on the staff of the Office of the Chief of Naval Operations, where he was heavily involved in the requirements generation process and the PPBS. He also was a program manager for a commodity line at the Naval Air Systems Command. After retirement, Mr. Vollmer was a Principal Staff Member at The BDM Corporation, where he managed several programs supporting international and domestic clients.

The reader who aspires either to be a successful Navy program manager (PM) or to be successful in dealing with Navy PMs must understand how the current Navy acquisition organization and management philosophy came to be. Although the Department of the Navy (DON) is fully compliant with current DoD acquisition policies, Navy PMs must understand and deal with the unique philosophies resulting from an evolving process that has provided weapon systems to the sailor over the last 200 plus years. This chapter will discuss the significant post–World War II organizational and responsibility changes within the Navy and the impact of the current organization on today's PM.

EVOLUTION AFTER WORLD WAR II

By the start of World War II, the Navy community responsible for the research, development, and acquisition of naval weapon systems had matured into a bilinear organization. The Chief of Naval Operations (CNO), through his headquarters staff (OPNAV), determined weapon system requirements which were given to material bureaus for resolution. The material bureau chiefs reported directly to the Secretary of the Navy (SECNAV) and controlled the research, development, acquisition, and in-service support of the resultant systems. Generations of Navy acquisition managers served under the philosophy that the bureau managers were free of OPNAV supervision once a system's requirements were established, although OPNAV retained responsibility for program sponsorship and funding.

The bilinear organization served the Navy well through World War II. However, as the Navy moved into the arena of nuclear power, missiles, jet aircraft, and generally more complex weapon systems, the cost and time required to develop and field new ships and weapon systems increased dramatically. The rapid advances in warfighting technology required a philosophical change in the Navy's management approach in order to maintain the fleet's warfighting capabilities at a high level. In addition, the more complex weapon systems under development crossed the traditional bureau boundaries, creating coordination issues that further stretched schedules.

In 1966 the SECNAV formally addressed the issue of weapon systems overlapping bureaus and eliminated the bilinear structure. The bureaus were required to report through a Chief of Naval Material, and the resultant organization was called the Naval Material Support Establishment. The next major change came in 1968 when Congress, responding to the Secretary of Defense's request, abolished the statutory basis for the material bureaus and relegated the authority and responsibility for their organization to the SECNAV. The Naval Material Command (NAVMAT), headed by a Chief of Naval Material, was established with six subordinate systems commands (SYSCOMs): air systems, ship systems, ordnance systems, electronic systems, supply systems, and facilities engineering. The traditional bilinear relationship was dropped and the CNO assumed responsibility for the Naval Material Command, except that the Navy Secretariat retained control of all business and contractual matters. This was a significant philosophical change for the former bureau personnel, who were used to a large degree of independence from the OPNAV staff. The next change occurred in the 1970s, when the Ship Systems Command and the Ordnance Systems Command were combined to create the Naval Sea Systems Command. This structure would remain in place until the mid-1980s.

Throughout this period, the Navy Secretariat had two key positions overseeing DON research, development, and acquisition. The Assistant Secretary of the Navy for Research, Engineering and Systems [ASN(RE&S)] oversaw the research and development of systems (with the exception of ships) until they were ready for full-rate production. The Assistant Secretary of the Navy for Installations and Logistics [ASN(I&L)], later renamed the Assistant Secretary of the Navy for Shipbuilding and Logistics [ASN(S&L)], assumed responsibility for the item at the production decision point and had oversight for the remainder of its service life. The ASN(S&L) also oversaw the entire ship research, development, acquisition, and in-service support effort. In addition, the SECNAV delegated senior procurement executive responsibility to the ASN(S&L), giving the incumbent contracting and business review authority over all research, development, and acquisition matters regardless of the phase of development. As a result, some PMs' programs were reviewed by one ASN, then later by the other, and possibly by both, depending on the type of project and its phase of development.

During the 1960s Secretary of Defense Robert S. McNamara imposed the planning, programming, and budgeting system (PPBS) on the services as he consolidated more power within the Office of the Secretary of Defense (OSD). OSD also directed the services to implement formal project offices and then imposed a formal review process on the service PMs. These actions had a profound effect on the CNO and OPNAV through lost authority and an expanded paper workload for OPNAV as staff offices responded to increasing OSD and congressional demands for information (Hone,[1] p. 82). CNO Admiral Elmo Zumwalt reorganized OPNAV in 1971, creating Deputy Chiefs of Naval Operations (DCNOs) for submarines, surface, and air. "This move represented a major change in focus because it shifted the OPNAV organization further away from functional lines and more toward warfare, or platform communities" (Hone,[1] p. 91). He also created and strengthened some directorates within OPNAV to better enable him to monitor and control what the newly created platform sponsors did.

The platform sponsors, also called program or resource sponsors, had program coordinators (PCs) who worked the headquarters issues (requirements, resource, priority, OSD, or congressional, for example) associated with their assigned weapon systems or acquisition programs. This led to a problem as the SYSCOMs informally aligned themselves with their OPNAV platform sponsors. As a result, weapon systems that did not necessarily fall within the force structure envisioned by the CNO could be funded and developed. When a 1979 paper "described OPNAV as divided into two camps—one organized around the platform sponsors (DCNOs for submarine, surface, and air warfare); the other, around the OP-090 directorate (Program Planning)" (Hone,[1] p. 108), Chief of Naval Operations Admiral Thomas B. Hayward moved quickly and, in 1980, changed the duties of OP-095, renaming it Director of Naval Warfare, and making it "responsible for implementing the CNO's policy for overall fleet readiness and modernization in regards to all phases of general purpose naval warfare" (Hone,[1] p. 112). To make OP-095 viable, he reassigned some of the best platform and program sponsor PCs to the new organization.

THE PIVOTAL 1980s

Reagan Administration Secretary of Defense Caspar Weinberger and his Deputy Secretary Frank Carlucci began holding the service secretaries and chiefs more responsible for managing their departments. Secretary of the Navy John Lehman quickly recognized this as an opportunity to greatly increase his control and influence over the DON. Lehman was also dissatisfied that the Chief of Naval Material reported to the CNO, for he recognized that the platform sponsors exerted a high degree of influence over the SYSCOM PMs. In 1985 he disestablished the Naval Material Command and directed the SYSCOM commanders to report directly to him. He renamed the Naval Electronics Systems Command the Space and Naval Warfare Command and assigned it the former NAVMAT responsibility for coordination among the three acquisition commands. Placing the SYSCOM commanders and their PMs under his purview reduced the authority and influence of the CNO and the OPNAV platform sponsors. Program decision authority for most projects was elevated to the assistant secretary level (Ref. 2, pp. 3–4).

Through the two assistant secretaries, Secretary Lehman personally oversaw and controlled many details of the Navy acquisition process. Examples of his personal review and approval included the development of program acquisition strategies and the preparation of requests for proposal. A specific example of his centralized control was the requirement for PMs to submit engineering change proposal requests for review through the CNO to him for his final approval.[3] Secretary Lehman also instituted a policy requiring competition and fixed-price-type contracts throughout the life-cycle process. This process became institutionalized in the SYSCOMs, and a new generation of acquisition professionals developed in this stringent environment. During this period, the CNO continued to exercise responsibility for both the requirements generation process and the PPBS.

In April 1986 President Reagan signed National Security Decision Directive (NSDD) 219, which instituted a streamlined reporting chain for PMs, allowing only two levels of authority between the PM of a major defense acquisition program and the defense acquisition executive (DAE). The two levels were to be a program executive officer (PEO), responsible for overseeing several programs and their PMs in a warfare area, and a service acquisition executive (SAE), who was responsible for all acquisition matters within the service. Secretary Lehman designated himself the Navy acquisition executive to comply with the SAE requirement. He was "assisted in this function by the ASN(S&L) for programs which fall under his cognizance and the ASN(RE&S) for programs which fall under his cognizance" (Ref. 4, p. 1). The Navy implemented the PEO requirement by assigning the additional PEO responsibilities to the SYSCOM commanders rather than following the Army's concept of creating PEOs independent of the existing acquisition command structure (Ref. 4, p. 2).

The Goldwater-Nichols Department of Defense Reorganization Act of 1986 (Public Law 99-433[5]) directed a fundamental realignment of power within the DoD. Congress transferred the weapon system acquisition process to civilian control through this act and also mandated that one office assume the responsibility for the process in each military department (Ref. 5, Title V, Part B, sec. 511, subsec. 5014). The Office of the Secretary of the Navy was assigned the responsibility for the research, development, and acquisition of weapon systems. Actually the Navy was the least compliant service, as the ASN(RE&S) and the ASN(S&L) continued to function as before, despite the act's requirement to have one assistant secretary in charge of research, development, and acquisition. In 1988 Secretary Lehman relinquished NAE responsibility to the Under Secretary of the Navy (Ref. 2, p. 2). The required transfer of the research and development authority from OPNAV to the secretariat was accomplished by simply "dual-hatting" the responsible OPNAV director to the secretariat. The legislation had little actual day-to-day impact on the acquisition community since Secretary Lehman's consolidation of power had already changed the Navy acquisition community from its prior independent status to one closely supervised by the secretariat.

Goldwater-Nichols also limited the number of OPNAV DCNOs to five (Ref. 5, Title V, Part B, sec. 512, subsecs. 5036–5037) and CNO Admiral Carlisle A. H. Trost changed the three platform sponsors (air, surface, and submarine) from DCNOs to Assistant Chiefs of Naval Operations (ACNOs). At the same time, the Director, Naval Warfare (OP-095) became the DCNO, Naval Warfare (OP-07). It is important to note that these changes reflected a shift back toward the functional organization the OPNAV staff previously had and was intended to reduce the influence of the platform sponsors.

In late 1989 the Vice CNO eliminated the title of OPNAV program coordinator and established requirements officers (ROs) "to preclude any semblance of noncompliance or conflict with Secretariat responsibilities, which the term 'Program Coordinator' may connote. . . . "[6]

The Secretary of Defense's defense management report (DMR) to the President, published in the summer of 1989, had a significant impact on the Navy's acquisition organization. First, the DMR reiterated the Goldwater-Nichols requirement that a single civilian assistant secretary responsible for all service acquisition functions be designated the SAE.[7] To comply, the Navy established an Assistant Secretary of the Navy for Research Development and Acquisition [ASN(RDA)] and eliminated the ASN(RE&S) and ASN(S&L) positions.[8] The second DMR impact was the requirement to follow the Army concept of PEOs, who were independent of the SYSCOM commanders. These two changes finally brought the Navy acquisition management organization into alignment with NSDD 219 and Goldwater-Nichols requirements. Implementation of the "independent" PEOs had a significant impact on the Navy acquisition community that has not been totally settled to date. The details have been formalized in charters signed by the principals and the ASN(RDA); however, resolving the details of making the system work is still ongoing.

THE 1990s: A TURBULENT DECADE?

In late July 1992 acting Secretary of the Navy Sean O'Keefe and CNO Admiral Frank B. Kelso II announced a far-reaching reorganization plan that would restructure the Navy to better meet its new roles in the emerging post–Cold War environment. Although the overhaul occurred primarily in OPNAV, it fundamentally changed the way the service allocates resources and responsibilities by eliminating the ACNOs for air warfare, surface warfare, and submarine warfare. The warfighting specialties will still exist on the OPNAV staff, but their power and influence has finally been diluted substantially.[9]

The demise of the Soviet Navy as a blue-water threat allowed the Navy to refocus on joint and combined operations in the littoral areas of the oceans. This emphasis on brown-water-type operations resulted in the establishment of a new OPNAV office for expeditionary forces that pulled amphibious-assault ships away from surface warfare sponsorship. The Navy now concentrates on six warfare mission areas: joint strike, joint littoral, joint surveillance, joint space and C4, strategic deterrence, and strategic sealift and protection.[10]

Although the OPNAV reorganization was to be completed by January 1, 1993, its full effect on the Navy will not be felt for some time. The first POM and budget assembled under the new structure will be an indicator of how successful the SECNAV and CNO were in breaking up the warfighting specialty fiefdoms.

In a little more than 25 years the Navy acquisition community has undergone several significant management and organizational changes: from dissolution of the long-standing bilinear organization with somewhat independent bureaus and PMs to Secretary Lehman's direct management and control of the SYSCOM commanders and PMs to the current reporting chain for major defense acquisition program PMs who are independent of the SYSCOMs from which they still receive matrix functional support. In light of Congress's desire to downsize the DoD significantly in response to the end of the Soviet threat, the reader must anticipate that further changes will occur in the near future.

CURRENT U.S. NAVY ORGANIZATION FOR ACQUISITION MANAGEMENT

Figure 13.1 is a general depiction of the current DON organization and chain of command in the DoD acquisition management structure. Both the Navy Secretariat and OPNAV have tremendous influence and control (indirectly in the case of OPNAV) over PMs. The ASN(RDA) and the CNO report to the SECNAV for all matters, with the ASN(RDA) also reporting directly to the DAE for acquisition management matters. As the chart shows, the CNO/OPNAV and the Commandant of the Marine Corps (CMC) and Headquarters Marine Corps (HQMC) are not in the official DON acquisition chain of command. Figure 13.2 provides an overview of the acquisition community within the DON. The duties and responsibilities of the major players in the acquisition process follow.

Navy Secretariat

The current structure of the Office of the Secretary of the Navy is displayed in Fig. 13.3. Even though PMs in the acquisition community report to the ASN(RDA)

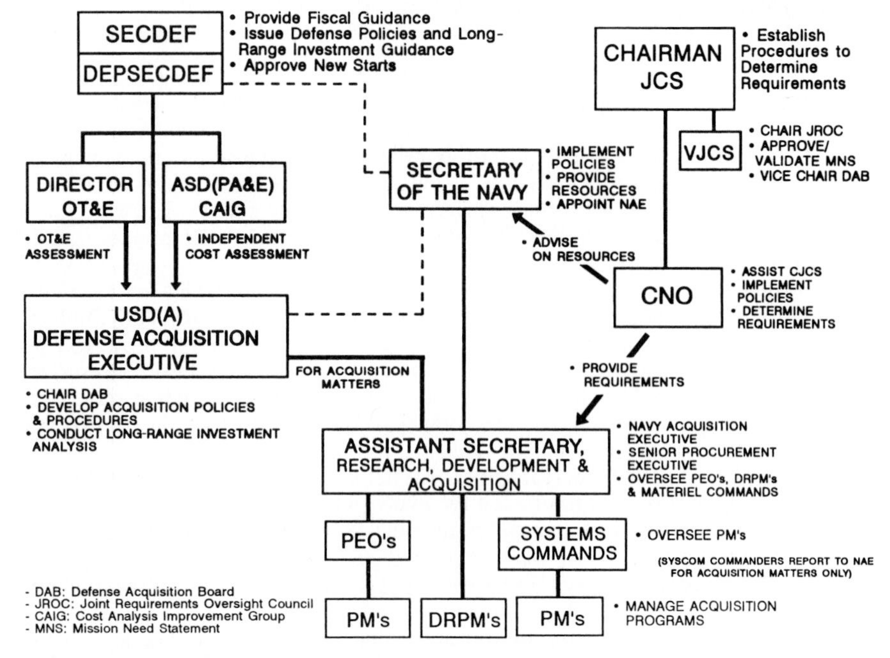

FIGURE 13.1 Overview of U.S. Navy acquisition management organization in relation to the Office of the Secretary of Defense and the Joint Chiefs of Staff.

through a PEO or SYSCOM commander, as appropriate, the other assistant secretaries can have a significant role in the weapon system acquisition process. Their acquisition-related duties are as follows.

Assistant Secretary for Financial Management [ASN(FM)]. The ASN(FM) is responsible for developing independent cost analyses for use in Navy program decision meetings and for OSD-level Defense Acquisition Board (DAB) review. The incumbent is also responsible for the finance, budget, and cost aspects of acquisition programs.

Office of the Comptroller of the Navy (NAVCOMPT). The ASN(FM) is also the Comptroller of the Navy. The Comptroller is responsible for formulating policies and prescribing procedures and systems which will exercise control over the financial operations of the DON through the application of accounting principles, PPBS development, and financial analysis of the DON. It should be noted that the Director, Office of Budget and Reports, and his staff in NAVCOMPT are dual-hatted there from their positions in N82, the Fiscal Division within OPNAV.

Office of Program Appraisal (OPA). Although OPA is not normally reflected on secretariat organization charts, it is the principal staff element used by the SECNAV to exercise direction, authority, and control over the DON. OPA assists the SECNAV in developing and implementing policies, evaluating plans and resource utilization, and monitoring programs. Acquisition-related responsibilities include the following:

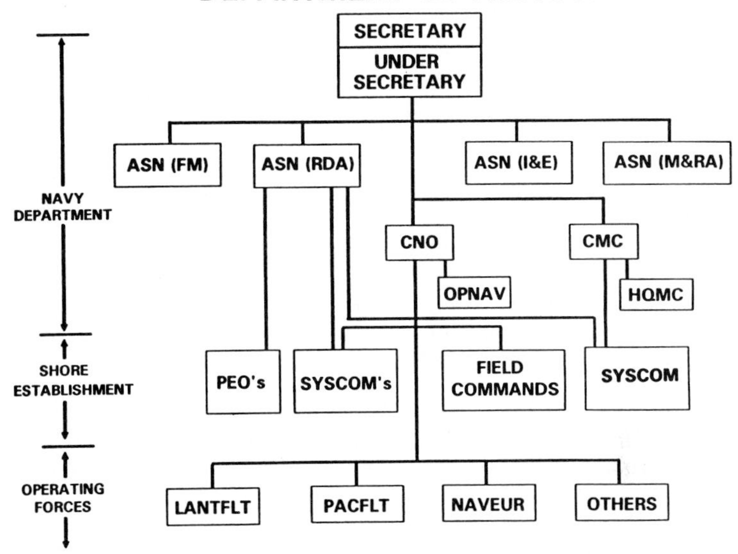

FIGURE 13.2 Overview of acquisition organization within the Department of the Navy.

Initiating actions, programs, and other tasks to ensure adherence to DON and DoD policies and objectives and ensuring that DON programs are designed to accommodate operational requirements

Advising the SECNAV on decision documents, policy statements, correspondence, and directives associated with issues impacting the DON

Managing the DON planning phase of the PPBS

Analyzing the validity and adequacy of programs in achieving the objectives of the DON

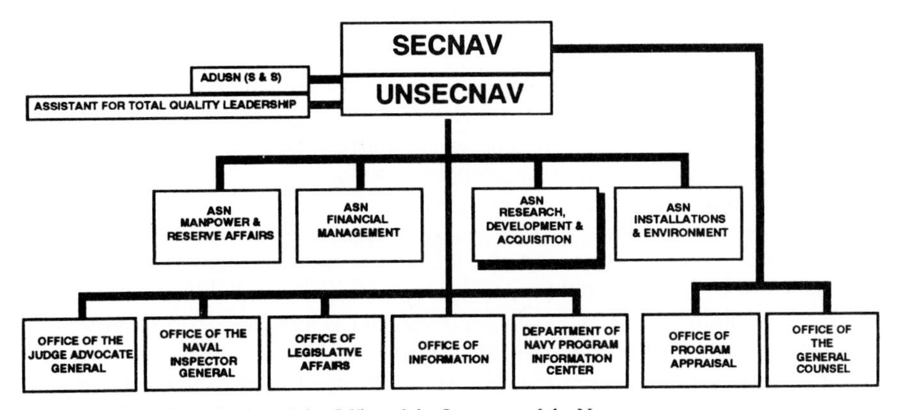

FIGURE 13.3 Organization of the Office of the Secretary of the Navy.

Briefing the SECNAV on the PPBS and, when appropriate, presenting policy and programming matters for action

Advising the SECNAV on policy and programming matters to prepare him for meetings of the Defense Planning and Resources Board

Resolving issues between offices in the secretariat, OPNAV, HQMC, and OSD, as appropriate

Assistant Secretary of the Navy for Manpower and Reserve Affairs [ASN (M&RA)]. The ASN(M&RA) is responsible for aspects of manpower and training affecting the research, development, and operational fielding of weapon systems.

Assistant Secretary of the Navy for Installations and Environment [ASN (I&E)]. The ASN(I&E) has responsibility over the impact of the weapon system on existing base facilities and any weapon system development and employment environmental issues.

Office of the General Counsel. The General Counsel advises the ASN(RDA) on legal issues which may surface due to the research, development, and acquisition process.

Assistant Secretary of the Navy for Research, Development and Acquisition [ASN(RDA)]. As the Navy acquisition executive (NAE), the ASN(RDA) is the single official within the DON responsible to the SECNAV and the DAE for providing the management and technical expertise over all aspects of Navy research, development, and acquisition to meet the funded requirements of the DON. The ASN(RDA) exercises line authority through the key positions in the Navy and Marine Corps acquisition organization, which are responsible for the performance, costs, and schedule of their assigned programs. The ASN(RDA) is supported by a small staff (Fig. 13.4), who are principal advisors for policy and procedures in their respective areas of expertise and assist in formulating guidance to the acquisition line managers. Figure 13.5 depicts the overall Navy acquisition management structure. Navy PEOs and their assigned PMs, direct reporting PMs (DRPMs), and SYSCOM commanders have continuous dealings with the ASN(RDA) staff. This interface addresses both major and less than major (ACAT I through ACAT IV) defense acquisition programs. The staff members are not in the line of command, but they evaluate information from the PM and other sources and advise ASN(RDA), who makes the decision.

The ASN(RDA)'s primary functions and responsibilities include the following (Ref. 11, pp. 4–5):

Serving as the DON's senior acquisition executive and senior procurement executive

Implementing DoD acquisition policy; establishing DON acquisition policy and managing all DON research, development, production, shipbuilding, and logistics support programs

Recommending milestone decisions on programs (ACAT ID) for which the Under Secretary of Defense (Acquisition) is the decision authority and serving as the milestone decision authority (MDA) for ACAT IC, II, and III programs

Supervising the PEOs, DRPMs, and SYSCOM commanders for all acquisition matters pertaining to their assigned programs

Establishing policy and providing oversight for the Navy's technology base, ad-

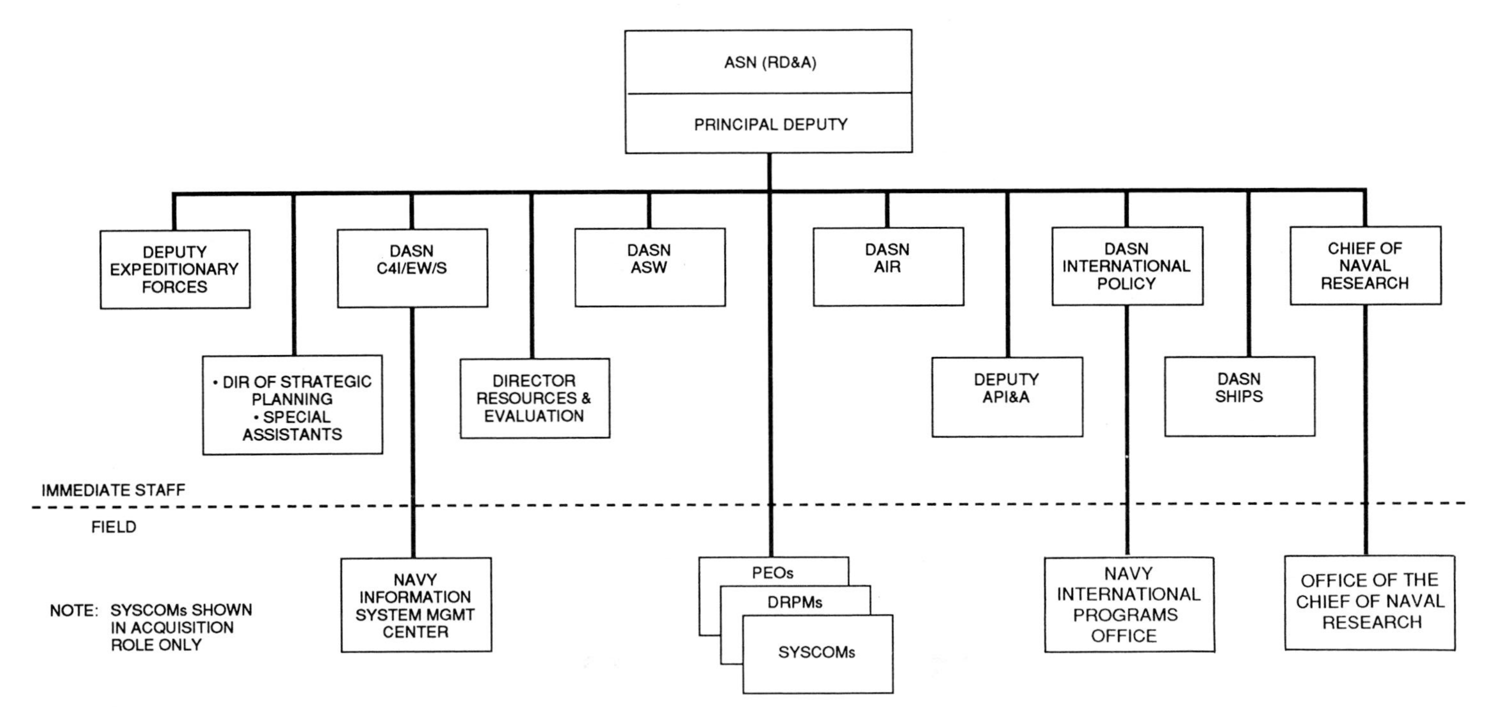

FIGURE 13.4 Organization of the Office of the Assistant Secretary of the Navy for Research, Development and Acquisition and reporting field activities.

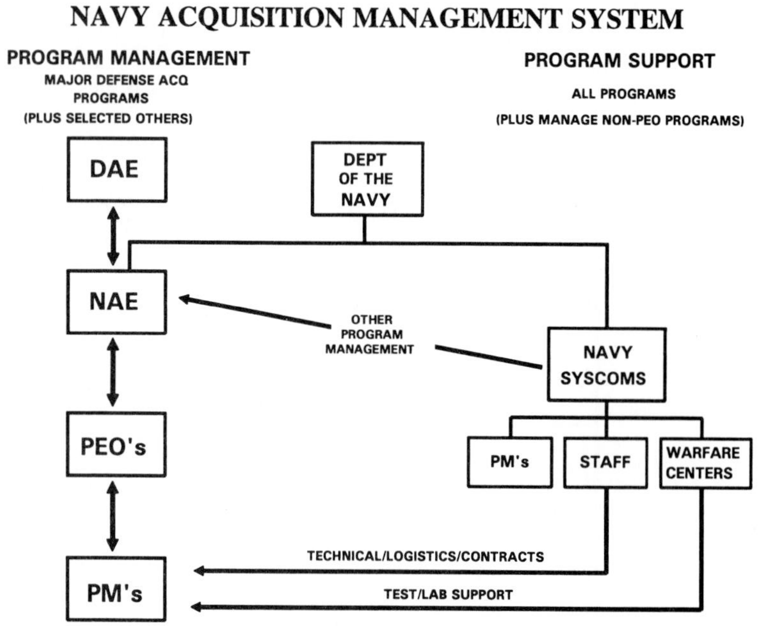

NAVY ACQUISITION MANAGEMENT SYSTEM

FIGURE 13.5 Overview of the Navy acquisition management system, showing relationship between various activities.

vanced development, laboratories and warfare centers, Navy international programs, competition, acquisition streamlining, nondevelopmental items, procurement integrity, and accountability

Serving as the MDA, establishing policy, and providing oversight for all matters related to information resource management systems and equipment

Acting for the SECNAV in establishing policy and providing oversight of logistics management and integrated logistics support

Establishing DON policy for, and overseeing, the education and training of the acquisition work force as required by the Defense Acquisition Workforce Improvement Act (DAWIA).

Office of the Chief of Naval Operations

Although Goldwater-Nichols prevents the CNO and the OPNAV staff from direct participation in the research, development, and acquisition management process, the CNO is still responsible for the planning and programming portion of the PPBS process, establishing requirements, setting priorities for programs, and for the operational test and evaluation of weapon systems prior to their introduction to the fleet (Ref. 12, pp. 1–10). OPNAV is composed of the CNO, Vice Chief of Naval Operations, Deputy Chiefs of Naval Operations, directors, special assistants, and the associated subordinate staff offices.

The OPNAV reorganization directed by Secretary O'Keefe and Admiral Kelso in July 1992 was intended to model the Navy staff along Joint Chiefs of Staff lines, rely more heavily on fleet input, centralize warfighting requirements, and reduce the OPNAV staff size by eliminating redundant functions. The most significant change was the reduction of platform sponsor power by reducing the platform sponsors from three-star positions to two stars and placing them under a new three-star position, N8.

The reorganization was "completed" in January 1993; however, no official documentation describing the new duties of the principals has been released. The following is the author's best attempt to describe the new duties, based on interviews with various officers on the OPNAV staff, briefing papers obtained during the interviews, and the old OPNAV organization delineated in OPNAVINST 5430.48C. The reader is strongly cautioned to obtain a copy of the new OPNAV organization manual (OPNAVINST 5430.48D) when it is released to be sure of the new duties and responsibilities within OPNAV.

The current OPNAV organization is shown in Fig. 13.6. Some key OPNAV offices the Navy PM must deal with follow. (Again, the reader is cautioned that these paragraphs are based on the author's opinion.)

Deputy Chief of Naval Operations (Manpower and Personnel), N1. N1 is responsible for managing the planning and programming of all manpower and personnel within the Navy. The duties are basically the same as for the old OP-01, except that the responsibility for training has moved to N7. In coordination with N8, the N1 staff develops the manpower portions of the Navy and DoD FYDP and implements the actions to comply with budgetary and congressional actions.

Deputy Chief of Naval Operations (Logistics), N4. It is assumed that N4 will have fundamentally the same duties as the previous OP-04. Under the direction of the ASN(RDA), N4 serves as the Navy's focal point for all logistics matters. The N4 staff is responsible for planning, determining, and providing for logistics research

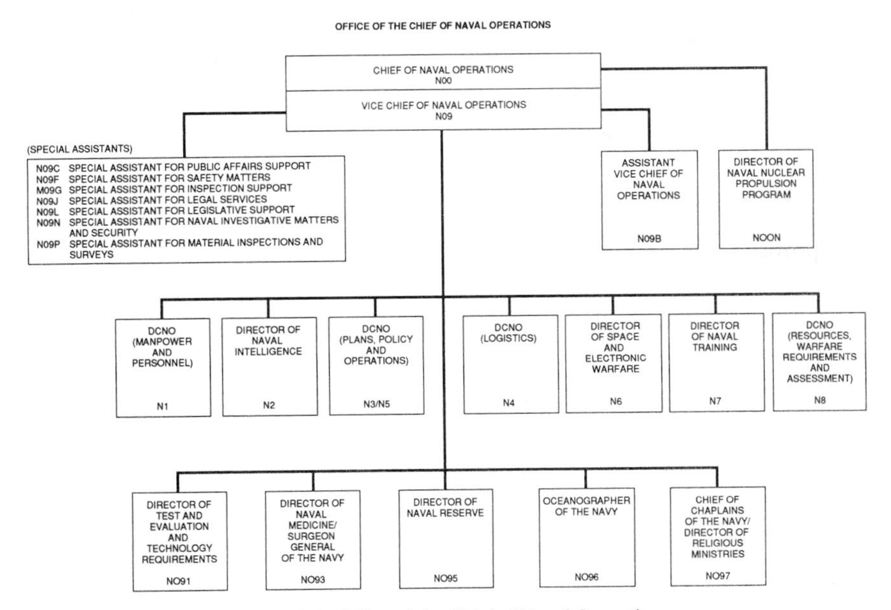

FIGURE 13.6 Organization of the Office of the Chief of Naval Operations.

and development and the support needs of the U.S. Navy. N4 sponsors the Navy's military construction appropriation and also represents the CNO to the Joint Logistics Commanders.

Deputy Chief of Naval Operations (Plans, Policy and Operations), N3/5. It is assumed that N3/5 will continue the basic duties of the previous OP-06 organization. N3/5 advises the CNO on JCS matters and is responsible for developing and implementing strategic plans and policies. N3/5 also counsels SECNAV and CNO on strategic planning, nuclear systems, and international politico-military matters. The incumbent is the principal CNO advisor on technology transfer, foreign disclosure, and security assistance.

Director of Space and C4 Systems Requirements, N6. The majority of the former OP-094 staff was transferred to this directorate. N6 is responsible for centralized coordination over policy, planning, and integration of requirements for space and command, control, communications, and computers (C4). To ensure that Navy space and C4 planning is compatible with the combat systems requirements developed by the platform sponsors in N8, the N6 staff reviews programming, reprogramming, and sponsor program proposals (SPPs) that involve space and C4 programs and recommends approval or disapproval to N8. N6 also validates research, development, and test and evaluation programs to ensure that the system will meet space and C4 requirements. Responsibility for platform-specific requirements is within N8.

Director of Naval Training, N7. It is assumed that the N7 organization continues the training-related responsibilities of the previous OP-01 and the actual training from the former Chief of Naval Education and Training (CNET). N7 is responsible for all training and education within the Navy as well as naval doctrine through the Naval Doctrine Command. N7 advises the CNO on training and doctrine matters.

Deputy Chief of Naval Operations (Resources, Warfare Requirements, and Assessment), N8. N8, depicted in Fig. 13.7, subsumed the previous OP-02, OP-03, OP-05, OP-07, and OP-08 organizations. N8 is responsible for the centralized coordination, assessment, and prioritization of naval warfare requirements and the application of

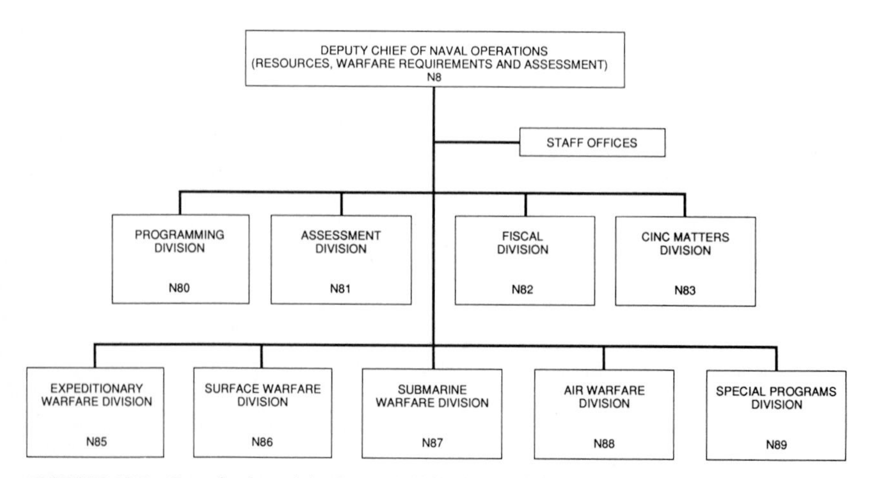

FIGURE 13.7 Organization of the Deputy Chief of Naval Operations for Resources, Warfare Requirements, and Assessment (N8).

resources against those requirements. The N8 staff integrates the operational needs of the fleet. It reviews research, development, and acquisition programs for the attainment of naval warfare requirements and advises SECNAV and CNO on matters involving the warfighting capabilities of naval forces. N8 also manages the mission need statement (MNS) and operational requirements document (ORD) staffing and approval process within the Navy. N8 provides OPNAV guidance on warfare system architecture and engineering to the acquisition SYSCOMs through liaison with the Space and Naval Warfare Systems Command (SPAWARSYSCOM). N8 is responsible for overseeing the development and implementation of plans, programs, and policies in the overall direction of the Navy. The staff reviews and evaluates programs (programmatic, financial, and manpower) to determine their impact on the overall balance of forces in the Navy. It recommends program adjustments or corrective actions to program sponsors and the Vice CNO to restore overall program balance. N8 provides the CNO with system analyses that evaluate the relative effectiveness of alternatives in Navy programs and sponsor program proposals. The staff arbitrates differences between OPNAV components, and between those components and external Navy activities. N8 coordinates the preparation of the Navy POM for the CNO and the administrative process which supports POM development. Figure 13.8 is a top-level depiction of the N8 decision-making process. A brief overview of the various directors in the N8 organization follows.

Director, Programming Division, N80. The former OP-80 staff was transferred to this directorate. N80 is responsible for developing policy on planning and programming matters as directed by SECNAV, appraising Navy programs and the programming system on a continuing basis to ensure balance within and between programs, and coordinating with NAVCOMPT and other offices to ensure that program actions are accommodated by, and integrated with, budget actions.

Director, Assessment Division, N81. The former OP-81 staff was transferred to N81, which is responsible for evaluating the effectiveness of alternatives in programs and program proposals and thereby assist in the decision-making process; to assess all major weapon systems at each milestone during the acquisition process; to implement the N8 responsibility for conducting scientific, analytical, and technical studies through the use of the Center for Naval Analysis; and to review and validate analytical models and methodologies used in program planning.

Director, Fiscal Division, N82. The Director of Budgets and Reports (NCB) in

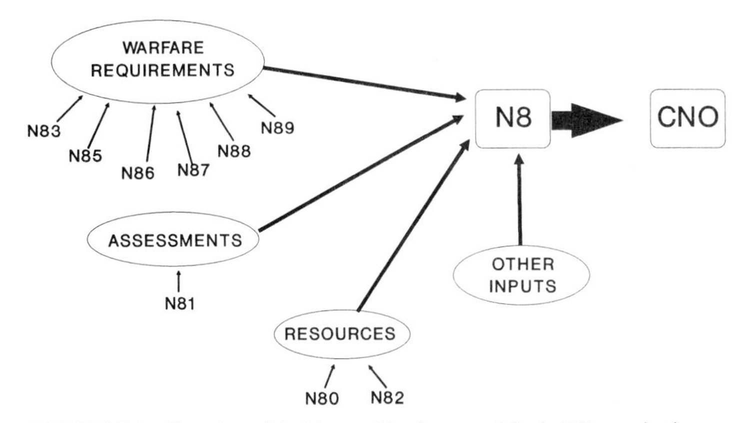

FIGURE 13.8 Overview of decision-making framework in the N8 organization.

the Comptroller's office is dual-hatted in OPNAV as N82. The former OP-82 staff was transferred to this directorate. N82 is responsible for developing, coordinating, and maintaining an integrated system of fiscal management to ensure effective management control of the funds and resources assigned to the CNO and supervising the resource management operations to meet the operating and support needs of the Navy.

Director, CINC Matters Division, N83. N83 is a new directorate created to foster a closer dialogue between the warfighting CINCs in the field and the OPNAV in Washington. It should serve as a conduit for requirements documentation, program assessments, program prioritization, and resource application between the Navy headquarters staff and the CINCs in the field.

Director, Expeditionary Warfare Division, N85. Those former OP-03 billets associated with amphibious warfare were transferred to this directorate. N85 assists N8 in setting policy, establishing requirements, setting priorities, and planning and programming for all aspects of expeditionary warfare. The N85 staff develops and coordinates the requirements documentation associated with expeditionary warfare to initiate the acquisition process and prepares the expeditionary warfare portion of the Navy and DoD FYDPs.

Director, Surface Warfare Division, N86. The former OP-03 staff was transferred to this directorate with the exception of the amphibious warfare billets, which were transferred to N85. N86 assists N8 in setting policy, establishing requirements, setting priorities, and planning and programming for all aspects of surface warfare. The N86 staff develops and coordinates the requirements documentation associated with surface warfare deficiencies to initiate the acquisition process and prepares the surface warfare portion of the Navy and DoD FYDPs. N86 sponsors the Navy's ship construction and weapons procurement appropriations.

Director, Undersea Warfare Division, N87. The former OP-02 staff was transferred to this directorate. N87 assists N8 in setting policy, establishing requirements, setting priorities, and planning and programming for all aspects of undersea warfare. With regard to program management, the N87 staff develops and coordinates the undersea warfare requirements documentation that initiates the weapon system acquisition process. It also prepares the portion of the Navy and DoD FYDPs dealing with undersea warfare.

Director, Air Warfare Division, N88. The former OP-05 staff was transferred to this directorate. N88 assists N8 in setting policy, establishing requirements, setting program priorities, and planning and programming for all aspects of naval aviation. The N88 staff develops and coordinates naval aviation requirements documentation that initiates the weapon system process and prepares the naval aviation portion of the Navy and DoD FYDPs. N88 sponsors the Navy's aircraft procurement appropriation.

Director of Test and Evaluation and Technology Requirements, N091. N091 is responsible for establishing and promulgating the policies and procedures for the conduct of operational test and evaluation (T&E) and acts for the CNO in resolving T&E issues. N091 is the research, development, test, and evaluation (RDT&E,N) appropriation sponsor and supports selected RDT&E field activities and the commander of the Operational Test and Evaluation Force (COMOPTEVFOR).

Commander, Operational Test and Evaluation Force

The commander of the Operational Test and Evaluation Force (COMOPTEVFOR) is head of the Navy's independent test agency and reports directly to the CNO for operational test matters. COMOPTEVFOR is responsible for providing an independent assessment of the operational test and evaluation test results, addressing operational effectiveness and suitability on ACAT I, II, III, and IVT programs (ACAT program levels are discussed in a later section of this chapter) directly to the PM and MDA prior to the milestone review at milestone III and at milestone II, when applicable (Ref. 2, enclosure 1, p. 5).

Program Executive Officer and Direct Reporting Program Manager

Program executive officers (PEOs) have management oversight authority over a number of similar interrelated acquisition programs and report directly to the ASN(RDA). Direct reporting program managers (DRPMs) have management authority over a single acquisition program and report directly to the ASN(RDA). Both have full authority over and are accountable for the cost, schedule, and performance matters of assigned acquisition programs. PEOs and DRPMs have small support staffs because they receive matrixed functional support from a designated SYSCOM. Major defense acquisition programs will normally be assigned to them on the basis of various factors. A formal operating agreement, approved by the NAE, has been negotiated between each PEO or DRPM and the supporting SYSCOM. The SYSCOMs or a PEO will manage the effort during the concept exploration and definition phase until just prior to the milestone I decision point when a PM is normally assigned. Mature ACAT I programs in stable production, not subject to significant product improvement or block upgrade, may be transferred to a SYSCOM commander (who assumes PEO responsibilities for the program) with the approval of the Under Secretary of Defense for Acquisition. Figure 13.9 displays the current PEOs, DRPMs, and SYSCOM commanders and their relationship with each other and with the ASN(RDA).

As the result of a realignment initiated in the summer of 1992, four PEOs and one DRPM are collocated with the Naval Sea Systems Command (NAVSEASYSCOM) from which they receive matrix support as needed. The PEO for Submarines is responsible for all assigned major and related nonmajor submarine programs. The PEO for Undersea Warfare is responsible for all assigned major and related nonmajor submarine combat and weapons systems. The PEO for Mine Warfare is responsible for all assigned mine countermeasure systems and equipments and for sea mines. The PEO for Ship Defense is responsible for all ship defense programs, including other related systems and technologies for the coordination and management of air defense by single and multiple ships against air threats. The DRPM for AEGIS is responsible for all programs supporting AEGIS ships and their associated combat and related support systems (Ref. 11, pp. 25–34).

There are three PEOs and one DRPM collocated with the Naval Air Systems Command (NAVAIRSYSCOM). They draw matrix support from NAVAIRSYSCOM as necessary. The PEO for Tactical Aircraft Programs is responsible for all assigned programs, platforms, and weapons for air superiority and strike interdiction. The PEO for Cruise Missile Project and Unmanned Aerial Vehicles (UAV) Joint Project is responsible for all assigned major and nonmajor programs for long-range cruise missile weapon systems and strike warfare mission planning systems. This PEO is also responsible for all joint service nonlethal UAV programs. The PEO for Air Antisubmarine Warfare, Assault, and Special Mission Programs is responsible for ASW-related fixed-wing and rotary-wing aviation platforms and their associated weapons, processors, and sensors; fixed-wing vertical-flight aircraft; airborne mine countermeasures; and the tactical jet trainer program. The DRPM for Advanced Medium Attack (A/F-X) is responsible for managing the development of a medium attack aircraft incorporating stealth technology (Ref. 11, pp. 35–46).

The PEO for Space, Communications, and Sensors is responsible for electronic

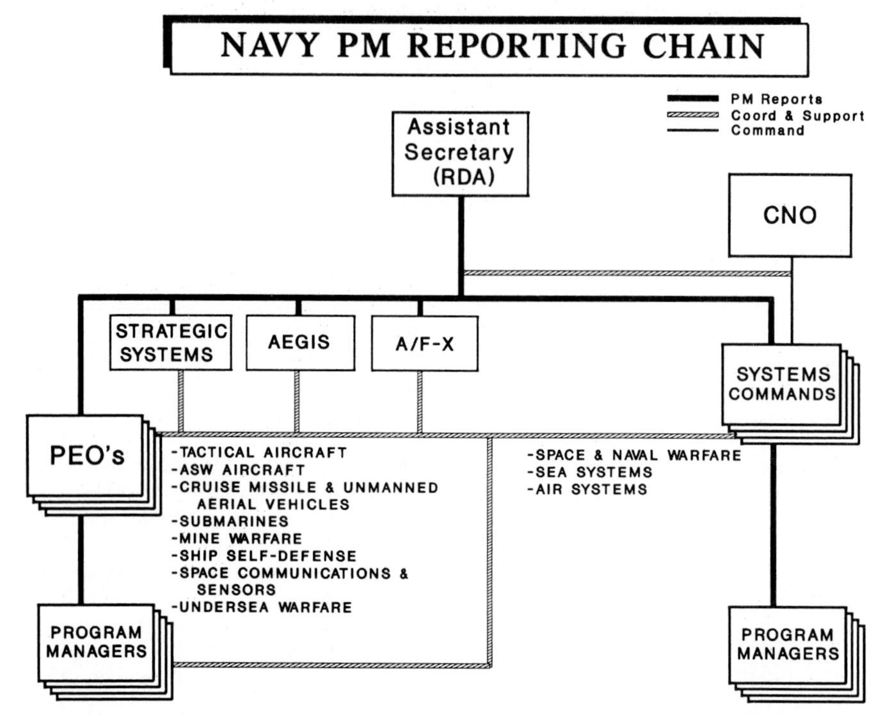

FIGURE 13.9 PM reporting chain showing relationship between PEOs, DRPMs, and the supporting SYSCOMs.

systems programs that use signal processing to receive, process, and transmit information to fleet units. The PEO is collocated with SPAWARSYSCOM and receives matrix support from that organization (Ref. 11, pp. 47–50).

The DRPM for Strategic Systems Programs (SSP) is responsible for managing all assigned major and nonmajor programs for submarine-launched strategic weapon systems, including the missiles, the platforms, associated equipment, security, training of personnel, and the installation and direction of necessary supporting facilities. The DRPM also reports to the CNO for matters involving fleet operations and support. SSP is a self-contained organization, controlling its funding and all support personnel. The PM for the Strategic Submarine Program (PMS-396), at NAVSEASYSCOM, is dual-hatted to the SSP for the design and construction of Trident submarines. The pre-milestone 0 Navy Theater Ballistic Missile Defense (TBM) Program, which is the Navy's part of the SDIO Global Protection Against Limited Strikes effort, is managed in SSP (Ref. 11, pp. 51–53). Once milestone 0 is passed, the Navy may well create a new PEO or DRPM to oversee the Navy's TBM Program.

The DRPM for Advanced Amphibious Assault (AAA) is responsible for management of the AAA program, currently the Marine Corps' only major defense acquisition program. The DRPM also manages other programs related to Marine Corps amphibious assault vehicles. The DRPM AAA receives support from Marine Corps Systems Command functional components, including legal, comptroller, contracting, and other matrix support (Ref. 11, pp. 55–57).

Some of the major functions and responsibilities of the PEO and the DRPM are (Ref. 11, pp. 19–22):

Reporting solely and directly to the NAE on all matters concerning cost, schedule, and performance of assigned programs and exercising authority, responsibility, and accountability for those programs

Controlling all funding allocated to their programs by the DON Comptroller

Chairing acquisition review boards (ARBs) for assigned programs and performing delegated review and approval authorities, including approval of acquisition plans and business clearances

Providing program management and oversight

Ensuring that readiness objectives are achieved through early support, performance, cost, and schedule trade-off analyses

Acting as source selection authority for assigned programs

Evaluating the performance of assigned PMs and preparing their fitness reports (military) or performance appraisals (civilian)

Acting as milestone decision authority (MDA) for assigned ACAT IV programs

Systems Command Commanders

Despite the transfer of the more highly visible acquisition programs to PEO and DRPM organizations, the SYSCOMs still manage a significant portion of the overall Navy acquisition budget. The missions of the acquisition SYSCOMs are (Ref. 11, pp. 21–22):

Managing those acquisition programs not assigned to the PEO or DRPM structure

Providing functional support services to the PEOs, DRPMs, and their assigned PMs without duplicating any of their management functions

Reporting directly to the NAE for their assigned acquisition programs with the same major functions and responsibilities as a PEO or DRPM

Acting as MDA for assigned ACAT IV programs

Providing supported PEOs and DRPMs with full contracting and comptroller support services and acting as the head of contracting activity (HCA)

Responsibility for life-cycle logistics support for all systems and programs unless the responsibility has been transferred to a PEO or DRPM, in which case they will provide matrix functional support to the PEO or DRPM for the accomplishment of the responsibility

In summary, each of the four weapon system acquisition SYSCOMs (Naval Sea Systems Command, Naval Air Systems Command, Space and Naval Warfare Systems Command, and Marine Corps Systems Command) has three principal duties: managing the assigned programs (including the mature, stable ACAT-I programs not assigned to a PEO or DRPM); providing necessary support services to the associated PEOs and DRPMs; and providing life-cycle logistics support to the fleet. For all assigned programs the SYSCOM commander has the same authority and responsibilities as a PEO. The incumbent also reports to the CNO (Commandant of the Marine Corps for Commander, Marine Systems Command) for matters involving fleet operations and support.

Despite the creation of PEOs and DRPMs, the SYSCOMs still play a significant role in the acquisition process and manage a large number of acquisition programs at all ACAT levels. Since the SYSCOM commanders are normally senior in rank to the PEOs and DRPMs they are supporting, issues can arise over management decisions on programs.

Program Manager

Unlike the Army, the Navy did not feel it necessary to assign different titles (that is, of program, project, or product) to managers of different ACAT level acquisition programs.

To comply with the Defense Acquisition Workforce Improvement Act of 1990,[13] the Navy created the Director, Acquisition Career Management (DACM), who is responsible for acquisition workforce matters and reports to the ASN(RDA). The ASN(RDA) chairs an acquisition workforce oversight council (AWOC), which develops policy guidance for, and oversight of, the entire Navy acquisition workforce. The DACM provides an executive secretary to support the AWOC as well as much of the information used during its deliberations. The AWOC approves all ACAT I PM assignments and any other senior acquisition workforce assignments.

All PMs of Navy ACAT I (ACAT ID and IC) level programs must be graduates of the Defense Systems Management College program management course (or an approved equivalent), have eight years of acquisition experience (two years in a system program office), and meet tenure requirements (serving until reaching the milestone closest to four years as PM) documented in a written agreement. PMs of "significant nonmajor" (ACAT II) programs must be graduates of the program management course and have six years of acquisition experience.

PMs, no matter what ACAT level they are managing, are responsible for the following (Ref. 11, pp. 19–22):

Managing a specific acquisition program, reporting to and receiving direction from a PEO or SYSCOM commander (PMs have full authority and responsibility and they are accountable for their programs. They are responsible for meeting all Navy acquisition policy requirements in implementing their programs.)

Managing their programs in a manner that is consistent with, and supportive of, the policies contained in the DoD 5000 series and Navy implementing instructions

Committing to a program baseline

Identifying shortfalls in personnel, functional management support, funding, and timeliness of funding that adversely affect achievement of MDA decisions and approved programs

Promptly reporting all imminent and actual breaches of MDA decisions and approved programs

Preparing and submitting periodic program acquisition information system (AIS) and defense acquisition executive summary (DAES) performance reports where applicable

REQUIREMENTS PROCESS AND DOCUMENTATION

The Navy still does not have an organization, such as TRADOC in the Army, which, externally to the service headquarters, deals with the requirements process, even though a naval doctrine command was established during fiscal year 1993. The new doctrine command has the mission of converting strategic policy and plans into naval doctrine and performing assessments of developed tactical doctrines, and is not involved in the requirements generation process.

Any component or agency which identifies a mission need, deficiency, or opportunity will forward it to OPNAV, which has been designated the Navy user representative (Ref. 12, p. 3). The following paragraphs describe the coordination and approval process within OPNAV.

The appropriate OPNAV sponsor will prepare a draft mission need statement (MNS) and forward it to N8 (DCNO, Resources, Warfare Requirements and Assessment), which will review the draft MNS and confirm that no nonmaterial alternatives are available to satisfy the deficiency. N8 will then circulate the draft MNS to the appropriate OPNAV offices and the fleet CINCs for review and comment. The sponsor will then consolidate the comments into a smooth MNS, assign a potential ACAT level, and recirculate it through N8 for final comment (Ref. 12, enclosure 6, p. 1).

The MNS will be submitted to the CNO for review, approval, and validation. A resources and requirement review board (R3B) is in the process of being established within OPNAV and detailed R3B procedures will be promulgated in a future revision to OPNAV Instruction 5420.2 (CNO executive board procedures). The functions of the R3B are to advise the CNO on warfare requirements and for the evaluation of warfighting effectiveness of existing or proposed warfare systems, force levels, and major cross-platform warfare matters. It will prioritize all ACAT I, and select ACAT II, MNSs, and ORDs with respect to warfare requirements (Ref. 12, enclosure 3, p. 1).

For potential ACAT I programs, the validated MNS is submitted to the Joint Requirements Oversight Council (JROC). After the JROC has reviewed and approved it, the validated MNS is forwarded to the Under Secretary of Defense for Acquisition [USD(A)] for a milestone 0 decision. The USD(A) will first determine whether the program will be an ACAT ID or IC. If it is to be an ACAT ID, the USD(A) will then sign an acquisition decision memorandum (ADM). For those programs designated ACAT IC, the ASN(RDA) will sign the ADM (Ref. 14, p. 2).

For an MNS representing a potential ACAT II, III, or IV program the validated MNS is sent to the staff of the ASN(RDA), which prepares a proposed ADM. The MNS and the proposed ADM are then submitted to the ASN(RDA), who, if in agreement, signs the ADM to document the milestone 0 decision, allowing the program to proceed into phase 0, the concept exploration and definition phase (Ref. 14, p. 2).

During phase 0 of the life-cycle process the OPNAV sponsor will prepare and forward a draft operational requirements document (ORD) to N8, which will review it and distribute it for review and comment (within Navy) in a process similar to that used for the MNS. The sponsor then consolidates the comments and prepares a smooth ORD. The remainder of the review and approval process basically replicates the MNS process (Ref. 12, enclosure 6, p. 2).

Preparation and coordination of the draft and smooth MNS and ORD are the responsibility of the requirements officer (RO) in OPNAV. The PM must work closely with the RO to ensure that the ORD contains achievable program threshold and objectives. The RO is also tasked with the preparation and coordination of all the documentation which initiates the sponsor's formal request for resources through the PPBS. Although the OPNAV staff is not in the PM's formal reporting chain, it is imperative that the PM keep a close working relationship with the RO. The RO must be kept appraised of the current program status as he or she shepherds it through the PPBS process (Ref. 12, enclosure 5, pp. 1–3).

ACAT LEVEL ASSIGNMENT FOR PROGRAMS

The Navy had been using the ACAT system prior to the OSD decision to incorporate it in the latest version of the DoD 5000 series. ACAT I and II program thresholds are specified in DoDD 5000.1 and discussed elsewhere in this handbook. The ASN(RDA) is the ACAT designation authority for all DON acquisition programs in categories II through IV (Ref. 14, p. 11).

Navy ACAT III programs do not have dollar thresholds. They are assigned to this category if the program affects the military characteristics of ships or aircraft or involves combat capability. This category applies to hardware and software, new systems, and modifications to existing systems (Ref. 14, p. 2).

ACAT IV is assigned to all Navy acquisition programs not otherwise designated ACAT I, II, or III. There are two categories of ACAT IV programs, IVT and IVM. ACAT IVT programs require operational test and evaluation and ACAT IVM programs do not (Ref. 14, p. 2).

PROGRAM REVIEW FORUMS

Program decision meetings (PDMs) are the DON forum for reaching Navy acquisition program milestone decisions and for MDA-directed program reviews. They are the Navy's equivalent of the Defense Acquisition Board (DAB) review process. PDMs are used for ACAT I through IV programs, and the documentation required to support a PDM for all levels is prescribed in DoDI 5000.2. Where the Navy has elected to tailor documentation requirements for ACAT II, III, and IV programs, this is reflected in SECNAVINST 5000.2A.[15]

Navy PDM procedures and the events leading to the formal meeting parallel those of the DAB outlined in the DoD 5000 series. PMs must allow sufficient time in their schedule to satisfy the procedures leading to a successful PDM. Following the PDM, the MDA will sign an acquisition decision memorandum (ADM) documenting the results of the PDM.

The ASN(RDA) as NAE chairs PDMs for all ACAT I, II, and III programs, although he or she may redelegate ACAT III MDA to the DRPM, PEO, or SYSCOM commander (Ref. 14, p. 2). The NAE convenes the PDM at each formal milestone to determine program status and readiness to enter into the next phase of development or production. The NAE holds a PDM for all ACAT ID programs before they are reviewed by the DAB. The NAE may also convene a PDM at any time to review the status of a program. Prior to the PDM, an acquisition review board (ARB) will be held by the PEO, DRPM, or SYSCOM commander, and a letter forwarding the ARB's recommendations will be sent to the ASN(RDA).

For ACAT IV programs (or those ACAT III with redelegated MDA), the PEO, DRPM, or SYSCOM commander responsible for management oversight has been designated the MDA (Ref. 14, p. 2). The MDA may provide separate guidance for the conduct and attendance of the PDM. However, adequate advance notice of the meetings shall be provided to the members of the ASN(RDA)-chaired PDM so that they may send representatives, if desired.

Navy PMs should be familiar with the specific MDA acquisition program policies discussed in enclosure 2 of SECNAVINST 5000.2A.[15]

In addition to the formal milestone reviews, the PM can expect the PEO (or SYSCOM commander) to hold periodic in-process reviews, usually on a semiannual basis. The PEO will normally chair the meeting with the commander of the SYS-

COM supporting the program present. Other attendees will include appropriate members of the PEO staff and supporting SYSCOM. Representatives from OPNAV, ASN(RDA), and other organizations may be included as deemed necessary.

RESOURCE MANAGEMENT AND THE PM

Obviously a PM must have resources if he or she is to execute a program. A Navy PM must have a basic understanding of how the PPBS is executed within the OPNAV and the secretariat. This section provides a very brief overview of the process within OPNAV and the key offices involved. The reader is again cautioned that, due to the ongoing OPNAV reorganization, no formal Navy PPBS policy reflecting the new structure had been released during the time this chapter was written. You must obtain copies of the appropriate POM serials to correctly understand the mechanics of how the Navy will prepare future POMs.

Each PPBS phase has an OPNAV coordination point. The planning phase is coordinated by N3/5, and N8 is responsible for promulgating PPBS policy, procedural guidance, and actual oversight of the OPNAV planning and budgeting phases. NAVCOMPT oversees the preparation of the DON budget and the actual budget execution.

The OPNAV offices which are responsible for assembling the data which eventually become a budget are called sponsors. There are two types of sponsors: resource sponsors and appropriation sponsors.

A resource sponsor is a DCNO or director responsible for an identifiable aggregation of resources that constitute requirements input to mission accomplishment. The requirements officers (ROs) in the resource sponsor organizations are the PM's OPNAV point of contact for all PPBS matters. The resource sponsors with acquisition-related resource areas are:

N1	Manpower and personnel
N2	Intelligence
N4	Logistics (including sealift)
N6	Space and C4 systems
N7	Training and education
N8/N85	Expeditionary warfare
N8/N86	Surface warfare (less expeditionary warfare)
N8/N87	Undersea warfare
N8/N88	Air warfare
N091	Research, development, test, and evaluation (science and technology)
N093	Medical
N096	Oceanography

Resource sponsors are the intermediaries between OPNAV and PEOs, DRPMs, and SYSCOM commanders and have the following responsibilities:

Providing guidance to the PEOs and SYSCOM commanders (and other claimants) to ensure the execution of program intent

Helping identify available resources from within the sponsor's resource base to reprogram for emerging program deficiencies

Ensuring that the program base for the POM is properly priced, documented, and supported

PEOs, DRPMs, and SYSCOM commanders have the following responsibilities in supporting their assigned programs:

Ensuring that the existing program base is priced accurately and provided to their resource sponsors

Providing cost estimates of program alternatives as requested by the resource sponsors

Ensuring that programs are adequately supported; imbalances should be reported to the resource sponsors

Appropriation sponsors are charged with the direction and supervisory control over an appropriation. Responsibility for Navy acquisition-related appropriations is as follows:

N1	Military pay, Navy
N8/N86	Shipbuilding and conversion, Navy
N8/N86	Weapons procurement, Navy
N4	Military construction, Navy
N8/N88	Aircraft procurement, Navy
N8/N82	Other procurement, Navy
N091	Research, development, test, and evaluation, Navy

While the resource and appropriation sponsors are important, N8 orchestrates the OPNAV PPBS process. N80 oversees the programming phase and promulgates PPBS guidance, data support, and decision coordination. This guidance is transmitted through a series of documents, called POM serials, which give the resource and appropriation sponsors the specific structure and guidance as to how the POM will be prepared.

Following the issuance of the initial POM serial memo, the CINCs submit their top maritime issues. Six joint mission area assessment teams and two support area assessment teams will take their input as well as input from the JCS, N8 (fiscal guidance), N3/5 (scenarios), N2 (threat assessment), and overarching Navy strategies. The teams will be made of core and mission area analysis groups and led by a one- or two-star flag officer. The CINCs and SYSCOM commanders will have representatives on the assessment teams. The six joint mission area assessment teams and two mission support area assessment teams are as follows:

- Joint mission areas
  1. Joint strike (lead N88)
  2. Joint littoral (lead N85/86)
  3. Joint surveillance (lead N87/88)
  4. Joint space and C4 (lead N6)
  5. Strategic deterrence (lead N87)
  6. Strategic sealift and protection (lead N87)
- Mission support areas
  1. Manpower, personnel, and training (lead N1/N7)
  2. Infrastructure (lead N81)

The outputs of these area assessments will include individual area "master" plans and other products to support key programmatic decision points. War games will be used to assess capabilities against the defense planning guidance and strategic concepts.

N81 will take the results of the eight individual area assessments and conduct an investment balance review (IBR). The IBR will include an assessment of Navy success in filling roles and missions, joint service efforts and contributions to Navy roles and missions, a "vision" of Navy doctrine of the future, and integrated analyses and tradeoffs of programs. The IBR is the bridge between the planning and programming phases and produces an integrated priority list (IPL) of all Navy programs.

The program and resource sponsors use the IBR and IPL to prepare their sponsor program proposal (SPP) inputs, which are consolidated and balanced throughout the programming phase by N80. Once the CNO and CINCs have decided on a tentative POM, it is reviewed in a DON program strategy board to arrive at a DON POM and submitted to OSD for review.

Following receipt of the program decision memorandum (PDM) from OSD, N82 incorporates PDM changes into the Navy budget through liaison with the appropriate claimant and resource sponsor. N82, in its dual-hatted role (described in the section discussing NAVCOMPT), also oversees the budget execution of programs for the Comptroller of the Navy.

CONSOLIDATION OF ACQUISITION-RELATED ACTIVITIES

In April 1991 Secretary of the Navy H. Lawrence Garrett III promulgated his intention to consolidate Navy research, development, test, and evaluation (RDT&E), engineering, and fleet support activities and facilities as part of the Navy post–Cold War restructuring.[14] The planned five-year consolidation formally got under way on January 2, 1992, when the Navy established four new warfare centers. These organizations will incorporate the activities of 36 research laboratories. When fully implemented in 1997, the consolidation is expected to save over $100 million per year and to have eliminated over 2500 jobs. A brief overview of the four centers follows (Ref. 14, enclosure 1, pp. 2–10).

The Naval Air Warfare Center (NAWC) is the full-spectrum (that is, RDT&E, engineering, and fleet support) center for air platforms, air warfare combat, and weapon systems. NAWC reports directly to COMNAVAIRSYSCOM. NAWC is organized into two major divisions: the Aircraft Division on the East Coast and the Weapons Division on the West Coast.

The Naval Surface Warfare Center (NSWC) is the full-spectrum center for surface platforms, surface warfare combat, and surface warfare weapon systems. It is also the focal point for all ship and submarine hull, mechanical, and electrical programs. The NSWC reports directly to COMNAVSEASYSCOM and is organized into four functional divisions: the Combat and Weapon Systems Research and Development Division, the Combat and Weapon Systems In-Service (ISE) Engineering Division, the Combat and Weapon System Engineering and Industrial Base Division, and the Hull, Mechanical and Electrical Research and Development and ISE Division.

The Naval Undersea Warfare Center (NUWC) is the full-spectrum center for submarine sensors and submarine combat and weapon systems. The NUWC reports directly to COMNAVSEASYSCOM and consists of two divisions: the Weapons and Combat Systems Division and the Weapons Systems ISE Division.

The Naval Command, Control and Ocean Surveillance Center (NCCOSC) is the full-spectrum center for maritime command, control, communications and intelligence, ocean surveillance technology, and fleet and shore support. The NCCOSC reports directly to COMNAVSPAWARCOM and is organized in three major directorates: the RDT&E directorate, the West Coast ISE directorate, and the East Coast ISE directorate.

REFERENCES

1. T. C. Hone, *Power and Change: The Administrative History of the Office of the Chief of Naval Operations, 1946–1986,* Naval Historical Center, U.S. Government Printing Office, Washington, D.C., 1989.
2. Secretary of the Navy Instruction 5000.1C, "Major and Non-Major Acquisition Programs," Sept. 16, 1988.
3. Secretary of the Navy Instruction 5000.33B, "Program Management Process," Jan. 12, 1987, p. 3.
4. Secretary of the Navy Instruction 4210.8A, "Acquisition Organization and Procedures," Aug. 12, 1987.
5. Public Law 99-433, "Goldwater-Nichols Department of Defense Reorganization Act of 1986," Oct. 1, 1986.
6. "OPNAV Program Coordinators," Memo, VCNO to OPNAV Principals, ser. 09/9U501181, Oct. 12, 1989.
7. R. B. Cheney, "Defense Management Report to the President," Washington, D.C., July 1989, p. 9.
8. "Establishment of the Position and Office of the Assistant Secretary of the Navy for Research, Development and Acquisition," Secretary of the Navy Notice 5430, Mar. 12, 1990, p. 1.
9. "OPNAV Staff Reorganization," Naval Message, Chief of Naval Operations, 221830Z, July 22, 1992.
10. ". . . From the Sea: Preparing the Naval Service for the 21st Century," Navy and Marine Corps White Paper, Oct. 1, 1992.
11. U.S. Navy, *Mission, Interfaces and Organization of the Office of the Assistant Secretary of the Navy for Research, Development and Acquisition,* Handbook, Oct. 1992.
12. Office of the Chief of Naval Operations Instruction 5000.42D, "Implementation of Defense Acquisition Management Policies, Procedures, Documentation, and Reports," Draft, Nov. 5, 1992.
13. Public Law 101-510, "Defense Acquisition Workforce Improvement Act," Nov. 5, 1990, Title 10, Subtitle A, chap. 87.
14. "Research, Development, Test and Evaluation, Engineering and Fleet Support Activities Consolidation," Memo, SECNAV to Various Navy Principals, Apr. 12, 1991.
15. Secretary of the Navy Instruction 5000.2A, "Implementation of Defense Acquisition Management Policies, Procedures, Documentation, and Reports," Dec. 9, 1992.

CHAPTER 14

PROGRAM MANAGEMENT IN THE U.S. AIR FORCE

Edward J. Trusela

Edward J. Trusela has been holder of the Air Force Chair in the Executive Institute, Defense Systems Management College, since December 1988. He is Chairman, Board of Directors, NATO AEW&C Programme Management Organization, and head of the delegation for the U.S./Danish Committee on Greenland Projects. As an Air Force civilian, Mr. Trusela has had assignments at the former Air Materiel Command and Aeronautical Systems Division (ASD) at Wright-Patterson Air Force Base, Ohio. He was Deputy Director of Procurement Policy in the Office of the Secretary of Defense. At HQ Air Force Systems Command, Andrews Air Force Base, he was Principal Assistant, Deputy Chief of Staff/Contracting and Manufacturing. In the Office of the Assistant Secretary of the Air Force he became Deputy for Acquisition and later Deputy for International Programs (Acquisition).

BACKGROUND

On February 28, 1908, the U.S. Army Signal Corps awarded the first contract for a military aircraft to the Wright Brothers of Dayton, Ohio. This began the acquisition for an air capability for the U.S. military. The Signal Corps remained the acquisition agency for the Army Air Corps procurements until the creation of the U.S. Air Force within the newly formed Department of Defense in 1947. Air Force acquisition from 1947 until 1961 was the responsibility of the U.S. Air Force Air Materiel Command (AMC), which managed the production and logistic support of the new systems. The basic research and development was conducted by the Air Research and Development Command (ARDC).

With the advent of the space program and the need to narrow the missile gap, in 1961 the Air Force reorganized its acquisition responsibilities. The logistic functions were split off from AMC and placed in the newly formed Air Force Logistics Command (AFLC). The remaining production functions were combined with research and development functions to be managed by a newly formed Air Force Systems Command (AFSC).

Under AFSC, the Air Force acquisition process evolved into its present form, beginning with the publishing of the 375 series regulation in the 1960s. Starting in the early 1960s, when Robert McNamara became Secretary of Defense, the services began to relinquish a good portion of their autonomy over their acquisitions. The Office of the Secretary of Defense (OSD) began to exercise more influence over the service-conducted acquisition process. OSD determined that independent planning and requirements generation was unacceptable. It was claimed that each of the services was preparing independent scenarios and that the services could not support each other in the event of a conflict. For example, neither the Navy nor the Air Force had the capability of moving the Army troops and equipment. As a result, the Assistant Secretary of Defense for Systems Analysis was created, which has evolved into the current Assistant Secretary of Defense for Program Analysis and Evaluation, with the responsibility to coordinate the DoD-wide military needs.

The first major program to be influenced by OSD's input was the TFX(F-111), which combined the Air Force and Navy requirement into a joint program, with the Air Force as executive agent. The program ran into difficulties in the planning and source selection process. For the first time the source selection authority, the Secretary of the Air Force, did not select the contractor recommended by the Source Selection Advisory Council. As a result, there were congressional hearings investigating how the contractors' technical and cost proposals were evaluated. The Secretary of the Air Force stated that the reason for not selecting the recommended contractor was that that contractor's cost estimates were unrealistically low. Even though, the contractors incurred little risk in underestimating the cost because they were not contractually committed to their prices at the time of source selection. During this period it was common to award a letter contract to the selected contractor with the firm requirements, contractual terms and conditions, and price subject to future negotiations. It was usually over a year before these were definitized. When the initial contract was finally settled, only the design development and testing were included, with the production of the inventory requirements left to be negotiated in a sole source environment. When the F-111 production quantities were negotiated, the cost was considerably higher than the initial estimates. This was true not only for the F-111, but for most programs not awarded as a result of meaningful competition.

To introduce greater competition in the process, in 1964 the Assistant Secretary of the Air Force, Installations and Logistics, Robert Charles, directed the consideration of total package procurement for Air Force systems. A simple explanation of the total package is contracting with a firm contract while in competition for as much of a program as can be described, usually on a fixed-price basis. The contract is awarded simultaneously with the announcement of the selected source.

In the middle 1960s the Air Force awarded total package contracts for the C-5A cargo aircraft, the TF-39 engine, the Maverick missile, and the SRAM (short-range attack missile). The problems with the C-5A, which caused the total package to fall into disfavor, are well documented. The Maverick and TF-39, however, were relatively successful. The SRAM total package had to be discarded because of the Air Force's inability to adequately define the interface requirements with the carrier aircraft—the F-111 and the B-52.

The main fallout of the total package experiment was changes in the source selection procedures, which resulted in a better description of the requirements and a completely negotiated contract with all competitors, so that a firm contract can be awarded simultaneously with the announcement of the selected source.

As the procedures changed and became more complex with many approaches to the acquisition of major systems, the Air Force Systems Command looked for a way to transfer lessons learned and experience between its many program offices.

In 1975 the command began conducting business strategy panels. One of the first panels conducted was for the lightweight fighter, which became the F-16. The purpose of the strategy panel was to gather together individuals from throughout the Air Force who could bring their expertise to the panels to discuss with the program office the pros and cons of different business approaches to managing a program. The panel was only advisory, and it was up to the program manager to decide what was the best strategy for his or her program. As a program progressed and circumstances changed, the program manager could request a reconvening of the panel. Through the years, the business strategy panels expanded into acquisition strategy panels to include all aspects of program management and not just the business portion. These panels have now become an integral part of acquisition planning for Air Force programs.

In July 1992 the Air Force reorganized to once again combine Air Force Systems Command and Air Force Logistics Command into a new Air Force Materiel Command. Under the new command, the Air Force will practice integrated weapon system management (IWSM). The definition of IWSM is to empower a single manager with maximum authority over the wide range of decisions and resources to satisfy customer requirements throughout the life cycle of that system.

ACQUISITION POLICY

The current Air Force acquisition policy is the result of the latest legislation and the implementation of the defense management review (DMR). The Air Force acquisition system encompasses all aspects of acquisition programs, from new start acquisition programs to the modification of existing systems. The Air Force acquisition executive (AFAE) is responsible for all Air Force acquisition. The Assistant Secretary of the Air Force (Acquisition) is the AFAE. As the AFAE, the assistant secretary is the senior corporate operating official for acquisition, responsible for overseeing all Air Force procurement activities and for implementing the AF information resource management (IRM) program.

The system is characterized by centralized policy development and decentralized execution. Authority and decision making are delegated to the lowest appropriate level consistent with the degree of oversight required to maintain the integrity of the process. Acquisition program management is accomplished through a streamlined chain to ensure short, direct lines of communication and no more than two levels of review. Accountability for program execution resides at each level of this chain.

To implement the streamlined chain of execution, program executive officers (PEOs) have been established in the command line between the acquisition executive and the program director (PD) for major and selected acquisition programs. Each PEO is responsible for a number of mission-related programs which collectively comprise the PEO's portfolio. The mission-related areas which have appointed PEOs are strategic, information systems, tactical/airlift, space, command control and communications, and tactical strike.

For other than major or selected programs, the commanders of the Air Force Materiel Command (AFMC) product centers or air logistics centers are specified as designated acquisition commanders (DACs). The DACs are in a direct reporting line between the subordinate program managers (PMs) and the acquisition executive. The DAC is responsible to the acquisition executive for assigned programs and will perform the same general functions as a PEO.

The PEO is directly responsible and accountable to the acquisition executive for the execution of a portfolio of programs. Except as may be expressly limited, the PEO exercises the authority of the acquisition executive. The PEO organization is not a part of the assistant secretary's acquisition staff, but is a field agency reporting directly to the assistant secretary in the role of acquisition executive. The PEO has direct access to the acquisition executive to ensure close and continuing communication. Day-to-day activities, however, will be conducted by direct and continuous interaction among the program offices, PEOs, acquisition command field activities and headquarters staffs, and the secretarial staff. These relationships form the basis of an acquisition team that collectively pulls together all aspects of program execution and support in a comprehensive, coordinated, and effective manner to ensure sound acquisition practices resulting in the required capability for the operating commands.

These principles apply equally to the DAC. The DAC is directly responsible and accountable to the acquisition executive for the execution of assigned system acquisition programs. These commanders, their subordinate PMs, their major command staff, and the secretarial acquisition staff will interact on a day-to-day basis to form the acquisition team under the acquisition executive, with responsibility for the execution of system acquisition efforts that are other than major and selected programs. Like the PEO, the DAC has a privileged line of communication to the acquisition executive.

The acquisition executive is designated by the Secretary of the Air Force as the senior acquisition official in the Air Force and is accountable to the secretary for all domestic and international service acquisition functions, including foreign military sales that require U.S. Air Force research, development, testing, and evaluation (RDT&E).

The acquisition executive establishes Air Force acquisition policy, advises the Air Force secretary on acquisition issues, and serves as a senior advisor on other issues as requested by the secretary. He or she nominates to the Secretary of the Air Force, with the advice of the Chief of Staff of the Air Force (CSAF), the PEOs and PMs for major and selected programs. He supervises and evaluates PEOs and decides programmatic issues by providing program direction and oversight. Oversight is provided through the PEO or DAC for all programs.

The acquisition executive chairs the Air Force System Acquisition Review Council (AFSARC), which is the Air Force corporate body that advises the executive on matters concerning the initiation, continuation of, or substantial changes to major defense acquisition programs. He or she responds to the defense acquisition executive on behalf of the Air Force and represents the Air Force at the Defense Acquisition Board.

The PEO or the DAC is directly responsible and accountable for the execution of an assigned portfolio of programs. As such, they exercise the delegated authority of the acquisition executive. The PEOs have been established in the command line between the acquisition executive and the PM for major and designated programs. Although the PEOs and DACs have a privileged line of communication to the acquisition executive, the day-to-day activities are conducted by direct and continuous interaction among the program offices, PEOs and DACs, acquisition command field activities, headquarters staff, and acquisition staff. These relationships form the basis of an acquisition team that collectively pulls together all aspects of program execution and support to ensure sound acquisition practices.

Among the major responsibilities of the PEOs and DACs are ensuring that the AFAE and acquisition staff are informed of all significant or sensitive problems of users, and assisting and advising the acquisition staff on programming and budgeting matters. The PEO directs PMs in all aspects of program execution to ensure that the program offices remain focused on satisfying the operational requirements and that programs are executed in terms of cost and schedule to meet the performance requirements within the approved baselines. The PEO or DAC ensures that program offices exercise contracting authorities and responsibilities as delegated by the head of the contracting activity (HCA). PEOs exercise below-threshold investment appropriation reprogramming authority for their portfolio of programs. This authority is a major tool available to assist PEOs in addressing cost schedule and performance objectives for their assigned programs.

The PM is the single operating official with the authority and responsibility for program execution within the approved acquisition program baseline (APB) and is accountable to the acquisition executive through the PEO. One specific responsibility of the PM is to develop the acquisition strategy and program baseline, including developing a management approach, providing budgetary estimates and alternatives, establishing program schedules, developing contracting strategies and structure, and establishing an interface framework with related programs. The PM executes the program within the guidelines and resources established in the approved program management directive, acquisition strategy, and acquisition program baselining, and conducts the day-to-day management of the program within the established policies and procedures.

The secretariat and AF headquarters acquisition staffs perform normal staff functions to support the program objectives and functions as the acquisition team's focal point and conduit for all interfaces in Congress, the Office of the Secretary of Defense, Joint Chiefs of Staff, other services, air staff major commands, and allied nation acquisition authorities. These staffs provide all acquisition input to the program planning and budgeting system (PPBS) and identify the program budget and reprogramming sources.

The AFMC is the key organization necessary to ensure effective support of all acquisition programs in which the U.S. Air Force participates. The command is accountable to the acquisition executive for maintaining the acquisition infrastructure so as to provide necessary program support. The responsibilities of the command are to implement military and civilian acquisition professional development programs, to maintain professional expertise in functional areas, and to provide advice and counsel to the program management personnel. The command makes sure that needed support is programmed, funded, and provided to the PM. Support includes technical assistance, infrastructure, test capabilities, laboratory support, professional education, training and development, and all other support for acquisition managerial personnel. The command, prior to milestone I, will support cost and operational effectiveness analysis performed by the operating commands and direct concept definition, evaluation, and integration studies, coordinating with the user and affected PEOs to develop alternative solutions to validated needs; and it will integrate life-cycle cost estimates to proposed alternatives for meeting these validated needs. The command will assist in the development of policy, implementation plans, and procedures within the guidelines established by the acquisition executive. Within these guidelines, the command will manage the science and technology (S&T) program and commodity acquisition.

CONTRACTING AUTHORITY

As agency head the Secretary of the Air Force establishes policies for, and directs and supervises the Air Force activities with respect to contracting and related matters. The Assistant Secretary of the Air Force (Acquisition) is the head of contracting activity (HCA) for major selected and other programs. The Secretary of the Air Force has delegated general contracting authority to the Assistant Secretary of the Air Force (Acquisition) and the Deputy Assistant Secretary (Contracting).

PEOs designated by ASAF(A) have delegated authority to enter into, execute, and approve contracts (including change orders, supplemental agreements, and other amendments to contracts), letter contracts, and other contractual actions, including terminations and settlements, for assigned major and selected programs. This authority may be redelegated to contracting officers within or supporting the program office under such terms, conditions, and limitations as may be deemed appropriate.

The commanders of the product centers and ALCs have delegated authority to enter into, execute, and approve contracts (including change orders, supplemental agreements, and other amendments to contracts), letter contracts, and other contractual actions, including terminations and settlements, for other programs and for major programs within their purview and not assigned to a PEO. This authority may be redelegated with or without the authority to make successive redelegations and under such terms, conditions, and limitations as may be deemed appropriate.

REQUIREMENTS PROCESS

The Air Force requirements process is governed by Air Force Regulation 57-1. The national strategy and the Joint Chiefs of Staff strategy guidance are the bases from which Air Force long-range plans for force modernization are derived. The planning process begins with an Air Force strategy review addressing key issues in preparation for the development of Air Force executive guidance. The Secretary and the Chief of Staff of the Air Force provide the executive guidance, addressing strategic environment, national strategy objectives, defense policy, national military objectives, and planning priorities. The planning process results in the publication of the Air Force Plan (AFP). The plan provides a summary of the executive guidance, fiscally constrained force structure levels, and assessments of forces. For those involved in identifying mission needs and operational requirements, the plan provides the link between planning priorities, fiscal reality, and potential Air Force programs.

The process begins with a mission area assessment, a "strategy-to-task" evaluation linking the need for certain military capabilities to the military strategy provided by the chairman of the Joint Chiefs of Staff (JCS). Air Force major commands continuously evaluate plans and JCS guidance for changes in assigned missions and objectives which must be accomplished. Once a task is identified, the major command must evaluate its ability to accomplish the task.

After the assessment is completed, a mission need analysis is accomplished. The objective of the analysis is to evaluate the Air Force's ability to accomplish identified tasks and missions using current and programmed future systems. If a major command identifies a shortfall in its ability to accomplish a task or mission, the first obligation is to determine whether a change in tactics, doctrine, or training (nonmateriel solutions) may solve the deficiency. If a materiel solution (new hardware or software) is required, the mission need is documented in a mission need statement (MNS).

Mission needs for major programs are reviewed and validated for the chairman of the JCS and the Secretary of Defense by the Joint Requirement Oversight Council (JROC). The JROC is chaired by the vice chairman of the JCS, and its members are the Vice Chiefs of Staff for the Air Force and Army, the Vice Chief of Naval Operations, and the Assistant Commandant of the Marine Corps.

Once a mission need is validated, the acquisition process begins. There are five phases within the process: phase 0—concept exploration and definition, phase I—demonstration and validation, phase II—engineering and manufacturing development, phase III—production and deployment, and phase IV—operations and support. Each of these phases is preceded by a milestone review.

When the milestone decision authority declares milestone 0, concept formulation, approval of the decision is documented in an acquisition decision memorandum (ADM) or an equivalent decision document. The document will identify a minimum set of alternatives and the source of funding. HQ USAF will prepare and issue the concept study program management directive (PMD). The PMD will direct the lead major command to identify, explore, and evaluate potential alternative solutions, accomplish a cost and operational effectiveness analysis (COEA), and prepare other documentation as required. For potential major programs the major command will determine the membership of and convene a concept action group (CAG) to manage the concept studies and prepare a COEA and other pre-milestone I documents. The major command commander uses the results of these studies and analyses to justify and select a preferred alternative and prepare a briefing to gain CSAF and SECAF approval. Once the Air Force position is established, the user implementer and the PEO or DAC prepare the documents required for the milestone I concept demonstration approved review. Historically the system program office (SPO) would prepare most of these documents, but under DoD 5000 series, a program office does not exist until a successful milestone I review. HQ USAF will assign in the PMD the implementing command for accomplishment of this documentation. The HQ USAF approved operational requirements document (ORD) is the basis for all follow-on documentation.

The initial ORD is prepared by the using command during phase 0, concept exploration and definition, following a successful milestone 0 decision. It is solution-oriented. It describes preliminary system-specific characteristics, capabilities, and other related operational variables. As program development matures, an updated and expanded ORD is required for each subsequent milestone. The ORD ensures that all participants articulate, develop, produce, and field military systems which meet the user needs in terms of intended mission and normal peacetime training requirements. A mandatory attachment to the ORD is the requirements correlation matrix (RCM).

The RMC tracks essential user needs and requirements over the program life cycle. It provides the basis for user-stated needs and requirements to be included in the additional required documentation. The RMC also correlates needs and requirements to contractual specification, depicts operational test criteria, and verifies that emergency capabilities under development continue to meet the needs of the user.

Senior-level requirements and acquisition program reviews (summits) are held for all ongoing major defense acquisition programs. Reviews for other programs may be directed by the CSAF. Summits are chaired by the CSAF. Membership consists of the AF acquisition executive, commanders of the operating implementing and supporting commands and the Air Force Operational Test and Evaluation Center, the mission area director, the PEO, the program director, the Assistant Chief of Staff for Plans and Operation, the Deputy Chief of Staff for Intelligence, the Assistant Chief of Staff for Logistics, and the Director of Operational Requirements.

Summits assess program progress in meeting user-stated needs and requirements. They consider operational concepts, the projected threat, and the capabilities of other supporting systems to ensure that technical solutions under development continue to meet user objectives. Summits are conducted to ensure that the user-prepared operational concept is correct and complete in order to overcome the system threat assessment and that the solutions under consideration are consistent with the mission deficiency; to confirm program funding levels, priorities, schedules, system baselines, available resources and overall management efforts are adequate; to establish and agree to realistic, achievable, and affordable performance and support goals and thresholds; and to resolve major operational or technical issues that could adversely affect the success of the program.

In summary, the Air Force acquisition process has evolved over the years, with the most significant changes since the defense management review. The most significant is in the area of requirements generation, which is now with the using commands and not the acquisition command. Once the requirements are established, the execution has been streamlined with the PM reporting to the PEO, who in turn reports to the AFAE, who has the ultimate responsibility for Air Force acquisition.

CHAPTER 15

NONTRADITIONAL APPROACHES TO MILITARY PROJECT MANAGEMENT

Robert R. Barthelemy and Helmut H. Reda

Robert R. Barthelemy became program director of the National Aero-Space Plane Program in 1987. As such, he is responsible for organizing and directing the Joint Program Office at Wright-Patterson Air Force Base. Earlier he served as technical director of the Air Force Wright Aeronautical Laboratories, a 2700-person, billion-dollar-a-year development activity. He previously served as deputy director of the Air Force's Aero Propulsion Laboratory and in numerous technical management positions. Born in Massachusetts in 1940, Dr. Barthelemy holds a bachelor's degree in chemical engineering and a master's degree in nuclear engineering, both from the Massachusetts Institute of Technology, and a doctorate in mechanical engineering from Ohio State University.

Helmut H. Reda became deputy director of X-30 System Development in 1989. As such, he is responsible for developing the program plan to design, build, test, and support the X-30. He was previously assigned to various flight test programs, including F-16, F-15E, F-4, A-10, and F-111. Born in Oregon, Major Reda holds a bachelor's degree in aeronautical engineering and a master's degree in business administration, both from Embry-Riddle Aeronautical University. Major Reda is a graduate of the Air Force Test Pilot School engineering course.

Necessity, they say, is the mother of invention. We have a lot of necessity in our society today and I think the time is here for inventing new approaches, new solutions to those various problems, so that we can indeed maintain America, and indeed the rest of the world as well, as the land of opportunity for all those that will be the achievers of the future.

Robert N. Noyce
Chief Executive Officer, SEMATECH
1989

INTRODUCTION

In military organizations, the management of large, complex projects is usually carried out in program offices. These program offices are supported by a full complement of staff, and their primary purpose is to deliver a product to the government on time, within budget, that meets specification. This goal is accomplished in a very complex, highly visible environment where almost every major decision is subject to scrutiny. Limited budgets and the need to promptly field weapon systems has motivated Congress and the administration to take a fresh look at the program acquisition and management process. Executive policy has become more lenient, allowing freedom and flexibility to execute procurements through smarter and simpler methods. Program offices can operate through traditional or nontraditional management approaches, depending on the opportunities and the constraints placed upon them. Traditional programs tend to have well-defined products, and they approach procurement from the classical sense: following existing or time-tested practices from previous programs. Nontraditional programs may have a general concept of the end product, and they exploit new or unique opportunities. Nontraditional programs also attempt to find creative solutions to external constraints or to obtain unusual flexibility in their acquisition approach.

This chapter discusses four nontraditional management alternatives available to program offices:

Joint government programs

Consortia

Contractor teaming

Skunk works

These alternatives warrant special consideration because of their potential for high payoff under the right circumstances. Joint government programs can be used when multiple customers or interests exist. Consortia and contractor teaming are viable alternatives when there is a clear advantage for competitive contractors to mutually attack difficult challenges. Finally, government and contractor organizations can apply Skunk Works operations if a quick, quiet, and cost-effective acquisition is required. These approaches generally apply to programs that are unrealistically constrained by cost or schedule, advance the state of the art, or seek to meet very challenging requirements. The method, or methods, to be used are primarily based on the final customer, the level of competition, and the distribution of existing talent and resources.

JOINT GOVERNMENT PROGRAMS

Whenever more than one service or agency stands to benefit significantly from a particular government project, the possibility of a joint program exists. If resources, management, information, or capabilities can be shared, a joint program might satisfy the requirements of each organization without any one of them having to shoulder the entire burden. While joint programs can be complex, the benefit to the taxpayer, as well as the government, warrant their consideration when the circumstances are appropriate.

There have been numerous, very successful joint government programs that are worth examining. Two interesting examples are the U-2 and the SR-71. The Air Force and the Central Intelligence Agency (CIA) formed a small joint program office and consolidated their requirements to procure a high-altitude strategic-reconnaissance/special-purpose research aircraft. The Lockheed Corporation was quickly awarded a contract and covertly developed the U-2 in minimum time and under budget while satisfying both customers' requirements. Several years later, Lockheed surprised the CIA again by offering an unsolicited proposal to develop a Mach 3+ reconnaissance aircraft with higher altitude and lower radar cross-section capability. Another contract was awarded to Lockheed for developing the SR-71. Both aircraft were successfully managed by small joint government program offices that provided this nation with many years of unopposed, reliable, and secure reconnaissance capability. During the 1960s and 1970s joint government programs provided the Air Force and Navy with mainstream fighter/bomber forces by procuring large quantities of A-7, F-4, and F-111 aircraft. During the 1980s and 1990s, the Strategic Defense Initiative Office (SDIO) and the National Aero-Space Plane (NASP) program were some of the more visible joint government programs.

The initial NASP effort began in 1984 after a Defense Advanced Projects Agency (DARPA) study concluded an aerospace plane was feasible. Shortly thereafter DARPA leadership decided that the most appropriate form for the NASP program was a joint government program. It was clear that an affordable and flexible space transportation system interested the Air Force, NASA, the Navy, SDIO, as well as DARPA. Each organization had technical expertise and knowledgeable individuals who could assist in conducting the program. Since the overall cost of the project was certain to be high, each organization found the joint program affordable by sharing this cost.

The joint program was initiated with each organization providing about 20 percent of the funding and contributing whatever expertise was needed to manage and conduct the technical aspects of the program. While DARPA retained overall program direction, the Air Force was declared the lead execution agency and assumed program management responsibilities. NASA led the technical management effort, while the Navy focused its hypersonic laboratories on the NASP program.

A NASP Steering Committee made up of senior leaders from all five organizations guided the program and the Joint Program Office (JPO) at Wright-Patterson Air Force Base. The JPO functioned with personnel from all five organizations. As the program matured, some aspects of the joint program have changed, but the fundamental joint nature of the program has been maintained due, in large part, to the merits and success of this approach to the NASP program.

If the conditions are right, joint programs can be a very successful program management technique. Sharing resources to attain a mutually important objective is the key criterion to assess their applicability. In a joint program the team can use the best resources from every organization to improve efficiency and minimize duplication. A joint program allows the government to show one "face" to the contractor, negating a need to respond to multiple customers by providing a single focal point for information exchange, technology transfer, program advocacy, and program direction.

Despite the obvious advantages, joint programs, like any collaborative endeavor, present some challenges. Since more than one master must be satisfied, the program office must be sensitive and responsive to the different demands of its sponsors. Productivity and efficiency may be compromised because of multiple management systems, and the need for consensus could result in delays and resource mismanagement. If one player tires of the game or withdraws its resources, the program may have to expend great efforts to keep on track or stay alive. As with all joint undertakings, leadership is critical. Managers who are capable of dealing with this complex, often political form of program management must be sought and empowered to act for the good of all parties. Despite these challenges, many joint government programs have been very successful, and their applicability to appropriate programs certainly should be examined.

CONSORTIA

Historically, consorting has been a popular technique when all partners stand to gain substantially from the process and not give up much to obtain the benefits. Men have consorted to achieve power, nations have consorted to stop aggression, and companies have consorted with and against each other for various reasons.

U.S. companies are collaborating to develop technologies to counter increased competition abroad or to bring new products to the market early. Industries are taking advantage of relaxed antitrust laws to pool research and share the results. Consortia are effective management tools to gain mutual collaboration on specific efforts. A consortium's advantage is achieved by leveraging each participant's capabilities while maintaining separate identities. Companies unite in generic efforts without feeling threatened from a proprietary standpoint. By eliminating duplication in technology development, consortia minimize each company's capital costs and investment of resources, while significantly reducing the industry's total expenditure to derive the technology. Consortia encourage mutual collaboration on high-payoff technologies too risky for a single company to pursue. Nevertheless, consortia have their own problems. Collaborative efforts diffuse authority, complicate communication, decrease the level of independence, and are managerially less efficient. However, these drawbacks can be overcome or minimized through effective management and clear contractual arrangements.

Recently organizations have formed consortia to concentrate sufficient resources on a problem whose solution is difficult to achieve, but is of mutual benefit to all of the contributors. The Japanese have used this technique to share information and technology among several organizations that could then develop internationally competitive products from the knowledge that they gained. In the United States several highly visible consortia formed to pursue specific electronic and manufacturing technologies. Each participant contributed relatively modest but equal resources, and all had access to the results of the endeavor. While input and output from recent consortia have varied, the collaborators generally contribute resources in the form of money, people, and information; the reward, equally shared, is more advanced information.

The Japanese have utilized the advantages of consorting for years. An interesting example is their international consortium called HYPR, aimed at developing a combined cycle, methane-fueled propulsion system focusing on the Mach 2.5 to 5 flight regime. The engine resulting from this program might power the third-generation supersonic civil aircraft. Interestingly enough, Pratt & Whitney, General Electric, Rolls Royce, and SNECMA are performing contract work for the Japanese HYPR consortium, adding a truly international flavor. These four western companies are also prime contenders to develop engines for the second generation of U.S. and European supersonic transports.

Semiconductor manufacturing is one of the most competitive businesses in the world. The global drive for smaller, more complex chips creates the necessity for extensive domestic research and development (R&D), and the continued evolution of more productive tools and manufacturing processes. Although the American semiconductor industry created, developed, and led the global semiconductor market for three decades with startling innovations, the U.S. semiconductor industry lost its leadership position to Japan in 1985. In 1989 Japan replaced the United States as the world's leading supplier of semiconductor equipment and materials for the first time in history.

Semiconductor Manufacturing Technology (SEMATECH) is a nonprofit research and development consortium created in response to national concerns about erosion of the domestic semiconductor industry and its strategic and economic implications. SEMATECH is a Department of Defense (DoD) and industry partnership dedicated to restoring America's leadership in semiconductor manufacturing technology. SEMATECH's 14 member industrial firms represent 80 percent of the semiconductor manufacturing capability in the United States and provide 50 percent of SEMATECH's funding.[1] The DoD matches industry's investment and also provides national laboratories for R&D. The contributing member firms, fiercely competitive in the open marketplace, put aside their differences in a spirit of precompetitive cooperation for the common good of the country, investing millions of dollars and the minds of their best engineers and scientists. SEMATECH does not concentrate on specific product manufacturing, but on the generic technology, standards, and specifications for use in broad applications of both commercial and military semiconductor manufacturing. By bringing together semiconductor manufacturers in precompetitive cooperation, SEMATECH defines and coordinates high-priority programs for the industry and facilitates the development of domestic manufacturing capabilities that can efficiently compete in global markets. Detailed plans are also under way to investigate the potential value of cooperating with the Joint European Submicron Silicon consortium by determining a thorough analysis of the European semiconductor industry and its infrastructure. SEMATECH is aggressively pursuing preeminence in semiconductor manufacturing by focusing limited resources for maximum impact. SEMATECH has already made important advances in semiconductor manufacturing and has effectively transferred that technology back into the industrial base.

The U.S. aerospace industry also has taken consortia to heart by establishing a nonprofit educational consortium called the National Center for Advanced Technologies (NCAT). Its goal is to pool information and resources from within industry to develop and mature technologies. NCAT technology demonstrations are generic and stop short of the critical stage at which companies compete to develop specific products. Their intent is to develop and demonstrate technologies to the point when they await application and help form strategic partnerships or joint ventures that will lead to the development of useful and enduring end products.

The largest current U.S. government-industry collaboration is the NASP Technology Maturation Program. This $250 million consortium is comprised of five NASP contractors (who were originally in direct competition with each other), 114 subcontractors, and 12 Air Force, Navy, and NASA laboratories. In the early stages of NASP, planners realized that the development of new materials to withstand the tremendous forces, heat loads, and temperatures of the hypersonic environment would present a great challenge. Although the government supported and funded materials and structures research in each of the competing aerospace companies, progress failed to keep pace with the needs of the program. Each of the five companies matched government funds in materials, but independent assessments by outside review teams indicated it would take more than 10 years to develop and incorporate the needed materials technology for the airframe and engine subsystems. Since the program schedule required those advances in only three years, a new approach was needed. The concept of a consortium was proposed almost simultaneously by both the contractors and the government.

The consortium pioneered an innovative, nationwide team-building approach to accelerate the development of critical-path materials, structures, and subsystem technology for the X-30 experimental demonstration vehicle. Because of the critical need and mutual acceptance of the concept, it was implemented in less than three months. Each company assumed responsibility for the development of one key material system using both government and internal company funds. The companies agreed to manage the separate development activities as a collaborative activity and to share all the resulting information with each other.

Although the overall competition for the NASP was still maintained, materials development was taken out of the competitive arena and became a shared resource among all five competitors and the government. Great progress was achieved. All five material systems were greatly enhanced, and each competitor used the technology derived in all five materials areas for its competitive approach. The government also benefited by having credible technologies and much more competitive contractors for the NASP. The materials progress originally estimated to take 10 years was achieved in three years with only a modest increase in resources to the government and the contractors.

While many hailed the NASP materials consortium as an unqualified success, the range of applicability for consortia is certainly not unbounded. All participants must see the benefits of the activity and share freely in the process. Equality and honesty are fundamental requirements, and a catalyst, such as the government, may need to start the process. If potential participants feel consorting with competitors will reduce or eliminate their competitive edge, their participation is certainly doubtful. The best situations involve meeting a challenging goal, or gaining on a common, superior enemy. The NASP consortium definitely serves as an example of the former, while SEMATECH has the latter characteristic.

Although consortia can be applied through many forms and structures, the usual rules for collaboration still apply. Precise goals, shared management, a commonly developed set of procedures and rules, clear commitment by participants (even if limited), and well-defined boundaries are necessary for a successful consortium operation. The key word in consortia is mutual: mutual sharing to achieve solutions to a mutual problem. It is a technique which can be used effectively in parallel with competition. In a very competitive situation, however, it may be impossible to achieve this kind of consensus. In less competitive situations, a consortium can lead to increased cooperation among the participants. In the case of the NASP, it eventually led to the formation of a joint-venture partnership among all of the competitors.

CONTRACTOR TEAMING

While consortia are useful in achieving contractor cooperation and synergism in specific areas, teaming, if it can be achieved, offers even greater possibilities.

Although contractors have used prime-subcontractor relationships for some time to achieve program objectives, teaming, as described in this chapter, refers to those arrangements that maintain equality between contractors while providing the potential for synergism through formal business agreements. While there are numerous approaches to achieve these objectives, the two most common forms of contractor teaming involve partnerships and joint ventures. Partnerships maintain the integrity of a contractor organization since they do not create a new legal entity for the government to deal with. Joint ventures generally take the form of a new organization which has to be capitalized, staffed, and legalized. A hybrid approach is the joint venture partnership, which creates a stronger bond than the simple partnership, but does not necessitate the formation of a new organization.

During the past two decades many partnerships and joint ventures have been formed to take advantage of the unique contributions of each participant and achieve a stronger competitive position or a synergistic initiative. Some national and international joint ventures have been extremely successful, capturing entire market segments. Many have failed, however, principally due to the managerial and organizational complexity of these arrangements.

After the success of the NASP Materials Consortium, the Joint Program Office and the NASP contractors formed two additional consortia, one in the subsystems development area and the second for technology transfer and program development. With three consortia under way it appeared logical to examine the possibility of one overall consortium to conduct the entire program. While the government program office warmed to the idea, the five prime NASP contractors initially met it with great resistance. The principal difficulty was associated with their perception of their competitive positions with respect to the overall program. Each felt they had the competitive advantage and none wanted to give it up. Some differences in capability certainly existed, but no one contractor had a monopoly on all of the good ideas and the challenge of the program required synergy rather than competition. When the government made this clear, while simultaneously declaring that teaming, rather than down selection, would occur within one year, the contractors moved in earnest toward collaboration. Despite this strong customer influence, it took more than one year to affect teaming, which eventually took the form of a joint-venture partnership. This form of teaming allowed the five prime contractor companies to stay involved in the program while still fostering future competition in operational derivatives of the experimental aerospace plane vehicle. A teaming agreement, based on principles created in conjunction with the government, was developed and eventually agreed to by all five corporations. With the teaming agreement, the joint-venture partnership became a legal entity and the government contracted with it for the remainder of the R&D phase of the NASP program.

The team set up headquarters at a neutral site and established a management structure that required all five players to participate in critical decision making. The team headquarters, called the National Program Office (NPO), was purposely kept small so that the majority of actual contract work flowed out to the five contractor home sites. The government set up a parallel structure and reinforced team behavior by focusing its attention on the NPO and away from the home sites.

While the joint venture partnership of the NASP contractors is less than one year old, it has overcome many problems associated with such a new approach. The contractors are now operating as a team, and significant synergy has been attained. If there has been a downside to the operation, it has been the loss of productivity while the contractors transitioned from a competitive group to a single team. Fortunately that inefficiency is decreasing as the team gels, and the collaboration obtained greatly offsets that deficiency.

Joint ventures, partnerships, and joint-venture partnerships certainly provide some rewards, but they are clearly not a panacea. When no single contractor (or contractor plus subcontractors) can handle the job alone, then this approach has great merit. If the reverse is true, then the awkwardness and potential inefficiency of teaming may negate its consideration. There are also certain legal restrictions on contractor teaming that must be considered. For purely R&D efforts the law allows contractor teaming. However, when the effort becomes more product-oriented, certain antitrust restrictions go into effect. If teaming is being considered, a thorough review of the law should be accomplished.

Teaming is much more formal than consortia, and it takes great effort to achieve a good team. The bureaucracy reacts strongly to innovative approaches. Everyone involved must deal carefully and sometimes slowly with the required details. Even more than in consortia, teaming requires a strong, persistent, and committed catalyst. If the government had not persisted, the NASP contractor team never would have formed. On the other hand, since teaming was attempted after three years of competition, it was probably more difficult to effect than if the government had pursued teaming from the onset. Nevertheless, any kind of nontraditional teaming requires extra effort, patience, commitment, and the ability to achieve consensus and some compromise.

SKUNK WORKS

Lockheed's supersecret advanced development projects organization is the aviation industry's premier leader for developing highly advanced aircraft. They have consistently delivered the next generation of aircraft well ahead of their domestic and foreign competition. Lockheed's management practices are so unique, original, and successful, that the U.S. Patent Office granted Lockheed the registered service mark "Skunk Works" to characterize their organizational procedures. While joint programs, consortia, and contractor teaming apply to external organizational teaming relationships, Skunk Works applies to an organization's internal team practices, relationships, and culture.

Skunk Works is a high-performance management philosophy that taps into an organization's reservoir of talent, creativity, and energy. It creates a culture that encourages its people to find new innovative ways of doing business while liberating them with the opportunity to make a real impact. Its success stems from the ability to develop small empowered teams entirely focused on one project with only one expected result: getting the job done right, on schedule, and within budget. This is primarily accomplished by creating an enthusiastic culture based on trust and nonadversarial customer-supplier relationships. All participants are positively committed to achieving success and jointly work as a team to overcome problems as they arise.

The Skunk Works method is not applicable to all programs. As in previous nontraditional approaches, these principles apply to select programs where all participants agree that the conditions are right and warrant this approach. Skunk Works should be considered when there is a need to deliver a special product in a quick, quiet, and cost-effective manner. The government must consider the need so vital, or generate a lead of such insurmountable advantage, that it justifies foregoing traditional acquisition methods. Skunk Works satisfies the unique case when the need justifies the means, as long as it follows legal and reasonable practices. There is also a misconception that Skunk Works applies to only covert aircraft programs. This is far from true; Skunk Works has been applied successfully to both classified and unclassified programs, involving both aircraft and ground-based systems.

Skunk Works performs best for small, time-constrained programs that seek high payoff in a short production run. Small Skunk Works organizations are desirable because they are relatively inexpensive, quick to execute, and help mitigate the risk of pursuing difficult projects. If a small team fails, the parent company will not go bankrupt. A program is kept small by having a limited core group assigned full time to the project. Short-term nonroutine tasks are accomplished by borrowing expertise from the parent organization. When tasks are completed, individuals are released back to the parent organization to work on other programs. This maximizes each person's utility to the Skunk Works project by focusing efforts on what must be accomplished, and minimizes the unnecessary bureaucratic activities that are designed to keep people busy while waiting for additional tasks. Within the small core group, communications are enhanced because people become acquainted and comfortable with each other. Communication occurs face to face, allowing information to be transferred freely and decisions executed promptly.

Clarence "Kelly" Johnson founded the original Skunk Works organization in July

1943, when the Army Air Corps asked Lockheed to design a prototype fighter around the British de Havilland-Halford H-1 turbojet engine within 180 days. Kelly quickly forged a team of 23 engineers and 103 shop mechanics by collocating them in an assembly shed near the Lockheed wind tunnel in Burbank, Calif. This allowed rapid turnaround between design modification and wind tunnel tests.

To stay on schedule, Kelly established a streamlined management structure and developed policies, principles, and practices to allow prompt execution of engineering and programmatic decisions. Kelly, an engineering and management genius, optimized his team's activities into quick, simple, and efficient processes. His team worked in a shroud of secrecy and completed the P-80 Shooting Star prototype 37 days ahead of schedule.

People became curious and noticed the beehivelike activity surrounding the assembly shed. In his autobiography, *Kelly—More than My Share of It All,* Kelly tells the story of how his team members responded to outsider's questions regarding what was happening inside the shed. On one occasion an engineer stated that Kelly was cooking up some kind of strange brew, and it quickly brought to mind a mysterious factory in the Li'l Abner comic strip called "Sk*o*nk Works" where hillbillies chopped up skunks for potions and other foul smelling brews. The name stuck, and it is still used today. Only now the term Skunk Works pertains not only to Lockheed's advanced development projects, but also to the unique streamlined management philosophy that Kelly inspired.

Kelly's successes continued after the P-80. The Saturn, XFV-1, F-104, C-130, X-7, U-2, and SR-71 aircraft were soon to follow. Kelly and his Skunk Works team received the Collier Trophy (aerospace industry's highest achievement award) for the F-104 and SR-71 aircraft. Through Kelly's staunch leadership, Lockheed's Skunk Works developed a reputation for building advanced and highly complex aircraft in a relatively short time span with limited funds. When Kelly Johnson retired from Lockheed, he left a legacy of success and the following 14 rules for Skunk Works operations (Ref. 2, pp. 4–8):

1. The head of the Skunk Works must have practically complete control of his program in all aspects.
2. The military and industry must provide strong but small project offices.
3. Restrict the number of people having any connection with the project.
4. Simple drawing and release systems with great flexibility for making changes must be provided.
5. There must be a minimum number of reports required, but important work must be recorded thoroughly.
6. There must be a monthly cost review covering not only what has been spent and committed, but also projected costs to the conclusion of the program.
7. The contractor must be delegated and must assume more than normal responsibility to get good vendor bids for subcontracts.
8. The Skunk Works' inspection system, which is approved by both the Air Force and Navy, should be used on all projects.
9. The contractor must be delegated the authority to test his final product in flight.
10. The specifications applying to the hardware must be agreed to in advance of contracting.
11. Funding a program must be timely so that the contractor doesn't have to keep running to the bank to support government projects.
12. There must be mutual trust between the military project organization and the contractor.
13. Access by outsiders to the project and its personnel must be strictly controlled by appropriate security measures.

14. Because only a few people will be used in engineering and most other areas, you must provide ways to reward good performance by pay not based on the number of personnel supervised.

Ben Rich became a senior engineer at Lockheed's advanced programs in 1950 and eventually took over as Executive Vice President and General Manager of Lockheed's Skunk Works in 1975. He continued the Skunk Works tradition by building several key aircraft, including the TR-1, YF-22, and F-117A. In 1989 Ben Rich and his Skunk Works team won the Collier Trophy for production and deployment of the F-117A stealth fighter. He also developed his own management philosophies concerning Skunk Works organizations (Ref. 2, pp. 8–9):

1. Leadership: First you need a strong leader—not a nitpicker, not someone who sticks his nose into everyone's business, and not a committee. You need one leader. Someone who sees the big picture, who isn't focusing on day-to-day details. He has to be the ultimate decision maker, but he can't do the research himself, he needs a dependable staff. His job is to weigh the alternatives and pick the best solution.
2. Teamwork: The leader must have full responsibility for selecting his own team.
3. Delegate: A leader must delegate both authority and responsibility to allow the team to get the job done with minimum interference.
4. Manage by charisma: The leader must give his team goals and objectives, but not all the step-by-step procedures to follow. People don't want to work for a leader who is directing every detail of a project.
5. Be practical: A leader shouldn't take too big of a step. He should walk before he runs. He should expect mistakes and correct them as quickly as possible.
6. Schedule tight: Time is money! So take no more time than is necessary to complete the project. In fact, give yourself less time than what you think you'll actually need. You'll miss a few deadlines, but on balance you'll come out ahead.
7. Demand and expect results.
8. Be ethical: No one is smart enough to lie.
9. Rewards and punishment: Reward those who do good work and punish those who don't. Take the time to say thank you. Give people credit for their successes. Don't tolerate mediocrity and nonperformance, give them a chance; if they repeat—get rid of them.
10. Enjoy your work! Take vacations, and if you work hard—play hard!

Skunk Works principles work, and they have been working for years (maybe that's why they call it—Skunk *Works*). In an attempt to emulate a Skunk Works organization, the NASP JPO continues to incorporate streamlined techniques into its daily operations. The JPO purposely was kept small to maximize productivity. All JPO members are keenly aware of the goals and objectives of the program and empowered to directly act to meet these goals. Communication is a key ingredient of the operation and teamwork (both within the government and with the contractor) is expected.

As the program moves closer to the X-30 airplane development phase, preparations are under way to transition the program to a true Skunk Works operation. The program office will be incorporated with the national team, and key decision authority will be vested with that joint organization. Shared management and streamlined functional matrixing will be used to minimize costs and improve efficiency. Most of the program activity (both contractor and government) will occur at one site, and a minimal amount of reporting will be required. Although all of the details of the X-30 program management have not been finalized, the NASP team is committed to utilizing the principles of Skunk Works. Besides being practical, these principles will allow this challenging undertaking to succeed with the shortest schedule and the minimum cost.

SUMMARY

This chapter defined traditional military program management as the classical, time-tested management practices that conventional programs have followed in the past. However, when a program cannot conform to conventional ways of doing business, or is unrealistically constrained by cost or schedule, or must advance the state of the art, a nontraditional approach to program management may be required.

Four nontraditional alternatives were discussed: joint government programs, consortia, contractor teaming, and Skunk Works. Joint government programs should be considered whenever more than one service or agency stands to benefit from an undertaking significantly. A key criterion to assess joint government applicability is the motivation and willingness to share resources to attain a mutually important objective. Consortia should be used if all parties stand to gain substantially from the process and not give up much to obtain the benefits. Consortia encourage mutual collaboration on high-payoff activities that are too risky for a single company to pursue, and they leverage each participant's capabilities while maintaining their *separate identity.* Contractor teaming can take the form of partnerships, joint ventures, or hybrids thereof. Like consortia, teaming also leverages the unique capabilities of each participant, but a *joint team entity* is formed to synergistically pursue a new initiative. Skunk Works is a high-performance management philosophy which focuses on the advantages of empowered teams by creating a culture which encourages smart ways of doing business. Skunk Works performs best for small, time-constrained programs that seek high payoff through a short production run.

Which management approach works best depends on the specific situation. The perceived threat, potential payoff, utility, and who the final customer will be influences how the government program office will organize, and the level of support it will receive. Potential payoffs must be balanced against external constraints, the resources available, and the risk involved (cost, schedule, and performance). Global and domestic market share, and the degree of technical readiness determine whether a contractor is motivated to collaborate against competitive threats. The level of collaboration is driven by each contractor's perceived competitive advantage, and their level of confidence to independently achieve the effort.

It has often been said, the best solution is achieved when all players win. Identifying each player's win scenario establishes the initial framework for constructing a successful acquisition and management approach. When facing a difficult challenge, be bold, smart, and creative. If a traditional management approach works best, use it. If not, seek a nontraditional approach to take maximum advantage of a team's synergistic benefits and its ability to utilize a large resource base.

A CASE STUDY: THE NATIONAL AERO-SPACE PLANE PROGRAM

Introduction

The National Aero-Space Plane is a look into the future. It is a vision of the ultimate airplane, one capable of flying at speeds greater than 17,000 miles per hour, 25 times the speed of sound (Mach 25). It is the attainment of a vehicle that can routinely fly from Earth to space and back, from conventional airfields, in affordable ways. It represents the achievement of major technological breakthroughs that will have an enormous impact on the future growth of this nation. Most of all, it is a projection of America at its best, at its boldest, at its most creative. It is more than a national aircraft development program, more than the synergy of revolutionary technologies, more than a capability that may change the way we move through the world and the aerospace around it. NASP is a revolutionary technical, managerial, and programmatic concept; it is a possibility of what can be in America.

The NASP program can be described from a technological and programmatic perspective. In each case, NASP has departed from the traditional evolutionary path. To achieve the NASP vision, innovative and revolutionary approaches are required. The technical challenges require the synergism of several major technology breakthroughs. The programmatic challenges require a fundamental change in the development, management, and implementation of this strategic high-tech program.

The Technical Challenge

The goal of the NASP program is to develop and demonstrate the feasibility of horizontal takeoff and landing aircraft that use conventional airfields; accelerate to hypersonic speeds; achieve orbit in a single stage; deliver useful payloads to space; return to Earth with propulsive capability; and have the operability, flexibility, supportability, and economic potential of airplanes. To achieve this goal, technology must be developed and demonstrated that is clearly a quantum leap from the current approaches being used in today's aircraft and spacecraft.

The NASP demonstration aircraft, the X-30 (Fig. 15.1), will reach speeds eight times faster than any other air-breathing aircraft. As it flies through the atmosphere from subsonic speeds to orbital velocities (Mach 25), its structure will be subjected to average temperatures well beyond anything ever achieved in aircraft. While rocket-powered space vehicles, like the Space Shuttle, minimize their trajectory through the atmosphere, the X-30 will linger in the atmosphere to use the air as the oxidizer for its ramjet and scramjet engines. The NASP aircraft must use liquid or slush hydrogen as its fuel, which present new challenges in aircraft fueling, storage, and fuel management. To survive the thermal and aerodynamic environment, the X-30 will be fabricated from a combination of highly advanced materials: refractory composites, metal matrix composites, and extremely high-temperature superalloys. Because no large-scale test facilities exist to validate aerodynamic and propulsion operation above

FIGURE 15.1 NASP demonstration aircraft X-30.

Mach 8, the design and operability of NASP aircraft must be carried out in "numerical wind tunnels" that use supercomputer-aided computational fluid dynamics (CFD). Propulsion systems based on subsonic and supersonic ramjet combustion will propel the X-30, and although these types of engines have been investigated in laboratories, there has been no significant flight testing. In the areas of aerodynamic design, flight control, thermal management, cooling systems, man-machine interface, and many other subsystems, NASP requires a major increase in capability in order to reach its objectives. The technical and system integration necessary to achieve single-stage-to-orbit aircraft operations will be more difficult than any yet attempted and will require a fundamentally new approach to aircraft design. In essence, NASP depends not on a single advance in technology, but on the synergism of breakthroughs in a number of major technical areas associated with aerospace vehicles (Fig. 15.2).

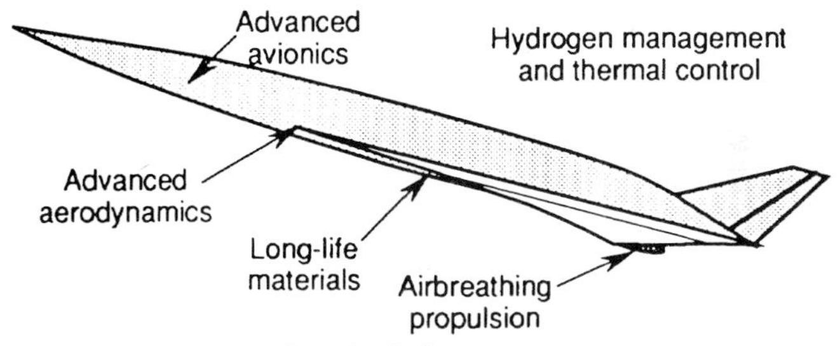

FIGURE 15.2 Key NASP technologies.

The Technical Response

The NASP program has been carefully orchestrated to achieve the technological advances and integration needed to attain the goals of the X-30. There are five key areas of technology that are the focus of the NASP development program: engines, aerodynamics, airframe-propulsion integration, materials, and subsystems. Significant development activities are under way and major advances have resulted. In the first three areas, approaches that were initiated at the start of the NASP program in 1986 are beginning to pay off. The work in materials and subsystems development was substantially accelerated in 1988, and there have been major breakthroughs since then in these critical technologies.

The feasibility and operability of the high-speed propulsion system are the key developments required in the NASP program, and those activities are receiving the greatest attention. The basic engine approach for NASP is a combined ramjet-scramjet air-breathing propulsion system (Fig. 15.3), which will provide high-efficiency thrust for much of the region between takeoff and orbit. Various low-speed systems and the use of rocket systems at very high Mach numbers and for orbital insertion are being investigated. Several key materials were identified as being critical to the feasibility of an air-breathing, single-stage-to-orbit aerospace vehicle. Because of the high-temperature, high-strength requirements of the NASP airframe and engine systems, most of the interesting configurations used combinations of high-temperature titanium aluminum alloys, carbon-carbon or ceramic composites, metal matrix composites, high-creep-strength materials, and high-conductivity composites. Although the development of these material systems has been under way for several decades, it was determined that the progress being made was insufficient to meet the requirements of the NASP program. Development of all five material systems was acceler-

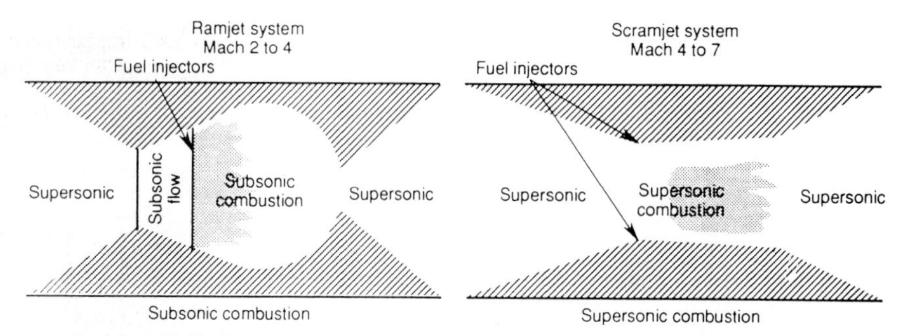

FIGURE 15.3 Schematic diagrams of ramjet and scramjet systems.

ated through the formation of a national materials consortium, focused on the five material types, with a greatly enhanced resource commitment (Fig. 15.4). The consortium fabricated, characterized, tested, and developed materials in each category, and significant progress had been achieved in the areas of super alpha-2 titanium aluminide, titanium-aluminide-silicon carbide metal matrix composites, and oxidation resistance coated carbon-carbon composites.

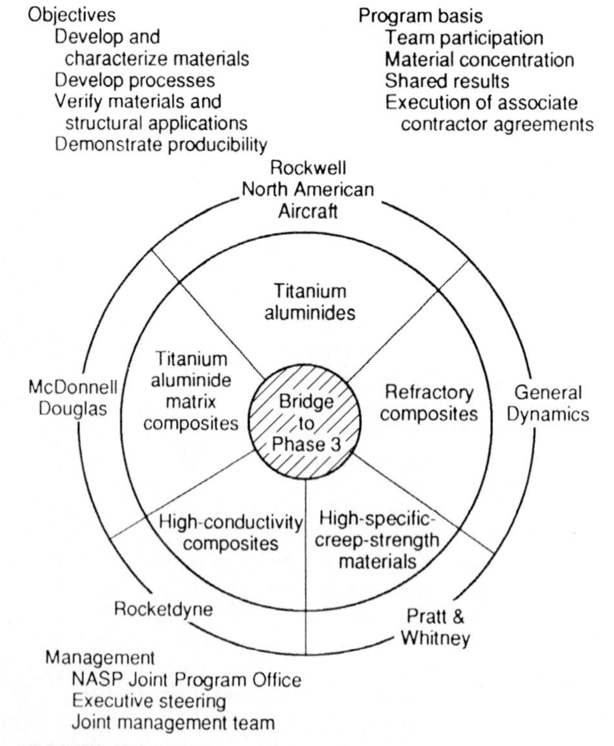

FIGURE 15.4 NASP materials and structures maturation program reduces risks and uncertainties.

The aerodynamics of hypersonic aircraft and aerospace vehicles has been the subject of considerable government, university, and industry attention for the past 30 years. Much is known about this subject, and the NASP program is taking advantage of the wealth of information available in the United States. The aerodynamic requirements of NASP vehicles, however, are extremely stressful and sensitive to small changes in vehicle configuration and performance. In addition, the specific flight regime of the X-30 has not been examined extensively through ground experimentation or flight testing. Because the X-30 itself will examine the aerodynamics of air-breathing aerospace vehicles, effort on the current development program has focused on developing detailed CFD models (Fig. 15.5) and on verifying them using several experimental tests. A massive CFD effort, using a significant fraction of the total U.S. supercomputer capability, is under way to develop experimentally valid models to predict the inlet, combustion, and nozzle operation of the NASP. Three-dimensional full Navier-Stokes codes that account for real-gas effects, chemical kinetics, and turbulent flow are being refined using shock tunnel, wind tunnel, and archival flight data to predict the critical NASP aerodynamic parameters.

Because the airframe and engine systems development for the NASP has been pursued by separate organizations, the level of airframe-engine integration required of hypersonic aircraft necessitated a major emphasis in this area (Fig. 15.6). Since the program began, this integration has commanded great attention and has received an enormous amount of government and contractor resources.

Although the previous four areas have demanded most of the resources of the NASP program, every subsystem of a hypersonic aircraft will be developed to the point where it will support the testing of an experimental vehicle. Major efforts are required to develop slush and liquid hydrogen systems, cryogenic tankage, fuel deliv-

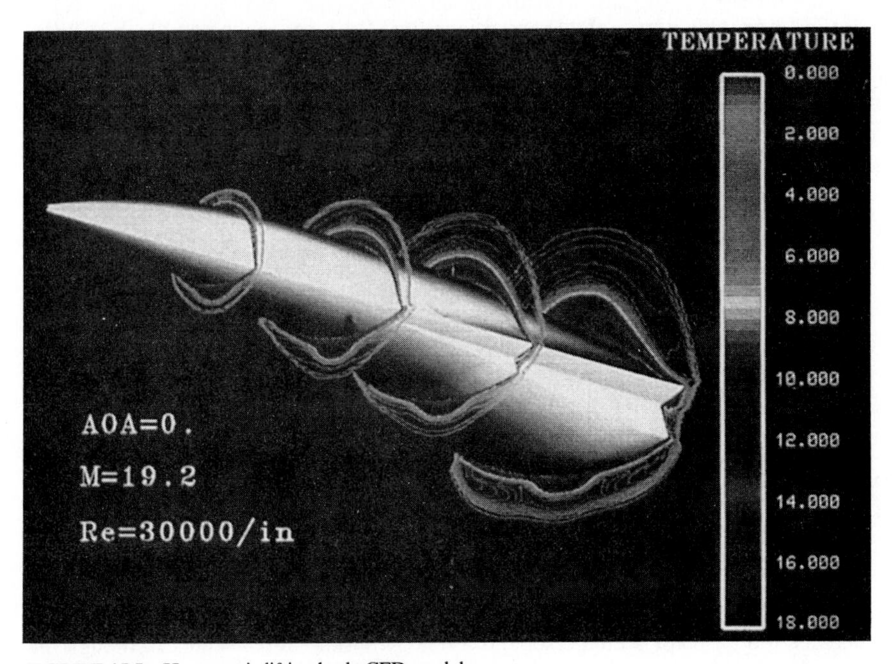

FIGURE 15.5 Hypersonic lifting body CFD models.

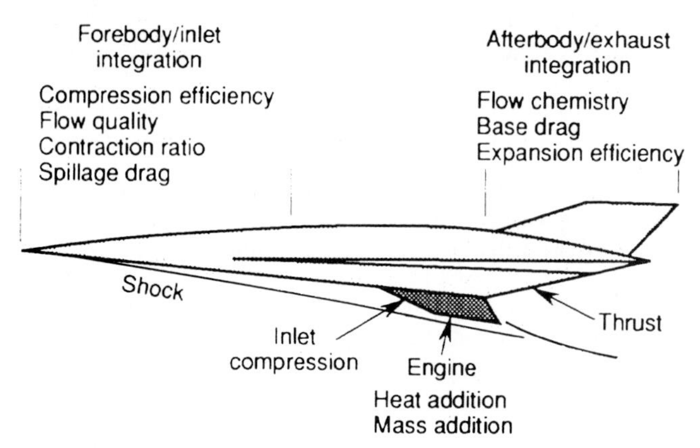

FIGURE 15.6 Airframe-engine integration.

ery systems, heat exchangers and turbopumps, avionics and cockpit systems, flight controls, and the instrumentation required to conduct the X-30 program.

The Programmatic Challenge

It has been over 80 years since humans first flew and over 40 years since aircraft flew supersonically. For the past 40 years, airplane speeds have advanced from Mach 1 to Mach 2, with only a few notable exceptions: the SR-71 was capable of Mach 3+ flight, and the rocket-powered X-15 achieved speeds around Mach 6. In general, however, it has taken us 80 years to go from Mach 0 to Mach 2. In contrast, NASP is attempting to increase the speed range of air-breathing airplanes to Mach 25 by means of a 10-year development and demonstration program. During the 1950s and 1960s much activity was aimed at the exploration of hypersonic vehicles and their possible configurations. Wind tunnels, shock tunnels, and experimental aircraft were fabricated and used to examine the key parameters of hypersonic flights. Unfortunately that activity prematurely ended in the 1960s, and the development of hypersonic aircraft virtually ceased until the NASP program began. A few government researchers and even fewer university and industry scientists kept the flame alive during those years, but progress in hypersonics has been extremely slow. Although research in the critical areas of materials, CFD, and combustion has progressed because of other demands, the national capability at the beginning of the NASP program was extremely limited, dispersed, and disorganized.

To conduct a challenging program such as NASP, an extensive, competent, well-integrated, and focused national team from industry, government, and academia had to be developed. A prime task of the development phase of the NASP program is not only to bring the key technologies to a point that will allow an X-30 airplane, but to form the team required to do the job. In 1990 there were over 5000 professionals working on the NASP program, as contrasted to 250 in 1985 (Fig. 15.7). Although the principal goal of the NASP program is to demonstrate an aerospace vehicle capable of aircraftlike operations while achieving single stage to orbit, the program has also become the basis for all hypersonic technology in the United States. Although the program must be focused on the goals of the NASP X-30 demonstrator, it must also

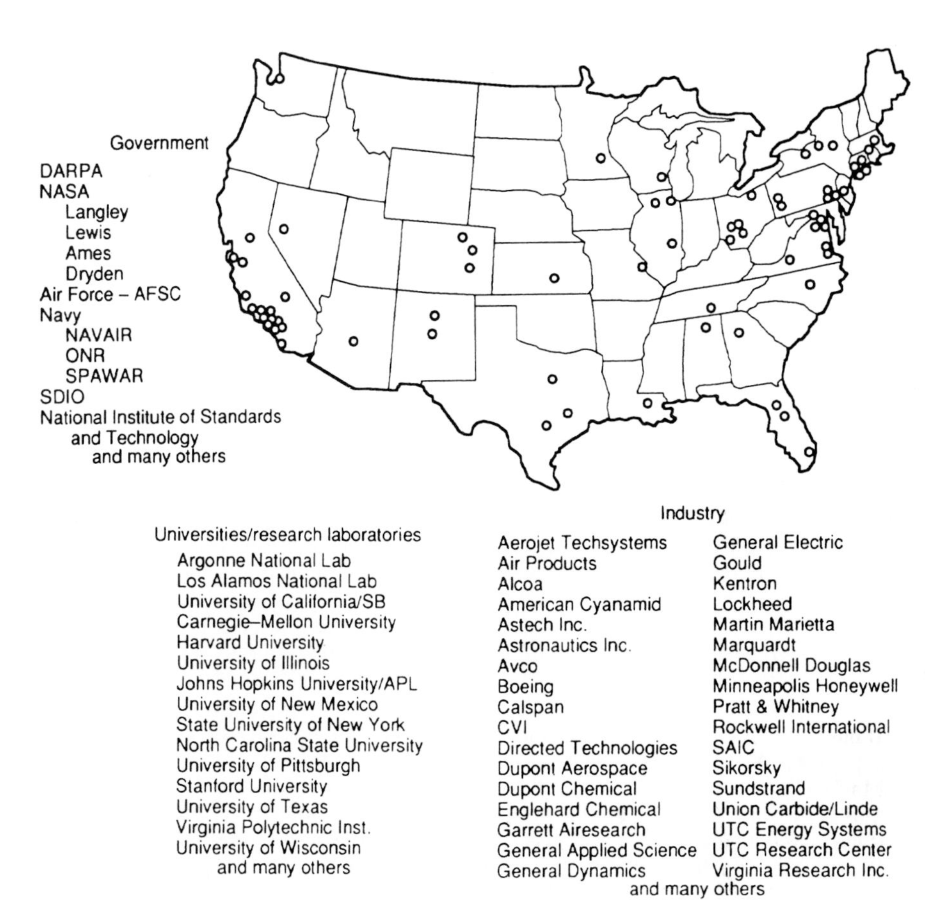

FIGURE 15.7 Participants in NASP program.

generate the technology that will allow a broad basis for future hypersonic vehicles and derivatives of the NASP demonstrator.

The Programmatic Response

The management of the NASP program has emphasized collaborative, participative approaches because the goal of the program is to develop a national team that can lead us into the aerospace era of the 21st century (Fig. 15.8). From the onset, the government laboratories and centers have been an integral part of the NASP team. Much of the initial expertise on the NASP program resided with government researchers, who will continue to play vital roles as consultants, contributors, and evaluators for the program. When the program began, only a few industrial companies and academic institutions had substantial efforts in the hypersonic area. These few not only had to be supplemented, but some of the leading aerospace companies had to be added to the field. Five major airframe companies (General Dynamics, Rock-

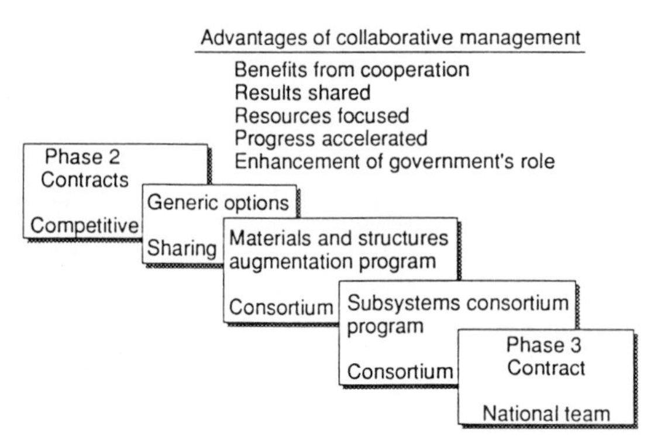

FIGURE 15.8 Collaborative Strategy in NASP program.

well International, McDonnell-Douglas, Lockheed, and Boeing) and three leading engine companies (Pratt & Whitney, General Electric, and Rocketdyne) received firm fixed-price contracts and were all heavily involved in the initial stages of the effort. Firm fixed-price contracts are rarely used for research and development efforts, but the JPO realized that intense competition existed between the contractors to become the industrial leader for developing future hypersonic aircraft. Whichever contractors were selected to design, build, and test the X-30 would have a virtual monopoly on the research data and a tremendous lead to design future hypersonic vehicles. In 1987 McDonnell-Douglas, General Dynamics, Rockwell, Pratt & Whitney, and Rocketdyne were selected to continue with the program.

Since the beginning of the NASP program, significant efforts were made to manage the program using innovative management concepts. Joint government-industry decision making has been a norm for the program. Consortia formation and generic government-industry technology development have been fostered, and very strong associate contractor agreements between all appropriate parties have been effected. In 1988 a materials consortium of the five major companies was formed to accelerate the development of NASP airframe and engine materials. The program was a complete success, with major materials advances and excellent cooperation between the companies. The success of the materials consortium, coupled with the need to develop a strong national industrial base for future hypersonic aerospace systems development, led in late 1989 to the consideration of a single NASP team. Progress and corporate contributions by the three airframe companies and both engine companies had been excellent, and a single team comprised of all five leading contractors seemed highly desirable. Although each company had pursued its own unique configuration approach, a national NASP team would allow a single synergistic configuration to emerge and all development efforts could focus on that concept. Another major advantage of a single team would be the guarantee of a broad industrial competitive base in the United States for future operational hypersonic and aerospace vehicles. Early attempts to foster such a national team paid off when the program schedule was extended in 1989 by two and a half years. With increased time for research and development and a spending rate which was essentially constant from 1988 through 1993, the idea of a single national NASP team took hold. In late 1989 the NASP JPO began procedures to form such a team. The five contractors were most responsive and agreed to form a joint venture partnership through a cost-plus-fixed-fee letter contract. This novel programmatic response was highly beneficial to the successful execution of the NASP phase 2 R&D program.

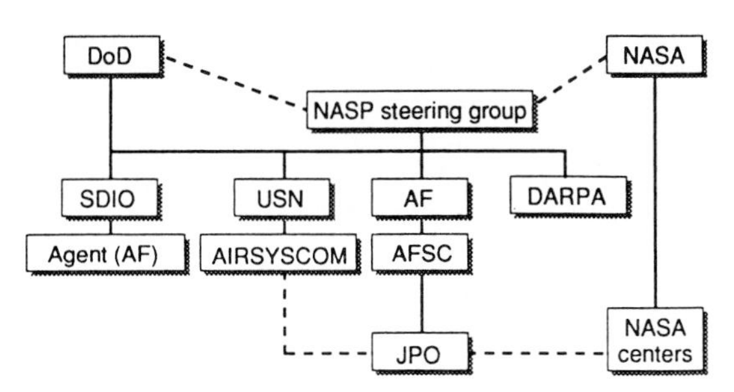

FIGURE 15.9 Management structure of NASP program.

On the government side, innovative and integrated management has been the program standard. The NASP program actually began in 1981 when Mr. Anthony duPont, an aerospace engineer, convinced Dr. Robert Williams at DARPA that an air-breathing hypersonic aircraft could fly all the way into a low Earth orbit and return. Initial research supported duPont's claims, and in 1984 DARPA conducted a larger study called "Copper Canyon" to validate duPont's concept. The Copper Canyon study agreed with duPont's concept and identified key technological challenges to be overcome. Realizing that a long-term financial commitment was required to achieve breakthroughs in multiple unproven high-risk technologies, DARPA enlisted the participation from potential beneficiaries through a unique five-part memorandum of understanding (MOU) between the Air Force, Navy, SDIO, NASA, and DARPA. Each agency and service branch pledged funding, and with the exception of SDIO, assigned people to create a single government program office, the NASP Joint Program Office at Wright-Patterson Air Force Base. The five-part MOU also created a NASP steering group comprised of high-level DoD, NASA, as well as other experts on aerospace, defense, and science (Fig. 15.9). The steering group advised and approved program direction and served as a source of advocacy to Congress and the administration.

To minimize bureaucratic aspects of a government-funded program, the NASP JPO maintained a manning level of approximately 75 people, used Skunk Works streamlined principles, and operated under specialized management practices authorized by Air Force Regulation 800-29. Several hundred government technical experts outside the JPO are being utilized principally to assist the contractors, rather than to evaluate them. About 20 percent of the program resources has been, and will be, spent in government R&D efforts. The JPO's principal role has been to focus the efforts of the thousands of personnel in hundreds of companies and universities toward the program goals. Executive direction is provided by a steering group of senior-level DoD and NASA officials, which meets biannually to guide the program. The JPO is manned with program and technical managers from the Air Force, Navy, and NASA and operates as a unified government organization with a strong total quality and high-performing team culture. The common vision for both government and industry partners is the X-30 and the experimental demonstration of the aerospace plane. It is this vision which drives the program and allows this unique programmatic response to be successful.

The NASP program is an experiment that tests the ability of the United States to work together to achieve revolutionary technology development and to effectively translate that technology into viable products. Because it is succeeding in meeting that goal, it has become an example of government-industry collaboration, effective technology utilization, long-range visioning, and focused national commitment.

These are the very principles that were at the core of the outstanding progress the nation achieved earlier in this century. They are the same principles that have been used so successfully by our economic competitors during the latter part of this century to capture a significant share of the markets and capabilities that once were ours exclusively. These are the principles for our nation's future growth, and NASP is the foundation for our aerospace leadership in the 21st century.

FUTURE TRENDS

A cultural change to the American work ethic. More value and utility per employee

Global alliances in certain niche markets (electronics, materials, subsystems)

New industrial policies/legalities to encourage national teaming of U.S. companies

Must ingrain a cultural change of the U.S. industrial base to take a new look at global competition

New streamlined acquisition processes

Higher reliance on subcontracting

Minorities assuming higher corporate responsibilities

Telecommunications at each person's workstation

Smaller government program office

Completely integrate rapid prototyping into the design process

Higher utilization of computational fluid dynamics in the design process

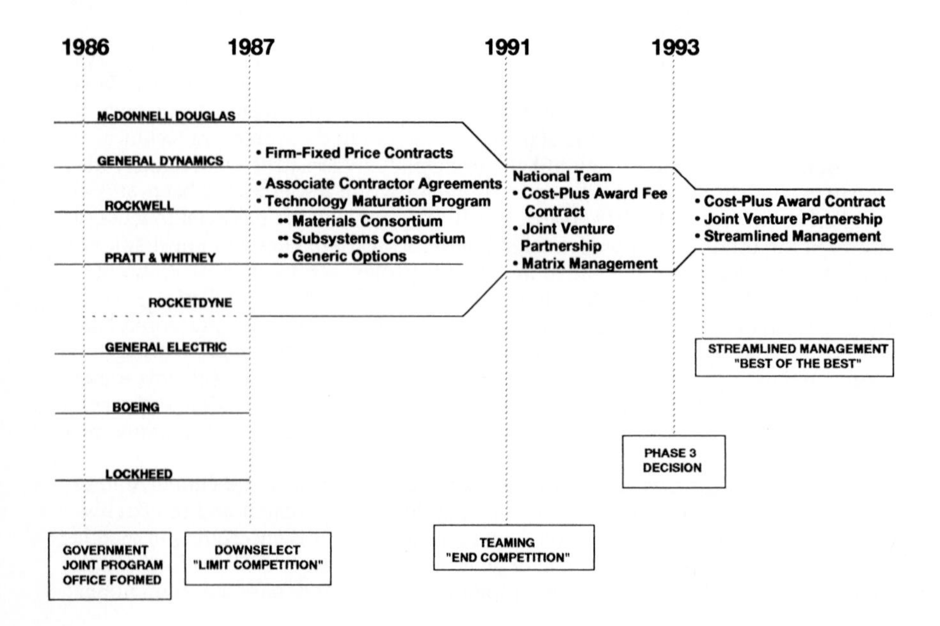

REFERENCES

1. "SEMATECH 1991 Update," SEMATECH Corp, Austin, Tex., Mar. 4, 1991, p. 9.
2. "Wright Brothers Leadership in Aeronautics: The Skunk Works' Management Style—It's No Secret," American Institute of Aeronautics and Astronautics 1988 Convention, Atlanta, Ga., Sept. 7–9, 1988.

CHAPTER 16

THE JOINT PROJECT OFFICE FOR UNMANNED GROUND VEHICLES: A CASE STUDY OF JOINT LEADERSHIP FOR THE FUTURE

Jeffrey D. Cerny

Jeffrey D. Cerny was the original Deputy Project Manager for the UGV JPO, from 1990 to 1992. In addition, he is an adjunct professor with the Florida Institute of Technology graduate programs. Upon graduation from West Point, Dr. Cerny served with the U.S. Army as a field artillery officer. He has commanded an armored artillery unit and has received Army qualifications in airborne, ranger, and pathfinder operations. He completed the army engineering internship in production engineering at Red River Army Depot. His other governmental assignments were at the Night Vision and Electro Optics Laboratory and the Research, Development and Engineering Center, U.S. Army Missile Command. Dr. Cerny has served as a corporate officer in General Dynamics' Washington operations. His duties as both a director and a manager have included support of the Land Systems Division with its M1 tank operations and the Defense Initiatives Organization with its new business development efforts. Dr. Cerny is currently a Visioneer with the Army Missile Laboratory's Advanced Systems Concepts Office.

VISION OF JOINTNESS

Project Leadership

Norm Augustine notes in this handbook's introduction that

> The job of project manager is among the most important and most difficult assignments in America's peacetime military.

Upon entering the 21st century, we will experience an increasing trend toward polarization, multiplicity, and change.[1–4] Because of this trend, Norm Augustine's observations in the foreword should be modified to state that the job of joint project leadership will be the most important and by far the most difficult function in America's changing commercial and military environments.

Project leadership is about change.[5,6] It involves the joint forces of implementing change as well as harnessing change. Both of these forces are constantly at odds with each other. Project leaders and their offices are the required balancing agents responsible for accomplishing change successfully. The project leader must not look only at his or her mission in light of fielding a particular item. Instead the project must be led as a process which is far larger and more complicated than just the item. In essence, the item is managed; the total process is led.

Concept of Jointness

The concept of jointness is the coordinated attempt to combat polarity and multiplicity within the environment. Jointness, as a management culture, is increasing. The world is experiencing a collision of dwindling industrial-age resources and increasing diversity of requirements and information. This trend is causing several changes. Today we use the word "unique" to describe joint projects. As the number of joint programs, ventures, and projects increases, we will come to refer to the "single-service" project manager (PM) as unique. The degree of jointness for a project management office (PMO) is displayed in Table 16.1.[7] It can range from the unique single-service project with no jointness to a fully integrated joint project office (JPO). The varying degrees of jointness are related by the amount of coordination established between the different individual services.

The question for all services, agencies, and activities is, when do these factors constitute the need for initiation of a joint project?

TABLE 16.1 Degree of Jointness

	Category	Characteristics
S-1	Single-service* PMO	Unique single-service program with no interest by others
S-2	Single-service PMO with product interest	Single-service program with interest by others
S-3	Single-service PMO with point of contacts	Single-service program with designated POC for liaison
S-4	Single-service PMO with on-site liaison	Single-service program with assignment of full-time liaison
S-5	Single-service PMO with senior representative	Single-service program with no formal coordination liaison assigned to PM
S-6	Fully integrated JPO	Multiservice program, staffed by services with formal coordination

*Single service implies one branch of military service.

Source: Adapted from Table 1-1, Joint Program Categories and Characteristics, Ref. 7, pp. 1–2.

INITIATION RATIONALE

Marriage

> ... the word joint does not necessarily mean togetherness. Most programs are the result of forced marriages ... Clearly joint programs require the very finest in management skills, particularly from the Program Manager.[8]

Joint projects, like commercial joint ventures, are marriages, whether they are forced or mutually agreed upon. Obviously those that are agreed upon will have a higher probability of success. An agreed-upon marriage reduces external conflict, provides a common vision, and assists in defining roles or mission. In today's changing environment we must recognize that the project is a subsystem of this environment. As such, the joint PM with his or her skills should strive to develop and enrich enduring relationships with "all" stakeholders of the joint project. The project's stakeholders, who are more than shareholders, include project personnel, customers, competitors, local community, industry, and governmental agencies. Each of these entities has a stake in the project's operation. All stakeholders, whether favorable or unfavorable, provide a reason for the project's existence. Recognition of these stakeholder relationships is the first step in the evaluation of joint project initiation. This recognition of stakeholder relationships forms the foundation for understanding the change process (Figure 16.1).[1,9] The next step is to weigh the apparent advantage offered by a joint project. This second step can only be fully evaluated if all the stakeholders have been identified and analyzed.

Advantages. The joint project is initiated because of a perceived advantage by the stakeholders to do so. During this second step, this advantage is analyzed within the

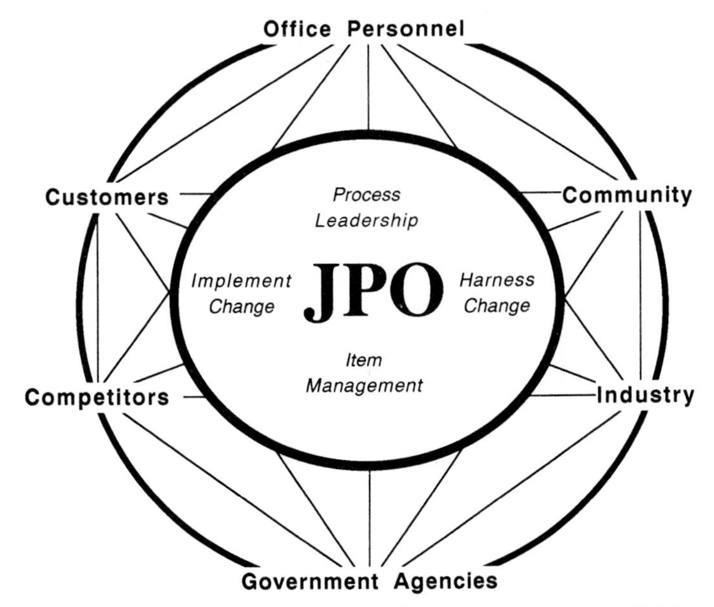

FIGURE 16.1 Joint project office stakeholders and culture: first step of initiation evaluation. (*Adapted from Cleland,[9] p. 161, and Hellriegel,[1] p. 165.*)

political, economical, and operational environments. Joint project initiation is contrasted and balanced to the ever-present uncertainties, costs, and risks of a changing environment. As outlined in Fig. 16.2,[4,10] each of the three environments provides its own pros and cons to the initiation of a joint project. In the economical environ, the owner stakeholders wish to reduce their development costs and individual financial risks by sharing them during the change process. In addition to this economic sharing, there is the desire to increase interoperability and capabilities within the mutual operational environs of the stakeholders. These two pros are the most obvious advantages of a joint effort. However, because of the marriage mechanics there is usually the less than obvious higher level of corporate commitment, or at least awareness, by the vesting parties. This higher-level vesting in the political environ helps in heightening the resolve of purpose during upcoming increased conflicts for joint efforts.

In essence, if the joint project has synchronized its resources and energies, it can

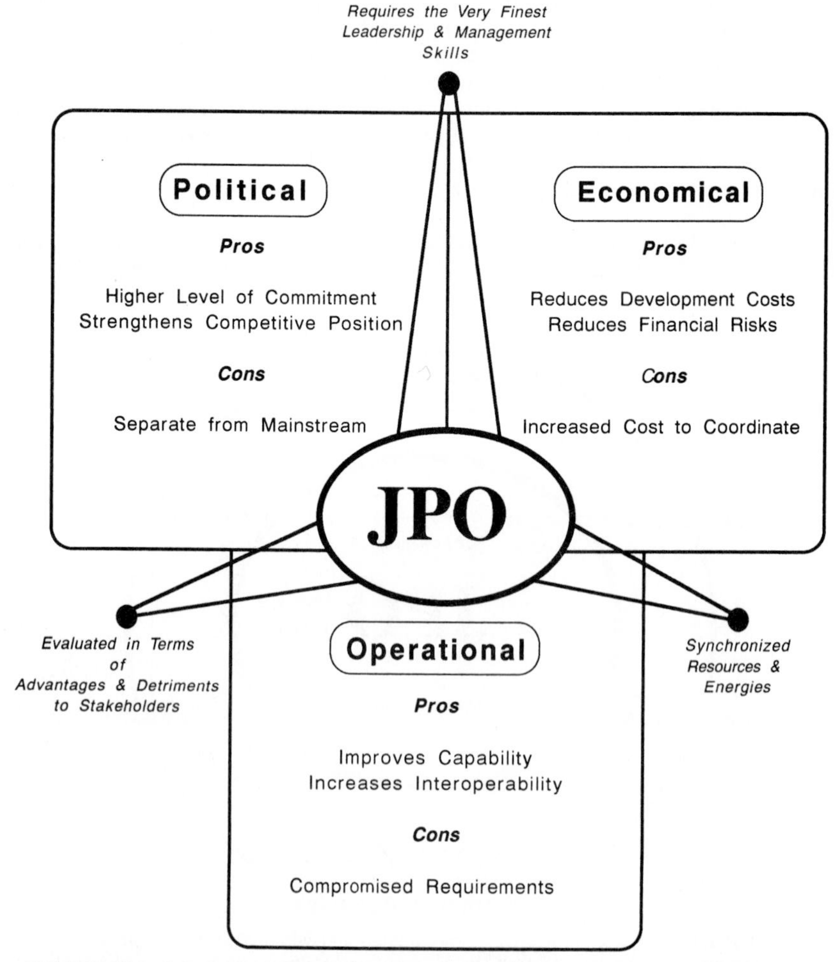

FIGURE 16.2 Joint project office environmental considerations: second step of initiation evaluation. (*Adapted from Kissler,[4] p. 250; Avots,[10] p. 41; and Appendix A.*)

strengthen the competitive advantage of its owner stakeholders. Ironically, this synchronization process, if not compensated, can provide often insurmountable obstacles or detriments to joint projects. It is this synchronization that requires the additional managerial and leadership skills of all joint PMs.

Detriments. To the detriment of joint efforts there is the burden of coordination among vesting services. This burden includes the increased cost to coordinate, in both time and resources, the potential of compromising the individual customer requirements, and the potential alienation of the PM Office because of the separate and unique structures of the core organizations or vesting services. It is because of these burdens that joint projects must be led by personnel with the very finest management and leadership skills. Each of these burdens is a major obstacle in its own right. Joint efforts do not allow for separation of these cons.

Inspiration and Persistence. To have project initiation, the pros must outweigh the cons. It is the inspired project leader's job to weigh and weight this evaluation. To continue past initiation, the project leader must ensure that the joint project has professionals that "outperform" in the political, economical, and operational environments of today's changing world. It is the persistence of the project's professionals, including the PM, that has to prevail during this more complicated journey through the world of jointness.

The successful joint leaders must recognize that they have more stakeholders than normal, have unique advantages over others, and have some detriments that must be overcome. All of these must be synchronized, developed, and anticipated. Joint projects give leaders the opportunity to be inspired and remain persistent. To simplify the understanding of this jointness phenomenon we can examine the old adage that you are not a true parent (or project leader) until you have at least two children with their own needs and individual wants.

Observation on Jointness

In this changing environment there has been a trend toward joint efforts. The most recent notable joint effort is that of the two major computer companies IBM and Apple.[11] Their partnership announcement on July 3, 1991, establishes a capstone of joint ventures. These two long-time rivals have obviously evaluated the pros and cons within their respective environments. More importantly, they have identified their stakeholders and analyzed the relationships thereto. An encompassing observation is that

> Joint ventures are inherently fragile. Each partner is pursuing a variety of objectives in its own dynamic setting. Consequently, change may occur in benefits sought, resources available or strategic objective of either partner.[12]

This statement effectively summarizes the fragile foundation of all joint ventures or projects. As stated earlier, project leadership is about change. Joint project leadership is about the compounded change of two or more partners. The leaders must continually strive to keep the jointness, objectives, and benefits in sync. All of this must be done within the allocated resources "ventured" to the effort.[13,14,3]

To be successful, a joint PM must:

- Understand and anticipate the dynamics of change
- Be a true leader of *all* stakeholders

- Overcome the detriments of jointness by synchronizing (this is a competitive advantage of joint projects)

CASE STUDY OF JOINTNESS: PP430 CASE

Situation

Four months ago you took over the U.S. Marine Corps (USMC) unmanned combat system (UCS) program. Your predecessor had eased the program into existence by getting approval of an initial statement of requirement (ISOR). Since the initial cost estimates put the program at the ACAT III level, there was tacit agreement by the user representative and the systems command that the ISOR's approval constituted passing the Marine Corps program decision meeting (MCPDM) 0, or milestone 0. Since taking over the program, you have concentrated on helping the USMC Warfighting Center translate the ISOR information into an operational requirements document (ORD), as directed by Department of Defense (DoD) Directive 5000.1. The ORD will be the basis for completing the other documents required for a MCPDM I decision. You have been working with the Office of Naval Research, surveying the technology available, and now have a summary of its work to date.

You are right on track for a MCPDM I in six months when the congressional liaison office at headquarters calls with the following news. The Senate Appropriations Committee has put language in the current DoD appropriations bill requiring DoD to "consolidate robotics programs into joint programs or joint development efforts." Your liaison officer is sure that the language will remain in the legislation when it is signed into the Defense Appropriations Act. You are about to go from a nonmajor program in an "umbrella" PMD to a joint major defense acquisition program.

The news has barely had time to sink in when you are given the following direction from your SYSCOM's chief of staff. As our program executive officer (PEO), the general has to brief the Joint Requirement Oversight Committee (JROC) on the Marine Corps' potential to be the lead service for this program. He really wants to be the lead service, even though the program will be moved outside the SYSCOM decision chain. He figures the best selling point will be a strong acquisition strategy. He needs you to put one together based on your UCS program and brief him on it next week. Be sure you cover all the points in Annex C of the integrated program summary (IPS) format in DoD Instruction 5000.2. Better be ready to spin him up on the pros and cons of joint programs, too.

Assignment

Given the ORD for the Marine Corps' UCS and the Office of Naval Research's technical survey, develop a brief to meet your PEO's needs. The brief should be 10 to 12 minutes long and cover the following broad areas of the acquisition strategy report on the IPS:

Program structure

Overview of the acquisition approach

Streamlining

Sources

Competition

Contract types
Use of fixed-price contracts
Major tradeoffs

Include in your brief a summary of the advantages and challenges of joint programs and how they may affect this program.

PP430 Observation

What has just been read is the situation and assignment required of some students at the Defense System Management College, Ft. Belvoir, Va. This PP430 lesson is derived from the actual JPO for Unmanned Ground Vehicles, Redstone Arsenal, Ala. According to a March 1991 communique to the JPO, the DSMC had finished a student section and the case has proven successful in portraying joint project concerns to the students. What follows is the actual situation and joint concerns for the Unmanned Ground Vehicle Office.

THE TACTICAL UNMANNED GROUND VEHICLE (TUGV) CASE (CALEB)

Situation

Background. For many years the individual military services of the United States and other U.S. government agencies have pursued technology to support the development of unmanned combat systems. Research continues in independent disciplines, but it has also produced transition projects in the areas of unmanned air vehicles (UAVs) and unmanned ground vehicles (UGVs). With respect to UGVs there have been two distinct efforts taken and two major stakeholders.

The Army had developed the teleoperated mobile all-purpose platform (TMAP) and the Marine Corps and the Navy had developed the ground-air telerobotic systems (GATERS) in the mid 1980s. GATERS was an integration of the airborne remotely operated device (AROD) and the teleoperated vehicle (TOV). The AROD as an aerial platform was separated from the TOV and placed under the Unmanned Air Vehicle JPO upon its formation. At the same time as this realignment, the Army and the Marine Corps corporate management independently investigated the advantages of forming an Unmanned Ground Vehicle JPO. This second step was entertained only after the stakeholders were identified and analyzed. This joint office became official with the signing of a memorandum of agreement (MOA) between the Army and the Marine Corps in April 1990 (see Appendix B). This project initiation was the result of perceived advantages outweighing perceived detriments. The project office was chartered as an S-6 fully integrated JPO. Its mission is to develop and acquire UGV systems and transition technology to fielded systems (see Appendix C). In the summer of 1990, according to the MOA, the Marine Corps office was moved from Quantico, Va., to Redstone Arsenal, Ala. The Army was established as the executive or lead service, with the Marine Corps providing the PM and another field-grade officer. The rest of the office has been staffed from Army personnel, both civilian and military (see Appendix D).

Marine Corps' Teleoperated Vehicle (TOV). The TOV is a large vehicle with extensive capabilities. The Marine Corps' draft requirement in 1983 intended to de-

TELE **O**PERATED **V**EHICLE

USMC RESEARCH DEVELOPMENT & ACQUISITION COMMAND

NAVAL OCEAN SYSTEMS CENTER

NOSC

REMOTE VEHICLE

CONTROL VAN (Exterior)

CONTROL VAN (Interior)

- **HMMWV, M998 PLATFORM**
  - DAY OPS, NIGHT OPS UPGRADE
  - STEREO "VISION", BINAURAL "HEARING"
  - NAVIGATION IN MILITARY GRID COORDINATES
- **MISSION MODULES**
  - RECONNAISSANCE, SURVEILLANCE, TARGET ACQUISITION
  - OFFENSIVE/DEFENSIVE WEAPONS
  - NBC RECONNAISSANCE/MONITORING
  - CONTROL OF SUPPORTING ARMS
- **HMMWV, M999 PLATFORM**
  - NSA C3I SHELTER
  - 3RV/MISSION MODULE CONTROL STATIONS
  - HEAD COUPLED VISION DISPLAY
  - STANDARD USMC COMMUNICATIONS
  - REALTIME, CONTINUOUS MAN-IN-LOOP, CONTROL ARCHITECTURE

TELEOPERATED VEHICLE (TOV)

REMOTE VEHICLE
- HMMWV, M998, 1.25 ton, 4x4 diesel platform
- Gross Veh. Wt: 6400 lbs, Payload: 3400 lbs
- Speed: 55 mph highway, 20 mph off road
- Range: 300 miles manual driving/30 km remote
- Remote driving controls - hydraulic
- Emergency abort - electric
- Mobility head module
  - ± 185° pan, ± 45° tilt. "Quick" head motions
  - Stereo camera "eyes" (COHU 4800 CCD)
  - Binaural "ears" (50-12000 Hz/15-146 dB)
  - 10-minute removal for manual ops
- Navigation
  - Dead reckoning with SATNAV update
  - UTM grid coordinate (6 bit)
  - Grid, speed, heading, altitude display
  - Accuracy: ± 200 M, 2% distance travel
- Electrical power
  - Onboard "quiet" diesel 1.2 kW DC genset
  - Draws fuel from HMMWV tank
  - Primary/auxiliary battery banks
- Command/data link
  - Fiber optic cable
  - Single mode, dispersion shifted
  - Full duplex digital for long distance opt S/N
  - Cmd link 100 Mb/sec, Data link 200 Mb/sec
- Command/control processor
  - CMOS low power, Std bus
  - 46-60 Hz throughput
  - Self-disgnostic
  - C language, ADA waiver
- Mission module interface ports
  - 24 V DC power
  - Analog/TTL binary digital instrumentation
  - RS 170 video
  - RS 422, 56 K baud serial
- "Quiet" mode operation - 4-hour duty cycle
- Endurance: 48-hour mission

The USMC/Naval Ocean Systems Center TeleOperated Vehicle (TOV) provides U.S. Army/U.S. Marine Corps "users" with needed equipment to assess the military worth of "battlefield" robotics and to develop concepts for their employment.

CONTROL VAN (CV)
- HMMWV, M998, 1.25 ton, 4X4 diesel, hvy duty suspension
- GICHNER Corp. Model GMS-1316 Shelter
- 3 RV Control Stations, Section Leader Station
- HALON fire fighting system, low O_2 warning
- MEP 003A mobile generator (120 VAC/60 Hz)

CONTROL STATION (CS)
- All controls, displays in standard 19" rack
- Duplicated HMMWV driving controls
- Helmet mounted remote driving displays
  - Model DH-132, Ground Control Crewman helmet
  - Stereo presentation "Binocular" adjustment
  - 40° field of view, color corrected (525 line 3:4 aspect ratio)
  - Polhemus 3 Space Isotrack head coupling sensors
- Joystick control of Mission Module motion
- Dual B&W COHU Mission Module displays
- Comms (std USMC gear)
  - Command Net - Dual AN/GRA 39B remote units
  - Landline connector port
  - Direct link to remote vehicle
  - Intercom to all stations
- Processor - VME bus

SECTION LEADER STATION
- Video switch access to all CS displays
- Data Acquisition System (DAS)

RECONNAISSANCE, SURVEILLANCE & TARGET ACQUISITION (RSTA) MODULE
- Extendable mast with pan/tilt platform
  - 6-12 ft. above ground, stable at max extension
  - Slow/high speed control <0.5 deg/sec to 30 deg/sec
- Daylight
  - COHU 4800 B&W video camera
  - Fujinon 10:1 zoom lens (40 to 4 degree FOV)
  - Autoiris, bright light limiter
- Night/Smoke
  - AN/TAS 4A thermal imager (4X/12X mag)
- Acoustic detection
  - USMC ADS system (6.2)/detect talk @ 400 M
- Laser Ranging
  - AN/PAQ3 MULE LDS-AN/GVS5 Laser Ranger
- Target Designation
  - AN/PAQ3 MULE on RSTA mast
- Enables RSTA from defilade

WEAPONS MODULES
- HELLFIRE missile launcher
- .50 CAL HMG heavy machine gun on universal mount
- Arm/charge/fire/de-arm/safe weapons remotely
- Engage targets
  - Max/min weapon ranges
  - Moving/stationary targets

NBC MODULE
- M8 Chemical detector, acoustic alarm

FIGURE 16.3 Teleoperated vehicle, Marine Corps single-service managed program.

T ELEOPERATED
M OBILE
A LL-PURPOSE
P LATFORM

U.S. ARMY MISSILE COMMAND

MARTIN MARIETTA

- 4WD - SKID STEER
- DIESEL POWER
- HYDRAULIC DRIVE
- QUIET SURVEILLANCE MODE
- JOYSTICK CONTROL
- LASER TARGET MARKER
- MANPORTABLE CONTROLLER POWER PACK (32 LBS. EA.)
- REAL TIME CONTROL SYSTEM ARCHITECTURE
- ACOUSTIC SENSORS

GRUMMAN

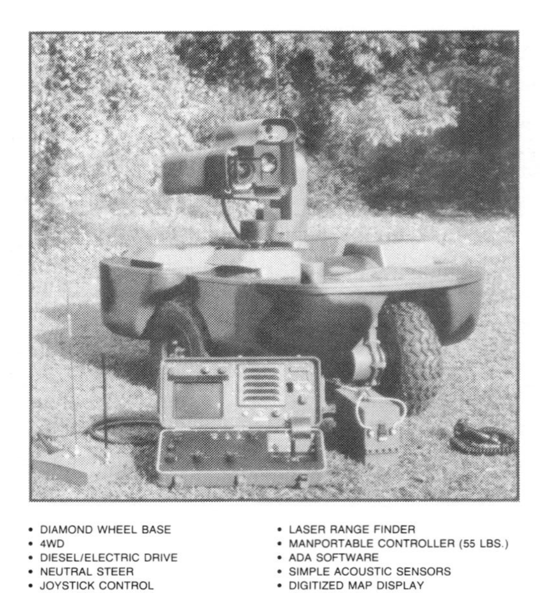

- DIAMOND WHEEL BASE
- 4WD
- DIESEL/ELECTRIC DRIVE
- NEUTRAL STEER
- JOYSTICK CONTROL
- QUIET MOBILITY SURVEILLANCE MODE
- LASER RANGE FINDER
- MANPORTABLE CONTROLLER (55 LBS.)
- ADA SOFTWARE
- SIMPLE ACOUSTIC SENSORS
- DIGITIZED MAP DISPLAY

TELEOPERATED MOBILE ALL-PURPOSE PLATFORM (TMAP)

The US Army remote control TMAP expands commanders' battlefield influence and minimizes casualties by offsetting the soldier from direct hazards. The platform is modular so it can accept any manportable mission package. It is a simple cost-effective "extension" of the soldier for battlefield missions in reconnaissance, forward observer, laser designator, sentry, explosive ordnance disposal, CBR reconnaissance, minefield detection/mapping and others where standoff capability is desired. TMAP--needed, simple, useful.

REMOTELY CONTROLLED
- o Removes soldier from direct line of fire
- o Operates beyond LOS in day and night conditions
- o Keeps soldier in the loop
- o Workload reduced by automatic target cuing
- o Enables surveillance or engagement of threat forces from concealment (defilade)
- o Color driving camera
- o LLLTV night camera

LIGHTWEIGHT
- o Platform (less than 800 lbs)
  - sized for easy HMMWV or self-transport
  - composite construction
  - size & weight mission dependent
- o Manportable Control Unit

MODULAR
- o Mission adaptable/reconfigurable

ALL-TERRAIN MOBILITY
- o Reliable four-wheel drive capability
  - negotiates 12" steps/trenches
  - remains stable on irregular terrain
  - climbs/descends 31° slope
  - turns in place
  - crosses water obstacles

PROVEN F/O CABLE & PAYOUT SYSTEM
- o Ensures fiber optic (F/O) cable battlefield survivability

PORTABLE CONTROL UNIT
- o 9" color monitor - high resolution
- o Simple, easy to read color displays provide accurate driving, status and navigation information

NAVIGATION
- o Overlays target coordinates, platform position & track
- o Accurate over entire mission profile
- o Dead reckoning navigation
- o Driver navigation aids

COMMUNICATIONS
- o Redundant links
  - fiber optic and RF

VERSATILITY
- o Designed for expansion to multiple platform operation by a single soldier

SPEED
- o 10 km/hr

FIGURE 16.4 Teleoperated mobile all-purpose platform, Army single-service managed program.

velop a system that would integrate current military vehicles. The remote platform was controlled from another sheltered vehicle. The design of the system was to include modular control and mission payload packages. The concept allows interchangeability of missions and specific capabilities. The TOV was installed on the fielded highly mobile multipurpose wheeled vehicle (HMMWV). The HMMWV allowed for a large payload and supportability of the chassis. The high mobility of the HMMWV was a plus, but the size created a considerable logistical burden to proposed users. Figure 16.3 shows a TOV.

The TOV operated with a 12-ft mast, which allowed remote sensing from defilade through a 14:1 zoom color-targeting camera, forward-looking infrared (FLIR) sensors, a laser designator and laser range finder, and acoustic sensors. It had a 30-km operational distance, which was limited by the length of the fiber-optic communications cable. Manned operation of the remote vehicle was possible when desired. The control of the remote platform was accomplished by a control shelter mounted in an HMMWV, containing three operator stations and a crew chief or supervisor's station. Stereo cameras slaved to the operator's head movements via a helmet-mounted display enhanced virtual driving.

Army's Teleoperated Mobile All-Purpose Platform (TMAP). The TMAP was required by draft documents to be small and transportable on the HMMWV. Unlike the TOV, which was developed by the Naval Ocean Systems Center, the TMAPs were developed by industry on contract from the U.S. Army Missile Command and Sandia National Laboratory (Fig. 16.4).

Initially built to carry antitank missiles, the TMAPs were modified to perform a strictly reconnaissance mission. One was built by Martin-Marietta and the other by Grumman Aerospace. Both were lightweight, about 320 kg, and had similar capabilities. They were much slower than the TOV, capable of speeds up to about 16 km/h, but both had FLIR or color zoom targeting cameras. The operator control units were man-portable. Both TMAPs were capable of being controlled by either fiber-optic or radio-frequency links. The unique feature of the Martin-Marietta version was a laser designator/range finder; the Grumman system had a quiet electric drive. The TOV and TMAPs had navigation information displayed to the operator.

The two TMAP designs as well as the three TOVs were built. Though these systems never matured out of basic research, all are currently being used as test beds for man-machine interface work, subsystem trials, and other technology exploration.

Synchronization

Customer Requirements. The Marine Corps and the Army draft requirements have gone through many changes prior to recent approval. The Army TUGV (Caleb) operational and organization plan was approved in November 1990. It is general enough to allow exploration of a wide variety of answers to the problems of fielding a remote combat vehicle. It was decided to require an initial mission of reconnaissance, surveillance, and target acquisition (RSTA), with possible follow-on requests of unspecified missions. The Marine Corps initial statement of operational requirement (ISOR), signed in December 1990, is similar in general requirements and also requires RSTA. Like the Army, the Marines stress modularity and capacity for growth into other missions, such as antiarmor, countermine, direct fire, and NBC reconnaissance.

Surrogate Equipment—A BETA Test Process. In an innovative approach, the surrogate teleoperated vehicle (STV) was procured as a test vehicle. This approach was to allow trades to be made by the JPO in conjunction with the user. The separate Army and Marine Corps requirements have been used to specify the development of a surrogate vehicle. The services will test the STV. Tests are designed to examine tradeoffs with respect to the desires of the two designated proponents or customers, the U.S. Army Infantry School (USAIS) and the Marine Corps Warfighting Center (WFC). Results of testing will also be used to further refine final requirements into the ORD. Many issues have been identified, such as vehicle size, mast height, speed, swimming parameters, administrative driving controls, the need for FLIR, laser designator requirements, types and capabilities of sensors, communications, propulsion type, telemetry, control architecture, battlefield controller type, and displays. Fourteen vehicles, built by Robotic Systems Technology (RST), Hampstead, Md., will represent the desired capabilities to allow nearly all the trades to be evaluated in a BETA model. Field testing will be conducted. After the field testing, analysis using gaming and simulation will be accomplished to match actual testing to simulations to further refine the objective requirements.

The STVs will have the following capabilities:

1. Maximum speed of at least 35 mi/h
2. Electric propulsion at low speeds for quiet positioning
3. A mast extending up to 14.5 ft
4. FLIR
5. Laser designator/range finder
6. Fiber-optic and radio-frequency control links
7. Man-portable controller
8. “Driver-in” capability
9. Advanced acoustic sensors
10. Color-targeting camera
11. Stereo day and night driving cameras
12. Flat screen and head-mounted or head-tracking display

The first STV is shown in Fig. 16.5. This is only a BETA test, and the objective or final TUGV system will be different in some of these capabilities. As in any BETA test, the intent is to get a working system into the hands of knowledgeable users.

Schedule. The initial product schedule supports fielding of the first-generation combat unmanned vehicle as soon as possible (see Appendix E). This schedule calls for a milestone I/II decision to start engineering and manufacturing development (EMD) in the 4th-quarter fiscal year 1993. All efforts of the project office are directed at providing the users with the capability to define their requirements and develop their doctrine and operational concepts. The early stages of the schedule include early user testing and evaluation (EUT&E) and concept of employment evaluation (COEE). This initial schedule of the TUGV is shown in Appendix E. The first equipped unit of these first-generation unmanned ground systems has been scheduled for 1998. With the defense budget changing so rapidly it is and will be a major leadership challenge to maintain the joint schedule.

Industry Support Groups. The JPO is attempting to openly interface with industry through the Association for Unmanned Vehicle Systems (AUVS). AUVS, as a professional nonprofit association, has established an industry support group (ISG) to support the unmanned ground efforts. The ISG is structured to address specific con-

Robotic Systems Technology

FIGURE 16.5 Surrogate teleoperated vehicle, joint project office managed program.

SURROGATE TELEOPERATED VEHICLE (STV)

The STV team includes Robotic Systems Technology, Polaris Industries, Martin Marietta, and Bio-Dynamics.

Our Advanced STV Design Allows for the Incorporation of a Variety of Modular Mission Payloads for Additional Applications.

The Robotic Systems Technology STV System Provides the U.S. Army/U.S. Marine Corps User With the Complete Capability to Evaluate All Potential Requirements for Unmanned Vehicles During Early User Field Testing.

REMOTE PLATFORM

General	6-Wheel Drive
Drive	Hybrid, Diesel/Electric
Steering	Ackerman
Remote Actuation	Back Drivable
Width	51 in.
Length	108 in.
Height	Variable 46-52 in.
Wheelbase	75 in.
Weight	800 lbs.
Turn Radius	162 in.
Engine	25 Hp Diesel 3 Hp Electric
Transmission	Automatic
Range	250 + Miles
Endurance	48 Hour Mission
Top Speed	35 mph
Front Slope	35 deg grade
Side Slope	25 deg grade
Water Crossing	Swim/Floatation
Payload Capacity	1200 lbs.
Tow Capacity	1500 lbs.
Power System	24 VDC
Batteries	100 Ahr

MOBILITY and RECONNAISSANCE, SURVEILLANCE & TARGET ACQUISITION MODULE

Mast Height	15 feet
Turret Motion Range	+/− 90 deg Tilt, +/− 270 deg Pan
Slew Rate	150 to 0.05 deg/sec
Day Driving Cameras	Color Stereo, 460 Line Resolution CCD, Auto Iris
Night Driving Cameras	Image Intensified, Stereo, 500 Line Resolution
Day Targeting Camera	460 Line Color CCD, 14:1 Zoom
Night Targeting Camera	Image Intensified, High Resolution, 10:1 Zoom
Acoustic Detection	20 Hz. to 25 kHz., Binaural Audio, Electronic Steering, Intruder Detection and Alert Systems
Navigation	GPS, 6 channel, 10 meter accuracy with Dead Reckoning Backup System
Designator/Ranger	LTM-86
FLIR	AN/TAS-4

COMMUNICATION LINK

General	Fiber-optic with Full-Capability Radio-Frequency Backup System, Dual Video Return with Bi-directional Audio and Data Lines
Fiber Optic	Primary Link, 10 km Payout System, Non-jammable
Radio Frequency	Full Backup Link, High Bandwidth, Multiple Frequencies Available

OPERATOR CONTROL UNIT

General	Man-Portable, Human Factor-Engineered
Displays	Dual color LCD's for Real-time Video and Navigation
Switches	Sunlight Readable, Water/Shock Proof
Controls	Dual Handlebar and Joystick Control of Driving and Targeting Functions

ROBOTIC SYSTEMS TECHNOLOGY

FIGURE 16.5 *(Continued)*

cerns and give feedback to the JPO from industry. This innovative mechanism allows for the free flow of pertinent communication required between industry and the government. This is another and very important means by which to synchronize the stakeholders.

Tactical Unmanned Case Observation

The development of UGVs offers a great number of difficult tradeoffs. All project tradeoffs deal with the three areas of cost, schedule, and performance.

The primary programmatic concern is to effectively marry the available technology with a reasonable requirement. Efforts of the project office are focused on delivering a mission-effective and cost-effective product to the user. Many requirements are written such that they can be met neither technically nor at a reasonable cost. Other systems have been built that do not answer the needs of the target requirements or provide no value to the unit supported. The JPO's intent is to synchronize the forces of need and cost. The project office is committed to the integration of a joint team of combat developers, material developers, and industry. This integration has meant seats on source selection boards for proponents, incorporation of other material developers' inputs, and inclusion of industry observations through an ISG. The users attend all meetings and support the project office with their perspectives at all times throughout the process. This process allows decisions to be balanced between necessary capabilities, affordability, and technical performance. This experiment or case study has and will have both its own time in history and its own organizational culture. What is needed to complete this experiment is the leadership during this change process. No two joint projects are the same. However, some common observations can be offered. To be successful, a JPO must:

Be led by a participating service PM (contrary to having a lead service PM).

Use BETA test items and simulations to identify areas of technical risk and uncertainty. These BETA tests also help solidify stakeholder requirements.

Recognize that each joint project is unique because of its compounded stakeholders' interests.

Enlist the support and feedback from industry through an industry support group.

Ensure that all members of the JPO have enlisted the vision or mission of the office and "live it" at work.

Become the "focusing point" of its represented technology. The JPO should accelerate and facilitate the flow of information to all appropriate stakeholders. (See Appendix F for an example of an information instrument.)

Summary and Conclusions

Leading the Joint Change Process. An executable project, which is long in coming, has finally begun in the DoD. Although changes will still be made, the capability to give the ground commander real-time information at long distances is now practicable. The joint programmatic responsibility is to integrate and coordinate perceived joint requirements with producible technology and design. It is anticipated that, upon acceptance of the first such system, many missions will be automated, keeping soldiers and marines out of harm's way. The joint project as planned maximizes the best use of combat developers, material developers, and industry. Robotics and unmanned vehicles are the future of many endeavors within the battle space. The DoD is committed to addressing this change issue intelligently and cost-effectively.

The project leader and office must be highly proficient in the technical procedures outlined in DoD Directives 5000.1 and 5000.2. Cochrane[15] gives a very good overview of these near-term transaction approaches for DoD organization. The project leader must also anticipate the future and the change it can have on the near-term technical approaches or procedures. By the time you have read this chapter the directives guiding JPOs will probably have changed. Change is inevitable. Your job is to implement it as well as harness it. It is this art of implementation that is difficult. Joint implementation is complex and even more difficult.

The joint project must continually evaluate its stakeholders as well as monitoring and, if necessary, influencing its environmental considerations, as outlined in Figs. 16.1 and 16.2. The joint project is a fragile system and therefore must balance its jointness, objectives, and benefits to the stakeholders. This changing system must be led by the PM anticipating, developing, and synchronizing activities within the vesting partner's cultures. Finally (Kerzner,[6] p. 886),

> Successful future enterprises will be differentiated by the quality of management planning and decision making. By virtue of broad experiences, flexible approach, and a "people-oriented" leadership style, tomorrow's project managers should be able to cope in the year 2000.

In the end we want to do more than cope; we want to jointly lead into this 21st century.

REFERENCES

1. D. Hellriegel and J. Slocum, *Management,* 6th ed., Addison-Wesley, Reading, Mass., 1992.
2. R. Carkhuff, *Empowering—The Creative Leader in the Age of the New Capitalism,* Human Resources Development Press, Amherst, Mass., 1989.
3. L. Hennebeck, "Acquisition Innovation for Today's Weapon Procurement," *Unmanned Systems,* pp. 19–21, Fall 1990.
4. G. Kissler, *The Change Riders: Managing the Power of Change,* Addison-Wesley, Reading, Mass., 1991.
5. M. C. Thomsett, *The Little Black Book of Project Management,* AMACOM, New York, 1990.
6. H. Kerzner, *Project Management,* 2d ed., Van Nostrand Reinhold, New York, 1984.
7. *Joint Logistics Commanders Guide for the Management of Joint Service Programs,* 3d ed., DSMC, Ft. Belvoir, Va., 1987.
8. R. Freeman, "Foreword," *Defense Systems Management Rev.,* p. 5, Spring 1979.
9. D. Cleland and W. King, *Management: A System Approach,* McGraw-Hill, New York, 1972.
10. I. Avots, "Project Management within the Systems Context," in H. Reschke and H. Schelle (Eds.), *Dimensions of Project Management,* Springer, New York, 1990, pp. 39–48.
11. D. Allen, "New Era of Cooperation," *BYTE,* p. 20, Feb. 1992.
12. W. H. Newman, "Focused Joint Ventures in Transforming Economics," *Academy of Management,* pp. 67–75, Feb. 1992.
13. J. Treece, K. Miller, and R. Melcher, "The Partners," *Business Week,* pp. 102–107, Feb. 10, 1992.
14. G. Bellman, *The Quest for Staff Leadership,* Scott, Foresman, Glenview, Ill., 1986.
15. C. Cochrane, "Defense Acquisition Policy," *Program Manager,* pp. 29–34, May–June 1991.

APPENDIX A

FACT SHEET
PROGRAM MANAGER'S NOTEBOOK

DEFENSE SYSTEMS
MANAGEMENT COLLEGE

Author: DSMC, Policy and Organization Management Department
Number: 8.2
Version: Update
Date: December 1988

I. TITLE

Joint and Multiservice Programs

II. REFERENCES

—*Joint Logistics Commanders Guide for the Management of Joint Service Programs*, Defense Systems Management College, Fort Belvoir, VA, Third Edition, 1987, especially Appendix C, "Memorandum of Agreement on Multiservice Operational Test and Evaluation and Joint Test and Evaluation"

—*The 31 Initiatives: A Study in Air Force-Army Cooperation*, R.G. Davis, Office of Air Force History, 1987, especially the appendices

—DOD Directive 5000.3

—AFLC/ASFC R 800-2/AMCR 70-59/NAVMATINST 5000.10 A, "Management of Multi-Service Systems/Projects/Programs"

III. POINTS OF CONTACT

Pre-Milestone I programs, call the Interoperability, Integration and Initiatives Division, J-7 (Joint Staff), (703) 694-4651; AV 224-4651

Milestone I and beyond, call the Systems Programs Evaluation Division, J-8, (703) 694-3681; AV 224-3681

Command, Control and Communications programs, call the Planning and Priorities Division, J-6, (703) 695-7168; AV 225-7168

IV. PURPOSE AND SCOPE

This fact sheet is designed to:

—Provide a description of joint programs
—Describe joint program initiation
—Describe the process of gaining approval of a joint PM charter
—Describe the contents of a joint PM charter
—Offer some program perspectives.

V. DOD POLICY

Joint programs are strongly supported and encouraged by the Office of the Secretary of Defense and the Congress. The reasons for advocating a joint program are many and varied but are ultimately reducible to some operational or economic advantage to DOD. Typically, one or more of the following factors is at work:

—Improving Combat Capability

—Coordination of Efforts. Coordination reduces duplication of effort, improves exchange of technical information, and channels individual Service efforts into mutually supportive programs.

—Interoperability of Forces and Equipment, especially in the areas of command, control and communications.

—Reduction in Development Costs. If the requirements of the combat components are compatible, and consolidation of programs does not increase risk unduly by closing out alternatives, it makes sense to fund one joint program, rather than multiple, single-Service efforts.

—Reduction in Production Costs. Consolidation of the Services' production requirements should lower unit price.

—Reduction in Logistics Requirements. Standardization across the Services offers potential for both reducing support costs and improving support provided to operating forces.

In 1973, the joint logistics commanders signed a memorandum of agreement that was subsequently promulgated as a joint regulation (a copy of the agreement is in Appendix A). That document sets forth principles of joint-program management that continue to provide a solid foundation to the establishment of a joint program management office. It introduces the concept of the executive (sometimes referred to as lead) and the participating Services, and establishes general responsibilities and authorities of both. It provides for use of executive service program management procedures in areas where common procedures do not exist and calls for multiservice program charters, program master plans, and joint operating procedures to be prepared as documentary instruments of joint program management.

Early every year, the Service R&D chiefs prepare and then review Joint Potential Designation Lists—the lists they use to keep tabs on programs with joint potential. These lists are sent to the Joint Requirements Oversight Council (JROC) every April for review. The JROC reviews all Acquisition Category I through IV programs (or their equivalents) coming up for Milestone 0, I and II reviews to assess the potential for joint Service development and/or production. THe JROC is composed of the Service Vice Chiefs and chaired by the Vice Chairman of the Joint Chiefs of Staff. Because the chairman of the JROC is also the vice chairman of the DAB, there is a direct link between the joint requirements review process and the milestone review process. This is beginning to have an impact on programs. In June 1988, for example, the JROC and Conventional Systems Committee of the DAB developed a "joint standoff weapons master plan" that reduced the number of air-to-surface standoff weapon programs from nine to five (and only two of the five will have extended production runs). This sort of executive action will increase if congressional pressure for joint programs grows.

The basic requirements document for a major acquisition program is the Mission Need Statement (MNS). A MNS identifies a specific military deficiency in a mission area, the priority assigned to correcting the deficiency, and the magnitude of resources needed to correct the deficiency. A joint MNS documents major deficiencies existing in two or more combat components. The DAB approval of an MNS is a prerequisite for initiation of a major system acquisition program. Operational requirements for less-than-major acquisitions continue to be stated in Service-peculiar requirements documents which tend to be more detailed and more weapon-system-oriented (vice mission-oriented) than an MNS. This same practice is likely to hold true for joint acquisitions: major acquisitions will be supported by a joint MNS; less-than-major acquisitions will be supported by a joint operational requirement (JOR), or similar document, which is more weapon-system-oriented than an MNS.

When a joint or OSD/Joint MNS is submitted, the Secretary of Defense (SECDEF) decision will be documented in a SECDEF Acquisition Decision Memorandum (ADM). The ADM may specify the lead DOD component and should provide explicit guidance on the responsibilities of the participating DOD components, including threat support. The lead DOD component will assign the program manager and may request the other participating DOD components to assign deputy program managers. The lead DOD component also will establish program objectives by promulgating a program charter after coordinating with the other participating DOD components.

VI. JOINT PROGRAM INITIATION

Few programs become joint without some initiative by the Joint Staff, the Congress or the Under Secretary of Defense for Acquisition.

Typically, the Under Secretary of Defense for Acquisition, with the assistance of the JROC, writes a memorandum designating one Service the executive or lead Service and directing it to charter a joint program. Less formal, but no less compelling, direction is given to the Services during program or budget reviews. The Vice Chairman of the Joint Chiefs of Staff is responsible for making sure that such reviews do not overlook the potential for joint development and that joint programs are not ignored in Service POMs.

The Services negotiate the ground rules of the joint program and agree to assignment of program authority and responsibility. The implementation of OSD direction is different in each of the Services. The Army and Navy simply forward by memorandum USD(A)'s direction to the appropriate development and acquisition activity via the chain of command. In the Air Force, the Assistant Secretary for Acquisition directs major command participation, either as lead or supporting elements, via program management directives (PMD). Further delineation of participation below major command level is promulated by Form 56 within AFSC, and program action directive (PAD) within AFLC.

Interservice negotiation and agreement on a joint program can be accomplished at any of several echelons in the Services' organizational hierarchies: the Service secretariats, the Service headquarters, the material development and logistics commands, or their commodity-oriented commands. Examples include the 31 initiatives agreed to in 1984 by the Army and Air Force. These agreements covered issues of doctrine, Service roles and missions, and joint development. Eventually expanded to 38, all but five of these initiatives had been resolved as of 1988. Their development and resolution show the Services must often reach agreement on their respective mission needs and responsibilities before they can move ahead with joint development.

If there is a general rule, it is to agree that the lowest level agreement is practicable, and that level varies from program to program. However, there are two advantages to agreements at the Service headquarters level: (1) it is the level at which operational requirements are validated and translated into equipment needs; and (2) it is the level at which funding priorities are established. Assuming Service headquarters staffs agree, PMs must be alert for any disagreement between the headquarters and Service secretariat levels. As all joint program managers soon learn, *nothing is more important to the success of a joint program than interservice agreement on requirements and funding.*

When agreement is reached at the Service headquarters and secretariat level, it is usually documented by a memorandum of agreement (MOA). There is no typical content of format for an MOA. It may be a long document defining all the ground rules for the joint program, much as would a charter. It may be very brief, covering only key areas of agreement, such as designation of the executive service and sharing of funding responsibility. Frequently, a program will have several MOAs associated with it, each covering a different topic.

VII. JOINT PROGRAM CHARTER PREPARATION

The charter, once promulgated, is the foundation of a joint program. It establishes the program and announces to all concerned the responsibility and the intended relationships among the participating Services. When possible, the individual selected to be the PM should be reponsible for the drafting of the charter. When OSD directs the initiation of a joint program, they can include in the directive provisions for an interim charter and staffing of the PMO. If OSD elects not to do this the lead Service should ensure the PM has an interim charter to use until an agreed-upon charter is obtained. The final charter should give the PM sufficient authority to accomplish the given responsibilities. As a minimum the PM must have adequate authority to:

—Make trade-offs among cost, schedule, supportability and performance within the established bounds for the program. It is difficult for a user in one Service to give this totally to a PM in another Service, but it must be done to have a successful program. The important thing is to define the bounds, and control the PM's action through reviews instead of limiting his/her power.

—Identify program funding needs and control funds allocated to the program. When Services "donate" funds to a joint program, there is a reluctance to give up total control of the funds, although the JROC is pressing for this. Getting agreement on the funding issue in the development of charter requires creative staffing to get a document.

—Determine and control hardware and software configuration.

—Communicate directly with the other Services and government agencies.

—Manage his/her military and civilian work force.

A good joint program manager will realize that to be successful, support from the participating Service is required. He/she can begin getting that support while drafting the charter. This is the time for the PM to "hit the halls" and find out the concerns of the Services and start lobbying for his/her proposed charter. First impressions are important and the PM must establish trust early. The more the Services trust a PM the more they will be willing to give up items in the charter.

Charters for joint programs are normally promulgated by the executive service. The JLC "Memorandum of Agreement on Management of Multi-Service Systems/Programs/Projects" calls for joint approval of joint program charters for major programs. However, such jointly signed charters are rare. Army charters are approved by the Secretary of the Army; Navy charters by the Navy Acquisition Executive; and the Air Force program management directives by the Assistant Secretary of the Air Force for Acquisition or the Deputy Chief of Staff (Logistics and Engineering). For non-major programs, the chartering is delegated to the materiel development or logistics commanders according to specific Service practices. On a few occasions, program charters have been promulgated directly by OSD, but that is unusual and has occurred primarily when OSD wanted to coordinate independent service programs in an active way. Even though there is no formal requirement to gain concurrence from the other Services, it is the best interest of the PM and the program to staff the charter with the partgicipating Services even if there has been an MOA signed.

If OSD retains approval authority, the lead Service is responsible for the submission of the charter. There are two ways the charter can be submitted:

—If OSD specifies the charter be submitted through the Joint Staff, the charter is submitted by the Service chief to the Joint Staff and a joint action is initiated to gain JCS recommendation for OSD. One thing to remember is that once the joint action is started *the responsibility for the action lies with the joint action officer and the lead Service reverts to being a voting player with the same status as the other Services*. Also, the Services that may not be a party to the program may be involved and will vote on the charter in the joint action.

—If there is no requirement for JCS recommendation, the charter will more than likely be submitted to OSD by the Service Secretary.

VIII. CONTENTS OF A JOINT PROGRAM CHARTER

Joint programs are exceptions to the Services normal acquisition practices. Thus, the joint program charter must include those elements essential to any charter and those needed to define specific relationships among the participating Services. The extent to which the latter must be defined in the charter depends on the circumstances surrounding establishment of the joint program. If, at the inauguration of a joing program, there exists a major issue involving responsibility, authority, or inter-Service relationships, it should be resolved in the charter, or *it will haunt the program throughout its life*. The following items are considered as essential items in the charter:

—*Designation of the Joint Program.*

—*Statement of the Program Objective.* It is extremely important that this section of the charter be well written and not open to interpretation. It is where the bounds are established.

—*Definition of the PM's Authority, Responsibility, and Accountability.* The accountability can be tricky because the participating Services will want some accountability by the PM to them. What must be avoided is having a joint PM answering to many people and organizations.

—*Specifications of Program Funding and Resources.* Again, this is probably the hardest item to get agreement on. The easiest thing is to have the lead Service fund the whole development project and let the participating Services fund for the procurement of their systems. It is the easiest way for the PM but difficult to get agreement on from the lead Service. If the funding is split among the participating Services, they will not agree unless they see it as a means of controlling the PM. The method that is best for a program will depend on the program and the environment when the charter is staffed.

—*Definition of the Services' Joint or Unilateral Responsibilities for Program Execution.*

—*Description of the Relationship of the Joint Program with Other Programs, Supporting Organizations, and Supported Organizations.*

—*Identification of the Chain of Command for Reporting and for Resolving Program Issues.* Every attempt should be made to keep the level for resolving issues as low as possible.

—*Reporting Requirements (Type, Format, and Frequency).* One thing often overlooked is a responsibility for the PM to keep the participating Service and the user, especially joint users, informed of program status. Provisions for this type of reporting should be included in the charter.

—*Project Office Organization and Initial Staffing.*

—*Requirement to Establish Joint Operating Procedures.*

The following items are "officially" optional elements but in reality *should be considered as essential:*

—*Assignment of the Deputy PMs from the Participating, Services, Definition of Their Responsibility and Authority, and Designation of Their Rating Officials.*

—*Methods of Resolving Conflicting Requirements or Objectives of the Services Involved.*

—*Creation of Joint Committees for Coordination or Approval of Key Aspects of the Program (i.e., requirements, funding, source selection, test and evaluation plans, and configuration).*

—*Performance Evaluation of Personnel.*

IX. A PROGRAM MANAGER'S PERSPECTIVE

At the outset of a joint program, the joint program manager should conduct a detailed technical requirements review that examines mission needs, operational concepts and environments, and performance parameters. He/she should ensure that requirements are understood, that conflicts are resolved, and that there is sufficient latitude to make the trade-offs essential to any program's success. This review should accomplish the following:

—Identify the similarities and differences in the Service's requirements and in the operation environments.

—Force a clear distinction between the "like to have" and "must have" requirements.

—Identify areas of technical risk or uncertainty.

—Identify the similarities and differences in the Services' logistic concepts, requirements, and procedures, including their approach to the implementation of the life-cycle cost concept.

Once the requirements of each Service are well understood, the joint program manager should define the set of essential requirements that is most demanding in terms of cost, schedule, and performance criteria. It will require determining the extent to which commonality of hardware and software, frequently an explicit or implied goal of a joint program, is a valid requirement and is achievable. Some joint programs will be considered successful only if they develop identical or nearly identical systems for use in all Services. The value of other joint programs however, may be only in sharing the costs of concept formulation and validation or in coordinating the engineering development of systems peculiar to each Service and ensuring their interoperability; trying to develop identical or nearly identical systems for all the Services may frustrate the program and lead to its failure.

The preparation for each milestone review should include a re-examination of the same items reviewed at the initiation of the joint program. This re-examination should determine not only that the participating Services' perceptions of the requirements have not changed, but also that the threat or other basis for establishing the system's need remains consistent with the initiating need.

In dealing with the contractor, the joint program manager must ensure that the interpretation of requirements within the scope of the contract comes *only* from the program office. There must be no other source, official or unofficial, stated or implied. This is the only way the joint program manager can maintain control of the program and hold the contractor accountable.

A joint program can be structured any way necessary to accomplish the program's goals. On the other hand, the base of experience for each type of joint program is small, and the advice and direction a new joint program manager receives (including that provided here) might have been formed from a joint program environment not at all similar to his own. Certainly the cost of a program, its importance, its urgency, and other factors which influence its visibility, will affect a joint program and its way of doing business. A joint mobile electric power (portable generator) program, for instance, will look different than a joint cruise missile program. The manager of each program will be influenced by different precepts even though both may be classified as "joint programs."

The DOD has established special arrangements for processing armaments and munitions requirements. An Armament/Munitions Requirements and Development (AMRAD) Committee has been established by the Deputy Secretary of Defense. The committee is staffed by members of the research and development directorates of the separate Services and reports to the Deputy Under Secretary of Defense (Tactical Warfare Programs). Although the objectives of most joint programs are outside the purview of AMRAD, the committee has more than 10 years experience in reconciling diverse requirements and has established a protocol for their harmonization. A program manager may find the committee's experience valuable.

Appendix A. MEMORANDUM OF AGREEMENT ON THE MANAGEMENT OF MULTISERVICE SYSTEMS/PROGRAMS/PROJECTS

1. PURPOSE

This Memorandum established policies for implementing multi-service systems, program/project management in accordance with DOD Directive 5000.1, "Acquisition of Major Defense Systems," 13 July 1971. It is the basic policy document for management of multi-service systems, programs and projects, and the framework within which, like DOD Directive 5000.1, acquisition management procedures must operate.

2. POLICY

The Service designated as the Executive Agent shall have the authority to manage the program/project under the policies and procedures used by that Service. The Program/Project Manager, the Program/Project Management Office, and, in turn, the functional elements of each Participating Service will operate under the policies, procedures, data, standards, specification, criteria and financial accounting of the Executive Service. Exceptions, as a general rule, will be limited to those where prior mutual agreement exists or those essential to satisfy the substantive needs of the Participating Services to accept certain deviations from their policies and procedures so as to accommodate the assumption of full program/project responsibility by the Executive Service. Demands for formal reporting as well as nonrecurring needs for information will be kept to a minimum.

3. RESPONSIBILITIES

a. The Executive Service will:

(1) Assign the Program/Project Manager.

(2) Establish an official manning document for the Program/Project Manager Office which will incorporate the positions to be occupied by representatives of the Participating Services; e.g., Department of the Army Table of Distribution and Allowances (TDA)/Department of the Navy Manpower Listing/Department of the Air Force Unit Detail Listing (UDL). The manning document developed from the Joint Operating Procedure on Staffing will also designate a key position for occupancy by the Senior Representative from each of the Participating Services.

(3) Staff the Program/Project Management Office with the exception of the positions identified on the manning document for occupancy by personnel to be provided by the Participating Services. Integrate the Participating Service personnel into the Program/Project Management Office.

(4) Be responsible for the administrative support of the Program/Project Management Office.

(5) Delineate functional tasks to be accomplished by all participants.

b. The Participating Services will:

(1) Assign personnel to the Program/Project Management Office to fill identified positions on the manning document and to assist the Program/Project Manager in satisfying the requirements of all participants. Numbers, qualifications and specific duty assignments of personnel to be initially provided by each Participating Service will be reflected in the Joint Operating Procedure.

(2) The Senior Representative from each Participating Service will be assigned to a key position in the Program/Project Management Office and report directly to, or have direct access to, the Program/Project Manager. This key position could include assignment as Deputy to Program/Project Manager. He will function as his Service's representative, with responsibilities and authorities as outlined in Paragraph 3.d of the Agreement.

(3) Provide travel funds and support necessary for the accomplishment of the responsibilities of their representatives in the management of the Program/Project.

(4) Accomplish Program/Project functional tasks as specifically assigned in the Charter, in the Master Plan, and Joint Operating Procedures (JOPs), or as requested and accepted during the course of the Program/Project.

c. The Program/Project Manager will:

(1) Satisfy the specific operational, support and status reporting requirements of all Participating Services.

(2) Be responsible for planning, controlling, coordinating, organizing and directing the validation, development, production, procurement and financial management of the Program/Project.

(3) Review, on a continuing basis, the adequacy of resources assigned.

(4) Assure that planning is accomplished by the organizations responsible for the complementary functions of logistics support, personnel training, operational testing, military construction and other facilities, activation or deployment.

(5) Refer to the appropriate authority those matters that require decisions by higher echelons. The following items will be referred to appropriate authority:

(a) Deviations from the established Executive Service policy except as specifically authorized by the Program/Project documentation (reference Paragraph 4 below).

(b) Increases in funding of the Program/Project.

(c) Changes to milestones established by higher authority.

(d) Program/Project changes degrading mission performance or altering operational characteristics.

(e) Participating Service Senior Representative(s) within the Program/Project Management Office will:

(i) Speak for his parent Service in all matters subject to the limitation prescribed by his Service. Authority of the Service Senior Representative is subject to the same limitations listed above for the Program/Project Manager.

(ii) Refer to his parent Service those matters which require decisions by higher echelons.

4. DOCUMENTATION

Management for particular Multi-Service Program/Projects shall be documented by:

a. **A Multi-Service Program/Project Manager Charter.** The responsible Commander in the Service having principal Program/Project management responsibility will cause the preparation, negotiation and issuance of a jointly approved Charter which will identify the Program/Project Manager and establish his management office. The Charter will define his mission responsibility, authority and major functions, and describe his relationships with other organizations which will use and/or support the Program/Project. The Charter will describe and assign responsibility for satisfying peculiar management requirements of Participating Services which are to be met in the Program/Project, and will be jointly approved for the Headquarters of each involved Service by persons officially appointed to approve such Charters.

b. **A Program/Project Master Plan.** This is the document developed and issued by the Program/Project Manager which shows the integrated time-phased tasks and resources required to accomplish the tasks specified in the approved statement of need/performance requirements. The plan will be jointly approved for each involved Service by persons officially appointed to approve such plans.

c. **Joint Operating Procedures (JOPs).** These will identify and describe detailed procedures and interactions necessary to carry out significant aspects of the Program/Project. Subjects for JOPs may include Systems Engineering, Personnel Staffing, Reliability, Survivability, Vulnerability, Maintainability, Production, Management Controls and Reporting (including SAR), Financial Control, Test and Evaluation, Training, Logistics Support, Procurement and Deployment. The JOPs will be developed and negotiated by the Program/Project Manager and the Senior Representatives from the Participating Services. An optional format is suggested in Attachment 1 to this Agreement. This action will be initiated as soon as possible and accomplished not later than 180 days after promulgation of the MultiService Program/Project Manager Charter. Unresolved issues will be reported to the Charter approving authorities for resolution.

d. **Coordination/Communication.** Where Participating Services are affected, significant program action, contractual, or otherwise, will not be taken by the Program/Project Manager without full consultation and coordination with the Participating Services while the matter is still in the planning stage. All formal communications from the Program/Project Management Office to higher authority in the Executive or Participating Services will be signed by the Program/Project Manager or his designated representative. Substantive change to the Charter, Master Plan, or JOPs will be negotiated with affected Participating Services prior to issuance as an approved change. No restrictions will be placed on direct two-way communications required for the prosecution of the Program/Project work effort, other than that required for security purposes.

We approve this Memorandum of Agreement and its implementing regulation.

/s/ HENRY A. MILEY, JR.
General, USA
Commanding General
US Army Materiel Command

/s/ I. C. KIDD, JR.
Admiral, USN
Chief of Naval Material
Naval Material Command

/s/ JACK J. CATTON
General, USAF
Commander
Air Force Logistics Command

/s/ GEORGE S. BROWN
General, USAF
Commander
Air Force Systems Command

20 July 1973

1. This memorandum of agreement is published as a joint regulation, AFLC/AFSC R 800-2.AMCR 70-59NAVMATINST 5000.10A.

APPENDIX B

MEMORANDUM OF AGREEMENT 25 APR 1990
BETWEEN
THE U. S. ARMY AND THE U. S. MARINE CORPS

SUBJECT: JOINT UNMANNED GROUND VEHICLE (UGV) PROGRAM

1. Purpose: This Memorandum of Agreement (MOA) defines the responsibilities of the U. S. Army and the U. S. Marine Corps in a joint program for the development, production and procurement of unmanned ground vehicles (UGV).
2. Mission: The Joint UGV Project Office manages the development, production, and procurement of UGV systems in accordance with DoD Directive 5000.1, Major and Non-Major Defense Acquisition Programs.
3. Conditions:
    a. Scope: This agreement applies to the U. S. Army materiel development organizations, including the U. S. Army Materiel Command and its subordinate elements and the Marine Corps Research, Development and Acquisition Command and its subordinate Command, the Marine Corps Tactical Systems Support Activity (MCTSSA).
    b. Period of Coverage: This agreement is effective for five years from date of signature. It may be renewed or terminated at any time by mutual consent or on order from higher headquarters.
    c. Changes to the Agreement: This agreement will be reviewed as needed for corrections and modifications. All changes to this agreement will be by mutual consent or on order from higher headquarters.
4. Organization:
    a. The U. S. Army is designated The Executive Agent. The Joint Project Office (JPO) will initially report to Director, MICOM Research, Development and Engineering Center and transition to the Army Program Manager Office for Unmanned Vehicles at Milestone II. The functional elements of each participating service will operate under the policies, procedures, data, standards, specifications, criteria, and financial accounting of the U. S. Army.
    b. The objective organization of the headquarters of the JPO will, as a minimum, include the positions listed below and be subject to changes in order to enhance its effectiveness as it progresses through the program milestones:

Title	Grade	Source	Location
Project Manager	COL	MCRDAC	MICOM
Deputy Project Manager	LTC/GM-14	AMC	MICOM
Operations Officer/APM USMC	Major	MCRDAC	MICOM
UGV Project Officer/APM USA	Major	AMC	MICOM
Tech Base Project Officer	Major/GM-13	AMC	MICOM
Chief Systems Engineer	GM-14	AMC	MICOM
Logistics Project Officer	Major/GM-13	AMC	MICOM
Plng/Programs/Budget Officer	GM-14	MCRDAC	MICOM
Administrative Officer	GS-09	AMC	MICOM
Secretary	GS-06	MCRDAC	MICOM
Secretary	GS-05	AMC	MICOM

    c. The objective organization will be fully established when the UGV program enters the PM at Milestone II. Prior to this juncture, while the program is 6.3-funded, the Project Office will be a subset of the objective organization. As a minimum, it will be comprised of the Project Manager, the Deputy Project Manager, and required support staff identified in paragraph 4.b above.
5. Funding:

6.3	DoD
6.4	Army/Marine Corps share proportionally to number of systems to be procured
Procurement	Army/Marine Corps according to number of systems procured

6. Responsibilities

a. The Joint Unmanned Ground Vehicle Project Office will exercise programmatic, technical and financial control of the UGV activities as follows:

(1) Manage and coordinate 6.3/6.4 (ADV/ENGR DEV), and procurement.

(2) Develop, maintain, and update a UGV management plan for all UGV development activities.

(3) Provide the Materiel Developer inputs to the Robotics Master Plans and Robotics Modernization Plans for the UGV Program.

(4) Serve as proponent for technology requirements and technology base efforts which support Unmanned Ground Vehicles.

(5) Provide UGV test beds and field test support to assist users in the definition of operational and maintenance requirements in terms of tactics, doctrine, and training.

b. The U. S. Army Materiel Command (AMC) will:

(1) Provide office space for the Joint Project Office.

(2) Provide administrative and matrix support to the Joint Project Office.

(3) Provide the personnel and force structure indicated in paragraph 4.b from existing AMC assets.

(4) Provide 6.4 and procurement funds for the Army.

(5) Transfer to the JPO, the two Teleoperated Mobile All Purpose Platforms (TMAPs) developed for the U. S. Army.

c. The Marine Corps Research, Development and Acquisition Command will:

(1) Provide personnel listed in paragraph 4.b.

(2) Provide 6.4 and procurement funds for USMC.

(3) Transfer to the JPO the Teleoperated Vehicles (TOV) developed by NOSC.

DONALD S. PIHL
Lieutenant General, GS
Military Deputy to the
Assistant Secretary of the
Army (Research, Development
and Acquisition)

J. R. DAILEY
Lieutenant General, USMC
Commanding General
Marine Corps Research,
Development and
Acquisition Command

APPENDIX C

PROJECT MANAGER

By direction of the Army Acquisition Executive, and by appointment of me, as the Commanding General, U.S. Army Missile Command, I hereby appoint

Lieutenant Colonel Robert J. Harper

as the Project Manager for the

Joint Project Office, Unmanned Ground Vehicles

in accordance with the Army Acquisition Management System.

As Project Manager (PM), you will perform as the Army centralized manager for your assigned Project reporting directly to the Commanding General.

You will, as the responsible management official, provide overall direction and guidance for the concept definition, development, acquisition, testing, product improvements, and fielding of the assigned project.

You will coordinate, integrate, and directly control your subordinate managers within the assigned mission area.

You will place primary management emphasis on cost estimating, planning, programming, budgeting, program integration, international and joint considerations, interoperability and oversight.

You are hereby delegated the full line authority of the Commanding General, for the centralized management of the assigned project.

Unless sooner terminated, this appointment will remain in effect so long as the Project Manager is assigned to MICOM.

William S. Chen
Major General, U.S. Army
Commanding

APPENDIX D

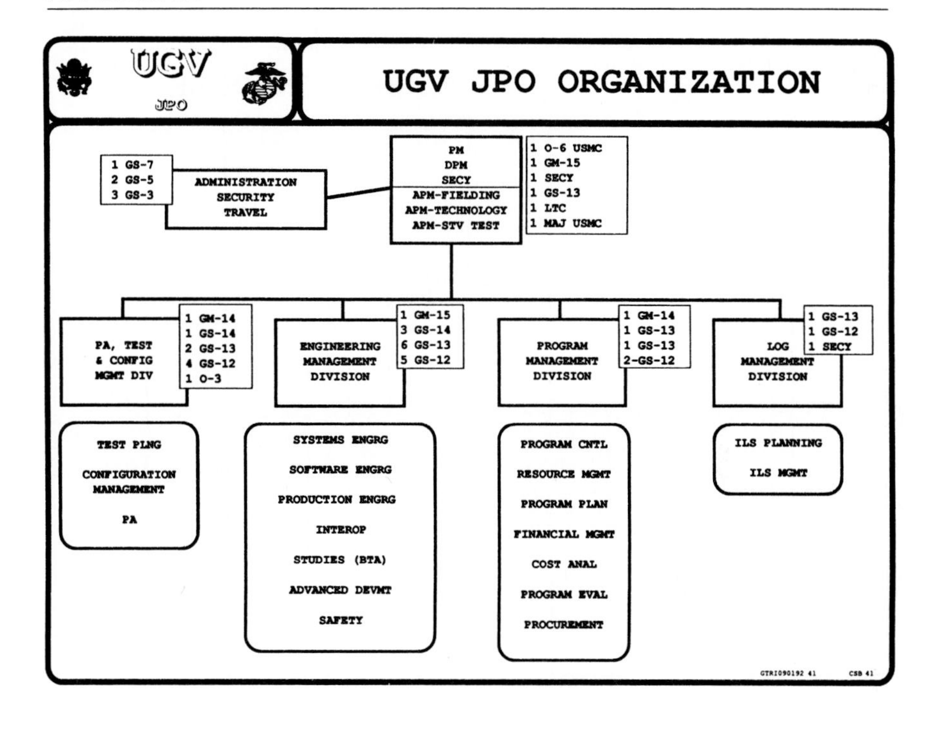
UGV
JPO
UGV JPO ORGANIZATION
PM
DPM
SECY
APM-FIELDING
APM-TECHNOLOGY
APM-STV TEST
1 O-6 USMC
1 GM-15
1 SECY
1 GS-13
1 LTC
1 MAJ USMC
1 GS-7
2 GS-5
3 GS-3
ADMINISTRATION
SECURITY
TRAVEL
PA, TEST
& CONFIG
MGMT DIV
1 GM-14
1 GS-14
2 GS-13
4 GS-12
1 O-3
ENGINEERING
MANAGEMENT
DIVISION
1 GM-15
3 GS-14
6 GS-13
5 GS-12
PROGRAM
MANAGEMENT
DIVISION
1 GM-14
1 GS-13
1 GS-13
2-GS-12
LOG
MANAGEMENT
DIVISION
1 GS-13
1 GS-12
1 SECY
TEST PLNG
CONFIGURATION
MANAGEMENT
PA
SYSTEMS ENGRG
SOFTWARE ENGRG
PRODUCTION ENGRG
INTEROP
STUDIES (BTA)
ADVANCED DEVMT
SAFETY
PROGRAM CNTL
RESOURCE MGMT
PROGRAM PLAN
FINANCIAL MGMT
COST ANAL
PROGRAM EVAL
PROCUREMENT
ILS PLANNING
ILS MGMT

APPENDIX E

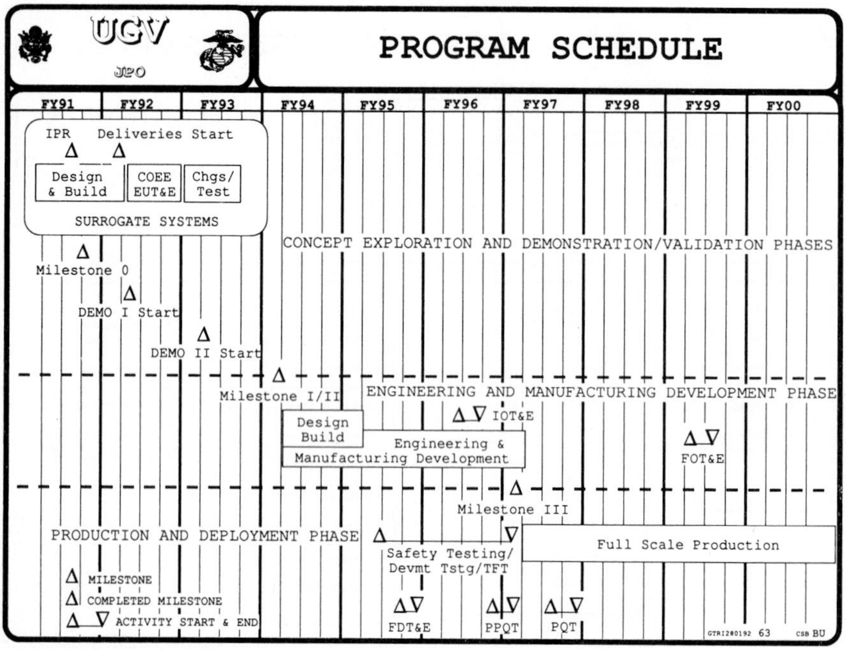

UGV
JPO

PROGRAM SCHEDULE

FY91 FY92 FY93 FY94 FY95 FY96 FY97 FY98 FY99 FY00

IPR Deliveries Start

Design & Build | COEE EUT&E | Chgs/ Test

SURROGATE SYSTEMS

CONCEPT EXPLORATION AND DEMONSTRATION/VALIDATION PHASES

Milestone 0

DEMO I Start

DEMO II Start

Milestone I/II

ENGINEERING AND MANUFACTURING DEVELOPMENT PHASE

IOT&E

Design Build

Engineering & Manufacturing Development

FOT&E

Milestone III

PRODUCTION AND DEPLOYMENT PHASE

Full Scale Production

Safety Testing/ Devmt Tstg/TFT

FDT&E

PPQT

PQT

Δ MILESTONE

Δ COMPLETED MILESTONE

Δ∇ ACTIVITY START & END

GTR1280192 63 CSB BU

APPENDIX F

UGS PROGRAM NOTES

An Industry Sponsored Publication of UGS Activities

UNMANNED GROUND SYSTEMS • IN OMNIA PARATUS • JOINT PROJECT OFFICE

Newsletter No. 4 January 15, 1992

What Exactly is a Tactical Unmanned Ground Vehicle?

Tactical Unmanned Ground Vehicle (TUGV) is the title given to the combat vehicle being developed by the Unmanned Ground Vehicle Joint Project Office (UGV JPO). The basic TUGV is composed of four major components: (1) a Mobile Base Unit (MBU), (2) an Operator Control Unit (OCU), (3) a Communications/Control Link, and (4) a Mission Module.

The Operator Control Unit will be one man portable. It will possess all the necessary controls and displays for both remote driving of the MBU and operation of the Mission Module.

Mission Module refers to the add-on package that is mounted on the MBU to give the TUGV a particular mission capability. The first generation TUGV is envisioned to have a Reconnaissance, Surveillance, and Target Acquisition (RSTA) capability.

The Mobile Base Unit is the vehicle platform which may carry the driving controls for manned operation and the actuators necessary for unmanned or remote driving. It will also be equipped with the necessary electronic and mechanical interfaces for mounting the Mission Modules and the required Communication/Control Link. This vehicle is not envisioned to be armored and maybe small enough to fit in the back of a standard HMMWV. It will possess the same level of mobility as the units which it supports.

Estimated Cost Percentages of TUGV Components

22.98%
12.89%
19.73%
44.40%

Operator Control Unit
Mobile Base Unit
Communication/Control Link
Mission Module

A Communication & Control Link is necessary for the operator to control the MBU and the Mission Module. Information and Commands will be transmitted via a fiber optic (FO) link and/or a wireless radio frequency (RF) link. The FO bandwidth allows non-line-of-sight capability. The tactical RF data link provides 2-way wireless communication.

NOTE: Estimates of total system cost by percentage for each component are illustrated in the pie chart.

~~ET~~ "TUGV" Phone Home

AT&T recently announced that they would have a video phone manufactured for public use. This advance in technology and commercial application leaves the UGV JPO with a very promising opportunity. Program adaptations of industrial technologies are always considered. The cost reduction from acquiring commercially tested and produced equipment would be significant.

Other commercial areas that may have an application for some of the technology being pursued in the JPO.

- Industrial Robots
- Space Exploration (NASA)
- Waste Management / Environmental Restoration (DOE)
- Transportation
  - — Intelligent Highway and Vehicle System (IHVS)
- Construction Industry
  - — Autonomous Trench Digger
  - — Autonomous Strip Mining
- Hazardous Events
  - — Transportation Accidents
  - — Bomb/Fire Incidents

Murphy's Laws of Combat

1. *If the enemy is in range so are you.*
2. *The easy way is always mined.*
3. *Teamwork is essential, it gives them someone else to shoot at.*
4. *The enemy diversion you have been ignoring will be the main attack.*

If these laws of combat apply to your program consider how a UGV could support your mission needs.

UGV Office is Growing

The Unmanned Ground Vehicle Joint Project Office (UGV JPO) is charged with developing, producing, and procuring UGV's. The UGV office continues to grow, both in responsibility for robotic programs and personnel. Program expansion has lead to more than doubling of the JPO staff. The current organization for the UGV JPO is depicted below.

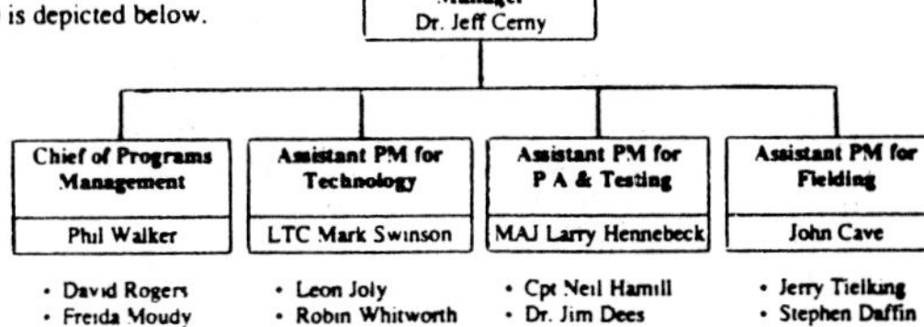

Now—hello and welcome aboard to the following people who have recently joined the UGV JPO staff.

Dr. Jim Dees - Human Factors/Test Engineer (Dec.)
Cpt Neil Hamill - Test & Evaluation Officer (Oct.)
Mr. Jerry Tielking - Logistics Specialist (Aug.)
Ms. Stephanie Daffin - Logistics Specialist (Dec.)
Ms. Frieda Moudy - Budget Analyst (Nov.)

Unmanned Ground Systems Program Notes

Newsletter No. 4 Page 2 January 15, 1992

RoboCop Presents

Sgt. Paul Hargrove of the New York City Police Department (N.Y.C.P.D.) will be the feature speaker at a dinner sponsored by the Pathfinder Chapter, Association for Unmanned Vehicle Systems (AUVS). The dinner/meeting will be at 6:00 pm on January 23 at the Redstone Arsenal Officers Club. Sgt. Hargrove's presentation will be on the use of robots by the N.Y.C.P.D. The department has used robots in hostage and other hazardous situations to reduce the danger to police officers and other personnel. For information/reservations: call Marilyn at (205) 922-7210.

For Your Information

The TUGV program is grandfathered under the old acquisition process—specifically the Army Streamlined Acquisition Process (ASAP). The TUGV program has completed the first requirement, the Operational and Organizational (O&O) concept, or Milestone Zero. The next requirement in our ASAP is the development of the Operation Requirements Document (ORD), for Milestone II.

Up Close and Personal -- The Operations Director

Major Lawrence M. Hennebeck was born in Ashland, Oregon. He graduated from Oregon State University, with a B.S. in Political Science and from Denver University, with a M.S. in Systems Management. Larry is married to the former Tomaline Conyers of Roseburg, Oregon and has two children.

Major Hennebeck was commissioned a 2d Lieutenant in the Regular Marine Corps in June 1973. He graduated officer basic school in December 1973, and after the necessary training, he was designated a Naval Aviator in Pensacola, Florida. He has performed various duties, including flight instructor and search and rescue pilot (SAR), accumulating over 4,000 hours of flight time.

He was assigned to the Marine Corps Research, Development, and Acquisition Command and worked with the USMC Ground-Air Telerobotics Program. The program later became part of the UGV JPO, where he is now Assistant Program Manager, U.S. Marine Corps and Operations Director.

Major Larry Hennebeck
Operations Director UGV JPO

Recently Major Hennebeck has been busy coordinating the Surrogate Teleoperated Vehicle (STV) activities to be conducted at Fort Hunter Ligget, CA. These activities have kept Larry on the move. Late last year he moved into a new office at the UGV JPO, and later this month he will be TDY at Hunter Ligget to schedule and oversee testing on the STV. GOOD LUCK!!

Surrogate Teleoperated Vehicle Activities

The Surrogate Teleoperated Vehicle (STV) is a small, lightweight, modular, teleoperated UGV. It can be transported in a HMMWV, towed by smaller vehicles, or driven either by an onboard driver or under remote command and control. The STV was built to formulate and evaluate concepts of employment and to engage the user in early test and evaluation. Robotic Systems Technologies, is delivering fourteen STV systems to the Government this month. The vehicles first stop is Aberdeen Proving Ground (APG) Maryland where the required safety, release, and acceptance testing will be performed.

Once the vehicles are released from APG, a series of development and operational tests and evaluation efforts will be conducted at Fort Hunter Ligget, CA (FY92). A Concept of Employment Evaluation (COEE) will use the STV's to test and refine operational and organizational concepts for both the Army and the Marine Corps. The COEE will include STV employment under controlled conditions and in field exercises that simulate operational scenarios. In essence, the COEE will be used to: develop how-to-fight tactics, techniques and procedures (TTPs); develop maintenance doctrine; and determine organizational mix. Additional user evaluations will be conducted at various military facilities (Fort Sill, Fort Knox, JTF6 and Fort McClellan). The COEE will be followed by a joint Early User Test and Evaluation (EUT&E). This hands-on user experience will clarify the operational role, value, and requirements of Unmanned Ground Vehicles (UGV's). Results will be used to develop the Operational Requirements Document (ORD) and to assess the combat value of TUGV's.

Based on the concepts developed within the COEE and EUT&E, selected technologies will then be integrated into the STV and tested. The resulting configuration, designated as STV(I), will reduce risk and provide a sound basis for Engineering Manufacturing and Development (EMD). In this respect, the COEE and EUT&E are designed to provide the operational experience required to field a first generation UGV system.

Conceptual drawing of STV

Coming Events

January
22-23 Project Review at Aberdeen
23 AUVS Pathfinder Chapter Dinner at the Redstone Arsenal Officers Club

February
10 COEE at Ft. Hunter Ligget, CA

March
3-4 Robotic Vehicle Working Group in Pasadena, CA

June
22-24 AUVS 92 Conference in Huntsville AL

Publication Notes

UGS Program Notes is published quarterly by W&G Sigma Services, Inc. Huntsville, AL.

The purpose of UGS Program Notes is to keep the UGS community abreast of program activities, and provide a center for the exchange of program-related requirements and information of general interest..

The Newsletter staff is listed below:
- Editor: Igor D. Gerhardt
- Staff Writers: William E. Powell
  Susan L. Gahagan

If you have comments or questions regarding the UGS Program Notes Newsletter or its articles, contact W&G Sigma Services, Inc., ATTN: UGS Newsletter, 3315 South Memorial Parkway, Suite 501, Huntsville, AL 35801 (Tel.: (205) 880-6910).

If you wish to publish an article of general interest to the UGS community, please forward your article to the above address.

APPENDIX G: GLOSSARY

AROD	Airborne remotely operated device
AUVS	Association for Unmanned Vehicle Systems
BAA	Broad agency announcement
BTA	Best technical approach
Caleb	Name of Army TUGV project
COEA	Cost and operational effectiveness analysis
COEE	Concept of employment and evaluation
DoD	Department of Defense
EMD	Engineering and manufacturing development
EUT&E	Early user testing and evaluation
FDT&E	Force development test and evaluation
FOT&E	Follow-on operational test and evaluation
GATERS	Ground-air telerobotics system
HMMWV	Highly mobile multipurpose wheeled vehicle
IPS	Integrated program summary
ISG	Industry support group
ISOR	Initial statement of requirements
JPO	Joint project office
JROC	Joint requirements oversight committee
MCPDM	Marine Corps program decision meeting
MOA	Memorandum of agreement
NOSC	Naval Ocean Systems Center
ORD	Operational requirements document
PEO	Program executive officer
PMO	Project or program management office
PPQT	Preproduction qualification test
RFP	Request for proposal
RSTA	Reconnaissance, surveillance, and target acquisition
STV	Surrogate teleoperated vehicle
TMAP	Teleoperated mobile all-purpose platform
TOA	Trade-off analysis
TOD	Trade-off determination
TOV	Teleoperated vehicle
TUGV	Tactical unmanned ground vehicle
UAV	Unmanned air vehicle
UCS	Unmanned combat system
UGV	Unmanned ground vehicle
USA	U.S. Army
USAIS	U.S. Army Infantry School
USMC	U.S. Marine Corps
WFC	U.S. Marine Corps Warfighting Center

CHAPTER 17
CRITICAL SUCCESS FACTORS IN DoD PROGRAM MANAGEMENT

James H. Dobbins

James H. Dobbins is an internationally recognized specialist in software quality and reliability with over 28 years of experience in these and related disciplines. He is currently Professor of System Acquisition Management at the Defense Systems Management College, Ft. Belvoir, Va. He is an attorney-at-law licensed to practice in Virginia and various federal courts.

INTRODUCTION

An objective of program management and of program management education is repeatable success as a program manager. It does little good if program managers are considered successful but do not know why they were successful and how to repeat that success. Success that is the result of luck is not really success, it is an illusion.

The use of critical success factors (CSFs) in the management of corporations has been the subject of several published studies. Very few CSF studies have been done of public-sector enterprises. Much of the private research conducted within individual companies, for obvious competitive reasons, has not been made public.

While profit-driven private-sector companies have virtual autonomy in their selection of suppliers and partners and almost absolute control over their expenditures, federal government program managers are required to engage in competitive bidding for almost every procurement. They also have a significant level of external control and oversight over schedules and expenditures due to the activities of the program executive officers (PEOs), the Defense Acquisition Board (DAB), and budget analysts working for Congress, the Government Accounting Office (GAO), and other oversight agencies. Program managers also have no profit motivation. In the absence of a clear profit motivation, researchers investigating CSFs have largely ignored the fundamental elements which determine success or failure of a federal government project.

The research conducted for this pilot study was accomplished using a questionnaire submitted to 130 DoD program managers responsible for the development and operation of systems involving large expenditures of taxpayer dollars. Most of the programs examined involved expenditures well in excess of $100 million. The results were grouped by whether the programs under development were embedded weapon system programs (group A) or information resource programs (group B). Based on analysis of the questionnaire responses, this pilot study identified 27 CSFs for group A and 24 for group B. There were 18 CSFs common to both groups, nine CSFs were unique to group A, and six CSFs were unique to group B.

The conclusion is that the CSFs, how they are identified and used, how they are measured, and how they are influenced should be a mandatory component of education for every federal government program manager or program manager selectee, as well as for all PEOs, federal budget analysts, members of congressional oversight committees, and members of the DAB.

THE RESEARCH ENVIRONMENT

The importance of CSFs in management first gained widespread attention through a publication by Rockart,[1] who showed the need among top executives for certain critical elements of information which were not being provided by the management information systems (MIS) personnel or the data analysis systems available. Executives suffered from data overload, but were starved for the right kind of data essential to making the decisions necessary to manage their enterprises effectively. This identified a need for both the determination of CSFs as well as the establishment of barometric indicators which can alert the executive when a CSF should be changed or when the assumption upon which a CSF is based is no longer valid. The initial Rockart paper was closely followed by the publication of a methodology for CSF identification developed by Bullen and Rockart.[2] The research conducted since then has been done either through the interview process, as described by Bullen and Rockart, or by the questionnaire method.

CRITICAL SUCCESS FACTORS

CSFs are identified as "the limited number of areas in which results, if they are satisfactory, will ensure successful competitive performance for the organization. They are the few key areas where things must go right for the business to flourish. If results in these areas are not adequate, the organization's efforts for the period will be less than desired".[1] If these activities are not done well, the project will not succeed except by accident. Clearly, the CSFs are "areas of activity that should receive constant and careful attention from management" (Rockart,[1] p. 85).

Identifying and managing the CSFs, and tracking them separately from the ever-increasing amount of data to which executives are subjected, has been the focus of significant private-sector research. Some of the research has limited the study to those activities over which the program manager has direct control,[3] while the majority of researchers have broadened the focus to include elements beyond the direct control of a project manager, but which are still within the sphere of things that he or she could influence, or which could exert significant influence on his or her activities.

Bullen and Rockart[2] have suggested that CSF identification be focused on whether there are identifiable CSFs which fall into one or more of several key areas. These key areas may be classified as follows:

1. *Global or Industry-Related.* Those activities essential to project success which would be true of any project or company operating in the particular environment (industry or business), not just this one company or project.

2. *External Influences.* The external factors which can influence the success of your endeavor significantly. Examples are Congress, higher authority outside your direct reporting chain, local government, or a major competitor.

3. *Internal Influences.* The internal factors which can influence project success significantly. This may include, for example, specific personnel skills, access to a particular technology, or essential funding.

4. *Current and Future.* The factors which are time-driven and which are essential to project success. They are those activities which must be done in the near future, that is, within the next two to three months, and those that are long-range, that is, they may not be required to be done until several months later. For the future activities, planning for their success may be an activity that requires immediate attention.

5. *Temporal and Enduring.* These are significant influences which either have a short-term duration or are present through most or all of a project. Loss of a critical-part supplier could create a temporal CSF. For a DoD program manager, having the prime contractor convicted of fraud can create both a temporal and an enduring set of CSFs. Continued availability of a product distribution network could be an enduring CSF.

6. *Risk Abatement.* These are the activities which, if not done, will pose a significant risk to project success. They are risk avoidance types of activities. Software quality assurance (SQA) should be viewed as a risk abatement activity, although it seldom is considered in that light, and the absence of an SQA activity at the contractor's location should be considered a major risk factor for a software-intensive development effort.

7. *Performance.* This is an identifiable level of performance or achievement which must be realized for the project to be successful. It may mean the achievement of a certain level of measurable reliability, the attainment of a critical speed, or a measurable level of availability of a system to the user community. For example, system availability could be a critical measure of performance when developing a system for air traffic control near major military airfields, or a system for controlling the speed and direction of smart bombs.

8. *Special Monitoring.* These are activities or events which require special monitoring, protection, or contingency planning in order to assure project success. They cannot ever be ignored.

9. *Quality.* These are quality requirements which, if not met, will mean the failure of the project. They can be anything from hardware tolerance limits to software processing characteristics, personnel skills, or any other process or activity which has critical quality requirements.

10. *Modification Management.* These are those activities or conditions which currently exist or are currently planned and which, if not changed, will cause the project to fail.

PRIOR RESEARCH

Prior research in the field of CSF has been focused primarily on the identification of CSFs for executive-level managers in specific industries, or heads of specific kinds of departments, principally MIS departments. This research has also focused attention on the diverse applications of CSFs. Some of the results follow.

1. *Use of CSFs as a Means for MIS Planning.*[4] An important outcome of this research was the recommendation that CSF identification not be driven by current information production capability within the organization, but rather by the actual information needs of management.

2. *Variation of CSFs over Stages in Project Life Cycle.*[5] In this research the authors hypothesized a set of CSFs and then conducted a validation study based on empirical evidence. The objective was to identify a set of CSFs, for each life-cycle phase, that were general rather than company- or industry-specific, and to determine the relative importance of the CSFs across life-cycle phases.

3. *Using CSFs as a Basis for Evaluating the Reliability of Information Systems.*[6] In this research the author developed a theory of reliability of an information system as a measure of the systems success based on CSFs.

4. *Use of CSFs as a Key Step in Overall Planning of Strategic Information Systems.*[7] In this research the author defines a nine-step process for information system planning, incorporating CSF identification and use as a major step in that process.

5. *Identifying the CSFs for Excellence in Project Management.*[8,3] In this area of research Kerzner identified six CSFs for project management excellence. In *Project Management Handbook,* Pinto and Slevin identified 10 CSFs for project management as follows:

1. Project mission
2. Top management support
3. Project schedule or plan
4. Client consultation
5. Personnel selection, recruitment, training
6. Technical tasks
7. Client acceptance
8. Monitoring and feedback
9. Communication
10. Troubleshooting

6. *Multiple Uses of CSFs.*[9] In this research, in addition to using CSFs for strategic planning purposes, the authors stress the use of CSFs for identifying criteria for strength and weakness assessment.

7. *Developing a Dual Look at CSFs.*[10] In this research the authors stress the importance of using CSFs to identify major causes of project failure as well as success.

8. *Using CSFs to Evaluate Effectiveness of Existing Information Systems.*[11] This research investigates the use of CSFs to evaluate the effectiveness of existing information systems.

9. *Determining the CSFs of Chief MIS Executives.*[12] This research identified seven CSFs which were felt to be universally applicable to MIS/DP executives.

10. *Comparative Analysis of the Differences between CSF Management Approaches and Process Management Approaches, and the Advantages of the CSF Approach.*[13] The author's view is that if the inquirer wants to know what management is, then the process view should be studied. However, if one wants to know why selected organizations are successful in highly competitive environments, then one must study CSFs.

There have been a few studies of public-sector strategies involving the application of CSFs. One such study was done in the social services field.[14] Here CSFs were identified for the abuse and neglect departments. In this example, application of CSFs is at a functional- or operational-level department as opposed to an executive-level application. The specificity of the CSFs indicates their limited scope of applicability to other public-sector organizations.

RESEARCH APPROACH

The research for this pilot study was conducted utilizing the recommended categories identified by Bullen and Rockart and adopted for a questionnaire approach. The questionnaire method was selected due to the number of and diverse locations of the program managers from whom information was desired. Along with the questionnaire, a package of instructional information about CSFs was provided, including the history of CSF development and how CSFs have been identified and used previously.

Strengths

The inherent strength of the questionnaire approach for this research lies in the number of program managers that can be accessed simultaneously, with the resultant savings in time and dollars. The questionnaire approach also allows the program managers to respond in several sittings as time becomes available, rather than having to take a contiguous block of valuable time for the interview on a subject of which they may have no prior knowledge. The questionnaire provides the basis for a more objective scoring and analysis of the results than would be possible using only personal interviews. It provides a neutral and standardized method of data collection and allows those responding to identify CSFs as well as giving them a view of the assumptions underlying the CSFs and ways in which the CSFs can be measured. The questionnaire also asked that those responding consider what, based on experience, they felt should be included as CSFs and as measures, regardless of what was actually practiced or actually available on their projects.

Weaknesses

The questionnaire was mailed just prior to the advent of activities in the Persian Gulf War. The primary overall weakness, therefore, was the need to elicit information from DoD program managers at a time of national crisis. This necessarily had a significant impact on the number of responses received.

With regard to the research method itself, one of the weaknesses of the questionnaire approach is the lack of opportunity to discuss a question with the program manager and explain the question more fully in case there is any doubt as to the meaning. It is subject to a certain degree of bias and does not allow for the pursuit of additional issues which might be important to the result and which are not a part of the questionnaire itself. In addition, there is no protection from the tendency either not to respond at all, or to respond hastily without giving the matter the intellectual time it requires. The questionnaire is lengthy and does take considerable time to answer. One other weakness is the possibility that the program manager to whom it is sent will have one of his or her staff personnel respond. This is not as disadvantageous as might be supposed since each level of management will have an identifiable set of CSFs, and the responses would be on behalf of the executive manager.

Participant Selection

Having selected the questionnaire approach, the set of program managers to whom the questionnaire was sent was obtained from lists maintained by the Information Resources Management (IRM) College of the National Defense University in Washington, D.C., and by the Defense Systems Management College (DSMC) at Ft. Belvoir, Va. This total list contained 130 program managers. Those from the IRM College are program managers for the development of nonweapon systems, previously identified as group B. They may include personnel systems, housing and relocation systems, payroll systems, foreign military sales, or other similar systems. Those from DSMC are program managers for the development of systems which are identified as mission critical computer resource (MCCR) systems, typically embedded weapon systems and major intelligence systems, previously identified as group A.

The effectiveness of the information system utilized in the management of the development of either group of systems is essential to the success of the development. Certain information systems may themselves be used to support or to provide information critical to the development of group A systems. These information support systems may include command and control systems, intelligence systems, or any other kinds of information systems. The questionnaire used:

1. Identifies the project parameters and state of completeness of the project
2. Guides the responder in the identification of CSFs
3. Assists the responder in iterating on the initial list to combine and pare the number of selected CSFs to 10 or fewer
4. Assists in the identification of specific measures for each CSF identified
5. Assists in the identification of specific data sources required for the CSF identification and measurement

The participants were then asked to group the CSFs identified by relative importance in each project life-cycle phase. The questionnaire is included in a DSMC research report.[15]

The questionnaire was mailed out to each identified program manager. Of the 130 to whom it was sent, there have been 34 returns to date, and of those, 21 are usable responses. Others from whom responses were sought have indicated that responses are still in preparation.

INITIAL RESULTS

Responses

Of the total 34 responses received, 13 were rejected as nonresponsive. The rejected responses included those that were returned unanswered due to lack of time to respond, one which indicated a failure to see the applicability of CSFs to the job of program management, and one in which the responder indicated that there were so many factors critical to someone in the chain of command (not necessarily the program manager) that he was overwhelmed by the thought of trying to identify them. The remaining 21 were analyzed for CSF identification.

The data were broken down between the two groups. Group A included 73 program managers. There were 20 group A returns, and 10 of those were usable. Group B included 57 program managers. There were 14 group B returns, and 11 of those were usable.

The returns were examined for identification of CSFs and for measures which might be or are being utilized. No program manager reported more than 10 CSFs. The program managers for group A reported a collective set of 37 CSFs; the program managers for group B reported a collective set of 29 CSFs.

The initial set of reported CSFs for group A are shown in Table 17.1, which lists

TABLE 17.1 Group A Initial Results

Number	Frequency	CSF
1	8	Stable and adequate funding
2	7	Clearly defined and stable requirements including interface
3	5	Risk management
4	6	Technically competent program office staff
5	2	Develop/execute program management strategic plan
6	1	Configuration management and control
7	3	Stable and adequate personnel resources
8	1	Thorough system documentation
9	5	Schedule management
10	3	Cost management
11	3	User involvement/support/acceptance
12	3	Strong and structured quality control
13	2	Clearly/objectively defined project goals
14	4	Continuous meaningful visibility using measures
15	2	Early and continued monitoring and evaluation
16	1	Other agency support for training and GFE
17	1	Adequate program office resources
18	3	Stable, qualified industrial base
19	4	Manage political influencing agents
20	4	Effective vertical/lateral communications
21	2	Contractor support
22	1	Leadership
23	3	Effective contractor review process
24	1	Test and evaluation master plan approval
25	2	Change management
26	5	Effective technical performance evaluations
27	1	RAM requirements
28	1	Program office teamwork
29	1	Establish system engineering competence
30	1	Open communication with contractor
31	1	Effective/timely decision making
32	1	Effective prototype performance
33	1	Foreign military sales
34	1	Requirements feasibility analysis
35	1	Measure and control ILS performance
36	2	Staff professional development
37	1	Initiation of new projects

TABLE 17.2 Group B Initial Results

Number	Frequency	CSF
1	7	Top management support
2	10	Stable and adequate budget
3	2	Clearly defined mission
4	2	Strong knowledge of life-cycle management
5	5	Stable project staff
6	5	Detailed requirements analysis
7	1	Objective economic analysis
8	2	Cost management
9	6	Schedule management
10	6	Strong quality control program
11	1	Achievable engineering objective
12	1	Stable, qualified industrial base
13	6	Effective lateral and vertical communications
14	3	Risk management
15	2	Common sense
16	5	Effective technical performance evaluation
17	1	Knowledge of government contracting process
18	9	User involvement and support
19	3	Incremental acquisition
20	8	Technically competent staff
21	3	Prototyping
22	1	Visibility
23	2	Adequate program office resources
24	5	Manage political influencing agents
25	1	Configuration management
26	4	Other agency support: training and GFE
27	1	Leadership
28	2	Requirements feasibility analysis
29	1	On-site team to prevent fraud, waste, abuse

the 37 CSFs identified and the frequency of reporting of each. Table 17.2 shows the same type of data for the 29 CSFs reported for group B.

Upon examination of the data it was determined that certain of the reported CSFs in each group could be combined. Even though they were not exactly the same, and there may be subtle differences between them, these CSFs were similar enough that for program management information system purposes, and in light of the measures required for each CSF, they could be combined. For group A,

CSFs 2, 27, and 34 were combined

CSFs 4, 29, and 36 were combined

CSFs 14, 15, 23, 26, and 32 were combined

CSFs 18 and 21 were combined

CSFs 20 and 30 were combined

For group B,

CSFs 4 and 17 were combined

CSFs 6, 11, and 28 were combined

CSFs 16, 21, and 22 were combined

Since there were 18 CSFs in the resultant list which were common to both groups, the CSFs were then renumbered to take into account the combinations and to apply the same number to both groups for those CSFs that were common to each. These data are shown in Tables 17.3 to 17.5.

It should be noted that a few CSFs were identified in addition to those reported here. These were very specific to a particular program and are therefore not included here.

Measures

The respondents were asked to identify measures they use, or would use, for the CSFs they had identified. The group A respondents identified a total of 41 different measures; each group A respondent identified at least six measures. The group B respondents identified a total of 27 different measures. Three of the respondents for group B were unable to identify any measures. One other group B respondent indicated that there were too many factors required to come together for success and the only real measure is the ultimate outcome.

Of the measures identified by the two groups, measures 1 through 14 are considered common to both groups. Although in some of the common 14 measures the wording differs between the two groups, the intent of the measure is felt to be the same. Those measures numbered 15 or greater in each group are unique to that group.

TABLE 17.3 Group A Modified List

Number	Frequency	CSF
1	8	Stable and adequate funding
2	9	Clearly defined and stable requirements, including interface
3	5	Risk management
4	9	Technically competent program office staff
5	1	Configuration management and control
6	3	Stable and adequate personnel resources
7	5	Schedule management
8	3	Cost management
9	3	User involvement/support/acceptance
10	3	Strong and structured quality control
11	2	Clearly/objectively defined project goals
12	15	Continuous meaningful visibility using measures
13	1	Other agency support for training and GFE
14	1	Adequate program office resources
15	5	Stable, qualified industrial base
16	4	Manage political influencing agents
17	5	Effective vertical/lateral communications
18	1	Leadership
19	2	Develop/execute program management strategic plan
20	1	Thorough system documentation
21	1	Test and evaluation master plan approval
22	2	Change management
23	1	Program office teamwork
24	1	Effective/timely decision making
25	1	Foreign military sales
26	1	Measure and control ILS performance
27	1	Initiation of new projects

TABLE 17.4 Group B Modified List

Number	Frequency	CSF
1	10	Stable and adequate budget
2	8	Detailed requirements analysis
3	3	Risk management
4	8	Technically competent staff
5	1	Configuration management
6	5	Stable project staff
7	6	Schedule management
8	2	Cost management
9	9	User involvement and support
10	6	Strong quality control program
11	2	Clearly defined mission
12	9	Effective technical performance evaluation
13	4	Other agency support: training and GFE
14	2	Adequate program office resources
15	1	Stable, qualified industrial base
16	5	Manage political influencing agents
17	6	Effective lateral and vertical communications
18	1	Leadership
19	7	Top management support
20	3	Strong knowledge of life-cycle management
21	1	Objective economic analysis
22	2	Common sense
23	3	Incremental acquisition
24	1	On-site team to prevent fraud, waste, abuse

TABLE 17.5 Group A/B Combined Common CSFs

Number	Frequency	CSF
1	18	Stable and adequate funding
2	17	Clearly defined and stable requirements, including interface
3	8	Risk management
4	17	Technically competent program office staff
5	2	Configuration management and control
6	8	Stable and adequate personnel resources
7	11	Schedule management
8	5	Cost management
9	12	User involvement/support/acceptance
10	9	Strong and structured quality control
11	4	Clearly/objectively defined project goals
12	24	Continuous meaningful visibility using measures
13	5	Other agency support for training and GFE
14	3	Adequate program office resources
15	6	Stable, qualified industrial base
16	9	Manage political influencing agents
17	11	Effective vertical/lateral communications
18	2	Leadership

The measures and the number of times each was identified by the group as a measure are listed in Tables 17.6 and 17.7.

Although some of the measures could be combined, such as 13 and 32 in the group A list, they are not. The differences in the measures themselves are considered useful, not only their intent. For example, measure 13 in group A lists specific quantitative measures for system reliability, some of which would usually require the availability of a computer-based reliability model. The measures identified in mea-

TABLE 17.6 Measures for Group A

Number	Frequency	Measure
1	1	Quantitative assessment of requirements
2	7	Number and frequency of requirements changes
3	4	Changes to budget
4	8	Deviation from cost
5	10	Deviation from schedule
6	7	Number of unique trouble reports
7	2	Number of customer complaints
8	4	Results of tests of independent systems
9	3	User/contractor walkthroughs and reviews
10	2	Prime contractor productivity per 7000.2
11	1	Program quality targets
12	3	Funding level vs plan
13	1	VROC, MTBF, Pd, MTTR
14	1	System availability
15	1	Response to change (qualitative)
16	2	Program plan assessment (qualitative)
17	1	Number of issues requiring higher approval
18	1	Time taken for approval decisions
19	1	Number of acquisition protests
20	4	Cost of change vs cost of delay = cost to improve
21	2	Number of reworks/rewrites
22	1	Number of first-time approvals
23	2	Time between problem occurrence and problem ID
24	1	Time to process approved change
25	2	Time delay of GFE deliveries
26	1	Workload stability
27	1	Number/effect of congressional interactions
28	1	Effectiveness of visibility processes
29	1	Number of delay/disruption claims from contractor
30	1	Reject rates
31	1	Number of quality deficiency reports
32	1	RAM measures
33	1	Number of miscues per month (no coordinations, misunderstandings)
34	1	Number of technical surprises per month
35	1	Number of suggestions adopted by contractor
36	1	Number of delinquent action items-days late
37	1	Cost vs operational effectiveness
38	1	Number of risks identified per month
39	1	Number of risks resolved per month
40	1	Number of qualified staff vs need
41	1	Number of physical resources vs need

TABLE 17.7 Measures for Group B

Number	Frequency	Measure
1	1	Requirements review
2	2	Number of system requirements changes
3	2	Budget changes
4	4	Deviation from cost
5	4	Deviation from schedule
6	4	Number of unique trouble reports
7	1	User acceptance
8	1	Test result data reports
9	3	Program reviews
10	1	Productivity
11	3	Product quality
12	2	Funding level vs plan
13	1	System reliability
14	1	Down time, rate and duration
15	2	User feedback
16	1	Progress demonstration
17	1	Contractor product demonstrations
18	1	Personnel evaluations
19	2	External and internal IV&V
20	1	Milestone resource review
21	3	System throughput (performance)
22	1	System backlog
23	1	Analysis reports
24	1	Evaluation against oversight criteria
25	1	Number of support complaints
26	1	Number of software changes
27	1	Time to complete software change

sure 32 refer to Reliability/Availability/Maintainability (RAM) requirements for the particular contract, and therefore encompass more than just reliability; they are less specific in terms of identification of what quantitative values are measured. This measure also overlaps with measure 14, system availability.

System Size

The responses received for group A indicated program sizes of $45 million to $30 billion. In most cases, the responses were from the program managers, with some responses from project managers.

The responses received from group B indicated program sizes of $0.4 million to $3 billion. Responses were from program managers in most cases, with some responses from project managers.

Responder Rank or Position

Responses were received from both military officers and civil service individuals. The responses from the military were mostly from O-6 level officers (colonel or navy captain), with three responses from officers in the O-5 grade (lieutenant colonel or navy commander). The responses from civilians were from those at the GS/GM-15 and GS/GM-14 levels.

DATA ANALYSIS AND FINDINGS

Analysis

Group A CSFs. All references, unless indicated otherwise, are to the modified list of CSFs shown in Table 17.3. In group A by far the most frequently recorded CSF was 12, continuous meaningful visibility, with 15 occurrences. The next most frequently reported CSFs were 2, clearly defined and stable requirements, and 4, technically competent program office staff. A unique component of the group A response for CSF 2 is establishing a system engineering expertise within the program office itself, not just within the contractor. This aspect of technical competence was not present in the group B responses. Stable and adequate funding, CSF 1, was the third most frequently cited CSF.

Contractor evaluations were broad-based and included technical performance evaluations, management reviews, and other evaluations.

Group B CSF. All references, unless indicated otherwise, are to the modified list of CSFs as shown in Table 17.4. In group B, stable and adequate funding was the most frequently cited CSF. Its prominence in comparison to the other CSFs was not as dramatic as the most prominent CSF in group A.

Contractor evaluation was concentrated on technical performance evaluations.

Groups A and B Combined. All references, unless indicated otherwise, are to the combined list of common CSFs shown in Table 17.5. Of the 18 CSFs common to both lists, three (5, 15, and 18) were reported only once in group B, while four (5, 13, 14, and 18) were reported only once in group A. There were differences in emphasis evident, in terms of both frequency of reporting as well as the subtleties of their content.

Overall Observations. Given the publicity afforded to configuration management (CM), it was a surprise that CM occurred only once in each group. This could mean that those reporting:

1. Did not recognize the importance of CM, or
2. Recognized the importance of CM, but believe it is done well enough now not to be a prime candidate for program manager attention, or
3. CM has not been a significant source of problems requiring program manager attention

The questionnaire requested the program manager to list those activities that were felt to be critical to the success of the program. Issues reported will, in some way, reflect those areas which have required a significant degree of program manager attention.

Activities reflective of evolutionary acquisition (EA) or incremental acquisition (IA) appeared peripherally. In group A the issue of prototyping, which is not exclusively an EA or IA issue, was reflected in CSF 32 on the initial list of 37 CSFs (Table 17.1); IA and prototyping appeared on the original group B list as CSFs 19 and 21 (Table 17.2).

Another unexpected result was the order of prominence of risk management in the group B list, being 13th in the order of frequency of response. This same CSF is ranked 6th in the order of importance in group A.

With regard to the CSFs common to both groups, and combined into the single list of Table 17.5, the list reflects a strong belief that continuous meaningful visibility using measures is of primary importance to program success. Stable and adequate funding, clearly defined and stable requirements, and technically competent program office staff were next in order of importance and nearly equal in prominence. These appeared more prominently than either cost management or schedule management. This may reflect a belief that if these top four CSFs are achieved, cost management and schedule management are more easily accomplished. This ordering was also a surprise given the attention and publicity that both cost management and schedule management receive in the available program management literature, and the level of focus each maintains among the various oversight groups, such as the DAB, budget analysts working for the GAO, and Congress. This may suggest that the focus of the external groups is not on those activities which are most important to program success, at least not as viewed by those responsible for executing the mission of program management. It also indicates that a focus on cost and schedule is more hindsight than management. It is generally understood among program managers that a cost and schedule anomaly is the result of factors which happened much earlier and of which the program manager did not have visibility, or which were not managed adequately at that time.

One must also recognize that the CSFs identified are not necessarily disjoint. For example, continuous and meaningful visibility will be a necessary component of risk management. This is likewise true of strong and structured quality control and technically competent program office staff. All of these tend to be means to manage program risk.

Measures

Analysis of the measures is accomplished in light of the identified CSFs. It was anticipated that the most frequently mentioned CSFs should have measures reflecting them, and those with minimal mention may be expected to have the least number of measures.

Group A Measures

1. In group A, the most frequently mentioned measure is 5, deviation from schedule. The next most frequently mentioned measure is 4, deviation from cost. This may be reflective of either or both of the following: (a) ease of data collection, and (b) the need to respond to GAO, Congress, and the DAB, rather than focusing on those activities which the program managers clearly felt were of significantly superior importance to program success.
2. Continuous meaningful visibility was identified as the CSF of primary importance to the program managers. Given the importance afforded this CSF, those measures which can be considered reflective of this CSF are listed in Table 17.8.

In terms of sheer volume, these measures reflect the importance afforded continuous visibility. These measures were distributed across the spectrum of those reporting, and the majority were only identified once, an obvious cause for concern. The most frequently mentioned of all the visibility measures are those related to require-

TABLE 17.8 Measures for Continuous Meaningful Visibility

Number	Frequency	Measure
1	1	Quantitative assessment of requirements
2	7	Number and frequency of requirements changes
6	7	Number of unique trouble reports
7	2	Number of customer complaints
8	4	Test results
9	3	Walkthroughs and reviews
10	2	Contractor productivity
11	1	Program quality targets
13	1	VROC, MTBF, Pd, MTTR
14	1	System availability
28	1	Effectiveness of visibility processes
31	1	Number of quality deficiency reports
32	1	RAM measures
34	1	Number of technical surprises per month
38	1	Number of risks identified per month
39	1	Number of risks resolved per month

ments changes and to trouble reports. The next most frequently mentioned measure was 8, test results.

Walkthroughs and reviews, CSF 9, a widely publicized source of visibility during early life-cycle stages, was mentioned only three times. This is cause for concern both in terms of both what walkthroughs and reviews can provide that is not being utilized, and the need to assure that these visibility mechanisms are properly reflected in the acquisition process, particularly in the RFP and the contract.

For CSF 4, technically competent program office staff, the second most frequently mentioned CSF, there are virtually no measures reflected. Some of the more evident measures which might have been included are personnel evaluations, training costs, number of errors due to lack of technical skills, time to evaluate contractor technical deliveries, and number of unresolved technical action items.

It appears evident that there is a disconnect between those activities deemed critical to program success and the measures generally used or suggested on active programs. The number of measures related to the top-priority CSF, but which were only reported once, reflects a lack of consistency among those engaged in acquisition activities in terms of how commonly recognized critical issues can and should be evaluated and reported to higher authority and oversight agencies. Every program manager is left to invent the measures anew, and the oversight community is placing no demands for evidence of visibility in its requests for information. Their focus is almost entirely on cost and schedule.

Group B Measures. In general, (1) the lack of dominance of any one measure, and (2) deviation from cost, deviation from schedule, and number of unique trouble reports being the three most frequently named measures is reflective of a lack of familiarity with quantitative evaluation processes as well as a lack of consistency and standardization among those engaged in nonweapon system acquisition in terms of how those activities deemed critical to success can and should be evaluated. This may be partially the result of a dependence on regulations such as the 7920 series, and the absence of the application of standards such as DoD-STD-2167A, DoD-STD-2168, MIL-STD-1521, and other similar documents.

CSF 2, detailed requirements analysis, is noted eight times. Measure 1, requirements review, was mentioned only once; measure 2, number of system requirements changes, was mentioned only twice.

Of the 27 measures reported, 16 were mentioned only once. Five of the measures were only mentioned twice. Therefore only six of the 27 measures were mentioned more than twice, although 11 of the 24 CSFs for group B were listed at least five times. There is a clear disjoint between the activities considered critical to program success and the management information available by which the program managers can evaluate those critical factors.

The most frequently named measures are more reflective of responses to issues most often cited in GAO reports and DAB reviews than they are of the issues actually considered by the program managers as most critical to success.

General Concern

One clear concern is that if program managers recognize certain activities to be critical to program success, and the program manager's information network does not provide measures reflective of those critical factors, then their ability to manage those critical factors, or to know when a factor is not being met, is jeopardized. The information system providing the information content by which the program is being managed is not supportive of the program.

Findings

Based on the preceding, the findings are as follows.

1. The CSFs for DoD program management are identifiable, and requiring their identification would clearly assist the program managers in maintaining management focus on the factors most important to program success.

2. A significant number of CSFs are common to both group A and group B type programs.

3. The component assumptions and emphasis for a given CSF common to both groups A and B may be slightly different for the two groups. This difference is largely a function of the difference between the missions of the two groups, group A being more concerned with the complete development of total systems than group B.

4. The CSFs identified by the program managers as the most significant for program success do not appear to be those factors which require the most attention in order to be responsive to the oversight agencies, Congress, and GAO.

5. The measures identified most often by the program managers as those used or recommended are significantly more oriented toward the cost and schedule factors, which must be briefed to the oversight agencies, Congress, and GAO, than toward those factors felt by the program managers in the field to be most critical to the program success.

6. There is no widely recognized and generally utilized set of measures consistent with the most frequently reported CSFs. This leads to the conclusion that even though various factors are recognized as critical, the information network required to manage against those critical factors is not available to the program managers.

7. A commonly recognized set of CSFs, and a consistent measurement-based information network based on these CSFs, would be of significant benefit to both the program managers and the oversight agencies, Congress, and GAO. Such a management system would significantly improve the management success potential on programs across the board, and would provide the external groups a consistent way of evaluating and comparing different programs so that recommendations for future improvements could be based intelligently.

8. A CSF-based information network for program management would not only lend itself to increased visibility and awareness across the life-cycle phases for the program manager and staff, but would provide the base for the establishment of barometric measures for determining when the underlying assumptions for a given CSF may be changing.

9. A CSF-based information network would provide a common framework for productive discussions between the program manager and the external groups.

10. A CSF-based information network would significantly reduce the duplicative reporting and diversions required of the program managers, which are experienced under the present conditions.

RECOMMENDATIONS

Based on the information discussed previously, the following is recommended.

1. Educate program managers and their staff in the CSF identification process.
2. Educate the oversight agencies in the CSF identification process.
3. Educate the program managers and their staff in the development of information networks consistent with CSFs.
4. Establish oversight reporting mechanisms consistent with CSFs.
5. Develop and provide to program managers a measurement system consistent with the CSF information requirements and the oversight reporting requirements.
6. Minimize duplication of effort between the development of information required for management of the program and that required to support the requests of oversight agencies.
7. Utilize DoD and military standards, as well as the revised DoD Instruction 5000.2, on all programs, not just on weapon system programs.
8. Continue the research in this area to refine the CSF identification and information system requirements so that DoD programs can be managed more effectively.

REFERENCES

1. J. F. Rockart, "Chief Executives Define Their Own Data Needs," *Harvard Bus. Rev.*, vol. 57, pp. 81–93, Mar.–Apr. 1979.
2. C. V. Bullen and J. F. Rockart, "A Primer on Critical Success Factors," CISR WP 69, MIT Sloan School of Management, Center for Information Systems Research, June 1981.
3. J. K. Pinto and D. P. Slevin, "Critical Success Factors in Effective Project Implementation," in D. I. Cleland and W. R. King (eds.), *Project Management Handbook,* 2d ed., Van Nostrand Reinhold, New York, 1988, chap. 20, pp. 479–512.

4. M. E. Shank, A. C. Boynton, and R. W. Zmud, "Critical Success Factor Analysis as a Methodology for MIS Planning," *MIS Quart.*, pp. 121–129, June 1985.
5. J. K. Pinto and J. E. Prescott, "Variations in Critical Success Factors over the Stages in the Project Life Cycle," *J. Manage.*, vol. 14, no. 1, pp. 5–18, 1988.
6. F. Zahedi, "Reliability of Information Systems Based on the Critical Success Factors—Formulation," *MIS Quart.*, pp. 187–203, June 1987.
7. P. V. Jenster, "Using Critical Success Factors in Planning," *Long Range Planning*, vol. 20, pp. 102–109, Aug. 1987.
8. H. Kerzner, "In Search of Excellence in Project Management," *J. Syst. Manage.*, vol. 38, pp. 30–39, Feb. 1987.
9. J. K. Leidecker and A. V. Bruno, "Identifying and Using Critical Success Factors," *Long Range Planning*, vol. 17, pp. 23–32, Feb. 1984.
10. J. J. Walsh and J. Kanter, "Toward More Successful Project Management," *J. Syst. Manage.*, pp. 16–21, Jan. 1988.
11. F. Bergeron and C. Begin, "The Use of Critical Success Factors in Evaluation of Information Systems: A Case Study," *J. Manage. Inf. Syst.*, vol. 5, pp. 111–124, Spring 1989.
12. E. W. Martin, "Critical Success Factors of Chief MIS/DP Executives," *MIS Quart.*, pp. 1–9, June 1982.
13. K. H. Chung, *Management: Critical Success Factors*, Allyn and Bacon, Newton, Mass., 1987.
14. L. Garner, "Critical Success Factors in Social Services Management," *New England J. Human Serv.*, vol. 6, no. 1, pp. 27–31 1986.
15. J. H. Dobbins, "Critical Success Factors in Federal Government Program Management," DSMC Research Rep. DSMC-SET-01-91-0001.01, Defense Systems Management College, Ft. Belvoir, Va., Mar. 1991.

CHAPTER 18

INTEGRATED WEAPON SYSTEM MANAGEMENT: EVOLVING A NEW USAF PROGRAM MANAGEMENT STRATEGY

Fred Abrams

Fred Abrams served as the USAF project officer for defining, developing, and implementing integrated weapon system management (IWSM) as the foundation management approach for the formation of Air Force Materiel Command (AFMC) in 1991–1992. Prior to this he served as a fighter pilot, including a Viet Nam combat tour in F-100s; as an acquisition manager in Air Force Systems Command, where he headed the F-15 site activation and field support efforts and was later Chief of Test; and as a logistician, including duty as the F-15E Deputy Program Manager for Logistics, Vice Commander of the Logistics Operations Center, and Headquarters Air Force Logistics Command Director of Tactical Force Structure. He has an M.S. degree in systems management (with honors and election to Tau Beta Pi) from the Air Force Institute of Technology (AFIT) and holds Certified Acquisition Manager and Certified Professional Logistician (CPL) credentials. He is currently employed by Modern Technologies Corporation (MTC), continuing to apply the integrated management approach described in this chapter.

INTRODUCTION

In January 1991 the Secretary of the Air Force announced that the Air Force would deactivate Air Force Logistics Command (AFLC) and Air Force Systems Command (AFSC) and would activate Air Force Materiel Command (AFMC), a single USAF organization with life-cycle responsibility for technology, development, and support of military systems. Later in 1991 USAF announced that AFMC also would assume some Air Force Communications Command (AFCC) acquisition activities. The magnitude of AFMC responsibilities can be better grasped with a few statistics—one major air command of 128,000 people (including 40 percent of all USAF civilian employees) spending over half the entire USAF budget (over $40 billion per year).

Among the most significant challenges facing the architects of AFMC was the existence of two different cultures—one with a focus on front-end development activities that turn technology and customer requirements into delivered systems via contracts with industry, and one with life-cycle support and with modification and repair activities involving significant organic industrial capabilities. It became immediately obvious that AFMC faced two basic options—either create a single headquarters while retaining separate focuses on acquisition and logistics within the staff and field organizations, or build an integrated command from top to bottom (or even better, from bottom to top). Decision makers chose the latter, using the phrase "integrated weapon system management" (IWSM) to describe the new philosophy.

This chapter will describe two different aspects of the efforts to accomplish this fundamental change. First, we will explore the corporate re-engineering approach used to flesh out the definition of IWSM and plan its implementation, and then describe the nature of the IWSM strategy that forms AFMC's foundation. The first is important because it is a lesson in applying total quality management (TQM) precepts in undertaking the re-engineering of a management philosophy and organization. Understanding the approach to getting the answer is pivotal to understanding the result and may be of value to military project managers in forming or re-engineering their own organizations. Understanding IWSM is important to contractors and other services that must deal with program managers in AFMC. In addition, program or project managers outside AFMC may find value in elements of this TQM-based management philosophy.

ORGANIZING TO PLAN CHANGE

The commanders of AFLC and AFSC had a major advantage as they faced the task of planning the integration of the two commands—both were strong advocates of a TQM approach to management and were willing to apply it to all planning and organizing efforts. AFLC won the 1991 President's Award for Quality and Productivity (the public-sector version of the much sought Malcolm Baldrige Award), and both commands had embraced team building, customer focus, and worker empowerment as basic tenets of their operations. The integration effort represented an opportunity to apply TQM to the largest merger in U.S. history. Traditionally such reorganizations had involved a small cadre of senior leaders who sequestered themselves, came up with the answers, and then passed those answers to the rest of the organization to implement. The AFMC integration effort also started with a small cadre of senior folks, but in this case they were charged with coming up with the concepts and approaches to use in evolving the IWSM process. The senior leaders' role was to agree on a vision, mission statement, goals, and objectives for the new organization and to assure that the implementation approach was built from the bottom up by those who knew the business best—the managers immersed in day-to-day hands-on program management. The effort was divided into two basic parts, the merger of the two headquarters (which is only mentioned in passing in this paragraph) and the development of a new program management culture, IWSM. The headquarters aspect had a mix of straightforward tasks, such as merging the chaplains, medical staff, and civil engineers, and some "toughies," as merging the requirements staffs where the very word requirements meant different things to folks in the two cultures. It was clear from the outset that these latter types of headquarters changes would have to evolve with IWSM.

The commanders' stated integration objective for all participants was "to build a stronger Air Force by:

- Integrating the work force and infrastructure (talents and capabilities) of the two commands and synergistically employing the strengths of both
- Improving the current business practices by providing a completely integrated weapon system management process using a cradle-to-grave philosophy
- Providing a single face to operational commands that covers all aspects of integrated weapon system management and establishes a clear line of accountability that enhances responsiveness."

An IWSM concept definition team was established in January 1991 and charged with defining basic IWSM tenets and planning the re-engineering approach to evolve exactly what AFMC would implement on July 1, 1992. In forming the concept team it became immediately obvious that there were very few people who had a good grasp of what went on in *both* commands. By the time the seven-person team began work it was also obvious that cross-cultural training would be an imperative for AFMC. The IWSM concept team analyzed the work of both commands and categorized it into eight core process areas, which are described later. The team also laid out the strategy for involving the largest number of stakeholders in coming up with (and buying into) process and organizational results in which all had confidence. The efforts of the concept team were overseen by an executive committee comprising the general officer field commanders of both commands in a further effort to reflect broad-based experience and to gain up-front buy-in. The concept team's efforts culminated April 2, 1991, in a concept presentation by the AFLC and AFSC commanders to the Secretary of the Air Force and USAF Chief of Staff. They approved the concept and launched the major part of the IWSM process development effort.

IWSM—The Concept

The foundation upon which AFMC and its program management philosophy is based has three major components that will serve repeatedly to test the direction of IWSM process development results. First, the new command would offer a single face to the user through a single organization responsible for each military system (a generic term that includes weapon systems) or commodity. The makeup of the single organization would likely change over time and could comprise elements at different locations, but the USAF practice of program management responsibility transfer (PMRT), the handoff from AFSC to AFLC with all its negotiation of residual tasks, would end. The responsibility for each program would never leave the hands of the program manager, although he or she may physically change locations within the organization as the center of gravity of his or her efforts evolve. The second major attribute of IWSM was the application of this philosophy from cradle to grave. These two events were defined as "not later than milestone 1" through either the retirement or the cancellation of the system. The third major attribute was that the single manager would preside over a seamless organization based around eight critical or core AFMC processes that were to be integrated across the life cycle. After July 1, 1992, a fourth major element was added, that of integrated product teams (IPT) that cross functional and organizational boundaries.

IWSM—Arriving at a Definition

A short working definition of IWSM was very carefully crafted and continually serves as the most important test of any decision on program management approaches or organizational alternatives. There are two main principles that underlie the definition. First, in arriving at the definition the concept definition team concluded that the single most important variable in program or project management is authority. The authority over activities in each critical IWSM process area and the way that authority changes as the life cycle progresses form the pivotal definition needed for each program. Within any particular process at any point in the life cycle the program manager has a discrete amount of authority over the decisions and resources involved in doing a specific task. The amount of authority can range from absolutely none (the manager is told what to do and given specific fenced resources to do it) to absolute (there is no need to ask or coordinate with anyone else and the manager can move resources directly to the task). Whenever authority is neither zero nor absolute the manager must form partnerships with other managers and negotiate the sharing of authority to get the task done.

The second principle is the recognition that IWSM is based on establishing these partnerships of shared authority in all process areas. The focus of management is on products (be they airplanes, electronic circuit cards, or software) operated by the warfighting commands and managed within AFMC. Most of these products are viewed from two product management perspectives. First, they are a subsystem within a military (or weapon) system or a product in support of such a system. Second they are part of a group of similar products that are aggregated for commodity management due to their common attributes. The partnership of shared authority between the system and commodity managers forms a foundation of mutual support to our warfighting customers.

This management foundation is also one side of another partnership of shared authority, that with the overall AFMC/USAF/OSD environment. The overarching infrastructure of AFMC is responsible for many decisions that must be made in a centralized or corporate-level manner. Likewise the infrastructure has control of many of the resources upon which the managers depend. These include not just the personnel resources for organizations and the money for centrally funded efforts, but also the specialized support capabilities such as test, laboratories, inventory control, and repair or maintenance. This third dimension of partnerships also includes those areas where higher authorities up to and including Congress become involved in decision making and resource control involving the program or project. Figure 18.1 depicts this three-dimensional relationship.

Finally, then, the working definition of IWSM that resulted from the process development efforts and was captured in the April 1991 IWSM roadmap is as follows:

> Empowering a single manager with authority over the widest range of decisions and resources to satisfy customer requirements throughout the life cycle of that system.

It is both worthwhile and instructive to walk through the elements of this definition.

> "Empowering a single manager . . . " carries the requirement to have only one manager throughout the life cycle. Based on the central elements of the command integration objectives, including decentralized execution and TQM precepts, we must empower that manager.
>
> " . . . with authority . . . " carries the requirement to seek the greatest empowerment or maximum authority feasible, but does not demand total authority.

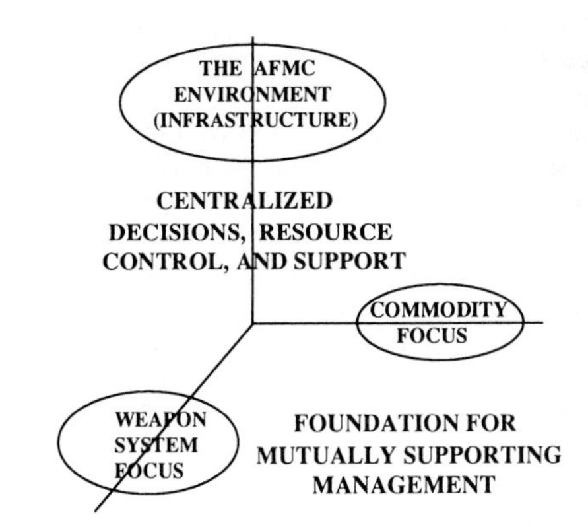

FIGURE 18.1 AFMC partnerships of shared authority.

" . . . over the widest range . . . " carries the requirement to look at all processes and all life-cycle phases.

" . . . of decisions . . . " carries the requirement to focus thinking on the decisions that must be made in each process across the life cycle and then to decide who is empowered to make each decision. This is the first key step in defining the essential program management partnerships. A program approaches full IWSM implementation when all such decisions have been defined and the authority to make them assigned.

" . . . and resources . . . " brings in the second major area of program manager authority. The program manager seeks two types of authority over resources: control to apply the resources where needed and flexibility to shift resources according to changing requirements and priorities. From an IWSM perspective, by the way, statements of priority among requirements are an essential part of stating requirements. Remember that the manager still may not have control over all essential resources.

" . . . to satisfy customer requirements . . . " assures we remember that while all IWSM process development actions (process changes and organizational re-engineering) should result in greater internal AFMC efficiencies, the ultimate test of process changes and single-manager organizations is improved service to the warfighting customers. The negotiation of agreed upon and feasible priorities between program managers and their customers is an essential part of the requirements partnership.

" . . . throughout the life cycle of that system" is a reminder that the empowered single manager's responsibilities span the total life cycle and that, since there will be no transfer of responsibilities, the manager must give greater weight to life-cycle implications of decisions.

IWSM—The Core AFMC Processes

Following an extensive review of the work performed by AFMC, the concept team categorized these activities into eight core process areas. The AFMC IWSM process areas represent a greater aggregation than the project management functions and military project management (MPM) framework processes that form the foundation of this handbook. This chapter will provide a cross reference between the IWSM and MPM functions and processes.

Requirements Process. This process encompasses all types of requirements from those sent to the program manager by external customers to those used internally to those that the program manager forwards to others for action. Among the different requirements included are customer statements of need and operational requirements documents; peacetime and wartime deployment, employment, and operations tempo; priority lists dictated by or negotiated with internal and external customers; computation of spares and support equipment buy requirements; planning and negotiation of depot workload requirements for repair, programmed depot maintenance, or modification; organic support infrastructure needs; modifications for incorporation in production and by retrofit; need for contractor or interim contractor support; data needed for maintenance and support concepts; sustainability analysis results that drive management action; resource needs stated both quantitatively and in dollars; support authorization levels as stated in spares and support equipment lists and manpower documents. The MPM framework processes included in the IWSM requirements process are role of the user and requirements.

Systems Engineering/Configuration Management Process. Among the different subprocesses and activities included here are system design, including design for support and tradeoffs; supportability engineering, including reducing and simplifying support requirements; systems requirements analysis; configuration baseline maintenance; configuration change and government-furnished equipment integration; modification development and proposal review; repair procedure development; engineering drawing management; application of computer-aided acquisition and logistics systems (CALS); standards and specifications application; support and training equipment design; manufacturing and producibility engineering; and quality assurance. The MPM functions and processes included here are engineering; configuration and data; and simultaneous engineering.

Financial Management. In an IWSM environment these subprocesses begin with the budget requirements and statements for priority application of resources that generate in the requirements process. Among the subprocesses included are program control; execution and expenditure control; funds propriety assurance; financial database management; life-cycle support funding profile development and maintenance; and translation of financial requirements into budget documents. The MPM function of program control is obviously included here.

Contracting Process. Competition and advertising; oversight of statement of work (SOW) preparation; request for proposal (RFP) preparation and release; proposal review and comments consolidation; contract fact finding and award negotiation; contract performance monitoring; warranty administration; and closeout are part of the IWSM contracting process. The following MPM functions and processes are also included here: contracting, manufacturing, contract administration; subcontract management; and contract law.

Technology Insertion Process. The overall technology process operates to a great extent outside the IWSM environment, but has essential interfaces with it in determining requirements and their priorities and moving the results to implementation. The major IWSM challenge recognized was the potpourri of stand-alone initiatives and lack of discipline with program management. The following technology subprocesses are in operation: Tech investment makes decisions to pursue technology development; tech transition moves from development/demonstrate into engineering development; tech insertion moves from engineering development into a product; tech transfusion moves beyond the first product; tech transfer moves to customers outside the USAF. The AFMC technology strategy involves all stakeholders in a coherent process encompassing all of these aspects.

Logistics Process. While it was the original intent of the IWSM concept team to avoid the use of the terms acquisition and logistics in defining AFMC processes, they encountered difficulty in reaching broad consensus on an omnibus process title that captured all of the support processes not otherwise included in requirements and system engineering. Sustainment is the most widely accepted term for those logistics processes generally considered to be operational support activities. Among the specific subprocesses included here are parts cataloging and procurement; integration of logistics support into life-cycle management; inventory control functions such as responding to customer requisitions and overall exchangeable and consumable item control; replenishment support equipment procurement and redistribution; depot maintenance scheduling, induction, and in-process management; surge repair and acceleration in response to wartime or contingency needs; depot repair assistance to the field; battle damage and crash damage repair response; and modification line operation. The MPM functions list also captures logistics as a broad category, but includes several aspects that IWSM placed elsewhere such as ILS and acquisition logistics, which become an inherent part of integrated product development in IWSM thinking, and logistics R&D, which is viewed as systems engineering under IWSM.

Test and Evaluation Process. This IWSM process includes all types of testing involving all types of test assets, from new technology development testing to verifying the function and serviceability of repaired or modified assets. Included among the subprocesses are test design; test conduct; results analysis; test resource management; test documentation management; and life-cycle test planning. The MPM functions list also recognize test and evaluation as a major project management function.

Program Management Process. The IWSM concept team recognized immediately that processes can be broken into two general categories—those done by functional area experts in support of and overseen by the program manager and those that form the heart of program or project management. The preceding seven IWSM processes all have functional disciplines associated with them. The program manager is responsible for assessing the degree of authority over those processes and defining the nature of the partnerships needed to conduct them effectively. Among the key subprocesses in the program management discipline itself are creation of baselines and their maintenance; creation, use, and maintenance of planning documents; planning for rapid response to wartime and contingency needs; conduct of external customer interface forums; conduct of program review activities; decisions concerning production line and manufacturing efforts; creation of organizations and delegations of authority; and assurance that program actions satisfy customer requirements. The MPM framework processes include the following major areas by comparison: information resource management; TQM; ethics; termination; project documentation and reporting; interpersonal skills; motivation; the political environment; strategic planning; risk management; project strategy formulation; team development; inspections and audits; modernization management; identifying key stakeholders; interfacing with professional organizations; and determining the need for specialized or streamlined management arrangements.

IWSM—The Process Development Methodology

Once the IWSM environment had been given a basic definition and process framework, it received the concurrence of senior Air Force leadership. Efforts were then focused on filling in the framework and involving the maximum number of stakeholders. While the concept team had all the characteristics of the sequestered insiders deciding and dictating, it is important to note that the major product was a process framework that others would be asked to fill in. All of this was captured in the key document that guided the process development effort, the IWSM roadmap. The IWSM process development effort was organized into four phases, and two major groups of stakeholders were identified—the process owners and the program management organizations where the real work of AFMC takes place. The mission here was to figure out how to come up with the best answers for AFMC.

The specific objectives contained in the roadmap to guide the process development effort were:

1. Develop and define integrated processes for commandwide implementation beginning July 1992.
2. Evolve and evaluate alternative IWSM approaches.
3. Define and establish partnerships with all internal and external customers.
4. Resolve geographic separation issues.
5. Eliminate distinction between acquisition and logistics activities in favor of integrated weapon system management processes.
6. Identify familiarization and training requirements needed to establish integrated management.
7. Identify and initiate solutions to problems in integrated management.
8. Evolve the policies and plans for integrated management.
9. Evolve AFMC strategies in specific technical areas (such as integrated diagnostics).

The basic strategy decision was to select a cross section of programs for early implementation of single management approaches. These programs were tasked to come up with their best answers, and to try them out while capturing both their process for reaching the answers and the issues raised in trying to implement them. The criteria for choosing these "selected programs" included involving all types of systems (aircraft, space, missiles, electronics, software, and communications), a cross section of commodities, systems of varying size and reporting chain, including both program executive officer (PEO) and designated acquisition commander (DAC) programs, and systems in each phase of the life cycle. It was also important to assure that all major AFSC, AFLC, and AFCC field organizations were involved. Twenty-one programs were selected to participate in the experiment. Their charter was to become expert in their processes, improve those processes, help AFMC figure out how to get quality answers, and develop improved processes for commandwide use.

The process owners were directed to form process action teams (PATs), a TQM-based approach that focuses on understanding how processes work and then improving them. A total of eight major PATs were formed, corresponding to the eight core IWSM processes. The roadmap told the PATs to draw most of their membership from the 21 selected programs. The objective was to assure that the process view included both how guidance said the processes should be done and how the programs actually did things. The goal was to evolve the ideal seamless processes for AFMC in a series of iterations between the programs and the PATs.

Four phases of IWSM process development were established to maintain a process focus while continuing to draw the overall effort together. Phase 1, program analysis, asked each of the selected programs to get the experts from both cultures together, undergo team training, and then work to gain a solid understanding of each process across the program's life cycle. The managers of each process area within each program were then to recommend an integrated approach. Phase 2, program integration, asked the two current program managers to continue team building in considering the recommendations of their program analysis teams. The result sought was a set of coherent and integrated processes that would lead to a logical single-manager organization and a team recommendation for the identity and location of the single manager. Not all programs followed this rigorous process-centered building-block approach. Several sat down and started to mesh wiring diagrams together to build a new organization. Their continuing efforts tended to be fraught with frustration and disagreements, while those who used a process approach had broad buy-in to the final recommendations.

As the focus of these two phases on the programs and processes within them gained momentum and single-manager recommendations started to emerge, the PATs began their phase 3 work in earnest. Phase 3 shifted the attention from individual programs to process analysis. The major focus was on defining the overall parts of each process, assuring that all involved understood the broader range of subprocesses inherent in life-cycle management. The selected programs had been tasked to provide issues that had been encountered during their program analysis and integration efforts. In most cases these were assigned to the PATs for resolution. The PATs worked these issues and were asked to identify the most important process changes necessary to resolve the issues and eliminate seams in merging both cultures' sets of processes. The resulting recommended changes were subjected to balloting by a wide population of AFMC stakeholders and were placed in priority order. The original issues had been categorized as those that each program could solve on its own (type 1), those that the IWSM team could solve (type 2), and those where the IWSM team could only influence a higher level of management authority (type 3). Many of these type 3 constraints bounded the realm of the possible. Naturally the recommended AFMC process changes took on these same characteristics, with some triggering multiyear initiatives reaching to the Office of the Secretary of Defense (OSD) and Congress. Many participants were frustrated by the realities of congressional appropriation policies that limit the degree of control and flexibility possible in the short and intermediate ranges.

Phase 4 of IWSM process development, process integration, turned the perspective back to the program manager by asking what integrated management processes needed coordinated attention from two or more of the PATs. A process integration team (PIT) was established to recognize that program managers focus on specific tasks to be done, such as developing, getting approval and funding, procuring, and installing a modification. The job of program management is accomplished through these activities that clearly involve several of the core processes. The program manager is interested and involved in each process and functional area, but the question he or she wants answered is, "How do I run a mod program in an IWSM environment?" The job of the PIT was to assure that these types of multiprocess questions were answered.

In addition to the selected programs and the process-centered activity, there were several other key groups of stakeholders who were brought into active involvement. The overall IWSM process development effort was overseen by a steering committee chaired by the vice commanders of AFSC and AFLC and comprising the commander of AFCC, the Air Force acquisition executive (AFAE), the Air Force deputy chief of staff for logistics, the commanders of the nine major product and logistics centers that made up AFSC and AFLC, and an IWSM project office. Several specific working groups were also established to address command policy, commodity management strategy, implementation plans, management information system "quick connects," and training. These working groups were assigned many of the issues that generated from the selected programs. In addition the steering committee placed a number of additional important stakeholders in an advisory group; membership included the Air Force PEOs, security assistance, legal, and logistics management systems senior staff.

The bottom line goals of all this activity and seemingly endless committees were to tap the expertise of the widest range of stakeholders and, having done so, to assure the broadest possible buy-in to the results. The final products sought were an implementation plan outlining the strategy for introducing the remaining several hundred program managers (beyond the first 21) to IWSM; an implementation guide explaining how they should go about integrating their program management; command guidance that reflected the AFMC processes; and training programs that would orient stakeholders of one culture to unchanged, but unfamiliar processes from the other in addition to educating everyone in the modified processes resulting from IWSM process development.

IWSM PROCESS DEVELOPMENT RESULTS

Before describing the results of all this activity it might be fun to quote from a document generated a third of a century ago (actually, December 31, 1959), when the USAF was trying to develop a weapon system management concept. At that time General S. E. Anderson, the USAF Air Materiel Command commander, wrote to Lt. General B. A. Schriever concerning efforts that would eventually lead to the establishment of Air Force Systems Command. The concept proposed was to "delegate to a specifically designated program director for each weapon system a maximum of authority to direct the implementation of the systems program." The amplifying narrative continued on to say that "... the total system must be integrated, time phased, produced and delivered to the user as an entity—a complete, operable and supportable system." It is sometimes humbling to see how long and hard we work toward objectives. Maybe this time our results will get us there.

The intent of process development was to build an AFMC framework of program management processes that fit into the existing USAF structure. It was recognized that issues would arise that could only be fully resolved by changes to processes and policies controlled outside of AFMC, that is, in DoD or Congress. Achieving such changes would serve to improve IWSM further, but were not essential to its implementation. The gratifying result was that stakeholders at all USAF levels began cooperating to enhance what these TQM-focused efforts were recommending. At the foundation of the results were better definitions of things many of the stakeholders thought they understood and the addition of some new terms upon which the management interfaces are built. The following paragraphs discuss several such "breakthrough" areas that were essential to bringing the cultures together.

Commodity Management. Perhaps the area where definition was most needed and certainly the one that took the longest (until November 1992) to clarify was commodity management. The existence of commodity groups was recognized as the focus of AFMC's expertise for various products, and these groups are the preferred source of management. There is not a revolution in commodity management as practiced, but the need to better define relationships between system and commodity managers became clear. Generally speaking, commodity management refers to technical and managerial practices exercised at the lowest level of indenture where items, subsystems, or systems possessing similar characteristics and applications are aggregated for management in a single organization. IWSM precepts apply to commodities just as they do to military (and weapon) systems. The major difference is that many management responsibilities for some commodity products may transfer from system managers to commodity managers during their life cycles.

The IWSM effort found it useful to categorize commodities as single-application, multiple-application, and stand-alone in defining these management responsibilities. Some single-application commodities are so highly integrated that the military system (their application) manager retains management for the life cycle while others, called federated, will transfer to a commodity manager once interface stability is achieved. Multiple-application and stand-alone commodities will normally be under a single manager within the appropriate commodity organization for their life cycles and application of IWSM precepts is straightforward. In all cases the partnerships of shared authority are key to supporting the warfighting customers, and participation on integrated product teams is essential. IWSM applies directly to the manager of each commodity product or item where the maximum authority should be concentrated in following the AFMC objective of decentralized execution. AFMC guidance places broadest aggregation of commodities in product and materiel groups.

The AFMC model divides items managed into three major categories. The identification of which items fall into which category is addressed in the AFMC program master list (PML). Systems are end items at the highest indenture that actually accomplish a military mission. These end items are managed by a system program director. Systems obviously comprise many items at lower indenture levels. Products are lower indentured items on which development activity is likely to continue. The management of products is aggregated in product groups with a single manager having authority over activity at both product and logistics centers. Materiel encompasses lower indentured items on which there is minimal, if any, development activity. The aggregated management activity is called a materiel group and is normally led from a logistics center.

The AFMC IWSM model released in November 1992 also made provisions for separate development activities (SDAs) to accommodate major development activities (perhaps needing PEO oversight) that might occur within a product group (such as electronic warfare). These SDAs would still depend on the expertise of the product group, and management would transition to the product group once the need for special oversight ended.

Acquisition versus Sustainment Activities

In defining the tasks the single manager performs in accomplishing life-cycle management, the IWSM effort stressed that they would be accommodated in an integrated organization. In other words, engineering is engineering without differentiating between transitioning a requirement into a new product and analyzing a field problem with equipment in operational use. The IWSM single manager is the single face to the operational customer. This differentiation did become important, however, because of the different responsibility chains within the USAF. The AFAE and the PEO/DAC chain is clearly responsible for all activities involving directed development and procurement. The Chief of Staff, USAF, and the AFMC commander are responsible for assuring that forces in operational use are prepared to carry out their wartime missions. This led to a distinction being drawn between acquisition and sustainment activities. In general terms, acquisition includes:

1. Initial weapon system or commodity development and procurement
2. Permanent modifications
3. Initial spares and support equipment

Sustainment, on the other hand, includes many of the activities that occur at the working level as well as those that do not involve the AFAE/PEO chain unless major issues are escalated:

1. Safety of flight and air worthiness activities, such as aircraft structural integrity program, system safety groups, and mishap investigations
2. Readiness and sustainability assessment, including control of mission capability (MICAP) impacting parts shortfalls and critical item programs
3. Maintenance support, including response to field problems and requests for assistance as well as overall maintenance requirements reviews
4. Replenishment spares and support equipment management, including requirements definition, buy, and repair
5. Modification requirements generating from product improvement working groups, maintenance data analysis, and materiel or quality deficiency report
6. Nondevelopment equipment buys
7. System retirement and usable asset reclamation

Corporate Level Activity

Both AFSC and AFLC recognized to differing degrees the need for certain activities to be performed only at headquarters level. In the transition of acquisition from AFSC to the AFAE, the headquarters (HQ) AFSC role became training, organizing, and equipping program managers and overseeing laboratory and test activities. AFLC likewise recognized the need for central decision making concerning the overall industrial infrastructure, but in addition there are some support processes (such as replenishment spares computation) and resources (such as depot-purchased equipment maintenance and the apportionment of stock fund obligation authority) that must retain central focus for at least the near future. While individual program managers may want absolute authority to assign repair or modification workload to a particular organic facility, such decisions must be made as part of a larger mission element posture planning process. The individual program managers should certainly perform analyses and make recommendations, but decisions need to fit a larger scheme. It was recognized that the mission element planning process assigns workloads where the existing capabilities can best be utilized while corporate planning determines where such capabilities will be established or maintained. There is also a still higher level of strategic planning performed at HQ AFMC and HQ USAF that determines acquisition and support doctrine and policy for the USAF.

Locating and Organizing the Single Manager

The earlier discussion on the process development methodology described how the individual programs were tasked to develop a single-manager recommendation. The roadmap also tasked them to come up with the single-manager organization that best suited their program. As the overall AFMC single-manager strategy evolved, there were a set of tenets developed for determining the program center of gravity and hence the location of the single manager. The single manager will be called a system program director (SPD) and will head an organization called a system program office (SPO), new terms for AFLC. There were also several pieces of organizational standardization guidance issued.

The location of an SPD depends primarily on the balance of risk-laden development activity with the level of operational support activity. Weapon system programs will usually begin their life at an AFMC product center because the major focus is transitioning laboratory technology in a risk abatement environment. AFMC Guidance issued in October 1992 emphasized that leadership remains at a product center as long as there is "significant development or integration risk." The role of the logistics center is recognized early in the life cycle and the SPD will assign a system support manager at a logistics center. When the preponderance of program activity has become operational support and there is no significant development still occurring, the single manager will normally be located at a logistics center. It is possible that in the later phases of a program's life cycle, after the SPD is located at a logistics center, significant risk-laden development work may recur. In this situation the SPD establishes a development system manager at a product center. This is different from the previous AFSC/AFLC arrangement of an SPD and a system program manager because under IWSM, when there are managers at two locations, one manager works for the other.

The location of product or materiel group SPD managers is usually at a logistics center in recognition of the focus of management and technical expertise that resides there. Product and materiel managers provide support to weapon system SPDs for single-application commodities and do the development and procurement of multiple-application and stand-alone commodities. When significant risk-laden development arises in a commodity program, the manager will usually call upon the development center of expertise of a product center.

GRADING THE RESULTS OF IWSM PROCESS DEVELOPMENT

Rather than just list all the changes to USAF program management that evolved under IWSM, it is more instructive to take a grade card approach and enumerate factors that experts say enhance or hurt program management. This section will then explain the impact of IWSM-driven changes with respect to each of these factors.

Air Force View—Program Success Factors and Inhibitors

The Air Force senior acquisition leadership has agreed on a set of factors that contribute to and inhibit program success. While many of these involve external influences, they do form a good set of criteria for measuring the impact of changes resulting from defining and implementing IWSM. In the following paragraphs we briefly review those factors and how the features of IWSM influence them.

Success Factors

Requirements. There is consensus that having a well-documented requirement that is strongly advocated, clearly articulated, and supported by a committed user, is key. This should result in a credible, realistic cost and delivery schedule that will be met as promised. A major step is the IWSM process development activities by the requirements PAT to assure a broader understanding of the requirements process and to give greater structure to premilestone 1 activities. The efforts of the IWSM manager PAT to make an integrated automated weapon system master plan (AWSMP) the key document to reflect the customer or acquirer requirements contract and to capture the evolution of program costs and schedules into budget requirements significantly support this factor. Efforts to reflect a wider range of program risks and resources (specifically support elements) in the acquisition program baseline (APB) and integrated program summary (IPS) portions of the AWSMP will help prevent support shortfalls. There is consensus that this well-defined and stable requirement should be reconfirmed periodically. By having a single AWSMP this should occur. The consensus is also that there needs to be a shared view of the baseline requirements. The systems engineering PAT is working to evolve a baseline concept description (BCD) and an improved systems requirements analysis (SRA) under IWSM to assure better requirements definition and to translate them into engineering requirements.

The acquisition leadership feels that program managers need to have a logical and valid story line and be able to provide sound explanations for program decisions and events. It is believed that the discipline of the AWSMP will help provide this when coupled with the basic IWSM philosophy that gives a single manager life-cycle accountability for decisions. The fact that a single manager (rather than two separate managers) will be providing status reports to senior leader and customers improves the "story line" quality.

There is consensus that conducting successful source selections to pick contractors who understand and will manage programs well, along with selecting a contract type and approach that is appropriate to the task and then keeping the contract in synch with the program decisions will help ensure success. The efforts of the IWSM contracting PAT in writing the AFMC Federal Acquisition Regulation (FAR) supplements have focused on trying to improve guidance in these areas. The PAT feels that the confidence in picking the right contractor will be improved by applying the most-probable-cost analysis technique that enables the government to assess the contractor's understanding of the requirement reflected in his or her bid.

Organization. The senior acquisition leadership has stressed that program office organization is crucial to success. Offices must be well led and well structured. Teamwork and trust must be built and maintained among the program office, customer, headquarters, OSD, Congress, contractor, and tester. SPOs need strong management and leadership with continuity and the ability to be innovative and responsive. Open communications are essential to avoid surprises. Program managers need clear lines of authority, responsibility, and accountability within the acquisition structure. The IWSM philosophy embodied in all progress review and approval efforts and upon which the IWSM roadmap was based repeatedly stressed these factors. IWSM process development emphasizes building organizations on process analysis that is founded in team-building efforts. Teamwork and maximum stakeholder involvement are key attributes of IWSM, as is the providing of maximum authority to a single manager through decentralized execution. IWSM seeks to clarify the lines of authority between the AFAE and the program manager and stresses the importance of both acquisition and sustainment activities to properly serving customers as well as the importance of partnerships of shared authority among system managers, commodity managers, and the overarching AFMC infrastructure that provides program support. The long-range initiatives by the financial PAT to provide managers with greater control and flexibility over money results in greater ability to be responsive and seize opportunities for innovation.

Resources. All managers recognize that to succeed, personnel and budget resources must be sufficient and properly phased, and reasonable stability in funding really helps a program. While IWSM cannot guarantee these, there will be a better definition of resources available, determining what can realistically be done and what must be pushed into program baseline deferred content. IWSM process development has worked hard to define the program manager resource support roles of the product and logistics centers. The greater funding flexibility sought will allow managers to preserve core functions better in periods of unstable funding. The clarification for all stakeholders of the realities of congressional constraints ensures that managers properly understand the "playing field" on which they manage their programs.

There is a consensus that an appropriate, phased mix of skilled (experienced and trained) manpower is essential. Strong DAC (center commander) matrixed support in people and technical expertise in processes is the key and should be done without trying to control program management decisions. IWSM in AFMC will be dealing with a single major air command (MAJCOM) and teams that cross prior product center or logistics center boundaries. Intensive efforts by the IWSM training working group have focused on providing all AFMC managers a broader exposure to, and understanding of, life-cycle management skills. The efforts of the systems engineering PAT to specifically define training to support a consistent AFMC systems engineering/configuration management process also will help.

External Influences. The final critical success factor pointed out by USAF leadership is the need for external influences to be minimized or managed by top leadership above the SPO, who should moderate inputs to make them more manageable. Top leadership should mitigate or neutralize "dumb or unreasonable" demands, hence buffering the core task. IWSM process development has forged a strong AFAE/HQ USAF/AFMC team that recognizes the importance of helping the program manager do his or her job. The AFMC headquarters role has been better defined in view of supporting the managers. Efforts to integrate more management information into single documents upon which all stakeholders agree, make it more difficult for some, but certainly not all, external influences to perturb the program. The AFAE/USAF/AFMC team has jointly worked initiatives to take forward to OSD and Congress that further strengthen the program manager's ability to succeed in managing his or her own program.

Inhibitors. The USAF acquisition leadership also identified a series of inhibitors to successful program execution and placed these in priority order.

Requirements. The most significant is the flip side of the most important success factor—in this case, the lack of quality requirements definition. The leadership felt that a particularly pervasive problem is inflated performance needs. As explained earlier in our discussion of success factors, the IWSM team placed particular emphasis on both the process of determining requirements and the tools (such as AWSMP) used to document them and lay out the plans to accommodate them. The focus of the requirements PAT on improving the development planning process and the engineering PAT on the BCD and SRA is particularly significant. The cost of operational performance requirements in terms of supportability risk and cost will be better captured through initiatives by the logistics PAT to place greater detail into baseline-type documents. This should aid the AFMC program manager in carrying out his or her sustainment role once the system is fielded. The PIT led an effort to provide clear definition of the USAF modification process and to assist the Secretary of the Air Force for Acquisition (SAF/AQ) and the USAF Deputy Chief of Staff of Logistics (USAF/LG) in defining their partnership with each other in directing, funding, and accomplishing modifications to systems and commodities. A particularly important outcome of IWSM was the definition of a single integrated technology insertion process by the PAT of the same name. This process is established for the USAF where only a mass of individually focused and bound programs existed before. It helps ensure that all stakeholders (operational customers, laboratories, industry, and the AFMC managers of systems, commodities, and infrastructure) work as a team to pick the right technologies to push and select only ready technologies for transfer into operational use. This should reduce the risk of trying to accommodate inflated performance with immature technology. These IWSM-driven improvements coupled with the USAF-wide move toward the more disciplined strategy-to-task approach for developing and defending requirements suggest major quality enhancements. Strategy to task will help ensure that the USAF better understands the evolution of a requirement and the impact to it as any factor between the overall strategy and the individual task changes.

Optimism. The second most prevalent inhibitor was felt to be excessive optimism by our people in setting cost, schedule, performance, and supportability objectives. Program managers too often set program objectives using best-case scenarios. This can be exacerbated by contractor buy-in. Once again the IWSM effort specifically addresses the means to obtaining and using realistic information. The greater role of supportability in cost and technical baselines and the implementation of an integrated product team approach bring a fuller range of disciplines together, improving communication and the sharing of information that deflates excessive optimism. Giving IWSM single manager's life-cycle accountability in a single office and providing better training on life-cycle processes and issues also suggests that managers will be more realistic in their views of program risks. The engineering PAT efforts to improve risk management techniques and the technology insertion PAT efforts to insert more discipline in the selection of risky technologies both reduce the propensity toward excessive optimism.

Self-Inflicted Wounds. AFMC senior management recognizes that we often corporately self-inflict manpower policy and financial policy wounds, hence becoming our own worst enemies. Policies and strategies for financing the pay of civilians, breaking our civilian force into groups, each paid by a different funding element, can lead to greater cuts among civilians funded by operations and maintenance (O&M). Decisions in this "dollar-driven workforce" environment need to be made by other than programmers and accountants. "Functional" elements of the Air Force often drive managers to do irrational things. IWSM implementation efforts are based on a process and product rather than having a functional focus. What previously could have been competition between acquisition and logistics managers now focuses on teamwork. What would have been an AFSC external customer relationship with AFLC is now one of internal customers, with a single boss who cares about the whole life cycle. The quality of the decisions concerning people and financial management is more likely to improve as they focus on the overall process of serving a warfighting customer. The whole area of self-inflicted wounds will likely become more visible as the total quality approaches foster the exposing of inhibiting policies and encourage the challenging of policies through a strong issues focus.

Contractor Buy-In. Contractor buy-in is viewed as the next major inhibitor. Our earlier discussion on AFMC contracting initiatives (notably most probable cost) pointed out where we hope to reduce occurrences as part of a best-value mentality.

People. Success is obviously inhibited by depending on inexperienced or untrained people who can lead a program astray or fail to recognize problems soon enough to prevent program breaches. However hardworking and well-intentioned folks may be, nothing beats experience and training. Programs must have balanced talent. The single AFMC team goes a long way to providing a large pool of broader experience where more cross flow is available. The application of an acquisition professional development program (APDP) to all of AFMC and the intensive emphasis on training in IWSM implementation will reduce this problem. We expect AFMC managers to be conversant in a wider range of life-cycle disciplines. We want to assure both cross-functional knowledge and a full understanding of the new IWSM processes, which are often easier to understand due to the attack on the seams that separated AFSC and AFLC activities. (The engineering efforts are a good example.)

The acquisition senior leadership pointed out that nothing beats a good team with a professional approach to their jobs. If that does not happen when a program office forms up, it can be a handicap difficult to overcome. Programs routinely languish because they establish neither a team nor a businesslike approach. The IWSM philosophy communicated to all stakeholders is a team mentality where there is a broader understanding and appreciation of the jobs all persons do. There is a better chance of forming effective teams when a process-driven organization is used which depends on the integrated product team (IPT) philosophy.

Excessive Oversight. The final inhibitor recognized was excessive auditor oversight. While not as significant as previous problems, auditors and inspectors can become a distraction to successful program management. Everyone recognizes the push to avoid front-page embarrassments for USAF programs. This dilemma is tough for IWSM or AFMC to overcome, but AFMC is focusing its own inspector general (IG) activities on program management assistance rather than compliance and fault finding.

In summary, we believe AFMC and our IWSM efforts have made superb progress at fostering an environment that promotes program execution in success while reducing the impact from these inhibitors. In the same way in which we use the IWSM definition itself as a success criterion, we can also continue to use these factors and inhibitors to drive continuous improvement of the IWSM approach, its core processes, and the overarching infrastructure that serves the program managers.

The Norm Augustine View

In the introduction to this handbook Mr. Norm Augustine makes some powerful points about project management, which help serve as a scorecard for how well the IWSM process development effort addressed what is really important. He makes the point that project managers have responsibility for the control of substantial financial resources—the IWSM financial PAT had as its cornerstone objective increasing the control and flexibility given managers over their financial resources. He stresses the challenge of meeting technical goals—the IWSM systems engineering and technology insertion PATs focused on improving the process of setting realistic technology transition goals and translating user requirements into engineering specifications. He points out the environment of intense scrutiny—the manager PAT looked particularly hard at the way that program managers report their progress and document their baselines. Mr. Augustine summarizes the job of program management as taking technology from the labs and getting it to the forces in the field. He cites the managerial judgment in transitioning technology as one of the major challenges that managers face. The technology insertion PAT gave the USAF for the first time a coherent technology process focused through technical planning teams on user requirements. The objective is to identify the technologies where accelerated maturation is most needed and those most ready for immediate use. For the first time all stakeholders work together in the transition and transfer process.

The second major challenge facing managers in Mr. Augustine's opinion is finding a substitute for market forces when dealing in a military acquisition environment. The selection of a program's acquisition strategy is where this must be embodied.

The IWSM contracting PAT took special aim at giving the program managers across AFMC greater authority over their own acquisition strategies by raising many of the thresholds that brought higher headquarters into the decision and approval process.

Mr. Augustine points out that project managers operate with highly constrained management flexibility, suffering from a diffusion of authority within government and the existence of multiple regulations to prevent problem recurrence. The IWSM philosophy stated in the definition seeks to increase the manager's authority, and the in-depth review of all processes (and the regulations that implement them) seeks to streamline them to the maximum extent possible. Having the selected programs as part of the process review ensured that we brought the line manager's view of excessive constraint into the analysis and the results. Much of the "zero basing" that Mr. Augustine calls for is inherent in the activities of the eight PATs and the various working groups. It was particularly encouraging to have the USAF/LG and the AFAE directly involved. The comments of the AFAE concerning IWSM say it all: "No one should question the value of IWSM as the foundation for improvements. It is forcing us to look at all the right issues and ask all the right questions." The higher headquarters' view was that USAF-level activities and organizations had to be examined to ensure that they supported the IWSM approach properly.

Mr. Augustine refers to four horsemen of the apocalypse that work to prevent good program management. AFMC believes that good progress has been made in harnessing these. First, to assure that only the best individuals hold responsible positions in program management, AFMC has rigorously applied the APDP criteria in all areas to demand and ensure that training and experience requirements for each job are evaluated and that individuals meeting those qualifications are assigned. The different categories of APDP ensure that both program managers and functional specialists are well trained and have solid experience. Second, Mr. Augustine calls for giving program managers reserves or margins to accommodate uncertainties. By working to give each program manager the maximum authority over decisions and resources AFMC feels that we are giving him or her the opportunity to keep the program on track. The third major area is preventing program turbulence. The IWSM efforts to refine the requirements process itself and to establish greater rigor in documenting and describing the translation of requirements into technical and financial baselines, coupled with the overall USAF move toward the "strategy-to-task" technique, should reduce the major nonpolitical sources of turbulence. The final "horseman" cited is the tendency to terminate a program and start another whenever problems are encountered. The consolidation of USAF life-cycle responsibility into a single command brings greater experience to bear on dealing with risk and the inevitable problems it spawns. More corporate memory is resident in the program management organization of the USAF.

In conclusion, we shall look at Mr. Augustine's "checklist for an acquisition adventure—a formula for failure" as a final grade card on where IWSM gives AFMC program managers a better shot at a passing grade in the tough business of program and project management.

- *Settle for less than the best people.* AFMC and IWSM stress rigorous application of qualification and training standards and give particular emphasis to establishing the right training courses using an occupational review team (ORT) approach that draws its subject matter experts from among those who really do the work and understand the processes. AFMC is dedicated to improving the expertise of its work force so that they are the best.
- *Build an adversarial relationship.* By blending acquisition and logistics into seamless organizations one of the USAF's most significant "us and them" relationships evaporates. By focusing on team building in a TQM environment to build new

IWSM organizations the focus is clearly on breaking down traditional adversarial relationships. The good news is that the process development experiment showed that those who focused on process before organization felt like members of the same team.

- *Change management frequently.* By eliminating the USAF program management responsibility transfer (PMRT) event and focusing on life-cycle program continuity IWSM brings much greater stability to programs.
- *Include all features anyone wants.* The greater rigor in requirements, use of "strategy to task" in the USAF, and implementation of technology oversight centers mitigate the chances of features slipping in that add risk without attendant value added.
- *Divide management responsibility among several individuals.* The basic tenet of IWSM, a single manager/single face cradle to grave, recognizes that splitting responsibility adds risk. Where split responsibility is unavoidable, the emphasis is on defining and forging the necessary partnerships—rather than fighting them, do them right.
- *Give reliability low priority.* The efforts of the IWSM logistics PAT to place life-cycle cost considerations into the baseline and to expand the evaluation of support risk in the IPS put action in place of good intentions. The use of integrated product teams places logisticians on the design teams up front. These efforts plus the change in mindset that results from having one command responsible for development and support help assure a higher priority to factors that influence supportability.
- *Develop underlying technology and end product concurrently.* The changes resulting from the IWSM technology insertion process definition place greater emphasis on assuring that technology is sufficiently mature before it is being transitioned.
- *Create as many interfaces as possible.* The IWSM move to a single manager with maximum authority seeks to limit the number of interorganizational agreements to a minimum essential number.
- *Focus on the big picture—the details will take care of themselves.* The IWSM efforts to promote a fuller understanding among program managers of a wider range of disciplines across the life cycle coupled with training programs that support that goal should result in managers that consider a wider range of details.
- *Disregard seller's track record.* AFMC efforts to identify blue-ribbon contractors with proven track records ensure that good performance gives a competitor an edge.
- *Ignore the users.* The development of an integrated weapon system master plan (WSMP) where chapter 1 is the requirements contract between user and program manager ensures that the user is not ignored.
- *Share authority with staff advisors.* The AFMC goal of decentralized execution is reflected in the concerted efforts to move decision making out of the headquarters except where it involves corporate-level planning and resource allocation for centrally managed accounts.
- *Minimize managers' latitude for judgment.* Efforts to give the IWSM single manager the maximum authority over decisions and resources does just the opposite.
- *Never delegate.* Once again the goals of decentralized execution and maximum manager authority are a major push toward maximum delegation.
- *Don't waste time communicating.* The IWSM effort sought to involve the maximum number of stakeholders in evolving IWSM, and its MIS working group fo-

cused on increasing the communications potential through tools such as program office videoteleconferencing.

- *Eschew strong systems engineering.* Strengthening the systems engineering and configuration management process is a major IWSM focus. Bringing greater discipline to the development and use of technical baselines and the interface with technologists was a major step.
- *Delay establishing configuration control until the last minute.* The IWSM efforts seek to bring greater discipline to this area.
- *Always pick the low bidder.* The contracting PAT efforts to establish a most-probable-cost technique helps spot contractors that have missed the true technical content or are attempting to buy in.
- *Don't worry about the form of contract.* The opportunity under IWSM to rewrite and simplify the Air Force FAR supplements was an opportunity to include better guidance on choosing the right type of contract for the task at hand.

SUMMARY

The foregoing explanations may have left the reader looking for the quick and easy explanation of exactly what IWSM means. The force of IWSM is captured in three main tenets—cradle to grave, single face to the user, and organization of seamless processes. These tenets are the heart of guidance given to the hundreds of USAF programs tasked to implement IWSM as their way of doing business. As explained earlier in this chapter, the definition itself is a criterion against which efforts are measured. The "IWSM in Ten Easy Steps" summary contained in Fig. 18.2 is designed as a more lighthearted view of IWSM for the stakeholders. When asked to capture the overall IWSM approach in a single sentence, we offered the following. IWSM is:

- A life-cycle
- Process-and product-focused,
- Customer-oriented,
- Empowered single-manager
- Philosophy dedicated to
- Partnerships of shared authority,
- Decentralized execution, and
- Continuous improvement

CONCLUSION

No one in AFMC is claiming that IWSM is the panacea for all the problems that have plagued acquisition and support for decades. What has happened in the USAF is a once-in-a-career chance to reassess everything we do, and to run wide open at finding better ways. Taking a process approach and involving the managers who would have to implement the results minimized the chances that functional influence would keep us doing it the way we always did.

No one is claiming that IWSM as implemented in July 1992 was the best answer

1. Think Process
   —Then Organization
2. Those Who Do the Job Know Best
   —Ask Them
3. Goal Is to Serve the Customer
   —Not to Protect Turf
4. One Product
   —One Boss
5. No Resource or Decision Authority
   —No Manager Needed
6. Partnerships of Shared Authority
   —Not Autonomy
7. Accountability
   —To Those Who Direct You
   —To Your Customers
   —Cradle to Grave
8. All Issues Are Solvable
   —If Floated High Enough
9. Authorized to Get Smarter
   —No Mistakes, Just Learning
10. IWSM Is Here
    —When We Stop Saying IWSM

FIGURE 18.2 IWSM in ten easy steps.

available. Even as final plans were being made for implementation, a roadmap of future improvement initiatives was being prepared and action was already under way to make IWSM better. The success of IWSM as a program management process is based in the TQM principle of continuous process improvement. The flurry of activity in the first six months of AFMC bore out the dedication to continually improving and refining IWSM. So long as a process focus is maintained and we continue to involve those who have to make it work, we will get better and better at this job.

GLOSSARY

AFAE	Air Force acquisition executive
AFCC	Air Force Communications Command
AFLC	Air Force Logistics Command
AFMC	Air Force Materiel Command
AFSC	Air Force Systems Command
APB	Acquisition program baseline
APDP	Acquisition professional development program
AWSMP	Automated weapon system master plan
BCD	Baseline concept description
CALS	Computer-aided acquisition and logistics systems
DAC	Designated acquisition commander
DoD	Department of Defense
FAR	Federal Acquisition Regulation
HQ	Headquarters
IG	Inspector general
IPS	Integrated program summary

IPT	Integrated product team
IWSM	Integrated weapon system management
MAJCOM	Major air command
MICAP	Mission capability
MPM	Military project management
ORT	Occupational review team
OSD	Office of the Secretary of Defense
PAT	Process action team
PEO	Program executive officer
PIT	Process integration team
PMRT	Program management responsibility transfer
R&D	Research and development
RFP	Request for proposal
SAF/AQ	Secretary of the Air Force for Acquisition
SOW	Statement of work
SPD	System program director
SPO	System program office
SRA	System requirements analysis
TQM	Total quality management
USAF/LG	USAF Deputy Chief of Staff for Logistics
WSMP	Weapon system master plan

CHAPTER 19

THE POLITICAL PROCESS AND PROJECT MANAGEMENT, OR WHO ARE THOSE GUYS?

Thomas C. Hone

Thomas C. Hone is Professor of Acquisition Management Policy at the Defense Systems Management College, Ft. Belvoir, Va. He has taught at the Naval War College and several universities. He has also worked in industry, most recently with Booz, Allen and Hamilton, Inc. His papers have appeared in journals such as *Armed Forces and Society,* and he is the author of *Power and Change: The Administrative History of the Office of the Chief of Naval Operations, 1946–1986.*

Mark Twain is supposed to have said that the members of Congress were America's only native criminal class. A nice line. However, there's more to Congress than the scandals of money, power, and sex that periodically crowd other forms of nasty news off the front pages of leading newspapers. Any major program manager—to say nothing of the senior civilian leaders of military acquisition—willing and eager to ignore the legitimate role Congress plays in acquisition is asking for trouble, *serious* trouble. The purpose of this chapter is to introduce prospective program managers to the ways Congress affects program management. The method of the chapter is to raise the usual questions program managers ask, and then answer them, tying all the issues together at the end in a brief review.

Question 1. How come Congress is involved so much?

Answer. Because the Constitution requires it. Article 1, Section 8 of the Constitution gives Congress the power and responsibility for "the common defense and general welfare of the United States." To fulfill this responsibility, Congress can "declare war," "raise and support armies, but no appropriation of money to that use shall be for a longer term than two years," "provide and maintain a navy," "make rules for the government and regulation of the land and naval forces," and "make all laws which shall be necessary and proper for carrying into execution the foregoing powers."

Section 9 of Article 1 states clearly that no public money can be spent "but in consequence of appropriations made by law," and that Congress must issue periodic financial statements. To execute a program, you—as a program manager—must have money and authority. Congress grants both. Congress also writes the laws (in Title 10, United States Code) on which the Defense Department regulations governing your behavior as a government official are based.

Question 2. Where does the authority of Congress leave off and that of the President begin? Who's really in charge of me?

Answer. They both are. Article 2, Section 2 of the Constitution makes the President, the chief executive, "commander in chief of the Army and Navy of the United States." Section 2 also gives the President authority to demand reports from "the principal officer in each of the executive departments, upon any subject relating to the duties of their respective offices." Section 2 also enjoins the President to share his power to appoint senior government officials (ambassadors, Supreme Court justices, and "all other officers of the United States") with the Senate. Section 3 requires the President periodically to "give to the Congress information of the state of the union, and recommend to their consideration such measures as he shall judge necessary and expedient."

The Constitution separates executive and legislative powers, but cleverly forces the executive and the legislature to work together. Congress appropriates, but the President is commander in chief. Congress makes the rules for the regulation of our military forces, but the President applies the rules, directs executive branch officials as they carry out their duties, and nominates key executive branch officials. Does this forced linking of executive and legislative actions and powers invite conflict? Yes. *But it was meant to be that way.*

The Constitution was not written with administrative efficiency in mind. No. The primary concern of the authors of the Constitution was accountability—making sure that Congress and the President acted openly and in accordance with rules, so that citizens could judge their performances. Yet ordinary citizens could not watch the actions of the President and his subordinates, or the actions of Congress, closely. Ordinary citizens, watching government from a distance, could not impose accountability on a day-to-day basis. To protect the ordinary citizen, the authors of the Constitution divided power between Congress and the President and forced the two branches to interact constantly. To make the pot boil even more, the Constitution left both branches open to pressure from groups or individuals affected by government action. Can this make *your* job harder? You bet. But it was meant to be that way. For the reasons why, read James Madison's *Federalist Number 10,* written to defend the divided but shared powers concept.

Question 3. So the Constitution gives me multiple masters?

Answer. Yes. Congress *must* be involved in military program management. The Constitution requires it.

Question 4. So all politically visible programs are shaped by partisan or "political" influences?

Answer. Yes, *but that does not mean issues such as technical performance do not matter.* They do. But *how* they matter to members of Congress may be very different from how they matter to you. Your focus is on meeting the needs of users. Members of Congress understand that, but they must also focus on the desires of their constituents, or on the need to gain publicity in order to stay in Congress, or on their efforts to gain influence over other members of Congress. These other goals can come before technical performance—and often do.

And why not? Constituents often care little about technical performance. Instead, they worry about jobs, or about taxes, or about educating their children. Members of Congress are supposed to be sensitive to these needs. Even if these constituent needs do not matter, members are still supposed to consider the *nation's* needs, however they define them. Moreover, members may not understand the technologies they are being asked to support. Did the Navy really need a "stealthy" attack plane (the A-12, now AX) that it could launch from aircraft carriers? How can members of Congress know for sure? They usually cannot. So members use their staffs to dig into programs and look for potentially embarrassing problems. If they are lucky, members will benefit from the work their personal and professional staff do. *At worst,* members will look like they are trying to protect the taxpayers. In fact, what they are really doing is finding out if the story your service has told them about its R&D and production programs "hangs together."

Question 5. Do members of Congress do an honest day's work?

Answer. In spades. Fifty years ago, you could literally retire to Congress. The pay was low, but the hours and "perks" were good, and many members found it easy to win reelection. Not now. Though almost all members of the House who ran for reelection in 1990 won, most had to raise serious money and spend time and energy campaigning. They also had to respond to the expectations of their constituents (a real time eater and staff builder) and do the work of Congress.

In 1964 there were 47 influential chairmanships in the House and Senate combined. Twenty years later, there were over 300. Put another way, only about 10 percent of the members held real positions of influence in 1964. In 1984 about 40 percent occupied such positions. More "leaders" means more committee hearings (where "leaders" exercise visible "leadership"), and more preparation time, and more disputes among committees over jurisdiction. It means more negotiation among members, and these negotiations have added to the length of the workday.

So has the increase in the number of floor votes. In 1960 the House of Representatives assembled to vote 180 times. In 1980 representatives did the same thing almost 1300 times. Members also get more mail. The Speaker of the House got five *million* letters, cards, phone calls, and telegrams in just one day in 1984. Members have responded to this increased workload by hiring more staff to work in their Washington offices and at home (the "personal" staff). Committees have reacted by hiring more professional staff and advisory help (including specialists in military matters in the Legislative Reference Service of the Library of Congress). But hiring more staff means the staff will have to be directed and supervised, and doing that takes yet more time.

In short, members are busy, sometimes because they are very ambitious, and always because they must deal with constituents, campaign and solicit donations, attend an average of four committee or subcommittee hearings a day when Congress is in session, and try to keep up with issues. If a member seems influential, he or she will have to deal with the insistent and often aggravating attention of the media and of literally hundreds of lobbyists. (There are at least 20,000 lobbyists in Washington.) Because members are often too busy to take the time to master complex technical issues, they rely on staff help and on the advice of colleagues and "friendly" lobbyists. Many members have complained in public about this situation, but it persists—and it directly affects you.

Members may evaluate your program on the basis of false or inaccurate impressions. Staff may misunderstand your program, or they may fasten on information which is only a part of the picture. An illustration: in the early 1980s several young staff investigators working for the defense subcommittee of the House Appropriations Committee wrote a classified report—later leaked to the press—charging that the first few AEGIS antiair warfare cruisers of the *Ticonderoga* class were top-heavy. In technical terms, the value of the first ship's metacentric height was lower than the Navy anticipated and wanted. To increase it, the Navy ordered the builder, Ingalls Shipbuilding, to ballast the ship's keel (lowering its center of gravity). The staff investigators were supposed to be looking at cost issues, not potential technical problems. They did not know that the first ship of a new type usually varies—sometimes over, sometimes under—from its projected weight (and weight distribution, too), and that *the Navy expects this.*

No matter. Their boss, the late Representative Joseph Addabbo, head of the defense subcommittee, did not care. He just wanted something to use as a lever against the Navy. This focus on gaining influence over someone at the expense of the "truth" is very common in politics, and you had better be prepared for it. You find a lot of it in Congress on defense issues because, first, most members cannot or will not master complicated technology and, second, staff members have an interest in keeping their bosses "happy." There are two points to keep in mind. First, if there is a perception of a problem regarding your program in the minds of influential members of Congress, then you have a problem, *period.* Second, because you may not be able to track the contest for influence within one legislative house or the other, you may not be able to guess *why* some member is after your program.

Question 6. Why are members of Congress and their staffs often so fixed on gaining and using this kind of influence?

Answer. Because they cannot get anything done without it. A member of Congress can get other members to accept a proposal only if those other members can see that the proposal has value for the country, for their own constituents, or for themselves *as members of a legislative organization.* In most cases, members do not have just one proposal concerning a particular issue to think about. Instead, they have several, and members must often choose between or among these proposals, *each of which has some merit* and for each of which there are strong arguments.

A classic illustration was the division in Congress over the kinds and numbers of nuclear weapon systems which the United States should field. Starting in the late 1970s, members were divided over how best to maintain the strategic "triad," or mixed force of manned bombers, land-based ICBMs, and sea-based ICBMs. The sea-based "leg" seemed secure enough, and the Navy's Trident program received steady support in both houses. But the House Armed Services Committee locked horns with three successive presidents over how to modernize the bomber and land-based missile forces. The result was over a decade of confusion and competition, with members arguing among themselves about how reliable and secure systems such as the rail-garrison MX were, how trustworthy service testimony about these systems was, and what the economic effects of building or canceling these systems might be.

Sensible arguments can be (and have been) produced to support a number of potential mixes of strategic nuclear weapon systems, which means that there is no obvious solution to the problem of what to develop and field. *In such a situation,* where there is no clear "right answer," the issue is often resolved on the basis of who has the most influence *in that particular case at that particular time.* The "cheapest" form of influence you can have is when you get people to share your point of view so that they will decide issues just as you would. This is why the dispute over what kind of nuclear forces the nation should have provoked a lot of testimony before congressional committees, plus barrels of official studies and unofficial books and articles—all designed to win influence in the minds of members of Congress or with the members of the public to whom members of Congress listen and respond. All this "partisan" analysis and rhetoric must be heard and weighed, which takes time. Because the analysts (including those in the Pentagon) themselves cannot agree on what to do, there is no consensus in Congress on what to do, and so the issue becomes one of influence—who has got it, and on what side of the question will they use it?

As various "outsiders" try to affect the issue, members work on one another, trying to influence one another's perceptions and votes. Sometimes they do this because they are concerned about national defense. Sometimes they do it because promoting a particular program will aid their constituents. Sometimes they do it because, if they are seen by other members as taking the lead on so important an issue, they will gain influence in the House or Senate *beyond* this matter, into areas which are of deep concern to their constituents. In most cases, all these motives are active at the same time in every member of Congress. Moreover, while this arguing, considering, persuading, and bargaining over one issue goes on, there are other issues—many just as complex—demanding attention.

One reason why members try hard to gain influence over other members (and over their own constituents, and over lobbyists, and over departments of the executive branch) is so they can reduce the turmoil of each legislative session. Having influence is a means of keeping lobbyists (whether in the government or outside the government), constituents, and even other members at bay, under some control. Without influence, a member finds that his or her agenda is set by others. Without influence, it is difficult to get things done, and difficult to gain respect within Congress itself.

Members often pursue influence because they like having it. Most are ambitious. They want to make their marks. But they also *need* influence in order to succeed as legislators. This need to be and appear influential is what pushes them to take actions which you might regard as unfairly hostile or foolish. Members of Congress often have two faces—public and legislative. They may attack your service or your program before the media and then return to a hearing room to ask some very serious and thoughtful questions away from the microphones about what you are doing. Be prepared. Do not assume the public face is the only face. Just remember that the members of Congress who wrestle with the funding and the fates of these programs probably are not having fun, either. They are simply trying to do their jobs, and part of doing "the job" of Congress is competing for influence, even if, in the process, the actual competition harms particular programs.

Finally, look what happened to the whole debate. Changes in the Soviet Union led President Bush to cancel the rail-garrison MX and the mobility part of the small ICBM program. Over a decade's worth of political conflict within Congress and between presidents and Congress was overtaken by events. Programs that were up, then down, then up again, then down again, are just gone. That sort of thing happens in politics, and you have to be ready for it.

Question 7. You have talked about authority and motives, but what are the mechanics of the congressional process that affect my program?

Answer. The congressional process is very involved. All the details of it cannot be explained in this chapter. In fact, one reason why each service has two legislative affairs offices and OSD and the joint chiefs each have one is so that you will have someone to turn to and work with as Congress deals with your program. Talk with and listen to your legislative affairs people. *Never try to deal with Congress on your own.* It's too complicated an institution.

Lawrence Monaco, a staff attorney with the law revision counsel's office in the House of Representatives, is fond of saying that the term "legislative process" is misleading. He thinks it is a miracle that Congress ever gets anything done, let alone on time. Why? Partly because the rules under which the House and Senate work are complex and sometimes cumbersome. Partly because, every year, Congress must work with the President to produce a budget, and the pattern now is one of routine conflict between the President and Congress over the budget. Because of the size of the national debt, they will fight a lot. But remember, Congress supported the defense buildup which started at the urging of former President Jimmy Carter and grew under former President Ronald Reagan, despite warnings in the early 1980s that such spending, coupled with no tax increase, might create a large budget deficit.

Because the budget process is so important, it deserves first crack at bat. In 1974 Congress passed the Budget and Impoundment Control Act in order to tie together its procedures for setting revenue and spending targets. Congress did so because, by the early 1970s, members had found it harder and harder to complete the government's budget on time. The federal government was entering each fiscal year on continuing resolutions which maintained spending and taxing levels temporarily at ceilings set for the previous year. This was unsettling to both the executive and the legislative branches. There was an obvious need for reform. Accordingly, the 1974 law (1) set a definite timetable for budget and budget-related legislation, (2) created a new pattern for developing that legislation, and (3) established new budget committees and the nonpartisan congressional budget office to oversee and implement the new process.

Here is how the new process was supposed to work. First, the White House would send the new budget to Congress. This budget, combined with a congressional assessment of the nation's economy and needs, would be reviewed by the new congressional budget committees (one in the House, the other in the Senate) and by the nonpartisan congressional budget office. After their reviews, the budget committees would set tentative spending and taxing ceilings. These ceilings would be reviewed by the whole House and Senate. Then both houses would pass nonbinding budget resolutions to guide the committees responsible for raising and spending funds. All this activity would take about four months. Once the budget resolutions were passed, the authorizing (permitting you to obligate) and appropriating (telling you how much you can obligate) committees would do their work, *guided by the ceilings set in the budget resolutions.* Before the beginning of the new fiscal year, the House and Senate would have to compare what the committees had done with what each house's budget resolution said and then make any final adjustments in the laws passed permitting federal agencies to spend and tax through the next fiscal year.

It was a neat idea, but it did not quite work. The authorizing and appropriating committees were jealous of their authority and regarded the budget committees as threats. Many members who joined the new budget committees believed that the authorizing and appropriating committees were too independent. The opposed perspectives made it hard to implement what was supposed to be a cooperative, collegial process. Congressional political party leaders were less than helpful because they thought that the new process robbed them of influence over the budget process. The authors of the Budget and Impoundment Control Act of 1974 believed that passage of the law would allow Congress to complete the budget process on time and balance a budget that was becoming chronically unbalanced. Events did not proceed as they had anticipated.

In 1985, in frustration, Congress established a procedure (through the Balanced Budget and Emergency Deficit Control Act) to eliminate gradually the federal deficit. This procedure, usually referred to as Gramm-Rudman-Hollings after the names of its Senate sponsors, required deliberate reductions in the federal deficit each year until the budget was balanced in fiscal year 1991. It also established a method of enforcing those target reductions called *sequestration*. Under the Budget and Impoundment Control Act, the House and Senate could use a process called *reconciliation* to bring the taxing or spending decisions of an individual committee into line with the budget resolution passed by a whole chamber. In the opinion of many members, however, reconciliation—which was essentially a process of negotiation—had not worked. Sequestration was more dramatic. The Gramm-Rudman-Hollings law *required* across-the-board reductions in spending if its federal deficit targets were not met. These reductions in discretionary federal spending (which excluded entitlements such as Social Security) were to be automatic and applied equally to domestic agencies and the Defense Department. Every eligible program was to be reduced by the same percentage.

Gramm-Rudman-Hollings slowed the growth of the federal deficit briefly, but the whole budget process threatened to come apart in 1990 when congressional Democrats and President George Bush failed to agree on how to deal with the spending crisis brought on by the collapse of many federally insured savings and loans. Before 1990, disputes between Congress and the White House had been handled by adjusting the Gramm-Rudman-Hollings targets and through relatively minor sequestrations. In 1990, however, those tactics could not cover the anticipated deficit, and Congress and the President could not agree on either taxing or spending levels. After much posturing and bluffing by both sides, the President and Congress finally agreed on a FY 1991 budget, and Congress passed the Budget Enforcement Act of 1990.

The Budget Enforcement Act is, in theory, a lot like the Gramm-Rudman-Hollings law which it replaced, but the new law places ceilings separately on discretionary appropriations categories, not just on total federal spending, and these ceilings were set in November 1990 for fiscal years 1991 through 1993. Under the old law, Congress could move funds from one appropriations category to another so long as it met the overall deficit reduction target for that fiscal year. Now Congress cannot "borrow" from one category to "help" pay for unanticipated expenses in another. Moreover, Congress must balance any new entitlement (Social Security is the best example) with an increase in taxes.

The only "loophole" in this process is the opportunity for the President to designate a program an "emergency" one. If Congress agrees, then all bets are off—the ceilings in the appropriation categories can be exceeded. This has already been done to finance the recent war against Iraq. Critics of the new process charged that the "loophole" would become a wider opening for exceptions once any exception were made. The issue came to a vote in the fall of 1991, when President Bush vetoed an "emergency" increase in unemployment benefits and his veto was sustained. That action may have set a pattern. "Emergencies" may be limited, and, if so, Congress will be able to avoid the issue of modifying the budget process until 1993.

The Budget Enforcement Act eliminated the requirement that Congress annually reconcile its spending laws with deficit-reduction targets. It did that because Congress and the President almost closed down the federal government in 1990 when they could not agree on how to cover the losses of the federally insured savings and loans. But the basic problem is still there: there is no consensus between Republicans and Democrats on spending and taxing priorities. Public opposition to tax increases is high and growing. At the same time there is a need for more spending in some areas, such as supporting the Federal Deposit Insurance Corporation. The result, from *your* perspective, is more pressure on the defense budget and more instability in defense programs.

Face it: the budget process is a killer. The Defense Department wants Congress to accept a two-year defense budgeting cycle. Congressional authorization committees agreed to go along in 1989; the appropriations committees refused. As a result, program managers must plan for a two-year PPBS cycle *and* for a one-year "adjustment" process during the off years of the two-year official PPBS cycle. To make matters worse from the program office point of view, the whole budget process—between Congress and the President and within Congress—is unstable and very hard to predict and influence. Finally, the transformation of the Soviet Union into a loose confederation led by the Russian Republic has so transformed the "threat" that effective long-range R&D planning (the basis for the planning portion of the PPBS process) seems impossible. Good luck!

Question 8. I understand that the rules governing what Congress does and when are complex, but why is that?

Answer. One reason is tradition. Congress is an 18th- and 19th-century institution somewhat uncomfortable with the 20th century. The other reason is that an understanding of the rules and procedures is a source of influence. Those who have mastered the rules are not eager to make things simpler and easier for newcomers, or for the public.

Question 9. How do the rules affect me?

Answer. By establishing a number of points where people outside your service can affect your program. *All* legislation starts in House and Senate committees. Committees are divided into subcommittees, and the split in seats between the political parties is carried down in some form through the committees into the subcommittees. What this means is that disputes within and between Republicans and Democrats in Congress are reflected in congressional committees and subcommittees. For example, in the House, the majority party (the Democrats) decides how many Republicans will sit with its members on the House Armed Services Committee. Because the House Armed Services Committee is large (55 members in 1991), its seven subcommittees and five panels really do the committee's business, and how the members of each political party are distributed among these groups affects how that business is done.

Numbers tell a story. In 1991 there were 56 Democrats in the Senate and 44 Republicans. In the Senate Armed Services Committee there were 11 Democrats and 9 Republicans. The percentages were as close as they could get: 44 percent of the Senate was Republican; 45 percent of the Senate Armed Services Committee was Republican. In the House, the split between the parties was 267 Democrats and 167 Republicans (with two empty seats). Republicans made up 38 percent of the House, and they held 40 percent of the seats on the House Armed Services Committee. The actual split was 33 Democrats and 22 Republicans.

These 55 representatives must fill seats on seven subcommittees and five special panels. This means, first, that some members will serve on more than one panel or subcommittee and, second, that members will rely on professional staff for assistance. They will have no choice. Members of the Senate Armed Services Committee have the same problem. There are 20 members spread across six subcommittees. There is overlap, and members must rely on staff support to do their work. To make sure that both Republicans and Democrats have staff they can trust, professional staff appointments are divided proportionately between the ranking Democrats and

Republicans on the Senate Armed Services Committee. The House Armed Services Committee does not do that.

The identity of the Democrats who sit on the House Armed Services Committee is determined by the House Democratic Party caucus. The caucus can also remove a committee chairperson. Disputes among the Democrats in the House may find their way into the composition of the Democratic membership of the Armed Services Committee. You may know nothing of these disputes, but they may affect whether and how members of the committee review your program. Consider this: when the chairperson of the House Armed Services Committee was selected from the majority party by seniority, his focus was on the House leadership, on the executive branch, and on pressure groups concerned about defense issues. He did not have to spend a lot of time "working" his own committee. Now the chairperson must also consider his status within the committee. He must "wheel and deal" (and the phrase is not used in a negative sense) inside the committee as well as outside it. Only your legislative liaison office will be close enough to the process to track this side to Congress.

Question 10. I understand that, but you said that procedures in Congress mattered because they gave people and groups not in the services opportunities to change service programs. Can you give some examples?

Answer. The classic example is the committee *markup.* Bills submitted to Congress are sent first to the relevant committee, which may decide to hold hearings on a bill if it deals with a subject of interest to the full committee. Otherwise committee chairs refer each bill to a subcommittee. The subcommittee studies the bill and, if necessary, holds hearings. Subcommittees may then schedule a markup session. In the Senate, markups are sometimes done in whole committees. A markup is where committee members and their staffs actually put the provisions of a bill together, paragraph by paragraph. It is the first major test of a bill.

To an outsider, a markup session can be astounding. Members and staff will seal their final bargains, make their final trades, and sometimes find that they can only agree to carry their disagreement to the next decision level (the whole committee). Markup sessions can be quiet and methodical, with a few knowledgeable members or a committee chair exercising control over the pace of the meeting. Or they can be like the floor of a commodities exchange, with lots of simultaneous discussions, rapid action, and raised voices. The point is that the markup process is one characterized by give and take, not by analysis. Markup is where the often important *nontechnical* issues surrounding a program emerge.

Once markup is done, the committee issues the bill and an accompanying report.

Question 11. I have heard that reports matter a great deal. Why?

Answer. A report describes the intent of a committee in issuing a bill. It explains what the bill means. If the bill is accepted on the floor of the House or Senate, its report guides negotiators at the *conference,* where House and Senate versions of the same bill are reconciled. This is very important for the Defense Department because House and Senate defense authorization bills are often different. The differences can only be ironed out in a closed conference attended by delegations from each chamber. The reports accompanying each bill set the starting points for the negotiations in the conference.

Reports are also a means for Congress to communicate with the executive branch. Reports justify, for example, the requests Congress makes each year for special studies from the Defense Department. Reports also are signals. Committees can "send a message" in a report. In general, reports are a means for congressional committees to say what they may not (yet) wish to say in a bill or a law.

For example, in the report attached to the House Armed Services Committee FY-91 Defense Authorization Bill markup, the committee told the Defense Department that the following standards should govern the decision whether to develop new weapons systems: (1) There "must be a demonstrated need based on a documented threat." (2) A service must show that no existing system—even a modified one—can meet "the threat." (3) The new "system must work as advertised." (4) It must be a "technological breakthrough." (5) It must save money over its life cycle when compared with existing systems. But what does "work as advertised" mean? Meeting performance thresholds? It is precisely that kind of question which your service legislative liaison office may be able to answer. However, there may not be agreement even within the committee about the proper interpretation of its words.

Federal judges have given congressional committee reports legal standing. Especially important is the report of the conference committee. To determine the meaning of a law, federal courts will begin by reading both the law and the report of the committee that drafted or approved the bill in its final form. To understand a law, you must read the reports of the committees which handled and approved it as it moved from introduction separately in the House and Senate to final signing by the President.

Matters are even more complex in the House of Representatives, where a special committee, called the Committee on Government Operations, can, if it wishes, send the Armed Services or Appropriations Committee its own report on what its members think the implications for the government will be of appropriations or authorizations. The *Rules of the House of Representatives* give the Committee on Government Operations jurisdiction over the "overall economy and efficiency of government operations and activities, *including federal procurement*" (italics added). That means Government Operations may hold hearings on how your service is handling certain types of programs, or negotiating certain types of contracts. Government Operations may then place its findings in the report of the Armed Services Committee. Of course, the jurisdiction of the Government Operations Committee could change with the next Congress because the House can change its rules every two years.

The important point is that the congressional process, based on committee assessments, is designed to give the opponents of any given piece of legislation several chances to derail it. There are subcommittee reviews and hearings and—usually—subcommittee markup sessions. Then there are full committee reviews, hearings, and markups. Then the committee's bill is brought to the floor for debate and a vote. Then if the Senate and House bills are not the same, there is a conference. The results of the conference are reviewed and voted on—again—in the Senate and House, and then the bill is finally sent to the White House for signing. If it is not signed but vetoed, the bill comes back for conference action and votes in the full House and Senate. The whole process can be compared to the ordeal of Pacific salmon struggling back up "fish ladders" to their home rivers to spawn. Like the salmon, the advocates of legislation often feel like giving up, and many participants in the process—especially congressional staffers—are "burned out" by the long hours and seemingly endless debate, meetings, and arguments.

Question 12. I understand reports, but what are "rules"?

Answer. You need to worry about "rules" only in the House. As you might expect, Congress has rules and "rules." The "ordinary" rules tell members how to conduct business. In the House, however, you need a special *rule* (permission, granted by the members of the House Rules Committee) to take a bill from a committee and carry it to the floor of the House for action. The Rules Committee of the House is a kind of traffic cop. The "rules" the committee prepares specify what kinds of amendments can be offered to a bill when it reaches the floor. This can be very important to the fate of a bill.

Imagine you are one of the sponsors of a bill. You may want, from the Rules Committee, a "rule" limiting the types of amendments which can be offered to your bill when it reaches the House floor. You may suspect that opponents of the bill will try to kill it by loading it with amendments that some members will not accept. If your opponents can attach harmful amendments to your bill, the other members who said they would support the bill in its original form may actually oppose it in its final form. So you could find yourself "winning" at the subcommittee and committee levels but "losing" once you got to the whole House. This has happened often enough that a bill's managers must watch carefully the proceedings of the House Rules Committee. Those proceedings are yet another point where opponents of a bill can weaken its chances of becoming a law.

Supporters of a bill in Congress must build and maintain voting coalitions. To turn their bills into laws, legislative managers (who are actually members, usually aided by one or more committee heads and political party leaders) must preserve their voting coalitions as they move across a series of legislative mine fields. Support in a subcommittee must be turned into support in a committee, and a committee's endorsement must be seen by influential members of the House as strong enough to ward off any efforts to defeat the bill on the House floor. Opponents of a bill may avoid a confrontation in a committee and instead work on the members of the Rules Committee. They may try to get a "rule" that will allow them to use floor debate to their advantage. A bill's managers must prepare for this possibility. As you can guess, "preparing" means a lot of communication—phone calls, notes, personal visits to other members, and lots of meetings among staff concerned with the bill. And that means more time, more energy, and perhaps more delay. It also means that you have to rely on your service's congressional liaison office. Only they will be able to follow all this complicated activity.

Question 13. How do the members of Congress keep track of all this activity—of who is doing what and why?

Answer. Staffers do a lot of it. And, as you might guess, the increase in legislative activity has been accompanied by an increase in the number of congressional staff. For example, in 1960 the House Armed Services Committee had 15 professional staff. By 1990 the committee had 74. Similarly, the Senate Armed Services Committee's professional staff increased from 23 to 49 over the same period. And that is just the *professional* staff. Personal staff, employed directly by members and working in members' offices, also work defense issues.

Question 14. How do these staffers get information on my program?

Answer. In five ways—from you, from formal reports and briefings, from specialists employed by congressional agencies such as the General Accounting Office, from the defense press, and from informal networks. One of the most revealing formal reports is the annual selected acquisition report, which the Defense Department is required to send to Congress for every major defense acquisition program starting engineering and manufacturing development and a few still in demonstration and validation. Congress has made the requirement that the department submit these reports, called SARs, part of Title 10, U. S. Code (Section 2432). SARs contain a wealth of information on programs, including program schedules, system performance data, and cost summaries. Department of Defense Manual 5000.2-M, "Defense Acquisition Management Documentation and Reports," has a whole section (Part 17) which describes what SARs must present. The important point for a program manager is that the data in a SAR will give professional congressional staff the kind of information they can analyze to question *intelligently* the way a program is being managed.

SARs are also reviewed by the analysts at the General Accounting Office. In fact, Charles Bowsher, Comptroller General of the United States and head of the GAO, has said that his agency did not have enough data to monitor Major Defense Acquisition Programs closely until the SARs started rolling in. But he has also noted that Defense Department management of technology is often of high quality, despite the technological obstacles which military service program managers face. The point is that GAO analysts and auditors (nearly 5000 strong, with half in Washington and the rest in field offices) are often very skilled, and they can provide professional congressional committee staff with insights into your program. Though not the "enemy," GAO is *out there,* and out there in sufficient force to catch you if you make serious errors or suffer from unanticipated problems.

A SAR is usually not prepared for the Office of the Secretary of Defense and Congress until after a program's budget for engineering and manufacturing development is approved at the service headquarters. A program office will complete a SAR annually after that, until the program's production schedule is almost or fully complete. Programs which cannot stick with their baseline estimates of cost, schedule, and performance will have to prepare quarterly updates to their annual SARs. The services nominate programs for the SAR requirement every November, but final review is in the hands of the staff of the Under Secretary of Defense for Acquisition and the Secretary of Defense. The congressional military authorization committees screen the final list sent over from the Pentagon and actually accept the SARs. In 1991 there were approximately 100 SAR programs.

Do not think that GAO analysts are tied to SARs *alone.* The General Accounting Office has a very broad mandate from Congress, which means, in practice, that only about half of the studies done by the GAO every year are actually commissioned by congressional committees. The rest are cooked up within the GAO itself. This *can* give GAO managers an opportunity to "show off" with Congress. That is, division directors within GAO can use studies and audits to gain favor with congressional committee staff and—more importantly—with members of Congress.

Washington is a place to make a career, and a lot of analysis is done with a mind to gain visibility for the "author" (whether an individual or an office). People (like members of Congress) who can help analysts or GAO division directors make careers are often willing to entertain studies which might help *them* make an impression in Congress. So the temptation to meddle in program management exists. Analysts are looking for ways to promote their careers. Members of Congress and their staffs want their views to prevail in committee. The danger is that, in pursuing their individual career and political interests, the analysts will produce and members will then publicize studies which lack sufficient depth. This does not *appear* now to be a widespread problem, but the potential is always there.

GAO is not the only source which Congress can quickly tap for expertise. Program element descriptive summaries, prepared within the services, are a major source of information. Perhaps the most important source is industry, whose voice you cannot control (and often cannot anticipate, either).

The Congressional Research Service of the Library of Congress has several dozen skilled, knowledgeable analysts in the areas of defense and foreign policy. They tend to focus less on acquisition than on issues of force structure and military operations. They will not come to your office (as GAO does) to look through your files, but they can fill members of Congress in on complex defense issues, *making it virtually impossible for you or your service to "tap dance" around them.* Congress is also supported by the Office of Technology Assessment, an organization of over 150 analysts, most with technical degrees and training. The specialists in Technology Assessment have focused on issues such as the Strategic Defense Initiative and the defense industrial base. Their work may affect how members of Congress make decisions about your program, but you may never even be aware that these specialists are studying what you or your service have done.

These formal organizations are supplemented by networks of informal contacts. For example, when the Navy AEGIS cruiser *Vincennes* shot down an unarmed Iranian airliner in 1988, members of Congress asked GAO analysts who had reviewed AEGIS operational test reports whether the accident revealed a flaw in the AEGIS system itself. The GAO analysts did not think so, but they got on the phone to others who were familiar with the system to seek their views. The author of this chapter was one of those consulted. This example is typical. Analysts get to know one another. Those working for Congress try to build a collection of contacts whom they can talk with "off the record." Analysts in the executive branch will often talk with their counterparts on the legislative side because they, too, want Congress to have the best, most accurate data available.

Do not ever underestimate the power of these networks. Lots of people are watching every Major Defense Acquisition Program. As a program manager, you will not even know who all these people are. Many will know almost as much about your program as you do. Others may know more than you do about how your program fits into the force structure, or how your system affects the natural environment. The networks provide congressional staff with information which supplements material, such as SARs, sent formally to Congress by the Defense Department. *You are not being spied on, even informally.* But you are being evaluated and critiqued, not only by the people whom you report to and brief, but even by specialists whom you do not know and may never see.

Defense acquisition resembles professional baseball. Like the manager of any major league team, you—as program manager or program executive officer—are judged by your statistics. You face an inquisitive, clever press. The television media may tape and then broadcast your most embarrassing mistakes. There is a network of interested, critical, energetic boosters and fans, and they—along with the sports reporters—communicate with one another all the time. As a result, *you rarely have secrets.* Even Congress may intervene, changing the rules under which you work. Front office (that is, headquarters) management often has a very weak sense of loyalty to you or to your program. Money managers (that is, comptrollers) are heartless. Players, like defense contractors, have their own agendas, and some are prima donnas.

So why play? First, there *are* ways to win. One is to create a national constituency for a particular system by spreading the work and by convincing members of Congress and interest groups that the system is needed. The Air Force did this with the B-1B. Second, *the game matters.* Rear Admiral Wayne Meyer, the manager of the Navy's $30 billion AEGIS shipbuilding program, once said that future wars were won on the drawing boards of today's engineers. What he did not say—but implied—was that future victory depended on the success of today's program managers.

Question 16. But even baseball has the World Series. . . .

Answer. So does acquisition. It is the annual defense authorization and appropriations acts. Everything comes together there. When the two houses of Congress complete their budget resolutions in the spring (hopefully, by April 15), the authorization committees consider the programmatic requests (and programmatic performance) of the services and the defense agencies. While their subcommittees are doing that, the subcommittees of the appropriations committees are going over your program element entries to see what you have spent, at what rates, and in which funding categories. (You will have other types of expenditures in addition to RDT&E.) The appropriations committees will, again hopefully, complete their efforts by the end of June, then reconcile their appropriations with the budget ceilings, and then take any House and Senate differences to joint conferences (one for authorizations and one for appropriations).

The joint conferences are the equivalent of the World Series. Each "team" (one from the Senate, one from the House) comes armed with instructions ("game plans") from its home institution. But each "team" is also divided into squads of Republicans and Democrats. The Secretary of Defense is an important "player," but the influence of the White House comes from the sidelines. The President does not attend the conference, but his influence has been a major factor all along, from the beginning of the congressional budget process. The teams toss their respective House and Senate committee reports at one another and try to strike a deal. A "win" is a bill which both houses can endorse (thereby making it an act) and which the President can sign (thereby turning the act into a law).

In 1990 the fiscal impact of the savings and loan industry failures almost brought the whole process to a halt. Members of Congress were unable to compromise on a budget bill because the President promised to veto any overall deal which the House Democrats were likely to accept. In retaliation, House Democrats stalled. They knew that any budget which the President vetoed would not survive because House Republicans stood with the President and would support his veto. It was a political standoff, and the clock just kept ticking. What eventually allowed congressional Democrats and the White House to strike a deal was an agreement which stretched out over time, reducing the immediate shock of expenditure reduction and giving each side something it wanted.

Yet the basic dispute between the political parties was not resolved. The President and most Republicans in Congress want to reduce the federal deficit by cutting discretionary spending, but discretionary federal spending (that part of the annual budget *less* funds for entitlements and interest on the national debt) is a shrinking piece of overall federal expenditures. Moreover, the President has supported certain tax reductions which his Democratic opponents claim would reduce that already shrinking piece of overall spending even more. The solution reached in the fall of 1990 was only temporary. It did nothing to solve the fundamental problem of the decline in discretionary federal spending.

As a program manager, you cannot do a thing about this larger issue of how best to manage the economy, but the lack of money has already affected lots of programs. What you must understand is that the dispute between Republicans and Democrats in Congress, and between the Democratic leaders in Congress on the one hand, and the President on the other, is *real.* It is not a "fake" issue. The two sides disagree strongly about how to manage the economy, and their disagreement cannot be settled by a simple compromise (such as increasing expenditures while simultaneously increasing taxes). A lot is at stake. The Republican Party's leaders (including President Bush) think they can wrest control of Congress (or at least of the congressional budget process) by tightly constraining spending and reducing taxes. Democrats oppose them because Democratic leaders have a very different view of national priorities and because many Democrats in Congress think promoting those different priorities will pay off politically. The two sides are not cynical, just determined. You are in the middle, and the problem will not go away until the economy grows substantially, or until one political party controls both the White House and Congress.

Question 17. So my world gets better fast if one of those things happens? There is nothing hiding in the weeds?

Answer. You are not catching on. Almost everything is in the weeds; for example, the conflict between the appropriations and authorizations committees. The authorizing committees give executive branch agencies their obligational authority. The appropriating committees *supposedly* determine the specific levels of that authority *under* the ceilings set by the authorizers. Before World War II you could get an authorization and it would carry over from one session of Congress to the next. The Navy, for example, was authorized some destroyers in 1916 that the appropriations committees did not fund until 1932. Not a bad system. Once you got by the authorizing committees, all you had to do was return every year to the appropriators.

No more! In 1959 Congress passed a law (PL 86-14) requiring annual authorizations of appropriations for the procurement of aircraft, missiles, and ships. In 1962 Congress extended this requirement to RDT&E associated with aircraft, missiles, and ships. And so it went. The snowball turned into an avalanche. Congress covered tracked combat vehicles with the requirement in 1963, "other weapons" in 1969, torpedoes in 1970, military construction of ammunition facilities in 1975, all ammunition programs in 1982, and working capital funds in 1983.

What is the point? The point is that Congress has increasingly focused on *programs* since 1959. Moreover, the requirement that the authorizing committees review programs each year has led to conflict with the appropriating committees. Thirty years ago, the authorizing committees would ask, "*Why* do you want this thing?" and the appropriators would ask, "*Why* do you want *this particular amount* for the next fiscal year?" Now the questions asked by the two camps are often the same, and they often deal with the details of program finance and management.

In effect, both kinds of committees have the authority to get down in the weeds, to "micromanage," *and they do.* Because they do not have giant staffs, they can only target a few programs for intense scrutiny, but your program—even if you think it is relatively small—may be one of the targets. In 1970, for example, 180 changes were made to Defense Department programs during the authorization process; 650 changes were made to Defense Department program elements during the appropriations process. In 1985 those numbers were 1315 and 1848, respectively. *That is micromanagement.* Moreover, you may be pulled in opposite directions by two different committees. The authorizers may tell you to proceed while the appropriators may not give you the money to do it. Then the authorizers will take you to task for not doing what they told you to do.

Question 18. This sounds like a bad dream. Any examples?

Answer. Unfortunately, yes. One is the Army's heavy expanded mobility tactical truck, or HEMTT. In 1980 Representative Marvin Leath of Texas, a member of the House Armed Services Committee, persuaded his colleagues on the committee to accept an amendment to the fiscal year 1981 Department of Defense authorization bill which granted the Army the authority to buy new, heavy cross-country trucks. The Army responded. In 1984 the service awarded a five-year contract to the Oshkosh Truck Company to produce 13,000 HEMTTs. In 1985, however, the Army chief of staff witnessed a demonstration in England of something called a palletized loading system, or PLS. The PLS allows one man to do the work of many in loading and unloading a large vehicle like HEMTT, and so the Army decided to pursue the PLS concept.

But what was the requirement? To put PLS on the new HEMTT, or solicit a completely new truck design from industry while limiting HEMTT production? At the end of 1984 the Army's transportation school—charged with coming up with the requirement—recommended mating PLS with the HEMTT. In 1987 the transportation school reversed itself, explaining that the HEMTT and PLS combination would not allow the Army's forces in Europe to meet NATO interoperability standards.

Representative Leath responded by persuading his fellow Democrats on the House Armed Services Committee to back the HEMTT/PLS combination. The Senate Armed Services Committee, however, sided with the Army, and supported a new competition for a new vehicle to carry the PLS. The issue went to conference, and the conference report (released in November 1987) approved the purchase of about 2800 HEMTTs in fiscal years 1988 and 1989 (1388 each year), and ordered the Army to halt HEMTT production at 4373 models. The conference report also authorized $17 million in research and development funds for a competitive evaluation of PLS on several different chassis, but the conferees directed the Under Secretary of Defense for Acquisition—*not* the Army—to conduct the evaluation and select a source for the new chassis for the PLS.

This was a classic case of congressional compromise. Every primary party to the dispute got something. The Oshkosh Truck Company did not get its total of 13,000 HEMTTs, but it got over a third of the total production it had anticipated. The Army got money to compare different chassis-mounting PLSs, but the conferees shifted the authority for source selection to the Office of the Secretary of Defense.

Did the compromise satisfy the needs of the Army? In one sense, *it does not matter.* Congress has its own needs. Furthermore, a representative such as Marvin Leath can use the potential leverage that comes from holding a seat on the House Armed Services Committee to force some part of his views on a service, even if the leaders of that service oppose him. But do not consider Leath's intervention completely unwise. After all, the Army did not appear to know what it wanted, leaving the door open for *needed* congressional direction.

Did the decision by the House-Senate conference on defense authorizations in 1987 affect the management of the HEMTT program? Yes, in a very direct way. The manager of the HEMTT program had his long-term acquisition strategy trashed. The conference gave him and Oshkosh an "out" by promising him relatively high numbers of vehicles for fiscal years 1988 and 1989, and it did not exclude Oshkosh from competing for the new chassis/PLS combination, but it did change the whole nature of the HEMTT program. *Program managers must plan for this sort of disruption when they put together their acquisition strategies.*

Question 19. How can we do that? It sounds impossible.

Answer. Yes and no. Take one last case. It is the summer of 1990. The Navy Department has decided that it will present the following case to Congress in the 1991 legislative session (working on the fiscal year 1992 authorizations and appropriations). Among carrier *attack* aircraft development programs, the A-12 stealth plane ranks first, followed by the modified F-14 (the F-14D) and then a modified F/A-18 (the F-18E/F). The concerned program offices within the Naval Air Systems Command (NAVAIR) have set their priorities and developed their plans for fiscal year 1992 accordingly.

Two events upset this applecart. First, the Secretary of Defense tells the Navy that it cannot have three carrier attack plane programs *and* ask Congress for money for long-lead items for the next heavy carrier. In response, the Secretary of the Navy, the Chief of Naval Operations (CNO), and the Assistant Chief of Naval Operations for Air Warfare (OP-05) decide to cancel the F-14D and accelerate the development of the F/A-18E/F (while keeping the A-12 as their first aircraft development priority). In the fall of 1990, however, the Secretary of Defense, learning that the A-12 stealth program is failing, cancels the A-12. In the space of about three months, three Navy attack aircraft programs have become one, and that one survivor was not supposed to have been the front-runner in the first place.

You are in the F/A-18C/D program office in NAVAIR, planning for the eventual transition to the F/A-18E/F. It is February 1991. The budget, just sent to Congress, does not have the A-12 or the F-14D. The armed services committees' staffs are concerned that the Navy does not have *any* effective attack aircraft programs in NAVAIR. The congressional delegation on Long Island is actively lobbying for the F-14D, produced by Grumman (who employs their constituents). Former Navy fighter ace and California Representative Randy Cunningham is crying foul, and saying publicly that the Navy is supporting the F/A-18E/F because it has decided to save McDonnell-Douglas at the expense of Grumman. The Navy is under a lot of pressure to show that the F/A-18E/F is a workable program, despite the fact that it was not anyone's first choice a year ago.

The pressure from Congress works its way down. The Office of the Chief of Naval Operations (OPNAV) orders the program executive officer (PEO) for tactical aircraft programs in NAVAIR to come up with a workable schedule for F/A-18E/F development that will also meet an initial operational capability (IOC) deadline of 1997. The PEO passes it on to your office. But your office is (as usual) busy with an ongoing modification of current production aircraft (the F/A-18C/Ds). Moreover, it is not clear that you can shift McDonnell-Douglas from the C/D model to the E/F model in a cost-effective way that will not involve some serious technical risk. To meet the IOC date, you think you will have to mix development and production. As one of your staff reminds you, however, that was done years back in order to meet the IOC set for the first F-18 models—and with near-disastrous consequences. One of those early models came apart during flight testing, despite Navy Department assurances to Congress that development had proceeded without major problems. You do not want to go through *that* again.

So, here you are. Stuck in the middle. OPNAV wants an airplane—by 1997. Your own engineers and test experts tell you that you cannot "get there" without concurrent testing and production. Yet the Assistant Secretary of the Navy for Research, Development and Acquisition [ASN(RDA)] has stated openly that the Navy's policy is to discourage or forbid concurrent development and production. While you are trying to balance the demands of the operators' representative (OPNAV) against the judgment of your staff, the Senate Armed Services Committee digs into your tentative timetable. The result is a report, prepared by the committee to accompany its version of the FY-92 defense authorization bill, which states that the committee thinks your initial program schedule is too optimistic. Congressional champions of the F-14D jump on that report and use it to press for a reconsideration of *their* solution to the Navy's future attack aircraft needs.

Great! Program planning and scheduling are supposed to be done carefully and conservatively, with an appreciation for technical and cost problems. Here, however, political issues are also driving program planning. The conflict over which firm will get the contract (Grumman versus McDonnell-Douglas), the pressure on OPNAV to find a suitable replacement for the obsolete A-6, and the lobbying by individual members of Congress such as Representative Cunningham are pushing you, the person closest to the technical problems, toward a schedule which is riskier than one you would otherwise choose. What do you do?

There is not much you can do. As the defense budget declines and the "defense industrial base" contracts, people are going to lose jobs and Congress will want to squeeze the greatest possible benefit from every defense dollar. Congressional "micromanagement" of high-ticket programs will increase, and program managers will see their plans shaped or altered by political decisions made by people who do not really know what those decisions will do to a program. In your case, you have little choice but to put together an acquisition strategy which makes clear the consequences of these "political" decisions. Then you must brief it to your PEO and his boss, the service acquisition executive. These senior officials *must* face the consequences of essentially political decisions forced down to your level.

Question 20. Sounds tough. How do you sum it all up?

Answer. Remember that military acquisition has *always* been influenced a lot by partisan politics. That is obvious in Congress. But it is also true for the executive branch. The Secretary of Defense and his senior deputies are political appointees. So are the service secretaries and their senior deputies. Moreover, as the case of the modified F/A-18 shows, military headquarters officers also get involved in politics. They are responsible for developing legislative strategies. Lobbyists, members of Congress, and congressional staff appointees know that, and so they treat what you might regard as technical or military decisions as negotiable or partisan decisions.

The world you live in as a program manager is not the world these political people (from both the executive and the legislative branches) live in. They do not see the world the way you do, and maybe they cannot. How many of them are engineers or pilots or production managers? So you need to understand *their* perspective. Otherwise, what they do will always catch you by surprise.

There is a kind of logic to partisan politics, but it is not a technical or even business logic. Instead, it is built around compromise, bargaining, trading *for political advantage,* personal ambition, the interests of constituents, and personalities. The better you understand it—and understand that it will not ever go away—the easier and more successful your professional career will be.

ACKNOWLEDGMENTS

The author wishes to thank the following for their very helpful comments and criticisms of his ideas and drafts of this chapter: Captain James Hollenbach, USN, Captain Max Current, USN, and Captain John Fedor, USN (ret.). Thanks are also due to *Armed Forces Journal International* (January 1988) for alerting the author to the story of HEMTT, and to former Representative Marvin Leath, who has spoken to Department of Defense program managers on congressional involvement in acquisition.

P · A · R · T · 3

THE HUMAN ELEMENT OF MILITARY PROJECT MANAGEMENT

It is often stated that the most important resource of any project is the human resource. Such is the case in the military project environment. With the continued reduction in the DoD acquisition budget, the military project manager must do more with less—and that includes staffing, education, and training of the project team. The human resources base is established by reviewing the ethical and motivational aspects of military project management. Team building provides a cohesiveness that is needed in the complex organization environment the military project manager operates. Training and adult learning for the project work force are also addressed in this section. Finally, the military project manager can turn to certain professional societies and associations for additional high-quality support and assistance.

Donald S. Fujii sets the stage in Chap. 20 by addressing the important subject of ethics in military project management. The chapter discusses a number of important aspects of this timely subject, including ethical values and ethical decision making. The latest executive order, public law, and DoD directives concerning ethics are also addressed.

In Chap. 21 Daniel G. Robinson discusses motivation. Starting with individual motivation, the author turns to group and organizational motivation and then addresses the specifics of motivation in the military project office.

Once the ethical standards are established and the project team is appropriately motivated, it is time for the project manager to build the team. Kevin P. Grant, in Chap. 22, addresses team building and the project manager and covers the complexities encountered by the military project manager in dealing with developing project teams to include personnel from matrix offices.

Considering the 1990 Defense Acquisition Workforce Improvement Act, the subject of training and development for the military acquisition work force addressed by J. Gerald Land in Chap. 23 is timely. The requirement for a defense acquisition university structure to provide education and training for the acquisition work force is discussed.

In Chap. 24 Michael J. Browne covers the subject of adult learning in the DoD acquisition environment. This thoughtfully prepared chapter addresses the theory of adult learning, discusses the process, and then provides a current prospectus with a focus on how the defense project management professional can use adult learning techniques.

In Chap. 25 coeditor James M. Gallagher discusses the role of professional and technical societies and associations in supporting the military project manager. Ten representative societies and associations are reviewed, highlighting the specific contributions that they can make to the military project manager.

In Chap. 26 Michael G. Krause presents an analysis of the role of interpersonal skills in the management of military programs. He concludes that since the program manager manages the integration of the various aspects of a program, it is essential to establish and maintain effective interpersonal relationships in the execution of the program management activities.

CHAPTER 20

ETHICS IN MILITARY PROJECT MANAGEMENT

Donald S. Fujii

Donald S. Fujii was born and raised in Hawaii. He attended the University of Michigan, where he was awarded degrees in zoology and psychology. He received the M.S. degree in human factors engineering and the Ph.D. degree in industrial organizational psychology from Purdue University. He retired as a colonel after a 26-year career in the U.S. Air Force as a human factors engineer, an instructor at the Air Force Academy, and a scientific manager in the Air Force technology base. He is a professor in the managerial development department at the Defense Systems Management College, Ft. Belvoir, Va.

WHY ETHICAL CONDUCT IS IMPORTANT IN MILITARY PROJECT MANAGEMENT

The Consequences of Unethical Behavior

How would the members of your family, the Browns, feel if they looked at the front page of your hometown paper and saw the headline, "Colonel Brown Convicted of Procurement Fraud." Would your spouse be embarrassed to face your friends? Would your children be taunted by their classmates at school? Would you be in trouble with your boss? These are a few of the many consequences that can easily arise if you are not proactive and think about how you should behave as a military project manager when you lead and manage your project. The purpose of this chapter is to give you a basic understanding of the fundamental ethics concepts and the congressionally mandated principles of ethical conduct; to have you reflect upon the relationship between your ethical values and your behavior as a military project manager; and to show you how to develop a personalized approach to deal with ethical dilemmas.

The Current Situation

Within the Department of Defense (DoD) the ethical behaviors of project managers, contracting officers, and others from the defense acquisition community have been the targets of extensive media attention, as Operation Ill Wind[1] has yielded approximately 50 individual and corporate convictions and more than several hundred million dollars in fines[2] and continues to grind to a conclusion. Some of the cases have involved blatant greed and acceptance of bribes by key defense acquisition personnel. However, in other cases the extent of unethical behavior has not been very clear or has centered around the *perception* of wrongdoing.

How extensive is the involvement of DoD personnel in unethical behavior? Alstott[3] reports that the DoD acquisition work force of some 200,000 military and civilian personnel process roughly 15 million contract packages each year with a total value of approximately $150 billion. Within the context of these large numbers, the 50 Ill Wind convictions for crimes that involved the DoD acquisition and defense contractor communities represent a miniscule portion of the total acquisition work force. However, the fact that the convictions made the front pages of the major newspapers in the nation makes ethics a major concern to the taxpayers and the defense acquisition community.

Ethical Conduct and Government Employment

Since only a small fraction of the total acquisition work force is guilty of unethical behavior, why should the military project manager be concerned with his or her ethical behavior? The answer is directly linked to the fact that the military project manager is a public servant. And, as a public servant, the project manager renders a public service, which, in turn, is a public trust. Thus the military project manager is required to conduct himself or herself in such a manner that every American citizen can have complete confidence in the integrity of the Department of Defense. Executive Order 12731[4] mandates that every military project manager comply with the following principles of ethical conduct.

PRINCIPLES OF ETHICAL CONDUCT FOR GOVERNMENT OFFICERS AND EMPLOYEES

Note that the term "employees" when used in the following principles refers to government officers and employees of the federal government.

1. Public service is a public trust, requiring employees to place loyalty to the Constitution, the laws, and ethical principles above private gain.

2. Employees shall not hold financial interests that conflict with the conscientious performance of duty.

3. Employees shall not engage in financial transactions using nonpublic government information or allow the improper use of such information to further any private interest.

4. An employee shall not, except pursuant to such reasonable exceptions as are provided by regulations, solicit or accept any gift or other item of monetary value from any person or entity seeking official action from doing business with, or conducting activities substantially affected by the performance or nonperformance of the employee's duties.

5. Employees shall put forth honest effort in the performance of their duties.

6. Employees shall make no unauthorized commitments or promises of any kind purporting to bind the government.

7. Employees shall not use public office for private gain.

8. Employees shall act impartially and not give preferential treatment to any private organization or individual.

9. Employees shall protect and conserve federal property and shall not use it for other than authorized activities.

10. Employees shall not engage in outside employment or activities, including seeking or negotiating for employment, that conflict with official government duties and responsibilities.

11. Employees shall disclose waste, fraud, abuse, and corruption to appropriate authorities.

12. Employees shall satisfy in good faith their obligation as citizens, including all just financial obligations, especially those—such as federal, state, or local taxes—that are imposed by law.

13. Employees shall adhere to all laws and regulations that provide equal opportunity for all Americans regardless of race, color, religion, sex, national origin, age, or handicap.

14. Employees shall endeavor to avoid any actions creating the appearance that they are violating the law or the ethical standards promulgated pursuant to this order.

The last principle of ethical conduct emphasizes the word "appearance." This makes the translation of the principles into actual behaviors a complicated, subjective, and often confusing task for the military project manager. The bottom line is that a project manager's behaviors will be *perceived* as ethical or unethical, depending primarily on the values and beliefs of the perceiver—the citizen or taxpayer.

The 14 principles of ethical conduct or "thou shall nots" are only a small, but fundamental part of a large body of public laws, statutes, directives, regulations, and so on, which are intended to ensure that federal employees behave ethically so that every citizen has complete confidence in the integrity of the federal government.

COMMON ETHICAL SITUATIONS FACED BY MILITARY PROJECT MANGERS

Of the 14 principles of ethical conduct for government officers and employees, let us focus on three which are frequently related to situations encountered by military project managers in their jobs. The material for the situations was abstracted from the pamphlets "Defense Ethics"[5] and "Acquisition Alerts for Program Managers."[6]

Financial Interests

Based on the second ethical principle, a military project manager or any of his or her personnel cannot participate personally and substantially in a particular matter in which they have a financial interest if their actions will have a direct and predictable effect on these interests. There is no minimum value or control that constitutes a financial interest. As a military project manager this prohibition applies to you if any of the following individuals and organizations have a financial interest in the matter: (1) your spouse; (2) your minor child; (3) your partner in a business enterprise; (4) an organization in which you serve as an officer, director, trustee, partner, or employee; and (5) a person or organization with which you are negotiating for prospective employment or have an arrangement for prospective employment.

As an example, your daughter has a single share in General Dynamics, which was given to her by your father on her last birthday. Since General Dynamics is the contractor on your project, and your role as project manager requires you to be substantially and personally involved with your counterpart from General Dynamics, you could be in violation of the law. In similar cases, the military project manager has been advised to sell his or her daughter's single share in General Dynamics or to disqualify himself or herself from the project.

Job-Related or "Insider" Information

According to the third ethical principle, as a military project manager you may not engage in personal, business, or professional activity, nor enter into any financial transaction that involves the direct or indirect use of "inside" information for personal advantage to yourself or others. This prohibition against the use of inside information gathered while with the DoD continues even after you leave government service.

As an example, one of your counterparts, the manager of one of the competing contractors, has reviewed the preliminary statement of work in the request for proposal package and has asked you, the project manager, a question about several assumptions that are made when overall system reliability is calculated. Because the assumptions were overlooked in the haste of putting the entire request for proposal together, and because the contractor, who is on the telephone, needs the information immediately in order to meet the deadline for submitting his company's proposal, you explain the underlying assumptions to him. Are you violating the third ethical principle? Or are you using common sense? *Answer:* You are in violation of the third ethical principle. Revealing information about the procurement to one contractor that is not revealed to all other contractors is a violation.

Accepting Gifts, Entertainment, and Favors

The fourth ethical principle means that neither you nor your family may solicit or accept any gifts, entertainment, or favors from anyone who has or seeks business with the DoD or anyone whose business interests are affected by DoD operations. There are some exceptions that are applied narrowly. This prohibition encompasses organizations that contract with or are regulated by a DoD component.

As an example, as the project manager you are at the contractor's plant for the critical design review on the laser inertial navigation gyroscope, a subsystem that has been holding up the entire project. Before the review you notice a tie tack with a miniature model of the new supercruise fighter aircraft the company is building. You did not ask for the tie tack; however, as a naval aviator you would be proud to wear it. You estimate the tie tack to have a retail value of approximately $9.85. Can you accept the tie tack without violating the fourth ethical principle? *Answer:* Yes! One of the exceptions allows you to accept unsolicited promotional items that are less than $10.00 in retail value.

DEFINITION AND PURPOSE OF ETHICS

At this point it is necessary to answer the question: "What is ethics and what does it deal with?" Michael Josephson of the Government Ethics Center of the Joseph and Edna Josephson Institute of Ethics (hereafter to be referred to as "Josephson") defines ethics as "the standards of conduct which indicate how one should behave on moral duties and obligations."[7] He says ethics deals with two aspects: the first involves one's ability to distinguish right from wrong, good from evil, and propriety from impropriety; the second involves the commitment to do what is good, right, and proper.

CORE ETHICAL VALUES

Values are core beliefs which guide or motivate attitudes and behaviors. They are the established ideals of life that members of a given society regard as desirable. Few values are ethical in nature; the majority are nonethical.

Ethical values. These are the values that are directly related to our beliefs concerning what is right, good, and proper. They impose moral obligations. They are concerned with our sense of moral duty.

Nonethical values. These are the values that deal with things we like, desire, or find personally important. Examples of nonethical values include wealth, status, happiness, fulfillment, pleasure, personal freedom, being liked, and being respected. Nonethical values are ethically neutral.

Now, let us focus on 10 major ethical values which Josephson[7] says form a core group of consensual ethical principles. By means of surveys and discussions, Josephson has confirmed the existence of the following core ethical values.

1. *Honesty.* One of the primary ethical values, honesty is associated with people of honor. As a concept, honesty requires a good-faith intent to be truthful, accurate, straightforward, and fair in all communication so people are not misled or deceived. Honesty has three basic dimensions: (a) Truthfulness—the obligation not to engage in intentional misrepresentation of fact, intent, or opinion (lying). (b) Sincerity or nondeception—the obligation not to engage in acts, including half-truths and out-of-context statements, that are intended to create beliefs or impressions that are untrue, misleading, or deceptive. (c) Candor—in relationships that involve legitimate expectations of trust, honesty may also require candor, forthrightness, and frankness, which imposes an obligation to affirmatively volunteer information that the other person considers important to know.

2. *Integrity.* This value is a cornerstone of ethics. It embraces more than honesty. Integrity refers to the ethical principle of wholeness, of consistency between principle and practices. Integrity operates on both the personal and the professional levels. On a personal level, it requires us to adhere to and honor personal convictions, especially those about right and wrong. Integrity means we do not choose whether to live by our principles of right and wrong, we are obligated to do so. It requires us to stand up and be counted, to fight for our beliefs, and to demonstrate the courage of our convictions. At the same time, integrity requires us to avoid self-righteousness. We violate the ethical principle of respect for others when we claim moral superiority whenever there is a clash of opinion on matters of values. On a professional level, the DoD is committed to being an institution representing the highest ethical standards.

3. *Promise Keeping.* When we make promises or other commitments which create a legitimate basis for another person to rely upon us to perform certain tasks, we undertake moral duties that go beyond legal obligations. The ethical dimension of promise keeping imposes the responsibility of making all reasonable efforts to fulfill our commitments. In an organizational context, this means DoD employees have a moral obligation to perform not only their personal promises and commitments, but also those made in the name of the Department of Defense.

4. *Loyalty.* Loyalty involves a special moral responsibility to promote and protect the interests of certain persons, organizations, or governments. This duty extends beyond the normal obligation of concern for others. It arises from relationships that create an expectation of allegiance, fidelity, and devotion. The ethical principle of loyalty does not justify violating other ethical principles such as integrity, fairness, or honesty. This is especially true for public servants acting in an official capacity since they must abide by the principles of public-service ethics which demand that all decisions be made on the merits, free of favoritism.

5. *Fairness.* This value embodies the notions of justice, equity, due process, openness, and consistency. Fairness is one of the most elusive ethical principles since it often refers to a range of morally justifiable outcomes rather than the discovery of *the* fair answer.

6. *Concern and Caring for Others.* At the core of many ethical values is concern for the interest of others. Persons who are totally self-centered tend to treat others simply as instruments of their own ends and rarely do they feel an obligation to be honest, fair, loyal, or respectful. The moral obligation of caring and concern does not preclude decisions which harm others. An ethical manager's decision to terminate a substandard employee may be construed as harmful by the subordinate. The key to the ethical value of care and concern is that one should consciously cause no more harm than is reasonably necessary to perform one's duties and comply with other ethical principles, including fairness and accountability.

7. *Respect for Others.* Respect is related to, but different than concern and caring. Respect focuses on the recognition of the intrinsic right to dignity and autonomy of all individuals.

8. *Responsible Citizenship.* As both a citizen and a public servant, DoD employees have a two-part obligation to adhere to the letter and spirit of the law, and to uphold and protect the national security of the United States.

9. *Pursuit of Excellence.* When a person is in a position of responsibility and others rely upon his or her knowledge, ability, or willingness to perform tasks effectively, the pursuit of excellence has an ethical dimension. It is not necessarily unethical to make mistakes or to be less than excellent, but individuals, and especially public servants, have a moral obligation to do their best, to be diligent and prepared.

10. *Accountability.* Ethical persons accept responsibility for their decisions. An accountable person does not shift blame or accept credit for another's work.

What are your core ethical values? Do they coincide with the previous 10 values that appear repeatedly in surveys conducted by Josephson? Or, do you have other unique ethical values? Before you can develop an approach to ethical decision making, you should invest some time and effort and attain a solid understanding of your core ethical values.

ETHICAL DECISION MAKING

Ethical Perspectives of Decisions

Josephson[7] states that there is an ethical dimension to every decision that can be evaluated in terms of its adherence to the previously discussed core ethical values. Thus any of your decisions which affect other persons have ethical implications, and virtually all of your important decisions reflect your sensitivity and commitment to ethics. In summary, in your role as a military project manager, what are the ethical dimensions as you deal with your superiors; the contractors; the functional personnel who provide support to your project; and the taxpayers, who because of your role as a public servant, expect you to honor their trust in you?

The Process of Ethical Decision Making

Ethical decision making involves the process by which a person evaluates and chooses among alternatives in a manner consistent with his or her core ethical values or principles. Thus as a military project manager, when you make an ethical personal or professional decision you:

1. Perceive and eliminate unethical options
2. Select the best from several competing ethical alternatives

Josephson[7] stresses that ethical decision making requires more than a belief in the importance of ethics. It also requires sensitivity to perceive the ethical implications of your decisions; the ability to evaluate complex, ambiguous, and incomplete facts; and the skill to implement ethical decisions without jeopardizing your career. Ethical decision making requires three things: ethical commitment, ethical consciousness, and ethical competence.

Ethical Commitment. This is the strong desire to act ethically, to do the right thing, especially when ethics imposes financial, social, or psychological costs. Surveys by the Josephson Institute reveal that, regardless of profession, nearly all people believe they are and should be ethical. While most are not satisfied with the ethical quality of society as a whole, they believe their profession is more ethical than others and they are at least as ethical as those in their profession.

Unfortunately our behaviors do not consistently conform to our self-images and moral ambitions. As a result, a large number of decent people who are committed to ethical values regularly compromise these values—often because they lack the strength to follow their conscience. As a military project manager, you will be confronted with a continuous stream of situations in which your ethical commitment will be constantly tested.

Ethical Consciousness. Josephson states that while weakness of will explains a great deal of improper conduct, a much greater problem arises from our failure to perceive the ethical implications of our conduct. Many of us simply fail to apply our moral convictions to our daily behaviors.

Some public employees do not always see ethical issues that are likely to trouble outsiders. Military project managers may not be aware that perfectly legal conduct often appears to be improper, inappropriate, or downright sleazy to the taxpayers who expect public servants to avoid even the appearance of impropriety.

Ethical Competence. Being ethically conscious and being committed to act ethically is not always enough. In complex situations (which are frequently faced by those involved in public service, especially those who work in the military project management environment), the following reasoning and problem-solving skills are also necessary.

Evaluation. The ability to collect and evaluate relevant facts and to know when to stop and how to make prudent decisions based on incomplete and ambiguous facts.

Creativity. The capacity to develop alternative means of accomplishing goals in ways which avoid or minimize ethical problems.

Prediction. The ability to foresee potential consequences of conduct and assess the likelihood or risk that persons will be helped or harmed by an act.

The Stakeholder Concept

A person concerned with being ethical has a moral obligation to consider the ethical implications of all of his or her decisions upon others.

Each person, group, institution, or constituency that is likely to be affected by a decision is a "stakeholder" with a moral claim on the decision maker. This stakeholder concept provides a systematic way of perceiving and sorting out the various interests involved in our ethical decision making. Within the defense acquisition environment, the stakeholders include your immediate boss, the key personnel at the service headquarters level, the program executive officer, the service acquisition executive, the defense acquisition executive, members of congress and their staffers, functional personnel who support your project, the taxpayers, and others.

The stakeholder concept reinforces our obligation to make all reasonable efforts to foresee possible consequences and take reasonable steps to avoid unjustified harm to innocent stakeholders—an ethical decision maker would never inadvertently cause harm.

A Model for Ethical Decision Making: Golden Kantian Consequentialism (GKC)

Josephson has developed a model of ethical decision making that avoids the shortcoming of the traditional approaches such as the Golden Rule (do unto others as you would have them do unto you) and Kant's categorical imperatives (higher truths impose absolute moral obligations which must be obeyed regardless of the consequences). Josephson's model can be practically applied to common problems found in competitive and stressful military project management situations. The three steps to the Golden Kantian Consequentialism (GKC) ethical decision-making model include:

1. All decisions must take into account and reflect a concern for the interest and well-being of *all* stakeholders.
2. Ethical values and principles *always* take precedence over nonethical values and principles.
3. It is ethically proper to violate an ethical principle only when it is *clearly necessary* to advance another *true ethical principle* which, according to the decision maker's conscience, will produce the greatest balance of good *in the long run.*

STAGES OF MORAL DEVELOPMENT

The extent to which we use the GKC approach to ethical decision making to guide our behaviors in project management situations will vary from person to person. Kohlberg[8] offers a handy framework for delineating the stage each of us has reached with respect to personal moral development. The examples were adapted from Manning and Curtis.[9]

Stage 1. Physical consequences determine moral behavior. At this stage of personal moral development, the individual's ethical behavior is driven by the decision to avoid punishment or by deference to power. Example: "I won't hit him because he is bigger than me and he may hit me back."

Stage 2. Individual pleasure needs dictate moral behavior. At this stage, a person's needs are the person's primary concern. Example: "I will help him because he may help me in return."

Stage 3. Approval of others determines moral behavior. This stage is characterized by decisions where the approval of others determines the person's behavior. The good person satisfies family, friends, and associates. Example: "I will go along with you because I like you."

Stage 4. Compliance with authority and upholding social order are a person's primary ethical concerns. "Doing one's duty" is the primary concern. Example: "I will comply with the General's instructions because it is wrong to disobey a senior officer."

Stage 5. Tolerance for rational dissent and acceptance of rule by the majority becomes the primary ethical concern. Example: "Although I disagree with her views, I will uphold her right to have them."

Stage 6. What is right or good is viewed as a matter of individual conscience, free choice, and personal responsibility for the consequences. Example: "There is no external threat that can force me to make a decision that I consider morally wrong."

What is your stage of personal moral development? Be honest with yourself and recall the decisions you made in recent ethical situations. The six stages are valuable landmarks as they tell you approximately where you are and what changes you will have to make in your behaviors and decisions to move to a higher level of moral development. The ultimate goal is to engage in ethical decision making at stage 6. However, the level that you do reach will depend on your ethical commitment, your ethical consciousness, and your ethical competence.

CONCLUSION

Understand and Comply with Standards of Conduct

As a minimum, you should understand and comply with the standards of conduct that are contained in DoD Directive 5500.7, "Standards of Conduct,"[10] in Public Law 100-679, Section 6, "Procurement Integrity,"[11] and in other executive orders and the relevant U.S. codes. In most military project management offices the standards of conduct are reviewed each year at mandatory training sessions and everyone is asked to certify that he or she understands the "thou shall nots" and asked to conduct himself or herself in an ethical manner. Thus you should be able to distinguish between right and wrong in most everyday ethical situations.

Take the High Road

Although this level of compliance is commendable, military project managers should do more in terms of their ethical decision making and ethical behaviors. Deputy Secretary of Defense Donald J. Atwood put it nicely when he said[12]:

> Ethics is more than a set of rules of what to say and what to do, or of what to avoid and what to overlook. It involves judgments that only those facing those decisions can make.
>
> Therefore, ethical government has to mean more than simply laws prohibiting certain actions. It must be a spirit, an imbued code of conduct, an ethos.
>
> There must be a climate in which everyone understands some conduct is correct and other conduct is clearly unacceptable.

As a military project manager, as a person who is responsible for fulfilling the public trust, and as a person who is fundamentally good and ethical, it is your duty to make your ethical decisions at the highest possible stage of moral development. Once the majority of project managers operate at Kohlberg's stage 6 of personal moral development, the frequency of headlines that blare the misdeeds of military project managers will be close to zero. Then the American public will have total trust and confidence in those who develop and procure the weapon systems that assure world peace.

REFERENCES

1. S. J. Crawford, "Ill Wind and Well-Learned Lessons," *Defense 90,* p. 15, July–Aug. 1990.
2. R. F. Howe, "Defense Procurement Fraud Figure Pleads Guilty," *The Washington Post,* p. A12, Aug. 23, 1991.
3. J. D. Alstott, "The Search for Honor: An Inquiry into the Factors that Influence the Ethics of Federal Acquisition," in J. A. Petrick, W. M. Claunch, and R. F. Scherer (eds.), *Institutionalizing Organizational Ethics Programs: Contemporary Perspectives,* Wright State University, Dayton, Ohio, 1991, pp. 182–194.
4. Executive Order 12731, "Principles of Ethical Conduct for Government Officers and Employees," presidential documents, *Federal Register,* vol. 55, no. 203, Oct. 19, 1990.
5. Inspector General, Department of Defense, "Defense Ethics: A Standards of Conduct Guide for DoD Employees," IGDG 5500.8 AFU, U.S. Government Printing Office, Washington, D.C., 1989.

6. Inspector General, Department of Defense, "Acquisition Alerts for Program Managers," IGDH 4245.1 AFU, U.S. Government Printing Office, Washington, D.C., 1988.
7. Government Ethics Center of the Joseph and Edna Josephson Institute of Ethics, "Ethics at the IRS: A Quest for the Highest Standards," Workshop and Resource Materials, Internal Revenue Service Management Training Program, Marina del Rey, Calif., 1991.
8. L. Kohlberg, "Stages of Moral Development as a Basis for Moral Education," in C. M. Beck, B. S. Crittenden, and E. V. Sullivan (eds.), *Moral Education: Interdisciplinary Approaches,* University of Toronto Press, Toronto, Ont., Canada, 1971, pp. 86–88.
9. G. Manning and K. Curtis, *Ethics at Work: Fire in a Dark World,* South-Western Publishing, Cincinnati, Ohio, 1988.
10. DoD Directive 5500.7, "Standards of Conduct," May 6, 1987.
11. Public Law 100-679, Sec. 6, "Procurement Integrity," Nov. 17, 1988.
12. D. J. Atwood, "Living up to the Public Trust," *Defense Issues,* vol. 5, no. 27, p. 1, 1990.

CHAPTER 21
MOTIVATION

Daniel G. Robinson

Daniel G. Robinson is Professor of Acquisition Management at the Defense Systems Management College. He served almost 22 years in the U.S. Air Force. His military career included duty in operational, manufacturing, and program management positions. He also spent 5 years in the defense industry as a management and organization development specialist and as manager of management and organization development. He is currently the Special Assistant for Operations Improvement, assisting the leadership of DSMC in implementing total quality principles and practices.

INTRODUCTION

The question of how to motivate people has been a popular topic for many years. From Maslow's hierarchy of needs to Herzberg's motivator factors and hygiene factors and beyond, many people have struggled with this issue. The issue is an important one. Our modern-day world is becoming more and more competitive. We are moving very rapidly to a world economy. There is an urgent need for us to coordinate our actions more effectively. If we in America are to regain our competitive edge, we will require the effort of dedicated and motivated teams.

Since motivation is a complex issue, this chapter will look at it from a number of perspectives. The personal, interpersonal, group, and organizational aspects of motivation will be explored. Some questions to be answered are: How do you motivate individuals? How does motivation affect interactions between individuals? How are groups motivated? What are the motivational factors affecting organizations?

INDIVIDUAL MOTIVATION

Much of the interest, and most of the literature on motivation, addresses how to motivate other individuals. In fact, the way how that is stated represents much of the problem. It stems from a stimulus-response (S-R) view of human behavior. In other words, if one applies the proper stimulus to or for another, the other will respond in the way the first person intended, that is, the person being stimulated will be motivated. So if managers could just learn to do the right things to or for their subordinates, the subordinates would be motivated to perform in productive ways.

Recent advances in our understanding of human behavior call into question the usefulness of the S-R view of motivation. Path-breaking work done by such researchers as William Powers, William Glasser, Humberto Maturana, Francisco Varela, Michael Gazzaniga, Fernando Flores, and others gives us another view of human cognition and behavior. There will be no attempt to summarize their work in this chapter, but it will serve as the basis for the view of motivation outlined here.

This new view of human behavior sees humans less as manipulable vessels and more as structurally determined entities that respond to interactions with others in their environment based on the nature of their structure rather than based on the nature of the interaction. This view also holds that the structure of the individual can modify itself over time, based on the nature of the interactions it has with others.

Put in plain language, individuals choose to respond to the attempt of others to motivate them based on their own makeup and not based on whatever it was that was done to or for them. Consequently, people motivate themselves, they are not motivated by others. Since people differ from one another, what is done to or for one individual may result in his or her motivating himself or herself. When that same thing is done to or for other individuals, they may not motivate themselves.

This creates a dilemma for managers who must effectively coordinate the actions of others. If managers cannot do things to or for others that will motivate them to perform in certain ways, what can they do? If motivation springs from within, not from without, what role can the manager play in all this? The answer is both simple and complex. It says that managers need to create an environment in which people can and will motivate themselves to perform effectively. It is both simple and complex, because it is easy to say and not all that easy to do. We will explore the group and organizational implications of this solution later. For now, let us turn to some of the aspects of individual structure to determine some roots of differences between individuals.

There are a number of perspectives from which to view individual differences. They include values, traits, attitudes, skills, and personality. For purposes of this work we shall explore the effects of an individual's personality type or temperament as determined by the Myers-Briggs type indicator (MBTI). There are many other tools available for this effort, but we will stay with the MBTI here.

The concept of personality type as determined by the application of the MBTI is based on the work of the Swiss psychiatrist Carl Jung. This concept posits that we all have a core to our personality. That core consists of two parts. The first is how we experience the world, or how we find out what is going on around us. The second is how we decide what to do with what we have experienced, or how we make decisions. Two other aspects of our personality surround the core. The first is how we orient ourselves to the world. The second is the attitude we take toward the world. Each of these aspects, our experiencing and deciding processes, our orientation, and our attitude have two separate and distinct modes of operation. These relationships are shown in Fig. 21.1.

One of the ways of using this knowledge is to look at four separate and distinct classes of people from a temperament perspective. The four temperaments can be seen as SJ, SP, NT, and NF. People in each of these classes have certain characteristics in common. They think, act, and behave in predictably similar ways. That is important to know as one sets out to coordinate action with the other. It is important because, if one is to understand the nature of the structure of others, one has to know what the possibilities are. Remember, in our interactions we are attempting to set a climate, an environment, in which the others will be able to motivate themselves.

When we understand and accept the existence of individual differences, we must

		EXPERIENCE	DECISION		
	E	S	T	J	
ORIENTATION					ATTITUDE
	I	N	F	P	
E = EXTROVERSION I = INTROVERSION		S = SENSING N = INTUITION	T = THINKING F = FEELING	J = JUDGING P = PERCEIVING	

FIGURE 21.1 Personality-type concept.

recognize that there is no one best way, no magic formula, for creating climates in which individuals can motivate themselves. Consequently, the astute manager will recognize that individual differences exist, take steps to understand the nature and extent of the differences that exist among the subordinates, and develop a strategy for dealing with that. This gets us away from looking for single methods for "motivating" people.

Each of the different sets of temperaments has a preferred way of operating. They share differing likes and dislikes. Some prefer structure; others prefer no structure. Some focus on the details of the moment; others on the possibilities of the future. The key to understanding and being able to use this is the term "preference." People have preferences for certain ways of living and working. They can live and work in different ways than those they prefer, but they pay a personal price for doing so. Organizations, also pay a price when they ignore the preferences people have and try to mold them into alien ways of working. Managers who can understand this can be more effective in working with others to maximize their potential than can managers who do not understand it, or worse, who ignore it.

For example, SJs prefer to work in orderly, structured environments. They tend to be loyal to the organization and believe rules and regulations are meant to be followed to the letter. They value hard work, and they value work for work's sake.

Another group, SPs, prefer openness and freedom of action. They tolerate rules but do not let them get in the way of getting the job done the best way it can be done. For them, rules are only useful as long as they do not get in the way. SPs also crave action. They are rarely satisfied with a calm and routine work environment.

A third group, NTs, prefer creative, dynamic, changing environments. They value conceptual work, the world of ideas, and organizational growth. They get excited with discussions of an organization's mission and the possibilities that are open to it. They constantly ask "why?" The world of work is for them an outlet for their creative energies, not necessarily for getting things done at this moment.

Finally, the NFs provide the catalyst for understanding and improving the extent and the nature of the human interactions in the organization. They prefer a harmonious working environment. Their interests are not in the work itself, but in how to help people get the work done more effectively and productively.

In summary, we have four groups of people whose interests range from stability to action, to concepts, to people. An effective manager is one who is aware of this, who knows how to identify the interests and preferences of the various people in the group, and who knows how to create an environment in which people can thrive and grow and be productive workers.

Thus the issue of motivation becomes not "what is it that I can do to motivate that person?" but rather, "what does that person prefer and what kind of an environment can I arrange that will be most comfortable for him or her?" This does not mean that a manager will always be able to create environments in which all the people can thrive all the time. It does mean that managers can be alert and aware of the preferences of their people and can help the people adapt to environments in which they are not comfortable.

For example, a task that requires attention to detail, closely following a prescribed, predetermined pattern of activity, and which must be done within a tight schedule, would be done most comfortably by an SJ. That is great if there is one available to perform the task. But what does a manager do if there is only one of the other types available to do the task? One strategy would be to talk to the person who is assigned the task and explain that it is not one that he or she may be comfortable doing. Explain the necessity of completing the task and offer help in structuring the work in ways that may make it more palatable for the person. Acknowledge their preferences and explore with them methods and processes that will appeal to their interests. In effect, when attempting to force a round peg into a square hole, you try to do so without damaging either the peg or the hole. The manager can help modify the shape of the hole while helping the peg adapt to the need to change for the duration of the task.

While the person is doing the task, it may be helpful for the manager to check occasionally to see how the person is doing. The key is to provide support, not to direct a specific process or to demand a specific outcome. Allow the persons to adapt the task to their own preferences as much as possible rather than insisting that they only adapt their own preferences to the needs of the task.

A trap for anyone here is that we all view the world through our own eyes. Each of us has our own set of preferences. We tend to think that the way we view the world is the way the world is. This is especially critical for managers. SJ managers will try to mold their subordinates to their own ways of thinking. They will expect others to see the world the way they do. When people see it differently, they are seen as "wrong."

Manager need to understand themselves first. They need to understand what their preferences are and how that affects the way they view the world. They need to clearly understand that their way is not the only way to see the world. Then they need to understand how their subordinates see it. That then leads to determining how to structure the work environment so that the greatest number can operate as close to their preferred way of operating as possible.

It is important to understand a very important caveat at this point. The issue here is not having happy, satisfied employees. The issue is creating a work environment in which the maximum number of employees can motivate themselves in ways that will allow them to be as productive as possible. The bottom line for all this is quality, productivity, and profit.

A very basic thesis is that people are most happy in a work environment when they can produce high-quality work that delights their customers. Consequently the stress on understanding people's preferences is to optimize the opportunity for them to excel at their work. Therefore if we understand the basic needs and preferences of people, design an environment that optimizes their comfort with the demands of their work, and give them opportunity to excel and take pride in their work, they will be productive and, consequently, motivated.

So, one more time, the goal is not to motivate people, but to create an environment in which people can become motivated through their work.

INTERPERSONAL MOTIVATION

A complex subject, motivation of the individual, becomes even more complex when you look at the next step. It is one thing to understand the issue of personal motivation, and yet another to understand motivation from an interpersonal perspective.

Once we understand how individuals motivate themselves, we can move to looking at how motivation can work when two people are attempting to coordinate their actions.

We will explore how to carry on effective conversations that lead to action. We will assume the goal of the conversation is to achieve some mutually desired result or outcome. There are a number of motivation factors at play in this scenario, but we will focus on the need to reach understanding rather than achieving personal success.

Reaching understanding in a conversation requires one to focus on the cares and concerns of the other, rather than on one's own ego or need to "win." The most effective way to accomplish this mode of operation is to use a questioning strategy rather than a telling strategy. This runs counter to our culturally derived need to compete and talk. To use this strategy successfully, one must be willing to collaborate and to listen.

From very early in our lives we are taught to compete. As infants we compete for our parents' attention. We move to competing with our siblings and friends. As we go on to school, we compete for the attention of the teacher and for grades. We then move to athletic or other competitive pastimes as we continue to mature. After a while it is difficult for many of us to recognize when it is effective to compete and when it is counterproductive. Competition is embedded in our bones by this time and becomes the naturally accepted way of life.

However, in normal conversation, competition tends to get in the way of understanding. We become so enthralled with our own ideas and our own voices that we end up listening to ourselves rather than to those with whom we are conversing. Consequently there is little understanding either of what the other is saying or of how the ideas of the other might be synergistically combined with our own ideas.

Because the human need to be listened to and to be understood is so strong, people tend to try to overwhelm others when they are engaged in conversation. When one recognizes the power of this need in others as well as in oneself, he or she can create an environment in which the others can motivate themselves toward collaboration and away from competition. As before, the motivation comes from within, not from without. However, the motivation to collaborate is suppressed when one is talked to rather than listened to. So we can allow others to surface their motivation to collaborate by listening to them and showing an interest in their cares and concerns.

Why the emphasis on collaboration in a discussion on interpersonal motivation? Collaboration is the key to mutually coordinated action between or among individuals. Through collaboration we reach consensus and understanding. Through consensus and understanding we reach coordinated action. Through coordinated action we achieve desired outcomes and obtain results. Therefore to be effective at what we do, we are best served by creating an environment in which those with whom we are coordinating our actions can motivate themselves to collaborate with us.

This discussion is based on the recognition that in organizational settings there are no independent performers. People who work in organizations are dependent on one another for all that they do. People get ideas, direction, resources, guidance, and support from others in order to do their jobs. They provide the same to others in return. They do this through speaking and listening to one another.

How can we speak and listen to one another more effectively? We can do so by understanding the nature of the conversations in which we engage when we speak and listen. We can recognize that we are talking about speaking and listening in the broadest sense; we are including writing and reading as well as using and interpreting nonverbal gestures and any other form of communication.

We next need to devise a strategy for bringing those we are conversing with into a collaborative mode with us. We can do this most effectively by recognizing their need to be listened to. As we listen and develop a level of trust with them, they can motivate themselves better to join us in a spirit of sharing, collaborating, and understanding.

The key to effective listening is to learn to ask questions rather than give answers. We do this by asking an interrelated series of open-ended questions. Open-ended questions are those starting with who, what, where, how, and when. Again, we tend to be culturally hampered here, because most of us have learned to ask close-ended questions, that is, those requiring a yes or no response. We do this primarily because we want to convince or persuade people to see things our way.

So we ask questions that tend to "box" people into certain restricted areas and that restrict their ability to think too carefully. We say things like, "Don't you think that . . . ?" or "Isn't it true that . . . ?" When we do ask an occasional open-ended question, it is usually "why?" The problem with that is that it tends to put people in a "justify yourself" mode and causes them to become defensive.

To develop a sense of trust in those with whom we are conversing, we have to steer away from making them feel defensive. We also have to give them the feeling that we truly care about what they think and feel. We want to show them by our actions and behavior that their cares and concerns, their objectives and objections are important to us as well as to them. To do this we must listen to them and ask questions that show that we are listening. We can do this best when we learn to listen to what it is that they are listening to and listening for.

How do we do that? We start by entering into the conversation with the express goals of understanding what it is that the other's cares, concerns, and objectives are and how those are related with our own. We first ask some probing questions to get a better understanding of the other's position. We then follow up with questions that help both of us clarify what is possible. One of the keys here is to help the other think through the issue with our help. Our help comes from our skill in asking questions that are relevant and that show the other that we are listening.

Our ultimate goals are simple. We want to understand the other's position, we want to see how it is similar to or different from our own, we want to show the other through this process that we are listening, and through all of this we want to develop a trusting relationship with the other. When we have developed trust, our chances of moving to a collaborative environment are enhanced. We then increase the likelihood that the other will be motivated to continue the relationship in that vein.

When we attain a collaborative environment, it enhances our ability to mutually coordinate action toward desired results. By adroitly understanding the commonalities and differences in our respective positions, we can move to a synergistic position. This is enhanced by the trust that has been developed during our conversation.

We can move through a conversation from disagreement to agreement in the following manner. When we find ourselves in disagreement with another, after stating our position we listen carefully to the other's position. To understand his or her cares and concerns, we begin asking open-ended questions. We continue that process by searching for areas of agreement and disagreement. We can then explore areas for possible agreement. We can do this by suggesting areas or by asking the other to suggest some. We continue this process until we have either reached agreement or have mutually agreed to disagree. An added benefit of this process is that we can much more quickly discern when it is unlikely that we will ever be able to agree on the issue in question.

The issue of learning to agree to disagree is an important one in organizational life. Disagreement usually is based on either goals, methods, factors, or values. Careful listening can help you determine which is the case. When values are the issue, it serves little purpose to argue interminably. When value differences exist, the issue becomes coexistence, not agreement. By identifying value differences we can agree to disagree and begin a new conversation over possibilities for coexistence, given our value differences. We follow the same strategy of listening, questioning, and searching that we used in the previous conversation.

In summary, we are attempting to create a climate of collaboration and trust in our interpersonal relationships with others. We do this by listening and questioning. We stay out of competitive or defensive situations. We search for mutual understanding of issues and a mutually agreed upon course of action. In this way we can achieve mutually compatible results.

GROUP MOTIVATION

There are a couple of modes of group motivation that will be explored here. The first pertains to a group being motivated to perform. The second pertains to groups being motivated to interact effectively and productively. Both of these are important for productive quality output in an organization.

Groups are interesting entities. As all of us know, groups can take on a life of their own, sometimes seemingly separate and distinct from the lives of their members. Group formation and operation has been the focus of many studies and theories. Tuckman wrote of the forming, storming, norming, performing stages of group life. Bion wrote about work groups and basic assumption groups. Janis brought us the term "groupthink," and many others have expounded at length about group dynamics. Much has been written and spoken about groups and their nature, yet we continue to observe suboptimized effort from groups on a daily basis.

This discussion will focus on the characteristics and behavior of high-performing groups. These factors have been culled from observations of hundreds of groups over the past 15 years. A majority of the groups observed in action were engaged in project management activities. Most of the groups observed were not assessed to be particularly effective. Some of the groups stood out, however, as being substantially more effective than the others. It is from those groups that these factors were derived.

The first characteristic of the high-performing teams that stood out quite clearly was that they all had a very clear understanding of what they were to accomplish. They knew what was expected of them. This was clear to all the members of the groups and was explicitly talked about. There was no ambiguity about the goals and objectives of the group.

The second characteristic was that all the members of the group has specific, clearly understood objectives that were aligned to the purpose and intended outcome of the group as a whole. They each knew that it may be necessary to suboptimize their own objectives in order to accomplish the goals of the group.

Also, each member knew what they needed from the others and what the others needed from them. There was no question that they were part of a team and that it was necessary that they had to support each other if the group as a whole was to succeed.

Fourth, there was a clearly understood system of accountability for all members. They knew that they would be held accountable by both their peers and their leader for fulfilling their commitments and attaining their objectives.

Fifth, the lines of communication were clear and complete. There was an active network of conversations that supported the completion of the work, the sharing of problems and solutions, and the discussion of possibilities.

These factors led to another two which were outcomes of the first five: (1) clear understanding of roles and responsibilities and (2) a high level of trust among all team members.

Groups that were observed to be operating under these conditions were judged to be much more highly motivated to perform productively and with high quality. There was no overt effort to "motivate" the group itself or the individuals in the group. The environment that existed for the group led to the group "motivating" itself.

It is one thing to have a group motivated and performing at the top of its capability, yet quite another to have two or more groups effectively interacting to produce high-quality results. The key to this scenario is the ability of the different groups to communicate effectively with one another. They must know each other's needs and expectations and have a strategy for fulfilling them. This will enhance their individual ability to motivate themselves to coordinate their joint actions effectively.

How might this be accomplished? One way is to have each group answer three questions about each other, namely:

"What do we get from you that we like?"

"What do we get from you that we do not like?"

"What do we not get from you that we need?"

Each group then shares its answers with the other. They take the answers and develop a list of what they think the other group is asking them to do differently, or to start to do. These lists form the basis for a group-to-group conversation about how to improve the nature and quality of their interactions in the future.

There are a number of motivational forces at play in this interaction between the groups. First, it forces each group to be aware of what its own needs and expectations are. Second, it allows each group to understand what the other needs and expects from it. Third, it leads to an increased understanding of their common objectives. Finally, it opens the possibility of increased and continued communication between the groups as they work together through the future.

The ability of each group to motivate itself is thus enhanced. They each have a better understanding of those aspects of their functioning that are common as well as of those which are not. Awareness and understanding are very powerful forces for enhancing quality and productivity in the interactions of groups. Those forces lead to the attainment of mutually desired results, which is what intergroup interaction and collaboration is really all about.

To summarize, in this section we have looked at group motivation from the perspective of the performance of individual groups and from the perspective of the interaction between different groups. We have examined the characteristics of high-performance groups and a process for improving communication, and consequently intergroup understanding, and performance between groups. The underlying theme remained constant—motivation does not occur from the application of stimuli from external sources, but springs from within when an environment is established which allows it to happen.

ORGANIZATIONAL MOTIVATION

From a macro sense, the issue of motivation at the organizational level can be quite intriguing. One may ask if motivation can be discussed at this level. After all, aren't people motivated, not organizations? One perspective is that just as with groups, organizations are made up of individual people who pool their talents, skills, and abilities for some common purpose. Not coincidentally, that is the key to understanding motivation at the organization level—common purpose. Highly motivated, high-performing organizations have a clear overriding purpose, which is understood and shared by all the people in the organization.

There are two perspectives from which organizational motivation can be explored. First is the issue of how the leadership sets the environment internally to assure the desired level and direction of motivation. The second is how an external entity can get what it desires from the other organization by understanding what motivates it.

For example, the leadership of a defense electronics firm has a vision of where they want to lead that organization. To do so, they must understand the purpose that drives the effort of the people. If that purpose is not clear or understood, or accepted across the organization, the organization as a whole will be less than optimally motivated.

For one firm in particular, there was a common purpose of technological excellence shared by the engineering community but not by the rest of the organization. This led to a high degree of motivation by those involved in technical activities that was not shared by the other functional areas of the organization. The task of leadership thus became one of either transferring an understanding of and need for that purpose to the other areas, or modifying or expanding the perception people held of the common purpose.

Needless to say, the motivation of an organization is closely tied to the culture of that organization. This particular firm had been started by a group of engineers who developed an engineering-oriented culture. The engineers in that organization shared a common purpose of striving for technological excellence. As the company grew and became a more diverse company with a strong manufacturing emphasis, the culture did not shift with that change. The striving for technological excellence hindered the firm's ability to produce products that met the expectations of its customers.

Considerable friction existed among the marketing, engineering, and manufacturing divisions of the company. There was no cohesion between the individual divisions and they had differing views on what the purpose of the company was. Without a common purpose the overall motivation of the organization was reduced and its capability was severely suboptimized. The company's leadership was faced with a considerable challenge to develop and promulgate a common purpose around which all of the company could rally.

Notice that here, as in the previous sections, it is the environment in which people work that either helps or hinders their ability to motivate themselves to excel. Another way to say this is that the system within which people work affects the level of motivation they are able to achieve. Of course, this is individual-dependent. Each individual differs in his or her preferences for different working environments. Collectively, however, the emphasis shifts from an individual's preferences to group preferences. At the organizational level the individual preferences become more diffuse and a cultural collectivity takes control. In spite of that, individuals still choose whether or not to work in certain cultures. The kinds of people who work in an organization help determine the culture which in turn helps individuals determine whether that is the kind of culture in which they would prefer to work.

An individual's motivation will depend on his or her level of comfort with the culture in which the individual works. That culture will also determine what kinds of activities the organization will be motivated to pursue. That is important for anyone to be aware of when attempting to conduct business with the organization.

Whether you are purchasing goods or services from the organization or selling them, it is important for you to be aware of the organizational preferences when dealing with it. The individuals with whom you deal with reflect the values and culture of the organization. Consequently you must be aware of both those individuals' preferences as well as the culture of the organization to understand what motivates them. This should help shape the manner in which you interact with them and what you should expect from them.

Returning to the example firm discussed earlier, we can explore what motivated excellence in that organization and how that affected the way in which those in the organization reacted to external stimuli.

Since this was a defense electronics firm, most of its business was done with the government. The company was most successful when working on technically complex, difficult to solve quick-reaction-capability problems. When the government did not know how to solve the problem and left that work up to the contractor, they excelled. The engineers were allowed the freedom of opportunity they craved and enjoyed. They worked long hours and developed creative, albeit costly, solutions.

However, when the needs of the government were less complex, and there was less of a challenge to the engineers, they continued to look for technically elegant solutions, even though that was not required or desired by the government. Even though the government was the customer, the engineering community considered the customer to be ignorant and decided to give it what it "needed" rather than what it had asked for. Had the government been more aware of this culture and the accompanying motivation, it could have been more selective in the types of contracts it entered into with the company. At a minimum, the government could have better understood the propensities of the company and managed the contracts with it more effectively. Had it recognized the cultural imperatives and motivations, it may have been able to work with the company to help overcome the effects of the prevailing motivations.

The leadership of the company on the other hand did not do its part in managing this problem internally or externally. Since the leadership had come out of the same culture and had the same motivation, they were blind to the potential consequences. They continued to work to expand the size and diversity of the company without working to adapt the culture to the changing demands being placed upon it.

In summary, organizational motivation is rife with complexity. At this level there is an interaction of forces stemming from individual and group preferences, cultural drives and manifestations, and leadership awareness and understanding of the effects of those issues. However, it is possible to understand motivations at this level and to deal with them whether your vantage point is from inside the company or outside it.

MOTIVATION AND THE PROGRAM OFFICE

So far, with the exception of organizational motivation, we have discussed the issue of motivation generically. What about some specific examples from program office settings? All of the motivation issues and factors discussed are at play in any program office at any time. Let us look at a few of them.

First consider individual differences between supervisor and subordinates. When a program manager or functional manager has a preference indicative of a particular temperament, he or she will tend to favor and even create an environment that is conducive to the manager's own comfort. The manager will then expect the subordinates to operate in ways that may not match their preferred way of functioning.

For example, an ENTP (NT) program manager once worked for an ISTJ (SJ) product-line director. The NT program manager was not very attentive to details and he was not overly concerned with intermediate deadlines. His boss, the SJ, was a stickler for both details and timeliness. The NT kept telling his boss that all was well, everything was in hand, and the program would come in on time and under budget.

However, he had little or no data to back up his claims. As it turned out, he was correct, but that did not satisfy his boss. The boss operated only on facts and was constantly upset with the program manager because he rarely had the facts available when they were asked for. Meanwhile the program manager was highly stressed and extremely nervous because of the demands placed on him by the boss.

In this case they were able to work out a compromise that was helpful to both of them when they became aware of the causes of the differences between them. They were able to work out an arrangement where the program manager was given a deputy who handled the details and the deadlines for him. This freed the program manager to work on the creative side of the program while his deputy kept the boss satisfied with the detailed facts of the situation.

An important point to note here is that the boss was able to get what he needed based on his temperament, without forcing the program manager to work in an environment that affected his motivation. It also required that the program manager recognize that his deputy had to have access to the facts and details that were needed to keep the boss satisfied. By understanding the differing needs and preferences of one another the product-line manager and the program manager were able to work together much more productively.

An interesting facet of this situation was that the program was in a state that required the skills and talents of the NT program manager to work through some especially knotty problems that it was experiencing. It required creative solutions and those are best worked by NTs. Had the program manager been replaced by the SJ deputy, there is a high likelihood that the program would not have been completed successfully. Had the product-line director kept pressuring the program manager to be more detailed, he would have continued an environment in which the program manager would have been less and less motivated to continue performing productively on the program.

On an interpersonal level, the two men were only able to come to the conclusion when they were coached in how to listen to each other's cares and concerns. Prior to this intervention they had been in a demanding mode of each other in their conversations. Only when they began to listen and to ask open-ended questions of one another were they able to move from the impasse they had created with one another. As they came to understand the nature of their differences, as well as the different strengths and weaknesses they brought to the situation, they were able to find a common ground. They came to understand clearly that they needed each other's strengths, and each was motivated by different kinds of behavior.

Another example that may be helpful is that of two program offices, one government and one industry, working on a common program. The program was an aircraft program and there was considerable friction between the two program offices prior to a jointly agreed upon intervention. The intervention was designed to allow each group to understand the needs and expectations of the other, and to find common ground from which they could improve their interaction and the productivity of the program.

The differences between the groups were extreme. All the senior members of the industry program office were SJs, and most of those from the government office were NTPs. Consequently the industry people preferred to work with details, facts, deadlines, and quick-decision processes, and they got very upset when schedules and agreements were not followed to the letter. On the other hand, the government people were focusing on possibilities, the bigger picture, and they felt that schedules and agreements were starting points for further discussion. The industry people were very concerned with the day-to-day needs of the program, while the government representatives were more concerned with longer-range issues and political considerations. They were constantly talking past each other rather than talking with one another.

The intervention was designed to help each group understand the kind of environment in which the other was most comfortable operating. They each answered the three questions mentioned earlier and used them as the basis for an extended series of conversations to find ways to allow them to support each other.

An important point here is that they each preferred specific environments that allowed them to maintain a high level of motivation. Those environments were quite different. When they were trying to force each other into an alien way of operating, there was constant aggravation and all concerned admitted that it was difficult to stay motivated.

Once they understood each other's needs and found ways to thrive in differing environments, their individual and joint motivation levels rose considerably. The quality and effectiveness of their interactions improved dramatically over time.

These examples are intended to show that the previous discussions were more than just academic in nature. They are taken from actual situations and the outcomes are the ones that actually occurred. By understanding the effects of organizational climate, people's motivation can be improved.

SOME FINAL WORDS

Motivation is but one of an interrelated complex of variables that affect how well we get our work done and how well we interact with others. This chapter has tried to show how motivation can be viewed from personal, interpersonal, group, and organizational perspectives. It has also tried to show that these perspectives are interrelated. To understand motivation and its effect on performance, we must look at all the interrelationships and search for a balance among them.

As individuals, we work both independently and interdependently in a system with other individuals. We can choose to become aware of and understand what it is that helps us motivate ourselves. We can also understand what it is that helps others motivate themselves. With this understanding we can choose to create environments and to interact with those others in ways that are more effective than the ways we have chosen in the past. The goals of doing this are for us to become more productive, to increase the quality of our goods and services, and to improve our competitive position in the world marketplace.

BIBLIOGRAPHY

Gazzaniga, M., *The Social Brain,* Basic Books, New York, 1985.

Gazzaniga, M., *Mind Matters,* Houghton Mifflin, Boston, Mass., 1988.

Glasser, W., *Control Theory,* Harper & Row, New York, 1984.

Keirsey, D., and M. Bates, *Please Understand Me,* Gnosology Books, 1984.

Maturana, H., and F. Varela, *Autopoiesis and Cognition,* Reidel, Dordrecht, The Netherlands, 1980.

Maturana, H., and F. Varela, *The Tree of Knowledge,* Shambhala, Boston, Mass., 1987.

Myers, I., *Gifts Differing,* Consulting Psychologists Press, Palo Alto, Calif., 1980.

Searle, J., *Expression and Meaning,* Cambridge University Press, Cambridge, England, 1979.

Winograd, T., and F. Flores, *Understanding Computers and Cognition,* Ablex Publishing, Norwood, N.J., 1986.

CHAPTER 22

TEAM BUILDING AND THE PROJECT MANAGER

Kevin P. Grant

Kevin P. Grant is an Assistant Professor and currently the director of the Graduate Systems Management Program at the Air Force Institute of Technology. He has previously served as a project manager in the A-10 System Program Office of the Aeronautical Systems Division and as a program manager in the Aeromedical/Casualty System Program Office of the Human Systems Division. He holds a Ph.D. degree in industrial engineering from Texas A&M University, an M.S. degree in systems management from the Air Force Institute of Technology, and a B.S. degree from the United States Air Force Academy.

INTRODUCTION

Rome was not built in a day, nor was it built by an individual. People have been working together to achieve results since the earliest days of civilization. In recent years, many organizations have realized significant improvements in productivity and quality through the use of work teams.[1,2] These reported benefits, along with the cultural changes which accompany the quality revolution, are fueling increased reliance on work teams in the acquisition arena. The emergence of project teams in the system acquisition environment will naturally result in an increased demand for team building.

"Team building is the process of taking a collection of individuals with different needs, backgrounds, and expertise and transforming them by various methods into an integrated, effective work unit. In this transformation process, the goals and energies of individual contributors merge and support the objectives of the team."[3] Further, team building is a process which must continue throughout the life of the project. It should not be viewed exclusively as a one-time organization development intervention. Project teams which are currently working in the acquisition environment face many significant challenges, including inconsistent organizational structures, geographical separation, excessive oversight, and frequent changes in project staffing.[4] These challenges intensify the need to view team building as an integral responsibility of the project manager for the duration of the project.

Team building is appropriately conducted at many levels within an organization (Fig. 22.1). It can be conducted with a relatively small project team which consists of

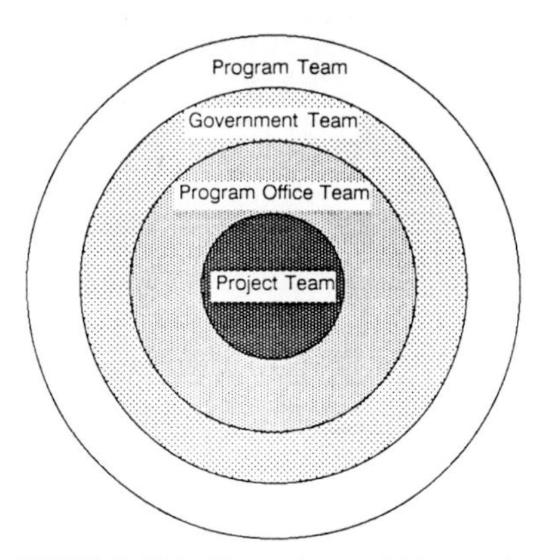

FIGURE 22.1 Teams in acquisition environment.

five individuals, or it can be directed toward a program office team which consists of 300 people. Team-building activities can also extend beyond the rigid boundaries that define acquisition organizations to include external partners and customers. Team building can be conducted to cultivate the collaborative efforts of team members representing several government organizations, and it can foster the cooperation essential to the success of the program team, which includes the contractors.

This chapter begins with a description of the nature of teams within the context of acquisition organizations. It then identifies team-building strategies for the directors or managers of program offices that may include many projects or smaller programs. Next it describes team-building activities which can be used by the project manager within the program office. Finally, the chapter stresses the importance of developing vital partnerships. This discussion focuses on opportunities to build the government team, which includes the operating commands, intelligence organizations, and laboratories in addition to the program office. It concludes by identifying opportunities to build the government-contractor team.

ACQUISITION PROGRAM OFFICES

Before embarking upon a discussion of team building it is important to describe the organizational structures which provide the environment within which project teams and project managers must perform.

The Matrix Organization. While projects are managed in a large variety of organizational structures, acquisition projects are most frequently managed in a matrix organization known as the program office or system program office (SPO). There are many factors which influence the structure of a program office. However, it remains useful to begin this discussion with a description of a typical program office which is structured as a matrix organization (Fig. 22.2). The program office is managed by a program director or program manager. The program director reports either to a program executive officer (PEO) or to a designated acquisition commander (DAC)[5] in the case of nonmajor programs.

The program office typically includes a projects division which comprises project managers. In a large program, these project managers may be responsible for specific subsystems, integration projects, or system modification efforts. If the program office is a "basket SPO," the project managers will typically be responsible for one or more of the many smaller programs managed within the program office. The program office will also include many functional divisions, such as contracting, program control, systems engineering, logistics, configuration management, test, and manufacturing.

The Project Organization. While a significant proportion of defense acquisition projects are managed under the auspices of a matrix organization, a second organizational form is also commonly found, the project organization (Fig. 22.3).

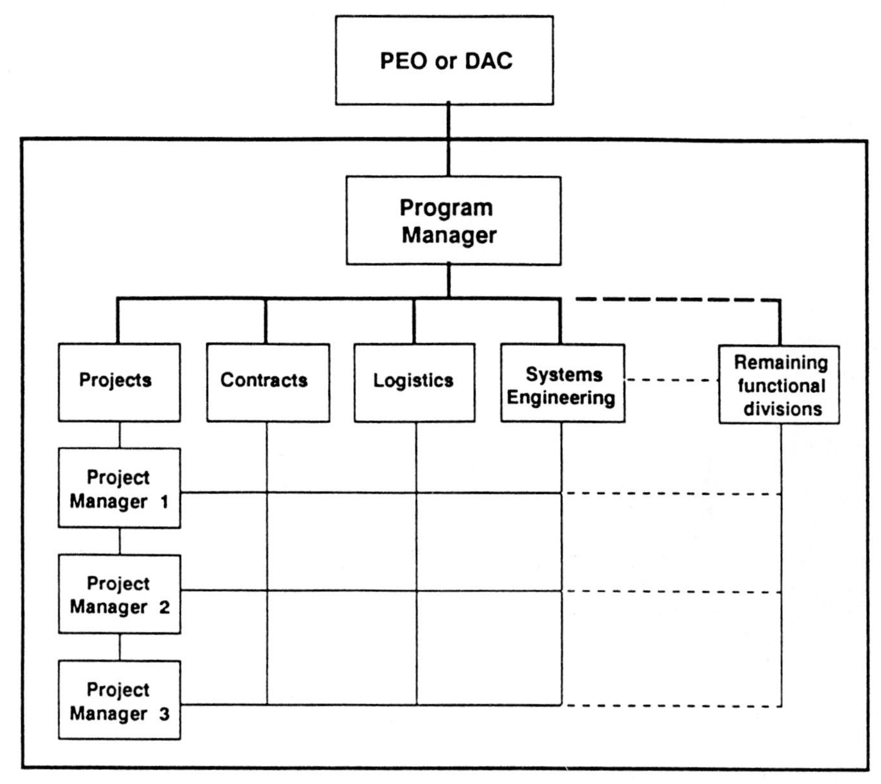

FIGURE 22.2 Typical program office.

In a pure project organization, the project team members report directly to the project manager rather than to a functional division chief. Frequently the team members will be collocated in the same office. Here team members from a variety of disciplines are integrated into a single, autonomous unit and are not required to support a large variety of competing projects. This organizational form is commonly found in laboratories as well as in advanced development program offices (ADPOs).

While Fig. 22.2 purports to describe a typical program office, in practice very few are "typical." Organizations are alive and vibrant. They flourish as requirements increase; they ebb as requirements wane or budgets decline. Program offices must constantly evolve to improve the manner in which they satisfy their customers' needs. Despite efforts to develop a standard SPO structure, there are many factors which will continue to influence the organization of the program office. Some are described in the paragraphs that follow.

The needs and characteristics of program customers will influence the structure of the program office. For example, a program which is deeply involved in foreign military sales may form a division to host representatives of the participating governments. Likewise, a program office may develop an internal liaison office to facilitate the partnership with the major commands primarily responsible for the operation of the systems managed by the program office.

The quality revolution is rapidly leaving its mark on the structure of program offices and will likely effect changes in the years to come. As more and more managers seek to institutionalize continuous improvement, they are erasing the lines and

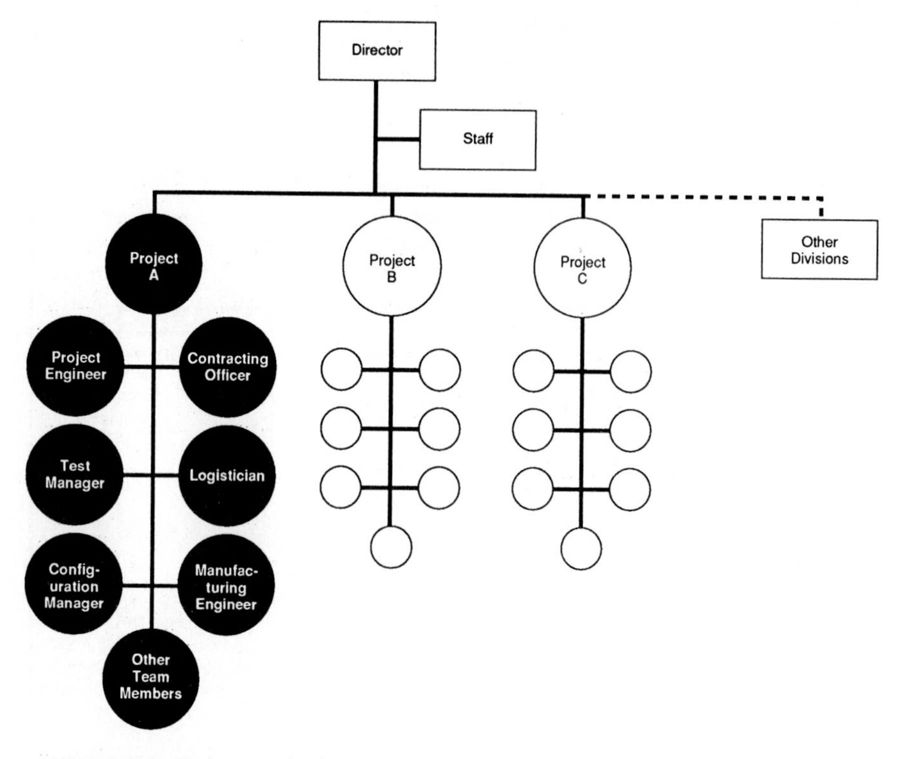

FIGURE 22.3 Project organization.

boxes of traditional matrix organizations and replacing the core of the organization with process action teams. These changes will increase dramatically the need for effective team building in the program offices of the future.

The size, complexity, and nature of the program will necessarily dictate key attributes of the program office. Large, complex, and enduring programs will likely exist as separate entities. On the other hand, smaller programs may be managed under the auspices of a "basket SPO." Moreover, some programs may remain separate entities for many years and then become absorbed by a new parent after many of the critical issues have been resolved and the requirements and resources no longer mandate high visibility.

The nature of contractor relationships will also continue to shape the program office. During the demonstration and validation phase a program office may develop separate divisions to manage the efforts of competing contractor teams. As the program progresses into engineering and manufacturing development, the program office may alter its form to deal more effectively and directly with the contractors selected to continue the program. In addition support contractors have augmented program offices for many years. Reliance on system engineering and technical assistance contractors as well as federally funded research and development centers has grown substantially throughout the acquisition community in recent years. The scope and nature of contractor support activities will inevitably affect the allocation and organization of the resources of the program office, especially its most vital resource—people.

Program strategy acts as a key determinant of the program office structure. For example, a program which adopts concurrent engineering may organize many multidisciplinary teams to integrate design, manufacturing, and support activities. In this case, the program office may reduce the staffing of the traditional functional divisions.[4] Or a program office may elect to pursue and qualify a second source. Once again, it is likely that adoption of this strategy will lead to changes in the structure of the program office.

Finally, program offices evolve simply as a function of the phases of the acquisition process. A test division may grow as a program approaches testing. A manufacturing division may grow as the program approaches low-rate initial production. The reliance on external sources of manpower may grow as a program enters source selection. As these examples illustrate, the nature of the work to be performed varies throughout the life of a program. Consequently the structure of the program office must change over time to confront these ever-changing demands.

BUILDING THE PROGRAM OFFICE TEAM

The program manager or program director of a large acquisition organization plays a crucial role in team-building activities. First, he or she is chiefly responsible for cultivating teamwork throughout the organization. Second, the program manager directly influences the success of team-building activities conducted at the project level. The following discussion identifies several strategies the program manager can use to cultivate the program office team and to facilitate project team-building activities.

Organize for Success. As the previous discussion revealed, there are many factors which impact the organizational structure of the program office. The program manager works in a very dynamic environment. One of the most fundamental steps the program director can take to set the stage for team building is to organize for success. The program director or program manager must proactively shape the structure of the program office so that it will be postured to succeed even in a turbulent environment.

Formulate a Vision. "Where there is no vision, the people perish. . . . " It is highly unlikely that this excerpt from the Old Testament book of Proverbs was intended to address modern-day strategic planning. However, the passage aptly suggests the importance that a vision may portend for the progress of an organization. A vision gives people throughout the organization a clear understanding of the direction in which the organization is heading. It is inspirational and describes the ultimate success of the organization. It should visually conjure images of an ideal state which motivates the collaborative efforts of all members in the organization. Through the establishment of a vision for the program office, the program manager can influence the success of the program as well as the many project teams within the program office.

There are many examples of program visions in the arena of government program management. Most Americans remember watching Neil Armstrong take the giant leap for mankind. This accomplishment was the fulfillment of the vision for the Apollo project—to place a man on the moon before the end of the decade. A second example is taken from the air force simulator SPO at the Aeronautical Systems Division. This program office is responsible for the development and production of simulators and rehearsal devices. While the following vision was never formally adopted, it remains a very colorful example: " . . . to make flight simulators so realistic that aircrews will stand in line with rolls of quarters to use them!!!"

Define the Mission. While the vision illuminates and guides, the mission describes the very reason the organization exists. It must be broad enough to include the contributions of all the members of the organization. It should be specific enough to distinguish any particular program office from others with similar missions. In the case of acquisition program offices, the mission frequently is oriented toward developing, providing, and sustaining a system or systems which will contribute to a required operational capability. The mission statement will define the boundaries of organizational activities. More importantly, it will focus on the customer.[6] Perhaps the greatest opportunity to lead an organization to internalizing the importance of customer satisfaction is to ensure that the mission statement is directed toward the customer.

Mission statements have been developed at various levels throughout the defense acquisition community. Each of the services has a mission, major commands have missions, product divisions or centers have missions, program offices have missions, and many project teams have missions. It follows then that workers on a project team are involved in the simultaneous pursuit of a complete hierarchy of superordinate missions. And while in many cases the project mission will function as the primary catalyst for team-building activities, there are several circumstances in which team members may find that the project mission provides little impetus for collaboration. In these cases it becomes imperative that the team members be able to understand and commit to the program office mission. The program office mission may prove particularly salient to program office members who spend the vast majority of their time working on functional activities, dedicate their efforts to several different projects, or work on a project that centers on a subsystem or component which is unlikely to contribute to operational capability independent of the complete system.

Establish Balance of Power. The matrix organization is very prevalent among acquisition organizations in the Department of Defense as well as in the defense industry. One of the most frequently cited difficulties associated with the matrix structure is the struggle for power, which arises between functional managers and project managers.[7] Unfortunately this delicate balance of power and authority can result in problems for project team members, who are torn between supporting the project manager and the functional division chiefs.[8] This conflict can result in dilemmas regarding priorities, commitments, and allegiances. The responsibility rests with the program director to determine the extent to which authority in a program office will reside with the project managers and the extent to which functional managers will retain authority for functional activities. The ability of the project managers to build teams at the project level will clearly be influenced by the support and authority which the project managers receive from the program director.

Commit Management Support. Senior management support is an absolute necessity for the project manager faced with effectively managing program office people and resources.[3] Through direct support to project team leaders, senior management facilitates the team building conducted at the project team level. This support can take many forms. The program director may strive to protect internal project teams from external threats or influences, excessive oversight and interference, and rapidly changing requirements. The program director may initiate internal process improvements which help every project team overcome the daily hurdles encountered in an acquisition program office. There are a multitude of actions which a program director can take to demonstrate commitment to project teams. The most important step is to commit personally and conspicuously to support all of the project teams in the program office.

Reward Team Performance. Rewards have long provided senior managers with a constructive means by which to encourage the behavior desired by members of the organization. Senior managers who are committed to encouraging teamwork within their organizations must develop rewards for team performance. It is not uncommon in military organizations to reward individuals. And individual rewards will continue to play an important role in any formal recognition program. However, an important catalyst for the cultural change to implement effective project teams is to develop measures and rewards for team performance.

Staffing and Selection. Senior management can also strongly influence the success of team-building activities within their organizations by paying careful consideration to the selection of project managers and project team members. There are a variety of skills which characterize successful project managers, including communication skills, technical competence, leadership skills, organizational abilities, and team-building skills.[9,10] The program manager can improve the likelihood of project success significantly by selecting project managers with the appropriate skill mix. In addition the program manager should also consider the composition of the project team when selecting the project manager. It is not uncommon that team characteristics can influence the relative importance of many project manager skills.[11] Finally, the program manager must provide education and training opportunities to project managers who are in the process of developing the skills essential to effective project management.

PROJECT TEAMS IN ACQUISITION ORGANIZATIONS

Before a project manager can initiate team-building activities, it is important to define the project team. In a project organization, team constitution is readily apparent. It is the very basis of the organization. However, in a matrix organization it is frequently difficult to identify team members. In a matrix organization, team members are typically assigned to a variety of functional offices and called upon to support project requirements whenever the need dictates. Consequently the nature and composition of the team is constantly changing. Figure 22.4 depicts a project team in a matrix organization. The shaded circles represent team members, each of whom dedicates varying amounts of time and effort to the project throughout the project life cycle. One of the advantages of the matrix organization is that functional team members can dedicate their time and efforts to a particular project when needed, and then return to their functional division to complete tasks which originate within the functional division or to support different projects. The matrix supports this flexible application of human resources without imposing a need to hire, terminate, or frequently reassign employees. However, building a team in a matrix organization can prove to be a formidable endeavor for many reasons. In a matrix organization the team members are usually assigned directly to the functional divisions and work directly for the functional division chief. Consequently their performance will typically be evaluated by the functional division chief rather than the project manager. Further, their assignment to the project team will frequently occur at the discretion of

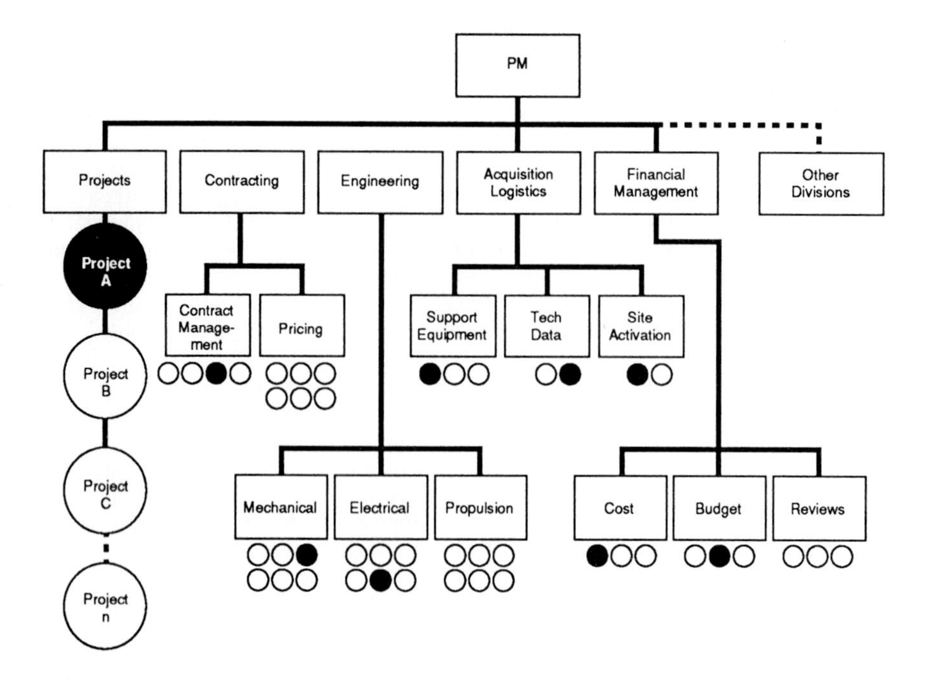

FIGURE 22.4 Teams in matrix organization.

the functional chief. In addition their time will be split between project responsibilities and functional responsibilities. Consequently team members are forced to balance their priorities. Of course, it is not uncommon for project managers to conclude that project priorities should prevail. Hence team building in a matrix organization is frequently characterized by conflicts between dedicated professionals. These are but a few of the challenges to team building in a matrix organization.

BUILDING THE PROJECT TEAM

The preceding sections have described a "typical" program office as well as several factors which impact the formation of the project team within the program office. It is within the framework of these organizations that the project manager must perform, that project managers must harness the resources of their project teams. While the specific structure of a program office may vary from program to program and over time, the project manager remains responsible for the successful completion of all project objectives, on time, and within the project budget.

Fortunately the project manager is not alone in contending with these challenges. There is the project team. Research conducted by Caron and Roderick led them to conclude that "the single most important element of management for a project manager is the building and motivating of the project team."[12] In a more recent study of project managers in industry, Posner reported that team building was one of the most frequently identified project management skills.[10] These research efforts, along with recent changes in the acquisition environment, strongly suggest increased emphasis on team building in project management.

Team building is the key to synergy. It enables diverse individuals with personal goals and perspectives to collaborate and accomplish more than the sum of their individual efforts could achieve otherwise.[13] This section first describes several activities which are important during the early stages of team performance, including creating a team environment, establishing a leadership role, defining the mission, and charting the course to success. Next this section will identify team-building activities which sustain the team throughout its life, namely, communicating regularly, rewarding deserving team members, and resolving the conflicts which occur between team members.

Create a Team Environment. One of the very first steps to team building is to create a positive team environment. The project manager must establish an atmosphere where team members are encouraged to participate freely, consider alternative viewpoints, act naturally, and—most importantly—work together.[14] It is also important for the project manager to properly set the stage for team building. The project manager needs to help the team understand that by defining the team mission, conducting integrated planning, setting goals, and establishing team member roles and responsibilities the team is beginning its journey to improved team productivity and performance.[15]

Take Charge. Leadership is frequently identified by experienced project managers as one of the most important project management skills.[10] The project manager is the team leader, and he or she is ultimately responsible for the performance of the project team. It is imperative that the project manager clearly establish this role. It is also equally vital that the project manager fulfill the leadership expectations of the project team members. Success in these regards will reduce team member competition for the leadership role.[3] Credibility is also a crucial ingredient to the formula for project leadership. Through integrity, sound decision making, technical competence, and positive relationships with key project stakeholders the project manager can establish the credibility with project team members which is essential to meeting their expectations.[3]

Define the Mission. The team mission is the very reason the collection of individuals has been assembled. It is why the team exists and specifically "what" the team is supposed to do. For acquisition project teams determining the team mission is usually straightforward. The team is formed to design, test, deliver, and sustain a critical capability. The mission will almost always derive from a formal operational requirement, or as a subset of a superordinate program mission. It must be stressed that even when the mission is readily apparent or formally directed, it is vitally important that all team members participate in defining the mission. Team member participation in the development of the team mission will foster team commitment toward completion of the mission, and it will also facilitate team understanding of the mission. It is difficult to overstate the motivational benefits accrued by team members who realize that their efforts are purposeful and directed toward a meaningful outcome. If relegated to the pursuit of purely functional objectives, the configuration manager may very understandably languish in the pursuit of the status of an engineering change proposal. However, as part of a team with a mission, this vital team member may find it very liberating to be correcting a flight safety issue or improving combat mobility. Examples of project mission statements are provided in Table 22.1.

Chart the Course to Success. Project planning is a critical activity which aims to identify specific project objectives as well as the ordered activities required to meet

TABLE 22.1 Project Mission Statements

The mission of the XYZ team is to design, test, and deliver medical shelters and support systems to conduct emergency medical operations in a chemical or conventional warfare environment, ultimately saving lives and returning our servicemen and servicewomen to duty.
The mission of the ABC team is to assess the ability of emerging electronic technologies to enhance the capability of the Joint Military Combat Command to detect and identify relocatable targets.
The mission of the Inertial Upper Stage (IUS) team is to provide an operational launch vehicle capable of deploying a tracking and data relay satellite to a geosynchronous orbit from the Space Shuttle.
The mission of the JSTARS radar team is to design, develop, and produce an integrated radar system for the E-8 aircraft that is interoperable with the Army's ground system module (GSM) in order to support the JSTARS mission of battle control.

these objectives. In addition project planning identifies the resources necessary to accomplish all project activities.[13]

The first step is to identify key project objectives. These project objectives should be specific, achievable outcomes which motivate and guide the performance of project team members. In addition they should be measurable and provide a vehicle the project team can use to assess team progress. From a team-building perspective, the formulation of project objectives is an excellent opportunity to secure team commitment and understanding. Several examples of project objectives are provided in Table 22.2.

The effective project manager will ensure that project objectives are explicit and understood by the project team members. To this end it may be useful to explain the objectives in terms of the system to be developed, deployed, or sustained. In addition the project manager should strive to develop consensus on the project objectives and hold project team members accountable for the successful completion of all objectives.[16] Success becomes far more likely when team members subordinate their personal or functional objectives to the project objectives. Creation of team objectives through the participation of the team members is a critical step toward securing the team member commitment essential to project success. Finally it should be noted

TABLE 22.2 Project Objectives

Award the contract for the engineering and manufacturing development phase no later than 15 November 199X.
Resolve all action items from the preliminary design review no later than 1 May 199X.
Deliver all items required to support operational test and evaluation no later than 31 October 199X.
Reduce the number of user-generated service reports by 50 percent within six months.
Improve the mean time between failures (MTBF) from 500 hours to 1000 hours within 12 months.

that an initial list of project objectives need not be cast in iron. Project objectives should be reviewed regularly and changed if warranted by significant changes in requirements, resources, or available information.

The second element of project planning is to identify and order the project activities. The work breakdown structure (WBS) is a very useful vehicle to identify project activities. Frequently acquisition project managers view the WBS solely as an instrument to define, monitor, and control contractor activities. This is understandable. The WBS provides a framework to estimate the costs of a contract, write the statement of work, and structure any cost performance reports. However, a WBS can also be used to identify those project activities which will be completed in-house. In a more general sense, a WBS is simply a hierarchical delineation of the work to be performed in a project. It begins with the entire project and breaks down the work into meaningful subunits one level at a time until discrete activities, of relatively short duration, have been identified. These work units can then be used as a basis for project planning, scheduling, resource allocation, and control.[8] Actively involving project team members in the definition of the project WBS will give team members a sense of ownership. Further, it will foster commitment to the project objectives and to the accomplishment of the project work units.[15]

Once the project activities are identified, the next step is to order the activities and develop the project schedule. Again, team participation in this activity is a vital element of project team building. Involving the team members who will be responsible for the activities is involving the team members who best understand what is involved with completing each activity. Consequently program schedules prepared in a team forum will be more accurate. More importantly, the project manager will be better able to hold team members accountable for performing their duties in accordance with a schedule which they helped prepare. A tool that can be used to facilitate this team activity is the precedence diagram (Fig. 22.5). It accomplishes thorough planning through the systematic delineation of task interdependencies and the logical sequencing of project activities.[17] When team members participate in the preparation of a precedence diagram it is easier for them to see how their individual contributions fit into the overall project. In addition they can see the detrimental impacts which will occur if they fail to complete their particular activities on time. It should also be noted that the information included in a precedence diagram can be used to generate a Gantt chart for the presentation of the project schedule.

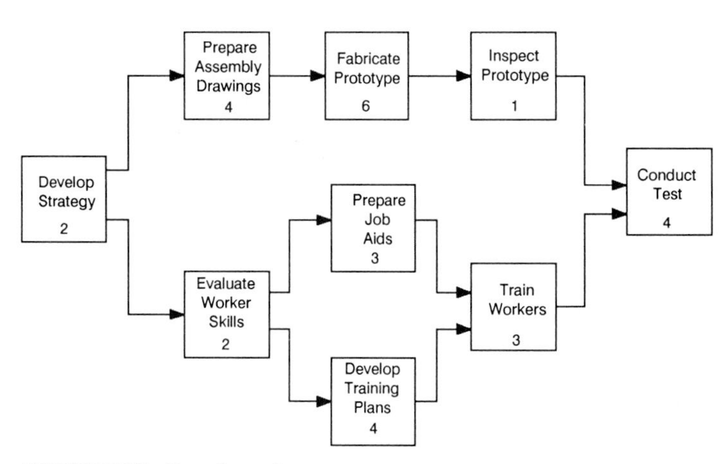

FIGURE 22.5 Precedence diagram.

	Project Manager	Engineer	Contracting Officer	Data Manager	Logistics
Acquistion plan	1	2	2		2
CBD announcements	3		1		
Source list	1	2	3		
Specifications	2,3,4	1,4			2
Statement of work	1,4	2,3,4	2,3,4	2,3	2,3
WBS	1	2,3	2,3	2,3	2,3
CDRL	2,3	2,3	2,3	1	2,3
Proposal preparation instructions	1	2	2	2	2
Evaluation factors	1	2	2	2	2

1	primary responsibility	3	must review
2	must contribute	4	must receive copy

FIGURE 22.6 Linear responsibility chart.

Finally, after the activities have been determined and scheduled, specific roles and responsibilities should be clearly identified and assigned. A convenient tool which can be used for this team process is the linear responsibility chart (LRC). It helps the team identify the team members who are primarily responsible for the completion of project tasks as well as the supporting roles which each of the team members may play. As with any tool, the LRC is flexible and can be tailored to meet the specific needs of the team. The LRC presented in Fig. 22.6 is an example of a chart which might be used to identify team roles involved with the preparation of a request for proposals (RFP). In this example the chart identifies who is responsible, who must contribute to each task or product, who must review each task or product, and who should ultimately receive a copy of each product. Through the use of this tool, each team member can identify his or her particular needs and expectations. The preparation of the LRC may involve several negotiations as team members determine who must contribute, or who really needs to review a document. However, by involving the team members in this process early in the project, the project leader can circumvent conflicts downstream.

Most of the team-building activities discussed to this point are especially crucial during the early stages of a project. The following team-building activities are more important to sustaining the continued performance of team members. Specifically these activities include communicating, rewarding performance, and resolving conflicts.

Communicate Regularly. Communication is the glue that binds an effective project team. The project manager must ensure that good communication prevails on the team. To this end, the project manager should make every effort to explain new policies and management decisions to team members.[14] In addition the project manager should frequently meet with team members to discuss their progress as well as their needs and concerns.[3]

There are several vehicles available to the project manager to facilitate team communications. Team meetings provide an opportunity to discuss what the team is doing well, and to identify areas which need further attention. Regular team meetings keep a team well informed about progress in diverse project areas.[3]

In recent years e-mail has proved to be a valuable medium for sending project information to the entire team. There are several advantages to be realized through the use of e-mail. First, it is quick. Second, e-mail messages will be stored until the team member is able to read them. This capability is very useful in the program office environment where many team members have to divide their time between several projects, or spend a considerable amount of time TDY. Another useful feature of e-mail is that it can confirm receipt of messages sent.[18] This feature provides a record of communication between team members. In some respects e-mail will never eliminate the need for face-to-face interaction between team members. However, when used in concert with team meetings, e-mail can contribute significantly to effective team communications.

Reward Deserving Team Members. "People are motivated in direct proportion to the value which they feel is being placed upon them; and the only demonstrated indication of this value is the rewards given them."[19] It is extremely important that project managers recognize team members for their good work and ensure that team members receive proper credit.[20]

Unfortunately it is not always easy for a project manager to provide direct rewards to a team member. Frequently the team members are evaluated by their functional division chiefs. Further, recommendations for awards, step increases, and merit promotions are also usually controlled by the functional division chief. Under these circumstances it is important that the project manager provide positive performance feedback to the functional supervisors when a team member deserves credit.[16] Actively providing this feedback will enable team members to receive the recognition they have earned. It also demonstrates that efforts dedicated to project success are no less worthy of recognition than efforts applied to functional objectives. Cultivating this understanding on the part of team members is a fundamental step of team building.

It is also important to recognize and reward team performance. Celebrating team success can be a very effective method to build the esprit de corps which is characteristic of effective teams. There are many methods to celebrate success. These include "milestone parties," public recognition, and team awards. There is no rigid formula for dispensing credit and recognition, or for providing rewards to deserving team members. What is important is that the project manager remain ever vigilant to ensure that team members receive the rewards they have earned through their efforts as vital team members.

Resolve Conflicts. Conflict is inevitable. The different professional identities and orientations characteristic of project team members who are drawn from several functional backgrounds frequently become ingredients for conflict.[21] Moreover, conflict can arise simply from the nature of project management. Thamhain and Wilemon identified seven sources of conflict which are prevalent in the project management arena. These include conflicts over schedules, priorities, manpower, technical issues, administration, personalities, and cost objectives.[16] So what should the project manager do to resolve conflicts on a project team? First, the project manager must be sensitive to the potential causes of project conflict.[16] Second, the project manager must be alert to the development of conflict on the team and take prompt, effective action to resolve the conflict before it gets out of control.[14] Finally, the project manager needs to confront conflict when it appears, and use compromise and encouragement to abate it.

In *The Managerial Grid,* Blake and Mouton identified five methods of conflict resolution: withdrawal, smoothing, compromising, forcing, and confrontation. Withdrawal was characterized as retreating from potential conflicts or disagreements. Smoothing involves the emphasis of commonalities and deemphasis of differences or conflicting issues. Compromising attempts to bring some degree of satisfaction to both parties through bargaining and negotiation. Forcing on the other hand is characterized by a win-lose situation wherein one party exerts his or her viewpoint at the potential expense of the other. Finally, confrontation involves a resolution of a problem through cooperative teamwork which directly addresses the area of disagreement.[16] Research has suggested that smoothing, compromising, and confrontation are associated with low conflict on project teams.[16]

The approach selected by each project manager will vary based on the source of the conflict, the styles of the manager, and even the phases of the project life cycle. The important implications for team building are that the project manager must be aware that conflicts will arise, must monitor team performance and relations, and must take immediate remedial actions when a conflict occurs.

DEVELOPING VITAL PARTNERSHIPS

The Government Team. Team building is "the process of influencing a group of diverse individuals, each with their own goals, needs, and perspectives, to work together effectively for the good of the project such that their team will accomplish more than the sum of their individual efforts could otherwise achieve." [13] This definition describes team building for the project team. However, with a slight modification, this definition becomes very useful in the context of building the government team. To effectively build the government team, the program manager must influence individuals positively, each representing various organizations with unique goals, needs, and perspectives, to collaborate in a partnership for the good of the program such that the government team will accomplish more than the sum of the organizations could achieve otherwise. The members of the government team represent a variety of organizations, including the acquisition program office, the ultimate users of the systems, and organizations which are responsible for threat analysis, test and evaluation, training, and system support. Team building in this context may largely center on those efforts dedicated to establishing and sustaining a viable and productive partnership. In some respects, it may mean less emphasis on a one-way customer relationship between the program office and the user, and more emphasis on the potential for a synergistic partnership. From this perspective it may not be as important to identify specific team-building strategies as it is to identify important opportunities for the organizations to work together to produce a result which is substantially better than the result that would have been generated by the organizations acting independently. This, after all, is the essence of teamwork.

Requirements Definition. The primary objective of requirements definition is to identify the performance capabilities of the system to be developed, produced, and deployed. It begins with the definition of broadly stated mission requirements and evolves to very specific system capabilities.[5] Throughout this process it is crucial that operators, developers, and supporters collaborate to identify options, alternatives, and implications. Analysts must share threat assessments and keep the members of the government team apprised of changes. The research and development community must communicate the availability and capabilities of emerging technologies. Operators must keep developers aware of changes in strategy, tactics, organization, or training. The developer must assess the affordability of the system and communicate life-cycle cost profiles to the remainder of the team. It is only through a highly integrated effort characterized by teamwork that acquisition programs will be able to effectively define requirements and remain responsive to the mission.

Technology Insertion. There are many challenges associated with inserting emerging technologies into either new or existing systems. These barriers include technical risk, lack of operational test data, undefined requirements, and a general lack of technical awareness.[22] Each of these barriers represents an opportunity to capitalize on teamwork. When the user is involved from the beginning and participates in demonstrating the technology and evaluating early prototypes, the user is better able to appreciate both the strengths and the limitations of the technological advance.[22] Moreover, the user develops a better understanding of the technical risks and the alternatives available to reduce the risks. Ultimately the government team is better able to deploy an enhanced capability successfully and to operate and maintain the capability with realistic expectations.

Trade Studies. Trade studies are conducted as an integral part of the system engineering process. During the design phase many decisions are made which effect total system performance as well as system costs and life-cycle costs. It is not uncommon that contractors will identify alternative strategies to meet system performance requirements throughout the design process. In some cases the contractor may recommend changes to requirements in an effort to improve total system performance, or to reduce cost or risk.[23] This decision process is an excellent opportunity to cultivate an effective partnership with the user and other government organizations. When the user participates in the process of weighing alternatives, assessing risks, and making design decisions, the user is better able to understand the evolution of the system and to plan for the operation and maintenance of the system. Once again, building this vital partnership can contribute significantly to the success of the program.

Programming Resources. The success of any program is clearly contingent on the receipt of funding provided through the federal budget process. Teamwork between the using commands and acquisition organizations is an essential part of this process. The acquisition commands must develop accurate estimates of system costs throughout the life cycle and share these estimates with the using commands. Further, acquisition organizations need to involve the using commands in the acquisition process to enable them to advocate the programs effectively. The using commands need to advocate programs essential to fulfilling operational requirements. Together, they must present and defend affordable and realistic program funding profiles. In the end, the very life of a program may depend on the successful collaboration between these vital partners.

This section has described several examples of opportunities for building a collaborative government team in the acquisition arena. Opportunities for team building occur regularly throughout the life of a program. Consequently team building is really an essential ingredient of management. It is an important part of program management each and every day.

The Government–Contractor Team. For all too many years, the link between the government and the contractor has been viewed by many as an adversarial relationship. In recent years there has been an increased recognition of the benefits to be realized through development of the government-contractor team. As with the discussion of the vital partnership between members of the government team, this discussion will focus on opportunities to foster collaboration between the members of these organizations.

Presolicitation Opportunities. There are many initiatives which the government can undertake to build relationships with contractors prior to the release of the solicitation. First, the government can invite contractors to discuss the anticipated contractual effort in an attempt to identify risk areas or strategies that might unnecessarily contribute to a higher contract price. The government may also elect to issue a draft RFP in an effort to solicit contractor recommendations to improve the RFP.

Throughout the preparation of the proposal the government should recognize that "most major contractors have been developing, producing and marketing their products for commercial, industrial and Government use for years." [24] This experience has enabled the contractors to develop significant expertise in the performance of acquisition contracts. Efforts directed at telling the contractor how to perform are frequently counterproductive. Success in today's acquisition environment comes from recognition that providing the contractor latitude for ingenuity is more likely to result in a quality, cost-effective system than arbitrarily specifying unnecessarily detailed requirements.[24]

Source Selection Opportunities. Clearly, the opportunities for collaboration are justifiably restricted once the solicitation is released and prospective offerors are in the competitive mode. However, there are some strategies the government should consider which can help establish a strong foundation for continued cooperation after the award of the contract. For example, the government may conduct a bidders' conference to give contractors an opportunity to clarify the requirements of the solicitation or the requirements of the anticipated contractual effort. The government may also establish a library of relevant information which the contractors can consult during the proposal preparation phase. In addition, the government may provide the contractor a reasonable amount of time to prepare the proposals. One of the more frequent criticisms of the government contracting process levied by contractors is that the government takes many months to prepare the solicitation and then attempts to streamline the schedule by restricting the contractor to a 30-day proposal preparation period.

Postaward Opportunities. Many government organizations have been conducting team-building sessions with the contractor shortly after contract award. These "kick-off" meetings have generally provided the contractor with the opportunity to clarify the requirements of the contract. They have also served as get-acquainted sessions for the members of both organizations. Kick-off meetings have frequently included specific team-building activities designed to establish group norms or communication patterns. In several instances, consultants have participated to determine the psychological types of the project participants and to discuss the implications of which psychological type may portend for future interactions between team members.

Postaward team-building opportunities continue throughout the performance of the contract. A mutual commitment to the success of the contract effort and ultimate success of the program provides a common ground for members of both the government team and the contractor team. Both teams can celebrate the same successes—together. In addition a free flow of information between organizations can greatly facilitate problem-solving activities. Greater emphasis on accountability may provide contractors the latitude needed to generate the best solutions. Recent initiatives to streamline acquisition have shifted the balance of responsibilities between the government and the contractor. The government is providing the contractor less-detailed guidance, more discretion, and greater incentives and rewards. At the same time, the contractors are assuming greater responsibility for developing producible and supportable systems on time and within budget.[24] These are the fruits of government-contractor partnership and teamwork.

CONCLUSION

The management of a technically complex acquisition project in a resource-constrained environment is an ominous undertaking. Teamwork provides the best hope for success. Team building is the means by which the program director can secure the contributions of the many varied members of the program office. It is the means by which the project manager can promote the synergistic collaboration of project team members. It is also the vehicle for the cultivation of vital partnerships between government organizations and between the government and contractors. In sum, team building is a way of life in project management.

REFERENCES

1. E. J. Poza and M. L. Markus, "Success Story: The Team Approach to Work Restructuring," *Organ. Dynamics,* vol. 8, pp. 3–25, Winter 1980.
2. R. E. Walton, "Work Innovations at Topeka: After Six Years," *J. Appl. Behav. Sci.,* vol. 13, no. 3, pp. 422–433, 1977.
3. D. L. Wilemon and H. J. Thamhain, "Team Building in Project Management," *Project Manage. Quart.,* pp. 73–81, June 1983.
4. Air Force Systems Command, APEX White Paper on SPO Team Building, 1990.
5. DoD Directive 5000.1., "Defense Acquisition," Feb. 23, 1991.
6. M. Hardaker and B. K. Ward, "Getting Things Done: How to Make a Team Work," *Harvard Bus. Rev.,* pp. 112–120, Nov.-Dec. 1987.
7. E. W. Larson and D. H. Gobeli, "Matrix Management: Contradictions and Insights," *Calif. Manage. Rev.,* vol. 29, no. 4, 1987.
8. J. R. Meredith and S. J. Mantel, Jr., *Project Management: A Managerial Approach,* 2d ed., Wiley, New York, 1989.
9. D. S. Kezsbom et al., *Dynamic Project Management: A Practical Guide for Managers and Engineers,* Wiley, New York, 1989.
10. B. Z. Posner, "What It Takes to Be a Good Project Manager," *Project Manage. J.,* Mar. 1987.
11. C. R. Baumgardner, *An Examination of the Perceived Importance of Technical Competence in Acquisition Program Management,* master's thesis, AFIT/GSM/LSY/91S-4, School of Systems and Logistics, Air Force Institute of Technology (AU), Wright-Patterson AFB, Ohio, Sept. 1991.
12. P. F. Caron and B. Roderick, "The Challenge of Project Management: Building and Motivating Teams," *Program Manager,* pp. 6–9, July-Aug. 1979.
13. "Project Management Body of Knowledge," Standards Committee, Project Management Institute, 1987.
14. D. D. Acker, "Managing Conflict at Work: The Art of Putting Away the Boxing Gloves and Living Together in the Professional Ring," *Program Manager,* pp. 6–12, July-Aug. 1980.
15. D. S. Kezsbom, "Are You Really Ready to Build a Project Team?" *Industrial Engineering,* vol. 22, no. 12, pp. 50–55, October 1990.
16. H. J. Thamhain and D. L. Wilemon, "The Effective Management of Conflict in Project-Oriented Work Environments," *Defense Management Journal,* pp. 29–40, 1975.
17. M. M. Obradovitch and S. E. Stephanou, *Project Management: Risks and Productivity,* Daniel Spencer Publishers, South Bend, 1990.
18. K. J. Farkas, *An Analysis of Factors Affecting the Electronic Delivery of Contractor Data,* master's thesis, AFIT/GSM/LSY/91S-9, School of Systems and Logistics, Air Force Institute of Technology (AU), Wright-Patterson AFB, Ohio, September 1991.
19. L. C. Stuckenbruck and D. Marshall, *Team Building for Project Managers,* Project Management Institute, Drexel Hill, Pa., March 1990.
20. R. M. Hodgetts, "Leadership Techniques in the Project Organization," *Academy of Management Journal,* pp. 211–219, June 1968.

21. R. E. Hill, "Managing Interpersonal Conflict in Project Teams," *Sloan Management Review,* pp. 45–61, Winter 1977.
22. S. J. Guilfoos, "Bashing the Technology Insertion Barriers," *Air Force Journal of Logistics,* pp. 27–32, Spring 1989.
23. R. K. Barry, *The Advanced Tactical Fighter Early, Structured, and Continuous User-Interface Process: A System Program Office/User Team-Building Concept,* master's thesis, AFIT/GSM/LSY/91S-3. School of Systems and Logistics, Air Force Institute of Technology (AU), Wright-Patterson AFB, Ohio, September 1991.
24. Department of Defense Handbook 248B, *Acquisition Streamlining,* 15 October 1987.

CHAPTER 23

TRAINING AND DEVELOPMENT FOR THE MILITARY ACQUISITION WORK FORCE

J. Gerald Land

J. Gerald Land is currently the Associate Dean of Faculty and Professor of Systems Acquisition Management in the Funds Management Department, Faculty Division, Defense Systems Management College (DSMC), Ft. Belvoir, Va. For the previous two years he served also as Deputy Director of the Education and Curriculum Support Department at DSMC. In addition to his position at DSMC, Colonel Land is a part-time Professor of Management in the School of Business, Florida Institute of Technology.

INTRODUCTION

Background of the Underlying Issue

Numerous investigations, commissions, congressional hearings, and studies have indicated that there are serious problems with the manner by which the Department of Defense (DoD) acquires its weapon systems and other major items required for the national defense and that action is necessary to correct problems which may be systemic to the acquisition process.

Periodically, during the past several years, there have been numerous examples of negative publicity concerning the defense acquisition process. Weapon systems have been canceled after millions of dollars had been spent on a weapon that does not perform as required. Weapon systems have taken longer in development than originally expected by either the government or the contractor. Weapon systems have grossly exceeded approved budgets and show excessive cost overruns. Some contractors have overcharged the federal government for such common items as spare parts, hammers, and coffee pots procured as part of a major acquisition. Some contractors have been found guilty of fraud in their federal government contracts and others have paid large fines rather than go to court. Several government employees have either admitted guilt or have been found guilty of accepting illegal payments from contractors in exchange for "insider" type information.

Such publicity raises several concerns: the manner in which the federal government spends public funds; the ability of acquired materiel to perform; the competence and trustworthiness of people involved in acquisition; and the actual inner workings of the defense acquisition system. There is a generally negative public perception associated with how DoD acquires weapons and other materiel intended to provide for the national defense. Congress has shared this public concern for several years. Various committees have held hearings on acquisition-related issues, and individual members have initiated many bills on different aspects of this public-policy issue. Past presidents and DoD senior officials have been concerned with the public perception as well as the actual problems experienced in defense acquisition.

While there may be general agreement that problems exist and improvements could be made to the acquisition system, opinions differ as to specific problems and what should be done to correct them. Defense acquisition is an extremely complicated system which, according to a 1990 congressional report, consists of three distinct elements (Ref. 1, p. 1): "(1) the policies, procedures, and processes which govern the operation of the acquisition system; (2) the organization of the resources (people, management structure, capital, and facilities) that execute the policies and procedures; and (3) the people within the organization that make the system work."

Although the reason for problems associated with defense acquisition often depends as much on the perspective and the perceptions of the individual expressing the problems as it does on reality, it should be obvious that there is not a single but rather a combination of causes for these problems. Because of its complexity, the system could break down for a multitude of reasons. I suggest that one basic reason, among several other potential reasons, concerns the people element of the acquisition system—the individuals involved in the day-to-day business of defense acquisition.

In this chapter I will address one aspect of the people element: I will consider the professionalism of the defense acquisition work force in general and, specifically, the training, education, and career development of the individuals who compose that work force. The examination of various aspects of training, education, and professional development of the defense acquisition work force will be made primarily from an historical perspective. The term "military" in the chapter title is intended to include the civilian acquisition professional as well as the active-duty acquisition professional service member.

Key Definitions in Defense Acquisition

Defense acquisition is defined as "the planning, design, development, testing, contracting, production, introduction, acquisition logistics support, and disposal of systems, equipment, facilities, supplies, or services that are intended for use in, or support of, military missions." Broadly defined, the defense acquisition work force is "the personnel component of the acquisition system. The acquisition work force includes permanent civilian employees and military members who occupy acquisition positions, who are members of an Acquisition Corps, or who are in acquisition development programs." The acquisition corps is "a subset of a DoD Component's acquisition work force, composed of selected military and civilian personnel in grades of Lieutenant Commander, Major, General Schedule and/or General Manager (GS/GM) 13 and above, who are acquisition professionals. There is one Acquisition Corps for each Military Department and one for all the other DoD Components (including the OSD and the Defense Agencies)." (Ref. 2, pp. vi–vii.)

Professionalism in Defense Acquisition

Generally accepted among professionals is the concept that true professionalism results from an individual possessing proper levels of training, education, and experience in the chosen field, combined with a measure of specific intrinsic traits or characteristics critical to that profession. The individual is accepted as a member of that profession when he or she acquires and demonstrates those factors. Among those factors, education and training may be obtained through attendance at formal or informal schools or through other types of course offerings. The individual may acquire experience by working and living the profession and, through this experiential process, can often also develop those intrinsic traits to become a professional. Proper development of the professional occurs when there is a systematic and managed combination of training, education, and experience. Demonstration of those factors occurs when the individual actually puts into practice the results of that training, education, and experience.

Career or professional development, as defined by DoD, is "the professional development of employee potential by integrating the capabilities, needs, interests, and aptitudes of employees participating in a career program through a planned, organized, and systematic method of training and development designed to meet organizational objectives. It is accomplished through . . . work assignments, job rotation, training, education, and self development programs." (Ref. 2, p. viii.)

According to an extensive study developed primarily through the efforts of Dr. James Edgar, an American Political Science Association fellow on assignment to the House Armed Services Committee of the second session of the 101st Congress, which was published in 1990 as a committee report: "Although it has been recognized in studies and commissions . . . for over thirty years that the quality and professionalism of the defense acquisition workforce should be improved, the great majority of reform efforts have focused on changes in policies and procedures, or organization (of the resources that execute the policies and procedures)." (Ref. 1, p. 1.)

Recent congressional action increased the overall awareness of the need for a truly professional defense acquisition work force. Following several years of interest in the broad area of defense acquisition, including the specific aspect of government acquisition personnel, in 1990 Congress passed a law (P.L. 101-510) containing a separate portion that pertains specifically to the defense acquisition work force.[3] That separate portion of P.L. 101-510, the Defense Acquisition Work Force Improvement Act, mandated specific requirements relative to the defense acquisition work force community.

EXECUTIVE BRANCH INQUIRY AND ACTION

Within the federal executive branch, concern about the acquisition work force was shown as early as 1949 with the establishment of the first Hoover Commission. This presidential task force has been followed by several others: the second Hoover Commission, 1955; the Fitzhugh Commission (President's Blue Ribbon Defense Panel), 1970; the Commission on Government Procurement, 1972; the Grace Commission, 1983; and the 1986 Packard Commission (President's Blue Ribbon Commission on Defense Management). Each of these executive-level groups recognized the importance of personnel to the government's acquisition process and commented on the need for a qualified professional work force in order to achieve a quality governmental acquisition program.

By Executive Order 12352 (1982), the President required each executive department to establish "career management programs, covering the full range of personnel management functions, that will result in a highly qualified, well-managed procurement work force." (Ref. 1, p. 63.) An executive committee was formed to assist in the implementation of that executive order. An element of that executive committee issued 38 recommendations pertaining to establishing procurement career management programs. Some of those recommendations were implemented, others were not (Ref. 1, p. 63).

The specific topic of education and training for individuals involved in a primary area of acquisition (that is, contracting) was addressed as early as February 1984 in a DoD Inspector General (IG) report based on an internal audit. The primary purpose of that audit was to determine whether senior- and intermediate-level contracting personnel had received mandatory training required by DoD directives. The resultant report stated that 67 percent of senior- and intermediate-level professional contracting personnel at the 24 activities visited had not taken all mandatory training and recommended that training provided to defense procurement personnel be improved (Ref. 4, p. 2).

In August 1985 the Deputy Secretary of Defense initiated "a comprehensive review designed, among other things, to identify actions needed to promote a more professional contracting, quality assurance, and program management work force." (Ref. 4, p. 1.) One outcome from this review was the creation of a study group, known as the Acquisition Enhancement I (ACE I) Program. The 1985 ACE I report set forth the following: (1) prerequisites of experience and training requirements for 15 job functions; (2) drafts of new directives and instructions to promulgate those prerequisites; and (3) a recommendation for a follow-on study of the DoD's acquisition training base.

In November 1985 the Assistant Secretary of Defense for Acquisition and Logistics published a document known as the "Wade White Paper on Acquisition Improvement." This document addressed several aspects of defense acquisition, including the need to revitalize the DoD acquisition and logistics work force through cooperation between the executive and legislative branches. One specific recommendation in that document was to " . . . establish an umbrella Defense Acquisition University encompassing all the existing acquisition related Defense schools. The University should include separate colleges of specialization such as contracting and acquisition, logistics, quality, program management, systems engineering, production and manufacturing, all offering accredited degrees." [5]

In May 1986 DoD tasked the Defense Systems Management College (DSMC) to conduct a follow-on study to the 1985 ACE Program. The study's final report, published late 1986, included numerous recommended changes pertaining to procedural matters and a specific recommendation to establish the DoD University of Acquisition Management. The basic concept of this recommendation for a DoD-level university stated that currently existing facilities providing education and training for DoD acquisition employees would be left in place, with their activities coordinated through a consortium of those educational facilities led by a university president and small staff (envisioned to be colocated with DSMC). If considered necessary, more extensive consolidation and direct control of those facilities would be accomplished at a later date (Ref. 4, p. vi).

President's Blue Ribbon Commission

In July 1985 President Reagan established his Blue Ribbon Commission on Defense Management, chaired by Mr. David Packard, to " . . . study issues surrounding defense management and organization, and report its findings and recommendations."

(Ref. 6, p. xi.) The President's Blue Ribbon Commission on Defense Management (hereinafter called the Commission) issued an interim report (February 1986) and a final report (June 1986). The latter presented the Commission's full findings and recommendations. Because of the pervasive nature of acquisition within DoD, a significant portion of the Commission's final report deals with the topical area of acquisition and a minor portion thereof with the education and training of acquisition personnel. Some of the more pertinent findings of the Commission pertaining to enhancing the quality of personnel in the defense acquisition work force follow (Ref. 6, pp. 65–69):

> Our study convinces us that lasting progress in the performance of the acquisition system demands dramatic improvements in our management of acquisition personnel at all levels within DoD.
>
> DoD must be able to attract and retain the caliber of people necessary for a quality acquisition program. . . . Federal regulations should establish business-related education and experience criteria for civilian contracting personnel, which will provide a basis for the professionalization of their career paths . . .
>
> The defense acquisition work force mingles civilian and military expertise in numerous disciplines for management and staffing of the world's largest procurement organization . . . compared to its industry counterparts, this work force is undertrained, underpaid, and inexperienced. Whatever other changes may be made, it is vitally important to enhance the quality of the defense acquisition work force—both by attracting qualified new personnel and by improving the training and motivation of current personnel.
>
> The caliber of uniformed military personnel engaged in program management has improved significantly of late. . . . Each of the Services has established a well-defined acquisition career program for its officers. . . .
>
> By contrast, much more remains to be done concerning civilian acquisition personnel generally. . . . We recommend federal law permit the Secretary of Defense to include other (i.e., besides scientists and engineers) critical acquisition personnel in such a system (i.e., an experimental Navy recruitment and retention plan to correlate pay, incentives and advancement with performance), and facilitate greater professionalism among civilian acquisition employees through government sponsorship of graduate instruction in acquisition management.
>
> Among acquisition personnel, contract specialists have an especially critical role. More than 24,000 members of DoD's acquisition work force specialize in the award and administration of contracts. . . . Contract specialists must master the extensive, complex body of knowledge encompassing materials and operations management, contract law, cost analysis, negotiation techniques, and industrial marketing. Yet, the Office of Personnel Management designates the Contract Specialist personnel series (GS-1102) as an administrative and not a professional series under Civil Service Title VIII. This administrative designation prohibits the establishment of *any* business education requirement for contract specialists. As a result, only half of DoD's contract specialists have college degrees, which may or may not be business-related. We recommend establishing a minimum education and/or experience requirement for the Contract Specialist series. . . .

Considering the importance of acquisition to DoD and to the national defense, it would seem that a maximum effort would be made to create the most efficient and effective education and training program for individuals employed in that functional area. The Commission reached a similar conclusion and recommended that " . . . such training (i.e., for contract specialists and all acquisition personnel) should be centrally managed and funded. . . . to improve the utilization of teaching faculty, to enforce compliance with mandatory training requirements, and to coordinate overall acquisition training policies." (Ref. 6, p. 69.)

Following receipt of the interim report to the President from the Packard Commission in February 1986, the President directed (by National Security Decision Directive 219) that appropriate elements within the Executive Department begin implementing " . . . virtually all of the recommendations presented to him in the interim report of the Blue Ribbon Commission on Defense Management (i.e., the Packard Commission)." This included changes in the national security planning and budgeting system, in the military organization and command structure, and in the acquisition organization and procedures (such as creation of a new level II position of Under Secretary of Defense (Acquisition); a strengthening of "personnel management policies for civilian managers and employees having contracting, procurement or other acquisition responsibilities"; and new procedures relating to government and industry accountability.[7] This was followed by the August 1986 National Security Decision Directive 238 requiring compliance with the Packard Commission recommendations.

LEGISLATIVE BRANCH INQUIRY AND ACTION

Congressional interest in the professionalism of government personnel involved in the defense acquisition process has also been demonstrated for several years. Early laws addressed key aspects of the defense acquisition work force. As early as 1984, Congress passed two laws that contained provisions pertaining to such employees. P.L. 98-369 (Section 2721) stated that each executive agency was to " . . . develop and maintain a procurement career management program in the executive agency to assure an adequate professional work force." [8] P.L. 98-525 (Section 1243) required that program managers of major defense weapon systems remain in their assignment for a minimum of four years or until that weapon completed a significant milestone in its development.[9] In 1985 Congress passed P.L. 99-145 (Section 924), which required an individual designated as program manager of a major defense weapon system to have completed the 20-week program management course at DSMC and have at least eight years' experience in one or a combination of the acquisition disciplines.[10] Several subsequent laws contained significant provisions affecting the defense acquisition system and its associated work force; these are now described further in greater detail.

Public Law 99-661

The National Defense Authorization Act for Fiscal Year 1987 (P.L. 99-661),[11] passed in November 1986, included language which set forth two specific requirements pertaining to training, education, and the professional development of defense acquisition personnel.

The first requirement of P.L. 99-661 was contained in Sec. 932, which required that the Secretary of Defense develop a plan for a personnel initiative designed to enhance the professionalism of, and the career opportunities available to, DoD acquisition personnel, and that DoD provide Congress a report describing that plan. The Deputy Secretary of Defense responded to that section of the public law by a May 13, 1987, letter to the chairman of the House Armed Services Committee in which the following points were set forth: "The plan being pursued consists of two sets of initiatives. The first is a legislative change that would allow the establishment of a formal alternative personnel management system. The second involves specific actions exclusively addressing educational qualification requirements and training opportunities for acquisition personnel." [12]

The first initiatives proposed by DoD in that early 1987 letter included such features as pay banding (wherein one band would encompass pay ranges for two or more grade levels of the general schedule); a simplified classification (to eliminate many administrative efforts of the classification procedures); pay for performance (in which an individual's progression in the pay band would be based on the concept that better performance would result in higher pay); and market sensitivity (which would allow DoD to hire a new employee at a pay rate, within the appropriate pay band, at a level consistent with local market conditions for an individual of that skill and ability).

The second set of DoD proposed initiatives in response to requirements in P.L. 99-661 covered two aspects pertaining to education for civilian acquisition personnel. The first pertained only to educational standards for civilian contracting specialists (GS-1102) of the defense acquisition work force. The letter informed Congress that DoD had contacted the Office of Personnel Management (OPM) about establishing firm educational requirements as a prerequisite for entry into that job series, and that OPM had indicated such requirements are precluded for the GS-1102 occupational series by the U.S. Code (that is, 5 U.S.C. 3308). DoD suggested that it would support a change to that code which, as a minimum, would permit OPM, in coordination with DoD, to establish minimum educational standards for the GS-1102 series and, as an optimum solution, would state that a college degree was a required prerequisite for entry into the GS-1102 series. The second aspect recommended a change to another provision of the U.S. Code (5 U.S.C. 4107) to permit DoD to pay for training expenses of civilian acquisition personnel who take such courses for the primary purpose of obtaining an academic degree. (Payment for college courses was specifically restricted to only those courses considered to be job-related.)

The second requirement of P.L. 99-661 directly affecting defense acquisition personnel was contained in Sec. 934, which required that the Secretary of Defense "... submit to the Committees on Armed Services of the Senate and the House of Representatives a report containing a plan for the coordination of educational programs managed by the Department of Defense for acquisition personnel of the Department."[11] The Assistant Secretary of Defense (Force Management and Personnel) responded to that section of the public law by similar letters dated March 2, 1988, to the chairman of the House Armed Services Committee and the chairman of the Senate Armed Services Committee that provided a copy of the DoD plan to improve the education and training of the acquisition work force and relevant directives that were to be revised in the near term to implement that plan. The letters also stated that the mission of DSMC had been expanded to require that school to coordinate the high-quality mandatory acquisition training and education courses intended to prepare military officers and civilian personnel for assignments in the acquisition career fields.

An enclosure to the March 1988 letters from the Assistant Secretary of Defense (Force Management and Personnel) to the two committee chairmen was a February 4, 1988, memorandum from the Under Secretary of Defense (Acquisition) to various internal DoD addressees in which the former requested support and action by the latter to implement specific recommendations of the ACE Program Action Group. Those ACE recommendations constituted the DoD plan for near-term improvements in the education and training of the acquisition work force.[13]

Hearings before the Senate Armed Services Committee, 100th Congress, Second Session

During July and August 1988 the Senate Armed Services Committee (SASC) of the 100th Congress held a series of hearings on the defense acquisition system. These hearings were the direct result of allegations of fraud in the DoD's source selection procedures (that is, the method by which the government selects the specific contractor to be awarded a contract to develop or produce a weapon system). As stated in the hearings proceedings (p. 2), the primary reason for the hearings was the fraud allegations, but the overriding reason was the committee's concern " . . . about the inability of the acquisition system to provide our men and women in uniform with quality equipment on a timely and cost effective basis."[14]

During the four days of hearings the committee heard from nine senior individuals who were either currently or previously involved in defense acquisition. Of the nine witnesses, seven discussed the professionalism of government acquisition personnel and the need for such employees to receive increased education and training in various aspects of acquisition.

Hearings before the House Armed Services Committee and the Acquisition Policy Panel, 100th Congress, Second Session

During June through October 1988 the House Armed Services Committee (HASC) and the Acquisition Policy Panel of the second session of the 100th Congress held a series of hearings on the defense acquisition system. As with the hearings before the SASC, these hearings were the direct result of allegations of fraud and bribery involving defense contractors, consultants, and certain government officials in DoD. As stated in the hearings proceedings, the primary reason for the hearings was the fraud allegations, but the main focus was their concern about " . . . the current integrity of the DoD [Department of Defense] acquisition system . . . " and the impact of current investigations of procurement fraud on the United States national security.[15]

During the nine actual days of hearings held during the five-month period, the committee heard from 16 senior individuals who had either a current or a prior involvement with the defense acquisition system. Five of the nine witnesses who appeared before the SASC also provided testimony to the HASC. Of the 16 witnesses, at least 10 commented on the professionalism of government acquisition personnel or the need for such employees to receive increased education and training in some aspects of acquisition.

Report of the Investigations Subcommittee of the House Armed Services Committee, 101st Congress, Second Session

Based on an extensive investigation of the quality and professionalism of the defense acquisition work force, which was based on data collected during the January to June 1989 period, the Investigations Subcommittee published its detailed findings May 8, 1990. According to the letter of transmittal from the subcommittee chairman, the Honorable Nicholas Mavroules, to the chairman of the full House Armed Services Committee, the Honorable Les Aspin, these investigation findings were to be included in a report of legislation proposed to improve the acquisition work force. As stated in the report, the investigation's primary objectives were to " . . . assess the qualifications and professionalism of the acquisition work force—both present and past, military and civilian; to review the efforts of the Department of Defense and the Military Departments to establish and manage the career development of that work force; and, where appropriate, provide recommendations for improving the quality and professionalism of that work force." (Ref. 1, p. 65.)

The resultant 776-page report addressed multiple facets of the defense acquisition work force: (1) organizational environment within which the work force operates; (2) operational personnel management systems which control work force members' careers; (3) detailed data pertaining to the work force characteristics; (4) work force professionalism, with emphasis on its education and training; (5) the mix of civilian and military personnel who compose the work force; and (6) compensation of work force members. In addition, the report provided an excellent historical background of the continuing issue of professionalism in the defense acquisition work force and the acquisition system. The report also provided information from DoD concerning actions accomplished to date as well as those planned, which were intended to enhance professionalism and educational opportunities for acquisition personnel.

Public Law 101-510

Earlier it was stated that findings from the Investigations Subcommittee on the Quality and Professionalism of the Acquisition Work Force were to be included in a report of legislation proposed to improve the acquisition work force. This was done. The findings and recommendations became a bill (H.R. 5211), introduced by Nicholas Mavroules, chairman of that subcommittee. Provisions of the Mavroules Bill were later incorporated into another bill (H.R. 4739), which was also under consideration by the House of Representatives. That latter bill was subsequently enacted as P.L. 101-510, signed by the President in November 1990.[16] A portion of that law, the Defense Acquisition Work Force Improvement Act, added a new Chap. 87 (entitled "Defense Acquisition Work Force") to that part of the U.S. Code which covers the various activities and functions of the DoD (that is, Title 10, U.S.C.).

Because of the ultimate positive impact this public law will have on training and development of the acquisition community, the cited Defense Acquisition Work Force Improvement Act and the newly created Chap. 87, Title 10, U.S.C., will undoubtedly prove to be a significant turning point not only in training and education, but also with regard to the overall professionalism of the entire defense acquisition work force.

Among other pertinent provisions and requirements contained in the Defense Acquisition Work Force Improvement Act portion of P.L. 101-510, the following were included: (1) recognition that the acquisition work force must include both civilian and military personnel, and that both categories require adequate training, education, and a career development program; (2) establishment of an acquisition corps for each military department and one or more corps for other DoD components; (3) specification of minimum educational and experience requirements for membership in the respective acquisition corps and provision of special education and training programs for member personnel; (4) establishment of policies and procedures for effective management (including the accession, education, training, and career development) of personnel serving in DoD acquisition positions; (5) specification of a career development process for acquisition personnel; (6) establishment of a defense acquisition university structure; (7) establishment of a coherent framework for educational development of personnel in acquisition positions from the basic level through intermediate and senior levels; and (8) requirement that the senior level of educational development be a senior course equivalent to existing senior professional military education school courses.[16]

This law required the Secretary of Defense to provide Congress (specifically, the Senate and House Committees on Armed Services) an implementation plan for the university structure no later than October 1, 1991, and to carry out that implementation plan by August 1, 1992.

THE DEFENSE ACQUISITION UNIVERSITY

Although this chapter is not intended to be a policy research study, there do appear to be sufficiently valid policy issues involved in the basic subject of the university to warrant a discussion of public-policy-related nature of the congressional action relative to the creation of the Defense Acquisition University (DAU). Policy research, according to Majchrzak,[17] concerns social-related research intended for policy makers to use in making decisions relative to addressing the basic social issue upon which the research was based. The fundamental social problem involved here relates to the professionalism of the defense acquisition work force and the potential negative implications to the national good if that work force is not adequately educated and trained to perform its assigned responsibilities. However, because Congress has already initiated action intended to alleviate the stated social problem, this discussion centers on congressional action instead of the more traditional approach whereby recommendations are made with the intention to alleviate the stated social problem.

The primary public-policy issue involved in the congressional creation of a defense acquisition university pertains to the fundamental question of whether such an action is a proper role of the federal government. Should the federal government establish an institution of higher education in which individuals involved in the defense acquisition professions receive training and education for those professions? Is this a proper function of the federal government? Is the federal government unfairly competing with existing public and private universities in establishing such an educational institution? Would defense acquisition professionals receive the same or a higher level of training and education if they were to attend nongovernment universities or courses offered by commercial educational and training providers?

My contention is that the establishment and operation of an institution of higher education in which individuals involved in defense acquisition receive training and education for those professions is a proper function of the federal government. Creation of a university for the purpose of providing education and training for defense-related skills and knowledge is not without historical precedence. The federal government, as early as 1802, established its first federal institution of higher education: the U.S. Military Academy at West Point. In addition to the four service academies, which offer baccalaureate degrees, there are other federal educational institutions from which graduate degrees may be earned. These include the Naval War College, the Naval Post Graduate School, and the Air Force Institute of Technology. (The latter two also offer courses appropriate for the defense acquisition professional.) Although recent legislation does not specify that DAU will offer academic degrees, there is nothing in the public law to preclude such a future action.

Will the government be in unfair competition with existing public and private universities with this acquisition university? While other universities may offer educational programs and academic degrees in subject areas important to the defense acquisition profession (such as business administration, engineering, general management, and even project management), most civilian universities tend to approach such disciplines from the civilian perspective rather than from the government perspective. However, there is nothing to preclude any civilian university from developing specific courses, or even formal degree programs (at either the master's or the doctoral level), in defense system acquisition management. The handbook of which this chapter is a part would seem to be a meaningful start to such a formal program. The best approach for the acquisition professional would appear to be getting an academic degree from a civilian university in one of the basic subject areas critical to defense acquisition and then receiving follow-on education from the defense perspective at DAU or other DoD educational activities, such as DSMC, the Air Force

Institute of Technology, the Naval Post Graduate School, or the Army Logistics Management College. While a form of competition might exist between DAU and civilian universities, it would not necessarily be to the detriment of the good civilian universities that offer meaningful academic programs.

DEPARTMENT OF DEFENSE IMPLEMENTING ACTIONS

Critical to improving training, education, and professional development of the defense acquisition work force is the manner in which DoD actually implements the findings and recommendations made by internal inspections, audits, and studies; recommendations made by senior-level commissions and boards; directions contained in presidential executive orders; and laws passed by Congress. Obviously, a lack of action by individuals who have the authority to act would result in no changes and, in the perception of some individuals, no improvements in the defense acquisition work force professionalism. On the other hand, immediate action in response to every recommendation made by an external source would result in the defense acquisition process, and the professionals executing that process, being in a mode of continuous change. Neither extreme would seem to be in the best interest of the defense acquisition community.

The DoD has made some significant changes in the training, education, and professional development of its acquisition work force in the recent past. While most of those changes had a greater direct and more immediate impact on its military members than on its civilian employees, the improvements that were implemented created a base that could be expanded to cover both categories of personnel.

As indicated earlier in this chapter, one of the significant outcomes of the 1984 Defense Inspector General's audit report relative to the then current training status of senior- and intermediate-level contracting personnel was a comprehensive review to determine how to achieve a more professional contracting, quality assurance, and program management work force. That review led to the creation of a 1985 study group, which was formalized as a permanent interservice action office (the ACE Program Office) in early 1988, and charged it with the responsibility for oversight of defense acquisition training and education programs. That office, which became part of DAU upon its formal activation October 1992, coordinated the writing and publication of several DoD-level directives and manuals pertaining to the training and education of acquisition personnel.

Significant among these publications is DoD Directive 5000.52[18] (October 25, 1991), which addresses the defense acquisition training and education program, sets broad policy statements on the subject, establishes responsibilities for implementing that policy, and provides procedures relative to that training and education program. Another important publication is the DoD Manual 5200.52M[2] (November 15, 1991), which provides detailed information on establishing the Defense Career Development Program for Acquisition Personnel, including its operation and administration. The manual established mandatory and desired training courses for the 13 acquisition career paths. The information is provided for the three certification levels within each of the 13 career fields.

Also of interest to the acquisition work force, the DAU ACE Office publishes an annual catalog specifying mandatory courses required for specific job series within the acquisition field, which are offered by the major defense training sources. During fiscal year 1992 there are 24 mandatory courses in the Defense Acquisition Education and Training Program.[19] That number is to be increased in the near future. Starting in fiscal year 1993, the mandatory courses will be expanded by adding another 32 courses. Planning documents indicate that the mandatory acquisition courses will be further expanded by nine more courses the following year.[20] Assuming the course-related planning in a December 1991 memorandum from the Office of the Under Secretary of Defense (Acquisition) is implemented, members of the various career fields of the acquisition work force will be offered 65 mandatory training and education courses effective in fiscal year 1994.

While many of the fiscal year 1992 mandatory courses apply to individuals in the contracting occupational series, others are required for individuals in manufacturing and production, quality assurance, acquisition logistics, business and finance, and for program managers. The newly designated mandatory courses for 1993 and 1994 are in various occupational (career) fields, such as those previously identified as well as test evaluation; system planning, research, development, and engineering; and auditing. DoD directives allow for the designation of certain courses to be equivalent to designated mandatory courses (that is, the equivalent course has been judged to contain the appropriate level of knowledge as contained in the mandatory course). There are currently seven such equivalent courses offered by training activities not under the auspices of the Defense Acquisition Education and Training Program. In addition to the equivalent courses offered by other training activities, DoD has recognized certain college courses offered by 20 universities and colleges to be equivalent to mandatory acquisition courses. Equivalent courses for the fiscal year 1993 and 1994 added courses have not yet been identified.

The first edition of DoD Manual 5200.52M (dated September 1990) constituted the first known attempt at a coordinated display of (1) "... the general requirements for merit placement, assignment, and career management of members of the acquisition work force and ... [(2) a listing of] minimum education, training, and experience requirements for specific acquisition work force job series career fields or specialties."[21] In addition to providing general guidance and responsibilities in the acquisition community, the manual addressed the recruitment, selection, assignment, and advancement of personnel in the acquisition work force; the career development of those personnel; and a master training and development plan (with a separate appendix for each job series expressed by level, job position, experience, education, and training) for acquisition personnel in specific job series in the acquisition field.

The second edition of DoD Manual 5200.52M[2] (dated November 1991) restated and updated general guidance and responsibilities of individuals and organizations in the acquisition community. It also confirmed the mandatory education, training, and experience standards for the acquisition work force as established in the original 1990 version of the manual. The 1991 edition of DoD Manual 5200.52M, which specifies procedures for the effective career development of all persons serving in DoD acquisition positions, is an extremely valuable reference document for anyone currently involved in defense acquisition or who plans to become a member of the acquisition community.

With regard to laws pertaining to acquisition training, education, and the professional development of its personnel, DoD has reacted positively to specific provisions of the various laws passed during recent years by changing its regulations to the extent required by those laws or by providing Congress specific information required by such laws.

In response to a requirement set forth in 1984 P.L. 98-369, there is a DoD-level procurement (namely, contracting, GS-1102) career management program, and that program is one of the better structured career programs in the acquisition community. However, as stated previously, there is an ongoing disagreement between executive agencies of the federal government concerning the educational level of civilian members of this career program. The OPM considers the GS-1102 occupational series to be in the administrative rather than the professional career field, and, in accordance with OPM standards, a college degree may not be required as a condition of employment in such administrative career fields. OPM states that firm educational requirements are precluded for the GS-1102 series by the U.S. Code (namely, 5 U.S.C. 3308). However, DoD has indicated that a college degree is the preferred minimum educational level for personnel in this occupational series.

Research findings are not consistent relative to the value of a college degree to the performance of a government employee in the contracting career field. One of the findings of a 1992 special study conducted by the U.S. Merit Systems Protection Board (MSPB) to determine the quality of government contract specialists and their work indicated that there was not a significant correlation between a college degree and work quality[22]:

> A second purpose of this study was to determine whether there is, in fact, a statistically significant relationship between the potential quality indicators and actual performance. Although the sizes of the relationships are small, there are a few general indicators of quality which might be of some limited use in tracking gross changes in the quality of the Federal procurement workforce.
>
> The single best indicator is the educational level of the workforce. In general, the more education completed by a worker, the higher the quality of his or her work. This is not to say, however, that a person must possess a college degree in order to be a high-quality contract specialist. The relationship between education and performance is not large enough to indicate that possession of a college degree should be a minimum qualification for admission to the field.

On the other hand, in an independent research effort conducted by this author[23] in conjunction with a doctoral course, the following findings were made:

> There was general agreement among the respondents that civilian employees in the contracting job series who had a college degree were better performers than such employees without a degree. At least it appeared to the respondents the degree was a contributing factor until the employee had approximately twenty years of experience.
>
> Among the most repeated responses was the opinion that a college degree provided the individual with a group of basic skills critical to a good contracting employee. Those skills, considered to be difficult to teach in follow-on professional development courses, included the following: an understanding of the English language; a mental aptitude (i.e., ability to think); reading and writing (composition skills); oral communication; interpretative ability; organizational skills; an awareness of how things fit together; questioning skills (i.e., curiosity); a global perspective; social skills; and initiative.
>
> Although there was not as wide an agreement, most respondents believed a degree should be a condition of employment into the contracting series (GS-1102). While a degree was considered "an absolute necessity" by some respondents, the specific field of study was not considered significant. Some respondents agreed that while a business degree would provide the individual an advantage in the business related aspects of the job, other specialized degrees were considered equally valuable when the individual did contracting work related to that specialty (e.g., a contracting officer with biology degree who worked with scientists and medical doctors was better able to understand the terminology of the specialty).
>
> Several respondents discussed the thought that the "best" among the new employees was an individual who had a college degree and then spent three years in a formal intern program wherein the person alternated between specialized courses and short (several months) assignment doing a variety of different jobs in the contracting community.

In response to a requirement set forth in 1984 P.L. 98-525, the DoD regulation pertaining to the tenure of program managers of major defense weapon systems required these individuals to remain in their assignment for a minimum of four years or until that weapon completed a significant milestone in its development. It should be noted, however, that the matter of program manager tenure was not a new concern when Congress enacted the 1984 law. As indicated in the previously mentioned 1990 report of the Investigations Subcommittee of the House Armed Services Committee, which pertained to quality and professionalism of the defense acquisition work force, inadequate tenure of DoD program managers had been a long-standing problem, dating as early as 1955. The second Hoover Commission report, published that year, reflected a two-year average tenure of military program managers. Thereafter similar findings were indicated in publications or reports dated 1962, 1970, 1979, and 1981 before the stated requirement of a four-year tenure was set forth in the 1984 P.L. 98-525 (Ref. 1, pp. 307–309). The 1990 report also contains detailed statistics indicating that as of mid-1988 the average tenure of military program managers of "major programs" ranged between 21 and 39 months, depending on the specific service and other circumstances.

In response to a requirement set forth in 1985 P.L. 99-145, DoD has required that the program manager of a major defense weapon system complete the 20-week program management course (PMC) at DSMC and have at least eight years' experience in one or a combination of the acquisition disciplines. That same 1990 report of the Investigations Subcommittee of the House Armed Services Committee provided detailed statistics relative to the training and experience of currently serving program managers as of mid-1988, when data for the report were being collected. Depending on the service and other circumstances, the percentage of serving military program managers who had attended PMC ranged between 29 and 96 percent; between 71 and 97 percent of those same program managers had at least eight years of appropriate experience (Ref. 1, pp. 334, 366, 397).

Requirements set forth in the 1986 P.L. 99-661 pertained to DoD preparation of (1) a plan to enhance the professionalism of, and career opportunities available to, DoD acquisition personnel and (2) a plan for the coordination of educational programs managed by DoD for acquisition personnel of the department. The first plan entailed several legislative changes to establish a formal alternative personnel management system and specific actions concerning educational qualification requirements and training opportunities for acquisition personnel. Elements of that first plan required congressional action; and Congress has provided many of the requested elements in the previously discussed portion of P.L. 101-510 pertaining to the defense acquisition work force (that is, the Defense Acquisition Work Force Improvement Act). Concerning the second plan, DoD provided for the coordination of educational programs with the publication of the previously described DoD Directive 5000.52[18] and DoD Manual 5000.52M.[2]

With regard to the specific requirements set forth in the 1990 P.L. 101-510, it may still be too early to determine the extent to which DoD has taken action to implement all those requirements. However, as stated by Land, in a series of memoranda dated November 20, 1990, the Under Secretary of Defense (Acquisition) created an implementation board whose members were to advise him on the implementation of the requirements contained in the Defense Acquisition Work Force Improvement Act. One subcommittee was to address the DAU structure and another the other provisions of the act. Creation of these structures only two weeks after the President signed the law containing the Defense Acquisition Work Force Improvement Act indicates DoD commitment to begin the task of improving the education, training, career development, and overall management of defense acquisition personnel (Land,[24] p. 22).

DoD has taken action to implement two of the significant requirements set forth in P.L. 101-510. This action was detailed in DoD Directive 5000.57, dated October 22, 1991. One requirement was to establish a defense acquisition university structure which would provide for professional educational development and training of the acquisition work force, and would conduct research and analysis of defense acquisition policy issues from the academic perspective. The other requirement was to establish a senior-level acquisition course that would be a substitute for, and equivalent to, existing senior professional military educational (PME) school courses.

The named directive established the DAU and provided for a senior acquisition course designed specifically for civilian and military personnel serving in critical acquisition positions. A significant decision relative to the DAU was that the internal core of the university structure would not be built around any one existing defense acquisition educational activity; rather, all existing activities would be included in the organization in a consortium structure.

As stated in DoD Directive 5000.57, the new DoD-level university, which will be under the direction, authority, and control of the Under Secretary of Defense (Acquisition), will be structured and operated as an education consortium consisting of those DoD component education and training institutions, organizations, and activities that provide the courses considered necessary to satisfy the acquisition education and training requirements specified by the USD(A). As specifically required by the cited public law, the DSMC is included in the DAU education consortium. The operational concept of the university consortium is that the participants will remain a part of their existing command structures and will retain control of their faculty and day-to-day activities.[25]

The university's mission shall be "to educate and train professionals for effective service in the defense acquisition system; to achieve more efficient and effective use of available acquisition resources by coordinating DoD acquisition education and training programs and tailoring them to support the careers of personnel in acquisition positions; and to develop education, training, research, and publication capabilities in the area of acquisition." (Ref. 25, p. 1.)

The DoD decision relative to the senior acquisition management course was that it would be conducted by the Industrial College of the Armed Forces, a subordinate academic element of the National Defense University. Early plans of ICAF indicate that the existing senior professional military education (PME) course conducted by that educational activity will be modified as necessary to provide for the appropriate level of education in defense acquisition management functional areas.

The implementing DoD Directive 5000.57 states that the Under Secretary of Defense (Acquisition) is responsible to coordinate with the chairman of the Joint Chiefs of Staff on this topic and then establish "... the acquisition content of a senior course at the Industrial College of the Armed Forces (ICAF) as part of the Senior Acquisition Education Program for acquisition professionals. The senior course will be the preeminent course for civilian and military members of the Acquisition Corps." (Ref. 25, p. 3.) In this regard, the chairman of the Joint Chiefs of Staff is responsible to "(1) provide representation on the policy guidance council established by the USD(A) to provide advice in relation to the university; (2) ensure that the President, National Defense University, and the Commandant, ICAF, provide the senior course for acquisition personnel serving in critical positions; and (3) ensure, in coordination with the USD(A), the relevance of the ICAF acquisition curriculum to the educational needs of senior acquisition professionals." (Ref. 25, p. 4.)

A longer passage of time may be necessary to put into better perspective the specific manner in which DoD has implemented these provisions of P.L. 101-510. While the basis for achieving many of the objectives contained in the law relative to the university had been in place as early as 1988 (when the mission of the DSMC was expanded, including the coordination of all mandatory defense acquisition training and education courses), and that it might have been the intent of Congress, in specifying in Sec. 1745 of the cited law that "... the term 'defense acquisition university' includes Defense Systems Management College . . . ,"[16] that DSMC would become the core educational activity for the mandated university, the actual intent of Congress is not known. However, it is not the purpose of this chapter to address congressional intent, to answer that question, or to question DoD decisions.

CONCLUSION

As stated in the previously described 1990 congressional report pertaining to the defense acquisition work force, the defense acquisition system consists of three distinct elements: "(1) the policies, procedures, and processes which govern the operation of the acquisition system; (2) the organization of the resources (people, management structure, capital, and facilities) that execute the policies and procedures; and (3) the people within the organization that make the system work." (Ref. 1, p. 1.) Without highly trained and professional people, the acquisition system will not work properly. This fact—and the need to improve the quality and professionalism of that work force—have been recognized in numerous studies and reports for the past 30 years.

Congress has acted to address the public-policy issue related to the professionalism of the acquisition work force through passage of the 1990 Defense Acquisition Work Force Improvement Act as part of P.L. 101-510. A significant provision of the law is the requirement for a defense acquisition university structure that provides for education and training courses from the basic level through intermediate and senior levels. Decisions made by senior DoD officials, the manner by which the department writes implementing regulations for that law, and the manner in which the service components actually implement defense directives will determine the ultimate effectiveness of the congressional action. The legal framework to increase professionalism of the acquisition work force through education and training has been provided by Congress. It is incumbent upon DoD to create and implement a comprehensive education and training structure to help achieve the high level of professionalism in the acquisition work force necessary to resolve the basic public-policy issue associated with the defense acquisition process. A highly educated, trained, and professional acquisition work force, working within a well-organized acquisition structure, will be better able to effectively and efficiently implement well-developed policies and procedures of the defense acquisition system. History will reveal the degree of success actually achieved by Congress, DoD, and the members of the defense acquisition work force in this endeavor.

REFERENCES

1. U.S. Congress, House, Committee on Armed Services, Investigations Subcommittee, *The Quality and Professionalism of the Acquisition Work Force,* 101st Cong., 2d Sess., 1990, Committee Print 10.
2. DoD Manual 5000.52M, *Department of Defense Career Development Program for Acquisition Personnel,* Government Printing Office, Washington, D.C., 1991.
3. Title XII, "Defense Acquisition Work Force Improvement Act," *National Defense Authorization Act of Fiscal Year 1991,* Statutes at Large, P.L. 101-510, 1990.

4. Acquisition Enhancement (ACE II) Study Group, *The Acquisition Enhancement (ACE) Program Report II,* vol. I, DoD, Ft. Belvoir, Va., 1986.
5. U.S. Congress, House, Committee on Armed Services, Acquisition Policy Panel, "DoD Acquisition Improvement—The Challenges Ahead. Perspectives of Assistant Secretary of Defense for Acquisition and Logistics," in *Integrity of Department of Defense Acquisition System and Its Impact on U.S. National Security,* 100th Cong., 2d Sess., 1988, HASC 100-89, p. 9.
6. President's Blue Ribbon Commission on Defense Management (David Packard, Chairman), *A Quest for Excellence: Final Report to the President,* The Commission, Washington, D.C., 1986.
7. President, *National Security Decision Directive 219,* 1986.
8. *Omnibus Deficit Reduction Act of 1984,* Statutes at Large, P.L. 98-369, 1984.
9. *Omnibus Defense Authorization Act of 1985; Defense Spare Parts Procurement Reform Act,* Statutes at Large, P.L. 98-525, 1984.
10. *Department of Defense Authorization Act of 1986,* Statutes at Large, P.L. 99-145, 1985.
11. *National Defense Authorization Act for Fiscal Year 1987,* Statutes at Large, P.L. 99-661, 1986.
12. Deputy Secretary of Defense, letter, May 13, 1987.
13. Under Secretary of Defense (Acquisition), "Initiatives to Improve the Education and Training of the Acquisition Work Force," memo, Feb. 4, 1988.
14. U.S. Congress, Senate, Committee on Armed Services, *Defense Acquisition Process,* 100th Cong., 2d Sess., 1988, Hrg. 100-963.
15. U.S. Congress, House, Committee on Armed Services, Acquisition Policy Panel, *Integrity of Department of Defense Acquisition System and Its Impact on U.S. National Security,* 100th Cong., 2d Sess., 1988. HASC 100-89, p. 1.
16. *National Defense Authorization Act for Fiscal Year 1991,* Statutes at Large, P.L. 101-510, 1990.
17. A. Majchrzak, *Methods for Policy Research,* Sage Pub., Beverly Hills, Calif., 1984, p. 12.
18. DoD Directive 5000.52, *Defense Acquisition Education and Training Program,* Government Printing Office, Washington, D.C., Oct. 25, 1991.
19. FY-92 Catalog, *Department of Defense Acquisition Education and Training Courses,* Acquisition Enhancement Program Office, Ft. Belvoir, Va.
20. Under Secretary of Defense (Acquisition), "Data Call for FY 1993/1994 DAU/ACE Programming and Budget Meeting," memo, Dec. 19, 1991.
21. DoD Manual 5000.52M, *Department of Defense Career Development Program for Acquisition Personnel,* Government Printing Office, Washington, D.C., Sept. 1990, p. 1-1.
22. U.S. Merit Systems Protection Board, *Workforce Quality and Federal Procurement: An Assessment,* MSPB, Washington, D.C., July 1992, p. 49.
23. J. G. Land, "Educational Requirements for Federal Government Civilian Contracting Work Force," University of Maryland, College Park, Md., 1991.
24. J. G. Land, "Defense Acquisition University Coming," *National Defense,* Jan. 1991.
25. DoD Manual 5000.57, *Defense Acquisition University,* Government Printing Office, Washington, D.C., 1991, p. 1.

CHAPTER 24

ADULT LEARNING IN THE DoD ACQUISITION ENVIRONMENT: PRACTICE AND APPLICATION

Michael J. Browne

Michael J. Browne, U.S. Air Force, is currently Professor of Acquisition Management at the Defense Systems Management College (DSMC), Ft. Belvoir, Va. His academic credentials include B.S. (chemical engineering), M.S. (systems management) and M.E. (educational leadership) degrees, and an Ed.D. (supervision and administration) degree from Northeastern University. Lt. Col. Browne has performed duties in various systems program offices, including being a project manager of one system, and has taught at the secondary, high-school, and adult levels. As the primary action officer responsible for implementing competency-based education at DSMC, Lt. Col. Browne is familiar with the process of converting a course from a knowledge-based to a competency-based curriculum.

Scholae sed vitae discimus (Not for school but for our lives we learn)

INTRODUCTION

Definitions

Pedagogy Traditional process of teaching or instructing in an authoritative manner, with emphasis on structuring the content to be learned rather than the process of learning.

Andragogy Alternative process of teaching or instructing, with emphasis on the set of procedures used to facilitate the acquisition of content by the learner. The instructor leans toward a collaborative process of guiding, advising, or joining with the students to acquire the content to be learned.

Knowledge Category of usable information, such as data, facts, insights, or models, organized around a specific content area or task-related activity.

Skill Demonstration of a set of behaviors or processes effectively and readily in execution or performance.

Ability Possession of knowledge and a complex of skills, coupled with the capability to successfully focus, integrate, and apply these to a specific task.

K-S-A Grouping of knowledge, skills, and abilities that are prerequisite for the development of a capability or competency to perform a specific task or activity.

Historical Perspective

It may be instructive to recall that the great teachers of ancient times taught adults, not children. This historical who's who in education includes Confucius, Jesus, Socrates, Plato, Aristotle, Cicero, and Euclid. These historic adult educators perceived learning as a process of active inquiry by the learners. So they developed specific techniques that would involve learners in active inquiry. These techniques included the *case method,* developed by the Chinese and the Hebrews, where "one member of a study group (not necessarily the leader) would present a paradox—often in the form of a parable—and the group would examine its background and explore possible resolutions." (Knowles,[1] p. 6.3.) *Socratic dialogue* was developed by the ancient Greeks, and in this technique "a member of the study group would pose a question, and the group would pool their resources to arrive at an answer." (Knowles,[1] p. 6.3.)

Over the millennia many of the ancient inquiry techniques lay dormant, buried in the archives of religious monasteries and libraries. But in 1926 Edward C. Lindeman again attempted to describe the unique characteristics of the adult learner. *The Meaning of Adult Education* is still recognized today as one of the "most insightful and inspiring works in the literature of adult education." (Knowles,[1] p. 6.4.) Lindeman captured what can be described as the true meaning of adult learning in the following[2]:

> I am conceiving adult education in terms of a new technique for learning, a technique as essential to the college graduate as to the unlettered manual worker. . . . It represents a process by which the adult learns to become aware of and to evaluate his experience. To do this he cannot begin by studying "subjects" in the hope that some day this information will be useful. On the contrary, he begins by giving attention to situations in which he finds himself, to problems which include obstacles to his self-fulfillment. Facts and information from the differentiated spheres of knowledge are used, not for the purpose of accumulation, but because of need in solving problems. In this process the teacher finds a new function. He is no longer the oracle who speaks from the platform of authority, but rather the guide, the pointer-outer who also participates in learning in proportion to the vitality and relevancy of his facts and experiences. In short, my conception of adult education is this: a cooperative venture in nonauthoritarian, informal learning, the chief purpose of which is to discover the meaning of experience; a quest of the mind which digs down to the roots of the preconceptions which formulate our conduct; a technique of learning for adults which makes education coterminous with life and hence elevates living itself to the level of adventurous experiment.

From this start, adult learning has evolved over the years and "andragogy" is now the term that is most commonly used for adult education and learning.

Current Perspective

The general theory of adult education and training has evolved over the years from Knowles' initial theory of andragogy to what is now known as "situational andragogy." Practitioners have discovered that participative, learner-centered techniques that tap the learner's experience are not always the best techniques because much depends on what is to be learned and who is doing the learning. Therefore the most commonly mentioned reason for turning from an andragogical method of instruction to a more pedagogical approach is the specific subject matter to be taught. For technical subjects which have an established "best way" to perform the task or tasks, pedagogical methods are usually more appropriate. It would not make sense for the student to identify options, define those options, and incorporate their various experiences in the process if there already exists a "best practice." So in those areas where the individual has few options, the use of an andragogical methodology for learning is usually not appropriate.

PRACTICE

There are numerous methods being applied in the education and training of adults today. Experiential learning, problem solving, and cognitive apprenticeship will be briefly discussed because the characteristics of these methodologies have potential for specific application to the education and training in the Department of Defense (DoD) acquisition or program management career field.

Experiential Learning

Experiential learning is defined by F. Gerald Brown as "learning how to perform a specific act or operation by doing it ('how to' learning); (or) learning the complexities of a professional role by experiencing the milieu in which the role is performed and attempting to perform parts of the role (role socialization)."[3] The essence of experiential learning is the use of concrete experiences as a catalyst for student observation, reflection, and action. One of the problems with traditional education is that the knowledge learned cannot usually be personalized. Experiential learning helps link the specific learner to the knowledge that is learned in two ways according to Kolb and Lewis[4] (pp. 99–100):

> First, by encouraging reflection on the meaning of abstract concepts in the light of shared personal experiences, the techniques allow concepts to become "real," that is, learners find examples and applications in their experience that illustrate concepts. Second, by encouraging personal action on concepts, the techniques allow learners to commit themselves to the idea; they accept responsibility for the choice of that idea and for learning the skills necessary to use it.

Experiential learning can bridge the gap between experience and concept by providing students a way to learn from their own and others' experiences and to develop a capability to translate concept to practice by using these experiences.

An experiential learning model was proposed by Kolb and Lewis[4] (p. 100) in 1986 and describes learning as a four-stage cycle (Fig. 24.1):

> In this cycle, immediate concrete experience is the basis for observation and reflection. The observations are assimilated into an idea or theory from which new implications for action can be deduced. These implications or hypotheses then serve as guides in acting to create new experiences. An effective learner needs four different capabilities: concrete experience (CE) skills, reflective observation (RO) skills, abstract conceptualization (AC) skills, and active experimentation (AE) skills. That is, he or she must be able to get involved fully, openly, and without bias in new experiences (CE), to reflect upon and interpret these experiences from different perspectives (RO), to create concepts

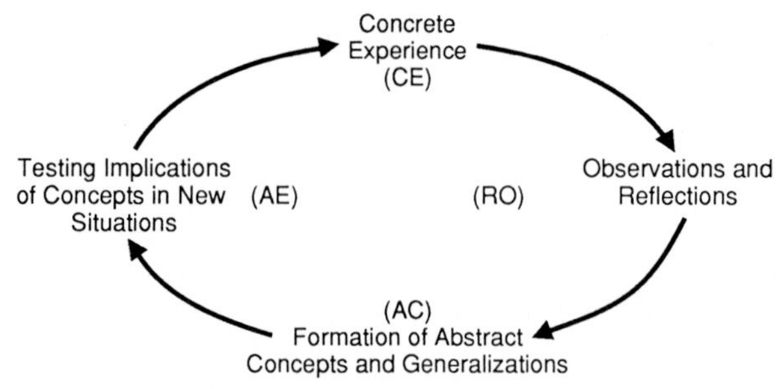

FIGURE 24.1 Experimental learning model. (*After Kolb and Lewis.*[4])

that integrate these observations in logically sound theories (AC), and to use these theories to make decisions and solve problems (AE) leading to new experiences.

Therefore all experiential models begin with the "experience" designed for classroom application. Students then share their reactions and observations concerning the experience. Next the pivotal step of group dynamics occurs where the experience is reconstructed in terms of specific patterns and interactions by "talking through" the experience. It is critical that the facilitator ensure that students understand the dynamics of the experience and not make premature generalizations. But sooner or later the "so what?" of the experience relative to real-life application must be addressed. Practical inferences must therefore be made that take learning from the structured classroom experience to the reality of everyday life. Translating generalizations to individual practical applications is the whole purpose of the structured experience in the first place. It is in the application where the inferences and generalizations are tested. Students develop an enhanced capability to infer from their practice in application. This step completes a typical experiential learning cycle.

Certain perceptions and the existing environment can make the implementation of experiential learning difficult. Experiential learning to the uninformed can be thought of as a game or a bull session or, in the worst case, the "blind leading the blind." Also, certain characteristics of the traditional education model can impact implementation and experiential learning negatively. For example (Kolb and Lewis,[4] p. 105):

> *Class schedules* are oriented to the lecture method. Exercises are difficult to process in fifty minutes or even two hours. Two- to three-hour or day-long sessions are more appropriate but hard to schedule.
>
> *Large fixed-seat classrooms* make student-to-student interaction difficult. Meeting places are needed for project or game teams, and open space is needed for exercises.
>
> *Traditional teacher-student role expectations* encourage one-way dispensing of information rather than coequal dialogue among learners with different perspectives on the focal experience.
>
> *Assessment or grading methods* emphasize fact retention rather than personal skill acquisition.
>
> *Limited budgets* fail to reflect the need for funds to conduct research, develop techniques, support simulation activities, and sponsor field-based expenses.

Chickering acknowledges these difficulties associated with the implementation of experiential learning, but describes the value and the viability of the methodology by stating[5]:

> There is no question that issues raised by experiential learning go to the heart of the academic enterprise. Experiential learning leads us to question the assumptions and conventions underlying many of our practices. It turns us away from credit hours and calendar time toward competency, working knowledge, and information pertinent to jobs, family relationships, community responsibilities, and broad social concerns. It reminds us that higher education can do more than develop verbal skills and deposit information in those storage banks between the ears. It can contribute to more complex kinds of intellectual development and to more pervasive dimensions of human development required for effective citizenship. It can help students cope with shifting developmental tasks imposed by the life cycle and rapid social change.

According to Kolb and Lewis[4] (pp. 86–89), "adults demand relevance and seek opportunities to test ideas against their own accumulated experience." Most adults question the utility or "value added" of traditional teaching methods which seem remote from the realities of problems they experience each day. Experiential learning "teaches people how to acquire, use, and evaluate information; arouses motivation and involvement; develops social skills and social behavior, and allows for mistakes without retributions." (Kolb and Lewis,[4] p. 106.)

Problem Solving

Erickson[6] (p. 96) states that problem solving, the second methodology to be described, is the

> . . . essence of learning how to think independently, but students can only think for themselves. Efforts to find solutions to problems require a relevant fund of knowledge, diversity of attack, and an overt, active search for relations between means and ends. By encouraging such diversity and pressing students to acquire open attitudes and beliefs, a teacher is helping students to sustain independent intellectual inquiry and thereby to become somewhat a gadfly within a chosen field of study or in an otherwise complacent society.

In general, problems are identified by individuals when they perceive that a relevant set of knowledge, skills, and abilities (K-S-A) is not included in their "bag of tricks." Normally, "our repertoire of verbal chains, conceptual pigeonholes, procedural habits, and value stereotypes are sufficient to take us around and over most conflicts and obstacles." (Erickson,[6] p. 93.) When learners come to the realization that certain problems cannot be packaged neatly into existing solution models, they experience what Erickson[6] (pp. 61–63) calls "proactive interference." At this point, learners have a great tendency to rush quickly to problem solutions without addressing the critical processes of problem identification and definition. Nowlen[7] (p. 27) states that "the emphasis on problem solving, as opposed to problem definition, is a natural consequence of the positivist model of knowledge." Schon[8] continues this argument by stating that the positivist

> . . . invariably ignores problem setting, the process by which we define the decision to be made, the ends to be achieved and the means which may be chosen. In the real world, problems never appear as such without the intervention of the problem setting or defining mind. It is a process in which interactively we name the things to which we will attend and frame the context in which we will attend to them.

Learners can develop a debilitating dependence by being "provided the problem to solve" in the typical classroom environment. When challenged by the real world, where they must identify and define the problem in the context of a complex set of interrelated issues, symptoms are often defined as problems. Valuable resources have been wasted by solving symptoms but never really addressing the problem. For example, Schoenfeld has found that students rely on their knowledge of (Collins et al.,[9] p. 2)

> . . . standard textbook patterns of problem presentation, rather than on their knowledge of problem-solving strategies or intrinsic properties of the problems themselves, for help in solving mathematics problems. Problems that fall outside these patterns do not invoke the appropriate problem methods and relevant conceptual knowledge. In other cases, students fail to use resources available to them to improve their skills because they lack models of the processes required for doing so.

Bransford and Stein[10] approach problem solving with five interrelated steps. The first step is to identify the problem. They state that problem identification is "overlooked in many problem-solving courses and books because the latter emphasizes solutions to ready-made problems . . . this is a serious omission, because problem identification or problem finding is often the most significant part of problem solving." Also, as the environment you are dealing with becomes more complex, the more difficult it becomes to identify the "real" problem. Therefore the importance of this first step, problem identification, is directly related to the complexity of the environment. The more complex the environment, the more critical problem identification becomes. This is certainly true when the environment is weapon system acquisition, which is extremely complex, with many interrelated issues.

The second step to problem solving is defining and describing the problem as carefully as possible. Once symptoms are identified, the specific problem is more precisely defined by attempting to discover the exact reasons for the symptoms. Different definitions of the identified problem will result in different actions. For example, a patient whose high blood pressure is caused by hardening of the arteries may need different types of treatment from one whose high blood pressure is due to job stress. Robertshaw, Mecca, and Rerick[11] state:

> . . . we produce an initial definition of the problem, we will begin to consider the generation and/or evaluation of some alternatives. As we do so, we may find that we must redefine the problem. An improved definition may in turn lead to new alternatives which may lead to a further improvement in the definition of the problem. As we complete a number of these iterations and improvements, we may tend to become satisfied that we have "zeroed in" on the problem, that we have completed the definition. This satisfaction, however, will eventually be replaced with a growing sense of frustration as we realize that the problem definition, by its very nature, will always be incomplete . . . our definition of the problem can only be complete if we are able to choose consistently between any two given alternatives. But we can choose consistently only if the complete set of alternatives is known. However, in the practical situation the complete set of alternatives can never be known. Thus the problem will never be defined completely. While this is frustrating and irritating to the rational perfectionist, we find that we must accept it as a fact of reality and proceed in spite of it.

The last three steps in the problem-solving process (namely, exploring possible strategies, acting on those strategies, and evaluating the effects of those actions) are where most problem-solving classes begin and end. But the importance of the first two steps, identification and definition of the problem, cannot be overemphasized.

Wales and Stager propose a general strategy they call "guided design"[12] for teaching problem solving and decision making. The application of this strategy can vary from highly structured programs requiring printed material, standard exercises, and regular evaluations to programs that are mere suggestions to the instructors. According to Resnick,[13]

> Central to all of the programs is extensive practice on actually solving problems, or designing and carrying out experiments. Various forms of social interaction are used (teacher modeling, working in pairs or groups), both to make visible certain aspects of the problem-solving process that normally remain covert, and to increase students' self-conscious monitoring and management of their thought processes. Particular attention is often paid to the uncertainties of problem solving and to the process of making and correcting—rather than avoiding or denying—errors.

Erickson[6] (p. 93) says it is "vital for students to know something about a field before engaging in the heuristic (trial and error) scramble for solutions to its problems." Students acquire this basic background in many ways, but knowing about a problem is only the initial step of a process for being able to identify, define, and ultimately develop viable alternatives to solving that problem. Therefore the background information and the guides relative to problem solving are all important, but sooner or later each student—with "hands on" or with "mouth open"—actively tries to solve problems. It is in the act of doing where learners restructure their problem-solving process beyond habitual modes of analysis and synthesis . . . with clarification and restatement of the problem very key steps. (Erickson,[6] p. 94.)

Learners, as they develop an ability to solve problems, expand their learning thresholds. And according to many authors, learning is enhanced when learners operate beyond their individual thresholds. But there is much evidence from the laboratory and the traditional classroom indicating that "beyond the threshold" learning is beneficial only if the learner has a good chance of succeeding from the start. If success is not likely, then Elshout[14] states:

> Learning by doing and discovery may have disastrous consequences if a solid groundwork of conceptual understanding is not laid, if the student is not prevented from slipping into the mode of unsystematic muddling through, and if informative feedback is not given when manifestly incorrect beliefs intrude. Without such support, students will tend to adopt an unsystematic working system, skipping a thorough analysis of the task environment and trying almost at once to reach the answer by some superficially attractive method.

Much of learning is focused on individual problem solving. Schoenfeld[15] (p. 16) advocates small-group problem solving for several reasons:

> First, it gives the teacher a chance to coach students while they are engaged in semi-independent problem solving: he cannot really coach them effectively on homework problems or class problems. Second, the necessity for group decision making in choosing among alternative solution methods provokes articulation, through discussion and argumentation, of the issues involved in exercising control processes. Such discussion encourages the development of the metacognitive skills involved in, for example, monitoring and evaluating one's progress. Third, students get little opportunity in school to engage in collaborative efforts; group problem solving gives them practice in the kind of collaboration prevalent in real-world problem solving. Fourth, students are often insecure about their abilities, especially if they have difficulties with the problems. Seeing other students struggle alleviates some of this insecurity as students realize that difficulties in understanding are not unique to them, thus contributing to an enhancement of their beliefs about self relative to others.

Schoenfeld[15] (p. 17) also places a unique emphasis on the specific sequencing of problems to "achieve four pedagogical goals: motivation, exemplification, practice, and integration." Learners must appreciate the criticality of the experimental or "trial and error" (heuristic) methods used for successful problem identification, definition, and solution. The capability to apply these methods becomes even more essential in solving complex problems requiring the integration of multiple problem-solving methods.

Cognitive Apprenticeship

The third method for adult education and training does not involve didactic teaching but observation, coaching, and successive approximation while carrying out a variety of tasks and activities and is described as "cognitive apprenticeship." One concern relative to traditional teaching that this approach addresses is that "skills and knowledge taught in schools have become abstracted from their uses in the world . . . (cognitive) apprenticeship embeds learning of skills and knowledge in the social and functional context of their use." (Collins et al.,[9] p. 1.) Cognitive apprenticeship is based on this concept—situated cognition. Collins et al.[9] (p. 3) describe cognitive apprenticeship as a sequence of activities:

> In this sequence of activities, the apprentice repeatedly observes the master executing (or modeling) the target process, which usually involves a number of different but interrelated subskills. The apprentice then attempts to execute the process with guidance and help from the master (coaching). A key aspect of coaching is the provision of scaffolding, which is the support, in the form of reminders and help, that the apprentice requires to approximate the execution of the entire composite of skills. Once the learner has a grasp of the target skill, the master reduces his participation (fades), providing only limited hints, refinement, and feedback to the learner, who practices by successively approximating smooth execution of the whole skill.

According to Rogaff and Lave,[16] observing a model of the target process aids learners in the development of a conceptual model of the target task or process prior to attempting to execute it. Having a conceptual model is critical to success in teaching complex skills. A conceptual model provides a focus for execution of the target process. It also provides the learner a structure for interpreting feedback from the instructor for effective implementation of needed changes. And finally, since the conceptual model is internalized, it provides a "benchmark" for independent practice using a process of successive approximation.

Another significant characteristic of cognitive apprenticeship is its application, which requires externalization of processes that are usually executed internally. This provides the learner an in-depth view of the process, so that a more comprehensive conceptual model can be built. In this way, learners have access to the mind of the instructor and can observe the unique cognitive and metacognitive characteristics of the target process.

Cognitive apprenticeship is an iterative process with a number of interrelated steps. The initial step distinguishes this method from all others—modeling. The modeling of a "real-life" activity allows the learner to experience reality as the first step rather than teaching prerequisite knowledge, skills, attitudes, or competencies with the assumption the learner will comprehend the relevance of each element and ultimately be able to apply those necessary to do the "real thing." The instructor models the actual process to be learned and applied by the student. The process must be modeled to include not only the extrinsic activities, which can easily be observed, but more importantly the intrinsic or "internal to the mind" type activities. The articulation of what the instructor is thinking, including the "tricks of the trade," will assist the learner in the application of the "domain-specific" methods needed to address all complex processes.

The instructor must first assess the progress of the learner in order to provide the appropriate support technique. If the diagnosis identifies a specific problem the learner is having in performing the task or activity, then the instructor should provide "scaffolding." The instructor would actually perform the activity in question while articulating what is being done and why it must be done a particular way. Otherwise, if the learner can accomplish the task or activity successfully, but only at a novice or basic level, then the support technique of coaching would be more appropriate. Successful coaching occurs while the learner performs the target activity and the instructor or coach provides feedback relative to performance and suggestions for improvement.

While the instructor supports the process with scaffolding or coaching, learners should reflect about the differences between their performance and the "model." This reflection process will assist the learner's development of self-monitoring and self-correcting skills. Once the learner gains some degree of confidence in performing the target activity with the scaffolding and coaching of the instructors, the next step in the process, fading, can then begin.

The instructor begins the fading segment of the process as the learner completes each successful approximation of the target activity. The gradual removal of the supports continues as the learner demonstrates increased ability to do the "real thing."

The last step in the process is described as exploration and begins when the learner can satisfactorily approximate the "real thing." The learner is now able to practice the target activity on his or her own as a self-directed learner.

Experiential learning, problem solving, and cognitive apprenticeship were described in some detail because each facilitates the development of higher-order thinking abilities—critical thinking, problem solving, and decision making. In each of the three teaching methodologies certain kinds of higher-order thinking recur. Problems or issues are analyzed and then restructured and redefined. Proposed solutions are evaluated against the specific criteria that were developed or provided. The implications of the solutions are determined and appropriate modifications are made rather than resorting to a quick solution based on one's initial analysis. Higher-order thinking abilities are critical in the complex and interrelated DoD weapon system acquisition environment.

APPLICATION

Requirements Defined

The DoD established a program in 1986 to enhance the professionalism and the effectiveness of its acquisition work force. A comprehensive review was made of the actions needed to promote a more professional and effective work force in the areas of contracting, quality assurance, and program management. This three-month review, led by the Defense Systems Management College (DSMC), included representatives from all of the services and the Defense Logistics Agency and was designated the Acquisition Enhancement (ACE) Program.[17] The ACE study addressed 15 acquisition career fields (Table 24.1) comprising approximately 56,000 federal government civilian employees and military personnel. This very significant education and training requirement demanded a comprehensive coordination effort that crossed individual service and agency lines. The reality of constrained resources dictated an implementation strategy with maximum benefit from funds, instructor and student

TABLE 24.1 Job Functions and Official Titles/Series Addressed in ACE Study

ACE job functions	OPM official titles/series
1. Program manager	1. a. Engineer/800 b. Program manager/340
2. Deputy program manager	2. a. Engineer/800 b. Program manager/340
3. Business/financial manager	3. a. Program analyst/345 b. Budget analyst/560
4. Contracting officer	4. Contract specialist/1102
5. Contract negotiator	5. Contract negotiator/1102
6. Contract specialist	6. Contract specialist/1102
7. Contract administrator	7. a. Contract administrator/1102 b. Contract termination specialist/1102
8. Procurement analyst	8. Procurement analyst/1102
9. Price analyst	9. Contract price/cost analyst/1102
10. Quality assurance specialist	10. Quality assurance specialist/1910
11. Procurement clerk	11. Procurement clerk/1106
12. Procurement assistant	12. Procurement assistant/1106
13. Purchasing series	13. Purchasing agent/1105
14. Industrial specialist	14. Industrial specialist/1150
15. Property administrator	15. a. Industrial property management specialist/1103 b. Industrial property clearance specialist/1103

Source: *ACE Program Report*,[17] p. 3.

time, and facilities as they are applied to the educational and training task. The report concluded that if competency-based learning concepts were vigorously applied with some consolidation of courses, the acquisition education and training requirement could be satisfied (Ref. 17, pp. v–vi).

The report concluded that competency-based education or instruction is both effective and efficient because it provides the specific knowledge, skills, and abilities required for professional and effective job performance. A competency-based model curriculum was proposed for each of the seven job functions plus an additional area titled "Supervision and Management" (Table 24.2). The proposed model curriculum consists of 13 mandatory functional courses taken from the following functions: contracting, industrial property management, purchasing, industrial specialist, quality assurance, business/financial management, and program management. Each course was designated to be competency-based and made mandatory for the indicated career field (Table 24.3).

DoD Directive 5000.52,[18] dated October 25, 1991, updated policy and responsibilities for a career development program for acquisition personnel. It stated (p. 2) as policy that "the Under Secretary of Defense (Acquisition) shall establish education, training, and experience standards for each acquisition position based on the level of complexity of duties carried out in that position."

The policies of DoD Directive 5000.52 were implemented by the reissuance of DoD Manual 5000.52M,[19] dated November 15, 1991. The purpose of this manual was (p. i) to "provide uniform procedures for a Department of Defense Career Development Program for Acquisition Personnel." The manual established (p. 2-1) "specific education, training, and experience standards for each position category and career field by career level for the DoD Career Development Program." The manual defined 12 job functions as acquisition positions (Table 24.4) and described typical as-

TABLE 24.2 Job Functions

Program management (includes program manager, deputy program manager, and other key professionals working in program management office)
Business/financial manager
Contracting (1102 series) (includes contracting officer, contract negotiator, contract specialist, contract administrator, procurement analyst, price and cost analyst)
Industrial property management
Purchasing (1105 series)
Industrial specialist
Quality assurance
Supervision and management

Source: *ACE Program Report,*[17] p. 37.

signments and the experiences, education, and training that are either mandatory or desired for each career path. The program management career path is shown in Table 24.5.

The focus on learner competency and the capability to demonstrate a task or activity is also integral to the U.S. Department of Education's *America 2000* strategy. One of the primary educational objectives of this strategy is for learners to "leave grades four, eight, and twelve having demonstrated competency . . . to reason, solve problems, apply knowledge . . . and for college graduates to demonstrate an advanced ability to think critically, communicate effectively, and solve problems."[20]

Competency Defined

Before discussing *what* competency-based education is, we need to provide a *working* definition of a competency. A competency encompasses many identifiable individual behaviors or skills that can be specified in behavioral terms. A specific competency statement should describe the observable demonstration of a "composite" of specific behaviors or skills. A competency defines performance based on the acquisition, integration, composite building, and application of a set of job- or task-related skills and knowledge. The value of learning a volume of skills and knowledge is limited; the real utility is found in combining and interrelating these, resulting in a capability for an integrative performance by the learner which exceeds the sum of the individual skills and knowledge. This integrating activity entails a growth process or gestalt building in addition to the development of individual skills and knowledge (Hall and Jones,[21] p. 29).

McAshan[22] (p. 45) defines competency as "demonstrable composite knowledge, skills, abilities, characteristics, or traits related to effective task performance on the job." Competency was also defined by McBer[23] in studies of professional and executive performance:

> Generic knowledge, skill, trait, self-schema or motive casually related to effective and/or outstanding performance in a job.
>
> - knowledge, a category of usable information organized around a specific content area, for example, knowledge of math;
> - skill, an ability to demonstrate a set of behaviors or processes related to a performance goal, for example, logical thinking.

TABLE 24.3 Comparison of Current and ACE Recommended Mandatory Training

Job function	Level	Current job requirement	ACE proposed requirements
Program management (several GS/GM occupational series and military equivalents)	I	N/A	1 course, 4–6 weeks (mandatory for key professionals in program management offices)
	II	N/A	Same as Level I
	III	N/A	1 course, 10–14 weeks (only mandatory for PMs and deputy PMs)
	IV	N/A	1 course, 20 weeks (only mandatory for PMs of major programs)
Business/financial management (several GS/GM occupational and military equivalents)	I	N/A	N/A
	II	N/A	1 course, 2–4 weeks
	III	N/A	N/A
Contracting (series 1102 and military equivalents)	I	3 courses, 8 weeks	1 course, 6–8 weeks
	II	2–3 courses, 5–9 weeks (depends on specialty)	1 course, 6–8 weeks
	III	1 course, 1 week	1 course, 1 week
Property management (series 1103 and military equivalents)	I	3 courses, 8 weeks	1 course, 3–4 weeks
	II	2 courses, 4 weeks	1 course, 2–3 weeks
	III	1 course, 1 week	1 course, 1 week
Purchasing (series 1105 and military equivalents)	I	N/A	1 course, 4–6 weeks
	II	N/A	1 course, 4–6 weeks
	III	N/A	N/A
Procurement clerk/assistant (series 1106 and military equivalents)	I	N/A	N/A
	II	N/A	N/A
	III	N/A	N/A
Industrial specialist (series 1150 and military equivalents)	I	2 courses, 10 weeks	1 course, 6–8 weeks
	II	2 courses, 5 weeks	1 course, 3–4 weeks
	III	1 course, 1 week	1 course, 1 week
Quality assurance (series 1910 and military equivalents)	I	N/A	1 course, 4–6 weeks
	II	2 courses, 5 weeks	1 course, 3–4 weeks
	III	N/A	1 course, 1 week
Supervisory and managerial training (training for supervisors and managers in all functions)	Supervisor	N/A	1 course, 3–5 weeks
	Manager	N/A	1 course, 3–5 weeks
	Executive	1 course, 2 weeks (series 1102, 1103, and 1150 only)	1 course, 2–3 weeks

Source: *ACE Program Report,*[17] pp. 38–39.

TABLE 24.4 Job Functions

Job Functions
Program management
Communications computer systems
Contracting
Purchasing (procurement assistant)
Industrial property management
System planning, research, development, and engineering
Test and evaluation engineering
Manufacturing and production
Quality assurance
Acquisition logistics
Business, cost estimating, financial management
Auditing

Source: DoD Manual 5000.52M,[19] pp. A-1–G-8.

trait, self-schema or motive casually related to effective and/or outstanding performance in a job.

- knowledge, a category of usable information organized around a specific content area, for example, knowledge of math;
- skill, an ability to demonstrate a set of behaviors or processes related to a performance goal, for example, logical thinking.
- trait, a consistent way of responding to an equivalent set of stimuli, for example, initiative;
- self-schema, a person's image of self and his/her evaluation of that image, for example, self-image as a professional;
- motive, recurrent concern for a goal, state or condition which drives, selects and directs behavior of the individual, for example, the need for efficacy.

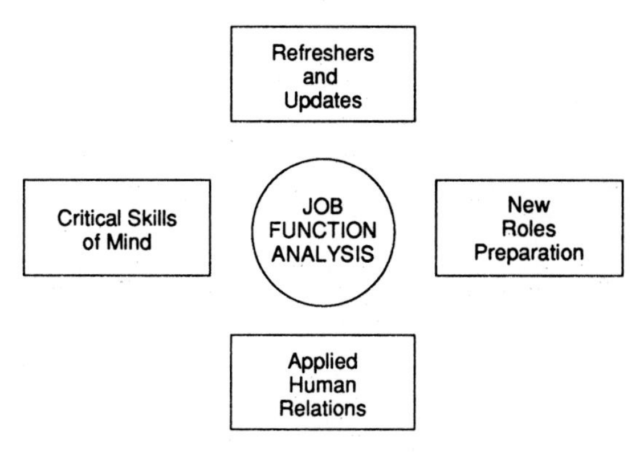

FIGURE 24.2 Competence model.

TABLE 24.5 Career Path Program-Management

Level or typical grade	Assignments	Experience	Education	Training
Level I GS 5–7 01--03	Program office Program management staff Engineering staff officer Contracting specialist Laboratory staff Test and evaluation Engineering staff	(M)* One year of acquisition experience	(D)* Baccalaureate degree in business or technical discipline	(M) A basic course in systems acquisition [The fundamentals of systems acquisition management course (DSMC, 1 week) satisfies this requirement] (D) An additional systems acquisition course
Level II† GS 9–12 03/04	Program office Staff officer Branch chief Division chief Project management Acquisition command HQ Program management, engineering, test, acquisition logistics, staff Broadening (operational assignment) Education with industry (EWI)	(M) At least 2 years of acquisition experience (D) At least two additional years of acquisition experience, preferably in a systems program office or similar organization	(D) Master's degree in engineering, systems management, or other appropriate field	(M) Intermediate systems acquisition course [Acquisition basics course (DSMC, 4 weeks) satisfies this requirement] (D) DSMC program management course (PMC), or management program at an accredited educational institution determined by the USD(A) to be comparable to DSMC PMC (D) Management and leadership training

Level III‡ GS 13 and above 04 and above	PEO Program manager/deputy program manager Service HQ staff Acquisition command HQ Director Division chief Acquisition staff Systems command HQ staff	(M) At least 4 years of experience in acquisition, at least 2 years of which were performed in a systems program office or similar organization (D) An additional 4 years of acquisition experience	(D) Master's degree in engineering, systems acquisition management, or other appropriate field	(D) DSMC program management course (PMC) or management program at an accredited educational institution determined by the USD(A) to be comparable to DSMC PMC (D) Management and leadership training

*(M)—mandatory; (D)—desired.

†Concurrent with achieving level II, a person should meet the criteria for the Acquisition Corps. Criteria for selection into the Acquisition Corps for civilian personnel includes serving in a position graded GS-13 and above, and in the case of officers it includes serving in the grade of 0–4 and above. Criteria further include 4 years acquisition experience; a baccalaureate degree or certification by an acquisition career board (ACPB); *and* at least 24 semester credit hours from among the following disciplines: accounting, business finance, law, contracts, purchasing, economics, industrial management, marketing, quantitative methods, and organizational management; or at least 24 semester credit hours in the person's career field and 12 semester hours in the disciplines listed.

‡For a general/flag officer at least 10 years experience in an acquisition position, at least 4 of which performed while assigned to a critical acquisition position. *Note:* Critical acquisition position may only be filled by members of the Acquisition Corps; see level II career path.

Source: DoD Manual 5000.52M,[19] pp. A.1-7–A.1-9.

Nowlen also addresses a potential flaw in the approach most educators take when describing competency. There is a tendency to assume that performance is totally dependent on the individual. Competency models focus on the individual and are (Nowlen,[7] p. 36)

> . . . not sensitive to the organizational context of business and professional activity. Some of the key influences on performance are found in the professionals' interior life, the critical higher order skills of mind, specialized knowledge and skill, motivation for excellence, the maturity to manage personal affairs well, and the grit to overcome setbacks. There are other influences, however, and they stem from the quality of the relationship individuals have to one another in the organizational setting; the ensemble of peers, subordinates, superiors, and systems. It is this ensemble that can cripple or enhance individual effectiveness. Performance is as much a function of the ensemble as it is of the individual.

Competency-Based Education (CBE) Defined

Nickse[24] proposes that competency-based education is "basically an instructional strategy that assumes different programmatic forms in different contexts." McAshan[22] (p. 38) defines competency-based education in terms of three major components, which are explained to the learner as elements of the introduction of the learning activities: (1) the specific competencies; (2) the enabling objectives and strategies to facilitate the achievement of the competencies; and (3) the assessment or evaluation techniques to determine learner success or progress in achieving the required level of learning or mastery. Houston and Howsam,[25] who were later supported by Collins,[26] propose three basic elements that characterize competency-based education[25]:

- Precise learning objectives defined in behavioral and assessable terms must be known to the learner and the teacher. They are viewed as means to a specific end, rather than the objectives of the learning experience. This suggests that the learning objectives are merely a guide, a way of getting there, and should not be the sole basis for the evaluation. This fits with the theory of goal free evaluation.
- The learners accept responsibility and know they are accountable for meeting the agreed upon criteria. They must be able to demonstrate the specified competencies to the required level and in the established manner.
- Programs are "individualized," "self-paced" and "personalized." This does not mean that each learner's activities are independent of other learners'.

The last element, individualization, seems to be the most common theme in all definitions. Finch and Crunkilton[27] provide a model with the focus on the individual that interrelates instructional content, media, environments, and strategies.

After defining exactly what competency is, the second element of competency-based education is the set of enabling objectives or strategies to facilitate the achievement of the competencies. Enabling objectives or strategies are means to the end—competency demonstration. The distinguishing aspect of competency-based education is that the learner is held accountable, not for the acquisition of a specific competency, but for the demonstration of that competency or capability. The enabling objectives provide the learner the knowledge base required, the "how to" learning, the expert demonstration of "how to," and the practice which will ultimately result in satisfactory competency demonstration by the learner. The determination of "satisfactory" competency demonstration by the learner is the third and last element of competency-based education—evaluation. (McAshan,[22] pp. 38–39).

The last element of competency-based education, and in many cases the most difficult, is evaluation and the resultant feedback to the learners. The two measurement systems in use today are norm-referenced and criterion-referenced. The norm-referenced system is designed to compare individuals with one another. According to Guliano[28] (p. 8.17), "this type of measurement system is not suitable for adult learners, especially if an instructor wants to build a learning environment that is free of anxiety and frustration. Criterion-referenced systems are designed to compare an individual with a given performance standard, and not with the other individuals in a class. Guliano also considers the backgrounds of the adult learners (p. 8.18):

> Adult learners enter a learning environment with various abilities and learning speeds. This does not mean that they cannot learn but only that they require a different time span for learning. If a normative system is used, it would compare slower learners against those who can grasp the material more readily. Some of the slower learners could be rated fairly low or could be eliminated from the program completely. The advantage of using a criterion-referenced system is that it compares each individual learner to the skills delivered by the course. This enables you to determine whether a learner can perform the skills and at what level of proficiency. The criterion-reference system takes into account the slow learner who can master the skills but needs more practice to become proficient at them. Criterion testing should be an essential part of any adult training program.

Hall and Jones[21] (p. 35) addressed how success is achieved in competency-based education and the types of grading schemes that apply when they said:

> . . . the student demonstrates that he can perform the required objectives. The student does not receive a letter grade on the objective, but instead his achievement is recorded with a "yes," a "pass," a "completed," a "finished," or some other notation. If he cannot demonstrate the attainment of an objective, that fact is recorded with a "no," a "fail," a "not yet," an "incomplete," or something; he gets the opportunity to try again, perhaps after another prescription from an instructor . . . Such "yes–not yet" records on individual skills, objectives, or subcompetencies or competencies do not preclude the use of grades in CBE efforts. In a number of institutions, the elimination of grades from a system will be next to impossible. Grades can still be computed by identifying a lower limit of acceptable skill as a B or a C, with higher grades being reserved for more proficient behavior. Similarly, since courses will consist of a number of subcompetencies or skills, course grades can be calculated by averaging "grades" on each of the enablers. The basic difference in grading with CBE is that the student knows in advance how grades will be determined, and he is aware that his grade will not be determined by comparing his behavior with that of the other students.

Collins[26] (p. 17) cites four choices for evaluating competencies:

- Actual performance. This is the ideal but it is expensive, takes years and the results are returned too late for meaningful help to the learner or the program.
- Simulated performance. The problems associated with simulation are that it is not "real" performance, it is expensive and time consuming and there are not many simulations available.
- Evaluation of products and performance. Problems with this method are that students often have had help, the pressures of testing are missing (is this so bad?) and results are hard to score.
- Pencil and paper test. These are cheap and available and are the most common.

Therefore the focus of competency-based education is on the learners' acquisition and demonstration of a predetermined set of job-specific capabilities or competencies. Specific learning objectives are determined and used as observable evaluation indicators relative to competency achievement. A minimum level of learning for each objective or competency is established as a criterion for success. Learning experiences are designed to optimize student achievement of the minimum-level competency and to facilitate assessment of student progress. Objectivity in the assessment of achievement is imperative; therefore competencies are stated so that assessment can be made through direct observation of student behavior.

Implementation of CBE—Background

In 1986 the Under Secretary of Defense (Acquisition and Logistics) mandated the implementation of CBE at many service schools within DoD, including the Defense Systems Management College (DSMC). The transition to CBE at DSMC is still in progress. This section describes the implementation to date and then proposes the ideal actions to conclude the process. The discussion attempts to tie in the theory, process, and practice discussions from earlier parts of this chapter.

DSMC is located at Fort Belvoir, Va. It was established in 1971 as an educational center for DoD civilian and military personnel involved in the complex process of acquiring military weapon systems. The principal offering at DSMC is the 20-week program management course (PMC). One of the strong, influential supporters of DSMC is the U.S. Congress. That body has directed that an individual assigned as the program manager (that is, the senior person of a major weapon system acquisition program) must be a graduate of the PMC. Most PMC students, however, are mid-level managers in the systems acquisition career field and do not immediately become the program manager of a major program. Approximately 840 students will graduate from PMC during 1992.

Curriculum material within the PMC is divided into three phases. The phases are designed to assist the student in his or her transition from the knowledge domain to the comprehension domain, followed by transition from the comprehension domain to the application domain. For the first few weeks of the PMC, students are presented with a wide range of facts in order to build a knowledge base regarding the various aspects of the acquisition process. The knowledge base is the essential foundation for developing skills and abilities to operate within the acquisition process. During this first phase, the standard lecture process (pedagogy) is the primary means used by the faculty to transmit the information.

The second phase or segment of instruction consists of a number of integrated subjects, lessons, and case studies. These pose problems or planning requirements of sufficient breadth to demonstrate the interrelationships among the elements of information in the knowledge base. The principal purpose of this second phase is to support the student's transition from the retention of myriad unrelated facts to a comprehension of the ways those facts can be used productively in the acquisition process. At this point the student begins to use the universal problem-solving process, with some assistance from the instructor who now takes on the role of team leader and collaborator. The instructor uses elements of andragogy with some lapses into pedagogy, while the student develops a deeper comprehension of the acquisition management arena and the managerial actions required within that arena.

The third phase consists of a multiscenario six-week classroom simulation, entailing representative issues involved in acquiring a defense weapon system. The simulation spans a nine-to-ten-year time period, from the initiation of military justification for a new weapon system to the decision point approving full-rate production. Students organize themselves into a government program management office structure (as well as a contractor team), taking turns at the roles of program manager, business manager, technical director, test and evaluation director, and so on. The class instructor interacts primarily as a facilitator in this third phase, attempting to use all the tools of andragogy to assist the students. Faculty members interject themselves at appropriate times to play additional roles as senior decision makers and staff officers in the scenarios. The students, in turn, use their knowledge, comprehension, and prior experience to accomplish the problem-solving or planning requirements of each scenario independently.

Implementation of CBE—An Example

The initiation of competency-based education at DSMC began with a specification process to define the competencies critical to graduates of the PMC. The 1986 ACE report provided an initial list of 785 PMC tasks or competencies. Through an iterative process, utilizing the DSMC faculty, the acquisition community, and a network of functional experts in program office, staff, and support positions, a list of 23 PMC capstone competencies was developed (Table 24.6). This list of 23 competencies is currently being validated through a process of interviewing acquisition executives and surveying recent graduates and their supervisors.

The next element of the implementation process is critical because weapon system acquisition is such a dynamic process. Competencies must continually be reviewed and validated by the DSMC faculty, the acquisition community, and a network of functional experts relative to the ever-changing acquisition environment. Based on the results of this process, the PMC curriculum, including all enabling objectives and evaluation procedures, must be revised or added to as necessary. This methodology will ensure the validity of the PMC curriculum relative to the needs and constraints of the acquisition community and DSMC.

The implementation process continues with an audit of the PMC curriculum to determine whether the existing instructional lessons and processes support the newly specified capstone competencies. The audit function must be "institutionalized" and applied to the PMC curriculum each time the capstone competencies are modified or when competencies are added to the curriculum. The audit process must analyze each enabling objective and the associated desired learning outcomes (DLOs) from the appropriate lessons to determine whether and how they relate to each other and support the capstone competencies. Each lesson plan must be reviewed to ensure that the content and the focus of the lesson are consistent with the educational requirements of the capstone competency(ies) they support. An individual learning program for each PMC student provides the architecture for implementing competency-based education. The critical elements of the individual learning program include preassessment instruments, the individual learning plan (ILP), a faculty advisor for each student, modules for individualized learning, feedback methodologies, and postassessment instruments.

A typical individual learning program could be described using the following scenario. A student would arrive for PMC, and during the first week of class he or she would take a preassessment instrument, which would evaluate his or her capability relative to each of the capstone competencies. The student would then meet with the faculty advisor to develop an initial ILP. The faculty advisor would use the preassessment results, special interests of the student, and any post-PMC assignment data available to recommend the ILP. The ILP will track the student through each experience he or she will have during the 20-week PMC. In this way there will be both a rationale and utility for each element of the ILP. The ILP will also provide the student the opportunity, based on expertise and interest, to play varied roles in the classroom. Potential roles include functional expert, evaluator, researcher, observer, or even instructor for certain elements of the curriculum. Each of these roles must first be agreed to by the faculty advisor and the student and then coordinated with the appropriate instructors. The student ILP will be reviewed periodically by the faculty advisor on a prescheduled basis or as requested by the student or any other faculty member. The ILP

TABLE 24.6 Capstone Competencies (Complete Statement)

1	Able to develop an acquisition strategy that is the framework for integrating functional activities essential to fielding a defense acquisition program and is the basis for developing specific program management documents.
2	Comprehends how to translate operational requirements into design requirements which are explicit functional quantitative ranges or point values of performance and is able to do a trade-off analysis.
3	Comprehends how to analyze an escalating threat relative to an existing system to determine viable courses of action (such as major system improvement, new start).
4	Able to analyze the technical adequacy of the existing design to meet known technical requirements.
5	Able to review and evaluate program documentation to ensure that the needs of an individual program are consistent with sound business practices, common sense, and the degree of risk involved.
6	Comprehends the processes of cost and budget reviews and how they interrelate with the program review/decision process and associated reporting. Able to prepare and analyze a program cost estimate and POM/budget input.
7	Able to develop, analyze, and update performance, schedule, cost, and program baselines.
8	Able to analyze contract cost/schedule performance data to determine contract and programmatic impacts, and recommend actions.
9	Able to assess a program's readiness to exit from a given phase of the acquisition process, pass the required review, and enter into the next phase.
10	Able to prepare a solicitation that effectively communicates the government's requirements, acquisition strategy, and evaluation criteria.
11	Comprehends both the competitive and noncompetitive award processes. In the noncompetitive process, able to apply the techniques of pricing, fact-finding, evaluating cost and technical proposals, and negotiating from both a government and a contractor perspective.
12	Comprehends the financial and technical systems, processes, and practices used by defense contractors to manage weapon system acquisitions, including the motivations and constraints in their implementation.
13	Comprehends how to develop and implement solutions to management issues associated with international considerations in defense programs.
14	Comprehends the philosophy, techniques, and tools of total quality management and how total quality management should be applied in the program office and the contractors' organizations.
15	Able to apply quantitative and qualitative methods to analyze an issue and present a decision briefing to higher authority.
16	Comprehends how congressional activities impact acquisition management.
17	Able to formulate tailored, human-skills approaches based on individual differences, the interpersonal communication process, and small group dynamics to maximize interpersonal effectiveness in the program management environment.
18	Comprehends how to identify and plan for the personal and functional management support required to execute the acquisition strategy of a defense acquisition program.
19	Comprehends how to develop, document, and staff a test and evaluation program in DoD.
20	Able to analyze program office and contractor status/plans for mission critical computer resource (MCCR) development, integration, management, and support.
21	Comprehends the DoD logistics policy and related management tools, such as logistic support analysis (LSA), the logistic planning and contracting processes, and how these should applied to program office and contractor logistic efforts during the acquisition life cycle.
22	Comprehends the principles of manufacturing management and how these principles should be applied to influence the design process, transition to production, and execution of the manufacturing plan. Able to analyze program office and contractor status/plans for transition to production.
23	Able to develop plans and review program status for deployment.

attempts to tailor the PMC experience to the needs, interests, and capabilities of each student while still supporting the demands of the acquisition community.

The curriculum will be designed as competency modules. In certain situations a student will have the option to complete part of a module independently, or as in most cases as part of a small group. The group dynamics is an important factor, and individual cooperation will be reinforced and encouraged. The progress of each individual as a member of the group will be emphasized rather than having individuals compete with each other. This type of group work and focus will provide students experience in teamwork, which will be essential for future team situations. The modular concept will also provide the opportunity for the student to "test out" of certain lessons or possibly whole modules. This will allow the student to include activities such as research, consulting, electives, or field trips in his or her ILP. Therefore the ILP has the flexibility to be tailored in many ways to accommodate individual differences and interests.

The final element required in the implementation of competency-based education for the PMC is the development of valid and reliable assessment procedures. The student will be accountable for demonstration or performance of each of the competencies either individually or as a member of a small group. In general there must be a criterion-type formative evaluation to help students during the learning process and identify required remediation.

The focus of evaluation must be on what the student "can do" relative to the competencies (that is, satisfactory demonstration of predetermined competencies). The evaluation instruments must test the student's ability to apply processes resulting in a specific capability or competency. It must, therefore, be *Performance*-based. The types of evaluative methodologies that should be used include essay exams, term papers, simulations, field experiences, critical incidents, and case studies.

Therefore both the PMC students and the curriculum must be assessed continually, using both internal and external resources, to ensure that the PMC optimizes the students' performance in the real world of the defense acquisition community.

CONCLUSION

The education and training of acquisition management personnel is a vital function. This process includes enhancing performance on existing jobs, educating for future jobs, and developing ways for individual growth as organizations evolve. As the defense budget decreases, it will be crucial to the defense posture of our nation to optimize the utilization of even more limited resources. One significant way to accomplish this is to develop better acquisition management personnel. This chapter has described ways to enhance the capability of any organization's most limited resource—its people.

REFERENCES

1. M. S. Knowles, "Adult Learning: Theory and Practice," in L. and Z. Nadler (eds.), *The Handbook of Human Resource Development,* 2d ed. Copyright © 1990 by L. and Z. Nadler. Reprinted by permission of John Wiley & Sons, Inc., New York.
2. R. Gessner (ed.), *The Democratic Man: Selected Writings of Edward C. Lindeman,* Beacon Press, Boston, Mass., 1956, p. 160.
3. F. G. Brown, "Three Types of Experiential Learning: A Nontrivial Distinction," in *New Directions for Experiential Learning: Developing Experiential Learning Programs for Professional Education,* vol. 8, Jossey-Bass, San Francisco, Calif., 1980, p. 14.

4. D. A. Kolb and L. H. Lewis, "Facilitating Experiential Learning: Observations and Reflections," in L. H. Lewis (ed.), *Experiential and Simulation Techniques for Teaching Adults,* New Directions for Continuing Education, no. 30, Jossey Bass, San Francisco, Calif., 1986.
5. A. Chickering, *Experience and Learning: An Introduction to Experiential Learning,* Change Magazine Press, New Rochelle, N.Y., 1975, pp. 86–87.
6. S. C. Erickson, *The Essence of Good Teaching,* Jossey-Bass, San Francisco, Calif., 1985.
7. P. M. Nowlen, *A New Approach to Continuing Education for Business and Professionals—The Performance Model,* Macmillan, New York, 1988.
8. D. A. Schon, *The Reflective Practitioner,* Basic Books, New York, 1983, p. 21.
9. A. Collins, J. S. Brown, and S. E. Newman, "Cognitive Apprenticeship: Teaching the Crafts of Reading, Writing, and Mathematics," in L. B. Resnick (ed.), *Knowing, Learning and Instruction: Essays in Honor of Robert Glaser,* Lawrence Erlbaum, Hillsdale, N.J., 1989.
10. J. D. Bransford and B. S. Stein, *The Ideal Problem Solver: A Guide for Improving Thinking, Learning, and Creativity,* W. H. Freeman, New York, 1984, p. 12.
11. J. E. Robertshaw et al., *Problem Solving: A Systems Approach,* Petrocelli Book, New York, 1978, p. 82.
12. C. E. Wales and R. A. Stager, *Guided Design,* West Virginia University Center for Guided Design, Morgantown, W. Va., 1977, p. 7.
13. L. B. Resnick, "Instruction and Cultivation of Thinking," in *Learning and Instruction: European Research in an International Context,* vol. 1, Pergamon Press/Leuven University Press, Oxford, U.K., 1987, p. 427.
14. J. J. Elshout, "Problem Solving and Education," in *Learning and Instruction: European Research in an International Context,* vol. 1, Pergamon/Leuven University Press, Oxford, U.K., 1987, p. 267.
15. A. H. Schoenfeld, "Problem Solving in the Mathematical Curriculum," Mathematical Assoc. of America, MAA notes no. 1, 1983.
16. B. Rogaff and J. Lave (eds.), *Everyday Cognition: Its Development in Social Context,* Harvard University Press, Cambridge, Mass., 1984, p. 3.
17. Acquisition Enhancement (ACE II) Study Group, "The Acquisition Enhancement (ACE) Program Report II," vol. 1, DoD, Ft. Belvoir, Va., 1986.
18. DoD Directive 5000.52, *Defense Acquisition Education, Training, and Career Development Program,* Government Printing Office, Oct. 25, 1991, p. 2.
19. DoD Manual 5000.52M, *Career Development Program for Acquisition Personnel,* Government Printing Office, Washington, D.C., 1991.
20. *America 2000: An Education Strategy,* U.S. Department of Education, Washington, D.C., 1991, pp. 38–40.
21. G. E. Hall and H. L. Jones, *Competency-Based Education: A Process for the Improvement of Education,* Prentice-Hall, Englewood Cliffs, N.J., 1976.
22. H. H. McAshan, *Competency-Based Education and Behavioral Objectives,* Educational Technology Publ., Englewood Cliffs, N.J., 1979.
23. *Understanding of Competence,* McBer and Co., Boston, Mass., 1978, p. 2.
24. R. Nickse (ed.), *Competency-Based Education: Beyond Minimum Competency Testing,* Teachers College Press, Columbia University, New York, 1981, p. 43.
25. W. R. Houston and R. B. Howsam (eds.), *Competency-Based Teacher Education: Progress, Problems and Prospects,* Science Research Assoc., Chicago, Ill., 1972, p. 29.
26. M. Collins, *Competence in Adult Education,* University Press of America, Lanham, Md., 1987.
27. C. R. Finch and J. R. Crunkilton, *Curriculum Development in Vocational and Technical Education,* Allyn and Bacon, Boston, Mass., 1989, p. 245.
28. D. Guliano, "Instructing," in L. and Z. Nadler (eds.), *The Handbook of Human Resource Development,* 2d ed. Copyright © 1990 by L. and Z. Nadler. Reprinted by permission of John Wiley & Sons, Inc., New York.

CHAPTER 25

THE ROLE OF PROFESSIONAL AND TECHNICAL SOCIETIES AND ASSOCIATIONS IN SUPPORTING THE MILITARY PROJECT MANAGER

James M. Gallagher

James M. Gallagher is the director of the Center for Executive Development and Project Management in the College of Business and Administration at Wright State University, Dayton, Ohio. He joined the university in April 1986 as executive assistant to the president. Previous to his position at Wright State, he served in a number of progressively responsible project management positions with the U.S. Air Force, retiring from the Air Force and the Senior Executive Service in March 1986. He is also an adjunct faculty member in Wright State's management department, and a Project Management Professional, certified by the Project Management Institute.

INTRODUCTION

Professional and technical societies and associations, hereafter referred to as associations, can play a unique role in supporting the military project manager. Much like the project manager, the association must walk a very thin line in terms of being a project advocate (see Chap. 43). In fact, some of the associations discussed in this chapter preclude specific project advocacy. In addition, the society and association can provide valuable technical and professional support and training to the project manager's staff.

At this early point I would like to acknowledge that I inevitably did not solicit *all* of the associations that impact and assist military project managers. Perhaps a second edition of this handbook will allow me to explore other associations identified by distraught readers who learn that their "favorite" has been neglected. I would also like to thank Wendy M. Claunch for her help in researching and editing this chapter.

I solicited a representative number of associations with different perspectives on military project management:

Aerospace Industries Association (AIA)
Air Force Association (AFA)
American Defense Preparedness Association (ADPA)
Dayton area Chamber of Commerce
Defense Systems Management College (DSMC) Alumni Association
National Contract Management Association (NCMA)
National Security Industrial Association (NSIA)
Project Management Institute (PMI)
San Antonio area Chamber of Commerce
Society of Logistics Engineers (SOLE)

While a number of citations on sources of information are given, all specific information included in this chapter can be found in the literature of the organizations discussed.

In this chapter I will discuss the role that each of these associations plays in assisting the military project manager and his or her staff and, when appropriate, I will reference specific military projects. My conclusions and recommendations address what associations can do to improve their support of military projects and the military project manager.

AEROSPACE INDUSTRIES ASSOCIATION

The Aerospace Industries Association (AIA) has a rich history dating back to 1919. While its original focus was on the unification of the air industry, public education, and technical concerns of the industry, the focus when the association became the Aircraft Industries Association in 1945 was on developing the industry's commercial aviation and related trade and commercial associations. The name was changed to the Aerospace Industries Association in 1959 to reflect the broader industry product line and space involvement.

The AIA is considered the premier national trade association of the aerospace industry. It is an aggressive and effective voice that speaks for the industry on such issues as competitiveness in the world market, the financial health of the industry, procurement and finance policies, research and technology, and defense and aerospace policies. The association enlists senior aerospace executives to carry industry's message to congress, the executive branch, the media, and the public.

The objectives and purpose of AIA are to foster continually the advancement of those aeronautical, astronautical, and related sciences, arts, technology, and industries which contribute to the public and private welfare of local communities, the nation, and the international community. Specifically, the continuing goal of the organization is:

1. To support industry in its responsibility for continued national security
2. To foster the peaceful conquest of space for the benefit of all humankind
3. To encourage safe and economical commercial and private air transportation

4. To promote the scientific, management, and manufacturing skills and techniques that will enhance the social, cultural, and economic well-being of the nation.[1]

The AIA consists of over 50 member companies and a volunteer board of governors. The board elects an executive committee from its members to exercise the necessary power when it is not in session. The AIA staff, headed by Don Fuqua, president, is located in Washington, D.C. The staff supports committees made up of member company representatives by keeping up with administrative and technical developments and relaying that information to members through regular and special meetings, workshops, seminars, special reports, and regular publications.

The association consists of nine major departments organized into councils, including two that should be of particular interest to the military project manager: procurement and finance, and technical and operations. The procurement and finance council monitors legislative and regulatory changes and initiates actions for improvement in procurement and procurement-related issues, including patents and data rights. The procurement and finance council includes the following committees: cost principles, legal, finance, procurement techniques, tax matters, economic advisory, facilities and property, and intellectual property. The technical and operations council focuses on all aspects of technology, operations, and engineering efforts to advance all aspects of project management, the industrial base, engineering, development, test, manufacturing, quality, material management, product support, and information to better address issues stemming from the production of aircraft, missile, and space vehicles. There are 18 committees under the technical and operations council and the product support, spare parts, and technical management committees should be of particular interest to the military project manager. For example, the product support committee is concerned primarily with the formulation and review of overall policies pertaining to product maintenance and support to assure maximum operational capabilities of aerospace equipment systems. The spare parts committee is a working committee of the product support committee concerned with the policies, procedures, and practices for the selection and ordering of spare parts, special tools, test and aerospace ground equipment, training aids, and training equipment. Each year the spare parts committee of AIA meets with its companion committee of the Electronic Industries Association (EIA) and sponsors a high-level meeting of industry and government representatives. In 1990 and 1991 this author had the honor of serving as the moderator for a cross-fire panel of senior government and industry representatives at which time a nonattribution, "no-holds-barred" discussion of issues was conducted. This forum provides an excellent opportunity for the military project manager to express concerns to his or her industry counterparts in a proactive manner.

AIA annually identifies the top 10 issues of importance to the aerospace industry. In 1992 these included the following:

- *Aerospace Technology Base.* To remain competitive into the 21st century, the United States must maintain a strong technology base. Industry, government, and academia must devise strategies for technology development while using commercial or military innovations.
- *Space Policy.* The U.S. aerospace industry supports a broad-based, balanced U.S. space program. Government investment in space launch technologies and infrastructure are crucial to the long-term competitiveness of U.S. aerospace.
- *Improved Contract Financing.* Policy changes to improve the financial health of the defense industry include a simplified progress payment process and higher rates, recoupment of nonrecurring costs limited to foreign sales of major defense equipment, and elimination of fixed-price contracts for R&D.

- *Streamlined Acquisition Policy.* Actions to streamline acquisition include eliminating redundant government oversight; industry participation in negotiated rule making and in developing standards and regulations; and fewer certification, disclosure, and reporting burdens.
- *Defense Industrial Base.* Congress, DoD, and industry must cooperate in bringing about an orderly transition to reduced spending levels. We must also ensure a viable industrial base that will meet emergency surge requirements and develop, produce, and support U.S. national defense.
- *National Industrial Security Program (NISP).* To assist the full implementation of the NISP, AIA will provide input on an NISP operating manual, FAR change package, NISP compliance review, and changes to laws, regulations, and policies.
- *Small Disadvantaged Businesses (SDB) Subcontracting Awards.* AIA's SDB database supports industry's goal of increased subcontract awards to SDBs. Incentives in the implementing legislation and regulations of DoD's SDB mentor-protege program would maximize industry participation.

AIR FORCE ASSOCIATION

The Air Force Association (AFA) represents a somewhat peculiar, but important, association that can be of assistance to the military project manager. The AFA provides an organization by which free people in the United States may unite to address the defense responsibilities of our nation imposed by the dramatic advance of aerospace technology; educates the members and the public at large as to what that technology could contribute to the security of free people and to the betterment of humankind; and advocates military preparedness of the United States and its allies adequate to maintain the security of the United States and the free world.[2] The association is organized into 12 regions around the United States with regional officers, state officers, and a number of local chapters. AFA is principally a membership organization with nearly 200,000 members. It is organized with a national board of directors which includes a chairman and 75 national directors, and it is supported by a staff at national headquarters in Arlington, Va., headed by Executive Director Monroe W. Hatch, Jr.

AFA at the national and local levels offers a number of programs and workshops intended to educate its members and the general public about the mission of the association. It publishes a bimonthly *Air Force* magazine as well as regular legislative updates, internal newsletters, fact sheets, and other printed material. Through the vehicle of its meetings and publications the AFA becomes a strong advocate for a number of Air Force programs. Thus it can become a valuable resource to the Air Force project manager. As a minimum, the Air Force project manager should encourage his or her project team to become members of AFA and should also use the AFA, when appropriate, as an advocate for his or her program.

AMERICAN DEFENSE PREPAREDNESS ASSOCIATION

Under the leadership of President Lawrence F. Skibbie and the volunteer board of directors who are prominent scientific, management, and industrial leaders, the

American Defense Preparedness Association (ADPA) provides a unique service to the military and the military project manager. ADPA is the voice of the defense industrial base and is dedicated to supporting a strong national defense and the industrial base necessary to make that defense credible. ADPA is a nonprofit education membership society devoted to national security. It encourages government-industrial dialogue to achieve its objectives under the following arrangements:

- Approximately 32,000 members who join as individual citizens and corporate representatives.
- 750 corporate members who recognize the need for a viable and responsive industrial base. Corporate members include companies from all aspects of support for national security, from hardware to professional services, from energy to the environment, and from communications to intelligence. Many of the corporate members are small businesses who are sensitive to the fact that their support of the association is a significant contribution to our national defense.
- 57 chapters carry the ADPA message to their local areas.
- 24 technical divisions, each devoted to a special area of industrial technology, are run by volunteer members.[3]

The National Defense magazine, the journal of the ADPA, is a vital educational tool of the association. The magazine is distributed not only to ADPA members, both corporate and individual, but to other influential leaders in the defense and legislative departments. The articles cover a wide array of defense and industry topics and provide another forum for the expression of the goals of the association. There is a small staff at the ADPA national headquarters in Arlington, Va., to carry out the central operations of the association.

Recent changes in the world situation and the needs of industry and government have been the catalyst for ADPA to enter into the following new initiatives:

- The establishment of a legislative affairs office to project ADPA's presence to Capitol Hill and to promote and sustain communications from the membership to Congress. It serves an educational and liaison function consistent with the association's charter.
- The promotion and participation in more overseas activities acknowledging the global marketplace.
- The addition of new technical subject areas as defense technology evolves.

The ADPA helps the military project manager in a number of ways, both directly and indirectly. Perhaps the most direct involvement is the endowment of an ADPA industry chair at the Defense Systems Management College (DSMC) to assist in the orientation of project managers who are completing the programs at DSMC with the perspective of various industry issues.

Chapters carry out association objectives on a local basis. The Wright Brothers Tristate Chapter, serving the southwest Ohio, northern Kentucky, and eastern Indiana areas, is an example of a chapter which has made a concerted effort to reach out to the community and involve, both on its board of directors and by its programs, a number of community leaders who are not directly tied to the defense project management establishment. For example, the chapter board of directors includes local politicians from both major political parties, the editor of the *Dayton Daily News,* and a number of business leaders whose businesses do not have a defense orientation. Involvement in the board helps to make these influential members of the community advocates of the needs of the defense and industrial base and has been a real bonus to both local and national defense projects.

As mentioned, the new ADPA legislative affairs department has the responsibility of carrying the defense preparedness message to Capitol Hill. This department also brings the congressional views on problems and possible solutions to ensure the solidity of the industrial base and the maintenance of our national security.

The final activity of ADPA with a direct relationship to the military project manager is the highly acclaimed technical forums carried out under the technical services function of the association. Each year, at more than 50 events, government employees, both military and civilian, and industry counterparts meet under an environment that fosters open communication and allows for the exchange of ideas between the producers and managers of government products and services. These meetings provide a legal, ethical forum for interchange in an increasingly legalistic business environment. ADPA meetings cover areas of interest and importance to national security. All forums strive to develop a common understanding between government and industry regarding the defense management process.

The military project manager can reap the benefits of ADPA in a number of different ways. The manager and staff can attend chapter meetings, national technical services division activities and government-industry forums, as well as roundtables providing an opportunity to update their management skills and knowledge on a number of specific issues. Through the vehicle of the chapter and the national leadership of the association, the military project manager can voice his or her concerns and solicit the support of the association in either a broad or a very specific way.

DEFENSE SYSTEMS MANAGEMENT COLLEGE ALUMNI ASSOCIATION

The preeminent institution for DoD acquisition education, research, and information is the Defense Systems Management College (DSMC) at Ft. Belvoir, Va. Through its various programs it offers advanced courses and information in defense acquisition management for officers and civilians from DoD and for a limited number of foreign officers and defense industry executives. Most current military project managers and essentially all future military project managers must attend the 20-week capstone project management course (PMC). All graduates of the PMC, DSMC faculty and staff members, graduates of DSMC short courses, and others holding key defense acquisition management positions are eligible to become regular or associate members of the DSMC Alumni Association. There is also a special membership category for honorary members. The DSMC Alumni Association was established in October 1983 by a group of PMC graduates representing virtually every PMC class and its predecessor course at the Defense Weapons Systems Management College at Wright-Patterson Air Force Base, Ohio. Current membership includes a mix of both civilian and military personnel involved in defense systems acquisition. The objectives of the DSMC Alumni Association, as stated in its constitution, are to provide:

- A members' forum for continuing professional growth of the defense acquisition community
- A source of defense acquisition management expertise for the Defense Systems Management College and the Association[4]

The association is managed by a volunteer board of advisors and a board of directors. A number of committees have been established in various functional and professional areas. Administrative support is provided by a small staff in the Washington area. There is also a provision for establishing local chapters of the association, and a number of chapters are in existence across the country.

The DSMC Alumni Association publishes a quarterly newsletter providing status information and an update of activities. Another significant activity of the association is the annual symposium, typically held in the summer, at Ft. Belvoir, Va. This provides an excellent opportunity for members to be updated on the current status of programs and activities at the college.

The synergism between the DSMC Alumni Association and the military project manager should be obvious. Since most military project managers are alumni of the college, they are automatically eligible for regular membership. The rare individuals who are military project managers or staff members but are not alumni are typically eligible for associate membership by nature of their holding key defense acquisition management positions. The Alumni Association provides another valuable resource to the military project manager in terms of contacts and interfaces with colleagues who are fellow military project managers. By the vehicle of the quarterly newsletter, the annual symposium, and through various chapters, the association can provide a valuable technology and management resource to the military project manager.

NATIONAL CONTRACT MANAGEMENT ASSOCIATION

The National Contract Management Association (NCMA) is somewhat different from the other associations and societies being addressed in this chapter inasmuch as it is a functional and technical association for the contracting and procurement profession. The association and its various activities can be an extremely valuable resource to the military project manager.

The NCMA was established in 1959 and was an outgrowth of three other contracting and procurement associations that were formed in the early 1950s. Unlike some of the other associations discussed in this chapter, from the beginning the emphasis was placed on the local chapters, and today, although the national office publishes the monthly magazine *Contract Management* and produces the national education and training programs and administers NCMA's professional certification programs, association activities are still focused on the chapter and the volunteer member. There are over 22,500 individual members of NCMA located in 141 chapters throughout the United States and in several foreign countries. The chapters are the strength of NCMA, and the chapter events focus on what the local membership wants and needs. Chapters support speakers for regular meetings, organize scholarship programs, act as liaisons along with local universities, and sponsor seminars and conferences to meet the needs of the chapter members. Each chapter belongs to one of eight geographic regions. A regional vice president represents the region's interests on the national level at meetings of the executive council and the board of directors.

The governing body of NCMA is the board of directors, which is composed of one representative from each chapter. There is also an NCMA board of advisors composed of top professionals in contracting. That board meets semiannually and addresses timely issues affecting the functions of the organization. The national president appoints national functional directors for a variety of areas, including certification, legislation, education, ethics, and program management. A small national office staff is located in Vienna, Va., a suburb of Washington, D.C., under the leadership of Jim Goggins, executive vice president. The national office is responsible for publishing the *Contract Management* magazine as well as related educational training material. This office also manages the day-to-day operations of the association.

In keeping with the important goal of advancing the professionalism of its members, NCMA administers two professional certification programs—the certified professional contract manager (CPCM) and the certified associate contract manager (CACM). An intensive examination based on the body of knowledge of the contracting and procurement career field along with experience determines the qualifications of an individual to become certified by the association.

A unique aspect of the NCMA is its annual national education seminar. This is a one-day course designed with the busy professional in mind and is offered in many locations in an effort to minimize participation expense. The topic for calendar year 1991 was "Managing Contracts for Peak Performance."

How, then, can the military project manager utilize the NCMA? The NCMA is a professional technical association for the contracting and procurement discipline. Since so many of the decisions that a military project manager must make involve the contracting and procurement career field, it behooves the project manager to encourage his or her professional contracting and procurement staff to participate in and reap the benefits of the NCMA.

NATIONAL SECURITY INDUSTRIAL ASSOCIATION

The National Security Industrial Association (NSIA) was founded in 1944 at the insistence of James Forrestal, then secretary of the Navy and later our first secretary of defense. It is a not-for-profit, nonpolitical association of some 400 manufacturing, research, and service companies of all sizes and types. The leadership of NSIA consists of a volunteer board of trustees made up of executives from the member companies, and a national headquarters staff located in Washington, D.C., headed by President James R. Hogg.

NSIA is a limited, special-purpose association of industrial and associated firms with a tradition of patriotic dedication to the national interest. Historically, NSIA has followed a policy of "no involvement" in political issues, including no support or opposition to any particular military policy or strategy, governmental budget level, or military force level; nor to any governmental material program or military weapon system. The association's province is the business and technical aspects of the government-industry relationship, encompassing government policy and practice in the entire procurement process, research and development, logistic support, and many other areas.

The association's primary objectives, as set forth in the NSIA bylaws, are:

- To maintain effective communication and a close working relationship between industry and those establishments of the government, including Congress, whose functions relate to the national security
- To consider and resolve problems through cooperation
- To provide industry views and recommendations to the government
- To assist the government with technical information and advice relative to industrial experience, capabilities, and practices
- To encourage scientific research[5]

Further, NSIA has continued to stress the areas of government legislation, policy, regulations, and procedures; military requirements, plans, programs, and problems;

and scientific and engineering matters—providing advice in many technical areas, such as antisubmarine warfare, automatic testing, research, and engineering.

Concerned with industry's contribution to national security, NSIA provides an effective forum for the exchange of ideas and information between its corporate members and government agencies through a network of committees, groups, chapters, national meetings, and conferences and visits to DoD installations. NSIA's principal vehicle in working with government is its committee program. NSIA committees have become an institution in American defense-industry relationships. From the earliest days of the association, it has provided sound, constructive counsel to the defense establishment, which it was created to serve. The views of industry, as expressed through the NSIA committees, are listened to and given consideration. Through the committee structure, interested member companies contribute ideas and recommendations to government and receive helpful information on government policies, programs, and problems.

The purpose of an NSIA committee is to coordinate the industrial resource of the association in providing the government with comprehensive and optimum assistance, information, and experience data. From objective study and analysis, it makes recommendations on policies, programs, regulations, procedures, and problems. All of the NSIA committees maintain continuing close contact with representatives of the DoD, the military services, and other government agencies, and many of the committees have identified liaison individuals with the services. If a liaison person is not already identified, the NSIA welcomes the interest and interface with the appropriate government organization.

The committees include logistics management, manpower and training, procurement, research and engineering, environment, communications, legislature, international, testing, amphibious warfare, antisubmarine warfare, space, software, manufacturing, and quality and reliability assurance. A special group covers industry computer-aided acquisition and logistic support (CALS).

The NSIA has a limited number of chapters principally located in major metropolitan or large defense-based communities. The chapters promote national security by developing and fostering mutual understanding and maintaining a close working relationship between industry interested in and concerned with national security and those elements of government responsible for national security at the geographic level represented by the chapter. The NSIA chapters are Dayton, Greater Hampton Roads, Greater Los Angeles, New England, New York, Philadelphia, Rocky Mountain, San Diego, and Washington, D.C. The chapters play a powerful and important role in carrying out the mission of the NSIA. For example, the Dayton chapter is extremely active with the local defense project management establishment. The chapter hosts bimonthly luncheons throughout the year, where a senior defense official speaks to the membership. Another important initiative is the chapter support of the Aeronautical Systems Center's "Chief Engineers Day," at which 25 company vice presidents of engineering/technology attend a full-day meeting with ASD's key command and engineering personnel. The chapter has also regularly sponsored technology briefings on such subjects as the source selection process.

NSIA also funds a chair (the James Forrestal Memorial Chair) at the Defense Systems Management College at Ft. Belvoir, Va. The purpose of this funded industry chair is to provide the insight and vision that the NSIA can offer to the students of the various courses and programs offered by the Defense Systems Management College.

Turning to the specific subject of how the association assists the military project manager, perhaps the best example is the calendar year 1991 program guide. In this program guide, which provides a focus of NSIA's activities for the forthcoming year, the emphasis is on acquisition. Section I of the guide gives a comprehensive look at

"The Acquisition Environment," recognizing that the defense procurement landscape has experienced a number of significant changes in recent years. The association has succinctly identified a number of items covered in Section II of the guide, titled "Important Acquisition Issues," including the following:

- Reasonable down-sizing of the defense industry
- Arresting the proliferation of unproductive policies
- Improving mechanisms for allocating risks
- Defense–industrial base
- Manufacturing technology
- Technology transfer
- Shaping the evolving relationships among DoD, industry, Congress, and the public
- Procurement integrity
- Overregulation and excessive oversight
- Total quality management
- Defense management review (DMR)
- The powers of the DoD inspector general
- Program stability
- Maximizing availability of information on defense acquisition[6]

As can be noted from this "shopping list" of items in NSIA's program guide, all of these issues are extremely important to the military project manager. The concept of working together as a team, while not unique to NSIA, seems to be epitomized by the various initiatives of the association, both its committees and its chapters.

PROJECT MANAGEMENT INSTITUTE

A relatively new association to military project management is the Project Management Institute (PMI). The mission of PMI, a nonprofit technical organization, is to be the leading recognized professional and technical association in advancing the state of the art of program and project management, and the development and dissemination of the theory and practice of effective management of resources in reaching project goals. PMI's objectives are to:

- Foster professionalism in the management of projects
- Advance the quality and range of project management
- Identify and promote the fundamentals of project management and advance the body of knowledge for managing projects successfully
- Provide a recognized forum for the free exchange of ideas, application, and solutions to project management challenges
- Stimulate the application of project management to the benefit of business/industry and the public
- Provide an interface between users and suppliers of hardware and software project management systems
- Collaborate with universities and other educational institutions to encourage appropriate education and career development at all levels in project management

- Encourage academic and industrial research in the field of project management
- Foster contacts internationally with other public and private organizations which relate to project management and cooperate in matters of common interest

When relating PMI to military program or system acquisition management, it is important to understand the accepted PMI definition of "project management." Project management in the general sense implies the broad conceptual approaches used to manage nonrepetitive work within the constraints of time, cost, and performance targets. Other terms which imply the use of project management techniques and methods include new venture, product, matrix, and program management. They differ principally in management's focus. Project management originated in the "advanced technology" areas, but it has recently been recognized that the application of project management methods cannot be limited to advanced technology work. Project management methods have proven very useful in "soft" sciences such as sociology, psychology, education, and management. Project management has been demonstrated to improve the chances of completing work efforts successfully on time, within cost targets, and of meeting performance criteria. It can be profitably applied to any nonrepetitive, clearly definable task. When we talk about project management, PMI's definition clearly includes military programs, systems, subsystems, and military program and system acquisition management. In the discussion of how PMI as a professional association can assist the military project manager, I will address two major areas. The first is the Defense Workforce Acquisition Improvement Act and the second is how a number of PMI initiatives, either existing or in the planning stages, can assist the project manager.

The Defense Workforce Acquisition Improvement Act mandates certain training, education, and professionalism for the defense acquisition work force. There are a number of steps being taken by the DoD to determine the best way to accomplish these mandates. But why "reinvent the wheel," when PMI, with its established professional certification program and its body of knowledge, with minor changes, might meet many of the needs of the defense acquisition community?

PMI has recently established a new special-interest group identified as the Defense/Aerospace Specific Interest Group (D/ASIG), which is being chaired by myself. PMI special-interest groups provide the following benefits: (1) A forum to provide PMI members a place to readily interact and network on project management issues and applications specific to their area of concern. (2) Special tracks based on interest groups can increase attendance at annual seminars/symposia. (3) The development of special-topic issues of the *Project Management Journal* and *PMNetwork* educational products are aided by interest group participation. (4) Corporate recognition of project management is enhanced by interest groups. For the Defense/Aerospace Interest Group the following are specific areas of interest and involvement: (1) It serves government members, businesses, and industries that develop, produce, and support defense and aerospace systems. (2) If appropriate, subfocus groups are developed at a later date to satisfy specific areas needing attention. (3) Information and idea exchange will assure a current, inclusive defense and aerospace body of knowledge that complements the project management body of knowledge (PMBOK). (4) Defense and aerospace unique project management practices and techniques are researched and developed. (5) Research topics, sponsors, and researchers can be identified and results disseminated to interest group members.

Each year, as part of its annual seminar/symposium, PMI sponsors a student paper competition, and the winner is invited to present that paper at the seminar/symposium. At the 1991 seminar/symposium in Dallas, Tex., the winner was Capt. Korina L. Kobylarz, USAF, an Air Force officer who prepared her paper while a graduate student in the systems acquisition management program at the Air Force

Institute of Technology. Her paper was entitled "Establishing a Department of Defense Body of Knowledge."[7] This paper generated considerable interest from among those present at the seminar/symposium, who were professionals in the DoD and related aerospace industries, and an initiative is under way to establish a DoD body of knowledge subgroup of the D/AIG.

How can PMI specifically assist the DoD project manager? As noted, PMI is a professional society of project managers with an established and recognized professional certification program. The DoD project manager can turn to PMI for professional support, not only for him- or herself, but for his or her project team. The new Defense/Aerospace Specific Interest Group (D/ASIG) will provide a vehicle for the defense project manager to identify and coordinate matters of professional interest. It gives DoD program and project managers another avenue of professionalism that they previously did not have access to. The various training programs that both PMI and its chapters offer to their membership and other interested project managers can also be a valuable tool for improving the professionalism of the DoD project manager's work force.

ROLE OF THE CHAMBER OF COMMERCE IN SUPPORT OF THE MILITARY PROJECT MANAGER

One of the last things that the military project manager should have to worry about is the future of the project office location and the quality of the facilities in which the project office is housed. Yet, in reality, there are always a number of cross currents working on the project—funding changes, mission changes, realignments, and so on. All of these factors can be very disruptive to the military project manager, especially to the civilian work force who does not typically have the same mobility requirements as the military members of the project office work force. An effective chamber of commerce can be of great assistance to the project manager in alleviating these concerns.

This section will review two chambers of commerce and how they approach the relationship with the military, with a specific focus on the military project manager. These are San Antonio, Tex., and Dayton, Ohio. While San Antonio has much larger overall military community, with five major installations, the Dayton area has more direct military projects with both the Aeronautical Systems Center and the information resource management activities of the Air Force Materiel Command.

San Antonio Chamber of Commerce

The Greater San Antonio Chamber of Commerce is the "granddaddy" of the chamber of commerce–military establishment relationship in this country. It has nearly a 100-year history of solid backing to its military missions and personnel, and it has established the following position statement on military relations in around 1991:

> We realize that local military missions could be affected due to the downsizing of the DoD, and we will support those restructuring decisions that make sense and do not undermine the security posture of our country. At the same time, we will zealously pursue the opportunity to obtain additional missions for our military community that make sense and fit into the present structures and facilities available on local military installations. We will also oppose local cuts that are a result of political considerations rather than what is best for our military.
>
> Throughout our 97-year history we have demonstrated solid backing for our military missions and personnel. We will continue that record throughout this decade and into

the 21st century. We are proud of our military. They are part of the San Antonio family, and we care for them as we do for our own families.

San Antonio has five major installations, Ft. Sam Houston, Brooks Air Force Base, Kelly Air Force Base, Lackland Air Force Base, and Randolph Air Force Base. The total economic impact of these installations is estimated at about $3.7 billion per year, with much of that impact coming from military and civilian payrolls, local construction projects, and contracts with San Antonio firms for support services. The less visible impact, though, is almost as important. Military people bring a strong sense of volunteerism to the community and the military citizen is usually highly educated, motivated, dedicated, and highly skilled.

John Williams is the vice president–military affairs of the Greater San Antonio Chamber of Commerce, and under the Military Affairs Council, the chamber has a very proactive set of programs in support of the military installations. Its largest single activity each year to support the military missions and personnel is its annual trip to Washington, D.C., which it calls "SA to DC." They visit the Pentagon and Capitol Hill for three days to thank both the military and the congressional leadership. On the first day they visit the Pentagon for a series of briefings to help better educate their contingent on matters of national defense. On the second day of the trip, the citizens break up into two- or three-person groups and schedule courtesy visits with members of Congress. In addition to the members Congress from Bexar County, the two Texas senators are also visited. Other calls are made on members of the House and Senate Armed Services Committees and other committees such as Appropriations and Military Construction. Clearly, when "SA to DC" arrives, Washington listens, and while San Antonio may not include a large number of DoD project management activities, the fringe benefits and the precedent that the Greater San Antonio Chamber of Commerce has established have clearly been beneficial to the military project manager.

One of the most significant activities of the San Antonio Chamber was the issue of reprogramming of fiscal year 1990 funds, which was vigorously supported by the chamber. Military leaders did not seek sequestration of military manpower from the FY 90 budget cuts, because the military wanted the flexibility to reprogram funds from the weapons and vehicles accounts to prevent drastic reductions in military personnel. In the past, the military was able to routinely shift money between accounts. The chamber actively supported the reprogramming issue because of the potential circumstances which could affect the military establishments in San Antonio. Without more dollars in the personnel account, for example, the Air Training Command might have had to stop its basic training at Lackland Air Force Base. The San Antonio Chamber launched a mailing effort to support the reprogramming, which was subsequently approved.

Dayton Area Chamber of Commerce

Although similar in mission, the thrust of the Military and Veterans Affairs Committee of the Dayton Area Chamber of Commerce is somewhat different from the Greater San Antonio Chamber of Commerce. Even though both chambers support major defense establishments with the related contractor support activity, the major responsibilities of the defense community in Dayton—the Air Force Materiel Command, the Aeronautical Systems Center, the Wright Laboratories, a number of other support organizations at Wright-Patterson Air Force Base, and the Defense Electronics Supply Center at Gentile Air Force Station, in Kettering, Ohio—are very much project management oriented.

As in the case of San Antonio, the Dayton area defense and veterans affairs installations provide nearly 50,000 jobs in the area and have an economic impact of well over $10 billion annually to the State of Ohio. A specific provision of recent contracts awarded by the information resource management activities of the Air Force Materiel Command is that the technical work is to be accomplished within a 25-mile radius of Wright-Patterson Air Force Base. This brings a considerable economic and technology impact on the region.

Under the leadership of chamber vice president Ron Wine, the Military and Veterans Affairs Committee is comprised of over 200 community leaders, aerospace contractors, and others who recognize the importance of the military community to the Dayton area. The purpose of this committee is to foster supportive relationships between the business community and area government installations. While the chamber staff coordinates the day-to-day activities of the committee, a steering committee, which serves as the coordinating organization for activities and as the principal advocacy organization at the state and federal levels, meets quarterly to review the progress of a number of initiatives. The steering committee consists of 25 members representing a wide cross section of defense establishments, the Department of Veterans Affairs Medical Center, State of Ohio, a number of national professional and technical organizations, and a number of local, regional, and State of Ohio technology-based organizations.

The activities of the chamber and the committee are varied. Ron Wine and other chamber staff members travel monthly to Washington, D.C., and Columbus, often accompanied by various community leaders, to coordinate support at all government levels for legislative and federal priority issues concerning important defense-related projects. While the greater Dayton area community has always supported the defense establishments (for example, the land which WPAFB occupies was purchased by community leaders and donated to the federal government in the 1920s), the Chamber of Commerce has, just within the last decade, intensified its focus on military and veterans activities. Similar to San Antonio, in 1985 the first community leaders' trip to Washington, D.C., was planned in response to the need to show more visible support for the area's federal installations and to increase the awareness of key defense issues affecting the Dayton region. Initially the visit targeted the Miami Valley congressional delegation and the Pentagon, but over the years it was expanded to include meetings with other members of Congress, the Department of Veterans Affairs, and the Defense Logistics Agency. In May 1991, 45 community leaders traveled to Washington to set the stage for a number of activities relating to the defense establishments in the Dayton area. Shortly before that visit, a major announcement was made concerning the location of the Air Force Materiel Command, an integration of the former Air Force Systems Command, headquartered at Andrews Air Force Base, Md., and the Air Force Logistics Command, headquartered at Wright-Patterson Air Force Base. The assignment of this new command to Wright-Patterson has significant positive impact to the greater Dayton area community, so the principal mission of the Pentagon visit in 1992 was one of appreciation of this decision. Meetings were held with the Air Force Chief of Staff, the Secretary of the Air Force, and the commander of the (then) Air Force Systems Command. Discussions on the Air Force Logistics Modernization Program and the Corporate Information Management Program, both projects managed at Wright-Patterson Air Force Base, were held at the Assistant Secretary of Defense level. Meetings on Capitol Hill included all of the Ohio congressional delegation as well as Senators Glenn and Metzenbaum. A lengthy discussion with Senator Sam Nunn, Chairman of the Senate Armed Services Committee, was also held.

In addition to the Washington and Columbus visits, the Military and Veterans Affairs Committee holds monthly "Commanders Call" breakfast meetings, at which time "flag rank" officers are featured to dialogue with members of the committee on items of mutual interest. Other activities include one or two annual conferences with a focus on project management and procurement issues, and a number of social activities, both receptions and luncheons, help the military feel a part of the Dayton area and provide opportunities for them to meet with members of the business community to discuss areas of mutual interest. Finally, the Military and Veterans Affairs Committee of the Dayton Area Chamber of Commerce sponsors a military outplacement program. Bimonthly breakfast meetings and one or two job fairs a year assist retirees, separatees, and spouses in securing employment in the region. The Military and Veterans Affairs Committee considers the military project manager and his or her staff to be a valuable resource to the community and all the activities described are focused on helping them to accomplish their jobs more efficiently and to be welcome in the community.

The technology base in the Dayton area has spawned a number of technology-based organizations with a linkage among the defense technology activities, the State of Ohio, and the business community. These include the Edison Material Technology Center, the Ohio Advanced Technology Center, the Ohio Computer Technology Center, and the Center for Artificial Intelligence Applications. The Dayton Area Technology Network is an informal yet dynamic organization established to coordinate and integrate the technology base that exists in the greater Dayton area. This network receives administrative support from the Dayton Area Chamber of Commerce and is responsible for maintaining a professional relationship with the defense establishments, aerospace, high-technology firms, and universities to enhance new business opportunities, encourage technology transfer between government and private industry, and encourage the linkage between business and area colleges and universities. Principally because of the unique technology base at the Wright Laboratories, the subject of technology transfer becomes vitally important to not only the community, but also the military project manager. Technology transfer is a two-way street for military project managers—they can reap the benefits of technology from other sources and they can provide technology to the private sector from their specific projects.

SOCIETY OF LOGISTICS ENGINEERS

The Society of Logistics Engineers (SOLE) was founded in 1966 as an international nonprofit professional organization of individuals with a primary purpose of engaging in educational, scientific, and literary endeavors to advance the state of logistics technology in management. There are over 100 SOLE chapters in 25 districts, including a number of foreign countries. Chapters and districts conduct and sponsor technical meetings, seminars, and workshops designed to provide the SOLE members with opportunities for professional advancement. SOLE also sponsors an annual international logistics symposium, which brings together hundreds of members from all over the world in a three-day technical meeting. Technical papers are available in a bound volume of proceedings. SOLE also publishes a quarterly professional journal, *The Logistics Spectrum,* which is distributed to all society members. Members receive a monthly *SOLEtter* and a scholarly technical journal entitled *The Annals of the Society of Logistics Engineers.* SOLE has a professional certification program designated as the Certified Professional Logistician (CPL) Program. This was established to recognize outstanding professional logisticians. The examination process is challenging and stimulating, and there are currently over 1200 logisticians who have achieved a CPL designation.

SOLE believes that the maximum interchange of knowledge among all elements of the logistics community is vital to the continued evolution of the logistics profession. Accordingly, SOLE works closely with other professional societies and is cosponsor of many important technical symposia. SOLE is managed by a volunteer board of directors, which is responsible for establishing policy for the association. The executive board is an element of the board of directors, consisting of nine members who are responsible for implementing the policies and the day-to-day operations. In addition, a high-level board of advisors provides guidance to the board of directors on policy matters and assists the board of directors chart its long-range course of action to achieve its goals of advancing the profession of logistics. A full-time executive director, Norman Michaud, and a small staff are responsible for the day-to-day operations of the association. International headquarters are located in New Carrollton, Md.

SOLE has four technical divisions established for two reasons: (1) to ensure that SOLE's international programs and publications remain in balance when considering the entire scope of logistics, and (2) to permit coverage of specific career fields within the broad spectrum of logistics. These four divisions are the technology division, product support division, management division, and material operation and distribution division. While all of these divisions focus on one or more aspects of the military project management activity, the management division has the most direct applicability in that it addresses such specific items as business logistics management, life-cycle cost management, project management, contract management, and supplier/subcontractor management.

How, then, can SOLE assist the military project manager? As with the National Contract Management Association, SOLE represents an important functional discipline of military project management. The project manager can turn to SOLE, either through its national staff or through its various chapters, for technical assistance. Clearly, the military project manager should encourage the logisticians on his or her project staff to affiliate with this important association.

SUMMARY

This chapter has addressed the role that a number of professional and technical societies and associations play in supporting the military project manager. In the scheme of organizational relationships, these societies and associations can be considered external stakeholders to the military project management organizational team. It is difficult to summarize and generalize what these societies and associations can bring to the military project management team, as each one tends to play a somewhat unique and differing role.

The Aerospace Industries Association uses its voice to carry its message to Congress, the executive branch, the media, and the public. In a similar vein, area chambers of commerce provide much needed support to projects themselves. Associations providing education and research opportunities, such as AFA, ADPA, DSMC, and SOLE, are vital to the development of the project manager. Those organizations that offer forums for discussion and advancement of the profession, PMI, NCMA, and NSIA, strengthen the project manager's knowledge base. Suffice it to say, each can bring to the military project manager a very special perspective on the project.

The societies and associations can often bring to the military project manager additional high-quality resources to work on specific project issues and problems. It is the perception of this author that the typical military project manager probably does not fully utilize the resources of these societies and associations, whether it is a "not invented here" attitude or just an awareness problem. The military project manager and his or her staff should indeed call on the valuable resources of the societies and associations discussed in this chapter as well as others to become external stakeholders on the project management team. As evidenced by the specifics discussed in this chapter, they can provide a valuable perspective to the military project manager and his or her team. Similarly, I encourage the societies and associations to continue to reach out to specific military project managers to assure that they are aware of the resources available to them.

REFERENCES

1. Bylaws, Aerospace Industries Association of America, rev. Nov. 15, 1990.
2. National Constitution and Bylaws, Air Force Association, Feb. 3, 1991, p. 4.
3. 1991 Report, American Defense Preparedness Association, Apr. 1991.
4. *Member Directory and Handbook,* DSMC Alumni Association, Feb. 1989, p. 1.
5. "Industry's Expertise and Experience for America's Security," Annual Report and Directory, National Security Industrial Association, Jan. 1, 1991, p. 1.
6. *Program Guide, Calendar Year 1991,* National Security Industrial Association, Washington, D.C., Dec. 12, 1990.
7. *Proceedings,* Project Management Institute 1991 Seminar/Symposium, Dallas, Tex., Sept. 27–Oct. 2, 1991, pp. 275–279.

CHAPTER 26

INTERPERSONAL SKILLS: THE KEY ESSENTIAL

Michael G. Krause

Michael G. Krause is the principal of Krause Associates, a management consulting firm established in 1976. He received an M.P.A. degree from the University of Southern California in organization development and training; an M.A. degree from the University of Maryland in public administration; and a B.S. degree from the U.S. Air Force Academy.

INTRODUCTION

Large organizational systems are complex and not understood very well. Department of Defense (DoD) efforts to reduce the cost and enhance the performance of weapon systems have focused on reporting requirements, organization structure, program stability, smaller staffs, and better communications. Currently the DoD philosophy is to centralize policies, procedures, and standards, but to decentralize their execution. Thus program management skills take on an even greater measure of importance. With this proliferation of power centers, program success is even more dependent on the support and cooperation of people or organizations outside of the formal authority of the acquisition manager.

A given part of the program manager's job is working with various "players" and "teams" (subordinates, functional personnel, service headquarters, users, DoD, other commands or services, and industry). The real job of a government program manager is integration.

The implicit nature of a human organization structure results in the program manager working for and with people. Intergroup myopia often paralyzes organizational progress when all groups with an interest in the program feel they have the given right to comment or veto concepts, plans, procedures, or activities. Self-interest can also work to the detriment of the program organization. Thus it is essential to establish effective interpersonal relationships, develop effective oral and written communications and listening ability, and influence skills which will enable people to share the vision of the program.

People are the program manager's resource for achievement. As the manager works toward achieving program goals and objectives, he or she must work with people to make decisions which enhance the likelihood that the cost, schedule, supportability, and performance objectives will be realized. Time and money can solve most technical problems. This is not the case for people problems. Effective teams take time to develop. Without people skills, the program manager's job can be very difficult and frustrating.

Effective interpersonal relations are based on trust. This begins with the assumption that people will work best if they believe the work they are doing is important and satisfying and provides for growth. Unfortunately trust is slow to develop, takes extensive time to nourish, and can be very fragile. Human nature can plan havoc with the balance. Building trust requires an open, supportive, and problem-oriented work environment, which values a learning process. As people are accepted for who they are, safeguards and facades are dropped for a more candid relationship, in which attitudes and feelings become known. With a deeper level of communications there is greater openness to give and receive negative information related to the program. This results in decisions based on more data which have not been filtered. However, there is a constant need for the program manager to ensure that there is compatibility between program needs and individual needs.

It is difficult to apply objective measures to determine a program manager's effectiveness. This is due to the complexity of defense programs, the involvement of a multitude of stakeholders, the relatively short period of time a government program manager is tasked to direct the program, the level of visibility or importance of the program, and the level of technological risk. Not to be overlooked is that the program manager's perceived success or failure may rest on the decisions made by predecessors.

ROLES OF PROGRAM MANAGER

While dealing with the technical and financial aspects of government programs, the psychological and social aspects of the role are often overlooked. To be successful, a broad range of behaviors needs to be developed to handle each role in a consistent and congruent manner. Like a true star actor in Hollywood, the program manager feels comfortable in each role that he or she must play. This helps to build a fan club of the various people and organizations associated with the program. In moving from role to role, predictability is the key. When role behavior is predictable, people know what to expect. Conflicting role behavior causes uncertainty.

The program manager has many interpersonal roles such as leader, manager, figurehead, spokesperson, negotiator, politician, peacemaker, cheerleader, entrepreneur, or disciplinarian. These roles are briefly described.

Leader. As the leader, the program manager has received a program charter which describes what he or she is expected to accomplish. Leadership implies responsibility for creating a climate that will enable the people associated with the program to be highly motivated and give their best efforts as they move the program ahead in the direction desired by the leader. Leaders have the knack of attracting the best people who find it highly desirable to be associated with the program; and of creating a climate where all work teams have mutual respect.

Manager. In this role the manager balances the tension between short-range tasks and objectives and long-range plans and objectives. This is accomplished by focusing employee resources toward viable activities and goals. Internal management activities such as planning, priority setting, decision making, agenda setting, scheduling, staffing and work assignments, personnel development, resource allocation, and monitoring systems are used to determine where there are problems and to initiate corrective action.

Figurehead. As the figurehead, the program manager is the symbolic top and the most visible representative of the program management office. He or she performs numerous ceremonial duties, which are required by law, expected socially, or desirable to maintain harmonious relationships. Examples include retirement, promotion, and award ceremonies; contract announcements; roll-outs or ribbon cuttings; banquets; and visits to the operational user.

Spokesperson. As the spokesperson, the program manager represents his or her program executive officer (PEO) and subordinates when communicating with external organizations in both public and private forums. In addition to providing information on the program, he or she is in an advocacy position and is seen as the prime program representative by the PEO, users, service, and Congress. This is one role from which the program manager should not shirk.

Negotiator. As the negotiator, the program manager deals with people who allocate resources; with contractors to reach technical, schedule, and cost parameters; and with functional managers to acquire staff; and represents the program in other give-and-take activities. Skill is needed to assure that a mutually beneficial agreement is reached.

Politician. In the political role, the program manager builds coalitions with the various stakeholders. As the politician, the program manager melds individual values into a common frame of reference and shared purpose. This requires social grace and political savvy, and several years of network building which began in the past and will continue in the future. Political decisions are made to minimize barriers to achieving the primary program goals.

Peacemaker. The other roles may place the program manager in the role of peacemaker. In this role, he or she works to eliminate or minimize disruptions, especially those with significant program impact.

Cheerleader. When the going gets rough, there is a special need for cheerleading or nurturing. In this role the needs of individuals are recognized and responded to. Confidence is built to enhance individual and team desires to continue contributing their best efforts to the program. Cheerleading can also be combined with other roles when the manager is away from the program office.

Entrepreneur. With the life of many programs ranging over several decades, entrepreneurship is often overlooked as a role required of government managers. This role functions to identify opportunities and to take action which translates into new advantages. In some situations this can result in the program turning into an entirely different direction.

Disciplinarian. Finally there may be a need to take prompt disciplinary action when there is a willful violation of law, contract, safety, policy, procedures, or regulations.

Effective program managers are able to adapt their interpersonal role as the situation demands. In order to do this, they need to have a clear vision of the outcomes they desire during their tenure as the program's manager. To obtain the vision, answer the following questions:

- What is my mission?
- What do I want to happen?
- What will success look like?
- How will I achieve it?
- Do I have the resources I need?

PROGRAM MANAGER COMPETENCIES

The Defense Systems Management College (DSMC) conducted a study of DoD program manager competencies.[1] The study identified 16 competencies which were placed in four clusters, namely, managing the external environment, managing the internal environment, managing for enhanced performance, and proactivity. Eight of the 16 competencies speak directly to the interpersonal skills of the most competent managers. They include the following:

Political awareness. Knows the influential players, what they want, and how best to work with them.

Relationship development. Spends time and energy getting to know program sponsors, users, and contractors.

Strategic influence. Builds coalitions and orchestrates situations to overcome obstacles and obtain support.

Interpersonal assessment. Identifies specific interests, motivations, strengths, and weaknesses of others.

Assertiveness. Takes or maintains positions despite anticipated resistance or opposition from influential others.

Managerial orientation. Gets work done through the efforts of others.

Results orientation. Evaluates performance in terms of accomplishing specific goals or meeting specific standards.

Critical inquiry. Explores critical issues that are not being explicitly addressed by others.

From the study it was noted that program managers at the beginning of the acquisition life cycle demonstrated a significantly greater degree of relationship development. As the program grew larger and transitioned into later phases, managers depended more on their staff.

As presently organized, the government program manager deals with numerous individuals on a wide range of organizational levels in stakeholder organizations both within and outside of government. Successful programs generally result when the program manager has established open, trusting, and cooperative working relationships among and between the program staff, matrix personnel, contractor personnel, and others who have an interest in the program. Relationships must be nurtured across a broad range of values and cultures inhabited by diverse professional groups (engineers, lawyers, political appointees, military officers, and so on). The program may be the only point of commonality between all these people.

DIFFERENCES IN PEOPLE

Over time our society has come to recognize that people must work together to accrue benefits for everybody. Anyone who has worked in a group or lived in a family is aware that differences in people can be either beneficial or disruptive. The socialization process helps shape individuals so they can be contributing members of society. As the person goes through various life phases, socialization and individualization can cause tension or conflict as the person strives to meet his or her needs and achieve his or her life goals. Conformity to organizational norms may not satisfy individual needs or preferences.

Myers-Briggs Type Indicator

The Myers-Briggs type indicator (MBTI) is one of the most widely used measures of personality preference. It is a psychological instrument based on Jungian typology and does not measure intelligence or abilities. Feedback is provided on four bipolar scales. The scales are based on how people use their psychic energy to deal with the environment; how they gather information; how they make decisions with that information; and how they relate to their environment in terms of control versus adaptation.

One aspect of the MBTI is that each of the 16 types has a different gift. There is no need to say that one type is better than another. In the acquisition environment, this is very useful. Using the MBTI as a framework, the program manager enhances his or her ability to look at differences and to have a model for assessing people's preferences and potential behavior. (This is especially critical if the person fully supports the program, but appears to be taking actions to hinder its forward movement.)

A person who takes the MBTI receives feedback on his or her type preference. For each scale, the letters are paired to form opposite ends of a continuum as follows: E–I, S–N, T–F, and J–P. (It is estimated that 75 percent of the population who have taken the MBTI have an E or S preference; 50 percent have a T, F, J, or P preference; and 25 percent have an I or N preference. On the T/F scale, 60 percent of the female population is likely to score F compared to 40 percent of the male population.[2] The four letters which reflect type are the focal point for understanding. They stand for the following:

E (extraversion). A person probably relates with more ease to the outer world of people and things

I (introversion). A person probably relates more easily to the inner world of ideas

S (sensing). A person probably would rather work with known facts based on the senses of sight, sound, smell, taste, and touch

N (intuition). A person probably would rather look for possibilities and relationships based on feelings, meanings, or hunches

T (thinking). A person probably would base his or her judgments more on impersonal analysis and logic, and where the decisions might lead

F (feeling). A person probably would base his or her judgments on personal values, and be concerned with how they matter to others

J (judging). A person would probably like a planned, decided, and orderly way of life which he or she controls

P (perceiving). A person would probably like a flexible, spontaneous way of life which he or she adapts to

The majority of the acquisition work force are sensors and thinkers. As sensors, they generally like to solve immediate problems, pattern their actions on what has worked before or based on what others are doing, and they do not like to fix things which are not broken. The thinking preference results in the task being more important than relationships, ideas are presented logically with plenty of back-up evidence, errors and inconsistencies are pointed out, and conflict may be allowed to drag on.

Using the MBTI to look at people's differences helps us understand why the intuitive type may be characterized as independent, and be seen as less predictable than the sensing type, who is generally more structured and conforming.

By combining the two middle letters which reflect mental functions, an awareness can be gained as to how a person might handle his or her job. ST types are practical and matter-of-fact, and as leaders they are the traditionalists. They like to use their technical skills with facts and objects in areas such as business, production, and applied science. The NT types are logical and ingenious, and their leadership is seen as visionary. They like to use their abilities for theoretical and technical developments, and they are most likely found in research, the physical sciences, management, and analytical work.

On the people side there are the SFs and NFs. The SF is sympathetic and friendly and likes to be helpful and provide services to people. SFs are found in patient care, teaching, sales, and community service. Intuition makes the NF enthusiastic and insightful. NFs like to use their ability to understand and communicate with people, and are found in the behavioral sciences, research, and teaching.

Another combination of the letters is useful for predicting individual or organizational tendencies. For example, the IS people will want to retain the status quo; the IN people will want to look at problems and operations from several different vantage points; the ES people will want to jump right in and lead the charge to get the job done; and the EN people will want to make changes. By recognizing individual preferences, it may be possible to minimize conflicts which are caused by a lack of understanding of where a person with a different type might be coming from.

Organizationally, the judging-perceiving scale has the potential to have the greatest impact. Judging types (Js) like to follow plans, reach closure as soon as possible, may make quick decisions, dislike project interruptions, get right to the point, like clear decisions, are deadline-oriented, and are sensitive to time pressures. Perceivers (Ps) adapt to, and are energized by, changing situations, are open to changes, may not make decisions, start many projects, do not live by the clock, and feel time pressures late. When Js work with Ps, they need to be flexible, listen for new information, modify their thinking, and stay open as long as possible. From the other end, when Ps work with Js, they should plan ahead, be timely, establish and live by deadlines, make decisions, and avoid diversions.

Recognizing Differences

Individual egos of functional experts, staff personnel, operational users, or contractors can challenge even the best program managers. To varying degrees, people have a need for autonomy. Generally, highly trained professionals expect autonomy. They want to be held accountable for their results. They may resist formal authority or informal cajoling or influence so they can do it their way. As functional experts, they have some degree of power. Also, they may feel they have to protect their profession's body of knowledge. Thus they may be very determined to maintain a position despite its adverse impact on a program.

How does a manager continue to get new and creative ideas? What happens to motivation when a decision is made against one person and in favor of another? While both views might be objective from an outside vantage point, each side sees the other as being inflexible, parochial, head-strong, or overly defensive. The program manager's art comes into play by making a decision which creates a balance where all the players see the decision as being a winner for everybody. To have this happen, the team has to be exposed continually to the manager's long-range vision.

There are three basic assumptions a manager must recognize in order to enhance his or her interpersonal relationships. They are:

1. People have differences in personality, attitude, and perceptions which should not be automatically classified as good or bad.
2. In the interpersonal arena, there is no cookbook which provides step-by-step instructions and a list of ingredients to create the perfect program management work team. Thus there is no right way to deal with differences.
3. People can change over time. Education, experience, life transitions, or new roles can change previously held attitudes and perceptions. Needs for achievement, control, affiliation, influence, or dependence are subject to change. The field of psychology helps us understand that conflicting needs, values, and attitudes inside a person can be a major source of personal problems if not well managed. Personal conflict may exhibit itself as sleepless nights, irritability, restlessness, depression, or lack of energy.

When significant differences occur, communication frequently becomes indirect. People talk at or about each other, rather than to each other. For example, a person is concerned that an overly optimistic schedule will impact on family time. Instead of stating discomfort that if overtime is required to achieve the schedule family time will suffer, the person is more likely to bring up risk or other programmatic reasons for schedule stretch-out.

WORKING WITH SUBORDINATES

As the size of the program office increases, the program manager can be several managerial levels away from the lowest level subordinate. With intense travel, he or she may have few opportunities to meet face to face. Thus much of the work with subordinates is done through intermediate managers, many of whom might be responsible for high-budget programs or projects themselves.

Information is filtered to prevent communication overload. As a result, different people are aware of different pieces of relevant information, which they are free to accept or reject, and may or may not place them in their brain's memory so that they can be recalled. To help subordinates understand that decisions are based on program needs, and not on individual personalities, the program manager must develop an internal process which allows information to flow to the lowest organization level relatively unfiltered, and lower-level perceptions and differences to percolate up. To enhance success, (1) assure that there is mutual understanding of the program objectives, plans, constraints, and resources; and (2) take extraordinary efforts to communicate with everybody by establishing relationships which minimize surprises.

There is a range of leadership behavioral patterns which can be used. Directive behavior would be used when there is a need to tell people what to do, when to do it, or how to do it. At the other extreme is sympathetic behavior. Here the manager listens, provides support, encourages, acts as a sounding board, and may facilitate action indirectly. The style would be dependent on the nature of the task and the employee's stage of development. Generally the newer the employee, the greater the direction that is needed. As the person gains competence in the job, understanding of the program and the world of acquisition, and self-assurance, the manager can shift to provide the individual with the maximum autonomy.

Repression of Differences

At times there is a need to repress differences. This may be done to present a single or harmonious viewpoint to people outside of the organization; to reduce or eliminate internal dissension; to force movement; or to gain control over the people through rewards, punishments, and indoctrination.

However, there are several risks inherent in a repression strategy. If the thrust is toward fostering loyalty and teamwork, differences may be suppressed. At some point the simmering pot may boil over and cause a mess. Personal stress may increase. If not worked out through effective office communications, the individual might take the stress home. In the domestic setting, hostility from the job can result in misdirected anger. Problems can be magnified and then brought back to the work setting, creating an endless loop which gets bigger and bigger.

Those who feel they are not in control may find some relief in resisting good ideas. In this way they take the spotlight and can use the rejection as a means to feel important by "bad mouthing" the boss with their peers or with friends away from the job.

Effective Listening

I often ask workshop participants what makes an effective boss. A vast majority of the lists mention the ability to listen. This includes:

- Letting the individual speak without interruption
- Responding with descriptive language rather than evaluative or judgmental language
- Asking leading questions, and following up with probing questions which will dig below surface issues
- Asking questions for clarification
- Restating what has been heard to verify
- Summarizing the discussion (often in long conversations)

Also, when the discussions are infrequent or conducted with a younger or less experienced team member, there is a need to use language which encourages the speaker. Body language is also important and includes sitting or standing posture, eye contact, proximity to the other person without "invading" his or her space, gestures, and facial expressions.

Effective listening enables the program manager to determine the nature of problems faced by program and contractor personnel. A determination should be made whether they are based on facts, methods, outside pressures, goals and objectives, personality, perceptions, or values. A brief discussion of each of these follows.

Methods. Individual educational and work experiences often lead to a person having a unique view on how work can be done best; and the short- and long-term activities which will enable the person to contribute to the achievement of both organizational and individual goals.

Outside Pressures. People respond to power, influence, and authority differently. Outside pressure can be real or imaginary, and it can include higher levels of authority, such as people who have power or rank but are not in the chain of command (Congress, GAO, OSD staff personnel, contractor executives, and so on).

Goals and Objectives. While program documentation should make the program's goals and objectives clear, the level of risk or uncertainty may lead to disagreement about what actions need to be taken to achieve them.

Personality. As was discussed, there is a wide range of individual personalities which are encountered. The extent to which a person is guarded or open has an impact on the information that will be shared.

Perceptions. We all have a consistent or inconsistent framework for viewing the world around us. Thus our response to a set of information may be expected or surprising, depending on our relationship with the person or people involved.

Values. Differences can stem from values. These can be based on a person's view of the American mission (or any other nationalistic mission), the nature of justice, Judeo-Christian or other religion-based beliefs, or the pioneer spirit of individualism. Organizationally there may be a set of public administration values at work. They include efficiency, responsibility, social equity, professional competence, civil service neutrality to political party differences, centralization or decentralization, and intergovernmental competition or cooperation. What is fraud, waste, or abuse to one person may not be to another. The use of power can also be seen as a value issue. A person in charge may see the use of power as one method for accomplishing program goals, whereas another may see it as an abuse of human dignity.

Once the category of problem is identified, the program manager needs to identify the cause. This can be done by asking subordinates to examine the assumptions on all sides of an issue, and can require individual or group meetings, research to gather information, or the use of a third party. Assuming there is trust, integrity, and a team orientation, it is possible to discuss the various viewpoints. With clarification it is possible to understand how everybody involved feels, and what they would like to accomplish. Once there is understanding, it is generally possible to make a decision which all people can support.

One means of checking for complete agreement with a decision is the use of the unqualified "yes." Once a decision is reached, the manager turns to each individual and asks whether he or she fully supports the decision. Any answer less than a quick, firm yes results in the manager's immediate use of a question such as:

"There seemed to be some hesitation in your response, what else is on your mind?"

"'I think so' is not what I want to hear. If you cannot give me an immediate yes, tell me why not."

"Why do you say, 'I don't think so?'"

MANAGING UP

Getting the job done is only part of management. To minimize the impact of the PEO's position and power, program managers need to develop their capability to manage up. It is possible to misread the boss just as easily as the subordinate. A misunderstanding can have a significant impact on the program. One factor is the relationship which has been developed with the boss. Ideally there is the need to become the PEO's trusted subordinate. Each has a responsibility to nourish it.

Legal requests of the PEO must be obeyed. However, the cost can be the loss of self-esteem. Thus it is essential to have the psychological willingness to leave the program manager position should the PEO demonstrate a lack of confidence in you as an individual or for the program. The willingness to depart can be the key to maintaining a dynamic posture as the subordinate.

How do you nourish the PEO–program manager relationship? First, remember your accountability. This requires discussions of progress, actions being taken to resolve problems, future activities, potential risks and consequences, and resource needs. Second, be the loyal opposition or the devil's advocate. The PEO must balance the needs of many program managers. Help the PEO understand the relationship of your program to the other programs, and support his or her decisions when they are made. Third, ask for feedback and information. Learn how the PEO views the world. Maintain two-way trust by not breaching the confidence placed in you. If your needs are not being met, let them be known. Finally, do not forget that the PEO is also a figurehead. Invite him or her to visit the facilities of your program and program contractors. Help the PEO become a shining star.

CONFLICT

Competition and interdependence are inherent in the American culture. When combined with the need to achieve, the seeds for conflict are firmly rooted in the workplace. Reflect for a second. Conflict originally meant to fight or do battle with an enemy. Over time it has also come to mean antagonism or sharp differences between people. Program managers need to recognize the source of conflict in themselves, their subordinates, their organizations, and their environment.

Program management allows technical and professional personnel to be used across functional areas. To be successful, the manager needs a long-range view. Conflict will be inevitable. He or she needs to be able to take the heat for his or her staff or program. Look for win-win alternatives, and know when to retreat and not face a frontal assault.

Causes of Conflict

Conflict occurs when two or more parties do not agree on something. It can occur between people within the same or different work groups, between work groups within and outside of the program office, or between organizations. In program management, conflict can result over resources, work tasks, differing personal opinions or goals, schedule pressures, differing priorities, tradeoffs, differing perceptions, narrow-minded functional thinking, short tempers due to long work hours, or external pressures or crises. The need for consensus often fuels the fires of conflict since professional viewpoints may be sacrificed in the trade-off process. The list which follows includes many causes of program conflict broken out by major category.

Competition for Scarce Resources. Resources include budget, time, equipment, materials, and people. Competition can be triggered by contraction, expansion, individual needs, or a stagnant situation.

Organization Design. This area includes program authority, the chain of command and reporting channels, organizational layering, personnel location (including social isolation), program organization (matrix, product, traditional), work group assignments, and line-stafff relationships.

Organization Procedures. This area includes management control of operating systems evolving from DoD or service directives; conflicting directives; and personnel systems established to create a framework for compensation, promotion, rewards, or performance reviews.

Organization Ambiguity. This area includes unclear goals, misunderstood roles and relationships, inconsistent administration or improper implementation of policy, poorly defined work or areas of responsibility, perceptual differences, communication breakdowns, managerial competition, or misuse of authority.

Goal Differences. This area includes professional-clerical relationships (such as the professional wanting to meet a deadline, and the secretary wanting to follow procedures), host-tenant relationships, differing life styles (such as military or civilian), or team pressures.

Values. Values and beliefs based on religion, philosophy, and so on, are difficult to change since they are ingrained. Functional values such as research versus production or labor versus management may be easier to change.

Societal Concerns. Conflict might arise from the concern over the rate of change in the world, nuclear war, global warming, energy shortages, terrorism, collapse of the monetary system, epidemics, work force diversity, escapism (for example, the use of drugs), or overcrowding.

Individual Concerns. These might arise from new life styles, incompatible expectations, work overload, unsettling conditions or uncertainty (more issues than answers), level of trust, few work contacts, unmet needs, criticism, being disadvantages (perceived and real), status or position, or emotional state. If role expectations are not congruent with experience, capabilities, or values, there will be internal conflict.

Dealing with Conflict

DoD does not have any specific policies directing managers on how to deal with conflict. However, as was described, procedures and directives can lead to conflict. Also, new world relationships are developing. In the past there was a common external enemy. It was easy to wave the flag as a patriot and put conflict aside in the name of moving the program ahead toward operational capability.

While some people say that conflict is conflict, there is a difference when the conflict is between two people, and when there are three or more people who can be involved in coalitions with different goals for the ultimate solution. There are some situations where the conflict may be ambiguous. For example, the budget process may result in a presidential budget submission which is integrated and makes strategic and tactical sense. When considered by Congress, there may be a coalition which is organized to reshape the President's program significantly. It is possible that the people on the hill may not agree among themselves on the specific line items, but they reach a compromise.

The lack of clear program objectives or the failure to communicate revised objectives will lead to conflict. It is clear that the program manager must determine the program's objectives (preferably with the input of the entire team), reaffirm them, and clearly communicate them to the program office and other stakeholders. The major objectives must be clear, understood, and agreed upon so that the whole team is pulling in the same direction. This requires that decisions be explained to assure that they have not been made in a vacuum or by popularity contest. Priorities might be established on the basis of urgency, closeness to a major milestone, technical risk, expected impact, PEO direction, user interest, or impact on other program areas or personnel.

Conflict between groups exists when the activities in one area are at odds (disagreements among people) with the activities of the other. This could be due to real or perceived interference, injury, and so on. The range of the conflict can be from disagreement to outright hostility. If organizational conflict is considered normal or natural, then a person's anxiety will be reduced related to the occurrence of conflict. Pretending that conflict does not exist (a variation of withdrawal, which will be discussed) may not be appropriate, although it could be a strategy in some situations. Suppression of conflict is like pouring water on a fire—if the fire is not drowned, it smolders.

Conflict is usually a sign of a healthy organization. A total lack of conflict may be a sign that an organization has some potential or actual morale problems. While conflict makes many people uneasy, it is essential for a dynamic organization. It empowers them to have different viewpoints aired, discussed, refined, and acted upon. A program manager should understand the difference between conflict management and conflict resolution.

Knowledgeable program managers know that conflict can create stress for the parties involved as well as for bystanders. For the bystander there can be anxiety, annoyance, fear, a desire to enter or withdraw from the conflict arena, or a beginning to doubt the manager's capability. Thus the manager must think beyond the conflicting parties.

Organizational productivity is enhanced under managers who have the capability to handle interpersonal and organizational conflict effectively. Those managers who are unable to handle the conflict may be damaging their programs. They may not be aware that delays result from conflict over minor issues, that the team sees itself as fragmented because the manager has not provided the decisions required for a cohesive unit, or that there is increased distance between people as a result of polarization, distrust, or self-preservation.

One part of our value system honors the notion of getting along with other people and respecting authority figures. As a result we might believe that conflict of any type is bad. This can cause the valuable aspects of conflict to be overlooked. For example, confronting conflict can settle old problems, clarify viewpoints, and increase understanding.

All conflict is not bad. Conflict can increase the tension in a program management office. It is important to identify the cause of conflict, and to redefine it if necessary. When handled in a constructive manner, problem identification, decision making, and problem solving can be enhanced. This is because the conditions are established for new ideas to surface, be explored, and be acted upon to solve problems. Thus the program manager should seek to create an environment in which team members work toward the success of their program through mutual achievements of the interim objectives.

Studies of creativity and innovation suggest that there be enough divergence and conflict among individuals to produce a stimulus that shakes them out of their personal comfort zones and forces them to take new viewpoints and make new assumptions.

Another concern in the conflict situation is whom the person is working for. For example, after the recent Pershing explosion in Germany, the program manager was dealing with technical questions related to the cause of the explosion and procedures to eliminate future problems. In other forums, which often overlapped, there were the political questions of nuclear safety and tactical implications to the NATO alliance. In resolving this conflict, there were significant issues as to who was working for whom.

In some situations conflict can develop over several identifiable stages. Schmidt and Tannenbaum[3] described five stages: anticipation, conscious but unexpressed differences, discussion, open dispute, and open conflict. They suggest that the sooner a manager becomes involved, the greater the potential he or she has to influence the final resolution of the differences.

To increase your competence, you need to develop the capability to assess the parties in conflict without judgment and to get them to discuss the issues, rather than competing for the win. Prior to facing the people, you will have thought about personalities and developed a tentative mental script of how to handle them. You learn to depersonalize individual differences whenever feasible; to remain objective; and to demonstrate interpersonal competence by hearing people out, asking questions to draw out their position, and focusing energy on the points of agreement and potential areas for compromise. When feasible, you allow for wide participation in goal setting and decision making, and you delegate if appropriate.

There are some people who have a greater tendency to rub people the wrong way. While it may be very difficult to change these people, awareness of their behavior's impact on others and the program office can improve your chances of assuring that the behaviors are not excessively disruptive. Likewise there are others who use power to their own advantage rather than to the best advantage for the program. You should know who these people are and make the decisions that minimize power vacuums which enable this abuse to occur.

The longer a conflict simmers, the greater the potential for increased organizational impact due to distrust, parochialism, or lack of decision making. Specific impacts might be noted such as:

- Lengthy decision-making processes
- Increased documentation needs
- Avoidance of confrontations if perceived to have an impact on careers
- Formation of coalitions or cliques
- Increased dysfunctional competition
- Increased psychological distance
- Decrease or destruction of sensitivity
- Distortions of reality
- Diversion of energy from the main issues
- People dropping out

If the conflict situation stimulates aggressive feelings or behavior, there will be increased misperception and misunderstanding. A we-they attitude will develop. People will seek to place blame, seek to undermine other people's credibility, or be emotionally ready to explode. Faults or mistakes will be denied, and conversation will not be as candid.

One factor must be kept in mind. As the stakes to the organization and program are higher, and the level of conflict is at a higher stage, more time will probably be required. Since time for most program managers is at a premium, the initial reaction may be to make a quick decision and move on. As has been described, this can be the wrong approach. To smooth out differences effectively takes time and thought. Several rounds of discussion, fact finding, negotiation, mediation, arbitration, or decision making may be required to satisfy all people involved and to eliminate the conflict permanently, rather than sweeping it under the rug. The end goal is channeling conflict constructively to enhance the program.

In a conflict situation, prepare a mental or written script prior to facing the parties. Define your outcomes for expected situations. Determine if the situation impacts mission, objectives, organization structure, authority, or communication systems. Know what your fallback might be. This means that you need to establish priorities and look at the various risks. Listen for understanding—save evaluation for later.

Develop procedures which facilitate problem solving, such as brainstorming, nominal group technique, or use of an electronic decision support system. The key is to select a process which will free people from having to defend their ideas and enable them to buy into and fully support the result. The proper group process avoids the one-on-one management style, which reduces information flow to a trickle, creates rivalries, and increases game playing.

Learn to recognize and accept that feelings are a legitimate part of organizational life. Even you, the program manager, may become emotionally involved in the situation and should be aware at what level—such as caring for the people involved, fear for the program, and so on.

Unresolved conflict can play havoc on the human psyche. The way a person perceives the conflict situation determines his or her emotional response. No-win situations are probably the worst. Stress from the conflict can result in negative self-talk, which acts on brain chemicals that signal a need to fight or flee. The person may respond with behavior patterns which do not fit the organization, such as misplaced anger or rage, or mood swings; or with physical symptoms, such as high blood pressure, irregular heart beat, poor digestion, or tense muscles. Feedback on this behavior then triggers another round of brain response, which may lead to even worse behavior. To deal with this spiral, the individual, or a helper, needs to be able to suspend judgment or remove him- or herself from the immediacy of the situation. This can be done through self-talk from any one of three views:

- Acceptance of what is and being willing to move on in life
- Resistance to what is and being willing to fight until victory is achieved
- Tolerance of what is and being willing to drift with the current

Another way to control one's emotions is by engaging in aerobic activity or using relaxation techniques. Two of the easiest are deep breathing and mental vacationing. For deep breathing take seven to ten slow, deep breaths and hold them for four to five seconds before slowly exhaling. You know you are doing this exercise correctly when you feel your chest expanding as you inhale. For the vacation, close your eyes for a minute and picture your favorite vacation location. In your mind look at the panorama of sight, listen to the sounds, smell the air, and feel the movement of your body.

Should the conflict have escalated to aggressive acts, it is essential that you separate the two parties using extreme caution or by calling in the proper authorities, and then engage in "shuttle" negotiations to obtain information and commitment which will enable you to take action to remove the conditions which led to the conflict.

Classical Methods

There are several classical methods for dealing with conflict. They are avoiding, smoothing, forcing, compromising, and collaborating.[4] An effective conflict manager knows that the initial method used will be based on the specifics of the situation and includes the parties in conflict, the length they have been in conflict, the stakes for both the program and the individuals, the intensity of the conflict, and the time available to make a decision or work with the parties. The seasoned manager also knows that if the first method does not work, a second method should be used as a follow-up. Prior to taking action, the following questions should be answered:

What is my objective?

Who are the conflicting parties?

What is the position of the conflicting parties?

Is the position value-based?

What are the interests of the conflicting parties?

How long have they been in conflict?

What is the payoff for resolving the conflict?

Who can resolve the conflict? (If not me, how will I get them involved?)

What information is available?

Do I need more information?

Can the conflict be depersonalized?

Are there areas of agreement?

What are the areas of disagreement?

What will it take for the conflicting parties to fully support the final agreement?

What is the range of potentially acceptable alternatives?

What are the consequences of delay?

Do I need to be involved? What are the consequences if I am not?

Do I need a third party?[5]

Avoidance

When avoidance is used, the manager withdraws from the conflict issue either intentionally or through denial. This is appropriate when one side needs to cool off, when more facts are needed, when the issue is of minor importance, when it is appropriate not to rock the boat, or when there is more to be gained with the parties resolving their differences. It is essential that the manager monitor the situation to assure that the parties do not see the conflict as a major issue. If they do, then the manager may have to step in and make something happen so that the parties come to closure on their problem.

An avoidance strategy can be a survival instinct. It provides a way out either emotionally or physically. If done to defuse a situation, this strategy may keep the door open for future resolution. However, if physical violence is a possibility, immediate action is called for.

Compromise

Compromising sounds like a fair method for settling disagreements in that it means that the conflicting parties enter into a process which will bring some satisfaction to both parties. It involves giving and getting, and it assumes that there is some good will between the conflicting parties. This can be done through bargaining or mediation. Compromises are often based on perceived areas of strength and weakness. With this method the parties can meet periodically to assess previous agreements or to establish new agreements.

Without further search for other solutions, compromising may be less than optimum. Over time it is possible that parties learn to begin the process with extreme positions, knowing that the end result will be somewhere toward the middle. However, compromising can be valuable when resources are limited and their distribution is noncritical to program success. Compromising lends itself readily to the use of a neutral third party who can assist in issue clarification, finding areas of common ground, and helping the parties move to a win-win position.

Collaboration

Collaboration works best when both sides are interdependent, have mutual self-interests, and do not flaunt an inequality of power. By clarifying the cause of the conflict it is possible to work in a joint problem-solving mode based on the integration of individual expertise, needs, values, attitudes, and perceptions in a mutually satisfactory manner. Both parties work together to surface differences in order to determine how the larger organizational goals can be met. When this process is successful, both are most likely to be satisfied with the results. Perhaps the only disadvantage with this method is that it generally takes the most time, and both parties must be motivated to be open and expend energy.

However, one party may be playing a win-lose game while pretending to be in a collaborative mode. This could be due to perceived self-interest, the thrill of victory, an "I'll get you back someday" attitude, or an inability to spend the time and effort required to achieve consensus through collaboration. When one or more of these factors are recognized, it may be possible to switch out of this competitive mode to a more constructive mode such as forcing, withdrawal, or compromising. This might preclude the open and trusting party from being hurt when the information is used against its position by the party playing a winner-take-all game.

Forcing

Deadlines can impact the manager's choice of methods. If a deadline is near, he or she is more likely to resort to forcing a decision through the use of power. As a program manager, forcing or using a power play should remain a last resort because it is seen by the subordinate as *you win* and *I lose.* This method can lead to information distortion, hostile feelings, subversive or diversive activities, or displaced energy channeled away from the program's goals.

Forcing works best when authority is exercised so that both parties feel it has been used with concern for their own welfare, not the self-serving needs of the manager.

Finally it is important that the manager have an understanding of the terms "conflict management" and "conflict resolution." As has been described, conflict is an inevitable part of program management. Managers must recognize those situations where immediate action must be taken to resolve the conflict so that it is not detrimental to the program. Ordinary conflicts are managed in relation to other program activities.

To minimize conflict, the key is open communications. This can be a very slow process for a program manager who inherits a program which had poor communications based on a competitive model. Only time can allow an environment to evolve where objectives, problems, failure, and successes are known to all. Control techniques such as the various plans, work breakdown structure, and cost data enable various people and organizations to know their responsibilities and work relationships. Informal activities also enhance the communication flow. This can include sports, cookouts, parties, and award ceremonies.

Conflict should be expected and planned for. To minimize the destructive consequences of conflict, you should confront conflict whenever it will hurt your program. In order for you to handle conflict successfully, you must know where you are going. By knowing what is to be accomplished (both short-range to end the conflict and long-range) as to how the conflict decision relates to the entire program, you can clarify the issues; assess possible outcomes as they relate to the individuals, organization, and program; decide upon the most acceptable solution (preferably as a group); and plan for implementation and evaluation of the results.

USING YOUR POWER

Power is the ability to unilaterally get a person or group to do what you want done. In the program environment it can be based on the formal authority vested in the program manager; on the ability to influence people's behavior through the use of rewards, punishments, and other organizational resources and incentives; or through expert or personal power.

People who become program managers have high power and achievement needs. The outstanding program managers rarely resort to directive management. In their quest to get the job done they know that their persuasion skills help maintain an organizational climate which is conducive for a productive environment, especially when the program is under pressure. They also have informal influence, which enables them to have people behave in a desirable manner without having formal authority over them. As an example they gather ideas in an open and neutral manner and build them into concepts which other people in their network can use.

There is another aspect of power. Depending on the organization structure, the program manager may have leverage based on his or her boss's rank. This often enables him or her to speak softly, knowing that people perceive that he or she has the full support of the senior person.

Effective program managers do not hesitate to initiate conversations involving something unpleasant. They use their power sparingly to surface issues and they recognize that they do not want to impose a solution if they do not have to. By following these seven steps, you can enhance your chances for influencing people so that they understand your position.

1. *Set a time and place for the discussion.* Let the person know what the issue is and that it concerns both of you. If possible, allow the person to select the time and

place for the meeting. The two key factors are that enough time is allowed for preparation, and that the meeting is held in a location where there is privacy.

2. *Open the meeting.* The best way to open the meeting is to candidly state your expectations for the meeting's outcome.
3. *Describe the events and your feelings.* To begin to establish a common ground upon which to reach closure, you need to describe the event as you saw it, or as it was described to you. With open communications it is possible to express also how you feel about the act.
4. *Obtain the other person's view.* Once you have spoken, ask the other person how he or she viewed the event and felt about it. However, remember that this may only be a surface issue. The person's real issue may be either at the conscious or at the unconscious level. It may take several attempts before the time is right to explore below the surface.
5. *Assess the event and gather additional information.* If the opening discussion does not provide a common basis to resolve the situation, discuss other factors of the situation. Make sure that both of you understand what is being said. This is done with good listening techniques and paraphrasing to assure there is understanding. Both the factual and the feeling content should be summarized.
6. *Propose a tentative solution.* Once it appears that all the information is available, propose a solution based on equality or fairness. By asking whether the person can support your proposition, his or her answer will provide insights as to the extent to which the person might be committed to the solution. If the commitment is not there, ask what solution would satisfy him or her. Then work toward resolving the differences. Do not impose a solution. This will only back the person into a corner or create an attitude problem.
7. *Reach agreement.* Agree on the solution and how it will be implemented and evaluated.

MATRIX MANAGEMENT AND THE PROGRAM OFFICE

With the concept of total quality management, the idea of serving the customer has surfaced. Unfortunately the functional personnel in the matrix serve several customers. Matrix team members often have more than one loyalty. In the simplest case, they can have two customers—their functional boss and the one program they have been assigned to work with. High-quality technical people are often needed by many of the programs assigned under one or more program executives. This can cause problems for the program manager who has total responsibility and accountability for a program's success.

How can they learn to please all customers, especially when due dates are nearly impossible? It pays for the program manager to have a thorough understanding of the goals and objectives of the functional units who serve him or her. With this knowledge, the program manager can communicate his or her program's priorities and seek to minimize the potential for conflict as other programs make their demands.

As a first step, objectives must be clearly established by the PEO or the PEO's boss. From here there must be communication that will enable program managers and functional managers to understand how the various program priorities relate to each other. Ideally a human resource planning system will be in place, which looks at each person's skills and how they fit into the program's plans. Without such a tracking system it is difficult to determine where problems will arise and cause conflict.

An alert program manager will be aware of key personnel, and influence them to appreciate the importance of his or her program. In this manner he or she may avoid conflict for his or her program, but may cause conflict for another program. Of course, this can result in conflict for the program manager when called in by the PEO and asked whether he or she can spare that key person for the other program who also needs the person's services.

By understanding more than his or her own program, the program manager can try to minimize problems for other programs by working with the functional manager and determining which other programs may need the same functional help that he or she requires. By looking ahead it may be possible to schedule people so that there is no conflict, or to hire the right people to get the job done for everybody. (Hiring can include the use of consultants if there is not enough work to occupy a new hire.)

Actions Based on the Situation

During the early phases of a program, objectives and outcomes may be defined loosely. If the program manager or personnel in the program office have extensive experience with the program's technology, then it is essential for them to use an integration and coordination approach. This requires that all people and organizations involved with the program work cooperatively, and that most actions be coordinated so that the proper integration is achieved. On the other hand, if the program manager and personnel have limited experience with the program's technology, then there is the need for more formality. This is accomplished through plans and more frequent presentations, reports, and reviews.

Internally, as the team is built, the program manager should select people for their technical knowledge as well as their experience and work records on large projects involving many organizations. The ideal would be to build a team of people who have worked together before. Even at this early stage, the government program manager should take all feasible steps to involve industry executives in the program. By doing this the door is open if problems occur at the program manager level.

By its very nature, DoD program management entails working relationships between military officers, enlisted personnel, and civilians who may or may not be government employees. It is critical that the program office establish and maintain effective military-civilian relationships. With different compensation and employment systems, this can be tricky, not to mention that individual motivations differ significantly. Joint training and development is one means for developing strong relationships. Mixed work groups is another method used to assure that each group understands the other.

As the team grows there is a need to be explicit about group roles and expectations of how individuals and organizations will work together. Due to the nature of the work, the majority of people are likely to have engineering backgrounds. This means that they may look at problem solving as a function of time and money. Given enough of both, physical laws can be used to solve a problem or challenge. However, they may fail to realize that program management requires tradeoffs to be made or numerous alternatives to be considered. It is often the budget which is the program manager's major uncertainty, or the contractual time schedule which drives events.

Individual initiative and flexibility should be encouraged. As people's capabilities become known, and they understand the program, there is greater likelihood that the people can be self-directed. This is when management by exception works very well. If challenged, the functional specialists will do a super job.

Meetings

In program management there is no way to escape meetings. Several factors are essential to ensure that your meetings are effective and efficient, and that they contribute to better interpersonal relationships. They include the following:

Meeting Intervals. If there is no new information, there is no need for the daily or weekly informational meeting. The interval between meeting times should be determined by the need and purpose of the meeting.

Meeting Time. The time of day and day of week are important factors for calling a meeting. Once the purpose and attenders have been determined, decide on the time and day. Meetings before the weekend may result in unnecessary work over the weekend, whereas the Monday morning meeting might result in setting the activities that will take place during the week. If meetings are not timed properly, job dissatisfaction may occur over time, the end result being the loss of an effective and maybe critical team player.

One interviewee from the DSMC's program manager competence study described how he called a crisis meeting on the weekend after receiving notification of a potentially lethal problem in an operational system. He called his team together to determine how the problem could be resolved. Due to the nature of the situation, all participants were highly motivated to solve the problem, and did not worry about a lost weekend and naps at the office.

Meeting Location. The location of the meeting is also important. If a meeting is needed, the purpose and participants will suggest the location for the meeting. If classified information is to be discussed, the range of meeting locations will be limited.

Meeting Minutes. The minutes of the meeting must summarize the meeting accurately and should be distributed to all interested parties. They become a record which can be referred to when there is doubt about a commitment to action.

SUMMARY

Large organizational systems are complex and not understood very well. With this proliferation of power centers in DoD, program success is even more dependent on the support and cooperation of people or organizations outside of the formal authority of the acquisition manager. As the government program manager manages the integration of the various aspects of the program, it is essential to establish effective interpersonal relationships. People are the program manager's resource for achievement. Time and money can solve most technical problems. Effective interpersonal relations are based on trust, which is slow to develop, takes extensive time to nourish, and can be very fragile. Building trust requires an open, supportive, and problem-oriented work environment which values a learning process.

Conflict is a given in acquisition management. An effective manager knows that the initial method used will be based on the specifics of the situation and will include the parties in conflict, the length they have been in conflict, the stakes for both the program and the individuals, the intensity of the conflict, and the time available to make a decision or work with the parties.

REFERENCES

1. "A Competency Model of Program Managers in the DoD Acquisition Process," Defense Systems Management College, Ft. Belvoir, Va., Feb. 1990.
2. I. Briggs Myers and M. H. McCaulley, *A Guide to the Development and Use of the Myers-Briggs Type Indicator,* Consulting Psychologists Press, Palo Alto, Calif., 1985, p. 45.
3. W. H. Schmidt and R. Tannenbaum, "Management of Differences," *Harvard Business Rev.,* Nov.–Dec. 1960.
4. R. R. Blake and J. S. Mouton, *The Managerial Grid,* Gulf Publ., Houston, Tex., 1964.
5. R. E. Walton, *Interpersonal Peacemaking: Confrontations and Third Party Consultation,* Addison-Wesley, Reading, Mass., 1969.

P · A · R · T · 4

MILITARY PROJECT MANAGEMENT FRAMEWORK

The framework of military project management is the basis on which the project environment, human resources, and functions are built. While Part 1 of this handbook addresses the basic functions that occur in the execution of a typical military project, Part 2 introduces the environment under which the military project manager operates, and Part 3 discusses the human element, this part presents a wide range of technical areas that contribute to the success (or failure) of a military project.

Defining the interrelationships in a part of a handbook that contains 15 chapters of separate subject matters presents a challenge. There are two common threads, (1) the diversity of both the subject matters and (2) the backgrounds and experience levels of the authors addressing these subjects.

The subject matters of the chapters cover a spectrum ranging from contract law and the legal aspects of military project management to technology transfusion and transfer, project documentation and reporting, strategic planning, risk management, and managing modernization projects.

The second common thread is the breadth and depth of the experience of the various chapter authors. Experience levels range from NCO to lieutenant general, with appropriate civilian equivalents, and academic backgrounds including doctoral degrees to a few graduates of "the school of hard knocks."

The stage is set for this part in Chap. 27 by Timothy J. Kloppenborg and Margaret G. Cunningham, who discuss total quality management for system projects. TQM is a timely subject and has been adopted by the Department of Defense (DoD) as well as the defense industry as a "way of life" in managing individual projects. In this thoughtfully prepared chapter the authors provide background on quality concepts, then address quality in projects. A seven-step process for diagnosing quality problems is presented, and the chapter concludes with a discussion on statistical process control and continuous improvement.

In Chap. 28 Earl D. Cooper covers the role of the military project manager in strategic planning, starting with a review of a generic model of classical strategic planning, then addressing the DoD process and its similarities with the classical process. The concept of strategic visioneering is addressed, and the applicability of this concept to the military project manager and how it can affect his or her strategic planning is discussed.

An important technology consideration in any project organization is the subject of technology transition and transfer. In Chap. 29 Steven M. Shaker addresses this area by covering the DoD technology base and critical technologies, and describing a unique technology business intelligence system available as a relatively new management tool which can assist decision makers in managing the massive amounts of data relating to technology that can impact their specific program.

In Chap. 30 Tom Bucher cites the example of the Advanced Tactical Fighter (ATF), F-22, and its innovative features used during the demonstration validation phase of the program to cover the subject of new approaches to project strategy development.

An important stakeholder in military project management is the user. Charles B. Cochrane addresses in Chap. 31 the role of the user, identifying the user for each of the military departments as well as the unified and specified commands within DoD. The chapter further covers the cost, schedule, and performance aspects of the military project and how the user interrelates to these important elements. Finally the acquisition phases and milestones are addressed from the perspective of the user.

David A. Yosua examines in Chap. 32 risk management in military acquisition projects. Because of the great amounts of money, the high degree of public visibility, the significant technical challenges, and the number of frequent changes in the military project management environment, the military project manager must deal with an inordinate amount of risk and uncertainty. This chapter provides background on understanding risk and how it should be treated in the planning process. It addresses risk identification and provides tools and techniques for accomplishing risk analysis. The chapter also discusses the important aspect of responding to the risks and implementing the risk management process.

The next four chapters address the contracting and legal aspects of the military project management framework. A thorough treatise on the subject of government contract law is provided by John A. Ciucci in Chap. 33. After presenting background on basic contract principles, including such topics as the legal elements of the contract, contract types, methods, and contract modification, the author discusses the timely subjects of ethics in contracting, fraud, and environmental issues.

In Chap. 34 Carl R. Templin covers subcontracting management and focuses on the defense subcontracting environment, stressing the impact on project cost, schedule, and performance. The controls placed by the government on subcontracting are unique to military project management and are discussed in some detail. The author concludes with a discussion of the management of DoD-supplier relationships.

Completion is the fate of all projects. At real issue are the circumstances at which project completion, whether planned or forced, takes place. In Chap. 35 Michael E. Heberling describes project termination and addresses several issues related to this subject. The major factors and causes that can play a role in a decision to terminate a program—political, cost, technical, and obsolescence—are addressed. Project termination from a contractual standpoint is also covered.

In Chap. 36 James H. Dobbins covers the subject of inspections and audits of DoD programs. The chapter explains the difference between inspections and audits for both hardware and software development activities and covers the separate procedures and controls for each. The governing documentation for the implementation and conduct of formal government inspections and audits is reviewed.

Bud Baker covers current perspectives on project documentation in Chap. 37. After describing the many categories of project documentation, the author addresses the current and emerging issues, including an interesting discussion on documentation in "black world" acquisition programs. The conclusions of this chapter are worth noting from the perspective of "are we really making progress?" from a documentation standpoint, as a result of decades of acquisition streamlining.

Coeditor David I. Cleland looks at concurrent engineering in Chap. 38 and discusses the background and impact of simultaneous rather than sequential engineering as a relatively new way of doing business in the management of military projects. The author brings an interesting perspective to this subject, having studied, at some depth, this subject in the non-DoD sector.

Managing information resource management projects is discussed by John F. Phillips in Chap. 39. The author cites the experience and success stories of managing the many projects under the aegis of the Logistics Management Systems Center (LMSC) of the (then) Headquarters Air Force Logistics Command (AFLC). Shortly after submittal of this chapter, the author was named commander of the newly established DoD Joint Logistics Center, so many of the concepts discussed here should have even wider DoD application.

Military projects age, project requirements change, and limited funding is available to initiate new projects. Modifying existing projects presents a major challenge. In Chap. 40 Jerry D. Schmidt addresses the subject of managing modernization projects: the process of upgrading, enhancing, or refurbishing military equipment. The adaptation of many project management techniques for new programs to modification programs is reviewed in this chapter.

In Chap. 41 C. Michael Farr describes international project management and asks the provocative question: do other countries do it better?

In conclusion, the reader should take the opportunity to explore each of the subjects treated in this part on the military project management framework. By the vehicle of the printed page you have the unique opportunity to "pick the brains" of some real giants in the business of military project management framework.

CHAPTER 27
TOTAL QUALITY MANAGEMENT FOR SYSTEMS PROJECTS

Timothy J. Kloppenborg and Margaret G. Cunningham

Timothy J. Kloppenborg received the Ph.D. degree in operations management from the University of Cincinnati. He is an Assistant Professor of Decision Sciences at Xavier University, Cincinnati, where he spearheaded development of an M.B.A. concentration in Quality Improvement. Dr. Kloppenborg is a major in the USAF Reserve, where he serves as a quality assurance officer. He is a Certified Project Management Professional and has published over a dozen articles and papers on quality and project management.

Margaret G. Cunningham received the Ph.D. degree in operations management from the University of Cincinnati. She is chair of the Information and Decision Sciences Department at Xavier University in Cincinnati. Dr. Cunningham teaches quality management in Xavier's M.B.A. program and has been involved in developing the M.B.A. concentration in Quality Improvement. Her current research interests include the practice of quality improvement in academia and writing management science cases.

INTRODUCTION

The crucial importance of quality in military systems projects is obvious and can hardly be overstated. Because quality problems can cause ineffective systems, unnecessary cost overruns, and dangerous risks to individuals and even to national security, it is vital to solve existing quality problems and to prevent future quality problems whenever possible. Beyond solving and preventing overt problems, the continuous improvement of various processes within a system can contribute to the development of systems with superior quality and effectiveness.

The next section of this chapter provides the military project manager with an overview of the concepts and principles of modern approaches to quality, including total quality management and the relationship between cost and quality. This is followed by a section on quality considerations that are specific to projects. The remainder of the chapter is devoted to techniques for maintaining and improving quality.

This includes sections on the seven simple tools for diagnosing quality problems, statistical process control, and an actual example that illustrates both. These tools have been proven to be quite effective and are becoming widely used by businesses in virtually every industry as well as in various branches of the military.

IMPORTANT QUALITY CONCEPTS

What Is Quality?

Although most people think that they intuitively know quality when they see it, developing a more formal operational definition is not a trivial task. There are several approaches to defining quality that are widely known and used.

Fitness for Use. One approach to defining quality is in terms of "fitness for use." According to this definition, a product has high quality if it is fit for use by the customer (actual user) for its intended purpose over a specified period of time. This definition requires that we have identified the actual user in the field or fleet, the intended use of the product, and the appropriate useful life of the product. Furthermore, in order to make the definition operational, we must develop some specified means of determining whether or not the product is fit for use. Therefore while this definition is conceptually useful, it is often difficult to apply in practice because many of our defense systems are designed by one command, purchased by another, the contractually defined quality is interpreted by a third organization, and they are used by troops in diverse operating environments which may not be identifiable at the design stage.

Quality of Conformance. Another approach to defining quality is to concentrate on conformance to product specifications developed for the measurable characteristics of the product. The product is considered to be of high quality if the specifications are met and of poor quality if they are not met. Of course this is only true if the specifications capture all of the desirable characteristics of the product adequately. This is the approach that has traditionally been used by the military and many other organizations for items supplied by outside vendors. This approach relies on extensive inspection to verify quality. It also assumes that meeting the specifications really signifies a useful product. However, with the complexity of our defense systems and rapidly changing threats this seems like an overly simplistic assumption.

Design Quality. It does little good for a product to conform to specifications if these specifications are based on a faulty design in the first place. Therefore quality of conformance by itself is inadequate. Attention must first be given to the quality of the product design. Since this may require a great deal of costly research for which there will be no direct or immediate payback, the temptation often exists to cut corners at the design stage and go as quickly as possible into production. This is a shortsighted strategy doomed to frequent failure in the longer term. Although it is often hard to justify beforehand, spending time and money on improving quality at the design stage is not only worthwhile but often essential.

What the Customer Wants. In the last few years, many consumer-oriented companies have been focusing their quality improvement efforts on market research activities. These efforts are intended to obtain customer input on the design of their products and services and to determine which product characteristics are most important to customer satisfaction. While the market research approach is less applicable to military projects, the general concept of obtaining user input and striving for user satisfaction is just as important when dealing primarily with internal customers. Frequent contact and cross assignments between operating and designing commands are helpful in determining field (customer) needs.

What Is Total Quality Management?

Only in the last few years has it become widely recognized by many major U.S. corporations that extensive use of acceptance sampling and statistical process control alone is not enough to achieve quality. In addition, the firm must be managed in such a way that quality management is facilitated. The term total quality management (TQM) has been coined to describe this approach to quality. Key features of TQM include:

Top management commitment
Input from customer (or internal user)
Involvement of workers at every level
Emphasis on design quality and process improvement
Decisions based on information instead of opinion
Continuous improvement through reducing variability

Current quality efforts in many firms deal with the proliferation of quality improvement activities throughout the firm. Although TQM does not necessitate an association with any particular quality consultant, it is compatible with what is currently being taught by the most prominent leaders in the field. World-reknowned quality leaders W. Edwards Deming and Joseph Juran, both in their nineties, now deal primarily with managerial issues in their public seminars and their work with individual firms. Although a thorough discussion of the contributions of prominent individuals is beyond the scope of this chapter, every person involved in quality improvement should have some familiarity with Deming's 14 points for management and Juran's seven-step breakthrough sequence.

Deming's 14 Points. W. Edwards Deming believes that many U.S. firms are in crisis and unable to compete with foreign competitors because of poor quality. He has developed 14 points for management which summarize his approach to quality management. These 14 points are discussed extensively in his book *Out of the Crisis*[1] (pp. 23–24):

1. Create constancy of purpose toward improvement of product and service, with an aim to become competitive and to stay in business, and to provide jobs.
2. Adopt the new philosophy. We are in a new economic age. Western management must awaken to the challenge, must learn their responsibilities, and take on leadership for change.
3. Cease dependence on inspection to achieve quality. Eliminate the need for inspection on a mass basis by building quality into the product in the first place.
4. End the practice of awarding business on the basis of price tag. Instead, minimize total cost. Move toward a single supplier for any one item, on a long-term relationship of loyalty and trust.
5. Improve constantly and forever the system of production and service, to improve quality and productivity, and thus constantly decrease costs.

6. Institute training on the job.
7. Institute leadership. The aim of supervision should be to help people and machines and gadgets to do a better job. Supervision of management is in need of overhaul, as well as supervision of production workers.
8. Drive out fear, so that everyone may work effectively for the company.
9. Break down barriers between departments. People in research, design, sales, and production must work as a team, to foresee problems of production and in use that may be encountered with the product or service.
10. Eliminate slogans, exhortations, and targets for the work force asking for zero defects and new levels of productivity. Such exhortations only create adversarial relationships, as the bulk of the causes of low quality and low productivity belong to the system and thus lie beyond the power of the work force.
11. a. Eliminate work standards (quotas) on the factory floor. Substitute leadership.
    b. Eliminate management by objective. Eliminate management by numbers, numerical goals. Substitute leadership.
12. a. Remove barriers that rob the hourly worker of his right to pride of workmanship. The responsibility of supervisors must be changed from sheer numbers to quality.
    b. Remove barriers that rob people in management and in engineering of their right to pride of workmanship. This means, abolishment of the annual or merit rating and of management by objectives.
13. Institute a vigorous program of education and self-improvement.
14. Put everybody in the company to work to accomplish the transformation. The transformation is everybody's job.

These 14 points form the basis for Deming's seminars and his consulting work with individual companies. Deming believes that all 14 points are related and should be adopted and implemented together, along with statistical process control, in order to be successful.

Juran's Seven Steps. Joseph Juran has long been a proponent of continuous process improvement. One of Juran's major contributions to the quality field is the development of a seven-step procedure for attacking chronic quality problems.[2] This procedure is known as the breakthrough sequence and is general enough to be applicable to almost any kind of process. The seven steps are as follows (pp. 100–101):

1. *Convince others that a breakthrough is needed.* Convince those responsible that a change in quality level is desirable and feasible.
2. *Identify the vital few projects.* Determine which quality problem areas are most important.
3. *Organize for breakthrough in knowledge.* Define the organizational mechanisms for obtaining missing knowledge.
4. *Conduct the analysis.* Collect and analyze the facts that are required and recommend the action needed.
5. *Determine the effect of proposed changes* on the people involved and find ways to overcome the resistance to change.
6. *Take action* to institute the changes.
7. *Institute controls* to hold the new level.

This series of steps can be used to improve virtually any kind of process. It generally involves the use of the seven simple tools presented later in the chapter.

Commitment to Quality

In order for quality improvement efforts to be effective, an organization needs to have a strong commitment to quality. This commitment needs to begin with the people at the top levels of the organization and must extend to all levels. It must also be expanded to include the organization's suppliers. In any large organization, this is difficult and cannot be expected to occur overnight, because it involves a major change in the culture of the organization. Experience has shown that once a commitment to quality is expressed at the top levels, there is a tendency for people at the middle levels to be suspicious that quality improvement will be a passing fad and therefore to take a wait-and-see attitude as to whether the commitment is permanent. Many military departments have been concentrating on the TQM approach long enough now that employees and suppliers recognize that TQM is here to stay.

Quality and Cost

It is especially difficult to establish an organizationwide commitment to quality when the organization is simultaneously engaged in short-term (and sometimes shortsighted) cost reduction efforts, because these two goals often come into conflict with one another. This is an especially difficult problem for military organizations since the military has come under extensive public criticism in recent years for what is perceived by some as wasteful spending. Another complicating factor in the military and aerospace industries is the boom and bust cycle. Constantly hiring and laying off workers to match fluctuating requirements adds greatly to the difficultly of achieving better quality because employees fearing termination are not motivated to be team players and because process knowledge leaves with experienced employees.

It is important to remember, and to constantly remind others, that in the long run quality improvement efforts will result in lower costs because the costs of poor quality will be reduced or eliminated. Ineffective systems, excessive scrap and rework, dissatisfied users, and other consequences of poor quality are very expensive.

The Cost of External Failures. The costs associated with the failure of a product to perform its intended function once the product is in the hands of the user are known as external failure costs. The worst possible scenario in the case of weapon systems would be for the system to malfunction during use. The consequences of such a malfunction could include impairment of military effectiveness and possible loss of life in addition to extremely high monetary costs. Obviously it is even more imperative to avoid external failures and the associated monetary and nonmonetary costs in military systems than it would be with most other kinds of products.

The Cost of Internal Failures. The costs associated with rectifying defects in a product that are discovered during production or through inspection and testing before the delivery of the product are known as internal failure costs. These costs include the cost of scrapping and replacing defective items that cannot be repaired and reworking items that can be repaired. Many defense contractors have elaborate material review board systems to decide the fate of defective items. Internal failures as well as external failures can be the result of either a faulty design or failure of the process to produce an item that meets the design specifications. Although not as serious as external failures, internal failures can be very expensive and can also result in nonmonetary consequences such as delayed delivery of a crucial weapon system. Producing better quality to begin with can prevent most failures and the associated costs.

The Cost of Inspection and Testing. Costs associated with testing and inspection in order to assess conformance to specifications are known as appraisal costs. These costs include the cost of inspection labor, testing equipment, and possibly the cost of replacing parts destroyed during destructive testing. For many years, military organizations concentrated quality assurance efforts on inspection and testing. In fact, the extensive tables of inspection plans developed by the military became widely used in other industries. It is now widely recognized that inspection alone does not improve product quality, but merely converts external failures into internal failures. As organizations, including the military, learn how to concentrate quality improvement efforts on product design and process planning and improvement, inspection and testing are becoming a smaller part of much more comprehensive quality improvement efforts. However, this does not mean that inspection and testing are unnecessary or will become obsolete. Final inspection as well as preliminary inspections for the purpose of process control will always be necessary in many instances, especially for such critical products as weapon systems.

The Cost of Quality Improvement. To improve quality it is necessary to engage in activities directed toward diagnosing and correcting existing process problems and preventing future quality problems. Costs associated with quality planning and quality improvement activities are known as prevention costs. These costs include the cost of establishing process control systems and engaging in process improvement activities and may also include the cost of certain kinds of research and development activities, market research, and employee training. Although these activities may be expensive, it has been the experience of many organizations that if they are effective in improving quality, these costs are more than offset by reductions in appraisal costs and often dramatic reductions in failure costs. Unfortunately it is usually impossible to use traditional cost-benefit analysis to justify quality improvement efforts since there is no way of knowing what failure costs would be incurred if the prevention efforts are not undertaken. Also, prevention costs are incurred immediately and savings from reduced failures occur over a period of time. Budgetary constraints often make it difficult to spend money now in exchange for possibly larger savings later. This is one reason why a strong commitment to quality improvement at the top levels of the organization is necessary for success.

QUALITY IN PROJECTS

Planning and Organizing for Quality

A quality product or service begins with development of a design that will meet the needs of the customer. This important portion of the quality effort establishes the design quality for the product. Next a process must be developed that can consistently meet the design specification. Finally a process control system must be established that will monitor the process and provide an early warning if problems begin to develop once the process is in operation. The process and its control system provide quality of conformance to the specifications established during the design phase.

In the case of a project, quality management must extend through the entire project life cycle. The firm's quality program needs to be flexible enough to be tailored to each project. The project manager must take overall responsibility for quality on the project, although obviously some of the work will need to be delegated. The role of the project manager is to ensure that the quality function is well organized and assigned appropriate project staff. Each person performing quality functions should have both the ability to identify quality problems and recommended solutions and the authority to implement those solutions, including the authority to temporarily stop further processing if necessary. Ideally all project team members should be involved in the quality effort in some way. One way to involve workers is through the use of quality improvement teams.

Quality Improvement Teams

Some organizations, including military organizations, have established special groups with the purpose of analyzing various processes within the organization and finding causes for quality problems or opportunities for process improvement that may prevent future problems. Such groups go by many names such as process action teams or employee involvement groups, and usually consist of production workers as well as employees from a wide variety of other jobs within the organization. Among the most commonly used problem-finding techniques are brainstorming and cause-and-effect diagrams, although the specific methods used can vary widely.

Usually the people working on a project are already organized as a team or perhaps as several interacting teams. This is an advantage in that team members are already familiar with the team form of organization, and the principles of teamwork. It is hoped that workers have already developed the ability to work in a constructive manner with other team members. Existing teams that are accustomed to working together can also serve as quality improvement teams, or special teams can be formed to undertake specific quality improvement tasks as the need arises.

Achieving Quality

Traditionally most military organizations have relied heavily on the use of inspection and acceptance sampling to achieve quality. While inspection can verify conformance quality or discover lack of quality, inspection alone does nothing to contribute to the improvement of quality except to hold the supplier accountable. It has thus become widely recognized in recent years that a much broader approach to the achievement of quality must be taken. If a quality problem is observed, its cause must be identified and removed to prevent recurrence of the problem. It has been found that if processes are monitored on an ongoing basis, many potential problems can be anticipated and thus prevented. Obviously this is a more desirable approach since it results in cost savings and time savings as well as better quality in the long run.

DIAGNOSING QUALITY PROBLEMS—SEVEN SIMPLE TOOLS

A number of tools and techniques have been developed for the identification and diagnosis of quality problems. It has been found through experience with a wide variety of processes in many industries that among the most effective tools are a set of simple graphic techniques, known collectively as the seven simple tools. These include:

1. Flowcharts
2. Check sheets

3. Pareto charts
4. Histograms
5. Fishbone diagrams
6. Scatter plots
7. Run charts

Flowcharts

A flowchart is a tool often used by systems analysts to provide a visual representation of a process or system. A flowchart helps clarify and communicate to others the steps in a process and how they are related. This is especially useful if the process is complex or includes more than one possible series of steps. Although flowcharts are primarily a documentation tool, they can also be used to help determine at what point in the process or system a problem may be occurring. These points become logical points to gather data in order to pinpoint problems. Many quality improvement efforts in the military start with constructing a flowchart to help understand the production process.

Check Sheets

A check sheet is used to tally the frequency and magnitude of various kinds of problems as they occur. This method is often used to keep track of the different kinds of defects an inspector may find in an item. This procedure provides more information than an undifferentiated count of defects. Check sheets can take many forms and can be used for virtually any type of process.

Pareto Charts

Often the results of a check sheet tally are displayed in a special kind of bar chart known as a pareto chart or pareto diagram. Each bar on a pareto diagram represents a particular category of defect or problem. This makes it easy to see which problems are occurring most frequently. Pareto charts can also be used to identify which shift, machine, supplier, and so on, is responsible for the most problems.

Histograms

The histogram is a graphic representation of the frequency distribution of a variable. It can provide information about the patterns of variation displayed by the variable. Understanding the pattern of variation is often an effective starting point in reducing that variation.

Fishbone Diagrams

The fishbone diagram or cause-and-effect diagram is a technique developed by Kaoru Ishikawa for identifying the possible causes of a problem.[3] The problem being considered is drawn as a box (the head of the fish) with a central horizontal line (the spine of the fish), and the possible causes are shown as slanting lines (the large bones of the fish) with arrows pointing toward the central line. Factors that may contribute to a particular cause are shown on additional lines (the small bones of the fish) pointing toward the appropriate cause. It is important to identify as many potential causes and contributing factors as possible, so this technique works best when used by a small group in an atmosphere similar to a brainstorming session.

Scatter Plots

Once a quality improvement team has some ideas about what may be causing a problem, a scatter plot can be used to take a quick look at possible cause-and-effect relationships among variables. If there does appear to be a relationship between certain variables, additional statistical analysis can be undertaken to confirm this and gain more specific information. A scatter diagram is also useful in identifying outliers, or nontypical data points. Investigation of the circumstances under which outliers occur often leads to important insights about the process that has generated the data.

Run Charts

Run charts show the value of a variable plotted over time. This can be useful in pinpointing when process problems are occurring, which sometimes provides a clue to the cause of the problem. A run chart can also help the user spot trends or patterns that are occurring over time.

Used together, these seven simple tools can help the project team diagnose quality problems and thus provide guidance for possible design or process improvements. Once existing quality problems have been overcome and a stable process has been achieved, statistical process control can be used to monitor the process on an ongoing basis.

STATISTICAL PROCESS CONTROL

Statistical process control is one of the most common quantitative approaches used to support a company's efforts toward maintaining quality of conformance. The use of statistical process control has become widespread in many industries over the past few decades. Various kinds of control charts are used to monitor production processes, and they are increasingly being used to monitor service processes as well. It should be emphasized that in practice these techniques play only a supporting role in producing a quality product or providing a quality service, since a good design and a process capable of meeting its specifications must already be in place before a process control system can be of value.

The first step in preparing to use statistical process control is to determine which specific aspects of a process need to be dealt with in this manner. In some cases the process is best monitored by measuring physical characteristics of the product at a certain point in the production process, for example, the diameter of a hole punched in a metal bracket. In other cases it may be more important to monitor a characteristic of the process itself, such as the temperature used in heat treating, which has an impact on the quality of the product or service. The selection of the particular attributes or variables to be measured and controlled calls for a thorough knowledge of the production process or service delivery system, especially the impact of each portion of the process on the key quality characteristics of the final product or service.

Next it is necessary to determine how these characteristics of the product or process are to be measured. Some characteristics, such as length, weight, and temperature, are relatively easy to measure as long as properly calibrated equipment is available. Other characteristics, such as the color of a fabric or the cleanliness of a restroom, may be important but largely subjective and therefore may require considerable effort in order to devise an adequate means of measurement. It is also necessary to make at least a preliminary decision on how large a sample size to use, how frequently to sample, and how a sample is to be chosen. Again, these decisions require a thorough knowledge of the particular process in question.

Once the data collection procedures have been established, preliminary data can be collected from the process in order to develop a trial control chart. In developing a control chart, it is important to use data from a period of time when the process is operating in statistical control. This means that only common causes of variation in the process are present. Common causes are those that are an integral part of the process and cannot be removed without changing the process itself. If special causes of variation, also known as assignable causes, are present, then the process is not operating under statistical control. This would indicate that some factor that is not always present is affecting the variability of the process. Unfortunately it is not always easy to distinguish common causes from special causes or to identify what particular special causes may be in operation at a specific point in time. In fact this is one of the purposes for which control charts are used. Until it can be established statistically that the process is indeed operating in control, the data collected and the trial control limits calculated from them must be considered preliminary. Nevertheless, chances for success are best if the preliminary data are collected during a period of time when to the best of the manager's knowledge no special causes are in operation.

Before actually developing a trial control chart, it is also necessary to decide which kind of control chart would be most appropriate for the situation at hand. This will be determined in part by whether the manager is dealing with attributes data or variables data. If the available data are attributes data, the manager will ordinarily use either a p chart or an np chart. If the data involve variables, the manager will probably need to use an $\bar{X}$ chart in conjunction with either an R chart or an s chart. These are the most common kinds of control charts; other kinds of control charts are available for special situations.

Attributes Data

Sometimes the data available to the manager are in the form of attributes data. This is the case when the product or process characteristic under consideration cannot be measured but can only be classified. Usually items are classified by whether or not they conform to established product or process specifications. Thus each item is identified as either a conforming item, which meets the appropriate specifications, or a nonconforming or defective item, which fails to meet the specifications.

Constructing a* p *Chart. If the data that have been collected are attributes data, we can use these data to construct a p chart, also known as a fraction defective or fraction nonconforming chart. Alternately, we could choose to construct an np chart for the number nonconforming from the same data.

Assuming to begin with that we prefer to construct a p chart, the first step is to express the number of defective items in each sample as a fraction defective. The next step is to calculate $\bar{p}$, the average fraction defective:

$$\bar{p} = \frac{p_1 + p_2 + \cdots + p_M}{M} \tag{27.1}$$

where M is the number of samples.

The average fraction defective is used as the centerline on the p chart. Ordinarily, the control limits are set three standard deviations above and below the centerline. The standard deviation can be estimated by:

$$s_{\bar{p}} = \sqrt{\frac{\bar{p}(1-\bar{p})}{n}} \tag{27.2}$$

where n is the number in each sample. The upper and lower control limits can then be calculated as follows:

$$\text{UCL}_p = \bar{p} + 3s_{\bar{p}} \tag{27.3}$$

$$\text{LCL}_p = \bar{p} - 3s_{\bar{p}} \tag{27.4}$$

When the lower control limit is calculated, a value less than zero is often obtained. Since a fraction defective less than zero is impossible, zero is used instead in constructing the control chart.

Constructing an* np *Chart. Instead of a p chart, we may prefer to construct an np chart, which shows the number of nonconforming items in each sample instead of the fraction nonconforming. The np chart requires equal sample sizes and therefore cannot always be used. But when it is appropriate, it is often preferable because it provides equivalent information in an easier to understand form.

The first step is to calculate $n\bar{p}$, the average number of nonconforming items per sample:

$$n\bar{p} = \frac{n_1 + n_2 + \cdots + n_M}{M} \tag{27.5}$$

This is the average number of nonconforming items per sample, and is used as the centerline on the np chart.

The standard deviation can be estimated using:

$$s_{n\bar{p}} = \sqrt{n\bar{p}(1-\bar{p})} \tag{27.6}$$

Again setting the control limits three standard deviations from the centerline, the upper and lower control limits can be calculated as follows:

$$\text{UCL}_{np} = n\bar{p} + 3s_{n\bar{p}} \tag{27.7}$$

$$\text{LCL}_{np} = n\bar{p} - 3s_{n\bar{p}} \tag{27.8}$$

Again, if the lower control limit is calculated to be below zero, zero is used instead.

Variables Data

Often the data available to the manager will be in the form of variables data rather than attributes data. When a product or process characteristic is measured on any kind of a continuous scale, the resulting data are variables data. If there is a choice, it is preferable to collect data in this form since they provide the manager with more information.

Constructing* $\overline{\mathbf{X}}$ *and* R *Charts. When variables data are used, it is necessary to develop control charts for both the mean of the process and also some measure of the variability of the process. The most common approach is to use an $\overline{X}$ chart, which shows the centering of the process as measured by the sample mean, along with an R chart, which shows the variability in the process as measured by the sample range. Another alternative would be to replace the R chart with an s chart, which shows variability in the process as measured by the sample standard deviation.

The first step in developing $\overline{X}$ and R charts is to compute the sample mean $\overline{X}$ and the sample range R for each sample. The sample mean is computed as follows:

$$\overline{X} = \frac{X_1 + X_2 + \cdots + X_n}{n} \tag{27.9}$$

where n is the sample size. The sample range is obtained by subtracting the lowest value in the sample from the highest value in the sample:

$$R = \text{highest value} - \text{lowest value} \tag{27.10}$$

The next step is to calculate $\overline{\overline{X}}$, the average mean or grand mean, which will be used as the centerline for the $\overline{X}$ chart, and $\overline{R}$, the average range, which will be used as the centerline on the R chart. To find $\overline{\overline{X}}$, simply take the average of the sample means:

$$\overline{\overline{X}} = \frac{\overline{X}_1 + \overline{X}_2 + \cdots + \overline{X}_N}{N} \tag{27.11}$$

where N is the number of samples. Similarly, to find $\overline{R}$, take the average of the sample ranges:

$$\overline{R} = \frac{R_1 + R_2 + \cdots + R_N}{N} \tag{27.12}$$

Finally, the control limits for both the $\overline{X}$ chart and the R chart can be computed. Ordinarily control limits that are three standard deviations from the centerline are used. Factors can be obtained from widely available tables to calculate such control limits from $\overline{X}$ and $\overline{R}$. These factors are based on the assumption that the sample means are normally distributed, and are tabled according to sample size. Using these factors, which can be found in Table 27.1, upper and lower control limits can be obtained for the R chart as follows:

$$\text{UCL}_R = D_4\overline{R} \tag{27.13}$$

$$\text{LCL}_R = D_3\overline{R} \tag{27.14}$$

TABLE 27.1 Control Chart Factors

	Factor			
n	A_2	D_2	D_3	D_4
2	1.880	3.686	0	3.267
3	1.023	4.358	0	2.574
4	0.729	4.698	0	2.282
5	0.577	4.918	0	2.114
6	0.483	5.078	0	2.004
7	0.419	5.204	0.076	1.924
8	0.373	5.306	0.136	1.864
9	0.337	5.393	0.184	1.816

where D_3 and D_4 are table values obtained from Table 27.1. Similarly, upper and lower control limits for the $\overline{X}$ chart can be obtained as follows:

$$\text{UCL}_{\overline{X}} = \overline{\overline{X}} + A_2 R \tag{27.15}$$

$$\text{LCL}_{\overline{X}} = \overline{\overline{X}} - A_2 R \tag{27.16}$$

where A_2 is a table value obtained from Table 27.1.

Interpretation of Control Charts

Once a control chart has been developed, it is used on an ongoing basis to monitor the process. There are several danger signals that may indicate that a formerly stable process may no longer be in statistical control. The most obvious of these is a point outside the control limits. While this happens occasionally by chance, any point outside the control limits should be investigated to see whether there is reason to attribute it to a special cause. A special cause can be any factor that is not an intrinsic part of the process.

In addition to a point outside the control limits, any other apparently nonrandom pattern should serve as a warning sign that the process may be out of control. Operators must be alert for such patterns. In most organizations using control charts, management provides operators with a specific set of guidelines. There are a number of similar standard lists. One commonly used list, adapted from *The Memory Jogger,*[4] (p. 55) is as follows:

- 2 points, out of 3 successive points, more than 2 standard deviations from the centerline in the same direction (may indicate that the process average has shifted)
- 4 points, out of 5 successive points, more than 1 standard deviation from the centerline in the same direction (may indicate that the process average has shifted)
- 9 successive points on one side of the centerline (may indicate that the process average has shifted)
- 6 consecutive points increasing or decreasing (may indicate an upward or downward trend)
- 14 points in a row alternating up and down (may indicate cyclical variation)

- 15 points in a row within one standard deviation of the centerline (may indicate that the variation in the process has been reduced)

It is possible but *extremely unlikely* for any of the foregoing patterns to occur strictly by chance. If any of these patterns appear to be occurring, the process should be investigated for possible special causes. This is usually best accomplished through the use of the seven simple tools presented earlier. Once the problem has been found and corrected and the process returned to statistical control, the control chart can again be used to monitor the process for new problems.

Determining Process Capability

A process may be under statistical control and still not capable of meeting specifications consistently. Determining process capability involves comparing the specification limits (also called the tolerance limits) to the observed variability of the process.

The process capability index is a standard way of measuring the ability of a process to meet specifications consistently. This index is based on the assumption that the variable of interest follows a normal distribution. Before using the normal distribution, it is desirable to do a goodness-of-fit test to determine whether the data conform to this distribution. Often the normal distribution is considered to be a reasonable approximation if a histogram of the data appears bell-shaped. If the normal distribution is appropriate, the standard deviation can be estimated from the range using the following formula:

$$\sigma = \frac{R}{D_2} \tag{27.17}$$

where D_2 is a value found in Table 27.1. The process capability index C_p can then be calculated as follows:

$$C_p = \frac{\text{USL} - \text{LSL}}{6\sigma} \tag{27.18}$$

where USL and LSL are the upper and lower specification limits, respectively.

If the process is not centered midway between the USL and the LSL, there will be more difficulty meeting one specification limit than the other. An alternate measure of process capability, C_{pk}, takes this factor into consideration by calculating both upper and lower one-sided process capability indices, and using the lower of the two as a measure of the capability of the process,

$$C_{pk} = \min\{C_{pu}, C_{pl}\} \tag{27.19}$$

where

$$C_{pu} = \frac{\text{USL} - \mu}{3\sigma} \tag{27.20}$$

$$C_{pl} = \frac{\mu - \text{LSL}}{3\sigma} \tag{27.21}$$

It is desirable to have a process capability index high enough that it is extremely unlikely that the process would produce any observations above the upper specification limit or below the lower specification limit. If a value less than 1 is obtained, the process is clearly not capable of meeting specifications consistently. The usual rule of thumb in military organizations is that C_{pk} should be at least 1.33. However, for processes where the variable being measured is extremely critical, as may be the case with weapon systems, a C_{pk} of 2.0 or even higher may be desired.

If the process capability index indicates that the process is not capable of meeting the specifications consistently, either the process must be changed or the specifications revised. Although it is sometimes possible to alter the specifications, especially if they were arrived at in an arbitrary manner to begin with, it is more often necessary to alter the process in some way to reduce variability. The seven simple tools can be used to find some of the common causes of variability in the process. Common causes of variability are those that are an intrinsic part of the existing process. The process must be changed in order to remove common causes of variability. Of course it is always possible that a process change will introduce new sources of variability as well.

EXAMPLE: WEAPON SYSTEM TIMING DEVICE

This example is a simplified version of an actual case history. It illustrates how several of the simple tools and control charts may be used to improve the quality of the same system. The authors thank Dan Zint from the Defense Logistics Agency for providing the information on which this example is based. The data are real. They represent the length of time, measured in milliseconds, that it takes a timing device to set. A similar timing device is part of many weapon systems. More parts and more processes are involved than are shown here; the example has been simplified to illustrate the use of quality tools better, without distracting the reader with needless detail.

Flowchart

One good starting point in quality improvement is to understand the process we are using. This can be accomplished by constructing a flowchart. A flowchart not only shows each step in the process, but provides a means of visualizing the process as a whole and how the steps are related to one another. This is especially helpful if the process is highly complex or if not everyone on the quality improvement team is equally familiar with the process. The flowchart for the timing device is shown in Fig. 27.1.

Check Sheet

Once we have a flowchart we can decide where to gather data. In this case we chose to collect the times in operation 80 (arm spin test). The specifications call for the timing device to arm in between 500 and 700 ms. Since this device regulates when the weapon system detonates, arming in a consistent, predictable amount of time is crucial. A check sheet was developed to collect the actual arm times. A sample of five units was tested from each lot; 20 samples were recorded in Fig. 27.2.

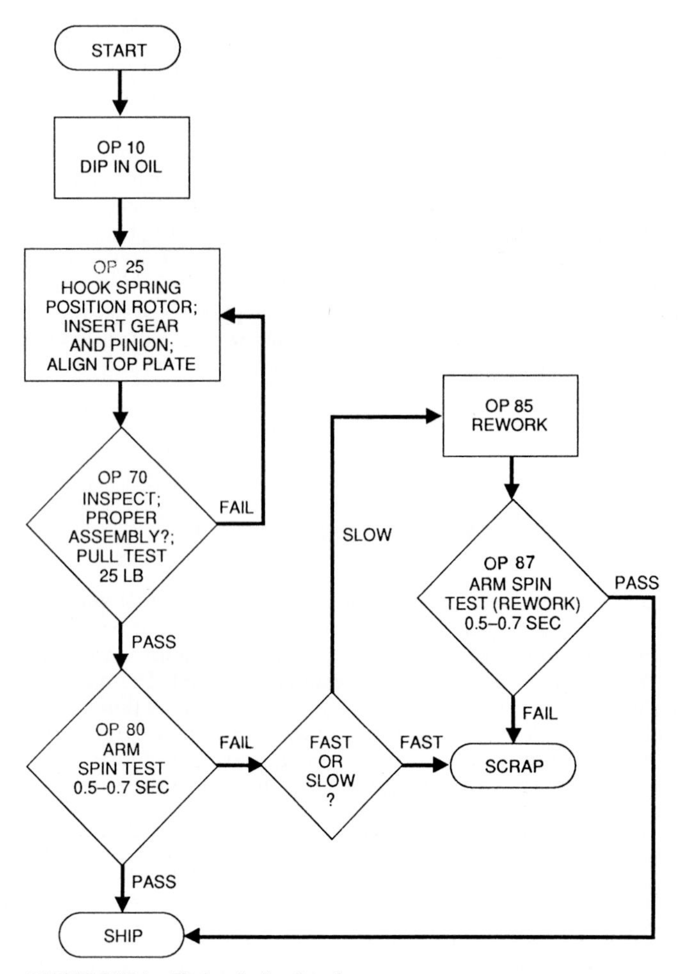

FIGURE 27.1 Timing device flowchart.

Pareto Chart

One immediate observation from the check sheet is that there appear to be many devices that require 852 ms to arm. That is because the test equipment registers 852 as a maximum value. To understand the patterns in which the times occurred more clearly, a pareto chart was developed. This chart is shown in Fig. 27.3. From observing this pareto chart it is obvious that almost half (46 percent) of the timing devices had unusually slow arming times. This tells us that we need to speed up the arming times of many devices dramatically. Note that the pareto chart could also be used to compare time of day, day of week, material suppliers, machine used, and many other factors in an initial effort to spot possible causes of the problem for further investigation.

PART NAME: TIMING DEVICE
SAMPLE SIZE: 5
TOLERANCE: 0.5 TO 0.7 SEC

OPERATION: 80 (ARM SPIN TEST)
CHARACTERISTIC: ARMING TIME
NOTE: SECONDS SHOWN IN THOUSANDTHS (852 =0.852 SEC)

DATE	TIME	SAMPLE #	DEVICE # 1	2	3	4	5
8-12-91	0900	1	852	605	608	852	852
	0905	2	604	605	599	852	852
	0910	3	852	574	852	852	604
	0925	4	619	583	574	612	852
	0940	5	620	852	599	852	852
	0950	6	600	603	852	852	852
	1005	7	582	609	852	852	852
	1015	8	604	620	613	590	597
	1025	9	664	603	636	603	613
	1035	10	595	601	598	605	852
	1045	11	852	852	852	583	852
	1100	12	852	852	852	601	852
	1115	13	852	852	583	852	852
LUNCH	1125	14	604	852	610	852	852
LUNCH	1210	15	619	852	852	601	611
	1220	16	852	852	603	606	602
	1230	17	594	852	596	625	852
	1240	18	611	600	609	852	608
	1255	19	852	852	587	852	607
	1310	20	597	852	606	734	852

FIGURE 27.2 Timing device check sheet.

Histogram

Another tool that could be used to help understand the variation in arm times is a histogram. The histogram in Fig. 27.4 shows clearly that there appear to be two distinct groups of arming times. The question is how can the slow group be sped up without speeding up the group that is already doing well? It takes a detailed understanding of the process to answer such a question.

Fishbone Diagram

The cause-and-effect diagram, or fishbone diagram, can be used to collect thoughts generated in a brainstorming session concerning possible causes of slow arm time. The cause-and-effect diagram for this example is shown is Fig. 27.5. A cause-and-ef-

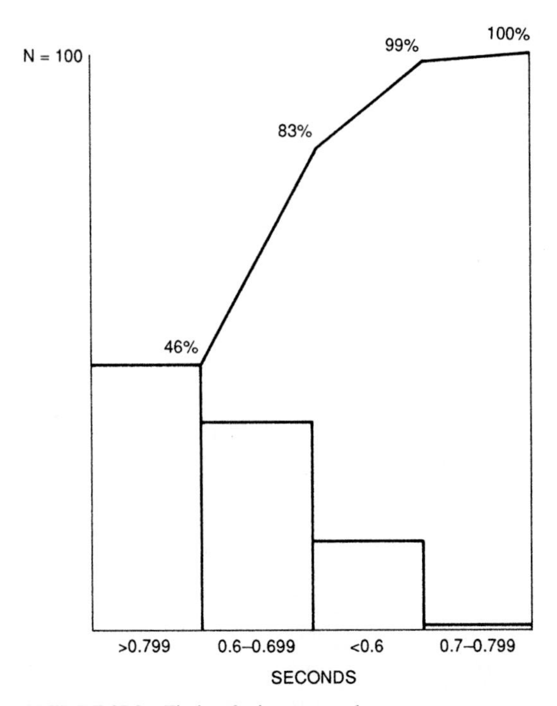

FIGURE 27.3 Timing device pareto chart.

fect diagram can be expanded to include considerably more detail by continually asking the question "why?" for each branch and incorporating the group's responses as another level of detail on the chart. Once an apparently exhaustive set of possible causes is generated, one or more potential causes that seem likely are selected for testing. In this example the number of turns of the spring was selected. It was thought that by tightening the spring one-quarter turn in operation 25 (see flowchart, Fig. 27.1) the arm times would be quicker. This was tested, and a new check sheet (Fig. 27.6) was used to collect data. A new pareto chart (Fig. 27.7) and a new histogram (Fig. 27.8) were developed to display the data. Clearly there is an improvement with only half as many slow times and still no fast times. There is still much room for further improvement. This could be pursued using the same tools or some additional tools.

Scatter Plot

A scatter plot (Fig. 27.9) was developed to compare the time of day to the average time for each sample. If there is a strong relationship between two variables, the points will appear to form a line or to show some other pattern. In this example the points appear to be quite random, so time of day does not seem to be relevant. We should investigate something else.

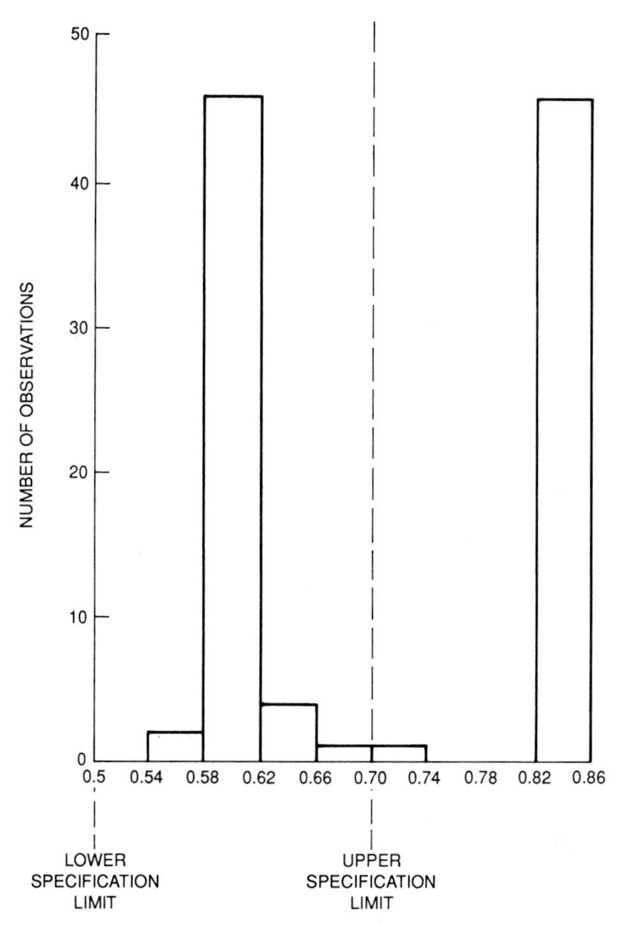

FIGURE 27.4 Timing device histogram.

Run Chart

A run chart can also be constructed (Fig. 27.10). In this case the percent acceptable in each lot was shown in the order in which they were produced. Vertical lines with comments can be drawn to show where different events take place. This can be used to help understand what conditions existed at the same time that improvements or degradations occurred in the product. For example, after the spring was tightened, more units were acceptable.

Control Charts

The final technique illustrated is the control chart. Often project activities are not repeated enough times to warrant the use of a control chart. An active research area in statistical process control is the development of special techniques for short produc-

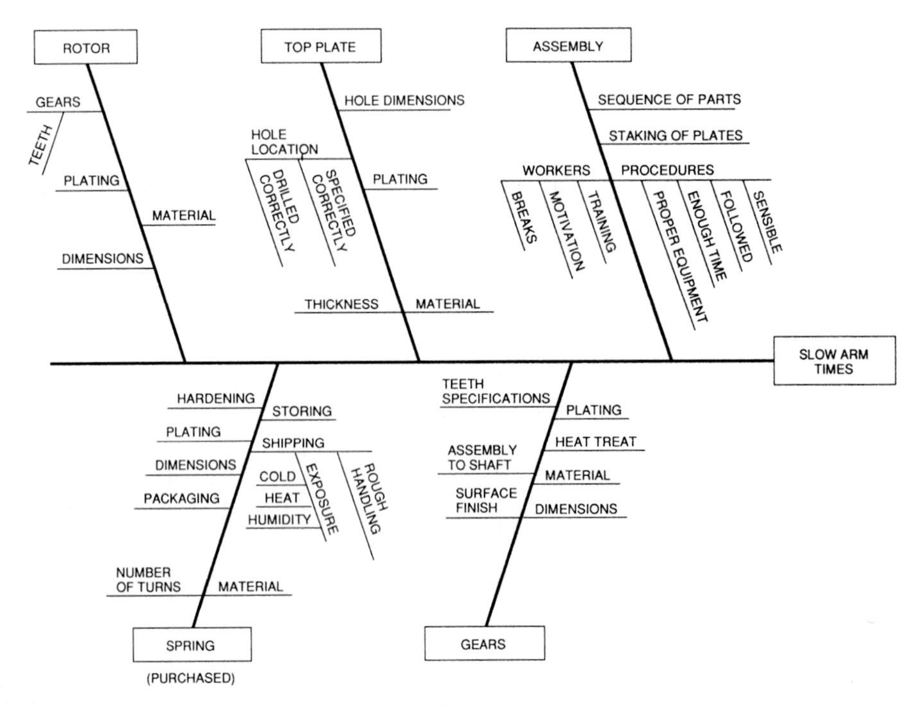

FIGURE 27.5 Timing device cause-and-effect diagram.

tion runs. Such special techniques are beyond the scope of this chapter. However, for processes that are performed repeatedly, standard types of control charts are a useful device.

In this example we took equal sample sizes (five) and measured a variable (arming time). The control charts that are appropriate here are the $\overline{X}$ chart to detect variation between samples and the R chart to detect variation within samples. These charts are shown in Figs. 27.11 and 27.12, respectively. This process is not in statistical control. On the $\overline{X}$ chart the sixth point is above the control limit. The R chart also shows the process to be out of control. In the first part of the R chart four of five consecutive points are at least one standard deviation above the mean. In the last part of the chart four of five points are at least one standard deviation below the mean, and two of three points are at least two standard deviations below the mean. Both charts indicate that while the process is getting better (the average is moving closer to the middle of our specified range and the variability is decreasing), it is not yet stable. Further use of the seven simple tools should be performed, and then the process should once again be checked with control charts for stability.

Process Capability

It is easy to see that this process is not capable of meeting the specifications consistently. We can calculate C_p as follows:

PART NAME: TIMING DEVICE
SAMPLE SIZE: 5
TOLERANCE: 0.5 TO0.7 SEC

OPERATION: 80 (ARM SPIN TEST)
CHARACTERISTIC: ARMING TIME
NOTE: SECONDS SHOWN IN THOUSANDTHS (852 =0.852 SEC)

DATE	TIME	SAMPLE #	DEVICE # 1	2	3	4	5
8-26-91	0900	1	594	593	852	594	561
	0930	2	600	852	852	607	852
	0955	3	592	617	587	852	594
	1015	4	583	566	852	572	571
	1035	5	582	584	852	852	585
	1100	6	852	600	852	852	687
	1115	7	852	743	696	579	595
	1145	8	596	852	585	613	593
8-29-91	0900	9	580	569	580	591	593
	0925	10	594	852	852	590	575
	0950	11	615	592	852	577	603
	1010	12	600	600	561	588	566
	1045	13	580	551	577	598	561
	1105	14	852	583	577	575	609
	1120	15	573	580	609	573	601
	1140	16	587	577	586	586	597
9-4-91	0905	17	589	852	613	852	852
	0930	18	567	647	584	600	599
	0945	19	571	592	574	576	597
	1005	20	852	778	663	565	624

FIGURE 27.6 Timing device check sheet no. 2 (after change in spring rotation).

$$C_p = \frac{\text{UTL} - \text{LTL}}{6\sigma} = \frac{700 - 500}{6(36.66)} = 0.9093$$

Although C_p is less than 1, indicating an incapable process, it is interesting to note that C_{pk} is much worse because of the fact that the process is not centered midway between the tolerance limits,

$$C_{pu} = \frac{\text{UTL} - \mu}{3\sigma} = \frac{700 - 650}{3(36.66)} = 0.4546$$

$$C_{pl} = \frac{\mu - \text{LTL}}{3\sigma} = \frac{650 - 500}{3(36.66)} = 1.3639$$

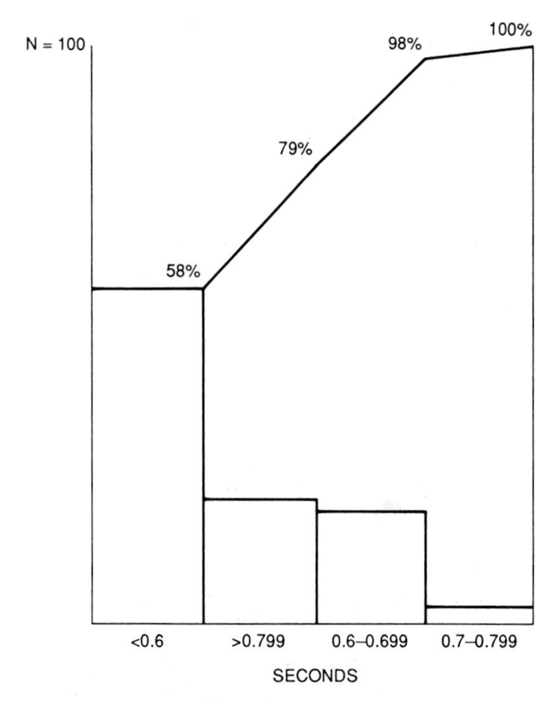

FIGURE 27.7 Timing device pareto chart (after change).

$$C_{pk} = \min\{C_{pu}, C_{pl}\} = C_{pu} = 0.4546$$

Therefore we can conclude that in addition to having too much intrinsic variability in the process, the process is not centered midway between the upper and lower tolerance limits. More process improvement is clearly necessary. The team should now look at some of the other process factors identified in the fishbone diagram. It may be necessary to use the seven simple tools repeatedly before an acceptable level of quality is achieved.

CONTINUOUS IMPROVEMENT

Since it is not possible to identify and remove all sources of variation, there will always be room for further quality improvement. The goal must be to improve both design quality and the processes that are used to produce the product continuously. Improving design quality involves obtaining input from the actual user of the product in order to develop appropriate product specifications. It is then necessary to design processes that can meet these specifications consistently, and to develop process control systems, including statistical process control, to monitor these processes on an ongoing basis.

When quality problems occur, quality improvement teams should investigate for possible process problems using the seven simple tools. Such groups should include

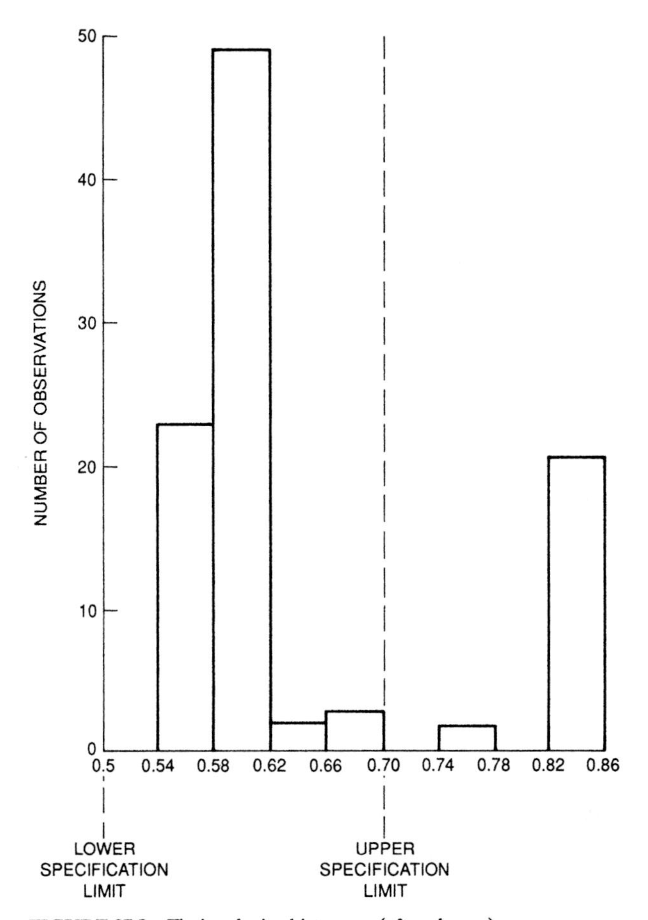

FIGURE 27.8 Timing device histogram (after change).

workers at all levels, especially workers at the operator level, because they are often the most familiar with the day-to-day fluctuations of the process and thus are in the best position to detect special causes that others might miss. Obtaining the cooperation of lower-level workers necessitates that they be treated with respect and in such a manner that they do not fear any adverse consequences.

Even when there are no overt quality problems, quality can often be improved still further by seeking ways to improve processes through finding ways to reduce variability. This should be done on an ongoing basis, thus emphasizing the prevention rather than the correction of quality problems.

Continuous quality improvement can be achieved only if a strong commitment to quality exists at all levels of the organization. When this commitment exists, it is possible for all members of the organization and its various constituents to work together to improve quality.

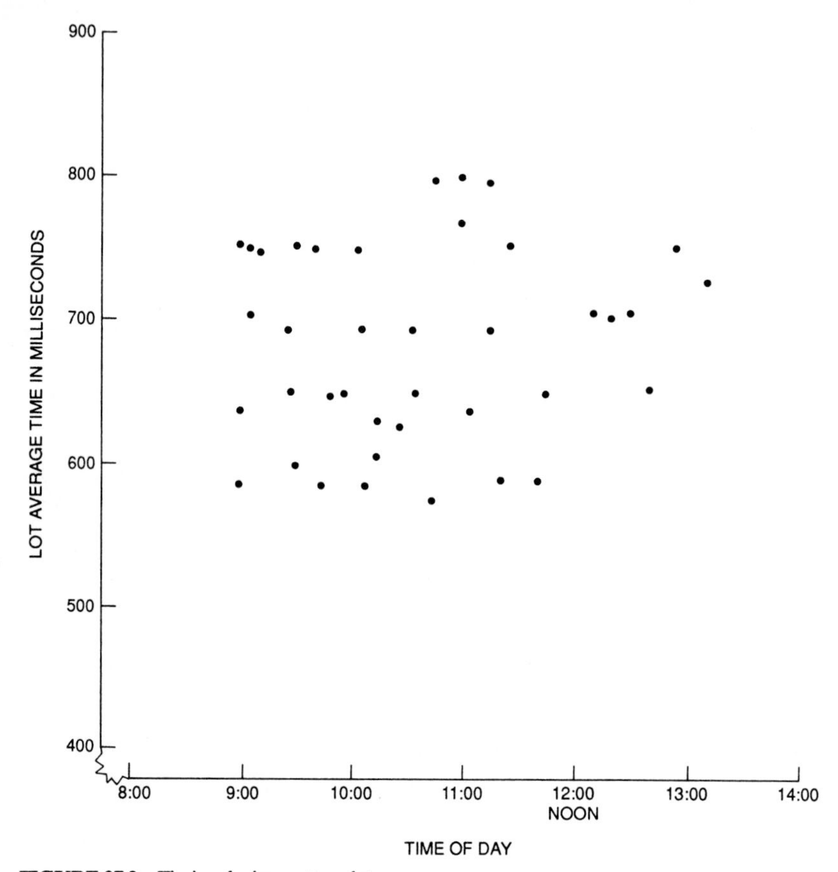

FIGURE 27.9 Timing device scatter plot.

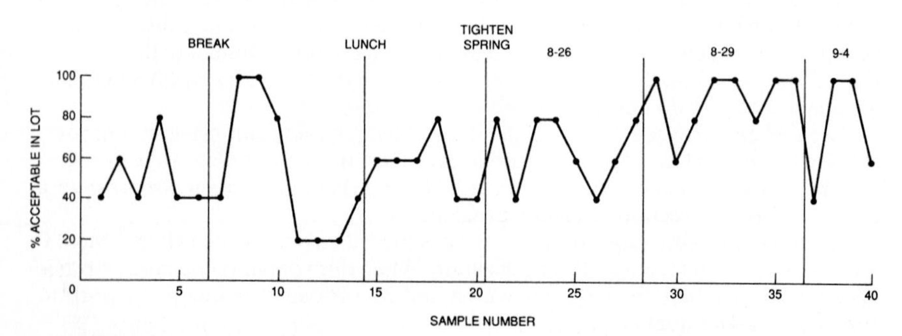

FIGURE 27.10 Timing device run chart.

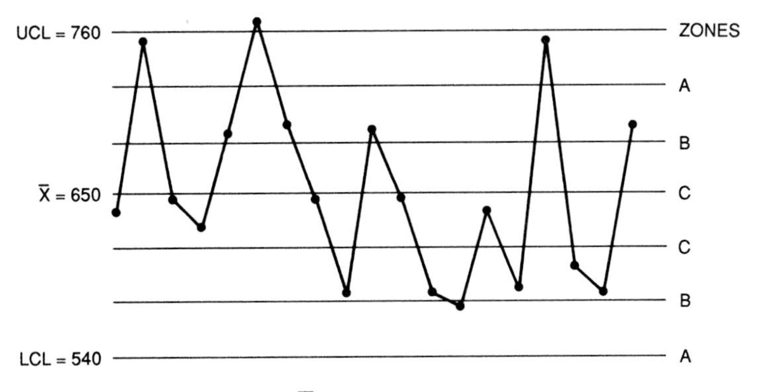

FIGURE 27.11 Timing device $\overline{X}$ chart.

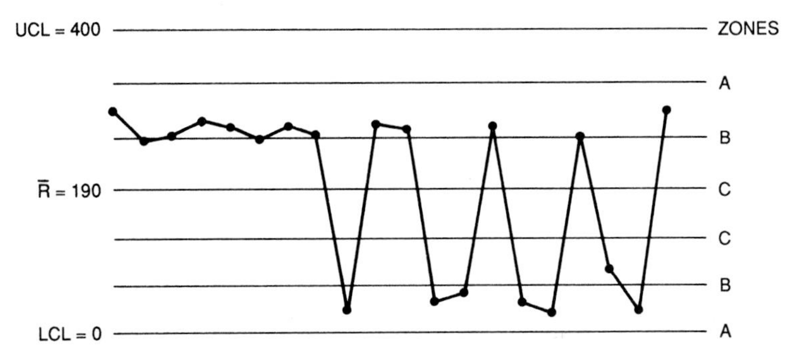

FIGURE 27.12 Timing device *R* chart.

REFERENCES

1. W. E. Deming, *Out of the Crisis,* M.I.T. Center for Advanced Engineering Study, Cambridge, Mass., 1986.
2. J. M. Juran and F. M. Gryna, *Quality Planning and Analysis,* McGraw-Hill, New York, 1980.
3. K. Ishikawa, *Guide to Quality Control,* Asian Productivity Organization, Tokyo, Japan, 1976.
4. M. Brassard, *The Memory Jogger,* Goal/QPC, Methuen, Mass., 1988.

BIBLIOGRAPHY

W. E. Deming, *Quality, Productivity, and Competitive Position,* M.I.T. Center for Advanced Engineering Study, Cambridge, Mass., 1982.

A. J. Duncan, *Quality Control and Industrial Statistics,* 3d ed., Irvin, Homewood, Ill., 1965.

J. R. Evans and W. M. Lindsay, *The Management and Control of Quality,* West Publ., St. Paul, Minn., 1989.

G. Taguchi, *Introduction to Quality Engineering,* Asian Productivity Organization, Tokyo, Japan, 1986.

CHAPTER 28

THE ROLE OF THE MILITARY PROJECT MANAGER IN STRATEGIC PLANNING

Earl D. Cooper

Earl D. Cooper is currently Program Director of the Florida Institute of Technology's graduate management programs in the national capital region and a consultant to various aerospace firms and governmental agencies. He holds the academic rank of Associate Professor and has a doctorate in public administration. As a member of the Senior Executive Service, he served in various senior management positions in naval aviation, including that of the Deputy Project Manager of a designated major defense acquisition program and Technical Director of Research and Technology for the Naval Air System Command.

INTRODUCTION

The planning system used in the Department of Defense (DoD) is defined in considerable detail in DoD Directives 5000.1, "Defense Acquisition," and 5000.2, "Defense Acquisition Management Policies and Procedures," both dated February 23, 1991. This system has evolved over many years and has been fine-tuned on numerous occasions in order to achieve cost savings, increase product quality, minimize graft and corruption, enhance competition, increase the degree of fair treatment among industry competitors, and so on. This system, although laborious and constraining to the project manager, reflects a response to the many, many special-interest forces that comprise our nation, and the checks and balances of our governmental system design.

Figure 28.1 is a generic model of a classical strategic planning process, which can be used by any organizational entity. Figure 28.2 is a generic model of the DoD process, in which a defense acquisition project progresses through the various acquisition phases, proceeding from one milestone to another. Figure 28.2 is very similar to Fig. 28.1 in that the basic strategic questions of the classical model are asked again, and the answers thereto are generated at each milestone.

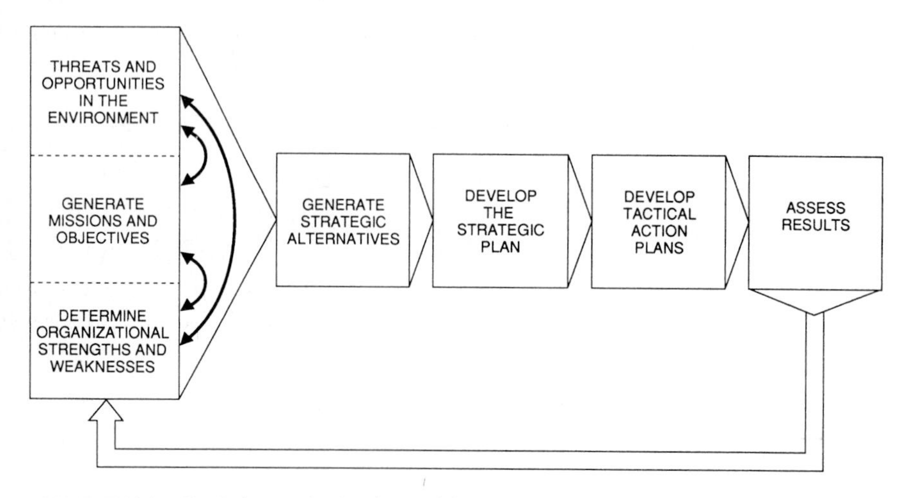

FIGURE 28.1 Classical strategic planning model.

The DoD project manager must adhere to the DoD 5000 series of directives in carrying out a project and must follow the mandatory policies and detailed procedures set forth therein. As such there is very little room left for innovation in the planning process and in the design of the planning system the project manager must use. The one area, however, wherein the project manager does have considerable

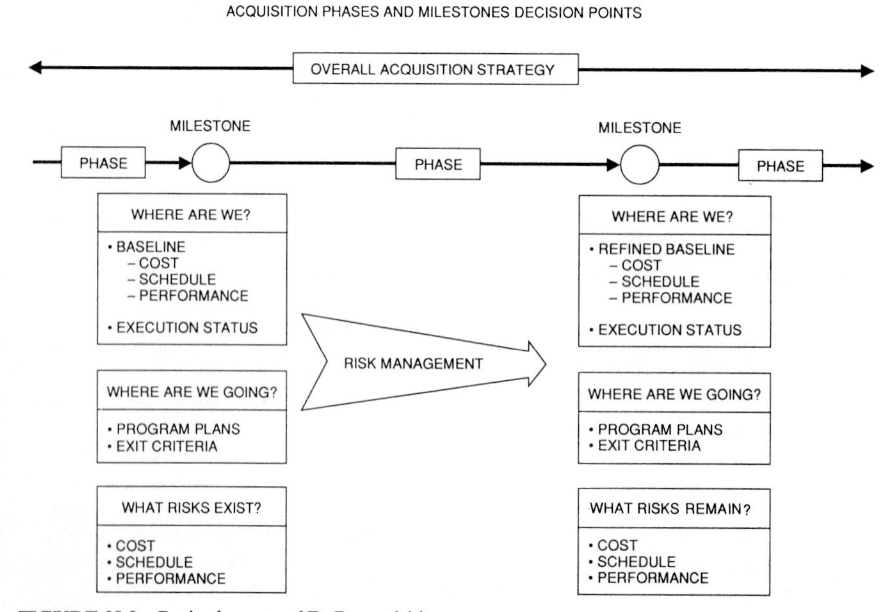

FIGURE 28.2 Basic elements of DoD acquisition process.

room to innovate and can perhaps make the biggest impact, is the "people area." A simple definition of management is "getting things done through people." For the project manager to be successful, his people must be successful. It has been said that the greatest contribution a manager can make to an organization is to unleash the potential of the people in that organization, which is its greatest asset. Thus the prime focus of the project manager and the area where he has the greatest leverage is people. He must actuate himself, the people on his team, his up-line hierarchy, and his cross-line associates or peers.

The project manager is the chief "visionary" for his project. This means that he, more than any other individual, must have a clear vision of project success. He must be able to define what project success is, be able to articulate and translate this definition to each project team member and members of all supporting groups in their terms and language. He is the "keeper of the vision." The project manager must be able to show his team the pathway to achieving his vision of project success, and at the same time warn them of the threats and negatives in the environment along the way. In articulating the vision of project success and showing how to get there, the project manager actually performs the basic steps in strategic planning. The establishment, maintenance, continual assessment of vision integrity, and vision modification as necessary can be mechanized into a structured process that blends in well with and supports the defined DoD acquisition system as defined by DoD directives 5000.1 and 5000.2. One such DoD-compatible advanced management concept is "strategic visioneering," which has been termed a new management tool for a new age. The strategic visioneering concept was developed by the author and first articulated formally at the World Future Society Professional Members Forum held in Salzburg, Austria, on June 5, 1990.

STRATEGIC VISIONEERING: WHAT IS IT AND WHAT WILL IT DO?

Strategic visioneering (SV) can be defined as the application of unique blends of behavioral science, psychology, and management theories coupled with advanced technology by which the properties of old management systems are modified or new management systems are created to be useful to humans in goal definition and achievement.

SV utilizes a systems approach and can be used effectively at all levels of activity, ranging from the corporate level down to the personal level. It is most effective when it is applied to all simultaneously. At the corporate level it is useful to define the organization's future visions of success and generate the pertinent, enabling plans and actions, and to link all of these elements. In addition, SV provides a continuous focus on the vision that is needed to ensure its realization.

At the individual level it can be used to develop challenging personal visions of career achievement and to link career planning and enabling actions to these visions. Simultaneously, as the process ensues, individual motivation is continually increased, and all the factors involved in the system propel the subject toward the goal with an ever-increasing crescendo.

At the corporate level the SV process mirrors that for the individual, except that the person is replaced by the corporate entity, which becomes the totality of all of the individuals involved. The ideal relationship of the corporate and employee body occurs when the visions of both are identical or mutually supporting. It is management's job to work to meld the two visions together in order to gain the ultimate employee motivation. This is not unlike Chris Argyris' "fusion theory" remedy for dealing with the evils of bureaucracies.

SV is simple in concept. This concept is similar to that which formed the basis for the theme song of the General Electric Company's exhibition at the New York World's Fair in 1964, which was called the Carousel of Progress. As it rotated about a series of stages, each stage displaying GE appliances for a particular time period (or GE visions realized) and moving with chronological ascension, the audience was treated to an inspirational song, the words to which are presented in Fig. 28.3. The song has three focal points. First is the "dream," or vision, which is obviously a challenging goal, having great rewards if achieved. Second is "follows . . . with mind and heart." This infers that all the thoughts and actions are dedicated to and focused continuously in the direction of the dream, resulting in a confluence of enabling forces, which act together to make the dream come true. Third is the "beautiful tomorrow," which represents the achievement of the goal and thereby reaping the rich rewards inherent in such success.

There's a great big beautiful tomorrow
Shining at the end of every day
There's a great big beautiful tomorrow
And tomorrow is just a dream away

Man has a dream and that's the start
He follows his dream with mind and heart
And when it becomes a reality
It's a dream come true for you and me

FIGURE 28.3 GE's Carousel of Progress.

An example of some of the dynamics involved can be seen in the example of a family contemplating and planning for a "dream" vacation. Their dream becomes more focused after a search for knowledge (that is, building a database) on such pertinent areas as the scenic places in the world, places where people can go to have fun, means of transportation, and lodging costs, with all the family taking part. Then a number of enabling actions follow, such as participative decisions on the best alternative location, determining the best route to travel, buying airline tickets, reserving a rental car, and so on. With more and more "talking it up" in family meetings, the excitement, anticipation, and database build. The family team members are motivated to do more information searches to further refine the vacation planning—and so it goes until finally the dream is achieved.

Basic to this discussion are the answers to two questions: (1) What qualifies an individual or organization to have or develop a legitimate vision? (2) What is required to completely engulf individuals in (or cause them to follow "with mind and heart") an endeavor to accomplish its purpose? The answer to the first question is superior knowledge. When we want to know the "ultimate" about a subject, we turn to an expert. An expert is defined as a person with superior knowledge. One of the most respected approaches to future forecasting is the Delphi method which, of course, features a consensus of experts. In a highly competitive marketplace the company with the "best vision" (in terms of realism and most pertinent knowledge) will probably be the winner. Hence it is incumbent upon the organization to have a superior knowledge which, when translated into today's terms, is a superior database and the ability to know how to use it fully. The answer to the second question is a system of continuous motivation such that the individuals involved use all of their conscious moments in pursuit of the vision and are simultaneously guided in the very same direction by the invisible hand of their subconscious minds. As such, management and all the team (or at the personal level, the individual himself) must develop an insatiable desire to achieve the vision which is borne of the belief in the great value of such an accomplishment and the continuous reinforcement of this belief. Levels of performance approaching Abraham Maslow's self-actualization or Friedrich Nietzsche's "Overman" must be achieved.

In summary the SV concept can be expressed in terms of Chester Barnard's famous book, *Functions of the Executive,* which specifies the project manager's prime tasks: (1) the establishment, articulation, and maintenance of organizational purpose (the vision); (2) the creation and maintenance of the willingness of the employees to cooperate (effective motivation); and (3) the establishment and maintenance of effective communication systems (to provide for effective coordination and continual feedback for motivation reinforcement). SV's uniqueness is found in its emphasis on a superior knowledge and stress on continual, positive motivational reinforcements.

SIMPLIFIED STRATEGIC VISIONEERING MODEL

Figure 28.4 is a model of the SV process. Its major elements consist of "the vision," motivating forces, reinforcement of the motivating forces, building a superior knowledge base, action planning, implementing actions, progress toward the vision, and a continuous evaluation of the process.

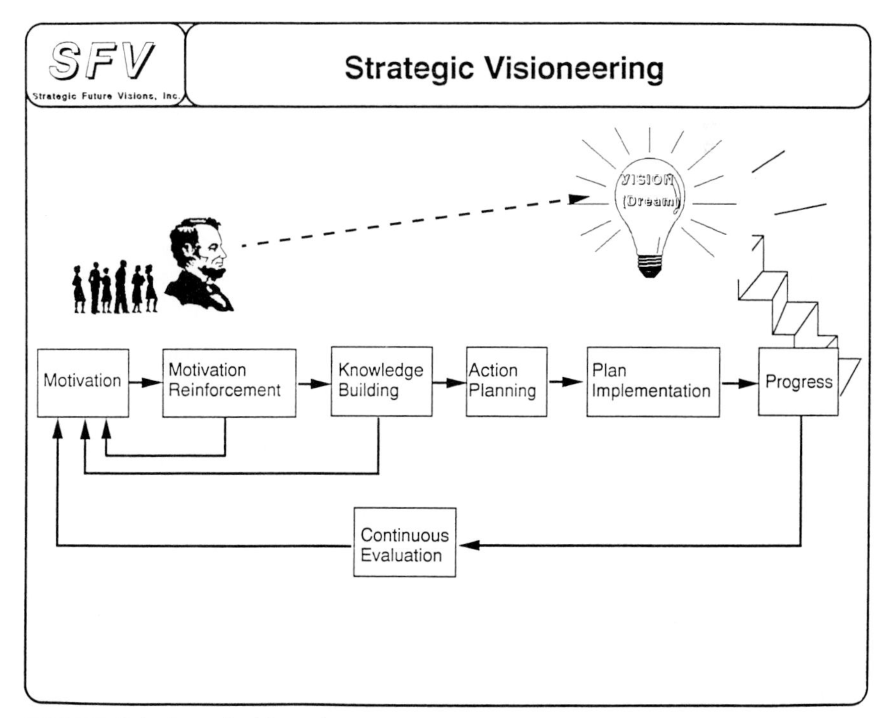

FIGURE 28.4 Strategic visioneering.

In practice it is difficult to pinpoint where the process begins. In general it starts with a vision (or dream) that is borne of a person's biological needs or instincts, or of new knowledge which in effect provides the person a footstool to stand on to see further and clearer and to recognize a vision, or which gives the person the building blocks needed to create one. The vision may be recognized in the individual's mind or the corporate mind, and may lay dormant until something perturbs the system to disturb the equilibrium and create the unbalanced forces necessary to get the process rolling. In the case of a defense acquisition, this perturbation is a military operational requirement articulated by a mission need statement.

One famous vision on a national scale provides an example of the process. It is President John F. Kennedy's vision "to land a man on the moon." Obviously many people of many generations prior to Kennedy had the vision of humans traveling to the moon. Prior to Kennedy's grand announcement, American scientists were already conducting research with government grants on space travel. It was the Soviet Sputnik satellite, however, beeping as it passed over the heartlands of America, and the threat implications embodied therein that provided the unbalanced forces necessary to get the process started in earnest. Of course there followed a series of comprehensive enabling plans and actions that made the vision become reality. It is significant to note that the increasing Soviet threat provided a continuing, reinforcing stimulus to the process. Also, as the project proceeded and space research yielded more and more "fallout" benefits to humankind, support for the space program became more widespread among the populace, and the space scientists experienced a motivational reinforcement.

An example on a personal level is a young U.S. Air Force Academy graduate who dreams of being a top Air Force general. After she graduates from the academy she is motivated to investigate and define the knowledge she needs and actions she must take to reach her goal. She determines that having a master's degree is a requirement to be promoted beyond captain, so she sets about to accomplish this intermediate goal, and so the process goes toward fulfilling her dream.

Motivation

The goal of the motivational programs is to create a conation within each individual involved and an environment to promote nisus (or a state of conation). If successful, each participant will have an instinctively motivated biological striving for the vision that may appear in consciousness as volition or desire, or in behavior as action tendencies. An individual exhibiting conatus is a high performer approaching Maslow's self-actualized level of activity.

Motivation comes about by the unbalancing of the forces of equilibrium at work within an organization or an individual. Its basis can be found in and articulated by human needs theories, some of the most famous being Abraham Maslow's hierarchy of needs, Frederick Herzberg's two-factor theory, and David McClelland's learned needs theory. Figure 28.5 compares these three classical content theories. Although in Maslow's hierarchy it is necessary not to neglect any of the levels, it is the top two on the pyramid that develop the greatest forces within individuals for action toward vision achievement. Esteem is the need of the individual to have a feeling of great self-worth and the appreciation and positive recognition from others. Self-actualization is the individual's need to fulfill himself in his ideal self-image by making the maximum use of his skills, abilities, and potential. In this sense management must understand precisely what each employee's needs are (and they are different for each) and try to strike a positive match or relationship with corporate needs.

Accordingly, it is necessary for the program manager to take a systems approach

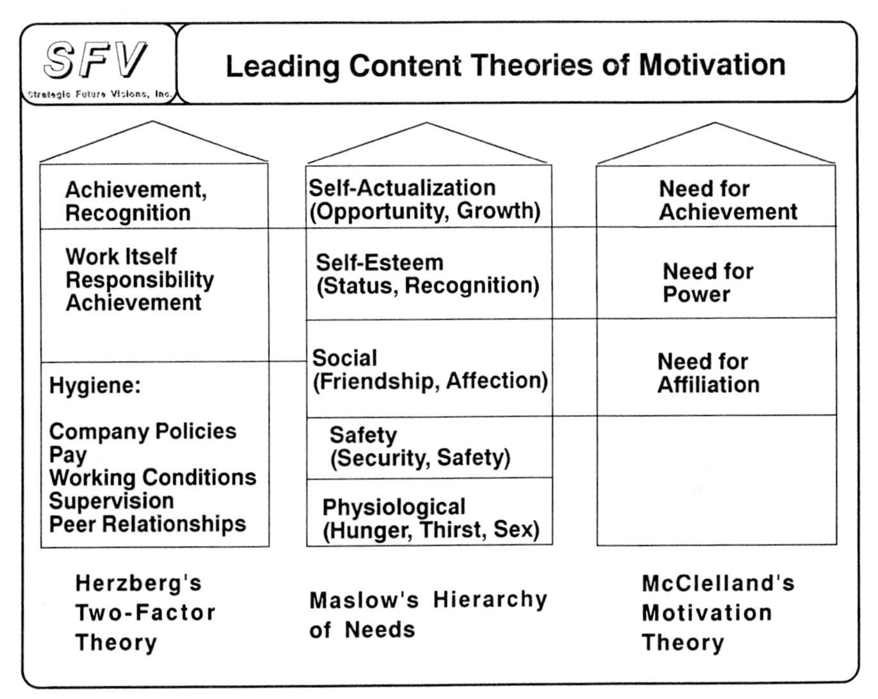

FIGURE 28.5 Leading content theories of motivation.

with his team members. In other words, he must consider the whole person, or "person system," that is the 24-hour, seven-days-a-week person, not just the person present during working hours. Figure 28.6 shows the person system in the milieu of a sampling of his basic concerns, which are motivational force generators. Management must work to ensure that any forces generated in these areas are positive, and that they remain in a positive direction such that they will converge to produce a conatus. It is obvious that an individual employee cannot provide an optimum output if he or she is worried about a sick child at home or the well-being of an elderly parent left at home alone. The provision of day care facilities for both the employee's preschool children and elderly parents is an example of innovations used by some companies to remove negative (worry) forces.

Reinforcement

The purpose of reinforcement is to maintain a continuous imbalance of the individual and corporate forces of equilibrium, which is necessary to keep individual and corporate eyes on the "dream" throughout the process, until it is realized. An analogy is given in Fig. 28.7 with the firing of the old Bullpup guided missile from an airplane toward a ground target. After launch the missile flew out ahead of and faster than the aircraft. The missile's guidance was remotely (radio) controlled by the pilot. In order to hit the target successfully it was necessary for the pilot to guide the missile down his line of sight to the target (that is, to keep the missile lined up with the

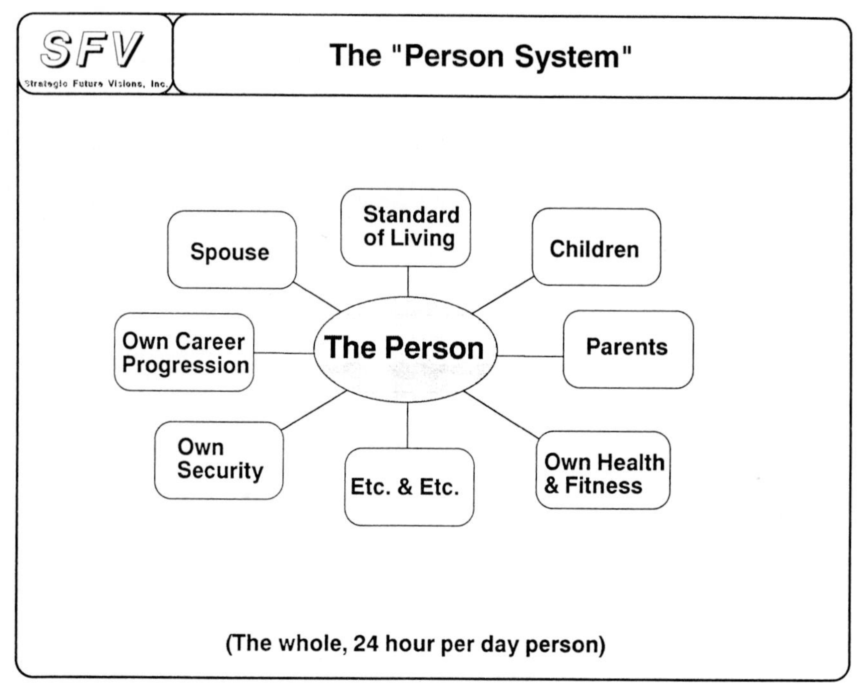

FIGURE 28.6 The "person system."

target). As such it is obvious that the pilot always had to keep his eyes both on the missile and on the target (as well as flying the airplane) until the mission was completed. What's more, the closer the aircraft came to the target, the more hectic the task became. In this analogy, hitting the target is the vision and the aircraft, pilot, and missile are the corporation in operation. Success was achieved only by continuously keeping the vision in view and all facets of the system concentrated on and working toward it. To every individual involved this means the whole person—the person system—mind plus heart.

A prime focus of SV is the human mind. Although it is difficult to obtain a consensus definition of the mind, having been variously defined as a process, a spirit, a complex of faculties, and an individual's total adaptive activity, it is nonetheless the entity that presides over and directs the person system. The mind uses two amazing person subsystems, the brain and the central nervous system, to translate its mandates into action. The brain serves as a supercomputer, while the central nervous system can be likened to a master servo control system.

The brain reflects an advanced computer technology which today is still far beyond human understanding and comprehension. The human three-pound brain is estimated to contain on the order of 12 billion neurons, each being capable of effecting up to 50,000 connections with other cells. As such it is capable of accomplishing one hundred trillion connections. This gives it a byte capacity well beyond any supercomputer ever conceived of. It is controversial as to whether or not humans use the full capacity of their brain power. Research, for example, has determined that the average human is capable of speaking between 150 to 200 words per minute and reads

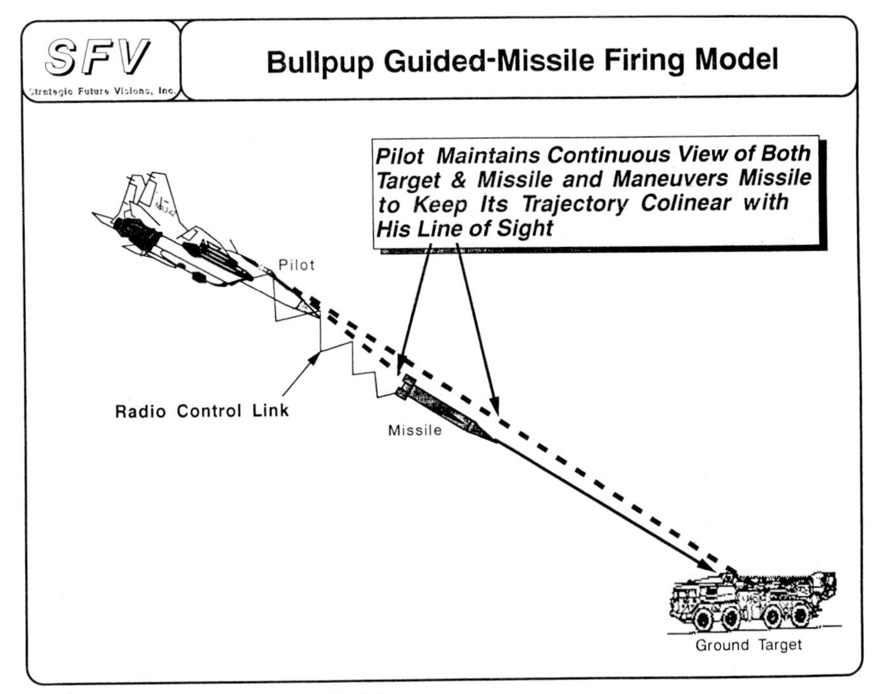

FIGURE 28.7 Bullpup guided missile firing model.

on the order of 300 to 500 words per minute. The brain on the other hand can process 2000 or more words per minute. As such it is obvious that excess brain capacity is being unused if an individual is simply listening to the spoken word. This, for example, presents a significant challenge to the college professor to command the complete attention of the student body during a lecture. The message here is that it is necessary for the professor to fill the void with more pertinent information, such as pictures or other charts, and to eliminate distracting negative views or sounds. This is why audiovisual presentations are so effective and why television is such a powerful communication tool, further giving credence to the old saying that "a picture is worth more than a thousand words."

Figure 28.8 is a person-system operational model, showing the human sensor-mind-body response relationships. This figure reflects one definition that describes the mind as the totality of the conscious and subconscious processes of the brain and the central nervous system that directs the mental and physical behavior of an organism having sense perception. In our conscious state we are generally aware of our mental activities and physical actions. However, our subconscious mind operates below the level of awareness and works continuously to guide our thoughts, feelings, and behavior. Hence during our waking moments we are directed both by our conscious and by our subconscious minds simultaneously. When we go to sleep, our conscious mind turns off, but our subconscious mind continues to operate.

That the subconscious mind exists is beyond question. We have all had the experience of trying unsuccessfully to recall a person's name from memory. Minutes, hours, maybe days later the name will pop into our conscious mind. Obviously we

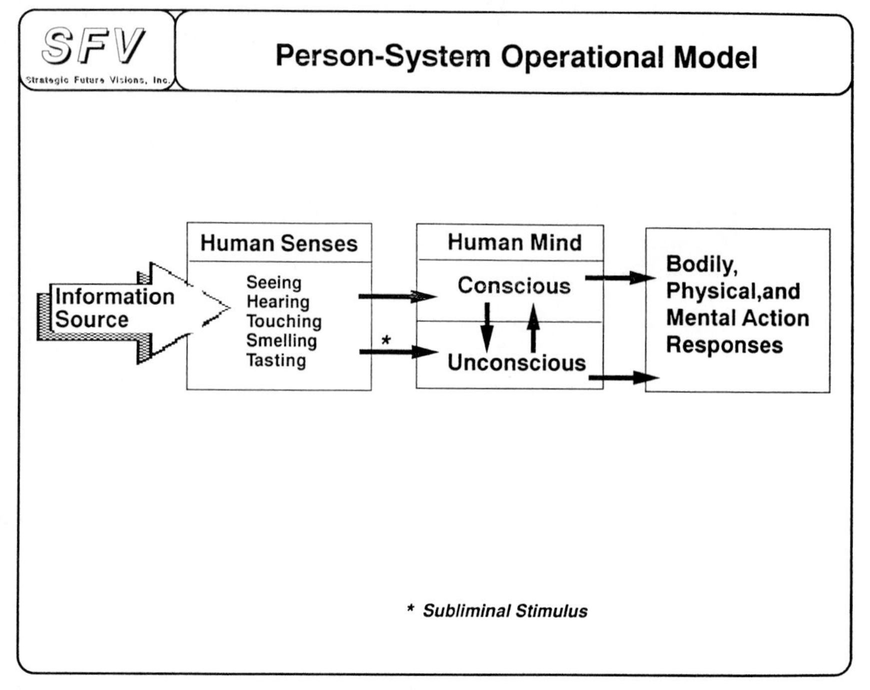

FIGURE 28.8 Person-system operational model.

had it stored within us all the time, but we just could not recall it from the data bank in our subconscious mind when we consciously desired to do so. The human mind has a great storehouse of knowledge, but this great storehouse and the knowledge therein are not located in our conscious mind. We must retrieve it from the subconscious when we need it. Research indicates that we can generally most easily recall information from the subconscious that has been reinforced by repetition or by a great emphasis (trauma, excitement, and so on) while it was being received. Techniques such as hypnosis have been used to unlock the knowledge storehouse of the subconscious. Just recently the murder of a Maryland state trooper was solved when another police officer working on the case was able to provide under hypnosis additional minute detail description of the murderer's automobile. That our subconscious mind directs our thoughts and actions can be seen in so many ways. It controls most of our bodily reactions and our metabolism. Once we learn to perform a task such as swim, drive a car, play a sport, or walk, this knowledge is stored in our subconscious data bank, and we perform it automatically without consciously thinking about the process or the various steps or techniques involved. The mind is continuously collecting information and storing it in the human database, or memory. This information may flow through the conscious mind into the subconscious or directly to the subconscious from the source. As we walk into a large room filled with many objects, our conscious mind will focus on a few of the objects that for some reason may be of greater interest and store this information in the memory with emphasis. The conscious mind will ignore the uninteresting objects. However, the subconscious observes all of the objects sensed by the body's sensors and stores this information in the subconscious directly. Later the individual may be able to recall some of the uninteresting objects by bringing information out of the subconscious into the conscious. If not, techniques such as hypnotism may be needed to unlock the data storehouse. An amazing aspect of this thinking is that we have been storing data on all of our life experiences in our subconscious data bank ever since birth or since our body's sensors were formed and became operative. This human storehouse contains all of our beliefs, habits, memories, emotions, attitudes, and culture. This subconscious mind plays a major role in directing us to be glad or sad, excited or bored, motivated or demotivated, and so on. Hidden in this subconscious is the basis for all of our fears, phobias, habits, and self-imposed limitations. As this human data bank is the sum total of all of the information inputs to our lives, it houses all the teachings and admonitions from out parents, relatives, friends, educational systems, and so forth. It reflects the environment that we grew up in. If then we have grown up with some negative, self-limiting, false, outdated ideas or other hangups, which are stored in the subconscious, we will probably tend to continue to create more self-limiting, negative, false thoughts, leading to more experiences of the same kind. If while growing up we have been told repeatedly that we will never amount to anything, that we are slow learners, or that we are ignorant, chances are that our subconscious mind will guide us to believe that we will not progress far in life, and such will be reflected in a lack of positive motivational forces and no interest in a positive vision of the future.

From an SV perspective it is important that the project manager work to improve the image of each of his team members. He must replace the false, self-limiting, negative outdated ideas with positive attitudes, beliefs, and thoughts. In other words, it is necessary for him to replace the old script in the subconscious mind with a new one that focuses on a challenging future vision and extols the positive and super capabilities of the individual concerned to achieve the goal and make the dream a reality. While more and more evidence points to the conclusion that humans do in fact use most of their brains' capacity and that it works for both the conscious and the subconscious mind simultaneously, the SV focus is on using this brain efficiently and effectively.

Hence the reinforcement phase of SV could be called a continuous team building/organizational development, which draws heavily upon Kurt Lewin's classical change process of unfreezing, converting, and refreezing. However, in the SV concept Lewin's three phases are somewhat modified in principle and are conducted simultaneously and continuously until the goals are accomplished and the vision becomes a reality. An analogy of the process is seen in a farmer who produces a new crop in his fields. First, he plows the field to remove the remnants of an old crop and to make the soil fertile and rich. Unfreezing, or the plowing of individual minds, means removing support for the false notions and self-imposed limitations harbored by the individual and making the mind receptive to positive new ideas. Second, the farmer plants new seeds or seedlings, which is likened to converting the human mind by introducing (or planting) and articulating new ideas of expanded horizons, new challenging visions, and increased self-capability and worth. Third, in refreezing, the new ideas are solidified to ensure that they stick. In the farmer analogy he nurtures his new crop continually with applications of water and fertilizer and protects the new plants from encroachment by weeds (interference or threat from nonproductive competitor plants).

The SV approach is to plow the minds, plant new ideas, and nurture and justify the new ideas continuously and simultaneously. As such, programs are developed at both the corporate and the individual levels to accomplish a continuous unbalance of the forces of equilibrium at play in the milieu, keeping the resultant force directed toward the vision. Such programs must involve communication with all levels of the mind—the conscious and the subconscious.

New ideas can be communicated to the subconscious by going through the conscious mind using repetition or strengthening of the new idea with some specially designed effect to force the desired idea into the subconscious. An example of instilling loyalty into the subconscious mind through repetition can be found in the gathering of the workers at a Japanese manufacturing plant at the beginning of each workday to sing the company song. Another example is the assembly each day of American school children in their classrooms and their joint pledge of allegiance to their country's flag and their singing of the national anthem together. In these examples communication is directly to and through the conscious mind of each person in the group, as each is aware that he or she is performing on an individual basis, with the experience being strengthened as the result of the group dynamics involved.

In order to communicate directly with the subconscious mind (bypassing the conscious) it is necessary to utilize subliminal communication designs and techniques. Subliminal communication involves sending and receiving messages at a level below conscious awareness. Although these subliminal messages can be received by all five of the body's sensor systems, it is the audio and visual sensors that are the most efficient receivers, with olfactory sensors being a distant third. Messages designed for the audio and visual channels are transmitted in frequencies which are not detectable by the conscious mind and yet are received by the subconscious mind and processed by the brain. The subconscious mind for example can, through the body's audio system, hear from 200 to 500 short messages per minute (depending on their length) or on the order of 3000 words per minute. Thus subliminal tapes can be prepared which send positive affirmations at these rates to reinforce the visions or objectives which are desired of achievement. This is much greater positive programming than could be accomplished by the individual himself repeating these messages verbally. Subliminal video tapes using the same principles as the audio tapes can be prepared that permit the subconscious mind to receive subliminal messages through the eyes directly, bypassing the conscious mind. As in the case of subliminal audio tapes, subliminal videos send messages at much greater data rates than the conscious mind can handle. It is then obvious that the most effective subliminal messages are those that use both audio and video channels simultaneously. Using the olfactory channel to transmit subliminal messages involves such a high level of sophistication and complexity in design that it is impractical at present. This is because no olfactory language has been developed. However, from our own experiences we can all recall being motivated by pleasant odors and demotivated by foul odors. Use of subliminal messages in management is a very sensitive subject, hence their employment, if at all, must be handled with extreme caution.

The bottom line is that any organization, be it the military project management team, a profit or nonprofit firm, or a government, religious, or other social organization, must develop effective programs to provide the necessary level of motivation and continually reinforce it in order to create and maintain the "super team." At the personal level, if working alone and not supported by other organizational assets, an individual must design, develop, and implement his or her own motivational programs to assure continuous motivational levels to create and maintain these "superstar" levels of activity and accomplishment. A caring management can go a long way to help. SV is a tool that is basic in creating and maintaining such programs. These programs must be designed to relate to each person system on the organization's team and not be limited to just working hours. They must respond to the full spectrum of individual needs. They must communicate with and impact both the conscious and the subconscious mind. The level of attention and concern which must be given to achieve success requires that the stimulus and direction of such programs emanate from high levels in the organization. At the personal level the individual must be concerned especially about self-reinforcment programs, which must be structured such that the person will not succumb to lethargy.

Knowledge Building

This component of SV reflects the old adage that "knowledge is power," and suggests that the organization or person with superior knowledge has superior power and as such has a competitive edge relative to all competitors. At the same time, however, today's organizations and individuals have to operate in increasingly sophisticated and complex environments, which require them to have new dimensions of knowledge in order to compete. If organizations are successful in acquiring the additional dimensions of information needed, they find that their problems are further compounded by difficulty in processing, analyzing, assessing, and interpreting orders of magnitude larger amounts of data, further translating them into an intelligible form and then getting the results to the right people at the right time to permit effective managerial decision making. It is important to distinguish between the terms "knowledge" and "data" as used in SV. Data refers to basic information or facts void of any understanding of their impact. Knowledge, on the other hand, refers to data that have been acquired, processed, analyzed, interpreted, and mastered to the extent that they impact vision accomplishment.

The increasingly sophisticated and complex environments which are fostering the requirement for new dimensions of knowledge attainment are the result of the following:

1. *Opportunities from Advanced Technology.* Technology holds the key to success in defense acquisition in many different ways. Manufacturing technology advances hold potential for reducing costs and increasing production rates of weapons presently being produced. The modification of present production designs to incorporate advanced materials and use advanced testing equipment could greatly improve the performance and reliability of future production items relative to weapons that are being produced today. It is obvious that advancing technology will also aid your competitors, so it is important to keep tabs on their technology. Technology increases will continue to create the basis for and suggest new weapon systems. The needs to solve military problems will continue to stimulate technology to increase. It is necessary for military organizations and individuals to have superior knowledge of advanced technology as it is pertinent to their vision achievement.

2. *Needs of the Person-System Employee.* A system focus on employees requires adoption of new management concepts and programs not heretofore associated with classical government or industrial firms. The trend to provide day care for both children and elder parents, physical fitness programs, professional development, and career progression programs requires new expertise and knowledge to be generated to construct the program designs which are the most performance- and cost-effective. Continual knowledge of emerging demographics is necessary to permit personnel programs to be developed in order to use this changing work force effectively. For example, future work force compositions will reflect more women. The future population will include increasingly larger numbers of able-bodied and experienced older people. As more employees engage in international travel, more knowledge relative to employee travel security (such as against terrorism) must be obtained.

3. *Opportunities within an Ever-Increasing Complex Environment.* Most democratic nation states have pluralistic societies, featuring a plethora of competing special-interest groups. Governments in such countries become the agencies through which mutual interests can be explored and preserved and competing interests adjusted. As a result, complex systems of laws and regulations have evolved. As more and more countries in the world choose democratic forms of government and as the mores and cultures of populations change, even more factions are forming which must be heard and satisfied, causing governments to design governing mechanisms of still further complexity.

Coupled with these changes is an increasing world concern about the physical environment of the planet and social inequality among the races and ethnic groups about the globe. These concerns are generating political forces that can be expected to produce dynamic and continuing change in existing laws and government regulations, and create new ones that could greatly impact the success of military projects. These same political forces will greatly influence public opinion, and hence public support for the project. As a result, the project manager and his team must reflect these changes and display a social and environmental conscience. Such a pro bono publico stance is not just altruistic, it makes good business sense. A strategy with this orientation can minimize the potential of harmful contractor employee strikes, fines and lawsuits for polluting the atmosphere, costly facility modifications, alienation of the tax-paying public, and so on.

Inherent in having concern for the social and physical environments is pertinent knowledge and requires a whole new dimension of data collection and understanding, and organizational strategies. Such social and environmental postures should not be considered as passive. In effect they are proactive if they look for and seize the many opportunities that are generated in such a dynamic atmosphere of change. Going still further, the project organization may need to mount a local, national, or international educational and lobbying campaign to correct false information, improve unfounded poor images and perceptions, and so forth. The feedback loop in the SV model of Fig. 28.2 from the knowledge building block to the motivation block should be noted. Positive knowledge is in itself a motivator.

Action Planning and Plan Implementation

Managers in organizations, whether they be in the military, profit or nonprofit, public, or private, share a consensus that we are entering an era of strategic planning. There is growing recognition that organization success or even survival depends on the longer-term perspectives. The Japanese have been very successful in using longer-term considerations. In effect they have been using the SV model to some degree. They have targeted various advanced technology areas wherein they planned to become the best in the world. After formulating such visions, Japanese industry, in concert with and supported by their government, has generated longer-term strategic plans and taken the necessary follow-through actions, and has achieved world market domination in so doing.

The SV approach to action planning recognizes and uses many classical strategic and tactical planning techniques that have been developed. SV does, however, concentrate on the creation and articulation of the visions, mechanisms for achieving high levels of motivation and superior knowledge. The SV concept emphasizes participative management utilizing modified forms of management-by-objective (MBO) schema. Good MBO designs create clear, effective top-down, bottom-up, and sideways communication channels and involve everybody in the organization, inputting their visions into the process and participating in deciding what their roles will be in achieving the company vision. MBO ties the organization's strategic and tactical planning together. The establishment of yearly objectives and success criteria for each person provides still further motivation. The great benefits of a properly structured MBO-type system far outweigh the time involved to create and implement it.

SV gives special emphasis in planning implementation to Herzberg's hygiene factors to assure that all of these negatives are avoided. Further, emphasis is given to using various classical "operations management" concepts, such as quality management and control, inventory management and control, locations and design of facilities, and choices of implementation technologies. DoD's total quality management (TQM) concept is used and is most effective to foster, ensure, and link high productivity with high quality. TQM is a philosophy which must be applied and tailored in each case, and, hence, its detailed definition is unique in each instance.

Progress and Continuous Evaluation

A measurement of progress generally involves assessing an individual's or a corporation's present status relative to a series of milestones or steps which have been defined to bring them closer and closer to vision accomplishment. Every employee under an MBO regimen will at least yearly have an understanding with his boss as to how far he has gone toward the overall vision or pieces of it that are his part of the overall effort.

It is necessary, however, to communicate to all the stakeholders the "big-picture" progress as well as to speak to their individual concerns. As such it may be necessary to use different levels of communication and define the progress in different terms to different stakeholders (such as immediate project team member versus supporting subcontractor employee). An example can be seen in a country at war. Although various citizens are performing many different, unique tasks to help the war effort, the government must give them big-picture progress reports on how the war is going. Progress reporting must be as often (continuous) as practical to ensure that every eye remains focused on the vision. Continuous evaluation involves a constant looking at progress in such a way as to provide early warning of potential weak areas so that remedial actions can be taken as appropriate.

REFLECTIONS OF STRATEGIC VISIONEERING CONCEPTS NOW IN PRACTICE

There is mounting evidence of a growing acceptance, especially by private industry, of two of SV's cardinal precepts, motivating the person system and building a superior knowledge base. Examples of this evidence can be seen in the following paragraphs.

Fitness and Health

Tenneco opened in 1982 an $11 million exercise complex at its Houston, Tex., headquarters to attract employees and to motivate them to stay on board in spite of having to run the gauntlet of downtown traffic-jammed streets in order to get to work. Tenneco's wellness program also includes employee self-help programs to combat nicotine and alcohol addiction, stress, and obesity. The company subsidizes the cost of healthy breakfasts for employees who arrive early in the morning for workouts before the workday begins. Even employee spouses are permitted to join them in using the fitness facilities. Tenneco is so concerned with wellness that it has established a corporate management position entitled Vice President for Health, Environmental Medicine and Safety, which is staffed by a medical doctor.

American Telephone and Telegraph Company started experimenting with wellness programs in 1983. Its corporate Vice President for Health Affairs reports that since the beginning of the program 70,000 of its 300,000 employees have participated. AT&T has projected that its employees who participated in the program would have well over 2000 fewer heart attacks and 245 fewer cases of cancer for each 100,000 employees over a 10-year period. AT&T has been so pleased with these results that it began a Health Risk Management Program in June 1990, which will feature giving 60,000 employees a free health appraisal each year which includes evaluating blood cholesterol, hypertension, and overweight problems.

International Business Machines, Ford Motor Company, and the Campbell Soup Company are other examples of major corporations who believe in the value of wellness programs not only as motivators, but who also appreciate these programs as cost savers in health benefits and reduced absenteeism, and as increasers of company productivity. The emergence of a whole new wellness industry with new companies (such as Staywell Health Services, Travelers Corporation's Center for Corporate Health Promotion, and Johnson and Johnson), which offer health promotion products and services to a wide range of organizations (from public to private in nature), serves to show that the wellness concept is catching on.

Frequent business travel is a well-known inhibitor to the continuance of health and fitness programs. Wellness-oriented companies are now booking their employees into hotels which have the best fitness amenities available. These amenities include well-equipped exercise rooms, exercise bikes in the rooms, indoor swimming pools, and jogging tracks, for example. This policy not only serves to reduce the tensions associated with travel, but serves to make the employee more productive and alert while conducting business. Typical responses by the hotel industry are seen in the provision of "health suites" at 93 Embassy Suites hotels across the United States, with each room featuring treadmills or exercise bikes, circuit weight-training units, and a refrigerator filled with health snacks. Another example is Holiday Inn's Sportshotel concept and their Holidome indoor recreation centers.

Day Care for Children and Elderly Parents

While the need for day care for preschool children is already a fact of life and is burgeoning, *Fortune Magazine* predicts that elder care assistance will be "the emerging employee benefit of the 1990s." Such trends and projections are supported by demographic studies and forecasts which indicate that the number of women in the workplace relative to men is increasing, and that this trend will continue over the next few decades. The need to hire more women, coupled with a rapidly growing population of single parents, has made it necessary for many companies to consider provision of day care facilities for the preschool children of their employees. Further demographic changes reflect the fruits of advancing medical science, which permits employees' parents to live much longer. The bad news is that in some cases these elders cannot care for themselves and, hence, become a burden physically, mentally, and financially on the employee. As a result, currently over 300 major U.S. companies have already established or are considering providing elder care programs. IBM, one of the first companies to enter this arena, is spending $3 million on elder care assistance programs. This IBM program includes hiring and training in-home social-service and health specialists, developing respite care and intergenerational programs. Examples of big-name companies actively involved in care giving are Johnson & Johnson, Atlantic Richfield, Honeywell Inc., Pitney Bowes, Stride Rite, Ciba-Geigy, Pepsico, American Express, and Colgate-Palmolive. Typical benefits offered are flexible work schedules, counseling information services, unpaid leave to permit care for any family member, and so forth.

As a result of the aforementioned increasing numbers of old people and single parents with young children, a whole new class has emerged in society called the sandwich generation, referring to those people who must care for both their children and elderly parents simultaneously. In reponse to the needs of their sandwich-generation employees, Stride Rite has opened an intergenerational day care facility which is designed to accommodate up to 55 children and 20 dependent elderly persons.

Security

The new emerging economic centers in Western Europe and the Pacific Rim, changes in governments in Eastern Europe, and the condition of the U.S. economy are creating great pressures for the establishment of joint or cooperative international military projects around the globe, which in turn are generating significant increases in international travel. At the same time forecasts indicate an increase in terrorist activity as new nations emerge from the old Soviet Union and borders of the new European community become more invisible. Further projections (and present experience) indicate great increases in crime in many countries (especially Eastern Europe). The net of this discussion is that the threat to the security of the U.S. government and business traveler (and the tourist as well) has increased dramatically.

Training their employees in antiterrorist tactics is not new to some multinational companies who have been operating in the few high-threat areas of the world. But now the few are becoming many. The growing number of companies and U.S. government organizations who are new to the international arena are becoming concerned and must gain a new dimension of knowledge and provide new levels of training in antiterrorist and crime prevention to their staff. While some companies, who have had experience in thwarting terrorism, have established pertinent intelligence links and private databases, the new entries must either build their own in-house capability or subscribe to new emerging anticrime and antiterrorist databases and assessment services. In addition, they must contract out necessary training to a new emerging commercial intelligence industry which is being spawned in this new environment. An example of the new commercial intelligence breed is International Travel Security Consultants, Inc. (ITSC), which advertises "intelligence and security for today's traveler." ITSC's focus is on terrorism, political unrest, and crime in particular countries of interest. In addition to offering subscriptions to its analyzed database, ITSC has custom-designed services in the areas of executive protection, crisis planning, crisis management, product contamination, diversion and counterfeit, aviation security, communications and computer security, industrial counterespionage, and customized in-depth analyses for inputs into decision making relative to entering developing overseas markets.

Still More Reflections

As a means of fulfilling the needs of the person system for career progression, many public and private organizations fund tuition for university courses leading to advanced degrees. Some of the military services even require their officers to have an advanced degree in order to qualify for promotion. In addition, a number of organizations have developed professional development programs which provide select groups of employees with specialized advanced learnings in the form of short courses, seminars, intensified courses, and so on. Professional development is considered to be preparation for the next higher-level positions in the career ladder while training is aimed at helping the employee to do his present job better. The Naval Air Systems Command conducts a showcase professional development program through its Naval Aviation Executive Institute (NAEI), which provides advanced learning to its 6112 institute members. The institute comprises all of the command's GS-13 civilian employees (in grades GS-13 and above) and naval officers (in ranks LCdr and above). In addition to these advanced learnings and support of graduate study, NAEI sponsors a senior executive development program to prepare junior executives for the federal senior executive service and a civilian material professional career program which is a systematic approach to the certification and career development of its employees interested in the acquisition management profession.

THE FUTURE OF STRATEGIC VISIONEERING

SV is a natural outgrowth of the explosive increases of information-age technologies, astounding technology advances in other pertinent areas (automation, telecommunications, robotics, and so on) and research to better understand the workings of the human mind and its supporting knowledge and control mechanisms (that is, the brain and the central nervous system). It is a stimulus that will bring about a revolution in management theory and concepts. Its focus is on building the "superteam" or individual "superstars" by generating high levels of motivation or enthusiasm within them and giving them a superior knowledge base to work with. SV significantly enhances creativity and innovation because of its supporting symbiotic relationship with knowledge, idea generation, and the acquisition of conceptual skills.

SV's emphasis on exploiting the whole mind (both the conscious and the subconscious) and more efficient use of the human "megabrain" will gain wider and wider acceptance as more and more understanding of the brain and how and why it works is gained. Already, for example, amazing results in curing human diseases (such as cancer) have been achieved through the application of psychoneuroimmunology (PNI). PNI is a self-healing process that is the result of, in effect, programming the subconscious mind to tell the brain to direct all of the appropriate human body internal forces to join together to rid the body of the problem. If this concept is successful in fixing existing problems in the body, it surely can be successful in preventing the problem from happening in the first place. PNI training will become a basic part of future organizational wellness programs and personal health and fitness regimens (with an obvious consequential positive stimulus to motivation).

Another SV focus, the subconscious mind, will command growing attention when the understanding of its importance to human motivation, action tendencies, and knowledge building becomes widely accepted. Use of various techniques to communicate with the subconscious to cause replacement of the old negative, self-limiting scripts with positive, limitless scripts can be expected to increase rapidly. Employment of subliminal audio and video programs to generate motivation and enthusiasm, increase knowledge, stimulate self-healing, overcome bad habits, and so forth will flourish. Use of hypnotic techniques to expand the knowledge base in the conscious will be common in the workplace. Training in self-hypnotism will become a standard part of many future training and professional development programs.

Ethical concerns over techniques used to communicate with the subconscious mind will naturally arise as accounts of brainwashing, negative mental programming, and psychological warfare are resurrected from the history files. These concerns will fade away as results such as those from Dr. Benjamin Lebet's research on the mind and brain (at the University of California at San Francisco) indicate that "free will" is, indeed, "alive and well" and can override and negate spontaneous intentions that arise from the subconscious. Brainwashing per se, such as it has been used in the past cases of prisoners of war, involved other dimensions. It had been conducted in an environment of physical or mental torture, wherein the needs of the individuals involved had been reduced to the very bottom level on Maslow's hierarchy and the issues were life or death.

The person-system concept will receive wide recognition as the relation of motivation and the whole person becomes understood. It will receive even greater emphasis as advanced technology provides opportunities for new innovative ways to satisfy human needs. For example, advancing telecommunications, robotics, and automation technologies are the basis for forecasts that the flexi-workplace concept will flourish in the future. As such, many workers could work at home, effectively utilizing high-tech data links with their office or plant. Such a decentralized approach, for example, could permit the worker to still do his assigned task and provide for child and elder care simultaneously while avoiding losses of travel time, fuel, and vehicle costs and the frustrations of traffic snarls in driving to and from the workplace.

SV, while emphasizing participative management and positive thinking, suggests other management concepts that go far to stretch organic management concepts further away from the mechanistic by using metapsychological approaches. Much further out in the future there will be attempts to employ intuitive management concepts using managers trained to enhance their dormant yet existing personal capabilities to employ parapsychology, clairvoyance, telekinesis, and the metaphysical.

Already corporate vice presidents for wellness or the equivalent exist in private industry. SV thinking suggests further corporate-level positions to capitalize on other new potentials in both government and industry. This thinking suggests that future organizations will feature vice-president-level positions (here expressed generically) or their equivalents in government for motivation/enthusiasm, communications, corporate knowledge, person-system support, social/environmental conscience, and so on. One approach is to design these positions into a matrix organizational format superimposed on a classical functional-type organization.

Henri Fayol, one of France's early management pundits, was the first to introduce the notion of the universality of management (namely, it is a body of knowledge that is useful for and can be applied to all endeavor). SV, which is a management tool, exhibits much of this universality in its potential application. SV is a powerful concept. By employing it, old cliches, such as "shoot for the stars," "the sky's the limit," become respected affirmations again. The SV concept is bound to command wide attention as its superteam or superstar end products begin to achieve success in making their dreams become realities by following them with a mind and heart borne of "strategic visioneering." It suggests a whole new dimension to the role of the military project manager and to his strategic planning as well. The net result is that military project managers, along with their manager counterparts in private industry, will shift their focus more and more on *people* as our nation shifts away from the information age to an age of caring.

BIBLIOGRAPHY

Brice, L., "Gateways to Self Discovery," Gateways Institute, Ojai, Calif., Winter-Spring 1990.

Brice, L., "Listen Your Way to a New Life," Jonathan Parker's Gateway Institute, Ojai, Calif., 1987.

Burt, B., "How to Choose Hotels with the Best Fitness Amenities," *Washington Flyer*, pp. 47–52, Mar./Apr. 1990.

Cole, A., "The Memory Puzzle," *AARP Newsl.*, Apr. 1990.

Crowley, S., "Firms Help Workers Who Help at Home," *The Washington Post,* Mar. 20, 1990.

Fanning, D., "How to Retire and Stay Alive Too," *The New York Times,* Apr. 1, 1990.

Locke, S., and D. Colligan, *The Healer Within: The New Medicine of Mind and Body,* New American Library, 1986.

McAuliffe, K., "Get Smart: Controlling Chaos," *Omni,* pp. 44–92, May 1990.

Miller, C., "Intelligence Systems," *Marketing News,* May 9, 1988.

Restak, R., "Your Brain Has a Mind of Its Own," *The Washington Post,* Mar. 27, 1990.

Rossi, E. L., *The Psychobiology of Mind-Body Healing,* W.W. Norton, New York, 1988.

Shapiro, J. P., "When Workers Choose between Careers and Taking Care of Aged Parents," *The Washington Post,* Mar. 20, 1990.

CHAPTER 29
TECHNOLOGY, TRANSITION, AND TRANSFER

Steven M. Shaker

Steven M. Shaker is a program manager with Global Associates Ltd., Arlington, Va. He is a noted futurist specializing in technology forecasting and the future applications of robotics, unmanned vehicles, and manned system automation. He is author of *War without Men: Robots on the Future Battlefield* (Pergamon, 1988).

INTRODUCTION

The transfer of sophisticated technology from one organization into another is not a simple feat. Successful technology transfer goes far beyond taking a formula, design specification, or research note and passing it on to another scientist or engineer in a different locale. It involves more than having a different organization replicate a scientific technique, method, or process.

Government and corporate executives are often unpleasantly surprised when they learn of the immense difficulty that a new company or manufacturing plant has in building a product which is readily manufactured elsewhere. What they soon learn is that there are intangibles encompassed in the technical know-how that require experienced people from the originator (transferor) organization to assist in hands-on operations at the replicator (transferee) organization. The successful program manager must master and facilitate this flow of technology know-how.

Technology transfer has varied connotations for different organizations. For the intelligence community, the Department of Justice, and much of the Department of Defense (DoD), as well as for the mass media and the public at large, technology transfer has meant the illegal acquisition of western technology by the former Soviet Union or by other adversaries, which ends up being incorporated in their weaponry used against us.

To the program manager technology transfer has typically meant the transfer of technology from one production source to another to attain competitive production sources or to enable multinational coproduction. The Defense Systems Management

College has published "A Program Office Guide to Technology Transfer" to assist the program manager in developing a technology transfer plan. The goal of the plan is to facilitate the qualification of a second source through a series of steps, including:

- Component verification
- Interchangeability demonstrations
- Process validation, fabrication, and assembly demonstrations
- Performance testing
- Configuration audits

Technology transfer may also infer to the program manager the Domestic Technology Transfer Program. A series of executive orders and congressional acts, such as the Stevenson-Wydler Technology Innovation Act of 1980 (P.L. 96-480), the Federal Technology Transfer Act of 1986 (P.L. 99-502), and Executive Order 12591 of April 1987, have developed programs to facilitate the flow of DoD-developed technology into civilian applications, and to provide for cooperative development of technologies of importance to the DoD and the civilian economy. With our nation's increasing economic difficulties and concern with competitiveness, program managers are being tasked with giving more importance to the Military Civilian Technology Transfer and Cooperative Development Program.

THE DoD TECHNOLOGY BASE

The collapse of communism in the Soviet Union and Eastern Europe combined with the growing economic might of Japan, the Pacific Rim, and the European Common Market is resulting in a transformation of the U.S. defense posture and the resultant allocated resources. Policies, budgets, and weapon programs are being dramatically altered. Likewise, the role of the military program manager is evolving and adapting to the changing world.

Although East-West tensions have dissipated, the competition for scarce dollars and expertise within the United States is escalating. During the periods of defense growth when dollars were plentiful, such as in the mid-1980s, the emphasis of the program manager, as the title implies, was on management. Resources being somewhat more stable allowed the program manager to devote greater effort toward the internal functioning of the organization. In periods of major cutbacks and overall defense decline, such as we are currently experiencing, a program manager must shift toward being a leader and advocate for the program. More time is devoted to influencing external decision makers and outside processes which ultimately determine the program's fate.

Key to the program manager's success is being able to monitor technological developments, and to transfer the relevant scientific and engineering know-how to the program. To do so effectively in today's world, where much of the leading state-of-the-art technologies are being derived from commercial applications, and often occur outside of the United States, requires the use of some innovative managerial tools and approaches.

Since World War II the United States has relied on a strategy of countering the numerical superiority of potential adversaries, principally the Soviet Union, with technically superior weapon systems. The fall and breakup of the Soviet Union has deflated some of this rationale. With the possible exception of the Peoples Republic of China, few potential adversaries can match the U.S. war machine in size. Many foreseeable Third World engagements, however, envision adversaries who, at least at the onset of hostilities, have a numerical superiority in tanks, armored vehicles, and troops. In some instances their weaponry will also represent the latest in western technology or in proven Soviet design.

Many political analysts also maintain that the U.S. populace, in today's media age, has a very low tolerance for casualties. U.S. forces can be victorious on the field, but what price is the public willing to pay? Technology is thereby one means, with the advent of low observable (stealth) technology, smart and brilliant munitions, automated systems, and unmanned vehicles, in which survivability of U.S. systems is enhanced and losses can be kept to a minimum. Having superior weapon systems and technology therefore continues to be essential to the United States in order to counter larger forces and reduce U.S. casualties. A major role of the program manager is to ensure that U.S. weapon systems incorporate the necessary technology to meet these goals.

Program managers have relied on the DoD science and technology (S&T) base to furnish the necessary technology. With many of the key emerging technologies being developed outside the United States and by commercial enterprises, it is highly questionable whether the DoD S&T base will in itself suffice for the needs of the program. Therefore new strategies and modus operandi are needed to utilize these nontraditional sources of technological expertise.

Traditionally the program manager's technological needs are supported throughout the acquisition process from the DoD S&T community. In 1991 the DoD budget for S&T activities was approximately $9 billion. This supported some 50,000 scientists and engineers, operating in 70 service laboratories and commands. These facilities comprise more than 1.3 million acres, including 38 million square feet of laboratory space, and are valued at more than $65 billion.

The DoD S&T community focuses on the front end of the research and development process encompassing three basic categories: basic research (6.1 program category), exploratory development (6.2), and the initial stages of advanced development (6.3A). Basic research involves advancing science through theorizing and experimentation without any direct linkage to application. Although most DoD laboratories have research departments, much of the basic research is contracted out to colleges and universities. This enables the labs to hire and retain some of the best scientific minds in academia.

Exploratory development is focused on applying scientific advances to the military environment. The output of 6.2 funding often results in databases and new methodologies, state-of-the-art design tools, and specifications for future systems.

Advanced technology development is aimed at furnishing proof of principle and concept demonstrations, which verify the feasibility of a system or approach, and advances a technology toward a future military capability. The funding supports the development of special components, subsystems, and feasibility prototypes called advanced technology transition demonstration (ATTD) projects.

CRITICAL TECHNOLOGIES

Key technologies and expertise required for many U.S. programs lie outside the country. The 1991 joint military net assessment states that although currently no country is ahead of the United States in any overall area of technology, the United States often lags behind foreign countries in incorporating the latest technological advantages in its fielded products. U.S. leadership in defense-related technologies is being increasingly challenged by the Pacific Rim countries, the European community, and the Soviet Union. The move toward the manufacture of defense products by international consortia is resulting in an increased use in the United States of foreign technology.

The Department of Commerce issued a 1990 report entitled "Emerging Technologies: A Survey of Technical and Economic Opportunities." It examined 12 emerging technologies that were considered to have a major impact on the economy. The report found that if current trends continue, the United States would lag behind Japan in most of these technologies and trail the European community in several of them.

Of the 12 emerging technologies, the trends reveal that in terms of research and development, the United States is losing to Japan in five of the technologies (advanced materials, biotechnology, digital imaging technology, sensor technology, and superconductors). There is no R&D area where the U.S. lead is increasing. In terms of product introduction, the U.S. standing via Japan is even more dismal, with the United States losing in nine of the technologies (advanced materials, advanced semiconductor devices, biotechnology, digital imaging technology, high-density data storage, high-performance computing, medical devices and diagnostics, optoelectronics, and superconductors).

Even in current technology areas the United States is becoming increasingly dependent on foreign sources for application and manufacture. During Desert Storm the U.S. government required the assistance of 30 foreign governments to ensure that supplies of critical components were kept in the logistics pipeline. This included components for the F/A-18 ejection seat, M1 tank optics in the gunner's primary sight, and microcircuits in the computer that aims and fires the tank's main gun.

A recent study conducted by the General Accounting Office maintains that "DoD does not know the impact or significance of its foreign dependency problems; DoD's awareness of dependencies is limited; previously identified dependencies still exist; and DoD's efforts to develop adequate information on dependence have been slow in coming and inadequate."

The program manager must be alerted to these real-world facts, namely, that the U.S. DoD S&T base may be insufficient in meeting the program's needs. The 1991 joint military net assessment states that much of this dependence (to foreign sources) is irreversible. Thus we must monitor foreign content carefully in order to ensure that dependence does not become a vulnerability.

A TECHNOLOGY BUSINESS INTELLIGENCE SYSTEM

The old adage "knowledge is power" has never been more true than in the so-called information age of today. Success for corporations, government agencies, and program offices is increasingly dependent on a timely awareness by key executives of rapidly changing events outside their organizations. Yet most organizations, including the program office, have become caught in a catch-22. To raise their knowledge base, program offices have become so involved with gathering information that they are swamped by facts and figures. They cannot readily separate the "wheat" from the "chaff" to determine what is important in order to manage this deluge of data.

A new management tool or, more aptly phrased, a process has been developed to assist decision makers in managing the massive amounts of data relating to technological developments and other factors which can impact the program. Called "business intelligence," it utilizes a similar structured process as practiced by the national security intelligence community in obtaining essential information in a timely manner, although modified for the acquisition and business world. Business intelligence is not industrial espionage. It is a legal and ethical approach to collecting, analyzing, and disseminating information which influences decision making. It is the smart way of doing business.

Corporations such as AT&T, Kodak, Motorola, General Dynamics, Phillips Petroleum, Southwestern Bell, and Nutrasweet have instituted their own business intelligence systems. Government organizations such as the National Institute of Standards and Technology and the National Science Foundation as well as several DoD program offices have set up similar mechanisms to monitor technological developments and to alert key decision makers to events as they occur, which can impact programs.

Developing a business intelligence system for the program office involves a form of intelligence engineering. If one envisions a pyramid hierarchy of information, the bottom level would be considered data. This is information in the public realm, usually published in the open literature. This information often is well known and "stale," that is, too late to capitalize on. The next level is information which has added value by being analyzed data. At the apex of the pyramid is intelligence. Intelligence is information which is "actionable" by the program manager and has foresight. It responds to the program manager's informational needs and leads to a decision. Currently, program managers have massive amounts of data and public information, small amounts of value-added information, and relatively little intelligence. Intelligence engineering involves shifting the flow of information from data to intelligence to support the program manager.

The intelligence is not the same type as gathered by the CIA, DIA, or other elements of the intelligence community. This intelligence relates not to technological developments of potential adversaries, but rather to technology which may be of value to our own programs and the acquisition business. It furnishes the program manager with an early warning of external events that may impact the program or office. It identifies new technological opportunities that can enhance the program, or technological threats which can threaten the program's current approach. Ultimately it would enhance decision making and the planning process. Business intelligence raises the program managers' awareness to external changes, thus preparing them for action. The program manager can be proactive rather than reactive to the course of events.

A business intelligence system would monitor technologies relating to the program. This would include intelligence on U.S. and foreign advanced systems, subsystems, components, and technologies similar to, or relating to, that of the U.S. programs. Commercial developments as well as basic technological breakthroughs would be monitored, which can be applied to the military program. In addition, intelligence relating to domestic and international industry structures and trends affecting the program, as well as political, economic, military, and social forces of change influencing the program, would be collected, analyzed, and disseminated to the program manager and other key decision makers. The business intelligence system can either be positioned within the program office with contractor assistance, or perhaps be a new service offered by the DoD laboratories to assist each program office.

In this period of dramatic changes within DoD and its acquisition programs, new approaches for conducting business are required, and the opportunities increase for innovative program management techniques.

BIBLIOGRAPHY

"A Program Office Guide to Technology Transfer," Defense Systems Management College, Ft. Belvoir, Va., Nov. 1988.

Cooper, E. D., and S. M. Shaker, "A Strategic Planning Methodology for Smart Munitions Programs," USAAMCCOM, June 15, 1989.

"Emerging Technologies—A Survey of Technical and Economic Opportunities," Technology Administration, Department of Commerce, Washington, D.C., Spring 1990.

"Manufacturing Technology: The Key to the Defense Industrial Base," American Defense Preparedness Association, Arlington, Va., Oct. 1989.

"RDT&E/Acquisition Management Guide," Department of the Navy, NAVSO P-2457, Washington, D.C., Jan. 1989.

U.S. Congress, Office of Technology Assessment, "The Defense Technology Base," OTA-isc-374, Government Printing Office, Washington, D.C., Mar. 1988.

Warren, R. A., and C. R. Cooper, "Science, Technology and the Program Manager," *Program Manager,* pp. 20–27, July–Aug. 1991.

CHAPTER 30

ADVANCED TACTICAL FIGHTER (ATF) ACQUISITION STRATEGY DEVELOPMENT

Wallace T. Bucher

Wallace T. Bucher has served in a wide variety of systems acquisition positions during his Air Force career. He began his career as an avionics maintenance officer and served as an advisor to the Vietnamese Air Force. His systems acquisition experience includes tours as systems program manager and deputy program manager for logistics for the AIM-9L sidewinder missile at Naval Air Systems Command. He also established the integrated logistics office for the Munitions Systems Division and was the deputy director for program control for the airlift and trainer program office. He was program director for the C20 and MC130H programs and deputy director of the F-22. He currently serves as the director of acquisition development for the space and missiles systems center.

INTRODUCTION

It appears to be traditional in the U.S. Air Force that when a new fighter is designed, a new management system is put in place to oversee its development. This was seen in the F-15 program, when the program director was given direct access to the highest levels of the Department of Defense (DoD) and several unique contractual features were instituted. The tradition was continued in the Advanced Tactical Fighter (F-22) Program, where an integrated management system was developed for the engineering and manufacturing development (EMD) phase of the program. This management system, as well as the total acquisition strategy of which it was a part, was carefully crafted and prototyped during the demonstration and validation (Dem/Val) phase of the program and has been called the most innovative and well-constructed program management technique in the DoD today. Many of the new concepts developed for the ATF are being adopted by other programs and used in the planning for future development efforts. This chapter will show how that management system and acquisition strategy were developed as integral parts of the total ATF Dem/Val effort. It will delineate the innovative features of the strategy by showing the objectives and the processes behind their development and use. In order to put the acquisition strategy in perspective, it is necessary to describe the entire Dem/Val effort.

ATF DEM/VAL PROGRAM OVERVIEW

The purpose of the ATF Program is to develop a weapon system which will maintain air superiority well into the 21st century. At this time, the F-15 will be at the end of its useful life and the United States must have a weapon system that has the characteristics required to dominate the emerging global threats. Specifically, the major characteristics required in the ATF are low observability, long range, large payloads, supercruise capability (the ability to cruise supersonically without the use of afterburners), high maneuverability, and excellent supportability. The Dem/Val phase of the program is the development period where the final decisions are made on the balance of requirements and technologies that will be taken into EMD and production.

The purpose of the ATF Dem/Val program was threefold. First, it was to define a set of potential weapon system characteristics based on demonstrated technology. This was crucial to assure that eventual EMD and production of the selected weapon system were based on proven technology, thus minimizing potential technical setbacks which result in cost or schedule problems. Second, it was to define and validate with the rest of the Air Force the requirements necessary to defeat the threat and dominate the air superiority arena. This activity was aimed at specifying the optimum balance of all potential choices. The third objective, although not clearly articulated at the beginning of the program, was to develop the structure of the EMD program by capitalizing on this unprecedented understanding of what was needed and could be done.

The Dem/Val program began after a short concept definition phase in which seven of the major weapon system contractors performed analytical studies to examine the feasibility of developing a weapon system with the desired characteristics. Initiated in 1986, the Dem/Val program built upon these studies and further refined and demonstrated the feasibility of the weapon system. The Dem/Val contracts were awarded to two prime contractors for the weapon system—the Lockheed Aeronautical Systems Company (LASC) and the Northrop Corporation (NAD). The two prime contractors established commercial agreements with other aerospace firms in order to share the costs of the program and to take advantage of significant technologies and areas of expertise that these firms, now as team members, could bring to bear on the program. NAD teamed with the McDonnell Douglas Corporation and LASC teamed with the Boeing Corporation and the General Dynamics Corporation. The government contracts themselves were firm-fixed-price efforts with no mention of cost sharing in the contracts. Throughout the Dem/Val program, the press and other sources referred to the ATF Program as a cost-shared program, which, in fact, was not the case. Government funding was considered adequate to meet the minimum requirements of the contract. Large contractor investments by all of the parties involved resulted from the force of competition and their desire to win the program.

In addition to the weapon system contract, two existing Air Force contracts for the ATF engines were brought into the program. These contracts were with the General Electric Corporation (GE) and the Pratt & Whitney Corporation (P&W) and were originally part of the Joint Advanced Fighter Engine Program begun in 1983.

This early start was necessary to provide sufficient lead time for developing the engines in advance of the weapon system. These contracts were also best effort, firm-fixed-price contracts subject to the same type of investment strategy on the part of these contractors. Under these contracts, both engine contractors focused on new centerline engine designs and technologies to support the ATF weapon system design (such as stealth, supercruise, supportability).

The significant characteristics of all of these contracts were that they called for the development of two prototype air vehicles (PAV) from each weapon system contractor and six prototype engines from each engine contractor. The weapon system contracts also called for the development of avionics ground prototypes (AGP) and avionics flying laboratories (AFL). All of the trade studies, analyses, and demonstrations required to develop full system specifications were also included.

The nature of the Dem/Val program was to allow the contractors the freedom to explore solutions and to demonstrate technology in a competitive environment under government direction. As a consequence, the statement of work (SOW) for the Dem/Val contract was kept very general in nature. It was identical for both the engine and the prime contractors and was kept that way throughout the entire Dem/Val program. There were no hardware deliveries required under the contract, as the only output of the Dem/Val program was to be data in the form of technical reports, trade studies, and analyses. Specifications of the program were goals documents, termed initially the system requirements document (SRD) and later the preliminary system specification (PSS). These specifications and all program goals were continually evaluated and updated by both the contractors and the Air Force. Specifications never reflected any one particular contractor's design, but rather those technical characteristics that the Air Force deemed feasible to fit within the technical risk, schedule, and cost goals of the program.

REQUIREMENTS REFINEMENT

While putting the EMD contracts in place, the ATF system program office (SPO) developed a unique methodology to refine the Air Force's requirements, based on the technologies to be demonstrated and matured throughout the Dem/Val program. This process is illustrated in Fig. 30.1, which shows that, at the beginning of the program, the Air Force had many wide-ranging and extremely demanding requirements for the weapon system, but very little data to determine what was, in fact, achievable. The contractors at the beginning of the program also had stated that they could fulfill all of those requirements. Consequently, as engineering data were generated throughout the Dem/Val program, the requirements were refined to meet the technologies agreed upon by the Air Force and the contractors.

Figure 30.1 also illustrates that, at the beginning of any program, the desires of the user generally exceed the current state of the art. Requirements are based on the user's assessment of his needs and of the status of the technologies to fulfill those needs. In general these technologies hold high promise but many times fail to achieve this promise—often by significant margins. Consequently a successful program must continually examine to the maximum extent practical the true state of the art so that the customer, in the ATF's case, the tactical air forces, has a clear understanding of what can be delivered. These assessments evolve into a very specific set of requirements at the start of EMD and should be reexamined throughout the entire life cycle of the program to assure that the requirements of the user can be developed within the allocated resources. The process of defining those requirements is

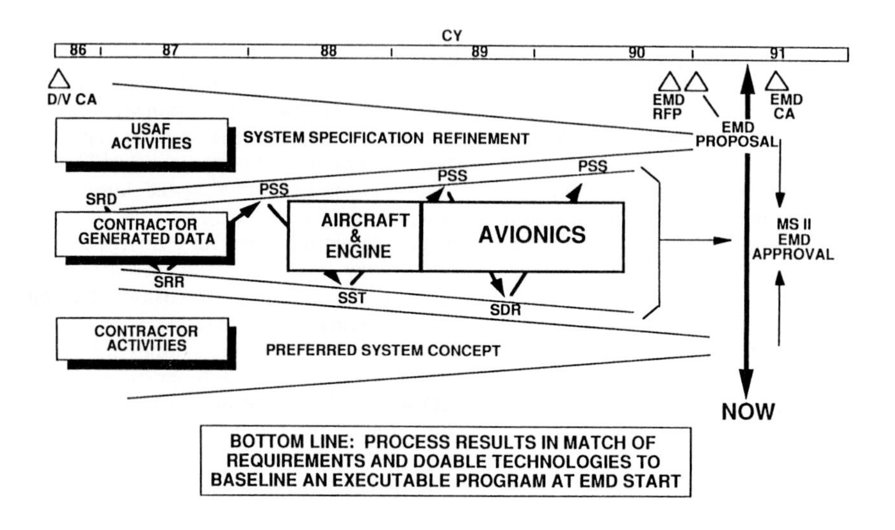

FIGURE 30.1 Dem/Val process to define requirements.

one of tradeoffs and compromises to ensure that the needs of the user are consistent within the context of other program goals, such as affordability.

In order to define these requirements, the user must be willing to participate and accept compromises as initial technical feasibility is determined. This process must include all of the customers of the system, the users, the maintainers, and industry, to name a few. The process has to be a well-documented, logical, complete track of all requirements, and tradeoffs must be rigidly maintained. The delicate balance between what is needed versus what is achievable and affordable must be assessed continually to ensure achievement of the optimum solution.

In the ATF Program the requirements provided at the beginning of the Dem/Val program were general in nature. Trade studies were conducted to refine the requirements against a standardized threat. The user was continuously involved in this disciplined requirements trade process, and the leadership of the Air Force was regularly briefed on the results. Periodic assessments of alternative solutions for air superiority were also analyzed to ensure that the baseline aircraft was clearly the best product to meet the needs of the Air Force. This balanced approach to the definition requirements assured commitment at every management level of program activity.

MATURATION OF TECHNOLOGIES

Just as requirements must be defined, so must the technologies which will be applied to the program. This is obviously an interactive process in which a disciplined evaluation of the technical risks must be compared with the absolute need for the incorporation of a technology into the program. As a consequence, the SPO needed to identify and manage all potential sources of technology evolution and to provide optimum levels of maturity for every part of the weapon system. A strong contractor and laboratory interface was established in which the SPO acted as facilitator and mediator. Before any technology was applied to the program, it was proven in terms of its achievability and its contribution to the program. As with all things in this process, the customer was involved in this evaluation. A fundamental approach the ATF SPO followed was to provide periodically a "menu" of viable options to its customer based on demonstrated results and showing what cost and effectiveness (against the threat) could be achieved. The customer then picked the option, and the process of refinement continued.

In the ATF Program a specific technology transfer program was established and reviewed every six months in order to focus emerging technologies on the ATF program and to expose the contractor to its benefits and risks. All sources of emerging technologies, including Air Force and Navy laboratories and other DoD agencies, participated in this program. All of these were focused through the SPO on the emerging system.

The purpose of this technology transition program was to ensure that, at the start of EMD, risks were refined to a manageable level. Stated another way, the ATF had to be a doable do. Consequently, during Dem/Val all of the technologies that would be used in EMD were demonstrated. Everything earned its way onto the ATF Program as part of this disciplined process. Also, a careful documentation track was always maintained and decisions were made in a team environment.

This entire maturation process was structured to provide the necessary options for the customer to make the hard choices of what the ATF should be. By being provided with validated information, the customer had confidence that his requirements could be met, significantly strengthening his support of the program.

CUSTOMER DEFINITION AND SUPPORT

There is one significant and common feature to the development processes described so far. It is the approach taken to involve all of the parties in the program. In order to make an integrated, interactive program approach work, the first order of business is to identify all customers of the program. This had to be done early in the program and included a broader view of customers than is traditionally taken. In the Air Force the traditional user is Tactical Air Command (TAC), Strategic Air Command (SAC), or Military Airlift Command (MAC). This classical definition had to be expanded to include support agencies, training agencies, reviewers (such as OSD), security interest, intelligence sources, and virtually any other agency that would have any involvement in the program. It was important to let each of these agencies know that their inputs were valuable, and that they were considered as integral to the development of the system. This team must also have the right types of expertise with the appropriate authority to seek out and resolve all issues. Team members must be willing to compromise as program trades are conducted. Consequently frequent and focused communication was essential to ensure that all team members had the necessary factual information to define and support the total program.

In the ATF Program team work was emphasized at all levels, from the beginning of the Dem/Val program. This was fostered and exemplified to all of the team members through the establishment of working groups. These working groups (which today are called integrated product teams) focused on virtually every aspect of the program and brought together the expertise and decision makers from throughout the Air Force, Navy, and the rest of DoD. They covered such diverse issues as armament integration, common avionics, security management, test, and cost estimation, to name but a few. Active participation by all members of the acquisition team was sought and issues were surfaced and brought to immediate resolution at the proper levels of authority. This process focused the entire team, ensured that all team members knew and understood the rationale behind the decisions, and proved that their point of view had been considered. This fostered a wide and diverse range of program support. This process is identical to the integrated product development approaches taking root in much of American industry today.

Another example of how this integrated team process was implemented was the general officer review (GOR). The GOR, which was conducted in the final year of Dem/Val, brought together general officers from throughout the Navy and the Air Force and represented every command and agency involved in the program. The purpose of the review was to look at the final draft request for proposal (RFP), which contained all the specifications and requirements of the government. Once the approval of this group was obtained, the draft RFP was given to industry, and all of the comments, discussions, and changes were then recoordinated with the staffs of each of the general officers who were present at the review. This effort brought together the technical and business communities and coordinated their efforts prior to the release of the EMD RFP. This activity sent a strong message to industry that the entire Air Force and Navy had a solid position and that the program was well supported.

DESIGN PROCESS DOCUMENTATION

The design of the ATF was documented through a series of design reviews and reflected in the Air Force preliminary system specification (PSS). The program office met continually with the contractors throughout the Dem/Val program to examine the results of their analyses and demonstrations. On an annual basis, as depicted in Fig. 30.1, the SPO would update the PSS and present it to the highest levels of the Air Force and contractor community to gain consensus and agreement on the need for and achievability of Air Force requirements. PSS became the document which told both contractors what the Air Force thought was feasible. The contractors maintained a companion document, entitled the preferred system concept (PSC), which showed the Air Force what their system could do in relation to the PSS. This allowed all parties within the Air Force and the industry to know where they stood relative to each other and to identify areas in which further work was required. The final RFP reflected this iterative process so that the competitors knew the weapon system requirements well in advance.

ACQUISITION STRATEGY DEVELOPMENT OBJECTIVES

In concert with defining requirements and maturing technologies, a business plan or acquisition strategy for the next phase of the program was concurrently constructed. This facet of the program took the same integrated team approach, which required, first of all, that the contractor be actively involved with the entire government team in the development of the acquisition strategy. This strategy had several key objectives. First, it had to be fair to all involved. It could not appear to the contractor as an assumption of undue financial or programmatic risks. The development of the acquisition strategy followed the same rules which were applied to the previously mentioned portions of the program, that is, the contractor must be committed to an executable program, and that program must have an open, measurable criteria and a system to track progress. The contractor must also be motivated to manage the plan through his continual involvement in the process. The contractor must be sure that all requirements are identified and achievable, that there is no duplication between the government's and the contractor's efforts, and that the resources of the program are viewed as a whole. He must also be sure that the government is willing to pay a fair and reasonable price.

In the ATF Program the acquisition strategy was evolved through extensive discussion with industry and through the employment of innovative management systems which fully integrated the contractor into the planning process. The first step that the SPO took was to establish an acquisition strategy working group which, over the course of Dem/Val, examined all aspects of the business arrangement in light of the executability of the program. This open discussion started by examining business concepts, evolved into actually providing several draft RFPs, and allowed the contractors to comment on them. This process permitted the contractors to examine the business aspects of the program while at the same time the Air Force was examining the technologies to be applied and the requirements the system would have to meet. This ensured a better understanding of the total risks of the program on the contractor's part.

Several innovative management "tools" were also introduced into the program during the acquisition strategy working group meetings. These tools constitute the integrated management system which would be used in EMD to meet the objectives of the program. The first tool is the integrated master plan (IMP). This plan, which is generated by the contractors, lays out their entire program for the development and production of the weapon system. It allows the contractor to define the maturation of the system and the criteria to measure his own progress. The IMP is placed on contract and the government uses it as one of its primary tracking tools. This positively motivates the contractor to manage the development effort according to his own plan, while permitting the government to reward his performance. Another benefit of the IMP is that it was prepared before the start of EMD, and consequently achieved the contractor's commitment much earlier in the program.

To complement the IMP, and to demonstrate to the contractors that the government was truly allowing them to define and manage their program, the ATF SPO allowed the contractors to write their own statements of work (SOWs). These contractor-generated SOWs laid out all of the program tasks required to accomplish the IMP. The SOW calls for unique contractor tracking tools which provide both the contractor and the SPO with the visibility to measure progress. These include integrated management schedule (IMS), technical performance measures, and a fully integrated cost-reporting system. Together these tracking tools and the IMP permit the contractor to lay out and track his own program with full and open communication with the government.

This integrated management approach is essential in the current environment of cost-plus contracting. In the ATF Program a cost plus award-fee contract was selected for EMD to provide the contractors with the financial risk protection needed for this effort while providing the government with the leverage to ensure their performance. The award fee will be monitored and given to the contractor on the basis of his performance against his own plan, as measured by the tools which he and the government are applying through the contract.

ATF ACQUISITION STRATEGY DEVELOPMENT PROCESS

The ATF acquisition strategy is a natural follow-on to the entire Dem/Val process. Examined within this context, the Air Force first decides what to buy and defines this in a functional system specification which is the result of the PSS described previously. The second step is to determine how to buy it through the construction of the acquisition strategy. Allowing these two steps to become disconnected is a formula for disaster.

Consequently, one of the primary objectives of the ATF acquisition strategy was to structure a program consistent with the risk of the final product. This is not just technical risk, but also cost and schedule risk. The acquisition strategy therefore had to take a long view of the program, that is, the acquisition strategy must not concentrate on the EMD program and the risk attendant to it, but it has to look at the production as well. After all, the purpose of the entire ATF program was to put a capable weapon system into the hands of the users in quantity. In doing that, the SPO wanted each of the contractors to plan his own program and then to commit contractually to the executability of that program. The government also wanted to provide the contractors the flexibility to manage their programs while it evaluated their performance, guided their efforts, and rewarded their progress. In addition there was one other primary objective to the acquisition strategy—it had to be a fair arrangement for both parties. The contract had to present a win-win situation for both the government and the contractors. The ATF SPO believed that if the contract were not perceived by the contractors as fair, the program would not achieve the level of commitment and team work required to execute the EMD and production programs successfully.

With these objectives in mind, the ATF SPO conducted an extensive process aimed at ensuring an open communication and thorough understanding by both the government and industry of the business arrangements included in the EMD contract. This resulted in the formation of the acquisition strategy working group, as mentioned previously. This group comprised key members of the ATF SPO and their industry counterparts. The purpose of the group was to examine every aspect of the contract, plans, and specifications. This group met quarterly from early 1988 until it was converted into a more intense monthly process immediately before the RFP was released in November 1990.

This effort was paralleled by a thorough review throughout all of government. Three acquisition strategy panels, which are independent, high-level reviews conducted by Air Force Systems Command (AFSC), were convened. The first, in January 1988, looked at the initial concepts to be used in the program; the second, in April 1989, examined the progress made in the implementation of the concepts; and the third, in August 1990, approved the final RFP. In order to achieve consensus at the highest levels of both government and industry, annual meetings were also conducted by the service acquisition executive (SAE), the chief of staff of the Air Force, and the chief executive officers (CEOs) of all the major corporations competing in the program. These CEOs meetings examined all aspects of the program, both technical and business-related.

The documentation associated with this effort was released in several phases. The first was the publication of two preliminary acquisition packages in January and April 1989. These acquisition packages showed the contractors the early requirements of the RFP and were released through and discussed in the acquisition strategy working group. These acquisition packages were not complete draft RFPs, but formed the basis of the draft RFP. In August 1989 the first of two draft RFPs was released. This permitted the contractors to see exactly what the government would be requiring them to bid on a full 16 months before release of the final document. After extensive discussions, guided by the acquisition strategy working group, the second draft RFP was released in April 1990. At that point the SPO conducted monthly in-process reviews (IPRs) in which every facet of the program was examined formally. Formal minutes were taken at each of these meetings, and all actions were documented so that the contractors knew precisely what would change in the RFP at any point in time. About 700 such action items were generated and resolved before the final RFP was released on November 1, 1990.

There were additional steps taken to ensure that full government participation and understanding of the acquisition strategy and the RFP were achieved. A GOR group, as mentioned previously, was formed and conducted in March 1990 to look at the draft RFP and the acquisition strategy. This GOR consisted of members from HQ USAF, TAC, NAVAIR, OPNAV, ATC, AFLC, HQ AFSC, AFOTEC, AFFTC, and ESD. All actions were documented and a continuing review was held with the action officers from these agencies, most of whom participated in the IPR when their portion of the RFP was discussed. Finally, in late August 1990 all of the staff officers coordinated on the final RFP.

Consequently all of the concerns and issues of the government and the contractors were resolved satisfactorily before issuance of the RFP on November 1, 1990. This extensive coordination and communication process was necessary due, in part, to the fact that the ATF had departed in many significant ways from the traditional acquisition approaches.

The ATF program was trying to achieve a thoroughly integrated management system for the execution of the program by focusing the program on the products to be developed and produced during the EMD and production phases. The objectives of the integrated management system were, first of all, to maximize the contractors' involvement by forcing the program to be well documented and planned by the competitors before the proposal was written. It was also to create a very measurable and trackable program description that would be used by the contractors and government to manage the program in a partnership environment. The ATF SPO was attempting to maximize the contractors' flexibility and creativity so that they could emphasize the strengths of their systems and utilize, to the maximum extent possible, their existing corporate processes and systems. This reduced the amount of data and paperwork required and focused on the familiarity of the government with the contractors' practices achieved during the Dem/Val program. The ATF SPO also wanted to focus the source selection on the contract rather than on proposals.

Figure 30.2 illustrates the relationship of the ATF integrated management process and reflects the construction of the ATF RFP and contract. The requirements of the contract in terms of specifications and the SOW were included in the contract. The integrated master plan (IMP) was also included in the contract. The IMP is the contractors' plan for the execution of the EMD program and includes all of their accomplishments, the criteria for the successful completion of those accomplishments, and the events or maturation milestones by which those accomplishments are to take place. The IMP is a modification of the system engineering master schedule currently in use on other Air Force programs. This event-based plan does not contain dates by which tasks are to be completed, but rather shows the overall maturity of the system by each event. The ATF SPO specified a few minimum events and tasked the contractors in their responses to the RFP, to expand the IMP to fully describe the activities they would accomplish in EMD. Also included in the contract is an extensive tracking system with four main features. First is an extensive technical performance measurement (TPM) system, which will show significant performance and technical characteristics of the program over time so that any changes or reestima-

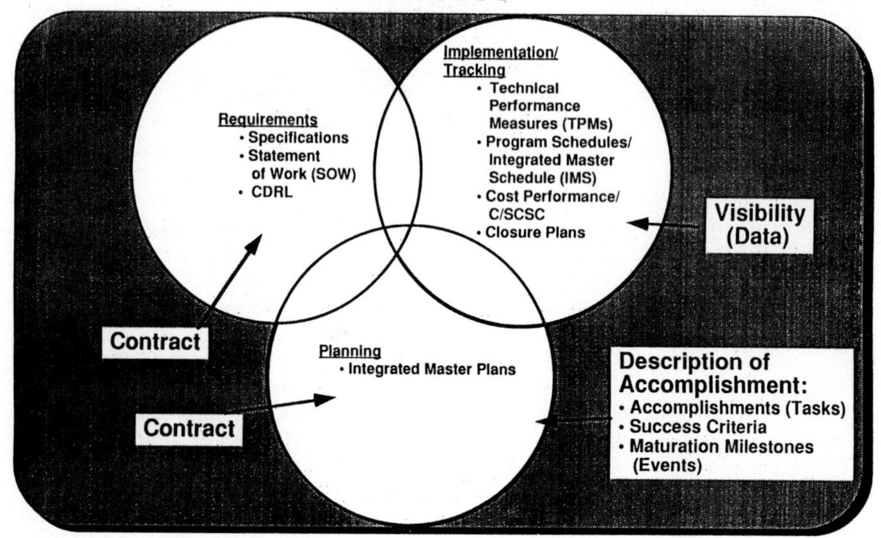

FIGURE 30.2 Integrated management process.

tion of the program's technical content or capability can be reviewed and tracked. Second is the integrated master schedule (IMS). This tiered scheduling system takes the events and accomplishments from the IMP, lays them out over time, and provides additional lower-level detailed tasks to complete the contractors' overall efforts. The third tracking tool is standard cost-schedule control systems criteria (C/SCSC) tracking via cost performance report and other data items. The fourth feature is closure planning, a formal system between the contractors and the government, which is instituted as problems arise or significant actions are required on the program.

Figure 30.3 illustrates the relationship between the principal documents of the ATF contract and states the objectives of the overall process. First the technical requirements of the program were documented in the weapon system specification. They were always kept at a functional level, that is, the specification describes to the contractors how the product is to work rather than telling them how to make it. The RFP called for, and the contractors delivered, 15 weapon system segment specifications which further broke that functionality down into the major constituent parts of the proposed system. This provided the Air Force with a much clearer picture of how the system would perform and the risk attendant to each consistent part of the system.

All specifications are arranged according to the outline of the program. This outline of the program, or work breakdown structure (WBS), uses a single integrated numbering system for the entire program. It is tailored to the development and manufacturing process and is used for cost reporting, planning, and organizing throughout the life of the ATF Program. The WBS is carried over to the SOW, and each SOW paragraph utilizes this numbering system. The ATF SOW is unique in that it

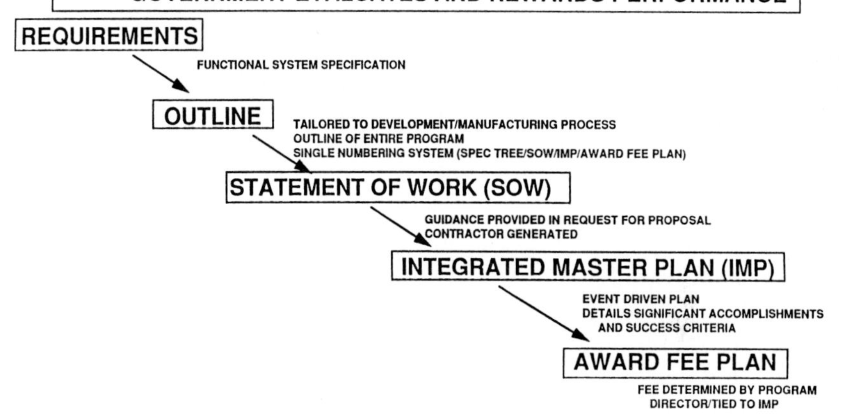

FIGURE 30.3 Integrated management structure.

was contractor-generated. The Air Force, in the RFP, provided the contractor with examples of SOW paragraphs containing the minimum essential requirements of the program. The contractors responded by providing their own unique, tailored SOWs. The SOW, in turn, is complemented by the IMP, which is an event-driven plan detailing the significant accomplishments and criteria for the contractors' success in the development of the ATF. This has all been tied to the award-fee plan, which is not part of the contract, but is provided to the contractors prior to the RFP release. The award-fee program, which will be discussed later, is the principal method by which the Air Force will reward contractors' performance.

The IMP was incorporated into the contract so that no portion of the ATF Program plan could be changed without the mutual consent of the government and the contractors. Consequently the IMP is event-based, that is, no dates appear in the IMP. All of the scheduling detail appears in the IMS. This permits the contractors the flexibility required to manage their programs without constantly modifying the contract. It also provides the government with the oversight necessary to ensure that the contractors are progressing according to their plans while maintaining the visibility of progress according to schedule.

Figure 30.4 shows the relationship of the IMP to the IMS. As can be seen in this example for the landing gear subsystem, the preliminary design will be completed by the preliminary design review (PDR). This design is complete when the duty cycle has been defined, preliminary analysis has been completed and reviewed, and preliminary drawings have been released. Examples of significant accomplishments which will detail the entire development of this subsystem are also shown. The IMS in this figure then takes the same accomplishment and shows what is required and when it is required in order to complete the accomplishment criteria. Consequently

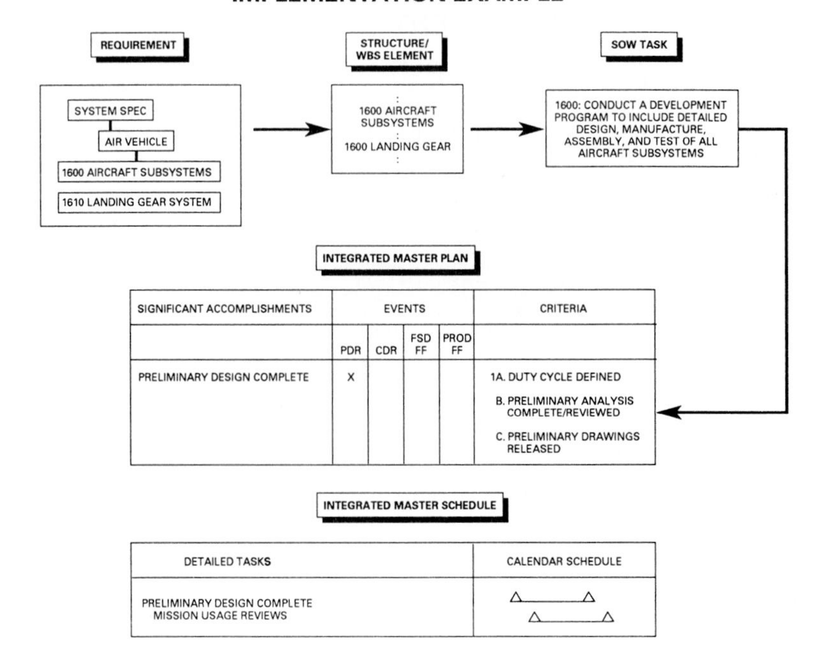

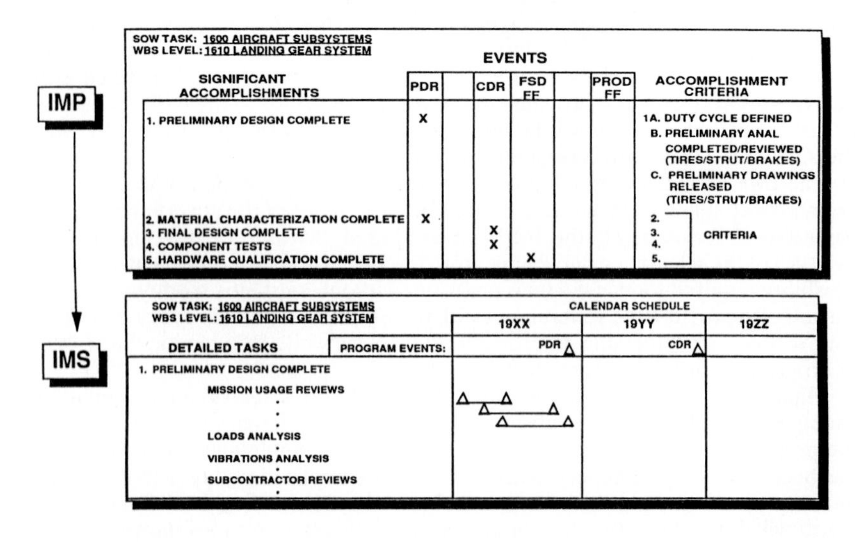

FIGURE 30.4 Implementation example.

these two systems provide the detail necessary for a thoroughly planned and scheduled program.

The IMP and the IMS provide all of the detail necessary to form work packages against which the resources of the program can be allocated. This system then forms the basis for all cost reporting through the C/SCSC. The work packages which are defined by these two elements are, because of the single numbering system of the program, all related directly to cost accounts. The work packages are then tracked via cost performance reporting, and can be tied directly to the IMP, thus tying the plan, the schedule, and the resources of this program together for increased management awareness and problem-solving ability.

The ATF award-fee program is tied directly to this integrated management system. The evaluation criteria contained in the award-fee plan is based on the contractors' execution of the total program as viewed through the integrated management system. The areas of evaluation in the award-fee plan measure the system maturity as reflected in progress through the IMP, as well as the ability to meet schedules as reflected in the IMS, control of cost, and progress toward meeting specification requirements. This offers both the contractors and the government a method for openly judging performance under the contract and for rewarding this performance.

With this integrated management system in place and the contract focused on the products of the ATF Program, the ATF SPO wanted to send an additional message that the total focus of the program was to be placed on the products. Consequently the philosophy of integrated product development was adopted as the organizational principle for the SPO. Integrated product development organizations were also required by the RFP to be put in place at each of the contractors' facilities.

Figure 30.5 shows the organization of the ATF SPO. Four primary integrated product development teams are shown at the bottom of this organizational chart.

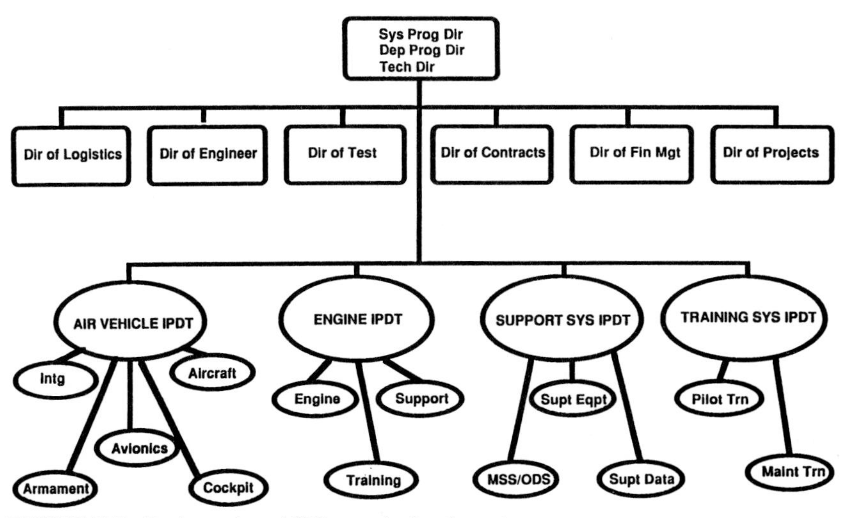

FIGURE 30.5 Product-oriented SPO organizational structure.

These teams correspond directly to the major products of the program—air vehicle, engine, support subsystem, and training system. Additional subteams were formed to concentrate on the major portions of the program. The functional offices listed across the top of Figure 30.5 have four primary responsibilities. First, they provide the manpower resources to each of the integrated product teams. Second, they act as an interface to their functional counterparts throughout the rest of the Air Force and contractor communities. Third, they are charged with the technical excellence of each of their respective disciplines. And last, they are responsible to promulgate policies and procedures for the execution of their disciplines within the program.

SUMMARY

In summary, the success of a program such as the ATF Program is due to two fundamental ideas. First, make absolutely sure, through a rigorous and disciplined process, that what is being asked for is truly needed and can be achieved with the time and resources allocated. Second, plan the work to be done in a manner that makes the contractors accept true ownership of the task. In establishing the ATF acquisition strategy, these two fundamental principles were always adhered to. The integrated management system, which is the centerpiece of the acquisition strategy, contains all of the elements of the program thoroughly integrated and tied to the organizations of both the contractors and the government.

CHAPTER 31
THE ROLE OF THE USER

Charles B. Cochrane

Charles B. Cochrane is Professor of Systems Acquisition Management, Acquisition Policy Department, Defense Systems Management College, Ft. Belvoir, Va. He retired from the U.S. Army with over 25 years of active service. While on active duty, Mr. Cochrane (then Lieutenant Colonel) was the principal staff officer at Headquarters, Department of the Army, for implementation of the Packard Commission recommendations for streamlining the program manager's reporting chain. He also played a key role as the acquisition member of the study group charged with implementing the DoD Reorganization Act of 1986.

WHO IS THE USER?

The user is generally thought of as the final recipient of the products or weapon systems produced by the defense acquisition system. The role of the user in defense system acquisition and the user's interface with military project management have not been well documented. Department of Defense (DoD) Instruction 5000.2,[1] *Defense Acquisition Management Policies and Procedures,* specifically recognizes the role of the user in requirements generation. However, the user's role extends well beyond the development of the requirement for a new weapon system.

To place the role of the user in perspective, it is helpful to understand the definitions dealing with the DoD acquisition system. DoD Instruction 5000.2 defines this system as "a single uniform system whereby all equipment, facilities, and services are planned, designed, developed, acquired, and disposed of within the Department of Defense."

The Federal Acquisition Regulation, subpart 2.101, states in part: "Acquisition begins at the point where agency needs are established and includes the description of requirements to satisfy agency needs . . . " The Defense Systems Management College defines acquisition as "the conceptualization, initiation, design, development, test, contracting, production, deployment, and logistic support, modification, and disposal of weapon and other systems, supplies, or services (including construction) to satisfy DoD needs, intended for use in or in support of military missions."

Clearly, the user is involved in requirements generation. The acquisition program managers and defense industry take the user's requirements and design, develop, and produce the system. Modifications are made as necessary, usually based on input from the user. What may not be well understood is the role the user, or the user's representative, plays in the total acquisition process depicted by the preceding definitions, including design, development, test and evaluation, deployment, and logistic support. This role will be expanded on in this chapter.

User and user representatives are defined by the Defense Systems Management College as:[2]

(1) That command, unit or element which will be the recipient of the production item for use in accomplishing a designated mission.

(2) User representatives are: Army, Training and Doctrine Command; Navy, OPNAV (the staff of the Chief of Naval Operations); Air Force, Operational Commands (such as Air Combat Command, Air Mobility Command, *et al.*); Marine Corps, Headquarters, USMC, and the Marine Corps Combat Developments Command.

(3) The operator and maintainer of the system.

Based on this definition, what may not be evident is that the ultimate users of weapon systems are the soldiers, sailors, marines, and airmen primarily serving in or under the operational control of combatant commands. The combatant commands are the unified and specified commands that have warfighting missions. The military departments (Army, Navy, and Air Force) are charged with maintaining a system for research, development, and acquisition of weapon systems to support the combatant commands and to provide a worldwide support structure for systems when fielded to those commands.

Users are also found in support roles within the military departments, or in the reserve force structure. Further, these users are represented within the defense acquisition management structure by major commands and headquarters staff agencies within the military departments, and by the Chairman, Joint Chiefs of Staff.

The unified and specified commands have operational control of all U.S. combat forces. National Guard and Reserve forces are earmarked for contingency to one or more of these commands. A unified command consists of forces from two or more military services, while a specified command is normally made up of forces from a single service. These commands report directly to the national command authorities, the President and the Secretary of Defense. Orders are transmitted to these commands through the Chairman, Joint Chiefs of Staff (CJCS). The relationship of these "using commands" to the military departments, the CJCS, and the office of the Secretary of Defense is illustrated by Fig. 31.1.

THREE LEVELS OF USERS

The role of the user in military project management must be discussed on three levels. First, there is the role of user representatives within the military departments. User representatives may be dedicated full time to this task, although in some cases, particularly at a headquarters staff, it may be a part time job. Then there is the Chairman, Joint Chiefs of Staff, supported by the joint staff, who is in continuous contact with the commanders of the combatant commands. Finally (but first in priority) are the actual users, the commanders in chief (CINCs) and the soldiers, sailors, marines, and airmen of the unified and specified commands. At each of these three levels the user brings a warfighter's perspective to the acquisition process.

The Military Departments

Within the military departments there are user representatives who provide the link between the warfighting requirement and the acquisition system. Each military department is organized for this task in a slightly different way.

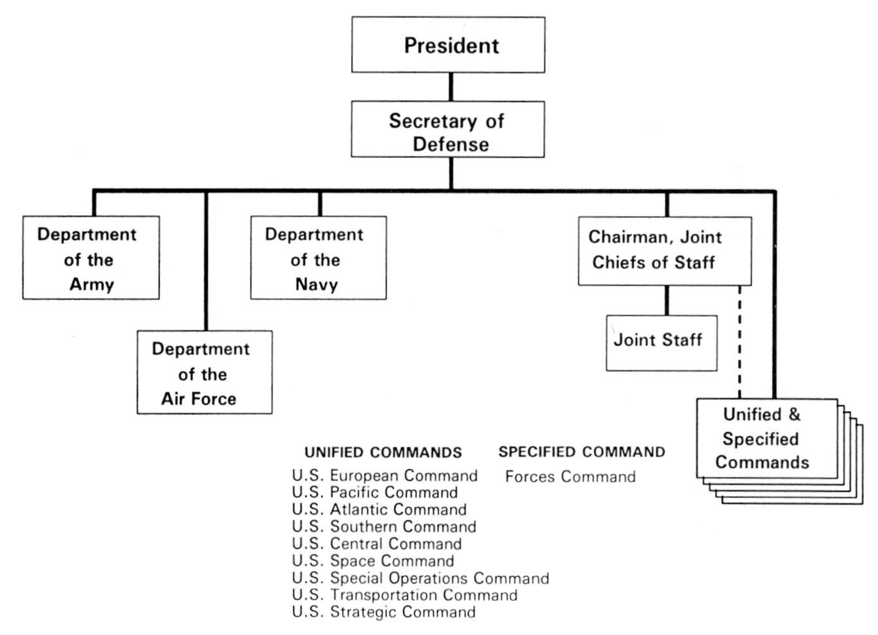

FIGURE 31.1 Users of defense weapon systems.

U.S. Army. Army operational users are located in the ground component commands of the combatant commands, such as U.S. Army Europe, U.S. Army Pacific, U.S. Army South, the Special Operations Command, and the Military Traffic Management Command. Forces Command, a specified command, also has operational users.

The Training and Doctrine Command (TRADOC), headquartered at Fort Monroe, Va., is the user representative within the Army. TRADOC maintains combat development directorates at its schools and centers throughout the United States. These combat development directorates receive input from Army combat units, government and industry technology base activities, and the Army leadership as to future needs of the Army. Working closely within this framework, and assisted by the Army acquisition work force, the combat developments staff officers identify warfighting deficiencies and produce the requirements documents necessary for the acquisition work force to modify existing systems or acquire new systems to correct these deficiencies.

At the Army staff level in the Pentagon, the Deputy Chief of Staff for Operations and Plans (DCSOPS), a lieutenant general, represents the user in dealings with the Army acquisition executive (AAE) (the Assistant Secretary of the Army for Research, Development, and Acquisition) and the other members of the Army staff and secretariat involved in the acquisition process. The DCSOPS also prioritizes, for the chief and the secretary, TRADOC's list of battlefield deficiencies and the acquisition programs necessary to correct these deficiencies. The DCSOPS has an assistant for force development, a major general, who oversees a number of warfighting mission area divisions headed by colonels. These divisions include aviation, fire support, intelligence and electronic warfare, command and control and signal, combat maneuver (infantry, armor, and engineer), future combat systems, and air defense.

Within DCSOPS mission area divisions, systems integration staff officers, or SIs, are the day-to-day link with the user. The SIs perform a series of critical functions as user representatives. They are the HQDA point of contact for requirements, force modernization, and fielding. They monitor all research, development, and procurement programs and provide input to the program objectives memorandum (POM) and budget. SIs help defend the program to the Office of the Secretary of Defense (OSD), the President's Office of Management and Budget (OMB), and Congress, and they play a role in determining the overall priority of the program for the Army. They are also a link between the user and the Army acquisition executive's (AAE) action officers.

U.S. Navy. Primary users are the major naval operating forces which make up the naval components of the combatant commands, such as the Pacific and Atlantic fleets; Naval Forces Southern Command; Naval Forces, Europe; and the Military Sealift Command. Other Navy users are located in the Navy shore establishment.

The staff of the Chief of Naval Operations, referred to as OPNAV, is the user's representative in the Navy. OPNAV, located at the Pentagon and other offices nearby, performs much the same function as the Army DCSOPS. However, the Navy does not have a formal structure for combat developments. Within OPNAV the Deputy CNO for Resources, Warfare Requirements, and Assessments, a vice admiral, is responsible for Navy warfare mission areas. This DCNO has directors, sometimes referred to as platform sponsors, who provide the link with Navy users in the operating forces for surface warfare, air warfare, submarine warfare, expeditionary warfare, and special programs. Within these directorates, requirements officers provide the daily link between the user community, OPNAV, and the office of the Assistant Secretary of the Navy for Research Development and Acquisition—the Navy acquisition executive (NAE).

U.S. Marine Corps. Although part of the Department of the Navy, the Marine Corps, like the Army, has a formal structure for identifying user requirements. The Marine Corps Combat Developments Command (MCCDC), located at Quantico, Va., performs a role similar to that of the Army's TRADOC. MCCDC develops requirements documents, operational concepts, and warfighting doctrine. MCCDC's Warfighting Center responds to needs identified by the users in the fleet Marine forces (Atlantic and Pacific), the Marine Air-Ground Task Force Commands, and other Marine users.

At Marine Corps headquarters, near the Pentagon, the Deputy Chief of Staff (DC/S) for Requirements and Programs, a major general, reviews and validates requirements; ties requirements to the planning, programming, and budgeting system (PPBS); and looks after combatant command priorities for Marine Corps assets. While the Warfighting Center and the Marine Corps Systems Command (MARCORSYSCOM) (also at Quantico) cooperate on user requirements from the fleet Marine forces, the DC/S for Requirements and Programs functions as an honest broker to ensure compliance with Marine Corps and Department of the Navy needs. The DC/S also acts as an interface with the OPNAV staff for Marine Corps requirements.

Marine Corps aircraft requirements and initiatives are supported by Navy funds, and are processed within the Navy system. Within headquarters, Marine Corps, the Deputy Chief of Staff for Aviation, a three-star general, is dual-hatted as the Deputy Chief of Naval Operations for Marine Aviation and sponsors aviation initiatives within OPNAV.

U.S. Air Force. Air Force users are located in the air component commands of the unified combatant commands, such as Air Forces Europe, Pacific Air Forces, and the

Air Force Special Operations Command. Other users are in the Air Mobility Command, the Air Combat Command,* and the Air Training Command.

The Deputy Chief of Staff for Plans and Operations, a member of the Air Staff located at the Pentagon, is the Air Force user representative. This organization has a directorate of operational requirements led by a major general with divisions for global operational requirements and contingency operational requirements. Staff officers within this directorate are the primary point of contact with the using commands, and provide the user a link with the Air Force acquisition executive, the Assistant Secretary of the Air Force for Acquisition.

Within each of the military departments, representatives of the users perform the following functions, which may vary slightly depending on the military service (Army, Navy, Marine Corps, and Air Force):

Conduct studies and analyses to determine warfighting deficiencies. This activity should consider the threat and warfighting mission areas of the service and take a long-term view, that is, look 10 to 20 years into the future.

Assess warfighting deficiencies to determine if nonmateriel solutions will correct the deficiencies. DoD Directive 5000.1 requires that nonmateriel solutions be addressed first. These nonmateriel approaches include changes in doctrine, tactics, training, or organizations (Ref. 3, p. 2-2).

If a nonmateriel solution is the appropriate course of action, develop and publish the changes in doctrine, tactics, training, or organizations for the user to follow.

If a nonmateriel solution will not resolve the warfighting deficiency, document the deficiency as a mission need in a mission need statement (MNS) and forward the MNS through operations channels to the appropriate service acquisition executive for action.

Participate in the defense acquisition process throughout each phase. This includes, but is not limited to, studies and analyses, development of operational and logistic doctrine to support fielding of the new weapon system, participation in operational test and evaluation activities, and assisting the program manager (PM) by updating and justifying the warfighting need (the requirement for the system) throughout the acquisition process.

Chairman, Joint Chiefs of Staff

The CJCS represents the user at Defense Planning and Resources Board (DPRB) reviews of defense planning guidance, the military departments' POMs, and review of major budget issues. The Vice Chairman, Joint Chiefs of Staff (VCJCS) represents the user at milestone decision reviews and program reviews conducted by the Defense Acquisition Board (DAB) of major defense acquisition programs. The VCJCS is the vice chair and only military member of the DAB.

The VCJCS also chairs the Joint Requirements Oversight Council (JROC), the single most important review body with respect to assessing user requirements for acquisition programs. The JROC reviews and approves the mission need statement (MNS) at milestone 0, prior to the DAB review. It also reviews each DAB (acquisition category ID) program prior to milestones I, II, III, and IV to assess the opera-

*Many readers are familiar with Tactical Air Command (TAC), Strategic Air Command (SAC), and Military Airlift Command (MAC). On June 1, 1992, these three commands were deactivated and two new commands were activated. The Air Mobility Command (AMC) includes intertheater airlift assets and most of the theater airlift and tanker forces. The Air Combat Command (ACC) consists of fighters based in the continental United States; bombers; intercontinental ballistic missiles; reconnaissance aircraft; command, control, and intelligence platforms; and some theater airlift and tankers.

tional performance baseline and the operational requirements document (ORD). The Under Secretary of Defense for Acquisition, as the chair of the DAB, relies heavily on the VCJCS and the JROC for their continuing assessment of the user's requirement (Ref. 1, pp. 13-D-1–3). JROC members, in addition to the VCJCS, include the vice chiefs of the Army and Air Force, the Vice Chief of Naval Operations, and the Assistant Commandant of the Marine Corps.

Unified and Specified Commands

The CINCs have a shorter planning horizon than the military departments, that is, they are primarily concerned with fighting the next war, which could come on short notice. The CINCs' staff is therefore generally concerned with those activities that must be accomplished in the short term, such as developing or updating war plans. They also conduct periodic training exercises to ensure combat readiness of assigned forces. For this reason the unified and specified command staffs may have limited resources to look 10 to 20 years into the future and uncover warfighting deficiencies. When these commands want to bring deficiencies to the attention of the acquisition community, they will normally do so through their ground, naval, and air component commanders. These subordinate component commanders have a direct link to their respective service chiefs of staff who have the assets to support their requirements. The role of these commands in defense systems acquisition can be summarized as follows:

Bring warfighting deficiencies to the attention of their supporting services, or if necessary directly to the attention of the Chairman, Joint Chiefs of Staff.

Consult with the Chairman, Joint Chiefs of Staff, during the preparation of the joint military net assessment (JMNA).* Of specific importance to defense acquisition are the U.S. technology and industrial base, and the research and development investment strategy portions of the JMNA.

Inform the Chairman, Joint Chiefs of Staff, of any concerns with the defense planning guidance, the military departments' POMs, and/or budgets during the PPBS process. The chairman is a member of the DPRB, which is used by the Deputy Secretary of Defense to manage the PPBS.

Support acquisition programs in testimony to Congress during the congressional budget process.

Significant deficiencies in the integration of weapon system acquisition programs with national military strategy were identified in the final report to the President by the President's Blue Ribbon Commission on Defense Management (the Packard Commission).[5] In light of the Packard Commission's findings, supported by the Center for Strategic and International Studies and others, Congress stated in the National Defense Authorization Act for Fiscal Year 1989[6] that:

> . . . the office of the Secretary of Defense, the Joint Staff, and the headquarters of the unified and specified combatant commands should more clearly define the necessary operational capabilities and concepts of operations as part of the requirements process and should explicitly consider alternative acquisition programs based on probable levels

*Section 113, Title 10, United States Code,[4] as amended by the National Defense Authorization Act for Fiscal Year 1989, requires annual joint military net assessments to assess the forces and capabilities of the U.S. armed forces and allies against those of potential adversaries.

of resources likely to be approved by Congress and trade-offs among the acquisition programs of the military departments.

Congress restated its concerns on this issue in the National Defense Authorization Act for Fiscal Year 1991, specifically directing the Secretary of Defense to determine the priorities assigned to major weapon system acquisition programs in order to fulfill the needs of the combatant commands.* As users and user representatives, the CINCs, the joint staff, and the Chairman, Joint Chiefs of Staff, play an increasingly important role in determining the requirements and priorities for defense acquisition programs. Acquisition managers at all levels, PMs, program executive officers (PEOs), and Service Acquisition Executives (SAEs) should all be knowledgeable of the annual joint military net assessments (JMNAs)† and the national military strategy reports for insight on the direction these documents may provide to the future of acquisition programs.

COST, SCHEDULE, PERFORMANCE, AND THE USER'S REQUIREMENT

PMs deal with the cost, schedule, and performance of their programs on a daily basis. Funding levels, schedule drivers, technical and supportability performance characteristics are all directly related in some fashion to the user's requirement. Successful acquisition programs reflect a close relationship between the user community and the program office. The influence of the user must be carefully balanced by the PM in terms of overall program objectives.

The user has two overriding roles throughout the acquisition process: requirements identification and requirements verification. It would be naive to think that requirements are not going to evolve throughout the process. The fast pace of emerging technology, changes in the political environment, and changes in the threat, coupled with the lengthy development of a complex weapon system, are likely to produce changes in the requirement. As a program progresses through the life cycle of development and production, just the process of clarification/understanding the user needs may result in derived requirements which may have a significant impact on the program. As the system evolves, the user is involved in requirements identification/clarification and then verification that the system meets the stated requirement, all while working hand-in-hand with the PM under the constraints of cost, schedule, and performance.

In DoD there are two basic requirements documents, the MNS and the ORD. The MNS, mentioned previously, documents a warfighting deficiency for the milestone 0 (concept studies approval) decision review. After a favorable milestone 0 decision by the acquisition executive, phase 0, concept exploration and definition, is initiated based on information contained in the MNS, supported by a valid threat assessment. The ORD translates the warfighting deficiency found in the MNS to system-specific requirements. It is developed by the user concurrent with the phase 0 activities of the PM and industry; is required at milestone I, concept demonstration approval; and is

*Section 901, National Defense Authorization Act for Fiscal Year 1991[7] directs the Secretary of Defense to submit to Congress a national military strategy report for fiscal years 1992, 1993, and 1994. It also defines the role of the Chairman, Joint Chiefs of Staff, in the development of each report.

†The JMNA is both a qualitative and a quantitative assessment of all types of forces, from nuclear to conventional, against both global and regional concerns. It is published in both classified and unclassified versions. The 1991 JMNA[8] covered the Fiscal Year 1992–1997 time frame of the future years defense program (FYDP).

TABLE 31.1 Comparison of MNS and ORD Contents

MNS	ORD
• Defense planning guidance	• Description of operational capability
• Mission and threat analysis	• Threat
• Nonmateriel alternatives	• Shortcomings of existing systems
• Potential materiel alternatives	• Capabilities required
• Constraints	• Integrated logistic support
• Joint potential designator	• Infrastucture support and interoperability
	• Force structure
	• Schedule considerations

Source: DoD 5000.2-M,[9] pp. 2-1-1, 3-1-1–3-1-3.

updated for all subsequent milestones. Contents of the MNS and the ORD are summarized in Table 31.1.

Understanding the relationship of these requirements documents to cost, schedule, and performance and to the acquisition program baseline is essential for both the user and the PM. Problems arise when the user does not understand the impact of changes in requirements to these three critical areas. Further, some users do not appreciate the implication of acquisition program baseline breaches that could occur if the PM does not conduct tradeoffs as necessary to keep the program on track. On the other hand PMs must provide the user a fair hearing. The PM and the user negotiate cost, schedule, and performance tradeoffs throughout the acquisition process. They discuss emerging technologies, the impact of evolving requirements on the program, and alternative acquisition strategies for program success.

Cost

Research and development, procurement, operations, maintenance, and construction costs, indeed the life-cycle costs of a system, are the result, to a large extent, of what the user specifies in the requirements document. The required performance of the system, and the acquisition strategy needed to design, develop, test, and produce the system and to field it in accordance with the user's requirement, set the stage for the life-cycle cost estimate. Development of the final cost estimate is the PM's responsibility. This cost estimate is validated by comparison with a service-independent cost estimate at milestone I and is updated for each subsequent milestone. The PM should consult with the user if cost estimates are too high and discuss tradeoffs.

Schedule

Program schedule is initially based on the user's timing of an enemy threat driven need, which first appears in the MNS at milestone 0. In the ORD at milestone I, prior to entering phase I, demonstration and validation, the user defines what actions, when complete, constitute attainment of an initial operational capability (IOC) and a full operational capability (FOC) for the system. IOC and FOC are defined by DoD Instruction 5000.2, as follows (Ref. 1, pp. 15-6 and 15-7):

> *Initial Operational Capability.* The first attainment of the capability to employ effectively a weapon, item of equipment, or system of approved specific characteristics, that is manned or operated by a trained, equipped, and supported military force.

Full Operational Capability. The full attainment of the capability to employ effectively a weapon, item of equipment, or system of approved specific characteristics, that is manned and operated by a trained, equipped, and supported military force.

Based on these definitions an example may be the fielding of a new field artillery cannon to a battalion. A battalion is the smallest Army unit typically capable of conducting independent combat operations. Once the new cannon is fielded to the battalion, the soldiers have been trained to operate it, and the support structure (maintenance capability, spare parts, and so on) is in place, the Army will declare: "an IOC has been achieved." When the new cannon has been fielded to all the field artillery units within the combatant command, the Army will declare that an FOC has been achieved.

Target dates may be specified in the ORD for IOC and FOC if achieving an operational capability within a certain time frame is important. The user will also describe the impact if the capability is not achieved within this time frame. The time frames for IOC and FOC are used by the PM to develop and update the program schedule.

In the past the user's IOC date was "chiseled in stone," and the PM's schedule was oriented toward that date, often without regard to schedule and technological risk. In the future this may not be the case. The IOC has been used as a benchmark requirement to counter enemy capabilities during a specific time frame. With the collapse of the Warsaw Pact and the Soviet Union, the enemy threat is no longer that easy for the intelligence community to predict. Further, current policy clearly states that "schedule shall be subject to trade-off as a means of keeping risk at acceptable levels" (Ref. 3, p. 1-5).

Performance

Technical and supportability performance requirements are almost totally user driven. These are the operational performance parameters which first appear in the ORD at milestone I. Technical performance parameters, such as speed, weight, range, probability of kill, and supportability parameters, such as reliability, availability, and maintainability, are expressed by the user in terms of thresholds and objectives. Performance thresholds are the user's minimum acceptable requirement for the system when fielded to an operating or combat unit. Performance objectives must reflect an operationally meaningful, measurable, cost-effective, affordable increment above the threshold, or may be the same parameter as the threshold in some cases.

Acquisition Program Baseline

This document is prepared by the PM for the milestone I decision review and is updated for all subsequent milestones. The baseline contains quantified targets for key cost, schedule, and performance parameters of the acquisition program. The baseline is signed by the PM, the PEO, the SAE, and, for programs reviewed by the DAB, the defense acquisition executive. The user's influence on cost, schedule, and performance, as discussed previously, is reflected in this important document. The baseline has two components for each parameter—an objective and a threshold. The user's ORD provides performance objectives and thresholds. The ORD also provides the user's requirement for both IOC and FOC, which have schedule implications. Cost objectives and thresholds in the baseline reflect the funding required for the program to meet the user's performance objectives.

SCIENCE AND TECHNOLOGY AND THE USER

The DoD science and technology (S&T) base includes RDT&E projects in basic research, exploratory development, and advanced technology development. These projects are typically managed by a laboratory or a research and development center, not a PM. A key element of the DoD strategy for S&T base activities is user involvement. To ensure early and continuous participation in the S&T world, users must have formal access to the process. Considering that S&T activities are not governed by the rules for traditional acquisition programs (DoDI 5000.2), this is not an easy task.

The Director of Defense Research and Engineering (DDR&E), in the office of the USD(A), has the lead for ensuring that technological superiority continues to support national security priorities and national military strategy. The DDR&E formulated a new S&T strategy for 1993 and beyond with "early, intensive, and continued involvement of warfighters" as one of its core elements.* The new S&T strategy recognizes that with the collapse of the Soviet Union and Warsaw Pact the threat to national security has changed. This strategy anticipates that regional powers will have increasing access to both Russian and western technology, and that the United States will have to continue to work hard to maintain a technological edge over potential adversaries. While the changing threat has forced the user to shift the analysis of warfighting mission areas from the traditional Soviet/Warsaw Pact threat in central Europe to regional threats all over the world, a similar shift is taking place to focus the user on an assessment of the S&T base. Historically, the S&T base has been relatively free of user involvement, preferring to link directly with materiel development commands or PM offices.

In addition to the early and continuous involvement of the user, the two other core elements of the S&T strategy are specifically related to satisfying user warfighting needs: the explosion of information technology, and the use of advanced technology demonstrators (ATDs) to demonstrate the military utility of new technology in as realistic an operational environment as possible.

ATDs are not new. The Air Force Have Blue program demonstrated stealth technology was feasible prior to development of the F-117 stealth aircraft. The Army Missile Command's assault breaker program in the mid-late 1970s demonstrated radar technology for the Joint Surveillance and Target Attack Radar System (JSTARS), and guidance and propellant technology for the Army Tactical Missile System (ATACMS). ATD efforts by the Army's Human Engineering Laboratory and users at the Field Artillery School led to the development and fielding of the M981 Fire Support Team vehicle (FISTV). Each of these systems saw action in the Persian Gulf during Operation Desert Storm in 1991.

The current S&T strategy is intended to expand the breadth and scope of future ATDs so that potential improvements to warfighting capabilities receive a rigorous assessment to provide better and more mature technologies for traditional acquisition programs. It is hoped that well-planned and executed ATD programs supported by user needs and priorities will reduce technical risk during subsequent

*DoDI 5000.2,[1] Part 5C has a short discussion of S&T and the transition from an S&T project to an acquisition program. Also, the DDR&E has issued guidance on the new S&T strategy in a booklet, *Defense Science and Technology Strategy.*[16] This booklet also discusses seven S&T thrust areas: global surveillance and communication, precision strike, air superiority and defense, sea control and undersea superiority, advanced land combat, synthetic environments, and technology for affordability. These seven areas represent the DDR&E's assessment of the demands placed on the S&T program by the user's most pressing military needs.

prototyping of weapons systems during the demonstration and validation phase of development.

The rapid advancement of information technology will provide the user with more effective command, control, communications, and intelligence structures. Emerging information technology will also assist the user in requirements generation and provide the user with the opportunity to better influence the development process. A major S&T area of user interest is "synthetic environments," or advanced simulation technology to create electronic battlefields where users from the operational forces can synthetically project technological solutions to warfighting deficiencies. This technique can be used to try out materiel solutions in a synthetic operational environment to enhance the requirements-generation process. Operators on an advanced electronic battlefield should be able to conduct simulated force-on-force combat operations and assess technical and cost implications prior to the start of an expensive development program. The same methodology could be used to assess the cost and operational benefit of upgrading current weapon systems by inserting new or emerging technologies.

Although the S&T strategy recognizes the user's importance, the actual process for user involvement is not well articulated.* To have an effective influence over S&T efforts the user must have some sort of a formal link into the laboratory systems and into the R&D centers. The user must be able to influence the priorities of the S&T efforts within each warfighting mission area so that only those that clearly support technologies that will enhance future capabilities are resourced. The military departments and the DDR&E have yet to cleanly define this process.

ACQUISITION PHASES AND MILESTONE DECISION REVIEWS

The defense acquisition process consists of five major milestones and five phases. The role of the user actually starts before milestone 0 in determining the mission need and assessing new technology (discussed earlier). The role of the user has been described by various members of the acquisition community as ranging from that of an "innocent bystander" to a "worrisome meddler." The truth is, the user is an integral part of the acquisition process. The user is part of the program management team, participating in the process from requirements development throughout the development phases and on into production and fielding. Figure 31.2 depicts acquisition milestones and phases and summarizes the role of the user during each.

Milestone 0, Concept Studies Approval

The user has an active role to play during each phase of the defense acquisition process and for the milestone decision reviews. Since it is the user who conducts mission area analysis, determines warfighting deficiencies, and writes the MNS, it is the user who must support the program at the milestone 0 decision review. At this early point in time, a PM may not have been assigned. The service headquarters operations staff, supported by the using command, supports the MNS at the JROC review. If the

*However, TRADOC, the Army's user representative, has established a series of battle labs that have the potential for influencing the S&T base using a combination of synthetic environments and prototyping to identify warfighting ideas, exploit technology, identify requirements, and seek solutions (technology insertions or new requirements) to warfighting capability needs.

Acquisition Milestones & Phases

	PHASE 0	PHASE I	PHASE II	PHASE III	PHASE IV
DETERMINATION OF MISSION NEED	CONCEPT EXPLORATION & DEFINITION	DEMONSTRATION & VALIDATION	ENGINEERING & MANUFACTURING DEVELOPMENT	PRODUCTION & DEPLOYMENT	OPERATIONS & SUPPORT
Conduct MAA. Identify Warfighting deficiencies. Evaluate S&T. Prepare MNS.	Help select/select best alternative for development. Prepare ORD. Prepare COEA. Input to: ILSP TEMP IPS	Update ORD. Update COEA. Input to: ILSP TEMP IPS	Update ORD. Update COEA. Input to: ILSP TEMP IPS Establish training base. Participate in IOT&E.	Update ORD & COEA. Input to ILSP, TEMP, & IPS. Operate training base. Establish support base. Operate & maintain system. Prepare requirements for modifications/product improvements.	

Identify and verify requirements. Evaluate technology and operational enhancements. Negotiate cost, schedule and performance trade-offs with PM. Participate in test & evaluation coordination groups, ILS management, configuration management, and attend technical reviews. Conduct affordability analysis. Evaluate S&T projects.

MS 0	MS I	MS II	MS III	MS IV
Concept Studies Approval	Concept Demonstration Approval	Development Approval	Production Approval	Major Modification Approval

FIGURE 31.2 User participation in acquisition process.

JROC review supports the effort, the same user representatives will support the MNS to the DAB.* The issues at milestone 0 are those dealing with the threat and with the mission need. Although affordability is always an issue, the acquisition community only plays a peripheral role at this milestone. After a favorable milestone 0 decision, the acquisition commands will be brought in to heavily participate in or conduct phase 0 and solicit alternative concepts from industry to resolve the warfighting deficiency identified in the MNS.

Phase 0, Concept Exploration and Definition

The user will participate with the lead organization conducting this phase to assist in the evaluation of materiel alternative concepts proposed by industry. The military departments approach this phase in a slightly different manner. In the Army and Air Force, there is a great deal of user influence and oversight during this phase. In the Navy, the acquisition systems commands manage phase 0, with the OPNAV staff in the role of the user.

In the Army, phase 0 is conducted by a special study group (SSG) chartered by the Training and Doctrine Command, or a special task force (STF) under the supervision of the HQDA Deputy Chief of Staff for Operations and Plans. The SSG or STF will manage the concept exploration and definition effort, including preparation of all required documentation to support a milestone I decision.[10] The Army

*For purposes of discussion here, the acquisition process for major defense acquisition programs will be used. These are the programs reviewed by the JROC and the DAB. However, the same basic procedure applies to nonmajor programs, except that the level of milestone decision review is at the service. A major difference is that, while the Vice Chairman, JCS, represents the user at DAB reviews, the actual user or a direct user representative will participate in the service-level reviews.

materiel commands have membership on the SSG/STF to assist in oversight of the contractors and to help prepare for the milestone I review.

The Air Force Deputy Chief of Staff for Plans and Operations issues a program management directive to the operating command for the conduct of phase 0. The operating command leads the study effort and establishes a concept action group (CAG) to explore materiel alternatives. The Air Force Materiel Command (called the implementing command) develops the documentation to support the milestone I review.[11]

During this phase the user must develop the ORD (mentioned earlier). An ORD must be developed for the most promising system concept. The user will also participate during the identification of opportunities for cost, schedule, and performance tradeoffs within and among the proposed alternatives.

A cost and operational effectiveness analysis (COEA) must be prepared during this phase. The COEA shows the relative advantages and disadvantages of the alternatives being considered during phase 0, as well as the sensitivity of each alternative to possible changes in key assumptions, such as the threat. The PM is not responsible for this analysis; in fact, the COEA must be prepared independently of the PM (Ref. 1, p. 4-E-1). In the Army, the Training and Doctrine Command is responsible for the COEA. In the Air Force, the operating commands are responsible. For Navy programs, the COEA is conducted by a study group reporting to an oversight board cochaired by officials from OPNAV and the Navy acquisition executive's office. COEAs for Marine Corps programs are conducted under the supervision of this same board, except that the Marine Corps Deputy Chief of Staff for Requirements and Programs participates instead of OPNAV. So this important analysis is accomplished by users in the Army and Air Force, while Navy and Marine Corps users heavily influence the oversight board.

The user should identify readiness objectives and thresholds for both technical performance and supportability, and should conduct an affordability analysis during phase 0. The acquisition program baseline (APB), the integrated logistics support plan (ILSP), and the integrated program summary (IPS) will be heavily influenced by the user's decisions in these areas. The APB was discussed earlier. The ILSP documents the planning for logistics support of the system once it is fielded. The IPS is the primary decision document for milestone decision reviews. It provides a summary of program status and the recommendations of the PM and the PEO. An affordability assessment is one of the required annexes to the IPS.

Although the IPS is prepared by the materiel developer (PEO/PM/materiel command), in this author's opinion the affordability assessment annex must be prepared by the service headquarters staff. This assessment must compare the proposed new system emerging from phase 0 studies to other new start programs already approved, and to ongoing programs in development and production. This information is not available to the PEO/PM. It is the user representative at the service headquarters who has access to the data. Obvious sources are the staff agencies of each service chief who are responsible for programming and prioritization of acquisition programs for the PPBS. The affordability assessment is broader than one PM's program. It must consider all programs, and if some must be cut or service priorities change, only the service headquarters should make these decisions, based on the warfighting needs of the user.

The user participates in system engineering technical reviews throughout the phases of development. These technical reviews are designed to enhance communication among the functional disciplines, such as engineering, logistics, quality assurance, cost control, and others. The first technical review, the system requirements review (SRR), occurs late in phase 0. This review is conducted to ensure that the preferred alternative is sufficiently defined to scope a development program in terms of cost, schedule, and performance. It is important for the user to be present at this review as the following items of user interest will be addressed: mission and requirements analysis, trade-off analysis, logistics support analysis, manpower requirements, and the schedule of program milestones.* From this review a draft systems-level specification will emerge to guide the contract for phase I. The user should ensure that this specification agrees with the operational performance requirements as outlined in the ORD.

A formal test and evaluation program is normally not conducted during phase 0; however, an initial test and evaluation master plan (TEMP) must be prepared. Each service will form an advisory group for test and evaluation coordination (the test integration working group in the Army, the Air Force test planning work group, and the Navy test and evaluation coordinating group). The user must be heavily involved with the test planning advisory group and in the writing and staffing of the TEMP. The user's representative must concur in and sign on the signature page of the TEMP before it is forwarded to higher authority. Of the two types of testing, developmental and operational, the user is most heavily involved in operational testing. Developmental testing is conducted by the PM to ensure that the system meets contract specifications. Operational testing is conducted to ensure the system meets the user's requirement as stated in the ORD. During the development of critical issues for test and evaluation, the user is a key source for critical operational test issues, and also provides input to planning for developmental testing.

Milestone I, Concept Demonstration Approval

This milestone is critical. Approval at this milestone constitutes "program initiation." While it is the PM's responsibility to inform the milestone decision authority of the results of phase 0 and of the acquisition strategy for proceeding with the program, it is the user's responsibility to confirm that the requirement for the system still exists. At milestone I, and at all subsequent milestones, the user should help the PM prepare for the review. Major programs will go through a series of reviews, both at the service headquarters and at the office of the Secretary of Defense.

Of particular importance is the review conducted for acquisition category I programs by the JROC prior to the DAB. The JROC will review the PM's proposed APB and the user's ORD to ensure that the user and the PM agree on operational performance thresholds and objectives. The user must be available to assist the PM if a JROC briefing is required. At a minimum, the user can be expected to brief the service member of the JROC. The Vice Chairman, Joint Chiefs of Staff (VCJCS), the chair of the JROC, represents the user at the DAB review. If there is a significant time lag between the JROC and DAB reviews, the user and the PM may provide the VCJCS an update prior to the DAB milestone decision review.

Phase I, Demonstration and Validation

After a successful milestone I decision review, demonstration and validation is the first phase of hardware and software development. During this phase the user updates the ORD, updates the COEA as appropriate, and participates in the continuing review and update of test, evaluation, and logistics support planning. The user attends the technical system design review (SDR) during this phase. The SDR considers total system requirements: hardware, software, operations, maintenance, test-

*See MIL-STD-1521B[12] for a complete discussion on technical reviews.

ing, and training. Engineering design and technology demonstration activities during the demonstration and validation phase may cause some changes to the user's operational performance objectives. If this is the case, the PM must coordinate these changes with the user prior to changing the baseline document. If the ORD is changed, the requirement must be revalidated by the JROC.

If combat forces are required for developmental or operational testing during this phase, the using commands will provide them. This phase may also involve competitive prototyping, that is a "flyoff," or "shoot-off" between two or more industry competitors. The winner may receive the contract for the next phase of development and subsequent production. The user should be involved in the source selection process to pick the winning design.

The PM will establish a number of coordinating and advisory groups to help manage the acquisition process and control the configuration of the system. Some of these groups may be established as early as phase 0, but all will be in place during phase I. These groups include the integrated logistic support management team (ILSMT), test and evaluation advisory/coordinating groups (mentioned earlier during the discussion on phase 0), configuration and change control boards for both hardware and software, interface control working groups, and a computer resources working group. These groups develop and coordinate plans for logistic support, test and evaluation, physical and functional configuration management, and interface control for those characteristics that serve as a common boundary between two or more components of the system. The user is a member of all these coordinating groups, but may not be a voting member depending on the group's charter or service policy.

Milestone II, Development Approval

The user provides essentially the same support at this milestone as before, at milestone I. There will be another JROC review of the updated ORD and the performance baseline thresholds and objectives. Again, the user assists the PM in preparing for the milestone review presentation, both at the service headquarters and at the DAB. After a successful milestone II decision review, the acquisition executive will authorize the program to proceed into engineering and manufacturing development (EMD).

Phase II, Engineering and Manufacturing Development

This phase of development is the most intense in terms of effort and the largest in terms of resources required for both the user and the PM. During this phase the design must be matured for production. Developmental and operational testing will be conducted on production-representative prototypes or on low-rate initial production items. Live fire testing will measure lethality of missiles and munitions and determine crew member survivability when attacked by enemy weapons. Typical users must operate and maintain the system under conditions that simulate combat conditions. User commands will also provide forces to simulate the enemy during testing whenever possible (Ref. 1, p. 8-9).

The user must update the ORD and the COEA in preparation for the next milestone decision review. The ILSP must be revised during this phase, and user input is especially important. The user must plan now to establish a support base for the new system during the production and deployment phase. The operations and support phase overlaps the production and deployment phase, and the user must ensure that required support is in place when the first production items are fielded. Required support includes spares, repair parts, maintenance facilities, and trained maintainers. The user must program and budget for this capability as early as possible so that it is in place when required.

The PM contracts for logistic support analysis (LSA) from industry during EMD. There are a number of logistic support analysis tasks that provide important data to the user. For example, training requirements and recommendations, transportability analysis, provisioning requirements, sources of manpower and personnel skills, combat resource requirements, and postproduction support analysis are all important user-related LSA tasks. The user provides input to the logistics support analysis report (LSAR) database and uses this database to help determine skill levels required for maintenance personnel, training, and other requirements.*

During EMD the user must plan for establishment of a manpower training base for both operator and maintainer training. Depending on the program schedule, construction of training base facilities may be started during this phase. If facilities for training are required, but will not be ready in time to train personnel for early system fielding, the user must consider how training activities will be accommodated within existing assets. Manpower skill level requirements may even create a demand for an entirely new skill, which may impact on recruiting efforts as well as training.

The user should attend two technical reviews at a minimum during EMD. The first is the preliminary design review (PDR), held to look at a configuration item or a number of configuration items.† The PDR is held early in EMD to evaluate the preliminary design of the system prior to the start of detailed design. Later in EMD, a critical design review (CDR) will be held to determine readiness for fabrication. Depending on the complexity of the system and the number of configuration items, more than one PDR and CDR may be held. The user should be there to answer questions on operational requirements, the expected warfighting scenario in which the system will be used, and the tactics and doctrine for the employment of the system. Users are critical to this process. Their role at the PDR and CDR is to verify that requirements are being accurately satisfied, thereby reducing the chance of rework downstream. User participation can save millions of dollars in cost and schedule delays.

Since it is extremely unlikely that a program will be approved for EMD but not for production, it is during this phase that an acquisition program generates the most interest. This interest comes from the service headquarters, the OSD staff, the Congress, and of course the news media. PMs must watch this interest carefully as many of these agencies will try to force changes into the program. Since the design of the system is being finalized, and prototypes that look, feel, and smell like the real thing are being produced, the user also tends to want changes to the configuration. In a 1972 paper for the Defense Systems Management College, Major David Jens Teal (now USAF Lieutenant General Teal) pointed out that "individuals in the using commands do not necessarily share the SPO's (System Program Office) goal or cost consciousness. Their personal crusade items are often made to appear as command requirements when in reality they are not."[15]

This view of the user community is prevalent today throughout the acquisition community. In many cases it is still an accurate reflection of users at various levels.

*MIL-STD-1388-1A[13] describes the LSA/LSAR process.

†Configuration items are components of a total system, or an aggregation of system elements, that satisfy a specific end-use function and are designated for configuration control, for example, an engine or an air frame. Their performance parameters and physical characteristics must be separately defined and controlled to provide management the insight needed to meet overall system performance requirements. See MIL-STD-499A.[14]

However, to be fair, it has been this author's experience that to prevent requirements creep, particularly in EMD, both the using commands and the materiel commands must be fended off. Now that the program enjoys the priority it takes to achieve the production, there will be many vendors of fantastic widgets visiting with the holders of the ORD (the user). The purpose of these visits is to convince the user to "slightly modify the requirement" so that this new widget will find a home with a successful program. It is not just the defense industry that approaches the user at this stage, government laboratories and research and development centers are just as anxious to "sell" their favorite technology and attach it to a successful program.

A very important role of the user is to work with the PM to stabilize the requirement and prevent gold plating. However, there may be changes due to emerging technology that could add to operational performance and supportability, while at the same time reducing life-cycle cost. Also, enemy research and development is not stagnant, and an increased threat may also force the user to change requirements.

Milestone III, Production Approval

This milestone results in approval of full-rate production for the system. To prepare for the milestone III review, the user updates the ORD and the COEA and assists the PM as appropriate to help ensure a successful review. Approval by the acquisition executive authorizes the program to proceed into phase III, production and deployment. If low-rate initial production (LRIP) started concurrent with EMD, prior to the end of phase II, more than likely there was a program or milestone review prior to milestone III. This is typically referred to as a milestone IIIA review.*

Phase III, Production and Deployment, and Phase IV, Operations and Support

As pointed out earlier, these two phases overlap. During production and deployment the system is produced and fielded to operational combat units. Also, quantities for fielding must reflect requirements of the training base. The user participates on the PM's new equipment training (NET) team. The NET team is designed to arrive at the operational unit with the new system and train the operators and maintainers at field locations. This gives the training base time to catch up to the requirement to produce and send trained operators and maintainers to the operating forces.

A support base must be ready for the new system when fielded. An initial support package will accompany the system; however, any new facilities required should be constructed, or existing support facilities modified, prior to the arrival of the new system. Planning for these facilities was conducted during phases I and II. The user representatives on the service headquarters operations staffs, mentioned earlier, will work closely with the combatant commands to ensure a smooth transition from the old to the new system for both operations and support.

During the early months of phase III, the users in the combatant commands must closely monitor the operation of the system in the field. Problems with operation or support should be brought to the immediate attention of the PM. Later, as fielding progresses, problems are worked out, and the support structure stabilizes, support

*DoDI 5000.2[1] does not address any requirement for LRIP milestone approval, or milestone IIIA review. At milestone II LRIP quantities must be approved, and the LRIP schedule may be approved as one of the acquisition strategy elements for phase II. However, prior to the award of the LRIP contract the acquisition executive will more than likely want to review the program. DoD 5000.2-M[9] refers to a milestone IIB for LRIP approval (p. 14-1-6).

problems will be handled through normal service supply and maintenance channels. The user must also keep the training base informed so that lessons learned in the field can be incorporated into classroom instruction. Further, as the operating commands gain experience with the new system, organizational or doctrinal changes may be required. Although there has been extensive test and evaluation during development, there is simply no substitute for experience gained from prolonged use of weapon systems by operational combat forces.

During the operations and support phase the user monitors the fielded system closely to assess the effects of aging on system capabilities. Acquisition management responsibility for the system may have transitioned from program management to functional management under a materiel command. Nevertheless, information provided from using commands is needed for decisions on whether or not to implement research and development efforts to extend the service life of the system. The user should attend postfielding supportability reviews conducted by the materiel commands to identify and resolve operational and supportability issues. These reviews may be held as joint reviews hosted by either the materiel or the using command.

While a system is still in production, field experience may bring about the need for a major modification. A major modification is defined as a modification that meets the criteria of an acquisition category I or II program. In recommending a major modification effort, the user considers changes in the threat, policy changes directed by the Secretary of Defense, deficiencies discovered during follow-on operational testing, and technological opportunities to reduce the cost of ownership. These major modifications are subject to a milestone IV decision review, major modification approval. For systems no longer in production, deficiencies resulting from threat, policy, or technology changes must be documented by the user in a new MNS and compete with other alternatives during a new phase 0.

Milestone IV, Major Modification Approval*

User responsibilities for this milestone are similar to the previous milestones. The ORD must be updated, and the ILSP, TEMP, and IPS will contain information to support the entry of the major modification effort into one of the acquisition phases. For example, an effort that requires major design work may be directed to enter phase I, demonstration and validation, after a successful milestone IV review. A modification that is ready for engineering development and testing may be directed to enter phase II, engineering and manufacturing development.

CONCLUSION

It is obvious that the user plays a very important role in the management of research, development, and acquisition programs. Primary user activities throughout the acquisition process include requirements identification, clarification, and verification; negotiation with the PM for cost, schedule, and performance trade-offs; and assess-

*Major modifications are those that exceed the dollar threshold for an ACAT II program: $75 million in total RDT&E costs or $300 million in total procurement costs, both in FY 1980 constant dollars. Modifications to weapon systems still in production require a milestone IV review. Modifications to weapon systems no longer in production are called *upgrades* and must compete with other technological opportunities during a new phase 0. The upgrade decision would be made at milestone I. Modifications and upgrades with a cost of less than that of an ACAT II program are treated as engineering change proposals or product improvements and are not subject to the milestone and phase constraints of DoDI 5000.2.[1]

ment of S&T projects for technological opportunities for new systems and technology insertion to upgrade current systems.

The PM depends on the user for continuous input to many of the required documents for milestone decision reviews, and for many of the program plans used by the PM office to guide the conduct of the program during the phases of the life cycle. The user writes the ORD that has a direct impact on the APB, plans for test and evaluation, life-cycle costs estimates, and the contract with industry. The user is responsible for all doctrinal aspects of the system, including logistics support. The ILSP cannot be developed without close and continuous user participation. The user is responsible for the conduct of the cost and operational effectiveness analysis, and participates in the selection of the preferred system alternative for development. Using commands provide operational forces for the conduct of operational test and evaluation. The establishment and operation of a training and support base for the fielded system is the responsibility of the user. The user plays a significant role in determining the priority of the acquisition program within the force modernization plans of the military department and service, and this priority determines what programs will be funded. Simply put, the PM cannot have a successful acquisition program without full user participation.

REFERENCES

1. DoD Instruction 5000.2, *Defense Acquisition Management Policies and Procedures,* Feb. 23, 1991.
2. *Glossary, Defense Acquisition Acronyms and Terms,* 5th ed., Defense Systems Management College, Ft. Belvoir, Va., Sept. 1991, p. B-118.
3. DoD Directive 5000.1, *Defense Acquisition,* Feb. 23, 1991.
4. Section 113, *Secretary of Defense,* Title 10, Armed Forces, United States Code.
5. *A Quest for Excellence,* Report of the President's Blue Ribbon Commission on Defense Management, June 1986, pp. 16–20.
6. Section 732, *Linkage of National Military Strategy and Weapon Acquisition Programs,* National Defense Authorization Act for Fiscal Year 1989.
7. Section 901, *National Military Strategy Reports,* National Defense Authorization Act for Fiscal Year 1991.
8. *1991 Joint Military Net Assessment,* Joint Chiefs of Staff, Mar. 1991.
9. DoD Manual 5000.2-M, *Defense Acquisition Management Documentation and Reports,* Feb. 23, 1991.
10. Memorandum for the Acquisition Community, Subject: Interim Operating Instruction (Army Regulation 70-1) for Army Acquisition Policy, April 17, 1992, p. 17.
11. Air Force Regulation 57-1, *Air Force Mission Needs and Operational Requirements Process,* Nov. 8, 1991, pp. 11, 12.
12. MIL-STD-1521B, "Technical Reviews and Audits for Systems, Equipments and Computer Programs," June 4, 1985.
13. MIL-STD-1388-1A, "Logistic Support Analysis," Apr. 11, 1983; also, see MIL-STD-1388-2B, "DoD Requirements for a Logistic Support Analysis Record," Mar. 28, 1991.
14. MIL-STD-499A, "Engineering Management," May 1, 1974, soon to be replaced by MIL-STD-499B, "Systems Engineering."
15. "USAF F-15 Fighter Program Management Innovations and Lessons Learned," Study Rep. Program Management Course 72-2, D. J. Teal, Nov. 1972.
16. Director of Defense Research and Engineering, *Defense Science and Technology Strategy* (booklet), July 1992.

CHAPTER 32

RISK MANAGEMENT IN MILITARY ACQUISITION PROJECTS

David A. Yosua

David A. Yosua manages the cost risk/acquisition analysis section for The Analytic Sciences Corporation (TASC). He has provided risk management services to the Federal Aviation Administration, NASA, U.S. Army, and U.S. Air Force. He is a Certified Cost Estimator/Analyst and a Certified Program Management Professional.

INTRODUCTION

Few of us in the military project management business would argue that projects in this environment are stable, well defined, and easily managed. In fact, most would argue just the opposite. While the regulations and procedures clearly delineate project management requirements and related expectations, it is not possible for them to account for the dynamic nature of any individual project. And if nothing else, a military development project *is* dynamic. Even with all the excellent guidance that is provided, project management in the military environment is difficult because:

1. There are great amounts of money involved and a corresponding high degree of public visibility.
2. Significant technical challenges are faced frequently—doing things that have never been done before.
3. It involves managing never-ending changes.

These items all contribute to a common management challenge—dealing with inherent risk and uncertainty. Managing these projects successfully means that a project manager is effectively managing the related risk and uncertainty. So, to a great extent, managing a military development or acquisition project is a business of managing risk.

Before starting this discussion on managing risk, a couple of caveats are necessary. Risk exists everywhere. It has many forms and occurs in many different circumstances. Much has been written on risk as it applies to the environment, investments, the insurance industry, and other disciplines. While there is common ground within this vast subject area, there are also large differences. This is particularly true in the application of risk management techniques. Bear in mind that we are focusing on risk management only as it applies to military project management. This is a unique environment and "more or less" establishes the rules by which a manager must "play the game."

When risk management is mentioned, people often think of large security investments or particularly hazardous events that require the constant attention of highly skilled people. Few people view managing a project as essentially a task of managing risk. Yet in project management there are clearly large investments, and they require constant attention by highly skilled people. Project management is not thought of as a business of managing risk, because the management processes have become so well established that they seem completely natural. It has become much like driving a car. Every time we start the engine and venture out into traffic, we are assuming risk, and we practice risk management. Most of this is done subconsciously. For example, we use our turn signals to indicate our intentions to others so that they may prepare for our turn. We apply judgments based on experience at intersections in determining when to cross; and when we grow impatient, we sometimes assume more risk and become somewhat more aggressive in crossing the intersection. When a car crosses our direct path, we take evasive action in some form—braking, turning, and so on. Few of us view this as risk management. The similarities in managing risk on a project are logical extensions of the items just described. Actions are taken to prevent or control some unwanted future event that can affect the program or the project's direction. By making risk management a more formal, systematic process, rather than a subconscious activity, program managers can better anticipate what *might* happen on their project and can then better prepare for upcoming events.

Chapter Overview

This chapter presents risk management as a systematic process. The early portions present the important concepts regarding risk, without specific application to project management. After establishing some baseline definitions, the discussion is turned toward a process which can be applied generically to all projects. Each step of this process is then examined and discussed as to the various methods available for the project manager to execute that step of the process. Within the discussion of each of the steps are some specific tools that have evolved for risk analysis, and some suggestions for obtaining data useful in risk management. The chapter concludes with some suggestions to make the process work in a project office.

A Brief History

Risk management in military procurement has been around for quite some time. Even before it was formalized in regulation, most program managers practiced risk management in some form. The recurring interest in this subject by policy makers is clearly visible through recent history. In 1969 Deputy Secretary of Defense David Packard wrote a memorandum to the military services that listed inadequate risk assessment as a major problem area. Later Frank C. Carlucci III also noted risk management as a problem, and as part of the Carlucci initiatives, the Department of Defense (DoD) was required to increase the visibility of technical risk inherent in weapon system budgets. Then in 1986 the General Accounting Office (GAO) cited inadequate technical risk assessment as a problem area within the military services in a report released to the chairman of the Committee on Government Affairs, U.S. Senate.[1] Interest in risk management will continue to grow as program managers and policy makers recognize that risk can effectively be planned for and controlled. As a result, program managers should anticipate greater emphasis from senior management in this area. Clearly, the need for more effective risk management will be felt even more strongly in the future as we face more challenging technical problems coupled with declining military budgets, and, more importantly, as we face more challenging business and environmental world problems.

Acquisition Regulations and Risk

Within DoD Directive 5000.1[2] risk management is explicitly called out for all major and nonmajor acquisitions. Part 1, sec. C, item 2, details the requirements and explicitly calls for a risk management plan. This plan must address how risks will be identified, assessed, and eliminated or reduced. It is important to note that the regulations do *not* require the risk management plan to be contained in a single document. Rather, it is an integral part of the other plans which are submitted during the planning and execution of the project. The companion documents, DoD Instruction 5000.2[3] and DoD Manual 5000.2M,[4] set out more specific requirements and formats for risk management. Throughout DoD Instruction 5000.2, risk management concepts are mentioned as a required part of the project management process. Part 5, sec. B, is dedicated solely to risk management of the project and within part 6 it is tightly interwoven into the engineering and manufacturing requirements.

These latest versions of the regulations place more emphasis on risk management than any of the previous sections. A key change in the regulations is in the degree of formality with which a program manager is required to deal with risk. Discipline in the systematic application of the methodologies developed by program managers for risk management will be a key ingredient in determining which programs continue at full funding, and which continue in another form. Considering the increased size, complexity, dollar value, and unknowns faced in the development of today's systems, this emphasis on risk management is for good reason.

Any military acquisition program manager preparing to undertake a risk management program should consider the reading of 5000.1 and 5000.2 as an essential ingredient in the initial formulation of the risk management process.

UNDERSTANDING RISK

Risk means different things to different people. It involves a wide variety of subject areas from psychology to probability. The following few paragraphs are intended to introduce some of the complexities that can enter into the management of risk on any project. These complexities sometimes cause program managers to ignore risk. This results in the assumption of more risk than necessary because little or no preventive actions are taken. Yet examining these elements of project risk reveals the fact that it can be broken down into easily understood and manageable concepts.

Types of Risk

Within the broader project management community (including non-DoD), risk is viewed as being one of two types—business risk or insurable risk. Insurable risks offer only a chance for loss while business risks offer both a chance for loss and a chance for gain. The bulk of this chapter deals with business risks, not because they are more important than insurable risks, but because they are most often the risks military system program managers must face. Insurable risks include such items as direct property damage, indirect consequential loss, legal liability, and personnel. While these are important project management considerations, they are definitely more critical to running a business as a whole, and to program managers in other environments such as construction. Typically they are dealt with outside the management of any single military project.

Distinction between Risk and Uncertainty

To a program manager caught up in the day-to-day management of a project, distinguishing between risk and uncertainty seems hardly worthwhile. Nevertheless, it is important to understand that they are *not* the same. The difference is highlighted in Fig. 32.1.[5] Within this figure, a range of "certainty" is represented. At the far left is the absolute certainty point (meaning everything is known about the project), and at the far right is complete uncertainty (meaning nothing is known about the project). Each of these situations presents unique circumstances to program management. If a condition of certainty exists, then there is no need for management reserve funds (costs are known), the schedule is set, and the product performance is definitely achievable. In short, there is no need for management since the result will be as predicted. On the other hand, total uncertainty implies that not enough is known about the project as to predict either schedule, cost, or performance. This implies that management of the project is nearly impossible (except to ensure the proper use of resources). At this extreme, further research is needed to define the project and related goals better. The real management of risk and uncertainty lies within these two extremes. It is within these bounds that the program manager must use history, judgment, and analytic evaluations to control the project.

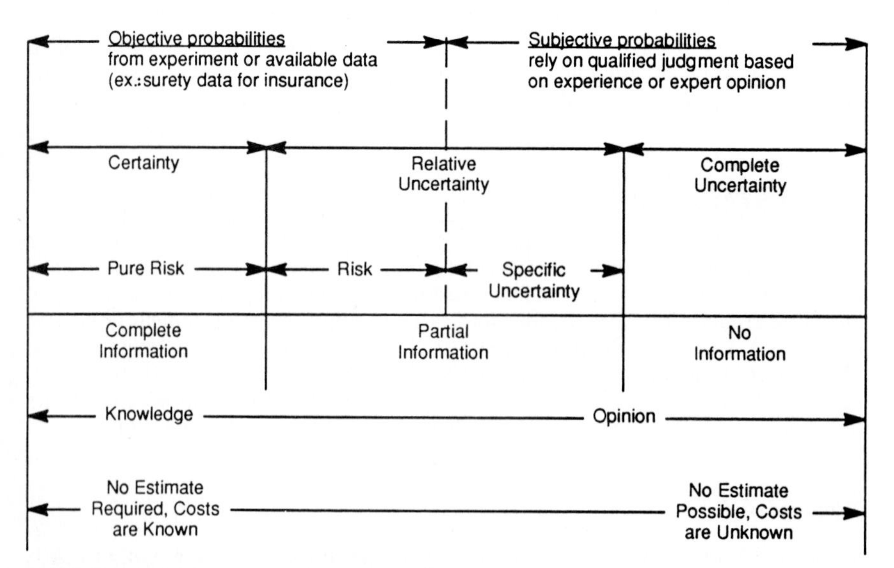

FIGURE 32.1 Risk and uncertainty spectrum.[5]

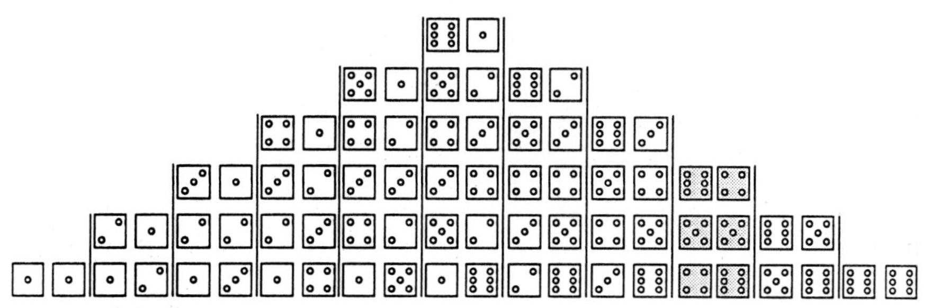

FIGURE 32.2 Probability (pure risk). Here 36 possible outcomes exist. The probability of throwing a 10 is 3/36 = 0.0833.

Information is a key differentiator between risk and uncertainty. For the program manager in military acquisition, the conditions related to having only partial information are the norm. The program manager is really dealing with both risk and uncertainty at the same time across the entire project. The technical definitions for these two terms are as follows.

1. *Risk.* A situation where the outcome of an event is not known with certainty, but can be bounded by a known probability distribution representing all possible outcomes. The classic example is the role of two dice, as illustrated in Fig. 32.2. In this figure all possible outcomes are known and can be described by the likelihood of any given outcome. This is also equivalent to "pure risk," as illustrated in Fig. 32.1. It differs from certainty in that a range of outcomes is known versus a single outcome.

2. *Uncertainty.* A situation where the outcome of an event is not known with certainty and there is insufficient information to describe the possible outcomes and their related probability (unknown probability distribution; far right spectrum of Fig. 32.1). This differs from complete uncertainty in that some of the more likely outcomes are known as opposed to no outcomes being known.

Theoretically this distinction is extremely important. However, in application the difference between risk and uncertainty is indeed a factor that must be dealt with, but is somewhat less meaningful. For example, trying to decide whether or not a technically challenging piece of avionics hardware needed for an airframe platform has a high degree of risk or a high degree of uncertainty may matter very little to the program manager. What is of great interest, however, is whether or not that risk or uncertainty is going to cost the program more time or dollars, and whether or not it will affect the performance of the equipment. So while understanding the distinction between risk and uncertainty is necessary to quantify the degree of "unknowns," it frequently gets lost in the overall evaluation with regard to what is at stake. Throughout the remainder of this chapter, use of the term risk encompasses both risk and uncertainty unless otherwise specified.

A Program Manager's View of Risk

Figure 32.3[6,7] portrays the three key elements of risk from a program manager's viewpoint: (1) the existence of an event that can cause change, (2) the event has some likelihood of occurrence, and (3) the event has some undesirable consequence. Note that there are two ways to control risk when viewed in this manner. A program

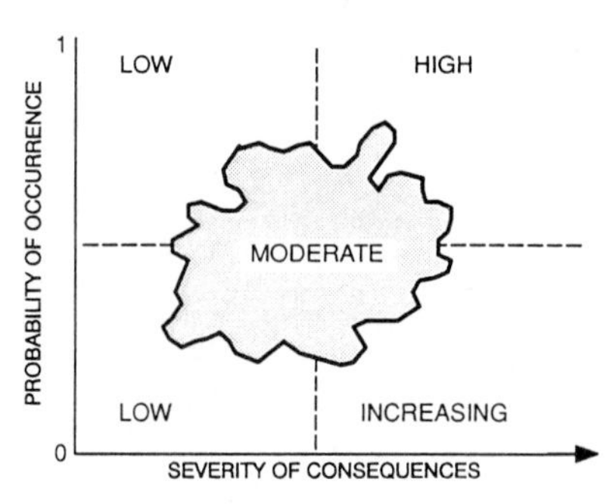

FIGURE 32.3 Risk management.[6,7]

manager can either focus on controlling or reducing the likelihood that a certain risk event will occur, or he or she can focus on minimizing the impact of the event should it occur, or both. An item of importance illustrated in Fig. 32.3 that can save a program manager time is assessing where a given risk event would be plotted on the graph. It would be of little value for a program manager to spend a great deal of time and effort managing a risk event that had a high likelihood of occurrence, but small consequence. By evaluating where each risk event lies within this graph, a program manager can quickly determine on which areas to focus all too often limited resources. In effect, this is a convenient mechanism to rank potential problem areas.

Another factor important to managing risk on a program is in interrelationships that exist between cost, schedule, and performance. Program managers are constantly forced into a balancing act between these three parameters. This sometimes requires that a relatively sophisticated methodology be used in the evaluation of risk, which considers three key measures: cost, schedule, and performance. Frequently actions taken to reduce one type of risk result in an increase in another risk area. An example of this can be seen in taking action to accelerate a project schedule. This is often accomplished by moving many events in parallel, producing a highly concurrent schedule. This can present both higher technical and cost risk: higher technical risk because a technical problem will ripple through many dependent activities very quickly, resulting in many more "fixes," and higher cost risk because of the technical problem impact, and because nearly anytime schedules are changed, costs go up.

Definition Problems

Most disciplines develop a common language of their own. Unfortunately the subject of risk has developed a language that is not very common at all. The terms risk management, risk assessment, risk analysis, risk mitigation, risk handling, risk planning, and so on, mean many different things. There are several versions of the set of terms developed to communicate in this field. The fact that there is not a uniform set of terms is not important in itself. What is important is that miscommunication can occur as a result. Take, for example, the term risk analysis. Within the DSMC risk management guide,[7] risk analysis is characterized by the quantification of uncertainty, the quantification of consequences, and performing sensitivity analysis on risk input variables to evaluate changes in consequences. In Rowe's *An Anatomy of Risk*[8] the term risk analysis is the top-level term encompassing all risk assessment and risk management activities. Within the Project Management Institute's body of knowledge,[9] the term risk analysis does not even appear. As can be seen, a program manager requesting someone to perform a risk analysis of a given situation must supply a more detailed description of the expected results than might first be thought. The problem is compounded when you consider that there are in excess of 20 risk "something" terms that are commonly used. This requires program managers to make clear the terms that they will use and the related definitions for communication and execution of risk management. This is best done at present within one of the program plans where risk is addressed. Note that the definitions should be in consonance with DoD Directive 5000.1 and DoD Instruction 5000.2.[2,3] The plans where these definitions should appear are the program management plan or the systems engineering management plan.

Risk and the Utility Function

The risk preference function is another concept essential to a program manager's understanding of risk. Recall from basic statistics that utility is a measure which describes the relative value of various outcomes of decisions to a decision maker. This measure of utility can be used in risk management to measure a decision maker's attitude toward risk. Figure 32.4 illustrates a decision maker's possible positions relative to risk. The risk averter graph depicts that when additional dollars are added to a game, the decision maker obtains lower and lower amounts of incremental gains in utility (meaning increasing amounts of satisfaction as the stakes are raised, but at a lower and lower rate). At the opposite extreme, the risk seeker graph shows gains in utility at an increasing rate as dollars at stake are raised. Note that the consequences do not always have to be monetary in nature. The actual calculation of these curves is beyond the scope of this chapter. Readers interested in finding out more about this subject are referred to any applied statistics handbook.

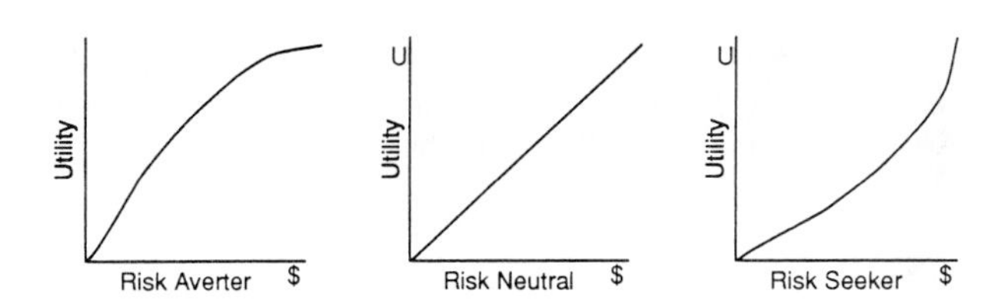

FIGURE 32.4 Risk preference and utility function.

Theoretically, this concept is very useful. But in practice it is very difficult to ascertain an individual's attitude toward risk with regard to the management of any single project. Much of the data are subjective in nature, and any decision maker's utility function can in fact vary somewhat from day to day. What is useful to a program manager, in addition to understanding his or her own risk utility function, is to gain a *general* understanding of both the supervisors' and the subordinates' attitudes toward risk so that the program manager can adjust his or her management style and approaches accordingly.

RISK MANAGEMENT AND THE PLANNING PROCESS

There are two aspects to the planning function and risk management. The first relates to the quality of the project planning in general. This is a prerequisite to effective risk management. In fact, if the project planning has not been addressed adequately, a major risk is introduced because of the absence of defined, documented direction. The second relates to describing how risk management will be carried out on the project. If this plan is insufficient, then there is some likelihood that risk will not be addressed with the proper emphasis, involvement, and understanding of the project personnel.

Project Planning and Risk

Risk management by its very nature is a proactive form of management. This implies that management actions are in the prevention or preparation for an event—not in reaction to it. Thorough planning is absolutely critical. If the planning on a project is not thorough, then examining what "could go wrong" on the project will be incomplete. The baseline plan is essentially the departure point for risk identification.

The key elements that need to be described in this baseline plan include the cost, schedule, and performance or technical requirements to be fulfilled on the project. Once these items are defined adequately, the process of identifying events that can cause changes to the plan is undertaken. It is easy to see that a crystal ball would be very beneficial at this point. Perfect information is certainly desired, but impossible to get. So a plan to gain information on these "potential plan-changing" events becomes a necessity.

Risk Management Planning

Before actually starting any risk management activities, there should be some guidelines produced to establish overall responsibilities and the risk management process. Gaining information about risky items is a necessity to plan for and manage risk effectively. As mentioned previously, a risk management plan should serve as a basis for the responsibility of how, when, and by whom this information is to be gained. Some important items to include in this plan are:

- Approach or process for risk management
- Risk identification techniques
- Focal points for risk management
- Techniques to be applied in risk management
- Definitions
- Rating schemes
- Frequency of analysis
- Formats for reporting

DEVELOPING AN APPROACH FOR DEALING WITH RISK

Within DoD Instruction 5000.2[3] there is a five-step process that is closely related to that set forth in the DSMC risk management guide, as illustrated in Figs. 32.5 and 32.6,[7] the major difference being the combination of risk identification and risk assessment into one activity within the DSMC approach.

FIGURE 32.5 DoD regulation risk management process.

The process set out in DoD Instruction 5000.2 starts at the top of Fig. 32.5 with the risk planning process and then moves clockwise to succeeding steps. Once the risk reduction phase is completed, the process begins again. This implies that the risk management process is never ending in a project. To some extent that is true. However, depending on program phase, remaining effort, complexity, and other factors, the level of detail and frequency in executing the process may vary as a whole and by

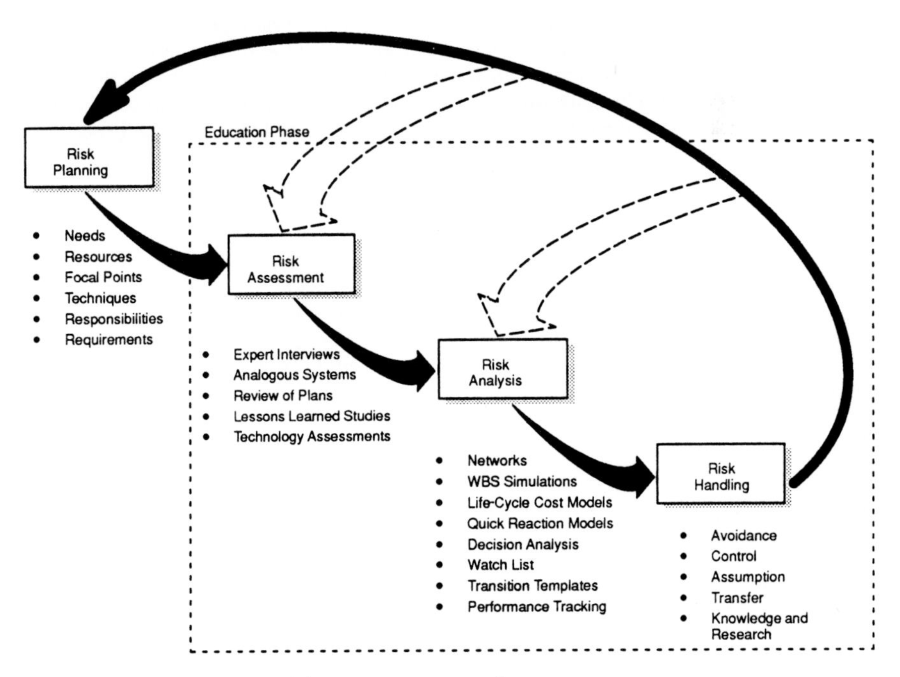

FIGURE 32.6 DSMC guide risk management process.[7]

individual process steps. At each milestone review a program manager must be prepared to brief (in detail) on each step of the process in terms of methodology and status.

The logic of the process is readily apparent: *plan* for managing risk, *identify* what the risks are, *assess* the magnitude of the identified risks, *analyze* the impacts and alternatives for the assessed risks, and *reduce* by management action the project impact of the risks. Since we have touched on the importance of planning for managing risk, the remainder of the chapter will focus on the techniques and tools used to carry out the rest of the risk management process. Specific attention is warranted in four areas:

- Risk identification and quantification
- Specific risk analysis tools
- Responding to risks
- Risk management in a program office environment

RISK IDENTIFICATION

One of the most difficult questions to answer in risk management is "how do I minimize the potential of making an error in the risk identification process?" The heart of the question relates to forecasting unknown events. We frequently speak of *known* unknowns and *unknown* unknowns, where the known unknowns are the events that have been identified as having the potential for unwanted change and the unknown unknowns are truly unforeseen events. The latter are the particularly worrisome events that must be actively sought out in the risk identification process. There is no best solution to this problem, but there are some techniques that can be used to minimize the frequency in which a potential risk area is overlooked. Program managers need to pay close attention to this process since their only real confidence that all risks have been uncovered can come from the knowledge that the process is valid and has been executed with discipline.

Sources of Risk

The first step in the risk identification process is to classify the sources of risk on a given project. There are several common classification schemes that can be used. The following two are samples that have worked well in the past, but may not be applicable to all projects. The first (from the DSMC risk guide) segregates sources of risk into five separate areas:

1. Technical
2. Programmatic
3. Supportability
4. Cost
5. Schedule

Technical risk is associated with efforts to evolve an increased performance level of some equipment, or the same performance level subject to some new constraints.

Programmatic risks are those risks that are outside of the program's control, but can affect the program's direction. These risks are primarily related to obtaining resources.

Supportability risks are the risks identified and associated with the operations and support phase related to the equipment. These risks comprise both technical and programmatic factors.

Cost risk is the potential financial exposure of the project.

Schedule risk is the potential time exposure of the project.

Cost and schedule tend to be somewhat different than their counterparts in this structure. These two items tend to be the measuring sticks that most people use to determine how well a project has been managed. In reality, cost and schedule are often driven by the technical task at hand and can only be managed by making the most effective use of the time and money allocated to the project. In this context there are only two possible sources of schedule and cost risk:

1. The risk that the time and money allocated to the project are unreasonably low and out of the range of being feasible
2. The risk that the program will not be executed in an efficient manner, thereby wasting the time and dollars allocated

Figure 32.7[6,7] shows some typical risks in each of the classifications, which should be helpful in understanding the categories.

The Project Management Institute has published a body of knowledge that con-

TYPICAL TECHNICAL RISK SOURCES	TYPICAL PROGRAMMATIC RISK SOURCES	TYPICAL SUPPORTABILITY RISK SOURCES	TYPICAL COST RISK SOURCES	TYPICAL SCHEDULE RISK SOURCES
Physical Properties	Material Availability	Reliability & Maintainability	Sensitivity to Technical Risk	Sensitivity to Technical Risk
Material Properties	Personnel Availability	Training	Sensitivity to Programmatic Risk	Sensitivity to Cost Risk
Radiation Properties	Personnel Skills	O&S Equipment	Sensitivity to Supportability Risk	Sensitivity to Programmatic Risk
Testing/Modeling	Safety	Manpower Considerations	Sensitivity to Schedule Risk	Sensitivity to Supportability Risk
Integration/Interface	Security	Facility Considerations	Overhead/G&A Rates	Degree of Concurrency
Software Design	Environmental Impact	Interoperability Considerations	Estimating Error	Number of Critical Path Items
Safety	Communication Problems	Transportability		Estimating Error
Requirement Changes	Labor Strikes	System Safety		
Fault Detection	Requirement Changes	Technical Data		
Operating Environment	Political Advocacy			
Proven/Unproven Technology	Contractor Stability			
System Complexity	Funding Profile			
Unique/Special Resources	Regulatory Changes			

FIGURE 32.7 Typical sources of risk.[6,7]

tains information on risk management which is also useful.[9] It classifies sources of risk in a different manner, as follows (see Fig. 32.8):

1. External, predictable
2. External, unpredictable
3. Internal, nontechnical
4. Technical
5. Legal

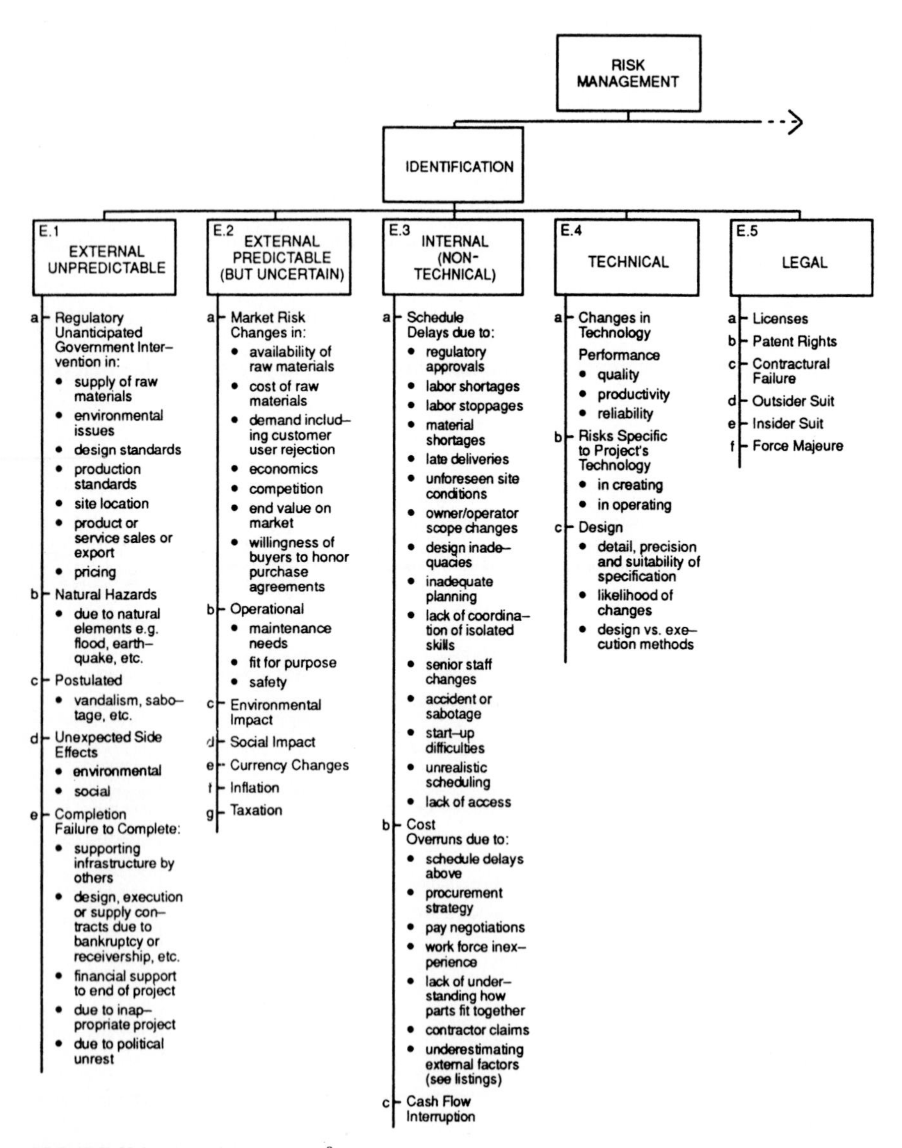

FIGURE 32.8 PMI risk structure.[9]

External predictable risks are risks inherent in doing business in a given environment. Generally they comprise events outside of the project's control, but impact the project.

External unpredictable risks are basically the same as predictable risks, but contain a higher degree of real uncertainty.

Internal nontechnical risks comprise events within the control of the project manager, but not directly related to technical challenges.

Technical risks relate to the use of state-of-the-art systems, processes, or products that may impact the project's direction and support.

Legal risks refer to potential problems with licenses, patent rights, lawsuits, and so on.

Note that both of these classification schemes focus on the cause of a risk, not its consequence. Completing this structure for classifying risk is only one-half of the ingredients necessary to identify project risk. The other half is the work breakdown structure (WBS). Since the WBS encompasses all of the work to be performed on the project, it serves as a great mechanism to systematically comb through the entire program looking for risk. A commonly held notion regarding the management of risk is that it is so difficult to get your arms around, that the only way to effectively deal with it, is to "smash" a grid over it, thereby breaking it into pieces, and then examine each piece individually. We now have a grid in place, as shown in Fig. 32.9.

Now the effort focuses on completing the blanks within the matrix set out in Fig. 32.9. Each WBS element is examined individually for inherent risk in each of the risk source classes. This matrix can be completed at any point within a program's execu-

	Technical Risks	Programmatic Risks	Supportability Risks	....
Space Vehicle				
Integration, Assembly and Test				
Software				
Spacecraft				
Integration, Assembly and Test				
Software				
Structure				
Integration, Assembly and Test				
Complex Mechanisms				
Other Structure				
Attitude Determination and Control				
Integration, Assembly and Test				
Software				
Attitude Determination				
Integration, Assembly and Test				
Software				
Digital Electronics				
Sensors				

FIGURE 32.9 Risk assessment matrix.

tion, but at least two particular times are very appropriate for this activity. The first is when the overall program planning is done. Frequently a program schedule is developed at this point in time, which reflects the overall WBS. During the development of the schedule an in-depth look is taken at each activity to be accomplished. This is a convenient time to consider the risks involved via the matrix of Fig. 32.9 and the schedule being developed. The other time is during the annual estimate process. Typically estimators and engineers together evaluate the entire program and often examine cost risk as a secondary issue. Again, this is a good time to evaluate more than just the impact to cost of possible risk events. It is a good time to thoroughly look for risks within the program. No matter when the risk identification process takes place, it requires information on difficulties that may be encountered in accomplishing the specific WBS task.

Techniques to Gain Information Relating to Risk

There are a large number of risk analysis tools in existence today that are beneficial to program managers facing decisions. All of these tools have one thing in common: they require information in the form of quantified variables. This information can be objective or subjective in nature. Objective information generally stems from databases on past experience on the current project, or similar previous projects. Subjective information is usually in the form of expert judgment. Of the two, objective information is the easiest to deal with since potential biases stemming from outside personal experiences do not enter into the data. For this reason it is important to glean as much useful information from past project histories, such as documented cost estimates, cost performance reports, or lesson learned studies, as possible. A subtle but important point beneath this need for objective information is that there should be a plan to record information on the current project for use in future risk management efforts. By so doing, a program manager can help ensure the success of future similar projects by documenting the history and progress of today.

Common Sources of Subjective Information. Subjective information can be gained by using variations (both in methodology and in application) of the following two common techniques: (1) expert interviews and (2) independent technical assessments.

Expert Interviews. Expert interviewing is probably the oldest method to collect information about risk on a program, and it is the most commonly used due to the fact that most programs contain uncertainty rather than true risk. Information on a particular area can come from a single individual expert or a panel of experts. Many people advocate the use of the Delphi technique, but in practice it can be difficult to apply (frequently because of time restrictions). In any case, a structured approach for extracting the information from the individual(s) is necessary. The following five steps have proven effective in interviewing experts for the purpose of extracting information on program risks.

1. *Identify the right individual.* Selecting the most knowledgeable individual is extremely important since judgment, or educated guesses, is involved in identifying risk areas. Up front, preliminary screens for relative expertise have been beneficial from the standpoint of saving time.
2. *Prepare for the interview.* Some thought should be given as to what areas will be covered during the interview. It is much easier to maintain control and direction during the interview if there is an agenda or list of areas that must be covered.
3. *Target the interest area.* The first portion of the interview should focus on the expert's area of expertise. It is very easy to let the interview gravitate toward the

more obvious risk areas, thereby glossing over some of the other important issues that may arise.

4. *Solicit judgment and general information.* Most people are willing to share their views on more than just their area of expertise. This can be valuable information to start another expert interview in the risk identification process. Frequently many possible risks are identified within a few interviews. This information is then refined as additional experts are interviewed.
5. *Quantify the information.* This is the most sensitive aspect of any risk analysis. Once the risk areas have been identified, an estimate of their potential impact on the program must be made. This requires that the expert consider the probability of the given event occurring, and what the potential impact may be in terms of cost, schedule, or performance. This aspect of the risk assessment process merits its own discussion and will be presented shortly.

Independent Technical Assessments. Independent technical assessments require people other than those under the direction of the program manager. Basically it is an application utilizing expert judgment in a totally independent manner. Here experts outside of the program organization (usually very senior personnel) are requested to evaluate certain aspects of a program and deliver a briefing on their view of the risks that exist in the various aspects evaluated. This is most commonly used when a program is perceived to be in trouble and critics have called attention to the program. If the trouble is real, the program manager can call on the experience of the independent team to assist in developing the corrective action plan. If the trouble is not real, this review will give the program manager additional credibility and will hopefully quiet critics. These reviews typically occur more frequently in development than in production.

Common Sources of Objective Information. As mentioned previously, objective information typically comes from historical data from the current or previous projects. Within the military project management environment, the following four sources of information have been valuable for risk identification: (1) program documentation evaluations, (2) current project performance data, (3) analogy comparisons and lessons-learned studies, and (4) transition templates.

Program Documentation Evaluations. This technique is aimed at identifying risks within a program by highlighting disparities in the program planning. More than anything else, it is a concentrated effort to seek out voids and contradictions in the baseline documentation. This documentation includes, among others,

- Program management plan
- Systems engineering management plan
- Acquisition plan
- Test and evaluation master plan
- Manufacturing plan
- Integrated logistics support plan
- Work breakdown structure
- Specifications
- Statement of work
- Schedules
- Cost estimates
- Contractual documentation

The objective is to ensure that the work to be performed has been fully accounted for and is adequately communicated to the appropriate parties. Each WBS element should be fully traceable to both the statement of work (SOW) and the specifications, and the program plans should provide the direction and coordination to facilitate accomplishing the tasks.

Current Project Performance Data. Most of the larger military projects have extensive performance reporting requirements. These data can often be used in the risk identification process. For example, if the cost schedule control systems criteria and the associated cost performance report (CPR) data are part of the contract, then this information can be used as a solid predictor of future performance. CPR data have in the past proven to be very good indicators of future performance based on progress to date. The cancellation of the Navy A-12 program was, in fact, based on CPR data to a large degree. The calculation of an estimate-at-complete derived from an index which considers the efficiency to date is an excellent example of using objective historical data from the current project. For more detail on the use of these data see Chap. 39. Another source of useful risk information is test data. Sometimes test results are classified "partially" successful or "the test would have been successful if the failure to part X had not occurred." The realistic evaluation and use of test data can help in planning more effectively and anticipating potential field problems. These data can be used in much the same way as CPR data.

There is a great amount of other data which can be used in a similar fashion for risk identification. Much of the data lie within the functional areas of engineering, manufacturing, logistics, and quality. It is incumbent upon the functional personnel receiving the deliverable items to perform an analysis of these data with an eye toward risk identification.

Analogy Comparisons and Lessons Learned Studies. These two sources of historical data are based on experiences gained from other projects. Here there is no particular methodology to follow other than to examine the records of past programs that may be valuable in terms of potentially identifying pitfalls that may exist on the current program and of which the program manager may not be aware. The main item to be cautious of in using this technique for risk identification is to fully understand the similarities between the current program and the historical program. Frequently program offices maintain a lessons-learned file just for the purpose of documenting past program experiences so that they may be referenced in the future.

Transition Templates. This is basically a collection of lessons-learned studies that have been accumulated over the past several years. The transition templates are actually elements called out in DoD Manual 4245.7-M.[10] This document is referenced several times within DoD Instruction 5000.2 as a recommended approach for risk management. The document uses the framework illustrated in Fig. 32.10 to highlight common program pitfalls and provides some solutions for dealing with them. One of the major benefits of this approach is that it views the project as a whole from development through production, and as such provides guidance for the total program and not just one phase.

Quantification of Identified Risks

Some form of risk identification is necessary to fully understand the existing risks, and to provide a mechanism for relating those risks to other parties within and outside the program. There are two general avenues followed to quantify risks within a program. The most commonly used avenue is that of determining a series of classes in which to place a risk, which is described qualitatively and then equated to a numerical level. These can then be integrated into a total program risk rating at a later

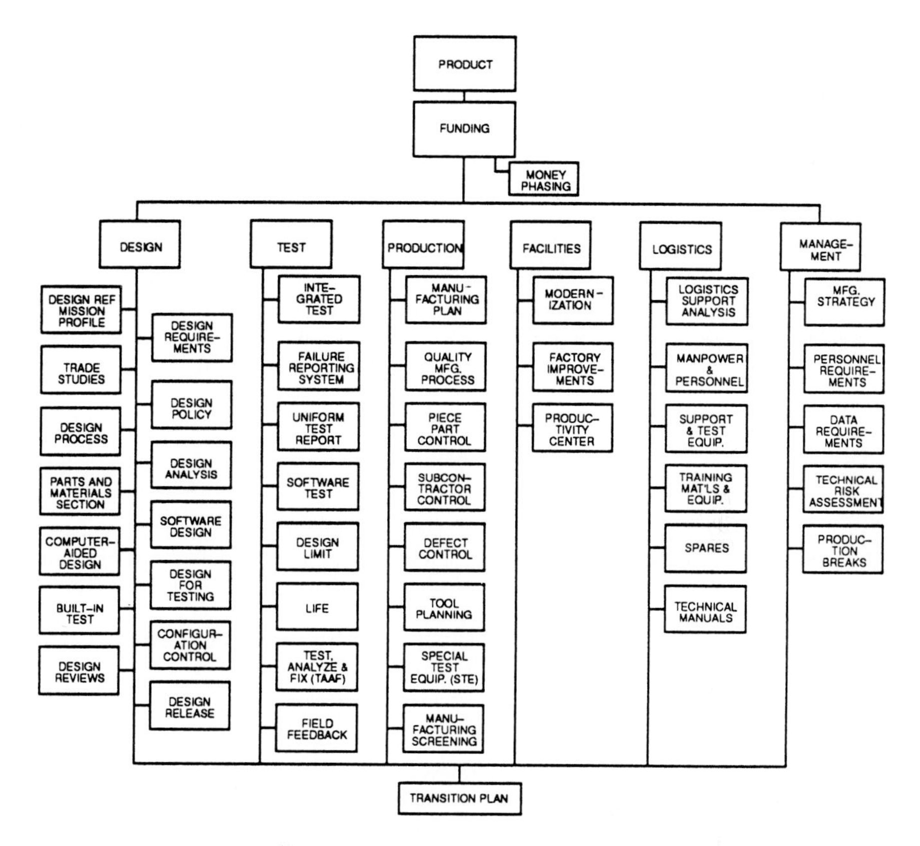

FIGURE 32.10 Templates.[10]

point. The less frequently used, but more meaningful, are the strict mathematical techniques involving the use of probability. The majority of the time, these two variations are used together in a modified mathematical approach. Before discussing these techniques, there are some points to consider.

As mentioned earlier, obtaining information from experts can introduce biases into the risk assessment process that may not be appropriate. This is particularly true when attempting to use expert judgment in the quantification of risk. These biases result directly from the experiences of the "expert" on other programs, and from his or her life in general. Many of these are difficult to detect and require awareness to them prior to interviewing the expert. The most common biases are illustrated in Fig. 32.11.

As can be seen from this figure, many of these biases are "built in" to the expert's opinion and must be actively compensated for if detected. There are techniques available to minimize the impact of these biases, which can be found in the DSMC "Risk Guide."[7]

Placing Risk into Classes. As mentioned previously, this is one of the more common techniques for assessing risk levels. It is much quicker, and in some cases more appropriate, to use this approach rather than an elaborate time-consuming modeling

Anchoring	Tendency toward keeping estimated distribution endpoints toward the starting value.
Availability/Common Occurrence	Tendency to recall more common events and neglect rare/infrequent events.
Controllability	Tendency to discount risks based on the belief that they are easily controlled.
Insensitivity to Sample Size	Tendency to incorrectly assume small sample sizes are representative to the population.
Involvement Biases	Tendency to have stronger belief in the accuracy of point estimates due to involvement in the development of the point estimate.
Overconfidence	Tendency to state overly narrow confidence intervals given the information available and the expert's knowledge base.
Proximity to Program	Tendency to view risks more favorably because of program advocacy.
Recent Event	Tendency to recall recent events as representative of the population of total events.

FIGURE 32.11 Common expert judgment biases.[14]

approach. Here the object is to develop a classification scheme on which to judge a risk, once identified. Normally it consists of a set of characteristics that may present risk and a corresponding numerical scale with descriptions of the varying degrees of risk. An example is shown in Fig. 32.12.

Note that the type of data contained in this figure have more of a flavor of being nominal or ordinal versus interval or ratio. Basically this means that the differences between pairs of numeric ratings are not necessarily equal. For example, under producibility, a rating of 6 indicates a feasible task. Apparently this is based on a valid data point from some previous experience with the risk area. A rating of 7 may in fact not be feasible—this is only a difference of 1 on our scale. This difference of 1 may not be equivalent to the difference in going from "established," a rating of 2, to "demonstrated," a rating of 3, also a difference of 1. In effect, the numerical data are used to develop a relative ranking of the risks that can be manipulated mathematically to integrate different aspects of the program. This technique is commonly used to assist in the development of probability distributions for more in-depth modeling as well. It can be modified easily to accommodate almost any program. Note that even though the data are not interval type, the precision is greatly enhanced by the characteristics described in each of the blocks. Many people fall into the trap of de-

RISK CATEGORIES	RISK SCORES (0=Low, 5=Medium, and 10=High)				
	0–2	3–4	5–6	7–8	9–10
1 Required Advancement	Existing	Minimum Advancement Required	State of Technology	Significant Advancement Required	New Concept Required
2 Technical Status	In Use	Prototype	Development	Design	Concept
3 Reliability	Historically High	Average	Known Problems	Serious Problems	New Concept Required
4 Producibility	Established	Demonstrated	Feasible	Serious Problems	No Known Production Experience
5 Criticality of Meeting Specification	Minimal Impact	Existing Alternatives	Potential Alternatives Available	Possible Alternatives Identified	New Concept Required

FIGURE 32.12 Technical risk categories and score criteria.[14]

scribing risk as low, moderate, or high without adequately defining the rating schemes.

Risk Quantification through Probability Distributions. Most of the more sophisticated tools available for risk analysis require the use of information expressed in terms of probability. This in turn requires that the basic rules of probability be followed, that is, that the probabilities of all possible events sum to 1, that the probability of any single event be between 0 and 1, and so on. These rules are important to remember when developing probabilities for events extracted from expert judgment. Several techniques can be used to assist in this process. The key point for program managers to realize is that the development of these probability distributions is one of the most important factors (if not *the* most important) in executing any of the risk tools covered below. Unfortunately it is difficult to determine how accurate the distributions really are. Therefore a disciplined, formal methodology to develop the distributions is necessary to minimize potential error.

TOOLS AND TECHNIQUES USED IN RISK ANALYSIS

The tools developed for risk analysis seem to be aimed at theoreticians, cost and schedule analysts, risk analysts—in short, almost everyone *but* the program manager. Bear in mind that many of these tools were specifically developed to address the risk issue. They were never intended to replace the existing program management tools currently used. Most were designed as a supplement to the array of program management tools, although some of the more recent software is being designed more and more as an integrated package which includes a significant risk analysis capability. The advantages of using a mathematical analysis tool is that it provides deeper insight into the interrelationships of the various program elements than other approaches. The disadvantage of using a model is that the results can look very impressive, yet still be very wrong, particularly if the input variables are not dealt with properly. These variables generally consist of the quantification of the identified risks, as discussed previously. However, assuming a logical process has been developed and implemented for quantifying the risks, these tools can be very beneficial. Figure 32.13 highlights the techniques that program managers found most useful from a survey conducted in 1988.

TECHNIQUE CLASS	NUMBER OF TIMES ANALYSIS RESULTS FOUND OF SOME OR SIGNIFICANT USE							
	DESIGN GUIDANCE	POM/BES PREPARATION	SOURCE SELECTION	ACQUISITION STRATEGY SELECTION	PROGRAM STATUS REPORTING	PROGRAM PLANNING	DAB MILESTONE BRIEFINGS	TOTAL
Network analysis	2	4	2	5	13	14	8	48
Decision analysis	1	0	2	2	3	3	2	13
Estimating relationships	2	3	1	2	4	5	2	19
Risk factors	0	2	0	0	1	3	1	7
Watch list	1	6	4	3	9	9	5	37
Independent cost estimates	5	3	2	4	6	6	6	32
Cost risk/WBS simulation	0	0	1	0	1	1	0	3
Other	3	1	2	4	3	5	1	19
Total	14	19	14	20	40	46	25	178

FIGURE 32.13 Usefulness of risk technique survey results.[12]

The following descriptions of common risk management tools is far from exclusive, nor are they very detailed. They are intended to give program managers just a taste of the variety of approaches possible, and to present a quick description of the more popular and useful tools. Before undertaking an exercise using any of these techniques, further research on the candidate techniques is an absolute must. The techniques are broken into two sections, (1) techniques that can be used to evaluate cost, schedule, and performance risks, and (2) techniques used only to evaluate cost impacts of risks.

Techniques with Broad Application

The following group of techniques are useful for examining risk from other than a pure cost impact viewpoint.

Network Analysis. One of the more common program planning techniques also used for risk analysis is network-based scheduling. (For more information on network-based scheduling see Chap. 39.) The advances in computing technology and power have provided very good extensions for using networks to examine risk within the program. Construction of the basic schedule network is the same for risk analysis as it is for general planning. The difference is in determining the time allocated to any (or all) activities within the network. Here the time is in the form of a probability distribution reflecting the likelihood of a range of possible outcomes. This distribution typically results from the activities described earlier under risk identification and quantification. Many of the more sophisticated network packages also provide for cost risk distributions to be developed, as well as schedule, for each of the activities. The most sophisticated provide for cost, schedule, and performance probability distributions to be developed and for modeling relationships between the distributions. A common use of this is where a relationship between time and cost can be identified, such as where there is a fixed cost of an activity independent of the time to complete (such as material) and a cost which is time-dependent (such as direct labor). This can be modeled as illustrated in Fig. 32.14.

The next step in the process is to execute a Monte Carlo simulation. This is essentially a sampling of 1000 iterations or more of the possible outcomes of the project, given the distributions specified. The result is an overall distribution which can be

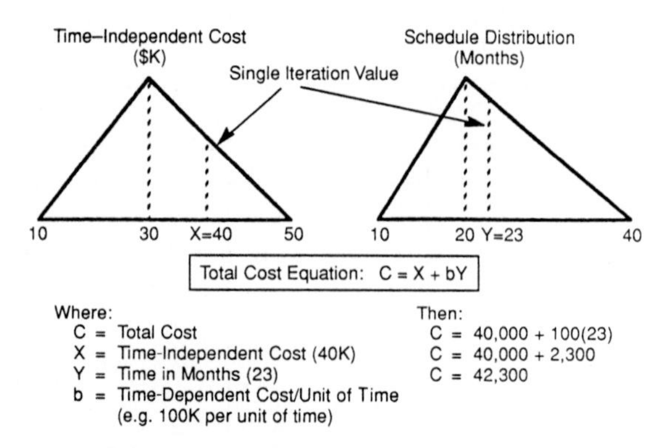

FIGURE 32.14 Example of cost and time sampling process.

used for decision making in terms of the amount of risk a program manager is willing to assume for project completion time and cost, and an understanding of the risk inherent in the activities within the network.

Decision Trees and Analysis. Most program managers are familiar with the concept of decision trees and analysis. Figure 32.15 shows a simplified decision tree constructed to assist in evaluating the economic consequences of a facility improvement decision. While this technique is fairly commonplace, its utility for complex decisions involving multiple uncertain decision criteria is limited. Most often it is used in situations that involve only economic decision criteria (versus technical characteristics combined with other variables). There is some very good software in existence for evaluating decision tables and trees. Many of these packages are relatively inexpensive and readily available.

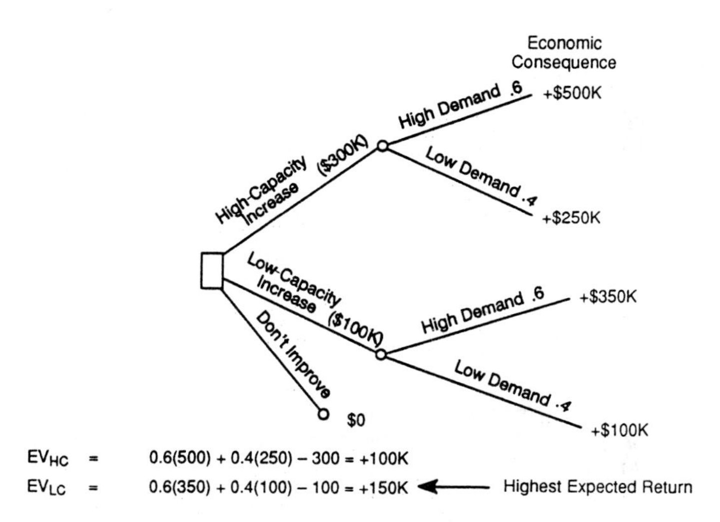

FIGURE 32.15 Decision tree for facility improvement.

Performance Tracking. This technique is focused on managing the technical risks within a project. It is basically a monitoring process designed for technical characteristics with clearly defined performance projections and preset alert criteria. Some standard criteria are shown in Fig. 32.16,[7] and an example of a control chart is presented in Fig. 32.17.[7] This technique, though not widely used, has had successful applications in a variety of forms and can be used by almost any program. The key to successful implementation is to adequately identify the correct criteria and alert thresholds.

Analytic Hierarchy Process. The analytic hierarchy process is a technique that allows a great deal of flexibility in the structure of a particular problem. It has some characteristics from decision trees, but can be tailored to a much broader array of problems. With AHP, the user selects those attributes necessary to the decision and fashions them in a hierarchical structure. This allows for the integration of a wide variety of low-level decision criteria. An example of an AHP tree is shown in Fig. 32.18.

AREA OF RISK	TECHNICAL RISK INDICATOR (TYPICAL UNIT OF MEASURE)	APPLIES TO						SOURCE			
			SUBSYSTEM								
		SYSTEM	SUBSYSTEM A	SUBSYSTEM B	SUBSYSTEM C	SUBSYSTEM D	SUBSYSTEM E	STATEMENT OF WORK	CONTRACT SPECS	CONTRACTOR PLANS	PRIOR EXPERIENCE
DESIGN	WEIGHT (POUNDS)	X	X	X	X	X	X		X		
	SIZE	X	X	X	X	X	X		X		
	CENTER OF GRAVITY (INCHES FROM REF. POINT)	X	X	X	X	X	X		X		
	THROUGHPUT (CLUSTERS PER MINOR CYCLE)	X							X		
	MEMORY UTILIZATION (PERCENTAGE OF CAPACITY)	X							X	X	
	DESIGN TO COST (DOLLARS)	X	X	X	X	X	X	X			
	DESIGN MATURITY (NUMBER OF DESIGN DEFICIENCIES)	X	X	X	X	X	X	X		X	X
	FAILURE ACTIVITY (NUMBER OF FAILURE REPORTS SUBMITTED)	X	X	X	X	X	X	X		X	X
	ENGINEERING CHANGES (NUMBER OF ECOs)	X	X	X	X	X				X	X
	DRAWING RELEASES (NUMBER OF DRAWINGS)	X	X	X	X	X				X	X
	ENGINEERING MANHOURS (MANHOURS)	X	X	X	X	X				X	X
TEST	CRITICAL TEST NETWORK (SCHEDULED DATES FOR CRITICAL TEST EVENTS)	X	X	X	X	X				X	X
	RELIABILITY GROWTH (MEAN TIME BETWEEN FAILURES)	X	X	X	X	X		X	X	X	X
PRODUCTION	TRANSITION PLAN (SCHEDULED DATES FOR CRITICAL PRODUCTION EVENTS)	X	X	X	X	X				X	X
	DELIQUENT REQUISITIONS (NUMBER OF DELINQUENCIES)	X	X	X	X	X				X	X
	INCOMING MATERIAL YIELDS (PERCENTAGE OF ACCEPTABLE MATERIAL)	X	X	X	X	X				X	X
	MANUFACTURING YIELDS (PERCENTAGE YIELD)	X	X	X	X	X				X	X
	UNIT PRODUCTION COST (DOLLARS)	X	X	X	X	X		X			
	UNIT LABOR & MATERIAL REQUIREMENTS (MANHOURS UNIT & MAT'L COST UNIT)	X	X	X	X	X				X	
COST	COST AND SCHEDULE PERFORMANCE INDEX (RATIO OF BUDGETED AND ACTUAL COSTS)	X	X	X	X	X		X			
	ESTIMATE AT COMPLETION (DOLLARS)	X	X	X	X	X				X	
	MANAGEMENT RESERVE FUNDS (PERCENTAGE REMAINING)	X								X	
MGMT	SPECIFICATION VERIFICATION (NUMBER OF SPECIFICATION ITEMS)	X	X	X	X	X			X		
	MAJOR PROGRAM RISK (RANKED LISTING)	X	X	X	X	X					X

FIGURE 32.16 Standard indicators.[7]

A typical hierarchy would be set up in a fashion that would start out at the top level with the focus of the problem (that is, choosing a design). The next levels would contain the criteria on which the decision would be based, followed by the alternatives available. If there are multiple facets to the problem, each facet could be analyzed individually in a separate hierarchy (as in a cost-benefit analysis where costs are analyzed separately from benefits and then compared in a ratio format). The structure of the tree is totally dependent on the problem to be solved.

The next step in the process is establishing information about how a particular decision maker (or decision makers) views the importance of each of the criteria, and how well each of the alternatives satisfies the criteria. This may be an iterative

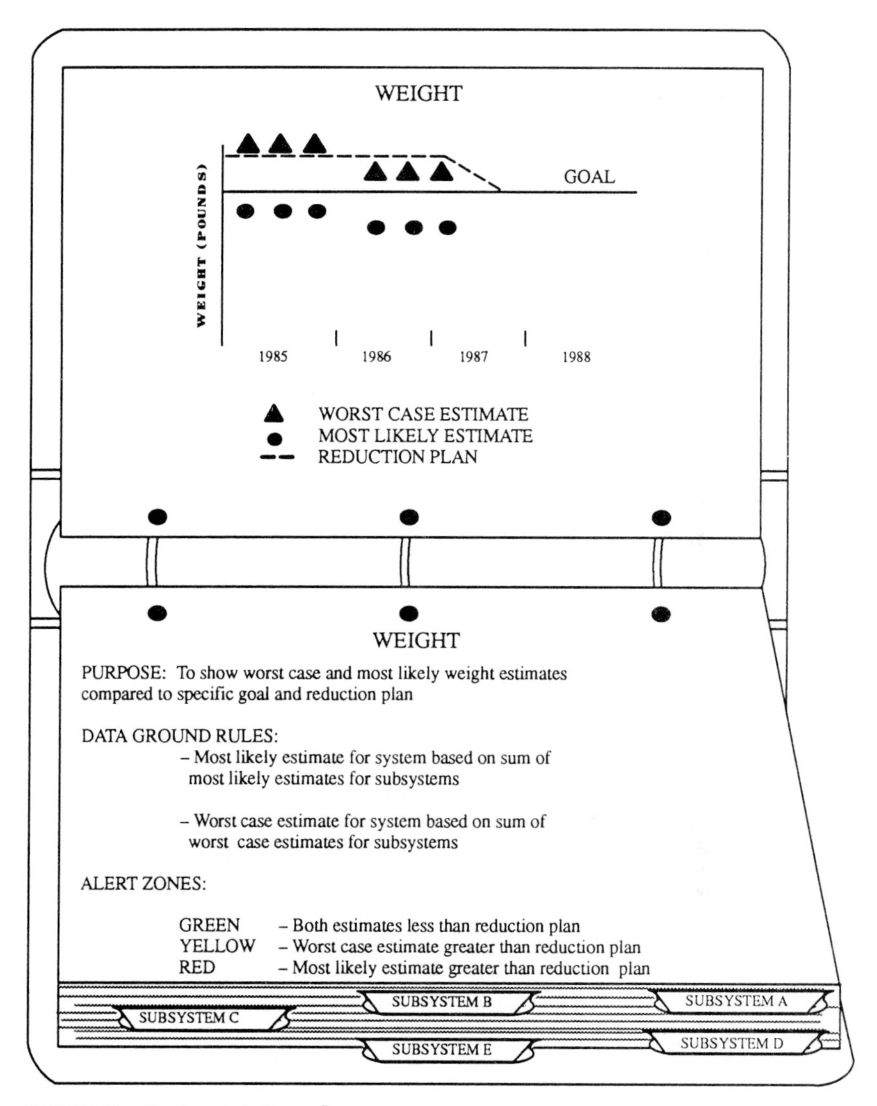

FIGURE 32.17 Sample indicator.[7]

process. The hierarchy does not have to remain constant. One of the benefits of AHP is that it is very flexible and allows a decision maker to quickly view a problem from different perspectives and different inputs. Delving into the matrix calculations that use the subjective judgments of the decision maker is beyond the scope of this chapter. The specific calculations used to develop the quantified solution of a hierarchy can be found in Saaty.[15] An important benefit of using AHP is that the user can also get a feel for how consistent his or her judgments are with respect to the criteria established for making the decision. The more consistent the judgments, the better a decision maker can feel about the certainty of the judgments made.

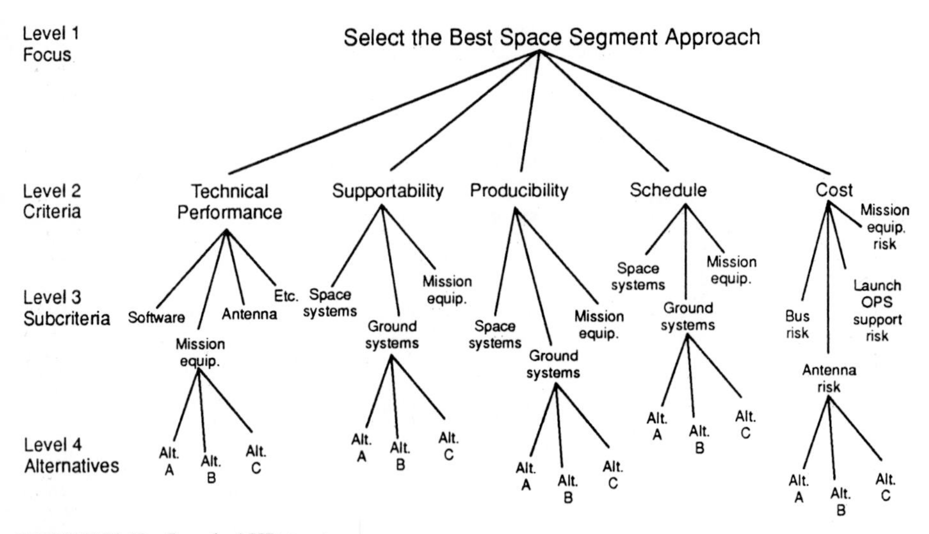

FIGURE 32.18 Sample AHP structure.

Watch Lists. This technique is almost fully described by its title. A watch list is nothing more than a prioritized ranking of risk areas with impact statements and management actions. Typically a watch list is the result of one of the other risk techniques described here, or the result of an intense risk identification process. Watch lists are one of the most common mechanisms to prioritize and focus risk management activity. Maintaining a top 10 list and reviewing progress on reducing or eliminating the top risks focuses the necessary resources in a proactive manner. This list should be prepared and maintained by a person designated by program management. Inputs to the list come from functional management. A typical watch list is shown in Fig. 32.19.[11]

AREA OF RISK	AREAS IMPACTED	ALTERNATIVE HANDLING TECHNIQUES
Loss of IV&V	Software Cost and Supportability	Obtain additional funds for V & V Do in-house (people??) Beef up QC Get Warranty
Inadequate Spares	User Availability	Obtain additional funds Get commodity CMD to budget more contractor support

FIGURE 32.19 Typical watch list.[11]

Techniques Focusing on Cost Risk

As might be expected, many of the risk techniques that have evolved focus specifically on quantifying the impacts of project risks in terms of dollars. This is logical in that dollars are the yardstick which is often used to evaluate performance. Unfortunately, unless the analyst performing the effort is conscientious enough to document the driving forces behind the cost risk thoroughly, visibility into the potential problem areas can be lost. Many of these techniques are particularly applicable to developing project cost estimates.

Risk Factors. The risk factors method for accommodating potential cost exposure due to program risk is simple to apply. The only requirements are that (1) there be a baseline cost estimate broken down by WBS elements in existence and (2) there be a systematic and scientific process to capture inputs from program personnel. The concept of the technique is to determine multipliers for each WBS element with which to increase the cost estimate for anticipated risk events. Typically these factors range from 1.0 to 2.0, with 1.0 representing no increase for risk and 2.0 representing a doubling of the cost estimate for risk. Each WBS element would have its own risk factor, which would be applied to the costs for that WBS element. Obtaining solid risk factors is the key ingredient to making this technique work. This technique is particularly useful if a cost estimate has been prepared which did not account for risk by element, and the program is currently being budgeted. This technique is only beneficial when examining cost as a function of risk.

Estimating Relationships. The estimating relationship method of calculating contingency funds for risk is probably one of the more difficult, but more defendable techniques for evaluating cost risk. The basic premise of the technique is that costs are correlated to other technical variables, such as design, performance characteristics, and maintenance concept. The key challenge in applying the technique is to uncover the variables in past programs that have influenced contingency funds (management reserve). Through regression analysis, an estimating equation can be developed that represents past experiences of these key program characteristics and their effect on management reserve. This equation can then be applied to the current program, substituting the new technical characteristics into the equation to calculate a predicted contingency fund. The regression may have to be run on many different variables to determine the characteristics which drive contingency funds. Once the equation is developed, the application to the current program is quick and easy to apply. An example of this application is illustrated in the following equation:[13] $Y = (0.0192 - 0.037X + 0.009X^2) \times 100$. Here X is the summation of engineering complexity (0 to 5), contractor proficiency (0 to 3), degree of system definition (0 to 3), and multiple users (0 or 1). Substituting values for these characteristics would yield a management reserve or contingency fund percentage that is reasonable based on previous experience.

WBS Simulation. This technique uses Monte Carlo simulation to develop a total program risk probability distribution. Here again, a program estimate by WBS elements (or by cost element structure) must exist. The difference between this and the risk factors method is that in WBS simulation, probability distributions are developed to reflect the cost risk versus a simple factor. Program completion is then simulated several hundred (or more) times to yield a range of probable results. Network models often incorporate the essence of this technique, but they typically require many more inputs. The output of this technique is a cumulative probability distribution from which a program manager can select an estimated completion cost at various confidence levels.

Technique Summary

The number and variety of techniques for analyzing risk are limited by individual creativity in problem solving. As mentioned previously, there are many other techniques readily available and useful in risk analysis, such as influence diagramming, method of moments, multiattribute utility modeling, life-cycle cost simulation, and CPR analysis. The important items to consider in selecting a technique are the applicability to the problem, the resources available to perform the analysis, and the desired format of the output.

RESPONDING TO RISKS

The risk identification, assessment, and analysis process means very little if there are no preventive actions taken which result from those processes. These management actions differ from what we typically think of as management actions because they are often aimed at preparing for or influencing the future versus reacting to the present. Since they are future-oriented, and not in response to an immediate need, the program manager has a great deal of flexibility in preparing for the management of the program's risk. (For example, the problem does not have to be "fixed" by tomorrow.) It would be impossible to list all management actions since they are dependent on the individual circumstances of each project. However, most actions can be categorized as described hereafter.

Categories of Risk Responses

After a program manager has been confronted with risk, he or she has several options to pick from. Typically these actions fall into one of the following response categories:

Risk assumption. In this case the program manager acknowledges the existence of the risk, but chooses to take no actions to change the current course of the project.

Risk avoidance. This implies a change of direction in some aspect of the project, or an alternative implementation of an aspect of the project in order to avoid the potential unwanted consequences of the current direction.

Risk control. This avenue focuses on controlling the impact of the risk, should it occur. Here contingency planning to provide for additional resources is a common practice.

Risk transfer. This involves a mechanism to either share risk with another party or completely transfer the risk to another party (like insurance policies and different contractual arrangements).

Combinations of the above. Risks are often best handled by a combination of actions falling into more than one of the categories described.

The choice of the response obviously depends on the unique circumstances of individual projects. For instance, the project phase would impact the type of risk problems being faced as well as the level of program advocacy, technology, funding, and so on. If the assessment and analysis of the risk have been done correctly, the program manager should have the opportunity to pick from several alternatives.

Contracting and Risk

One of the more effective and challenging risk-reducing strategies is developing a contractual arrangement which allocates risk equitably between the contractual parties. There are a great number of potential contractual arrangements possible which shift the risk from buyer to producer, as illustrated in Fig. 32.20. Selecting the "best" contractual arrangement is dependent on many factors, of which risk is only one. Since risk is such an integral part of nearly all other aspects of a project though, it should be given a considerable weight in the selection of the contractual approach. Some of the considerations include the following:

- Degree of work definition
- Product resulting from contract (research, hardware, and so on)
- Complexity of requirements
- Performance period
- Extent of competition
- Urgency of requirement
- Overall degree of cost, schedule, technical uncertainty

The degree of risk is a significant driver in the type of contractual arrangements, as represented in Fig. 32.21. (Note that cost and schedule distributions are typically skewed.) Although financially it is in the best interest of both parties to try to shift more risk onto the other party, this is not always the most effective way to manage a project in the purest sense. Some other considerations in trying to distribute the risk equitably include an assessment of who is responsible for creating the risk, who can best manage the risk, who should be involved in the management of the risk, and what other consequences might result to each party from the risk.

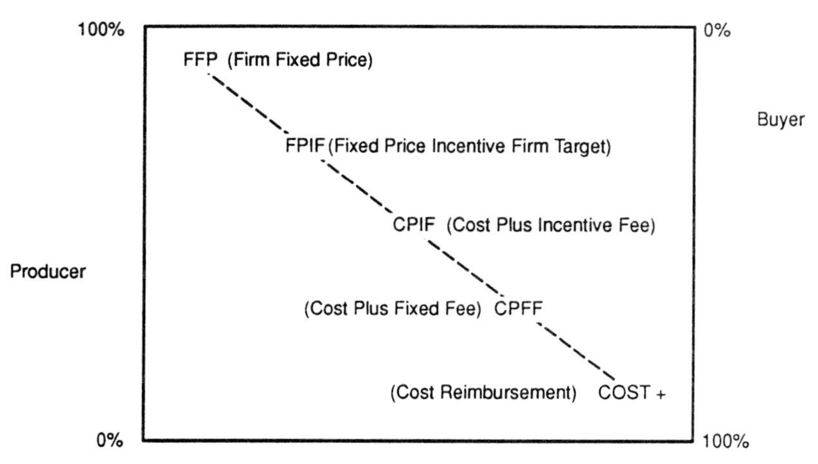

FIGURE 32.20 Contract types and risks.

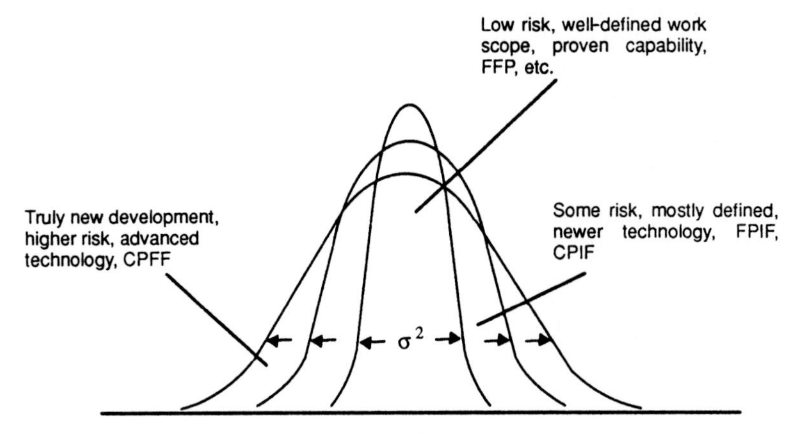

FIGURE 32.21 Sample risk distributions and contract types.

IMPLEMENTING THE RISK MANAGEMENT PROCESS

More often than not, program managers in the defense business find themselves with limited time, money, and skilled workers to meet ever-increasing demands for increased performance from the systems being developed and procured. This, in itself, poses substantial risk of failing to meet at least one of either cost, schedule, or performance goals as set for the project. It is critical that the program manager choose a risk management approach that is consistent with the existing organization, and with the other planning and control mechanisms in place on the project, is executable based on existing skills, time, and money available, and provides real value in terms of preparing for the future of the project. Thinking through how the risk management function will be executed, and where the responsibilities for executing the function will lie is a relatively easy task that will yield great benefits. The following five steps are intended to serve as thought generators for implementing a formal risk management process.

1. *Plan for risk management.* The program manager is ultimately responsible for managing risk. Most program managers require the establishment of a systematic process which provides the summary level information necessary for decision making. This process, the organization, focal points, the techniques to be used, and the general requirements are established at the initiation of a risk management program, and are communicated to project personnel. Thorough planning and proper emphasis increase the chances for a beneficial risk management program.

2. *Identify the known risks and prepare for the unknown risks.* Utilize the experts available to provide information regarding what the current risks are and their potential impacts. Rely on historical experience to indicate the possible extent of cost or schedule problems that could develop beyond the risks presently known. Update the list of known risks as the project evolves. Use the WBS to look at the total program for potential risks.

3. *Analyze the options.* Establishing a series of alternative plans is an excellent way to minimize the impact of decisions handed down from higher levels of authority. Programs are constantly being reshaped as a result of changes in requirements, funds, and so on. Preparing for these as well as identified technical risks can greatly benefit a program manager. The depth of the required analysis depends on the size and complexity of the program. An effective risk management program does not have to be expensive or use elaborate and costly techniques to achieve some good results.

4. *Make decisions based on long-range benefits.* Most program managers believe they do this. However, some people would argue that the current process for developing and procuring military systems fosters a short-term perspective. This means that life-cycle cost should become an even more important criterion in evaluating systems than before, and program managers need to look at the total picture very closely.

5. *Update the planning.* Periodic review of the risk management process put in place is a necessary activity to maintain usefulness over the life of a project. As the program proceeds, risks are reduced and there is less uncertainty than at startup. This suggests that the approaches to manage risk also change and evolve with the program. Figure 32.22 illustrates some key elements for risk management in each pro-

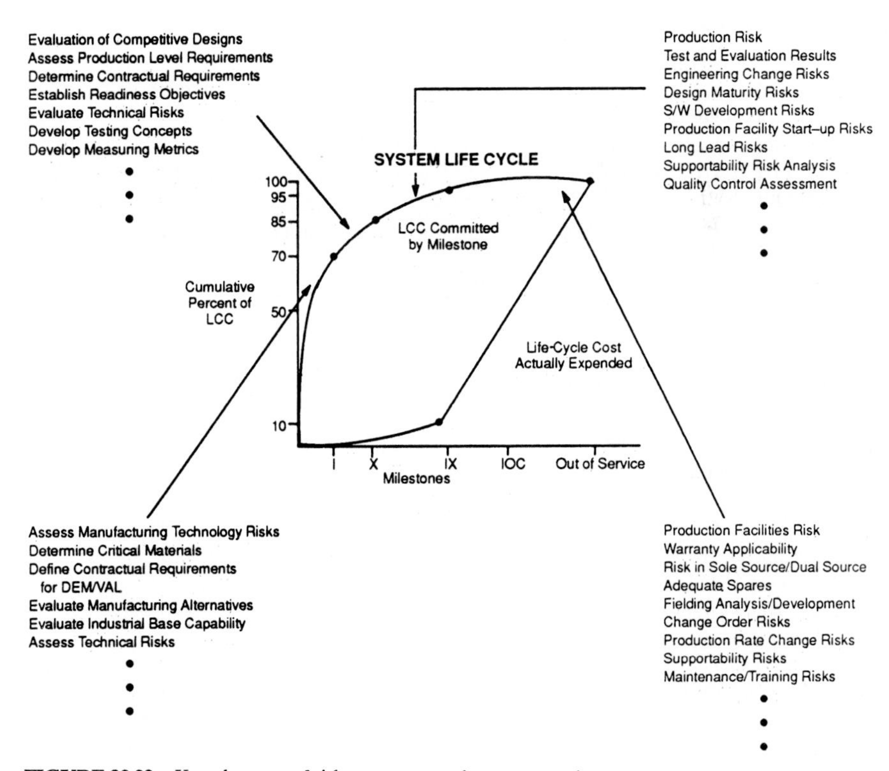

FIGURE 32.22 Key elements of risk management by program phase.

gram phase. Note that the nature of the risk problem changes as the program progresses.

SUMMARY

There are a couple of familiar sayings that are very appropriate in concluding this brief section on the subject of risk. The first is "no amount of planning will ever replace dumb luck." While we all have experienced this from time to time, we should never forget that "luck" is really nothing more than what results from "preparation meeting opportunity." Which, of course, brings us back to the requirement for preparation or, in other words, for good project planning. It is the project planning which serves as the departure point for examining all risk—cost, schedule, technical, or otherwise. And it is this examination of risk which allows a project manager to "prepare" for those possible but unlikely events.

REFERENCES

1. "Technical Risk Assessment: The Status of Current DoD Efforts," General Accounting Office, Washington, D.C., Apr. 1986.
2. Dod Directive 5000.1, "Defense Acquisition," Feb. 23, 1991.
3. Dod Instruction 5000.2, "Defense Acquisition Management Policies and Procedures," Feb. 23, 1991.
4. DoD Manual 5000.2-M, "Defense Acquisition Management Documentation and Reports," Feb. 1991.
5. Project Management Institute, Risk Workshop material presented by George H. Perino, DSMC, PMI National Symp., Calgary, Alta., Canada, 1990.
6. The Analytic Sciences Corp. (TASC), TR5541-1-7, June 30, 1988.
7. "Risk Management Concepts and Guidance," Defense Systems Management College, Ft. Belvoir, Va., Mar. 1989.
8. W. P. Rowe, *An Anatomy of Risk,* Robert E. Krieger, Malabar, Fla., 1988, p. 45.
9. "Project Management Body of Knowledge," Project Management Institute, Drexel Hill, Pa., 1987.
10. DoD Manual 4245.7-M, "Transition from Development to Production," Sept. 1985, pp. 1–8.
11. T. V. Caver, "Risk Management Workshop for Program Managers," Defense Systems Management College, Ft. Belvoir, Va., Nov. 1987.
12. The Analytic Sciences Corp. (TASC), "Program Management Office Survey," TR5541-1-6, May 1988.
13. M. Evrivaiades, "Management Reserve Cost Estimating Relationship," Cost and Analysis Division, Directorate of Cost Analysis, Hanscom AFB, Mass., Mar. 1980.
14. The Analytic Sciences Corp. (TASC), "Cost Risk Methodology," TR9042-2, May 1990.
15. T. L. Saaty, *Decision Making for Leaders,* RWS Publ., University of Pittsburgh, Pittsburgh, Pa., 1990.

BIBLIOGRAPHY

Adams, J. R., and M. D. Martin, "Professional Project Management, A Practical Guide," Universal Technology Corp., Dayton, Ohio, 1987.

Air Force Systems Command, USAF, "The AFSC Cost Estimating Handbook," Dec. 1985.

Caver, T. V., "Risk Management as a Means of Direction and Control," Fact Sheet, Program Manager's Notebook, No. 6.1, Defense Systems Management College, Ft. Belvoir, Va., Apr. 1985.

Rudwick, B., "Cost Estimating Workshop," Defense Systems Management College, Ft. Belvoir, Va., Nov. 1987.

CHAPTER 33

GOVERNMENT CONTRACT LAW

John A. Ciucci

John A. Ciucci is corporate secretary and general counsel to the Logistics Management Institute, Bethesda, Md. He also provides insights into contract law in some of the institute's acquisition and logistics tasks. As a federally funded research and development center, the institute is the Department of Defense's prime center for studies and analyses of logistics and acquisition issues. Colonel Ciucci retired from the U.S. Air Force in 1989 after 26 years, most of which was served with the Air Force Systems Command and the Air Force Logistics Command. His more recent assignments included General Counsel, Air Force Commissary Services; Staff Judge Advocate, San Antonio Air Logistics Center; and Vice Commander, Air Force Contract Law Center, Wright-Patterson Air Force Base. At Wright-Patterson he participated in source selection of many varied weapon systems. He holds a B.S. degree in business administration from Penn State University, an M.B.A. degree from Auburn University, and a J.D. degree from Georgetown University.

CAVEAT

Time and space limitations prevent our treating all the legal issues that a program management office can expect to encounter in the development of major DoD systems. We try to provide background information on certain fundamental contract policies, authorities, and procedures and to discuss selected current real issues. We do not offer this section as a substitute for the sound advice and counsel of the program manager's legal staff. The writer's observations while serving on active duty in the U.S. Air Force compel the conclusion that able, competent legal talent is available to program offices.

Some program managers request that a full-time lawyer be assigned to their office. In the Air Force we seldom honored such a request, even though we understood the perceived advantages. In our view, the disadvantages were too great. Having the lawyers assigned to the host installation legal office and dedicated to certain major acquisition programs, we were convinced, was the better relationship. We also provided a back-up lawyer for large programs. Our approach had a serendipitous effect in legal matters that is not available at the program office. In the final analysis, this relationship invariably proved the most effective.

BASIC CONTRACT PRINCIPLES AND POLICIES

In the United States, private parties have full power to contract as they please as long as the object of the contract is legal and not against public policy. Government contracting, on the other hand, faces more stringent limitations, primarily because of the government's unique status as a contracting party. First, while the United States is a sovereign power and has special immunities and unusual powers, it is a government of enumerated and delegated powers. For example, the Department of Defense (DoD) or the military departments are restricted to the contracting authority that the Congress or the President chooses to delegate.

Many attorneys on their initial assignment in support of a program office are amazed to learn that general common law principles of contract formation do not always apply to government contracting. In addition, the Uniform Commercial Code (UCC), often referred to as the benchmark in contracting in the private sector, does not apply to government contracts. While the UCC is a tremendous aid in contract formation and administration, public-sector courts and boards seldom turn to its principles in resolving disputes between DoD and contractors. Government contract law is a creature of federal statutes, implementing regulations* and decisions of U.S. courts and boards. Most boilerplate clauses in government contracts are grounded in this body of law. However, knowing that body of law is not enough; one must also be familiar with decisions of the courts and boards. As an example, the standard Changes clause in a supply contract requires the contractor to submit a request for an equitable adjustment within 30 days of a contracting officer's change that results in an increase in the contract price. The Armed Services Board frequently ignores that time requirement. On the other hand, the Disputes clause requires a contractor to appeal the *final decision* of a contracting officer within 90 days, and the Armed Services Board has upheld that requirement almost without exception.

Many clauses required in government contracts have no counterparts in the commercial sector. For instance, the Truth in Negotiations Act requires certain contractors to provide cost and pricing data and to certify under penalty of law that the data are accurate, current, and complete. Another clause, the standard Disputes clause, requires the contractor to continue performing the contract while engaged in a dispute with the government and while the case is before a court or board of contract appeals.

Two basic procurement statutes—the Armed Services Procurement Act of 1947 and the Federal Property and Administrative Services Act of 1949—govern the procurement of most government supplies and services. The former controls procurement of all supplies and services for DoD, which includes the departments of the Army, Navy, and Air Force, and for the Coast Guard and NASA. The 1949 Act controls purchases of the General Services Administration (GSA) and the other executive agencies. Among the many other laws that affect one phase or another of government procurement are the Defense Production Act, Buy American Act, Services Contract Act, Contract Disputes Act, and Small Business Act. Labor statutes permeate the process. For example, the Services Contract Act requires payment of minimum wages and benefits and imposes other conditions of work on most services

---

*Primary among those regulations are the *Federal Acquisition Regulation* (FAR) and the *DoD FAR Supplement* (DFARS).

contracts; the Davis Bacon Act, now more than 60 years old, requires that federal construction projects pay a federal minimum wage that is the prevailing labor rate in the geographic area; and the Walsh-Healy Act specifies the types of contractors eligible to enter into supply contracts with the government. Many other socioeconomic laws that apply to government procurement have no commercial counterparts and complicate the process for contracting with DoD. The process is further complicated when we consider that other federal agencies—the Department of Commerce, the Labor Department, the Small Business Administration, and the Environmental Protection Agency, to name a few—have jurisdiction to administer these socioeconomic laws. The EPA, relatively new on the scene, is responsible for publishing and enforcing a whole host of recent and changing regulations implementing environmental laws.

In 1984 Congress enacted the Competition in Contracting Act (CICA), which made extensive and sweeping changes to both procurement acts. When CICA became effective in April 1985, it was the first overhaul in the two basic procurement statutes since their passage. This important law is discussed in the section titled "Methods to Procure."

The Office of Federal Procurement Policy (OFPP), established in 1974, embarked on a mandated effort to establish a unified system of procurement regulations, a recommendation made by the Commission on Government Procurement. Its efforts culminated in a drastic change in the regulatory framework for government procurement when on April 1, 1984, the FAR came into effect.

Before 1978, DoD used its own system of procurement regulations, called the Armed Services Procurement Regulations (ASPR). The DoD, by directive in March 1978, changed the name of the ASPR to the Defense Acquisition Regulation (DAR). The federal civil agencies' counterpart regulations were known as the Federal Procurement Regulations (FPR). Since it became effective, the FAR has been used for most government supply and service procurement. DoD and other agencies have narrow authority to supplement the FAR with their own regulations (the DoD supplement is the DFARS). Both the FAR and DFARS, when appropriately published, have the force and effect of law, even though they are not statutory.

In addition to the cited acts and regulations, government program offices have to monitor the following sources that affect procurement law:

- Executive orders
- Judicial decisions
- Boards of Contract Appeals decisions
- The Comptroller General
- The Attorney General
- Department of Commerce
- Department of Labor
- Small Business Administration

ELEMENTS OF A CONTRACT

Government contracts are formed from the same four elements as private-sector contracts: capacity, offer and acceptance, consideration, and lawful purpose. In practice, however, we see significant differences. The *Restatement of Contracts** defines a

**Restatement of Contracts* is an authoritative compendium that addresses the current state of contract law in the United States.

contract as "a promise or a set of promises for the breach of which the law gives a remedy, or for the performance of which the law recognizes a duty." In a technical sense, a contract and an agreement differ in that an agreement is broader and encompasses some promises that a court will enforce and others that it will not.

To be enforceable, a contract must include all four elements. Each party to the contract must have the legal *capacity* to contract, and an *offer* must be made and *accepted.* The parties must each promise or give something of value, known in the law as *consideration,* and finally, neither the objective of the contract nor its subject matter must be in contravention of law or public policy, that is, they must be *legal.* The contract must also be in a form required by law, often in writing and signed. We explore these four elements in the following subsections.

Capacity

Capacity in the private sector usually devolves to whether the parties are able to contract, that is, whether they are of legal age, sane, sober, and so on. Corporations must stay within the "objects and purpose" granted by their corporate charter or certificate of incorporation; otherwise their contracts are considered *ultra vires* and not enforceable. In the government sector, litigation most often involves the authority of the government representative or the contractor's responsibility. Contractors often have difficulty in ascertaining the extent of the authority of the many government people who participate in a major procurement. The FAR is very clear on this point. FAR 1.602-1 states:

> Contracting officers have authority to enter into, administer, or terminate contracts and make related determinations and findings. Contracting officers may bind the government only to the extent of the authority delegated to them. Contracting officers shall receive from the appointing authority . . . clear instructions in writing regarding the limits of their authority. Information on the limits of the contracting officers' authority shall be readily available to the public and agency personnel.

More specific guidance is given at FAR 43.102:

> Only contracting officers acting within the scope of their authority are empowered to execute contract modifications on behalf of the government. Other government personnel shall not
>
> **(1)** Execute contract modifications;
>
> **(2)** Act in such a manner as to cause the contractor to believe that they have authority to bind the government; or
>
> **(3)** Direct or encourage the contractor to perform work that should be the subject of a contract modification.

Since the U.S. government operates under the principle of delegated powers, contracting officers, to bind the government, must operate within the bounds of their express authority, that is, their contracting officer's warrant. As a general rule, government officials, including contracting officers, do not have apparent authority (as in the private sector) to bind their principal. Thus contractors must ascertain the actual authority of government representatives. In a landmark case in 1947, the Supreme Court ruled on this authority issue:

> It is too late in the day to urge that the Government is just another private litigant, for purposes of charging it with liability, whenever it takes over a business theretofore conducted by private enterprise or engages in competition with private ventures. Govern-

ment is not partly public or partly private, depending upon the governmental pedigree of the type of a particular activity or the manner in which the government conducts it. The government may carry on its operations through conventional executive agencies or through corporate forms especially created for defined ends. . . . Whatever the form in which the government functions, anyone entering into an arrangement with the government takes the risk of having accurately ascertained that he who purports to act for the Government stays within the bounds of his authority. The scope of this authority may be explicitly defined by Congress or be limited by delegated legislation, properly exercised through the rule-making power. And this is so even though, as here, the agent himself may have been unaware of the limitations upon his authority. . . .

That case law prevails today.

The contractor's capacity is determined by an even more curious practice, for the government contracting officer is obliged by law to determine affirmatively that the prospective contractor is responsible. FAR 9.104 establishes the guidance for the contracting officer's determination:

(a) Have adequate financial resources to perform the contract, or the ability to obtain them;

(b) Be able to comply with the required or proposed delivery or performance schedule, taking into consideration all existing commercial and governmental business commitments;

(c) Have a satisfactory performance record;

(d) Have a satisfactory record of integrity and business ethics;

(e) Have the necessary organization, experience, accounting and operational controls, and technical skills, or the ability to obtain them . . . ;

(f) Have the necessary production, construction, and technical equipment and facilities, or the ability to obtain them; and

(g) Be otherwise qualified and eligible to receive an award under applicable laws and regulations.

Therefore, unless the government contracting officers find that the contractor is responsible, the contractor does not have the requisite capacity to contract and will be denied the award, even though it may be the low bidder. This responsibility determination is many times based on a rather subjective standard, such as poor past performance, but the General Accounting Office (GAO) or the Appeals Board defers to the contracting officers' findings unless they have abused their discretion or engaged in bad faith.

Offer and Acceptance

It has often been said that a contract is formed when an offer is accepted. An offer is a bid or proposal made by an entity, known as the offeror, to an offeree that a contract be formed. If it is acceptable, the offeree properly communicates a valid acceptance and a contract is formed. Notice in the *Commerce Business Daily* or otherwise posted for a government need does not constitute an offer. It is akin to an advertisement in commercial practice and is considered an invitation for offers. The same is true for an invitation for bids or a request for proposals. The responding bid or proposal is the offer, and if the offer is acceptable, the contracting officer is the proper person to accept it. An offer continues to exist until accepted or until its stated time expires (if no time is stated, until the expiration of a reasonable time). In the commercial world, an offeror's right to terminate or revoke an offer at any time prior to acceptance is virtually absolute unless it amounts to a paid option. Such is the case in offers to the government in a sealed bidding context. After bids are open, however, the Firm-Bid Rule becomes operative and the bid cannot be withdrawn for the time specified in the solicitation. Courts and the GAO routinely enforce the Firm-Bid Rule.

The acceptance of an offer must be absolute and unequivocal; it cannot be conditional. After an offer expires, it may not be accepted; an attempt to do so constitutes a counteroffer, which starts the offer-acceptance process again. What is required is the mutual assent of two separate entities. Some writers refer to this assent as a "meeting of the minds." A mutual mistake prevents this "meeting of the minds."

Consideration

Consideration is not generally a problem in the initial contract formation stage. Rather, the government program manager will more often encounter consideration problems when a contractor asks for an extension of the delivery schedule or a relaxation of a specification or makes a similar request. In such situations the government must also receive consideration for waiving some right to which it was entitled, that is, strict contract performance. The law distinguishes between "adequacy" of consideration and "sufficiency" of consideration. Courts and boards will not second guess what might be the appropriate value of the exchange, but are interested only in the existence of consideration—that what is given in exchange has value in the eyes of the law. A tangible item of monetary value does not have to be exchanged to constitute legal consideration. Often, forbearance of a legal right is the basis of consideration. Consideration questions do not occupy a lot of time and effort in the typical program office.

Lawful Purpose

Although the right to contract is fundamental, if the contract violates a statute, it is deemed to be void and will not be enforced. The same is true if the contract's intended purpose is contrary to public policy or "shocks the conscience of the community." In such cases the courts take a hands-off policy and are said to "leave the parties where they find them." An example would be a case in which a contractor agrees with another party to pay a contingent fee if the contractor is awarded a contract as a result of the other party's undue influence. Not only would the agreement be unenforceable under the covenant against contingent fees but it could violate the criminal antikickback statute if it were some arrangement between a prime contractor and a subcontractor.

Form

In addition to the four primary elements of a contract, care must also be given to the contract's form. Problems with form, however, seldom arise in a program office. 31 U.S.C. § 1501 requires that all obligations of U.S. government funds be supported in writing by documentary evidence of a binding agreement. The law was originally enacted after the Civil War to curb reported abuses that certain agencies were recording obligations based on oral contract purportedly to prevent unobligated funds from reverting to the Treasury at the expiration of the period for appropriation. The law has been amended several times since then, and the Supreme Court, in its interpretation, has refused to enforce several contracts that were not in writing. Instances of failure to commit a contract to writing for major acquisitions are rare.

CONTRACT TYPES

The government uses various types of contracts to procure its required goods and services. The type chosen is driven largely by the government's ability to determine the exact quantity it needs and to define the work it wants precisely. If the government is certain about the quantity and work, it will use a contract that fixes the obligations and rights of both parties precisely. In basic research, for example, if the quantity and work are not certain, the government will use a contract that provides for greater flexibility—both in the product delivered (perhaps, level of effort) and the cost to the government (perhaps, an arrangement to reimburse the contractor for its direct costs plus other customary burdens such as overhead, general and administrative expenses, or independent research and development accounts). Government contracts are usually referred to by their pricing arrangement. The contracting officer has broad latitude in selecting the type of contract. This authority is well stated in the FAR at 16.103:

> **(a)** Selecting the contract type is generally a matter for negotiation and requires the exercise of sound judgment. Negotiating the contract type and negotiating prices are closely related and should be considered together. The objective is to negotiate a contract type and price (or estimated cost and fee) that will result in reasonable contractor risk and provide the contractor with the greatest incentive for efficient and economical performance.
>
> **(b)** A firm-fixed-price contract, which best utilizes the basic profit motive of business enterprise, shall be used when the risk involved is minimal or can be predicted with an acceptable degree of certainty. However, when a reasonable basis for firm pricing does not exist, other contract types should be considered, and negotiations should be directed toward selecting a contract type (or combination of types) that will appropriately tie profit to contractor performance.

This contracting officer authority is not absolute. The contractor may attempt to influence the type of contract. The old saying that "everything is negotiable" is perhaps true in this process. Section 15.102 of the FAR states the following:

> Negotiation is a procedure that includes the receipt of proposals from offerors, permits bargaining, and usually affords offerors an opportunity to revise their offers before award of a contract. Bargaining—in the sense of discussion, persuasion, alternation of initial assumptions and positions, and give-and-take—may apply to price, schedule, technical requirements, *type of contract* or other terms of a proposed contract. [Emphasis added.]

In broad terms, the government uses two contract types: the fixed-price contract and the cost-reimbursement contract. It uses many variations of these types, ranging on a continuum from the firm-fixed-price contract under which the contractor bears the full responsibility for performance and costs (affecting its profit or loss) to the cost-reimbursement contract under which the contractor's performance obligation and the government's obligation to pay are related to the actual costs incurred.

1. *Firm, fixed-price contracts.* This type of contract is used most frequently. The price is firm and places maximum risk on the contractor. On the other hand, it

provides maximum incentive on the contractor to control costs and perform effectively to obtain a higher profit. It contains no formula for a variation or adjustment to either price or property or service delivered. It works well, of course, in purchasing commercial or off-the-shelf items. It generally is used in sealed bidding.

2. *Fixed-price contracts with economic price adjustment.* This type of contract provides for an adjustment, upward or downward, of some element of price, and is used when the labor or material marketplace are not stable. The adjustment most often allowed is for an increase or decrease in actual costs of labor or materials. At other times, particularly in contracts in which performance extends for a long period (usually exceeding a year), the parties may use an adjustment based on established cost or material indices.
3. *Fixed-price, incentive contracts.* This type of contract normally includes a predetermined formula for sharing overrun or underrun of the target cost. The formula consists of four elements: target cost, target profit, share formula, and ceiling price. It contains a ceiling price above which the government will not share cost; exceeding the ceiling does not excuse the contractor from strict compliance in performing the contract. It is common, although not necessary, to make the share line symmetrical on both sides of the target cost, and a common sharing is 70/30 (government-to-contractor) although the proportions should vary according to the uncertainties involved in contract performance. For instance, if the contract is for a first-article requiring limited development, the program manager will probably want to change the share line to assume some risk of the development from the contractor. Therefore you can expect to see contracts with 50/50 or 60/40 shares under target and perhaps go to 80/20 over target. An informed, commonsense approach to the risks involved is what is necessary. This risk analysis should include such things as engineering, production, financial, and management considerations. This analysis should then become part of the negotiation strategy. If these risks are evaluated well, then the share lines and target should fall easily into place. This split should create a win-win result, as far as practicable.
4. *Fixed-price, redetermination contracts.* This type of contract allows the parties to adjust profit and set the firm price based on an established formula at or near the end of contract performance. This type lends itself well to R&D contracts. Two kinds of redeterminable contracts are authorized:
   a. *Fixed-price contract with prospective redetermination.* This type of contract provides for a fixed price for the initial contract period and a prospective redetermination of the price for the remaining period or periods at stated times during contract performance. In the Air Force we routinely used this contract type to purchase aircraft engines, particularly with contracts whose accounting systems used periodic, plantwide prospective pricing methods.
   b. *Fixed-price contract with retroactive price determination.* This type contains a fixed ceiling price and allows a price determination after contract completion. The FAR allows use of this type of contract for R&D contracts estimated at $100,000 or less when it is established at the outset that a fair and reasonable firm fixed-price contract cannot be negotiated at that amount and a short performance period makes the use of any other fixed-price contract impractical (see FAR 16.205 and 16.206).

In the other major contract type, cost reimbursement, the government agrees to pay the contractor for the allowable costs incurred in performing the contract to the extent provided for in the contract. The contract generally includes an estimated cost, which is the figure the government uses legally to obligate funds and fix the ceiling that the contractor should not exceed unless it is willing to assume a substantial risk of not being reimbursed. However, the contractor is not obligated to perform or incur additional costs beyond that discrete ceiling. Of course, the government may increase the ceiling or later ratify it if it is exceeded and the contracting officer is so inclined. In a cost contract environment, the contractor assumes less risk but accepts much more government oversight. The program office must be vigilant in this effort, for a contractor has less motivation to control its costs when operating under this type of contract. The cost-reimbursement contract is to be used only after the government determines that it is likely to be less costly than another type of contract and finds that purchasing the required goods or services is impracticable without using a cost-type contract.

Like the fixed-price contract, the cost-reimbursement contract comes in varying forms.

1. *Cost-plus, incentive-fee contract.* This arrangement injects a sharing formula to permit the government and the contractor to share in efficient contract performance. It apportions cost overruns and underruns just as the fixed-price, incentive contract does.
2. *Cost-plus, award-fee contract.* Used extensively by NASA, this type contract provides for a fee consisting of a base amount fixed at the beginning of the contract and an amount based on a judgmental evaluation by the program officer deemed sufficient to motivate the contractor and provide for excellent performance of the contract. Please note that award fee determinations are usually within the sole discretion of the program office and not subject to the Disputes clause of the contract.
3. *Cost-plus, fixed-fee contract.* A widely used type, this arrangement pays the contractor its actual and allowable costs and a fixed fee for contract performance. The fees allowed are subject to a maximum specified by statute. The contract also may be awarded with no fee. It is most commonly used for research projects with universities, nonprofit organizations, or federally funded research and development centers or for facility contracts.
4. *Cost-sharing contract.* This contract provides no fee to the contractor and reimburses only for a portion of the contractor's allowable costs. The sharing arrangement is agreed upon in advance, and contract performance usually begins without either party knowing what amount is to be expended on the program. Therefore the Anti-Deficiency Act is a factor that causes the government to use a strict Limitation of Fund clause in the contract. Contractors who exceed the limitation do so at their peril.

CONTRACTING METHODS

The two principal statutes setting forth the legal methods to form government contracts were discussed earlier; they are the Armed Services Procurement Act of 1947 and the Federal Property and Administrative Services Act of 1949. Other statutes, especially the annual authorization and appropriation acts, not only authorize certain departmental supplies and services but also establish requirements in contract formation that add to those already enacted in the two basic procurement laws. While, as noted earlier, the Competition in Contracting Act drastically amended the basic procurement statutes of 1947 and 1949 and established a new regimen in contract formation, we can gain a better understanding of what we are doing today by reviewing the legal requirements for contract formation prior to the CICA.

Competitive Contracting

The two basic procurement statutes required the government to strive for full competition in its procurement. To achieve that competition, agencies used one of two methods for procurement—formal advertising (sometimes referred to as price competition) and negotiation. Formal advertising was the preferred method to procure supplies and services, and negotiated procurement was a method permitted only by statutory exceptions. Formal advertising required the buyer to specify a minimum need, solicit bids, publicly open and evaluate the bids, and award the contract to the lowest responsive, responsible bidder without negotiations. For complex items such as major weapon systems, the government had to find an exception to the formal advertising scheme and justify the use of negotiation by a fact-specific determination and finding (D&F). The D&F had to be approved by the agency head or his or her delegate.

Negotiation provided the flexibility for complex procurements. It allowed the contracting officer to "deal" with respect to the terms, conditions, and price with all potential contractors in the competitive range, that is, with all contractors whose proposals were good enough to be deemed responsive to the basic specifications of the contract even in light of known deficiencies. Further, contracting officers did not have to award to the contractor offering the lowest price. They were given the discretion to procure what was in the best interest of the government. If the difference between the price of the awardee and that of the low offeror was justified by enhanced technical performance or by a lower life-cycle cost, for example, the contract could be awarded to other than the low offeror. All that was needed was a memorandum in the contract file justifying procurement of the higher-price items. In almost all cases in which offerors protested to the Comptroller General, the protest was routinely denied on the grounds that the Comptroller General did not have the technical expertise to second guess the agency.

As stated earlier, to use the negotiated procedures, agencies had to present a procurement plan that justified negotiating a contract by fitting the procurement into 1 of the 17 exceptions to formal advertising. The majority of justifications invoked the so-called (a)(10) exception—"competition is impracticable." That exception was probably used because it fit better than any of the others. Nevertheless, it ignored the fact that a program office might have half the country's aircraft producers engaged in a head-to-head competition for a new fighter aircraft. Use of that exception led to a perception in Congress that negotiation was not a legitimate competitive procedure. In addition, the Commission on Government Procurement found that when negotiation was appropriate, agencies had to engage in an "expensive, wasteful, and time-consuming" procedure to justify its use. It recommended further that competitive negotiation should be authorized and established as an "acceptable and efficient alternative" to formal advertising. With the enactment of CICA, this contractual regime was changed drastically.

The Competition in Contracting Act

The CICA directed that contracts be awarded on the basis of full-and-open competition, which means that all responsible sources are eligible to bid or propose on government procurements. CICA established the competition advocate to ensure maximum competition for purchases. It actually raised the negotiated procurement method to the level of the advertised method, and gave the latter a new name—"sealed bidding." It recognized the need for tighter control on sole-source contracting and established tighter control for awarding noncompetitive (other-than-full-and-open competition) contracts—the justification and approval process (J&A).

The FAR's implementation of CICA requires contracting officers to use sealed bidding in certain instances:

> **(a)** Sealed bids. Contracting officers shall solicit sealed bids if—
> **(1)** Time permits the solicitation, submission and evaluation of sealed bids;
> **(2)** The award will be made on the basis of price and other price-regulated factors;
> **(3)** It is not necessary to conduct discussion with the responding offerors about their bids; and
> **(4)** There is a reasonable expectation of receiving more than one sealed bid.

But note how the CICA raised the heretofore "negotiation" procedures on a par with sealed bidding:

> Competitive Proposals.
> **(1)** Contracting Officers may request competitive proposals if sealed bids are not appropriate under paragraph (a) above. [cf. FAR 6.401]

In some situations sealed bidding and competitive proposal (negotiation) are not appropriate and open competition is not available. In such situations, sole-source contracting is necessary.

Sole-Source Contracting (Other-than-Full-and-Open Competition)

Sole-source contracting, now called other-than-full-and-open competition, is sanctioned by the CICA. It provides seven exceptions to full-and-open competition:

1. *Only one responsible source.* The supplies or sources required are available from only one responsible source and no other type of supplies or services will meet the agency's requirements.
2. *Unusual or compelling urgency.* The agency's need is of such unusual or compelling urgency that the government would be seriously injured unless the agency is permitted to limit the number of sources.
3. *Industrial mobilization or experimental, developmental, or research work.* The contract must be awarded to a particular source for the following reasons:
   - **a.** To maintain an essential industrial capability in case of a national emergency
   - **b.** To establish or maintain an essential engineering, research, or development capability to be provided by an educational or other nonprofit institution or by a federally funded research and development center
4. *International agreement.* The contract award is required by the terms of an international agreement or treaty or at the written direction of a foreign government.
5. *Authorized or required by statute.* A statute requires acquisition to be made through another agency or from a specified source, such as the Federal Prison Industries or Government Printing Office, or the agency needs a brand-name commercial item for authorized resale.
6. *National security.* The disclosure of the agency's needs would compromise the national security unless the agency is permitted to limit the number of sources.

(Merely because the acquisition classified is insufficient to obtain the use of this authority.)

7. *Public interest.* The head of the agency determines that it is not in the public interest for the particular acquisition concerned and notifies the Congress of such determination not less than 30 days before award of the contract. This authority may be used if no other exception applies, and may not be delegated.

A contracting officer before beginning the process of negotiating a contract without a full-and-open competition must *justify* the action in writing, certify the accuracy and completeness, and obtain the necessary approvals. This "justification and approval" process is based on 10 U.S.C. § 2304(f)(3):

> **(f)(3)** The justification . . . shall include
>
> **(A)** a description of the agency's needs:
>
> **(B)** an identification of the statutory exception from the requirement to use competitive procedures and a demonstration, based on the proposed contractor's qualifications or the nature of the procurement, of the reasons for using the exception;
>
> **(C)** a determination that the anticipated cost will be fair and reasonable;
>
> **(D)** a description of the market survey conducted or a statement of the reasons a market survey was not conducted;
>
> **(E)** a listing of the sources, if any, that expressed in writing an interest in the procurement; and
>
> **(F)** a statement of the actions, if any, the agency may take to remove or overcome any barrier to competition before a subsequent procurement for such needs.
>
> **(4)** The justification . . . and any related information shall be made available for inspection by the public consistent with the provisions of § 552 of Title 5 U.S.C. [The Freedom of Information Act]

The statute further prohibits using other than competitive procedures to avoid advance planning or because of concerns about the amount of funds available to the agency for procurement functions.

After the project officer satisfies these statutory requirements, the negotiation process may begin. Experience indicates that the contracting personnel in a major project office are expert in the steps leading to negotiation, including the preparation of the request for proposals, evaluation of proposals, source selection criteria, and competitive range determination. In fact, we doubt that discussion of these subjects is necessary. However, in a recent National Contract Management Association (Washington Chapter) newsletter, Joseph Bolos presents the following sound and sage advice on "negotiations":

> Negotiations. What comes to mind? Do you think of two parties sitting or standing across from each other, each with their own "bottom line" and each with an all consuming desire to reach an agreement at any cost that satisfies all their requirements? Or do you see two parties, both having something to bring to the arrangement, and both willing to reach an agreement that has provisions that are mutually beneficial?
>
> All too often, both in Government and in industry, negotiators have a "win at any cost" attitude. I'm not certain if there is a single cause for this attitude. But often the system of absolute competitiveness breeds a combative approach.
>
> But there is a better way of achieving agreements that satisfy the mutual concern of the parties. It is not just starting high then splitting down the middle. It's not just recognizing that you have a superior position and then digging in with a "take it or leave it" position. What it involves is searching for and identifying the needs of the other side and then attempting to balance those needs against your truly important needs in coming to

> an agreement. To some degree, based upon a balance of needs, both sides come out winners. The underlying assumption is that there is an environment and a set of attitudes on both sides that will allow this balanced type perspective and agreement to occur.
>
> Negotiation is not a single event; a contract, a new car, a house, a job. It's a complicated process with principles, rules, customs, and regulations all coming together to satisfy needs. Once this process is learned, it can be used for any type negotiation, personal or professional.

We think Mr. Bolos is "right on target."

Contracting officers now have statutory authority to award a contract without negotiating with the offerors (other than discussing for purposes of minor clarifications) provided the solicitation included a statement that the government intended to make award without discussions.* This authority has to be used wisely. In November 1991 the Air Force awarded a contract for desktop computers to a small northern Virginia firm using this procedure. In the process, it eliminated 20 of 22 companies in the competition. Early in 1992 the Air Force canceled the award because of a rash of protests from those eliminated contractors.

CONTRACT MODIFICATIONS

Definition

The contract amendment that restructured the C-5 aircraft contract and perhaps saved the Lockheed Corporation was Mod-1000, the one-thousandth change to the original contract. I recall assisting on a change to an F-111 aircraft contract with the General Dynamics Corporation—the modification was 3000 plus. Modification, made possible mainly by the Changes clause, is the program manager's most important tool. Simply stated, a modification is a new agreement that varies the terms and conditions of the agreement originally struck by the government and the contractor. The change in the contract is usually driven by such conditions as an urgent need to accelerate the delivery schedule, some change (up or down) in the program funding profile, a technological breakthrough enabling some desirable, increased performance in the weapon system, the incorporation of a value engineering change proposal, or overcoming defects discovered in the specification.

Contract Changes

Modification occurs through a change order, whether unilateral or bilateral, issued solely by the contracting officer. The FAR states: "A change order is a written order, signed by the contracting officer, directing the contractors to make a change authorized by the Changes clause." At common law, one party may not unilaterally change the obligation of the agreement; in the government arena, however, the right to do so is based on the advance agreement between the government and the contractor authorizing certain changes within the general scope of the contract. The particular Changes clause prescribed by the FAR is slightly different from one contract type to another; that is, the clause in the fixed-price supply contract is slightly different from the clause in a cost-reimbursement supply contract. However, all such clauses require three conditions:

*Section 802, DoD Authorization Act of 1991. The Department sought this authority for several years.

1. The change must be within the scope of the contract.
2. The order must be written.
3. The change must be ordered by an authorized government official.

The Changes clause prescribed in fixed-price supply contracts allows the contracting officer to make only the following specified changes. (Note the consequences of adjustment in price.)

> **(a)** The Contracting Officer may at any time, by written order, and without notice to the sureties, if any, make changes within the general scope of this contract in any one or more of the following:
>
> **(1)** Drawings, designs, or specifications when the supplies to be furnished are to be specifically manufactured for the Government in accordance with the drawings, designs, or specifications.
>
> **(2)** Method of shipment or packing.
>
> **(3)** Place of delivery.
>
> **(b)** If any such change causes an increase or decrease in the cost of, or the time required for, performance of any part of the work under this contract, whether or not changed by the order, the Contracting Officer shall make an equitable adjustment in the contract price, the delivery schedule or both, and shall modify the contract.
>
> **(c)** The contractor must submit any "proposal for adjustment" (hereafter referred to as proposal) under this clause within 30 days from the date of receipt of the written order. . . .
>
> **(e)** Failure to agree to any adjustment shall be a dispute under the Disputes clause. However, nothing in this clause shall excuse the contractor from proceeding with the contract as changed.

Authority for Changes

Obviously, the Changes clause grants broad authority to the contracting officer to change the contract. Contractors should be aware that not all government officials who appear at the plant have authority to issue changes. Overreaching inspectors and engineering personnel often implore a contractor to undertake efforts that increase the cost or delay delivery of the product. If the contracting officer has not properly delegated authority to that official, the change is unauthorized, and if the change is not subsequently ratified by the contracting officer, the contractor will probably not be compensated for the additional costs or schedule delay. If the circumstances amount to a constructive change, then the contractor can recover. We discuss constructive changes later in this section and in the section on contract performance.

Unpriced and Undefinitized Changes

Although the program office cannot always do so, it should attempt to avoid unpriced change orders and undefinitized changes. If time does not permit pricing the changes, a ceiling should be agreed upon by the parties. Since these ceilings are usually upheld if a dispute arises, the not-to-exceed price (ceiling) should be negotiated carefully, especially if the contractor is to undertake added performance risks. Undefinitized change orders can build up quickly and create chaos in the orderly administration of long-term contracts for major systems. In attempting to definitize change orders after the work is completed, both parties run the risk of violating the law, which makes the cost-plus-percentage-of-cost-type contract illegal. In such cases the contractor may jeopardize being awarded an otherwise well-deserved profit. A huge backlog of undefinitized change orders is detrimental to both parties. For some reasons contractors are often simply tardy in submitting cost proposals. "Jawboning" by the program manager is probably the most efficient method to overcome this neglect. However, I have seen cases in which program managers had to resort to a rather severe measure—withholding progress payments—to correct this inattention. Such action does not foster cooperation between the program office and the contractor, and that cooperation is essential to the long-term success of the program.

Changes within Scope

The only changes permitted are those *within* the general scope of the contract. This requirement creates problems at times and is responsible for many disputes before the Armed Services Board of Contract Appeals (ASBCA). A change that cannot be made "within scope" should be considered a new procurement, and all parties should follow the established procedures for awarding another contract from "scratch." Thus one can readily see the desirability of judiciously taking advantage of the convenient Changes clause.

The Supreme Court defines the scope of the contract as that which is "considered to be fairly and reasonably within the contemplation of the parties at the time of contract award." The scope is essentially the same work that the parties bargained for when the contract was entered into. The number of changes is not nearly as important as the quality or nature of the changes ordered. A change increasing the total contract price two and a half times has been upheld by the ASBCA as being within scope. On the other hand, a change that decreases the cost by much less could be deemed outside the scope of the contract and would have to be considered a partial termination under the Terminations clause of the contract. A contractor who refuses to perform the ordered change would be in breach of the contract if the change is within scope. The contractor could, however, legally refuse to proceed if the work is beyond the scope of the contract, that is, the change in the work is of such a fundamental nature and character to be actually outside that which both parties contemplated at the time of award.

Adjustments and Disputes

Of course, the contractor is entitled to an equitable adjustment in the contract price if the change increases the cost or schedule. Conversely, a decrease in cost or schedule should result in a decrease in the contract price. The Changes clause requires the contractor to submit a "proposal for adjustment within 30 days from the date of receipt of the written order." The 30-day time limitation is rarely met in major procurements, and in fact, courts and boards routinely ignore the requirement. Contracting officers are free to make the adjustment at any time before final payment of the contract price. As usual, the burden of proving the increase or decrease in the contract price is on the party seeking the increase or decrease.

Many times, the ultimate decision on pricing comes from the ASBCA because one of the parties disputes the amount, pursuant to the Disputes clause of the contract. Disputes often develop when the parties are unable to set a price on the change in work before performing it and fail to agree on the price during performance or after the work is completed. The correct measure of an equitable adjustment, based on leading court precedent, is the reasonable cost to the contractor in performing the changed work. If contractors do not segregate the additional costs—and often they do not—arriving at the correct adjustment may be difficult. If the contractor segregates the additional costs or is able to retrieve them accurately from its general accounting records, they are presumed to be reasonable. Many times, however, contractors are unable to retrieve the added costs successfully. In such instances they will attempt to recover them on a so-called total-cost basis. That cost is calculated as the difference between the original contract price and the total cost of performing the contract as changed. Contractors should not rely on the use of the total-cost approach to prove their added costs, for the courts and boards frown on it, although they sometimes accept it. The reasons are obvious for its general nonacceptance: the contractors may have "bought-in" the original contract or the changed work may have been done inefficiently.

If the equitable adjustment is satisfactory to both the contractor and the contracting officer, they must execute a release in the related supplemental agreement. This writer, during his government tenure, generally attempted to tailor the release so that it would include the circumstances giving rise to the specific cost adjustment required for the changed work and also would eliminate all prior claims for adjustments for the total time the contract existed and constitute a release from those claims. This approach was often unacceptable to contractors who, while they may not have had additional claims in mind, understandably did not wish to commit to a release of unknown meritorious claims. We submit, however, that this method may have great merit when dealing with a troublesome contractor who is particularly litigious, especially in presenting doubtful or frivolous claims.

Cardinal Changes

A change that is outside the scope of the contract is known as a *cardinal change,* and this determination is often the subject of litigation. One court described the cardinal change as "a change which has the effect of making the work as performed essentially not the same work as the parties bargained for when the contract was awarded." A cardinal change, then, actually results in the government breaching the contract, and the contractor is usually entitled to a recovery for breach. That recovery is often more favorable than recovery under an equitable adjustment pursuant to the Changes clause.

Constructive Changes

Ordinarily, writers couple a discussion of constructive change orders with a discussion of contract modifications. A constructive change is a legal fiction that occurs frequently when a contractor is required to expend additional effort because of the conduct of a government official who, *inter alia,* directs, instructs, or interprets changes to the contract without using the official Changes clause formalities. This concept includes such things as defective specifications, impossible performance, government's superior knowledge, or government interference with contractor performance. We discuss the constructive changes doctrine in the next section under contract performance.

CONTRACT PERFORMANCE

Specifications

We have discussed authority, methods, and procedures. But the heart of the contracting process is really performance, and performance is based on the contract specification or statement of work. A program's ultimate success or failure is predicated upon the attention and effort expended in drafting a realistic statement of work that is both clear and precise. The statement of work is the basis for preparing proposals, conducting negotiations, and measuring performance. It drives inspection and acceptance. The majority of litigation in government procurement is directly or indirectly about contract performance.

If a federal or military specification is published for a product in common use, the department is obliged to use that product when it is procured. The program office may find various items in the *DoD Index of Specifications and Standards* that suit its needs and find it convenient and helpful to use those specifications. When purchasing major items or systems, the program office will develop its own peculiar specifications, statements of work, or both. These may be design specifications, performance specifications, or some combination of both. Design specifications provide a detailed description of the item to be built; specify the material, size, features, and shape; and are usually accompanied by exact drawings. Sometimes, design specifications describe a process, such as "heat the inconel alloy at 2000°C for a 2-hour period." If a reliable contractor carefully follows the specifications, the end item should be acceptable. In fact, the government makes an implied warranty of the adequacy of the specifications and can be held responsible for their breach. At other times the program office, in its sound discretion, may choose to use a performance specification for which the design detail, materials, or fabrication processes are determined by the contractor. Performance specifications express requirements in terms of speed, range, payload, or other operational needs. In such instances the government makes no implied warranty as to the adequacy of the specification; this risk is, therefore, usually assumed by the contractor.

In major system procurements we rarely see a pure design specification or a pure performance specification. Rather, some mix of both is used. However, for convenience the specification usually is referred to by the dominant type. In litigation, however, courts and boards look behind this feature in determining such issues as whether implied warranties exist or which party assumed the risk of performance.

To determine whether a contract is strictly performed so that the contractor is discharged of its obligation, the contract as a whole must be read, not merely the statement of work or the specification or, in sealed bidding, the item description. The contract usually has many other specifications and standards, such as quality or testing, usually incorporated by reference. The specifications incorporated by reference are just as much a part of the contract and as legally binding as though they were set forth verbatim in the contract itself. This also is true for the contract data requirements list (CDRL) and its implementing data item descriptions. I believe the government pays for far too much data that it never receives or are never delivered in satisfactory form, content, and accuracy to comply with the enforcement standards usually incorporated in a contract by reference. Therefore, more attention should be directed to data management, especially staffing with qualified people.

Many times contractors get somewhat carried away in their quest to win a contract award and promise too much in the proposal. Many view that practice as nothing more than "sales puffing," and after award, during contract performance, quickly allude to the fact that the proposal was not part of the contract. A good tactic that has proved useful for the government is to incorporate the proposal into the contract by reference. That practice tends to have a salutary effect in achieving a measure of credibility in the proposal process. It also assists the evaluators during source selection. With what appears to be extreme compartmentalization when piecing together a government contract (that is, incorporating the myriad specifications and standards for major system buys), conflicts and discrepancies are bound to arise among the various requirements imposed on the contractor. Therefore an order-of-preference clause is necessary to establish, for example, whether the specification takes precedence over the drawings should they be in conflict. That clause cannot be written by the contracting officer without assistance from the project engineer, among others. When the government incorporates the contractors's proposal, it probably should be placed near the bottom of the order-of-precedence listing.

Government-Contractor Relationship

One cannot overemphasize the need for attention to detail in formulating the work expected from a contractor (specifications, standards, schedule, data items, order of preference, and so on) during contract performance. All members of the procurement team should be constantly mindful of the *contra proferentem* rule, which means that ambiguities and conflicts will be construed against the person who drafted the language. Courts and boards turn their decisions upon this rule quite frequently.

Either contracting party has a right to expect strict compliance with quantitative contract specifications. Courts and boards enforce strict compliance as a general rule. In recent years, however, in a few meritorious cases, the Armed Services Board has recognized the Doctrine of Substantial Compliance. When this occurs, the news of the decision usually "makes the professional journals." Therefore contractors should strive for strict compliance in performing the contract, for to do less opens the contractor to default termination or breach of contract.

Constructive Changes Doctrine

Contract performance might appear to be a one-way street running against the contractor. My experience dictates that the government-contractor relationship has to be something like a close partnership. Of course, the government has a fundamental legal duty to cooperate with the contractor during contract performance. Yet, surprisingly, it often takes a court decision to drive that point home. Government officials, without ill will or less-than-honorable motives, often find that they have made the contractor's performance more onerous in such a way that the government is liable for the unanticipated result. Examples might be withholding critical knowledge, overzealous inspecting, or erroneous contract interpretation. The following discussion emphasizes some instances of these shortcomings that very frequently occur. The program office needs to be very sensitive to such occurrences.

1. *Superior knowledge.* This occurs when the government has critical or vital knowledge that the contractor does not have and that is necessary to perform the contract. This condition usually becomes an issue when the court or board is attempting to decide who assumed the risk of performance. For example, the Armed Services Board recently granted a contractor additional costs in attempting to produce a pistol part to a standard specification. The government knew that no previous contractor could produce the part in accordance with specification requirements without a waiver but did not share that knowledge with the contractor, and the contractor had no reason to be aware of it. Therefore, the board held that the government breached its "implied duty to communicate."

2. *Impossibility.* A contractor who agrees to perform a contract having a performance specification generally assumes the risk to achieve the required result and assumes the further risk that it knows the "state of the art" and that such result is possible. A contractor that fails to meet the performance requirements will often advance the argument that performance is impossible. If the government has drafted the specification and the contractor overcomes the burden of proof, the contractor will be given relief. Legal theory recognizes two types of impossibility: actual and practical.

- *Actual impossibility.* Sometimes called physical, technical, or absolute impossibility, it means that based upon an objective standard, neither the contractor nor any contractor can perform to the specifications as written.
- *Practical impossibility.* Sometimes called legal, commercial, or economic impossibility, this definition is based on a more subjective standard, and although the contractor is able to perform the contract, it can do so only at a cost that is unconscionable. Of course, the high cost should not be the result of the contractor's inept performance but rather of an efficient, objective, good-faith effort to perform. Some years ago a small high-tech contractor agreed to supply the Air Force with an avionics component to retrofit a series of fighter aircraft. A new "high-voltage power supply" was needed, and the contractor began performance. The mean time between failures (MTBF) required by the specification was 300 hours. The contractor was unable to obtain first-article approval because the power supply experienced arcing at high altitudes, thereby causing an MTBF of 50 to 60 hours. The contractor then employed expert consultants and established clean rooms to assemble the units. These extra efforts resulted in a projected doubling of the contract costs and added 18 months to the delivery schedule. The specified MTBF was determined to be an unreasonable requirement, and although the contractor could probably perform the contract strictly with tremendous additional cost and extension of schedule, the Air Force recognized a good case for economic or practical impossibility and the contractor recovered the additional expenses with an appropriate adjustment in the delivery schedule. The lesson from this example is that project engineers should be familiar with the state of the art in the particular equipment they are dealing with and should establish realistic requirements. In the case cited, the contractor was forced into financial difficulty. But because of a finding that it and the desired avionics equipment were critical to the national defense, the company was rescued from bankruptcy by an award under Public Law 85-804.

3. *Interference with contractor performance.* As noted, the government has a duty to cooperate with the contractor during the period of contract performance. It also has a duty not to interfere with the contractor's performance. This writer listened to discussion of an interference case at a recent procurement conference. As described, the government suffered a contractor to perform (although not completely pleased with the contractor's performance) during a period when the work being supplied was critical to an operational effort. After the critical need no longer existed, the contractor alleged that more than 60 government officials visited the plant to audit the contractor's handling of government-furnished property. The visits continued on and off for a period of several months, disrupted the efficient production schedule, and occupied time of key personnel. After the examination, the contractor filed a claim for about $1 million, which was denied by the contracting officer's final decision. The contractor appealed to the Armed Services Board. While the appeal was pending, the contractor declared bankruptcy and placed his claim, which by then had grown significantly, before the Bankruptcy Court. The judge decided in favor of the contractor. The government believed it had no choice but to appeal the decision to the appropriate U.S. District Court on a jurisdictional issue. The government won in the District Court; the contractor appealed to the U.S. Circuit Court of Appeals and lost. The contractor again lost when the Supreme Court denied his writ of *certiorari.* By now, the amount claimed had grown to about $20 million, a lot of which was legal expenses. Later, in an attempt to continue pursuit of the claim, the contractor sought assistance from the local congressional representative.

A private bill for relief was considered. Both the contractor and the government were put to a tremendous amount of legal expense plus the time and effort diverted from valuable, productive ventures. The sad point was to learn that the case could have been settled early on for not more than $250,000—a settlement that was blocked by *one* government official and no other government official was willing or disposed to raise the question to the next higher level in the chain of command. The lesson from this example is crystal clear for both the government and the contractor: you cannot afford to harden your position and become intransigent at all costs. This cause should have been raised in government channels at least to a level at which it could have been evaluated in a disinterested fashion. A rare case, indeed, of government personnel neglecting to solve the problem in house and early on.

4. *Accelerating the delivery schedule.* Accelerating most often occurs when by reason of government action, namely, erroneous interpretation of specification, late delivery, or delivery of defective government property, the contracting officer holds the contractor to the original delivery schedule when the delivery schedule should have been extended. In these cases, usually the contracting officer will be deemed to have changed the delivery schedule by operation of law and the contractor is able to recover the additional costs plus an extension of the delivery schedule.

In other instances a government official may unwittingly expose the government to liability. Other examples frequently encountered involve situations of overinspection and furnishing defective specifications.

To permit contractors to recover for their unexpected additional effort in these types of situations, courts and boards have developed a legal theory, known as the *Constructive Changes Doctrine,* to find a basis to compensate a contractor who has been wronged by the kinds of government conduct described. (We defined constructive change at the end of the previous section.) Courts and boards say the effect of a constructive change is the same as though the contracting officer had issued a formal change order under the Changes clause of the contract. Constructive changes provide the basis of a large number of cases currently at the courts and boards. All program office personnel who deal with contractors should know this doctrine thoroughly and its resulting legal consequences.

Commercial Products

Before we leave this topic and recognizing the importance of specifications, we should emphasize that Congress has directed the DoD to reduce its reliance on newly developed items in favor of commercial products (also called nondevelopmental items, or NDI). 10 U.S.C. § 2325 states the following:

> **(a)** PREFERENCE.—The Secretary of Defense shall ensure that, to the maximum extent practicable—
>
> **(1)** requirements of the Department of Defense with respect to procurement of supplies are stated in terms of—
>
> **(A)** functions to be performed;
>
> **(B)** performance required; or
>
> **(C)** essential physical characteristics;
>
> **(2)** such requirements are defined so that nondevelopmental items may be procured to fulfill such requirements; and
>
> **(3)** Such requirements are fulfilled through the procurement of nondevelopmental items.

The following year Congress criticized DoD for not making sufficient effort or progress in procuring off-the-shelf nondevelopmental items. It is difficult to predict whether the congressional pressure on this issue will continue. We suspect it will. However, recent experience in operations Desert Shield/Desert Storm would indicate that this policy might need to be reexamined. It seems to me that off-the-shelf items may not be as suitable to harsh environmental conditions as the items procured to military specifications. The penchant of the policymakers for commercial items, however desirable for economic reasons, may have to give way to the needs of the operators. If so, specification problems will remain with us for the long run.

Warranties

Although warranties are not generally required in supply contracts, the government often uses the Warranty clause to expand the rights and remedies to which it is already entitled under the Inspection clause. In commercial transactions, the law of warranties is contained in the Uniform Commercial Code, a body of law adopted by all states except perhaps Louisiana, and is designed to codify and simplify the laws governing commercial transactions in the private sector. On occasion the Armed Services Board has turned to the Uniform Commercial Code for guidance in novel cases involving warranty law where the federal procurement laws have not preempted state laws.

The Uniform Commercial Code specifies two types of warranties: express (described by the seller, usually in writing) and implied (established by law). The Uniform Commercial Code states that "all statements, affirmations of fact, or promises made by the seller that became part of the basis of the bargain between the parties create an express warranty that the goods shall conform to the statement, affirmation, or promise." Express warranties can include oral or written representations, descriptions of the goods, samples or models, or plans and blueprints. While you may similarly rely on an oral express warranty, proof can become difficult. The law distinguishes between oral warranties and "sales pitches" or "sales puffing," especially if a reasonable person might have cause to doubt their believability. If the representation is inherently believable and is relied upon by the buyer, it will be enforced as an expressed warranty.

Traditionally buyers' rights in sales contracts were controlled by the maxim, *caveat emptor,* or "buyer beware." Courts began to recognize the harshness of this rule and began "raising up" certain minimum implied requirements in the goods sold. The Uniform Commercial Code has therefore codified the following two implied warranties: (1) a warranty of merchantability and (2) a warranty of fitness for a particular purpose. The Uniform Commercial Code, at Section 2-314(2), details the minimum guarantees of "merchantability"; but in general, it means that the supplies will be fit for sale in the usual course of trade at the usual selling price. The implied warranty of "fitness" means the supplies will be fit for a particular purpose if the seller has reason to know of any particular purpose for which the goods are required or that the buyer is relying on the seller's skill or judgment to select or furnish suitable goods. The Uniform Commercial Code provides for excluding both of these implied warranties, providing the language is explicit and conspicuous. If not properly excluded, both warranties will apply by operation of law.

The government's approach to warranties, although not mandatory (except in certain weapon systems contracts), favors express warranties and excludes the implied warranties when an express warranty is used. The FAR defines warranty as "a promise or affirmation given by the contractor to the Government, regarding the nature, usefulness, or condition of the supplies or performance of services furnished under the contract." It says the purpose of a warranty is to delineate the rights and obligations of the contractor and the government for defective items and services and to foster quality performance. It also makes it clear that in supply contracts the use of warranties is "not mandatory." Contracting officers should follow the five criteria for use of warranties in FAR 46.703. The government uses five different Warranty clauses, depending on the complexity of the supplies or the services procured. Each specifies different remedies for defective supplies. Therefore the contracting officer needs to become familiar with the different rights and remedies and to seek assistance of counsel in enforcing them. The remedies range from accepting the defective supplies with a reduction of the contract price to rejecting the supplies and demanding a replacement.

With respect to weapon system acquisitions, sec. 794 of the 1984 DoD Appropriations Act caused a significant expansion in the warranty arena. It was widely reported that a member of Congress believed that DoD should obtain the same kind of guarantees for its weapon systems that the member enjoyed for a farm tractor. Called the Weapons Systems Warranty Act, it caused a rash of protests from DoD and contractors alike because, if implemented as written, it was thought to have shifted untoward risk upon weapon systems contractors. As a result, the Defense Procurement Reform Act of 1985 modified the original law and eased its harsh and inflexible coverage. Of course, this law affects DoD contracts only.

The DFARS defines a weapon system as "a system or major subsystem used directly by the armed forces to carry out combat missions." Included are such systems as tracked and wheeled combat vehicles, naval vessels, bombers, fighters, and missiles. (See DFARS 246.770-1 for a more detailed definition and specific examples.) These warranties are required by statute; the contracting officer lacks discretion to include them or exclude them as is the case of warranties in supply contracts. The requirement may, however, be waived at the Assistant Secretary level.

Weapon system warranties when required apply to all production contracts and modifications, whether the contract is fixed-price, cost-reimbursement, or another type. The program office must obtain the three warranties specified at DFARS 246.770-2:

(a) Under 10 U.S.C. 2403, departments and agencies may not contract for the production of a weapon system with a unit weapon system cost of more than $100,000 or an estimated total procurement cost in excess of $10 million unless—

(1) Each contractor for the weapon system provides the Government written warranties that—

(i) The weapon system conforms to the design and manufacturing requirements in the contract (or any modifications to that contract),

(ii) The weapon system is free from all defects in materials and workmanship at the time of acceptance or delivery as specified in the contract; and

(iii) The weapon system, if manufactured in mature full-scale production, conforms to the essential performance requirements of the contract (or any modification to that contract); and

(2) The contract terms provide that, in the event the weapon system fails to meet the terms of the above warranties, the contracting officer may—

(i) Require the contractor to promptly take necessary corrective action (e.g., repair, replace and/or design) at no additional cost to the Government;

(ii) Require the contractor to pay costs reasonably incurred by the Government in taking necessary corrective action, or

(iii) Equitably reduce the contract price; or

(3) A waiver is granted under 246.770-8.

The government may exercise any of three remedies if the weapon system fails to meet the required warranties: it may require the contractor to correct the deficiency at no cost, it may make the necessary corrections and pass the costs to the contractor, or it may accept the system and reduce the contract price accordingly.

A common problem in warranty implementation is the difficulty the program office faces in performing a valid cost-benefit analysis as required. This problem is even more acute when the weapon system contract has concurrency of development and production. In 1989 the General Accounting Office studied DoD's effectiveness in implementing the Weapon System Warranty Act.* One shortcoming, according to GAO, was the lack of the cost-benefit analysis or the inadequacy of it. DoD admitted to difficulty in completing good sound analyses for many reasons and believed its shortcomings would take considerable time to correct. The GAO, curiously, recommended that DoD should be more actively seeking waivers to the requirement for warranties in weapon system purchases.

At any rate, the program office will find it difficult to prove a breach of warranty. There has not been a lot of litigation since the law requiring warranties became effective—about a half-dozen each year. The government has failed to prove breach in more than half the cases. Our advice to program offices is that they appraise the cost of warranties carefully, and if they are not cost-effective, request a waiver. The contractor should do its own cost-benefit analysis and offer it to the government.

CONTRACT DISPUTES

Contract Disputes Act of 1978

Disputes or disagreements over contract terms, conditions, payment, performance, and so forth are controlled by the Contract Disputes Act (CDA) of 1978 and the standard Disputes clause in government contracts. Actually, the CDA is applicable even in the absence of the Disputes clause in the contract since it is a self-executing statute. Litigating government contracts has become a specialty in law practice. Many large law firms, especially those in the Washington, D.C., area, have established separate divisions in the speciality. It is a lucrative practice. Although a contractor may be self-represented before the court or board, most often the contractor will find it necessary to involve an attorney early in the disputes process. The ASBCA, the board that handles administrative appeals for DoD contract disputes, has an average annual docket of 2300 cases involving hundreds of millions of dollars.

Disputes Process

The disputes process begins when one party files a claim against the other party. Sometimes the government files the claim for such items as liquidated damages, excess reprocurement costs, or a price reduction because of a contractor's defective cost or pricing data. When the government files such a claim, it takes the form of a contracting officer's final decision. If the contractor wishes to dispute the claim, it may file an appeal. Most often, however, claims are initiated by the contractor for a broad range of damages, including allowable costs, breach of contract, suspension of work, and a whole host of reasons, as discussed previously in the section entitled "Constructive Changes Doctrine."

*GAO/NSIAD-89-57, *DoD Warranties*, Sept. 1989.

No form has been established for contractors to submit a claim for damages to the contracting officer. For a simple claim, a short letter is sufficient. Most of the time, though, the contractor should submit a professional "claims package." The claim should demand payment of a specific sum with supporting accounting data and the legal reasoning as to why the contractor is entitled to relief. Complete documentation should be available for backup in case of litigation before a court or board. If the amount claimed exceeds $50,000, it must also be certified, as is discussed subsequently in this section. In certain instances the party asserting the claim may find it advantageous to inform the other party of a willingness to settle the claim amicably.

In any event, once the claim is filed with the contracting officer and is not settled, the contracting officer must issue the final decision within 60 days if the claim is for $50,000 or less; if the claim exceeds $50,000, the contracting officer must, within no more than 60 days, issue the final decision or inform the contractor when a decision will be issued. The final decision must meet the following criteria:

- It must be in writing.
- It must be delivered to the contractor.
- It must state that this is the final decision of the contracting officer.
- It must present the reasons for the decision.
- It must notify the contractor of the right to appeal.

In some situations the contracting officer refuses to issue a final decision or is derelict in issuing one. In those instances the contractor has the right to appeal—to the court or board—on the grounds of "refusal to render a final decision." The CDA empowers the board to order the contracting officer to render a final decision. If a final decision is rendered and is defective because it fails to meet one or more of the criteria mentioned, the contractor should appeal it just to be on the safe side.

The Disputes clause states: "the contractor shall proceed diligently with performance of this contract pending final resolution of any request for relief, claim, appeal, or action arising under or relating to the contract, and comply with any decision of the contracting officer." This is a rather onerous requirement, but it should not be ignored except for very compelling reasons. Only when the government is guilty of a *material breach* may the contractor stop work, and it should never do so without advice of competent legal counsel.

Because the CDA provides for payment of interest on any amount paid as a result of a claim, some contractors file their claim as early as possible. Interest begins on the date the contracting officer receives the claim until payment is made. A policy of early filing, however, is somewhat risky since the contractor must certify the following under penalty:

- The claim is made in good faith.
- Supporting data are accurate and complete to the best of the contractor's knowledge and belief.
- The amount requested accurately reflects the contract adjustment for which the government is liable.

The company's certification must be made by a senior company official in charge at the company's plant or location involved or an officer or general partner of the contractor having overall responsibility for the conduct of the contractor's affairs. A recent federal court decision held that Grumman Corporation's financial officer, the person signing the certificate, was not a proper person to certify the claim. At present it is only safe to advise that the chief executive officer/chief operating officer (CEO/

COO) is the proper official to certify the claim. It is incredible that after 12 years, the claim certification issue remains in total disarray. Even as this is being written, many different parties are working to solve the issue of "who may certify"!*

The contracting officer's final decision must be appealed to a court or board within specified time limits. The contractor must file its appeal (for DoD contracts) with the Armed Services Board within 90 days after receiving the final decision. It is well settled that the 90-day time limit is jurisdictional, which means that the board is without power to decide a late appeal and it cannot waive this requirement. Therefore time is a most important condition since an appeal, even one day late, can produce a harsh result. No special form is required to file the "notice of appeal." The simplest of letters has been held to be legally sufficient, although the contractor should follow the instructions in the final decision.

The CDA provides for a new direct appeal to the claims court.† That provision, which is somewhat of a safety net, requires that an appeal be filed within 12 months of receipt of the final decision. Thus if a contractor misses the 90-day filing time for the board, it may appeal to the claims court within a year. When that provision became law in 1979, many experts believed contractors would take their appeals to the claims court rather than the boards. That has not happened. During the summer of 1991, a news leak in the Washington area sent out the rumor that the Justice Department would seek legislative reform to reduce the 12-month appeal period to align it with the 90 days required for an appeal to the boards. Assistant Attorney General Stuart Gerson, in a speech to the Public Contract Law Section of the American Bar Association (August 1991), all but confirmed the rumor. Contracting personnel must carefully watch congressional action on this change, if it happens.

At any rate, once an appeal is submitted, things become very legalistic. For instance, some combination of the following phenomena (if not all) will occur:

- Complaint
- Subpoenas
- Amendments to complaint or answer
- Discovery
- Discovery and search of records
- Depositions
- Prehearings
- Prehearing briefs
- Board hearings
- Posthearing briefs

What does this mean to the program management office? At this stage the program office can expect its department's expert trial attorneys to take over the case in a fashion. It means some of the program manager's best people are going to spend undue time and energy assisting the attorneys in preparing for ultimate trial. The attorneys and contracting personnel, engineers, and so on, should constantly evaluate

*Public Law 102-572 (Section 907), enacted October 29, 1992, is designed to correct the confusion over certification. Whether the law has its intended effect remains to be seen. It is interesting to note that the American Bar Association Section on Public Contract Law is already urging its repeal, because it imposes a "knowledge" requirement not imposed on claims under contract with civilian agencies.

†Renamed the U.S. Court of Federal Claims under recent legislation which also expanded the Court's jurisdiction.

the strengths and weakness in the government's case during this period and *if feasible,* attempt continually to settle.

My experience indicates that this effort is not as glamorous as an ongoing major system procurement. Going through many boxes of old files, some of which have already been archived, is tedious, difficult work and not always interesting, to say the least. Most program offices look on claims and appeals as "old business" and a distraction from exciting new programs. No doubt about it, preparing for a hearing or trial and appearing at the trial or hearing all tend to sap productivity of an organization. Contracting personnel, government engineers, and so on, usually look back upon the trial or hearing as an unpleasant experience. Many do not wish ever to appear again. My only advice is that one who gets "gun shy" from such an experience may find that his or her subsequent decisions are affected. For the government employee, the uppermost question has to be the impact on his or her ability to protect the government's interest at all times.

After the hearing or trial, the board or court will usually render its decision within a year, although we have waited as long as 2 to 3 years for a decision in novel, controversial, or complex cases. If either the government or the contractor is not satisfied with the board or claims court decision, either may appeal the decision to the Court of Appeals for the Federal Circuit within 120 days.

Disputes Settlements

Government policy continues to be to settle disputes at the lowest possible level and as early as possible. The contracting officer is "key" to settlement, and should attempt settlement in appropriate cases during all stages of the disputes process. The contracting officer's authority to settle extends to those claims within his or her warrant but does not extend to areas in which another agency has statutory power, such as waiving claims for liquidated damages or claims involving fraud. A former supervisor had difficulty in convincing this writer that some claims generated by major systems acquisitions are just *too complicated to litigate,* but as I gained experience, I became somewhat of a convert. Many times administrative law judges or court judges will simply delay the proceedings and instruct the parties to try to settle the case. A recent example is a claims court judge suggesting that the government and McDonnell Douglas Corporation (with General Dynamics Corporation) begin negotiations to settle the multibillion dollar disagreement for the U.S. Navy's canceled A-12 aircraft program. An assistant attorney general commenting upon the decision to begin settlement talks said it was "entirely consistent with common sense." He noted further that the effort was consistent with an October 1991 Presidential Executive Order* directing government agencies to avoid litigation whenever possible.

A middle ground, which in my opinion is not used enough in government contract disputes but used widely in the commercial sector, is alternate dispute resolution (ADR). Broadly defined as any alternative to full-scale litigation, ADR as used here is noncoercive, voluntary, and usually nonbinding. It is generally thought to be preferable to litigation because it is less expensive, quicker, and less intimidating to the parties. The parties usually have a voice in choosing third-party neutrals to decide the disagreement. It is particularly good in resolving disputed factual issues rather than legal issues. The size of the case is not usually of overriding importance. Its use has produced satisfactory results in very large claims situations. The nonbinding decision is advisory; however, the parties may agree to binding arbitration. Although

*Executive Order 12778 attempts to implement some of the civil law reforms, enunciated in Vice President Quayle's speech before the American Bar Association's House of Delegates in August 1991.

ADR may take many forms, some common techniques involve minitrial, arbitration, and mediation. The simplicity or complexity desired by the parties to the dispute may be built into the ADR agreement.

The chairman of the Armed Service Board of Contract Appeals states: "Parties should consider using the ADR option more frequently. Too often, parties litigate because of habit. . . . " The Armed Services Board has adopted ADR methods, which are consensual and voluntary and may be used after the board has jurisdiction of the dispute. The board simply suspends the hearing while the parties engage in an alternative method of resolving their dispute, and if that fails, the hearing is reinstated. The ADR methods range from using one of the board administrative judges as a mediator (settlement judge) to using the highly flexible minitrial, to a summary trial with binding decision (decision not appealable, except for fraud). The board generally permits any method that is suggested by the parties so long as it is reasonable and aids in resolving the appeal at issue. In addition, the board encourages the parties at any stages of litigation to use other ADR techniques not requiring the board's participation, such as settlement negotiations, procedures mediation, minitrials, or a fact-finding conference between the parties.

The U.S. Army Corps of Engineers uses ADR (minitrials) very successfully. Its first effort in a case pending before the ASBCA for over $630,000 was resolved in about three days for about $375,000. Shortly thereafter a $55 million claim was similarly resolved in just four days. The corps uses ADR successfully in environmental cleanup contract disputes.

CONTRACT TERMINATION

The government may legally terminate its contracts in two ways without incurring liability for breach of contract: it may terminate for convenience or it may terminate for default.

Termination for Convenience

The Termination for Convenience clause is a unique provision in government contracting. It gives the government the right to terminate the contract without cause, and at the same time it limits the contractor's recovery to costs incurred for the work done and profits earned only to the point of termination. Profits anticipated by the contractor are not allowed.

The government may terminate contracts—in whole or in part—for its convenience any time during contract performance. Courts and boards have interpreted that right so broadly as to render it virtually unlimited. The need for this tool arose out of the government's need to wind down the huge procurement effort following mobilization. Even before adopting the Termination for Convenience clause, the U.S. Supreme Court (in 1875) recognized the special need to terminate and said "it would be a serious detriment to the public service if the power . . . did not extend to providing for all possible contingencies by modification or suspension of the contracts, and settlement with the contractors."

The program manager will encounter about as many reasons to terminate a contract as one could imagine. The law is pretty well settled by now. About the only problems that arise deal with a reasonable settlement and payment to the contractor. Part 49 of the FAR has detailed coverage—almost a cookbook approach—on convenience terminations. A word of caution is in order, however. The contracting officer should only terminate for convenience after considering all the circumstances and finding the termination to be in the best interest of the government. The contracting officer should effect a no-cost settlement instead of issuing a termination notice when it is known that the contractor will accept one, government property was not furnished, and all contractor obligations are satisfied on the contract.

The more important question concerns a contractor's remedy for wrongful termination. Usually the question is about the contractor's right to recover a profit for the work terminated, that is, legally called, anticipatory profits. The right to recover anticipatory profits will prevail if the termination is wrongful and results in the government's breaching the contract. The government runs this risk if its termination is based on bad faith (which requires a specific "bad" intent) or if the contracting officer abuses his or her discretion (which requires no specific intent standard). Once the contract is terminated for convenience, the action is final and the contractor is discharged from any obligations respecting the terminated portion.

In the landmark case, G.L. Christian and Associates versus United States, 312 F 2d 418 (1963), the court held that even though the contracting officer omitted the termination for convenience clause from the contract, the government's right to terminate was binding by "operation of law." The Supreme Court refused to review the case, and today the law holds that those clauses that are deemed to reach the status of *public procurement policy* will be incorporated by law even if inadvertently omitted from the contract. The result of this case has become known as the Christian Doctrine. Contractors beware! Government beware also, because the clauses vary with the different types of contract involved. Therefore all parties should study the applicable clause to determine precisely their rights and obligations.

Termination for Default

Contractors universally use all means to avoid a default termination because, *inter alia,* it may have a deleterious effect on future government awards, especially in cases in which past performance becomes a criterion in selection.

The government is generally entitled to strict performance of the contract. The Termination for Default clause allows the contracting officer to terminate the contract in whole or in part if the contractor fails to deliver or perform within the time specified or fails to comply with other contract provisions or fails to make progress in such a degree as to endanger overall contract performance. Again, each particular default termination clause must be carefully examined to ascertain the respective parties' rights since the clauses do differ.

When the contractor fails to perform according to the contract terms, it is in breach and the government gains the right to terminate for default, although that right does not have to be exercised. This option permits the government to view its total contract needs and what is in its best overall interest. The government may accept nonconforming items and require a downward adjustment pursuant to the Inspection clause in the contract. Default termination is severe; not only is the government relieved of liability for costs of unaccepted work, it is entitled to have returned unliquidated progress payments and advance payments. In addition, the contractor is liable for excess cost, if any, to reprocure the property or service, plus other proven actual or liquidated damages. Insofar as terminating contracts for default is concerned, Part 49 of the FAR contains an adequate and detailed treatment of default terminations. Such is beyond the scope of this section, but several concepts should be emphasized, namely, anticipatory breach, forbearance, and waiver and cure notices.

Ordinarily, a contract is not breached until the time for its performance has passed. By utilizing the modern legal concept called anticipatory breach, breach may occur prior to the time for performance completion when by a contractor's words, behavior, or inaction it is clear and convincing that the contractor does not intend to perform as promised. This failure to perform is sometimes called anticipatory repudiation; a clean, clear example is when the contractor informs by letter that it will "abandon the job" or will "cease performance." In that situation the contractor is not entitled to a 10-day notice period called for in the Default clause—called a cure notice. If the contracting officer intends to terminate a contractor for *failure to make progress so as to endanger performance of the contract,* then the 10-day cure notice must be strictly observed. Courts and boards faithfully enforce this requirement.

As mentioned, the government may forbear to exercise its right to terminate for default. However, it must be careful not to take any action during its period of forbearance that would amount to a waiver of the right to forbear. Some examples are issuing changes, accepting deliveries, or even imploring the contractor to perform. Likewise, the contracting officer should be sure before terminating for delay that the contractor is not excused from the failure to perform because of an "excusable delay." Excusable delays occur frequently and generally are occurrences that are outside the contractor's control and occur without the contractor's being guilty of negligence. Grounds for excusable delay are listed in the Default clause. They include such things as fires, floods, strikes, and acts of God. Much litigation occurs over excusable delays.

The program manager can expect most contractors to "fight the good fight" if defaulted. The Default clause provides that if a default termination is in error, it will be changed to a termination for convenience. The contractor is entitled to appeal a default termination under the Disputes clause and often has compelling reasons to have the default termination overturned. Damage to a contractor's business reputation is one. Another real reason is government's right to reprocure the item or service against the defaulted contractor's account. Other damages include the cost of moving government furnished property, if any, extra freight charges, cost of additional inspection, and so forth. It is no surprise that default terminations, especially on the issue involving excusable delays, is a very busy practice before the Armed Services Board of Contract Appeals. Contracting officers should never terminate a contractor for default without coordinating with their trial attorney's office.

ETHICS—THE PROCUREMENT INTEGRITY ACT

After a brief tenuous and tortuous history, the Procurement Integrity Act is back in full force and effect. It provides new rules affecting government officials and contractors alike with respect to offers of employment, gratuities, and the protection of proprietary and source-selection information. It specifies both civil and criminal sanctions for violations. The law applies broadly to the governmentwide procurement process and overlaps, in some instances, with existing conflicts-of-interest laws. The Procurement Integrity Act initially was in effect from July 16, 1989, until November 30, 1989, when it was suspended for a 1-year period. It was automatically reinstated December 1, 1990, with the exception of a revolving-door provision, which we shall discuss later.

The act applies equally to competing contractors and procurement officials during the conduct of a federal agency procurement, which begins with the decision to satisfy an agency need by contract. In most instances a procurement will begin when an agency official directs that a specification or statement of work be drafted to meet a stated requirement. It ends when the contract is awarded or the procurement canceled. Any of the following actions starts a procurement:

- Drafting a specification or a statement of work
- Reviewing and approving a specification
- Computing requirements at an inventory control point
- Developing a procurement or purchase request
- Preparing or issuing a solicitation
- Evaluating bids or proposals
- Selecting sources
- Conducting negotiations
- Reviewing and approving the negotiations or the awarding of a contract or modification

A competing contractor is any entity that is reasonably likely to become a competitor for, or the recipient of, a contract or subcontract under a procurement. It also includes any person acting on behalf of such entity. Competing contractors includes sole-source contractors. During the conduct of a procurement, a competing contractor is prohibited from the following activities:

- Discussing employment or offering employment to a procurement official, unless the official has been properly recused from the procurement
- Offering a gratuity or anything of value to a procurement official
- Soliciting or obtaining proprietary or source-selection information, directly or indirectly from an agency official. (An agency official includes a procurement official and other government officials.)

A procurement official is any civilian or military officer (including contractor employees) who participates personally and substantially in any of the following activities for a particular procurement:

- Drafting or reviewing and approving a specification or statement of work
- Developing or preparing the purchase request
- Issuing the solicitation
- Evaluating the bids or proposals
- Selecting sources
- Negotiating price, terms, and conditions
- Reviewing and approving the award

During the conduct of a procurement, procurement officials may not take the following actions:

- Discuss employment or solicit employment from a competing contractor, unless the procurement official is properly recused
- Receive gratuities from a competing contractor
- Disclose proprietary or source-selection information to competing contractors

Source-selection information as defined by the implementing regulations (whether marked by a legend or not) consists of the following:

- Bid prices submitted in response to a solicitation for sealed bids
- Proposed costs or prices submitted in response to a solicitation for other than sealed bids
- Source-selection plans
- Technical evaluation plans
- Cost or price and technical evaluation of competing proposals
- Competitive range determinations
- Rankings of bids, proposals, or competitors
- Records and evaluations of source-selection panels, boards, or councils
- Any other information marked "Source selection information—see FAR 3.104" (provided the contractor is legally correct in so marking it)

As mentioned earlier, the Procurement Integrity Act places postemployment restrictions on government officials. These provisions, in addition to having been suspended for the above-mentioned one-year period, were further suspended until May 31, 1991, purportedly to allow the Senate to hear objections from government and industry on its application. Government and industry showed no interest in the planned hearings, and the postemployment provision resumed effect on June 1, 1991, as scheduled.

The postemployment provision states that a procurement official on a particular procurement may not do the following for a period of two years:

- Participate in any manner on behalf of a competing contractor regarding that procurement
- Perform work on the resulting contract

The act also applies to certain subcontractors.

Procurement officials are allowed to obtain permission to withdraw from further participation in a procurement to discuss future employment with a competing contractor, but a procurement official who has participated personally and substantially in evaluating bids or proposals, selecting sources, or negotiating the contract may not be recused. Employment discussions may not occur until after written approval of the recusal request. Disapproval of the request is not an adverse personnel action.

The act was amended during its suspension to institute a legal concept known as *ethics advice.* Under that concept, an employee or former employee who is uncertain about whether future specific conduct would violate the law is allowed to request an advisory opinion from an ethics advisory official. An employee who provides all relevant information reasonably available may rely on the opinion and not be deemed to have violated the act, provided the future conduct is based on good-faith reliance on the ethics opinion.

The law requires various certifications:

1. Procurement officials must certify the following:
   - **a.** They are familiar with prohibited conduct, recusal, and certification requirements.
   - **b.** They will not engage in conduct prohibited by the act.
   - **c.** They will report violations to contracting officers.
2. Prior to award, modification, or extension of an existing contract (exceeding $100,000), competing contractors must certify the following:

   a. They have no information on a violation or possible violation of the act (and do not know of the existence of such information).
   b. They have disclosed the details of such possible violations in the certificates, and all such information has been disclosed.
3. Contractor employees who participate formally and substantially in formulating an offer or proposal must certify the following:
   a. They are familiar with the act.
   b. They will report all information regarding any violations.

The criminal sanctions in 18 U.S.C. § 1001 apply to those individuals and companies who are found guilty of making false certifications.

The act establishes other criminal and civil penalties for violations. An individual may be sentenced to a maximum of five years in prison or fined, or both, for violating the proprietary or source-selection provisions. Contractors may be fined up to $1 million in a civil action for engaging in prohibited conduct. Any employee may, likewise, be fined up to $100,000. Administrative and contractual remedies are available against the contractor and range from profit recapture to debarment.

FRAUD—QUI TAM *ACTIONS*

One of the most useful tools in combating fraud is the False Claims Act, 31 U.S.C § 3729–3733, which prohibits false and fraudulent claims for government payment. The act, originally passed in 1863, was designed to prevent profiteering during the Civil War and to curb such actions as mixing sawdust with gun powder or delivering sand as a substitute for sugar. Both civil and criminal penalties may be imposed for violating the law. Anyone who knowingly and willfully makes a false statement or submits a false claim to the government and is convicted may be fined not more than $1 million or imprisoned up to five years, or both.

The False Claims Act has been used quite successfully in combating wrongdoing in a broad range of situations affecting not only the original contract pricing but also all aspects of the contract throughout its life. Some frequent applications include prosecution for production substitution, submitting false testing data, mischarging or overcharging labor costs, and submitting unacceptable overhead costs.

The civil fraud provision imposes severe penalties on those found in violation of the act: up to $10,000 for each false claim plus treble damages sustained by the government and forfeiture of payments due under the contract. Actually the amounts can turn out to be huge. Consider, for instance, a contractor overcharging the government $2 million in an ongoing billing arrangement by submitting 60 invoices. The contractor could face a civil liability of $6.6 million ($10,000 for each of the false vouchers) plus $6 million in treble damages.

Recently, on January 15, 1992, Eleanor Spector, Director, Defense Procurement, distributed an interim rule implementing sec. 836 of the 1991 DoD Authorization Act (P.L. 101-570), which permits agencies to reduce or suspend payments to a contractor when the agency head determines that the contractor's request for payment is based on fraud.

The False Claims Act was essentially rewritten by the False Claims Amendment Act of 1986, which prescribes civil liability for any of the following actions:

- Knowingly submitting a false claim to the government
- Knowingly submitting a false statement to support a claim to the government

- Conspiring to defraud the government on a claim
- Delivering property in inaccurate quantities or accepting inaccurate receipts for property with the intent to defraud the government
- Knowingly obtaining property from persons in the government who cannot lawfully sell or pledge the property
- Knowingly submitting a false record or statement to conceal, avoid, or decrease an obligation to pay or transmit money or property to the government

Although the law incorporates a knowledge standard, civil liabilities can be imposed if the government can prove deliberate ignorance or a reckless disregard for the truth.

The impetus for the 1986 amendment was the perception by Congress that fraud in government contracts was out of control. That perception was occasioned by activities such as the spare-parts "fiasco" and behavior that resulted in the Ill-Wind investigations. What the amendment actually did was to invigorate private law suits in behalf of the government. That result was one of the stated purposes of the amendment when Congress rewrote the *qui tam* provision that made it easier for individuals with knowledge of wrongdoing to sue based on that knowledge and rewarded them for bringing the suits. *Qui tam* is an abbreviated Latin phrase that liberally translates to "bringing an action for the king and for himself." In our context, it means the right of any person or company (referred to as the relator) to sue to recover damages from a person or persons who have perpetrated a fraud against the government, even though the person filing the suit has suffered no damages. Without doubt, Congress's intended purpose was met, for *qui tam* suits started immediately after the 1986 amendment and have increased dramatically ever since. The suits have all been interesting, many controversial, and some surprising—surprising to observe both government and contractor employees coming forward with knowledge of wrongdoing: instant millionaires have been made.* The rest of this section discusses *qui tam* actions.

Qui tam actions are filed in the name of the government. The government has the right to take over the lawsuit, and to help it decide whether to take over the lawsuit, the relator must furnish the Department of Justice or the United States Attorney with a copy of the proposed complaint along with extracts of the relevant supporting evidence.

The government generally has 60 days to decide to take over the suit or decline to do so. Meanwhile the complaint remains under seal with the court. If the government declines to prosecute the action, the relator can pursue the lawsuit in its own right. In either event, the relator is entitled to a percentage of the judgment or settlement: usually between 15 percent and 25 percent if the government prosecutes the action and between 25 percent and 30 percent if the government does not join in the action. In either instance, the relator is also entitled to recover reasonable attorney fees, costs, and related expenses from the defendant.

The statute has a provision to encourage employees to report information about fraud by protecting whistle blowers from reprisals by employers such as discharging, demotion, or other forms of harassment for pursuing *qui tam* actions; other remedies are also available in case of reprisals.

The statute imposes limitations on bringing *qui tam* lawsuits. It provides that if the complaint is based on public information, that is, information from a court or public hearing, for instance, the relator must be an *original source,* that is, a person

*In July 1992, the *qui tam* plaintiff's share of the government's recovery arising from the case of United States *ex rel* Taxpayers Against Fraud versus General Electric was over $14 million. The Department of Justice is seeking to reduce plaintiffs' share to something less than $5 million.

with direct and independent knowledge of the circumstances (apart from the public disclosure) who makes the information available to the government prior to filing action. That limitation is supposed to prevent the "parasitic-type" lawsuits that were common in the 1930s, when relators filed lawsuits based on information copied from government files and indictments. The statute also prohibits a relator from bringing a lawsuit based on the facts that already form the basis of an existing *qui tam* lawsuit. That prohibition is aimed at preventing class action or multiple suits founded on the same factual base. Other practical reasons provide limitations. In cases in which the government does not choose to intervene, the relator might find it literally too expensive to pursue the action privately. Should a defendant prevail in a private action and be able to convince a court that the suit was frivolous, vexatious, or brought to harass, the relator can be held liable for the defendant's attorney fees, costs, and expenses. This rule provides a certain chilling effect against suits intended to harass a former employer.

A *qui tam* lawsuit can expose a government contractor to huge liability. Many times a contractor will not know the lawsuit exists until the seal is removed and it is served with a summons and complaint and then is required to file an answer. The law provides the *qui tam* defendant with a method to decrease exposure by the so-called come-clean provision. That provision requires a series of disclosures and cooperation by the person alleged to have committed the violation. Of course, the usual defenses—statute of limitations, original source, attack the merits, and so on—are available to the *qui tam* defendant. Contractors have been singularly unsuccessful in having the law declared unconstitutional, although several cases are now pending.

As stated earlier, the *qui tam* provision of the False Claims Amendment Act of 1986 has stirred considerable controversy. Even government officials become critical when government employees become plaintiffs in *qui tam* actions. In such cases the Department of Justice believes that government employees are jurisdictionally barred from bringing suit based on information acquired in the course of their employment because of the public disclosure exclusion of the law. In the past 15 months, Justice has argued its position in the 1st, 9th, and 11th federal circuit courts. Its legal position prevailed in the 1st Circuit, although the court stated that the decision was very fact-specific. The court's opinion in that case relies on tortuous logic, in my opinion and that of many others. The other two circuit courts rejected the Department of Justice jurisdictional argument and criticized the 1st Circuit Court decision.

It seems clear now that lawsuits based on information gained in the course of government employees' work are not automatically barred. The judges sympathized with the Justice Department's argument, but stated that the court is obliged "only with interpreting the statute . . . not amending it to eliminate administrative difficulties." They noted that it is Congress that can change the law. Meanwhile more suits by government employees are expected.* It would seem that the Justice Department will follow the court's suggestion for a legislative solution to the government employee issue. In a recent speech at the American Bar Association convention in Atlanta, Assistant Attorney General Stuart Gerson stated that the Department of Justice has discussed remedial legislation with Congressional staff members.

The moral is clear. Program managers and contractors alike must create an environment to foster good communication with employees; complaints of fraud must be taken seriously and investigated; and reprisal actions against whistle blowers are foolhardy.

---

*In one of the cases mentioned in the appeal, the suit was brought by a civilian attorney who worked in a legal office in Japan. He complained to his superiors and when no action was taken, brought a *qui tam* suit. The suit will probably be settled for an amount exceeding $30 million.

ENVIRONMENTAL ISSUES

After more than 50 years of intense effort in producing a war machine—sometimes without much regard to safeguarding the environment—DoD has shifted its emphasis over the past several years to "cleaning up" some potentially dangerous conditions at its installations. Although environmental laws have been in effect for at least a dozen years, the Environmental Protection Agency (EPA) has been very slow to provide implementing regulations. In addition, when it does provide regulations, they are somewhat of a moving target. In the past few years, however, the regulatory pattern has become more stable and now DoD program managers can expect to spend more time on cleanup and remedial matters. Someone has posited the question: "Will it take longer to clean up Rocky Flats than it did to create the waste that is there?"

Companies doing or contemplating cleanup work or offering new technology to the government are facing a decision on whether they wish to "bet their company" in the interest of doing the work because firms seeking to enter the field face the strict and absolute liability standard in most of the environmental laws. Some of the laws also utilize a cradle-to-grave concept, which enforcement officials have interpreted to mean that joint and several liability attaches to all parties who have been in the chain from creation of the hazard through its cleanup. The owners of the property and the operators—past and present—face potential liability for breaking the law.

The DoD recognizes the importance of the environmental issue and reflects that awareness in its cleanup budget, which is one of its few budget areas that is still growing. Environmental issues affect all government contractors. The DoD program managers' problems are perhaps more acute for government-owned, contractor-operated (GOCO) facilities, but they are also pervasive with the contractor-owned, contractor-operated (COCO) facilities, especially those committed principally to government work. Utilizing the Defense Production Act to invoke rated orders or providing government property is thought to provide a nexus sufficient to invoke government liability.

The liability question is real. The Comprehensive Environmental Response, Compensation, and Liability Act (CERCLA, or Superfund) imposes strict liabilities and requires that the EPA be notified before hazardous materials are disposed of. Liabilities also arise from the Clean Air Act, the Clean Water Act, and the Resource Conservation and Recovery Act (RCRA). Fines and penalties could bankrupt a company in short order. Some fines specify an amount for each day the person or firm is in violation of the law. The laws apply to both contractors and the agencies, and in some cases, agency officials. Criminal liabilities confront both the companies and the individuals. In fact, several government employees have been tried and convicted. Individuals who owned, operated, or disposed of waste at a particular site are liable for cleanup of that site, whether they caused the release of a hazardous substance or not.

Congress has unequivocally waived its sovereign immunity from being sued on some of these statutes; judicial decisions are interpreting the extent of the waiver as each day passes. All efforts to change the law to grant government officials personal immunity when acting within the scope of their employment have been unsuccessful. In fact, some officials have tried to insure against the potential threat to no avail.

Contractors have an added concern. The Justice Department insists on the *unitary executive theory,* which maintains that the EPA is prohibited from formal enforcement actions against other federal agencies. This Justice Department position is based on the premise that the Constitution precludes such judicial resolution because since the EPA and the involved agencies are part of the executive branch, this would amount to nothing more than the government suing itself. This interpretation has led some contractors to allege that the EPA has increasingly become aggressive in directing enforcement toward government contractors and operators of GOCOs. The problem for the GOCO contractor becomes more acute when we consider the position of a contractor who maintains and operates a GOCO facility. The contractor is not really in control of its destiny since the agency owns the facility, controls the facility's budget generally, and is involved in the daily management of the facility.

Add to the equation the fact that EPA is methodically delegating enforcement authority to the various states, some of which have regulations more onerous than the federal laws, and the uncertainty becomes more apparent. The writer has been involved in hazardous wastes spills and cleanups when it was necessary to deal with three regional EPA offices, the Justice Department, the state authorities, the Occupational Safety and Health Administration (OSHA), and local government officials. Throw in an interested citizens' action group and progress becomes exceedingly difficult. RCRA and the Superfund Acts permit public interest groups to serve the role of a *private attorney general* and file civil actions for enforcement. Such actions are becoming more frequent. The Sierra Club, for example, compelled a contractor and the Department of Energy (DOE) to go through the permit process to operate incinerators at a Colorado facility. A contractor convicted of a criminal violation of these laws can be barred from contracting with the government.

In any event, cost recovery in remedial environmental actions is by no means a sure thing for contractors. Fines and penalties for violations of the law are not allowable costs under the FAR. Contractors' caution or reluctance to enter the cleanup industry is understandable. DoD and NASA contractors have some protection under the FAR Clause 52.228-7, "Insurance-Liability to Third Persons." That clause states that contractors "shall be reimbursed for certain liabilities . . . to third persons not compensated by insurance or otherwise. . . . " This reimbursement theory is limited, however, since it is subject to the Anti-Deficiency Act. Therefore protection is limited to the availability of funds appropriated. Commercial insurance against this peril is rapidly becoming less available or its cost is increasing to unreasonable amounts. Insurance companies—in the face of unbounded liabilities—are limiting coverage under the policy terms and conditions, and even placing dollar limitations on per-occurrence incidents.

At a time when we most need additional contractors and new technology, we lack the incentives to attract them. In a survey by the GAO, 60 percent of the defense contractors stated that a recent statutory mandate to expand DoD's independent research and development program will have little or no effect on their investment in environmental technologies. DoD, NASA, and DOE are putting forth a good-faith effort to share the burden with contractors. In September 1991 the Assistant Administrator for Procurement at NASA asked the FAR Council to reopen its environmental cost principle case, claiming, "I don't personally believe we can put the burden 100 percent on the contractor. I think that's unreasonable." DOE uses the Price-Anderson Act and its nonstatutory general contract authority to indemnify its contractors against "liability for uninsured risks" [see Department of Energy Acquisition Regulation (DEAR) 950.7001(c)].

The DoD has proposed an environmental cost principle. In addition, contractors have proposed that DoD use its general contract authority under Public Law 85-804 (Extraordinary Contract Relief) to promise to indemnify them for unusually hazardous risks. FAR 52.250-1 is the clause implementing P.L. 85-804. It provides indemnification to the contractor "to the extent that the claim, loss or damages (1) arises out of or results from a risk defined in this contract as unusually hazardous or nuclear and (2) is not compensated by insurance or otherwise." We believe this authority is a good sound basis, when used judiciously, for agreeing to indemnify contractors in some environmental situations. Of course, contractors should not be indemnified when the damage or loss is caused by their own willful or wanton behavior or their negligent acts.

One beauty of P.L. 85-804 is that it is not subject to the Anti-Deficiency Act, which effectively prohibits open-ended conditions in government contracts. However, after payment, certain amounts must be reported to Congress. There are well-publicized instances of meritorious usage of P.L. 85-804 in critical circumstances. For example, the Secretary of the Army recently approved P.L. 85-804 indemnity for GOCO ammunition plant contracts to protect the contractors against certain environmental liabilities. During Desert Storm, when the Civilian Reserve Air Fleet (CRAF) was activated and the aircraft's commercial insurance coverage was either terminated automatically or ineffectual in a war zone, the Secretary of the Air Force authorized the use of P.L. 85-804 to indemnify owners of the CRAF aircraft against loss or damage from operations in the theater.

In the past some of us who relied on this indemnification authority would look ahead, in some practical way, to a source of funds in the unlikely event government liability should attach under the indemnification agreement. We often found a reliable source of funds in the so-called M account—an expression for treasury memorandum account. Even though agreements made pursuant to P.L. 85-804 are not subject to the Anti-Deficiency Act, nonetheless, good program managers inevitably try to budget for potential liabilities. The M account provides a needed flexibility since once the funds expired and entered the account, their expenditure need not legally be tied back to the purpose of their original appropriation. This flexibility is quite useful for the program manager, but Congress, concerned with reported abuses of the M account, abolished it through P.L. 101-570 of November 1990. The unexpended balance is being phased out over a period of years.

The law (P.L. 101-570) provides the following in pertinent part:

> . . . on September 30th of the 5th fiscal year after the period of availability for obligation of a fixed appropriation account ends, the account shall be closed and any remaining balance (whether obligated or unobligated) in the account shall be canceled and thereafter shall not be available for obligation or expenditure for any purpose.

Thus, under the public law, some M account funds have already been canceled.

In the future separate accounts will have to be maintained for each fixed account with its fiscal-year identity for five years. The funds remain available for adjusting and paying proper charges during that five-year period, after which it will be necessary to charge current appropriations. The program manager then runs the risk of an unanticipated shortfall from the current account. Without doubt, with the loss of the M account, a very practical tool has disappeared from the program manager's tool kit.

CHAPTER 34

SUBCONTRACTING MANAGEMENT

Carl R. Templin

Carl R. Templin is Assistant Professor of Contracting Management and Director of the Graduate Contracting Program at the Air Force Insititute of Technology, where he teaches graduate courses in acquisition and contracting management. He received a B.A. degree from Brigham Young University, an M.B.A. degree from the University of Wyoming, and a Ph.D. degree from Arizona State University. He has been involved with defense contracting and manufacturing since 1979.

Subcontracting plays a significant role in the development and production of a weapon system or any other military project. As weapon systems have increased in complexity and sophistication, prime contractors have come to depend heavily on subcontractors for significant portions of the contracted effort. The term subcontractor refers to any supplier, distributor, or vendor that furnishes supplies or services to a prime contractor or another subcontractor.[1] The terms supplier and subcontractor will therefore be used interchangeably. This chapter will help project managers understand the importance of subcontracting management and their responsibilities in dealing with subcontracting issues. This will be done by first discussing the impact subcontracting can have on the development and production of a weapon system, especially in regard to cost, schedule, and performance. Next the unique environment that surrounds defense subcontracting will be explored. Project managers must understand the relationship between the government, the prime contractor, and subcontractors at various levels since decisions and actions at each level can seriously impact the others and affect the performance of the overall project. Next, specific regulatory and contractual issues will be addressed. That section will describe the government controls and policies that impact the subcontracting decisions made by the prime contractor. Finally, managerial issues will be addressed, primarily focusing on how to conduct subcontracting to effectively achieve many of the benefits of commercial style purchasing while still complying with government guidelines.

IMPORTANCE OF SUBCONTRACTING

In only the simplest of military projects is it even feasible for the prime contractor to perform the entire contracted effort without the support of suppliers. At a bare minimum, raw materials must be obtained. In most cases there is an extensive amount of materials, parts, and services that must be subcontracted since the prime contractor lacks the capacity or capability to produce them. In other cases the prime contractor chooses to subcontract in order to achieve lower costs, higher quality, better delivery schedules, needed technical expertise, compliance with government policies, or some combination of these factors. Whatever the reason, most military projects have substantial amounts of subcontracted activity, which can significantly impact project costs, schedule, and performance.

Impact on Costs

Subcontracted items account for a significant portion of a project's cost. Figure 34.1 charts the ratio of material costs to the value of finished products over the last two decades for U.S. industries as a whole and for five industry groups with the largest sales to the federal government. Purchased materials account for just over half the price of finished products for U.S. industries as a whole. The transportation industry group, which includes production of motor vehicles, aircraft, ships, railroad equipment, missiles and space vehicles, and related parts has one of the highest ratios with about 60 percent of the total effort subcontracted out. On the low end, the instruments and related products group (producing navigation equipment, measuring and controlling devices, medical instruments, and photographic equipment) subcontracts for about a third of the value of the finished product. The other three industry groups, fabricated metal products, industrial machinery and equipment, and electronic and electric equipment, have ratios hovering around 50 percent. Of course, these figures include sales to both the government and private sectors.

DoD estimates suggest that there can be an even greater cost impact on DoD

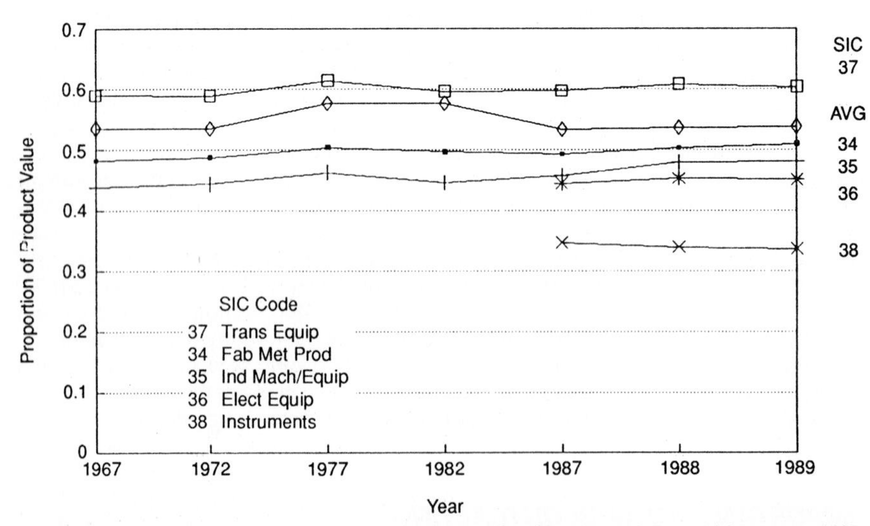

FIGURE 34.1 Material costs in proportion to product value. (*Sources: U.S. Dept. of Commerce, 1987 Census of Manufacturers, General Summary, MC87-S-1, Mar. 1991, Tables 1 and 3; and U.S. Department of Commerce, 1989 Annual Survey of Manufactures, Statistics for Industry Groups and Industries, M89(AS)-1, June 1991, Table 2.*)

programs. According to one estimate, the percentage of procurement dollars going to subcontracting has grown from only 9 percent in 1950 to 41 percent in 1980 and 53 percent by 1990.[2] Other estimates suggest that between 40 and 60 percent of a weapon system's cost is subcontracted by the prime contractor or suppliers,[3] with some going as high as 80 percent.[4] Thus the magnitude of the value of subcontracted efforts alone warrants serious attention by both government and industry project managers.

Impact on Schedule

Subcontracts also have a significant impact on a project's schedule. A 1980 congressional study into the state of the defense industrial base concluded: "The defense industrial base is unbalanced; while excess production capacity generally exists at the prime contractor level, there are serious deficiencies at the subcontractor levels."[5] Such capacity problems at the subcontractor level create bottlenecks in obtaining required materials and supplies, which results in longer lead times or increased inventory requirements or both. Studies have found that the procurement lead time for defense contractors is considerably longer than that for their private sector counterparts. One study found the administrative lead time required to establish the subcontract to be three times longer for defense contractors and the actual production lead time more than double that of private sector firms.[6] Another study investigating the growth of DoD procurement lead times determined that the single most important factor contributing to a defense contractor's production lead time is the time it takes to acquire material from suppliers, accounting on average for 70 percent of the total production lead time.[7]

Impact on Performance

The significance of subcontracted efforts on project performance cannot be overstated. Prime contractors often depend heavily on the technological expertise of their subcontractors to solve technical problems and achieve desired levels of system performance. This is particularly true if the primary role of the prime contractor is that of system integrator, where the prime contractor relies on subcontractors for their expertise in developing and producing critical subsystems which can be crucial to the success and performance of the weapon system as a whole. Consider the impact, for example, of a faulty flight control system or defensive avionics system on a newly developed fighter, or a defective guidance control system on a missile. In such cases subcontract performance could make or break the success of the program.

The quality of purchased materials and components also has a direct impact on the production line. Variations in quality, due to changing suppliers or inconsistent quality control of the subcontracted materials, introduce variation into the production process and make it difficult to keep the production process under control. Subcontractor quality deficiencies also take time to resolve due to the waiver or deviation process or the requirement to fix or remanufacture the deficient materials. Such delays are detrimental to the prime contractor, who is held responsible for any schedule delays resulting from its subcontractors' performance, and to the government, which in turn must endure schedule slippages that impact mission accomplishment.

Since subcontracting can have a major impact on project performance, schedule, and costs, project managers must ensure that subcontracting responsibilities are carried out in a responsible manner. This includes consideration of subcontracting issues in preparing acquisition plans, conducting design and program reviews, and performing day-to-day project management and oversight. It also includes determining that subcontracting actions and decisions are in the best interest of the project and are in compliance with applicable government regulations. In order to properly oversee, manage, and evaluate subcontracting activities, government and industry project managers must understand the defense subcontracting environment which defines the freedoms and limitations, the opportunities and constraints that project managers face in dealing with subcontracting issues.

THE DEFENSE SUBCONTRACTING ENVIRONMENT

Almost every aspect of military project management is governed to some degree by the force of government laws, regulations, or policies, and the subcontracting area is no exception. Defense subcontracts are contracts between private contractors and therefore are subject to state common law and the Uniform Commercial Code. However, because of the sovereign nature of the U.S. government, defense subcontracts are also governed by extensive controls arising from federal laws and defense regulations, policies, and practices. This gives rise to several dichotomies in the defense subcontracting environment. First, the DoD holds its prime contractors responsible for subcontract management and the performance of their subcontractors, disavowing to a large degree any direct responsibility in this area. Yet, at the same time, the government wields a large amount of control over both the prime contractor and subcontractors through oversight and contract flow-down provisions. This section explores this environment in terms of the contractual relationships of the various parties involved and the nature of the tiers of the defense industrial base in which most subcontractors operate.

Contractual Relationships

DoD managers must walk a tightrope when dealing with subcontracting issues to protect the government's interests on the one hand and to avoid overstepping their authority and usurping the prime contractor's responsibility on the other. In order to do this, everyone involved in subcontracting management should understand the contractual relationships, rights, and responsibilities of the parties involved.

Figure 34.2 provides a graphic, though somewhat simplified, diagram of the relationship of government and private-sector parties involved in subcontracting. A buying office enters into a contract with a prime contractor. This contract specifies the terms and conditions to which both parties have agreed and, depending on the type of contract and dollar amount, may contain provisions giving the government certain rights and controls over the prime contractor's subcontracting activities. The details of these controls will be discussed later. The point here is that the government often retains extensive rights over subcontracting. However, the government's direct contractual relationship exists only with the prime contractor, and the government holds the prime contractor responsible for the performance of the subcontractors involved in a particular project.

In order to perform the contracted work, the prime contractor enters into contracts with its subcontractors, generally referred to as first-tier subcontractors. There is usually a chain of such subcontracts, resulting in second tiers, third tiers, and so

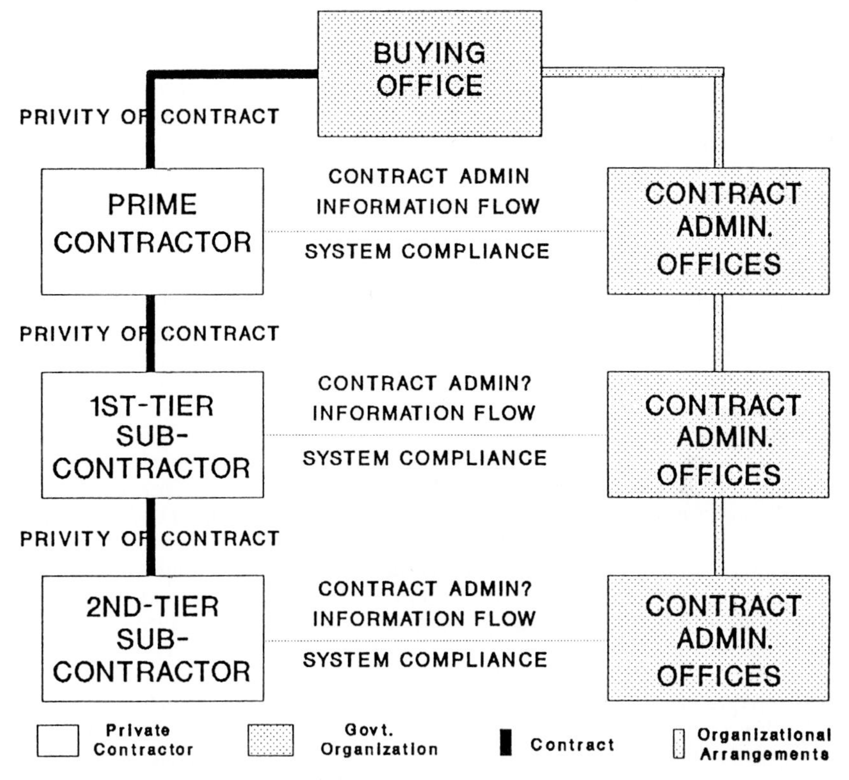

FIGURE 34.2 Subcontracting relationships.

forth. The parties to each prime contract or subcontract are said to have *privity of contract*:[8]

> The term privity refers to the relationship which exists between two contracting parties: historically, a party could not bring a law suit concerning a contract if he was not in privity with the party being sued. That is an important factor in subcontracting: in general, a subcontractor cannot sue or be sued by the government on matters based on either the government contract (to which the subcontractor is not a party) or the subcontract (to which the government is not a party).
>
> The fact that the prime contract may require advance government approval of a subcontract does not confer privity on the subcontractor. In fact, there is no privity between the subcontractor and the government even where the subcontract is subject to all the terms and conditions of the prime contract.

The lack of privity of contract between the government and subcontractors places the prime contractor in the role of intermediary. The government buying office, for the most part, deals only with the prime contractor, who in turn deals with its subcontractors. That means that even though poor subcontract performance may be placing the project in jeopardy, the buying office works the problem through the prime contractor.

Another way the prime contractor serves as an intermediary between the government and the subcontractor involves governmental actions that have caused damages to a subcontractor, such as defective government specifications or property. Subcontractors do not have the right to seek redress directly from the government for such damages.[9] However, if the subcontractor is affected by a dispute between the government and the prime contractor, the prime contractor can seek redress with the government for and in behalf of its subcontractors or the subcontractor can appeal in the name of the prime contractor.* However, where the dispute does not involve any obligation of the government, but rather is limited to the prime contractor and the subcontractor, the government does not get involved in any adjudication or arbitration. The prime contractor and the subcontractors must seek remedy through appropriate state and federal court systems or through arbitration.

Even though the government holds the concept of privity of contract sacrosanct, that does not translate into an entirely hands-off policy relative to government involvement with subcontractors. As Fig. 34.2 indicates, both the prime contractor and the subcontractors deal frequently and directly with government contract administration personnel.

When the buying office awards the contract, it determines who will perform the contract administration activities. Some contract administration functions can be retained by the buying office or, as is usually the case, delegated to a Defense Contract Management Command (DCMC) office. In addition to the contract administration activities that deal directly with subcontracting, which will be discussed in a later section, nearly every contract administration function has subcontracting overtones. For example, contract quality assurance requirements, in addition to monitoring the prime contractor's in-plant quality assurance program and performing required quality inspections, also involves the evaluation of the prime contractor's control of subcontractor quality. This includes incoming inspection, flow down of contractual quality requirements, and analysis of subcontractor quality programs. Throughout the entire spectrum of contract administration, the contact administrator provides oversight and feedback to the buying office concerning trends in subcontract management and performance that impact project performance.

Government contract administration offices also become directly involved with subcontractor operations. This occurs in several ways. The government becomes involved at the request of the prime contractor. Contractors frequently need government assistance in dealing with their subcontractors. For example, if a subcontractor will not give the prime contractor access to its operations and records, the prime contractor can request government support through its contract administration office, which then arranges for the appropriate office to obtain the required information. For example, the Defense Contract Audit Agency can audit the subcontractor's books and the contract administration office with cognizance over that subcontractor provides technical and pricing evaluation of the subcontractor's proposal. That information is supplied to the prime contractor through its contract administration office. Similar support could also be provided in other areas, such as quality inspection or production surveillance. Thus specific contract administration activities can be delegated to government contract administration offices located at or near the subcontractor's operations.

Contract administration offices also become involved in subcontractor operations by providing system surveillance and approval. A DCMC contract administration office provides reviews of the subcontractor's purchasing system, quality system, cost accounting system, configuration control, and so on, and either approves or disapproves them. Approved systems may be required before the subcontractor can be qualified to receive subcontracts, such as having a MIL-Q-9858A qualified purchas-

*For a detailed discussion of the right of indirect appeal see R. J. Wehrle-Einhorn,[8] pp. 15-9 to 15-10.

ing system or having a cost accounting system that is in compliance with government Cost Accounting Standards. A third situation in which government contract administration offices become involved in subcontractor operations is at the request of the buying office. In such cases the contract administration office provides surveillance and information. Recent trends have been for the government to become more and more involved in such surveillance. However, care must be taken not to provide management direction to subcontractors when there is no privity. General system deficiencies related to compliance with government regulations which impact government approval of the subcontractor's systems, of course, are appropriately handled by the contract administration office. However, for contract-specific issues, corrective action should come through the prime contractor for contract-specific deficiencies.

Now that the contractual and organizational relationships of the parties have been explored, it is important to focus on the nature of the industrial base as it impacts subcontracting.

Defense Industrial Base Concerns

Traditionally the industrial base is classified as being made up of three major tiers: (1) prime contractors, which deliver complete systems to the government; (2) subcontractors, which sell major subsystems to the government or to a prime contractor; and (3) parts suppliers, which sell component parts or materials, again either directly to the government or to another defense contractor.* While companies categorized as prime contractors also serve as subcontractors, it is primarily the bottom two tiers which are of interest in assessing the subcontracting portion of the industrial base.

A 1980 study by the Defense Industrial Base Panel of the House Committee on Armed Services concluded that the industrial base as a whole was deteriorating and that there were particular concerns about serious deficiencies at the subcontractor levels. It noted that subcontractors are hit the hardest by defense program instability, burdensome government paperwork, and inability to raise capital.[5] According to those subtier contractors (Ref. 5, p. 13),

> . . . the main factors contributing to the failing health of their sector of the defense industrial base include excessive government administration, on-again, off-again procurement practices, restrictive documentation and specification requirements levied by many primary contractors, critical shortages of trained manpower, and lack of sufficient "flow-down" of contracting benefits from prime contractors.

It further concluded (Ref. 5, p. 14):

- Subcontractors prefer doing business in the commercial sector.
- They charge up to double the price for comparable commercial products.
- Prime contractors do not shield subcontractors from administrative burdens and even add administrative burdens of their own.
- Subcontractors are less able to withstand manpower shortages and therefore have increased their dependence on foreign sources.

This study suggests that the lack of capacity at the subcontractor level is partly caused by the level of administrative burden placed on the subcontractors by both the government and the prime contractors.

As a result of these problems plaguing the lower tiers of the defense industry, the

*For a more complete and historical perspective on these tiers, see Peck and Scherer[10] and Gansler.[11]

number of aerospace suppliers has declined dramatically such that in many segments there are only a few suppliers remaining. Jacques Gansler, one of the few researchers who has focused on problems of the lower tiers of the defense industrial base, indicates that prime contractors, the DoD, and Congress have exacerbated the problem. The prime contractor, whose contract is adjusted according to program risk, does not do the same for its suppliers, even when they have higher technical risk. Prime contractors are also reluctant to pass down favorable clauses and benefits to their suppliers, such as progress payments or multiyear contracts, in spite of the fact that subcontractors face greater market uncertainty and have greater borrowing needs and less borrowing power than the prime contractors. The government invests in equipment and technology at the prime contractor level but not at lower levels. Prime contractors tend to have and use negotiation leverage against subcontractors which have relatively weak bargaining positions. Prime contractors impose even more stringent specifications on their suppliers, even though small subcontractors are at a disadvantage in handling detailed defense regulations, procedures, and requirements. Subcontractors and parts suppliers are less profitable than prime contractors and do not receive the government bailouts often available to prime contractors during periods of financial difficulty (Gansler,[3] pp. 260–261).

Given the impact of subcontracting on project costs, schedule, and performance discussed earlier, project managers should be concerned at the state of the lower tiers of the defense industrial base. For the most part this is a result of years of neglect by both the government and the prime contractors relative to the health and well-being of defense subcontractors, in spite of the fact that the performance of the industrial base as a whole is constrained by the lack of capacity at those lower levels. While major improvements would demand DoD-wide industrial planning and investment, project managers should not overlook the impact of their subcontracting decisions regarding contractual requirements, specifications, and flow-down provisions on subcontractors.

GOVERNMENT SUBCONTRACTING CONTROLS

There are varying types and levels of controls placed by the government on subcontracting efforts. Factors that impact the level of control include whether or not the contractor is a small business, the nature of the product (for example, commercial), the degree that competition was used to award the contract, dollar thresholds, and the type of contract used. In general, greater levels of controls are placed on high dollar contracts with large contractors, especially if the contract was not awarded competitively and is a cost-reimbursement contract.

Some controls deal directly with the contractor's subcontracting decisions and process. These include review and approval of the contractor's purchasing system, make-or-buy program, and subcontracts prior to award to ensure that the prime contractor conducts its subcontracting activities in accordance with government policies and directives. In addition, there are other requirements whose benefits are primarily socioeconomic in nature. These include approval of subcontracting plans to ensure that contractors are giving small businesses an appropriate share of their business and evaluation of contractor efforts to meet small and disadvantaged business goals. There are also required contractual clauses and flow-down provisions that contractors must include in their subcontracts. This section briefly reviews the major government controls that apply to the prime contractor's subcontracting efforts and in some cases to subcontractors as well.

Subcontracting Plan

One way the government exerts control over subcontracting is by promoting the use of small businesses and small and disadvantaged businesses. For contracts (and subcontracts) expected to exceed $500,000 ($1,000,000 for construction) which have subcontracting possibilities, contractors are required to submit a subcontracting plan describing the efforts that will be made to ensure that such businesses have an equitable opportunity to compete for subcontracts. It also requires separate percentage goals for using small businesses and small and disadvantaged businesses as subcontractors.[12] This plan must be submitted and approved prior to contract award. The plan is reviewed by the contracting officer, after obtaining the advice and recommendations from the local Small Business Administration representative and the agency's small and disadvantaged business utilization representative. Approval is based on the plan's adequacy and the contractor's past performance. The contract administration office monitors the contractor's performance, and if the contractor fails to make a good-faith effort to comply with the plan, there are contract provisions that require the contractor to pay liquidated damages to the government.[13] Personal services contracts, those performed entirely outside the United States or its territories, or contracts resulting from small business set asides or accomplished under the 8(a) program are exempt from the requirement.

Make-or-Buy Program

The government also exerts a measure of control over the contractor's make-or-buy program. Contractors are required to submit a proposed make-or-buy program for all negotiated contracts with an estimated value of $5 million or more, unless the work is not complex, involves research and development with no follow-on production, is priced on the basis of adequate price competition, or involves commercial items with established catalog or market prices. Contractors must categorize each major item of work as "must make," "must buy," or "can either make or buy." The contractor must also provide the rationale for each categorization and the justification for each make-or-buy decision when a choice is possible. The contracting officer evaluates the proposed program based on the evaluation factors described in the solicitation and the recommendations of advisors before negotiating an agreement with the contractor. Advisors generally include technical and program management personnel, small and disadvantaged business representatives, and Small Business Administration representatives. The contractor's recommendations are usually accepted, unless they are inconsistent with government interests or policy. Some areas of government interest are the impact of decisions on price, quality, delivery, performance, and technical or financial risk. However, socioeconomic factors also play a role. In some cases, such as contracts for major systems, subsystems, or components or for certain cost reimbursable contacts for supplies and services, the make-or-buy program is incorporated into the contract. In cases where major items or work efforts cannot be fully identified prior to contract award, a clause (FAR 52.213-21) is incorporated into the contract which specifies how to make changes to the agreed upon make-or-buy program. Changes must be approved by the contracting officer, and an equitable adjustment in contract price might be required.

Consent to Subcontract

Under certain conditions, the government requires prime contractors to obtain government approval prior to awarding subcontracts. Consent is required when the subcontracted effort "is complex, the dollar amount is substantial, or the government's interest is not adequately protected by competition and the type of prime contract or subcontract."[14] It is generally required for certain types of contracts, specifically (1) fixed-price incentive and redeterminable contracts, (2) subcontracts resulting from unpriced modifications to firm-fixed-price or fixed-price with economic price adjustment provisions, (3) cost-reimbursement contracts, and (4) letter contracts.[15] Various clauses are used, depending on the type of prime contract.* The consent requirement does not generally apply to all subcontracts, but rather to those that meet specific thresholds, such as dollar amounts (for example, $100,000 under fixed-price contracts), percentage of total estimated cost of the prime contract (for example, fixed-price subcontracts exceeding 5 percent of the total estimated cost of a cost-reimbursement contract), types of subcontracts (for example, cost-reimbursement subcontracts under a cost-reimbursement prime contract), or other criteria (such as experimental, developmental, or research work under a cost-reimbursement contract).[15]

When consent is required, contractors must provide details of the proposed subcontract to the contracting officer. Depending on the particular subcontract clause, the contractor must furnish the following information: (1) a description of the supplies or services, (2) the type of subcontract proposed, (3) identification and justification of the proposed subcontractor, (4) proposed price with associated cost or pricing data and price analysis, (5) the subcontractor's status relating to Cost Accounting Standards (if required), and (6) a negotiation memorandum describing the significant points of the negotiation.†

The contractor cannot award the subcontract until the contracting officer has consented to the subcontract. The FAR provides 13 criteria to be used by the contracting officer. They include, among others, (1) whether the decision to subcontract is consistent with the contractor's make-or-buy program; (2) compliance with contract socioeconomic requirements; (3) the basis for selecting the subcontractor, including determining subcontractor responsibility and checking the consolidated list of debarred, suspended, and ineligible contractors; (4) the basis for determining that the price is fair and reasonable (use of competition, cost or price analysis, and so on); and (5) the appropriateness of contract type. The FAR instructs the contracting officer to pay especially close attention if (1) the prime contractor's purchasing system is inadequate; (2) there exists an unusually close supplier relationship which may result in higher prices due to lack of competition; (3) noncompetitive subcontracts have unreasonable prices, and (4) cost-reimbursement, time and materials, or labor-hour subcontracts are used.[17]

The contracting officer's consent to subcontract does not imply government approval of the terms and conditions of the contract nor that it has any responsibility regarding the subcontractor's performance. Consent requirements can be relaxed or waived if the contractor has a government-approved purchasing system.

Contractors' Purchasing System Reviews

The previous areas have described government involvement in subcontracting issues related to a specific prime contract. The contractor purchasing system review (CPSR), as the name implies, is a review of the purchasing system as a whole. The review is conducted by DCMC personnel when a contractor's sales to the government

---

*FAR 52.244-1 for fixed-price contracts, FAR 52.244-2 for cost-reimbursement and letter contracts, FAR 52.244-3 for time-and-materials and labor hour contracts, and FAR 52.244-4 for fixed-price architect-engineer contracts.[16]

†See FAR 52.244-1(c) or FAR 52.244-2(b)(2).[16]

(prime contracts, subcontracts under government prime contracts, and modifications), using other than sealed bidding procedures, are expected to exceed $10 million over 12 months and every three years thereafter.[18] The review provides a basis for the Administrative Contracting Officer (ACO) to grant, withhold, or withdraw approval of the contractor's purchasing system.[19]

The objective of the review is to evaluate the efficiency and effectiveness of the purchasing system, as well as compliance with government policy in subcontracting efforts. There are nine areas of special government interest: (1) the extent of price competition used; (2) pricing policies and techniques (such as methods of obtaining cost or pricing data and required certifications); (3) evaluation of subcontractor responsibility, including making sure subcontractors have not been debarred or suspended; (4) treatment of affiliates or suppliers with close working arrangements with the contractor; (5) compliance with socioeconomic policies and procedures (such as labor surplus area concerns, small business concerns); (6) management of major subcontract programs; (7) awarding contracting in compliance with Cost Accounting Standards; (8) appropriateness of use of contract types; and (9) the administration and control of progress payments to subcontractors.[20]

The ACO approves the contractor's purchasing system when the review indicates that its policies and practices are efficient and safeguards the government's interests. Between reviews, the ACO conducts surveillance of the purchasing system and on an annual basis either continues approval or, if significant deviations from approved policies and practices become apparent, initiates a special review or requests a CPSR.[21] The ACO withholds or withdraws approval when there are serious deficiencies or there has been a deterioration of the contractor's purchasing system, especially in areas relating to cost or pricing data, Cost Accounting Standards, consent to subcontracts, or small business contracting.[22]

Flow-Down Provisions

One other type of governmental control over subcontracting involves contractual requirements that must be included in subcontracts or can be included at the discretion of the prime contractor. There are also contract provisions that cannot or should not be placed on the subcontractor.* The scope of this chapter will permit only a general discussion of requirements with examples.

Mandatory flow-down requirements are requirements or clauses in the prime contract that must be included, in whole or in part, in subcontracts. In some cases the wording must be verbatim, in others only the substance must be included. In some cases mandatory flow-down provisions apply to all subcontracts, while in other cases there are thresholds that must be met. Some mandatory flow-down provisions are required to protect the government's rights and interests. For example, subcontracts requiring access to classified information must include the substance of FAR clause 52.204-2, "Security Requirements." Clauses dealing with the acquisition of patent rights (such as Ref. 25) must be included in all subcontracts involving experimental, developmental, or research work. Such clauses define the rights and relationships of the government, the prime contractor, and the subcontractors with regard to patent rights connected with research efforts. There are also a large number of clauses that deal with socioeconomic and safety issues that often apply equally to subcontractors as to prime contractors. For example, subcontracts over $500,000 require the same type of subcontracting plans that the prime contractor must submit.[26] Similarly, sub-

*For a more detailed discussion of flow-down requirements, see DSMC[23] and Arnavas and Ruberry.[24]

contractors must also comply with FAR 52.222-26, "Equal Opportunity," and FAR 52.222-36, "Affirmative Action for Handicapped Workers," to name a few.[27]

Some clauses place contractual obligations on the prime contractor, but are not mandatory flow-down provisions. Prime contractors should include these provisions in subcontracts to protect their own interests. For example, prime contracts always have some form of a Changes clause (such as Ref. 28), which if not included in subcontracts would put the prime contractor at risk of default if the government should change the prime contract and the subcontractor refused to comply. Prime contracts also contain a clause that gives the government the right to terminate for convenience.[29] Without that same right to terminate subcontracts, the prime contractor could find itself terminated by the government and facing breach of contract law suits from its subcontractors. If the prime contract includes FAR 52.210-5, "New Material,"[30] the clause should also be placed in subcontracts. Otherwise the prime contractor could find itself in default if subcontractors supplied subsystems or parts that are not in compliance with that clause. Similarly, the prime contractor should tailor FAR 52.215-22, "Price Reduction for Defective Cost or Pricing Data,"[31] to apply to subcontracts in order to have recourse against suppliers should the prime contractor be held for defective pricing due to a subcontractor's actions. If a prime contractor fails to judiciously apply nonmandatory flow-down provisions to its subcontractors, it may find itself unable to properly fulfill its contract requirements or to protect itself should the government exert some of its rights that are not automatically available to prime contractors in dealing with their subcontractors.

Finally there are contractual provisions that apply only to the prime contractor and should not be put in subcontracts. For example, the Disputes clause (FAR 52.233-1[32]) provides the prime contractor with the right to pursue claims against the government. However, contracting officers are specifically precluded from permitting subcontract clauses that give subcontractors the right to deal directly with the government in the event of disputes or that make the results of arbitration, judicial settlements, or any voluntary settlement between the prime contractor and the subcontractor binding on the government. However, in some situations the prime contractor can give the subcontractor the right of indirect appeal to an agency board of contract appeals. A subcontract clause can also bind the prime contractor and the subcontractor to a contracting officer's or board's decisions.[9] Therefore the prime contractor must create a new clause rather than use the Disputes clause. Other examples of clauses which should not flow down include "Gratuities" (FAR 52.203-3) and "Covenant Against Contingent Fees" (FAR 52.203-5), which prohibit unauthorized payments to obtain government contracts.[33]

Determining the appropriateness of flow-down provisions requires considerable knowledge of government contracting. It also demands some management attention and judgment to minimize, wherever possible, the adverse impact of tremendous amounts of regulation that accompany defense-related contracts. These decisions require a balance between protecting the government's interests through insertion of mandatory provisions, protecting the prime contractor's interests through the use of optional and special provisions, and protecting subcontractors from unnecessarily restrictive requirements. These and other management decisions regarding subcontract management are necessary to nurture the kind of supplier relationships required to compete successfully in today's highly competitive defense market.

MANAGING DEFENSE SUPPLIER RELATIONSHIPS

This section explores issues that the government and the prime contractor face relative to defense supplier relationships. The government has important decisions at the outset of an acquisition that will determine whether critical subsystems will be contracted for by the government or the prime contractor and how to best take advantage of defense supplier capabilities during all phases of the acquisition process. Both the government and the prime contractor also determine the degree to which innovative commercial-style buying practices can be used to improve subcontracting performance.

Government Supplier Decisions

The government makes many decisions at the outset of an acquisition that define to some degree the supplier relationships that form later on. A few of the most significant decisions will be discussed.

Component Breakout. This is one of the first supplier decisions that must be made. At the beginning of an acquisition, the government must determine which subsystems or component parts, if any, should be acquired directly by the government and provided to the prime contractor as government furnished property (GFP). Component breakout will require the government to obtain, if it has not already done so, a complete data package with government data rights from the developer. It makes sense if it will result in reduced costs over the life of the system and improved logistics support. However, there are risks that must be considered. The government assumes the risk that poor performance (quality, reliability, and delivery) of the GFP will adversely impact the delivery and performance of the end product. That risk is lowest when the government has experience in acquiring the component, especially if it is in government stock. If not, the government must determine whether it can manage the development, production, and delivery of the component better than the prime contractor and whether the benefits outweigh the risks. That decision warrants careful deliberation before deciding on the appropriate strategy.*

Other Issues. There are many other important issues that must be considered very early in a program. One of the most important involves the government's attitude toward change and improvement. There are established avenues for change currently available to facilitate product improvement and cost savings at the prime- and subcontractor levels. Such programs include Total Quality Management (TQM), Value Engineering (VE), the Industrial Modernization Improvement Program (IMIP), and Preplanned Product Improvement (P^3I). The government establishes the environment for change (enthusiasm, support, cooperation, and so on) under which these programs operate. This environment is important because government attitudes and procedures can either encourage and facilitate change or create barriers that all the improvement programs in the world cannot get past. In a similar vein, the government also determines to a large degree the extent to which it will use innovative commercial buying practices and the extent to which prime contractors and subcontractors will be permitted to also use those practices.

Application of Commercial Buying Practices

The general trend of buyer-supplier relationships in the private sector has been toward establishing supplier partnerships that help both improve their competitive positions. Such relationships are used to achieve consistently high quality of incoming products, reduced lead times, reliable delivery schedules, and technical exchange of

*For a more complete discussion, see DSMC,[23] pp. 10-4–10-5.

information to improve product performance. Numerous studies have called for the government to pursue commercial-style buying practices and management approaches (see Refs. 34–37) and preliminary research has indicated that defense industries can achieve the productivity, quality, and lead time benefits enjoyed by their commercial counterparts in both their production and their purchasing systems (see Refs. 38, 39). The challenge is to find ways to achieve the benefits of commercial buying practices while still complying with government requirements. One of the most important steps is to establish a method of evaluating suppliers' performance in order to identify and select only the best performers to be part of the supplier base.

Supplier Performance Rating System. The importance of a supplier rating system cannot be overemphasized. It is essential to properly determine subcontractor responsibility in the awarding of subcontracts and it is also an important element when past performance is to be used as an evaluation factor in competitive proposals. Contractors are generally free to determine the type of rating system used, though the contract administration office has the prerogative to review and evaluate it. For the rating system to assist in the determination of subcontractor responsibility, it should provide information to help assess the subcontractor's financial capability, past delivery performance, past quality performance, integrity, and the capacity and capability to produce the product or service in question, as well as any specialized requirements necessary for the subcontract in question.[40] In addition to determining contractor responsibility, the supplier evaluation system should help identify the best suppliers. In order to do that, companies must identify the most important performance criteria and standards. These criteria vary by company and even by product, but could include such factors as quoted price relative to other producers, willingness to share information, product reliability, technical capability, and reliability and consistency in meeting delivery, quality, and performance standards. In addition to determining the evaluation factors, companies also must determine how to evaluate the factors. Approaches range from simple to complex. A relatively simple approach is to score each performance variable either categorically or numerically and then rank suppliers. A more complex approach would be to weight the performance variables and compare the weighted point totals for each supplier under consideration. An even more complex approach would be to impute a cost to the performance criteria to permit quotes to be evaluated on a cost basis while also considering past performance and the impact of that performance on the buyer.* Proper selection and application of evaluation criteria can help identify the best performers as well as motivate suppliers to better performance.

Reducing the Supplier Base. In conjunction with a supplier rating system, many companies are aggressively reducing the number of suppliers in their supplier base. This is done in several ways. First, only the best suppliers are considered for contracts. Companies are also concentrating their purchase of similar items into one contract and awarding that to the most qualified source. In so doing, companies reduce the numbers of total suppliers they must track and have fewer contracts to administer, permitting them to focus management attention on developing good supplier relationships and achieving good performance. Defense contractors are free to take action in this area, but must take care to not pare the supplier base so low as to jeopardize the use of competition.

*For additional information consult any good purchasing text. See also Timmerman[41] and Thompson.[42]

Developing Supplier Partnerships. One of the prominent trends in the commercial sector over the last decade has been the move to develop close, long-term supplier partnerships to improve the buying firm's competitive posture.[43] Supplier partnerships can achieve standardized quality and delivery performance that permits great savings in reduction of safety stocks, discrepant material, production-line problems due to variation of material characteristics, and so forth. In addition, it provides incentive for both sides to work closely to achieve product improvement and cost reduction necessary to improve market competitiveness. Achieving such relationships in the defense environment is somewhat challenging due to government pressure to use multiple sources and competition whenever possible, as well as suspicion of overly close supplier relationships. Yet defense contractors have been able to achieve quasisupplier partnerships by consolidating requirements and competing one-year contracts with several one-year options. If the competition is based on price and other factors, such as quality history and delivery performance, so that the award goes to the best performer, such initiatives would satisfy the government's requirement for competition while also producing a relatively long-term arrangement with a proven supplier. The use of options limits the contractor's liability to one year, which is necessary if the prime contractor does not have a multiyear contract. Prudent use of termination for convenience provisions could protect the prime contractor further. Of course, this type of arrangement is especially suitable for common items used on several different products and on contracts that have long production runs or that are multiple-year in nature.

Supplier Development. In cases where conditions do not favor the use of long-term supplier partnerships, companies can still take action to work with suppliers to develop the required capabilities, especially in the areas of improving quality, on-time deliveries, and cost. Supplier development aims to take competent suppliers and make them even better by involving teams of functional specialists which work with suppliers to develop plans and strategies to achieve the desired improvement in technical, quality, delivery, and cost capabilities.[44] Prime contractors frequently have good technical interaction with their suppliers. However, there are other important areas that must also be emphasized.

Quality has always been recognized as important on government programs. Traditionally, however, this has been done primarily through inspections, either at the source or at the prime contractor's facility. Supplier development calls for an effort to help suppliers attain quality at the source, that is, to ensure that suppliers build quality into the product through the use of statistical quality control and improved production methods that will increase the ability of the process to consistently yield high-quality products. Once prime contractors have mastered the use of such programs in their own facilities, they could export that knowledge to suppliers.

Another important area for development in the defense sector is lead time reduction. As has already been shown, lead times in the defense sector not only are longer, but also have greater variation than those in the commercial sector. This leads to greater schedule uncertainty and also to increased costs due to higher levels of safety inventories. Both commercial and defense firms have had considerable success in reducing production cycle times through the use of JIT manufacturing techniques. Defense contractors that have applied JIT in their own operations have experienced dramatic reductions in cycle time, work-in-process inventories, and lot-size reductions as well as significant improvements in quality. However, while defense contractors have been willing to work with suppliers in terms of improving quality, they have not been very aggressive in helping suppliers incorporate JIT techniques in their production facilities.[39] If defense contractors could successfully interest subcontractors for their most critical parts in pursuing JIT, the length and variability of lead times for those items could be reduced. However, it is unlikely that this could be done without working toward the establishment of long-term supplier partnerships with those companies.

Improving performance in the subcontracting area could yield great benefits to the health and profitability of the prime contractor and to the performance of the military project. The generally poor condition of the subcontractor tiers of the defense industrial base suggests that the government and the prime contractors have not been very proactive in improving this vital area. Commercial firms have looked at the purchasing function as one of the last great frontiers to improving product performance and increasing profitability. The same could be done in the defense sector as well. However, it will take the attention and cooperation of managers at the governmental, prime contractor, and subcontractor levels to undertake the required changes and achieve the desired results.

REFERENCES

1. FAR 44.101.
2. E. V. Mooney, Jr., "Subcontract Management: A Key Function in Acquisition Process," *Program Manager,* pp. 12–13, July-Aug. 1991.
3. J. S. Gansler, *Affording Defense,* M.I.T. Press, Cambridge, Mass., 1989, p. 258.
4. DoD Manual 4245.7-M, "Transition from Development to Production," Sept. 1985, pp. 5–12.
5. U.S. Congress, House, Committee on Armed Services, Defense Industrial Base Panel, *The Ailing Defense Industrial Base: Unready for Crisis,* 96th Cong., 2d Sess., Dec. 31, 1980, pp. 11–13.
6. J. H. Perry, "Procurement Lead Time: The Forgotten Factor," *Natl. Contract Manage. J.,* vol. 23, no. 2, pp. 15–24, 1990.
7. J. H. Perry, I. Salins, and L. B. Embry, "Procurement Leadtime: The Forgotten Factor," Rep. DTIC AD-A176 086, Logistics Management Institute, Defense Technical Information Center, Alexandria, Va., Sept. 1986, pp. 3–10.
8. R. J. Wehrle-Einhorn (ed.), *Government Contract Law,* Air Force Institute of Technology, Dayton, Ohio, 1988, p. 15–8.
9. FAR 44.203.
10. M. J. Peck and F. M. Scherer, *The Weapons Acquisition Process: An Economic Analysis,* Harvard University, Boston, 1962, pp. 114–115.
11. J. S. Gansler, *The Defense Industry,* M.I.T. Press, Cambridge, Mass., 1980 pp. 3–7, 36–50, 128–161.
12. FAR 19.704.
13. FAR 19.705-7.
14. FAR 44.102.
15. FAR 44.201.
16. FAR 52.244.
17. FAR 44.402-2.
18. FAR 44.302.
19. FAR 44.301.
20. FAR 44.303.
21. FAR 44.305-1; DoD FAR Supplement 244.304.
22. FAR 44.305-3.

23. Defense Systems Management College, *Subcontracting Management Handbook,* 1st ed., Government Printing Office, Washington, D.C., 1988, pp. 9-1–9-2, app. C.
24. D. P. Arnavas and W. J. Ruberry, "Terms and Conditions," in *Government Contract Guidebook,* Federal Publ., Washington, D.C., 1987, chap. 22, pp. 22-1–22-8.
25. FAR 52.227-11, "Patent Rights-Retention by the Contractor" (short form) or FAR 52.227-12 (long form); FAR 52.227-13, "Patent Rights-Acquisition by the Government."
26. FAR 52.219-9, "Small Business and Small Disadvantaged Business Subcontracting Plan"; FAR 52.220-4, "Labor Surplus Area Subcontracting Program."
27. FAR 52.222-26; FAR 52.222-36.
28. FAR 52.243-1, "Changes—Fixed Price"; FAR 52.243-2, "Changes—Cost-Reimbursement"; FAR 52.243-3, "Changes—Time-And-Materials or Labor-Hours"; FAR 52.243-4, "Changes."
29. FAR 52.249-1, 52.249-2, 52.249-3, 52.249-4, 52.249-6, or 52.249-7.
30. FAR 52.210-5.
31. FAR 52.215-22.
32. FAR 52.233-1.
33. FAR 52.203-3; 52.203-5.
34. The Commission on Government Procurement, *Report of the Commission on Government Procurement,* Government Printing Office, Washington, D.C., Dec. 1972.
35. President's Private Sector Survey on Cost Control, *Report to the President,* Government Printing Office, Washington, D.C., Jan. 15, 1984).
36. President's Blue Ribbon Commission on Defense Management, *A Quest for Excellence: Final Report to the President,* Government Printing Office, Washington, D.C., 1986.
37. Office of the Under Secretary of Defense for Acquisition, *Final Report of the Defense Science Board, 1986 Summer Study,* Government Printing Office, Washington, D.C., Jan. 1987.
38. L. Besser et al., "Effects of Just-In-Time Manufacturing Systems on Military Purchasing," *Program Manager,* vol. 17, pp. 39–43.
39. C. R. Templin, "JIT and TQM: Implications for Defense Industries," in *Proc. 76th Annual NAPM International Purchasing Conf.* (New Orleans, La., 1991), pp. 148–153.
40. FAR 9.104.
41. E. Timmerman, "An Approach to Vendor Performance Evaluation," *J. Purchasing and Materials Manage.,* vol. 22, pp. 2–8, Winter 1986.
42. K. N. Thompson, "Vendor Profile Analysis," *J. of Purchasing and Materials Manage.,* vol. 26, pp. 11–18, Winter 1990.
43. R. Landeros and R. M. Monczka, "Cooperative Buyer/Seller Relationships and a Firm's Competitive Posture," *J. Purchasing and Material Manage.,* vol. 25, pp. 9–18, Fall 1989.
44. C. K. Hahn, C. A. Watts, and K. Y. Kim, "The Supplier Development Program: A Conceptual Model," *J. Purchasing and Material Manage.,* vol. 26, pp. 2–7, Spring 1990.

GLOSSARY OF TERMS

Approved purchasing system A contractor's purchasing system that has been reviewed and approved in accordance with FAR 44.101.

Consent to subcontract The contracting officer's written consent for the prime contractor to enter into a particular subcontract (FAR 44.101).

Contractor purchasing system review (CPSR) The evaluation of a contractor's purchasing system, including the purchase of material and services, subcontracting, and subcontract management from requirement development up through subcontract completion (FAR 44.101).

Privity of Contract The direct contractual relationship that exists between the parties of a contract.

Subcontract Any contract entered into by a subcontractor to furnish supplies or services required for the performance of a prime contract or subcontract, including purchase orders and changes or modifications to purchase orders (FAR 44.101).

Subcontractor A supplier, distributor, or vendor that furnishes supplies or services to a prime contractor or another subcontractor (FAR 44.101).

CHAPTER 35
PROJECT TERMINATION

Michael E. Heberling

Michael E. Heberling is head of the Graduate Acquisition Management Department at the Air Force Institute of Technology. He was the contracting officer on the A-10 acquisition program and the acquisition staff officer for both the national aero-space plane and the ground-launched cruise missile. Lt. Col. Heberling received his Ph.D. degree in commercial procurement from Michigan State University. He also had a one-year education-with-industry assignment with the General Electric Company.

INTRODUCTION

Termination is the fate of all projects. At issue is under what circumstances will the termination take place. Will it be the result of a successfully completed project within performance, cost, and schedule parameters? Or will the project end prematurely through termination for convenience or the more severe termination for default?

Project managers continually face the prospect of some form of termination throughout the project's life cycle. Project terminations are frequently only partial in nature. A partial termination will result when the project's scope or quantities are reduced.

This chapter will discuss several issues surrounding termination. These include causes, when they most likely occur, and the contractual issues.

TERMINATION PRIOR TO COMPLETION

Out-and-out termination of major defense programs before completion has been exceedingly rare. However, due to the compounding effects of budgetary deficits and the political events in the Soviet Union, termination of major programs may become commonplace.

The recent termination of the Navy's $57 billion, 620-aircraft A-12 Avenger II program after a $3.1 billion investment was a watershed event. This was the most costly weapon program ever canceled. The significance is illustrated by comparing it to earlier major program terminations. Before the A-12, the demise of the Army's Sergeant York division air defense (DIVAD) gun was the best known major pro-

FIGURE 35.1 T-46A next-generation trainer. (*Source: U.S. Air Force.*)

gram cancellation. It ended after a $1.8 billion investment.[1] The Air Force's $3.0 billion, 650-aircraft T-46 next generation trainer (NGT) met the same fate in 1986, after four years of work and an investment of $600 million.[2]

There are several factors that play a role in the decision to terminate a project. These factors fall into four broad areas: (1) political, (2) technical performance, (3) cost and schedule, and (4) change in requirements or obsolescence.

The decision to terminate a program is rarely attributable to any single factor. It is the collective negative impact of many factors that prove fatal. For example, the Air Force canceled the T-46 NGT (Fig. 35.1) program due to poor performance, schedule delays, and cost overruns. These factors together made the T-46 vulnerable to the politics of shifting budget priorities.[3]

Politics. Project termination frequently occurs for political reasons. The politics can be external or internal to the service of the program. Often it has nothing to do with the enemy threat or the service requirement. In spite of media rhetoric to the contrary, Congress abhors outright project cancellation. It prefers the more subtle approach of partial termination. The standard solutions include reducing the scope of the project, stretching out the program, or just lowering the quantities. These alternatives are far more palatable to constituents than the headline grabbing *Program Canceled.*

The primary problem is that Congress will focus exclusively on the budget for the next year. This short-term perspective frequently leads to the expedient recommendation to reduce quantities to save money. Unfortunately, as annual quantities decline, unit costs rise. This occurs because defense systems are produced at uneconomical rates. All of the fixed costs of plant, equipment, and management must be absorbed by fewer units (Gansler,[4] p. 124). From fiscal year 1981 to fiscal year 1986 the average production program had a quantity decrease of 17 percent. Ironically these actions resulted in a cost increase of 22 percent (Gansler,[4] p. 129).

This reduce-quantity–stretch-out-program approach worked well during the

1980s because of the large defense budget, which lasted through much of the decade. However, in the 1990s this approach will not work due to a declining defense budget and a changing enemy threat. When there is inadequate funding, defense planners are left with two difficult options. The choices are (1) program cancellation or (2) continue the vicious cycle of reducing quantities and stretching out programs (Gansler,[4] p. 103). Option (2) was common in the eighties; option (1) will be common in the nineties.

Technical Performance. Whenever a defense system is unable to meet the contracted performance specifications, termination is a likely option. The Army's DIVAD gun illustrates this point (Fig. 35.2). Approved in 1977, this program was viewed as a low-risk venture. It was considered simply an integration project using established technology. It combined the chassis from the M-48 tank, an existing gun system in production in Europe, and a derivative of the F-16 APG-66 radar for guidance.

After a "shoot off" between General Dynamics and Ford Aerospace in 1980, Ford won the development and production contract. Technical problems quickly hampered the progress of the program. The principal difficulty centered on integrating the aircraft radar with the ground vehicle portion.

Initial production tests found that the system failed 22 of 163 contract specifications. The Army's operational testing and evaluation agency concluded that the gun could not meet reliability and maintainability requirements. DIVAD's availability during tests was only 33 percent when the requirement was for 90 percent. The OSD's operational test agency found that the Sergeant York needed too much maintenance to be operational. Based primarily on these tests, Secretary of Defense Weinberger canceled the program on August 27, 1985.[5]

Another example, the Air Force's Skybolt missile (Fig. 35.3), illustrates that even

FIGURE 35.2 U.S. Army DIVAD gun system. (*Source: DoD.*)

FIGURE 35.3 Skybolt missile. (*Source: U.S. Air Force.*)

with political support, poor performance will cancel a program. When Kennedy was running for president in 1960, he criticized the Eisenhower administration for canceling the B-70 program. He promised to support the B-70 if elected, citing the "bomber gap" as justification. However, once elected he and Secretary of Defense McNamara, agreed with Eisenhower that the bomber was not needed. To ease the loss of the B-70, Kennedy promised the Air Force that he would support the development of the Skybolt missile. This was a long-range nuclear missile that could be fired from the B-52 bomber. Unfortunately the Skybolt performed poorly in tests and the program was canceled in 1962 (Kotz,[6] p. 77).

Cost and Schedule. Whenever the cost or the schedule for a program deviates markedly from the contract, program termination becomes a consideration. Problems with cost and schedule occur hand in hand. This was true with the A-12. The Navy had a development contract for $6.2 billion with the teaming contractors of McDonnell Douglas and General Dynamics. By January 1991 the Navy had paid more than $3.1 billion in progress payments, yet had received only $1.2 billion in actual goods and services.[7] In addition, the program was at least a year behind schedule (Morrison,[1] p. 31). The cost and schedule overruns were the primary factors in the decision by Secretary of Defense Cheney to cancel the A-12 program.

Gansler concluded that of the three primary project considerations (performance, cost, and schedule), the Department of Defense (DoD) places its priority on performance, followed by schedule and then cost. Consequently, as a general rule, most defense systems come very close to the original performance specifications. Based on the order of priority, the deviations from schedule would be more severe than performance, but less than cost. This was, in fact, the case. Schedules on average run about 30 percent late. Many studies have shown that cost growth in the range of 100 percent is the norm for major systems (Gansler,[4] p. 177). Augustine came to a similar conclusion: " . . . past program cost estimates have been in error, on the average, by about 100 percent" (Augustine,[8] p. 306).

An important question is: How prevalent are program terminations that result from cost overruns? The 1979 General Accounting Office report "Inaccuracy of Department of Defense Weapons Acquisition Cost Estimates" (November 16, 1979) provides some insight. The report stated that after reviewing programs covering a 10-year period, it failed to find one example where the DoD accurately estimated (or overestimated) the cost of any major weapon system. Once initial funds were provided, it was found that programs were rarely terminated for cost overruns. Termination occurred only in the most extreme cases.

Obviously the cancellation of the A-12 program was a significant event. It represents a marked departure from the historical treatment of cost and schedule overruns. This action "sends a clear signal that the Pentagon, under intense pressure to cut spending, no longer is willing or able to expend the money or political capital with Congress required to save troubled weapons development programs."[7]

Change in Requirements or Obsolescence. Many weapon systems are designed to counter specific enemy threats. Consequently any change in the enemy threat, as in a technology breakthrough, will alter requirements. As a minimum, a change may require major modifications to a system. In the extreme case, they can make a system obsolete.

Consider the Air Force's B-70 Valkyrie bomber program of the late 1950s (Fig. 35.4). This program was for 250 bombers with a price tag of $6 billion. The bomber was designed to fly at an altitude of 70,000 feet and an air speed of Mach 3 (Kotz,[6] p. 72).

On May 1, 1960, a single event drastically undermined the B-70's reason for being. On that day the Soviets shot down the CIA's high-altitude U-2 spy plane piloted by Francis Gary Powers. The Soviets used a high-altitude surface-to-air antiaircraft missile (SAM) (Kotz,[6] p. 60). The mission of the manned penetration bomber shifted almost over night from high altitude to low level. Another bomber, the B-52, in production during this time frame, had its mission modified to counter the new threat. The B-70, on the other hand, simply became obsolete.

FIGURE 35.4 B-70 Valkyrie bomber. (*Source: U.S. Air Force.*)

Today, due to our cumbersome defense acquisition process, obsolescence is frequently built into our new systems. The following observations, one by Gansler and the other by Augustine, highlight this point. Gansler stated: "Equally serious is the unreasonably long acquisition cycle (10–15 years) for a typical major weapon system. This long cycle leads to unnecessary development costs, to increased 'gold plating,' and to the fielding of obsolete technology" (Gansler,[4] p. 148). According to Augustine: "Based in part on the fact that the half-life of most technologies has been determined to be on the order of three to ten years, it appears that if the current glacial development trend persists, most new systems will be obsolete only slightly before they are born" (Augustine,[8] p. 357).

Beginning in 1989, the threat to the United States and the accompanying requirements for defense have been altered drastically. The focus of our military strategy is no longer based on a global nuclear confrontation between the NATO nations of the West and the Warsaw Pact nations of the Soviet bloc. As a result of the war with Iraq, attention has shifted to potential regional conflicts involving third-world powers.

This trend does not bode well for programs specifically designed for the Cold War nuclear environment. Future terminations are likely for programs that fall in this class. This is even more probable in the current era of budgetary constraints.

WHEN IS TERMINATION MOST LIKELY TO OCCUR?

Up until very recently it has been a general rule that once past full-scale development and especially into early production, major weapon programs rarely get canceled. Joseph Campbell, a vice president for PaineWeber, observes: "Weapon systems generally die as people do—it's either infant mortality or old age. Very few get killed in middle age."[9]

Which Program Is More Likely to Be Terminated—A Development Program or a Production Program? Based on historic trends, it is possible to make some observations on the conditions that favor termination. Unfortunately politics plays a major role in the decision. In its simplest form, the issue is jobs. To a politician, existing jobs are far more important than potential jobs. The B-58 and the B-70 bomber programs illustrate this point.

Although Convair's B-58 Hustler bomber could reach speeds of up to 2.5 Mach, it had serious drawbacks, including reliability. It was unable to reach targets in the Soviet Union from bases in the United States. As a medium bomber, its weapon payload was extremely limited. For example, it could not carry any missiles. In 1959 DoD wanted to cancel the program. To do this, however, would have aggravated the already high unemployment rate in the Fort Worth area, home of the Convair plant. At that time the national unemployment rate was at 7 percent and increasing.

President Eisenhower felt that neither the B-58 nor the B-70 was needed for the nation's defense. Yet he canceled the B-70 development contract in favor of building 42 more of the B-58 bombers at a cost of $1 billion (Kotz,[6] p. 57).

In some cases termination is not in the best interest of the government. If a project is near completion, termination may be more expensive than completion. Termination actions should take place only when savings will result. The decision to terminate must take into account all settlement costs and the impact on subcontractors.

General Conclusion 1. When given the option of canceling one of two programs, one in production and the other in development (and neither of which is needed), the one currently in production with real (as opposed to potential) jobs will continue.

General Conclusion 2. A program with technical problems will be canceled before a program with schedule problems. Ironically, in spite of all the grandstanding to the contrary, a program with a cost overrun will be the least likely to be canceled.

General Conclusion 3. If you are associated with a defense program that (1) is in development, (2) is experiencing technical problems, (3) is behind schedule, and (4) is in a cost overrun condition, then you should begin working on your resume.

CONTRACTUAL ISSUES OF PROJECT TERMINATION

From a contractual standpoint there are three possible termination outcomes. These are successful project completion, termination for convenience, and termination for default. Since successful project completion poses few problems, the discussions that follow will only address termination for convenience and termination for default.

Termination for Convenience*

The government reserves the right to terminate a project (in whole or in part) at any time. The contracting officer may terminate a contract when this action is in the best interest of the government.

Contractors understand that this risk exists when they conduct business with the government. Termination may occur even though the contractor is performing within the negotiated cost, schedule, and performance parameters. Termination for convenience, frequently called T for C, may take place due to budgetary constraints, technology obsolescence, or as a result of changing priorities.

The termination for convenience process begins when the contracting officer provides the contractor with a notice of termination. This specifies the extent of the termination and the effective date. The contractor is then obligated to stop all work related to the termination. This includes all in-house as well as all subcontract work related to the terminated project.

Government Responsibilities. Once the government decides to terminate a project, the contracting officer must notify the contractor as quickly as possible. Swift notification is meant to minimize the government's liability.

The government must pay the contractor for all completed and accepted work as well as for the preparation work done on items still in process. The contractor will normally be reimbursed for the subcontract settlement costs. The government will also reimburse special tooling and test equipment. However, the contractor must show that this equipment could not be used on other work. Compensation is also made for additional costs that result directly from the termination. This would include the costs for removal, handling and shipping, preservation, storage, and the administrative costs of handling terminated inventory. It also would include the costs of preparing for and settling the termination claim.

The government will pay a reasonable profit (or fee) for work done up to the

*Federal Acquisition Regulation (FAR) Part 49 covers termination of contracts.

time of the termination. This assumes that the contractor would have made a profit had the contract been completed. If the contracting officer makes a determination that the contractor would have lost money on the contract, then the government will not pay any profit. The government will not, however, pay the contractor any anticipatory profits (that is, profits the contractor might have received, had it been allowed to complete the total project).

The government will not pay for common items that could be used on other contracts. Any excess materials that exceed the known requirements will not be reimbursed. However, if the contractor purchased all of the needed material at one time to receive a discount, then this would be reimbursable.

Also, the government is not under any obligation to pay consequential damages (such as the loss of future business or other contracts) that result from the termination.

Contractor Responsibilities. Upon receipt of the formal termination notice, the contractor will immediately initiate stop work actions. Simply identifying the work to be terminated can be complicated. Terminated work is frequently interspersed with nonterminated work. In addition, orders to subcontractors are frequently negotiated and priced on a minimum quantity. If part of the order involves terminated work, should the entire order be canceled?

The contractor must transfer title and deliver fabricated (and unfabricated) parts, work in process, completed work, supplies, and other materials to the government. The Termination clause (FAR 52.249-2) states that the contractor has one year from the effective date of the termination to submit a settlement proposal to the government. This proposal details the costs of the termination. In fixed-price contracts the settlement can be based on either total cost or inventory. The inventory method is preferred.

If the contractor fails to provide a termination proposal, then the contracting officer may unilaterally determine the amount that the contractor should receive. When this happens, the contracting officer's decision is final and it may not be appealed.

A partial termination, especially one that reduces the quantities, impacts the contractor's ability to amortize the overhead costs. There are now fewer items to absorb the fixed costs. This means that the unit costs for the remaining items must go up. The Termination clause recognizes this situation. It allows the contractor to submit an equitable adjustment proposal to reprice the remaining units. The contractor has 90 days to submit this proposal from the effective date of the termination. This is negotiated separately from the termination proposal.

*Termination for Default**

When a government contractor fails to perform, the government may terminate the contract for default. This type of termination occurs when the contractor fails to deliver on time, meet specifications, or make progress. Other reasons for default termination include fraud, knowingly employing incompetent key personnel, or the abandonment of effort [(FAR 52.246-3(h)].

Although the termination for default for major weapon systems has been rare, this may change. Touhey and Thornton note that " . . . in the face of a lessened military threat and a reduced defense budget, the military services have been more aggressive in terminating troubled major weapons programs for default." They cite the Navy's A-12 fighter program and the P-7 submarine hunter aircraft programs as vivid examples.[10]

*A more detailed discussion of termination for default can be found in FAR Subpart 49.4.

There are a sequence of events that lead up to the termination for default. These include the issuance of a show-cause notice, a cure notice, and the final notice of termination for default.

Show-Cause Notice. When the contracting officer is concerned that the contractor may breach the contract, the issuance of a show-cause notice is appropriate. The notice requests that the contractor provide information on why the contract should not be terminated for default. The contractor's failure to perform may be due to extenuating circumstances beyond the control of the contractor. The failure may even be attributable to government action (or inaction).

Although the show-cause notice is not mandatory, it does serve as a forum for the contractor to present any mitigating circumstances.

Cure Notice. Before a contract can be terminated for default, the contracting officer must give the contractor a cure notice. This notice states the government's assessment of the impending failure. The contractor then has 10 days to correct the failure or provide reasons why the contract should not be terminated. If the contractor does not rectify the situation within the 10-day grace period, then the contractor will receive one more notice: termination for default.

Consequences of a Termination for Default. A contractor that is terminated for default is held liable for the excess costs of reprocurement through another contractor and for damages. There are two types of damages, (1) accrued liquidated damages and (2) actual damages suffered by the government.

In some cases a contractor is erroneously terminated for default. If evidence later reveals that the contractor's failure was due to extenuating circumstances, then the default is in fact excusable. The default clause in FAR 52.249 stipulates that when a contract is erroneously terminated for default, the termination is converted into a termination for convenience.

The government must be careful not to jeopardize the termination-for-default option. Failure to take action on the part of the government may be construed as condoning the breach of contract.

As was stated at the beginning of this chapter, a premature termination of major programs has been exceedingly rare. Even more rare has been the termination for default of a major program. The A-12 program termination fell into this latter category. Secretary of Defense Dick Cheney terminated the A-12 contract for default, citing the failure of McDonnell Douglas and General Dynamics to perform. The default was based on "the inability of the contractors to design, develop, fabricate, assemble, and test the A-12 aircraft within the contract schedule and to deliver an aircraft that meets contract requirements."[7]

As part of the termination for default, the government is demanding a refund in the amount of $1.9 billion for overpaid progress payments. As was stated, the government must be careful that it is not an accomplice in the contractor's failure to perform. The two contractors insist that they do not owe the Pentagon anything. If fact, they claim that the government owes them $1.6 billion for overrun expenses caused by excessive and unforeseen Navy demands (Morrison,[1] p. 30).

CONCLUSION

Termination is the fate of all defense projects. Ideally this will occur after the successful completion within the original performance, schedule, and cost parameters. Unfortunately premature terminations (both partial and complete) are always a possibility. A number of major factors play a role in the decision to terminate a project. These include political, cost, technical, and obsolescence issues. From a contractual standpoint, a project can be terminated for convenience or for default.

REFERENCES

1. D. C. Morrison, "Deep-Sixing the A-12," *Government Executive,* pp. 30–35, Mar. 1991.
2. D. M. North, "Congressional Impasse on T-46A Prompts NGT Recompetition," *Aviation Week & Space Technol.,* p. 16, Oct. 27, 1986.
3. D. E. Fink, "T-46A—Dead or Alive?" *Aviation Week & Space Technol.,* p. 13, Aug. 18, 1986.
4. J. S. Gansler, *Affording Defense,* M.I.T. Press, Cambridge, Mass., 1989.
5. T. L. McNaugher, *New Weapons, Old Politics: America's Military Procurement Muddle,* Brookings Inst., Washington, D.C., 1989, pp. 103–104.
6. N. Kotz, *Wild Blue Yonder: Money, Politics, and the B-1 Bomber,* Pantheon, New York, 1988.
7. J. D. Morrocco, "Navy Weighs Alternatives after Cheney Kills Avenger 2," *Aviation Week & Space Technol.,* p. 18, Jan. 14, 1991.
8. N. Augustine, *Augustine's Laws,* Penguin, New York, 1986.
9. J. Kitfield, "Cancellations: Will They Really Happen?," *Military Forum,* pp. 26–33, Sept. 1989.
10. T. J. Touhey and D. W. Thornton II, "Terminations of Defense Contracts Revisited in 1991," *Contract Manage.,* pp. 4–9, July 1991.

CHAPTER 36

INSPECTIONS AND AUDITS ON DoD PROGRAMS

James H. Dobbins

James H. Dobbins is an internationally recognized specialist in software quality and reliability with over 28 years of experience in these and related disciplines. He is currently Professor of System Acquisition Management at the Defense Systems Management College, Ft. Belvoir, Va. He is an attorney-at-law licensed to practice in Virginia and various federal courts.

INTRODUCTION

Inspections and audits of Department of Defense (DoD) programs are assumed to mean audits of contractors, but they can also mean audits of DoD agencies developing systems, particularly software, as an internal government project. It is important to understand that inspections and audits are two very different activities. They are different for both hardware and software development activities and require separate procedures and control mechanisms. These two distinct disciplines have a unique relationship which, if properly understood, can serve to improve the overall quality of the products under development significantly. Without them, the likelihood of success on any project is severely diminished.

The fundamental document governing the implementation and conduct of formal government inspections and audits is MIL-STD-1521B.[1] It is augmented by other documents, which will be referred to herein. These implementing documents reflect the policy guidance for conducting configuration audits as specified in DoD Instruction 5000.2[2]

Definition of Inspections

Inspections are a means whereby a product, whether hardware, software, or documentation, at any stage of development, from requirements definition through production, is evaluated by analysis of data collected through physical examination of the product. This examination can be done by people or by automated support mechanisms, including computer-aided software engineering (CASE) tools, or both. The final evaluation of the data is done by the inspectors who make a professional judgment regarding the acceptability of the product in its current stage of development.

Definition of Audits

Audits are an activity conducted by personnel whose mission is to evaluate processes and the resulting products to determine whether or not these processes, and the resultant products, meet acceptable development standards. Audits can be done on a routine basis or they can be spontaneous. Audits are conducted internally within a development organization, or a supplier can be audited by a customer. The auditing agency can be the government or a prime contractor. From the government perspective, audits done internally by a contractor, or done by a contractor on a subcontractor, are considered informal, even though to that contractor, within the particular processes, they may be very formal. The government considers formal audits to be those conducted by the government in accordance with the provisions of the latest released version of MIL-STD-1521.

When prime contractors conduct internal audits, the audit reports are subject to government review. These informal audits are conducted in accordance with contractor internal standards, as opposed to the formal audits conducted in accordance with the provisions of MIL-STD-1521.

Routine audits are those informal audits conducted according to a predetermined schedule and are not triggered by a particular problem. They are conducted to assure that the development process and products are in accordance with the contractual terms and conditions or, if an internal government or contractor development project, with the agreed-to development requirements.

A spontaneous audit may be conducted at any time, without prior notice, and is usually in response to a particular problem situation. It will be conducted by a contractor internally, or by a contractor auditing a subcontractor, or by a government activity, such as the Defense Logistics Agency (DLA) representatives, conducting an audit of a contractor. These DLA audits will be discussed in more detail later in this chapter.

FORMAL GOVERNMENT AUDITS—MIL-STD-1521B

MIL-STD-1521B[1] describes and establishes the requirements for formal reviews and audits for configuration items. A configuration item is hardware or software, or an aggregation of both, which is designated by the contracting agency for configuration control. The requirements for the conduct of the two kinds of audits described in MIL-STD-1521B, the functional configuration audit (FCA) and the physical configuration audit (PCA), implement the policy as specified in DoD Instructions 5000.2, Part 9, sec. 3(h).[2] These two audits are also described in the draft MIL-HDBK-61.[3] The FCA and PCA are principally a paper chase, are labor-intensive, and may take several days each. These two audits are conducted very late in the development process, following the test readiness review (TRR). Consequently any serious problems discovered during an FCA or PCA are indications of a breakdown in the entire contract management process. Realistically these audits should confirm and validate that the products and processes are as they should be, and they are not the place or time to find major problems.

The contractor is responsible for conducting the formal reviews and audits described in MIL-STD-1521B as well as for assuring subcontractor participation. Usually routine informal audits will be conducted one or more times prior to a formal government review. These informal audits are conducted either by the government (program office or DLA) or internally by a contractor, and certainly prior to any formal government FCA or PCA. The formal reviews and audits are conducted at the contractor facility or a designated subcontractor facility. The contractor establishes the time, place, and agenda for the audits, subject to coordination with the contracting agency. A contracting agency representative acts as cochair for the audit teams.

Following the formal government FCA or PCA, there are three basic kinds of responses provided by the contracting agency:

1. *Approval.* Indicates that the activity was completed satisfactorily.
2. *Contingent approval.* Indicates that the activity is not considered accomplished until satisfactory completion of the resulting action items.
3. *Disapproval.* Indicates that the activity was seriously inadequate. This will generally be the case only when a proper program of regular audits and inspections has not been carried out beforehand. It is as much an indication of failure on the part of the government to manage properly as it is one of failure of the contractor to develop properly.

Functional Configuration Audit

FCAs are conducted in accordance with the provisions of Appendix G, MIL-STD-1521B.* The FCA is done prior to the PCA. In order to conduct an FCA, the contracting agency must have authenticated the functional baseline and the allocated baseline. Therefore an FCA cannot begin until the formal reviews at which the baselines are authenticated and allocated have occurred. The functional baseline is established as a result of the formal system requirements review (SRR). The allocated baseline for software is established following the formal software specification review, and for hardware, following the hardware preliminary design review (PDR).

The objective of the FCA is to verify that the performance of the configuration item complies with the hardware requirements specification, software functional requirements specification, and interface requirements specifications. It is not an actual test, but an audit of test result data. An FCA is a prerequisite to acceptance of the configuration item by the contracting agency. For software, it is required that a technical understanding be reached on the validity and degree of completeness of the software test reports, the software test procedures, the operator's manual, the user's manual, and the system diagnostic manual.

Progressive FCA. An FCA does not have to be a one-time event for any given configuration item, and usually is not for a complex configuration item. When specified by the contracting agency, a series of progressive FCAs can be conducted throughout the development period of the configuration item. The FCA is considered completed only after the completion of the qualification testing, which may re-

*Note that the FCA and PCA activities and requirements are also described in the new draft MIL-STD-983, dated April 22, 1991, secs. 5.6–5.6.4.6, pp. 64–79. Following full approval and dissemination of MIL-STD-983 in its final form, the PMO is cautioned to evaluate any potential conflicts between MIL-STD-1521B and MIL-STD-983 when both are invoked on a given contract.

quire full integration testing, and after review of all the current discrepancies at the final FCA meeting.

For hardware in particular, the FCA is done on that configuration which is considered representative of the production configuration, even though the FCA may be done on a prototype. If a prototype or preproduction article is not produced, the FCA is done on the first production article. It is a check that the risk in going to full production is acceptable. For software, the first article is the final article in the absence of an engineering change proposal (ECP), and therefore the focus on the FCA is different. For software it is like doing the audit after the production contract is complete instead of before it begins.

Acceptable FCA Testing. The testing results considered acceptable for use in conducting an FCA are those tests accomplished using contracting agency approved procedures and witnessed by the government or its designated representative. The conduct of such a test is considered sufficient to ensure the performance of the configuration item as set forth in the specification, and is also considered adequate to meet the quality assurance requirements.

In the event some performance requirements cannot be verified by testing, adequate analysis and simulations can be substituted. The analysis and simulation should be rigorous and thorough enough to ensure the configuration item performance as outlined in the specification.

Any item which fails to pass quality assurance testing or qualification requirements is analyzed for cause and corrected before the configuration item is subjected to requalification.

Software FCA. For computer software configuration items (CSCIs) there are additional requirements. For the selected CSCIs being audited, the contractor shall provide a briefing to the FCA team. This briefing shall include the test results and findings for each audited CSCI. The briefing shall include, as a minimum, the requirements which are not met, a proposed solution to each such item, an accounting of each ECP incorporated and tested, a briefing on each proposed ECP, and a general briefing on the entire test effort and results.

In addition, the FCA team shall audit the formal test plans, descriptions, and test procedures and compare these against the official test data. Test reports shall also be audited to validate that the reports are accurate and completely describe the CSCI tests. The FCA team shall also audit all approved ECPs to verify they have been incorporated. All updates to previously delivered documents shall be audited, as well as the minutes of the PDR and the critical design review (CDR). For each selected CSCI the interface requirements shall be audited as well as the testing of these interfaces. Operational characteristics such as database characteristics, storage allocation data, timing, and sequencing characteristics shall be audited for compliance with the requirements.

Obviously doing all this is a laborious and time-consuming process. For large systems, even though what is examined is a sample of the total, the job is still large and requires a careful selection of the personnel who comprise the audit team. Without the proper team makeup, the necessary results will not be forthcoming, and the entire cost of performing the audit will have been wasted. The excuse that the right kind of people are hard to find is both lame and nothing more than evidence of poor planning and management. As late in the process as these audits occur, there is no acceptable excuse for not having the proper team makeup. There has been too much planning time available to find the right team.

Government Responsibility in the FCA. Given the preceding, it is the responsibility of the government to assure that those participating on the FCA team be conversant with the program, the functional requirements, and the issues which are the focus of the audit. It will require both hardware and software expertise as well as system engineering expertise. If the contracting agency does not have ready access to this kind of expertise, the government must gain access through an independent verification and validation (IV&V) or systems engineering technical assistance (SETA) contract. The program manager must do whatever it takes to get the job done right, and that takes advance planning and preparation.

Physical Configuration Audits

A PCA is conducted, in accordance with the provisions of Appendix H of MIL-STD-1521B, after successful completion of the FCA. It is a formal examination of the as-built version of a configuration item against the design documentation. The purpose is to establish the formal product baseline. The PCA also determines that the acceptance testing requirements prescribed in the documentation are adequate for the acceptance of production units and meet the requirements of the quality assurance activities. A PCA is conducted on the first configuration item to be delivered by a new contractor, even if a PCA was previously accomplished on the first article of the selected configuration item delivered by a different contractor. Following successful PCA, all changes to the product must be by ECP.

A PCA is a detailed audit of engineering drawings, specifications, technical data, and tests utilized in the production of hardware configuration items (HWCIs), and a detailed audit of design documentation, listings, and manuals for CSCIs. For software, the software product specification (software requirements specification) and version description document (VDD) shall be a part of the PCA.

For software, a final review is conducted of all operational and support documents, including the computer system operator's manual, the software user's manual, the computer system diagnostic manual, the software programmer's manual, and the firmware support manual to check format, completeness, and conformance with the applicable data item descriptions (DIDs).

For a hardware PCA, if there are any differences between the physical configurations of the production unit and the development units used for the FCA, the contractor shall identify the differences and shall certify or demonstrate to the government that these differences do not degrade the functional characteristics of the selected units. The contracting agency cochairperson shall identify a representative number of drawings and associated manufacturing instruction sheets for each selected item. A valid sampling basis may be chosen for selecting the actual items audited. Any discrepancies between the instruction sheets, the design details, and the changes in the drawings will also be reflected in the hardware. Therefore the chain of documentation must be examined very carefully. All records of the baseline configuration for the HWCI are audited by direct comparison to the contractor's engineering release system and change control procedures. In addition, the contractor's engineering release and change control system is audited to assure that the changes are adequately and properly controlled.

Software PCA. In addition to any other provisions of MIL-STD-1521, a software PCA must include a review of the software product specifications for format and completeness. This review presumes that the specifications have long ago been verified for technical adequacy. In addition, the FCA minutes will be reviewed for any recorded discrepancies and any action items and action item results. The PCA team will also examine the design descriptions for conformance to proper form, symbology, references, and data descriptions. The PCA team also compares the top-level computer software component design descriptions to lower-level computer software component descriptions to assure consistency between the document levels. The PCA team then compares all lower-level design documents with the software listings for accuracy, completeness, and consistency. Discrepancies discovered in these multi-level examinations would signal a difference between the software tested and the software delivered, the software tested and the set of requirements as specified, or would signal a potential software support capability deficiency.

The PCA team, to assure that the support personnel can adequately maintain and enhance the delivered software, also checks the software user's manuals, the software programmer's manual, the computer system operator's manual, the firmware support manual, and the computer system diagnostic manual. The examination of these documents is for format, completeness, and conformance with the applicable DIDs. This is not the time, for example, to initially discover that a set of support equipment, such as a programmable read only memory (PROM) burner, will be required but for which there has been no planning. Finally, the actual CSCI delivery media and the annotated listings for the software are examined for compliance with the software requirements specification and the approved coding standards.

Clearly, these MIL-STD-1521B FCA and PCA audits are generally conducted late in development. This is especially so for software since software has no separate production process. These initial development items are the final product. Basically, all that is done after initial development is making copies of the first article. Also, the PCA, like the FCA, is not a technical correctness audit as much as it is a traceability and format and DID conformance audit. This creates a need for a technical process and product correctness audit much earlier in the life-cycle phases. One way this is accomplished is through the DLA in-process audits.

DLA IN-PROCESS AUDITS

The DLA, headquartered at Cameron Station, Va., through its DPRO offices, upon receipt of a letter of delegation from the government program manager, can provide in-process audits of a contractor. DPRO serves as the quality assurance arm of the government, and the representatives report technically to the program manager and organizationally to DLA headquarters. These DPRO audits can be for hardware, software, or both, although the level of software expertise among the DPRO quality assurance representatives (QARs) has traditionally not been extensive or widespread.

The basic documents upon which the QARs depend is the DLAM 8200.5[4] and the DLA handbook DLAH 8250.2.[5] The QAR also uses the provisions of other standards and specifications, such as MIL-Q-9858A, DoD-STD-2167A, and DoD-STD-2168 (which supersedes MIL-S-5279A), whether military or industry, for both hardware and software, when invoked on the contract. For software development efforts, the foreword of DoD-STD-2168 states that MIL-STD-2168 satisfies the provisions of MIL-Q-9858A for software.

The DPRO audits, which take place throughout the development life cycle of a system, follow a DPRO review of the quality requirements on the contract. The DPRO representatives and the contractor quality assurance personnel should recognize that both organizations have a common objective, that of assuring that a high-quality product is delivered to the government. The sometimes hostile relationships, which have historically developed between the contractors and DPRO, are ill-founded and unnecessary. Some QARs develop a knight-in-shining-armor attitude, assume that all contractors are out to take every possible advantage of the government, and it is their personal job to protect the government. Conversely, contractors sometimes feel that the QAR is incompetent and is trying to tell them how to do their jobs, and in the process are doing nothing but get in the way of progress on the contract. In some isolated individual cases both attitudes may have been justified, but these attitudes are not generally based on fact. Proper education of each to help them understand the other's tasks and responsibilities can alleviate this animosity. Also helpful is regular and routine two-way communication between the contractor and DPRO. The total quality management (TQM) philosophy has also helped considerably, and the basic themes of DLAM 8200.5 and DLAH 8250.2 are TQM-based.

Once the contract quality requirements are determined, the contractor's quality procedures manual is reviewed, and the quality assurance plans for the contract have been reviewed, DPRO develops an audit checklist which identifies each of the items (characteristics) to be audited.* Each audit requires that the resultant information be recorded on government forms or entered into a database established for that purpose. Any resulting action items must be reported and tracked.

The checklist is used to perform regular in-process audits, which can take place as often as every 30 days. QARs can also perform spontaneous audits when they feel the situation warrants such an audit. Spontaneous audits can also be conducted by personnel from the DPRO regional headquarters.

Each DPRO audit conducted, whichever type it might be, is called a program evaluation (PE). The fundamental premise of the PE is that for each contract quality requirement, there should be a contractor-published procedure to verify that the requirement is met, and a set of objective evidence procedures to validate that the evaluation has occurred. If not, the contractor's quality process is per se deficient. If the written process is not deficient, but is not properly implemented, then the contractor's implementation process is deficient.

The DPRO checklist has several different items based on the initial QAR quality review. Each of these checklist items are called characteristics. In an initial PE, all characteristics on the checklist are checked. After that, "the frequency of data collection and analysis is left to the judgment of the Government in-plant QA personnel based on specific circumstances in the facility."† PEs are required for newly developed and revised contractor procedures.‡ The frequency of PEs is also based on the degree and consistency of compliance with the requirements, the contractor's history of compliance or noncompliance, and the particular characteristic criticality. It is important to remember that the DPRO audits focus on quality, not cost. It is important to remember that the DPRO audits focus on quality, not cost. A quality requirement is a quality requirement, not a cost-effectiveness requirement, and therefore how much implementation might cost is not relevant to the DPRO audits. If there are cost concerns, they should have been part of the contract negotiation, or should be addressed by the program manager, contractor, and contracting officer. It is not a DPRO issue.

When the QAR reviews the contractor's written quality procedures, the result is never in the form of approval, but rather disapproval. It is a technical point, but the QAR has the right to *disapprove* the contractor's procedures if they do not conform to contract requirements§ but does not approve the procedures. In other words, the QAR may not like the procedures, but if it cannot be substantiated that the procedures are not in contract compliance, the QAR cannot classify them as disapproved.

---

*DLAM 8200.5,[4] sec. II, "IQUE Methodology," p. 2-1 et seq.
†DLAM 8200.5,[4] sec.2-102(c)4(c), p. 2-4. See also DLAM 8200.5, sec. 2-103(a)(5-10), pp. 2-4–2-5.
‡DLAM 8200.5,[4] sec. 2-102(b)(4), p. 2-2.
§DLAM 8200.5,[4] sec. 2-102(b), pp. 2-1–2-2.

QAR Nonconformance Reporting

QAR nonconformance reporting begins with a corrective action request (CAR). If not corrected, the issue is escalated to the contractor executive management using a series of processes called *methods.* Methods are categorized as methods C through E.* These methods differ in terms of problem severity, location, and the kinds of responses required from the contractor.

A CAR may be verbal or written. If the matter can be corrected on the spot, the CAR is verbal. If the matter is more serious, the CAR is written and a formal request for corrective action is issued. The written CAR is the initial type of permanently recorded written quality report which occurs during a PE. It requires specific contractor action for correction as well as a description of what action the contractor plans to take to ensure that the particular problem does not recur. It documents a quality requirement which is not being met and, in effect, is saying that the contractor is out of contract compliance. If the formal CAR is not properly responded to, or not responded to within the required period of time, it is escalated to executive-level contractor management. The implication is that the lower levels of management have refused, or have been unable, to meet the contract requirements. When this escalation takes place, it is elevated as a Method C.

Method C places a burden on the contractor top-level management to correct the discrepancy. It is a very serious situation which the contractor cannot well afford to ignore, since it says that there is both a quality deficiency and a management deficiency. If positive action is not taken, which means correcting the problem noted as well as the causative factors, the issue is escalated. The corrective action means also correcting the management practices and environment which resulted in the problem, which should have been solved at the first-line management level, at the executive management level. If the problem is not corrected at the executive management level, it is further escalated to a method D.

Method D has the result of figuratively shutting the contractor's door. The government stops accepting the product and the contractor receives no further payment for those products produced. This is a very serious situation for the program manager and the program. It is likewise serious for the subcontractors, and could possibly subject the prime contractor to legal action by the subcontractors since the products they produce must be paid for through the prime contractor and must be delivered to the government by the prime contractor. No government program manager should ever be so lax and unaware of what is going on as to allow this sort of condition to occur without continual and focused awareness and action. The DPRO QAR will not be initiating such action without making the program manager, as a well as DPRO headquarters, very aware of the actions and consequences.

Another way for the prime contractor to incur a method D is through the inaction of a subcontractor. If DPRO has a delegation to audit a subcontractor facility, or if there is an automatic delegation at a subcontractor site,† the DLA auditor may discover deficiencies which lead to the issuance of a method C to the subcontractor. When a method C is issued to a subcontractor, a *method E* is issued to the prime contractor. The method E states that the prime contractor has been deficient in managing the subcontractor and requires immediate and effective prime contractor executive management action. The method E is a lead-in to a method D being issued to the prime contractor because of the quality deficiencies existing at the subcontractor facility, coupled with the prime contractor's lack of awareness and management effectiveness in managing that subcontractor.

*DLAM 8200.5, sec. 2-104, p. 2-5 ff.

†DLAM 8200.5,[4] sec. 2-104(b)(2)(c), p. 2-7.

These DPRO audits are a means for the program manager to gain direct insight and information from the government perspective. It is important for the contractor to understand the role of DPRO and to be informed when DPRO has a letter of delegation for an acquisition. The program manager should strongly encourage regular and meaningful communication between the contractor and the QARs, and should require that any QARs assigned to the program have the level of competence necessary to perform the required tasks. This negotiation with DPRO can and should be completed before the letter of delegation is issued and the costs of DPRO oversight incurred. It makes little sense to incur the costs if the level of QAR competencies is less than the minimum necessary to effectively accomplish the desired task. Such a deficiency will simply exacerbate the program manager's task, not assist it.

From the contractor's perspective, the more knowledgeable the QAR, the better the level of communication will be and the more meaningful the relationship. The contractor, in the face of an inexperienced QAR, can try to take advantage of that lack of experience and keep the QAR in the dark, or can try to help educate the QAR to make the relationship between the two more effective. Assisting the QAR to become technically proficient is obviously not a requirement, but is an activity which many of the better contractors recognize and do to make their own jobs easier. An inexperienced or ignorant QAR can require more contractor time and attention responding to requests and quality deficiency reports, or QDRs, that are invalid than they will expend if the QAR is at a higher level of competence.

Conversely, a highly competent QAR can, while communicating with the contractor, and even while issuing QDRs, help guide and educate a contractor whose quality processes and methods are deficient. Such an action on the part of the QAR can help bring a contractor to a higher level of process competence and can thus aid the program manager more in this way than by simply doing audits and writing QDRs. To do this takes a combination of a highly skilled QAR and a contractor willing to listen and learn.

Together, DPRO and a contractor can aid a program manager in gaining the kind of visibility and control necessary to manage a program to a successful conclusion. However, no matter how good a DPRO organization might be, the quality requirement is on the contractor, not the government. It is the contractor who has the responsibility to develop acceptable quality processes, and to follow those processes and assure that the products being delivered to the government meet acceptable standards and operational requirements. The contractor must develop the quality processes, including audits and inspections, to assure that the issues and activities important to the acquisition, such as those described in Part 6 of DoD Instruction 5000.2,[2] are addressed and accomplished.

These vary from contract to contract. Section P of Part 6 of DoD Instruction 5000.2 covers the overall quality requirements for systems acquisition. It replaces, among other documents, DoD Directive 4155.1,[6] and covers both hardware and software. It requires that the quality effort on a contract be an integrated system engineering effort. The focus should be on preventing as well as identifying and reporting deficiencies.

Another Kind of Government and Contractor Software Audit: SPA/SCE

The Software Engineering Institute (SEI) at Carnegie Mellon University in Pittsburgh is a DoD federally funded research and development center (FFRDC), which is assigned software research tasks believed to benefit the software engineering process. One of its tasks was to find a way to evaluate the software process used by a contractor to help minimize risks in software acquisition. The result of this effort was realized by the creation of the SEI software process assessment (SPA) and the government counterpart, software capability evaluation (SCE). These two techniques are similar in many ways, but are quite different in other very fundamental ways.

The SPA is a questionnaire and interview driven assessment process, having the purpose of evaluating the process maturity of a contractor and finding a path of process improvement. The contractor is assessed by the SEI or is trained by the SEI to conduct self-assessments. The result of the assessment is a software process rating on a scale of 1 to 5.*

At level 1 the contractor is not process-oriented, has an ad hoc management practice, and the software engineering processes are not repeatable. Whether or not the program is a success is largely an accident, is highly dependent on the skill levels of specific people, and will be little influenced by what the government program manager does or does not do. Success, if there indeed is success, happens in spite of, not because of, the program manager.

At level 2 the contractor is able to learn from past experiences, has established a rudimentary software quality assurance activity, and has established configuration management.

At level 3 the contractor's software engineering process is defined, the contractor is using software metrics with the requisite level of understanding, and it has established a software process control group. The risk of which the government was not aware before the SPA was available is that enormous risk is incurred simply by having a level 1 contractor as either a prime or a major subcontractor. When, in 1987, the SEI performed its first general assessment of DoD contractors who volunteered to participate, approximately 86 percent of the DoD contractors evaluated were at level 1. This explained, at least partially, some of the procurement results that had been experienced on past contracts with a major software component. It pointed out, in very dramatic terms, that process maturity, and not cost, should be the major driver in software contract awards. The effect of process immaturity on actual software development life-cycle cost, and not on contract bid price, was the driving factor in whether or not a contract came in on schedule and at cost. Process maturity, combined with domain knowledge, are the two most important considerations when doing source selections. Accepting a level 1 contractor who submits the low, or near low, bid is a formula which carries with it almost insurmountable and unmanageable program risk.

The SCE is a process conducted by the government which is based on the same software engineering model as the SPA, but is conducted quite differently and for a different purpose. The principal objective of the SCE is its use in source selection. In a way, it picks up where the SPA leaves off. In the request for proposal (RFP) the contractors are all told that the SCE will be used in source selection. The government then uses a trained team of government personnel to conduct the SCE. The team asks each of the contractors to complete the SPA questionnaire for five to seven programs they have produced in the past. The questionnaire results are then evaluated by the SCE team and an on-site visit plan is put together. The SCE team visits the contractor site, interviews a number of preselected contractor personnel, and reads a large number of documents. This is a "proof-of-the-pudding" process designed to validate the questionnaire response of the contractor, and to probe various process areas to develop a set of contractor strengths and weaknesses. None of the results are communicated to the contractors, but the set of strengths and weaknesses for each bidding contractor are provided to the source-selection committee for their consideration in contractor selection.

*In the updated model going into effect in November 1992 it will be a determination of strengths and weaknesses, not a number rating.

After contract award, the SCE results for the selected contractor are communicated to that contractor. During the course of the contract the SCE can be used as a contract-monitoring process. When used during a contract, the objective is similar to the way an SPA is used, namely, to use the SCE as another kind of audit, and to use the results as a process improvement tool focusing on strengthening those areas deemed weak in the initial SCE. During source selection there was no objective of process improvement, just evaluation of strengths and weaknesses. During the contract, the objective is improvement and consequent risk reduction.

It is important to understand that the general trend for process cognizant developers is to do less physical inspection of product as a means of quality control, and more inspection and audit of the development processes and environments. It is being understood more clearly than ever that process improvement means resultant product quality improvement, and that the majority of the quality evaluation activity needs to be done early in a development program if it is to be effective and provide the right kind of information at the right time to make the technically correct and cost-effective decisions necessary. It is becoming increasingly evident that focusing our management attention on cost and schedule variances, instead of those elements which control or drive cost and schedule, is an activity which begs for the undesired conclusion. There are few genuinely pure cost or schedule overruns, just failures in management control of those elements, usually technical, which result in cost and schedule overruns. Proper use by the government of inspections and audits, including the use of SCE in source selection, can and should alleviate many of what are otherwise predictable program failures.

For hardware, the processes and techniques for quality assurance are generally well understood by program managers. Several government and industry standards, including MIL-Q-9858A and various IEEE standards, determine the requirements which the quality program must meet. But most program managers are not software conversant and do not consider themselves software development managers. This picture is changing as more and more large procurements become software-intensive. In today's environment nearly all new major weapon systems and information systems have a significant software component, and software is now providing the functionality for most or all of these systems. This became evident to the world at large during the Persian Gulf War. What the program manager can expect from the prime contractor in a software quality program is driven by the contractor's internal defect prevention and detection processes. To gain a better understanding of software audits and inspections, as conducted by the contractor and reported to the government directly or as part of objective evidence, the processes which the program manager can and should expect to see named, but not necessarily fully described, in proposal responses will be discussed. The level of detail included is not for the purpose of teaching how to perform inspections, but rather to get a firm understanding of what a good contractor means when saying that they do software formal inspections, and where to expect the cost loading and the subsequent derived cost benefits.

CONTRACTOR SOFTWARE INTERNAL AUDITS AND INSPECTIONS

Contractor Software Audits

Software audits are best accomplished only after careful planning. Prior to an audit, the internal company standards to which the contractor has agreed to adhere, and government standards, such as DoD-STD-2167A, DoD-STD-2168, MIL-STD-1521B, and others as appropriate, invoked on the contract, must be analyzed. These documents define the underlying requirements for the processes the contractor has agreed, or has been obligated, to follow. This set of process requirements form the basis for the routine audits conducted. From these documents, audit checklists for routine audits can be developed which examine the developer from any of several aspects, such as program management, software engineering, software quality assurance, and configuration management. These process areas are examined for internal operations as well as for relations with subcontractors. Although different provisions may apply to different contracts, there should be a generalized audit process tailored for each contract, not a new audit process creation for each contract.

If the audit checklist is constructed using a spreadsheet, where all the calculations can be established in advance by formulas, the scoring can be set up such that 1/0 entries in the yes/no column (the contractor area audited does or does not perform a given activity) will cause a weighted score to be computed for each item listed, and a total score computed by section and for the total audit. The entries are initialized to zero to ensure that the contractor does not get credit, by default, for a process not utilized. Such a software process audit checklist is also a software development risk analysis audit. The score is based on whether a given process or technique is present or used by the contractor. It reflects both the presence of that process or technique and the risk to the government if that process or technique is not present at the contractor being audited. The score computation algorithm, based on the weighting factor, is a risk measurement. The lower the score, the higher the risk. The raw score as well as a percentile score can be provided in the spreadsheet for each part and for the audit in its entirety.

An example of such an audit spreadsheet, which can be used by the government to perform software audits as well as by a contractor to perform internal or subcontractor audits, is included in the Appendix at the end of this chapter. In computing the score in the example spreadsheet, the result is based on the two columns to the left of SCORE. The SCORE value is computed using an IF-THEN algorithm. If the value in the first column is a 1, the score is usually 1 times the weight given in the second column. If a process is felt to be particularly good or value added, the score is sometimes computed as the weight multiplied by a number larger than 1. If the value in the first column is a 0, then the score is usually a 0. However, for cases where the absence of that item is considered particularly detrimental to the program, the score may be the weight times a negative number having absolute value equal to or greater than 1.

After the routine audit is conducted, the audit results should be communicated to the group or groups audited. Any action items which result from the audit must be documented, filed, and submitted to the audited department together with a time frame for resolution of the issues indicated.

When doing a spontaneous audit, the audit process and objective are different from routine audits. Spontaneous audits result from an identified or perceived problem which warrants a more focused investigation. They are more specific in level of detail and less general in terms of scope. They may be, but do not have to be, announced in advance, nor is it generally true that the information to be examined is identified in advance. Spontaneous audits require careful planning and, where checklists are used, specialized checklists. These audits have three fundamental objectives:

1. To determine whether a perceived problem exists and, if so, to assess the potential risk
2. To determine whether the problem—assuming it does exist—is correctable
3. To establish a measurable corrective action plan

For any audit conducted by the prime contractor, whether internal or an audit of a subcontractor, the audit report is filed and is available for review by the government. Reports of contractor-conducted audits become part of the objective evaluation evidence maintained by a contractor as required by DoD-STD-2167A and DoD-STD-2168, and are among the items evaluated during audits conducted by DLA.

The government should use the results of the audits to work with the contractor in improving the software development process. Using an audit as a club is archaic and an activity of last resort and little effectiveness. Likewise, contractors have a professional obligation to do that which is necessary to improve their process capability, and to use the audits as a gauge for what needs improvement and how well improvement activities are working. Audits, properly utilized, can be an excellent means for achieving the kind of communication and cooperation needed for the government and contractors to reach their common goals.

Software audits are generally conducted by the department which is generically referred to as software quality assurance or software quality engineering. This department will be a sister department to hardware quality assurance or a component of systems quality engineering. The audits are described in the contractor's software quality procedures manual and cover both internal and subcontractor audits.

Spontaneous audits, as previously noted, are usually conducted in response to a perceived concern. The result of either kind of audit is recorded as objective evidence and is subject to government review at any time, including during any DLA audit. These routine and spontaneous audits are conducted to bring to technical and program management's attention whether or not the internal development is of a level of quality required for contract compliance. They also serve to identify those areas where improvements can and should be made to the software engineering process. As such, they serve as one of the integral components of a total quality improvement process.

Subcontractor Audits

Subcontractor audits are some of the most important audits which a contractor can conduct. The ability to manage a subcontractor, and to obtain from a subcontractor the level of quality products necessary for program success, is today an all-important objective. On large programs today, a prime contractor can have as many as 20 or more different subcontractors, each having to be managed as if the entire program were depending on that one subcontractor, and as often as not, it is. Therefore subcontractor audits, just like internal audits, must be accomplished in accordance with a previously developed plan and a previously defined process, both of which are tailored for the contract and for each subcontractor. The audit checklist included at the end of this chapter can be tailored for subcontractor audits as well as for internal audits.

Among the items audited by a contractor are the types and results of the human testing conducted by the contractor and subcontractors. Human testing is that software testing which occurs prior to the start of computer-based testing. There are four types of contractor human testing: desk checking, peer review, walkthroughs, and formal software inspections. The effectiveness varies widely among these methodologies, with formal inspections being by far the most effective. Formal inspections can be a major element in software process quality improvement as well as a major source of visibility to both the contractor and the government program manager. For this reason, and because the program manager should understand what formal inspections are and the benefits to be expected from them, as well as consider requiring them of a contractor on contracts which are software-intensive, they are further described herein.

Contractor Software Inspections

The formal design and code inspection process originated with the publication of a paper by Michael Fagan in the *IBM Systems Journal.* Inspections are a marked departure from other similarly intentioned processes, such as reviews and walkthroughs. The inspection process is characterized by formality, rigor, scheduling, phasing, and associated documentation. The software inspection process particularly addresses the need for a manageable software process, and one for which the input, output, and internal processes are well described. It also satisfies the need for a process, or a series of process steps, for which there is an entry and exit criterion at each phase. This removes the ambiguity and allows for the establishment of definitive process control checkpoints.

Inspections are initiated upon the completion of software design, either high- or low-level, or upon the completion of the first clean compilation of code. Developers should not spend any time doing desk checking of the product if these conditions have been met. There is always the feeling that by doing desk checking, time will be saved, or exposure of one's mistakes will be minimized. In most instances, no time is really saved since the inspections must be held and the inspection team will still have to review the material. In addition, the only eyes, experience, and talent being applied to the product is that of the author. Often this is insufficient. The whole idea is to make good use of the variety of backgrounds, experiences, and talents of the entire team to get a full, rigorous, and thorough examination of the product. Doing it twice or more does not really accomplish much other than to alleviate some imaginary wounds to the pride of the author.

It is absolutely imperative that there be management support, both real and perceived, if the inspection process is to work. If the management team is not absolutely supportive, the lack of a supportive attitude will become known, will be communicated to the programmers, and will have a definite negative impact on the entire process. This means, among other things, that adequate time for inspections is included in schedules and that work at one level (such as code) does not begin until the successful completion, including rework, of the inspection for the most proximate previous level (such as design).

The software inspection process is described here to acquaint the DoD program manager with the technical advantages and cost benefits of this process, why this process should be considered as a mandatory requirement on software development contracts, to provide an understanding of the costs and schedule implications which will be reflected in the proposal, and to provide a sufficient knowledge base for the technical proposal evaluation team to recognize whether or not the contractor himself understands what inspections are.

The following definitions are applicable to software inspections.

Inspection A formal, rigorous examination of a software product by a small group of trained peers of the author of that product.

Peer A person who is a professional coworker of the author of a software product (design or code), who works on the same task or contract as the author, is trained in the inspection process, and is either in the software development, systems engineering, or software test organization.

Moderator The moderator is part of the inspection team and is responsible for the successful completion of all six phases of the inspection process. The moderator serves as the manager of that particular inspection. The moderator should be a more senior member of the inspection team.

Preparation time The amount of time each inspector spent preparing for the inspection meeting. This does not include the time spent by the author to develop the design or code.

Inspection time The amount of time spent by each team member at the inspection meeting.

Inspection meeting The time during which the inspection team members, after sufficient preparation, meet to discuss the defects in the software product.

Reinspection The reexamination of the software product, using the same process as the original inspection. The moderator makes the decision as to whether to reinspect or not. Reinspection is generally required if more than 5 percent of the software product must be changed to correct the deficiencies identified, if the defects require complex corrections, or if the requirements have changed significantly.

Purposes of Software Inspections. There are both primary and secondary purposes for conducting inspections. The primary versus secondary categorization has more to do with the process of conducting the inspections than with the net impact of having done inspections.

Primary Purpose. The inspection, whether design or code, has only one primary purpose, that is, to identify and remove defects as early as possible in the development process. The purposes of the inspection preparation and meeting are to:

1. Identify potential defects during preparation and validate them at the meeting
2. Validate the fact that identified items are actual defects
3. Record the existence of the defect
4. Provide a record to the developer to use in making fixes
5. Discover additional defects as a result of the group interaction during the inspection meeting

It is not the intent of the inspection process to find solutions to identified defects, although not providing solutions is extremely difficult for some people. The fact that a defect exists will be sufficient cause in the minds of most people to search for a solution. However, the time of the inspection process preparation and meeting is to be used to identify and record the existence of defects. It is the sole responsibility of the author to define the solution. It is also in the province of the author to request, outside the inspection process, assistance in finding a solution. Therefore one of the most difficult tasks many moderators face is keeping the discussion centered on finding defects and away from discussing solutions. In some cases the mere identification of the defect may be sufficient to also state the solution, but in most cases this is not true, and the inspection process will be unreasonably lengthened if the solutions are pursued when the defects are identified. It is also a usurpation of the right of the author to fix his or her defects.

Secondary Purposes. There are secondary purposes which result from the inspection process:

1. Provide traceability of requirements to design
2. Provide a technically correct base for the next phase of development
3. Increase programming quality
4. Increase product quality at delivery

5. Lower life-cycle cost
6. Increase effectiveness of test activity
7. Provide a first indication of program maintainability
8. Encouragement of entry and exit criteria software management
9. Shorten junior programmer learning curves

These secondary purposes are all part of the net effect of performing inspections properly and professionally. The fact that they are secondary purposes does not in any way diminish their importance to the overall software development effort.

If the inspections are performed properly and according to defined procedures, and limited to that, then these secondary benefits will be naturally contributed to by having done this. Some of these secondary purposes will be achieved directly as a result of the inspections, and others will be achieved by the inspections working in concert with other activities. Inspections are not the sole cause, but they are a significant, and in some cases primary, cause of the secondary purposes being achieved.

The moderator of an inspection is responsible for the entire inspection process for that software product. There are six distinct inspection phases, namely, planning, overview, preparation, inspection meeting, rework, and follow-up.

During the inspection meeting it is expected, and it is the responsibility of the moderator to assure, that the team members conduct themselves in a professional manner. This means, as a minimum, that the participants have only one discussion at a time, that they address problems in as objective, professional, and impersonal a manner as possible, and that they try to convey the feeling that the person whose product is being inspected is being helped, not criticized. The activity during the meeting is limited to finding defects, not solutions. It is the responsibility of the author to find the solutions.

Inspection Types. There are three software product inspection types typically performed. These are the high-level design inspection (I_0), the low-level design inspection (I_1), and the code inspection (I_2).

High-Level Design Inspection (I_0). The high-level design phase is that phase during which the overall design for a module or function is produced. This stage commences with the issuance of a preliminary software requirements specification (SRS) and the initiation of the interface design specification (IDS), and it ends with the successful completion of the I_0 inspection. In this stage the high-level architecture of the software is determined and recorded in the SRS material. This SRS information is examined during the I_0 inspection. For each function, the SRS will provide:

1. The source of the design (new, other contract, and so on)
2. A graphic presentation of the function allocation to hardware resources
3. A graphic presentation of function flow
4. A description of scheduling, timing, and synchronization
5. A definition of interfaces
6. The process of decomposition
7. The design definition
   a. Retained modules (reference to an existing SRS, if applicable)
   b. Modified modules (reference to an existing SRS, where applicable, and a narrative description of the changes)
   c. New modules (high-level description of interfaces and processing)

During the I_0, the SRS information is further expanded to include a description of each new task or module, including interfacing and processing. An I_0 is held for each function. For the retained and modified modules, an inspection plan is written by the programmer which defines the required further level of inspection. These requirements are based on the anticipated extent of modification. Resource utilization estimates are generated during this design phase and are collected and maintained by system engineering, with software development periodically providing input. The system engineering organization reviews and approves the program design specification (PDS) and the I_0 material to assure compliance with the baseline documentation. All six phases of the inspection process (with the possible exception of the overview phase) are conducted for each I_0.

Purpose of the High-Level Design Inspection (I_0). The purpose of the I_0 is to conduct a formal examination of the software product in order to verify that the functional design at the task level is a correct expansion of the SRS at the mode level. A mode level function may be narrow-band processing, classification, tracking, FICA tax computation, and so on. The verification is performed by identifying the allocation of SRS requirements to processes and tasks. A single I_0 is typically performed for each mode.

Low-Level Design Inspection (I_1). The low-level design, more commonly called detailed design or module design, reflects back on the overall design objectives. Key objectives considered in designing the software are:

1. Accuracy and performance requirements are met.
2. The software is reliable and fault-tolerant so that the system will continue to perform in the event of hardware intermittent failures or other unexpected occurrences.
3. The software is flexible and can accommodate change and growth.
4. The software can be tested easily.
5. The software can be maintained easily.

Detail module design is developed, and I_1 is held for all modules that meet any of the following criteria:

1. New module development.
2. Any change to the external interface or function of an existing module.
3. A structural change in an existing module.
4. A 40 percent or greater change in the source lines of code (SLOC) in an existing module. This percentage presumes that the size requirements as specified in the military standards are adhered to. In the case of unrestricted module size, where the module is large, this percentage may not be a proper criterion.

Purpose of the Low-Level Design Inspection (I_1). The objective of the I_1 is to refine the I_0 design stepwise to an intermediate level before translation to the target language code is authorized. All interfaces between processes, tasks, and procedures are defined to the field or bit level. The level of decomposition must be sufficient to show the highest level of control structure for each procedure and to show the operations performed on the inputs and outputs. This does not necessarily define all the control structures and all the internal data structures completely. All six phases of the inspection process (with the possible exception of the overview phase) are conducted for each I_1. Only after the successful completion of the I_1 and any required reinspections can the module coding process begin.

Managers have the responsibility to assure that no code begins on a module until successful completion of the design inspection for that module. This is a technical and product quality decision and should not be controlled by schedule.

The portion of the program design specification (PDS) for that software which has completed the I_1 phase should be complete by the end of that inspection.

Code Inspection (I_2). An I_2 is held for all new code or code from another task which is being modified to meet the requirements of the new task or contract. The code inspection is not performed until after there has been an error-free compile. All six phases of the inspection process (with the possible exception of the overview phase) are conducted for each I_2.

Code Inspection Activities (I_2). The code inspection will satisfy the following evaluation objectives before proceeding with program test functions:

1. Verification that the code conforms to the SRS, PDS, and IDS requirements for operational software
2. Confirmation that the design has been correctly converted to the target language
3. Early audit of code quality by the programmer's peers
4. Early detection of errors
5. Verification that the code meets level-to-level module interface requirements
6. Review of module test specifications, which are provided with the inspection materials package
7. Verification that the module test specifications (module test plan) are necessary and sufficient to test the requirements specified for that module and reviewed during the design inspections (I_0 and I_1)
8. Verification that the proper test tools and test environment have been identified and are available
9. Verification that the test dependencies are correct and the module is testable based on the dependencies
10. Decision about allowing the module test (unit test) to begin
11. Verification that the software products conform to the contract or internal standards and conventions

Design and code inspections for each software module are typical of the type of pass-fail events which serve as key milestones in the software development schedule. The result of the successful completion of the inspections should be a complete code which conforms to the high-level design, the low-level design, and the SRS. Only after successful completion of the code inspection can the module test begin.

Data Collection. The inspection process, if it is to be most beneficial to the government and the contractor, must be applied consistently across all programs in the facility. To achieve this consistency, there should be only one set of data collection forms. These forms should be designed for the inspectors to use during the preparation period and for the moderator to use when producing the summary of the inspection. The defects should be broken out by defect type, and the data reports produced focused on reporting defects by type, not by individual. By regular and routine analysis of defect-type data, plans can be implemented for improvement of the software engineering process by interjecting processes designed to prevent the types of defects most frequently discovered through the inspection process.

The summary data reports are provided to the software quality engineering (SQE) department or whichever department will maintain the inspection results database. The information is entered into the SQE database and is available for government review. From this database the SQE department publishes a set of statistical charts showing graphically, and in some cases digitally, the cumulative results and trends of all the inspections for a given project and, if requested, the results for a given department.

Software Requirements Inspections. The design and code inspection process is now being applied to the requirements generation process. The same basic philosophy, procedures, and objectives of the design and code inspections are present in the requirements inspection process. The most fundamental differences are the type of product inspected and the modification of the forms to meet the needs of the requirements documents, which fundamentally means looking at a different set of defect types.

Subcontractor Inspection Data Reporting. Subcontractors who initiate the inspection process and data reporting as a result of a statement of work (SOW) from a prime contractor are expected to follow the same procedures and initiate the same data collection and reporting described in this document. In some cases subcontractors have viewed the data collection and reporting as a way to keep score and, in a sense, to evaluate them as subcontractors. This view is incorrect. The inspection process is included in SOWs for the same reason it is used for prime contractors. The results have proven the process to be cost-effective and have made the software development itself more manageable and therefore less risky. The increased management visibility during development, and improved quality of the delivered product, which have resulted from this process, dictate that inspections be used on all product developed for or delivered from the prime contractor to customers.

Since the prime contractor is ultimately responsible for the final delivered product to the customer, the same quality requirements which the prime contractor expects of its own product are expected from the subcontractors. The data collection requirement is the same because of the advantages of a consistent set of data collected for all software developed on a given contract. Because of the advantages which have been seen from the use of inspections, the decision should be made to share this technology with all software subcontractors and assist them in incorporating this process. It is felt that this will be beneficial to the subcontractors and it will also achieve the prime contractor's goal of having all software delivered by the prime contractor to the customer go through the inspection process.

Advantages of the Software Inspection Process. The advantages of the inspection process, in addition to what has been already stated, are that it provides a significant degree of process control and consistency and will result in a minimum of 70 percent of the life-cycle defects in the software being removed prior to the first unit test. This means that the defect density at the beginning of integration test will be between 0.5 and 2.0 defects per 1000 source lines of code (KSLOC), instead of the normal 15 to 28 defects per KSLOC. This translates to orders-of-magnitude costs avoided in defect detection and removal compared to programs where inspections have not been done. This can be as much as $25 million in net costs avoided for a 1 million line program. The marked decrease in the number of defects which must be found and removed during the traditional test periods means that the program manager has considerably more flexibility in how the normal integration test phases are utilized. In fact, the test phase is transformed from one whose primary purpose is defect detection and removal, to one whose primary purpose is validation of what has been done prior to the test phase. The schedule and cost control achieved is remarkable.

If the inspection process is augmented by the use of CASE technology such as that incorporated into the Battlemap* family of tools, the fundamental advantages of inspections can be multiplied, with the program manager deriving extremely attractive visibility and control benefits throughout the development phases of the program.

Caution. The implementation of the software inspection process requires a high degree of process control and consistency as well as proper training. It requires management commitment and a willingness to make inspections a milestone in the software development schedules. It also means that the development contractor must understand process control and have the maturity to recognize the life-cycle technical advantages and cost benefits of inspections. Without this level of maturity, the contractor will be unable to achieve the benefits of inspections. This lack of achievement may likely be the case with contractors who are at a maturity level of 1 on the SEI assessment scale, leading the contractor to believe that the inspection process is ineffective instead of realizing that it is the contractor's own level of process immaturity that is the problem. The light at the end of the tunnel, however, is that if properly implemented, the inspection process can be a major first step in helping a level 1 contractor move up the process maturity chain and consequently become more competitive and of more value to the government.

REFERENCES

1. MIL-STD-1521B, *Technical Reviews and Audits for Systems, Equipments, and Computer Software,* AMSC F3631, June 4, 1985.
2. DoD Instruction 5000.2, *Defense Acquisition Management Policies and Procedures,* Feb. 23, 1991, Part 9, sec. 3(h), p. 9-A-3.
3. Draft MIL-HDBK-61, *Configuration Management,* sec. 7, pp. 106–114. (This is the companion handbook to draft MIL-STD-973, *Configuration Management.*)
4. DLAM 8200.5, *In-Plant Quality Evaluation* (*IQUE*), Oct. 1990.
5. DLAH 8250.2, *Software Quality Assurance* (*SQA*), Dec. 1990.
6. DoD Directive 4155.1, *DoD Quality Program,* Aug. 10, 1978.

BIBLIOGRAPHY

Dobbins, J. H., *Software Quality Assurance and Evaluation,* ASQC Quality Press, Milwaukee, Wis., 1990, chap. 4.

Fagan, M. E., "Design and Code Inspection to Reduce Errors in Program Development," *IBM Syst. J.,* vol. 15 no. 3, 1976.

Kohli, O. R., "High Level Design Inspection Specification," IBM Tech. Rep. TR21.601, July 21, 1976.

Kohli, O. R., J. Ascoly, M. J. Cafferty, and S. J. Gruen, "Code Inspection Specification," IBM Tech. Rep. TR21.630, May 3, 1976.

Kohli, O. R., and R. A. Radice, "Low Level Design Inspection Specification," IBM Tech. Rep. TR21.629, Apr. 15, 1976.

McCloskey, H. A., "Programming Publications Inspection Procedures" (IBM internal use only), IBM Tech. Rep. TR21.612, Dec. 8, 1975.

*Battlemap is a product of McCabe & Associates, Columbia, Md.

Larson, R. R., "Test Plan and Test Case Inspection Specification," IBM Tech. Rep. TR21.586, Apr. 4, 1975.

Schulmeyer, G., and J. McManus (eds.), *Handbook of Software Quality Assurance,* Van Nostrand-Reinhold, New York, 1987, chap. 9.

APPENDIX: SOFTWARE AUDIT SPREADSHEET EXAMPLE

SW AUDIT CHECKLIST DATE: PROJECT: COMPANY:

Note 1: Plans, processes, or standards must be written in order to be considered existent or applicable.

Note 2: PRIME is prime contractor; SUB is subcontractor.

Category	Y(1) N(0)	Weight	Score
1. General for prime contractor			
A. SW risk management process in place	0	3	0
B. SQA function exists	0	5	–5
C. SQA is organizationally independent	0	5	0
D. SW CM function exists	0	5	–5
E. SW CM is organizationally independent	0	5	0
F. SW test department exists	0	5	0
G. SW test is organizationally independent	0	5	0
H. Procurement is SW intensive	0	5	3
I. Experience with DOD-STD-2167A	0	3	0
J. Experience with DOD-STD-2168	0	3	0
K. Experience with MIL-STD-1521	0	4	0
L. Has Ada experience (>100K)	0	3	–3
M. May subcontract SW	0	4	0
N. Has SW subcontractor management plan	0	5	–5
O. Evaluates using metrics	0	4	–4
P. SW management uses metrics	0	4	–4
Q. COTS will be part of system	0	2	0
R. GFE SW will be part of system	0	3	0
S. Has good domain knowledge	0	5	–10
T. Has SW system engineering function	0	4	–4
U. SW system engineering on SW design team	0	4	–4
V. Has SW CCB for each project	0	4	–4
W. Has SW project management training program	0	4	–4
X. Has SW inspection process training	0	4	–4
Subtotal			–53
Maximum score			104
Percentile score			–50.96%
2. Software development engineering: prime contractor has or does			
A. SW process maturity model	0	2	0
B. SW process improvement plan (for inspections or walkthroughs)	0	3	0
C. Performs formal inspections:			
Of requirements	0	5	0
Of design	0	5	0

Category	Y(1) N(0)	Weight	Score
Of code	0	5	0
Of test documents	0	5	0
D. Performs walkthroughs:			
Of requirements	0	3	0
Of design	0	3	0
Of code	0	3	0
Of test documents	0	3	0
E. Uses case tools:			
For requirements generation	0	5	0
For requirements analysis	0	5	0
For design generation	0	5	0
For design analysis	0	4	0
For code generation	0	3	0
For code analysis	0	3	0
F. Procedures limiting CSU size	0	3	0
G. Does SW development using:			
Structured programming	0	4	0
Top-down design	0	4	0
Object-oriented design	0	4	0
H. Designs software for:			
Reuse	0	3	0
Maintainability	0	5	–5
I. Uses internal standards for SW development	0	5	–5
J. Uses measures as tool to manage:			
Requirements generation	0	4	0
Design	0	4	0
Code	0	4	0
Test	0	4	0
Percent of erroneous fixes	0	2	0
K. Keeps SW documentation current	0	4	–4
L. Has quality control process	0	5	–5
M. Has SW virus control process	0	4	0
N. Has SW security management process	0	5	0
O. Has internal SW IV&V	0	3	0
P. Performs rigorous testing:			
Modules (unit test)	0	5	–5
Integration test	0	5	–5
SW system test	0	5	–5
Using read-only library	0	5	–10
Q. Has uniform SW error process:			
Error reporting process	0	5	–5
Data collection process	0	4	0
Error analysis process	0	4	0
R. Correlates error analysis to:			
Reliability requirements	0	3	–3
Quality requirements/goals	0	4	–4
SW safety	0	3	0
SW cost analysis	0	3	0
Schedule plan	0	3	0
Program risk elements	0	5	–5
SW improvement process	0	5	0
S. Has SW safety analysis process	0	3	0
T. Has SW TQM process	0	5	0
U. Does SW prototyping	0	3	0

Category	Y(1) N(0)	Weight	Score
V. Correlates data between:			
Errors across development phases	0	3	0
Costs to error profiles	0	3	0
Schedule to error profiles	0	3	0
Subtotal			–61
Maximum score			253
Percentile score			–24.11%
3. SW quality assurance			
A. Performs routine internal audits:			
Using internal procedures	0	4	–4
Per with government standards	0	5	0
Per internal standards	0	4	0
Issues audit reports	0	5	0
Tracks audit action items	0	4	–4
B. Contributes to subcontractor SOW	0	3	0
Performs:			
C. Spontaneous internal audits	0	3	0
D. Routine audits of sub	0	3	0
E. Spontaneous audits of sub	3	3	0
F. Regular evaluations of:			
SW development engineering	0	5	0
Programmer unit development folders	0	4	0
SW CM	0	4	0
SW test	0	4	0
Corrective action process	0	5	0
SW development libraries	0	5	0
G. Activity begins at project initiation	0	4	0
H. Has written procedures	0	5	–10
I. Follows procedures	0	5	–2.5
J. Sits on SCCB	0	4	0
K. Evaluates inspection results	0	5	0
L. Produces inspection data reports	0	5	0
M. Evaluates SW test results	0	4	0
N. Develops SW trend analyses	0	4	0
O. Maintains objective evidence of:			
Human testing evaluations	0	5	0
Computer testing analysis	0	4	0
All SQA evaluations	0	5	0
P. Reviews all SW CDRL deliveries	0	5	–5
Q. Approves all SW CDRLs	0	3	0
R. Maintains SW master library	0	3	0
S. Witnesses acceptance test	0	3	0
T. Does metric analysis of:			
SW development process	0	4	0
SW test process	0	3	0
Subtotal			–25.5
Maximum score			154
Percentile score			–16.56%
4. SW CM			
A. SW CM has:			
SW document configuration ID	0	5	0
SW product configuration ID	0	5	–5

Category	Y(1) N(0)	Weight	Score
SW document configuration control	0	5	0
SW product configuration control	0	5	–10
B. SW CM maintains:			
SW configuration status accounting	0	5	0
ID of SW deliveries	0	5	–5
Minutes of SCCB meetings	0	3	0
C. Chairs the SCCB	0	3	0
D. SW CM process:			
Begins with SW development	0	4	–4
Subtotal			–24
Maximum score			50
Percentile score			–48.00%
5. Project management			
A. Project management has:			
SEI SPA rating	0	5	0
SEI SPA rating ≥ 3	0	5	-10
Subcontractor management process	0	5	0
Open communication with customers	0	5	0
Open communication with subs	0	5	0
Traceability of requirements to design	0	5	0
B. Project management uses:			
Standard SW development process	0	5	0
Formal management review process	0	5	0
SW subcontractor management process	0	5	0
SW development folder standards	0	5	0
Formal SW size estimating method	0	5	0
Formal SW scheduled estimating method	0	5	0
Formal SW cost estimating method	0	5	0
Formal process for SW requirements	0	5	0
Formal requirements validation process	0	5	0
Man-machine interface standards	0	5	0
Statistics to help analyze:			
SW requirements errors	0	5	0
SW design errors	0	5	0
SW code errors	0	5	0
SW test errors	0	5	0
SW risk identification process	0	5	0
Metric-based information system	0	5	0
Early SW error removal process	0	5	0
Error data for process changes	0	5	0
Root cause error analysis	0	5	0
Productivity analysis process	0	5	0
Productivity measurement process	0	5	0
Error prevention process	0	5	0
Regular routine SW reviews	0	5	0
C. Coding standards applied to each project	0	5	0
D. SW size measured against target system	0	5	0
E. Action items tracked to closure	0	5	0
F. Critical success factors:			
Identified	0	5	0
Measured	0	5	0
G. Risks tracked and measured	0	5	0

Category	Y(1) N(0)	Weight	Score
I. SW first line managers track cost/schedule	0	5	0
J. Standard process used to control changes:			
To requirements	0	5	0
To design	0	5	0
To code	0	5	0
To approved test procedures	0	5	0
To integration libraries	0	5	0
K. Formal process used to determine:			
If prototyping appropriate	0	5	0
Applicability of SW reuse	0	5	0
Adequacy of regression test	0	5	0
Test procedures completeness	0	5	0
Completeness of requirements	0	5	0
Consistency of requirements	0	5	0
If requirements meet user needs	0	5	0
L. Regularly evaluates adequacy of SW tools	0	5	0
M. SW process improvement is standard practice	0	5	0
Subtotal			–10
Maximum score			260
Percentile score			–3.85%

Total audit score	–173.5
Total maximum score	821
Total percentile score	–21.13%

CHAPTER 37
CURRENT PERSPECTIVES ON PROJECT DOCUMENTATION

Bud Baker

Bud Baker is currently Assistant Professor of Management at Wright State University, Dayton, Ohio, and also coordinates Wright State's M.B.A. concentration in project management. He has extensive experience in government service, most recently serving as a program manager for a major Air Force acquisition effort. In addition to Wright State, Dr. Baker has served on the faculty of the U.S. Air Force Academy, Regis University, and the University of Dayton. He holds an M.B.A. degree from the University of North Dakota and M.A. and Ph.D. degrees from the Peter F. Drucker Center of the Claremont Graduate School.

The subject of project documentation in defense acquisition is a vast and amorphous one. Every engineering technical report, every cost analysis, every "what if?" generated in response to a budget exercise properly falls within the realm of project documentation. So, too, do cost performance reports, project baselines, system operational concepts, even annual and end-of-project histories. Add to all this the ever-growing demand of a burgeoning defense bureaucracy—Acquisition Executive Monthly Reports, Defense Acquisition Executive Summaries, System Maturity Matrices, and the like—and one can begin to appreciate the scope of the topic of documentation in the defense acquisition business. At times it appears that the sheer volume of project documentation threatens to bring the defense acquisition business to a grinding halt.

If there is any good news amid this jungle of documentation and reporting, it is that the requirements for these myriad documents are spelled out—often in excruciating detail—in various publications. So it will not be necessary here to fully duplicate those manuals and regulations. Instead, after reviewing the major types of reports and documentation, this chapter will look at a number of current or emerging issues in the project documentation field. Through this approach we hope to highlight a number of areas that merit attention by project managers, acquisition executives, and oversight bodies.

For the general purposes of this discussion, program documentation can be divided into two parts: (1) documentation in support of a milestone review and (2) documentation that is part of a separate system of periodic review.

MILESTONE DOCUMENTATION

This area can, in turn, be subdivided into three major segments: requirements documents, the Integrated Program Summary with Annexes, and stand-alone support documents.

Requirements Documents

The three major requirements documents called for in DOD 5000.2-M, the manual which governs DoD reporting and documentation, are:

1. *Mission Need Statement.* A non-system-specific statement of required operational capability. The first document prepared in the normal programmatic sequence of events, the Mission Need Statement (MNS) should not exceed five pages. The basic parts of the MNS are Mission and Threat Analyses, Exploration of Alternatives, and Constraints. An approved MNS is the primary document for initiating a Milestone 0 (Concept Studies Approval) review.

2. *Operational Requirements Document.* Contains key performance parameters for a proposed system. A required document for a Milestone I (Approval for Concept Demonstration) review, the Operational Requirements Document must contain system-level parameters (such as, range, survivability, availability) with specific objectives for each. The document is updated and expanded for Milestone II (Approval for Development) and then *should* be modified only in exceptional situations. (In reality, though, this type of stable requirement is rarely found: since the user is responsible for developing the document, there is often a strong tendency to add new "requirements" throughout the life of the program.)

3. *System Threat Assessment Report.* Prepared by the Service's intelligence agency, the System Threat Assessment Report (STAR) is the primary threat document used at milestone review. It is prepared prior to Milestone I for all major programs and updated for every subsequent milestone for those programs (Category ID) requiring Under Secretary of Defense for Acquisition approval. Major components of the STAR include system description, threat assessments (out to Initial Operational Capability plus 10 years), and potential reactive (that is, adaptive) threats that an enemy might develop.

Integrated Program Summary

After requirements documents, the second major category of milestone documentation involves the Integrated Program Summary (IPS) with its many annexes. The IPS provides a comprehensive assessment of program structure, status, plans, and recommendations to support decisions by Service and DoD executives. The IPS summarizes status and plans, assesses risk management efforts, and establishes the thresholds for cost, schedule, and performance which will become the "exit criteria" for moving to the next acquisition phase. Initially prepared prior to Milestone I, the IPS is updated at every subsequent milestone.

The IPS is composed of two major sections. The first, the Executive Summary, has nine segments:

1. Execution Status
2. Threat Highlights/Shortfalls of Existing Systems

3. Alternative Assessment
4. Most Promising Alternative, with Rationale
5. Acquisition Strategy
6. Cost Drivers and Major Tradeoffs
7. Risk Management
8. Affordability
9. Recommendations

Then seven annexes are appended to the IPS. They include:

1. Program Structure (typically a Gantt-type chart)
2. Program Life-Cycle Cost Estimate Summary
3. Acquisition Strategy Report (acquisition approach, streamlining efforts, competition, contract types, and major tradeoffs between cost, schedule, and performance)
4. Risk Assessment (looks at threat, technology, logistics, cost, and schedule)
5. Environmental Analysis (alternatives, environmental effects, rationale for alternative chosen, mitigation of environmental impacts)
6. Affordability (cost projections—constant and then-year dollars—for each year of procurement, contrasted with the costs of any current systems; also compares life-cycle costs for a 15-year period, beginning at IOC)
7. Cooperative Opportunities Document (assesses advantages of cooperative development program with Allied nations)

"Stand-Alone" Reports

The third category of milestone documentation, after requirements and the IPS, is something of a catch-all, a collection of independent reports. There are eight individual reports, most of them imposed by Congressional direction.

1. *Manpower Estimate Report.* Documents the total number of people (military, civilian, and contractor) needed to operate and support a system upon deployment. Specific effects of the system on fiscal-year "end-strength" numbers must be addressed, that is, will the system mean a net increase or a net decrease in total personnel?

2. *Test and Evaluation Master Plan.* Of all the areas of Defense documentation, Test and Evaluation (T&E) is the one most bogged down in bureaucracy. Even the foreword of DOD 5000.2-M is indicative of the problem: it has not one, but *two* authorizing signatures. One belongs to the Acting Under Secretary of Defense for Acquisition, with the caveat: "For all matters in this Manual except operational test and evaluation." But there is a second signature, that of the Director of OT&E, and that has a complementary caveat: "For all matters in this Manual relating to operational test and evaluation."

Once inside the rules and regulations concerning the Test and Evaluation Master Plan (TEMP), the bureaucratic obstacles intensify. Submitting just one copy of the TEMP to support a Milestone I decision is not sufficient:

> Fifteen copies of a preliminary Test and Evaluation Master Plan are to be submitted to the Deputy Director of Defense Research and Engineering (Test and Evaluation) 45

days (draft) and 10 days (final), prior to the Defense Acquisition Board Milestone I Committee Review of the program.

Fifteen copies of a "*draft* preliminary report"? Followed by 15 *more* copies of a "*final* preliminary report," 35 days later? Clearly, the concept of the "paperless office" has yet to reach the T&E community.

The actual format of the TEMP contains much of the same information already included elsewhere: yet another mission description, another summary of the operational performance thresholds already contained in the requirements documents, another system description. Part II finally begins to actually address testing, with integrated test schedules and tasks for all the participating test organizations. Part III addresses Developmental Test and Evaluation, while Part IV concerns Operation Test and Evaluation. A summary of required test resources is at Part V, and several appendices, annexes, and attachments close out the TEMP.

3. *Cost and Operational Effectiveness Analysis.* This is another report largely redundant to those already mentioned. The 14 pages of detailed instructions have a very different tone than the rest of 5000.2-M, almost academic in nature (for example, "Recognize that grossly overestimating or underestimating the threat can lead to the formulation of inferior alternatives."). Unfortunately it also raises a question as to whether or not the authors of this "stand-alone" document were aware that the vast majority (all, really) of what they request is already available in previously mentioned reports. For example, we see here another summary of need, yet another threat assessment, and more statements of operational requirements and thresholds. Also, still more in the way of alternative analysis, costs, and tradeoff analyses are mandated. Frankly, from a "value-added" perspective, the Cost and Operational Effectiveness Analysis would seem to be of limited marginal utility.

4. *Low-Rate Initial Production Report for Naval Vessels and Satellites.* This report, prepared by the program manager, is submitted to Congress at Milestone II (Approval for Development). It identifies the minimum quantity and production rate necessary to preserve the industrial base for that program. It is interesting to note that while this report is now limited to ships and satellites, the logic behind it could easily be applied to other high-value, low-quantity programs as well. With, for example, $800 million B-2 bombers being produced at the rate of two or three per year, minimum production rates and industrial base preservation take on special importance.

5. *Live-Fire Test and Evaluation Report.* The Secretary of Defense must certify to Congress that weapon systems which could reasonably be expected to be exposed to combat can in fact stand up to "live-fire" testing. There are two aspects to this: *survivability* (can a system absorb a hit and still survive?) and *lethality* (can munitions or missile systems perform as advertised, when actually fired at a target in a combat configuration?). Obviously there are some situations in which live-fire testing is impractical. No one, for example, has yet proposed flying the aforementioned $800 million B-2 (and crew) against an SA-10 surrogate, to see whether the B-2 can absorb a real missile strike. For cases in which live-fire testing is impractical or excessively expensive, the Secretary of Defense can submit a waiver to Congress prior to Milestone II.

6. *Acquisition Program Baseline.* The Acquisition Program Baseline (APB) is developed by the program manager and submitted as part of the Milestone I decision package as a "Concept Baseline." It is then updated at Milestone II (as the "*development baseline*") and at Milestone III (as the "*production baseline*"). The APB is *intended* to focus on "the key parameters that define the system," and should be limited to one or two pages in length.

In reality, as we shall see later, this sort of stable, top-level baseline is often not adhered to. Instead, the baseline can become a rolling-update document, serving as the user's avenue for continually adding requirements to the system.

7. *Program Office and Independent Cost Estimates.* Milestone decisions require both program office and independent costs reviews. For major programs, these typically are briefed to OSD's Cost Analysis Improvement Group (CAIG). Basic assumptions include the following: focus is on full life-cycle cost, not just procurement or Six-Year Defense Program periods: all cost categories are included, even items procured for other purposes. (For example, test time on an existing test range would require an estimation of cost for the use of the range.) All alternatives, including those not selected, must be priced. Specific documentation and methodologies are briefly covered in DoD 5000.2-M.

8. *Beyond Low-Rate Initial Production Report.* Before designated large programs can proceed beyond Low-Rate Initial Production (LRIP), the Director for Operational Test and Evaluation must submit a written report to Congress. The report must address adequacy of OT&E, including operational effectiveness and suitability for combat use.

This wraps up the documentation needed to support milestone reviews. Now it is time to turn to that category of documentation which is prepared on a periodic basis, regardless of the program's stage of development.

PERIODIC REPORTING

The previous section dealt with documentation which supports milestone decisions. But there is another major category of reporting, for the most part unrelated to program phase or progress. These reports are developed and submitted on a periodic basis, in response to Congressional or DoD mandate.

Three of these reports are exceptionally significant: the Defense Acquisition Executive Summary, the Selected Acquisition Report, and the Unit Cost Report. In addition, there exist a number of lesser reports which merit attention, and these will also be discussed briefly in the following pages.

1. *Defense Acquisition Executive Summary.* Prepared in eight sections, the Defense Acquisition Executive Summary (DAES) is designed to provide, on a quarterly basis, advance warning of problems and potential issues. The quarterly reporting is on a sort of rotating basis: the Under Secretary of Defense for Acquisition assigns a "reporting month" to a program, and the program manager then reports every three months thereafter. Exceptions to the quarterly system occur in the event of potential or actual baseline deviations or cost breaches, and in the submission of the DAES Section 8 (Annual Program Objective Memorandum/Budget Estimate Submission Summary), which occurs in the same month each year for all systems, regardless of quarterly cycle.

In addition to a highly formatted cover sheet, the DAES' eight sections include:

Section 1, Executive Summary. Includes issues, changes since last report, and data on the acquisition program baseline.

Section 2, Assessments. Addresses performance against program objectives with a color-code system. The program manager rates nine characteristics (performance, test and evaluation, logistics, cost, funding, schedule, contracts, production, and management structure) on a "red-yellow-green" basis.

Section 3, Program Manager's Comments. Permits the program manager to amplify on Section 2's color ratings and to comment on other areas as well.

Section 4, Program Executive Officer/Service Acquisition Executive (PEO/SAE) Comments. Provides an opportunity for the PEO and the SAE to give their opinions on program issues.

Section 5, Approved Program Data. Displays, in tables, key programmatic parameters in terms of technical performance, cost, and schedule. Thresholds come from the APB, IPS, and other sources. Demonstrated performance levels (where available) and the program manager's current estimates are compared to both initial and current program thresholds.

Section 6, Program Background Data. Contains detailed total life-cycle costs, including MILCON and operations and maintenance costs.

Section 7, Supplemental Contract Cost Information. Gives detailed information for each contract on the program. A specific format governs the 28 items of information required.

Section 8, Annual Program Objective Memoranda/Budget Estimate Submission Funding Summary. Enables all program offices to provide, at the same time, POM or BES data, as required. This is the third section of the DAES (in addition to sections 5 and 6) to require detailed total cost data.

2. *Selected Acquisition Report.* The purpose of the Selected Acquisition Report (SAR), according to DOD 5000.2-M, is "to provide standard, comprehensive summary reporting of cost, schedule, and performance information for major defense programs within the Department of Defense and to Congress." If that sounds familiar, it's because much of what's called for in the SAR already exists in other reports, including the DAES. For example, the SAR requires not only tremendously detailed cost and budget data, but also separate sections on program history, developments since the last report, threshold breaches, schedule milestones, performance characteristics, and so forth.

If there's any good news about the SAR, it's that it is *possible* for the report to be only required once a year, as of December 31. But, practically speaking, there exist quite a number of conditions which, if they develop, trigger quarterly SARs. For example, a 15 percent increase in then-year cost, or a six-month delay in any schedule milestone will mandate a quarterly report. In addition, other changes—funding turbulence, baseline adjustments, and the like—may also generate more frequent reports.

The effort required to prepare an SAR is very substantial. To get a feel for the detail required, one needs to review the procedures in Part 17 of DOD 5000.2-M. It is there that the true scope of the SAR becomes apparent, with *12* pages of general directions, followed by a full *29* pages of specific preparation instructions and definitions and an additional *15* pages of examples.

3. *Unit Cost Report.* When the first SAR is submitted, Unit Cost Reporting (UCR) begins. The data required by DOD 5000.2-M—detailed cost analysis—are reported quarterly, actually becoming Sections 6 and 7 of the Defense Acquisition Executive Summary. But the UCR rules mandate additional requirements, including immediate reporting by the program manager of cost increases over 15 percent, either programwide or for any major contract. The instructions for the UCR are extremely difficult to wade through. The following two sentences illustrate the point:

> When a report has been submitted in accordance with paragraphs 3.b. or 3.c., above, showing an increase of 15 percent or more in the program acquisition unit cost, current

procurement unit cost, or cost of a major contract and the Program Manager has reasonable cause to believe that an additional increase of 5 percent or more since the most recent report submitted under paragraphs 3.c. or 3.e. has occurred in the program acquisition unit cost, current procurement unit cost, or cost of a major contract, the Program Manager will again immediately submit a report to the DoD Component Acquisition Executive containing the same unit cost information as required in the quarterly reports; i.e., Defense Acquisition Executive Summary. This requirement reverts back to 15 percent at the beginning of each fiscal year for the program acquisition unit cost and current procurement unit cost only.

Still later in the UCR instructions, a third breach threshold—25 percent—is discussed, along with sanctions, including suspension of financial obligations in the event of late reporting.

4. *Program Deviation Report.* Covered in Part 19 of DOD 5000.2-M, this provides standard reporting procedures for breaches of acquisition program baselines. In addition, this Part describes the format used in developing acquisition program baselines and the procedures for changing them.

5. *Cost Management Report.* Part 20 of DoD 5000.2-M addresses standard contractor cost-reporting procedures and formats. Specific subparagraphs of Part 20 discuss Contractor Cost Data Reporting (CCDR), the Cost Performance Report (CPR), the Cost/Schedule Status Report (C/SSR), and the Contract Funds Status Report (CFSR).

6. *Multiyear Procurement Contract Certification.* Part 21 of DOD 5000.2-M governs the procedures involved in submitting necessary multiyear procurement certifications. It also references those areas of public law which govern multiyear procurement.

7. *Fixed-Price-Type Contract Certification.* A major lesson learned from some notable development programs of the recent past (such as the Navy's A-12 and the Air Force's YF-22/YF-23 ATF competition) concerns the proper allocation of risk between government and contractor. To prevent the inappropriate use of fixed-price contracts in high-risk development situations, Under Secretary of Defense for Acquisition approval is required in order to use either (1) a fixed-price research and development contract of $10 million or more, or (2) a fixed-price contract for the lead ship of a class. These procedures are set forth in Part 22 of DOD 5000.2-M.

With this we conclude our look at the specifics of documentation, both for milestone reviews and for regular periodic reporting. In the following section we shall address some current and emerging issues in project documentation.

THE BASELINE DOCUMENT AND THE ILLUSION OF STABILITY

The new program manager was impressed. Throughout her (limited) training, one thing had been stressed again and again: in program management, *instability kills!* The idea was simple: late changes to a design—for whatever reason—have a tremendously negative impact on program cost and schedule. Sometimes such changes were unavoidable, of course, but it would be essential that they be minimized at worst, eliminated at best.

So she was pleased to open the project baseline document to find these words on the front page:

> With the objective of enhancing program stability and controlling cost growth, we, the undersigned, approve this document to baseline the program. Our intent is that the program be managed within the programmatic and financial constraints identified herein . . . This baseline document . . . contains likely parameters, concepts, numbers and dates which are the basis for agreement among the participating commands. The program will be managed to satisfy the requirements defined herein.

Six signatures were affixed, the lowest-ranking of which was her boss, an Air Force major general. Clearly, here were some people who understood the importance of program stability.

The first hint of a problem was the date of the document: it had been published the month before. But this program had been under way for nearly 8 years—why was the baseline only being established now? Her second dose of reality occurred a week or so later, when she received a call from Captain Smith, "the baseline review officer" at her user's headquarters. "Baseline review officer?" Surely, Smith's job was not to just go around updating all his command's baseline documents? No, no, Smith assured her. He was only responsible for *this* baseline document, for *this* system. And this was his entire job, a sort of aeronautical version of the painters on the Golden Gate Bridge: just when they finally finish, it is time to start all over again.

This is not an isolated, or even an unusual, situation. Secretary of Defense Cheney's 1989 "Defense Management Report to the President" cited the need for greater program stability, and specifically pointed out that "reaching and adhering to baseline agreements on factors critical to a program's success . . . [had] been proven to yield substantial savings over the life of a system."

So, if the benefits of strong baseline documentation are so well understood, why then do we have so much trouble actually *doing* it?

In truth, the *reaching* of the baseline agreement is seldom the problem. In the warm glow so typical of the early stages of a project, when goodwill and good intentions remain relatively undisturbed by the laws of physics and economics, such agreements come fairly easily. No, the real problem lies not in *reaching* the agreement, but rather in *adhering* to it. Our acquisition systems—driven by such things as leadership changes, technological updates, mission revisions, and the constant pursuit of "better" at the expense of "good enough"—appear to lack the discipline necessary to adhere to the baseline agreements reached at previous stages of the project.

Examples of such malpractice abound. One particularly compelling case involves a current aircraft acquisition effort. Five years into the program, a new leader of the using command decided—on what might most generously be described as an intuitive basis—to increase the number of aircrew members provided for. This drove a costly major system redesign, in which a significant portion of the avionics suite was relocated to the only area available—one with an exceedingly negative environment in terms of heat, vibration, and accessibility. Three years later—same program—came another leader, and another change, this one altering the aeronautical ratings of the crew members. Again, the effect was significant: on software, training systems, and technical data. The worst example followed later, when the same commander initiated a "no smoking in the cockpit" policy, and demanded that the cockpit ashtray be removed. A trivial change at first, until one considers the hundreds and thousands of descriptions, drawings, schematics, technical manuals, and the like which were impacted.

Not all examples are so blatant. In fact, a prudent and realistic observer recognizes that with systems which regularly take more than a decade from inception to operational capability, it is pretty naive to think that changes—necessary, unpreventable, essential changes—*will not* occur. The enemy threat may change; certainly technology will move forward; business base erosion can be relied upon to eliminate critical members of the contractor team. These and other sources of turbulence will mandate *some* baseline revisions. The task for the program manager becomes how best to hold those revisions to an absolute minimum.

In addressing this issue, there are two principles which come to mind: education and planning. First, education. Acquisition specialists tend to assume that the dangers of instability—so clear to them—are equally clear to their users. But that is, in fact, rarely the case. Usually untrained in acquisition, and often lacking significant business or engineering background, users frequently fail to grasp the significance of baseline changes, and therefore tend to see acquisition managers as recalcitrant and lacking "customer orientation" when they resist change. Acquisition managers, for their part, do not see education of users as part of their mission. The resulting miscommunications and cultural gaps then become inevitable.

However, a proactive approach can pay dividends. A constant process of customer education can show users the impacts of their decisions. But this kind of education is rarely done. When, for example, was the last time a program office sponsored one of its *users* to attend even a *short* acquisition course, much less something as significant as Defense Systems Management College? Without the development of shared knowledge and values between acquisition partners, the mistrust, mistakes, and misperceptions so common today will continue.

The second area offering hope for improvement is *planning.* Rather than permitting situations in which staff members develop continuously rolling baseline changes, top leaders across the project need to plan for *scheduled, periodic* baseline updates just as they now plan for events like Preliminary and Critical Design Reviews. The importance of baseline stability is such that it can no longer be left to lower- and middle-level management to develop, negotiate, and then pass baseline changes to top managers for ratification. Senior leaders of all parties need to devote personal attention to the task, in keeping with Secretary Cheney's vision of stable acquisition management.

The moral of all this: project instability *does* kill. Proper documentation—beginning with a clear, coherent, and coordinated baseline document—can alleviate some of that instability, but *only* if the system has the necessary discipline to adhere to it. Otherwise the baseline document produces only the *illusion* of stability, and that illusion can be a dangerous one.

"EXIT CRITERIA" AND THE SYSTEM MATURITY MATRIX

Both Congressional and Executive Branch DoD reform initiatives have identified deficiencies in the process of the Defense Acquisition Board (DAB) and its predecessors. One of those shortcomings involves the observation that programs were sometimes permitted to transition from one stage to the next based on a predetermined schedule which did not in all cases relate to actual program achievements.

To remedy this, the 1989 Defense Management Report to the President called for greater discipline in the DAB process. Toward that aim, the Under Secretary of Defense for Acquisition was charged with using "exit criteria," program thresholds which, if *not* achieved, would deny a program's passage to the next acquisition phase.

The Cheney report was not so specific on exactly from where those thresholds would come. But in the months following the report's publication in July 1989, the armed services saw increasing use of the System Maturity Matrix (SMM), a device which had been previously mandated by Congress for certain extremely high-value programs. The basic process behind the SMM is eminently logical and is, in fact, the same process used to develop performance thresholds for other reports:

1. Identify critical performance parameters (such as speed, payload, range, survivability).
2. Using the DoD-approved Test and Evaluation Master Plan (TEMP), time-phase the achievement of specific thresholds in those critical areas.
3. Integrate planned milestone events.

A typical SMM for an aircraft development program—hugely simplified—is illustrated in Table 37.1.

The approved SMM then becomes a "contract" between the program and DoD and Congressional decision makers, operating in much the same way as the baseline document does between the user and its acquisition agency. As with all contracts, stability is crucial, and any regular juggling of the thresholds obviously defeats the purpose of the document. At the same time, changes are inevitable, such as those caused by new threats, unforeseen world developments, scientific breakthroughs, or technological barriers.

While the concept behind the SMM is sensible, in practice it presents challenges to the program manager. One is the bureaucratic workload: the SMM needs to be fully coordinated with all acquiring, using, and testing agencies, and it needs to be updated periodically as new and inevitable revisions are made. It is not uncommon to find the preparation, negotiation, and revision of the SMM to be a significant drain on the resources of a program office.

But by far the larger problems are more basic ones: identifying critical performance areas, developing achievable thresholds, and then keeping the test program on track to meet them. It is critical for the program manager to keep some advice in mind:

1. Not every parameter is critical. If basic Pareto analysis demonstrates that 20 percent of an organism is responsible for 80 percent of its results, then it stands to reason that only your truly critical parameters ought to be included. And 20 percent ought to be a ceiling. Fewer is better.

TABLE 37.1 Typical System Maturity Matrix

Date	31 Dec 92	31 Dec 93	31 Dec 94	31 Dec 95
Speed	Mach .8	Mach .8	Mach 1.4	Mach 1.5
Range	Confirmed by Analysis	Achieve 90% in Simulated Mission Profile	Achieve 95% of Range Specification	Achieve 100% of Range Specification
Weapons Carriage	Ground Compatibility Testing Complete	Captive Carry Testing Complete	Conventional Certification Complete	Nuclear Certification Complete
	▲ Milestone II		▲ Milestone IIIa	▲ Milestone IIIb

2. Keep the thresholds as "top level" as possible, so that a 2 percent shortcoming in one area can be offset by a similar surplus in another. Here's an example: rather than having six different range or altitude or payload thresholds, have *one* that is a composite of six different profiles. That way, when you run into some unforeseen anomaly five years down the road, you'll enjoy a built-in management reserve. It won't save a *bad* program, but it may keep you from losing a *good* program due to statistical flukes or one or two minor shortfalls.

In sum, a well-planned, well-executed SMM can facilitate a program's transition from one program phase to the next. It can also be used to satisfy increasingly frequent Congressional restrictions (such as "fences" used to withhold funds from a program) on major programs. But *poorly* planned, that same SMM has the ability to slow or even kill a program, and for relatively minor shortcomings.

PERIODIC REPORTING, AND WHY THERE IS LITTLE HOPE FOR IMPROVEMENT

Every major review of defensive acquisition has cited one theme: the stifling, stagnating effects of program reporting requirements. The 1989 Cheney report was no different:

> . . . the system is encumbered by overly detailed, confusing, and sometimes contradictory laws, regulations, directives, instructions, policy memoranda, and other guidance. Little room now remains for individual judgement and creativity of the sort on which the most successful industrial management increasingly relies to achieve higher levels of productivity and lower costs. Much of this stifling burden is a consequence of legislative enactments, and urgently requires attention by Congress. Much also has been administratively imposed and requires prompt corrective action by DoD.

Secretary Cheney went on to charter a task force to conduct a DoD-wide review of acquisition reporting. Although the task force was directed to complete its report in 1990, it is probably fair to say that the net effect of its efforts has yet to be felt in the field. Basically the same reporting requirements remain in effect, although there has been a consolidation of governing regulations with the 1991 publishing of DODD 5000.1 and related publications.

This section began with a pretty pessimistic title: "Periodic Reporting, and Why There's Little Hope for Improvement." What is it that makes the picture so bleak?

It has become axiomatic that—like the Cheney report—any report on defense reform has, as one of its themes, the tremendous costs and inefficiencies associated with redundant reporting requirements. The reform report can be relied upon to argue eloquently and vehemently against the evils of micromanagement, and then to recommend a task force to solve the problem. Sadly, one can equally well rely on the certainty that a year or two later, after the warm glow of the new report has faded, there will be not only no net reduction of reports, there will probably be a net *increase*.

This entire phenomenon is fairly clear, and probably inescapable. The reason for these repeated failures is that such task forces, special commissions, "blue ribbon" panels, and the like all focus on the same thing: the *process*. But the real focus needs to be elsewhere: on the *people*, the thousands upon thousands of "decision makers" involved in the acquisition process.

The Cheney report says that the actual number is even more discouraging, citing an acquisition work force of 582,131—not including those thousands outside DoD, such as Congress, rapidly growing congressional staffs, and other constituents. With these vast legions of "decision makers"—many of whom have the authority to say "no" but very few of whom have the authority to say "yes"—it can be seen that a form of Parkinson's law—work expands to fill the time available—is hard at work.

The problem revolves around people, in both the acquisition work force and Congress, and the problem is twofold: quantity and quality.

If, as Parkinson said, work really *does* expand to fill the time available—and who among us seriously doubts that?—it stands to reason that it also expands to occupy the *number of people* available as well. The huge size of our acquisition work force—over a half-*million* people—brings with it layer upon layer of middle management, and with that comes an ever-burgeoning demand for histories, reports, executive summaries, analyses, certifications, and the like.

It is here that the *quality* of our people comes into play as well, but it is not in the way that those who have made careers out of "DoD bashing" would have us believe. Rather, it is the very fact that we enjoy a high-quality, dedicated work force that in turn drives up the level of reporting requirements. As the legions of committed, conscientious DoD acquisition staff members all try to "get into the act," they find that they can't do it unless they're in the reporting loop. And despite generally high levels of skepticism regarding the value added of Congressional staffers, there is really no reason to think that most of *them* are any different: just honest, hard-working people who are trying to bring their personal expertise to bear on the situation. And the way they do that is by requiring ever-increasing numbers of reports.

What is the answer? Since it certainly is not to lower the *quality* of our people, it follows that we need to work on the *quantity* side of the problem. The current and future down-sizing of the federal government work force may help, but probably not to a meaningful degree. And so long as other non-DoD functions—most notably, Congressional staffs—continue to grow, so will the number and complexity of reporting requirements.

ACQUISITION STREAMLINING AND THE RISE AND FALL OF "BLACK WORLD" PROGRAMS

One characteristic of the post-1980 defense buildup by the Reagan administration was increasing use of "black" programs. A black program was one in which the detail (and, in some cases, the very existence) of an acquisition effort was screened from public view by use of security classification and "special access" compartmentalization. In all cases, black programs were sold on the basis of extraordinary need for secrecy, typically to protect high-technology weapons or sensitive intelligence-gathering techniques. And the programs developed in the "black world"—ranging from the 1950's development of the U-2 to the 1960's SR-71 and the 1980's F-117 stealth fighter—seem to have enjoyed extraordinary levels of operational success, indicating that the protection of the advanced technologies did in fact pay big dividends.

But there was another great advantage enjoyed by black programs, one that is rarely mentioned. It involves their absolute minimal levels of program documentation, and the resultant absence of the bureaucratic stagnation that so plagues more traditional programs.

This minimalist approach to documentation was the product of two things. First, program direction typically forbade the kind of detailed, redundant reporting laid out previously in this chapter. A typical Program Management Directive for a black program would have a passage like this:

> Reporting will be kept to an absolute minimum, consistent with the need to maintain security. Written reports, summaries, and documentation will be avoided where at all possible. Instead, reporting will be on an "exception only" basis, and will be face-to-face whenever practical.

And it was not unheard of to have directors of black programs brief their status quarterly, directly to the Secretary of Defense, dispensing with much of the bureaucracy described earlier in this chapter.

The second way in which documentation was streamlined successfully was through the limited-access nature of these programs. Each project was compartmented and assigned a "special-access" identifier. Only people with the requisite high-level clearances *and* a strictly defined "need to know" were permitted access. That access was controlled by senior DoD officials, with support of Administration and Congressional leaders. By ruthlessly restricting such access, two benefits were derived. First, national security interests were protected. Second, with thousands of the peripheral decision makers in the defense hierarchy out of the picture, more energy could be focused on actual programmatic work, and less on the task of keeping the army of acquisition bureaucrats "in the loop."

It is fair to say that the days of the black programs are on the wane, for at least three reasons. First, the decline—in the public perception, at least—of the Soviet threat makes such levels of secrecy more difficult to justify. Second, with defense budgets dropping in real terms since 1986, new starts of any systems—"black" or "white" world—are bound to continue to decline. Finally, some unfavorable experiences with certain special-access programs—the Navy's canceled A-12 comes first to mind—have given ammunition to the bureaucratic legions in their fight to continue bringing their wisdom and oversight to the acquisition process. Their main argument—that greater levels of oversight would have prevented embarrassments like the A-12—is a weak one: one has only to look at the myriad "white world" programs in trouble to see that the relationship between amount of oversight and program success is a tenuous one.

Three of the basic principles of Lockheed's "Skunk Works"—that premier organization which produced such aerospace success stories as the F-104, U-2, SR-71, and F-117—merit consideration here. Lockheed's rule number three is pertinent:

> The number of people having any connection with the project must be restricted in an almost vicious manner. Use a small number of good people (10% to 25% compared to the so-called normal system).

Two additional rules support this minimalist approach:

> Access by outsiders to the project and its personnel must be strictly controlled by appropriate security measures.
>
> There must be a minimum number of reports required, but *important* work must be recorded thoroughly.

On balance, these concepts—which underline the whole concept of black world acquisition programs—have served us well in the past. They make even more sense for the future.

CONCLUSION

The need for proper documentation in defense acquisition is clear and unassailable: in order to make rational and correct decisions, senior defense leaders need accurate and timely data. While the existence of such data does not guarantee *good* decisions, it is fair to say that its absence goes a long way toward driving *bad* ones. But even with so clear and compelling a need, the current system of documentation has marked shortcomings in terms of both efficiency and effectiveness.

The inefficiencies of defense documentation are readily apparent to the most casual observers. High levels of redundancy mean that great resources are expended reporting reams of similar—or even identical—information in various reports. This inefficiency, cited by every significant defense-streamlining effort in recent years, has proven to be totally resistant to the best intentions and efforts of the reformers. In fact, the system is more than just *resistant*: it actually seems to thrive and grow under even the most withering criticism.

Less obvious than the lack of efficiency—but no less real—is a resultant lack of *effectiveness* as well. If the endless and repetitive reports were actually improving decision making, their great cost might be justifiable. But products of the huge documentation bureaucracy produce not better decisions, but *worse* ones. This is true for two main reasons. First, the really important data tend to get lost amid the volumes of boilerplate and repetition. Second, the fact that many reports are built around the same data but are "sliced" differently for different constituencies ensures that many of the key actors in the decision process are forced to read their lines from very different scripts.

It would be nice to end this chapter on a positive note, citing reasons to hope for improvement. Nice, but not honest. The fact is that, with decades of acquisition streamlining efforts behind us, we are worse off now—from a program documentation perspective—than ever.

BIBLIOGRAPHY

"Basic Operating Rules of the Lockheed Skunk Works," handout from the Air Force Institute of Technology, undated.

DoD Directive 5000.1, "Defense Acquisition," Feb. 23, 1991.

DoD Instruction 5000.2, "Defense Acquisition Management Policies and Procedures," Feb. 23, 1991.

DoD 5000.2-M, "Defense Acquisition Management Documentation and Reports," Feb. 1991.

DoD Instruction 5000.33, "Uniform Budget Cost Terms and Definitions," Aug. 15, 1977.

DoD Directive 5134.1, "Under Secretary of Defense (Acquisition)," Aug. 8, 1989.

DoD 7000.3-G, "Preparation and Review of Selected Acquisition Reports," May 1980.

DoD Directive 7750.5, "Management and Control of Information Requirements," Aug. 7, 1986.

DoD 7750.5-M, "Procedures for Management of Information Requirements," Nov. 1986.

Office of Management and Budget Circular A-109, "Major System Acquisitions," Apr. 5, 1976.

Secretary of Defense Report, "Defense Management Report to the President," July 1989.

Title 10, United States Code, Sec. 2306(h), "Kinds of Contracts."

Title 10, United States Code, Sec. 2350 a.(e), "Cooperative Opportunities Document."

Title 10, United States Code, Secs. 2365, 2438, and 2502, all relating to competition and the Defense industrial base.

Title 10, United States Code, Sec. 2366, "Survivability and Lethality Testing Before Full Scale Production."

Title 10, United States Code, Sec. 2399(b)(1), "Operational Test and Evaluation."

Title 10, United States Code, Sec. 2400(c), "Low Rate Initial Production of Naval Vessels and Satellite Programs."

Title 10, United States Code, Sec. 2430, "Major Defense Acquisition Program Defined"; Sec. 2302(5), "Major Systems Defined."

Title 10, United States Code, Sec. 2432, "Selected Acquisition Reports."

Title 10, United States Code, Sec. 2433, "Unit Cost Reports."

Title 10, United States Code, Sec. 2434, "Independent Cost Estimates/Operational Manpower Requirements."

Title 10, United States Code, Sec. 2435, "Enhanced Program Stability."

Title 40, Code of Federal Regulations, Parts 1500–1508, "National Environmental Policy Act Regulations," July 1, 1986.

CHAPTER 38

CONCURRENT ENGINEERING—A MILITARY PERSPECTIVE

David I. Cleland

David I. Cleland is the author/editor of 22 books and many articles in the fields of project management, engineering management, and strategic planning. He is a Fellow of the Project Management Institute and was appointed to the Ernest E. Roth professorship in the School of Engineering, University of Pittsburgh, in recognition of outstanding productivity as a senior member of the faculty. He is currently Professor of Engineering Management.

INTRODUCTION

Changes in technology and military-political affairs around the world are altering the business of the development, production, and support of military systems. One of the more significant changes is found in the emerging concept of *concurrent engineering.* This chapter provides some generic information on this important aspect in the design of products and the supporting organizational processes such as marketing, manufacturing, procurement, and after-sales service. Before exploring information on concurrent engineering phenomena, some background on the global changes in military markets is in order.

The decline of the cold war and the quick win in the Persian Gulf along with the decline of communism in the world had many influences on defense global markets and in the global defense industry. In Europe the new market for the development and production of military systems has taken an unusual competitive turn. Defense companies are consolidating their work through joint projects between companies and countries. Rolls Royce, who heads Europe's largest engine maker, is proposing a wide range of projects with France's SNECMA, Turbomecca, Fiat-Avio of Italy, and Spain's Industria de Turbo Propulsores. GEC and British Aerospace are cooperating in a project to bid against IBM and British Aerospace. Germany's Messerschmitt-Bolkow-Blom are working with Raytheon Company of the United States to upgrade the Patriot missile. The Eurosam consortium of Aerospatiale, Thomson-CSF, Italy's Alenia, and the newcomer Inisel of Spain is readying a similar antimissile program for the late 1990s. Most ambitious of all is the four-nation European fighter aircraft, which could generate sales of 1400 throughout the world. NATO's plans for a new rapid-deployment force in Europe will bring about more changes, such as a strategy of high-tech, high-powered weaponry and mobility. European spending on research and development for defense is already growing at a double-digit clip.

With the removal of trade barriers in Europe, it will be tougher for the U.S. defense companies to compete, without joining in a cooperative project with European partners. Already discussions are underway between European and U.S. plane builders to come up with a new jointly developed military cargo aircraft.

A message is clear out of all of this: If defense contractors want to stay in business, they will have to put aside nationalist feelings and find ways to work together on joint projects with other companies and countries in the global market for defense products.[1]

Defense contractors and the Department of Defense (DoD) also face a more awesome challenge: the need for more efficiency and effectiveness in the development, production, and support of military weapon systems.

BETTER MANAGEMENT REQUIRED

Reduction of cost, higher quality, and the shortening of the development and production cycles of military systems have become strategic issues in the DoD. The typical major weapon system now requires 10 to 15 years to develop and put in its operational environment. In 1986 the President's Blue Ribbon Commission on Defense Management (the Packard Commission) addressed the issues of the time to develop, the high cost, and the inadequate performance of weapon systems. In global competition in nondefense products and services, similar problems of cost, schedule, and performance became painfully evident to many U.S. commercial producers—made vividly clear in a loss of market share and the producing of products and services that fall behind the prevailing state of the art in the marketplace.

In the development of weapon systems the classic approach of using major life-cycle phases serially is used. In general, these phases provided for a lock-step approach going from a requirement phase through a product development phase, then to a prototype, followed by the actual building of the product in a production phase. Finally the weapon system was turned over to a user. Because of the changes in technology that became available and the serial nature of the process, engineering changes and redesign of both the weapon system and the processes by which it came into existence during its long life cycle were carried out too far downstream into the production phase. All of this increased costs and extended the time when the user could put the system to work in supporting military objectives.

The serial life cycle has been in use for a long time. In today's competitive and global defense markets, costs and schedules have become increasingly beyond realistic military budgeting and military expectations. A more vexing problem arises when the technology tends to advance, as it inevitably does. Weapon system developers tried to remedy the technology lag through redesign and engineering changes—the clock kept running, reflecting cost and schedule overruns. All too often weapon system design or manufacturing deficiencies were discovered at the prototype stage, at which time much of the funds available for further R&D had been spent. If the prototype assessment validated poor design, or pointed to a lag in the manufacturing technology required to produce the weapon system, additional cost and schedule overruns were inevitable.

In the DoD continual efforts are under way to improve the understanding and efficiency of design, test, and production processes. Much effort is spent in ensuring that engineering strategies are implemented on schedule as an important way to reduce the risk in military programs. Efforts are continually put forth to shorten the acquisition process. Although there have been many strategies implemented to shorten the development-production schedule, the cycle has grown longer and the criticism stronger. The reason for shortening the cycle has been to reduce cost—but in the larger context it has been done to improve military readiness.

In the acquisition process many product-process problems do not become apparent until the program goes from full-scale development to production. These problems include technical reasons, poor design, inadequate testing, all leading to manufacturing problems which impact production schedules and costs. Evidence of the "hidden factory syndrome" is found during and after production with its redesign and rework costs, leading to impacted operational and training schedules as well as increased costs—and compromise of military readiness. Product changes have been a problem in the development of military weapon and support systems for a long time.

The management of engineering changes has always been challenging in the development of any product. In DoD weaponry programs dealing with such changes is particularly challenging because of the usual reviews and approvals that are required. The challenge of managing engineering changes can be facilitated by the process of concurrent engineering because manufacturing decisions and user decisions on the system can be made back in the design process, establishing firm support and maintenance systems much earlier.

CONCURRENT ENGINEERING—A PROMISE

The Institute for Defense Analysis (IDA) has defined concurrent engineering as (Ref. 2, p. v):

> . . . a systematic approach to the integrated, concurrent design of products and their related processes, including manufacture and support. This approach is intended to cause the developers, from the outset, to consider all elements of the product life cycle from conception through disposal, including quality, cost, schedule, and user requirements.

An aerospace executive has defined concurrent engineering as:[3]

> . . . the structuring of a product development organization such that significant contributors to the design process are encouraged and disciplined to operate as a team, working toward a common set of design goals, whose end product is a total "build package."

Concurrent engineering is the process of integrating the technology of the weapon system and the supporting organizational processes required to conceive, design, develop, produce, and place the system in its operational environment. In the history of warfare, technology has made possible decisive victories and has changed the direction of history. The stirrup, an enormous breakthrough for its time, allowed superior performance by the mounted English soldiers in the Battle of Crécy in 1346. The machine gun brought about changes in the deployment and use of troops on the battlefield. In the Battle of the Somme in World War I the British tactics of using infantry against entrenched German machine guns caused a loss of thousands of their best troops. The tank reduced the mastery of the battlefield by the machine gun. The technological advantage of the airplane and sophisticated "smart" bombs and missiles was demonstrated dramatically in the brief Persian Gulf War. The atomic bomb—and its later development of advanced nuclear weaponry—provides a technological superiority that will last for some time. But if the R&D on weaponry technology continues, we can expect to see more advanced technological improvements that will likely be as dramatic as the shift from the repeating rifle of the infantryman to the machine gun, and the consequent casualties.

The benefits of concurrent engineering in nondefense goods and services provide earlier commercialization of the product, reduction of costs, higher quality, greater profitability, and a more efficient and effective work force in part because of the development of a project team culture in the company which encourages participation, innovation, and creativity. By using product-process design teams, a means is provided for the continuous improvement of both products and processes. Of course, in the private sector there is likely to be less bureaucracy and less turbulence in the management of new products and processes. Once a new product-process development process is initiated, there is likely to be somewhat of a steady progress through the phases of the life cycle of the product development since the control of the resources required for the product is within the approval authority of the chief executive officer and the staff of the private enterprise. There may be delays, projects will be canceled as the promise of technology feasibility fades, or the market opportunity has been lost because a competitor got to the market first. In the private enterprise there are politics, competition for the scarce product-process development monies, and delayers, debaters, and other people who can, in many subtle ways, delay the development of new products and the supporting organizational processes. It is not that simple in managing projects in the defense business. According to Adelman and Augustine:[4]

> Projects are started, stopped, accelerated, or slowed; budgets are increased or decreased; schedules, objectives, designs, and even people are continually changed . . . the average development program today takes over eight years. It takes fourteen years, on average, to go from the beginning of the concept formulation stage to operational development.

(See Fig. 38.1.) In addition, Adelman and Augustine note that the budgeting process is one of the main culprits in promoting instability in the development and manufacture of defense products.

Yet in the face of the turbulence and instability of doing business in the defense industry, both the government and defense contractors continue to improve the management of the defense products and services. Concurrent engineering has helped in such improvements, even though it is done in an ambience in which many of the supporting processes required to get the product developed and delivered to the defense user are simply beyond the control of any product-process design team, and even beyond the control of senior contractor and defense executives. Even considering these constraints, the process of concurrent engineering in the defense community continues the promise of making meaningful contributions.

A ROSE BY ANY OTHER NAME

Other names have been given to the concept of concurrent engineering. At the Northrop Corporation it is called "parallel release," and at Boeing the term "de-

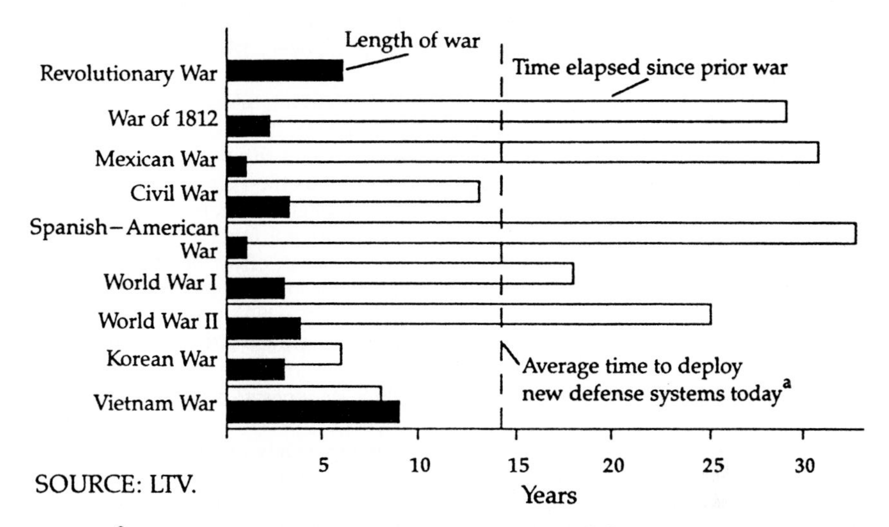

FIGURE 38.1 Length of U.S. wars and peace compared with average time to deploy new defense systems today. (*From Adelman and Augustine,*[4] p. 179.)

sign/build" describes the concept. In the Westinghouse Electric Corporation it is called "design for value." Whatever it is called, concurrent engineering provides for getting the product and organizational process designers together to work side by side in doing better designs and to help in making the product easier to build. The change brought about by concurrent engineering in the way people work together is so great that it will not likely evolve as a grass roots movement from the bottom of the design and manufacturing organizations. At the Northrop Corporation, the commitment of senior management was considered so essential to making concurrent engineering work that six vice presidents signed a document which committed their support in making the change to concurrent engineering a reality.

At Northrop parallel release concepts and processes were used with its work on the Advanced Tactical Fighter (ATF) Program. The initial efforts at parallel release were successful—part of the success was that the product-process design team was well on its way to realizing its goal of a 70 percent cut in engineering changes after a part had been released for production. Goals of labor costs and the time between design and production were established between 30 and 50 percent. Initial success in the parallel release process indicated that these goals were realistic and stood an excellent chance of being attained. Early success in the new approach to product-process design can be attributed in part to the collocation of the product and manufacturing designers, who work together first in reviewing the design strategies on computer displays, and then in finalizing the design that could meet both product and manufacturing process requirements.[5]

At McDonnell Douglas in the Delta II rocket plant in Pueblo, Colo., a product-process design project team shut the plant down and took a fresh look at how fairings (a kind of protective shroud that covers the tip of a rocket) were being manufactured. The plant was experiencing serious schedule and quality problems. Working together to eliminate waste and improve efficiency, the team not only caught up to schedule, but increased production dramatically from four fairings in an eight-month period to a current average of one fairing every 18 days; in the process it also improved quality.[6]

Concurrent engineering may be thought of as a building block in a continuous-improvement strategy for the enterprise. It is the integration—through a product-process design team—of the improvement of the work of the traditional serial life cycle resulting in a design process toward enhancing the efficiency and effectiveness of the downstream processes. The key to concurrent engineering is to integrate disciplines (reliability, software, electrical) and organizational functions (manufacturing, design, marketing, and so on) through a simultaneous effort of integrating the workings of people with the technical and management information available to the enterprise leaders.

Companies doing business with DoD have been exposed to the use of project (program) management techniques in performing on contracts for defense work. Thus the use of teams to bring about interdisciplinary cooperation should not be foreign to these producers. However, the use of product-process design teams early in the life cycle of a product-process design is not being widely practiced in many defense companies. The intensity of cooperation required among the organizational disciplines early in the life cycle of a new defense product idea under the concurrent engineering philosophy requires many changes in the strategies and practices of the companies.

DoD INITIATIVES

Concurrent engineering, which holds promise in being more successful in developing effective and timely weapon systems, has attracted the attention of policy and strategy executives in DoD. In early 1989 several new policies were issued which encouraged the use of concurrent engineering in weapon system development.

DoD has sponsored major strategic initiatives in concurrent engineering. The Defense Advanced Research Projects Agency (DARPA) has provided support for the Concurrent Engineering Research Center (CERC) at West Virginia University in Morgantown, WVa. Also, DARPA has sponsored the DICE (DARPA initiative in concurrent engineering) programs with GE Aircraft Engines in the concurrent engineering programs in Cincinnati, Ohio. This program, started in July 1988, is a five-year effort by a university-industry team led by General Electric Aircraft Engines. Advanced concepts in concurrent engineering technology are being developed, validated, and demonstrated under the program. The program is being sponsored by CERC at West Virginia University. The objective of the DICE program is to reduce the introduction time and life-cycle costs of advanced products and systems. The cost benefits to be expected from a realization of the objectives of the program are a reduction of (1) the time from concept to production by 30 percent, (2) the number of design- or manufacturing-caused engineering changes by 50 percent, (3) the cost of product support by 50 percent, and (4) sustaining engineering by 50 percent.[7] The mission of the DICE program[7]

> . . . is to create a Concurrent Engineering environment that will result in reduced time to market, improved total quality and lower cost for products or systems developed and supported by large organizations. This environment will enable all disciplines important in the life cycle of a product or system to cooperate interactively in its definition, planning, design, manufacture, maintenance, refinement and retirement from service.

TRADITIONAL DESIGNS

Many professional product developers, and managers as well, have a romantic view of product development: the lonely creative individual working in a highly unstructured, closeted place where innovative products come out once in a while. The lonely, solitary inventor is more a product of Hollywood than of reality. In fact, in the truly successful firms today, product development is a highly disciplined, organized activity.

Consider that it is generally known that about 80 percent of the manufacturing costs are locked in during the first 20 percent of the product design. Japanese designers believe that somewhere between 75 and 90 percent of a product's cost is locked in at the design stage. Consider further that:

- Poor design will contribute to downstream quality problems upwards of 40 to 50 percent of such problems.
- Overcomplicated designs can cause delays and lead to costly engineering changes.

Traditional product-process design results in increased costs, delayed schedules, rework, and engineering changes, and in general reflected unfavorably on the credibility of the entire product-process design function. The compartmentalized disciplines reinforced by separate sets of values among design engineers, manufacturing managers, and professionals heightened the lack of the establishment or understanding of common goals.

The traditional design of products and processes was done within the confines of functional and psychological walls between the people that were doing the design work. Design engineers resisted any critique of their designs.

Jack Reichert, CEO of Brunswick, the $3-billion-a-year sporting goods company, decided in the mid-1980s that developing new products—outboard motors, automatic bowling scorecards, fishing reels—took too long. He said:[8] "Product development was like elephant intercourse. It was accompanied by much hooting, hollering, and throwing of dirt, and then nothing would happen for a year."

In using traditional "over the wall" product-process design strategies, the firm is subject to the design dilemma—most of the product cost, quality, and manufacturability are committed very early in the design before more details about the product and its manufacture, marketing, quality, and vendor relationships have been developed. Using a product-process interdisciplinary design team permits the development of information leading to the intelligence needed to preclude people from becoming enamored of their approach without full assessment of how everything fits together, in the development of a product that meets all of the stakeholders' expectations. Differences that are bound to come up between design and manufacturing engineers are more easily reconciled.

Figure 38.2 compares a sequential approach to product development at the top with a concurrent approach in the lower half. The arrows in the figure indicate the flow of information—one direction of flow in the traditional approach vis-à-vis a bidirectional flow in the concurrent approach. The sharing of information in the concurrent approach requires both technological and organizational change.

Concurrent engineering makes an extraordinary contribution to the solution of some of the problems of traditional product-process design. These contributions include:

- Fostering of broad and more effective communication patterns among the principals concerned with product-process design.

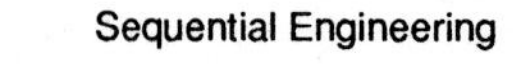

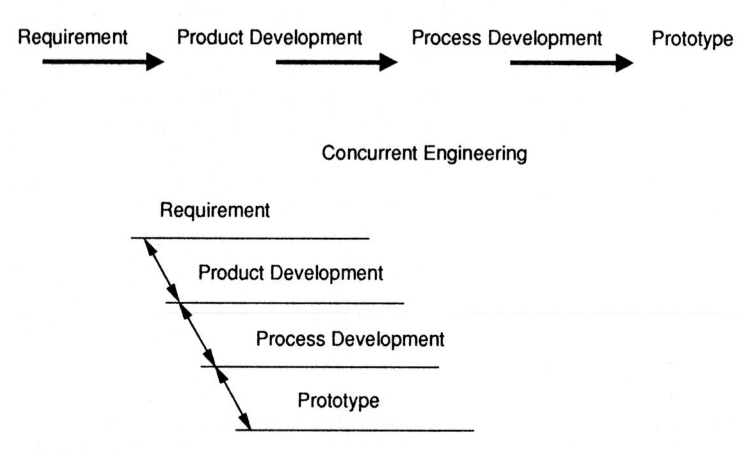

FIGURE 38.2 Comparison of sequential and concurrent engineering. (*From Winner et al.*,[2] p. 12.)

- Dissolving the disciplinary and organizational barriers among the professionals concerned with the aspects of designing, developing, manufacturing, and supporting the military system in its operational environment.
- Nurturing a common philosophy of values among the product-process design team members. These values are most often reflected in a group concern for cost reduction, reduced product-process development schedules, and enhanced customer-user satisfaction with the resulting weapon system.
- Greater results through a product-process means and ends to support DoD objectives and goals.
- A commitment toward total quality achievement in a manner unsurpassed by any other management technical strategy.

Design has always been an elusive and fleeting art. Today under the umbrella of concurrent engineering, design encompasses how a product is conceived, assembled, manufactured, and sold. It means studying the use of alternative materials, fabricating techniques, marketing strategies, and ease of service considerations.

A GROWING USE

In the material that follows additional examples of companies that use concurrent engineering in both defense and nondefense industries are given.

The 3M Company is broadly based in many products and technologies. About 50,000 different products are drawn from 100 different product technologies. There are 53 business units, each responsible for its own sales, marketing, laboratory, product development, and manufacturing. Today the 3M marketing strategy needs to be changed since shorter product life cycles and the requirement to commercialize products sooner are competitive realities. The company is moving into the use of product-process design teams in a concurrent engineering context to do a better and more timely job of product-process development. It is using the quality function deployment, which includes the development of an understanding of the customer's needs up front, and integrating those needs into the products and the supporting processes. By empowering these teams to take the action needed without interference (but with support) from functional management, it is expected that the product will be moved to manufacturing sooner and to the customers faster.[9]

At the Product Definition Development Center (PDDC) at Northrop Aircraft Division in Hawthorne, Calif., an example exists of a shift in the way companies design, plan, manufacture, and support defense systems. Through a concurrent engineering approach project team personnel are collocated, including designers, quality assurance personnel, material planners, engineers, and assembly planners. Under this arrangement the firm anticipates a 30 percent reduction in average project time and cost, a 75 percent drop in design changes after release, a 50 percent inventory reduction, and a 25 percent reduction in direct labor.[10]

McDonnell Aircraft Company used a simultaneous engineering approach to develop the night attack nose cone for the AV-8B jet. Members of the product design team were collocated, and through the full use of computer-aided design were able to emphasize design for producibility. The first nose cone was completed five months ahead of schedule with no tooling or part rework. Almost no assembly fit problems were encountered, and a significant drop in design changes was realized.[10]

Johnson & Johnson is the world's largest, most decentralized health care company. Operating in a broad range of technologies, the company has been dominated by marketing and financial people for the last couple of decades. But this is changing: general managers are trying to understand more about the management of technology—more specifically in the area of concurrent engineering. The company is moving strongly into much more customized products that require new materials and new technologies. To accommodate these changes the company has established umbrella projects that bring together the marketing, R&D, engineering, and production people as a team. The use of concurrent engineering is being done to reduce the time to go from raw material to the finished goods delivered to the customer. The company believes that the EDA and other governmental agencies have to figure out better ways to get through the entire approval process to be more responsive to the market needs.[9]

Boeing's Ballistic Systems Division reduced parts and materials lead times by 30 percent. ITT reduced the design cycle for an electronic countermeasures system by 33 percent. Transition-to-production time was reduced to 22 percent; time to produce a certain cable harness was reduced to 10 percent.

Texas Instruments studied how much concurrent engineering could save the company. By comparing "over the wall" serial engineering with concurrent engineering, it evaluated the time needed to review a drawing, produce a model, and make detailed drawings. The company also reviewed the cost reductions that needed to be made just to keep on budget once a product was in production. On these reduction programs it was able to go back and take out anywhere from 10 to 40 percent of the costs. These reductions provided the basis for setting new engineering objectives: reduce the cost of doing mechanical design by 40 percent, reduce product cost by 25 percent, and reduce the concept-to-market time from the conventional 18 to 36 months to 24 months.[11]

Major principles of concurrent engineering were first formally introduced at McDonnell Aircraft Company (MCAIR) on the AV-8B Harrier II Program. This program faced severely limited resources. In addition, the company's CAD/CAM technology had developed so that three-dimensional computer graphic systems could be used for the system's integration work. In August 1988 a senior-level multidisciplinary team was organized to head up this effort and was given the mission to: "Improve the quality of the MCAIR product definition so as to provide first time quality on schedule and at low cost in the development and production cycles." The organization of this multidisciplinary team is depicted in Fig. 38.3.

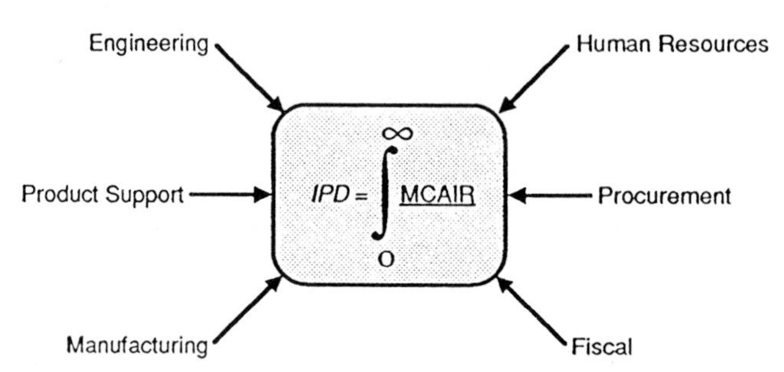

FIGURE 38.3 Multidisciplinary team. (*From McGovern and Grossman,*[12] p. 98.)

To facilitate having all of the organizational disciplines involved feel a responsibility for accomplishing the mission of the team, the term "integrated product definition" (IPD/CE) was used to connote concurrent engineering at MCAIR. At the company a total quality management system was in place which used empowered teams, disciplined systems and processes, as well as a supportive cultural environment in achieving customer satisfaction. Multidisciplinary teams called "product developers" worked cooperatively and simultaneously to create a single definition of the product and process. The output of the teams' effort was a documented "build to packages," which contained all of the information required to fabricate or assemble a first-time quality product. The IPD/CE process is depicted in Fig. 38.4. In using this process the following advantages were gained:

- The requirements of manufacturing, support, and business considerations were considered up front in the design process. This permitted early risk assessment as an inherent part of the IPD/CE.
- Conceptual layouts to include key process descriptions were released, which helped instill discipline in the configuration development process.
- Compatibility between structure and the subsystems of the product was checked through the use of computer graphics fixtures.
- Build-to-package planning and scheduling was done collectively through the team, which provided an integrated master schedule, thus eliminating the need for separate functional schedules.
- Fully sized assembly layouts were released prior to the build-to-packages release. This facilitated bringing forth the final objective of the team's effort, a build to package, the contents of which are shown in Fig. 38.5.

A second multidisciplinary team was formed at MCAIR to consider operational considerations as the team prepared a formal IPD implementation plan.

Success with the initial efforts of MCAIR in concurrent engineering has moti-

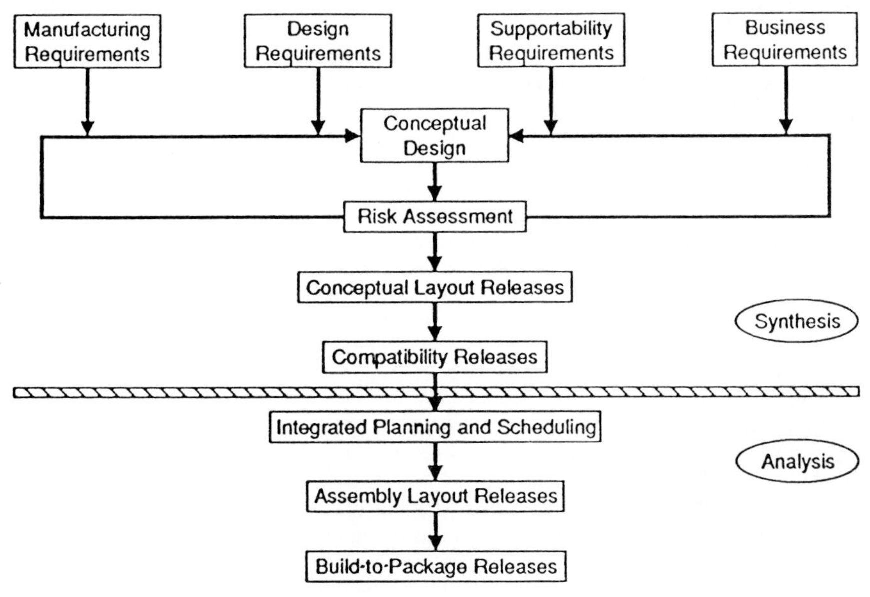

FIGURE 38.4 IPD/CE process. (*From McGovern and Grossman,*[12] p. 100.)

vated a series of future strategies to expand the company's application of concurrent engineering. These include (1) formation of a new quality process division, (2) use of two high-level teams to develop strategies for implementing concurrent engineering in the procurement of avionics and airframe systems equipment, (3) development of guidelines to extend concurrent engineering to logistics definition packages, (4) development of new computer graphics capable of fully integrating CAD, CAE, CAM, and CALS (computer-aided acquisition and logistics support) software, (5) considering the potential applications of artificial intelligence to IPD/CE, (6) training of

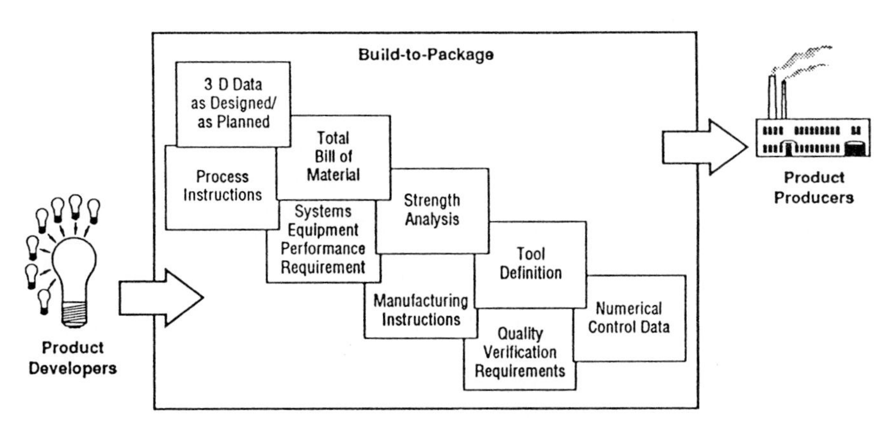

FIGURE 38.5 Contents of build to package. (*From McGovern and Grossman,*[12] p. 101.)

product and process personnel in Taguchi methods and quality function deployment, and (7) a commitment to implement IPD/CE to the maximum extent possible on future full-scale development efforts. Certainly the early successes in the development and implementation of concurrent engineering at MCAIR has had an immediate influence on the way the company develops its products and processes. This influence has also affected the culture of the company, and will doubtlessly influence the manner in which the company does business with its customers. One influence will be to encourage customer representatives to be collocated with the product-process development team. (McGovern and Grossman.[12])

There is an early precedent for concurrent engineering, the Skunk Works strategy. The famous Lockheed Skunk Works developed early high-performance aircraft and probably never heard the term "concurrent engineering." At the Skunk Works the design methodology incorporated many of the characteristics we write about today under the umbrella of simultaneous engineering. These characteristics included a sacred schedule, teamwork, concurrent design for manufacturing, and ongoing team review of product-process design.[13]

Since World War II, Lockheed's Advanced Development Projects (known as Skunk Works) has successfully led to top-secret, highly advanced aircraft for U.S. military and intelligence agencies. These aircraft were produced in an ambience of creativity and innovation, under the leadership of Clarence "Kelly" Johnson. The idea behind the Skunk Works operation was to encourage bright people to come up with creative ideas and make those ideas work around seemingly insurmountable problems. A few basic operating rules had been established for the Skunk Works. For example:

1. The Skunk Works manager must be delegated practically complete control of his or her program in all aspects. The manager should report to a division president or higher.
2. Strong but small project offices must be provided by both the military and industry.
3. The number of people having any connection with the project must be restricted in an almost vicious manner. Use a small number of good people.
4. A very simple drawing and drawing-release system with great flexibility for making changes must be provided.
5. There must be a minimum number of reports required, but important work must be recorded thoroughly.
6. There must be a monthly cost review covering not only what has been spent and committed, but also projected costs to the conclusion of the program. Don't surprise the customer with sudden overruns.
7. The contractor must be delegated and must assume more than normal responsibility to get good vendor bids for subcontract work on the project.
8. The inspection system as currently used by Advanced Development Projects (ADP), which has been approved by both the Air Force and the Navy, meets the intent of existing military requirements and should be used on new projects. Push more basic inspection.
9. The contractor must be delegated the authority to test the final product in flight. He can and must test it in the initial stages. If he does not, he rapidly loses his competence to design other vehicles.
10. The specifications applying to the hardware must be agreed to in advance of contracting.

11. Funding a program must be timely so that the contractor does not have to keep running to the bank to support government projects.
12. There must be mutual trust between the military project organization and the contractor, with very close cooperation and liaison on a day-to-day basis. This cuts misunderstanding and correspondence to an absolute minimum.
13. Access by outsiders to the project and its personnel must be strictly controlled by appropriate security measures.
14. Because only a few people will be used in engineering and most other areas, ways must be provided to reward good performance by pay not based on the number of personnel supervised.

A new R&D vice president initiated a "continuous improvement" strategy for his corporation. This included a way of ranking development projects based on customer needs, assigning teams of design engineers, materials purchasers, and manufacturing engineers to work on major design projects. More formal communication between the shop floor and the design center was carried out. Machine tools were, insofar as possible, standardized. The result: a 34 percent drop in average cost of design engineering for new parts; a 26 percent decline in the ratio of engineering overhead—including specification, database, quality assurance—to direct engineering; and a 40 percent drop in product development time.[14]

DESIGN FOR DISASSEMBLY

In the process of concurrent engineering companies are designing products so that they come apart more easily—in one sense to make the products more environmentally responsive. Called "design for disassembly," the fewer the parts, the easier it should be to dispose of the product after it has served its useful life. Square D Company's Trilliant original circuit-breaker metal box contained 173 parts. The redesigned injection-molded plastic Trilliant has only 42 parts. Fewer parts mean fewer parts to manufacture, reducing product cost.

Some key considerations in carrying out a design-for-disassembly strategy include:

- Design with ease of separating, handling, and cleaning of product components.
- Develop two-way snap fits or break points on snap fits.
- Identify break points in materials, such as molded-in logos.
- Apply tight-tolerance design principles so that parts are consolidated and fasteners are reduced substantially.
- Reduce energy usage in production and disassembly by eliminating unnecessary manufacturing, assembly, and disassembly steps.

Like concurrent engineering, design for disassembly is a long-range strategy through considering the full life cycle of the product, from conception to disuse and destruction at the end of its useful life. Sometimes getting rid of the product costs more than the product itself. Wooden pallets are an example. It costs $3 or $4 to build one, but it costs $1 to $5 to get rid of them in a landfill. The entire life-cycle costs approximate $7 to $9.[15]

The concept and process of concurrent engineering is still emerging. There are some misconceptions as well as benefits.

MISCONCEPTIONS AND OTHER CONSIDERATIONS

Concurrent engineering is not a magic formula for success. It is an approach that improves the efficiency of people, but is no guarantee of success. Concurrent engineering does not eliminate any of the phases of the sequential process of developing products and processes. Rather all of the downstream processes are codesigned toward an all-encompassing, cost-effective optimum design. It is not the simultaneous or overlapped design of production; it is the simultaneous design of the product and of the downstream processes. It is not beginning production of an item that has not completed its test, evaluation, and fix phase. Concurrent engineering is not the same as conservative design. Conservative design seeks robustness by using detachable parts, redundancy, close tolerances, and so on. Nor does concurrent engineering imply conservatism with respect to the integration of new technologies in the product. Although concurrent engineering is dependent on total quality management, it does not imply conservative inspection strategies. Instead it seeks to achieve manufacturing repeatability through product robustness and by designing a manufacturing process that includes the means for monitoring and controlling itself. (After Winner et al.,[2] pp. 21–22.)

Benefits from concurrent engineering reported in a study done by the Institute for Defense Analysis (Winner et al.,[2] p. vi) include:

- Through improvement in the quality of designs, dramatic reductions in engineering change orders (greater than 50 percent) were realized.
- Product development time was reduced by as much as 40 to 60 percent.
- Manufacturing costs were reduced by as much as 30 to 40 percent.
- Scrap and rework was reduced by as much as 75 percent.

The use of design-process teams in the design development of weapon systems has had both success and failure. The failures resulted in part from putting people on the teams who did not have the authority to commit the organizations. At one aerospace firm problems developed in putting the skilled people in enough teams to do 100 percent of the technical work. The approach now is to select the high-leverage areas—the areas where the most impact can be gained—and assign teams to these areas. By getting two strong executive vice presidents at the corporate level together, one on the engineering and one on the operations side, at off-site meetings a lot of barriers between these two organizational units have been reduced. A better working relationship has come out of these meetings, that is, a better understanding of each other's problems than had existed previously. Through the use of a CAD system and a more effective communications system, a much greater exchange of both engineering and manufacturing information was possible.

SOME DIFFICULTIES

The introduction of simultaneous engineering into an organization is not without its difficulties and pitfalls. Concurrent engineering is a team effort. Anything that detracts from the efficiency and effectiveness of the teams has the potential to create difficulties and impair the potential success of engineering. Some of the common difficulties that exist include the following:

1. Failure to properly prepare the people for the change. The team members will continue to have a loyalty to their functional departments. If they have not been adequately indoctrinated in the need for the change and have not been given the required training, the chances are good that they will resist the change to the detriment of an effective concurrent engineering effort.

2. Failure of the team to engage in open and intense communication about the technical and managerial aspects of the concurrent engineering effort.

3. Working toward unrealistic product-process development schedules. Going from the traditional to the new product-process development system simply takes time. If overly optimistic schedules are locked in at the beginning of the development process, and there is unusual pressure to adhere to those schedules, there is the chance that people will take risky shortcuts in order to meet the schedules.

4. Change in product performance specifications. Nothing is more frustrating to a team than to have the product specifications changed several times during the development effort. Some of these changes are unavoidable, such as changes in customer needs for features that are important to the customer. Many products continue to evolve during the development process; changing product specifications will cause the product design to change, which will cause the tooling, manufacturing, after-sales service, and supplier strategies to change. Changes should be kept to an absolute minimum. If the product specifications are tentative because of insufficient information on customer needs, then everyone should be informed that the team is working with a moving target, and the team should conduct its studies and work accordingly.

5. Failure to bring the vendors into the spirit and intent of the concurrent engineering processes. In preparing the vendors' people for the change, to include adequate training is essential to the success of the product-process development effort. The procurement function in companies has been subject to the growth and maintenance of bureaucracy just like other organizational functions. Reducing this bureaucracy, reducing the number of vendors, getting vendors to assume more responsibility for the design of parts and subassembly, and improving the quality of the supplies they furnish to OEMs, all require substantial change to their thinking and in their operations. This simply takes time.

6. Failure to have top management commitment and support to the simultaneous engineering effort. If top management announces the intent to change the product-process development approach, allocates the needed resources, and then goes back into their offices hoping that the change will come about in due course, they are in for a big disappointment. Senior management must develop and articulate the vision for concurrent engineering and then have ongoing visible presence to lead the organization to its new posture in product-process development. Without their leadership, the new initiative in product-process will not be successful.

If there is not effective communication among the people involved in concurrent engineering, difficulties will exist.

COMMUNICATION

There has always been some degree of effectiveness in communication in the development of products. Most of this communication has been more accidental than planned. Little has been done by way of specific strategies to improve communication between product and process designers. The finale of throwing the product design "over the wall" to the manufacturing engineers was a firm method of communication, but primarily one way. Informal chats at the coffee machine or during luncheon no doubt helped to facilitate better communication. Today when companies are rushing to get their products commercialized sooner, the informal or accidental means of communication will no longer suffice.

Communication has always been an important consideration in the culture of an organization. Poor or nonexistent communication can impair or even destroy the ability of any group of people to work together to produce a desirable result. It is no different on a product-process design team. Some companies have taken special care to provide an ambience which facilitates communication.

A chief project engineer for the New Airplane Division of Boeing Commercial Airplanes stated:[16]

> Although we have just implemented this design build (simultaneous engineering) process two observations are worth noting. The first is that co-location of the team members is essential to their success. Communication is most important and the traditional meetings, telephone calls and memos will not suffice. There is no substitute for sitting with the other team members. Facilities must be organized around product teams and not the traditional functional organizations.

Information is critical to the practice of concurrent engineering. Information flows are multidirectional, and decisions involving the product-process design are multidirectional as well. Sharing of information in the concurrent management of the organization will require both organizational and technological change. Advocacy and full support by senior management are essential.

Some companies deal with the need for enhanced communication by putting the manufacturing engineers in the design team's office where a formal product design team has not been organized.

Flynn advanced the hypothesis that the computerization of the design and manufacturing functions in organizations brings existing communication and coordination problems to light and exacerbates these problems.[17]

Communication is the essence of concurrent engineering. In the engineering and manufacturing community the management of engineering documentation is a critical matter. Engineering copiers, printers, plotters, associated peripherals, software, and other products for the management of engineering documentation are critical to the concurrent engineering process. The design and test of new products through the utilization of simulations on animated, three-dimensional digital workstations powered by supercomputers is a technology that is coming. By relying on simulations, a manufacturer can reduce the time and cost of getting a product from design drawing to the production line. Information technologists are working toward the building of a corporatewide information database for design. The linking of this information system with the minds and memories of design and manufacturing engineers will provide a real magic in the better design of products and manufacturing processes.

Communication has always been an important matter in the success of organizations. Today it is no different. The complex splintering of technology creates abundant opportunity for people to be isolated from members of their peer group who work in a different technology and in a different part of the organization. One of the clearest benefits of using multidisciplinary product-process design teams is the opportunity to bring the people together to talk directly to their counterparts in the same organization.

Concurrent engineering really means that management concepts and processes have to be rallied to bring about more effective purposeful *communication* among the functions of the organization, in particular, between the design and manufacturing functions of the enterprise. Communication is the heart and soul of concurrent engineering. The technology involved in concurrent engineering is mostly an enabler—people in their social and cultural settings have to make the process work.

Timely and accurate information is essential to the concurrent engineering process. Some companies are making concentrated efforts to innovate supportive information systems to support concurrent engineering. For example, at the Cornell Design Research Institute scientists and engineers are working to eliminate the need for physical prototypes. By creating a new design strategy, combined with a process for testing new products and components through simulations on animated three-dimensional digital workstations powered by supercomputers, it is hoped that further cuts in the time and cost of getting a product from design drawing to the production time can be made. In addition to research into finding ways to eliminate physical prototypes, Xerox and Cornell are devising ways to build a corporatewide information database for design, called a corporate memory of design. When operational, this memory will enable the designer to obtain instant on-line access to corporatewide data on product specifications, engineering drawings, previous design simulations, discarded designs, supplier price and specification lists, service histories, and updated status reports on all phases of product and process development. The corporate memory of design will enable everyone to talk with everyone else in the corporation. In other words, all corporate information will be available at the designer's fingertips to enable good design decisions to be made. Additional research complementing the corporate memory of design will include the exploration of how organizational structures will impact the design and engineering of products.[18]

Making the decision to use concurrent engineering is the easy part, implementing it is the real challenge.

IMPLEMENTING CONCURRENT ENGINEERING

One of the first things that a company needs to do when considering the implementation of concurrent engineering is to knock down the traditional functional departmental barriers, whether real or imagined. These barriers have become institutionalized in a company that has been around for a long time, particularly if no special effort has been undertaken to integrate interfunctional cooperation. In some companies an individual who is an expert in both product and manufacturing engineering serves as a "guru," resists the parochialism of the other functional specialists, and because of a senior engineering reputation, brings the functional specialists together.

The recognition of the need to change the way products and processes are designed through the use of product-process design teams has to start at the top, with a vision communicated by the senior leaders of the company. Once the vision has been established and general guidelines have been given on how the teams should be organized and operated, the implementation of the teams' work must come from these same senior managers—they must assist the teams' work with guidance and leadership from the top down. It will not and cannot start at the bottom of the organizational structure and seep up through the organizational hierarchy.

The parallel product-process design team yields more creative designs and more communications than does the "over the wall" approach. When concurrent engineering is instituted, nearly everything is influenced and has to be changed in the enterprise. The organizational structure, roles, authority, responsibilities, and review and reward systems have to be overhauled. Team members are shifted to be near each other. By having them work in close proximity rather than having them joined electronically, the level of communication and decision-making improves.

As many specialists as needed should be on the product-process design team. Members should be chosen for their "compatibility chemistry" as well as the technical expertise that they bring to the team. The team leader is the most common cause of success or failure. This leader has to lead the team to a meaningful conclusion in the design of the product-processes on time, on budget, and expressed in effective design of both the product and the processes. The leader leads the people by encouraging meaningful debate and adverse positions in reaching a decision. The activities of the product-process design team cannot be a "love-in"—it has to be an objective evaluation of product and process design decisions, properly evaluated and executed.

The product-process design implementation strategy usually consists of several distinct steps in the concurrent design of both products and the supporting processes. These steps include:

- The development of a strategic direction for the enterprise, which includes the probable and possible product development strategies that support the technological and business objectives of the firm.
- A more definitive selection of the product-process development projects that support existing and future customer needs. Description of these needs means that the product performance specifications have been determined and adequately defined to be set up as a product-process research project to support corporate purposes.
- The development of plans to carry out the product-process realization strategy, such as delineation of the needs for the evolution of the product beyond the design stage and through the entire product life cycle, including after-sales support to customers in terms of the "bundle of services" needed by them to support their operational needs.
- In all of these steps, assurance that the entire product life cycle, including distribution, support, maintenance, recycling, and disposal, is given full consideration in the planning and execution of the product-process realization process.

THE ORGANIZATIONAL DESIGN

In carrying out the concurrent engineering activity, the hard part is the organizational aspects. Years of engrained bureaucratic cultural values become threatened—the stakes that managers and professionals alike hold are challenged. Concern for the organizational design considerations and remedies for dealing with this challenge are found in the literature and in the teachings and counseling provided by experts in the field. But all of the literature and teachings do not do the job. People have to get used to the change to an end-item-oriented product-process development team which has the following characteristics:

- A matrix organizational design, providing for the management of a complete "turnkey" design-manufacture package to manufacturing.
- An interdisciplinary team project leader–manager who has the leadership skills and legal authority to manage the team and work with the many stakeholders who have vested interests in the product-process activities as well as the outcome of the team's work. Although the disciplines represented on the team vary depending on the particular nature of the weapon system, the following disciplines are usually

represented: the traditional engineering disciplines, integrated logistic support, reliability, maintainability, manufacturing engineering, tool design, tool fabrication, training, and spares.

Some of the internal barriers to concurrent engineering are lack of respect, understanding, and empathy among the various functions of the enterprise. Our egos get in our way. The designers' egos often get in their way—they like to do it all, complete and ready to turn over to the manufacturing people. We have difficulty working in a teamwork mode. These difficulties come in part from our heritage of individualism—reinforced by our early education—where cooperation and teamwork were discouraged, even penalized if the teamwork resulted in "cheating" in doing our school homework and in taking our examinations.

Although there has been some success in the use of product-process design teams, much needs to be learned about how to improve the use of such teams in the concurrent engineering context. Issues in the management of these teams relate to the use of project management concepts and processes, including the manner and style in the design and implementation of planning, organizing, motivation, leadership, and monitoring and control functions. The creation and use of cross-functional teams characterized by the matrix organization requires special consideration. Communication issues relating to the creation, filtering, transfer, and use of information among the team members requires special attention. The internal aspects of information both within the company and with customers and suppliers deserve the fullest mindfulness. Models need to be developed on how information is and should be exchanged, and how multidisciplinary teams are expected to operate in the concurrent product-process design activities.

The initiation of concurrent engineering processes in an organization influences the organizational culture.

CULTURAL CONSIDERATIONS

One distinguishing characteristic of the concurrently managed product-process design process is the breadth and depth of networks—both horizontal and vertical—that link information and people together in the making and implementation of decisions. For example, in the design of the Boeing 777 a huge computer network with international partners on line enables engineers to iron out bugs on video screens, where fixes are cheap, instead of on expensive life-size models, called mock-ups. The new concurrent engineering process brings together representatives from design, production, and Boeing's outside international partners and suppliers, with regular input from airline customers, maintenance, and finance. Boeing hopes this concurrent engineering approach will save as much as 20 percent of the 777's estimated $4 billion to $5 billion development costs.[19] When these networks function efficiently, decisions are made faster, although the decisions do not necessarily follow the formal decision structure of the enterprise. For example, the engineering design manager's role in conducting design reviews and in approving the final design will be principally preempted by the product-process design team, which will be reviewing designs by the team members on an ongoing basis. When the time comes for the final approval of the design, the design engineers and the manufacturing engineers will have largely made the required decisions through the networks and coalitions that they built. Although the formal authority may have rested with the engineering design manager, the de facto authority has come out of the deliberations carried out by the product design team within the networks and coalitions that have been built.

In the simultaneous organization ongoing education and learning become requirements for survival. Learning is required to upgrade the knowledge and skills of the people to better perform their jobs, as well as to teach them how to participate in a culture that is becoming more characterized by consensus and consent than the older traditional anachronistic command-and-control management style.

Organizational cultures can block change from the old design to the concurrent design process. Old habits can enter in and keep people from accepting the consensus-and-consent organization and wanting to know more about the customers, their suppliers, and the competitors' products. When this dissatisfaction appears, the time is right to work with the team in broadening their responsibilities in the organization.

One characteristic of the design process that is managed concurrently is that the managers seem to have easy jobs; they do not seem to be harried. Operational processes seem to flow smoothly. People work at exploiting opportunities more than they do at solving problems. Problem solving is done on an anticipatory basis rather than fighting fires and solving crises. Steady progress in the organization comes from the success of the product-process development work. When crises and problems arise, teams go to work to solve matters, and to review existing products and processes to see what remedial action is required to fine-tune policies and procedures in order to prevent the same problems coming up again. Organizational problem solving in concurrent engineering is facilitated by asking a few questions like these:

- What is the problem, and what are the underlying symptoms? Be sure and understand the differences. A perfect solution to a symptom does not solve the root problem.
- Has this problem occurred before? If the problem has come up before, why did our solution at that time not provide a remedy?
- Where is most of the effort being placed, on problem solving with suitable resolution through improved policies and procedures, or on solving problems that have come up before? If the same problems are appearing again and again, suitable solutions have not been designed and implemented.
- If the same problems keep coming up, are these problems a symptom of new or larger problems as yet unfathomed?

Several common characteristics have been identified with companies that have successfully deployed concurrent engineering (Winner et al.,[2] p. 20):

1. Upper management supported the initial change and continued to support its implementation.
2. Changes were usually substitutions for previous practices, not just additional procedures.
3. The members of the organization perceived a need to change. Usually there was a crisis to be overcome. Often the motivation seemed to center around retaining or regaining market share.
4. Companies formed teams for product development. Teams included representatives with different expertise, such as design, manufacturing, quality assurance, purchasing, marketing, field service, and computer-aided design support.
5. Changes included relaxing policies that inhibited design changes and providing greater authority and responsibility to members of design teams. Companies practicing concurrent engineering have become more flexible in product design, in manufacturing, and in support.

6. Companies either started or continued an in-place program of education for employees at all levels.
7. Employees developed an attitude of ownership toward the processes in which they were involved.
8. Companies used pilot projects to identify problems that were associated with implementing new concurrent engineering techniques and to demonstrate their benefits.
9. Companies made a commitment to continued improvement.

Much of the success of concurrent management is due to the quality of leadership provided to the product-process design team. Leadership of the team is about coping with change that the team needs to accommodate in order to survive. In coping with change, the team leader helps in finding the product-process design *vision* that is needed to provide the general pathway for change in the organization. Achieving the vision requires motivation and inspiration—keeping people moving in the right direction, overcoming obstacles, building networks with the team's stakeholders, and appealing to their need for accomplishment. The leader works at aligning the people to support the vision they hold. Aligning the people is a communications challenge, talking to all of the team members and the other stakeholders who have, or feel that they have, a vested interest in the team's work and its output. The leader's communication must be credible—people have to believe the message that the leader is sending forth. When people feel that they are properly aligned with the leader and team's vision, they feel empowered and develop a sense of belonging, recognition, self-esteem, a feeling of having one's work and life in control and the opportunity to reach one's objectives for self-fulfillment. This is real *empowerment* which provides the team members the strength and motivation to produce quality results (after Kotter[20]).

CUSTOMERS AND SUPPLIERS

The integration of customer input throughout the deliberations of the product-process design team makes considerable sense. Customers usually know what they want. Too many companies have relied too much solely on engineering in the design of products. Some design engineers forget about the customer's need for simplicity, and for a product that operates when the customer wants it to operate.

In concurrent engineering special efforts should be constantly under way to build a better supplier-customer relationship. Quality improvement and the reduction of costs are distinct advantages to be pursued in such a relationship. The relationship is characterized by mutual respect, trust, and benefit. Mutual respect and trust come out of careful planning and a constant working to build mutual understanding between the parties. Ideas have to be exchanged, information and processes have to be shared, and mutual knowledge of each other's processes and values helps to build a relationship of trust and mutual understanding. Some specific steps that can be taken to build a mutual understanding and trust include:

- Commit to and build toward a trusting relationship through mutual understanding of reciprocal problems, opportunities, and strategies.
- Establish clear accountabilities and responsibilities so that each party to the relationship understands these requirements that have to be met to provide mutual benefit to the parties.

- Select suppliers that have the knowledge and wherewithal to do quality work.
- Once the requirements have been established, work toward building a relationship that emphasizes conformance to the product and the relationships stated in specific conformance requirements.
- Develop a mutually acceptable system of measurement to determine whether both parties are carrying out their obligations and respecting the obligations of the other party.
- When deviations occur, correct these deviations and investigate what caused the deviations to see whether remedial action is required to realign the resources in order to preclude the deviation from coming up again.

Concurrent engineering is paralleling the pace of change in manufacturing.

CHANGES IN MANUFACTURING—TOWARD CONCURRENCY

The modern manufacturing business is becoming transformed into a systems design whereby the manufacturing functions are seen as one integrated process, starting with product-process ideas through production, delivery to the customer, and providing after-sales support—indeed providing the customer a "bundle of services" not just a product or service. To follow this systems concept requires that the entire business be redesigned and reorganized so that products and services provided by the enterprise can be managed as an integrated flow from internal producers to external customers.

Manufacturing does not stop when the product is produced; physical packaging and distribution, education of the customers, and providing of quality after-sale service are part of the overall production function. To emphasize this philosophy, some companies guarantee 24- and 48-hour replacement of spare parts anywhere in the world. Caterpillar and Deere, two giants in the construction industry, abide by this policy.

The emergence of new technologies in the design and processing of products and services creates the conditions for making the manufacturing company an integrated systems-oriented enterprise. The technologies of just-in-time inventories, total quality management, computer-aided design and computer-aided manufacturing, total information systems, flexible manufacturing, computers, and concurrent engineering have been the main contributors to the crusade to view the manufacturing organization as an integrated flow challenge. Some industries, such as petroleum refining and large-scale construction, practice integrated systems flow in the management of the enterprise.

The traditional factory has been managed mostly from a functional compartmentalized, "series" basis, where the organizational functions were carried out as successive and often isolated steps in producing goods and services. The use of project teams in the concurrent engineering activity, where product-process design teams are used to bring functions into a parallel and simultaneous process, has improved competitiveness through lower costs, higher quality, and earlier commercialization of products and processes. These product-process design teams have brought the various organizational functions together from the inception of a new product or process project. The use of such multidisciplinary teams has helped bring about the integration process of tying everything together. The technologies that have been mentioned have both caused and facilitated the shift from the series organization to the parallel and simultaneous organization. The impact of these technologies has been felt most acutely on the human and social systems of the manufacturing organization. Manufacturing managers and professionals have to change and think like business or total organizational people, since their technical decisions will impact the total organization in its integrated systems flow of activity, as products and processes are moved through their life cycle in the flow of activities in the enterprise.

THE FUTURE OF MANUFACTURING

Manufacturing employment is headed down, at least through the middle of the decade. This prediction is expressed by an analysis of the effects of scheduled cuts in defense spending on U.S. industry, released by the Manufacturers' Alliance for Productivity and Innovation (MAPI), a policy research organization representing some 500 major industrial companies. According to the MAPI prediction, some 405,000 defense jobs will be cut. In addition, labor markets will have to cope with the loss of about 250,000 defense-related nonmanufacturing jobs plus declines in DoD armed services and civilian payrolls averaging 119,000 full-time and 14,000 part-time jobs a year. Although defense accounted for only 6.7 percent of manufacturing output in 1990, some industries such as steel mills and foundries, electrical machinery, materials handling and metal working machinery, missiles and space vehicles, tanks, and shipbuilding will be impacted by defense spending. Growth in aircraft industries will also slow down significantly.[21]

In the face of continued DoD cutbacks, concurrent engineering and advance manufacturing are likely to become an increasingly important means for DoD to reduce manufacturing costs and get more weaponry and support systems for its investment. Higher-quality products will be demanded through manufacturing processes designed to produce defect-free products that provide the desired technical performance capabilities on time and within budget.

SUMMARY

More and more product-process development is being done through the use of product-process design teams. Such teams are one of the best ways to pull together the interdisciplinary efforts required to commercialize technology. These teams save time through reducing the resources required to redo work—such as through engineering changes—coordinating communication among the often disparate functional elements in the organization, facilitating reviews and approvals, and passing work on to other areas and people of the company. By having the multifunctional teamwork together there will be a reduction in the risk of poor decisions on both the product and the process. Successful teams tend to exhibit many important characteristics, such as a respect for the interdisciplinary effort that is required and a recognition of the equality of the disciplines. These teams also help, because they tend to take on many administrative tasks, to cut overhead, to reduce the product-process development costs as well as time, and to improve quality.

Concurrent engineering is an idea whose time has come, and is moving forward in its application in defense work. The changes being brought about by concurrency in product and process design will continue. In the not too distant future the concurrent engineering process will become the accepted and "traditional" way of developing, manufacturing, and supporting military weaponry. But as this innovation becomes institutionalized, there is little doubt that new innovations in the management of product-process technology will come forth and change things again.

One thing is certain about the use of product-process design teams in a concurrent engineering aproach: A more adaptive organization that values true participation and shared results coming about through more creative people will emerge.

REFERENCES

1. R. A. Melcher et al., "Europe's Weapons Makers Launch Operation Market Shield," *Business Week,* pp. 44–45, July 8, 1991.
2. R. I. Winner et al., "The Role of Concurrent Engineering in Weapons System Acquisition," Rep. R-338, Institute for Defense Analysis, Alexandria, Va., Dec. 1988.
3. J. Ayres, VP, Product Development, Aircraft Division, LTV Aerospace and Defense Company.
4. K. L. Adelman and N. R. Augustine, *The Defense Revolution,* ICS Press, Institute for Contemporary Studies, San Franciseo, Calif., 1990, p. 178.
5. T. M. Rohan, "Designer/Builder Teamwork Pays Off," *Industry Week,* pp. 45–46, Aug. 7, 1990.
6. "World-Class Quality: The Challenge of the 1990s," *Fortune,* p. 2, Jan. 21, 1991.
7. "DARPA Initiatives in Concurrent Engineering (DICE)," pamphlet, Concurrent Engineering Program, GE Aircraft Engines, Cincinnati, Ohio.
8. B. Dumaine, "How Managers Can Succeed Through Speed," *Fortune,* p. 56, Feb. 13, 1989.
9. "Concurrent Engineering, Global Competitiveness, and Staying Alive: An Industrial Management Roundtable," *IM Manage. Roundtable,* pp. 6–10, July-Aug. 1990.
10. P. M. Noaker, "How the Best Make It," *Manuf. Eng.,* pp. 52–54, Nov. 1988.
11. "Texas Instruments Has a Good Defense," *Industry Week,* pp. 56–62, June 17, 1991.
12. D. R. McGovern and D. T. Grossman, "Concurrent Engineering at McDonnell Aircraft Company," in *3d Annual Best Manufacturing Practices Workshop Proceedings* [San Diego, Calif., Sept. 1989, Office of the Assistant Secretary of the Navy (Shipbuilding and Logistics) Reliability, Maintainability and Quality Assurance Directorate], pp. 93–100.
13. B. R. Rich, "How the Stealth Fighter Got Off the Ground," *Wall Street J.,* June 4, 1990.
14. M. F. Blaxill and T. M. Hout, "The Fallacy of the Overhead Quick Fix," *Harvard Business Rev.,* pp. 93–101, July-Aug. 1991.
15. "Design for Disassembly," *Industry Week,* pp. 44–46, June 17, 1991.
16. R. Johnson, Boeing Commercial Airplanes, Seattle, Wash., personal communication, Aug. 27, 1990.
17. M. S. Flynn, in *The Integration of Design and Manufacturing for Deployment of Advanced Manufacturing Technology,* presented at the TIMS/ORSA Joint National Meeting, Los Angeles, Calif., Apr. 1986.
18. "Talk This Way," *Template—The Magazine of Engineering Systems and Solutions,* pp. 14–17, Winter 1991.
19. D. Jones Yang, "Boeing Knocks Down the Wall Between the Dreamers and the Doers," *Business Week,* pp. 120–121, Oct. 28, 1991.
20. J. P. Kotter, "What Leaders Really Do," *Harvard Business Rev.,* pp. 103–111, May-June, 1990.
21. "Defense Cuts Deliver a Body Blow to Factory Jobs," *Business Week,* p. 18, June 17, 1991.

CHAPTER 39

INFORMATION RESOURCE MANAGEMENT

John F. Phillips

John F. Phillips is the commander of the Joint Logistics Systems Center (JLSC), which was officially established at Wright-Patterson AFB, Dayton, Ohio, on January 1, 1992. The JLSC is the single organization responsible for Department of Defense information systems supporting logistics processes.

INTRODUCTION

The continuing advances in technology and the increasing complexity of projects have greatly magnified the amount of information a project manager must have available to do his job competently. The times when the project manager could personally "manage" the entire operation, when he could keep in his mind or, literally, on the back of an envelope all the pertinent information regarding the task at hand have long since passed. No longer are all of his staff within shouting or even walking distance, but rather they may be scattered in several buildings or even distant cities. The problem of "keeping on top" of everything is magnified by the introduction of engineering change proposals (ECPs), technical change proposals (TCPs), and a myriad of other factors which alter the original objectives, or at least the means of achieving those objectives. In addition to technical performance, he must also be equally concerned with schedule and cost. It has, therefore, become necessary for the manager to have available a storehouse of information and a tool box of systems or capabilities which enable him to maintain the required cognizance over everything that is transpiring. Whether he enjoys "hands-on" time at the keyboard or he prefers only, albeit reluctantly, to even acknowledge the existence of automated information systems, the successful manager achieves a reasonable level of computer literacy. He recognizes the need for information, quickly learns what is available, requests development of capabilities to acquire that which is not, and ensures that he and his staff use all relevant data on a timely basis.

The dilemma has been made less threatening with the development of a program management support system (PMSS) which, when appropriately applied, can generate a wealth of useful information for the program manager. Inherent in PMSS are the capabilities of planning, networking, scheduling and resource loading, configuration and data management, financial management, measurement of progress and performance efficiency, and reporting. Sophisticated data automation systems greatly facilitate the capture, manipulation, processing, and display of these voluminous data. Unfortunately some managers elect to ignore PMSS and its products, preferring instead to "fly by the seat of their pants."

Much of the guidance and direction relative to acquisitions has been directed at weapon systems since they constitute the lion's share of all acquisitions. Likewise, tools developed to assist the program manager in fulfilling his responsibilities have largely been targeted for this arena. The continual growth, however, in the acquisition of automated systems and the application of information resource management has resulted in the publication of specific guidance for this type program at both DoD and component levels. As headquarters Air Force Logistics Command (AFLC) initiated their Logistics Management Systems Center (LMSC) and a $1.7 billion effort to modernize outdated logistics systems with nine large information resources programs, they elected to apply the available acquisition guidance and tools and further decreed that PMSS would be a standard management tool to be used by each of the LMSC program offices. In this chapter we will zero in on these tools and the support system and see how available information resources can facilitate the job of the project manager.

RISK MANAGEMENT

Any effort which involves the design, production, and implementation of a new product or service encompasses considerable risk. Basically there are two types of risk, technical and business. Because technical information is generally visible and reasonably accessible, technical risk is normally recognized and consequently relatively easy to manage. Business risk, however, partly because it is not always acknowledged as an integral part of project control, is often overlooked. By overlooking or ignoring the role of business elements, management misses opportunities for reducing risk.

The effective management and use of information resources is a critical element in risk management and requires the establishment of standards which: (1) address the unique characteristics of a program, (2) ensure that adequate and timely communication exists between program participants, and (3) provide a series of checks and balances. Including the various business elements of a program in the decision-making process incorporates a business engineering approach into the program management philosophy. This coupled with the application of the PMSS tools, which will be discussed throughout this chapter, will not only lessen the opportunities for surprises but increase the chance of program success significantly.

PLANNING

Work Breakdown Structure—A Must for Planning Definition

Background. The concept of a work breakdown structure (WBS) has been in use within the military environment for several decades. In its infancy, however, and particularly as projects became more technically complex, a WBS was typically tailored to a functional manager's own effort, with little regard for the entire project. Consequently in 1968, MIL-STD-881A,[1] entitled "Work Breakdown Structure for Defense

Materiel Items," was published with an objective of achieving a single project WBS which would reflect the interrelationships of the several functional areas involved while maintaining and satisfying the individual manager's needs. In addition to explaining the concepts of a WBS and defining various elements, the Military standard also provides a summary WBS for the following systems: aircraft, missile, electronics, ordnance, ship, space, and surface vehicle.

What is a WBS? A WBS is a technique for segmenting a complete task or job into specified subcomponents which, when displayed, reflect the relationship of each component to one another as well as to the entire project. Commonly referred to as a family-tree-type structure, a WBS systematically divides a project into manageable pieces, each piece ultimately being assigned to a performing organization responsible for its accomplishment. These pieces can be thought of as products, and the sum of these products, be they hardware, software, services, data, or other work tasks, constitute the total effort. Thus level 1 of a WBS is the entire project or program (in the case of an LMSC software development program, the stock control and distribution system). Level 2 represents the major elements which comprise the total system (program management, system engineering, ADP hardware, software, data, and so on). Level 3 is a further breakout of subordinate elements under each of the level 2 parent items. These top three levels of the WBS are initially referred to as the preliminary, or summary, WBS. At LMSC, this preliminary WBS was often made part of the request for proposal (RFP), along with a requirement that offerors utilize the WBS in formatting their submissions. This greatly facilitated the source-selection process as all proposals by necessity followed a similar structured approach (Fig. 39.1).

Requirement. MIL-STD-881A has been mandated for use by all departments and agencies of the Department of Defense (DoD) and is applicable to each of the following types of projects (Ref. 1, p. i):

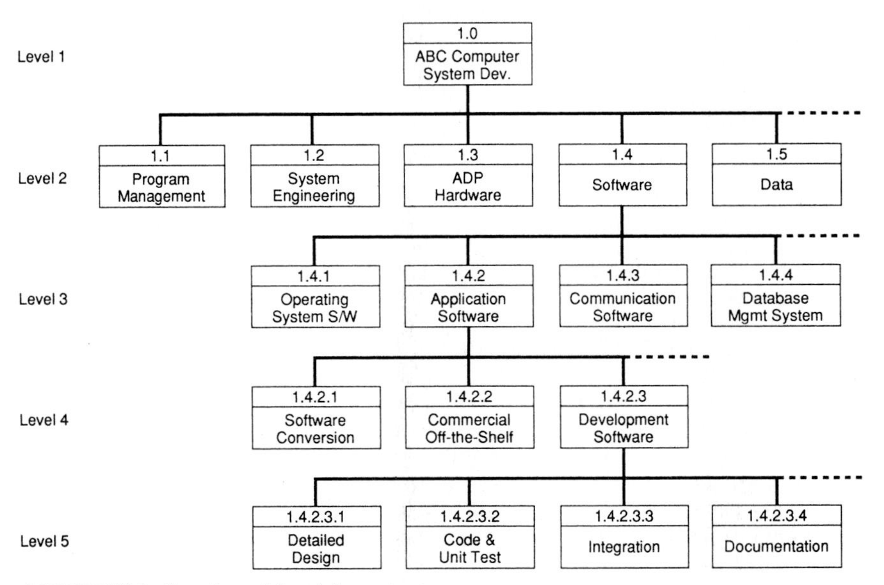

FIGURE 39.1 Sample work breakdown structure.

(1) Materiel items in the 6-year defense program.
(2) Materiel items or major modifications in projects which exceed $10 million in RDT&E financing.
(3) All production follow-on of (1) and (2) above.

A WBS may also be employed in total or in part for other defense materiel items at the discretion of the DoD component, or when directed by the Director of Defense Research and Engineering.

Although automated systems may not be thought of as materiel items, the acquisition of such a system frequently constitutes a major procurement effort, and the concepts of MIL-STD-881A apply. Thus the decision to use a WBS is not an arbitrary one, but rather one dictated by policy.

Other Reasons for Using a WBS. The WBS plays a significant role in just about every aspect of program management, literally from program conception until final operating capability. It is a key ingredient for program definition, solicitation, budgeting, scheduling, networking, performance measurement, and reporting. The WBS is a living document and is modified as required to reflect contractual changes approved by the responsible military component. It should be pointed out that the standard WBSs included in MIL-STD-881A are guides, for it is unlikely that any will exactly match your project, and it is entirely proper that you tailor your WBS to reflect what you want to see and how you want to manage. This is precisely what the LMSC community of AFLC did as they embarked on their massive software development effort. A standard WBS for software programs was developed. Each program management office (PMO) then tailored this standard WBS to conform to its unique program requirements. The WBS was expanded to the level required to acquire the desired visibility, sometimes to level 7, and then loaded into the computer. Each task and effort, regardless of size or duration, was now identified by code, thereby facilitating networking and scheduling. Budgeting, performance measurement, and reporting also keyed off the WBS. Everybody was speaking the same language, and traceability was improved significantly.

Summary. Because of the significant role a WBS plays in all aspects of program management, any attempt to manage a program of appreciable size without one is charting a course for disaster.

COST ESTIMATING

Certainly one question foremost in the mind of all parties involved in a development project is cost. This concern over cost is not a new phenomenon, one caused by continuing inflation, or, in the case of the military, a shrinking defense budget. Cost was a consideration some 2000 years ago when it was stated: "Suppose one of you wants to build a tower. Will he not first sit down and estimate the cost to see if he has enough to complete it?"[2] Regardless of how diligently this concept might be applied, the fact remains that with government development programs, cost overruns have become a way of life. Acquisition programs within the military, including software programs, are no exception.

"No sense in reinventing the wheel" is not an unfamiliar expression in the military community. To the contrary, it makes sense to use, whenever possible, a previously tried and proven technique. Following this logic, the LMSC community availed itself of several widely accepted models and methodologies in estimating the costs of its projects.

Methods for Cost Estimating

There are basically four techniques used to estimate the cost of various projects, namely, analogy, expert judgment, grass roots (bottom up), and parametric (top down), each with certain advantages, each with its own limitations. These techniques have generally stood the test of time. Their effectiveness and degree of accuracy have been established, and a range of error has been determined.

Analogy. The analogy method is the simplest of the four and involves comparing the new program effort to a previous program of similar nature. Whether the product under development is hardware or software, there is a chance that a similar effort has already been accomplished. Except for the Wright Flyer, there has always been a predecessor airplane for developers to use as a comparison or starting point in estimating aircraft costs. Thus the advantage of this methodology is that it is based on actual experience. The disadvantage is that pure similarity between programs is rare, particularly when talking about software programs. Another criticism of this technique is that costs due to inefficiencies or poor management of the analogous program are then built into the program being estimated.

Expert Judgment. This technique involves calling upon the "professional" judgment of recognized experts who have experience in the particular area involved. A problem arises, however, when there are no experts in that field, and there is the ever present danger of personal bias on the part of the individual.

Grass Roots. Also known as the bottoms-up technique, the grass-roots approach involves decomposing the task into small pieces, estimating the cost of those individual pieces, and then summing them up to determine the cost of each computer software configuration item (CSCI). A disadvantage, however, is that while the projected costs of the individual pieces may be fairly accurate, the cost of integrating these pieces into successively higher units does not share the same degree of accuracy. This technique can also be quite time-consuming, particularly in the early stages of an effort when the amount of detail is still somewhat limited. On the plus side, the estimating is often done by the individual responsible for accomplishing the actual effort and therefore feels a personal responsibility to "keep things on track."

Parametric. Just the opposite of the grass-roots method, the parametric, or top-down, approach estimates the overall project cost based on the design characteristics or parameters of the program. Compared to the other techniques, this method is relatively fast, and since it requires limited detailed information, is relatively easy to use. Likewise, statistical ranges can be established around point estimates. Accuracy may suffer somewhat, however, because of the high level at which the costs are estimated. Regardless, this method is the model of choice in many DoD activities because it has a low effort-to-estimate ratio.

Models in the LMSC

In the LMSC community a variety of parametric models, including COCOMO, SLIM, GE PRICE, and SPQR20, are used to estimate software development costs. All models have advantages and disadvantages. The key is to select one or maybe two models and be consistent in their application. LMSC has many different contractors, each using various models and techniques of choice to estimate software development effort. This in turn forces LMSC to use many different models in its analysis. "The Constructive Cost Model (COCOMO) is probably the most comprehensive, formal and exhaustive of the software estimating models."[3] With COCOMO offering three increasingly detailed models (basic, intermediate, and detailed), LMSC resorted chiefly to the intermediate version, which uses, as a primary input, size in terms of source lines of code. The derived size is then tempered with other parameters, such as personnel experience level, complexity, reliability requirements, and a dozen or so others. An added feature, one of significant interest to LMSC, is the ability of COCOMO to measure the annual levels of effort for software support.

Another model used quite extensively in LMSC is the software life-cycle investment model (SLIM). "This model, marketed by Quantitative Software Management (QSM), employs a mathematical relationship known as the QSM software equation to relate cost, or level of effort, to program size."[4] The primary inputs to SLIM include SLOC (source lines of code), a calibratable productivity factor, and a manpower build-up (staffing) factor.

A third model used in LMSC is GE PRICE-S. GE PRICE-S is the only timeshare cost-estimating model. It uses lines of code, a productivity factor, and schedule to estimate effort.

A fourth model used occasionally in LMSC is the SPQR/20. Principal outputs of this model are software productivity, quality, and reliability, and therein lies its acronym SPQR. The 20 refers to the number of questions to which the user must respond in providing information relative to the developer's experience, the development methodology involved, and the development environment. The driving input to all the parametric software cost-estimating models is the sizing of the project. While most are based on lines of code, SPQR/20 also uses function points. The function-point technique for sizing software was developed by A. J. Albrecht in 1979 while employed by IBM. This technique was developed to provide a unit of measure that did not penalize high-level languages when measuring cost per line of code. It is based on counting inputs, outputs, inquiries, data files, and interfaces. Function points are rapidly becoming the method of choice for sizing software and can be converted to lines of code for use in currently developed models. The decision on which one to use depends on the particular situation faced by the program manager and the availability of the required input data. The key is to be consistent in the model's application, use more than one model where feasible, and calibrate the models with historical data when possible.

NETWORKING

Background. As suggested earlier, the very success of a project depends heavily on the manager's understanding of the task at hand and his application of thorough and practical planning and control techniques. It is imperative that he be able to address the complexity of multiple activities that will exist in any large-scale project. Fortunately the process of computer networking has been developed to assist in establishing this level of visibility. A network is simply a visual display of the project plan. Networking can be thought of as a form of business engineering. It is an offshoot of earlier scheduling techniques. One of the first techniques was the Gantt chart, which was primarily a bar chart reflecting the activities of a project, each with an estimated start and completion date. As projects became more complex, the need for more sophisticated scheduling arose. The increasing capabilities of computers answered this challenge, and we saw the evolution of the program evaluation and review technique (PERT), the critical path method (CPM), and the precedence diagram method (PDM).

Networking as a Planning Tool. The very construction of a network is beneficial in that it requires the participation of managers and key workers at all levels of the organization. The project is literally decomposed into finite tasks. Task durations are determined and interdependencies and constraints identified. This necessitates communication within the organization as schedules are established and resources assigned. The resultant network provides management with an opportunity to assess the reasonableness and "doability" of various decisions before they are actually implemented. This is often referred to as a "what-if" capability.

Application. AFLC elected to use a PMSS precedence diagramming network capability resident on a minicomputer to provide this visibility for its LMSC software development endeavors. As each individual effort of a program is identified and made part of the WBS, the following information is ultimately provided: task description, planned start dates, duration, and an office of primary responsibility (OPR). As constraints are determined, preceding and succeeding activities are identified. Where appropriate, the network is subsequently resource-loaded, which permits a determination of total resource requirements as well as potential staffing conflicts.

Once established, the networks are continually monitored and updated. Actual start and finish dates are entered. Outputs are generated with the critical path appropriately identified. (A critical path is that sequence of activities that will take the longest time to complete, given the interdependencies that exist.) Various listings were also generated. For example, a manager can ask for a listing of all activities which are late in starting or that are behind schedule. As deviations from plan occur, visibility over the impact is readily available and the manager is able to use the "what-if" capability to evaluate possible courses of action.

Summary. The network is a powerful tool in the project planning and control process. With respect to planning, it provides the vehicle for the collection of all activities and events relative to program execution. It also instills discipline and the need for good communication. As a control tool, it provides a visual assessment of project performance and identifies problem areas, both existing and potential.

BASELINES

A baseline is an agreed to set of parameters which, once adopted, cannot be changed without a formal review and approval process. Baselines are imperative within programs to provide definition and to achieve stability.

There are three primary baselines established during the life cycle of a program. The first is a functional baseline generated at the conclusion of the concepts definition phase and established at the system requirements review (SRR), which authenticates the system specification against the functional requirement. This baseline sets the stage for all subsequent project activities and is the foundation for design and development efforts that will follow in successive phases of the project life cycle.

Next is the allocated baseline, which consists primarily of performance-oriented specifications and interface requirements. As its name implies, it allocates design efforts to each of the functional capabilities reflected in the functional baseline. Developed during the design phase, it is typically established in conjunction with the system design review (SDR).

Finally we have the product baseline generated during the development phase. This baseline is established by and with the authentication of the product specifications and in conjunction with the product verification review (PVR). Two audits are performed to validate the achievement of the development requirements. One, the functional configuration audit (FCA), is performed to validate the satisfactory development of each configuration item and ensure that it performs as required. The second, the physical configuration audit (PCA), establishes the product baseline and is used for the production and acceptance of configuration items.

The timeliness in establishing baselines is critical. Once established, a baseline can be changed only upon formal review and approval by the Configuration Control Board (CCB). Thus establishing a baseline too early may limit design flexibility. On the other hand, if there is a delay in establishing a baseline, discipline and product integrity may suffer.

In addition to the programmatic baselines just discussed, there are other baselines which must be maintained by the program office. The first is a funding baseline which reflects the degree to which the program has received funding as an integral part of the President's budget as well as the projected funding for future years. Obviously the most desirable situation is when the particular program is fully funded. When this is not the case, however, the program office must now manage in light of the shortfall. Further, the projected funding for future periods must be considered as work is planned and contractual effort evaluated for follow-on work.

The second is an acquisition program baseline (APB). Modified in mid-1990, the APB now includes provisions for three sets of program cost, schedule, and performance parameters, namely, the milestone baseline, the program manager's baseline, and the current estimate. Initially established at program startup and modified following major milestone reviews or to reflect other external events which modify the directed program, APBs must be approved by the designated acquisition commander or higher approval authority, depending on the size and nature of the program.

Baselines are a way of life in LMSC. They are on the agenda for each program review, the SRR, SDR, preliminary design review (PDR), critical design review (CDR), PVR, system validation review (SVR), as well as the quarterly program manager's review. Like the WBS, the baselines are indispensable tools that help ensure that everybody has a common goal.

CONFIGURATION MANAGEMENT

Configuration management (CM) is defined by DoD Directive 5010.19 as an engineering management procedure that includes the four functions of CM, namely, identification, control, status accounting, and audits. CM is a process for identifying the characteristics, both functional and physical, of a program during its life cycle. It is the means through which the integrity and continuity of the design, engineering, and trade-off decisions made between technical performance, development, operations, and support are recorded, communicated, and controlled by project and functional managers. The rewards of effective CM are stability, discipline, and technical and product integrity throughout the life cycle.[5]

Configuration Identification. Configuration identification describes what it is that is to be delivered. It "identifies" the performance requirements, qualifications, attributes, operational characteristics, and the criteria to be used in accepting individual configuration items.

Configuration Control. Changes are inevitable in any large development effort. To minimize the confusion that results from unexpected changes, the concepts of configuration control are applied. A configuration control board (CCB), normally chaired by the program manager, is convened to receive, process, approve, or disapprove any and all proposed changes to an existing baseline. Without such a formal process, chaos would soon reign supreme.

Configuration Status Accounting. It is imperative that positive control and traceability be maintained over any and all configuration changes. This is the task of status accounting. All activity resulting from the actions of the CCB, whether completed, on-going, or projected, must be documented. As changes are implemented, baselines are changed accordingly.

Configuration Audits. Audits are necessary to ensure that physical and functional requirements are satisfied and that program configuration corresponds to that spelled out in configuration identification. The FCA and the PCA are the vehicles for verifying such compliance.

As evidenced in the previous discussion, the various functions of configuration management necessitate the generation, update, and display of large amounts of data. During the design, and in particular during the development and test stages of a program, literally hundreds of design problem reports (DPRs) and software problem reports (SPRs) are generated. The corrective action required to remedy these problem reports frequently result in a baseline change request (BCR). Without the aid of an automated information processing system, keeping a handle on all of these transactions would be nearly impossible. In the LMSC world, and as a part of the PMSS, automated tracking systems were constructed to provide the required visibility over this voluminous problem reporting and resolution activity. Without these tools, the functions of configuration management either would have required an army of people or would not have been completed in timely fashion.

MANAGEMENT OUTPUT

In previous sections the discussion centered on various management tools available through PMSS. However, as in the case of a boat that never gets in the water, these tools are of little value until they are utilized. When employed, they are used to collect and manipulate management information and, on demand, generate various output in the form of listings, reports, schedules, networks, charts, and graphs. Everyone is familiar with the expression that a picture is worth a thousand words, and this adage certainly holds true in the arena of information resources. Sometimes the mere volume of information pertaining to a particular subject precludes its display in pictorial fashion and necessitates a tabular listing. When possible, however, a graphic or other visual display often facilitates "getting the message across" with the least difficulty. Certainly, there are times when there is a need for both. Given the scenario of a program that has had some 900 DPRs submitted against it, the programming staff would want detailed information on each, and this could only be provided with a tabular output. Management, on the other hand, is interested in the magnitude of the overall problem, and a graph showing the number and status (open, pending, being worked, closed) of the DPRs might suffice.

Figure 39.2 is a schedule example of a precedence diagram network, each box representing an event or milestone. On this type of output, the critical path would be identified by color coding and represent the sequence of events which would, because of their interdependencies, take the longest time to complete. Figure 39.3, a master schedule, is a subset of the network shown in Fig. 39.2 and reflects by month

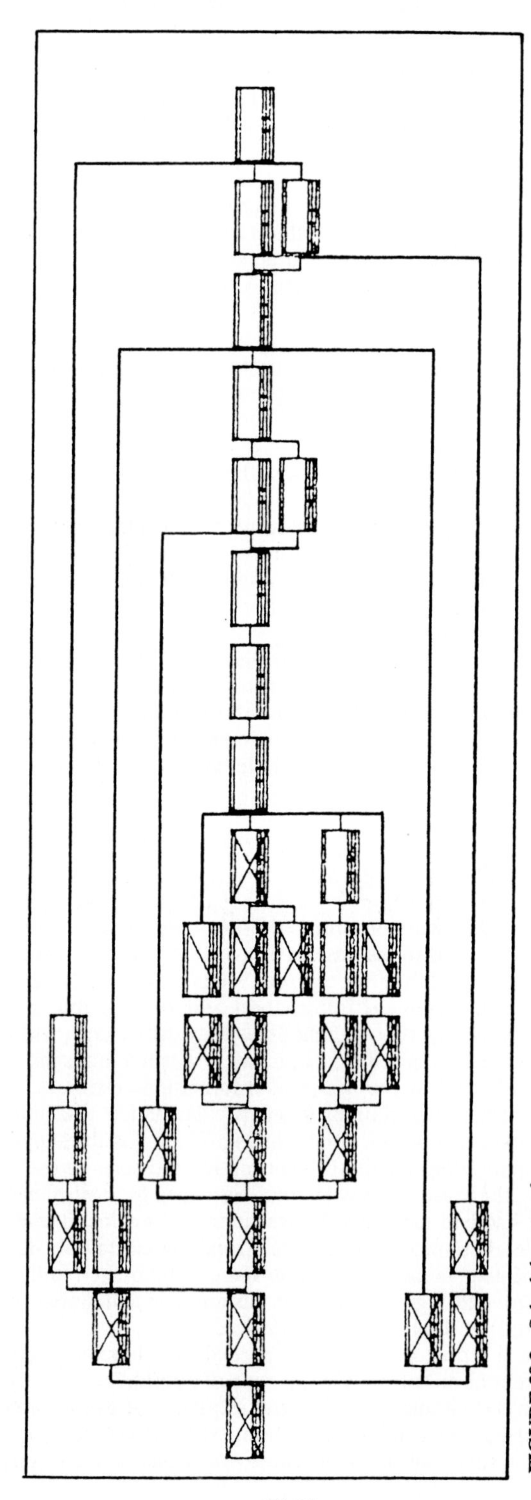

FIGURE 39.2 Schedule example

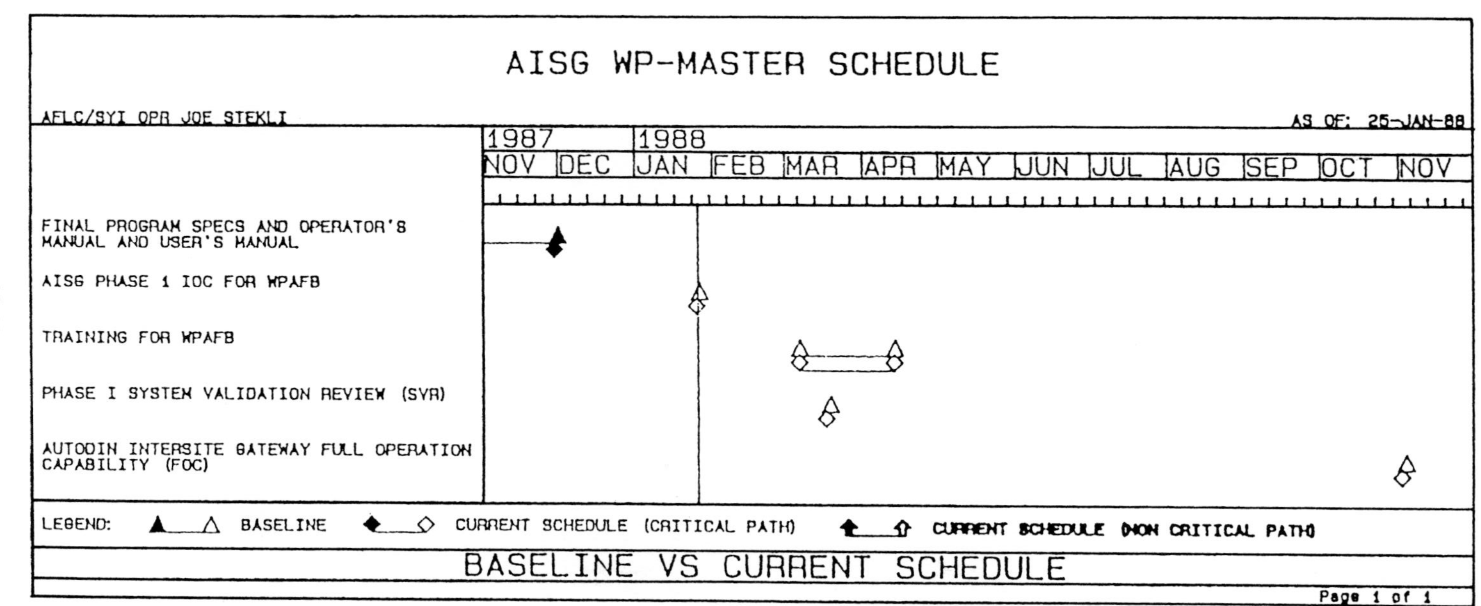

FIGURE 39.3 Master schedule.

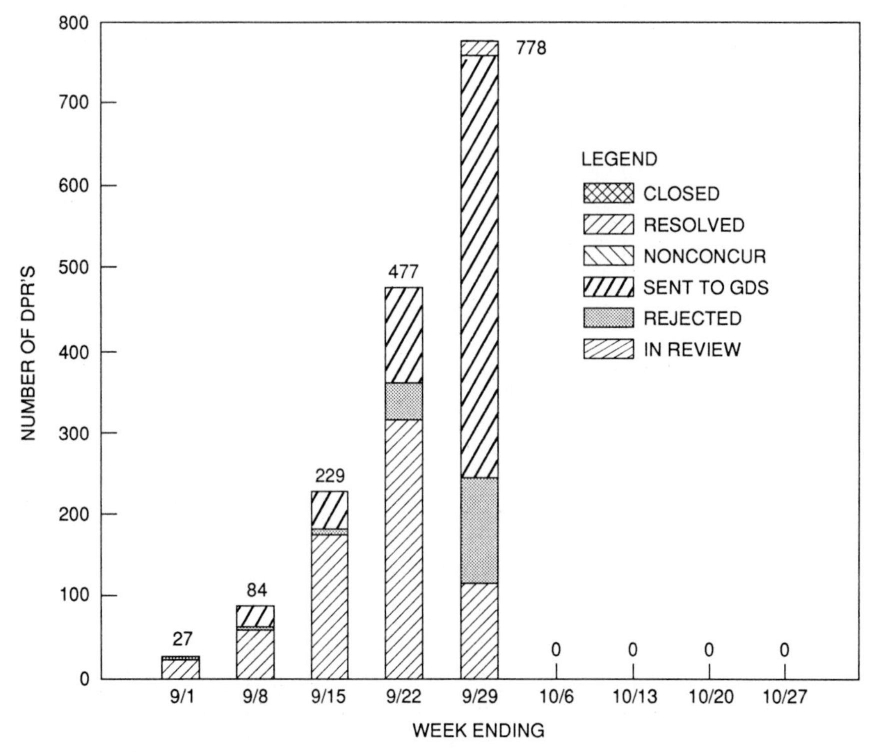

FIGURE 39.4 Current DPR status.

and activity the calender schedule for specific tasks and events. The example shown also reflects the baseline schedule, thereby permitting a comparison of the current schedule with the original plan.

The selection of the format for the output is very flexible. Generally systems are menu-driven, thus providing the user a variety of choices. Bar charts, pie charts, and

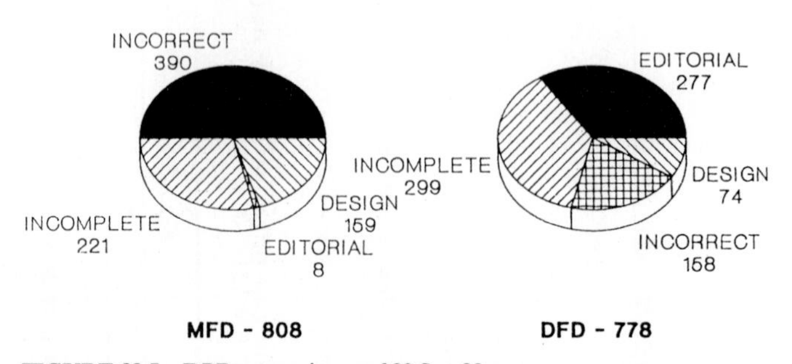

FIGURE 39.5 DPR categories as of 29 Sep 89.

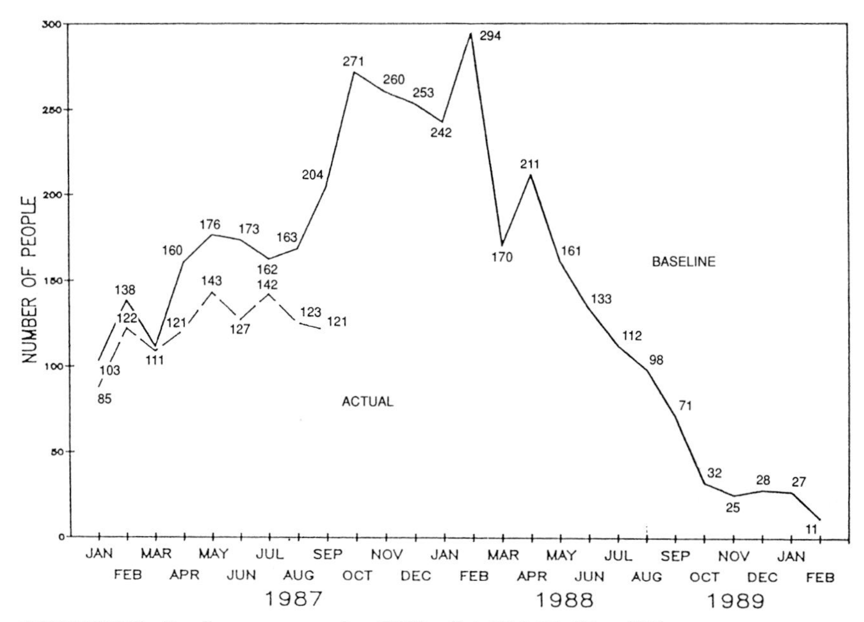

FIGURE 39.6 Baseline manpower plan. CPCIs: pilot, 4/5, 7, 8A, 8B, and 8C.

line graphs are the most commonly used output. Figure 39.4 is a bar chart reflecting DPRs, with the classes of DPRs segregated by fill pattern. Figure 39.5 is a pie chart, again reflecting DPRs by category.

Line graphs are another common way of portraying management data for both informational and comparison purposes. Figure 39.6 reflects a baseline manpower plan for several computer program configuration items (CPCIs) and quickly shows the peaks and valleys regarding personnel requirements. A second line tracks actual manpower, and in this example readily quantifies the shortfall of assigned personnel versus the plan. Figure 39.7 provides "heads-up" information relative to personnel available for a new start program. The solid line shows the projected manpower requirements to accomplish the planned work over an approximate two-year period. The dashed line reflects available manpower and alerts management that during the period from September 1987 to late December there will be insufficient labor resources to maintain the plan. Having this advance information permits management to look at and implement work-around plans.

As mentioned at the outset of this chapter, it is a rare manager who can keep in his mind all the necessary information on a large program. Information resources provide the solution for this dilemma, but even then, it too must be managed. With the availability of a variety of graphic packages and other applications which have similar features embedded, the form of data retrieval and display is limited only by the imagination of the user and system programmer.

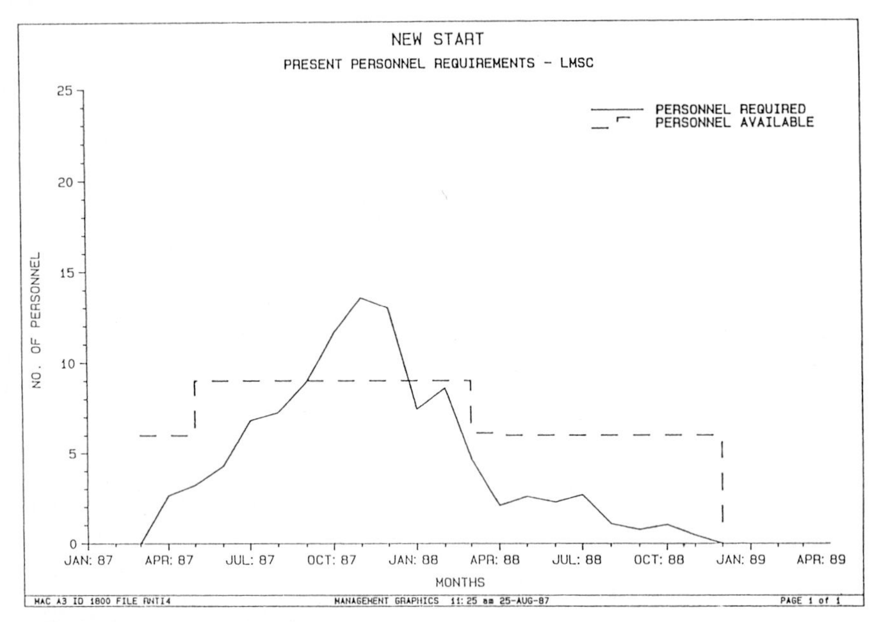

FIGURE 39.7 Personnel requirements, new start.

PERFORMANCE MEASUREMENT AND PROGRESS TRACKING

Updating Network Data

Need for Currency. As discussed in an earlier section, the size and complexity of the typical project will dictate the use of a networking capability. The effectiveness of a network, however, is directly influenced by the degree of currency with which it is maintained. The contractor and the government component share a responsibility to update, on a timely basis, the data file which generates the network. It is equally important that information (proposed start dates, durations, constraints, and projected completion dates) relative to other tasks also be updated as changes occur. Only in this fashion will a program manager have the necessary visibility over the progress of his program.

Schedule Performance. Regardless of how closely a plan is monitored, deviations to schedules will occur. Whether the variance is favorable or unfavorable, it is quite possible that adjustments will be necessary. Task B may be dependent on task A and cannot be initiated until task A is complete. Knowing this in sufficient time may permit the development of work-around plans or allow the use of the personnel resources in other areas. It is therefore imperative that a program manager have the most current data upon which to base his decisions.

While on the surface it may seem simple to maintain visibility over schedule status, in actuality it may be more complicated than one realizes. On tasks of short duration it is quickly obvious whether or not the effort is being completed on time.

However, on work efforts of a longer time span, the necessity of establishing intermediate milestones becomes apparent. Let us assume that as program manager we have a work effort for which we have allocated five months. Two months have elapsed and, in theory, we can conclude we are 40 percent complete (two out of five months). But suppose difficulties were encountered in those first two months. The task was more complex than anticipated, and progress was not as good as originally planned. What then is our schedule status? To be able to answer such a question, it would have been necessary to have established intermediate milestones. For example, we could have determined what would have been a reasonable amount of accomplishment to expect at the end of month 1, month 2, and so on, and established those accomplishments as intermediate milestones. Then, at the end of month 1, actual accomplishment could have been compared to the milestone. Rather than 20 percent complete (one month out of five), we might have determined we were considerably ahead (or behind) plan. The program manager now has a more realistic assessment of his schedule position.

Cost Performance. As in the case of schedules, cost variances are common in major programs. Although one might think that firm-fixed-priced (FFP) contracts are immune to cost overruns, such is not always the case when contract modifications and ECPs become a reality. Further, it is naive to conclude that because you have an FFP contract, you can rest assured that the final product will be delivered on time and at cost. Regardless of the type of contract, to provide visibility over cost deviations from plan, it is imperative that a time-phased budget be established against which contract performance can be measured. Some program offices are comfortable in comparing actual expenditures against this plan to reveal how they are doing costwise. Such a technique is of limited value in that it merely reflects how fast the funds are being expended versus the budget plan. What is missing is a measurement of achievement—what was accomplished for that expenditure of funds. We are now entering the arena of performance measurement.

Performance Measurement. DoD Instruction 5000.2, "Defense Acquisition Management Policies and Procedures," requires application of the cost/schedule control systems criteria (C/SCSC) on significant defense contracts ($60 million RDT&E, $250 million production), except for those which are firm-fixed-price. The purpose of this requirement is severalfold[6]:

1. Preclude the imposition of specific cost and schedule management control systems by providing uniform evaluation criteria to ensure that contractor cost and schedule management control systems are adequate.
2. Provide an adequate basis for responsible decision making by both contractor management and DoD component personnel by requiring that contractors' internal management control systems produce data that:
   (a) Indicate work progress
   (b) Properly relate cost, schedule, and technical accomplishment
   (c) Are valid, timely, and able to be audited
   (d) Provide DoD component managers with information at a practical level of summarization
3. Bring to the attention of DoD contractors, and encourage them to accept and install, management control systems and procedures that are most effective in meeting requirements and controlling contract performance.

Although the application of C/SCSC to firm-fixed-price contracts is not authorized, as indicated earlier, cost reporting may well be appropriate.

To achieve actual performance measurement, in addition to the preparation of a budget plan and the collection of actual costs discussed previously, a third element must be introduced. This element, known as earned value, is the budgeted cost for the work actually performed, the planned expenditure for what was actually accomplished. The difficulty arises in deciding which technique to use to measure accomplishment, choosing the one that will be the most objective. The use of discrete milestones is the most desirable, level of effort being the least. Level of effort is simply spreading the task equally over the period of performance and taking credit with the passage of time. This technique is certainly appropriate for measuring activity such as the program manager's time, which is devoted to all aspects of the program. It should not be used, however, on discrete tasks, which have a measurable output or product.

Having data available is of little value if they are not analyzed and used in making management decisions. Failure to do so can put a program in serious jeopardy, as indicated from the following quotes from a memorandum dated November 28, 1990, for the Secretary of the Navy concerning the A-12 full-scale development program.

> The . . . team failed to utilize the CPR information to identify to the Government the potential schedule and cost implications of the performance problems it encountered.
>
> We found no focused utilization of the CPR data by. . . .
>
> There were shortcomings . . . in integrating engineering, industrial, and C/SCSC oversight functions into a "program" perspective.
>
> Specifically, they (the government) failed to reconcile physical contract performance with costs incurred and charged to the Government despite the substantial cost and schedule variances being reported in the CPRs.

Obviously, the indifference to information resource management data had an adverse effect on the A-12 acquisition program.

Application. For those programs requiring application of the C/SCSC, the reporting vehicle is typically the cost performance report (CPR). Submitted on a monthly basis, the CPR consists of five formats, the first being a reflection of the contractor's performance by WBS. It provides data for both the current month and cumulative to date and includes the budgeted amount, the earned value (credit for work accomplished), the actual costs, and a projection of total costs at completion. The report also shows by WBS element the schedule variance, the cost variance, and the projected variance at completion.

Format 2 reflects similar information, but displays it by functional area. Format 3 is a baseline report, showing for the near term by month and for the long term by quarter (or other specified period) the incremental budget amount. All authorized changes to the baseline are also annotated. Format 4 is a manpower loading report, which shows by organizational or functional category the projected man-months of effort envisioned; and format 5 is a problem analysis of those variances which exceeded a preestablished threshold. A proper problem analysis report should address the nature of the problem, the reasons for the variance, the impact on the immediate task, as well as any impact on the total program, and, finally, any corrective actions taken or planned. The CPR can also be used on programs that do not require application of C/SCSC.

Designed particularly for nonmajor programs is the cost/schedule status report (C/SSR), which consists of only two formats. Format 1 is similar to format 1 of the CPR, but reflects only cumulative-to-date data. Format 2 is the problem analysis report addressing significant variances.

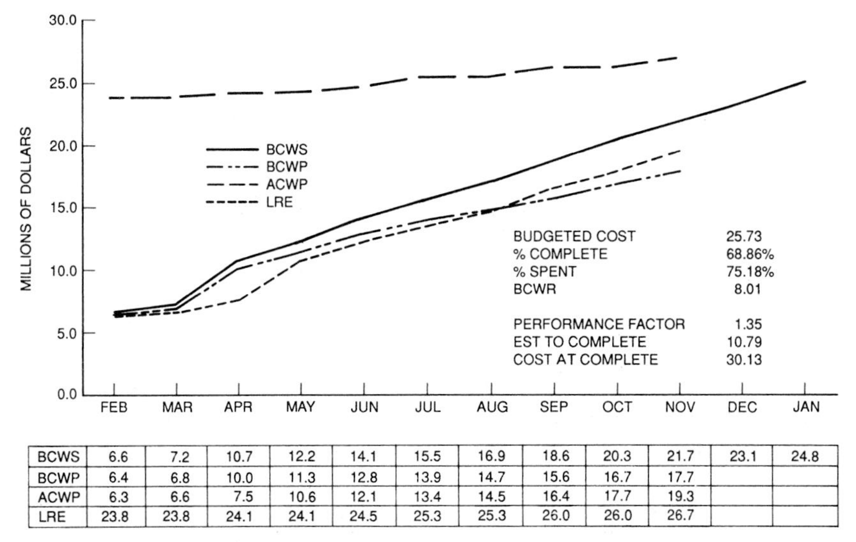

	FEB	MAR	APR	MAY	JUN	JUL	AUG	SEP	OCT	NOV	DEC	JAN
BCWS	6.6	7.2	10.7	12.2	14.1	15.5	16.9	18.6	20.3	21.7	23.1	24.8
BCWP	6.4	6.8	10.0	11.3	12.8	13.9	14.7	15.6	16.7	17.7		
ACWP	6.3	6.6	7.5	10.6	12.1	13.4	14.5	16.4	17.7	19.3		
LRE	23.8	23.8	24.1	24.1	24.5	25.3	25.3	26.0	26.0	26.7		

FIGURE 39.8 Cost and schedule status report.

The C/SSR was the report of choice within LMSC, primarily because of the initial size of the programs and the type of contracts involved. This report is also required on a monthly basis and is scrutinized by financial analysts in the program management office (PMO) as well as by senior analysts in the program control directorate. Pertinent data are extracted from the report and portrayed in graphic form for use by management and during program reviews. Figure 39.8 reflects by month the various performance elements, namely, budgeted cost of work scheduled (BCWS), budgeted cost of work performed (BCWP), actual cost of work performed (ACWP), and the latest revised estimate (LRE). These data, shown in both tabular and graphic form, can be extracted by WBS element or for the total program and are used to calculate schedule and cost variances from plan. Figure 39.9 is a graph showing schedule and cost variances by month and provides a quick assessment of trends.

A second report used by the LMSC community is the contract funds status report (CFSR). It is required on a quarterly basis and reflects funding requirements as well as open commitments, accrued expenditures, a forecast of billings to the government, and estimated termination costs.

Through the use of an automated system, a series of analytical reports are generated, showing various performance factors. Summary data from these reports are consolidated onto one chart, fondly referred to as the "birdleg" chart, which is frequently shown at quarterly program management reviews and provides top management with a quick summary of program status (Fig. 39.10).

Manpower Assessment. The ineffective use of manpower is often a contributing factor to cost and schedule problems. Several information resource management tools and techniques are available to maintain visibility over manpower utilization, both planned and actual. Utilizing a PMSS precedence diagramming networking capability, it is possible to resource load a complete project. Such a capability not only provides information on conflict avoidance but will clearly identify those areas where there is insufficient available manpower. Using a "what-if" analysis capability,

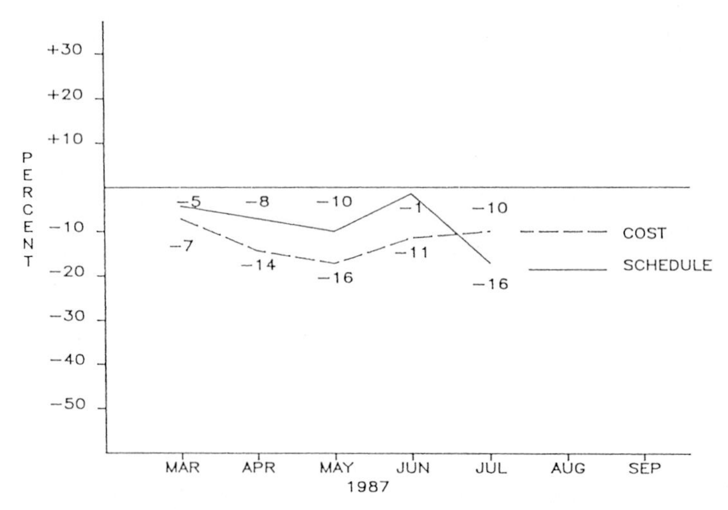

FIGURE 39.9 Cumulative C/SSR variance data (total program cumulative) as of 31 Jul 87.

FINANCIAL OVERVIEW

PROGRAM NAME: FLYING WIDGET
CONTRACT NO.: F33657–90–C–0001

WBS 1100

($000'S)

DATA AS OF: JUN 91

MONTH OF: JUNE					
BUDGETED COST OF WORK SCHEDULED	BUDGETED COST OF WORK PERFORMED	ACTUAL COST OF WORK PERFORMED	BUDGETED COST OF WORK SCHEDULED	BUDGETED COST OF WORK PERFORMED	ACTUAL COST OF WORK PERFORMED
1000	900	1030	4875	4650	4890

	SCHEDULE VARIANCE	COST VARIANCE	SCHEDULE VARIANCE	COST VARIANCE
$	−100	−130	−225	−240
%	−10	−14.4	−4.6	−5.2

MAJOR CAUSE OF VARIANCES:

SCHEDULE: LACK OF REQUIRED RESOURCES
COST: INCREASED LABOR RATES

TRENDS: CUMULATIVE

	SCHEDULE VARIANCE	COST VARIANCE
JUN 91	−4.62	−5.16
MAY 91	−3.23	−2.93
APR 91	−3.65	−3.61
MAR 91	−4.74	−6.08

	% COMPLETE	% SPENT	TARGET %
JUN 91	58.13	61.13	60.94
MAY 91	46.88	48.25	48.44
APR 91	34.63	35.88	35.94
MAR 91	22.63	24.0	23.75

AT COMPLETION

	CONTR.	AF
BUDGET AT COMPLETION	8000	8000
CONTR. LATEST REV. EST.	8380	
AF EST. AT COMPLETION		8412.9

FIGURE 39.10 Financial overview.

it is possible to introduce several differing scenarios and observe projected results. Further, by maintaining burn rates, a manager can develop a trend analysis of manpower utilization as well as future project costs and budget requirements.

Project Stability, a Reflection of Requirements Definition. Most managers want to know fairly precisely what their charter is, what is expected of them, before they begin an effort. Often when they ultimately fail to meet the goal, the expectations (requirements) have changed and schedule slippage or cost overruns, or both, result.

Although it may seem that it should not be a problem, a serious dilemma of any program manager is requirements uncertainty or instability. This can happen from one or more of the following causes and at varying times during the development and implementation processes[7]:

a. Lack of a clear definition of *the problem* prior to contract award. "We think this is the problem, so let's start to work it."
b. An evolutionary development of the functional description causing cost and schedule uncertainties. "We don't have the full picture yet, but let's start with what we do know."
c. User driven changes after contract award and the attendant cost and schedule growth resulting therefrom. "Now that development is under way, we're not sure that's what we want."
d. The restructuring of programs and/or the reduction of functionality to live within the cost baselines. "We can't afford it all, what can we give up? What trade-offs can we make?"
e. User impatience with the pace of development. "Maybe we can do it better ourselves." Also, maybe the user is reluctant to share departmental processes.
f. Lack of documentation and topology of systems being replaced. "We think we understand where we are going."

Further adding to the headache of a program manager are technology-driven issues such as:

g. Technology driven changes and technological promises. "Let's achieve state-of-the-art capability even though it is constantly changing."
h. Competing solutions to problems are offered by transient teckies choosing among correct solutions. "I think this other approach would work better. Let's do it my way."
i. Interoperability in a heterogeneous environment. "Let's all do it the same way, even though what we do is entirely different."
j. Inability to simulate an operational environment which leads to false starts at implementation. "Gee, it looked different way back when."

There are several personnel issues which can also factor into the requirements arena:

k. Problems may arise in achieving a fully staffed program office which possesses the requisite acquisition skills.
l. Changing-of-the-guard and the requirement to re-educate up and down the chain-of-command.

The key is to keep the objective in the clearest focus possible. Changes are inevitable, and the program manager must make the best of it. It is imperative, however, that regardless of the cause of whatever changes may occur, the manager must maintain strict configuration management and control.

PROGRAM REVIEWS

Thus far the discussion of this chapter has centered primarily on the information resource management tools available to assist the program manager. It should now be evident that a great number of data can be accumulated to provide visibility over technical, schedule, and cost performance. However, as evidenced by the lack of attention paid to such information within the now canceled Navy A-12 aircraft program, the mere existence of the data is insufficient to increase the chances of success. The data must be collected, displayed graphically or in reports, analyzed, and appropriate management action taken. There must be a continuous monitoring of what is happening and the progress being made (or the lack thereof) toward achievement of the stated goals. Even doing all of this will not guarantee a successful program, but the chances will be greatly enhanced.

Formal periodic reviews are a tried and proven way of "staying on top" of a program, and hopefully every DoD program manager uses such a technique. The author's past six years have been involved with AFLC and the acquisition of information systems, and we will cover program reviews from that aspect. The overall approach, however, should be similar to that used in other acquisition arenas.

There are two types of control a program manager must establish and maintain. The first is technical control, which includes the planning and controlling of tasks necessary to ensure that the user's requirements are ultimately satisfied by the development, implementation, and operation of the final system. The second is cost and schedule control, which endeavors to ensure that the final product is delivered on time at the agreed-to price.

Technical Control. Formal technical control is achieved by a series of technical reviews supplemented by configuration audits. Within AFLC the following reviews have been incorporated to verify that the design does, indeed, satisfy the requirement as established by the user. As part of the systems engineering process, these reviews ensure visibility and technical understanding.

System Requirements Review. As its name implies, SRR is accomplished to determine that the initial design approach will satisfy and be responsive to the stated requirement of the user. Normally conducted during the concepts development phase, it can be accomplished only after a significant portion of the functional requirements has been defined. It becomes a decision point for the establishment of the functional baseline.

System Design Review. SDR is held only after the design effort has progressed sufficiently to a point where the system characteristics and the design approach have been well defined and configuration items have been identified. Typically associated with this review is the allocated baseline, which assigns the various design efforts to the functional entities specified in the functional baseline.

Preliminary Design Review. During the PDR, each configuration item is evaluated relative to the technical adequacy of the selected design approach, the progress being made, and risk resolution. This review follows establishment of the allocated baseline.

Critical Design Review. Once the detail design has been completed and fabrication drawings have been prepared and are ready for release, the CDR is conducted. The purpose of this review is to ascertain whether the design definition as set forth in the allocated baseline has been satisfied by the completed detail design. Following a successful CDR, the contractor should be prepared to begin operational coding.

Product Verification Review. Conducted to approve the product baseline for each configuration item and to ensure that preparations for testing have been completed, PVR must be preceded by both an FCA and a PCA. Within the LMSC community of AFLC, there are actually two of each of these types of audits, namely, a preliminary FCA/PCA and a final FCA/PCA. The objectives of the preliminary FCA/PCA are to determine the state of readiness to begin validation testing and to verify performance against that specified in the system specification and functional description. In the event major problems are identified, consideration is given to postponing the PVR. Following PVR, the final FCA/PCA is performed with the major objectives including the validation of the tested performance against the system specification and the functional description, and the review of any BCRs and ECPs to verify that they have been incorporated and validated by the testing. Additional objectives of this review are to ensure that technical documentation accurately reflects the functional and physical characteristics of the configuration item and to determine the current status of open design problem reports and software problem reports.

System Verification Review. The objectives of the SVR are to confirm that the *system* performance meets the specified requirements, that it is technically acceptable for an operational environment, and that appropriate arrangements have been completed for implementation, operational support, and system management transfer (SMT).

System Transfer Review. The purpose of the STR is to ensure that all events planned in the SMT plan have been accomplished satisfactorily and that the system is ready for transfer to the ultimate user. Of equal importance is the determination that all resources, training, and equipment necessary to support the system are in place.

Schedule and Cost Control. As discussed earlier in this chapter, there are a number of PMSS tools which provide the manager with visibility over the schedule and cost elements of the program. The attentive manager will use this information to avoid surprises, and when appropriate will share such information with higher management. At AFLC, program managers accomplish a program assessment each month during which they evaluate the following 10 factors:

System performance
System readiness
Cost performance
Contracts
Key decisions
Schedule
Test and evaluation
Program funding
Reliability and maintainability
Training

Based on the assigned ratings of these factors, an overall program rating is determined. On a quarterly basis, the commander of LMSC holds a program review during which the program manager's assessment is critiqued. Comments and direction are provided as appropriate. Schedule and cost performance are also topics for discussion at this review, and the "birdleg" chart (Fig. 39.10) discussed earlier is typically shown and evaluated.

In addition to the aforementioned formal reviews, in-process reviews are frequently held to maintain visibility over progress in each task area. Finally, on an as-required basis, a technical interchange meeting (TIM) with contractors is convened to resolve open issues and questions.

The need for visibility of program status is unquestionable. The appropriate management of information resources will help to provide that visibility, and the timely application of program reviews will ensure that problem areas are addressed in a timely fashion. The combination of the two, while not a guarantee, is certainly a recipe for success.

CONCLUSION

Information—can we ever have too much of it? With the systems and capabilities available today to generate data, there exists a potential for data overload. A manager could conceivably be inundated with such voluminous information that it becomes difficult to distinguish between the "need to know" and the "interesting, but so what." The key is to be selective and to obtain and use that which offers the biggest payback. Within the LMSC community at AFLC, PMSS has become the tool of choice. PMSS has the capability of generating more data than any one manager could hope to use. But no two managers are exactly alike, and their requirements will differ as well. One individual's "must have" will be another person's "don't need." PMSS permits users to be selective in what they want to see.

The success of the LMSC program attests to the value of PMSS and the application of information resources. This was clearly substantiated in the House of Representatives Report of the Committee on Appropriations in the 1991 Appropriations Bill, which read in part: " . . . the Air Force Logistics Modernization . . . among the most successful computer systems in DoD history."

POSTLUDE

In an October 4, 1989, memorandum to the secretaries of the military departments and other addressees, the Deputy Secretary of Defense introduced the corporate information management (CIM) philosophy. This CIM initiative was established to reduce non-value-added work and costs and is perceived as one of the key ways of achieving cost reductions prescribed by the Defense Management Report given to the President the previous July.

The principal objective of CIM is to improve business processes while maintaining or improving the effectiveness of the DoD military missions. "DoD recognizes that information must be managed, just as capital, materiel, and people must be managed, to improve effectiveness and efficiency of operations."[8] Recognizing that there is duplication both in existing information systems and in new systems currently under development, the fundamental concept of CIM is to approach information management decisions on a business case basis. This simply means that current business practices will be examined and alternative ways of performing or even improving the particular function, while at the same time minimizing total expenses, will be considered. Secretary Andrews, in his report to the Defense Subcommittee of the House Appropriations Committee, went on to say: "The success of CIM hinges in large part on the ability to standardize processes and data and to install an open systems architecture as we move the Department into an era emphasizing information management."

The CIM effort was formally initialized when the comptroller of the DoD in early 1991 provided for the designation of an executive agent (EA) for each business function area and supporting system, where feasible. The EAs have been charged with the responsibility for developing a concept and a corresponding technical plan for achieving, within their respective business areas, enhanced capabilities with no loss of functionality and with a cost savings. Thus information management, and in particular the resources and processes used to achieve the required degree of management visibility, have assumed an even more important role in ensuring that the objectives are achieved at a reasonable cost. The responsibility for achieving the goals of the CIM initiative rests with every manager responsible for a given process and with the staff who provide the required management information.

ABBREVIATIONS

ACWP	Actual cost of work performed
AFLC	Air Force Logistics Command
APB	Acquisition program baseline
BCR	Baseline change request
BCWP	Budgeted cost of work performed
BCWS	Budgeted cost of work scheduled
CCB	Configuration control board
CDR	Critical design review
CFSR	Contract funds status report
CIM	Corporate information management
CM	Configuration management
COCOMO	Constructive cost model
CPCI	Computer program configuration item
CPM	Critical path method (of networking)
CPR	Cost performance report
CSCI	Computer software configuration item
C/SCSC	Cost/schedule control systems criteria
C/SSR	Cost/Schedule status report
DPR	Design problem report
EA	Executive agent
ECP	Engineering change proposal
FCA	Functional configuration audit
FFP	Firm fixed price
LMSC	Logistics Management Systems Center
LRE	Latest revised estimate
OPR	Office of primary reliability
PCA	Physical configuration audit
PDM	Precedence diagram method
PDR	Preliminary design review
PERT	Program evaluation and review technique
PMO	Program management office
PMSS	Program management support system
PVR	Product verification review
SDR	System design review
SLIM	Software life-cycle investment model
SLOC	Source lines of code
SMT	System management transfer
SPR	Software problem report
SRR	System requirements review
STR	System transfer review
SVR	System validation review
TCP	Technical change proposal
TIM	Technical interchange meeting
WBS	Work breakdown structure

REFERENCES

1. MIL-STD-881A, "Work Breakdown Structure for Defense Materiel Items," Apr. 1975.
2. *Holy Bible,* Revised Standard Version, Luke 14:28.
3. C. N. Moser, "Information Systems Acquisition," *Program Manager,* Defense Systems Management College, July–Aug. 1990.
4. D. V. Ferens, "Defense Systems Software Project Management," pp. 10-5, Jan. 31, 1990.
5. Robbins-Gioia, Inc., "A Systems Approach to Managing the Project Life Cycle," Configuration Management chapter, p. 1.
6. DoD Instruction 5000.2, "Defense Acquisition Management Policies and Procedures," Feb. 23, 1991, p. 11-B-1.
7. J. F. Phillips, USAF, 1989 Briefing on LMSC.
8. D. P. Andrews, Assistant Secretary of Defense, C3I, Statement before the Defense Subcommittee, House Appropriations Committee, April 24, 1991.

CHAPTER 40

MANAGING MODERNIZATION PROJECTS: THE PROCESS OF UPGRADING, ENHANCING, AND/OR REFURBISHING AGING MILITARY EQUIPMENT

Jerry D. Schmidt

Jerry D. Schmidt is the Air Force Deputy in the Joint Integrated Avionics Directorate (Army, Navy, Air Force organization). He has management responsibility for ensuring that Air Force avionics programs reinforce and complement Army and Navy programs and that duplication is eliminated.

INTRODUCTION

The military services are all victims of their tremendous successes in Desert Shield/Desert Storm and the crumbling of the Communist empire. A grateful American people (for the success in Desert Storm) will for many years praise the professionalism of our military and hold the military profession in high esteem. Its elected representatives, however, will examine each major military expenditure critically. No representative wants to reduce the nation's ability to project and use power. Each, however, according to his or her own beliefs, will attempt to make changes to the military budget that will allow the nation to remain the premier military power in the world. Most will maneuver to protect military-related jobs in his or her district or state, and many will continue to pursue the peace dividend. All will find increasing pressure to spend more money on domestic projects and many will exert more pressure to reduce military expenditures to pay for the ever-increasing domestic needs. One of the many results will be a reduction in new system procurement and an extension of the life of currently fielded systems. While future technology will dictate that we continue to procure new systems to maintain our technical leadership in military systems and technology, modernization through modifications will increase in importance as the world moves toward the western democratic values of our nation and the free world. The military won the "hot" war in the desert, and the nation, partially because the military was being prepared for 45 years, won the cold war. This chapter will discuss how the modernization programs will be conducted to ensure that we maintain technological currency in our weapon systems and remain the world's only true superpower.

OVERVIEW OF THE MODIFICATION PROCESS

Modification programs are acquisition programs and are covered by DoD Instruction 5000.1 and DoD Manual 5000.2M. Modification programs are subjected to milestone decisions similar to (and often identical to) new development programs. The length of time to proceed from one milestone to another is dependent on the complexity of the modification. The acquisition milestones are illustrated in Fig. 40.1.

Modification programs are placed in one of four categories. The categories are determined by the estimated program cost for research, development, test, and evaluation (RDT&E), or total program costs. The four acquisition categories (ACAT) are as follows:

ACAT I: RDT&E greater than $300 million or total program cost of $1.8 billion, both in fiscal-year 1990 constant dollars.

ACAT II: RDT&E greater than $115 million or total program cost of $540 million, both in fiscal-year 1990 constant dollars.

ACAT III: Programs that do not meet the criteria of ACAT I or II but have been designated as category III by the DoD component acquisition executive.

ACAT IV: All other modification programs.

The ACAT I or II program milestone decision authority is the service under secretary for acquisition or the service secretary. (The milestone decision authority is

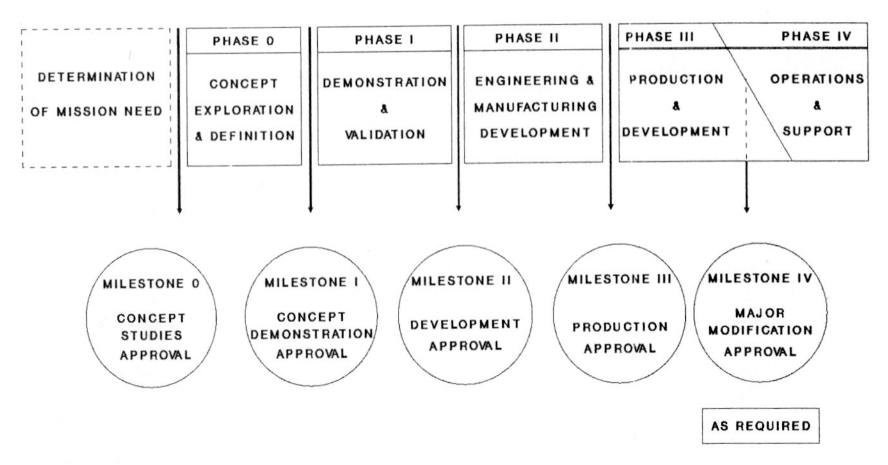

FIGURE 40.1 Acquisition milestones and phases.

the individual who can approve of the program moving into the next phase of its development or life cycle.) ACAT III programs are initially reviewed by the service under secretary for acquisition and at his or her discretion may be turned over to a program executive officer (PEO) or a designated acquisition commander (DAC). ACAT IV programs are generally delegated to a milestone decision authority below that of an ACAT III program, and for many small-dollar modifications the milestone decision authority may be placed in the program office. The levels of review and approval are thus tied to the complexity and probable cost to the nation.

Like for acquisition programs, the program manager (PM) reports directly to the milestone decision authority and bypasses the many levels of review that were required when the service staffs totally controlled the modification programs.

SOURCE OF MODIFICATION REQUIREMENTS

Modifications to military systems originate from many sources. These sources include the operating commands, the overhaul activity, the PM, defense contractors, and a host of other activities. Modification requirements generally fall into three major categories: (1) safety, (2) reliability and maintainability (R&M), and (3) operational capability.

Safety modifications are usually initiated as a result of problems found during inspections or to correct a deficiency found during an accident investigation involving the system. The safety modification generally receives the highest priority for obvious reasons: no one wants to place the health of the work force, system operators, or the population at large at risk because proper actions were not taken to correct a known system problem.

Modifications are also performed to prevent future safety problems. These modifications generally are structural in nature and through monitoring programs can be predicted far enough in advance to schedule the modification with depot maintenance prior to the safety problem becoming a reality.

R&M modifications are performed to increase mission availability while generally reducing life-cycle cost. Replacing problem systems, subsystems, or items can often result in many more missions being performed in a given time frame while fewer spares are being consumed and less maintenance is being performed on the system.

R&M modifications generally are initiated by the operating command, the PM, an unsolicited proposal from a contractor, or a standardization effort. When the reliability or maintainability of a system is such that the operators are being hampered from completing their mission, they will often request the PM to work an R&M modification. The PM, when reviewing data reported by the operators, may conclude that the R&M is below an acceptable limit because of operational availability impacts or that life-cycle costs (LCCs) could be reduced by modifying the system.

Many R&M modifications occur because contractors often provide unsolicited proposals which provide excellent solutions to known problems. If solutions are available from multiple sources (even though one contractor may have convinced the user or the PM of the requirement), the government is generally required to build a proposal which eventually can lead to the modification being competed. The fourth source of R&M modifications is standardization programs. Often offices are established to replace many items in different systems doing the same function with a common item. The item manager (IM) is responsible for the overall item modification package. This item modification package is provided to each PM, who programs the interfaces into his or her weapon system. Each PM must submit his or her requirements to the IM, who will then build the total package for submission to higher headquarters.

The operational commands normally initiate capability modification requirements by submitting a mission need statement (MNS) to the service headquarters. The MNS is the new DoD standard directed in DoD Instruction 5000.2 and DoD Manual 5000.2-M.

An MNS is a brief statement prepared by the commanders-in-chief (CINCs), service headquarters, or the using or operating major command that is normally one page (not more than five pages) in length. When it is practical and cost-effective, programs for several commands or several services will be consolidated. The objective is to prevent duplication of development efforts and improve commonality, standardization, and interoperability. When an MNS is designated multicommand or multiservice, a single command or service is designated to lead the effort for all commands or services. The MNS consists of the following elements:

1. *Defense Planning Guidance Element.* This element describes to what in the defense planning guidance or long-range investment plans this MNS responds.

2. *Mission and Threat Analyses.* The deficiency which is to be corrected or the need that must be satisfied is described here in terms of mission, objectives, and needed capabilities. New Defense Intelligence Agency (DIA)–validated threats are discussed as well as postulated future threat environments and shortfalls of existing systems.

3. *Nonmateriel Alternatives.* This element describes changes in U.S. or allied doctrine, operational concepts, tactics, organization, and training that were considered to satisfy this deficiency and the reason these changes were rejected as inadequate.

4. *Potential Materiel Alternatives.* This element identifies known potential solutions, either domestic or foreign, military or commercial, that are in development or in use. Discussed here is the potential for joint development, and initial preferences for emphasis are indicated.

5. *Constraints.* This element identifies key boundary conditions that may impact on satisfying the needs. Subjects to be addressed include mission planning, logistics support, transportation, mapping, charting, geodesy support, manpower, personnel and training constraints, command and control, communications, intelligence interfaces, security, standardization, interoperability, nuclear, biological, chemical, advanced technology, terrorism, sabotage, and natural weather. Mission capabilities desired in these environments should be described to the extent possible.

An MNS is required for all permanent modifications that exceed $10 million in total estimated costs, except for temporary modifications and approved safety modifications.

The operational capability modification is generally designed to respond to a threat change, to incorporate new technology and increase system effectiveness, to remove a capability no longer required, or to change the mission of the system. Capability modifications are initiated by several sources, but generally are driven by a requirements pull or a technology push. The operating command will submit a requirement for a modification when encountering a situation where a threat has appeared that the weapon system cannot counter. The operating command also submits the requirements to delete a capability from a weapon system. (An example is the deletion of LORAN bombing from the F-4 aircraft.) Many modifications are added to increase the effectiveness of a weapon system. (The modifications of most U.S. fighter and bomber aircraft to deliver guided weapons are excellent examples of increasing the lethality of current operational systems.) The mission of a weapon system is sometimes changed to alter its basic role in warfare. (A premier example is the change of the F-111A aircraft to the EF-111A. The mission was changed from an attack aircraft to a standoff or escort electronic warfare aircraft.) An area of modifications receiving increased emphasis is the service life extension program. Because of the very high costs of replacement systems, it is often more economical to extend the life of selected systems. Often the system is performing its mission well, but has life-limiting problems that would normally lead to retirement. By concentrating modifications on these selected problem areas, the system life can often be doubled or tripled. (The Air Force recently modified its infil/exfil PAVE LOW III MH-53J helicopter with an operating and cost improvement modification program designed to extend its service life beyond the year 2000. Engines were modified, selected airframe components were modified or replaced, the hydraulic and electrical systems were renovated, gearboxes were modified, the main landing gear was upgraded, and the intercom system was modified to be tempest-secure. The service life after modification completion is 25 years.) Increasingly, as costs continue to increase for replacement systems, the services are modifying existing systems for a fraction of the replacement cost and thus are simply extending their service lives.

BUILDING THE MODIFICATION MANAGEMENT PROPOSAL

Once the requirement for a modification is identified, the PM must take action to define the exact changes required and prepare the proposals [program management proposal (PMP) in the Navy; modification proposal and analysis (MPA) in the Air Force; and material change process in the Army] to gain approval and to execute the modification.

The PM has several avenues available to build the proposals. He or she will generally start with his or her own staff and lay out possible options. Often the staff, with assistance from the entire service staff, will complete the proposal without contractor assistance.

Inputs are received from the logistics support organizations for spares, support equipment, software support, provisioning, and reprocurement data. Engineering provides the performance requirements, R&M goals, and testing requirements. Training requirements are received from the training equipment organization and training instruction development organization. The PM must organize the inputs and structure the modification proposal from the Configuration Control Board (CCB).

Should the PM not have the technical expertise or the personnel available, he or she will use contractor assistance, generally through a sustaining engineering contract. Many times that contract is back to the prime contractor since it has the expertise and generally maintains a very complete set of all system drawings and data. When safety is involved, the prime contractor is generally called in to assist in finding a rapid solution. Should assistance be required by the PM on a modification proposal which will be procured competitively, a contractor is generally selected who will not compete for the contract.

PROGRAMMING FOR THE MODIFICATION

Biennially the services submit their modification requirements to the office of the Secretary of Defense (OSD) through the program objectives memorandum (POM) process. POM inputs are required 18 months to two years before funds are made available to begin the modification. These POM inputs cover the Future Year Defense Plan (FYDP). They are provided by the MAJCOM that operates the aircraft. The baseline for all POM inputs is the most current FYDP with approved amendments. Costs for out years are submitted in then-year dollars. POM inputs must include the research and development dollars required and the year of the requirement, the quantity and cost of the modification kits to be procured in each of the POM years, the number of kits to be installed in each fiscal year, and the installation costs. Spare parts must be priced out lead time away from the requirement. Test equipment must be planned for and plans for training equipment must be developed to modify it concurrently with the system. Technical data and any testing requirements must be included in the package.

Often safety modifications cannot wait for the POM process and must be started immediately. The PM has limited authority to move money from funded modifications to accomplish an approved safety modification. Often the PM is required to get approval of the service headquarters since funds required are beyond the PM's control. If the safety requirement is large enough, higher headquarters may be required to move money from other systems to cover the requirement. Beyond certain thresholds, these program changes must be coordinated with OSD and Congress.

Modifications, to the extent possible, are scheduled for installation during normal depot maintenance on the system. This generally results in greatly decreased installation costs since processing for the modification and for depot maintenance is very similar and thus is only required once. Often the area of the modification is opened up during depot maintenance, thus further reducing the installation cost of the modification. Another benefit is out-of-service time. Concurrent accomplishment of modifications and depot maintenance reduces the total downtime of the system and thus provides the equivalent of additional systems to the operators.

BLOCK MODIFICATIONS

Since it is often possible to do many modifications while the system is undergoing routine maintenance, modifications are grouped for installation. These modifications are designated block modifications. When the changes are extensive enough, the weapon system is often given a new number. (An example is the Marine HH-53s that were modified for Air Force special operational forces use. The Air Force renumbered the aircraft the MH-53J.)

BUDGETING FOR MODIFICATIONS

Modifications must be funded to procure a whole number of modification kits in any fiscal year. Installation costs are budgeted in the same year as the kit procurements and are now chargeable to the procurement account. Contractor Engineering Technical Services (CETS), which is required to support the modification, is either an O&M charge (for O&M-funded personnel) or a DMIF charge (for DMIF-funded personnel). Other items chargeable to the procurement account for the modification include the initial integrated logistics support plan (ILSP), trainer kits, training plans, trainer hardware, support equipment, and publications. Factory training, follow-on integrated logistic support (ILS), and updates to the ILSP are funded by the O&M account.

An exhibit is prepared for forwarding to OSD and Congress. This exhibit shows all models of the systems being modified, the quantity involved, a description, and a justification for the modification. All the types of funding are displayed (RDT&E, O&M, procurement), and the planned installation methodology is detailed.

PRIORITIZING THE MODIFICATION CANDIDATES

The system user together with the PM must define the criteria for establishing the priority ranking on the weapon system. Considerations include availability of technology for the modification, threats which justify the modification requirements, O&M costs before and after the modification, anticipated funding support for the modification, and the proper support from the service staffs to ensure a successful program.

All modifications to all similar systems are eventually rolled into one priority list by the major users. In the Air Force, Air Mobility Command (AMC) would prioritize all airlift modifications in conjunction with Air Force Material Command. Air Combat Command (ACC) and AFSOC (representing SOCOM) would do the same for their systems. NAVAIR, NAVSEA, and so on, working with OPNAV, would go through a similar process for the Navy. The Army Aviation and Troop Command, Army Tank Automotive Command, and so on, working with TRADOC, perform the same service for the Army. The modifications are then reviewed by the service staffs and the assistant secretary of the service for acquisition, and decisions are made on which modifications to proceed with.

A myriad of reviews take place with OSD, OMB, and the congressional staffs, with the services providing inputs and additional information as appropriate. Eventually the approved modifications are included in the congressional budget, and plans can begin to start the modifications the following October. The entire process up to this point has taken about 18 months. By the end of September, Congress authorizes programs and passes appropriation legislation. Often Congress is late in taking these actions, resulting in a continuing resolution, that is, continue to spend at last year's funding levels. A continuing resolution applies to ongoing programs only. New starts (modifications that have no previous approval) are not allowed until the appropriation legislation is passed and signed by the President.

MODIFICATION STAFFING

Most modifications begin with an engineering change proposal (ECP) since by definition a modification is a change to a configured system or item. The PM normally receives the ECP from contractors or his or her organic (service) engineering staff.

The ECP is reviewed by a team assembled by the PM in accordance with MIL-STD-480B. The engineering approach is evaluated for completeness and to ensure that the ECP satisfies its intended purpose. All ILS elements are addressed, and reasonability of cost and schedule are reviewed. All technical information changes (data, drawings, publications, and so on) are reviewed as well as requirements for government-furnished equipment and facilities. Spares are computed and included in the modification package. Training systems and training course material proposals are reviewed for adequacy. The functional managers for all areas are asked to coordinate on the modification package before it is submitted to the Configuration Control Board (CCB).

Once the PM is satisfied that the modification package is complete, he or she will assemble it for submission to the CCB. CCBs have various thresholds (dollar values) for modifications they look at in the different services, but all PMs must meet a CCB for programs above a service-established dollar value.

RECENT POLICY CHANGES AFFECTING MODIFICATIONS

Major changes have occurred recently in modification procurement. Modifications are acquisition programs and are now subject to the same milestone approval and documentation requirements as a major weapon system acquisition. All modifications must now follow the DoD 5000 series instructions and must have operational testing and a formal production decision before production quantities are ordered. Acquisition program baselines (APBs) are now required for all modifications in accordance with DoD Instruction 5000.2. If a modification's cost is estimated to exceed $10 million, the user must prepare and submit an MNS.

The time required to use a modification after completion has recently been changed from a two-year/four-year rule to a two-year/five-year rule. This change was brought about because of the uncertainty of phase-out dates for the weapon system. The new criterion states that a modification may not be planned, programmed, or budgeted unless two or more years of programmed life for the weapon system are forecasted to remain after completion of the modification.

A modification will not be installed unless five or more years of programmed life are forecasted to remain after installation. This five-year rule is mandated by congressional statutory language. Modification approval, funding, acquisition, and installation will be adjusted accordingly. System PMs are responsible for obtaining timely force structure data and enforcing the five-year rule. Headquarters USAF/XO is responsible for providing accurate and timely force structure data. This five-year rule may not apply to some commodity time compliance technical orders (TCTOs) which are programmed for multiple-host platforms scheduled to remain in the inventory. However, the decision to modify these commodities should consider the effects of items pulled from retiring platforms and added to the total inventory. Safety modifications are exempt from these rules.

ACQUISITION PLANNING

After the modification has been approved and funded, the entire package is forwarded to a contracting officer as part of a purchase request (PR) package. The heart of the PR package is the statement of work (SOW). The SOW details the technical requirements, including performance, R&M, delivery schedules, quantity, and testing. The contractor must translate the information provided into a contract that defines the roles and responsibilities of both the government and the contractors.

An acquisition plan must be written that addresses the method of acquisition, how competition will be used, criteria for selecting a winning contractor, how milestones will be used to arrive at a contract, and if competition is not used, justification for sole-source or selective-source contracting.

SOLICITING THE INDUSTRY

The government uses two methods to solicit industry for goods and services. They are the request for proposal (RFP) and the invitation for bid (IFB). The RFP is the government's solicitation for goods and services from industry used during competitive source selection. It describes what the government wants, when it is needed, and how it is to be delivered. How the acquisition is to proceed is described and industry is invited to submit proposals to fulfill the government's requirements.

An IFB is used in the case of sealed bidding procedures. This method is seldom used for sophisticated modifications. It is generally used where the item described is well defined and not subject to misinterpretation. For most sophisticated modification programs the competitive proposal method is used.

METHODS OF PROCURING THE MODIFICATION

Generally the modification, if it is a major structural modification, is procured from the prime contractor. Tooling to work on the structure is generally stored by the prime contractor and is only available at the prime contractor's facility. Structural design data and original test data are also only available from the prime contractor. In addition, few contractors feel competent to work on the structures of a major system designed and built by someone else. The usual result is a contract to the prime contractor, even though several contractors may be invited to bid.

Safety modifications often involve structures and thus are procured from the prime contractor. Installation may be performed by the prime contractor, a service maintenance facility, or a field team. Generally the maximum number of installations are accomplished in conjunction with regular overhaul maintenance. For safety modifications, however, urgency may dictate that contractor installations or installations by field teams are required.

Most modifications other than safety are accomplished through competitive procedures. Those that are not competed must be accompanied by a justification and authorization (J&A) package. The J&A describes the reason for the noncompetitive procurement and must be signed by an official in OSD. There are seven conditions that permit other than full and open competition. They are:

1. Only one responsible source and no other supplies or services will satisfy agency requirements. This generally implies that the source has demonstrated a unique and innovative concept; offered a concept or service not otherwise available to the government; and does not resemble the substance of a pending competitive acquisition.

2. Unusual and compelling urgency. When an agency's needs for supplies or services are such that serious injury could occur to the government, this authority may be used. Several procurements were made during Desert Shield/Desert Storm to procure critically needed materiel. Another example of this authority is when a major military system may be stood down because of the lack of some supplies or services.

3. Industrial mobilization; or engineering, developmental, or research capability. The government agency may award a sole-source contract to maintain a facility, producer, manufacturer, or other supplier for possible national emergencies, to achieve industrial mobilization, or to build and maintain essential engineering, research, or development capability to be provided by an educational institute, a nonprofit institution, or a federally funded research and development center.

4. International agreements. Competition can be limited or not provided when a foreign government or international organization requires in writing that the product or service be bought from a specific supplier and that government or organization is paying for the product or service, or when the proposed acquisition is for services to be provided, or supplies to be used, in the sovereign territory of another country and a treaty or agreement specifies or limits the sources to be solicited.

5. Authorized or required by statute. Competition need not be provided when a statute expressly authorizes or requires that the acquisition be made through another agency or from a specific source, or the agency's need is for a brand name commercial item for authorized resale.

6. National security. If the national security could be compromised because of full and open competition, the agency can limit the number of sources.

7. Public interest. Determination that full and open competition is not in the public interest can be made only by an agency head and cannot be delegated by the agency head. This authority applies only when none of the preceding exceptions are applicable and requires the agency head to notify Congress a minimum of 30 days before contract award.

Simple modifications for nondevelopmental items are often bought through price competition (IFB). Contractors must establish their ability to complete the contract requirements before their bids are accepted. Once the government has established which bidders are acceptable, bids are accepted and a contract is awarded.

Many modifications are sophisticated enough to require a formal source selection process. For these programs a source selection authority (SSA) is established. That SSA could be the assistant service secretary for acquisition, one of the service acquisition executives (SAE) established under the Goldwater-Nickols Act, or a designated acquisition commander (DAC) established in the same legislation. The acquisition category (I, II, III, or IV) will determine the level of the SSA. The SSA directs the PM to form a source selection evaluation board (SSEB). This board generally functions through several independent teams. Teams are generally assembled to evaluate contractor proposals in the areas of technical adequacy of the proposed approach, management, pricing, and production. The SAE and the SSA will also establish a source selection advisory committee (SSAC) from experts throughout the service. The SSAC meets with and reviews the SSEB's selection criteria and recommends acceptance or changes to the SSA. The SSAC will also provide advice to the SSA during the evaluation process. Scoring criteria are developed by the SSEB and approved by the SSA to ensure that the contractor adequately addresses all aspects of the modification, including nonrecurring engineering, modification kits, technical data, logistics, training systems, factory training, production readiness, adequate oversight of subcontractors, R&M, test equipment, government-furnished equipment (GFE) required, government-approved cost control system, installation planning, adequate progress review plans, and others as deemed necessary. Each modification will require many of the same items as previous modifications, but each will require tailoring to meet its intended purpose. Several methods can be used to acquire the modification. The entire modification package, including the kit and the installation, can be procured as one entity. This places maximum responsibility on the contractor, and because more risk is shifted to the contractor, it may increase the cost of the modification. In addition, contractor installation could result in fewer systems in the hands of the operators as opposed to installing the modification in conjunction with periodic maintenance. Generally, the modification is procured as a kit, and installation is done in conjunction with a major overhaul or periodic maintenance activity. This allows multiple sources to install the kits and work the system availability to meet the user's requirements.

Some modifications, like initial system procurements, are excellent candidates for multiple development contracts. The same is true for modifications which are replacements of common systems (such as updated radios) in many larger systems and systems which add a capability to current systems. The global positioning system, which is being added to aircraft, tanks, ships, and vehicles and is considered for every conceivable military use, is an excellent example of multiple-source development. It may also be desirable to maintain dual (or more) sources in procurement for those subsystems where the numerical requirements are very high and many years are required to achieve the desired equipage. By maintaining dual sources, competition is maintained and each contractor is motivated to improve the R&M and thus make its unit more attractive to the buyer.

MONITORING THE MODIFICATION CONTRACT

Once a contract has been awarded, MIL-STD-1521 and MIL-STD-2167 describe the design reviews that must be accomplished to ensure that the contractor is progressing according to the terms of the contract. Other functions of these reviews include reviewing the contractor's design for the modification and providing feedback on any problems detected in meeting the objectives on the modification.

SYSTEM REQUIREMENT REVIEW

The first major review is the system requirement review (SRR). This review ensures that all modification kit requirements and interface requirements are properly defined and understood. All the proposed logistics functions are reviewed for adequacy. Requirements are reviewed to ensure that the major system being modified and any other equipment needed to accomplish the modification are properly interfaced with the design.

The SRR is normally conducted at the beginning of the demonstration/validation phase of the modification. For modifications that do not include major development efforts, the SRR may not be required or included in the contract.

SYSTEM DESIGN REVIEW

Each configuration item (CI) is reviewed to ensure that the performance requirements for the modification are covered adequately in the draft hardware specification before the CI baseline is established. Once the PM is satisfied that all CI baselines are adequate, complete development specifications for the CIs that make up the total modification package can be finalized. Checks are made by the government team to ensure that the analysis data used by the contractor led to the best possible modification alternative to satisfy the operational requirements.

The government engineers establish an allocated baseline for each CI as it completes its system design review (SDR). The allocated baseline is a contract between the government and the contractor. Tests are designed to verify that each CI meets its required performance. The baseline provides the basis for the full-scale development of the modification. Each change to the modification baseline requires the contractor to submit an additional engineering change proposal. For many large new systems the number of ECPs runs into the thousands. For major modifications, they typically run into hundreds or less if the initial design was well engineered. The objective, of course, is to build a well thought out and engineered modification that will keep the number of ECPs as low as possible.

SOFTWARE SPECIFICATION REVIEW

The software specification review (SSR) performs similar functions for software as the SDR does for hardware. Analysis data are reviewed to ensure that an appropriate software alternative is selected to meet the software performance requirement. The SSR is where the software specifications are agreed to and the baseline for the software CI is established.

PRELIMINARY DESIGN REVIEW (PDR)

Once the government engineers are satisfied that they have adequately addressed all the functions required to accomplish the modification requirements and fully decomposed all the functions, MIL-STD-1521 requires that a preliminary design review (PDR) be conducted before detailed design begins. The PDR checks the CIs to ensure that the baseline requirements, including all interfaces internal to the design and external to the weapon system being modified, are properly addressed. It is normally accomplished during the early stages of full-scale development.

If the PDR reveals areas of concern to the government engineers, the contractor is requested to redesign those parts of the modification where weaknesses were detected. All areas of the design and the contractor's analysis data are reviewed, questions are asked, and, where necessary, corrections are initiated.

CRITICAL DESIGN REVIEW (CDR)

The critical design review (CDR) is the last major government review of the contractor's design before fabrication of the modification begins. The CDR ensures that the detailed design addresses the baseline requirements and can reasonably be expected to meet the modification requirements. Hardware design is generally contained in schematics and drawings which are all reviewed for each CI.

Software design review is also a major part of the CDR. DoD-STD-2167 defines the software products that must be available for the CDR. Software products required for the review include the software detail design documents (SDDD), the interface design documents (IDD), and the database design documents (DBDD).

The government engineers must be satisfied that there are no weaknesses in the contractor's design. This is the last major opportunity to stop an incorrect or technically deficient design from going to the fabrication shops. Correction of design problems after CDR is generally very expensive and can often result in modification schedule slips.

TEST READINESS REVIEW

The software documents presented at the CDR provide the baseline for the contractor to write the software code. Software has become such a major element of modern weapon system design that the test readiness review (TRR) is dedicated to it. It provides for code testing under conditions that are simulated to be as close to the expected operating conditions as possible.

Both the SSR and the TRR are required by DoD-STD-2167.

FUNCTIONAL CONFIGURATION AUDIT

A functional configuration audit (FCA) plan is generally submitted to the government 30 days before the formal FCA. The FCA is a formal audit to validate that the development of a CI has achieved the performance and functional characteristics specified in the functional or allocated configuration baseline.

The objective of an FCA is to verify that the CI's actual performance complies with its hardware development or software requirements and interface requirements specifications. Test data are reviewed to verify that the hardware or computer software performs as required by its functional or allocated configuration identification. For software, a technical understanding must be reached on the validity and the degree of completeness of the software test reports and the software user's manual.

Complete drawing packages must be available for service (Army, Navy, Air Force) review. The FCA audit consists of reviewing drawings on a sampling basis. For hardware CIs, drawings for the selected CI are compared directly to released engineering data. Any level of CI may be reviewed, including commercial equipment. The level, amount, and detail of this review is accomplished at the discretion of the service and coordinated between the service and the TSD audit chairperson.

At the conclusion of each day, all discrepancies will be made a part of the FCA minutes. Completion dates for all discrepancies shall be clearly established and documented. All FCA events are conducted according to contract or government requirements and the FCA plan.

The service acknowledges the accomplishment of a partial completion of the FCA for those CIs whose qualification is contingent upon completion of integrated systems testing, in accordance with para. 70.4.11 of MIL-STD-1521B.

The FCA documentation review includes, but is not limited to, an examination or evaluation of the following information:

Engineering drawings or documentation

Nomenclature

CI number

Specification tree

All deviations or waivers
Contractor test procedures or results
System test plan
System specification
Acceptance test procedures
Contractor test plans—SCT, environmental, burn-in
All engineering change proposals
Internal documentation (version description document)
PDR/CDR minutes and action items
Procurement specifications, source, or specification control drawings for each CI
Prime item development specifications
CI product fabrication specifications
CI item development specifications
Calibration measurement requirement summary
Commercial off-the-shelf specifications
Logistics support analysis documents
Noncomplex computer program specifications
Computer system operator's manual, or equivalent document
Software user's manual (includes firmware), or equivalent documents
Computer system diagnostic manual, or equivalent documents

The purpose of the documentation audit is to validate that the development of a CI has been completed satisfactorily and that the CI has achieved the performance and functional characteristics specified in the functional or allocated configuration identification.

The product specification for the CI under audit is reviewed against the shortage list. The product specification references a drawing number. The drawing for each CI is compared to the as-built configuration listing for revision status compatibility. Any difference between the listing and the drawing is resolved by examination of the change notices accompanying the drawing package. The drawing package is then reviewed. Where applicable, the acceptance test procedure (ATP) and the associated test data for each prime item are examined.

The objective quality evidence, such as inspection records for a given assembly, demonstrates proof of inspection at each level or step of assembly or test. It also verifies completion of actions or changes (see MIL-STD-1521B, para. 70.4.2e). The FCA would normally be done just prior to the completion of full-scale development.

PHYSICAL CONFIGURATION AUDIT

A physical configuration audit (PCA) is held to ensure that the design documents accurately describe what the contractor actually built. If the government engineers agree that the contractor has produced what it contracted to produce, they will sign the specification and establish the modification baseline. Often the PCA is called a first article test.

IMPLEMENTATION

The PM must ensure that all aspects of the modification are ready for installation and operational use. Items on the PM's checklist which must be addressed include:

1. Kits, government furnished equipment (GFE), and installations
2. Spares for peacetime and wartime
3. Kits, spares, and installations for simulators
4. Support equipment, support equipment qualification, and any needed modifications to existing support equipment
5. Software needs
6. New publications and publication updates
7. Trial installation and trial proof kits
8. Training requirements for operator and support personnel
9. Test and evaluation requirements
10. Adequate technical data
11. Installation planning, including negotiation of release of the systems for modifications by the operational command

Generally a trial installation is accomplished and the kit is checked out before full-scale production of the modification kits begins. Limited initial production rates are also being used to ensure proper operational test and evaluation before large-scale production release. The big exception to this "try before major buy" policy is safety modifications. The time urgency may necessitate concurrent kitproofing and production.

Once these concerns are satisfied, kitproofing has been accomplished satisfactorily, all of the reviews leading up to the production decision have been completed, and test and evaluation results validate the worth of the modification, the full rate of production is initiated and modification of all the weapon systems affected is accomplished.

SUMMARY

The modification of military weapon systems required hundreds of pages of explanations in service regulations and manuals. The entire process is still evolving as the service staffs continue to work with the service acquisition executives to define and clarify their respective roles and responsibilities. Although the review authorities may change slightly over the next few years, the efforts required for a successful modification as described in this chapter will still be required. From the initial establishment of modification requirements to their successful installation and operational use, all parties must understand their respective responsibilities and the process involved if the military services are to continue to hold the technology edge so vital to our position as the world's only military superpower.

BIBLIOGRAPHY

"A-7 Cast Functional Configuration Audit Plan," Allied Signal Aerospace Co., Teterboro, N.J., Nov. 1991.

AFR 57-1, "Air Force Mission Needs and Operational Requirements Process," Department of the Air Force, Washington, D.C., Nov. 1991.

AFR 57-4 (draft), "Operational Requirements, Modification Approval and Management," Department of the Air Force, Washington, D.C., Aug. 1991.

Aircraft Modification Process Training Course, Student reference handbooks, Department of the Navy, Arlington, Va., June 1991.

Federal Acquisition Regulation (FAR), FAC 90-5, July 1991.

Introduction to Acquisition Management, Air Force Institute of Technology, Department of the Air Force, Wright-Patterson AFB, Ohio, Apr. 1990.

CHAPTER 41

INTERNATIONAL PROJECT MANAGEMENT: DO OTHER COUNTRIES DO IT BETTER?

C. Michael Farr

C. Michael Farr is a graduate of the U.S. Air Force Academy and holds a Ph.D. degree in operations management from the University of North Carolina at Chapel Hill. He began his career as a Minuteman III combat crew member, where he gained experience as a user of military systems. He has been associated with the defense acquisition community since 1980 and is currently chief of the Systems Management Division at the Air Force Institute of Technology.

INTRODUCTION

In February 1989 Clifton Berry noted that European governments buy their weapons differently than the United States and posed the question "... but do they do it better?"[1] This question has been considered by several authors and has also resulted in studies by the General Accounting Office (GAO), the RAND Corporation, and the House Armed Services Committee. While many dimensions of project management have been examined, a recurring theme has been the centralized, civilian-controlled acquisition process used by many European countries.

Several authors have pondered whether the United States should adopt the European model of a "civilian superagency," and Congress has been no exception. In the wake of cost overruns and allegations of fraud, a reform-minded U.S. Congress repeatedly pressured the Department of Defense (DoD) to change the weapon acquisition process during the 1980s. During 1988 Senator William Roth and House Representative Dennis Hertel sponsored separate but similar legislation to create a civilian acquisition corps (Berry,[2] p. 75). Neither bill passed, but interest in the European model still exists.

Despite congressional interest, none of the previous studies have recommended U.S. adoption of the European centralized system. In fact, a 1989 report by the House Armed Services Committee[2] was quite clear in its recommendation *against* emulating the European system. However, several of the studies did find significant merit in various management approaches taken by our allies.

Therefore this chapter compares the project management approaches of selected international allies with management tendencies in the United States. The intent is to identify ideas and approaches that might be useful to U.S. project managers (PMs). This review begins with an overall look at acquisition processes and organizational structures, a look at "the system" and the environment in which it operates. The chapter then changes its focus more toward individual differences and human factors that affect daily management at the working level. Finally, the chapter concludes with a look at the impact of cultural differences on management styles.

COMPARISON OF ACQUISITION PROCESSES

This section first examines the debate over centralized procurement systems and whether they might offer significant advantages to the United States. The section then highlights other fundamental differences in the way various countries acquire weapon systems.

Centralized Procurement Systems: A Model to Emulate?

While there are some notable differences among the more centralized acquisition establishments found in Europe, they do have several common characteristics that stand in contrast to the U.S. approach. The pivotal difference in centralized systems is that the military services do *not* individually manage the weapon acquisition process. Instead there is a highly trained civilian acquisition corps that centrally manages research, development, and acquisition on behalf of the military services. There are comparatively few high-level political appointees, and the programs are managed by experienced, highly trained civilians with a clearly defined career in acquisition management.

By comparison, the military services play a dominant role in acquiring U.S. weapon systems. Also, most U.S. PMs are military officers and usually bring considerable operational experience to their jobs. While many people consider this operational experience to be a strength of the U.S. process, a directly related concern is that there is no guarantee that these managers have appropriate acquisition management education or experience. (However, it should be noted that the Defense Workforce Improvement Act of 1990 has resulted in initiatives by each military service to require specified education and experience for acquisition managers.)

The pressing question is whether centralized systems work better than the U.S. approach. Advocates of the centralized model believe that such an approach might eliminate duplication of weapon development programs among the military services, curb the potential for parochial and counterproductive competition among the services, and achieve greater efficiency in the acquisition process. "The expected outcome would be lower costs and more effective weapon systems for the same defense dollars."[3] While several qualitative studies have suggested that this might be so, quantitative comparisons of actual program outcomes suggest otherwise.

For example, in their analysis of attack and fighter aircraft programs, Gansler and Henning found that schedule and performance outcomes on U.S. projects were better than those on similar European projects.[3] Similarly, Berry[1] (p. 77) quoted the conclusion of a 1986 RAND report: "Weapon program outcomes in Europe are generally less satisfactory than those in the United States, especially in terms of schedule length and slippage during the development phase." Other authors also agree that, at best, the evidence that centralized procurement systems produce better project outcomes is inconclusive. However, there are other issues that led the House Armed Services Committee to conclude in 1989 that U.S. adoption of a centralized system was inadvisable.

In addition to the 1989 report by the Armed Services Committee, Gansler and Henning[3] and the General Accounting Office[4] have all noted at least two serious problems with the U.S. pursuit of a centralized acquisition system. First, the number of U.S. acquisition personnel and the size of the acquisition budget exceeds well over 10 times that of the European country most committed to defense expenditures, the United Kingdom. The ratio exceeds 15 or 20 times that of other European countries. The acquisition infrastructure of any single U.S. military service is three or four times larger than the entire acquisition establishment of any European country. Even a consolidated U.S. agency would be massive compared to its European counterparts. "Given the existing data, and without a great deal of additional study, it is not possible to determine whether such a large organization could work" (Ref. 2, p. 10).

So while previous studies question whether any efficiencies or other advantages would be gained from centralization, another serious obstacle has also been identified. The European centralized models imply more than just a central procurement agency. Any attempt to emulate them successfully would require a realignment of the entire defense establishment. For example, strong central coordination of military requirements and the defense budget would also be necessary. The United Kingdom has previously found that military control of operational requirements did not work effectively with a central procurement agency.

The 1989 House committee concluded that deeply rooted foundations of culture, national objectives, acquisition policies, and weapon system performance goals would have to change in order for the United States to actually copy the European model. U.S. commitment to worldwide mission requirements, attitudes about technology and risk taking, our history of detailed legislative oversight, and the frequently arm's-length (some would say adversarial) relationship between government and industry are all factors that would affect the success of a centralized defense establishment. The consensus of previous research has therefore expressed at least caution, if not outright opposition, to U.S. adoption of a centralized procurement process.

Other Fundamental Differences

However, there has also been a strong consensus that certain European policies and practices make a lot of sense irrespective of a centralized procurement system. Notably some of these ideas are aligned with recommendations made by the Packard Commission. This section identifies some of these other differences in acquisition philosophy and highlights those that, if adopted by the United States would improve our weapon acquisition process.

Interface of the Political and Acquisition Processes. It is no secret that the acquisition process has been subjected to extensive legislative oversight by the U.S. Congress. The process is intensely scrutinized, is highly legalistic and regulated, and receives an abundance of "free" advice. In their article "The Joint Chiefs of Congress," Hiatt and Atkinson link this detailed oversight to detrimental effects on the cost and schedule of several military projects.[5]

In addition to the voluminous and ever-changing array of legislation, U.S. acquisition programs are also subjected to a line-item review of the budget. Funding profiles have often been erratic, and U.S. industry finds it difficult to plan beyond one or two years into the future with any confidence.

Finally there are many high-level political appointees involved in the U.S. acquisition process. By comparison, Europeans achieve much greater program stability through their parliamentary process. Legislative oversight is much more limited, with the exception of Germany there is no line-item review of defense budgets, and there are very few political appointees involved in the weapon acquisition business. The result is an acquisition process that is somewhat insulated from politics.

If adopted in the United States, the less legalistic and less political European approach could facilitate a more stable budget environment and better long-range planning by industry.

Government-Industry Relationship. Former Assistant Secretary of Defense Jacques Gansler, among others, has noted that the United States is perhaps the only country in the world that fails to treat its defense industry as a valued national resource. The U.S. government promotes competition in the defense marketplace as if it were the same as any commercial market. In fact, there is only one buyer, there are significant barriers to entry and exit from the defense market, and the quantities purchased are frequently quite low. Yet R&D investments are typically high and the buyer sets unique specifications and requirements.

Governments around the world tend to work more cooperatively with their defense industries in several respects. First, Russell and Fischer note that in France "defense companies have been rationalized."[6] For example, there is one helicopter company in France compared to five in the United States. A 1986 GAO report[4] identified the number of prime producers per product type (aircraft, tanks, ships, and so on) in various countries around the world. Not surprisingly, the United States typically has more producers per product type than other countries. Given the U.S. government's insistence on promoting competition for a relatively small production volume, this larger number of producers has not always been economically sound. In fact, in a world where defense spending has cycled upward and downward repeatedly, Gansler and Henning[3] (p. 35) have noted that "The U.S. result is often an unhealthy defense industry—with considerable labor instability, program uncertainty and high-cost products."

Another concern related to the large number of suppliers and forced competition is highlighted by examining Japanese practices regarding suppliers. Japanese success with just-in-time inventory systems and quality derives in part from their preference for long-term, stable relationships with a limited number of carefully chosen suppliers. Close cooperative working relationships, improved quality, and a stable planning horizon have been achieved by the Japanese with this approach.

Finally, some additional comments about the arm's-length, legalistic, and often adversarial philosophy of the U.S. government are appropriate. Other governments work closely with their industries in developing and achieving technological and industrial goals, and in keeping the defense industry economically viable. In his 1989 book *Affording Defense,* Gansler noted that the Japanese Ministry of International Trade and Industry (MITI) is one example of a government organization steering a country toward a specifically planned industrial structure.[7] He also noted that MITI thrives on close working relationships among universities, the government, and industry, and that great emphasis is placed on compatibility between military and civilian technology.

With the exception of recent changes in the United Kingdom,* other countries promote competition much less aggressively than the United States. Whenever do-

*The United Kingdom has become very aggressive in promoting competition within its defense industry, one of the only countries in the world with policies similar to those in the United States. A major difference is that the United Kingdom is more comfortable whenever the competition has an international origin.

mestic quantities are too small to economically support more than one source, these governments assist defense companies with export sales and improving their international competitiveness. They contend that, even with a single supplier, the need to provide a good price on the export market is in itself a reasonable form of competition.*

Education and Experience of Acquisition Personnel. In its 1989 report the House Armed Services Committee identified three European practices it considered particularly important for the United States. One was "professionalism and training of acquisition personnel." The personnel system within the French Delegation General for Armaments has been cited by numerous authors as a particularly good model.

An elite group of approximately 1000 armament engineers comprise the top leadership. These individuals receive an intensive seven-year program of higher education and also gain hands-on experience in both R&D as well as manufacturing facilities. Entry into the program is highly competitive; candidates must pass a rigorous qualifying exam. Armament engineers hold military officer status and spend a portion of their early training with operational military units.

However, they have a separate but clearly defined career path that promotes them on the basis of engineering and management skills and compensates them on a par with their worth in the private sector. French PMs always come from this group of armament engineers. They operate in a much smaller bureaucracy with fewer layers of reporting, shorter chains of command, and are granted greater authority. They usually serve in a program management assignment for at least five years, which provides greater continuity and program stability.

The result of this system is appealing: PMs who understand the military but have no parochial interest in duplicating weapons, who have extensive education and experience before assuming senior leadership roles, and who stay in the job long enough to encourage planning and decision making for the best long-term results.

Cost-Performance Tradeoff. Russell and Fischer[6] (p. 6) noted that R&D expenditures in the United States are 10 times greater than in France. They further observed that U.S. equipment is "generally better, but not ten times better." A related observation, and perhaps a partial explanation, is that the United States starts many more projects than it has money to produce. Further, many of these multiple development projects are aimed at the same operational requirement.[8] Much of this duplicate R&D never produces any payback through production or export sales. Worse yet, the United States has withdrawn from several international partnerships because one of our "other" programs turned out to be a higher priority. By comparison, development does not usually begin in Europe until there is a reasonable expectation that the system will be funded for production.

The Gansler and Henning study revealed that both Europe and the United States generally achieve their program goals. Given our focus on maximum performance, U.S. equipment does generally perform better and is fielded a little faster. However, Europeans have also been successful in achieving lower costs and risks by settling for "acceptable" performance. The study supported this conclusion through an analysis of costs incurred per unit of performance, which revealed that U.S. and European outcomes were essentially equivalent when viewed in those terms.

Noting these results in his recent book,[7] Gansler calls for the United States to examine program affordability more aggressively and to consider a more balanced

*This sentiment was expressed during personal interviews conducted by the author during the summers of 1990 and 1991 with defense officials in Rome, Bonn, Paris, The Hague, London, and Brussels.

tradeoff between cost and performance. Given the dramatic changes around the world and the associated draw down in defense spending, these changes appear more essential than ever before.

Technology, Risk, and Mission Requirements. We have already noted that U.S. acquisition has traditionally relied on aggressive advances in technology to counter a known Soviet advantage in weapon quantities. While the issue is frequently debated, we have also accepted increased costs and risks in supporting this strategy. In the worldwide defense role that the United States has set for itself, we have also implicitly accepted operational performance requirements that exceed those of other countries. This is true in two respects. First, the United States requires advanced technology across the board in every weapon category, whereas other countries are more selective in their pursuit of advanced military technology. And second, the United States chooses to be capable of operating those weapons in any environment or location around the globe.

In the absence of a political decision to take less defense responsibility for the rest of the world, it would seem that the present situation is inflexible. However, Gansler and others have suggested that a more reasonable cost-performance balance is possible. To achieve this goal, a more rational and less duplicative requirements process is an absolute necessity.

Meeting this need does not require that the entire procurement process be centralized. However, it probably *is* necessary to have a strong, independent organization with the authority to guide the requirements process through the tough choices needed to eliminate duplication and conform to the cost realities of the 1990s.

THE HUMAN FACTOR: INDIVIDUAL DIFFERENCES

This part of the chapter looks more at the working level of acquisition management, that part of the business where some people would say "the rubber hits the road." First, fundamental differences in management philosophy and individual management tendencies are examined, especially in light of the differing environments in which PMs from different countries operate. Then the effects of cultural differences on management styles are examined.

Because of the acquisition environment, U.S. managers may not be completely free to adopt other practices that seem worthwhile. However, there may be some practices that U.S. managers would want to consider. That possibility aside, there is another reason for including this part of the chapter. Many U.S. managers find themselves involved in cooperative development and production ventures with other countries, and it is a great help to at least understand how your management counterparts tend to think.

Management Differences

PMs operate in a substantially different acquisition environment overseas. These differences exert considerable influence on management styles, and Gadeken has found that they even influence perceptions of training needs.[9] These environmental differences also partially explain why many U.S. management practices differ from those of our allies. Accordingly, some of the key environmental differences are recounted here.

U.S. managers work in much larger organizations than their allied counterparts, and perform much more detailed oversight of contractors in an environment characterized by lack of trust. Further, U.S. PMs are subjected to much more external oversight and are granted less authority and freedom in the day-to-day management of their projects. In *The Defense Management Challenge,* Fox offered observations from DoD managers, industry, and the 1986 Packard Commission regarding PM authority[10]:

> The program manager finds that, far from being the manager of the program, he is merely one of the participants who can influence it.

Fox goes on to describe a PM's vulnerability to a host of "special-interest advocates" who levy demands on the PM, yet the system allows the PM little discretion to use his or her judgment to balance these sometimes conflicting requirements.*

> In DOD there is nobody really in charge as is a contractor's program manager. Instead, we have many interest groups that can influence the system. There are so many checks and balances that decisions are very slow. Many can say no and very, very few can say yes.

Finally, U.S. PMs face many more layers of management and a political system that holds them accountable at each step of the way. The more parliamentary processes in other parts of the world interfere with PMs much less during the acquisition process and judge the results *afterward.*

In this environment, Gadeken's research revealed that U.S. PMs do not personally perform many of the analytical, hands-on management tasks typically undertaken by allied managers. Instead, U.S. managers are forced to focus on leading and coordinating the efforts of large groups of specialists. A complaint of U.S. PMs is that they sometimes lack sufficient control over these "functional" experts.

Allied PMs, who experience much less oversight themselves and who generally have close working relationships with industry, exercise less oversight of their contractors and are more comfortable in delegating certain responsibilities to industry. Gadeken's research, which focused on differences between the United States and the United Kingdom, indicated that PMs in the United Kingdom perceived much lower need for personal training in the areas of budgeting, test and evaluation, and production because they trust industry to manage these aspects of their programs.

The continual review of defense acquisition programs, which sometimes becomes both political and public, produces more uncertainty about whether U.S. projects will continue once they are under way. U.S. PMs, who are usually military officers with a strong action orientation and a brief tenure in their position, frequently develop a strong program advocacy and a comparatively short-term planning horizon in response to this environment. While the response is understandable and even predictable given the circumstances that often prevail, these practices have been challenged by various experts as suboptimal.

Cultural Differences

This section does not extol the virtues of copying other cultures. However, given the frequently international nature of business in the commercial as well as the defense

*These special-interest advocates represent legitimate concerns related to issues such as small and minority business utilization, competition, preference for domestic sources, maintainability, reliability, producibility, and so on. The difficulty is that, by comparison to allied counterparts, U.S. PMs have little freedom to exercise personal judgment in balancing these concerns.

sectors, it is vitally important to be aware of cultural tendencies which may be different than our own and which may exert significant influence on our business transactions.*

Hodgetts and Luthans at Florida International University have defined culture in the following manner:[11]

> ... acquired knowledge that people use to interpret experience and to generate social behavior ... this knowledge forms values, creates attitudes, and influences behavior. Culture can affect technology transfer, managerial attitudes, managerial ideology, and even business-government relations. Perhaps most important, culture affects how people think and behave.

During 1991 Anthony Amadeo conducted an award-winning study of cultural dimensions of international business.[12] Among other issues, Amadeo examined the effect of cultural differences on management practices on international armaments projects. The remainder of this section summarizes the key findings.

While many Americans approach decision making as an exact science, other countries often use a completely different style, which can prove frustrating. As compared with many international counterparts, Americans prefer quick factual decisions made by an individual who has been delegated the authority. Much of this tendency derives from American culture and its emphasis on time, individuality, and goal orientation. In contrast, other cultures place more importance on group decisions. Patience and flexibility are two mandatory characteristics for success in the international environment.

Two more management areas which are significantly affected by culture are the treatment and use of lawyers and written contracts. American ideas about the use of lawyers and contracts are not always consistent with those of other countries. Americans generally require the use of contracts with each aspect of the agreement stipulated in writing. If it is not in writing, then there is not an agreement. Also, lawyers are accepted members of a business team and are often used as a means to resolve disagreements. In fact, lawyers are often very influential members of the U.S. negotiation process. This is usually not the case internationally. Much more time is spent building relationships in other countries in an attempt to reduce the importance of written contracts and lawyers. Agreements in other countries may range from a simple handshake to a formal document. In some cases a verbal agreement can be more important than a contract. In some cultures a contract is merely an indication of work that is intended to be completed.[13] The implication of bringing a lawyer to an overseas business meeting is generally one of mistrust.

The significance of these two areas provides some insights into another important aspect of cultural dimensions, communications. It is essential that both parties fully understand exactly what agreements are being made. In many cases this must be accomplished without the use of complex contracts and lawyers.

Amadeo's analysis indicated that differing management practices or styles do create obstacles that PMs must overcome. The biggest problem associated with these differences was a general slowing of the entire process.

The analysis also revealed that developing personal relationships is somewhat more important in the international environment, however, they were not considered a major problem area. Respondents were split concerning whether lawyers should be actively involved in international transactions or negotiations.

---

*The research fellows program at the Defense Systems Management College produced studies two years in a row, in 1990 and again in 1991, on international armaments cooperation. The first report was "Europe 1992: Catalyst for Change in Defense Acquisition," by Cole, Hochberg, and Therrien. The second was "International Cooperation: The Next Generation," by Johnson, Engel, and Atkinson.

Also, several factors were identified that were considered important in determining the success of international transactions. The most important of these factors were preparedness and patience of the U.S. team. International dealings were generally considered to be more time-consuming; our international partners lack the same sense of urgency to "get the job done and move on" that typically characterizes U.S. managers.

Based on his research findings, Amadeo drew several conclusions concerning the effects of cultural factors on successfully negotiating and managing international programs. Those practices that are most relevant to program management are summarized in the following:

1. Several cultural factors can greatly influence the success of international negotiations. Factors associated with different managerial styles, negotiation tactics, legal systems, financial processes, and so on, presented the greatest problems. While it is impossible to completely resolve these differences, it is possible to lessen them with knowledge, understanding, and preparation. Surprisingly few problems were associated with any perception that there were deficiencies in the international counterpart team. U.S. team deficiencies such as insufficient planning, insufficient authority, and lack of experience were believed to affect success to a greater extent. The international negotiator controls these factors directly and can greatly enhance the chances of success by effectively managing them.

2. It is important to adjust your international negotiating style according to the culture with which you are dealing. An international negotiator must not only be aware of cultural differences which exist, but also of individual differences. He or she must be more flexible and patient than in U.S. negotiations. The international negotiation process is more time-consuming, with more emphasis on establishing personal relationships through informal conversations and entertainment.

Small, internationally experienced negotiating teams make this goal easier to achieve. This may require the use of a local culture expert or of an experienced international business traveler to accompany the team. The use of a core international negotiating team with additional technical experts partially addresses these concerns. Such a team should also have sufficient authority to make decisions locally without excessive calling or faxing back and forth to the United States.

3. As part of the planning and preparation for an international negotiation, the negotiator must take the time to study the counterpart's culture. While the importance of this cannot be overstated, there is a definite lack of cultural understanding by U.S. business people. The successful U.S. business person cannot simply transfer the same knowledge, tactics, and techniques overseas and expect to be successful. The importance of this cultural preparation must be emphasized through international training courses and literature.

4. As indicated, personal relationships are more important in the international arena. It is important to create an atmosphere of trust throughout the entire program. This trust can be built through social contact, honesty, patience, and understanding of different culture and customs. The U.S. business person must plan for additional time to be spent building these relationships. International transactions should not be approached with the typical U.S. attitude of "time is money." U.S. managers must also recognize that informal social events are often far more productive than the officially scheduled meetings.

5. While lawyers have a role in the international environment, U.S. teams traditionally place too much emphasis on their participation in the process. Internationally, lawyers signify a feeling of mistrust which greatly hampers the development of personal relationships. Therefore Amadeo concluded that lawyers should review U.S. positions, but should not usually accompany U.S. teams on international travel.

If it is necessary for a lawyer to be a member of the team, their role during face-to-face transactions with international counterparts should not be a prominent one. In addition, honesty and forthright explanations concerning U.S. business practices can go a long way in dismissing any feelings of mistrust.

6. Several management and organizational factors have important influences on the success of international transactions. These factors are directly controlled by the PM and therefore should be considered before venturing into an international program. Similar to factors which affect the success of negotiations, these factors (such as preparedness, patience, and technical expertise of U.S. team, familiarity with counterpart's business practices and customs, and personal ties built through the years) are important in determining the success of international transactions. Preparedness and proper planning are the keys to success. A large part of this planning is the understanding of your counterpart's culture and customs. Proper planning and understanding of differences can overcome most of the cultural barriers to successful programs.

SUMMARY

Based on the strong consensus found in previous studies, this chapter does not advocate copying the European centralized procurement process. However, also based on strong consensus, there are some international practices that deserve serious consideration by the United States, and as noted at the outset, some of these practices coincide with recommendations of the 1986 Packard Commission. Unfortunately, as one moves from the political environment through the DoD hierarchy and on down to the working level of project management, it becomes increasingly difficult for managers to implement new ideas based on their own judgment. Certainly PMs can become more aware of the potential for adopting good ideas from overseas and can achieve varying degrees of success in making some of those changes. However, full realization of the potential benefits will require cooperation between Congress and DoD that has not been forthcoming in the past. On a brighter note, however, it has been said that necessity is the mother of invention. In the face of dramatic changes around the world, the opportunity has probably never been greater for taking positive steps toward resolving some of the historical problems with the defense acquisition business.

REFERENCES

1. F. C. Berry, Jr. "Defense Procurement, European Style," *Air Force Mag.*, pp. 74–77, Feb. 1989.
2. "A Review of Defense Acquisition in France and Great Britain," Subcommittee on Investigations, House of Representatives, Committee on Armed Services, 101st Congress, 1st Sess. Government Printing Office, Washington, D.C., Aug. 16, 1989.
3. J. S. Gansler and C. P. Henning, "European Acquisition and the U.S.," *Defense and Diplomacy,* 1989.
4. U.S. General Accounting Office, "Weapons Acquisition: Processes of Selected Foreign Governments," GAO/NSIAD-86-51FS, Washington, D.C., Feb. 1986.
5. F. Hiatt and R. Atkinson, "The Joint Chiefs of Congress," *The Washington Post,* nat. weekly ed., Aug. 12, 1985.

6. T. B. Russell and C. K. Fischer, "How to Do Business with the French Delegation General for Armaments," Army Materiel Command Representative—France, Office of Defense Cooperation, APO New York 09777, May 1989.
7. J. S. Gansler, *Affording Defense,* M.I.T. Press, Cambridge, Mass., 1989, p. 311.
8. C. M. Farr, "Managing International Cooperative Projects: Rx for Success," in E. B. Kapstein (ed.), *Global Arms Production: Policy Dilemmas for the 1990s,* Center for International Affairs, Harvard University, Cambridge, Mass., 1992.
9. O. C. Gadeken, "Through the Looking Glass: Comparisons of US and UK PMs," *Program Manager,* Defense Systems Management College, Ft. Belvoir, Va., pp. 22–26, Nov.–Dec. 1991.
10. J. R. Fox, *The Defense Management Challenge: Weapons Acquisition,* Harvard Business School Press, Boston, Mass., 1988.
11. R. M. Hodgetts and F. Luthans, *International Management,* McGraw-Hill, New York, 1991, p. 35.
12. A. L. Amadeo, "Cultural Dimensions of International Business," M.S. thesis, Air Force Institute of Technology, Dayton, Ohio, Sept. 1991.
13. L. Copeland and L. Griggs, *Going International: How to Make Friends and Deal Effectively in the Global Marketplace,* Random House, New York, 1986, p. 94.

P · A · R · T · 5

STRATEGIC OUTCOMES

The coeditors have chosen to focus attention on three special subjects which can and do affect the strategic outcome of military projects. The first, source selection, could be considered an input, but in actuality it is an output—the critical issue of selecting the right contractors to support the government team in carrying the project to completion. The second subject, project advocacy, has a significant impact on the success or failure of the project. Finally, proper consideration of lessons learned from other projects can make a difference to the instant project. A legacy that the project manager can leave to others is the lessons learned from the instant project.

In Chap. 42 Curtis R. Cook and Vernon J. Edwards review source selection and the significant role that it can play in the military project. A thorough review of the source-selection process is contained in this chapter, and conclusions are reached as to the effectiveness of this concept.

In Chap. 43 Earl D. Cooper addresses the always controversial subject of project advocacy and analyzes the various forces, both within and outside the project office, that impact on the success of the project. The chapter addresses the question of whether or not the manager of a military project can be its chief advocate, ensure his or her career progression, and serve the public interest—all at the same time.

In Chap. 44 Daniel R. Vore and Norah H. Hill review the role of lessons learned in defense acquisition projects. The chapter addresses the importance of considering lessons learned and then discusses the peculiar programs of each of the services. Taking advantage of the successes and failures of other military projects should clearly be a goal of the military project manager, and the lessons learned program provides the vehicle for accomplishing this.

CHAPTER 42
SOURCE SELECTION

Curtis R. Cook and Vernon J. Edwards

Curtis R. Cook is Director of Project Management Research and Training for Educational Services Institute. Dr. Cook has taught project management and contracting courses to practicing professionals around the world. A member of the Project Management Institute and the National Contract Management Association, he is both a certified Project Management Professional and a Certified Professional Contracts Manager. Most recently Dr. Cook headed the Department of System Acquisition Management at the Air Force Institute of Technology, the graduate school for the U.S. Air Force, where he was responsible for professional continuing education and master-of-science degree programs in contracts management, systems/project management, information resource management, and software systems management. He holds a Ph.D. degree in logistics, operations, and materials management from the George Washington University and an M.B.A. degree in business administration.

Vernon J. Edwards is a senior instructor for Educational Services Institute. He has taught thousands of industry and government representatives a wide variety of courses, including source selection, federal contracting basics, construction contracting, competitive proposals contracting, and contracting for project managers and technical representatives. An accomplished author, Mr. Edwards has written texts on contracting, source selection, and a variety of other related topics. He holds a bachelors degree from UCLA and has extensive experience in major Air Force and DoD source selections.

INTRODUCTION

The Department of Defense (DoD) has long held the policy of fostering competition in its major acquisitions. This policy became law in 1984, when Congress passed the Competition in Contracting Act (CICA), making full and open competition mandatory for all federal acquisitions, with certain limited exceptions. CICA also eliminated the mandatory use of sealed bidding procedures, and established competitive negotiations as the appropriate method of awarding complex contracts when sealed bidding procedures are inappropriate. Source selection, the subject of this chapter, is nothing more than the process of conducting competitive negotiations.

Source-selection procedures vary among the military services, and even according to the dollar amount of the contracts. But in general the military follows a standard procedure we might describe as follows. A solicitation for offers is issued; an offer is tendered; the offer is analyzed and evaluated against the military's requirements; the offer is either accepted or negotiation ensues to reach agreement; counteroffers follow until mutually agreeable terms are established; an award is made by an official designated as the source selection authority.

While this process appears simple enough, it is in fact quite complex, sometimes taking a year or more to complete. Surprisingly, the complexity and mystery surrounding formal source selection is not the result of excessive Federal Acquisition Regulation (FAR) guidance. On the contrary, the objectives of the source-selection processes are clearly stated in the FAR at 15.603:

- Maximize competition
- Minimize the complexity of the solicitation, the evaluation, and the selection decision
- Ensure impartial and comprehensive evaluation of offerors' proposals
- Ensure selection of the source whose proposal has the highest degree of realism and whose performance is expected to best meet stated government requirements

These are rational goals. For example, competition is the law of the land. Therefore a process that increases competition is one which should willingly be adopted. Likewise, any measure that minimizes the complexity of the procurement process must be applauded. The third goal is of equal importance, for all offerors should be treated fairly. And finally, the government certainly has the right to receive best value for the money it expends. This idea of best value will be explored at length.

The FAR gives the military services wide latitude in implementing these objectives. DoD likewise has little to say in its Defense FAR Supplement (DFARS) other than to provide guidance on the alternative four-step process. However, this is where the simplicity ends. Each of the services has published detailed guidance on exactly how source selections should be conducted. As a result, the source-selection process has become quite complex and lengthy. Despite recent attempts at streamlining, the process remains cumbersome and confusing to those lacking experience in contracting with the DoD.

FUNDAMENTAL PRACTICES

While individual military services have heavily supplemented the FAR and DFARS with their own regulations describing service-unique processes, the same fundamental steps are followed in any successful source selection:

- Acquisition planning
- Solicitation preparation and issuance
- Preproposal conference
- Proposal preparation
- Proposal analysis and evaluation
- Decision to award on the basis of initial proposals or to conduct discussions
- Discussions
- Best and final offers
- BAFO analysis and evaluation
- Contractor selection and award

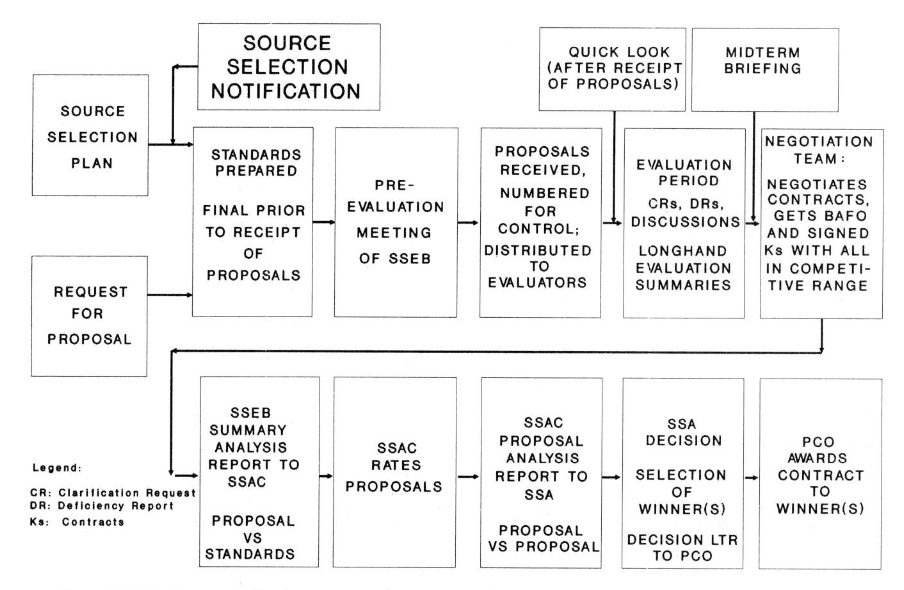

FIGURE 42.1 Source-selection process (conventional).

Figure 42.1 captures this sequence in large part and is meant to illustrate a typical conventional source selection in the military services.

What does the FAR itself say regarding source selection? Essentially, neither the FAR nor DFARS tells the services how to conduct source selection. Some guidance is provided, however, on the fundamentals that must be considered in every source selection. For example, the FAR states that every source selection must consider price or cost as an evaluation factor in the award. Other than this one factor, agency heads are free to tailor evaluation factors to suit their particular circumstances.

When must source-selection procedures be used? FAR 15.602 states that the procedures are applicable to negotiated contracts, when award is based on cost or price competition, or on the comparison of cost or price and other factors. Small purchases, those less than $25,000, are exempt. Note that less complex procurements using the sealed bidding technique, even if over the small purchase threshold, are not negotiated, and are therefore not subject to these procedures.

The FAR distinguishes between formal source selection and other less formal award decisions. For example, in formal source selections the official responsible for the source-selection decision must be designated in writing. This official is known as the source selection authority (SSA), and he or she is typically supported by a technical evaluation board, usually termed the Source Selection Evaluation Board (SSEB), and an advisory council, frequently known as the Source Selection Advisory Council (SSAC).

The FAR, at 15.612(b)(2), merely requires the SSA to establish an evaluation group structure "appropriate to the requirements of the particular solicitation." Anything beyond this simple guidance has been left up to individual agencies.

FAR 15.612(c) does require, in formal source selection, that a source-selection plan be developed which considers the following:

- A description of the organization structure
- Proposal presolicitation activities

- A summary of the acquisition strategy
- A statement of the proposed evaluation factors and their relative importance
- A description of the evaluation process, methodology, and techniques to be used
- A schedule of significant milestones

The FAR also requires that the evaluation factors used in source selections be set forth in the solicitation, and that their relative importance be stated. Evaluation of proposals—the assessment of the proposal itself and the offeror's ability to accomplish the contract—must be accomplished solely on the basis of the evaluation factors set forth in the solicitation.

How should proposals be "scored"? Despite the wide variety of scoring methods used by the military, FAR 15.608(a)(2)(iii) requires only that a "summary, matrix, or quantitative ranking of each technical proposal in relation to the best rating possible . . . " be established. The individual preferences of the services are described in their source-selection regulations. The Air Force, for example, forbids the use of numerical measures when evaluating proposals against primary selection criteria. Instead, a color scheme is used to summarize the relative merits of each proposal. This will be explained in a later section.

The FAR also provides general guidance on conducting oral and written discussions with offerors, and strictly forbids the practices of technical leveling, technical transfusion, and auction techniques. These will also be discussed later.

Source-selection decisions must be made relative to the factors set forth in the solicitation, must consider any rankings and ratings prepared by evaluation and advisory groups, and must be documented in writing, including the relative strengths and weaknesses of the proposals.

While the FAR and DFARS contain some additional guidance, most of the detailed instructions are set forth in individual agency regulations. For example, Air Force Regulation 70-15 (Apr. 1988) on the subject is about 36 pages long. NASA's handbook on source selection is 64 pages long. The Army and Navy have similarly lengthy documents. Why such detail? Most of the procedures now in use have evolved as a result of protests and appeals of source-selection decisions—in other words from case histories from the various agency boards of contract appeals and from courts of appeals decisions. The services have developed detailed procedures to protect themselves from protests and appeals of award decisions that inhibit the program manager's ability to maintain progress. A number of these case histories will be cited later in the chapter.

Of course, in addition to "protection," any process as important as source selection of contractors for multibillion dollar programs is inherently complex, and a certain amount of detail and standardization is necessary to ensure consistency of judgment across the huge organization known as the DoD.

We now turn our attention to the process of source selection, but rather than reiterate agency regulations, our exploration will proceed from the perspective of a major thrust in DoD today—that of selecting sources on the basis of best overall "value" for money spent.

BEST-VALUE CONTRACTING

The primary emphasis in source selection today is on best-value contracting. Best-value contracting literally means as integrated assessment of price, price-related factors, and other criteria to get best value when awarding contracts. Source-selection authorities and contracting officers make quality and price tradeoffs when evaluating competitive proposals to improve the quality of selection decisions.

The idea that we should be willing to pay more initially for better-quality products and services is a concept we are familiar with in Western culture. Better-quality products and services are more capable, durable, reliable, accurate, thorough, and so on. We may pay more for them initially, but over the long haul (the product life cycle) we will spend less—and be more satisfied.

Nevertheless, in government procurement there is a strong bias in favor of awarding contracts to the offeror with the lowest price. Many contracting officers are cautious about concluding that a higher-priced offer may represent a better overall value to the government than a lower-priced offer.

This reluctance may arise from fear that the exercise of judgment will expose the decision maker to second-guessing (not unusual in military procurement). It may also arise from a sense that the best-value approach is too subjective and, therefore, unfair to competing offerors in some way. It may be that some contracting officers are not quite sure how to go about making price and quality tradeoffs, or how to justify their decisions. Finally, some may feel that best-value selection or award processes are too complex and time-consuming, or that price-only decisions are simply easier to make.

To attenuate these fears, and to actually begin to practice best-value source selection, we need to examine the concept of best value. What exactly is it? It can be defined as the best combination of the following:

- Supplier reputation and capability
- Product or service features and characteristics
- Delivery terms and conditions
- Price or estimated cost and fee

A BRIEF HISTORY OF FEDERAL PROCUREMENT

For many decades federal contracts were, for the most part, awarded after the solicitation and opening of sealed bids.* The government awarded contracts to the lowest bidder. The best value was the lowest bid. However, sometime after 1861 the concept of "responsibility" was developed. Thereafter, following the opening of bids, an investigation of the bidder's responsibility was conducted. Only after the investigation confirmed the responsibility of the bidder could the award of the contract be made. The criteria for award then became not merely price, but price and responsibility. Best value was the lowest bid received from a responsible bidder.

Federal agencies rarely engaged in negotiated procurement—the award of contracts through procedures other than sealed bidding—before World War II, and then only with a special grant of authority. Until the end of World War II negotiations were usually noncompetitive. Any determination of a prospective contractor's responsibility was virtually inherent in the decision to do business with the contractor on a sole-source basis. Best value was a fair and reasonable price from the sole source.

The development of a competitive approach to negotiation followed the develop-

*Prior to the passage of the Competition in Contracting Act of 1984, this procedure was known in federal procurement as "formal advertising." The FAR has dropped the terms "advertised" and "formal advertising."

ment of the procedure for determining contractor responsibility subsequent to bid opening, and it created something of a procedural puzzle. How should the contracting officer proceed in a negotiated procurement? Should the contracting officer evaluate an offeror's proposal, rank it against the competition, determine it to be the best, and then determine if the offeror was responsible? Or should the responsibility determination be integrated with the determination of the acceptability of an offer before comparing it to other offers?

This question about procedure still lingers in the FAR and is reflected in the various approaches taken by different federal agencies. The question of whether a contracting officer can include "responsibility-type factors" among the evaluation criteria applied to competitive proposals has been, and continues to be, the subject of protests. The responsibility procedure is a "cultural artifact" of the days when sealed bidding was the normal procedure for awarding contracts.

Regardless of this confusion, the policy has long been established that contracting officers need not award competitively negotiated contracts strictly on the basis of price, or price and price-related factors. For at least four decades contracting officers have been free to make tradeoffs among a variety of factors pertinent to the question of which competing proposal represents the best value for the taxpayer. As stated in the FAR at 15.605(c):

> While the lowest price or lowest total cost to the Government is properly the deciding factor in many source selections, in certain acquisitions the Government may select the source whose proposal offers the greatest value to the Government in terms of performance and other factors. This may be the case, for example, in the acquisition of research and development or professional services, or when cost-reimbursement contracting is anticipated.

The policy that the government may pay more to get more is well established in the protest decisions of the Comptroller General of the United States. In *Litton Systems, Inc.*, B-239123, Aug. 7, 1990, 90-2 CPD para. 114, a protest against an Army award of a fixed-price contract to a higher-priced offeror, the comptroller affirmed his long-standing support of the best-value concept. In that procurement, offerors' proposals were evaluated on the basis of four factors: price, technical, product assurance and test, and production and management. The solicitation stated that the non-price factors were significantly more important than price. The Army awarded a contract to ITT, whose proposal was rated higher than Litton's, at a significantly higher price ($50 million higher).

In responding to Litton's protest that the Army's decision was unreasonable, the comptroller stated:

> Under a solicitation like the one here, which calls for award on the basis of best overall value to the government, there is no requirement that award be made on the basis of low price. Agency source selection officials have discretion in determining the manner and extent to which they will make use of the technical and price evaluation results. *Institute of Modern Procedures, Inc.*, B-236964, Jan. 23, 1990, 90-1 CPD para. 93. Technical and price tradeoffs are permitted but the extent to which one may be sacrificed for the other is governed by the test of rationality and consistency with the established evaluation factors. See *Grey Advertising, Inc.*, 55 Comp. Gen. 1111 (1976) 76-1 CPD para. 325. We will accord due weight to the judgement of selecting officials concerning the significance of the difference in technical merit of offers and whether that difference is sufficiently significant to outweigh the price difference. See *Institute of Modern Procedures, Inc.*, B-236964, *supra*.

FAR 15.602(a), which defines the applicability of FAR subpart 15.6, dealing with source selection, makes a distinction between cost or price competition between proposals that meet the government's minimum requirements stated in the solicitation; and competition involving an evaluation and comparison of cost or price and other factors.

It is this FAR passage that, through subtle wording, gives us a clue as to the distinction between the best-value approach to selection or award, and the approach in which award is made to the responsible offeror with the lowest price. This is a critical distinction, since both approaches involve the consideration of criteria other than price and price-related factors in making the selection or award. In both approaches the contracting officer must consider responsibility-type factors, but only in the best-value approach does the contracting officer make a determination of the relative merit of a proposal on the basis of "other factors."

EVALUATION OF BEST VALUE—A TWO-STEP PROCESS

Evaluation involves making comparisons. Evaluation is a matter of determining or setting the value of something, and is always a matter of comparing the thing being evaluated to some criterion, or standard, of value. Value is a matter of relative worth. For example, the value of gold may be measured by how much of some other commodity it can buy. Since an ounce of gold will buy more than an ounce of silver, we conclude that gold is worth more—that it has a higher value.

Our judgment of value depends on the criteria we use. Gold is not very valuable as food. And if we have a fixed amount of money, we can buy more lead than gold, which will make lead the better value if we intend to use it for ballast. We determine the value of something in light of some criterion of value.

The best-value concept implies a two-step assessment. The first step is to compare each competing proposal to criteria of value and to determine how well or how poorly the proposal measures up to those criteria. A score is then assigned. The score may be a number between 0 and 1.0, with 1.0 being best; it may be a choice of adjectives, such as excellent, good, average, or poor, or blue, green, yellow, and red; it may be a letter such as A for excellent, B for good, C for average, D for marginal, or F for unacceptable; or it may be any other expression of our judgment of the value of the proposal relative to the criteria.

This is the determination of the value of each proposal relative to the evaluation criteria. The second step is to determine the value of the proposals relative to one another. It is the determination made in this second step that forms the basis for the determination of which of the competing values is best. All of this is merely implied by the term "best value."

DEFINING QUALITY AND ASSESSING BEST VALUE

Quality is most often defined as meeting the customer's needs or providing customer satisfaction. Within this context we can also define quality as the set of functions or features of a product or service, and the properties and characteristics they possess, assuming these features and characteristics were defined by the customer. Examples of product features include functions, size, weight, material composition, and component design. Characteristics and properties include strength, durability, accuracy, reliability, and maintainability. Service features include functions performed, proce-

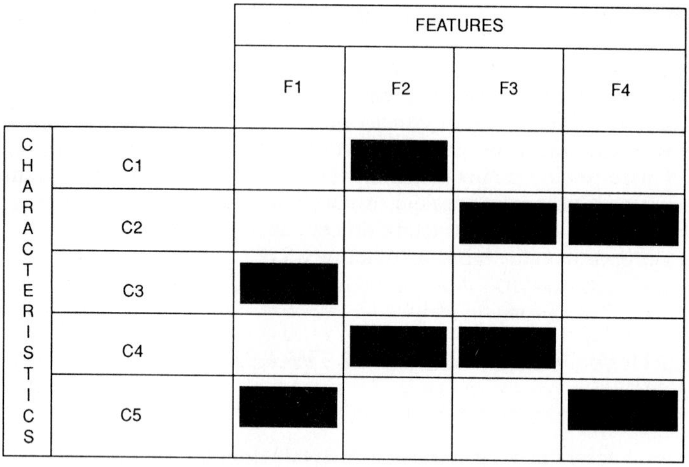

CHARACTERISTICS	FEATURES: F1	F2	F3	F4
C1		■		
C2			■	■
C3	■			
C4		■	■	
C5	■			■

The quality of a good or service is defined as its functions and features and their respective characteristics. Shaded areas represent the requirement that a feature possess a specified characteristic.

FIGURE 42.2 Quality matrix.

dures, location of performance, frequency of performance, and duration of performance. Service characteristics include accuracy, timeliness, and thoroughness.

Price is a feature. Reasonableness and realism are price characteristics. The price and its characteristics lend a proposal a certain quality. These relationships between features and characteristics can be portrayed in a matrix, as shown in Fig. 42.2.

The specification of a product or service is the description of the buyer's requirements for functions and features, the properties and characteristics those functions and features must possess, and the degree to which those properties and characteristics must be present.

For example, in writing a specification for a rope we might choose to describe the required features in terms of length, diameter, materials, and type of weave. We might further state that the rope must possess a characteristic strength, specifically that it must be strong enough to bear, without breaking, a force equal to a mass of 100 kilograms falling through a distance of 10 meters. We should include a description of the tests we will use to confirm the presence of these features and the strength characteristic. If these qualities are specified to be minimum, then the determination of the value of any proposed rope involves determining whether or not the rope possesses at least the minimum specified features and qualities. If it does, it has value. If it does not, then it has no value.*

In making a purchase of products or services, one should consider not only the quality of the products or services offered, but also the quality of the offerors themselves. Their offers constitute promises of satisfactory performance. The question arises whether we should believe their promises. When evaluating a proposal, we are evaluating the credibility of a set of promises.

*Properties and characteristics can, and frequently are, specified in terms of upper and lower limits.

Every source selection embodies a prediction that the promises of the selected supplier are likely to be kept. This involves evaluating the credibility of the promisor. This, in turn, involves making an assessment of the promisor's capability of, and reputation for, keeping its promises. Obviously we must expand our definition of quality to include the reputation and capability of the offeror.

The features relating to the quality of an offeror's reputation include its past performance, the nature of its relationships with its customers and its own suppliers, its labor relations, its credit standing, and so forth. The characteristics one might look for include reliability, cooperativeness, honesty, stability, and consistency. The features relating to its capability include its facilities, equipment, personnel, management systems, and financial capacity. The characteristics one might seek include adequacy, suitability, and reliability. The measure of the presence or absence of these features and characteristics will help us decide whether we can rely on an offeror's promises. The value of a proposal is determined in part by the presence or absence of these qualities of the offeror, as well as the qualities of its proposed products or services. The relative degree of credibility of each proposal is an important factor in identifying the proposal with the best value.

DESCRIBING FINDINGS

Value, as we have discussed it here, is a quality or a combination of qualities. How can these qualities be described in such a way as to facilitate comparison of competing offers to determine best value? A great debate rages over this question. Those with a quantitative bent believe that quantitative measures should always be used, while the "kinder and gentler" sect maintain that qualitative descriptions are more appropriate. In truth, a combination of measures is usually most meaningful.

Some aspects of quality lend themselves to quantification. Obviously, features such as size, weight, width, length, and thickness can be expressed quantitatively. Characteristics such as strength, reliability, and maintainability can also be expressed quantitatively. We can express strength in terms of the relative ability to lift a number of kilograms, or withstand a pressure of so many kilograms per square centimeter. Reliability can be expressed in terms of mean time between failures. Maintainability can be expressed as mean time to repair. What if we wish to evaluate the soundness of an offeror's plan to perform certain tasks? How can "soundness" be quantified meaningfully? How is an offeror's "understanding of a problem" measured?

Arguments in favor of numerical or verbal expressions of value are endless, and we will describe some of the scoring systems in use by federal agencies later. At this point it is sufficient to state that determinations of value are inherently subjective, and always entail a mix of fact and opinion. Subjective measures imply that reasonable people will differ as to the value of a proposal, and as to the relative value of two proposals. This is unavoidable, for value is an inherently fuzzy notion.

The Comptroller General has recognized that value determinations are subjective. In a discussion of numerical scoring in *Fox & Co.,* B-190507, 78-1 CPD para. 418 (1978), the comptroller stated:

> Numerical point scores, when used for proposal evaluation, are useful as guides to intelligent decision making, but are not themselves controlling in determining award, since these scores can only reflect the disparate, subjective and objective judgments of the evaluators. Whether a given point spread between competing offers indicates the significant superiority of one proposal over another depends on the facts and circumstances

of each procurement, and while technical scores must of course be considered by source selection officials, such officials are not bound thereby.

Regardless of how opinions are described, not all opinions are equally worthy. The opinion of a knowledgeable, trained, and experienced professional is more valuable than the opinion of a novice. Contracting officers and other selection officials must always ensure that judgments and opinions are properly documented such that an impartial observer can ascertain the reasoning supporting the decision. This is their best defense against future protests and the resulting scrutiny of their logic in making source selections.

THE PRICE/QUALITY DECISION

The challenge in military best-value source selection is twofold: how to make reasonable judgments that result in selection of the best overall source, and how to withstand public scrutiny of those judgments. Contracting officers have lost protests concerning use of their judgment. In *System Development Corp.,* B-213726, June 6, 1984, 84-1 CPD para. 605, the comptroller sustained a 1984 protest against an Army award of a cost-reimbursement contract for engineering support services on grounds that the Army had not justified paying a higher price than that offered by a technically acceptable, lower-cost offeror.

In that case the solicitation stated that award would be made to the offeror with the "best overall response," which was defined to be "superior technically with realistic estimated costs." Technical factors were numerically scored on a 100-point scale. Cost was not numerically scored. The winner's proposal received eight more points than did the protestor's, but the protestor's proposal was considered technically acceptable. However, the winner's proposed cost was $11 million, while the protestor's was $7 million. After reviewing the facts, the comptroller concluded that:

> While we recognize that the record does indicate that the [technical evaluation panel] determined [the winner's] proposal to be better technically than [the protestor's], the record contains no justification for paying $4 million more for a proposal only slightly better than [the protestor's] technically acceptable proposal. In fact, the record indicates that the Army made its award determination without considering cost.

This contracting officer's mistake is apparent. Best value cannot be determined without considering cost.

In *DLI Engineering Corporation,* B-218335, June 28, 1985, 85-1 CPD para. 742, the comptroller sustained a 1985 protest against a Navy decision to award a cost-reimbursement service contract to a lower-cost offeror in a procurement in which cost was the least important factor. In that case the protestor's technical proposal had received 96 out of 100 points and involved an estimated cost and fee of $1,467,175. Award was made to an offeror whose technical proposal received 76 out of 100 points and had an estimated cost and fee of $787,544. The Navy made adjustments to the protestor's estimated cost for "realism" and considered that a better estimate would be $923,175. After discussing the facts, the comptroller concluded:

> The contracting officer's stated position is that an award to [the protestor] would be justified if the firm's proposed costs were 30 percent or even 40 percent higher than ISA's, but that the technical merit of DLI's offer did not justify an award where the cost differential was more than 50 percent. In our view, such a distinction is arbitrary. Although the Navy heavily relies upon our decision in *System Development Corp.*, B-213726, *supra,* as

support for this position, we point out that our holding in that case was based on the fact that the awardee's cost was more than 50 percent higher, but that its technical proposal was scored only 8 points higher.

BEST-VALUE DECISION LOGIC

The best-value decision must be based on a comparison of the value of offered functions, features, and characteristics, including offered price or estimated cost and fee. Value is determined by comparing each offer to the selection criteria. Best value is determined by comparing the value attributed to each offeror to those of its competitors. Basically the decision requires a marginal analysis of the relative strengths, weaknesses, and prices of the competing proposals. This in turn involves three steps.

First, identify and set aside any final proposals that do not possess the minimum qualities specified. They are not given further consideration. Second, sort the remaining offers into those that only meet the minimum requirement and those that exceed it in some beneficial way. Third, compare those that meet the requirement with those that exceed it, and those that exceed it with each other. This is the process of ranking, which will establish which offer is best.

BEST-VALUE EVALUATION PROCEDURES IN GOVERNMENT

Our discussion of evaluation procedures will address the appointment of a decision maker, the organization and appointment of evaluation panels, and evaluation and scoring techniques.

Appointment of a Decision Maker

Progressive companies in the private sector emphasize a team approach to supplier selection. The American Society of Quality Control's supplier certification guidebook states that the ideal team is made up of representatives of purchasing, engineering, quality assurance, and the user. But interview and policy analyses indicate that in most private-sector firms the purchasing agent or purchasing manager makes the selection decision.

The FAR permits either the contracting officer or a higher-level official to make the source-selection decision. It seems to be the practice of most federal civilian agencies to have the contracting officer make the selection. However, the DoD commonly reserves the selection decision to senior technical officials or higher-level officials.

The Army supplement to the FAR (AFARS) states that the contracting officer is the decision maker for all procurements except for major systems or other specially designated procurements. The Navy subjects competitive range determinations and selection decisions for procurements in excess of specified dollar thresholds to "business clearance" review and approval. The Defense Logistics Agency's supplement to the FAR states that a senior manager must make a special appointment of a decision maker for all procurements in excess of $1 million, but does not require appointment at any particular management level. The Secretary of the Air Force is the decision maker for major Air Force programs, but usually delegates that authority to a general officer or senior civilian unless the program is of national significance. For other important programs the decision maker is a general officer, colonel, or senior civilian, usually a commander or someone with a technical background. For smaller procurements the decision-making authority may be delegated to the contracting officer.

The Comptroller General takes the position that any official in the procurement chain of command may overrule a decision made by a lower-level official, since all procurement authority is ultimately vested in the agency head. In *Bank Street College of Education,* B-213209, June 8, 1984, 84-1 CPD para. 607, the comptroller supported the action of an agency manager in overruling a decision made by a contracting officer, affirming his long-held position that:

> While it may be true that in most procurements the contracting officer ultimately makes the award decision, the contracting officer derives the power to bind the government from the general grant of authority to the agency head.... [A]s [National Institute of Education's] ultimate contracting authority, the Director has the discretion to exercise his contracting authority whenever he thinks that it will further NIE's statutory objectives.

The comptroller also held that this is true even when qualified evaluators disagree with the decision maker's judgment:

> The selection official, here the Director, is not bound by the recommendation of evaluators, and as a general rule our Office will defer to such an official's judgment, even when that official disagrees with an assessment of technical superiority made by a working level evaluation board or individuals who normally may be expected to have the technical expertise required for such evaluations.... The fact that the protester or the evaluators disagree with the selection official's conclusion does not in itself render the evaluation unreasonable.

Clearly, however, in best-value selection or award, the key qualification for the decision maker is competence, not position. A selection or award decision maker must be competent to recognize best value, explain his or her analysis of the relative merits of the competing proposals, and justify the selection decision. In many cases the contracting officer will possess the requisite knowledge, experience, and judgment, but in highly technical procurements those qualifications may be possessed only by the program manager or higher-level managers of requisitioning offices.

Organization and Appointment of Evaluation Panels

The FAR describes two approaches to team organization. One is formal source selection, which is defined in FAR 15.612(a) as follows:

> A source selection process is considered "formal" when a specific evaluation group structure is established to evaluate proposals and select the source for contract award. This approach is generally used in high-dollar-value acquisitions as prescribed in agency regulations. The source selection organization typically consists of an evaluation board, advisory council, and designated source selection authority at a management level above the contracting officer.

The Air Force provides a good example. The organization used by the Air Force in formal source selections is depicted in Fig. 42.3. In this organization the source-selection authority is a senior agency official. The source-selection advisory council is typically made up of middle managers from contracting, finance, law, program management, logistics, the user, and so on. The source-selection evaluation board is made up of working-level technical experts. Note that the evaluation board is subdivided into panels, which correspond to areas of technical specialization. The contract definitization team is headed by the contracting officer and is the group that meets with the offerors during discussions and negotiations. The source-selection authority is in charge of the entire selection and award process. The evaluation board is the group that actually reads and scores the proposals against the evaluation criteria, but they do not compare the proposals to one another. That process is the responsibility of the advisory council. The advisory council can accept or modify the scores assigned by the evaluation board. It compares the scores of the competing offerors and ranks them. This ranking and the detailed findings of the evaluation board are given to the source-selection authority as the basis for making the decision. The advisory council does not make an award recommendation unless asked to do so by the source-selection authority. As discussed, the source-selection authority is not bound by the findings or recommendations of the evaluation board and advisory council.*

The principal deficiency of military evaluation panels is in their sheer size. Literally dozens of experts are assigned to these ad hoc teams for lengthy periods of time. During the source selection, these individuals are not available to perform their day-to-day work, causing problems for ongoing programs. Much has been written on the need for the services to use smaller teams of highly qualified individuals, and to shorten the time they must spend evaluating proposals.

*An excellent inside view of a selection and award process using this organization was provided by the General Services Board of Contract Appeals in its decision in *Contel Federal Systems, Inc.*, GSBCA No. 9743-P, Dec. 14, 1986, 89-1 BCA para. 21,248, especially in findings of facts 6 through 22. See also Carnes.[1]

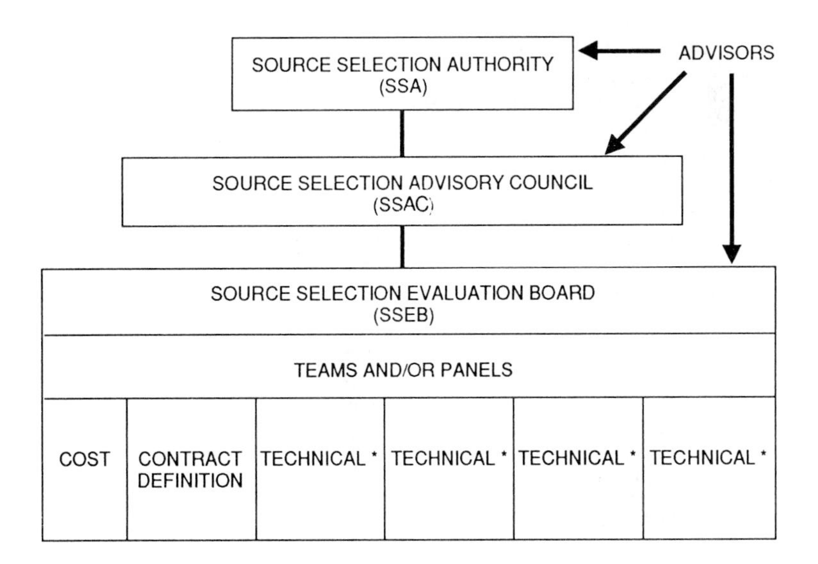

FIGURE 42.3 Air Force source-selection organization. (*From Air Force Regulation 70-15, Apr. 1988.*)

One observer described the difficulties that can be encountered when evaluators do not possess the requisite knowledge and experience (Carnes,[1] p. 5):

> None of the evaluators had the circuit design experience necessary to determine if the circuit proposed by one offeror to satisfy a given functional requirement was indeed superior to another offeror. While the technical evaluators had formal educations and had technical job descriptions, few of them had any currency in technology. Thus it was difficult to even have a superficial comprehension of an intricate circuit let alone attempt to pass critical judgment on the circuit.

On the other hand, experienced evaluators must keep an open mind[1]:

> The few [technical evaluators] who had maintained technical currency because their particular job function involved daily circuit design work in a government laboratory often advocated their own design approaches without assessing the offeror's. Usually these alternative circuits would be over designed and thus could not be considered by an offeror in a cost competitive environment.

Evaluation and Scoring Techniques

The decision to award a contract to a particular supplier always involves a wager. The decision maker is predicting that the supplier he or she selects will be fully successful. He or she is wagering the success of the program on the strength of the prediction. The determination of best value is simply a form of subjective probability assessment. The goal of proposal evaluation is to establish a reasonably sound basis for making that assessment.

Government agencies rely heavily on the evaluation of written proposals as the basis for determining best value. Offerors respond in writing to a government solicitation and an evaluation panel is convened to read, evaluate, and score the proposals. The selection or award decision is based on those scores.

A DoD solicitation has two parts. The first part is a model contract, which typically is complete except for blank spaces for the insertion of prices and the signatures of the parties. The specification is included in this part. The second part is the call for offers and includes a description of the evaluation criteria the agency will use and a set of proposal preparation instructions.

The proposal comes in three parts. The first part includes the model contract with the blanks for prices and signatures filled in by the offeror as appropriate. In a sense this is the actual offer, the promise to do what the contract says in return for the prices inserted in the blank spaces. Evaluation of this part usually involves checking to see whether the offeror filled in all the blank spaces appropriately, acknowledged any amendments to the solicitation, or took any exceptions to any of the terms and conditions of the solicitation.

The nature and content of the second part of the proposal depends upon the agency's proposal preparation instructions. Usually it is some sort of verbal-graphic-numerical description of the specific good or service offered in response to the agency's requirements. It may also include a discussion of the procedures the offeror will use to satisfy the requirement; a description of its resources and managerial control systems, such as cost, schedule, and quality controls; and a set of resumes for key managers and employees. This second part of the proposal is often labeled the technical volume or discussion, or it may be broken into separate technique and management volumes or discussions.

The third part of the proposal is usually referred to as the price or cost proposal. It may include a detailed breakdown of the offeror's prices or estimated costs. It may also include cost or pricing data if the procurement is subject to the requirements of the so-called Truth in Negotiations Act. This volume serves to explain and justify the prices inserted into the blanks in the model contract.

The technical and management sections of a proposal contain two types of information—facts, which can be determined to be true or false, and promises, which can only be determined to be believable or unbelievable. Facts include such information as the design documentation for a product which actually exists and has been produced; the results of product performance tests; the number of square feet of floor space in a factory owned or occupied by the offeror; lists of equipment owned or rented by the offeror; lists of current and former contracts and customers; and the education and experience of key personnel. Promises include the capability and performance of products that are still undergoing design; the efficacy of techniques which the offeror has not yet used; and the time required for performance of a task the offeror has never performed.

Evaluation of the factual information is relatively straightforward, but evaluation of the promises is more difficult. There is a higher degree of risk involved in selecting contractors based on their promises than on the facts. The risk also exists of being influenced by the quality of the presentation of the information, rather than by its content. This is especially true of an inexperienced evaluation board. The preparation of a persuasive written proposal is an art, and may depend as much on form as on content.

The purchase of services in general, and especially professional services, presents special challenges in this regard because services are intangible. There is a tendency among contracting officers to solicit proposals in the form of plans and descriptions of "approaches" to the work. A commonly seen proposal preparation instruction tells offerors to "address every task in the statement of work and provide a detailed description of your approach." Offerors are usually warned not to "merely parrot the statement of work," and that "mere promises of compliance" will be unacceptable. Offerors confronted by such instructions have no alternative but to engage in creative writing.

It is a mistake to use criteria which are not useful in identifying differences among the competing offerors. Criteria for management capability that depend on the analysis of written descriptions of an offeror's management systems often result in a conclusion that all offerors are equally acceptable. One observer of an evaluation described the difficulties one evaluation team encountered in this regard (Carnes,[1] p. 36):

> [T]he results of the management and logistics evaluations were inconclusive. This can be attributed to the fact that it is almost impossible to judge the adequacy of a management control system, configuration management system, parts control system, change control system, drawing release system, or technical order publication system based on the description of these systems in a written proposal.

The observer may have been wrong in concluding that all the management systems were equally acceptable. It may have been that only the descriptions in the proposals were equally acceptable. One or more of the offerors may have had problems with their systems that they did not describe.

A classic example of what can happen when evaluators use insubstantial, unverifiable criteria is illustrated in the General Services Board of Contract Appeals' protest decision in *SMS Data Products Group, Inc.,* GSBCA No. 8589-P, Dec. 3, 1986, 87-1 CPD para. 19,496. The Environmental Protection Agency used criteria such as "soundness of approach" and "demonstrated performance and reliability" to evaluate off-the-shelf commercial automatic data processing equipment. Two of the offerors proposed exactly the same items of hardware, but one was given a higher score than the other. The Board of Contract Appeals concluded that the lower score was attributable to the fact that the lower-scored offeror "had offered fewer words than [the higher-scored offeror] to explain the reasons for its product selection." In deciding to sustain the protest, the board wrote:

> [T]he record developed in these proceedings clearly demonstrates that much of the low score assigned by the greatest value panel to [the protester's] technical proposal was due to [the protester's] inability to write what can best be described as good ad copy.

Unfortunately good ad copy is too often a major element in government best-value awards. Insubstantial criteria such as "soundness of approach" and "understanding of the problem" encourage the submission of "fairy tales" and should be avoided if at all possible.

Inexperienced evaluators may use too many criteria. When this happens, no one criterion can have an effect on the outcome. In *Contel Federal Systems, Inc.,* GSBCA No. 9743-P, Dec. 14, 1988, 89-1 BCA para. 21,458, the General Services Board of Contract Appeals described an Air Force procurement to buy a computer system network designed to upgrade the word processing capability of the Office of the Secretary of Defense. The source selection plan listed hundreds of evaluation factors—so many that word processing accounted for only 1.7 percent of the total evaluation. The offeror's proposal was good enough in other factors to get the best evaluation score, even though the all-important word processing software was highly undesirable. As a consequence, the Air Force was forced to select the offeror whose word processing software it liked least.

The government's reliance on written proposals and insubstantial criteria is so pervasive it has been coined the "tyranny of the proposal."[2] In general, the government awards its contracts on the basis of data contained in offerors' proposals. This is unlike private-sector awards, which rely not only on written proposals, but also on a great deal of personal contact, demonstrations, and lengthy "handholding" sessions where the buyer and seller get to know one another and gain mutual confidence.

It is difficult to imagine how the government can predict a prospective contractor's enthusiasm, creativity in solving problems, or dedication to staffing the project with the best people by relying solely on the written proposal. Yet a selection or award decision is a prediction. And while it may be unrealistic to assume that the government will change its solicitation and award procedures substantially, evaluators should at least use criteria that involve analysis of information with high predictive value.

Facts have more predictive value than promises. An offeror's reputation has more predictive value than its promises to do well. An offeror's experience tells us more about an offeror's ability to do the work than a description of a "new and innovative" approach which the offer has written specifically to persuade us to hire them.

Proposal Evaluation

The first task in evaluation is the development of a source-selection plan which specifies the evaluation criteria. The evaluation criteria must identify those supplier capabilities and those product or service features that will be subjected to evaluation. The evaluators may conclude that a standard is met, not met, or exceeded. In the event the standard is not met, the evaluators should prepare a description of the deficiency in the proposal and the effect that deficiency would have on performance if not corrected.

If a standard is exceeded, and the excess will be helpful to the government by increasing quality, improving delivery, or reducing cost, then the evaluators should prepare a description of this strength in the proposal. When a proposal has been compared to every standard, and all deficiencies and strengths have been documented, the evaluation of that proposal is essentially completed. All that remains to be done is for the evaluators to summarize their findings and assign scores which express those findings.

Proposal Scoring

Scoring systems used by the military include the use of numbers, adjectives or colors, and other symbols and expressions. Numerical scoring is a common technique, and a variety of numerical schemes are in use. Probably the most popular are schemes which provide for the assignment of a number from 0 and 10 or from 0 and 100, with the highest number being best. While not a military service, it is interesting to note that NASA uses a 1000-point scale. Likewise, the Environmental Protection Agency has used a 0 to 5 scale, with each number accompanied by a brief description of its meaning.

One scheme also in use involves the development of a ratio of dollars per technical point. This is done by assigning a numerical score to the technical proposal, then dividing the proposed price by the technical score. For example, using a numerical scale of 0 to 400, a price of $28,500 divided by a score of 355 points results in a dollar per point score of $80.28. In this scheme the offeror with the lowest ratio of dollars per point is the best offeror. Table 42.1 illustrates this scheme.

Adjectival schemes usually involve four or five adjectival expressions, such as exceptional, acceptable, marginal, and unacceptable. Applicable regulations provide definitions of these expressions to guide evaluators in their application. Another scheme uses the adjectives excellent, very good, acceptable, susceptible of being made acceptable, and unacceptable.

Much has been made of the Air Force's color-coding method, but this is nothing but a technique for developing briefing slides in support of an adjectival scheme. The color blue is used to signify an exceptional part of a proposal; green for acceptable; yellow for marginal; and red for unacceptable. Color coding has not gained much acceptance outside the Air Force, where its use is mandatory for formal source selections.

In numerical scoring schemes, evaluators must understand that numbers are not objective expressions of truth. They are used as adjectives to describe opinions. Unsophisticated computations involving such numbers may have little if any statistical

TABLE 42.1 Application of Dollars per Point Scoring Systems*

Offeror	Price	Technical points	Dollars per point
A	$30,000	374	$ 80.21
B	$28,500	355	$ 80.28
C	$42,500	385	$110.38
D	$32,670	362	$ 90.24

*Each offeror's price is divided by its point score. Determination of best value is made on the basis of the resulting dollars per technical point.

validity. For example, in *Bank Street College of Education,* B-213209, June 8, 1984, 84-1 CPD para. 607, the Comptroller General reported a situation in which one offeror received a technical score of 82.1 out of a possible 100, while another received a score of 80.9. A poll of the evaluation board revealed that of the 10 evaluators four had ranked the offeror receiving the 80.9 as best, while only one had ranked the offeror receiving the 82.1 as best. The explanation almost certainly may be found in an outlying high or low score.

Army regulations prohibit averaging or voting as a means of establishing a team-wide score, and require instead the development of a consensus among the evaluators.* Those same regulations prohibit decision makers from selecting a source on the basis of the score alone. The regulation states: "The decision shall be made on the basis of an assessment of evaluation results as a whole."†

Award without Discussions

The FAR, at 15.610(a), permits the contracting officer to award a contract without discussions following the initial evaluation of proposals. Limitations on the use of this procedure require that the contracting officer ensure that he or she has obtained the best possible value, and that discussions and best and final offers would not improve any offer significantly. Among federal agencies, the DoD has more discretion to use this procedure than other agencies.

Government contracts are among the world's most complex business documents. The number and scope of terms and conditions make their analysis and interpretation truly challenging. In light of the limited opportunities for contracting officers and offerors to communicate during proposal preparation and evaluation, it is difficult to imagine a situation wherein award without any discussion would be a sound approach. Despite this, the authors are aware of the emphasis within the DoD on this strategy.

Award without discussions greatly increases the probability of postaward misunderstandings, claims, and disputes. Its only virtue is to shorten the selection or award process and eliminate the cost of discussions and preparation and evaluation of best and final offers. But that virtue comes at the expense of genuine understanding and agreement between parties who must depend on one another during contract performance.

COMPETITIVE-RANGE DETERMINATION AND DISCUSSIONS

FAR 15.609 requires the contracting officer to determine which of the evaluated offers has a reasonable chance of being selected for award. This group of offerors is said to fall within the "competitive range." Those who are considered to have no reasonable chance of being selected are eliminated from further consideration and are so notified. FAR 15.609(b) tells contracting officers that they are not limited to one competitive range decision, and that they may eliminate additional offers after discussions if it appears that these offers have no real chance of winning.

Although there are a number of protests about competitive range determinations, the concept presents contracting officers with no real difficulties. The most common mistake contracting officers make is to determine the competitive range without giving due consideration to all the evaluation criteria, especially price or cost.

*AFARS 15.608(a)(2)(iii).
†AFARS 15.605(e).

The purpose of determining a competitive range is to save money. Offerors with no real chance of being selected must be spared the additional expense of discussions and preparation of best and final offers. The government must be spared unnecessary expenses in conducting discussions and evaluating best and final offers. As a practical matter, however, the competitive range determination poses a dilemma for some contracting officers. Should they eliminate an offeror and face a potentially disruptive protest in the middle of the process, or should they keep all offerors involved in the competition, incurring unnecessary expenses thereby, and deal with any protest after the selection of a winner?

In *SMS Data Products Group, Inc.,* GSBCA No. 8589-P, Dec. 3, 1986, 87-1 CPD para. 19,496, the Board of Contract Appeals severely criticized a contracting officer for keeping a firm in the competition long after it became apparent that they had no reasonable chance of winning. The Environmental Protection Agency was forced to pay SMS $500,000 to compensate them for the needless expense SMS incurred after the date they should have been eliminated.

The contracting officer must engage in discussions with all offerors in the competitive range. The object of discussions is to improve the quality of the offers tendered to the government. This is accomplished by having the contracting officer notify the offerors of deficiencies and possible mistakes in their proposals. Improvements are actually made through the submission of best and final offers.

Discussions may be conducted orally, in writing, or both. Some agencies conduct discussions face to face, some by telephone, some entirely in writing, and some use a combination of the three methods. Some agencies conduct discussions in a single session, while others engage in several sessions.

When conducting discussions, contracting officers may not coach careless or incompetent offerors in order to raise their proposals to the same level of quality as those of their competitors. To do so would amount to "technical leveling." Government officials may not reveal to any offeror the contents of another offer's proposal. That would be "technical transfusion." Finally, officials may not conduct an "auction," which would entail revealing to an offeror the price of another competitor or its price standing with the purpose of making a price reduction a condition of selection or award. Each of these practices is expressly prohibited in FAR 15.610(d).

Perhaps the most frequent mistake contracting officers make is failing to conduct "meaningful" discussions by failing to fully disclose all deficiencies and possible mistakes in an offeror's proposal. This omission undermines the very purpose of discussions, which is to enhance the competition by giving offerors a chance to make changes that will improve their proposals, hence, the value of the government's choices.

Fear of breaking the rules against leveling, transfusion, and auctioning drives many contracting officers to severely curtail their bargaining behavior during discussions. In some agencies the scope of the proceedings is limited to reading a list of deficiencies and possible mistakes and asking whether the offeror understands them. Moreover, under the pressures of time and the need to conduct discussions with each of the competitors, contracting officers do not take the time to patiently discuss the solicitation terms and conditions in order to ensure mutual understanding and agreement with the offeror. As a result, contracts are too often awarded to contractors who have only a limited understanding of the nature and scope of their contractual obligations. This frequently leads to misunderstanding, claims, and disputes during performance, with a severe adverse impact on technical performance, schedule, and cost.

The Comptroller General clearly favors full disclosure, and protest decisions that sustain complaints of technical leveling are extremely rare. Even so, discussions do not have to be "all-encompassing." In *Bank Street College of Education,* B-213209, June 8, 1984, 84-1 CPD para. 607, the comptroller said:

> We have specifically rejected the notion, however, that agencies are obligated to afford offerors all-encompassing negotiations. The content and extent of meaningful discussions in a given case are a matter of judgment primarily for the determination by the agency involved and not subject to question by our Office unless clearly arbitrary or without a reasonable basis. *Information Network Systems,* B-2089009, March 17, 1983, 83-1 CPD para. 272. Where a proposal is considered to be acceptable and in the competitive range, the agency is under no obligation to discuss every aspect of the proposal receiving less than a maximum ranking. *Gould Defense Systems, Inc., et al.,* B-199392.3; B-199392.4, Aug. 8, 1983, 83-2 CPD para. 174.

In response to a complaint in the same protest that the contracting officer disclosed the government estimate to one offeror, but not to the protester, the comptroller said:

> We also do not believe that the agency's discussion of the government's cost estimate with Harvard without conducting similar discussions with Bank Street amounted to unequal treatment of offerors. An agency is not required to hold the same kind of detailed discussions with all offerors since the degree of weaknesses or deficiencies, if any, found in the acceptable proposals will obviously vary. *Pope Maintenance Corporation,* B-206143.3, Sept. 9, 1982, 82-1 CPD para. 218. Thus, an agency can discuss costs with one offeror without conducting similar discussions with another offeror, where, as here, it does not appear that the agency considers the other offeror's cost proposal to be deficient. *Tracor Jitco, Inc.,* B-208476, Jan. 31, 1983, 83-1 CPD para. 98.

BEST AND FINAL OFFERS AND CONTRACTOR SELECTION

When discussions with all offerors have been completed, the contracting officer must call for best and final offers. FAR 15.611(a) requires that contracting officers establish a common date for all offerors to submit best and final offers. The purpose of the best and final offer is to implement any changes the offeror wishes to make as a result of discussions. This includes the elimination of deficiencies and the correction of mistakes. The offeror is neither obligated to change anything, nor limited in what it may change.

The contracting officer must ensure that the best and final offer is evaluated against the same criteria that were applied to the initial proposals, unless those criteria were changed by amendment of the solicitation. Best and final offers are scored and then ranked, and the usual expectation is that the contracting officer will select a contractor upon completion of this final evaluation. However, if the contracting officer does not find any of the best and final offers to be acceptable, he or she may call for another round of discussions. The practice of multiple rounds of discussions and best and final offers is discouraged, but not prohibited. The DoD now requires approval by a senior agency official before a contracting officer may request multiple best and final offers. In general, contractors do not like multiple best and finals because of the pressure to cut prices to unreasonably low levels.

The best and final offer procedure presents an offeror with something of a dilemma. Solicitations typically warn offerors that the contracting officer may make an award without discussions, and encourage offerors to submit their best proposals initially. The dilemma for offerors is whether to submit their best prices with the initial proposal or leave something in to "give away" in their best and final offers. Many suppliers suspect that the government always expects a price cut in the best and final offer and that failure to make such a cut will hurt their competitive position.

CONCLUSION

The usual justification for the competitive range–discussion–best and final offer procedure is to maintain the pressure of competition through development of best and final offers. This policy reflects the unconditional faith placed in contestlike competitions as a means of obtaining best value. The costs and other drawbacks associated with the procedure may outweigh its benefits, especially in the age of improving relationships between buyers and sellers in the interest of improving quality and reducing overall costs.

The selection and award process entails both contractor selection and contract formation. In the authors' view, the FAR places undue emphasis on contractor selection and not enough on the development of genuine understanding and agreement between the parties. This may result in contracts between parties who hardly know one another for the performance of tasks they may not fully understand.

It is not realistic to expect that true understanding and agreement will be achieved during the discussion process that has arisen out of current policy. Without comprehensive and thorough discussion, full understanding and agreement cannot be achieved, especially if reliance is placed on a reading of the solicitation. Neither do limited-scope negotiations after selection provide a solution to this problem. More emphasis is needed on achieving a meeting of the minds during the selection and award process, not afterwards.

At the outset, the objectives of the source-selection process were stated to be to maximize competition; to minimize complexity of solicitation, evaluation, and the selection decision; to ensure impartial and comprehensive evaluation of proposals; and to ensure selection of the source who would provide "best value." While the last term, best value, is not explicitly stated in the FAR, the authors have attempted to show that best-value source selection is exactly what the writers of the FAR intended.

The military services have made some progress toward best-value contracting, although the current emphasis on limited discussions with offerors is seen as counter to the best-value philosophy. The relationship between risk and information is inverse. As more information is gained, the amount of risk involved in any project decreases. Government program managers and contracting officers are encouraged to enter into meaningful discussions with their potential contractors to reduce the unknowns that may later lead to problems.

Best-value contracting should be the goal of every source selection. By concentrating on an integrated assessment of all the criteria upon which the selection decision will be based, source selections can be carried out in a rational manner—one that is fair to all parties and results in the selection of contractors who offer the government the best value for the taxpayers' money.

REFERENCES

1. C. P. Carnes, *Participation in Source Selection: A Case Study,* No. AD-A028956, U.S. Department of Commerce, National Technical Information Service, Apr. 1976.
2. S. Kelman, *Procurement and Public Management: The Fear of Discretion and the Quality of Government Performance,* AEI Press, Washington, D.C., 1990, pp. 25–26, 39.
3. V. J. Edwards, *Competitive Proposals Contracting,* 2d ed., Educational Services Institute, Falls Church, Va., 1991.

CHAPTER 43
PROJECT ADVOCACY

Earl D. Cooper

Earl D. Cooper is currently Program Director of the Florida Institute of Technology's graduate management programs in the national capital region and a consultant to various aerospace firms and governmental agencies. He holds the academic rank of Associate Professor and has a doctorate in public administration. As a member of the Senior Executive Service, he served in various senior management positions in naval aviation, including that of the Deputy Project Manager of a designated major defense acquisition program and Technical Director of Research and Technology for the Naval Air System Command.

The project manager practices his trade in a complex, pluralistic environment that consists of many external and internal competing forces, any of which could constitute a fatal threat to his job and his project. On the other hand, any one of these forces could be a significant, positive supporting influence. These pluralistic forces consist of many and various special-interest groups inside or outside of government, or coalitions of private industry and government groups, and sometimes include citizen and community groups.

In order to achieve project success the project manager must deal effectively with these special-interest groups. His goal, in general, is to seek their satisfaction at various levels. The project-advocacy issue centers about the hows, whats, and whys involved as the project manager communicates with these special interests. The pertinent major questions that ultimately arise have ethical overtones. Simply stated, they are: Can the manager of a project be its chief advocate, avoid conflicts of interest, be objective, ensure his career progression, and serve the public interest all at the same time? How much of an advocate should the project manager be? Who should be the principal advocate for the project? What are the roles of the contractor and the various agencies involved? The focus of this chapter is to answer these questions while at the same time speaking to the aforementioned external and internal forces involved. In addition, project management decision making and communications as they bear directly on the subject of advocacy are treated.

PERTINENT DEFINITIONS

Advocate

Webster's *New International Dictionary* defines an advocate as: "One who pleads the cause of another before a tribunal or judicial court; a counselor; one who defends or espouses any cause by argument, a pleader; an intercessor."

Projects versus Programs

A project is an integrated collection of activities using both material and human resources to accomplish a specific end result at a specific time. In the Department of Defense (DoD) it is widely understood to be a specific weapon research, development, and production effort that is a one-time undertaking having both definite beginning and end points. At the conclusion of the project effort, the work of the project office and all the supporting organizations related to it is terminated, and peculiar organizations that were established to support and execute the project work, including the project office, are disbanded.

A program, on the other hand, is generally understood to be an area or collection of efforts that may include a number of related yet separate distinct efforts, some of which may even have small-project status. A program operates to fulfill broad goals, is continuous, and has no specific end point. A program is continuously spawning new efforts and ending others. As such, the successful project manager always works himself out of a job, whereas the program manager's job is never finished.

There are three categories of DoD weapon system acquisition efforts. The first is called a major defense acquisition. The principal characteristics of this kind of effort are that it is:

- Not a highly sensitive classified effort
- Formally designated by the Under Secretary of Defense for Acquisition as responding to an urgent need, has a development risk, has joint (interservice) funds, and has significant congressional interest or other special considerations
- Estimated to require expenditures for research, development, test, and evaluation (RDT&E) of more than $200 million or eventual expenditures for procurement of more than $1 billion (both in fiscal-year 1980 constant dollars)

The second type of defense acquisition is one that is not a highly sensitive classified effort, is not formally "designated" by the Secretary of Defense, and whose funding projections are of lesser values than those stated for the designated efforts described in the foregoing paragraph.

The third type is the highly sensitive classified effort, which receives special handling and most likely becomes a "black" program. In relating the foregoing project and program discussion to the pertinent and continually changing DoD directives and the DoD vernacular in general, it should be noted that the terms program and project are many times used interchangeably. For the purposes of this chapter, the terms project manager and program manager are defined as follows:

Project manager. Manager of a specific task having specific bounds, including a definite end point

Program manager. Manager of more than one task, each of which may or may not have specific bounds and which collectively have no explicit end point

System Project Offices versus Matrices

Two popular organizational designs for project management are defined by the classical system project office (SPO) and the matrix models. The SPO is generally a relatively large organization, consisting of members who are solely dedicated to the particular project and report to the project manager through a mechanistic hierarchy. The SPO is a self-contained group that has all the expertise and manpower needed to manage the project.

Advancing technology has permitted more and more capable weapons to be built. Modern "high-tech" weapons generally require a much higher degree of integration with the carrying platforms, which adds other dimensions of complication and a focus on the systems concept. As the degree of complication increased, functional divisional leaders found it more and more difficult to manage projects which cut across other divisional lines. One solution was the matrix organization. In a matrix the project manager generally has a small office consisting of a small staff. This project office draws support from the various functional divisions. The team members in his immediate office are dedicated solely to the work of the project and generally individually report directly to the project manager. By far the majority of expertise required to manage the project is provided by the various functional groups in the particular organization. The team members from the functional groups, although they report to the project manager for project matters, also report to the functional group heads for functional matters and sometimes even receive project technical direction from them as well. Although the matrix organization is deemed to be more efficient from a personnel-use point of view, in such cases where project team members are receiving direction from two bosses, the number of communication channels is increased significantly and the lines of authority are blurred. This increases chances for disputes and conflict regarding technical matters and priorities. Hence the matrix organizational design places pressures on the project manager to advocate for his project to his immediate superior (or higher in house) on matters of personnel priorities and internal funding matters.

Project Management Functions

A simple definition of management is, getting things done through people. Hence the manager is successful if his people are successful. A major system, as defined by pertinent DoD documents and the very nature of complex military weapon systems, involves the project manager with team members having diverse backgrounds and allegiances working to produce the specified product. He presides over and guides his project team, in many instances from cradle to grave. His success depends on his team members each being successful.

The project manager's job can be summarized as consisting of three major functions, as described by Chester Barnard in his famous book, *Functions of the Executive,* as follows:

1. *Maintenance of the integrity of organizational purpose.* In carrying out this function the project manager articulates the validity and importance of the work. This is the advocate function which includes his serving as the project's chief spokesperson and point of contact, the taxpayer's watchdog, and the government's smart buyer.
2. *Maintenance of the willingness of the team members to cooperate.* This is primarily a motivation function and overlaps the first function in that a clear under-

standing and acceptance of the real importance of the project can be a motivating factor in itself.

3. *Establishment and maintenance of an effective communication system.* This is the glue that holds the other two functions together. The team members at headquarters, in far-flung field activities and labs, at contractors' and subcontractors' plants, and so on, need to know what is going on. They continually need to hear motivating words and about the importance of their work, participate in the project advocacy, and so on.

THE PROJECT MANAGEMENT ENVIRONMENT

The major forces that impact the projects in the DoD are depicted in Fig. 43.1. These forces are treated individually in the following paragraphs.

Political Forces

Norman Augustine's excellent articulation of the importance, difficulty, and consequences of the job of today's military project manager serves to focus on and further sensitize the subject of project advocacy. His words are indicative of the "signs of the

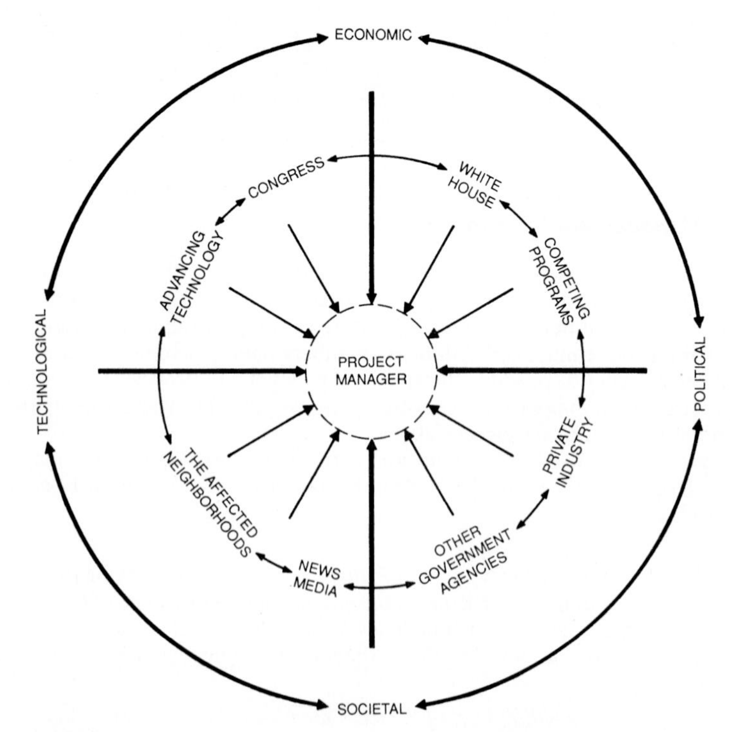

FIGURE 43.1 External forces impacting the project manager.

times," wherein the project manager (and the attendant project management concept) has emerged as the accepted best organizational model to best acquire military weapon systems.

Careers on the Line. Further signs can be seen in the creation of professional materiel acquisition career paths in each of the military services for both officers and civilians. In naval aviation, for example, in the past it used to be that a successful tour of duty as the commander of an aircraft carrier at sea (at the Captain level) was virtually a necessary prerequisite for promotion to Admiral. With the latter-day recognition of the importance of project management to the overall success of the Navy (operationally and financially), the names of many major system project managers are found on the promotion list to Admiral. Thus, in effect, promotions are being won on the procurement "battlefields" as well as in the classical battles at sea.

Mr. Augustine's comment that the job of project manager can be a career "buster" was recently borne out in the firing of the project manager for the Navy's A-12 stealth aircraft and two of his flag-level superiors due to significant cost overruns. On the other hand, the job of a project manager can be a career maker, and Mr. Augustine cites the project managers for the A-10 aircraft, the Tomahawk cruise missile, and the Abrams main battle tank as examples of success stories. The intensity of such positions, when "careers are on the line," and very large sums of taxpayer dollars as well as many civilian jobs in various (politically) key congressional districts are involved, further tends to heighten interest in the project manager's advocacy.

The Iron Triangle. Every project has its supporting constituency. This is a natural association of the external and internal special-interest groups who profit in some way by having the project succeed. This supporting constituency consists of a network of specific individuals or groups of individuals at multiple levels in the executive and legislative branches in government and in private industry. Constituency membership may also include state and local officials as well as citizen groups. They operate as vertical autocracies and are referred to as the "iron triangle" (see Fig. 43.2). In the executive branch membership includes the project office led by the project manager himself, partisan supporting government laboratories, and staff members in higher service operational echelons as well as staff in the pertinent ser-

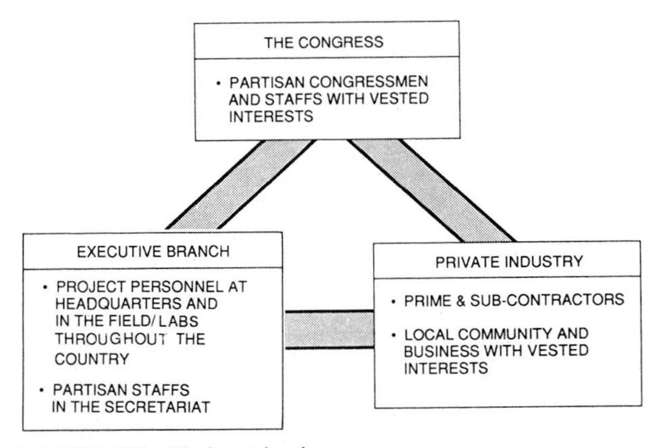

FIGURE 43.2 The iron triangle.

vice secretariat, all of whom are involved in the "purse string control" chain. Other iron triangle members include congressional staffers in various congressional committees, who have a vested interest in the project. These staffers may be attached to specific congressional committees (such as the House Appropriations Committee and the Senate Armed Services Committee) or to individual legislators (such as a member of Congress in whose district significant numbers of jobs would be lost should the project be canceled).

The private industry leg of the triangle consists of the weapon system prime contractor and all the supporting subcontractors. These firms can generate considerable advocacy forces through their representatives in the House and the Senate, state and local delegations, and various citizen groups. Some firms can support key influential congressional members outside their district and states with substantial financial resources.

These iron triangles can be very powerful. As such, the project manager can use these channels to bring advocacy pressures to bear on any decision maker at any point and level in the approval chain. Thus he can "end around" his immediate hierarchy of military superiors, his service secretary, and the Secretary of Defense. To activate these supporting forces, the project manager need not overtly go over the head of his superior, but by having developed close personal working relationships with all the players in this vertical autocracy, he can discretely inform them of his need for help. Such behind the scenes advocacy on the part of the project manager is obviously risky for him careerwise. On the other hand if his project is not completed successfully, his vision of rapidly rising up the career ladder vanishes.

There are many famous examples of managers who have worked the iron triangle over the heads of their bosses, such as Admiral Hyman Rickover, father of America's nuclear navy, and Dr. William McLean, father of the Sidewinder, the world's most successful air-to-air guided missile. These men believed in their programs and the programs' extraordinary need and usefulness to the security of our country. In these cases, elements of the iron triangle were powerful enough to protect these managers' careers.

Iron triangles can be negative forces as well. For example, a vertical network which is promoting project A can very well become a ready-made and strong force to lobby against the competing project B.

Black Projects. Project managers that operate under the cover of very high levels of security classification, or in the "black," still work with vertical autonomous networks. However, there are much fewer numbers of people in the network. In this special case the vertical networks can also be the chain of command. Security restricts the project manager from advocating for his project outside of this network of persons who are "read into" (given knowledge of) the program. If he encounters a negative internal member in the network in a key decision position, the project manager has little room in which to mount an advocacy campaign. If he goes public, he goes to jail.

On the other hand, black projects, by their very nature, receive much less oversight. Hence they are immune to many external negative forces, which greatly reduces threats to the project or the need for an advocacy program.

Associations. The skillful project manager can use various military service or military professional organizations as political action groups to gain support or publicity for his project.

Service-Peculiar Organizations. The U.S. Air Force Association and the Navy League are representative of organizations which are service-peculiar in scope and have a significant political clout. Each publishes a monthly magazine that gets wide circulation in the halls of Congress and in private industry. Each has political leaders in its membership and as major speakers at its functions.

Delving in more depth into the Navy League, one finds its stated objectives to include: "To foster and maintain interest in a strong Navy, Marine Corps, Coast Guard, and Merchant Marine as integral parts of a sound National defense and vital to the freedom of the United States."

The Navy League was founded in 1902. At the founding ceremonies, the League's biggest proponent at that time, President Theodore Roosevelt, said:

> It seems to me that all good Americans, interested in the growth of their country and sensitive to its honor, should give hearty support to the policies which the Navy League is founded to further. For the building and maintaining in proper shape of the American Navy, we must rely upon nothing but the broad and farsighted patriotism of our people as a whole; and it is of great importance that there should be some means by which this patriotism can find an effective utterance. Your society offers just the means needed.

Teddy Roosevelt was so interested in and supportive of the League that he donated a portion of his Nobel Peace Prize cash award to help get this new interest group started.

Today the Navy League claims a membership of about 50,000, coming from all the states of the union and some foreign countries. Military personnel on active duty are precluded from becoming members. The Navy League organization is structured on national, regional, state, and local levels. It is governed by national officers and a board of directors.

The Navy League advertises itself as a nonprofit, educational organization. Although in a benevolent mode, the league carries on a significant financial and legal assistance program to sea service personnel and their families. The educational program is by far the most influential stimulus on the non-Navy world. The league's monthly publication *Sea Power* (often quoted in the *Congressional Record*) contains strong pointed editorials. Seminars on Soviet maritime growth are held periodically in various cities throughout the country to provide the country with a periodic update on the relative strength of the Soviet and American navies and the implications thereof. The league sponsors an all-volunteer youth training organization (membership about 10,000 cadets, ages 14 to 17). The cadets gain an appreciation of our Navy's history, customs, traditions, and the significance and importance of a modern Navy. They are inculcated with a sense of patriotism, courage, self-reliance, and confidence, qualities contributing to high moral character, good citizenship, and the enhancement of the quality of our country's future leadership.

American Defense Preparedness Association. The American Defense Preparedness Association (ADPA) was founded in 1919 as the Army Ordnance Association. Subsequently it was reorganized as the American Ordnance Association to cover all three armed services, and then to the ADPA, still further increasing its scope to cover most aerospace industries. The membership consists of over 30,000 members, mostly representatives of various aerospace manufacturers. The staff is composed largely of retired military officers. Over 470 major corporations (covering all 50 states and Canada) are represented in the membership.

The key ADPA precepts are as follows:

1. ADPA is a nonprofit and nonpolitical organization. It is a membership association of American citizens. Nominees of corporate members participate as individuals.

2. ADPA strives to improve the effectiveness and efficiency of the government-science-industry relationship in the development and production of weapons and weapon systems. In fulfilling its objective, ADPA acts as a liaison at the management level between industry and government.

3. ADPA cooperates to every practical extent with other recognized defense-oriented technical and industrial associations in assisting the armed services of the United States. Its mission is to keep the nation's defense strong in peace and in war. Its functions are as important and as worthy of support in times of international quiet as in times of emergency.

4. ADPA keeps its members and the public informed on problems affecting defense preparedness. It does this through its publications, through its national and regional meetings, and through the activities of its local chapters in all parts of the country.

The main outward thrust of ADPA is through its monthly newsletter *The Common Defense* and bimonthly magazine *National Defense.* Both of these publications are widely quoted and feature pointed, prodefense editorials.

National Security Industrial Association. The National Security Industrial Association (NSIA) was organized in 1944 with the help of then Secretary of the Navy James Forrestal "to assure a clublike approach to industry's dealing with the military."

Unlike ADPA and the Navy League, membership is not open to individual citizens. Rather, NSIA membership is at the corporate industry level and, today, consists of over 250 major defense corporations.

NSIA serves two main functions: as a facilitator for technical and program information interchange (within) and as a powerful lobby (without). Evidence of the lobby aspect is seen in the following excerpts from the NSIA Program of Action (FY 1974):

> ... Programs will be broadened to include liaison with the Office of the Science Advisor to the President.... Within the limits of NSIA legislative policy, the Association will continue to devote special emphasis to improve liaison and communications with Congress, establishing or expanding relations with congressional elements in areas of mutual interest.... (NSIA) Advisory Committee will be encouraged to hold congressional seminars and peer group discussions in areas of mutual interest as a means of informing the Congress of industry's needs, problems and capabilities. The merit and feasibility of modifying current NSIA policy will be reviewed to permit discussion and support of the defense budget in the NSIA congressional relations program.

Military Retirees. Almost all companies in the military-industrial complex, as a matter of practice, hire retired military officers who bring military expertise, knowledge of the system, and friendships with people in the system. The higher-level flag officers (admirals, generals) can open key doors in the Pentagon or higher in the executive branch, depending on the positions they have held and high-level personal acquaintances they have established while in the service. In addition, these admirals and generals have had a close relationship with their previous subordinates who have succeeded them, and are now assigned to their old influential flag billets (these subordinates having been promoted).

Economic Forces

Budget and Funding Dynamics. The project manager finds himself continually reacting to changes in his funding structure. This may be the result of budget changes imposed by Congress or by a superior level in the executive branch of the government. As each of these funding changes is the result of competition for limited dollar resources, impact statements are generally required to be prepared as well as revised planning documents. These responses are most likely required on a "crash basis." The net result is that the potential error and the risk of overruns increase with each deviation from the original "well thought out" project plan.

An obvious and sensible solution is to provide for contingency "kitties" of funds to cover the increased risk and unforeseeable problems. However, any such kitties that are made visible in project funding documents are the first to be cut by higher-level reviewers in a budget squeeze. As a result many project managers "beef up" or inflate other line items in their budget plans to provide for such a contingency. In other instances, project schedule changes may cause a shortage or excess of funds during a specific time period. The project officer may solve this problem by swapping funds with another project manager who has an inverse type of problem. Both of these funding maneuvers to "make ends meet" are difficult and risky to articulate to cost-cutting budget reviewers looking for "fat" in DoD programs. The net is that this juggling is kept camouflaged. The pressure on the project manager to use such tactics is very great. For example, an unexpected delay in funding could cause a small subcontractor to have to lay off a team of skilled workers that find jobs elsewhere. When the funds finally become available for the particular task in question, a whole new team has to be trained up with consequent increased cost and schedule setbacks. There are past instances where resulting cash-flow squeezes forced some small companies out of business and new suppliers had to be found.

"Keep-on-Life-Support" Syndrome. There are many examples of weapon developments never having resulted in weapons being delivered to the fighting men and women on the battlefields on land, at sea, or in the air because of politically inspired budget cutbacks over the life of the project. In most of the cases the projects were so drawn out schedulewise and unit costs because so inflated as production quantities decreased that they became too costly and were canceled for cost reasons, or they were outmoded by advancing technology. Yet these projects never suffered any technical failures during their tenure. In an attempt to dramatize this political damage to the development process, this author helped prepare a presentation titled "Trash Heap of Technically Successful Naval Developments."

The pressures on the project manager to protect his project by "not telling all" are obviously great. It is easy for him (and the contractors involved) to promote the "keep-on-life-support" philosophy, ever hoping for the project to come back to life, instead of drawing a line and defining the point at which he recommends project cancellation because the effort has become economically unsound. Project cancellation could mean the loss of the project manager's job and perhaps his career. Also, it means the loss of many jobs and profits in private industry.

Technological Forces

Engulfed in "High-Tech." Advancing technology permits the building of new weapons today that are astoundingly more capable than those that were in the field 10 years ago ("smart weapons" versus "dumb weapons"). This trend and the use of even more advanced technology are expected to continue well into and through the next century. These new more capable weapons reflect the "high-tech" era where simple technology has been replaced by integrated sophisticated, more complex and expensive technologies. This increases the project manager's challenge in that the scope of knowledge has become so broad that he cannot personally comprehend all of it. He is therefore more dependent on his specialist team members and the contractors for explanations of the basis for technical performance progress or failures, to justify budget requests, and so on. This situation is further exacerbated if the project manager is placed in the "smart buyer" role and must advocate for the project based on significant high-tech explanations from the contractor.

Not Invented Here. Many project managers must deal with strong government laboratories on their teams, which may have invented the technology to be used in their particular projects or which may advocate a competitor technology to that decided upon for use in the project. Conflicts arise easily on matters of technical approach and on the program control mechanisms. Program control proposals range from detailed control by the government issuing detailed drawings to the contractor (favored by many government labs) to simple performance specifications (favored by the contractors). In such cases the project manager is continually involved with keeping the peace, motivating the laboratory-contractor team, and keeping lines of communication open between them. He must advocate in a manner and with content that respects all the team members and gives due appreciation for all vested competing concepts.

"My Project Forever" Syndrome. An ongoing successful weapon system project generally has built up a significant amount of what might be called supporting funding as well as political and technical inertia. Such an atmosphere of good feeling and project acceptance as well as the potential from rapid advances in pertinent technologies have motivated some project managers to advocate project continuation and modification to add new capabilities in order to satisfy new operational requirements. The supporting contractor team is obviously motivated to be a prime player in such thinking and may in fact be the originator of the new modification concepts.

The Navy's Bullpup air-to-surface guided missile manufactured by the Martin Company (later Martin Marietta) was a radio remote-controlled weapon powered by a rocket and carrying a large armor-piercing warhead. The pilot would aim his airplane downward at the target. He would launch the missile out in front of his aircraft, the missile traveling much faster than the aircraft. The pilot would then guide the missile via a joy stick control and try to keep the missile lined up with his line of sight to the target until its impact.

With the help of the Bullpup project officer, project funding, the Martin Company, and several naval laboratories, proposals to modify the basic Bullpup design to perform other missions were developed and many test vehicles were built and tested. These included:

Bullarm. An antiradar missile

Bulleye. A T.V.-guided missile (two versions—remotely controlled and automatic homing)

Air-to-Air Bullpup. An antiaircraft weapon

These other versions of Bullpup were in direct competition with approved ongoing system developments.

Societal Forces

Interfaces with the Community. Practically every military base and contractor's plant supporting a weapon project has some kind of interface with the local citizenry. Every airbase or large research and test facility borders on a town or countryside where people live. The attitudes of the citizens obviously reflect how much they have at stake in the particular military installation. If it is a relatively large employer in the area and responsible for a large number of sales in area shops, then the area member of Congress and a consensus of area citizens are likely to be supportive. However, the citizens who are closest to the interface and the related local civic associations may decry the base's existence (for example, if the airplanes are noisy or the fumes from one of the plants have a foul odor). The project manager's advocacy tailored to this group may be crucial to having a supportive relationship in order to keep down disruptive protests and strikes.

Burgeoning Bureaucracy. The results of actions by special-interest forces (sometimes backed up by political forces) have worked to complicate the governmental procurement system significantly by adding many levels of review and the mandatory treatment of societal issues. This has served to increase the difficulties the project manager must overcome and the times and levels at which the project manager must advocate for his project.

Table 43.1 is a chart (subsequently updated) that had been used by a former commander of the Naval Air Systems Command to illustrate the increasing approval levels and signatures over the years that are required to initiate the acquisition of a new naval aircraft. A typical reaction of the administration to problem solving in the acquisition of weapon systems is to form a committee to study the problem, which inevitably results in the procurement system being reorganized and made more cumbersome by adding more review layers and watch dogs to the system. In the recent Ill-Wind scandals in the Department of the Navy much more attention was given to changing the system rather than to focusing on the dishonesty of the "bad guys" or getting after their immediate bosses (some of whom were political leaders in the administration) for not doing their jobs to properly supervise the perpetrators and be knowledgeable of what their immediate subordinates were doing.

TABLE 43.1 Growth of Chain of Command in Naval Aviation over the Years

1921	1959	1963	1966–1986	1986–today
1. President	**1.** President	**1.** President	**1.** President	**1.** President
	2. Secretary of Defense	**2.** Secretary of Defense	**2.** Secretary of Defense	**2.** Secretary of Defense
2. Secretary of the Navy	**3.** Secretary of the Navy	**3.** Secretary of the Navy	**3.** Secretary of the Navy	**3.** Under Secretary of Defense for Acquisition
			4. Chief of Naval Operations	**4.** Secretary of the Navy
		4. Chief of Naval Material	**5.** Chief of Naval Material	**5.** Chief of Naval Operations
3. Chief of Bureau of Aeronautics	**4.** Chief of Bureau of Naval Weapons	**5.** Chief of Bureau of Naval Weapons	**6.** Commander, Naval Air Systems Command	**6.** Commander, Naval Air Systems Command

PRO BONO PUBLICO

Serving Self or Serving the Public

Dr. George A. Graham, renowned professor of public administration and advisor to U.S. presidents, says that "a public interest is developed as persons or groups find that they are affected by the acts of others with whom they have no adequate channels of communication. Government then becomes the common agency through which mutual interests can be explored and preserved and competing interest adjusted. Administration is the means by which these policy adjustments are made effective."[1] The term "administrator" in this case is directly synonymous with "project manager" as he proceeds to sort out and deal with competing interests within or on the outside of government.

George Graham's formula for serving the public interest when applied by the project manager directs him to set up a structure by which system values rather than personal values are the basis for legitimizing his decisions. In making his decisions he ensures that he obeys the law as well as administrative rules and regulations governing his agency. Further, throughout all phases of the decision-making process, he must act in an informed, fair, rational, and reasonable manner and use administrative due process.

A project manager who can make the case that he fully applied this formula in making his decisions will most likely be supported by the court when facing legal challenges. There may be times when the project manager believes that it is necessary to take actions "above the law," or lay aside his agency's rules and regulations or his superiors' orders in order to serve the public interest as he perceives it. In such instances (for example, Jefferson's unauthorized purchase of Louisiana) the project manager must be confident that his foresight will be in tune with the public's hindsight, or be prepared to suffer the consequences.

Project Manager Qualifications

Professor Graham also specifies the qualifications needed by the project manager that are pertinent to the manager ensuring that the public interest is served. They are enumerated as follows:

- Competence
- Good judgment
- Maturity
- Values reflecting the highest rules of society
- Open-mindedness
- Commitment to the public interest, the total welfare
- Courage

Avoiding Negative Group Dynamics

The project team or group can reflect the positive or negative attitudes of an individual. In many cases the team will reflect just about all the characteristics (good and bad) of a strong project manager. In other cases the group dynamics at play in the team may be such as to change the attitude and other characteristics of a weak project manager.

The negative group dynamics that the project manager must guard against especially is called "groupthink" or the "Watergate syndrome." Groupthink has been written about extensively by I. L. Janis, who defines it as referring to "deterioration of mental efficiency, reality testing, and moral judgment that results from in-group pressures."[2] The net effect on the group is that it overestimates its power and morality, it creates pressures for uniformity, and its members become close-minded. Some manifestations are the illusion of invulnerability and the encouragement to take great risks and to ignore the ethical or moral aspects of their decisions and actions. The group that approved, planned, and executed the famous Watergate break-in exhibited all of these characteristics in their actions.

This author has witnessed a close-mindedness on the part of several strong project managers which then permeated their teams. One project manager took this to the extreme and in effect defined his environment as consisting of two kinds of people, either friends or enemies. The friends were people who completely agreed with his favored solutions and supported his project. All others were enemies. Finally this project officer was fired as the result of his unbending position and his heated argument with and subsequent demonstration of a lack of respect for a high-level superior who disagreed with him.

In today's world of high-tech weapon systems the project manager must rely more and more on the individual and collective expertise of his team. In addition, he must motivate the team members such as to cause them to release all of their individual potentials and to direct these energies toward achieving project success. Hence he must use a participative management and decision-making style that makes full use of all of the talent on his team, and yet ensure that the team's output and the advocacy generated therefrom are objective and serve the public interest. Various group management decision-making techniques are available to the project manager, which are designed to eliminate or greatly reduce the threats to objectivity and to utilize each team member fully. Examples of such management mechanisms are the nominal group techniques, the Delphi technique, the Osbourne technique, and various other forms of brainstorming. In these cases the project manager either presides over these processes as an objective facilitator or he calls in an objective interventionist from outside to facilitate these sessions.

Career Success and the Public Interest

The relationship between long-term personal career success and project success places extraordinary pressures on the military project officer, whether or not he is a military officer or a civilian. These pressures are over and above those in his external and internal environment as discussed previously. This make or break atmosphere can impact the project manager's advocacy significantly. The classical slow-reacting, multilayered government bureaucracy is bound to place hurdles and roadblocks in the project manager's path and to threaten that things will not get done on schedule. Many times micromanagement by Congress can also threaten schedules as well as funding levels. The tendencies to bypass or "end around" these perceived roadblocks to success are great.

There is an old saying that there is at least one bad apple in every barrel. Likewise it is possible that a project manager may be confronted with situations involving fraud, waste, or abuse. The Ill-Wind scandals in the Navy serve as examples that wrongdoers can be found at any level of an organization. There are strong tendencies to avoid involvement in any way in such matters. In such situations, name association, even on the right side, can prove negative to career enhancement. Whistle blowing many times is perceived as antiestablishment-oriented and non-career-enhancing. However, in these cases the project manager must have courage and the public interest must be served.

PROJECT MANAGERS AS "SMART BUYERS"

A trend that has affected military managers over the last 30 years is still intact. The popular terms that characterize this trend implore the military to do "more with less." One manifestation of this trend is the contracting out of as much of government as possible. The desired outcome of these moves to privatize various heretofore government functions is to reduce the size of government and, at the same time, introduce the element of competition into the accomplishment of these functions. The net expected is a more efficient, higher-quality, and cheaper product or service, yielding significant cost savings to the taxpayer.

In general the impact of this trend on the military services is to significantly dilute their in-house technical expertise and facilities, which heretofore had been brought to bear in the acquisition of weapon systems to determine whether the military leaders in the field were going to obtain the weapons having the performance they need and to ensure that taxpayers are getting their money's worth.

This trend to "do more with less" has led to the "smart buyer" concept of project management. It reflects the condition of a project manager who does not have enough hands-on engineering expertise either personally, on his headquarters team, or in his in-house laboratory supporting team, to design and build the weapon he is responsible for acquiring. Rather, it is implied that the "smart buyer" (under these conditions) is astute enough to understand and evaluate the different system or component design alternatives offered by the contractor and is capable of selecting that which is the best buy and provides the "most bang for the buck."

The project manager modus operandi, which is considered to provide him with the ultimate control of the project endeavor, is that where he (the government) provides or controls the weapon drawings and detailed specifications. A very large supporting government headquarters and laboratory staff is needed in these cases. The "smart buyer" represents the "flip side," or other end, of the project manager's control spectrum. His controlling influence is focused much more on his personal powers of persuasion.

A project manager advocate who is in the "smart buyer" role must communicate that his decisions are based more on system values than on specific weapon development achievements. In other words the emphasis is more that good, knowledgeable people who have a good system and a good track record are bound to produce a successful, desirable product. This trend is expected to continue, requiring the smart buying project manager to seek new, innovative ways of advocating for his project.

SUMMARY AND FUTURE TRENDS

A parallel can be drawn between the tribune in the ancient Roman army and today's modern military project manager. A tribune was generally a well-educated person (an aristrocrat) who was chosen by the people to protect their rights against the "establishment" (the patricians) of those times. Today's military project manager is likewise a protector or champion of the people relative to their interest in the weapon system acquisition project for which he has been given responsibility. He is the people's advocate for his project. Accordingly he must play the role of an activist.

There are a number of trends in the acquisition of weapon systems (some are new and some have been in place for some time) that are reshaping the role and the desired characteristics of U.S. military managers. These trends, which impact on advocacy issues and which have symbiotic relationships to some degree, are listed in Table 43.2.

The cry to "do more with less" is one of the signs of our times, which from the defense perspective will be austere times for years to come. The perceived lessening of military threats to the United States coupled with severe difficulties in the U.S. economy have already caused huge cutbacks in defense expenditures and personnel. Many more reductions are yet to come as the nation fashions its much smaller, "lean and mean" armed forces and defense establishment. The military project manager will be thrust into the middle of an intense competition for fewer available dollars. In order to get his fair share of funding in this heightened competitive atmosphere, the project manager must improve his advocacy. He must increase his effective networks and communication channels throughout pertinent parts of federal and state governments and in private industry. Roland Schmitt, President of Rensselaer Polytechnic Institute, in speaking to the National Academy of Sciences, the National Academy of Engineering, and the National Science Board on academic research, used words that are directly applicable to the military project manager: " . . . Our political and economic systems are strongly biased toward the competitiveness of U.S. industry, regional economic development, education and human resources and governmental functions such as health, defense and environmental protection." In justifying our programs he states: "We must get better at dramatizing the linkages and aiming the arguments more precisely. We tend to assume that our enthusiasm for what we do is shared by the public. . . . We sometimes get generous support because what we ask for is politically attractive. . . . we need to become more focused and sophisticated on how we pursue federal funding. We need to target particular appropriations commit-

TABLE 43.2 Trends Reshaping the Role and Desired Characteristics of Military Project Managers

Trend	Impact
Increasing privatization of government functions.	Decreasing in-house expert support emphasizes the "smart buyer" role.
Downsizing of military establishment.	Smaller project management staffs.
Increased competition for fewer dollars within and outside of the services.	More emphasis on communicating linkages between military programs and other national needs.
Increased competition among junior officers for senior positions.	More stress on academic credentials as basis for promotion.
More emphasis on doing more with less will promote more joint-service and more cooperative international programs.	Project managers must be prepared to advocate from a purple-suit and global perspective.

tees with particular messages on why our enterprise should compete more favorably with other, often more understandable priorities. . . . each of these arguments has to be carried to particular agencies and particular congressional committees by informed people—preferably constituents—who appreciate the public interest." The bottom line of Dr. Schmitt's message, applied to the project officer, is that he must better demonstrate that his work is instrumental to the nation's goals. In pursuing such objectives the military project officer must walk a fine line to ensure that he has the complete support and understanding of his hierarchy of superiors and that he does not undercut the other projects and programs in his service which are also competing for finite resources. He must search for innovative arguments and ways of funding (such as joint service potential or technology transfer enhancements).

A climate of shrinking defense dollars stimulates thinking on joint military service programs. The new era of global relationships puts further emphasis on cooperative international programs. The future project manager must therefore be prepared to play a "purple-suit" role for joint service projects and his advocacy must be supportive of this enlarged scope of responsibility. Cooperative international projects have similar broader areas of concern, but may have cultural dimensions impacting advocacy as well.

More and more emphasis on the privatization of governmental functions can be expected to ensure that more of the shrinking defense dollars go into jobs in private industry. In contracting out these heretofore government functions, project leadership inside the DoD will generally be left with lesser and lesser amounts of in-house expertise. This has already led to smaller project offices, more use of matrix management concepts, reduced in-house laboratory support, and so on. In other words, project managers will increasingly be required to do their jobs with less help. This is leading to more lean and mean project management concepts and is placing more emphasis on the smart-buyer end of the project manager's spectrum of roles.

The swell of interest in and emphasis on defense acquisition professionalism is beginning to impact project managers greatly and can be expected to affect them even more so in the future. All of this change bodes well for the project manager himself and for the defense acquisition process as a whole.

First, there is the drive on the part of all the military services to upgrade the key acquisition personnel to professional status and to establish an infrastructure that supports professionalism and its further growth.

Second, there is the recognition of the significant importance of the military project manager's job and the sensitivity of his success relative to the value added to the protection of the citizenry and in view of the large number of taxpayer dollars at risk.

Third, there is the creation of definite career paths in the acquisition field, which can also lead to the top jobs in each service as have the heretofore accepted classical paths through the command of "big decks" at sea or through leadership assignments in major operational commands around the world. Also important has been the creation of similar career paths for civilians, leading to top-level civilian jobs, including the senior executive service.

Fourth, there has been an emphasis on formalizing and increasing the education credentials needed by the individuals in this new acquisition corps. This reflects the new and added visibility given the weapon acquisition function and the growing complexity of the environment in which these acquisition professionals, including the project managers, have to work. Many of the formal university programs required are designed to enhance the project manager's ability to effect his advocacy with emphasis on understanding and motivating people.

In summary, the project manager is the highest-level full-time government advocate for his project. His advocacy should only be bounded by the public interest, consistent with the overall program of the DoD and the programs of his particular military service. He must develop a network of advocates in all the federal, state, and local governments and in the private industry that has an interface with his project.

Finally, it can be said that the project manager is the government's full-time leader, chief spokesperson, and point of contact, and the taxpayer's top watchdog for the project. His work is accomplished in a complex and often hostile environment of competing economic, technological, political, and societal forces. His overall challenge is to lead the project to a successful conclusion and to completely serve the public interest in so doing.

REFERENCES

1. G. A. Graham, "Ethical Guidelines for Public Administrators," *Public Administration Review,* Jan.–Feb. 1974.
2. I. L. Janis, *Victims of Groupthink,* 2d ed., Houghton Mifflin, Boston, 1982.

CHAPTER 44

THE ROLE OF LESSONS LEARNED IN DEFENSE ACQUISITION

Daniel R. Vore and Norah H. Hill

Daniel R. Vore received the B.S. degree in chemical engineering from Michigan Technological University in 1984. He subsequently entered active duty as a TR-1 aircraft project officer. Capt. Vore worked in reconnaissance programs and managed TR-1/U-2 aircraft production, the TR-1 ground station, and the commanders' tactical terminal until his selection for the systems management program at the Air Force Institute of Technology. He is a September 1990 distinguished graduate of AFIT and currently serves as chief of the Targeting System Division of the LANTIRN program office.

Norah H. Hill received a B.S.E. degree in chemical engineering with honors from the University of Connecticut in 1983. She entered the Air Force in December 1982 and began as acquisition project officer for the ground mobile forces satellite communications terminal. Her job responsibilities expanded to include Air Force program management responsibilities for that Army production program as well as advanced planning for the development of an Air Force unique man-transportable version. In 1987 Capt. Hill was chosen to serve as executive officer for the Electronic Systems Division Deputy Commander for Tactical Systems. She then earned the M.S. degree in systems management from the Air Force Institute of Technology, where she is currently a professor of system acquisition management.

INTRODUCTION

"Those who do not learn from history are condemned to relive it" is an often cited quotation that first comes to mind when introducing a chapter on the role of lessons learned in defense acquisition. In the current environment of close scrutiny of defense acquisition by both Congress and the general public, the Department of Defense (DoD) cannot afford to appear to be unable to learn from its past experiences or to need to be taught a lesson more than once. Although humans naturally seek to learn from their own and others' past experiences, formal programs to encapsulate information regarding military acquisitions for later use just began in the 1980s. Certainly we want to learn from our collective previous mistakes in acquisition to avoid repeating them. Equally important, we want to repeat "tried and true" techniques that work to avoid "reinventing the wheel" on each new acquisition program.

A lesson learned can be defined simply as "a recorded experience of proven value in conducting future [activities]."[1–3] But exactly how we learn from our own experiences, and from those of others, and what is to count as fact, or right and wrong, is of critical importance in determining whether the results of applying lessons learned are truly beneficial.

The intended purpose of lessons learned programs, therefore, is to provide means to systematically access, scrutinize, and choose from past experiences those lessons we can apply in a new situation with a high probability that their use will result in a better course of action and results than would have been expected without their use.

With this understanding of the role of lessons learned in DoD acquisition, we will describe the lessons learned programs which currently exist within each service of the DoD. We will then assess current use of lessons learned and suggest guidelines for analyzing and applying lessons learned successfully. Finally, we will suggest potential changes to the current lessons learned programs to improve the application of lessons learned within DoD. The primary aim is to stimulate thinking regarding the most effective use of lessons learned and to identify some potential pitfalls to avoid when considering the use of a demonstrated course of action in a new situation.

CURRENT DoD LESSONS LEARNED PROGRAMS

Each service has a central agency responsible for the collection, validation, and dissemination of both positive and negative lessons learned. These agencies gather potential lessons recorded from past experiences, determine that they are correct and useful, and provide means for others to have access to this information. In accordance with a Joint Logistics Commanders' (JLC) joint agreement, each service maintains an independent database, since "each service's lessons learned are essentially hardware specific,"[4] and shares lessons learned with the other services on a regular basis. Also specified by the JLC was a standard single-page format for the submission of lessons to facilitate their exchange (see Fig. 44.1). Each service has developed a unique approach to its implementation of lessons learned programs.

Air Force Lessons Learned Program

The Air Force lessons learned program (AFLLP) was initiated in 1977 to manually collect, validate, and maintain lessons learned for the logistics community. The program began using an automated database the following year and has expanded ever since.[5] Located at Wright-Patterson Air Force Base in Ohio, the lessons learned program office (LLPO) is now the single Air Force manager for lessons learned.[6]

Collection. The AFLLPO does not develop the lessons itself. It solicits and accepts potential lessons learned from anyone in the Air Force and its contractors. Contributors provide potential lessons on their own initiative[5] or in response to requirements from leadership or regulation.[6,7]

Validation. Potential lessons learned are reviewed by volunteer subject-matter experts within the Air Force (not on the AFLLPO staff) to ensure that the potential

TOPIC	One or two lines which accurately describe the contents of the lesson
LESSON LEARNED	One or two sentences stating the single most important finding—cause and effect
PROBLEM	If the lesson learned is negative, a concise general statement describing what went wrong If the lesson learned is positive, state "none"
DISCUSSION	One to three paragraphs providing an account of the findings as they relate to the specific situation, procedure, or design
APPROPRIATE ACTION	A course of action (WHAT), time phase (WHEN), and vehicle (WHO, HOW) to implement the lesson

FIGURE 44.1 Air Force lessons learned format.[4]

lesson is "beneficial, valid, and applicable." A lesson is considered beneficial if it applies to a situation that a manager or designer is likely to face. The expert must determine that the lesson is both "factually and logically correct"[1] in order to declare it a valid lesson. Finally, a lesson must be specific in its recommended action and consistent with current laws and regulations to be applicable. If all three criteria are met, the lesson is entered into the database. Since lessons learned provided by the other services are validated by the originating service before they are sent to the AFLLPO, they do not undergo additional validation by the Air Force before they are entered into the database.[8] Additional validation is sought from database users. By requesting customer feedback, the AFLLPO obtains independent assessments of the usefulness and validity of individual lessons learned.

Dissemination. The database provides on-line access to more than 2000 validated lessons learned of the DoD to all authorized users. Interested parties can request online access directly from the AFLLPO or that AFLLPO conduct the research and send the results to the requestor. Current AFLLPO customers include all the major commands of the Air Force, defense firms contracted with the government, the Coast Guard, the Internal Revenue Service, the Federal Bureau of Investigation, NASA, and the Canadian Ministry of Defense. Users can identify lessons applicable to their programs or efforts by identifying the program phase (Fig. 44.2) or the subject of interest (Fig. 44.3), or specific words within the text.

The availability of the database is publicized widely, but whether and how it is used appears to be left up to the initiative of managers and their superiors.[6,8]

Navy Lessons Learned Program

The Navy lessons learned program began in 1982 to "support the Joint Services Advanced Vertical Lift aircraft acquisition."[2] A Naval Air Test Center team was formed to gather information from interviews with acquirers, supporters, operators, and maintainers of weapon systems of all three services from which lessons learned concerning reliability, maintainability, and supportability were derived and verified.

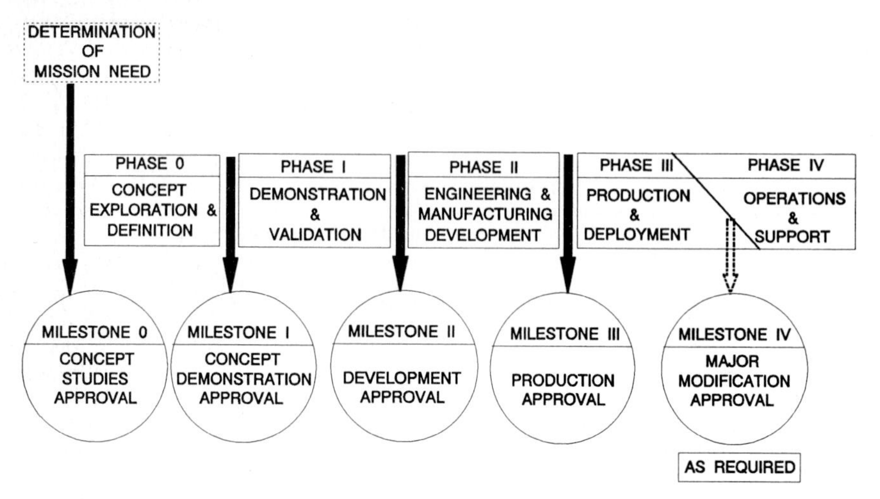

FIGURE 44.2 Acquisition milestones and phases.[9]

From this initial effort has evolved the present naval aviation lessons learned (NALL) program, which gathers, validates, and disseminates lessons learned in the areas of acquisition system design, logistics, program and contract management, and naval air operations.[2]

Collection. NALL are collected by members of a team of full-time researchers through a structured process. First, the researcher identifies a potential lesson learned through field visits, communication with contacts within the acquisition and operations communities, and government and contractor reports, periodicals, and other documents. In addition to self-initiated lesson gathering, researchers respond to specific requests. Next, the researcher queries the NALL database to ensure that a lesson has not already been documented in this area. If the lesson exists, the researcher determines whether the further research is needed on this lesson or not. If the potential lesson is determined to be worthy of pursuit, a lesson learned is drafted, accompanied by all source material, including a summary of the research and analysis conducted, and any recommended changes to official documentation.[2,10]

Validation. Once the researcher believes the lesson is accurate and complete, a review board consisting of all NALL team members, and engineers as required, verifies the lesson content for accuracy and consistency with the backup data and validates or rejects the lesson. All validated lessons are then sent to Naval Air Systems Command for approval before they are entered in the NALL database. Lessons in the database are reviewed once per year by the NALL team to ensure that they remain valid and do not require update.

Dissemination. Recommended changes to technical documents are sent to the document's issuing agency as soon as possible after the lesson is validated by the NALL organization itself. The NALL database includes a remote bulletin board system (RBBS) through which customers can obtain NALL, in a manner similar to the

1 Computer Resources (Support)
2 Energy Management
3 Engineering Data (Tech Data)
4 Facilities
5 Funding (Logistics Support)
6 Logistics Management Information Support
7 Provisioning
8 Maintenance Concept (Planning)
9 Modification Planning
10 Manpower Requirements
11 Ordnance
12 Reliability and Maintainability (R&M)
13 Safety
14 Supply Support
15 Support Equipment
16 Survivability
17 Technical Orders (Tech Data)
18 Test and Evaluation
19 Packaging, Handling, Storage, and Transportation (PHS&T)
20 Training and Training Support
21 Expert Systems/Artificial Intelligence (AI)
22 Propulsion Systems
23 Systems Integration (Hardware)
24 Systems Integration (Management)
25 Software
26 Software Management
27 Repair Techniques
28 Test Equipment
29 Corrosion Control
30 Configuration Management
31 Contract Administration
32 Contracting
33 Data Management
34 Engineering
35 Foreign Military Sales (FMS)
36 Human Factors Engineering
37 Life-Cycle Cost
38 Manufacturing
39 Operational Requirements
40 Program Control
41 Quality Assurance (QA)
42 Source Selection
43 Program Management Responsibility Transfer (PMRT)
44 Logistics Support Analysis (LSA)
45 Program Management
46 Environmental Management
47 Warranties
48 Hazardous Materials
49 Automated Information Systems (AIS)
50 Total Quality Management (TQM)
51 Personnel
52 Operations
53 Security
54 Composites
55 Trainers/Simulators
56 Avionics
57 Treaties
58 Identification
59 Maintenance Engineering
60 Materials
61 Logistics Assessment
62 TECHTIPs
63 TECHTAPs
64 Blue Two Visits
65 Program Managers Concerns
66 Facts (Fasteners, Actuators, Connectors, Tools, and Subsystems)
67 Desert Shield/Desert Storm

FIGURE 44.3 AFLLPO database impact areas.[8]

Air Force on-line access system, including search by impact area (Fig. 44.4), work unit code (Fig. 44.5), and aircraft/equipment type.[11]

Army Lessons Learned Program

The Army lessons learned program dates back to the mid-1980s when a number of archives and databases were designed and developed at various training centers with the goal of improving battlefield performance.[12] In 1989 the Center for Army Lessons Learned (CALL) was established by regulation to "collect, analyze, disseminate, and provide for the implementation of combat relevant lessons which impact on doctrine, training, organization, materiel, and leadership concepts and requirements."[13] One of CALL's main challenges is to integrate the various information gathering, processing, storage, and dissemination activities which are now in operation or planning. Current programs include the wartime Army lessons learned program (WALLP) implemented twice to date—during operations JUST CAUSE and

Acquisitioning	Ordnance
Automatic Test Equipment	Personnel
Computer Resources	Quality Assurance
Configuration Management	Packaging and Handling
Contract Administration	Provisioning
Contracting	Reliability
Corrosion Control	Repair Techniques
Data Management	Safety
Design-Engineering	Software
Facilities	Software Management
Funding	Source Selection
Human Factors	Supply Support
Identification	Support Equipment
Life-Cycle Cost	Survivability
Logistics Management	Technical Publications
Logistics Support Analysis	Test Equipment
Maintainability	Test and Evaluation
Maintenance Engineering	Training
Maintenance Plan	Training Support
Materials	Transportation
Operations	

FIGURE 44.4 NALL impact areas.[11]

DESERT STORM, the Army after-action reporting system, the classified joint universal lessons learned system maintained at the Pentagon, the Army lessons learned management information system, and the combat training center archive (Fig. 44.6).[14,15]

Collection. The Army Training and Doctrine Command (TRADOC) is responsible for the collection of observation reports from all sources, including historical analyses, joint, combined, and other combat operations and exercises, training center activities, simulations, war games, tests and evaluations, and terrorist incidents.[13] In addition, the combat training centers study various issues for possible solutions or improvements in what they call their "focused rotation" program.[14] An aspect of every combat experience and training exercise is to collect observations. The data to be collected are specified by Army regulation to include "doctrine and tactics employed, organization designs, materiel adaptations, and innovations . . . status of training, manning, logistics, morale and discipline of the combatants, event, or players."[13] Major command observers are tasked to maintain and document "operational plans, sequences of engagements, staff journals, and units after actions reports [and to] place emphasis on the systematic collections of information and statistical data [to provide for] a full evaluation."[13] After-action reports may also include field commanders' lessons learned.[14] After each event, these detailed observations are sent to CALL by each observing major command for analysis.

Validation. CALL personnel, in conjunction with schools, agencies, and contractors, as appropriate, analyze all observations to validate current policies, procedures, tactics, and other efforts; to derive lessons learned and issues requiring further study prior to resolution; and to recommend changes to current practice, as necessary.[13,14] Before validation is complete, a thorough assessment of the recommended solution is made.

Air Vehicle
11 Airframes
12 Fuselage Compartments
13 Landing Gear
14 Flight Controls
15 Helicopter Rotor System
17 Escape Systems
18 Training Equipment Flight Char./Force Generating Equipment

Power Plants
21 Reciprocating Engines
22 Turboshaft Engines
23 Turbojet Engines
24 Auxiliary Power Systems
26 Helicopter Drives/Transmissions
27 Turbofan Engine Power Installation
29 Power Plant Installation
32 Propellers
34 Rotary Wings
36 Ducted Fans

Utilities
41 Environmental Control/Pneumatic Systems
42 Electrical Power Supply/Distribution/Lighting Systems
44 Lighting Systems
45 Hydraulic System
46 Fuel Systems
47 Oxygen Systems
48 Ice and Rain Removal/Protection Systems
49 Miscellaneous Emergency Systems

Instrumentation
51 Instrumentation Systems
52 Autopilot Systems
53 Drone Guidance Systems
54 Telemetry Systems
56 Flight Reference Systems
57 Integrated Guidance and Flight Control Systems
58 In-Flight Test Equipment Systems
59 Target Scoring and Augmentation Systems

Communication
61 HF Communications Systems
62 VHF Communications Systems
63 UHF Communications Systems
64 Interphone Systems
65 IFF Systems
66 Emergency Radio Systems
67 COMM/NAV/IFF (CNI) Integrated Package Systems
69 Miscellaneous Communications System

Navigation, Bombing Fire Control, ECM and Photographic/Reconnaissance
71 Radio Navigation Systems
72 Radar Navigation Systems
73 Bombing Navigation Systems
74 Weapons Control Systems
75 Weapons Delivery
76 Countermeasures Systems
77 Photographic/Reconnaissance Systems
78 Training Equipment Instrumentations/Communications Systems
79 Training Equipment NAVI/Weapons Delivery, ECM, Reconnaissance System

Airborne Missiles/Rockets/Expendable Ordnance
81 Airborne Guided Weapons
83 Airborne Expendable Ordnance
84 Weapons Delivery Systems
85 Weapon Containers

Miscellaneous Equipment/Systems
91 Emergency Equipment
92 Tow Target/Towed Main Countermeasures Systems
93 Deceleration/Drag Chute System
94 Meteorological Equipment
96 Personnel Equipment
99 Training Equipment General-Purpose Computer System

FIGURE 44.5 NALL work unit codes.[11]

Dissemination. Since the Army lessons learned program's inception, 37 lessons learned bulletins have been published (Fig. 44.7), each focusing on a specific subject matter of interest to Army personnel. In addition to these periodic newsletters or pamphlets, CALL distributes videotapes, special reports, and briefings on topics of interest or concern to various units. Once disseminated, lessons learned are expected to be implemented by the commanders. Action plans are developed and published to implement specific lessons learned.[13]

Although the current Army lessons learned system is the most decentralized and

ARMY LESSONS LEARNED AUTOMATION SYSTEM (ALLAS)
A conceptual automated systems architecture which links multiple disparate lessons learned databases and which provides a feedback mechanism of measurable training performance to training developers.

INTELLIGENT GATEWAY PROCESSOR (IGP)
A software package developed by Lawrence Livermore National Laboratories which integrates multiple disparate databases, permitting access through a single point of entry to all linked systems.

ARMY LESSONS LEARNED MANAGEMENT INFORMATION SYSTEM (ALLMIS)
A database of subjective observations derived from combat training centers, worldwide exercises, historical experience, and actual combat which deal with doctrinal, training, organizations, materiel, and leadership issues.

COMBAT TRAINING CENTER ARCHIVE
A database of digital, papercopy, video, and audio training performance materials collected from the National Training Center, Ft. Irwin, Calif., the Joint Readiness Training Center, Ft. Chaffee, Ariz., and the Combat Maneuver Training complex, Hohenfels, Germany.

ARMY AFTER-ACTION REPORTING SYSTEM (AAARS)
A conceptual database to hold digitized unit and organization after-action reports as required under AR11-33, the Army Lessons Learned Program, System Development and Application.

JOINT UNIVERSAL LESSONS LEARNED SYSTEM (JULLS)
A database of subjective observations derived from joint operations with multiservice applications.

COMBAT OPERATIONS RESEARCH FACILITY (CORF)
An array of local microcomputer and workstation-based equipment and software, which provides access to the ALLAS and its allied databases, and supports graduated levels of data displays and replays.

FIGURE 44.6 Army lessons learned databases.[14]

least automated of the services' systems,[8,12] that is changing fast. Within the decade, CALL plans to provide on-line access to all databases through a single source from which information can be downloaded to a user's personal computer[12] to provide a more accessible, responsive system.

ESTABLISHING CRITERIA FOR THE APPLICATION OF LESSONS LEARNED

In order to discuss the application of lessons learned more clearly, a theoretical basis for establishing criteria for a successful application must be introduced. Without overburdening the reader with vocabulary or concepts, this section will serve as a backdrop for the remainder of the chapter's discussion.

The management cybernetician Stafford Beer defines variety as a measure of

Initial Issue	July 1986
Read Operations	November 1986
Lessons Learned (General)	April 1987
Light Infantry in Action I	April 1988
Light Infantry in Action II	June 1988
Deception	July 1988
RC Brigade to the NTC	April 1989
Commander's Memorandum—CG NTC	November 1986
Lessons Learned by/for Division Commanders	September 1986
MILES Checklist	June 1986
Fort Hood Leadership Study	December 1986
Commander's Comments—The CS Team	May 1987
Seven Operating Systems	January 1986
Intelligence	May 1986
Fire Support	September 1986
Command and Control	February 1987
Leadership	July 1987
Commander's Survivability	January 1988
Minefield Breaching	May 1988
Heavy Forces	Fall 1988
Non-Mechanized Newsletter	Spring 1989
Heavy-Light Lessons Learned	August 1989
NCO Lessons Learned	October 1989
Corps/Division Lessons Learned	November 1989
Commander's Casualty Evacuation System	November 1989
Fire Support for Maneuver Commander	February 1990
Reserve Component Deployment	March 1990
The Stone Forest	May 1990
Low-Intensity Conflict	May 1990
Fire Support Newsletter	June 1990
Musicians of Mars	June 1990
Winning in the Desert	August 1990
Winning in the Desert II	September 1990
Operation JUST CAUSE	October 1990
Inactivation	November 1990
Getting to the Desert	December 1990
Rehearsals	April 1991

FIGURE 44.7 Army lessons learned publications.[16]

complexity with which the manager is faced.[17] Management cybernetics, the science of effective organization, offers a methodology, known generally as the viable system model, for the manager to effectively organize and conduct operations to deal with the reality of proliferating variety.

Beer warns of the common management mistake of approaching situations within a high-variety system by arbitrarily reducing the system's variety via a subjective paradigm (model) or interpretation. Although the intention to simplify one's view of the situation as much as possible is understandable, if the resulting paradigm does not include all critical parameters pertinent to the system, any conclusions derived therefrom will of necessity be inaccurate or inappropriate.[17] No matter what the label, one always runs the risk of borrowing the "right answer" or method from another acquisition program only to find that this "answer" produces disastrous effects on one's own program. Those familiar with DoD acquisition are all too aware of the tendency of some to advocate the latest policy trend as a "miracle cure" applicable to every situation. A simple example was rampant in the mid-1980s in the contracting arena. With the admirable intent of controlling government costs, fixed-price contract types became the only "acceptable" means of contracting work. While fixed-price contracts are most desirable in many cases, when a system design is immature and the costs cannot be predicted accurately, a cost-reimbursement type contract is more appropriate.[18] If misapplied, a fixed-price contract, during early development of a system design, for example, is likely to threaten the industrial base by demanding that contractors absorb high losses in order to maintain their reputation and the opportunity for future work, or to incentivize the contractor to spend as little money as possible on early design, which will cost the government more in the long run in redesign, modification, and support.

The common cause of such misapplication of lessons learned is based in one's failure to fully understand one's own system or program and its output states—the results likely to be observed, given any change to the system—before recommendations for action are sought and applied to one's own situation.

The late Defense Systems Management College professor Fred Waelchli defines two basic approaches to problem solving: algorithmic and heuristic.[19] In each case one would consider the approach as one rooted in an implicit or explicit model of the situation under assessment in order that possible solutions could be identified, considered, and implemented. The algorithmic approach is essentially a "cookbook" approach, that is, a checklist or set of rules which, if followed, produces a solution to the problem. In contrast, the heuristic approach is described by Waelchli as "enlightened search."[19] In cybernetic terms, the algorithmic approach a priori reduces (constrains) the "solution space" (set of solutions considered) and therefore attenuates situational variety. In other words, an algorithmic approach may be based on a model of the system, problem, or situation which is incomplete, inaccurate, or oversimplified. If this is the case, any "answer" derived from this model will be likely to "fail." The heuristic approach, seeking "intelligent expansion of the solution space,"[19] a priori amplifies situational variety which the manager then can use to achieve requisite variety, or parity, with the problem at hand. If the variety is indeed requisite, the model upon which the heuristic is based accurately reflects the situation in which the solution or course of action to be derived from that model is to be implemented. Clearly, not all problems are inherently high-variety in nature—the algorithmic approach has its place in the manager's "toolbox." However, the highly complex, thereby high-variety, environment of defense acquisition breeds predominantly high-variety problems and situations. Therefore for a lesson learned to be truly applicable to a new situation, the situation in which it was originally learned must be the same, in all significant aspects, to the new situation. In order to make this determination one must have a full understanding of both situations.

To the extent that the process of system analysis and consideration of alternatives does not result in a complete understanding of the similarities and differences between the two situations, use of lessons learned parallels the algorithmic approach described by Waelchli. One simply finds a situation which seems similar to the problem at hand, reads how the problem was solved, and uses a parallel solution. A more thorough, systematic approach is more likely to provide beneficial results.

Of subtle, yet significant difference would be an approach that would use lessons learned as a means to expand the solution space and thereby amplify situational variety. Under this approach the lessons learned program would be expected to provide a repertoire of potential solutions, or summarily to define a part of the overall solution space, rather than to dictate a singular solution to apply methodically in many seemingly similar programs.

Through the use of times series, and discriminant analysis and bayesian statistics, Forrester and Beer, respectively, have developed predictive computer modeling techniques competent to simulate complex-probabilistic situations such as those faced within DoD acquisition.[19–21] If such models of DoD acquisition were developed using detailed observations as data with which the models could be refined over time to more and more accurately reflect reality and the broad scope of possible activities and outcomes, then such models could be useful tools to chose and assess potential courses of action. Recent modeling efforts sponsored by the Defense Science Board and the establishment in 1991 of a DoD simulation office may be the genesis of such an application.

POTENTIAL CHANGES TO IMPROVE THE CURRENT LESSONS LEARNED PROGRAM

Of the three programs described herein, the Army system appears best designed to capture the necessary array of parameters and to assess the applicability of a lesson learned in a training (simulated) environment prior to its actual employment. All three programs categorize data so that the thorough users, aware of all parameters which may affect their program's performance, may conduct a search and determine, after researching the source data for the lessons, one or more courses of action worthy of their implementation with the defendable expectation of beneficial results. However, three basic changes could drastically improve one's ability to derive information pertinent to a new situation from collective DoD past experiences and make less attractive an algorithmic approach to complex problem solving (determination of courses of action).

Collection

The current limitation to a single page to summarize the lesson and the resulting recommended action encourages algorithmic application. A more comprehensive documentation of the environment, circumstances, and details of the observed situation, and the analysis, activities, and changes that resulted in the identification of the recommended action, would vastly improve the likelihood that its true applicability to a new situation could be assessed accurately at some future time.

Maintenance

If the lessons were maintained in complex probabilistic simulation models, like those of Beer and Forrester, rather than in simple databases, any new lesson could be used to refine or expand this model, as appropriate, providing a more comprehensive solution space for the potential lessons learned user.

Validation

Whether a lesson is valid or not is dependent on the situation to which it will be applied. Consequently a better approach to validation would be conducted by the potential lessons learned user at the time of his or her query into the lessons learned system. Beginning with a thorough analysis of the programmatic and situational parameters (Fig. 44.8), the user could input these into the lessons learned model in a two-step approach. First, the solution space of possibilities "known" to the model would be chosen by the computer, based on the parameters of the new system or situation of the user. Then the probable results of the use of one or any combination or

Begin the search for a course of action to solve or avoid problems by asking yourself the following questions:

1. Do I have a credible model of my system which describes the interrelationships of the major elements?
2. Do I understand the relationship between inputs and outputs of my system?
3. Have I thoroughly defined the problem I am facing?
4. If I have knowledge of what appears a similar problem on another program and am thinking of borrowing that solution, do I understand the other system well enough to know why that solution worked?
5. What is different about the other program and how will those differences affect my application of the other program solution?

FIGURE 44.8 Guidelines for the application of lessons learned.

extrapolation of the possible solutions could be assessed using the simulation model. Armed with this information, the user could make an informed and rational decision regarding his or her courses of action.

SUMMARY

The necessity of viable lessons learned programs within the DoD is undeniable. Increasing senior manager interest in the programs is resulting in increased resources and use. However, the current and planned future databases may not be able to provide the expected results—the avoidance of repeated problems or unsuccessful results. Lessons learned programs need to be means to help DoD managers develop and use system models to make smart decisions about future courses of action based on reliable predictions of probable results.

As we consider how best to implement the expanding lessons learned programs, we must consider multilevel plans. The single best architects of a solution to the problem at hand are from within, not from outside. Each program, and the interaction of its elements, is unique. Each solution, therefore, to at least some degree, will be unique. However, each new program will not be able to start from "scratch." They must begin from the basis of existing knowledge and derive, over time, program-unique models which have better predictive value. Consequently one would expect each program to have an ever-evolving model of itself with which to consider alternatives and plan courses of action which would be a subset of one or more models within the service and DoD. Such an implementation would require careful design of interactive simulation models by a team of statisticians, computer scientists, and acquisition managers and informed use by knowledgeable program participants.

REFERENCES

1. ALD/LSE, *Air Force Lessons Learned Program,* Wright-Patterson AFB, Ohio, Feb. 15, 1991, p. 1.
2. *Naval Aviation Lessons Learned Preparation Guide,* Naval Air Test Center (RW83), Patuxent River, Md., Sept. 13, 1991, pp. 1, 2, 8, 9.

3. U.S. Army Combined Arms Command, *Center for Army Lessons Learned,* briefing charts, Ft. Leavenworth, Kansas, 1991.
4. *Joint Agreement on the Joint Logistics Commanders' Lessons Learned,* Mar. 16, 1989.
5. ALD/LSE, *Air Force Lessons Learned Program,* pamphlet, Wright-Patterson AFB Ohio, undated.
6. Air Force Supplement 1, DOD Instruction 5000.2 draft, *Acquisition Management,* Part 5, sec. H, "Air Force Lessons Learned," June 1991, p.1.
7. B. W. Cawthorn, *AFSC Program Manager Lessons Learned Information Network,* undated reprint of Gen. Ronald W. Yates' letter to Hon. John Welch regarding support for the Lessons Learned Program, electronic mail, Feb. 22, 1991.
8. R. Kerr, Air Force Lessons Learned Program, Acquisition Logistics Division, Wright-Patterson AFB, Ohio, personal communication, Aug, 7, 1991.
9. DoD Instruction 5000.2, *Defense Acquisition Management Policies and Procedures,* Part 2, "General Policies and Procedures," Feb. 23, 1991, p. 2-1.
10. S. Pistachio, Naval Aviation Lessons Learned Production Coordinator, Patuxent River, Md., personal communication, Sept. 23, 1991.
11. *Naval Aviation Lessons Learned Program Information Brochure,* Naval Air Test Center (RW83), Patuxent River, Md., July 1991, pp. 1–3, 8.
12. Becker, Center for Army Lessons Learned, Ft. Leavenworth, Kansas, personal communication, Sept. 23, 1991.
13. Army Regulation 11-33, *Army Lessons Learned Program: System Development and Application,* HQ Department of the Army, Washington, D.C., Oct. 10, 1989, pp. 1–5.
14. Center for Army Lessons Learned, standard briefing, charts, 1991.
15. Becker, *Army Lessons Learned Automation System,* annotated briefing charts, undated.
16. Center for Army Lessons Learned, "Rehearsals," *Newsletter,* 91-1, p. 29, Apr. 1991.
17. S. Beer, *The Heart of the Enterprise,* Wiley, New York, 1979, p. 32.
18. Federal Acquisition Regulation, Part 16, Government Printing Office, Washington, D.C., 1990.
19. R. Espejo and R. Harnden, *The Viable System Model: Interpretations of Stafford Beer's VSM,* Wiley, New York, 1989, pp. 51–73, 349–351.
20. J. W. Forrester, *Industrial Dynamics,* M.I.T. Press, Cambridge, Mass., 1961.
21. S. Beer, *Decision and Control,* Wiley, New York, 1966, pp. 270–495.

INDEX

ABOUT THE EDITORS

DAVID I. CLELAND is a professor of engineering management at the University of Pittsburgh and the author/editor of 19 books, including the forthcoming *Global Project Management Handbook.*

JAMES M. GALLAGHER is Director of the Center for Executive Development and Project Management at Wright State University. He previously served as a project manager in the Air Force, and held various management positions at Wright-Patterson Air Force Base.

RONALD S. WHITEHEAD is the Midwest regional representative for Vought Aircraft Company, and was formerly a lieutenant colonel in the Air Force.